Regions

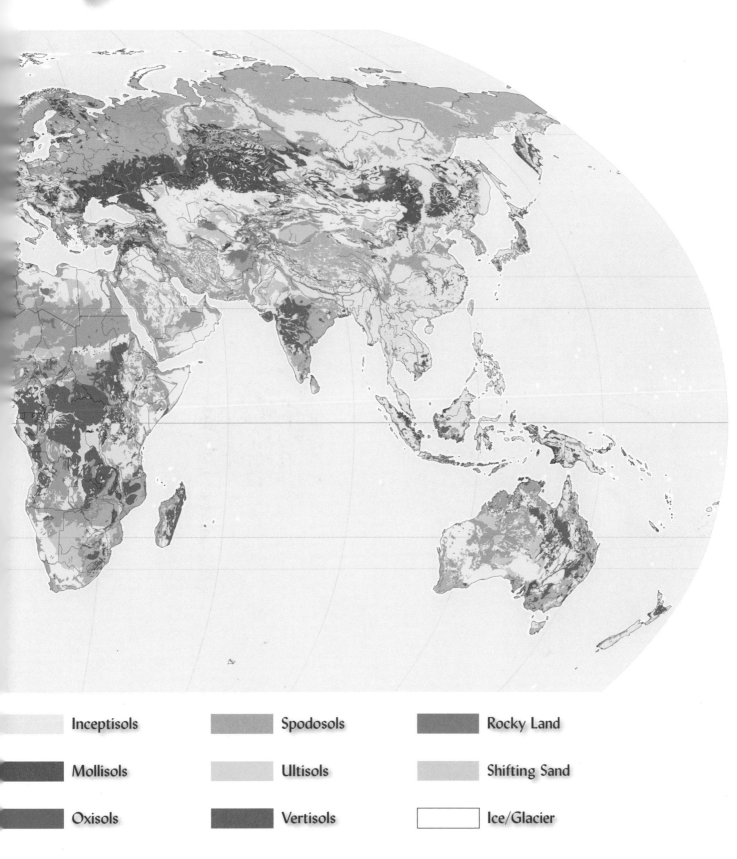

Inceptisols	Spodosols	Rocky Land
Mollisols	Ultisols	Shifting Sand
Oxisols	Vertisols	Ice/Glacier

The Nature
and Properties of Soils

The Nature
and Properties of Soils

REVISED FOURTEENTH EDITION

NYLE C. BRADY
PROFESSOR OF SOIL SCIENCE, EMERITUS
CORNELL UNIVERSITY

RAY R. WEIL
PROFESSOR OF SOIL SCIENCE
UNIVERSITY OF MARYLAND

PEARSON

Prentice
Hall

UPPER SADDLE RIVER, NEW JERSEY
COLUMBUS, OHIO

Library of Congress Cataloging-in-Publication Data

Brady, Nyle C.
   The nature and properties of soils / Nyle C. Brady, Ray R. Weil.—14th ed.
      p.   cm.
   Includes bibliographical references.
   ISBN 0-13-227938-X
   1. Soil science. 2. Soils. I. Weil, Ray R. II. Title.
   S591.B79 2008
   631.4—dc21

                                    2001034595

Editor-in-Chief: Vernon R. Anthony
Editorial Assistant: Sonya Kottcamp
Production Editor: Alexandrina Benedicto Wolf
Design Coordinator: Diane Ernsberger
Cover Designer: Thomas Mack
Cover Photo: by Ray Weil—Tree, soil, water and rock by Nile River.
Production Manager: Laura Weaver
Director of Marketing: David Gesell
Marketing Manager: Jimmy Stephens
Marketing Coordinator: Alicia Dysert

This book was set in Stone Serif by GGS Book Services PMG. It was printed and bound by Courier/Kendallville. The cover was printed by Phoenix Color Corp.

Copyright © 2008, 2002, 1999, 1996 by Pearson Education, Inc., Upper Saddle River, New Jersey 07458.
Pearson Prentice Hall. All rights reserved. Printed in the United States of America. This publication is protected by Copyright and permission should be obtained from the publisher prior to any prohibited reproduction, storage in a retrieval system, or transmission in any form or by any means, electronic, mechanical, photocopying, recording, or likewise. For information regarding permission(s), write to: Rights and Permissions Department.

Earlier editions by T. Lyttleton Lyon and Harry O. Buckman copyright © 1922, 1929, 1937, and 1943 by Macmillan Publishing Co., Inc. Earlier edition by T. Lyttleton, Harry O. Buckman, and Nyle C. Brady copyright © 1952 by Macmillan Publishing Co., Inc. Earlier editions by Harry O. Buckman and Nyle C. Brady copyright © 1960 and 1969 by Macmillan Publishing Co., Inc. Copyright renewed 1950 by Bertha C. Lyon and Harry O. Buckman, 1957 and 1965 by Harry O. Buckman, 1961 by Rita S. Buckman. Earlier editions by Nyle C. Brady copyright © 1974, 1984, and 1990 by Macmillan Publishing Company.

Pearson Prentice Hall™ is a trademark of Pearson Education, Inc.
Pearson® is a registered trademark of Pearson plc
Prentice Hall® is a registered trademark of Pearson Education, Inc.

Pearson Education Ltd.
Pearson Education Singapore, Pte. Ltd.
Pearson Education Canada, Ltd.
Pearson Education—Japan

Pearson Education Australia, Pty. Limited
Pearson Education North Asia Ltd.
Pearson Educación de Mexico, S.A. de C.V.
Pearson Education Malaysia Pte. Ltd.

13 12 11 10 9 CKV 16 15 14 13 12
ISBN-13: 978-0-13-227938-3
ISBN-10:   0-13-227938-X

*To all the students and colleagues in soil science who have
shared their inspirations, camaraderie, and deep love of the Earth.*

CONTENTS

xii *Contents*

NOTE: Every effort has been made to provide accurate and current Internet information in this book. However, the Internet and information posted on it are constantly changing, and it is inevitable that some of the Internet addresses listed in this textbook will change.

Soils are at the heart of terrestrial ecosystems. An understanding of the soil system is therefore key to the success and environmental harmony of any human endeavor on the land. The importance of soils and the soil system is increasingly recognized by business and political leaders, by the scientific community, and by those who work with the land. Scientists and managers well versed in soil science are in short supply and becoming increasingly sought after.

This book is designed to help make your study of soils both fascinating and intellectually satisfying. Much of what you learn from these pages will be of enormous practical value to help you meet the many natural-resource challenges of the 21st century. You will soon find that the soil system provides many opportunities to see practical applications for principles from such biology, chemistry, physics, geology, and other sciences.

This newest edition of *The Nature and Properties of Soils* explains the fundamental principles of soil science relevant to your interests. The text emphasizes the soil as a natural resource and soils as ecosystems. It highlights the many interactions between soils and other components of forest, range, agricultural, wetland, and constructed ecosystems. This book will serve you well, whether this is your only formal exposure to soil science or you are embarking on a comprehensive soil science education. It provides both an exciting, accessible introduction to the world of soils and a reliable, comprehensive, professional reference.

Each chapter has been thoroughly updated with the latest advances, concepts, and applications. Hundreds of new key references have been added. This edition includes new discussions of the pedosphere concept, subaqueous soils, ethnopedology, geophagy, soils and human health, organic farming, engineering properties of soils, nonsilicate colloids, inner- and outer-sphere complexes, nuclear contamination, radioactive waste storage, effective CEC, the proton-balance approach to soil acidity, cation saturation, acid sulfate soils, acid precipitation, arid region soils, irrigation techniques, biomolecule binding, soil food-web ecology, disease suppressive soils, soil archaea, forest nutrient management, phosphorus site index, lead contamination, indicators of soil quality, soil ecosystem services, charcoal and Terra Preta soils, plant production for biofuels, global climate change, and many other topics of current interest in soil science.

In response to their popularity in recent editions, I have also added many new boxes that present either fascinating examples and applications or technical details and calculations. These boxes both highlight material of special interest and allow the logical thread of the regular text to flow smoothly without digression or interruption. Examples of applications boxes include cases studies of how soils influenced the Hurricanne Katrina levee failures, amelioration of selenium pollution in wetlands, capillary barriers for nuclear waste, the invasion of North American forests by earthworms, runoff farming in ancient deserts, and the debate over nitrate toxicity. New boxes have also been added to provide detailed calculations for soil water content, profile water holding capacity, soil heat capacity, cation exchange capacity, and many other numerical problems.

Two extremely popular features have been the high quality figures that bring soils alive in the text, as well as the many Web universal resource locators (URLs) set in the margins. These Web sites developed by colleagues and organizations around the world expand and elaborate on certain topics in ways that would not be possible in a printed book. These features are made even more effective and convenient on the new and expanded companion Web site (**www.prenhall.com/brady**) where you will find beautiful full-color versions of most of the black and white figures in the book, as well as clickable links to the hundreds of web URLs printed in the margins of the book next to related text. The web site also offers practice quizzes for each chapter that provide interactive feedback. These quizzes will both challenge you and build confidence in your growing understanding of the topics and their interrelationships.

Dr. Nyle Brady, having been in retirement for a number of years, decided to stand aside for the writing of this revision—making this the first edition of *The Nature and Properties of Soils* since 1952 not to see his direct participation. However, he remains as first author in recognition of the fact that his vision, wisdom and inspiration continue to permeate the entire book. Although the responsibility for writing the 14th edition was solely mine, I certainly could not have made all the improvements just mentioned without the many valuable suggestions, ideas, and corrections sent in by soil scientists, instructors, and students from around the world. The fourteenth edition, like preceding editions, has greatly benefited from such contributions. The high level of professional devotion and camaraderie shared by so many students, teachers, and practitioners of soil science never ceases to amaze and inspire.

Special thanks go to Amy Kremen for her superb editorial and research assistance. I also thank the following colleagues (listed by institution) for generously reviewing portions of the text in detail and making valuable suggestions for improvement: Michéli Erika (Univ. Agricultural Science, Hungary), Duane Wolf (University of Arkansas); Tom Pigford (University of California, Berkeley); Chip Appel (California Polytechnic State University, San Luis Obispo); Thomas Ruehr (Cal Poly State University); J. Kenneth Torrence (Carleton University); Jessica Davis (Colorado State University); Pedro Sanchez and Cheryl Palm (Columbia University); Cristian P. Schulthess (University of Connecticut); Harold van Es, Susan Riha, and Martin Alexander (Cornell University); Dan Richter (Duke University); Owen Plank (University of Georgia); Robert Darmody, Colin Thorn and Michelle Wander (University of Illinois); Roland Buresh (International Rice Research Institute); Lee Burras (Iowa State University); Daniel Hillel (University of Massachusetts, Emeritus); Lyle Nelson (Mississippi State University, Emeritus); Jimmie Richardson (North Dakota State University); Rattan Lal (Ohio State University); David Munn (Ohio State ATI); Elizabeth Sultzman (Oregon State University); Darrell Schultze (Purdue University); Ross Spackman (College of Southern Idaho); Joel Gruver (Western Illinois University); Ivan Fernandez (University of Maine); Mark Carroll, Delvin Fanning, Robert Hill, Bruce James, Brian Needelman, Martin Rabenhorst, and Patricia Steinhilber (University of Maryland); Martha Mamo (University of Nebraska); Jose Amador (University of Rhode Island); Russell Briggs (State University of New York); Murray Milford (Texas A & M University); Mike Swift (UN Tropical Biology Program); Allen Franzluebbers, Jeff Herrick, Scott Lesch, and Jim Rhoades (USDA/Agricultural Research Service); Bob Ahrens, Mary Ellen Cook, Susan Davis, Bob Engel, Hari Eswaran, Paul Reich, and Sharon Waltman (USDA/Natural Resources Conservation Service); Fred Magdoff and Wendy Sue Harper (University of Vermont); W. Lee Daniels, S. K. de Datta, and Lucian Zelazny (Virginia Tech); Peter Abrahams (University of Wales); Luther Carter (Washington, DC); Clay Robinson (West Texas A & M University); Larry Munn (University of Wyoming); and Tom Siccama (Yale University).

Last, but not least, I wish to express my deep appreciation to Trish, my wife and spiritual partner, for her sense of humor (she likes to call this book "Dirts I Have Known"), understanding, patience, and encouragement, without which I could not possibly have completed this labor of love.

R. R. W.

Earth, unique for soil and water. (NASA)

1

THE SOILS AROUND US

For in the end we will conserve only what we love.
We will love only what we understand.
And we will understand only what we are taught.
—BABA DIOUM, AFRICAN CONSERVATIONIST

The Earth, our unique home in the vastness of the universe, is in crisis. Our planet is covered with life-sustaining air, water, and soil. However, great care will be required in preserving the quality of all three if our species is to continue to thrive.

Human activities are changing the very nature of the Earth's blanket of air. Depletion of the ozone layer in the stratosphere is threatening to overload us with ultra-violet radiation. Increasing concentrations of carbon dioxide and methane gases are warming the planet and destabilizing the global climate. Tropical rain forests, and the incredible array of plant and animal species they contain, are disappearing at an unprecedented rate. Groundwater supplies are being contaminated in many areas and depleted in others. In parts of the world, the capacity of soils to produce food is being degraded, even as the number of people needing food is increasing. It will be a great challenge for the current generation to bring the global environment back into balance.

Soils[1] are crucial to life on Earth. From ozone depletion and global warming to rain forest destruction and water pollution, the world's ecosystems are impacted in far-reaching ways by processes carried out in the soil. To a great degree, the quality of the soil determines the nature of plant ecosystems and the capacity of land to support animal life and society. As human societies become increasingly urbanized, fewer people have intimate contact with the soil, and individuals tend to lose sight of the many ways in which they depend upon soils for their prosperity and survival. Indeed, the degree to which we are dependent on soils is likely to increase, not decrease, in the future.

Soils will continue to supply us with nearly all of our food (except for what can be harvested from the oceans). How many of us remember, as we eat a slice of pizza, that the pizza's crust began in a field of wheat and its cheese began with grass, clover, and corn rooted in the soils of a dairy farm? Most of the fiber we use for lumber, paper, and clothing has its roots in the soils of forests and farmland. Although we sometimes use plastics

---

[1] Throughout this text, bold type indicates key terms whose definitions can be found in the glossary.

FIGURE 1.1 Environmental and economic imperatives suggest that we will become more dependent on soil to produce renewable materials that can substitute for increasingly scarce and environmentally damaging nonrenewables. Biodiesel fuel (*left*) produced from soybean and other oil crops is far less polluting and has less impact on global warming than petroleum-based diesel fuel. Soybean (*lower right*) and other oil crops can substitute for petroleum to produce nontoxic inks (*bottom*), plastics, and other products. Cornstarch can be made into biodegradable plastics for such products as plastic bags and foam packing "peanuts" (*upper right*). (Photos courtesy of R. Weil)

and fiber synthesized from fossil petroleum as substitutes, in the long term we will continue to depend on terrestrial ecosystems for these needs.

In addition, biomass grown on soils is likely to become an increasingly important feedstock for fuels and manufacturing, as the world's finite supplies of petroleum are depleted during the course of this century. The early marketplace signs of this trend can be seen in the form of gasohol and biodiesel fuels made from plant products, printers' inks made from soybean oil, and biodegradable plastics synthesized from cornstarch (Figure 1.1).

One of the stark realities of the 21st century is that the human population that demands all of these products will increase by several billion, while the amount of soil available to provide for them will not increase at all. In fact, the resource base is actually *shrinking* because of soil degradation and urbanization. It is clear that we must greatly improve our understanding and management of the soil resource if we as a species are to survive and if we are to leave enough habitat for the survival of the other creatures that share this planet with us.

The art of soil management is as old as civilization. As we meet the challenges of this century, new understandings and new technologies will be needed to protect the environment and, at the same time, produce food and biomass to support society. The study of soil science has never been more important for foresters, farmers, engineers, natural resource managers, and ecologists alike.

1.1 FUNCTIONS OF SOILS IN OUR ECOSYSTEM

In any ecosystem, whether your backyard, a farm, a forest, or a regional watershed, soils have six key roles to play (Figure 1.2). *First*, soil supports the growth of higher plants, mainly by providing a medium for plant roots and supplying nutrient elements that are essential to the entire plant. Properties of the soil often determine the nature of the vegetation present and, indirectly, the number and types of animals (including people)

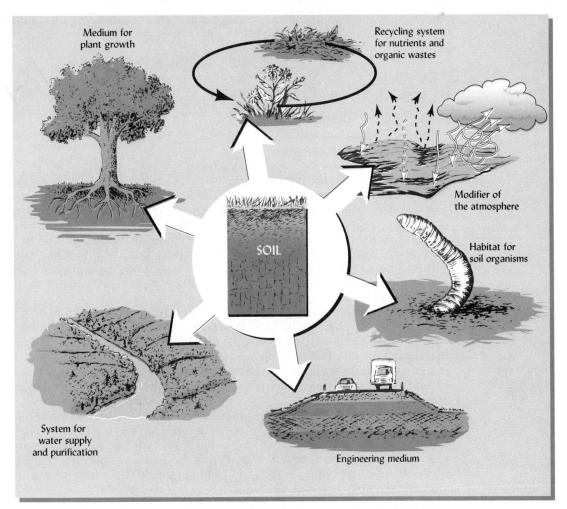

FIGURE 1.2 The many functions of soil can be grouped into six crucial ecological roles.

that the vegetation can support. *Second*, soil properties are the principal factor controlling the fate of water in the hydrologic system. Water loss, utilization, contamination, and purification are all affected by the soil. *Third*, the soil functions as nature's recycling system. Within the soil, waste products and dead bodies of plants, animals, and people are assimilated, and their basic elements are made available for reuse by the next generation of life. *Fourth*, soils provide habitats for a myriad of living organisms, from small mammals and reptiles to tiny insects to microscopic cells of unimaginable numbers and diversity. *Fifth*, soils markedly influence the composition and physical condition of the atmosphere by taking up and releasing large quantities of carbon dioxide, oxygen, methane, and other gases and by contributing dust and re-radiated heat energy to the air. *Finally*, in human-built ecosystems, soil plays an important role as an engineering medium. Soil is not only an important building material in the form of earth fill and bricks (baked soil material), but provides the foundation for virtually every road, airport, and house we build.

1.2 MEDIUM FOR PLANT GROWTH

Plant germination video: http://plantsinmotion.bio. indiana.edu/plantmotion/ earlygrowth/germination/ germ.html

When we think of the forests, prairies, lawns, and crop fields that surround us, we usually envision the **shoots**—the plant leaves, flowers, stems, and limbs—forgetting that half of the plant world, the **roots**, exists belowground. It is often competition among plant roots, rather than shoots, that plays the greater role in governing the coexistence of different plant species within an ecosystem. Because plant roots are usually hidden from our view and difficult to study, we know much less about plant–environment interactions belowground than aboveground, but we must understand both to truly understand either. To begin with, let's list and then briefly discuss what a plant obtains from the soil in which its roots proliferate:

- Physical support
- Air
- Water
- Temperature moderation
- Protection from toxins
- Nutrient elements

First, the soil mass provides physical support, anchoring the root system so that the plant does not fall over or blow away. Occasionally, strong wind or heavy snow does topple a plant whose root system has been restricted by shallow or inhospitable soil conditions (Figure 1.3).

Plant roots depend on the process of respiration to obtain energy. Since root respiration, like our own respiration, produces carbon dioxide (CO_2) and uses oxygen (O_2), an important function of the soil is *ventilation*—allowing CO_2 to escape and fresh O_2 to enter the root zone. This ventilation is accomplished via networks of soil pores.

An equally important function of soil pores is to absorb rainwater and hold it where it can be used by plant roots. As long as plant leaves are exposed to sunlight, the plant requires a continuous stream of water to use in cooling, nutrient transport, turgor maintenance, and photosynthesis. Since plants use water continuously, but in most places it rains only occasionally, the water-holding capacity of soils is essential for plant survival. A deep soil may store enough water to allow plants to survive long periods without rain (see Figure 1.4).

As well as moderating moisture changes in the root environment, the soil also moderates temperature fluctuations. Perhaps you can recall digging in garden soil on a summer afternoon and feeling how hot the soil was at the surface and how much cooler just a few centimeters below. The insulating properties of soil protect the deeper portion of the root system from extremes of hot and cold that often occur at the soil surface. For example, it is not unusual for the temperature at the surface of bare soil to exceed 35 or 40 °C in mid-afternoon, a condition that would be lethal to most plant

FIGURE 1.3 This wet, shallow soil failed to allow sufficiently deep roots to develop to prevent this tree from blowing over when snow-laden branches made it top-heavy during a winter storm. (Photo courtesy of R. Weil)

FIGURE 1.4 A family of African elephants finds welcome shade under the leafy canopy of a huge acacia tree in this East African savanna. The photo was taken in the middle of a long dry season; no rain had fallen for almost five months. The tree roots are still using water from the previous rainy season stored several meters deep in the soil. The light-colored grasses are more shallow-rooted and have either set seed and died or gone into a dried-up, dormant condition. (Photo courtesy of R. Weil)

roots. Just a few centimeters deeper, however, the temperature may be 10 °C cooler, allowing roots to function normally.

There are many potential sources of **phytotoxic substances** in soils. These toxins may result from human activity, or they may be produced by plant roots, by microorganisms, or by natural chemical reactions. Many soil managers consider a function of a good soil to be protection of plants from toxic concentrations of such substances by ventilating gases, by decomposing or adsorbing organic toxins, or by suppressing toxin-producing organisms. At the same time, it is true that some microorganisms in soil produce organic, growth-stimulating compounds. These substances, when taken up by plants in small amounts, may improve plant vigor.

Soils supply plants with the elements they need, mainly in the form of dissolved inorganic ions, or **mineral nutrients**. A fertile soil will provide a continuing supply of dissolved mineral nutrients in amounts and relative proportions appropriate for optimal plant growth. The nutrients include such metallic elements as potassium, calcium, iron, and copper, as well as such nonmetallic elements as nitrogen, sulfur, phosphorus, and boron. The plant takes these elements out of the soil solution and incorporates most of them into the thousands of different organic compounds that constitute plant tissue. Animals usually obtain their mineral nutrients indirectly from the soil by eating plants. Under some circumstances, animals (including humans) satisfy their craving for minerals by ingesting soil directly (Figure 1.5 and Box 1.1). Plants also take up some elements that they do not appear to use, which is fortunate as animals do require several elements that plants do not (see periodic table, Appendix B).

Interactive periodic table—look up the essential elements:
www.webelements.com

Of the 92 naturally occurring chemical elements, 17 have been shown to be **essential elements**, meaning that plants cannot grow and complete their life cycles without them. Table 1.1 lists these and several additional elements that appear to be quasi-essential (needed by some but not all plants). Essential elements used by plants in relatively large amounts are called **macronutrients;** those used in smaller amounts are known as **micronutrients**. To remember the 17 essential elements, try this mnemonic device:

C.B. HOPKiNS CaFé—
Closed Monday Morning and Night—
See You Zoon, the Mg

FIGURE 1.5 A mountain goat (in Glacier National Park) visits a natural *salt lick* where it ingests needed minerals directly from the soil. Animals normally obtain their dietary minerals indirectly from soils by eating plants. (Photo courtesy of R. Weil)

BOX 1.1 DIRT FOR DINNER?[a]

You are probably thinking, "dirt (excuse me, *soil*) for dinner? Yuck!" Of course, various birds, reptiles, and mammals are well known to consume soil at special "licks," and involuntary, inadvertent ingestion of soil by humans (especially children) is widely recognized as a pathway for exposure to environmental toxins (see Chapter 18), but most sophisticated residents of industrial countries, anthropologists and nutritionists included, find it hard to believe that anyone would *purposefully* ingest soil. Yet, a long history of documented research on the subject shows that many people do routinely eat soil, often in amounts of 20 to 100 g (up to $\frac{1}{4}$ pound) daily. **Geophagy** (deliberate "soil eating") is practiced in societies as disparate as those in Thailand, Turkey, rural Alabama, and urban Uganda (Figure 1.6). Immigrants from South Asia in the United Kingdom have brought the practice of soil eating to such cities as London and Birmingham. In fact, scientists studying the practice suggest that geophagy is a widespread and normal human behavior. Children and women (especially when pregnant) appear more likely than men to be geophagists. Poor people eat soil more commonly than the relatively well-to-do.

FIGURE 1.6 *Bars of clay soil sold for human consumption in a shop in Kampala, Uganda. (Photo courtesy of Peter W. Abrahams, University of Wales)*

People usually do not eat just any soil, but seek out a particular soil, be it the hardened clay of a termite nest, the soft, white soil in a particular riverbank, or the dark clay from a certain deep soil layer. People in different places and circumstances seek to consume different types of soils—some seek calcium-rich soils, others soil with high amounts of certain clays; still others seek red soils rich in iron. Interestingly, unlike many other animals, humans rarely appear to eat soil to obtain salt. Possible benefits from eating soil also vary and may include mineral nutrient supplementation (especially iron), detoxification of ingested poisons (by adsorption to clay—see Chapter 8), relief from stomachaches, survival in times of famine, and psychological comfort. Geophagists have been known to go to great lengths to satisfy their cravings for soil. But before you run out and add some local soil to your menu, consider the potential downsides to geophagy. Aside from the possibly difficult task of developing a taste for the stuff, the drawbacks to eating soil (especially surface soils) can include parasitic worm infection, lead poisoning, and mineral nutrient imbalances (because of adsorption of some mineral nutrients and release of others)—as well as premature tooth wear!

[a] This box is largely based on a fascinating book chapter by Abrahams (2005) and a review article by Stokes (2006).

TABLE 1.1 Elements Needed for Plant Growth and Their Sources[a]

The chemical forms most commonly taken in by plants are shown in parentheses, with the chemical symbol for the element in bold type.

Macronutrients: Used in relatively large amounts (> 0.1% of dry plant tissue)		Micronutrients: Used in relatively small amounts (< 0.1% of dry plant tissue)
Mostly from air and water	*Mostly from soil solids*	*From soil solids*
Carbon (CO_2)	*Cations:*	*Cations:*
Hydrogen (H_2O)	Calcium (Ca^{2+})	Copper (Cu^{2+})
Oxygen (O_2, H_2O)	Magnesium (Mg^{2+})	*Cobalt (Co^{2+})[b]
	Nitrogen (NH_4^+)	Iron (Fe^{2+})
	Potassium (K^+)	Manganese (Mn^{2+})
	Anions:	Nickel (Ni^{2+})
	Nitrogen (NO_3^-)	*Sodium (Na^+)[b]
	Phosphorus ($H_2PO_4^-$, HPO_4^{2-})	Zinc (Zn^{2+})
	Sulfur (SO_4^{2-})	*Anions:*
	*Silicon (H_4SiO_4, $H_3SiO_4^-$)[b]	Boron (H_3BO_3, $H_4BO_4^-$)
		Chlorine (Cl^-)
		Molybdenum (MoO_4^{2-})

[a] Many other elements are taken up from soils by plants, but are not *essential* for plant growth. Some of these (such as iodine, fluorine, barium, and strontium) do enhance the growth of certain plants, but do not appear to be absolutely required for normal growth, as are the 20 listed in this table. Still other elements (e.g., chromium, selenium, tin, and vanadium) are incorporated into plant tissues, where they may be used as essential mineral nutrients by humans and other animals, even though plants do not appear to require them. See periodic table in Appendix B.

[b] Elements marked by (*) are quasi-essential elements (*sensu*, Epstein and Bloom, 2005), required for some, but not for all, plants. Silicon is used in large amounts to play important roles in most plants, but has been proved essential only for diatoms and plants in the *Equisetaceae* family. Cobalt has been proved essential for only *Leguminoseae* when in symbiosis with nitrogen-fixing bacteria (see Section 13.10). Sodium is essential in small amounts for plants using the C_4 photosynthesis pathway (mainly tropical grasses).

The bold letters indicate the chemical elements in this phrase; finding copper (Cu) and zinc (Zn) may require some imagination.

In addition to the mineral nutrients just listed, plants may also use minute quantities of organic compounds from soils. However, uptake of these substances is not necessary for normal plant growth. The organic metabolites, enzymes, and structural compounds making up a plant's dry matter consist mainly of carbon, hydrogen, and oxygen, which the plant obtains by photosynthesis from air and water, not from the soil.

Plants *can* be grown in nutrient solutions without any soil (a method termed **hydroponics**), but then the plant-support functions of soils must be engineered into the system and maintained at a high cost of time, energy, and management. Although hydroponic production on a small scale for a few high-value plants is feasible, production of the world's food and fiber and maintenance of natural ecosystems will always depend on millions of square kilometers of productive soils.

1.3 REGULATOR OF WATER SUPPLIES

There is much concern about the quality and quantity of the water in our rivers, lakes, and underground aquifers. Governments and citizens everywhere are working to stem the pollution that threatens the value of our waters for fishing, swimming, and drinking. For progress to be made in improving water quality, we must recognize that nearly every drop of water in our rivers, lakes, estuaries, and aquifers has either traveled through the soil or flowed over its surface.[2] Imagine, for example, a heavy rain falling on the hills surrounding the river in Figure 1.7. If the soil allows the rain to soak in, some of the water may be stored in the soil and used by the trees and other plants, while some may seep slowly down through the soil layers to the groundwater, eventually entering the river over a period of months or years as base flow. If the water is contaminated, as it soaks through the upper layers of soil it is purified and cleansed by soil processes that remove many impurities and kill potential disease organisms.

---

[2] This excludes the relatively minor quantity of precipitation that falls directly into bodies of fresh surface water.

FIGURE 1.7 The condition of the soils covering these Blue Ridge foothills will greatly influence the quantity and quality of water flowing down the James River in Virginia. (Photo courtesy of R. Weil)

Contrast the preceding scenario with what would occur if the soil were so shallow or impermeable that most of the rain could not penetrate the soil, but ran off the hillsides on the soil surface, scouring surface soil and debris as it picked up speed, and entering the river rapidly and nearly all at once. The result would be a destructive flash flood of muddy water. Clearly, the nature and management of soils in a watershed will have a major influence on the purity and amount of water finding its way to aquatic systems. For those who live in a rural home, the purifying action of the soil (in a septic drain field as described in Section 6.8) is the main barrier that stands between what flushes down the toilet and the water running into the kitchen sink!

1.4 RECYCLER OF RAW MATERIALS

What would a world be like without the recycling functions performed by soils? Without reuse of nutrients, plants and animals would have run out of nourishment long ago. The world would be covered with a layer, possibly hundreds of meters high, of plant and animal wastes and corpses. Obviously, recycling must be a vital process in ecosystems, whether forests, farms, or cities. The soil system plays a pivotal role in the major geochemical cycles. Soils have the capacity to assimilate great quantities of organic waste, turning it into beneficial **humus**, converting the mineral nutrients in the wastes to forms that can be utilized by plants and animals, and returning the carbon to the atmosphere as carbon dioxide, where it again will become a part of living organisms through plant photosynthesis. Some soils can accumulate large amounts of carbon as soil organic matter, thus having a major impact on such global changes as the much-discussed *greenhouse effect* (see Sections 1.14 and 12.2).

1.5 MODIFIER OF THE ATMOSPHERE

The soil interacts in many ways with the Earth's blanket of air. In places where the soil is dry, poorly structured, and unvegetated, soil particles can be picked up by winds and contribute great quantities of dust to the atmosphere, reducing visibility, increasing human health hazards from breathing dirty air, and altering the temperature of the air

Chronology of climate
change science:
http://www.aip.org/history/
climate/timeline.htm

and the Earth. Moist, well-vegetated, and structured soil can prevent such dust-laden air. The evaporation of soil moisture is a major source of water vapor in the atmosphere, altering air temperature, composition, and weather patterns. Soils also breathe in and out. That is, they absorb oxygen and other gases such as methane, while they release gases such as carbon dioxide and nitrous oxide. These gas exchanges between the soil and the atmosphere have a significant influence on atmospheric composition and global warming.

1.6   HABITAT FOR SOIL ORGANISMS

When we speak of protecting ecosystems, most people envision a stand of old-growth forest with its abundant wildlife, or perhaps an estuary such as the Chesapeake Bay with its oyster beds and fisheries. (Perhaps, once you have read this book, you will envision a handful of soil when someone speaks of an ecosystem.) Soil is not a mere pile of broken rock and dead debris. A handful of soil may be home to *billions* of organisms, belonging to thousands of species. In even this small quantity of soil, there are likely to exist predators, prey, producers, consumers, and parasites (Figure 1.8).

Rangeland soil
communities:

http://www.blm.gov/nstc/
soil/index.html

How is it possible for such a diversity of organisms to live and interact in such a small space? One explanation is the tremendous range of niches and habitats in even a uniform-appearing soil. Some pores of the soil will be filled with water in which swim organisms such as roundworms, diatoms, and rotifers. Tiny insects and mites may be crawling about in other larger pores filled with moist air. Micro-zones of good aeration may be only millimeters from areas of **anoxic** conditions. Different areas may be enriched with decaying organic materials; some places may be highly acidic, some more basic. Temperature, too, may vary widely.

Hidden from view in the world's soils are communities of living organisms every bit as complex and intrinsically valuable as their counterparts that roam the savannas, forests, and oceans of the Earth. Soils harbor much of the Earth's genetic diversity. Soils, like air and water, are important components of the larger ecosystem. Yet only now is soil quality taking its place, with air quality and water quality, in discussions of environmental protection.

FIGURE 1.8   The soil is home to a wide variety of organisms, both relatively large and very small. Here, a relatively large predator, a centipede (shown at about actual size), hunts for its next meal—which is likely to be one of the many smaller animals that feed on dead plant debris. (Photo courtesy of R. Weil)

FIGURE 1.9 Built entirely from natural, locally occurring materials, this typical rural Tanzanian house has soil walls made by daubing mud onto a framework of sticks. Both its soil walls and grass thatch roof are shown in the process of being refurbished. For scale, note man standing left of the house. (Photo courtesy of R. Weil)

1.7 ENGINEERING MEDIUM

Modern and historic buidings made of soil: www.eartharchitecture.org/

Soil is probably the earliest and certainly one of the most widely used building materials. Nearly half the people in the world live in houses constructed from soil. Soil buildings vary from traditional African mud huts (Figure 1.9 and Plate 79) to centuries-old English homes to modern, environmentally friendly buildings built with cement-stabilized, hydraulically compacted "rammed-earth" walls (see Web link in margin).

"*Terra firma*, solid ground." We usually think of the soil as being firm and solid, a good base on which to build roads and all kinds of structures. Indeed, most structures rest on the soil, and many construction projects require excavation into the soil. Unfortunately, as can be seen in Figure 1.10, some soils are not as stable as others. Reliable construction on soils, and with soil materials, requires knowledge of the diversity of soil properties, as discussed later in this chapter. Designs for roadbeds or

FIGURE 1.10 Better knowledge of the soils on which this road was built may have allowed its engineers to develop a more stable design, thus avoiding this costly and dangerous situation. (Photo courtesy of R. Weil)

building foundations that work well in one location on one type of soil may be inadequate for another location with different soils.

Working with natural soils or excavated soil materials is not like working with concrete or steel. Properties such as bearing strength, compressibility, shear strength, and stability are much more variable and difficult to predict for soils than for manufactured building materials. Chapter 4 provides an introduction to some engineering properties of soils. Many other physical properties discussed will have direct application to engineering uses of soil. For example, Chapter 8 discusses the swelling properties of certain types of clays in soils. The engineer should be aware that when soils with swelling clays are wetted, they expand with sufficient force to crack foundations and buckle pavements. Much of the information on soil properties and soil classification discussed in later chapters will be of great value to people planning land uses that involve construction or excavation.

1.8   PEDOSPHERE AS ENVIRONMENTAL INTERFACE

The importance of soil as a natural body derives in large part from its role as an **interface** between the worlds of rock (the **lithosphere**), air (the **atmosphere**), water (the **hydrosphere**), and living things (the **biosphere**). Environments where all four of these worlds interact are often the most complex and productive on Earth. An estuary, where shallow waters meet the land and air, is an example of such an environment. Its productivity and ecological complexity far surpass those of a deep ocean trench, for example (where the hydrosphere is rather isolated), or the upper atmosphere (where rocks and water have little influence). The soil, or **pedosphere**, is another example of such an environment (Figure 1.11).

The concept of the soil as interface means different things at different scales. At the scale of kilometers, soils channel water from rain to rivers and transfer mineral elements from bed rocks to the oceans. They also remove and supply vast amounts of atmospheric gases, substantially influencing the global balance of methane and carbon dioxide. At a scale of a few meters (Figure 1.11b), soil forms the transition zone between hard rock and air, holding both liquid water and oxygen gas for use by plant roots. It transfers mineral elements from the Earth's rock crust to its vegetation. It processes or stores the organic remains of terrestrial plants and animals. At a scale of a few millimeters (Figure 1.11c), soil provides diverse microhabitats for air-breathing and aquatic organisms, channels water and nutrients to plant roots, and provides surfaces and solution vessels for thousands of biochemical reactions. Finally, at the scale of a few micrometers and smaller (less than a millionth of a meter), soil provides ordered and complex surfaces, both mineral and organic, that act as templates for chemical reactions and interact with water and solutes. Its tiniest mineral particles form micro-zones of electromagnetic charge that attract everything from bacterial cell walls to proteins to conglomerates of water molecules. As you read the entirety of this book, the frequent cross-referencing between one chapter and another will remind you of the importance of scale and interfacing to the story of soil.

1.9   SOIL AS A NATURAL BODY

You may notice that this book sometimes refers to "soil," sometimes to "the soil," sometimes to "a soil," and sometimes to "soils." These variations of the word "soil" refer to two distinct concepts—*soil* as a material or *soils* as natural bodies. *Soil* is a material composed of minerals, gases, water, organic substances, and microorganisms. Some people (usually *not* soil scientists!) also refer to this material as *dirt*, especially when it is found where it is not welcome (e.g., in your clothes or under your fingernails).

My friend, the soil; Hans Jenny interview:
http://findarticles.com/p/articles/mi_m0GER/is_1999_Spring/ai_54321347

A soil is a three-dimensional natural body in the same sense that a mountain, lake, or valley is. *The soil* is a collection of individually different soil bodies, often said to cover the land as the peel covers an orange. However, while the peel is relatively uniform around the orange, the soil is highly variable from place to place on Earth. One of the individual bodies, *a soil*, is to *the soil* as an individual tree is to the Earth's vegetation. Just as one may find sugar maples, oaks, hemlocks, and many other species of trees in a particular forest, so, too, might one find Christiana clay loams, Sunnyside sandy loams, Elkton silt loams, and other kinds of soils in a particular landscape.

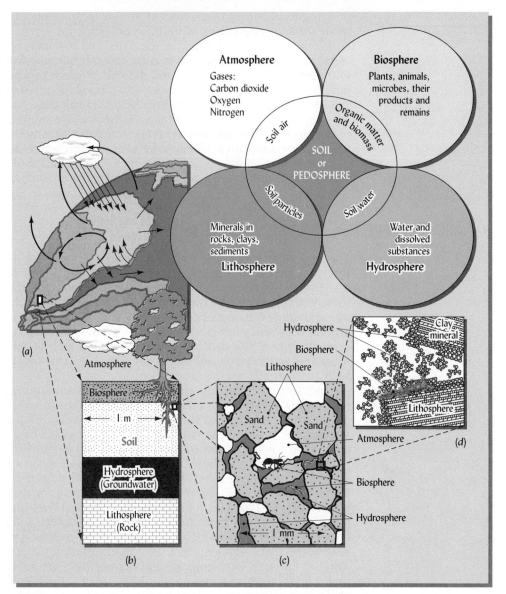

FIGURE 1.11 The pedosphere, where the worlds of rock (the lithosphere), air (the atmosphere), water (the hydrosphere), and life (the biosphere) all meet. The soil as interface can be understood at many different scales. At the kilometer scale (*a*), soil participates in global cycles of rock weathering, atmospheric gas changes, water storage and partitioning, and the life of terrestrial ecosystems. At the meter scale (*b*), soil forms a transition zone between the hard rock below and the atmosphere above—a zone through which surface water and groundwater flow and in which plants and other living organisms thrive. A thousand times smaller, at the millimeter scale (*c*), mineral particles form the skeleton of the soil that defines pore spaces, some filled with air and some with water, in which tiny creatures lead their lives. Finally, at the micro- and nanometer scales (*d*), soil minerals (lithosphere) provide charges, reactive surfaces that adsorb water and cations dissolved in water (hydrosphere), gases (atmosphere), and bacteria and complex humus macromolecules (biosphere). (Diagram courtesy of R. Weil)

Soils are natural bodies composed of soil (the material just described) *plus* roots, animals, rocks, artifacts, and so forth. By dipping a bucket into a lake you may sample some of its water. In the same way, by digging or augering a hole into a soil, you may retrieve some soil. Thus, you can take a sample of soil or water into a laboratory and analyze its contents, but you must go out into the field to study a soil or a lake.

In most places, the rock exposed at the Earth's surface has crumbled and decayed to produce a layer of unconsolidated debris overlying the hard, unweathered rock. This unconsolidated layer is called the ***regolith*** and varies in thickness from virtually nonexistent in some places (i.e., exposed bare rock) to tens of meters in other places. The regolith material, in many instances, has been transported many kilometers from the site of its initial formation and then deposited over the bedrock which it now covers. Thus, all or part of the regolith may or may not be related to the rock now found below it. Where the underlying rock has weathered in place to the degree that it is loose enough to be dug with a spade, the term **saprolite** is used (see Plate 11).

Through their biochemical and physical effects, living organisms such as bacteria, fungi, and plant roots have altered the upper part—and, in many cases, the entire depth—of the regolith. Here, at the interface between the worlds of rock, air, water, and living things, soil is born. The transformation of inorganic rock and debris into a living soil is one of nature's most fascinating displays. Although generally hidden from everyday view, the soil and regolith can often be seen in road cuts and other excavations.

A soil is the product of both destructive and creative (synthetic) processes. Weathering of rock and microbial decay of organic residues are examples of destructive processes, whereas the formation of new minerals, such as certain clays, and of new stable organic compounds are examples of synthesis. Perhaps the most striking result of synthetic processes is the formation of contrasting layers called **soil horizons.** The development of these horizons in the upper regolith is a unique characteristic of soil that sets it apart from the deeper regolith materials (Figure 1.12).

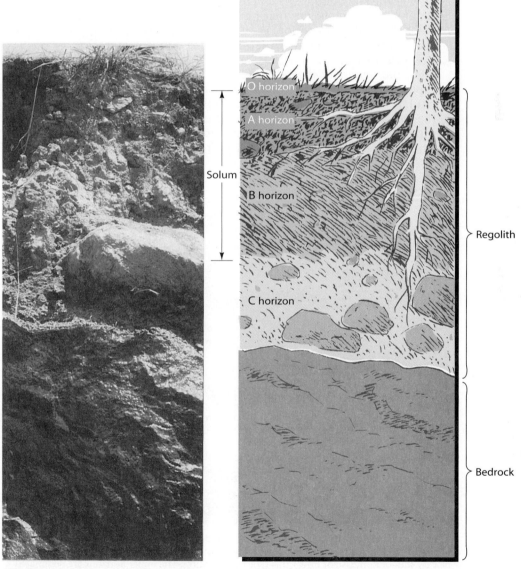

FIGURE 1.12 Relative positions of the regolith, its soil, and the underlying bedrock. Note that the soil is a part of the regolith and that the A and B horizons are part of the **solum** (from the Latin word *solum*, which means soil or land). The C horizon is the part of the regolith that underlies the solum, but may be slowly changing into soil in its upper parts. Sometimes the regolith is so thin that it has been changed entirely to soil; in such a case, soil rests directly on the bedrock. (Photo courtesy of R. Weil)

Soil scientists specializing in **pedology** (*pedologists*) study soils as natural bodies, the properties of soil horizons, and the relationships among soils within a landscape. Other soil scientists, called **edaphologists**, focus on the soil as habitat for living things, especially plants. For both types of study it is essential to examine soils at all scales and in all three dimensions (especially the vertical dimension).

1.10 THE SOIL PROFILE AND ITS LAYERS (HORIZONS)

Google "soil profile" then click on "Image results"

Soil scientists often dig a large hole, called a *soil pit*, usually several meters deep and about a meter wide, to expose soil horizons for study. The vertical section exposing a set of horizons in the wall of such a pit is termed a ***soil profile***. Road cuts and other ready-made excavations can expose soil profiles and serve as windows to the soil. In an excavation open for some time, horizons are often obscured by soil material that has been washed by rain from upper horizons to cover the exposed face of lower horizons. For this reason, horizons may be more clearly seen if a fresh face is exposed by scraping off a layer of material several centimeters thick from the pit wall. Observing how soils exposed in road cuts vary from place to place can add a fascinating new dimension to travel. Once you have learned to interpret the different horizons (see Chapter 2), soil profiles can warn you about potential problems in using the land, as well as tell you much about the environment and history of a region. For example, soils developed in a dry region will have very different horizons from those developed in a humid region.

Horizons within a soil may vary in thickness and have somewhat irregular boundaries, but generally they parallel the land surface (Figure 1.13). This alignment is expected since the differentiation of the regolith into distinct horizons is largely the result of influences, such as air, water, solar radiation, and plant material, originating at the soil–atmosphere interface. Since the weathering of the regolith occurs first at the surface and works its way down, the uppermost layers have been changed the most, while the deepest layers are most similar to the original regolith, which is referred to as

O horizons

A horizons

B horizons

C horizons (parent material)

FIGURE 1.13 This road cut in central Africa reveals soil layers or horizons that parallel the land surface. Taken together, these horizons comprise the profile of this soil, as shown in the enlarged diagram. The surface litter or O horizon may be very thin or nonexistent under this type of vegetation. The upper horizons are designated *A horizons*. They are usually higher in organic matter and darker in color than the lower horizons. Some constituents, such as iron oxides and clays, have been moved downward from the A horizons by percolating rainwater. The lower horizon, called a *B horizon*, is sometimes a zone in which clays and iron oxides have accumulated, and in which distinctive structure has formed. The presence and characteristics of the horizons in this profile distinguish this soil from the thousands of other soils in the world. (Photo courtesy of R. Weil)

the soil's *parent material*. In places where the regolith was originally rather uniform in composition, the material below the soil may have a similar composition to the parent material from which the soil formed. In other cases, the regolith material has been transported long distances by wind, water, or glaciers and deposited on top of dissimilar material. In such a case, the regolith material found below a soil may be quite different from the upper layer of regolith in which the soil formed.

In undisturbed ecosystems, especially forests, organic materials formed from fallen leaves and other plant and animal remains tend to accumulate on the surface. There they undergo varying degrees of physical and biochemical breakdown and transformation, so that layers of older, partially decomposed materials may underlie the freshly added debris. Together, these organic layers at the soil surface are designated the *O horizons*.

Soil animals and percolating water move some of these organic materials downward to intermingle with the mineral grains of the regolith. These join the decomposing remains of plant roots to form organic materials that darken the upper mineral layers. Also, because weathering tends to be most intense nearest the soil surface, in many soils the upper layers lose some of their clay or other weathering products by leaching to the horizons below. *A horizons* are the layers nearest the surface that are dominated by mineral particles but have been darkened by the accumulation of organic matter.

In some soils, intensely weathered and leached horizons that have not accumulated organic matter occur in the upper part of the profile, usually just below the A horizons. These horizons are designated *E horizons* (Figures 1.14 and 1.15).

The layers underlying the A and O horizons contain comparatively less organic matter than the horizons nearer the surface. Varying amounts of silicate clays, iron and aluminum oxides, gypsum, or calcium carbonate may accumulate in the underlying horizons. The accumulated materials may have been washed down from the horizons above, or they may have been formed in place through the weathering process. These underlying layers are referred to as *B horizons* (Figure 1.14).

Plant roots and microorganisms often extend below the B horizon, especially in humid regions, causing chemical changes in the soil water, some biochemical weathering

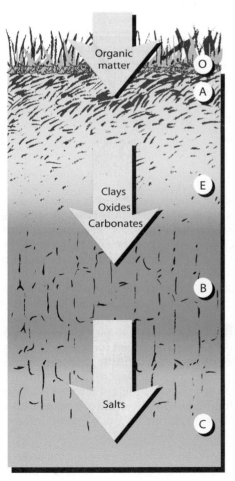

FIGURE 1.14 Horizons begin to differentiate as materials are added to the upper part of the profile and other materials are translocated to deeper zones. Under certain conditions, usually associated with forest vegetation and high rainfall, a leached E horizon forms between organic-matter-rich A and the B horizons. If sufficient rainfall occurs, soluble salts will be carried below the soil profile, perhaps all the way to the groundwater. Many soils (e.g., the soil in Figure 1.13) lack one or more of the five horizons shown here.

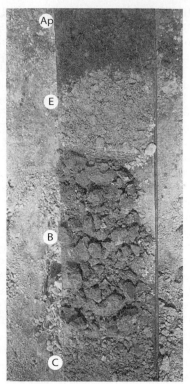

FIGURE 1.15 (*Left*) This soil profile was exposed by digging a pit about 2 meters deep in a well-developed soil (a Hapludalf) in southern Michigan. The top horizon can be easily distinguished because it has a darker color than those below it. However, some of the horizons in this profile are difficult to discern on the basis of color differences, especially in this black-and-white photo. The white string was attached to the profile to clearly demarcate some of the horizon boundaries. Then a trowel full of soil material was removed from each horizon and placed on a board, at right. It is clear from the way the soil either crumbled or held together that soil material from horizons with very similar colors may have very different physical properties. (Photos courtesy of R. Weil)

of the regolith, and the formation of *C horizons*. The C horizons are the least weathered part of the soil profile.

In some soil profiles, the component horizons are very distinct in color, with sharp boundaries that can be seen easily by even novice observers. In other soils, the color changes between horizons may be very gradual, and the boundaries more difficult to locate. However, color is only one of many properties by which one horizon may be distinguished from the horizon above or below it (see Figure 1.15). The study of soils in the field is a sensual as well as an intellectual activity. Delineation of the horizons present in a soil profile often requires a careful examination, using all the senses. In addition to seeing the colors in a profile, a soil scientist may feel, smell, and listen[3] to the soil, as well as conduct chemical tests, in order to distinguish the horizons present.

1.11 TOPSOIL AND SUBSOIL

The organically enriched A horizon at the soil surface is sometimes referred to as *topsoil*. Plowing and cultivating a soil homogenizes and modifies the upper 10 to 25 cm of the profile to form a plow layer. The plow layer may remain long after cultivation has ceased. For example, in a New England forest that has regrown on abandoned farmland, you will still be able to see the smooth boundary between the century-old plowed layer and the lighter-colored, undisturbed soil below.

In cultivated soils, the majority of plant roots can be found in the topsoil as that is the zone in which the cultivator can most readily enhance the supply of nutrients, air, and water by mixing in organic and inorganic amendments, loosening the structure, and applying irrigation. Sometimes the plow layer is removed from a soil and sold as topsoil for use at another site. This use of topsoil is especially common to provide a rooting medium suitable for lawns and shrubs around newly constructed buildings, where the original topsoil was removed or buried and the underlying soil layers were exposed during grading operations (Figure 1.16 and Plate 44 after page 112).

The soil layers that underlie the topsoil are referred to as *subsoil*. Although usually hidden from view, the subsoil horizons can greatly influence most land uses. Much of

---

[3] For example, the grinding sound emitted by wet soil rubbed between one's fingers indicates the sandy nature of the soil.

FIGURE 1.16   The large mound of material at this construction site consists of topsoil (A horizon material) carefully separated from the lower horizons and pushed aside during initial grading operations. This stockpile was then seeded with grasses to give it a protective cover. After the construction activities are complete, the stockpiled topsoil will be used in landscaping the grounds around the new building. (Photo courtesy of R. Weil)

the water needed by plants is stored in the subsoil. Many subsoils also supply important quantities of certain plant nutrients. The properties of the topsoil are commonly far more conducive to plant growth than those of the subsoil. In cultivated soils, therefore, productivity is often correlated with the thickness of the topsoil layer. Subsoil layers that are too dense, acidic, or wet can impede root growth. It may be extremely difficult and expensive to physically or chemically modify the subsoil.

Many of the chemical, biological, and physical processes that occur in the upper soil layers also take place to some degree in the C horizons, which may extend deep into the underlying saprolite or other regolith material. Traditionally, the lower boundary of the soil has been considered to occur at the greatest rooting depth of the natural vegetation, but soil scientists are increasingly studying layers below this in order to understand ecological processes such as groundwater pollution, parent material weathering, and geochemical cycles (Box 1.2).

1.12   SOIL: THE INTERFACE OF AIR, MINERALS, WATER, AND LIFE

We stated that where the regolith meets the atmosphere, the worlds of air, rock, water, and living things are intermingled. In fact, the four major components of soil are air, water, mineral matter, and organic matter. The relative proportions of these four components greatly influence the behavior and productivity of soils. In a soil, the four components are mixed in complex patterns; however, the proportion of soil volume occupied by each component can be represented in a simple pie chart. Figure 1.18 shows the approximate proportions (by volume) of the components found in a loam surface soil in good condition for plant growth. Although a handful of soil may at first seem to be a solid thing, it should be noted that only about half the soil volume consists of solid material (mineral and organic); the other half consists of pore spaces filled with air or water. Of the solid material, typically most is mineral matter derived from the rocks of the Earth's crust. Only about 5% of the *volume* in this ideal soil consists of organic matter. However, the influence of the organic component on soil properties is often far greater than its small proportion would suggest. Since it is far less dense than mineral matter, the organic matter accounts for only about 2% of the *weight* of this soil.

The spaces between the particles of solid material are just as important to the nature of a soil as are the solids themselves. It is in these pore spaces that air and water circulate,

BOX 1.2 USING INFORMATION FROM THE ENTIRE SOIL PROFILE

Soils are three-dimensional bodies that carry out important ecosystem processes at all depths in their profiles. Depending on the particular application, the information needed to make proper land management decisions may come from soil layers as shallow as the upper 1 or 2 centimeters or as deep as the lowest layers of saprolite (Figure 1.17).

For example, the upper few centimeters of soil often hold the keys to plant growth and biological diversity, as well as to certain hydrologic processes. Here, at the interface between the soil and the atmosphere, living things are most numerous and diverse. Forest trees largely depend for nutrient uptake on a dense mat of fine roots growing in this zone. The physical condition of this thin surface layer may also determine whether rain will soak in or run downhill on the land surface. Certain pollutants, such as lead from highway exhaust, are also concentrated in this zone. For many types of soil investigations it will be necessary to sample the upper few centimeters separately so that important conditions are not overlooked.

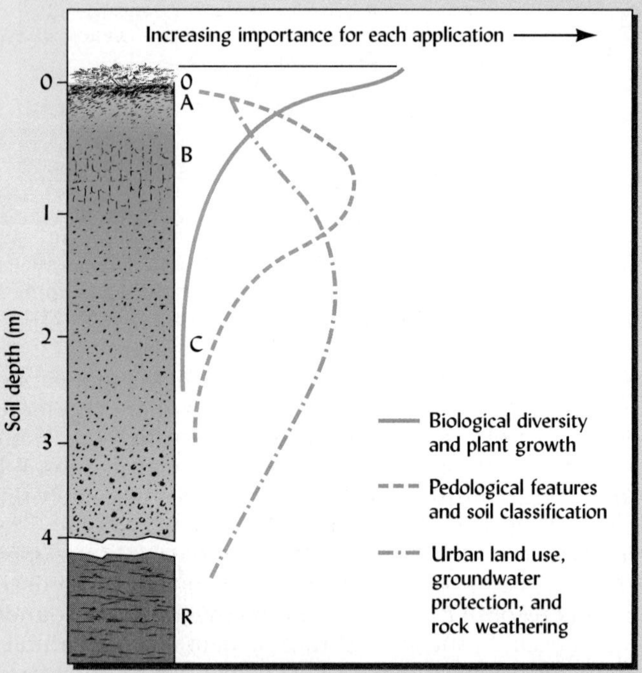

On the other hand, it is equally important not to confine one's attention to the easily accessible "topsoil," for many soil properties are to be discovered only in the deeper layers. Plant-growth problems are often related to inhospitable conditions in the B or C horizons that restrict the penetration of roots. Similarly, the great volume of these deeper layers may control the amount of plant-available water held by a soil. For the purposes of recognizing or mapping different types of soils, the properties of the B horizons are often paramount. Not only is this the zone of major accumulations of minerals and clays, but the layers nearer the soil surface are too quickly altered by management and soil erosion to be a reliable source of information for the classification of soils.

In deeply weathered regoliths, the lower C horizons and saprolite play important roles. These layers, generally at depths below 1 or 2 meters, and often as deep as 5 to 10 meters, greatly affect the suitability of soils for most urban uses that involve construction or excavation. The proper functioning of on-site sewage disposal systems and the stability of building foundations are often determined

FIGURE 1.17 *Information important to different soil functions and applications is most likely to be obtained by studying different layers of the soil profile.*

by regolith properties at these depths. Likewise, processes that control the movement of pollutants to groundwater or the weathering of geologic materials may occur at depths of many meters. These deep layers also have major ecological influences because, although the intensity of biological activity and plant rooting may be quite low, the total impact can be great as a result of the enormous volume of soil that may be involved. This is especially true of forest systems in warm climates.

roots grow, and microscopic creatures live. Plant roots need both air and water. In an optimum condition for most plants, the pore space will be divided roughly equally among the two, with 25% of the soil volume consisting of water and 25% consisting of air. If there is much more water than this, the soil will be waterlogged. If much less water is present, plants will suffer from drought. The relative proportions of water and air in a soil typically fluctuate greatly as water is added or lost. Soils with much more than 50% of their volume in solids are likely to be too compacted for good plant growth. Compared to surface soil layers, subsoils tend to contain less organic matter, less total pore space, and a larger proportion of small pores (*micropores*), which tend to be filled with water rather than with air.

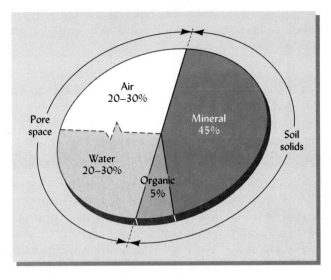

FIGURE 1.18 Volume composition of a loam surface soil when conditions are good for plant growth. The broken line between water and air indicates that the proportions of these two components fluctuate as the soil becomes wetter or drier. Nonetheless, a nearly equal proportion of air and water is generally ideal for plant growth.

1.13 MINERAL (INORGANIC) CONSTITUENTS OF SOILS

Except in organic soils, most of the soil's solid framework consists of **mineral**[4] particles. The larger soil particles (stones, gravel, and coarse sands) are generally rock fragments consisting of several different minerals. Smaller particles tend to be made of a single mineral.

Excluding, for the moment, the larger rock fragments such as stones and gravel, soil particles range in size over four orders of magnitude: from 2.0 millimeters (mm) to smaller than 0.0002 mm in diameter (Table 1.2). **Sand** particles are large enough (2.0 to 0.05 mm) to be seen by the naked eye and feel gritty when rubbed between the fingers. Sand particles do not adhere to one another; therefore, sands do not feel sticky. **Silt** particles (0.05 to 0.002 mm) are too small to see without a microscope or to feel individually, so silt feels smooth but not sticky, even when wet. **Clay** particles are the smallest mineral particles (<0.002 mm) and adhere together to form a sticky mass when wet and hard clods when dry. The smaller particles (<0.001 mm) of clay (and similar-sized organic particles) have **colloidal**[5] properties and can be seen only with the aid of an electron microscope. Because of their extremely small size, colloidal particles possess a tremendous amount of surface area per unit of mass. Since the surfaces of soil colloids (both mineral and organic) exhibit

Electron micrographs and other clay images:
http://www.minersoc.org/pages/gallery/claypix/index.html

---

[4] The word *mineral* is used in soil science in three ways: (1) as a general adjective to describe inorganic materials derived from rocks; (2) as a specific noun to refer to distinct minerals found in nature such as quartz and feldspars (see Chapter 2 for detailed discussions of soil-forming minerals and the rocks in which they are found); and (3) as an adjective to describe chemical elements, such as nitrogen and phosphorus, in their inorganic state in contrast to their occurrence as part of organic compounds.

[5] Colloidal systems are two-phase systems in which very small particles of one substance are dispersed in a medium of a different substance. Clay and organic soil particles smaller than about 0.001 mm (1 micrometer, μm) in diameter are generally considered to be colloidal in size. Milk and blood are other examples of colloidal systems in which very small solid particles are dispersed in a liquid medium.

TABLE 1.2   Some General Properties of the Three Major Size Classes of Inorganic Soil Particles

	Property	Sand	Silt	Clay
1.	Range of particle diameters in mm	2.0–0.05	0.05–0.002	Smaller than 0.002
2.	Means of observation	Naked eye	Microscope	Electron microscope
3.	Dominant minerals	Primary	Primary and secondary	Secondary
4.	Attraction of particles for each other	Low	Medium	High
5.	Attraction of particles for water	Low	Medium	High
6.	Ability to hold chemicals and nutrients in plant-available form	Very low	Low	High
7.	Consistency when wet	Loose, gritty	Smooth	Sticky, malleable
8.	Consistency when dry	Very loose, gritty	Powdery, some clods	Hard clods

electromagnetic charges that attract positive and negative ions as well as water, this fraction of the soil is the seat of most of the soil's chemical and physical activity (see Chapter 8).

Soil Texture

The proportion of particles in these different size ranges is described by **soil texture.** Terms such as *sandy loam, silty clay,* and *clay loam* are used to identify the soil texture.

BOX 1.3   OBSERVING SOILS IN DAILY LIFE

Your study of soils can be enriched if you make an effort to become aware of the many daily encounters with soils and their influences that go unnoticed by most people. When you dig a hole to plant a tree or set a fence post, note the different layers encountered, and note how the soil from each layer looks and feels. If you pass a construction site, take a moment to observe the horizons exposed by the excavations. An airplane trip is a great opportunity to observe how soils vary across landscapes and climatic zones. If you are flying during daylight hours, ask for a window seat. Look for the shapes of individual soils in plowed fields if you are flying in spring or fall (Figure 1.19).

Soils can give you clues to understanding the natural processes going on around you. Down by the stream, use a magnifying glass to examine the sand deposited on the banks or bottom. It may contain minerals not found in local rocks and soils, but originating many kilometers upstream. When you wash your car, see if the mud clinging to the tires and fenders is of a different color or consistency than the soils near your home. Does the "dirt" on your car tell you where you have been driving? Forensic investigators have been known to consult with soil scientists to locate crime victims or establish guilt by matching soil clinging to shoes, tires, or tools with the soils at a crime scene.

Other examples of soil hints can be found even closer to home. The next time you bring home celery or leaf lettuce from the supermarket, look carefully for bits of soil clinging to the bottom of the stalk or leaves (Figure 1.20). Rub the soil between your thumb and fingers. Smooth, very black soil may indicate that the lettuce was grown in mucky soils, such as those in New York State or southern Florida. Light brown, smooth-feeling soil with only a very fine grittiness is more typical of California-grown produce, while light-colored, gritty soil is common on produce from the southern Georgia–northern Florida vegetable-growing region. In a bag of dry pinto beans, you may come across a few lumps of soil that escaped removal in the cleaning process because of being the same size as the beans. Often this soil is dark-colored and very sticky, coming from the "thumb" area of Michigan, where a large portion of the U.S. dry bean crop is grown.

Opportunities to observe soils in daily life range from the remote and large-scale to the close-up and intimate. As you learn more about soils, you will undoubtedly be able to see more examples of their influence in your surroundings.

FIGURE 1.19   *The light- and dark-colored soil bodies, as seen from an airliner flying over central Texas, reflect differences in drainage and topography in the landscape. (Photo courtesy of R. Weil)*

FIGURE 1.20   *Although this celery was purchased in a Virginia grocery store in early fall, the black, mucky soil clinging to the base of the stalk indicates that it was grown on organic soils, probably in New York State. (Photo courtesy of R. Weil)*

Texture has a profound influence on many soil properties, and it affects the suitability of a soil for most uses. To understand the degree to which soil properties can be influenced by texture, imagine sunbathing first on a sandy beach (loose sand), and then on a clayey beach (sticky mud). The difference in these two experiences would be due largely to the properties described in Table 1.2.

To anticipate the effect of clay on the way a soil will behave, it is necessary to know the *kinds* of clays as well as the *amount* present. As home builders and highway engineers know all too well, certain clayey soils, such as those high in smectite clays, make very unstable material on which to build because they swell when wet and shrink when dry. This shrink-and-swell action can easily crack foundations and cause retaining walls to collapse. These clays also become extremely sticky and difficult to work with when they are wet. Other types of clays, formed under different conditions, can be very stable and easy to work with. Learning about the different types of clay minerals will help us understand many of the physical and chemical differences among soils in various parts of the world (see Box 1.3).

Soil Minerals

Minerals that have persisted with little change in composition since they were extruded in molten lava (e.g., quartz, micas, and feldspars) are known as **primary minerals.** They are prominent in the sand and silt fractions of soils. Other minerals, such as silicate clays and iron oxides, were formed by the breakdown and weathering of less resistant minerals as soil formation progressed. These minerals are called **secondary minerals** and tend to dominate the clay and, in some cases, silt fractions.

Most of the chemical elements essential for plant growth originate in the soil minerals and occur mainly as components of mineral crystalline structures. However, a small but important portion of these nutrient elements are found as charged ions on the surfaces of colloidal clay and organic matter particles. Thanks to several critical mechanisms (see Section 1.18), plant roots are able to use these surface-held nutrient ions and transfer them from the lithosphere to the biosphere.

Soil Structure

Sand, silt, and clay particles can be thought of as the building blocks from which soil is constructed. The way these building blocks are put together is called **soil structure.** The particles may remain relatively independent of each other, but more commonly they are associated together in aggregates of different-size particles. These aggregates may take the form of roundish granules, cubelike blocks, flat plates, or other shapes. Soil structure (the way particles are arranged together) is just as important as soil texture (the relative amounts of different sizes of particles) in governing how water and air move in soils. Both structure and texture fundamentally influence many processes in soil, including the growth of plant roots.

1.14 SOIL ORGANIC MATTER

Soil organic matter consists of a wide range of organic (carbonaceous) substances, including living organisms (the soil *biomass*), carbonaceous remains of organisms that once occupied the soil, and organic compounds produced by current and past metabolism in the soil. The remains of plants, animals, and microorganisms are continuously broken down in the soil, and new substances are synthesized by other microorganisms. Over time, organic matter is lost from the soil as carbon dioxide produced by microbial respiration. Because of such loss, repeated additions of new plant and/or animal residues are necessary to maintain soil organic matter.

Under conditions that favor plant production more than microbial decay, large quantities of atmospheric carbon dioxide used by plants in photosynthesis are sequestered in the abundant plant tissues that eventually become part of the soil organic matter. Since carbon dioxide is a major cause of the greenhouse effect, which is believed to be warming Earth's climate, the balance between accumulation of soil organic matter and its loss through microbial respiration has global implications. In fact, more carbon is stored in the world's soils than in the world's plant biomass and atmosphere combined.

FIGURE 1.21 Abundant organic matter, including plant roots, helps create physical conditions favorable for the growth of higher plants as well as microbes (*left*). In contrast, soils low in organic matter, especially if they are high in silt and clay, are often cloddy (*right*) and not suitable for optimum plant growth.

Even so, organic matter comprises only a small fraction of the mass of a typical soil. By weight, typical well-drained mineral surface soils contain from 1 to 6% organic matter. The organic matter content of subsoils is even smaller. However, the influence of organic matter on soil properties, and consequently on plant growth, is far greater than the low percentage would indicate (see also Chapter 12).

Organic matter binds mineral particles into a granular soil structure that is largely responsible for the loose, easily managed condition of productive soils. Part of the soil organic matter that is especially effective in stabilizing these granules consists of certain gluelike substances produced by various soil organisms, including plant roots (Figure 1.21).

Organic matter also increases the amount of water a soil can hold and the proportion of water available for plant growth (Figure 1.22). In addition, organic matter is a major source of the plant nutrients phosphorus and sulfur and the primary source of nitrogen for most plants. As soil organic matter decays, these nutrient elements, which are present in organic combinations, are released as soluble ions that can be taken up by plant roots. Finally, organic matter, including plant and animal residues, is the main food that supplies carbon and energy to soil organisms. Without it, biochemical activity so essential for ecosystem functioning would come to a near standstill.

Humus, usually black or brown in color, is a collection of very complex organic compounds that accumulate in soil because they are relatively resistant to decay. Just as clay is

FIGURE 1.22 Soils higher in organic matter are darker in color and have greater water-holding capacities than soils low in organic matter. The soil in each container has the same texture, but the one on the right has been depleted of much of its organic matter. The same amount of water was applied to each container. As the photo shows, the depth of water penetration was less in the high organic matter soil (*left*) because of its greater water-holding capacity. It required a greater volume of the low organic matter soil to hold the same amount of water.

the colloidal fraction of soil mineral matter, so humus is the colloidal fraction of soil organic matter. Because of their charged surfaces, both humus and clay act as contact bridges between larger soil particles; thus, both play an important role in the formation of soil structure. The surface charges of humus, like those of clay, attract and hold both nutrient ions and water molecules. However, gram for gram, the capacity of humus to hold nutrients and water is far greater than that of clay. Unlike clay, humus contains certain components that can have a hormone-like stimulatory effect on plants. All in all, small amounts of humus may remarkably increase the soil's capacity to promote plant growth.

1.15 SOIL WATER: A DYNAMIC SOLUTION

Water is of vital importance in the ecological functioning of soils. The presence of water in soils is essential for the survival and growth of plants and other soil organisms. The soil moisture regime, often reflective of climatic factors, is a major determinant of the productivity of terrestrial ecosystems, including agricultural systems. Movement of water, and substances dissolved in it, through the soil profile is of great consequence to the quality and quantity of local and regional water resources. Water moving through the regolith is also a major driving force in soil formation (see Box 2.1).

Two main factors help explain why **soil water** is different from our everyday concept of, say, drinking water in a glass:

1. Water is held within soil pores with varying degrees of tenacity depending on the amount of water present and the size of the pores. The attraction between water and the surfaces of soil particles greatly restricts the ability of water to flow as it would flow in a drinking glass.

2. Because soil water is never pure water, but contains hundreds of dissolved organic and inorganic substances, it may be more accurately called the *soil solution*. An important function of the soil solution is to serve as a constantly replenished, dilute nutrient solution bringing dissolved nutrient elements (e.g., calcium, potassium, nitrogen, and phosphorus) to plant roots.

When the soil moisture content is optimal for plant growth (Figure 1.18), the water in the large- and intermediate-sized pores can move about in the soil and can easily be used by plants. As the plant grows, however, its roots remove water from the largest pores first. Soon the larger pores hold only air, and the remaining water is found only in the intermediate- and smallest-sized pores. The water in the intermediate-sized pores can still move toward plant roots and be taken up by them. However, the water in the smallest pores is so close to solid particles that it is strongly attracted to and held on the particle surfaces. This water may be so strongly held that plant roots cannot pull it away. Consequently, not all soil water is *available* to plants. Depending on the soil, one-sixth to one-half of the water may remain in the soil after plants have wilted or died for lack of moisture.

Soil Solution

The soil solution contains small but significant quantities of soluble organic and inorganic substances, including the plant nutrients listed in Table 1.1. The soil solids, particularly the fine organic and inorganic colloidal particles (clay and humus), release nutrient elements to the soil solution from which they are taken up by plant roots. The soil solution tends to resist changes in its composition even when compounds are added or removed from the soil. This ability to resist change is termed the soil **buffering capacity** and is dependent on many chemical and biological reactions, including the attraction and release of substances by colloidal particles (see Chapter 8).

Many chemical and biological reactions are dependent on the relative levels of hydrogen ions (H^+) and hydroxyl ions (OH^-) in the soil solution, which are commonly determined by measuring the *pH* of the soil. The pH is a logarithmic scale used to express the degree of soil acidity or alkalinity (Figure 1.23). The pH is considered a master variable of soil chemistry and is of great significance to nearly all aspects of soil science. Therefore, carefully study Figures 1.23 and 1.24 to understand the meaning of pH and the range of pH levels commonly found in various soils.

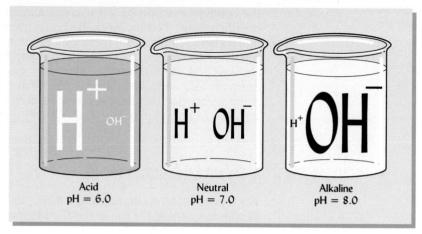

FIGURE 1.23 Diagrammatic representation of acidity, neutrality, and alkalinity. At neutrality the H^+ and OH^- ions of a solution are balanced, their respective numbers being the same (pH 7). At pH 6, the H^+ ions are dominant, being 10 times greater, whereas the OH^- ions have decreased proportionately, being only one-tenth as numerous. The solution therefore is acid at pH 6, there being 100 times more H^+ ions than OH^- ions present. At pH 8, the exact reverse is true; the OH^- ions are 100 times more numerous than the H^+ ions. Hence, the pH 8 solution is alkaline. This mutually inverse relationship must always be kept in mind when pH data are used.

1.16   SOIL AIR: A CHANGING MIXTURE OF GASES

Approximately half of the volume of the soil consists of pore spaces of varying sizes (refer to Figure 1.18), which are filled with either water or air. When water enters the soil, it displaces air from some of the pores; the air content of a soil is therefore inversely related to its water content. If we think of the network of soil pores as the ventilation system of the soil connecting airspaces to the atmosphere, we can understand that when the smaller pores are filled with water the ventilation system becomes clogged. Think how stuffy the air would become if the ventilation ducts of a classroom became clogged. Because oxygen could not enter the room, nor carbon dioxide leave it, the air in the room would soon become depleted of oxygen and enriched in carbon dioxide and water vapor by the respiration (breathing) of the people in it. In an air-filled soil pore surrounded by water-filled smaller pores, the metabolic activities of plant roots and microorganisms have a similar effect.

Therefore, soil air differs from atmospheric air in several respects. First, the composition of soil air varies greatly from place to place in the soil. In local pockets, some gases are consumed by plant roots and by microbial reactions, and others are released, thereby greatly modifying the composition of the soil air. Second, soil air generally has a higher moisture content than the atmosphere; the relative humidity of soil air approaches 100% unless the soil is very dry. Third, the content of carbon dioxide (CO_2) is usually much

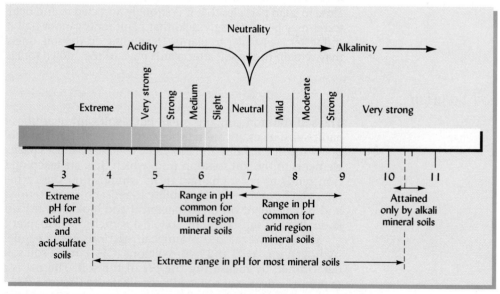

FIGURE 1.24   Extreme range in pH for most mineral soils and the ranges commonly found in humid region and arid region soils. Also indicated are the maximum alkalinity for alkali soils and the minimum pH likely to be encountered in very acid peat soils.

higher, and that of oxygen (O_2) lower, than contents of these gases found in the atmosphere. Carbon dioxide in soil air is often several hundred times more concentrated than the 0.035% commonly found in the atmosphere. Oxygen decreases accordingly and, in extreme cases, may be 5 to 10%, or even less, compared to about 20% for atmospheric air.

The amount and composition of air in a soil are determined to a large degree by the water content of the soil. The air occupies those soil pores not filled with water. As the soil drains from a heavy rain or irrigation, large pores are the first to be filled with air, followed by medium-sized pores, and finally the small pores, as water is removed by evaporation and plant use. This explains the tendency for soils with a high proportion of tiny pores to be poorly aerated. In such soils, water dominates, and the soil air content is low, as is the rate of diffusion of the air into and out of the soil from the atmosphere. The result is high levels of CO_2 and low levels of O_2, unsatisfactory conditions for the growth of most plants. In extreme cases, lack of oxygen both in the soil air and dissolved in the soil water may fundamentally alter the chemical reactions that take place in the soil solution. This is of particular importance to understanding the functions of wetland soils.

1.17 INTERACTION OF FOUR COMPONENTS TO SUPPLY PLANT NUTRIENTS

As you read our discussion of each of the four major soil components, you may have noticed that the impact of one component on soil properties is seldom expressed independently from that of the others. Rather, the four components interact with each other to determine the nature of a soil. Thus, soil moisture, which directly meets the needs of plants for water, simultaneously controls much of the air and nutrient supply to the plant roots. The mineral particles, especially the finest ones, attract soil water, thus determining its movement and availability to plants. Likewise, organic matter, because of its physical binding power, influences the arrangement of the mineral particles into clusters and, in so doing, increases the number of large soil pores, thereby influencing the water and air relationships.

Essential Element Availability

Cation exchange in action:
Univ. of New Engl.
http://www.une.edu.au/agss/
ozsoils/images/SSCATXCH.dcr

Perhaps the most important interactive process involving the four soil components is the provision of essential nutrient elements to plants. Plants absorb essential nutrients, along with water, directly from one of these components: the soil solution. However, the amount of essential nutrients in the soil solution at any one time is sufficient to supply the needs of growing vegetation for only a few hours or days. Consequently, the soil solution nutrient levels must be constantly replenished from the inorganic or organic parts of the soil and from fertilizers or manures added to agricultural soils.

Fortunately, relatively large quantities of these nutrients are associated with both inorganic and organic soil solids. By a series of chemical and biochemical processes, nutrients are released from these solid forms to replenish those in the soil solution. For example, the tiniest colloidal-sized particles—both clay and humus—exhibit negative and positive charges. These charges tend to attract or **adsorb**[6] oppositely charged ions from the soil solution and hold them as **exchangeable ions**. Through ion exchange, elements such as Ca^{2+} and K^+ are released from this state of electrostatic adsorption on colloidal surfaces and escape into the soil solution. In the example below, a H^+ ion in the soil solution is shown to exchange places with an adsorbed K^+ ion on the colloidal surface.

$$\boxed{\text{colloid}} \; K^+ + H^+ \text{ion} \longrightarrow \boxed{\text{colloid}} \; H^+ + K^+ \text{ion}$$

$$\text{Adsorbed} \qquad \text{Soil} \qquad\qquad \text{Adsorbed} \qquad \text{Soil}$$
$$\text{solution} \qquad\qquad\qquad\qquad \text{solution}$$

(1.1)

The K^+ ion thus released can be readily taken up (absorbed) by plants. Some scientists consider that this ion exchange process is among the most important of chemical reactions in nature.

---

[6] *Adsorption* refers to the attraction of ions to the surface of particles, in contrast to *absorption*, the process by which ions are taken *into* plant roots. The adsorbed ions are exchangeable with ions in the soil solution.

TABLE 1.3 Quantities of Six Essential Elements Found in Upper 15 cm of Representative Soils in Temperate Regions

Essential element	Humid region soil			Arid region soil		
	In solid framework, kg/ha	Exchangeable, kg/ha	In soil solution, kg/ha	In solid framework, kg/ha	Exchangeable, kg/ha	In soil solution, kg/ha
Ca	8,000	2,250	60–120	20,000	5,625	140–280
Mg	6,000	450	10–20	14,000	900	25–40
K	38,000	190	10–30	45,000	250	15–40
P	900	—	0.05–0.15	1,600	—	0.1–0.2
S	700	—	2–10	1,800	—	6–30
N	3,500	—	7–25	2,500	—	5–20

Nutrient uptake and recycling by plants:
http://fao.org/ag/magazine/spot3.html

Nutrient ions are also released to the soil solution as soil microorganisms decompose organic tissues. Plant roots can readily absorb all of these nutrients from the soil solution, provided there is enough O_2 in the soil air to support root metabolism.

Most soils contain large amounts of plant nutrients relative to the annual needs of growing vegetation. However, the bulk of most nutrient elements is held in the structural framework of primary and secondary minerals and organic matter. Only a small fraction of the nutrient content of a soil is present in forms that are readily available to plants. Table 1.3 will give you some idea of the quantities of various essential elements present in different forms in typical soils of humid and arid regions.

Figure 1.25 illustrates how the two solid soil components interact with the liquid component (soil solution) to provide essential elements to plants. Plant roots do not ingest soil

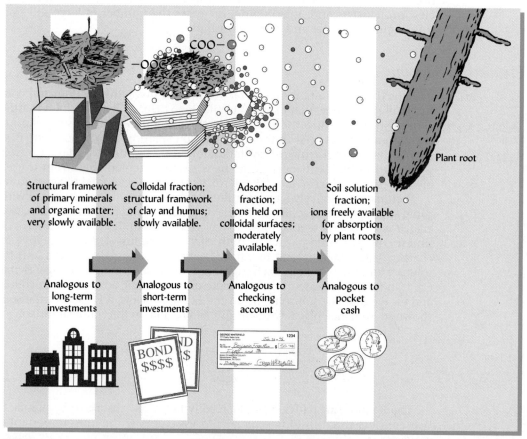

FIGURE 1.25 Nutrient elements exist in soils in various forms characterized by different accessibility to plant roots. The bulk of the nutrients is locked up in the structural framework of primary minerals, organic matter, clay, and humus. A smaller proportion of each nutrient is adsorbed in a swarm of ions near the surfaces of soil colloids (clay and organic matter). From the swarm of adsorbed ions, a still smaller amount is released into the bulk soil solution, where uptake by plant roots can take place. (Diagram courtesy of R. Weil)

particles, no matter how fine, but are able to absorb only nutrients that are dissolved in the soil solution. Because elements in the coarser soil framework of the soil are only slowly released into the soil solution over long periods of time, the bulk of most nutrients in a soil is not readily available for plant use. Nutrient elements in the framework of colloid particles are somewhat more readily available to plants, as these particles break down much faster because of their greater surface area. Thus, the structural framework is the major storehouse and, to some extent, a significant source of essential elements in many soils.

The distribution of nutrients among the various components of a fertile soil, as illustrated in Figure 1.25, may be likened to the distribution of financial assets in the portfolio of a wealthy individual. In such an analogy, nutrients readily available for plant use would be analogous to cash in the individual's pocket. A millionaire would likely keep most of his or her assets in long-term investments such as real estate or bonds (the coarse fraction solid framework), while investing a smaller amount in short-term stocks and bonds (colloidal framework). For more immediate use, an even smaller amount might be kept in a checking account (exchangeable nutrients), while a tiny fraction of the overall wealth might be carried to spend as currency and coins (nutrients in the soil solution). As the cash is used up, the supply is replenished by making a withdrawal from the checking account. The checking account, in turn, is replenished occasionally by the sale of long-term investments. It is possible for a wealthy person to run short of cash even though he or she may own a great deal of valuable real estate. In an analogous way, plants may use up the readily available supply of a nutrient even though the total supply of that nutrient in the soil is very large. Luckily, in a fertile soil, the process described in Figure 1.25 can help replenish the soil solution as quickly as plant roots remove essential elements.

1.18 NUTRIENT UPTAKE BY PLANT ROOTS

To be taken up by a plant, a nutrient element must be in a soluble form and must be located *at the root surface*. Often, parts of a root are in such intimate contact with soil particles (see Figure 1.26) that a direct exchange may take place between nutrient ions

FIGURE 1.26 Scanning electron micrograph (SEM) cross section of a barley root growing in field soil. Note the intimate contact between the root and the soil, made more so by the long, thin root hairs that permeate the nearby soil and bind it to the root. The root itself is about 0.3 mm in diameter. (Photo courtesy of Margaret McCully, CSIRO, Plant Industry, Canberra, Australia)

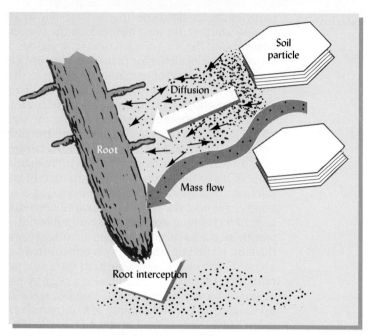

FIGURE 1.27 Three principal mechanisms by which nutrient ions dissolved in the soil solution come into contact with plant roots. All three mechanisms may operate simultaneously, but one mechanism or another may be most important for a particular nutrient. For example, in the case of calcium, which is generally plentiful in the soil solution, mass flow alone can usually bring sufficient amounts to the root surface. However, in the case of phosphorus, diffusion is needed to supplement mass flow because the soil solution is very low in this element in comparison to the amounts needed by plants. (Diagram courtesy of R. Weil)

adsorbed on the surface of soil colloids and H$^+$ ions from the surface of root cell walls. In any case, the supply of nutrients in contact with the root will soon be depleted. This fact raises the question of how a root can obtain additional supplies once the nutrient ions at the root surface have all been taken up into the root. There are three basic mechanisms by which the concentration of nutrient ions at the root surface is maintained (Figure 1.27).

First, **root interception** comes into play as roots continually grow into new, undepleted soil. For the most part, however, nutrient ions must travel some distance in the soil solution to reach the root surface. This movement can take place by **mass flow**, as when dissolved nutrients are carried along with the flowing soil water toward a root that is actively drawing water from the soil. In this type of movement, the nutrient ions are somewhat analogous to leaves floating down a stream. On the other hand, plants can continue to take up nutrients even at night, when water is only slowly absorbed into the roots. Nutrient ions continually move by **diffusion** from areas of greater concentration toward the nutrient-depleted areas of lower concentration around the root surface.

In the diffusion process, the random movements of ions in all directions causes a *net* movement from areas of high concentration to areas of lower concentrations, independent of any mass flow of the water in which the ions are dissolved. Factors such as soil compaction, cold temperatures, and low soil moisture content, which reduce root interception, mass flow, or diffusion, can result in poor nutrient uptake by plants even in soils with adequate supplies of soluble nutrients. Furthermore, the availability of nutrients for uptake can also be negatively or positively influenced by the activities of microorganisms that thrive in the immediate vicinity of roots. Maintaining the supply of available nutrients at the plant root surface is thus a process that involves complex interactions among different soil components.

It should be noted that the plant membrane separating the inside of the root cell from the soil solution is permeable to dissolved ions only under special circumstances. Plants do not merely take up, by mass flow, those nutrients that happen to be in the water that roots are removing from the soil. Nor do dissolved nutrient ions brought to the root's outer surface by mass flow or diffusion cross the root cell membrane and enter the root passively by diffusion. On the contrary, a nutrient is normally taken up into the plant root cell only by reacting with a specific chemical binding site on a large protein carrier molecule. These proteins form a hydrophilic channel across an otherwise hydrophobic lipid (fatty) membrane. Energy from metabolism in the root cell is used to activate this carrier protein so that it will pass the nutrient ion across the cell membrane and release it into the cell interior. This carrier mechanism allows the plant to accumulate concentrations of a nutrient inside the root cell that far exceed that nutrient's concentration in the soil solution. Because different nutrients are taken up by specific types of carrier molecules, the plant is able to exert some control over how much and in what relative proportions essential elements are taken up.

Since nutrient uptake is an active metabolic process, conditions that inhibit root metabolism may also inhibit nutrient uptake. Examples of such conditions include excessive soil water content or soil compaction resulting in poor soil aeration, excessively hot or cold soil temperatures, and aboveground conditions that result in low translocation of sugars to plant roots. We can see that plant nutrition involves biological, physical, and chemical processes and interactions among many different components of soils and the environment.

1.19 SOIL QUALITY, DEGRADATION, AND RESILIENCE

Global Population counter: http://www.ibiblio.org/lunarbin/worldpop

Soil is a basic resource underpinning all terrestrial ecosystems. Managed carefully, soils are a *reusable* resource, but in the scale of human lifetimes they cannot be considered a *renewable* resource. As we shall see in the next chapter, most soil profiles are thousands of years in the making. In all regions of the world, human activities are destroying some soils far faster than nature can rebuild them. As mentioned in the opening paragraphs of this chapter, growing numbers of people are demanding more and more from the Earth's fixed amount of land. Nearly all of the soils best suited for growing crops are already being farmed. Therefore, as each year brings millions more people to feed, the amount of cropland per person continuously declines. In addition, many of the world's major cities were originally located where excellent soils supported thriving agricultural communities, so now much of the very best farmland is being lost to suburban development as these cities thoughtlessly expand.

Finding more land on which to grow food is not easy. Most additional land brought under cultivation comes at the cost of clearing natural forests, savannas, and grasslands. Images of the Earth made from orbiting satellites show the resulting decline in land covered by forests and other natural ecosystems. Thus, as the human population struggles to feed itself, wildlife populations are deprived of vital habitat, and overall biodiversity suffers (Figure 1.28). Efforts to reduce and even reverse human population growth must be accelerated if our grandchildren are to inherit a livable world. In the meantime, if there is to be space for both people and wildlife, the best of our existing farmland soils will require improved and more intensive management. Soils completely washed away by erosion or excavated and paved over by urban sprawl are permanently lost, for all practical purposes. More often, soils are degraded in quality rather than totally destroyed.

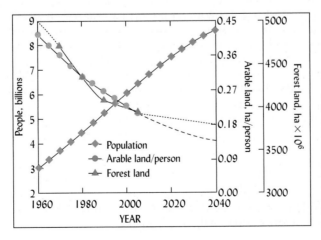

FIGURE 1.28 With more than twice as many human beings crowded on Earth in 2005 as in 1960, the per capita supply of land capable of growing food (arable land) has been cut nearly in half, shrinking from 0.41 to 0.21 hectare (ha) per person. Any expansion of arable land area has come largely at the expense of forestland, the absolute area of which continues to decline.
Data and projections based on FAOSTAT and FAO (2005).

Overview of soil quality concepts:
http://soils.usda.gov/sqi/concepts/concepts.html

Soil quality is a measure of the ability of a soil to carry out particular ecological functions, such as those described in Sections 1.2 to 1.7. Soil quality reflects a combination of *chemical, physical*, and *biological* properties. Some of these properties are relatively unchangeable, inherent properties that help define a particular type of soil. Soil texture and mineral makeup (Section 1.13) are examples. Other soil properties, such as structure (Section 1.13) and organic matter content (Section 1.14), can be significantly changed by management. These more changeable soil properties can indicate the status of a soil's quality relative to its potential, in much the same way that water turbidity or oxygen content indicates the water-quality status of a river.

Mismanagement of forests, farms, and rangeland causes widespread degradation of soil quality by erosion that removes the topsoil, little by little (see Chapter 17). Another widespread cause of soil degradation is the accumulation of salts in improperly irrigated soils in arid regions (see Chapter 10). When people cultivate soils and harvest the crops without returning organic residues and mineral nutrients, the soil's supply of organic matter and nutrients becomes depleted (see Chapter 12). Such depletion is particularly widespread in sub-Saharan Africa, where degrading soil quality is reflected in diminished capacity to produce food (see Chapter 20). Contamination of a soil with toxic substances from industrial processes or chemical spills can degrade its capacity to provide habitat for soil organisms, to grow plants that are safe to eat, or to safely recharge ground and surface waters (see Chapter 18). Degradation of soil quality by pollution is usually localized, but the environmental impacts and costs involved are very large.

While protecting soil quality must be the first priority, it is often necessary to attempt to restore the quality of soils that have already been degraded. Some soils have sufficient **resilience** to recover from minor degradation if left to revegetate on their own. In other cases, more effort is required to restore degraded soils (see Chapter 17). Organic and inorganic amendments may have to be applied, vegetation may have to be planted, physical alterations by tillage or grading may have to be made, or contaminants may have to be removed. As societies around the world assess the damage already done to their natural and agricultural ecosystems, the science of **restoration ecology** has rapidly evolved to guide managers in restoring plant and animal communities to their former levels of diversity and productivity. The job of **soil restoration**, an essential part of these efforts, requires in-depth knowledge of all aspects of the soil system.

1.20 CONCLUSION

The Earth's soil is comprised of numerous soil individuals, each of which is a three-dimensional natural body in the landscape. Each individual soil is characterized by a unique set of properties and soil horizons as expressed in its profile. The nature of the soil layers seen in a particular profile is closely related to the nature of the environmental conditions at a site.

Proverbs about soil and the land. Yoseph Araya:
http://www.iuss.org/bull103files/The%20Soil%20in%20Oral%20Culture.htm

Soils perform six broad ecological functions: they act as the principal medium for plant growth, regulate water supplies, modify the atmosphere, recycle raw materials and waste products, provide habitat for many kinds of organisms, and serve as a major engineering medium for human-built structures. Soil is thus a major ecosystem in its own right. The soils of the world are extremely diverse, each type of soil being characterized by a unique set of soil horizons. A typical surface soil in good condition for plant growth consists of about half solid material (mostly mineral, but with a crucial organic component, too) and half pore spaces filled with varying proportions of water and air. These components interact to influence a myriad of complex soil functions, a good understanding of which is essential for wise management of our terrestrial resources.

If we take the time to learn the language of the land, the soil will speak to us.

STUDY QUESTIONS

1. As a society, is our reliance on soils likely to increase or decrease in the decades ahead? Explain.

2. Discuss how *a soil*, a natural body, differs from *soil*, a material that is used in building a roadbed?

3. What are the six main roles of soil in an ecosystem? For each of these ecological roles, suggest one way in which interactions occur with another of the six roles.

4. Think back over your activities during the past week. List as many incidents as you can in which you came into direct or indirect contact with soil.

5. Figure 1.18 shows the volume composition of a loam surface soil in ideal condition for plant growth. To help you understand the relationships among the four components, redraw this pie chart to represent what the situation might be after the soil has been compacted by heavy traffic. Then draw another pie chart to show how the four components would be related on a mass (weight) basis rather than on a volume basis.

6. Explain in your own words how the soil's nutrient supply is held in different forms, much the way that a person's financial assets might be held in different forms.

7. List the essential nutrient elements that plants derive mainly from the soil.

8. Are all elements contained in plants essential nutrients? Explain.

9. Define these terms: *soil texture, soil structure, soil pH, humus, soil profile, B horizon, soil quality, solum,* and *saprolite.*

10. Describe four processes that commonly lead to degradation of soil quality.

11. Compare the pedological and edaphological approaches to the study of soils. Which is more closely aligned with geology and which with ecology?

REFERENCES

Abrahams, P. W. 2005. "Geophagy and the involuntary ingestion of soil," pp. 435–457, in O. Selinus (ed.), *Essentials of medical geology.* Elsevier, The *Hague.*

Epstein, E., and A. J. Bloom. 2005. *Mineral nutrition of plants: Principles and perspectives* (2nd ed.). Sinauer Associates, Sunderland, MA.

FAO. 2005. Global forest resources assessment 2005. Food and Agric. Organ. of the United Nations, www.fao.org/forestry/site/fra2005/en (verified 03 December 2005).

Stokes, T. 2006. The earth-eaters. *Nature* 444:543–554.

Sampling moon "soil" (NASA Apollo 14)

2

FORMATION OF SOILS FROM PARENT MATERIALS

It is a poem of existence . . . not a lyric but a slow epic whose beat has been set by eons of the world's experience. . . .
—JAMES MICHENER, CENTENNIAL

The first astronauts to explore the moon labored in their clumsy pressurized suits to collect samples of rocks and dust from the lunar surface. These they carried back to Earth for analysis. It turned out that moon rocks are similar in composition to those found deep in the Earth—so similar that scientists concluded that the moon itself began when a stupendous collision between a Mars-sized object and the young planet Earth spewed molten material into orbit around the planet. The force of gravity eventually pulled this material together to form the moon. On the moon, this rock remained unchanged or crumbled into dust with the impact of meteors. On Earth, the rock at the surface, eventually coming in contact with water, air, and living things, was transformed into something new, into many different kinds of living soils. This chapter reveals the story of how rock and dust become "the ecstatic skin of the Earth."[1]

We will study the processes of soil formation that transform the lifeless regolith into the variegated layers of the soil profile. We will also learn about the environmental factors that influence these processes to produce soils in Belgium so different from those in Brazil, soils on limestone so different from those on sandstone, and soils in the valley bottoms so different from those on the hills.

Every landscape is comprised of a suite of different soils, each influencing ecological processes in its own way. Whether we intend to modify, exploit, preserve, or simply understand the landscape, our success will depend on our knowing how soil properties relate to the environment on each site and to the landscape as a whole.

---

[1] The apt description of soil as "ecstatic skin" is from a delightfully readable account of soils by Logan (1995). Intriguing data from Mars Path Finder suggested that Earth may not be the only planet with a skin of soil. However, data from the Rover lander and the OMEGA orbiter (see Kerr, 2005, for details) suggest erosion and formation of secondary minerals like gypsum from the movement and evaporation of surface water, but almost no weathering and no clays. Scientists have concluded that the Martian surface was briefly flooded with water billions of years ago, but has been dry and cold since.

2.1 WEATHERING OF ROCKS AND MINERALS

The influence of weathering, the physical and chemical breakdown of particles, is evident everywhere. Nothing escapes it. Weathering breaks up rocks and minerals, modifies or destroys their physical and chemical characteristics, and carries away the finer fragments and soluble products. However, weathering also synthesizes new minerals of great significance in soils. The nature of the rocks and minerals being weathered determines the rates and results of the breakdown and synthesis (Figure 2.1).

Characteristics of Rocks and Minerals

Geologists classify Earth's rocks as igneous, sedimentary, and metamorphic. Igneous rocks are those formed from molten magma and include such common rocks as granite and diorite (Figure 2.2).

Igneous rock is composed of such primary minerals[2] as light-colored quartz, muscovite, and feldspars and dark-colored biotite, augite, and hornblende. The mineral grains in igneous rocks interlock and are randomly dispersed, giving a salt-and-pepper appearance if they are coarse enough to see with the unaided eye (Figure 2.3, *left*). In general, dark-colored minerals contain iron and magnesium and are more easily weathered. Therefore, dark-colored igneous rocks such as gabbro and basalt are more easily broken down than are granite, syenite, and other lighter-colored igneous rocks.

---

[2] Primary minerals have not been altered chemically since they formed as molten lava solidified. *Secondary minerals* are recrystallized products of the chemical breakdown and/or alteration of primary minerals.

FIGURE 2.1 Two stone markers, photographed on the same day in the same cemetery, illustrate the effect of rock type on weathering rates. The date and initials carved in the slate marker in 1798 are still sharp and clear, while the date and figure of a lamb carved in the marble marker in 1875 have weathered almost beyond recognition. The slate rock consists largely of resistant silicate clay minerals, while the marble consists mainly of calcite, which is much more easily attacked by acids in rainwater. (Photos courtesy of R. Weil)

Rock texture	Quartz / Light-colored minerals (e.g., feldspars, muscovite)		Dark-colored minerals (e.g., hornblende, augite, biotite)	
Coarse	Granite	Diorite	Gabbro	Peridotite / Hornblendite
Intermediate	Rhyolite	Andesite	Basalt	
Fine	Felsite / Obsidian		Basalt glass	

FIGURE 2.2 Classification of some igneous rocks in relation to mineralogical composition and the size of mineral grains in the rock (rock texture). Worldwide, light-colored minerals and quartz are generally more prominent than are the dark-colored minerals.

Animations of rock formation (click on Chapter 6):
http://www.classzone.com/ books/earth_science/terc/ navigation/visualization.cfm

Metamorphic rocks and how they form. California State University:
http://seis.natsci.csulb.edu/ bperry/ROCKS.htm

Sedimentary rocks form when weathering products released from other, older rocks collect under water as sediment and eventually reconsolidate into new rock. For example, quartz sand weathered from granite and deposited near the shore of a prehistoric sea may become cemented by calcium or iron in the water to become a solid mass called sandstone. Similarly, clays may be compacted into shale. Other important sedimentary rocks are listed in Table 2.1, along with their dominant minerals. The resistance of a given sedimentary rock to weathering is determined by its particular dominant minerals and by the cementing agent. Because most of what is presently dry land was at some time in the past covered by water, sedimentary rocks are the most common type of rock encountered, covering about 75% of the Earth's land surface.

Metamorphic rocks are formed from other rocks by a process of change termed "metamorphism." As Earth's continental plates shift, and sometimes collide, forces are generated that can uplift great mountain ranges or cause huge layers of rock to be pushed deep into the crust. These movements subject igneous and sedimentary rock masses to tremendous heat and pressure. These forces may slowly compress and partially remelt and distort the rocks, as well as break the bonds holding the original minerals together. Recrystallization during metamorphism may produce new (usually larger) crystals of the same minerals, or elements from the original minerals may recombine to form new minerals. Igneous rocks like granite may be modified to form gneiss, a metamorphic rock in which light and dark minerals have been reoriented into bands (Figure 2.3, *right*). Sedimentary rocks, such as limestone and shale, may be metamorphosed to marble and slate, respectively (Table 2.1). Slate may be further metamorphosed into phyllite or schist, which typically features mica crystallized during metamorphism.

FIGURE 2.3 Primary minerals are randomly interlocked in igneous rocks, as in the syenite on the left. High heat and pressure have deformed and reoriented the crystals and caused lighter minerals to separate from heavier ones, forming the light- and dark-colored bands typical of gneiss, a metamorphic rock (*right*). In this case, the primary mineral content of both rocks is similar, the light-colored minerals being mainly feldspars and the darker ones hornblende. Scale in inches and cm. (Photo courtesy of R. Weil)

TABLE 2.1 Some of the More Important Sedimentary and Metamorphic Rocks and the Minerals Commonly Dominant in Them

	Type of rock	
Dominant mineral	Sedimentary	Metamorphic
Calcite ($CaCO_3$)	Limestone	Marble
Dolomite ($CaCO_3 \cdot MgCO_3$)	Dolomite	Marble
Quartz (SiO_2)	Sandstone	Quartzite
Clays	Shale	Slate
Variable, silicates	Conglomerate[a]	Gneiss[b]
Variable, silicates		Schist[b]

[a] Small stones of various mineralogical makeup are cemented into conglomerate.
[b] The minerals present are determined by the original rock, which has been changed by metamorphism. Primary minerals present in the igneous rocks commonly dominate these rocks, although some secondary minerals are also present.

Metamorphic rocks are usually harder and more strongly crystalline than the sedimentary rocks from which they formed. The particular minerals that dominate a given metamorphic rock influence its resistance to chemical weathering (see Table 2.2 and Figure 2.1).

Weathering: A General Case

Animated overview of weathering:
http://www.uky.edu/AS/ Geology/howell/goodies/ elearning/module07swf.swf

Weathering is a biochemical process that involves both destruction and synthesis. Moving from left to right in the weathering diagram (Figure 2.4), the original rocks and minerals are destroyed by both *physical disintegration* and *chemical decomposition*. Without appreciably affecting their composition, physical disintegration breaks down rock into smaller rocks and eventually into sand and silt particles that are commonly made up of individual minerals. Simultaneously, the minerals decompose chemically, releasing soluble materials and synthesizing new minerals, some of which are resistant end products. New minerals form either by minor chemical alterations or by complete chemical breakdown of the original mineral and resynthesis of new minerals. During the chemical changes, particle size continues to decrease, and constituents continue to dissolve in the aqueous weathering solution. The dissolved substances may recombine into new (secondary) minerals, may leave the profile in drainage water, or may be taken up by plant roots.

Three groups of minerals that remain in well-weathered soils are shown on the right side of Figure 2.4: (1) silicate clays, (2) very resistant end products, including iron and aluminum oxide clays, and (3) very resistant primary minerals, such as quartz. In

TABLE 2.2 Selected Primary and Secondary Minerals Found in Soils Listed in Order of Decreasing Resistance to Weathering Under Conditions Common in Humid Temperate Regions

Primary minerals		Secondary minerals		
		Goethite	FeOOH	Most resistant
		Hematite	Fe_2O_3	
		Gibbsite	$Al_2O_3 \cdot 3H_2O$	
Quartz	SiO_2	Clay minerals	Al silicates	
Muscovite	$KAl_3Si_3O_{10}(OH)_2$			
Microcline	$KAlSi_3O_8$			
Orthoclase	$KAlSi_3O_8$			
Biotite	$KAl(Mg,Fe)_3Si_3O_{10}(OH)_2$			
Albite	$NaAlSi_3O_8$			
Hornblende[a]	$Ca_2Al_2Mg_2Fe_3Si_6O_{22}(OH)_2$			
Augite[a]	$Ca_2(Al,Fe)_4(Mg,Fe)_4Si_6O_{24}$			
Anorthite	$CaAl_2Si_2O_8$			
Olivine	$Mg,FeSiO_4$			
		Dolomite[b]	$CaCO_3 \cdot MgCO_3$	
		Calcite[b]	$CaCO_3$	
		Gypsum	$CaSO_4 \cdot 2H_2O$	Least resistant

[a] The given formula is only approximate since the mineral is so variable in composition.
[b] In semiarid grasslands, dolomite and calcite are more resistant to weathering than suggested because of low rates of acid weathering.

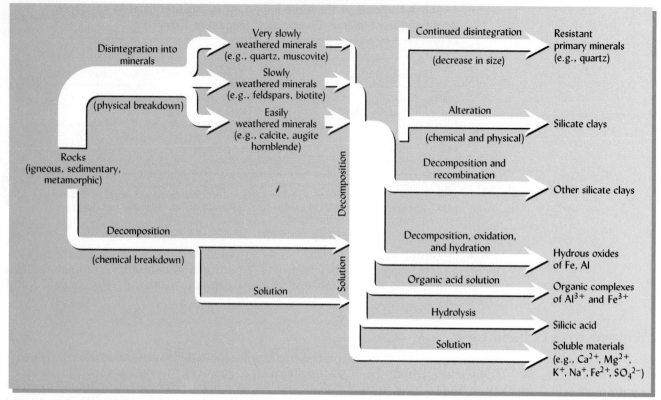

FIGURE 2.4 Pathways of weathering that occur under moderately acid conditions common in humid temperate regions. The disintegration of rocks into small individual mineral grains is a physical process, whereas decomposition, recombination, and solution are chemical processes. Alteration of minerals involves both physical and chemical processes. Note that resistant primary minerals, newly synthesized secondary minerals, and soluble materials are products of weathering. In arid regions the physical processes predominate, but in humid tropical areas decomposition and recombination are most prominent.

TABLE 2.3   Partial Elemental Analysis of a Granite Gneiss Rock and the B-Horizon of a Mature Soil Developed from That Rock under Forest Vegetation in a Warm, Humid Climate

Note that during weathering and soil formation there was a relative loss of calcium, sodium, potassium, and silicon, but a relative increase in iron, aluminum, and copper. A declining ratio of silicon to aluminum is considered an indicator of more complete weathering.

Element[a]	Analysis, mg/g		Change, %
	In rock	*In soil B horizon*	
Ca	27.2	0.184	−99
Na	36.2	0.197	−99
Mg	5.28	1.38	−74
P	0.496	0.383	−23
K	9.79	7.88	−20
Si	324	308	−5
Al	88.1	128	+45
Fe	20.8	40.1	+93
Cu	0.003	0.022	+633
Si/Al	3.7	2.4	−35

[a]Elemental analysis is given here, although many geologists report these values in terms of the oxides of the element: for example, 166.4 mg Al_2O_3, rather than 88.1 mg Al. [Data from Richter and Markewitz (2001) with permission of Cambridge University Press.]

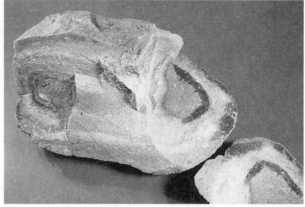

FIGURE 2.5 Two illustrations of rock weathering. (*Left*) An illustration of concentric weathering called *exfoliation*. A combination of physical and chemical processes stimulate the mechanical breakdown, which produces layers that appear much like the leaves of a cabbage. (*Right*) Concentric bands of light and dark colors indicate that chemical weathering (oxidation and hydration) has occurred from the outside inward, producing iron compounds that differ in color. (Right photo courtesy of R. Weil)

highly weathered soils of humid tropical and subtropical regions, the oxides of iron and aluminum, and certain silicate clays with low Si/Al ratios, predominate because most other constituents have been broken down and removed (see Table 2.3).

Physical Weathering (Disintegration)

TEMPERATURE.   Rocks exposed to sunlight heat up during the day and cool down at night, causing alternate expansion and contraction of their constituent minerals. As some minerals expand more than others, temperature changes set up differential stresses that eventually cause the rock to crack apart.

Because the outer surface of a rock is often warmer or colder than the more protected inner portions, some rocks may weather by ***exfoliation***—the peeling away of outer layers (Figure 2.5). This process may be sharply accelerated if ice forms in the surface cracks. When water freezes, it expands with a force of about 1465 Mg/m^2, disintegrating huge rock masses (Figure 2.6, *right*) and dislodging mineral grains from smaller fragments.

ABRASION BY WATER, ICE, AND WIND.   When loaded with sediment, water has tremendous cutting power (Figure 2.6, *left*), as is amply demonstrated by the gorges, ravines, and valleys around the world. The rounding of riverbed rocks and beach sand grains is further evidence of the abrasion that accompanies water movement.

FIGURE 2.6 Effects of water on weathering and the breakdown of rock. (*Left*) The V-like notch carved by water in this sandstone cliff in Montana is evidence of the cutting power of water laden with sediment. The cave below the notch was carved by eddies under the ancient waterfall. (*Right*) The expansion of water as it freezes has broken up these Appalachian sedimentary rocks into ever smaller fragments. (Photos courtesy of R. Weil)

Windblown dust and sand also can wear down rocks by abrasion, as can be seen in the many picturesque rounded rock formations in certain arid regions. In glacial areas, huge moving ice masses embedded with soil and rock fragments grind down rocks in their path and carry away large volumes of material.

PLANTS AND ANIMALS. Plant roots sometimes enter cracks in rocks and pry them apart, resulting in some disintegration. Burrowing animals may also help disintegrate rock somewhat. However, such influences are of little importance in producing parent material when compared to the drastic physical effects of water, ice, wind, and temperature change.

Biogeochemical Weathering

While physical weathering is accentuated in very cold or very dry environments, chemical reactions are most intense where the climate is wet and hot. However, both types of weathering occur together, and each tends to accelerate the other. For example, physical abrasion (rubbing together) decreases the size of particles and therefore increases their surface area, making them more susceptible to rapid chemical reactions.

Chemical weathering is enhanced by such *geological* agents as the presence of water and oxygen, as well as by such *biological* agents as the acids produced by microbial and plant-root metabolism. That is why the term **biogeochemical weathering** is often used to describe the process. The various agents act in concert to convert primary minerals (e.g., feldspars and micas) to secondary minerals (e.g., clays and carbonates) and release plant nutrient elements in soluble forms (see Figure 2.7). Note the importance of water in each of the six basic types of chemical weathering reactions discussed in the following.

HYDRATION. Intact water molecules may bind to a mineral by a process called *hydration*.

$$\underset{\text{Hematite}}{5Fe_2O_3} + \underset{\text{Water}}{9H_2O} \xrightarrow{\text{Hydration}} \underset{\text{Ferrihydrite}}{Fe_{10}O_{15} \cdot 9H_2O} \tag{2.1}$$

Hydrated oxides of iron and aluminum (e.g., $Al_2O_3 \cdot 3H_2O$) exemplify common products of hydration reactions.

HYDROLYSIS. In hydrolysis reactions, water molecules split into their hydrogen and hydroxyl components, and the hydrogen often replaces a cation from the mineral structure. A simple example is the action of water on microcline, a potassium-containing feldspar.

$$\underset{\text{(solid)}}{KAlSi_3O_8} + \underset{\text{Water}}{H_2O} \underset{\text{Hydrolysis}}{\rightleftharpoons} \underset{\text{(solid)}}{HAlSi_3O_8} + \underset{\text{(solution)}}{K^+ + OH^-} \tag{2.2}$$

$$\underset{\text{(solid)}}{2HAlSi_3O_8} + \underset{\text{Water}}{11H_2O} \underset{\text{Hydrolysis}}{\rightleftharpoons} \underset{\text{(solid)}}{Al_2O_3} + \underset{\text{(solution)}}{6H_4SiO_4} \tag{2.3}$$

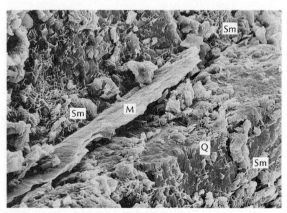

FIGURE 2.7 Scanning electron micrographs illustrating silicate clay formation from weathering of a granite rock in southern California. (*Left*) A potassium feldspar (K-spar) is surrounded by the silicate clays, smectite (Sm) and vermiculite (Vm). (*Right*) Mica (M) and quartz (Q) associated with smectite (Sm). (Courtesy of J. R. Glassmann)

The potassium released is soluble and is subject to adsorption by soil colloids, uptake by plants, and removal in the drainage water. Likewise, the silicic acid (H_4SiO_4) is soluble. It can be removed slowly in drainage water, or it can recombine with other compounds to form secondary minerals such as the silicate clays.

DISSOLUTION.   Water is capable of dissolving many minerals by hydrating the cations and anions until they become dissociated from each other and surrounded by water molecules. Gypsum dissolving in water provides an example.

$$CaSO_4 \cdot 2H_2O + 2H_2O \xrightleftharpoons{\text{Dissolution}} Ca^{2+} + SO_4^{2-} + 4H_2O \qquad (2.4)$$
$$\underset{\text{(solid)}}{} \qquad \underset{\text{Water}}{} \qquad \underset{\text{(solution)}}{} \qquad \underset{\text{Water}}{}$$

ACID REACTIONS.   Weathering is accelerated by the presence of acids, which increase the activity of hydrogen ions in water. For example, when carbon dioxide dissolves in water (a process enhanced by microbial and root respiration) the carbonic acid (H_2CO_3) produced hastens the chemical dissolution of calcite in limestone or marble, as illustrated when the following reactions go to the right:

$$CO_2 + H_2O \rightleftharpoons H_2CO_3 \qquad (2.5)$$

$$H_2CO_3 + CaCO_3 \xrightleftharpoons{\text{Carbonation}} Ca^{2+} + 2HCO_3^- \qquad (2.6)$$
$$\underset{\text{Carbonic acid}}{} \quad \underset{\substack{\text{Calcite} \\ \text{(solid)}}}{} \qquad \underset{\text{(solution)}}{} \quad \underset{\text{(solution)}}{}$$

Soils also contain other, stronger acids, such as nitric acid (HNO_3), sulfuric acid (H_2SO_4), and many organic acids. Hydrogen ions are also associated with soil clays. Each of these sources of acidity is available for reaction with soil minerals.

OXIDATION-REDUCTION.   Minerals that contain iron, manganese, or sulfur are especially susceptible to oxidation-reduction reactions. Iron is usually laid down in primary minerals in the divalent Fe(II) (ferrous) form. When rocks containing such minerals are exposed to air and water during soil formation, the iron is easily oxidized (loses an electron) and becomes trivalent Fe(III) (ferric). If iron is oxidized from Fe(II) to Fe(III), the change in valence and ionic radius causes destabilizing adjustments in the crystal structure of the mineral.

In other cases, Fe(II) may be released from the mineral and almost simultaneously oxidized to Fe(III). For example, the hydration of olivine releases ferrous oxide, which may be oxidized immediately to ferric oxyhydroxide (goethite).

$$3MgFeSiO_4 + 2H_2O \xrightleftharpoons{\text{Hydrolysis}} H_4Mg_3Si_2O_9 + SiO_2 + 3FeO \qquad (2.7)$$
$$\underset{\substack{\text{Olivine} \\ \text{(solid)}}}{} \quad \underset{\text{Water}}{} \qquad \underset{\substack{\text{Serpentine} \\ \text{(solid)}}}{} \quad \underset{\text{(solution)}}{} \quad \underset{\substack{\text{Fe(II) oxide} \\ \text{(solid)}}}{}$$

$$4FeO + O_2 + 2H_2O \xrightleftharpoons[\text{Hydrolysis}]{\text{Oxidation}} 4FeOOH \qquad (2.8)$$
$$\underset{\substack{\text{Fe(II)} \\ \text{oxide}}}{} \qquad \qquad \underset{\substack{\text{Goethite} \\ \text{[Fe(III) oxyhydroxide]}}}{}$$

The oxidation and/or removal of iron during weathering is often made visible by changes in the colors of the resulting altered minerals (see Figure 2.5, *right*).

COMPLEXATION.   Soil biological processes produce organic acids such as oxalic, citric, and tartaric acids, as well as the much larger fulvic and humic acid molecules (see Section 12.4). In addition to providing H^+ ions that help solubilize aluminum and silicon, they also undergo organic reactions with the Al^{3+} ions held within the structure of silicate minerals. By so doing they remove the Al^{3+} from the mineral, which then is subject to further disintegration. In the following example, oxalic acid forms a soluble complex with aluminum from the mineral, muscovite. As this reaction proceeds to the right, it destroys the muscovite structure and releases dissolved ions of the plant nutrient, potassium.

$$K_2[Si_6Al_2]Al_4O_{20}(OH)_4 + 6C_2O_4H_2 + 8H_2O \xrightarrow{\text{Complexation}} 2K^+ + 8OH^- + 6C_2O_4Al^+ + 6Si(OH)_4^0 \qquad (2.9)$$
$$\underset{\substack{\text{Muscovite} \\ \text{(solid)}}}{} \qquad \underset{\text{Oxalic acid}}{} \quad \underset{\text{Water}}{} \qquad \underset{\substack{\text{Potassium hydroxide} \\ \text{(solution)}}}{} \quad \underset{\substack{\text{Complex} \\ \text{(solution)}}}{} \quad \underset{\text{(solution)}}{}$$

FIGURE 2.8 First stages of soil development: biochemical weathering of a rock under the influence of mosses and lichen (a symbiotic combination of algae and fungi). Lichens are especially effective at pioneering inhospitable sites. The algal component provides energy-rich compounds from photosynthesis, while the fungal partner unlocks mineral nutrients from the rock. The fungus produces organic acids that break down the rock by hydrolysis and complexation reactions. Note the pitted surface of the rock where a lichen mat was pulled away to expose the weathering rock surface. The resulting loose mineral material and soluble nutrients, in conjunction with trapped dust and organic debris left by the lichen, will eventually provide a medium for the growth of higher plants. These in turn will further accelerate the process of soil formation. White bar is 1 centimeter. (Photo courtesy of R. Weil)

Had there been no living organisms on Earth, the chemical weathering processes we have just outlined would probably have proceeded 1000 times more slowly, with the result that little, if any, soil would have developed on our planet.

INTEGRATED WEATHERING PROCESSES. The various chemical weathering processes occur simultaneously and are interdependent. For example, hydrolysis of a given primary mineral may release ferrous iron [Fe(II)] that is quickly oxidized to the ferric [Fe(III)] form, which, in turn, is hydrolized to give a hydrous oxide of iron. Hydrolysis or complexation also may release soluble cations, silicic acid, and aluminum or iron compounds. In humid environments, some of the soluble cations and silicic acid are likely to be lost from the weathering mass in drainage waters. The released substances can also be recombined to form silicate clays and other secondary silicate minerals. In this manner, the biochemical processes of weathering transform primary geologic materials into the compounds of which soils are made (Figure 2.8).

2.2 FACTORS INFLUENCING SOIL FORMATION[3]

We learned in Chapter 1 that *the soil* is a collection of *individual soils*, each with distinctive profile characteristics. This concept of soils as organized natural bodies derived initially from late-19th-century field studies by a brilliant Russian team of soil scientists led by V. V. Dukochaev. They noted similar profile layering in soils hundreds of kilometers apart, provided that the climate and vegetation were similar at the two locations. Such observations and much careful subsequent field and laboratory research led to the recognition of five major factors that control the formation of soils.

1. *Parent materials* (geological or organic precursors to the soil)
2. *Climate* (primarily precipitation and temperature)
3. *Biota* (living organisms, especially native vegetation, microbes, soil animals, and increasingly human beings)
4. *Topography* (slope, aspect, and landscape position)
5. *Time* (the period of time since the parent materials began to undergo soil formation)

The five soil forming factors:
http://www.soils.umn.edu/
academics/classes/soil2125/
doc/slab2sff.htm

[3]Many of our modern concepts concerning the factors of soil formation are derived from the work of Hans Jenny (1941 and 1980) and E. W. Hilgard (1921), American soil scientists whose books are considered classics in the field.

Soils are often defined in terms of these factors as *dynamic natural bodies having properties derived from the combined effects of climate and biotic activities, as modified by topography, acting on parent materials over periods of time.*

We will now examine how each of these five factors affects the outcome of soil formation. However, as we do, we must keep in mind that these factors do not exert their influences independently. Indeed, interdependence is the rule. For example, contrasting climatic regimes are likely to be associated with contrasting types of vegetation, and perhaps differing topography and parent material as well. Nonetheless, in certain situations one of the factors has had the dominant influence in determining differences among a set of soils. Soil scientists refer to such a set of soils as a **lithosequence, climosequence, biosequence, toposequence**, or **chronosequence**.

2.3 PARENT MATERIALS

Geological processes have brought to the Earth's surface numerous parent materials in which soils form (Figure 2.9). The nature of the parent material profoundly influences soil characteristics. For example, a soil might inherit a sandy texture (see Section 4.2) from a coarse-grained, quartz-rich parent material such as granite or sandstone. Soil texture, in turn, helps control the percolation of water through the soil profile, thereby affecting the translocation of fine soil particles and plant nutrients.

The chemical and mineralogical composition of parent material also influences both chemical weathering and the natural vegetation. For example, the presence of limestone in parent material will slow the development of acidity that typically occurs in humid climates.

Parent material deposition: http://sis.agr.gc.ca/cansis/taxa/genesis/pmdep/ontario.html

The parent material may contain varying amounts and types of clay minerals, perhaps from a previous weathering cycle. The nature of the parent material greatly influences the kinds of clays that can develop as the soil evolves (see Section 8.5). In turn, the nature of the clay minerals present markedly affects the kind of soil that develops.

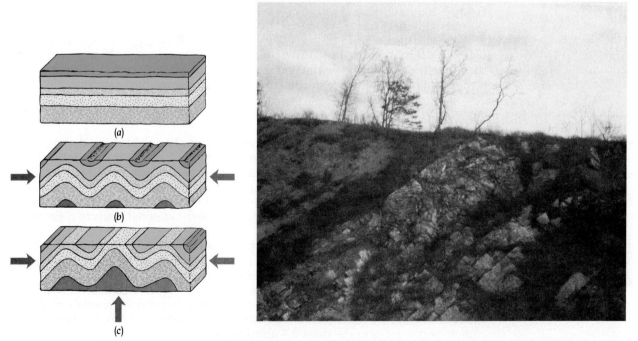

FIGURE 2.9 Diagrams showing how geological processes have brought different rock layers to the surface in a given area. (*a*) Unaltered layers of sedimentary rock with only the uppermost layer exposed. (*b*) Lateral geological pressures deform the rock layers through a process called *crustal warping*. At the same time, erosion removes much of the top layer, exposing part of the first underlying layer. (*c*) Localized upward pressures further reform the layers, thereby exposing two more underlying layers. As these four rock layers are weathered, they give rise to the parent materials on which different kinds of soils can form. (*d*) Crustal warping that lifted up the Appalachian Mountains tilted these sedimentary rock formations that were originally laid down horizontally. This deep road cut in Virginia illustrates the abrupt change in soil parent material (lithosequence) as one walks along the ground surface at the top of this photograph. (Photo courtesy of R. Weil)

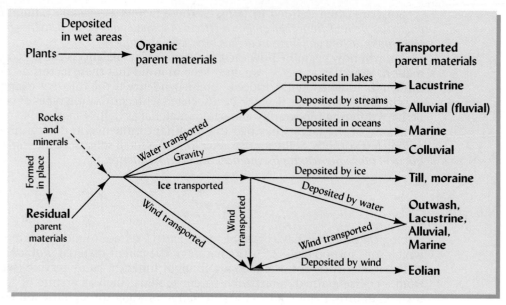

FIGURE 2.10   How various kinds of parent material are formed, transported, and deposited.

Classification of Parent Materials

Inorganic parent materials can either be formed in place as residual material weathered from rock, or they can be transported from one location and deposited at another (Figure 2.10). In wet environments (such as swamps and marshes), incomplete decomposition may allow organic parent materials to accumulate from the residues of many generations of vegetation. Although it is their chemical and physical properties that most influence soil development, parent materials are often classified with regard to the mode of placement in their current location, as seen on the right side of Figure 2.10.

Although these terms properly relate only to the placement of the parent materials, people sometimes refer to the soils that form from these deposits as *organic soils, glacial soils, alluvial soils*, and so forth. These terms are quite nonspecific because parent material properties vary widely within each group and because the effect of parent material is modified by the influence of climate, organisms, topography, and time.

Residual Parent Material

Residual parent material develops in place from weathering of the underlying rock. In stable landscapes it may have experienced long and possibly intense weathering. Where the climate is warm and very humid, residual parent materials are typically thoroughly leached and oxidized, and they show the red and yellow colors of various oxidized iron compounds (see Plates 9, 11, and 15). In cooler and especially drier climates, the color and chemical composition of residual parent material tends to resemble more closely the rock from which it formed.

Residual materials are widely distributed on all continents. The physiographic map of the United States (Figure 2.11) shows nine great provinces where residual materials are prominent (shades of *green* on the map).

A great variety of soils occupy the regions covered by residual debris because of the marked differences in the nature of the rocks from which these materials evolved. The varied soils are also a reflection of wide differences in other soil-forming factors, such as climate and vegetation (Sections 2.4 and 2.5).

Colluvial Debris

Colluvial debris, or **colluvium**, is made up of poorly sorted rock fragments detached from the heights above and carried downslope, mostly by gravity, assisted in some cases by frost action. Rock fragment (talus) slopes, cliff rock debris (detritus), and similar heterogeneous materials are good examples. Avalanches are made up largely of such accumulations.

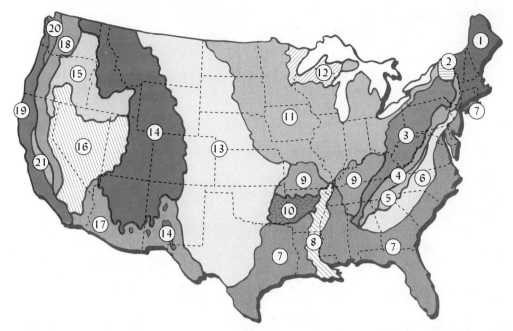

FIGURE 2.11 Generalized physiographic and regolith map of the United States. The regions are as follows (major areas of residual parent material are italicized and shown in shades of green on the map):

1. New England: mostly glaciated metamorphic rocks.
2. Adirondacks: glaciated metamorphic and sedimentary rocks.
3. *Appalachian Mountains and plateaus:* shales and sandstones.
4. *Limestone valleys and ridges:* mostly limestone.
5. Blue Ridge Mountains: sandstones and shales.
6. *Piedmont plateau:* metamorphic rocks.
7. Atlantic and Gulf coastal plain: unconsolidated sediments; sands, clays, and silts.

8. Mississippi floodplain and delta: alluvium.
9. *Limestone uplands:* mostly limestone and shale.
10. *Sandstone uplands:* mostly sandstone and shale.
11. Central lowlands: mostly glaciated sedimentary rocks with till and loess.
12. Superior uplands: glaciated metamorphic and sedimentary rocks.
13. *Great Plains region:* sedimentary rocks.
14. *Rocky Mountain region:* sedimentary, metamorphic, and igneous rocks.
15. Northwest intermountain: mostly igneous rocks; loess in river basins.

16. Great Basin: gravels, sands, alluvial fans; igneous and sedimentary rocks.
17. Southwest arid region: gravel, sand, and other debris of desert and mountain.
18. *Sierra Nevada and Cascade mountains:* igneous and volcanic rocks.
19. *Pacific Coast province:* mostly sedimentary rocks.
20. Puget Sound lowlands: glaciated sedimentary rocks.
21. California central valley: alluvium and outwash.

Colluvial parent materials are frequently coarse and stony because physical rather than chemical weathering has been dominant. Stones, gravel, and fine materials are interspersed (not layered), and the coarse fragments are rather angular (Figure 2.12). Packing voids, spaces created when tumbling rocks come to rest against each other (sometimes at precarious angles), help account for the easy drainage of many colluvial deposits and also for their tendency to be unstable and prone to slumping and landslides, especially if disturbed by excavations.

Alluvial Stream Deposits

There are three general classes of alluvial deposits: *floodplains, alluvial fans,* and *deltas.* They will be considered in order.

FLOODPLAINS. That part of a river valley that is inundated during floods is a floodplain. Sediment carried by the swollen stream is deposited during the flood, with the coarser materials being laid down near the river channel where the water is deeper and flowing

FIGURE 2.12 A productive soil formed in colluvial parent material in the Appalachian Mountains of the eastern United States. The ridge in the background was at one time much taller, but material from the crest tumbled downslope and came to rest in the configuration seen in the soil profile. Note the unstratified mix of particle sizes and the rather angular nature of the coarse fragments. (Photo courtesy of R. Weil)

with more turbulence and energy. Finer materials settle out in the calmer flood waters farther from the channel. Each major flooding episode lays down a distinctive layer of sediment, creating the stratification that characterizes alluvial soils (Figure 2.13).

If, over a period of time, there is a change in grade, a stream may cut down through its already well-formed alluvial deposits. This cutting action leaves **terraces** above the floodplain on one or both sides. Some river valleys feature two or more terraces at different elevations, each reflecting a past period of alluvial deposition and stream cutting.

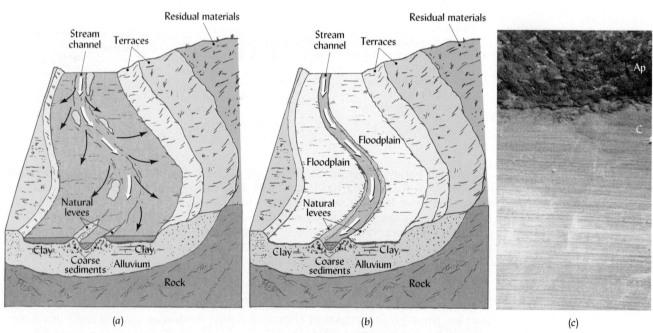

FIGURE 2.13 Floodplain development. (a) A stream at flood stage has overflowed its banks and is depositing sediment in the floodplain. The coarser particles are deposited nearest the stream channel where the water is flowing most rapidly, while the finer particles settle out where the water is moving more slowly. (b) After the flood the sediments are in place and vegetation is growing. (c) Profile of a soil on the Mississippi River floodplain showing contrasting thin layers of sand and silt sediments in the C horizon. Each layer resulted from a separate flooding episode. (Diagrams and photo courtesy of R. Weil)

Major areas of alluvial parent materials are found along the Nile River in Egypt and Sudan; the Euphrates, Ganges, Indus, Brahmaputra, and Hwang Ho river valleys of Asia; and the Amazon River of Brazil. The floodplain along the Mississippi River is the largest in the United States (area 8 in Figure 2.11), varying from 30 to 125 km in width. Floodplains of smaller streams also provide parent materials for locally important soil areas.

To some degree, nutrient-rich materials lost by upland soils are deposited on the river floodplain and **delta** (see following). Soils derived from alluvial sediments generally have characteristics seen as desirable for human settlement and agriculture. These characteristics include nearly level topography, proximity to water, high fertility, and high productivity. Although many alluvial soils are well drained, others may require artificial drainage if they are to be used for upland crops or for stable building foundations.

While alluvial soils are often uniquely suited to forestry and crop production, their use for home sites and urban development should generally be avoided. Unfortunately, the desirable properties of many alluvial soils have already led civilizations to found cities and towns on floodplains. As the many disastrous floods of recent years have illustrated, building on a floodplain, no matter how great the investment in flood-control measures, all too often leads to tragic loss of life and property during serious flooding.

In many areas, installation of systems for drainage and flood protection has proven costly and ineffective. Farmers and the general public pay high costs to keep such areas in agricultural or urban uses. Steps are therefore being taken to reestablish the wetland conditions of certain flood-prone agricultural areas that originally were natural wetlands. These and other alluvial soils can provide natural habitats, such as bottomland forests, which are very productive of timber and support a high diversity of birds and other wildlife.

ALLUVIAL FANS. Streams that leave a narrow valley in an upland area and suddenly descend to a much broader valley below deposit sediment in the shape of a fan, as the water spreads out and slows down (see Figure 2.14 and 19.16). The rushing water tends to sort the sediment particles by size, first dropping the gravel and coarse sand, then depositing the finer materials toward the bottom of the alluvial fan.

Alluvial fan debris is found in widely scattered areas in mountainous and hilly regions. The soils derived from this debris often prove very productive, although they

FIGURE 2.14 Characteristically shaped alluvial fan in a valley in central Nevada. Although alluvial fan areas are usually small and sloping, they can develop into productive, well-drained soils. (Photo courtesy of R. Weil)

may be quite coarse-textured. The Sacramento Valley in California and the Willamette Valley in Oregon are examples of large, agriculturally important areas with alluvial fan materials.

DELTA DEPOSITS.   Much of the finer sediment carried by streams is not deposited in the floodplain but is discharged into the lake, reservoir, or ocean into which the streams flow. Some of the suspended material settles near the mouth of the river, forming a delta. Such delta deposits are by no means universal, being found at the mouths of only a few rivers of the world. A delta often is a continuation of a floodplain (its front, so to speak). It is clayey in nature and is likely to be poorly drained as well.

Delta marshes are among the most extensive and biologically important of wetland habitats. Many of these habitats are today being protected or restored, but civilizations both ancient and modern have also developed important agricultural areas (often specializing in the production of rice) by creating drainage and flood-control systems on the deltas of such rivers as the Amazon, Euphrates, Ganges, Hwang Ho, Mississippi, Nile, Po, and Tigris.

Coastal Sediments

Streams eventually deposit much of their sediment loads in oceans, estuaries, and gulfs. The coarser fragments settle out near the shore and the finer particles at a distance (Figure 2.15). Over long periods of time, these underwater sediments build up, in some cases becoming hundreds of meters thick. Changes in the relative elevations of sea and land may later raise these marine deposits above sea level, creating a coastal plain. The deposits are then subject to a new cycle of weathering and soil formation.

A coastal plain usually has only moderate slopes, being more level in the low-lying parts nearer the coastline and more hilly farther inland, where streams and rivers flowing down the steeper grades have more deeply dissected the landscape. The land surface in the lower coastal portion may be only slightly above the water table during part of the year, so wetland forests and marshes often characterize such parent materials.

Marine and other coastal deposits are quite variable in texture. Some are sandy, as is the case in much of the Atlantic seaboard coastal plain. Others are high in clay, as are deposits found in the Atlantic and Gulf coastal flatwoods and in the interior pinelands of Alabama and Mississippi. Where streams have cut down through layers of marine sediments (as in the detailed block diagram in Figure 2.15), clays, silts, and sand may be encountered side by side. All of these sediments came from the erosion of upland areas,

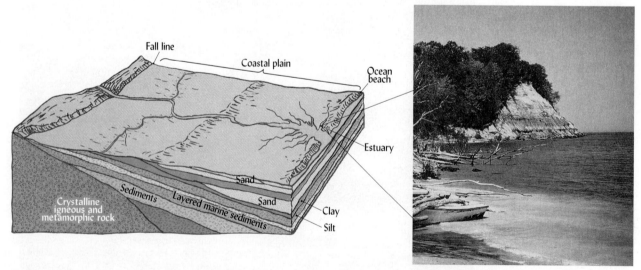

FIGURE 2.15   Diagram showing sediments laid down in marine waters and washed off the interior hills onto coastal areas. The diagram represents the coastal plain of the southeastern United States where such sediments cover older crystalline igneous and metamorphic rocks. Changes in the location of the shoreline and currents over time have resulted in sediment layers consisting alternately of fine clay, silts, coarse sands, and gravels. The photo shows such layering in coastal marine sediments along the Chesapeake Bay in Maryland. (Diagram and photo courtesy of R. Weil)

some of which were highly weathered before the transport took place. However, marine sediments generally have been subjected to soil-forming processes for a shorter period of time than their upland counterparts. As a consequence, the properties of the soils that form are heavily influenced by those of the marine parent materials. Because seawater is high in sulfur, many marine sediments are high in sulfur and go through a period of acid-forming sulfur oxidation at some stage of soil formation (see Sections 9.6 and 13.20 and Plate 109, after page 656).

Parent Materials Transported by Glacial Ice and Meltwaters

Debris transport by alpine glacier:
http://www.uwsp.edu/geO/faculty/lemke/glacial_processes/MoraineMovie.html

During the Pleistocene epoch (about 10^4 to 10^7 years ago), up to 20% of the world's land surface—northern North America, northern and central Europe, and parts of northern Asia—was invaded by a succession of great ice sheets, some more than 1 km thick (Figure 2.16). Present-day glaciers in polar regions and high mountains cover about a third as much area, but are not nearly so thick as the glaciers of the Great Pleistocene Ice Age. Even so, if all present-day glaciers were to melt, the world sea level would rise by about 65 m. Some scientists predict that if the current global warming trend continues, these present-day glaciers could partially melt, causing an increase in sea level, thus flooding many coastal areas around the world.

In North America, Pleistocene-epoch glaciers covered most of what is now Canada, southern Alaska, and the northern part of the contiguous United States. The southernmost extension went down the Mississippi Valley, where the least resistance was met because of the lower and smoother topography.

As the glacial ice pushed forward, the existing regolith with much of its mantle of soil was swept away, hills were rounded, valleys were filled, and, in some cases, the underlying rocks were severely ground and gouged. Thus, the glacier became filled with rock and all kinds of unconsolidated materials, carrying great masses of these materials as it pushed ahead (Figure 2.17). Finally, as the ice melted and the glacier retreated, a mantle of glacial debris or drift remained. This provided a new regolith and fresh parent material for soil formation.

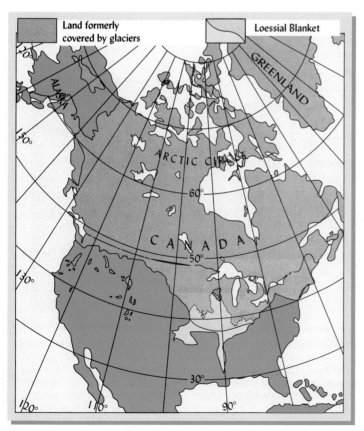

FIGURE 2.16 Map of North America showing the greatest extent of the glaciers of the Great Ice Age during the Pleistocene period. Most of the areas then covered by glaciers are now covered by glacial till. A large area in the central United States, partly over the glacial till and partly beyond the extent of the glaciers themselves, is now covered by a blanket of loess generated after the glaciers retreated. [Map adapted from Schlee (2000)]

FIGURE 2.17 (*Left*) Tongues of a modern-day glacier in Canada. Note the evidence of transport of materials by the ice and the "glowing" appearance of the major ice lobe. (*Right*) This U-shaped valley in the Rocky Mountains illustrates the work of glaciers in carving out land forms. The glacier left the valley floor covered with glacial till. Some of the material gouged out by the glacier was deposited many miles down the valley. (Left photo A-16817-102 courtesy of National Air Photo Library, Surveys and Mapping Branch, Canadian Department of Energy, Mines, and Resources; Right photo courtesy of R. Weil)

GLACIAL TILL AND ASSOCIATED DEPOSITS. The name **drift** is applied to all material of glacial origin, whether deposited by the ice or by associated waters. The materials deposited directly by the ice, called **glacial till**, are heterogeneous (unstratified) mixtures of debris, which vary in size from boulders to clay. Glacial till may therefore be somewhat similar in appearance to colluvial materials, except that the coarse fragments are more rounded from their grinding journey in the ice, and the deposits are often much more densely compacted because of the great weight of the overlying ice sheets. Much glacial till is deposited in irregular ridges called **moraines**. Figure 2.18 shows how glacial sheets deposited several types of soil parent materials.

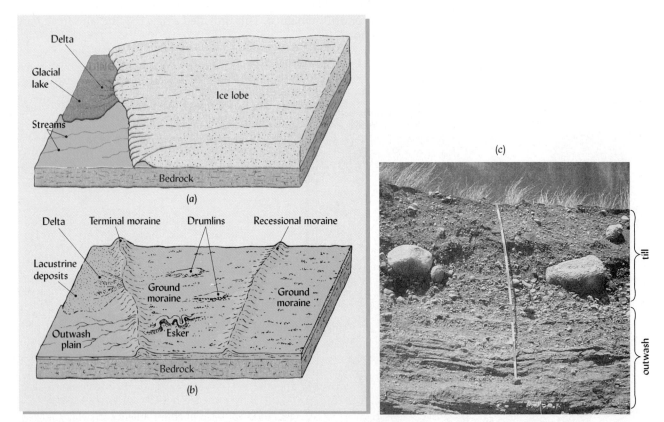

FIGURE 2.18 Illustration of how several glacial materials were deposited. (*a*) A glacier ice lobe moving to the left, feeding water and sediments into a glacial lake and streams and building up glacial till near its front. (*b*) After the ice retreats, terminal, ground, and recessional moraines are uncovered along with cigar-shaped hills (drumlins), the beds of rivers that flowed under the glacier (eskers), and lacustrine, delta, and outwash deposits. (*c*) The stratified glacial outwash in the lower part of this soil profile in North Dakota is overlain by a layer of glacial till containing a random assortment of particles, ranging in size from small boulders to clays. Note the rounded edges of the rocks, evidence of the churning action within the glacier. Scale is marked every 10 cm. (Photo courtesy of R. Weil)

GLACIAL OUTWASH AND LACUSTRINE SEDIMENTS. The torrents of water gushing forth from melting glaciers carried vast loads of sediment. In valleys and on plains where the glacial waters were able to flow away freely, the sediment formed an **outwash plain** (Figure 2.18). Such sediments, with sands and gravels sorted by flowing water, are common **valley fills** in parts of the United States. Figure 2.18c shows the sorted layering of coarse and fine materials in glacial outwash overlaid by mixed materials of glacial till.

When the ice front came to a standstill, where there was no ready escape for the water, ponding began; ultimately, very large lakes were formed (Figure 2.18). Particularly prominent in North America were those south of the Great Lakes and in the Red River Valley of Minnesota and Manitoba. The latter lake, called Glacial Lake Agassiz, was about 1200 km long and 400 km wide at its maximum extension.

The **lacustrine deposits** formed in these glacial lakes range from coarse delta materials and beach deposits near the shore to larger areas of fine silts and clay deposited from the deeper, more still waters at the center of the lake. Areas of inherently fertile (though not always well-drained) soils developed from these materials as the lakes dried.

Parent Materials Transported by Wind

Wind is capable of picking up an enormous quantity of material at one site and depositing it at another. Wind can most effectively pick up material from soil or regolith that is loose, dry, and unprotected by vegetation. Dry, barren landscapes have served, and continue to serve, as sources of parent material for soils forming as far away as the opposite side of the globe (Figure 2.19). The smaller the particles, the higher and farther the wind will carry them. Wind-transported (**eolian**) materials important as parent material for soil formation include, from largest to smallest particle size: **dune sand**, **loess** (pronounced "luss"), and **aerosolic dust**. Windblown **volcanic ash** from erupting volcanoes is a special case that is also worthy of mention.

DUNE SAND. Along the beaches of the world's oceans and large lakes and over vast barren deserts, strong winds pick up medium and fine sand grains and pile them into hills of sand called *dunes*. The dunes, ranging up to 100 m in height, may continue to slowly shift their locations in response to the prevailing winds. Because most other minerals have been broken down and carried away by the waves, beach sand usually consists mainly of quartz, which is devoid of plant nutrients and highly resistant to weathering action. Nonetheless, over time dune grasses and other pioneering vegetation may take root, and soil formation may begin. The sandy soils that extend for many kilometers east of Lake Michigan provide an example of this process. Some of the very deep sandy

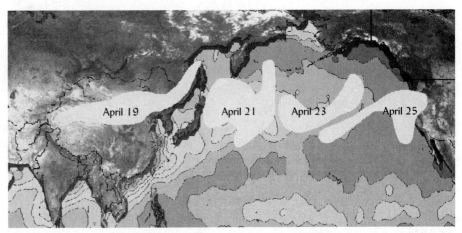

FIGURE 2.19 Map showing the path of a giant dust cloud that was generated by wind erosion of loess soils in the Gobi desert region of China. Within a week, the dust from China was contributing to soil parent material (and air pollution) on the west coast of North America. In a similar manner, dust from the Sahara desert in Africa rides the westerly winds across the southern Atlantic Ocean and contributes to soils in South America. (Courtesy of Rudolf Husar, School of Engineering and Applied Sciences at Washington University)

soils on the Atlantic coastal plain are thought to have formed on dunes marking the location of an ancient beach.

Desert sands, too, are usually dominated by quartz, but they may also include substantial amounts of other minerals that could contribute more to the establishment of vegetation and the formation of soils, should sufficient rainfall occur. The pure-white dunes of sand-sized gypsum at White Sands, New Mexico are a dramatic example of weatherable minerals in desert sands.

LOESS.   The windblown materials called *loess* are composed primarily of silt with some very fine sand and coarse clay. They cover wide areas in the central United States, eastern Europe, Argentina, and central China (Figure 2.20*a*). Loess may be blown for hundreds of kilometers. The deposits farthest from the source are thinnest and consist of the finest particles.

In the United States (Figure 2.20*b*), the main sources of loess were the great barren expanses of till and outwash left in the Missouri and Mississippi river valleys by the retreating glaciers of the last Ice Age. During the winter months, winds picked up fine materials and moved them southward, covering the existing soils and parent materials with a blanket of loess that accumulated to as much as 8 m thick.

In central and western China, loess deposits reaching 30 to 100 m in depth cover some 800,000 km^2 (Figure 2.21). These materials have been windblown from the deserts of central Asia and are generally not associated directly with glaciers. These and other loess deposits tend to form silty soils of rather high fertility and potential productivity.

AEROSOLIC DUST.   Very fine particles (about 1 to 10 µm) carried high into the air may travel for thousands of kilometers before being deposited, usually with rainfall. These fine particles are called *aerosolic dust* because they can remain suspended in air, due to their very small size. Although this dust has not blanketed the receiving landscapes as thickly as is typical for loess, it does accumulate at rates that make significant contributions to soil formation. Much of the calcium carbonate in soils of the western United States probably originated as windblown dust. Recent studies have shown that dust, originating in the Sahara Desert of northern Africa and transported over the Atlantic

Dust across the oceans. Click on "China during April of 1998." NASA:
http://toms.gsfc.nasa.gov/aerosols/dust01.html

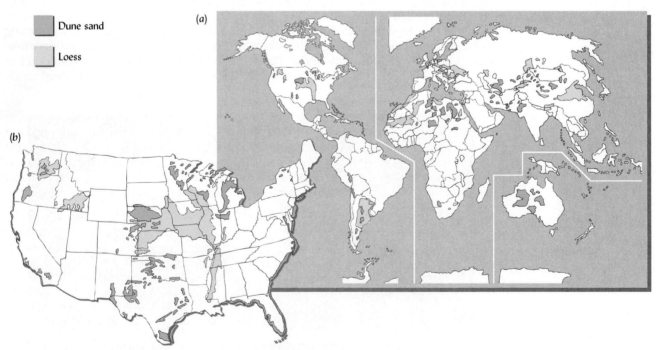

Dune sand

Loess

(a)

(b)

FIGURE 2.20   (*a*) Major eolian deposits of the world include the loess deposits in Argentina, eastern Europe, northern China, and the large areas of dune sands in north Africa and Australia. (*b*) Approximate distribution of loess and dune sand in the United States. The soils that have developed from loess are generally silt loams, often quite high in fine sands. Note especially the extension of the central loess deposit down the eastern side of the Mississippi River and the smaller areas of loess in Washington, Oregon, and Idaho. The most prominent areas of dune sands are the Sand Hills of Nebraska and the dunes along the eastern shore of Lake Michigan.

FIGURE 2.21   Villagers carve houses out of thick loess deposits in Xian, China. The loess consists mainly of silty materials bound together by small amounts of clay. The clay binder helps stabilize the loess when excavated, but only if the material is protected from rain, as in these vertical walls. Sloping excavations of this material would quickly slump and wash away when saturated with rain. Vertical road cuts are therefore a common feature of loessial landscapes around the world. (Photos courtesy of Raymond Miller, University of Maryland)

Ocean in the upper atmosphere, is the source of much of the calcium and other nutrients found in the highly leached soils of the Amazon basin in South America. Likewise, in the springtime, dust from wind storms in the loess region of China blows across the Pacific Ocean to add soil parent materials (and air pollution) to the western part of North America (Figure 2.19).

VOLCANIC ASH   During volcanic eruptions cinders fall in the immediate vicinity of the volcano, while fine, often glassy, ash particles may blanket extensive areas downwind. Soils developed from volcanic ash are most prominent within a few hundred kilometers of the volcanoes that ring the Pacific Ocean. Important areas of volcanic ash parent materials occur in Japan, Indonesia, New Zealand, western United States (in Hawaii, Montana, Oregon, Washington, and Idaho), Mexico, Central America, and Chile. The soils formed are uniquely light and porous and tend to accumulate organic matter more rapidly than other soils in the area (Section 3.7). The volcanic ash tends to weather rapidly into allophane, a type of clay with unusual properties (see Section 8.5).

Organic Deposits

Organic material accumulates in wet places where plant growth exceeds the rate of residue decomposition. In such areas residues accumulate over the centuries from wetland plants such as pondweeds, cattails, sedges, reeds, mosses, shrubs, and certain trees. These residues sink into the water, where their decomposition is limited by lack of oxygen. As a result, organic deposits often accumulate up to several meters in depth (Figure 2.22). Collectively, these organic deposits are called **peat**.

DISTRIBUTION AND ACCUMULATION OF PEATS.   Peat deposits are found all over the world, but most extensively in the cool climates and in areas that have been glaciated. About 75% of the 340 million hectares of peat lands in the world are found in Canada and northern Russia. The United States is a distant third, with about 20 million hectares.

The rate of peat accumulation varies from one area to another, depending on the balance between production of plant material and its loss by decomposition. Cool climates and acidic conditions favor slow decomposition, but also slower plant production. Warm climates and alkaline conditions favor rapid losses, but also rapid plant production. Enrichment with nutrients may increase the rate of organic production more than it does the rate of decomposition, leading to very high net accumulation rates (Table 2.4). As we shall see in Chapter 12, artificial drainage, used to remove excess water from a peat soil, lets air into the peat and drastically alters the balance between production and decomposition of organic matter, causing a reversal of the accumulation process and a loss or subsidence of the peat soil.

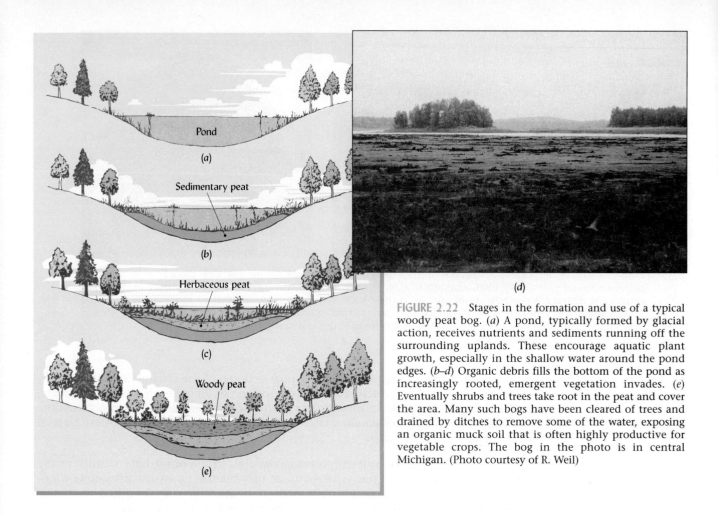

FIGURE 2.22 Stages in the formation and use of a typical woody peat bog. (a) A pond, typically formed by glacial action, receives nutrients and sediments running off the surrounding uplands. These encourage aquatic plant growth, especially in the shallow water around the pond edges. (b–d) Organic debris fills the bottom of the pond as increasingly rooted, emergent vegetation invades. (e) Eventually shrubs and trees take root in the peat and cover the area. Many such bogs have been cleared of trees and drained by ditches to remove some of the water, exposing an organic muck soil that is often highly productive for vegetable crops. The bog in the photo is in central Michigan. (Photo courtesy of R. Weil)

TABLE 2.4 Various Freshwater Wetlands, Their Rates of Accretion (Increasing Soil Thickness), and Accumulation of Carbon, Nitrogen, and Phosphorous in Their Peat Soils.

Type and location of wetland	Characteristics of wetland types	Accretion rate, mm/yr	Accumulation rate, $g/m^2/yr$		
			C	N	P
Bogs (Massachusetts)	**Bogs** are peat-accumulating ponds with no outflow and little inflow. They receive little calcium from the surrounding landscape and are therefore quite acidic.	4.3	90	1.2	—
Fens (Michigan)	**Fens** are peat-accumulating ponds with no outflow and little inflow. Calcium-rich mineral matter from the surrounding landscapes makes them relatively alkaline.	0.9	42	3.0	0.11
Pocosin swamps (North Carolina)	**Swamps** are periodically inundated with shallow, slow-moving water and dominated by shrubs and trees.	2.6	127	3.0	0.06
Okefenokee swamp (Georgia)		—	82	3.8	0.15
Everglades marsh (Florida), unenriched	**Marshes** are periodically inundated with shallow, slow-moving water and dominated by grasses and herbaceous plants. The Everglades is a very large wetland area in Florida with both marshy and swampy components.	1.4	65	4.7	0.06
Everglades marsh (Florida), enriched		6.7	223	16.6	0.46

[Data compiled by Craft and Richardson (1998)]

TYPES OF PEAT MATERIALS. Based on the nature of the parent materials, four kinds of peat are recognized:

1. Moss peat, the remains of mosses such as sphagnum
2. Herbaceous peat, residues of herbaceous plants such as sedges, reeds, and cattails
3. Woody peat, from the remains of woody plants, including trees and shrubs
4. Sedimentary peat, remains of aquatic plants (e.g., algae) and of fecal material of aquatic animals

Organic deposits generally contain two or more of these kinds of peats. Alternating layers of different peats are common, as are mixtures of the peats. Because the succession of plants, as the residues accumulate, tends to favor trees (see Figure 2.22), woody peats often dominate the surface layers of organic materials.

In cases where a wetland area has been drained, woody peats tend to make very productive agricultural soils that are especially well suited for vegetable production. While moss peats have high water-holding capacities, they tend to be quite acid. Sedimentary peat is generally undesirable as an agricultural soil. This material is highly colloidal and compact and is rubbery when wet. Upon drying, it resists rewetting and remains in a hard, lumpy condition. Fortunately, it occurs mostly deep in the profile and is unnoticed unless it interferes with drainage of the bog area.

The organic material is called **peat**, or **fibric**, if the residues are sufficiently intact to permit the plant fibers to be identified. If most of the material has decomposed sufficiently so that little fiber remains, the term **muck** or **sapric** is used. In mucky peats (hemic materials) only some of the plant fibers can be recognized.

WETLAND PRESERVATION. Wetland areas are important environmental buffers and natural habitats for wildlife. Drainage of these areas reduces the benefits of wetlands. While organic soils are the foundation for some very productive agricultural systems, environmentalists argue that such use is unsustainable because, once drained, the organic deposits will decompose and disappear after a century or so; therefore, these areas might be better left in (or returned to) their natural state (see Section 7.7).

Recognizing that the effects of **parent materials** on soil properties are modified by the combined influences of **climate**, **biotic activities**, **topography**, and **time**, we will now turn to these other four factors of soil formation, starting with climate.

Disappearing marshes. Cornelia Dean, New York Times. Also click on Marsh Mess video:

http://www.nytimes.com/2005/11/15/science/earth/15marsh.html?ex=1289710800&en=debebd7482392dcc&ei=5088&partner=rssnyt&emc=rss

2.4 CLIMATE

Climate is perhaps the most influential of the four factors acting on parent material because it determines the nature and intensity of the weathering that occurs over large geographic areas. The principal climatic variables influencing soil formation are *effective precipitation* (see Box 2.1) and *temperature*, both of which affect the rates of chemical, physical, and biological processes.

Effective Precipitation

We have already seen that water is essential for all the major chemical weathering reactions. To be effective in soil formation, water must penetrate into the regolith. The greater the depth of water penetration, the greater the depth of weathering soil and development. Surplus water percolating through the soil profile transports soluble and suspended materials from the upper to the lower layers. It may also carry away soluble materials in the drainage waters. Thus, percolating water stimulates weathering reactions and helps differentiate soil horizons.

Likewise, a deficiency of water is a major factor in determining the characteristics of soils of dry regions. Soluble salts are not leached from these soils, and in some cases they build up to levels that curtail plant growth. Soil profiles in arid and semiarid regions are also apt to accumulate carbonates and certain types of cracking clays.

Temperature

For every 10 °C rise in temperature, the rates of biochemical reactions more than double. Temperature and moisture both influence the organic matter content of soil through their effects on the balance between plant growth and microbial decomposition

BOX 2.1 EFFECTIVE PRECIPITATION FOR SOIL FORMATION

Water from rain and melting snow is a primary requisite for parent material weathering and soil development. To fully promote soil development, water must not only enter the profile and participate in weathering reactions, but also percolate through the profile and translocate soluble weathering products.

Let's consider a site that receives an average of 600 mm of rainfall per year. The amount of water leaching through a soil is determined not only by the total annual precipitation but also by at least four other factors as well (Figure 2.23).

a. **Seasonal distribution** of precipitation. The 600 mm of rainfall distributed evenly throughout the year, with about 50 mm each month, is likely to cause less soil leaching or erosion than the same annual amount of rain falling at the rate of 100 mm per month during a 6-month rainy season.

b. **Temperature and evaporation.** In a hot climate, evaporation from soils and vegetation is much higher than in a cool climate. Therefore, in the hot climate, much less of the 600 mm will be available for percolation and leaching. Most or all will evaporate soon after it falls on the land. Thus, 600 mm of rain may cause more leaching and profile development in a cool climate than in a warmer one. Similar reasoning would suggest that rainfall concentrated during a mild winter (as in California) may be more effective in leaching the soil than the same amount of rain concentrated in a hot summer (as in the Great Plains).

c. **Topography.** Water falling on a steep slope will run downhill so rapidly that only a small portion will enter the soil where it falls. Therefore, even though they receive the same rainfall, level or concave sites will experience more percolation and leaching than steeply sloping sites. The effective rainfall can be said to be greater on the level site than on the sloping one. The concave site will receive the greatest effective rainfall because, in addition to direct rainfall, it will collect the runoff from the adjacent sloping site.

d. **Permeability.** Even if the above conditions are the same, more rainwater will infiltrate and leach through a coarse, sandy profile than a tight, clayey one. Therefore the sandy profile can be said to experience a greater effective precipitation, and more rapid soil development may be expected.

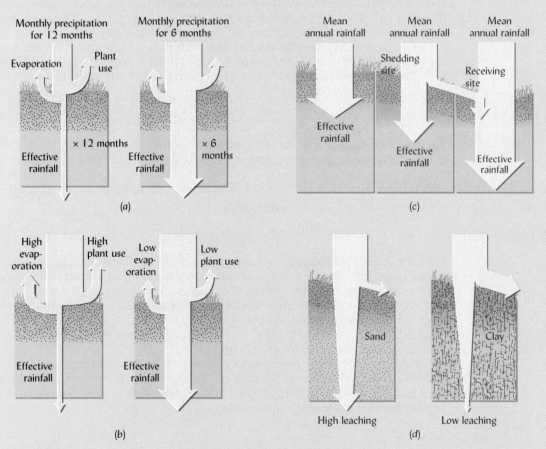

FIGURE 2.23 *The seasonal rainfall distribution (a), evaporative demand (b), site topography (c), and soil permeability (d) interact to determine how effectively precipitation can influence soil formation. (Diagram courtesy of R. Weil)*

FIGURE 2.24 A generalized illustration of the effects of two climatic variables, temperature and precipitation, on the depths of regolith weathered in bedrock. The stippled areas represent the range of depths to which the regolith typically extends. In cold climates (arctic regions), the regolith is shallow under both humid and arid conditions. In the warmer climates of the lower latitudes, the depth of the residual regolith increases sharply in humid regions, but is little affected in arid regions. Under humid tropical conditions, the regolith may be 50 or more meters in depth. The vertical arrows represent depths of weathering near the Equator. Remember that soil depth may not be as great as regolith depth.

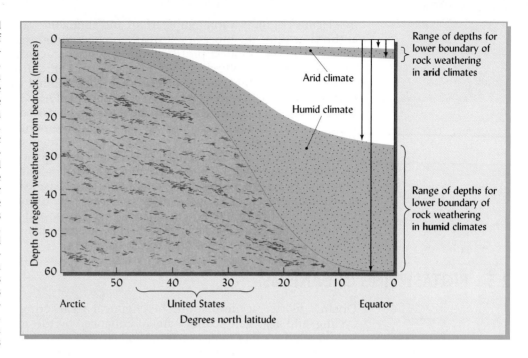

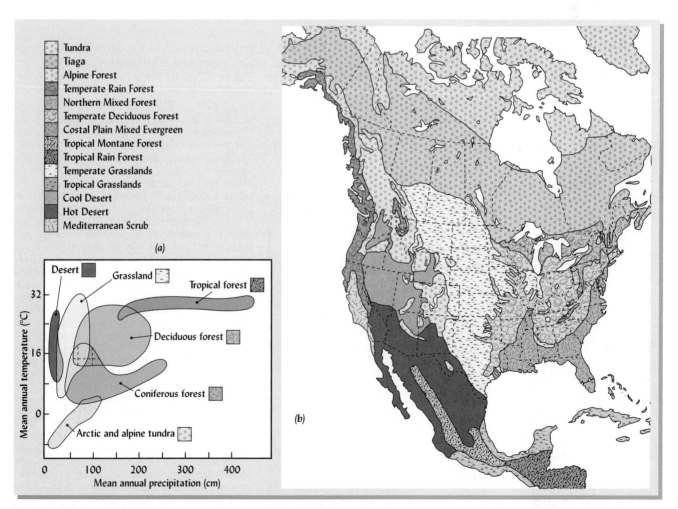

FIGURE 2.25 Effect of climate on vegetation. (*a*) General distribution of vegetation ecotypes (biomes) along gradients of mean annual temperature and precipitation. (*b*) Major biomes in North America. [(a) from NSF (1975); (b) from U.S. Dept. Energy Oak Ridge Lab]

(see Figure 12.20). If warm temperatures and abundant water are present in the profile at the same time, the processes of weathering, leaching, and plant growth will be maximized. The very modest profile development characteristic of cold areas contrasts sharply with the deeply weathered profiles of the humid tropics (Figure 2.24).

Climate also influences the natural vegetation. Humid climates favor the growth of trees (Figure 2.25). In contrast, grasses are the dominant native vegetation in subhumid and semiarid regions, while shrubs and brush of various kinds dominate in arid areas. Thus, climate exerts its influence partly through a second soil-forming factor, the living organisms.

Considering soils with similar temperature regime, parent material, topography, and age, increasing effective annual precipitation generally leads to increasing clay and organic matter contents, greater acidity, and lower ratio of Si/Al (an indication of more highly weathered minerals; see Table 2.3). However, many places have experienced climates in past geologic epochs that were not at all similar to the climate evident today. This fact is illustrated in certain old landscapes in arid regions, where highly leached and weathered soils stand as relics of the humid tropical climate that prevailed there many thousands of years ago.

2.5   BIOTA: LIVING ORGANISMS

Organic matter accumulation, biochemical weathering, profile mixing, nutrient cycling, and aggregate stability are all enhanced by the activities of organisms in the soil. Vegetative cover reduces natural soil erosion rates, thereby slowing down the rate of mineral surface soil removal. Organic acids produced from certain types of plant leaf litter bring iron and aluminum into solution by complexation and accelerate the downward movement of these metals and their accumulation in the B horizon.

Role of Natural Vegetation

ORGANIC MATTER ACCUMULATION.   The effect of vegetation on soil formation can be seen by comparing properties of soils formed under grassland and forest vegetation near the boundary between these two ecosystems (Figure 2.26). In the grassland, much of the

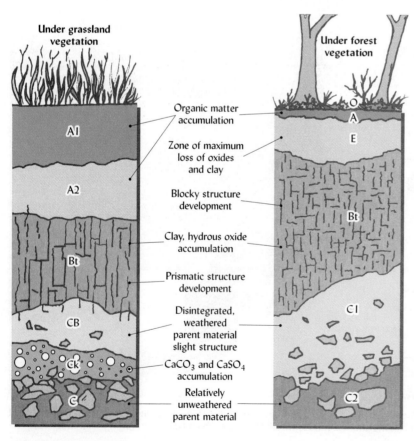

FIGURE 2.26   Natural vegetation influences the type of soil eventually formed from a given parent material (calcareous glacial till, in this example). The grassland vegetation is likely to occur in a somewhat less humid climate than the deciduous forest. The amount, and especially the vertical distribution, of organic matter that accumulates in the upper part of the profile differs markedly between the vegetation types. The forested soil exhibits surface layers (O horizons) of leaves and twigs in various stages of decomposition, along with a thin, mineral A horizon, into which some of the surface litter has been mixed. In contrast, most of the organic matter in the grassland is added as fine roots distributed throughout the upper 1 m or so, creating a thick, mineral A horizon. Also note that calcium carbonate has been solubilized and has moved down to the lower horizons (Ck) in the grassland soils, while it has been completely removed from the profile in the more acidic, leached forested soil. Under both types of vegetation, clay and iron oxides move downward from the A horizon and accumulate in the B horizon, encouraging the formation of characteristic soil structure. In the forested soil, the zone above the B horizon usually becomes a distinctly bleached E horizon, partly because most of the organic matter is restricted to the near-surface layers, and partly because decomposition of the forest litter generates organic acids that remove the brownish iron oxide coatings. Compare these mature profiles to the changes over time discussed in Sections 2.7 and 2.8. (Diagrams courtesy of R. Weil)

Labels on figure (grassland, left): Under grassland vegetation; A1; A2; Bt; CB; Ck; C

Center labels: Organic matter accumulation; Zone of maximum loss of oxides and clay; Blocky structure development; Clay, hydrous oxide accumulation; Prismatic structure development; Disintegrated, weathered parent material slight structure; CaCO₃ and CaSO₄ accumulation; Relatively unweathered parent material

Labels on figure (forest, right): Under forest vegetation; O; A; E; Bt; C1; C2

organic matter added to the soil is from the deep, fibrous, grass root systems. By contrast, tree leaves falling on the forest floor are the principal source of soil organic matter in the forest. Another difference is the frequent occurrence in the grasslands of fires that destroy large amounts of aboveground plant matter and surface litter. Also, the much greater acidity under many forests inhibits the action of certain soil organisms that otherwise would mix much of the surface litter into the mineral soil. As a result, the soils under grasslands generally develop a thicker A horizon with a deeper distribution of organic matter than in comparable soils under forests, which characteristically store most of their organic matter in the forest floor (O horizons) and a thin A horizon. The microbial community in a typical grassland soil is dominated by bacteria, while that of the forest soil is dominated by fungi (see Chapter 11 for details). Differences in microbial action affect the aggregation of the mineral particles into stable granules and the rate of nutrient cycling. The light-colored, leached E horizon typically found under the O or A horizon of a forested soil results from the action of organic acids generated mainly by fungi in the acidic forest litter. An E horizon is generally not found in a grassland soil.

CATION CYCLING BY TREES.[4] The ability of natural vegetation to accelerate the release of nutrient elements from minerals by biogeochemical weathering, and to take up these elements from the soil, strongly influences the characteristics of the soils that develop. Soil acidity is especially affected. Differences occur not only between grassland and forest vegetation, but also between different species of forest trees. Litter falling from coniferous trees (e.g., pines, firs, spruces, and hemlocks) will recycle only small quantities of calcium, magnesium, and potassium compared to those recycled by litter from some deciduous trees (e.g., yellow poplar, beech, oaks, and maples) that take up and store much larger amounts of these cations (Figure 2.27). Conifer tree roots take up less Ca, Mg, and K from minerals weathering deep in the profile and allow more of these

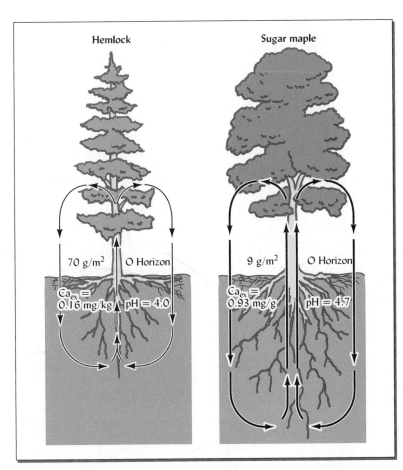

FIGURE 2.27 Nutrient cycling is an important process by which plants affect the soil in which they grow, altering the course of soil development and the suitability of the soil environment for future generations of vegetation. For example, hemlock (a conifer) and sugar maple (a deciduous hardwood) differ markedly in their ability to accelerate mineral weathering, mobilize nutrient cations, and recycle them to the upper soil horizons. Sugar maple roots are efficient at taking up Ca from soil minerals, and the maple leaves produced contain high concentrations of Ca. When these leaves fall to the ground, they decompose rapidly and release large amounts of Ca^{2+} ions that become adsorbed as exchangeable Ca^{2+} on humus and clay in the O and A horizons. This influx of Ca^{2+} ions may somewhat retard acidification of the surface layers. However, the maple roots' efficient extraction of Ca from minerals in the parent material may accelerate acidification and weathering in deeper soil horizons. In contrast, hemlock needles are Ca-poor, much slower to decompose, and therefore result in a thicker O horizon, greater acidity in the O and upper mineral horizons, but possibly less rapid weathering of minerals in the underlying parent material. [Data for a Connecticut forest reported by van Breemen and Finzi (1998)]

---

[4] For a book with intriguing papers on the ecological impacts and competing theories of how trees affect nutrient cycling and soil formation, see Binkley and Menyailo (2005).

FIGURE 2.28 The scattered bunch grasses of this semiarid rangeland in the Patagonia region of Argentina have created "islands" of soil with enhanced fertility and thicker A horizons. A lens cap placed at the edge of one of these islands provides scale and highlights the increased soil thickness under the plant canopy. Such small-scale soil heterogeneity associated with plants is common where soil water limitations prevent complete plant ground cover. (Photos courtesy of Ingrid C. Burke, Short-Grass Steppe Long-Term Ecological Research Program, Colorado State University)

nonacidic cations to be lost by leaching. Therefore, soil acidity often develops more strongly in the surface horizons under coniferous vegetation than under most deciduous trees. Furthermore, the acidic, resinous needles from conifer trees resist decomposition and discourage earthworm populations leading to the accumulation of a thick O horizon with distinctly separate layers of fibric (undecomposed) and sapric (highly decomposed) material. The leaves of deciduous trees generally break down more readily and form a thinner forest floor with less distinction between layers and with more litter mixed into the A horizon. In Chapter 8 (Box 8.2) we will see that nutrient cycling by plant roots can even alter the types of clay minerals found in a soil.

HETEROGENEITY IN RANGELANDS. In arid and semiarid rangelands, competition for limited soil water does not permit vegetation dense enough to completely cover the soil surface. Scattered shrubs or bunch grasses are interspersed with openings in the plant canopy where the soil is bare or partially covered with plant litter. The widely scattered vegetation alters soil properties in several ways. Plant canopies trap windblown dust that is often relatively rich in silt and clay. Roots scavenge nutrients such as nitrogen, phosphorus, potassium, and sulfur from the interplant areas. These nutrients are then deposited with the leaf litter under the plant canopies. The decaying litter adds organic acids, which lower the soil pH and stimulate mineral weathering. As time goes on, the relatively bare soil areas between plants decline in fertility and may increase in size as they become impoverished and even less inviting for the establishment of plants. Simultaneously, the vegetation creates "islands" of enhanced fertility, thicker A horizons, and often more deeply leached calcium carbonate (see Figure 2.28 and Section 10.1).

Role of Animals

The role of animals in soil-formation processes must not be overlooked. Large animals such as gophers, moles, and prairie dogs bore into the lower soil horizons, bringing materials to the surface. Their tunnels are often open to the surface, encouraging movement of water and air into the subsurface layers. In localized areas, they enhance mixing of the lower and upper horizons by creating, and later refilling, underground tunnels. For example, dense populations of prairie dogs may completely turn over the upper meter of soil in the course of several thousand years. Old animal burrows in the lower horizons often become filled with soil material from the overlying A horizon, creating profile features known as *crotovinas* (Figure 2.29). In certain situations, animal activity may arrest soil development by increasing the loss of soil by erosion. Such activity in alpine meadows is illustrated in Figure 2.30.

FIGURE 2.29 Abandoned animal burrows in one horizon filled with soil material from another horizon are called *crotovinas*. In this Illinois prairie soil, dark, organic-matter-rich material from the A horizon has filled in old prairie dog burrows that extend into the B horizon. The dark circular shapes in the subsoil mark where the pit excavation cut through these burrows. Scale marked every 10 cm. (Photo courtesy of R. Weil)

EARTHWORMS AND TERMITES.   Earthworms, ants, and termites mix the soil as they burrow, significantly affecting soil formation. Earthworms ingest soil particles and organic residues, enhancing the availability of plant nutrients in the material that passes through their bodies. They aerate and stir the soil and increase the stability of soil aggregates, thereby assuring ready infiltration of water. Ants and termites, as they build mounds, also transport soil materials from one horizon to another. In general, the mixing activities of animals, sometimes called **pedoturbation**, tends to undo or counteract the tendency of other soil-forming processes to accentuate the differences among soil horizons. Termites and ants may also retard soil profile development by denuding large areas of soil around their nests, leading to increased loss of soil by erosion (Figure 2.31).

HUMAN INFLUENCES AND URBAN SOILS.   Human activities widely influence soil formation. For example, it is believed that Native Americans regularly set fires to maintain several large areas of prairie grasslands in Indiana and Michigan. In more recent times, human

FIGURE 2.30 Pedoturbation by pocket gophers (family *Geomyidae*) can accelerate erosion and inhibit soil profile development in alpine tundra landscapes. In the winter the gophers make tunnels through the thick snow as they move about on the soil surface feeding on aboveground vegetation. Later, as they burrow through the soil, they pack these snow tunnels with excavated soil material. When the snow melts, the tubular cylinders of packed soil remain as snow casts like those shown in these scenes from two areas in the Rocky Mountains. The fine soil in the snow casts and that disturbed by the burrowing are highly susceptible to erosion by rain and snowmelt water. The gopher-disturbed area is thought to erode 10 to 100 times faster than comparable undisturbed tundra. See also Black et al. (1994) and Thorn (1978). (Large photo courtesy Robert Darmody, University of Illinois; inset photo courtesy of R. Weil)

FIGURE 2.31 How ants may influence profile development. (*Left*) Vertical mixing by bringing up from the C horizon bits of weathered granite rock the size of coarse sand (termed **grus**) in an eastern Colorado prairie (11 cm knife handle). (*Right*) Removal of vegetation from around another ant nest in formerly irrigated land in the Nevada desert leaves the soil unprotected and subject to crusting and erosion (clearing is 9 m across). (Photos courtesy of R. Weil)

destruction of natural vegetation (trees and grass) and subsequent tillage of the soil for crop production has abruptly modified soil formation. Likewise, irrigating an arid region soil drastically influences the soil-forming factors, as does adding fertilizer and lime to soils of low fertility. In surface mining and urbanizing areas today, bulldozers may have an effect on soils almost akin to that of the ancient glaciers; they level and mix soil horizons and set the clock of soil formation back to zero.

In other situations, people actually engineer new soils, such as those used in most golf greens and certain athletic fields (see Section 5.6 and Plate 78), the cover material used to vegetate and seal completed landfills (see Section 18.10 and Plate 64), and the plant mediums on rooftop gardens. Humans may even reverse the processes of erosion and sedimentation that normally destroy soils and counteracts soil formation (see Section 2.6 and Chapter 17). For example, in a recent project called *Mud to Parks* (Web link in margin), calcareous sediment dredged from the bottom of the Illinois River was placed as a thick layer of muddy parent material on barren, highly disturbed land. Within a year, the new parent material had dried out, was supporting lush vegetation, and began to develop such soil characteristics as granular and prismatic structure.

Illinois river muck becomes topsoil for new Chicago park, NPR, David Schaper: http://www.npr.org/templates/story/story.php?storyId=1919840

2.6 TOPOGRAPHY

Topography relates to the configuration of the land surface and is described in terms of differences in elevation, slope, and landscape position—in other words, the lay of the land. The topographical setting may either hasten or retard the work of climatic forces (as shown in Box 2.1). Steep slopes generally encourage rapid soil loss by erosion and allow less rainfall to enter the soil before running off. In semiarid regions, the lower effective rainfall on steeper slopes also results in less complete vegetative cover, so there is less plant contribution to soil formation. For all of these reasons, steep slopes prevent the formation of soil from getting very far ahead of soil destruction. Therefore, soils on steep terrain tend to have rather shallow, poorly developed profiles in comparison to soils on nearby, more level sites (Figure 2.32).

In swales and depressions where runoff water tends to concentrate, the regolith is usually more deeply weathered and soil profile development is more advanced. However, in the lowest landscape positions, water may saturate the regolith to such a degree that drainage and aeration are restricted. Here the weathering of some minerals and the decomposition of organic matter are retarded, while the loss of iron and manganese is accelerated. In such low-lying topography, special profile features characteristic of wetland soils may develop (see Section 7.7 on the soils of wetlands).

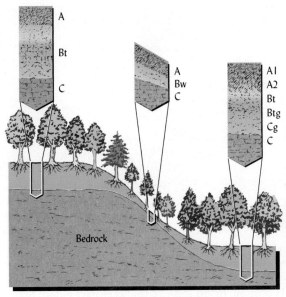

FIGURE 2.32 Topography influences soil properties, including soil depth. The diagram on the left shows the effect of slope on the profile characteristics and the depth of a soil on which forest trees are the natural vegetation. The photo on the right illustrates the same principle under grassland vegetation. Often a relatively small change in slope can have a great effect on soil development. See Section 2.9 for explanation of horizon symbols. (Photo courtesy of R. Weil)

Soils commonly occur together in the landscape in sequence called a **catena** (from the Latin meaning *chain*—visualize a length of chain suspended from two adjacent hills with each link in the chain representing a soil). Each member of the catena occupies a characteristic topographic position. Soils in a catena generally exhibit properties that reflect the influence of topography on water movement and drainage (see, for example, Figure 2.23c). A **toposequence** is a type of catena, in which the differences among the soils result almost entirely from the influence of topography because the soils in the sequence all share the same parent material and have similar conditions regarding climate, vegetation, and time (see, for example, Figure 2.32 and Plate 16 after page 112).

INTERACTION WITH VEGETATION. Topography often interacts with vegetation to influence soil formation. In grassland–forest transition zones, trees are commonly confined to the slight depressions where soil is generally wetter than in upland positions. As would be expected, the nature of the soil in the depressions is quite different from that in the uplands. If water stands for part or all of the year in a depression in the landscape, climate is less influential in regulating soil development. Low-lying areas may give rise to peat bogs and, in turn, to organic soils (see Figure 2.22).

SLOPE ASPECT. Topography affects the absorbance of solar energy in a given landscape. In the northern hemisphere, south-facing slopes are more perpendicular to the sun's rays and are generally warmer and thereby commonly lower in moisture than their north-facing counterparts (see also Figure 7.30). Consequently, soils on the south slopes tend to be lower in organic matter and are not so deeply weathered.

SALT BUILDUP. In arid and semiarid regions, topography influences the buildup of soluble salts. Dissolved salts from surrounding upland soils move on the surface and through the underground water table to the lower-lying areas (see Section 10.3). There they rise to the surface as the water evaporates, often accumulating to plant-toxic levels.

PARENT MATERIAL INTERACTIONS. Topography can also interact with parent material. For example, in areas of tilted beds of sedimentary rock, the ridges often consist of resistant sandstone, while the valleys are underlaid by more weatherable limestone. In many landscapes, topography reflects the distribution of residual, colluvial, and alluvial parent materials, with residual materials on the upper slopes, colluvium covering the lower slopes, and alluvium filling the valley bottom (Figure 2.33).

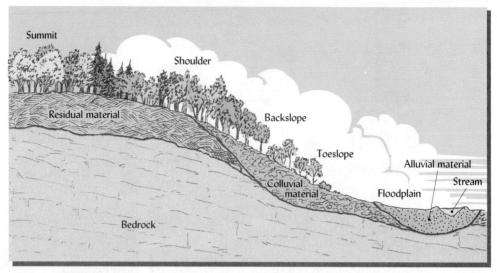

FIGURE 2.33 An interaction of topography and parent material as factors of soil formation. The soils on the summit, toeslope, and floodplain in this idealized landscape have formed from residual, colluvial, and alluvial parent materials, respectively.

2.7 TIME

Soil-forming processes take time to show their effects. The clock of soil formation starts ticking when a landslide exposes new rock to the weathering environment at the surface, when a flooding river deposits a new layer of sediment on its floodplain, when a glacier melts and dumps its load of mineral debris, or when a bulldozer cuts and fills a landscape to level a construction or mine-reclamation site.

RATES OF WEATHERING.   When we speak of a "young" or a "mature" soil, we are not so much referring to the age of the soil in years, as to the degree of weathering and profile development. Time interacts with the other factors of soil formation. For example, on a level site in a warm climate, with much rain falling on permeable parent material rich in reactive minerals, weathering and soil profile differentiation will proceed far more rapidly than on a site with steep slopes and resistant parent material in a cold, dry climate.

In a few instances soils form so rapidly that the effect of time on the process can be measured in a human life span. For example, dramatic mineralogical, structural, and color changes occur within a few months to a few years when certain sulfide-containing materials are first exposed to air by excavation, wetland drainage, or sediment dredging (see Plate 109 and Sections 9.6 and 13.20). Under favorable conditions, organic matter may accumulate to form a darkened A horizon in freshly deposited, fertile alluvium in a mere decade or two. In some cases, incipient B horizons become discernible on humid-region mine spoils in as few as 40 years. Structural alteration and coloring by accumulated iron may form a simple B horizon within a few centuries if the parent material is sandy and the climate is humid. The same degree of B-horizon development would take much longer under conditions less favorable for weathering and leaching. The accumulation of silicate clays and the formation of blocky structure in B horizons usually become noticeable only after several thousand years. Developing in resistant rock, a mature, deeply weathered soil may be hundreds of thousands of years in the making (Figure 2.34).

EXAMPLE OF SOIL GENESIS OVER TIME.   Figure 2.34 is worth studying carefully, as it illustrates changes that typically take place during soil development on residual rock in a warm, humid climate. During the first 100 years, lichens and mosses establish themselves on the bare exposed rock and begin to accelerate its breakdown and the accumulation of dust and organic matter. Within a few hundred years, grasses, shrubs, and stunted trees have taken root in a deepening layer of disintegrated rock and soil, adding greatly to the accumulation of organic materials and to the formation of the A and C horizons. During the next 10,000 years or so, successions of forest trees establish themselves and

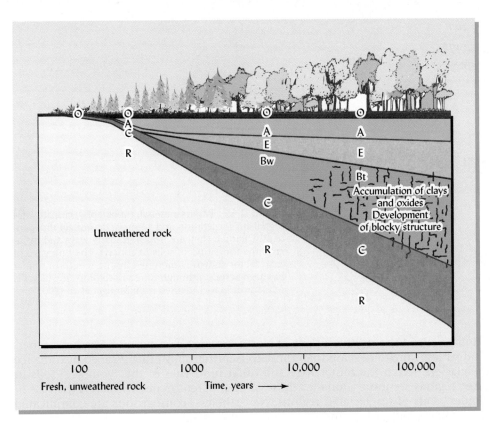

FIGURE 2.34 Progressive stages of soil profile development over time for a residual igneous rock in a warm, humid climate that is conducive to forest vegetation. The time scale increases logarithmically from left to right, covering more than 100,000 years. Note that the mature profile (right side of this figure) expresses the full influence of the forest vegetation as illustrated in Figure 2.26. This mature soil might be classified as an Ultisol (see Section 3.14).

the activities of a multitude of tiny soil organisms transform the surface plant litter into a distinct O horizon. The A horizon thickens somewhat, becomes darker in color, and develops a stable granular structure. Soon, a bleached zone appears just below the A horizon as soluble weathering products, iron oxides, and clays are moved with the water and organic acids percolating down from the litter layer. These transported materials begin to accumulate in a deeper layer, forming a B horizon. The process continues with more silicate clay accumulating and blocky structure forming as the B horizon thickens and becomes more distinct. Eventually, the silicate clays themselves break down, some silica is leached away, and new clays containing less silica form in the B horizon. These clays often become mixed or coated with oxides of iron and aluminum, causing the B horizon to take on a reddish hue. As the entire profile continues to deepen over time, the zone of weathered, unconsolidated rock may become many meters thick.

CHRONOSEQUENCE. Most soil-profile features develop so slowly that it is not possible to directly measure time-related changes in their formation. Indirect methods, such as carbon dating or the presence of fossils and human artifacts, must be turned to for evidence about the time required for different aspects of soil development to occur. In a different approach to studying the effects of time on soil development, soil scientists look for a **chronosequence**—a set of soils that share a common community of organisms, climate, parent material, and slope, but differ with regard to the length of time that the materials have been subjected to weathering and soil formation. A chronosequence can sometimes be found among the soils forming on alluvial terraces of differing age. The highest terraces have been exposed for the longest time, those in lower positions have been more recently exposed by the cutting action of the stream, and those on the current floodplain are the youngest, being still subject to periodic additions of new material.

INTERACTION WITH PARENT MATERIALS. Residual parent materials have generally been subject to soil-forming processes for longer periods of time than have materials transported from one site to another. Soils of the uplands in the southeast part of the United States, for example, have been developing for a much longer period of time than nearby soils in low-lying areas that are developed on marine or alluvial materials. Soils forming on

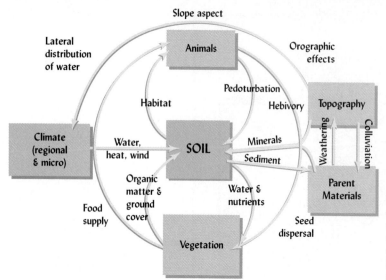

FIGURE 2.35 Parent material, topography, climate, and organisms (vegetation and animals) do not act independently. Rather, they are linked in many ways and influence the formation of soils in concert. The influence of each factor shown is modified by the length of *time* it has been acting, although time as a soil-forming factor is not shown here. [Adapted from Monger et al. (2005)]

glacial materials in the northern parts of North America, Europe, and Asia have generally had far less time to develop than those soils farther south that escaped disturbance by the glaciers. Often the mineralogy and other properties of the "younger" soils in glaciated regions are more similar to those of their parent materials than is the case in the "older" soils of unglaciated regions. Nevertheless, comparisons are complicated because climate, vegetation, and parent material mineralogy also often differ between soils in glaciated and unglaciated areas.

The five soil factors of soil formation act simultaneously and interdependently to influence the nature of soils that develop at a site. Figure 2.35 illustrates some of the complex interactions that can help us predict what soil properties are likely to be encountered in a given environment. We will now turn our attention to the *processes* that cause parent materials to change into soils under the influence of these interacting soil-forming factors.

2.8 FOUR BASIC PROCESSES OF SOIL FORMATION[5]

The accumulation of regolith from the breakdown of bedrock or the deposition (by wind, water, ice, etc.) of unconsolidated geologic materials may precede or, more commonly, occur simultaneously with the development of the distinctive horizons of a soil profile. During the formation (**genesis**) of a soil from parent material, the regolith undergoes many profound changes. These changes are brought about by variations in the four broad soil-forming processes (Figure 2.36) considered next. These four basic processes—often referred to as the soil-forming, or **pedogenic**, processes—help define what distinguishes soils from layers of sediment deposited by geologic processes. They are responsible for soil formation in all kinds of environments—maybe, as suggested in Box 2.2, even under water.

Transformations

Transformations occur when soil constituents are chemically or physically modified or destroyed and others are synthesized from the precursor materials. Many transformations involve weathering of primary minerals, disintegrating and altering some to form various kinds of silicate clays. As other primary minerals decompose, the decomposition products recombine into new minerals that include additional types of silicate

---

[5] For the classic presentation of the processes of soil formation, see Simonson (1959). In addition, detailed discussion of these basic processes and their specific manifestations can be found in Birkeland (1999), Fanning and Fanning (1989), Buol et al. (2005), and Schaetzl and Anderson (2005).

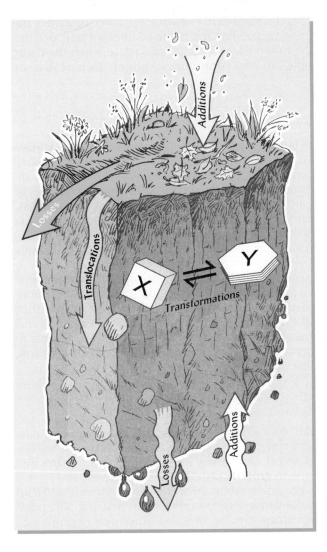

Soil-forming processes
animated:
http://www.environment.
ualberta.ca/soa/process2.cfm

FIGURE 2.36 A schematic illustration of additions, losses, translocations, and transformations as the fundamental processes driving
soil-profile development. (Diagram courtesy of R. Weil)

clays and hydrous oxides of iron and aluminum (see Figure 2.4). Other important transformations involve the decomposition of organic residues and the synthesis of organic
acids, humus, and other products. Still other transformations change the size (e.g.,
physical weathering to smaller particles) or arrangement (e.g., aggregation) of mineral
particles.

Translocations

Translocations involve the movement of inorganic and organic materials laterally
within a horizon or vertically from one horizon up or down to another. Water, either
percolating down with gravity or rising up by capillary action (see Section 5.2), is the
most common translocation agent. The materials moved within the profile include dispersed fine clay particles, dissolved salts, and dissolved organic substances.
Translocations of materials by soil organisms also have a major influence on soil genesis. Important examples include incorporation of surface organic litter into the A and B
horizons by certain earthworms, transport of B and C horizon material to the surface by
mound-building termites, and the widespread burrowing actions of rodents.

Additions

Inputs of materials to the developing soil profile from outside sources are considered
additions. A very common example is the input of organic matter from fallen plant
leaves and sloughed-off roots (the carbon having originated in the atmosphere).
Another ubiquitous addition is dust particles falling on the surface of the soil (wind

may have blown these particles from a source just a few meters away or across the ocean). Still another example, common in arid regions, is the addition of salts or silica dissolved in the groundwater and deposited near or at the soil surface when the rising water evaporates. Animals and people can also contribute additions, such as manure and fertilizers.

Losses

Materials are lost from the soil profile by leaching to groundwater, erosion of surface materials, or other forms of removal. Evaporation and plant use cause losses of water. Leaching and drainage cause the loss of water, dissolved substances such as salts or silica weathered from parent minerals or organic acids produced by microorganisms or plant roots. Erosion, a major loss agent, often removes the finer particles (humus, clay, and silt), leaving the surface horizon relatively sandier and less rich in organic matter than before. Organic matter is also lost by microbial decomposition (see Section 12.10). Grazing by animals or harvest by people can remove large amounts of both organic matter and nutrient elements.

These processes of soil genesis, operating under the influence of the environmental factors discussed previously, give us a logical framework for understanding the relationships between particular soils and the landscapes and ecosystems in which they function. In analyzing these relationships for a given site, ask yourself: What are the materials being added to this soil? What transformations and translocations are taking place in this profile? What materials are being removed? And how have the climate, organisms, topography, and parent material at this site affected these processes over time?

Soil-Forming Processes in Action: A Simplified Example

Consider the changes that might take place as a soil develops from a thick layer of relatively uniform calcareous loess parent material in a climate conducive to grass vegetation (Figure 2.37). Although some physical weathering and leaching of carbonates and salts may be necessary to allow plants to grow in certain parent materials, soil formation really gets started when plants become established and begin to provide *additions* of litter and root residues on and in the surface layers of the parent material. The plant residues are *transformed* by soil organisms into humus and other new organic substances. The accumulation of humus enhances the capacity of the soil to hold water and nutrients, providing positive feedback for accelerated plant growth and further humus buildup. Earthworms, ants, termites, and a host of smaller animals come to live in the soil and feed on the newly accumulating organic resources. In so doing, they accelerate the organic *transformations*, as well as promote the *translocation* of plant residues, loosening the mineral material as they burrow into the soil.

A-HORIZON DEVELOPMENT.   The resulting organic-mineral mixture near the soil surface, which comes into being rather quickly, is commonly the first soil horizon developed, the A horizon. It is darker in color and its chemical and physical properties differ from those of the original parent material. Individual soil particles in this horizon commonly clump together under the influence of organic substances to form granules, differentiating this layer from the deeper layers and from the original parent material. On sloping land, erosion may remove some materials from the newly forming upper horizon, retarding, somewhat, the progress of horizon development.

FORMATION OF B AND C HORIZONS.   Carbonic and other organic acids are carried by percolating waters into the soil, where they stimulate weathering reactions. The acid-charged percolating water dissolves various minerals (a *transformation*) and leaches the soluble products (a *translocation*) from upper to lower horizons, where they may precipitate. This combination of transformation and translocation create [eluvial] zones of depletion in the upper layers and [illuvial] zones of accumulation in the lower. The dissolved substances include both positively charged ions (cations; e.g., Ca^{2+}) and negatively charged ions (anions; e.g., CO_3^{2-} and SO_4^{2-}) released from the breakdown of minerals and organic matter. In semiarid and arid regions, precipitation of these ions produces horizons enriched in calcite ($CaCO_3$) or gypsum ($CaSO_4 \cdot 2H_2O$), designated as a Bk layer in Figure 2.37.

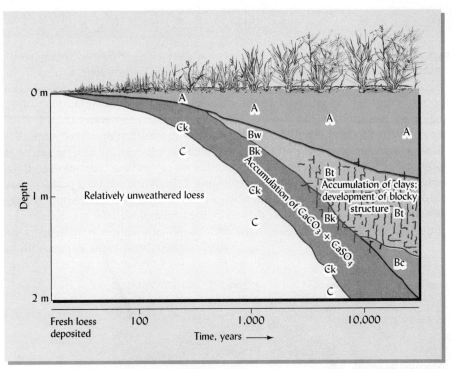

FIGURE 2.37 Development of a hypothetical soil in about 2 m of uniform calcareous loess deposits where the warm subhumid climate is conducive to tall grass prairie vegetation. The time scale increases logarithmically from left to right, covering about 20,000 years. In the initial stages, rainwater, charged with organic acids from microbial respiration, dissolves carbonates from the loess and moves them downward to a zone of accumulation (Bk horizon). As this happens, plant roots take hold in the upper layer and add the organic matter necessary to create an A horizon. Ants, beetles, earthworms, and a host of smaller creatures take up residence and actively mix in surface litter and speed the release of nutrients from the minerals and plant residues. Over time the carbonate concentration zone moves deeper, the A horizon thickens, and noncalcareous B horizons develop as changes in color and structure occur in the weathering loess above the zone of carbonate accumulation. Eventually, silicate clay accumulates in the B horizon (giving it the designation Bt), both by stationary weathering of primary minerals and by traveling there with water percolating from the upper horizons. Compare the stages of development and rates of change to those illustrated in Figure 2.34 for a different parent material, vegetation, and climate. Note that the mature profile (right side of this figure) expresses the full influence of the grassland vegetation as illustrated in Figure 2.26. This mature soil would be classified as a Mollisol (see Section 3.12). (Diagram courtesy of R. Weil)

Over time, the leached surface layer thickens, and the zone of Ca accumulation is moved downward to the maximum depth of water penetration. Where rainfall is great enough to cause significant drainage to the groundwater, some of the dissolved materials may be completely removed from the developing soil profile (*losses*), and the zone of accumulation may move below the reach of plant roots or be dissipated altogether. On the other hand, deep-growing plant roots may intercept some of these soluble weathering products and return them, through leaf- and litter-fall, to the soil surface, thus retarding somewhat the processes of acid weathering and horizon differentiation.

The weathering of primary minerals into clay minerals becomes evident only long after the dissolution and movement of Ca is well underway. The newly formed clay minerals may accumulate where they are formed, or they may move downward and accumulate deeper in the profile. As clay is removed from one layer and accumulates in another, adjacent layers become more distinct from each other, and a Bt horizon (one enriched in silicate clay) is formed. When the accumulated clay in the Bt horizon periodically dries out and cracks, blocklike or prismatic units of soil structure begin to develop (see Figure 2.37 and Section 4.4). As the soil matures, the various horizons within the profile generally become more numerous and more distinctly different from each other.

BOX 2.2   SUBAQUEOUS SOILS—UNDERWATER PEDOGENESIS?

By traditional definitions, soil stops at the water's edge. However, many well-drained soils are occasionally submerged underwater during flooding or heavy rain. Other soils, especially in wetlands (see Section 7.7) are saturated with water most of the time and support emergent aquatic vegetation—hydrophytic plants that are rooted in the soil and grow above the water's surface. So it should not be too surprising that soil scientists are beginning to study soil profiles that develop under shallow water (Figure 2.38) with only submerged aquatic vegetation (SAV—mainly aquatic grasses that are rooted, but do not emerge above the water).

Several teams of soil scientists studying sediment "landscapes" lying under about a meter of Atlantic Ocean water in estuaries and bays have suggested that we expand the concept of soils to include the natural bodies they find in these underwater environments (Figure 2.39). Working in the shallow water or from floating drill rigs, soil scientists have mapped distinct soil bodies in which the four basic soil-forming processes have created distinguishable horizons (Table 2.5). Note the striking similarities to the soil-forming processes pictured in Figure 2.36. Understanding the different soils in the underwater landscape may help estuary managers to inventory and conserve marine coastal resources in much the same way that soils information helps managers of dry land to protect and optimize terrestrial resources.

FIGURE 2.38   *Soil scientists using a long-handled soil auger (right) or a boat equipped with a vibration-corer and winch (left) bring up soil samples to aid in mapping "subaqueous soils" in Chincoteague Bay along the mid-Atlantic Coast of North America. (Photos courtesy of John Martin, left, and Tom McKinna, right)*

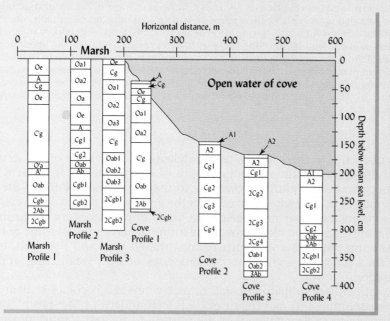
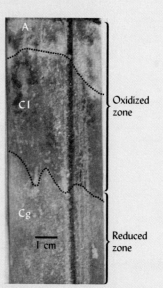

FIGURE 2.39   (Left) *A transect of soil profiles based on auger borings across a tidal marsh and out into the adjacent bay emphasizes the continuity of the soil landscape beyond the water's edge. (Right) A soil core from under the bay water (see Section 2.9 for explanation of the horizon symbols). The buried O horizons (e.g., Oab) were formed by accumulation of plant debris in ancient marshes and subsequently buried by sediment washed in over them. Organic matter from submerged aquatic vegetation (SAV) incorporated into the soil profile has created an A horizon in the upper few cm of the subaqueous soil. [Diagram based on Balduff and Rabenhorst (2005). Photo courtesy of Martin Rabenhorst, University of Maryland]*

TABLE 2.5 Soil-Forming Processes Observed in Sediments with Submerged Vegetation under Shallow Water in a Coastal Estuary

Process	Examples observed in subaqueous soils
Additions	Calcium carbonates added from clam and oyster shells.
	Organic matter added from roots and leaves of aquatic vegetation.
	Mineral sediments washed in by currents.
Losses	Carbon lost by decomposition of organic matter.
	Surface mineral material washed away by storm currents.
Translocations	Oxygen moved into the profile by diffusion from seawater.
	Oxygen and organic material mixed into upper 15 cm by burrowing clams and tubeworms.
Transformations	Formation of humus-like organic matter by microbial processing of plant residues.
	Formation of the mineral pyrite from sulfate in the seawater and iron in the sediments.

Summarized from Demas and Rabenhorst (2001).

2.9 THE SOIL PROFILE

At each location on the land, the Earth's surface has experienced a particular combination of influences from the five soil-forming factors, causing a different set of layers (horizons) to form in each part of the landscape, thus slowly giving rise to the natural bodies we call **soils**. Each soil is characterized by a given sequence of these horizons. A vertical exposure of this sequence is termed a **soil profile**. We will now consider the major horizons making up soil profiles and the terminology used to describe them.

The Master Horizons and Layers[6]

Six **master** soil horizons are commonly recognized and are designated using the capital letters O, A, E, B, C, and R (Figure 2.40). Subordinate horizons may occur within a master horizon and these are designated by lowercase letters following the capital master horizon letter (e.g., Bt, Ap, or Oi).

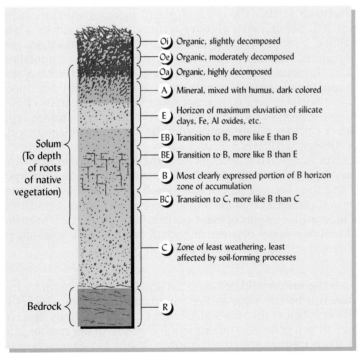

FIGURE 2.40 Hypothetical mineral soil profile showing the major horizons that may be present in a well-drained soil in the temperate humid region. Any particular profile may exhibit only some of these horizons, and the relative depths vary. In addition, however, a soil profile may exhibit more detailed subhorizons than indicated here. The solum includes the A, E, and B horizons plus some cemented layers of the C horizon.

---

[6] In addition to the six master horizons described in this section, **L** (limnic—Greek *limnē*, marsh) and **W** (water) are also considered to be master horizons or layers. The L horizon occurs only in certain organic soils and includes layers of organic and mineral materials deposited in water or by aquatic organisms (e.g., diatomaceous earth, sedimentary peat, and marl). Layers of water (frozen or liquid) found within (not overlying) certain soil profiles are designated as W master horizons.

O HORIZONS. The O group is comprised of organic horizons that generally form above the mineral soil or occur in an organic soil profile. They derive from dead plant and animal residues. Generally absent in grassland regions, O horizons usually occur in forested areas and are commonly referred to as the **forest floor** (see Plates 7, 19, and 70). Often three subordinate O horizons can be distinguished.

The **Oi horizon** is an organic horizon of **fibric** materials—recognizable plant and animal parts (leaves, twigs, and needles), only slightly decomposed. (The Oi horizon is referred to as the *litter* or *L layer* by some foresters.)

The **Oe horizon** consists of **hemic** materials—finely fragmented residues intermediately decomposed, but still with much fiber evident when rubbed between the fingers. (This layer corresponds to the *fermentation* or *F layer* described by some foresters.)

The **Oa horizon** contains **sapric** materials—highly decomposed, smooth, amorphous residues that do not retain much fiber or recognizable tissue structures. (This is the *humidified* or *H layer* designated by some foresters.)

A HORIZONS. The topmost mineral horizons, designated A horizons, generally contain enough partially decomposed (humified) organic matter to give the soil a color darker than that of the lower horizons (Plates 4, 7, and 20). The A horizons are often coarser in texture, having lost some of the finer materials by translocation to lower horizons and by erosion.

E HORIZONS. These are zones of maximum leaching or *eluviation* (from Latin *ex* or *e*, out, and *lavere*, to wash) of clay, iron, and aluminum oxides, which leaves a concentration of resistant minerals, such as quartz, in the sand and silt sizes. An E horizon is usually found underneath the A horizon and is generally lighter in color than either the A horizon above it or the horizon below. Such E horizons are quite common in soils developed under forests, but they rarely occur in soils developed under grassland. Distinct E horizons can be seen in the soils in Plates 10, 19, and 31.

B HORIZONS. B horizons form below an O, A, or E horizon and have undergone sufficient changes during soil genesis so that the original parent material structure is no longer discernable. In many B horizons, materials have accumulated, typically by washing *in* from the horizons above, a process termed **illuviation** (from the Latin *il*, in, and *lavere*, to wash). In humid regions, B horizons are the layers of maximum accumulation of materials such as iron and aluminum oxides (Bo or Bs horizons—see Plates 9, 10, and 31) and silicate clays (Bt horizons), some of which may have illuviated from upper horizons and some of which may have formed in place. Such Bt horizons can be clearly seen in the middle depths of the profiles shown in Plates 1 and 11. In arid and semiarid regions, calcium carbonate or calcium sulfate may accumulate in the B horizon (giving Bk and By horizons, respectively). See Plates 3, 8, and 13.

The B horizons are sometimes referred to incorrectly as the subsoil, a term that lacks precision. In soils with shallow A horizons, part of the B horizon may become incorporated into the plow layer and thus become part of the topsoil. In other soils with deep A horizons, the plow layer or topsoil may include only the upper part of the A horizons, and the subsoil would include the lower part of the A horizon along with the B horizon. This emphasizes the need to differentiate between colloquial terms (*topsoil* and *subsoil*) and technical terms used by soil scientists to describe the soil profile.

C HORIZON. The C horizon is the unconsolidated material underlying the solum (A and B horizons). It may or may not be the same as the parent material from which the solum formed. The C horizon is below the zones of greatest biological activity and has not been sufficiently altered by soil genesis to qualify as a B horizon. In dry regions, carbonates and gypsum may be concentrated in the C horizon. While loose enough to be dug with a shovel, C horizon material often retains some of the structural features of the parent rock or geologic deposits from which it formed (see, for example, the lower third of the profiles shown in Plates 7, 11, and 31). Its upper layers may in time become a part of the solum as weathering and erosion continue.

R LAYERS. These are consolidated rock, with little evidence of weathering.

Subdivisions within Master Horizons

Often distinctive layers exist *within* a given master horizon, and these are indicated by a numeral *following* the letter designation. For example, if three different combinations of structure and colors can be seen in the B horizon, then the profile may include a sequence such as B1–B2–B3 (see Plates 4, 7, and 9).

If two different geologic parent materials (e.g., loess over glacial till) are present within the soil profile, the numeral 2 is placed in front of the master horizon symbols for horizons developed in the second layer of parent material. For example, a soil would have a sequence of horizons designated O–A–B–2C if the C horizon developed in glacial till while the upper horizons developed in loess.

Where a layer of mineral organic soil material was transported by *humans* (usually using machinery) from a source outside the pedon, the "caret" symbol (^) is inserted before the master horizon designation. For example, suppose a landscaping contractor hauls in and spreads a layer of sandy fill material over an existing soil in order to level a site. The resulting soil might eventually (after enough organic matter had accumulated to form an A horizon) have the following sequence of horizons: ^A-^C-2Ab-2Btb, where the first two horizons formed in the human-transported fill (hence the ^ prefixes), and the last two horizons were part of the underlying, now buried soil (hence the lower case "b" designations).

Transition Horizons

Transitional layers between the master horizons (O, A, E, B, and C) may be dominated by properties of one horizon but also have characteristics of another. The two applicable capital letters are used to designate the transition horizons (e.g. AE, EB, BE, and BC), the dominant horizon being listed before the subordinate one (e.g., Plate 1). Letter combinations with a slash such as E/B, are used to designate transition horizons where distinct parts of the horizon have properties of E while other parts have properties of B.

Subordinate Distinctions

Since the capital letter designates the nature of a master horizon in only a very general way, specific horizon characteristics may be indicated by a lowercase letter following the master horizon designation. For example, three types of O horizons (Oi, Oe, and Oa) are indicated in the profile shown in Figure 2.40, which presents a commonly encountered sequence of horizons. Other subordinate distinctions include special physical properties and the accumulation of particular materials such as clays and salts. A list of the recognized subordinate letter designations and their meanings is given in Table 2.6. We suggest that you mark this table for future reference and study it now to get an idea of the distinctive soil properties that can be indicated by horizon designations. By way of illustration, a Bt horizon is a B horizon characterized by clay accumulation (t from the German *ton*, meaning clay); likewise, in a Bk horizon,

TABLE 2.6 **Lowercase Letter Symbols to Designate Subordinate Distinctions Within Master Horizons**

Letter	Distinction	Letter	Distinction
a	Organic matter, highly decomposed	n	Accumulation of sodium
b	Buried soil horizon	o	Accumulation of Fe and Al oxides
c	Concretions or nodules	p	Plowing or other disturbance
d	Dense unconsolidated materials	q	Accumulation of silica
e	Organic matter, intermediate decomposition	r	Weathered or soft bedrock
f	Frozen soil	s	Illuvial organic matter and Fe and Al oxides
ff	Dry permafrost	ss	Slickensides (shiny clay wedges)
g	Strong gleying (mottling)	t	Accumulation of silicate clays
h	Illuvial accumulation of organic matter	u	Presence of human-manufactured materials (artifacts)
i	Organic matter, slightly decomposed	v	Plinthite (high iron, red material)
j	Jarosite (yellow sulfate mineral)	w	Distinctive color or structure without clay accumulation
jj	Cryoturbation (frost churning)	x	Fragipan (high bulk density, brittle)
k	Accumulation of carbonates	y	Accumulation of gypsum
m	Cementation or induration	z	Accumulation of soluble salts

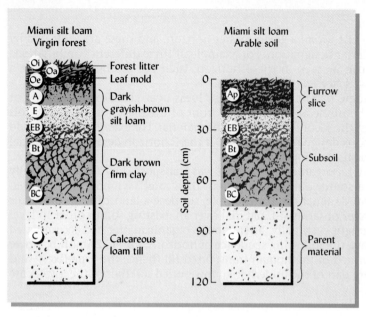

FIGURE 2.41 Generalized profile of the Miami silt loam, one of the Alfisols of the eastern United States, before and after land is plowed and cultivated. The surface layers (O, A, and E) are mixed by tillage and are termed the *Ap* (plowed) *horizon*. If erosion occurs, they may disappear, at least in part, and some of the B horizon will be included in the furrow slice.

carbonates (k) have accumulated. We have already used the suffixes i, a, and e to distinguish different types of O horizons. The significance of several other subordinate horizon designations will be discussed in the next chapter.

Horizons in a Given Profile

It is not likely that the profile of any one soil will show all of the horizons that collectively are shown in Figure 2.40. The ones most commonly found in well-drained soils are Oi and Oe (or Oa) if the land is forested; A or E (or both, depending on circumstances); Bt or Bw; and C. Conditions of soil genesis will determine which others are present and their clarity of definition.

When a virgin (never-cultivated) soil is plowed and cultivated for the first time, the upper 15 to 20 cm becomes the plow layer or Ap horizon (Plate 4). Cultivation, of course, obliterates the original layered condition of the upper portion of the profile, and the Ap horizon becomes more or less homogeneous. Another example of an Ap horizon can be seen in the upper 20 cm of the soil shown in Plate 1 (following page 112). In some soils, the A and E horizons are deeper than the plow layer (Figure 2.41). In other cases where the upper horizons are quite thin, the plow line is just at the top of, or even down in, the B horizon.

In some cultivated land, serious erosion produces a **truncated profile** (see Plate 65 after p. 656). As the surface soil is swept away over the years, the plow reaches deeper and deeper into the profile. Hence, the plowed zone in many cases consists almost entirely of former B horizon material, and the C horizon is correspondingly near the surface. Comparison to a nearby noneroded site can show how much erosion has occurred. Another, sometimes perplexing, profile feature is the presence of a buried soil (see Figures 2.39 and 2.42) resulting from natural or human action. In profile study and description, such a situation requires careful analysis.

Soil Genesis in Nature

Not every contrasting layer of material found in soil profiles is a **genetic horizon** that developed as a result of the processes of soil genesis such as those just described. The parent materials from which many soils develop contained contrasting layers *before* soil genesis started. For example, such parent materials as glacial outwash, marine deposits, or recent alluvium may consist of various layers of fine and coarse particles laid down by separate episodes of sedimentation (see Figure 2.42). Consequently, in characterizing soils, we must recognize not only the genetic horizons and properties that come into being during soil genesis, but also those layers or properties that may have been inherited from the parent material.

FIGURE 2.42 A soil profile forming in recent alluvium in the Atlantic coastal plain. The upper 80 cm of this profile has developed in recent alluvium laid down by occasional floods during the past 300 to 400 years. The A horizon is the result of soil formation, but the various C horizons differ because of geologic rather than soil-formation processes. Several of these layers show unique properties associated with the history of the past three or four centuries. The dark layer at about 50 to 60 cm depth contains abundant bits of charcoal. This charcoal was probably washed off the watershed slopes by heavy rains shortly after European settlers extensively deforested the area using slash-and-burn techniques to open the land for farming in the early 1700s. The layer just above the charcoal was laid down somewhat later and contains such artifacts as bits of antique glass from the late 1700s or 1800s. The layer below the charcoal was laid down prior to widespread European settlement. It contains numerous oyster shells that probably washed in from Native American encampments along the stream. Still deeper in the profile lies a buried soil. Part of a buried A horizon (Ab) and a buried Bg horizon (Bgb) are visible. These layers probably formed over a long period of time when the streambed was in a different location and the site was a poorly drained area with fine-textured materials. (Photo courtesy of R. Weil)

2.10 CONCLUSION

The parent materials from which soils develop vary widely around the world and from one location to another only a few meters apart. Knowledge of these materials, their sources or origins, mechanisms for their weathering, and means of transport and deposition are essential to understanding soil genesis.

Soil formation is stimulated by *climate* and living *organisms* acting on *parent materials* over periods of *time* and under the modifying influence of *topography*. These five major factors of soil formation determine the kinds of soil that will develop at a given site. When all of these factors are the same at two locations, the kind of soil at these locations should be the same.

Soil genesis starts when layers or horizons not present in the parent material begin to appear in the soil profile. Organic matter accumulation in the upper horizons, the downward movement of soluble ions, the synthesis and downward movement of clays, and the development of specific soil particle groupings (structure) in both the upper and lower horizons are signs that the process of soil formation is under way. As we have learned, soil bodies are dynamic in nature. Their genetic horizons continue to develop and change. Consequently, in some soils horizon differentiation has only begun, while in others it is well advanced.

The four general processes of soil formation (gains, losses, transformations, and translocations) and the five major factors influencing these processes provide us with an invaluable logical framework in site selection and in predicting the nature of soil bodies likely to be found on a particular site. Conversely, analysis of the horizon properties of a soil profile can tell us much about the nature of the climatic, biological, and geological conditions (past and present) at the site.

Characterization of the horizons in the profile leads to the identity of a soil individual, which is then subject to classification—the topic of our next chapter.

STUDY QUESTIONS

1. What is meant by the statement, *weathering combines the processes of destruction and synthesis?* Give an example of these two processes in the weathering of a primary mineral.

2. How is water involved in the main types of chemical weathering reactions?

3. Explain the weathering significance of the ratio of silicon to aluminum in soil minerals.

4. Give an example of how parent material may vary across large geographic regions on one hand, but may also vary within a small parcel of land on the other.

5. Name the five factors affecting soil formation. With regard to each of these factors of soil formation, compare a forested Rocky Mountain slope to the semiarid grassland plains far below.

6. How do *colluvium, glacial till,* and *alluvium* differ in appearance and agency of transport?

7. What is *loess,* and what are some of its properties as a parent material?

8. Give two specific examples for each of the four broad processes of soil formation.

9. Assuming a level area of granite rock was the parent material in both cases, describe in general terms how you would expect two soil profiles to differ, one in a warm, semiarid grassland and the other in a cool, humid pine forest.

10. For the two soils described in question 5, make a profile sketch using master horizon symbols and subordinate suffixes to show the approximate depths, sequence, and nature of the horizons you would expect to find in each soil.

11. Visualize a slope in the landscape near where you live. Discuss how specific soil properties (such as colors, horizon thickness, types of horizons present, etc.) would likely change along the toposequence of soils on this slope.

REFERENCES

Balduff, D., and M. Rabenhorst. 2005. "Subaqueous landforms and their associated soils in Chincoteague Bay, Maryland." Abstracts of the SSSA International Annual Meetings (November 6–10, 2005), Salt Lake City, Utah. Soil Science Society of America.

Binkley, D., and O. Menyailo (eds.). 2005. *Tree species effects on soils: Implications for global change.* Kluwer Academic Publishers, Dordrecht.

Birkeland, P. W. 1999. *Soils and Geomorphology,* 3rd ed. (New York: Oxford University Press).

Black, H. C. 1994. *Animal damage management handbook.* General Technical Report PNW-GTR-332. U.S. Department of Agriculture, Forest Service, Pacific Northwest Research Station, Portland, OR.

Buol, S. W., R. J. Southard, R. C. Graham, and P. A. McDaniel. 2005. *Soil genesis and classification,* 5th ed. Iowa State University Press, Ames, IA.

Craft, C. B., and C. J. Richardson. 1998. "Recent and long-term soil accretion and nutrient accumulation in the everglades," *Soil Sci. Soc. Amer. J.* 62:834–843.

Dalton, R. 2002. "California observatory sweeps the skies for springtime Asian dust." *Nature* 416:688.

Demas, G. P., and M. C. Rabenhorst. 2001. "Factors of subaqueous soil formation: A system of quantitative pedology for submerged environments." *Geoderma* 102:189–204.

Fanning, D. S., and C. B. Fanning. 1989. *Soil: Morphology, Genesis, and Classification.* John Wiley and Sons, New York.

Hilgard, E. W. 1921. *Soils: Their Formation, Properties, Composition, and Plant Growth in the Humid and Arid Regions.* Macmillan, London.

Jenny, Hans. 1941. *Factors of Soil Formation: A System of Quantitative Pedology.* Originally published by McGraw-Hill; Dover, Mineola, NY.

Jenny, Hans. 1980. *The Soil Resource—Origins and Behavior.* Ecological Studies, Vol. 37. Springer-Verlag, New York.

Kerr, R. A. 2005. "And now, the younger, dry side of Mars is coming out." *Science* 307:1025–1026.

Likens, G. E., and F. H. Bormann. 1995. *Biogeochemistry of a Forested Ecosystem*, 2nd ed. Springer-Verlag, New York.

Logan, William Bryant. 1995. *Dirt: The Ecstatic Skin of the Earth*. Riverhead Books, New York.

Marlin, J. C., and R. G. Darmody. 2005. *Returning the soil to the land: The mud to parks projects. The Illinois Steward*, Spring. Available online at: http://www.wmrc.uiuc.edu/special_projects/il_river/IL-steward.pdf. (verified 2 May 2007).

Monger, H. C., J. J. Martinez-Rios, and S. A. Khresat. 2005. "Arid and semiarid soils." In D. Hillel, ed., *Encyclopedia of soils in the environment*, pp. 182–187. Elsevier, Oxford.

National Science Foundation. 1975. "All that unplowed land," *Mosaic* **6**:17–21. National Science Foundation, Washington, DC.

Richter, D. D., and D. Markewitz. 2001. *Understanding Soil Change*. Cambridge University Press, Cambridge.

Schaetzl, R., and S. Anderson. 2005. *Soils—Genesis and geomorphology*. Cambridge University Press, Cambridge.

Schlee, J. S. 2000. *Our changing continent*. U.S. Geological Survey Information Services. http://pubs.usgs.gov/gip/continents/ (posted 02/15/2000; verified 10 April 2007).

Simonson, R. W. 1959. "Outline of a generalized theory of soil genesis," *Soil Sci. Soc. Amer. Proc.* **23**:152–156.

Soil Survey Division Staff. 1993. *Soil survey manual*. U.S. Department of Agriculture Handbook 18. Soil Conservation Service. http://soils.usda.gov/technical/manual/ (verified 18 February 2007).

Thorn, C. 1978. "A preliminary assessment of the geomorphic role of pocket gophers in the Alpine zone of the Colorado front range." *Geografiska Annaler* **60A**:3–4.

van Breemen, N., and A. C. Finzi. 1998. "Plant–soil interactions: Ecological aspects and evolutionary implications," *Biogeochemistry* **42**:1–19.

Roy Simonson investigates a soil (R. Weil)

3

SOIL CLASSIFICATION

*It is embarrassing not to be able to agree on what soil is.
In this the pedologists are not alone. Biologists cannot
agree on a definition of life and philosophers on philosophy.*
—HANS JENNY, THE SOIL RESOURCE:
ORIGIN AND BEHAVIOR

We classify things in order to make sense of our world. We do it whenever we call things by group names, based on their important properties. Imagine a world without classifications. Imagine surviving in the woods knowing only that each plant was a plant, not which are edible by people, which attract wildlife, or which are poisonous. Imagine surviving in a city knowing only that each person was a person, not a child or an adult, a male or a female, a police officer, a hoodlum, a friend, a teacher, a potential date, or any of the other categories into which we classify people. So, too, our understanding and management of soils and terrestrial systems would be hobbled if we knew only that a soil was a soil. How could we organize our information about soils? How could we learn from others' experience or communicate our knowledge to clients, colleagues, or students?

From the time crops were first cultivated, humans noticed differences in soils and classified them according to their suitability for different uses. Farmers used descriptive names such as *black cotton soils, rice soils,* or *olive soils.* Other soil names still in common use today suggest the parent materials from which the soils formed: *limestone soils, piedmont soils,* and *alluvial soils.* Such terms may convey some valuable meaning to local users but they are inadequate for helping us to organize our scientific knowledge of soils or for defining the relationships among the soils of the world.

In this chapter we will learn how soils are classified as natural bodies on the basis of their profile characteristics, not merely on the basis of their suitability for a particular use. Such a soil classification system is essential to foster global communications about soils among soil scientists and all people concerned with the management of land and the conservation of the soil resource. Soil classification allows us to take advantage of research and experience at one location to predict the behavior of similarly classified soils at another location. Soil names such as Histosols or Vertisols conjure up similar mental images in the minds of soil scientists everywhere, whether they live in the United States, Europe, Japan, developing countries, or elsewhere. A goal of the classification system is to create a universal language of soils that enhances communication among users of soils around the world.

Compared to most sciences, the organized study of soils is rather young, having begun in the 1870s when the Russian scientist V. V. Dokuchaev and his associates first conceived the idea that soils exist as natural bodies in nature. Russian soil scientists soon developed a system for classifying natural soil bodies, but poor international communications and the reluctance of some scientists to acknowledge such radical ideas delayed the universal acceptance of the natural bodies concept. In the United States, it was not until the late 1920s that C. F. Marbut of the U.S. Department of Agriculture, one of the few scientists who grasped the concept of soils as natural bodies, developed a soil classification scheme based on these principles.

The natural body concept of soils recognizes the existence of individual entities, each of which we call *a soil*. Just as human individuals differ from one another, soil individuals have characteristics distinguishing each from the others. Likewise, just as human individuals may be grouped according to characteristics such as height or gender, soil individuals having one or more characteristics in common may be grouped together. In turn, we may aggregate these groups into higher-level categories of soils, each having some characteristic that sets them apart from the others. Increasingly broad soil groups are defined as one moves up the classification pyramid from *a soil* to *the soil.*

Pedon and Polypedon[1]

There are seldom sharp demarcations between one soil individual and another. Rather, properties gradually change as one moves from one soil individual to an adjacent one. The gradation in soil properties can be compared to the gradation in the wavelengths of light as you move from one color to another in a rainbow. The change is gradual, and yet we identify a boundary differentiating between what we call green and what we call blue. Soils in the field are heterogenous; that is, the profile characteristics are not exactly the same in any two points within the soil individual you may choose to examine. Consequently, it is necessary to characterize a soil individual in terms of an imaginary three-dimensional unit called a **pedon** (rhymes with "head on," from the Greek *pedon,* ground; see Figure 3.1). It is the smallest sampling unit that displays the full range of properties characteristic of a particular soil.

Pedons occupy from about 1 to 10 m^2 of land area. Because it is what is actually examined during field investigation of soils, the pedon serves as the fundamental unit of soil classification. However, a soil unit in a landscape usually consists of a group of very similar pedons, closely associated together in the field. Such a group of similar pedons, or a **polypedon,** is of sufficient size to be recognized as a landscape component termed a **soil individual**.

All the soil individuals in the world that have in common a suite of soil profile properties and horizons that fall within a particular range are said to belong to the same **soil series**. A soil series, then, is a class of soils, not a soil individual, in the same way that *Pinus sylvestrus* is a species of tree, not a particular individual tree. The more than 20,000 soil series characterized in the United States are the basic units used to classify the nation's soils. As we shall see in Section 19.7, units delineated on detailed soil maps are not purely one soil, but are usually named for the soil series to which *most* of the pedons within the unit belong.

Groupings of Soil Individuals

In the concept of soils, the most specific extreme is that of a natural body called *a soil,* characterized by a three-dimensional sampling unit (pedon), related groups of which (polypedons) are included in a soil individual. At the most general extreme is *the soil,* a collection of all these natural bodies that is distinct from water, solid rock, and other natural parts of the Earth's crust. Hierarchical soil classification schemes generally group soils into classes at increasing levels of generality between these two extremes.

Many cultures have traditional names for various classes of soils that help convey the people's collective knowledge about their soil resources (see Box 3.1). Scientific classification of soils began in the late 1800s stemming from the work of Dokuchaev in

---

[1] Although widely used, the polypedon concept is not without its critics. See Ditzler (2005).

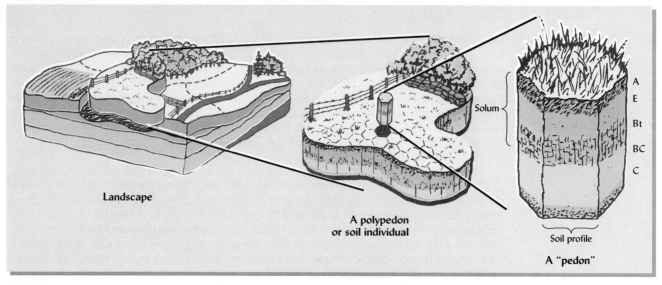

FIGURE 3.1 A schematic diagram to illustrate the concept of pedon and of the soil profile that characterizes it. Note that several contiguous pedons with similar characteristics are grouped together in a larger area (outlined by broken lines) called a *polypedon* or soil *individual*. Several soil individuals are present in the landscape on the left. (Diagram courtesy of R. Weil)

Links to online information about soil classification systems from around the world:

http://www.itc.nl/~rossiter/research/rsrch_ss_class.html#National

Russia (see Section 2.2). Australia, Brazil, Canada, China, the United Kingdom, Russia, and South Africa are among the countries that have developed and continue to use their own national soil classification systems.[2] To provide a global vocabulary for communicating about soils and a reference by which various national soil classification systems can be compared and correlated, scientists working through the Food and Agriculture Organization of the United Nations have developed a three-tier classification system known as the World Reference Base for Soils (see Appendix Table A.1). At the highest level, the world's soils are classified into 32 Soil Reference Groups that are differentiated mainly by the pedogenic process (such as the accumulation of clay in the subsoil) or parent material (such as volcanic ash) that is most responsible for creating the soil properties that typify the particular group. The system also uses an extensive list of prefix qualifiers that can provide a second tier of more specific classes within a Soil Reference Group. The prefix qualifiers include typical types found within a Soil Reference Group, as well as intergrades that indicate similarity to one of the 31 other Soil Reference Groups. A third tier of detail is provided by another list of suffix qualifiers that indicates specific soil properties and features.

In the United States, the Soil Survey Staff of the U.S. Department of Agriculture began in 1951 to collaborate with soil scientists from many other countries with the aim of devising a classification system comprehensive enough to address all soils in the world, not just those in the United States. Finally published in 1975, and revised in 1999, the resulting system, *Soil Taxonomy*, is used in the United States and approximately 50 other countries. This system will be employed throughout this text.

3.2 COMPREHENSIVE CLASSIFICATION SYSTEM: SOIL TAXONOMY[3]

Soil Taxonomy[4] provides a hierarchical grouping of natural soil bodies. The system is based on *soil properties* that can be objectively observed or measured, rather than on presumed mechanisms of soil formation. The system uses a unique nomenclature that gives a definite connotation of the major characteristics of the soils in question. It is truly international because it is not based on any one national language.

---

[2] See Appendix A for summaries of the World Reference Base for Soils and the Canadian and Australian Systems of Soil Classification. For more information on other national systems and their interrelationships, see Eswaran et al. (2003).

[3] For a complete description of *Soil Taxonomy* see Soil Survey Staff (1999). The first edition of *Soil Taxonomy* was published as Soil Survey Staff (1975). For an explanation of the earlier U.S. classification system, see USDA (1938).

[4] Taxonomy is the science of the principles of classification. For a review of the achievements and challenges of *Soil Taxonomy*, see SSSA (1984).

BOX 3.1 ETHNOPEDOLOGY: HOW LOCAL PEOPLE CLASSIFY THEIR SOILS[a]

For thousands of years, most societies were primarily agricultural, and almost everyone worked with soils on a daily basis. Raw survival, on a personal and community level, depended on the food that could be coaxed from the different soils that people found in their environment. By trial and error, people learned which soils were best suited to various crops and which responded best to different kinds of management. As farmers passed their observations and traditions from one generation to the next, they summarized their knowledge about soils by developing unique systems of soil classification. In some regions this local knowledge about soils helped shape agricultural systems that were sustainable for centuries. For example, formal Chinese soil classification goes back two millennia. In Beijing one may still visit the most recent (built in 1421) of a series of large sacrificial altars covered with five differently colored types of soils representing five regions of China (listed here with modern names in parentheses): (1) whitish saline soils (Salids) from the western deserts; (2) black organic-rich soils (Mollisols) from the north; (3) blue-grey waterlogged soils (e.g., Aquepts) from the east; (4) reddish iron-rich soils (Oxisols) from the south; and (5) yellow soils (Inceptisols) from the central loess plateau.

Local languages often reflect a sophisticated and detailed knowledge of how soils differ from one another. *Ethnopedological* studies carried out by anthropologists with an interest in soils (or soil scientists with an interest in anthropology) have documented many hitherto underappreciated indigenous classification systems for local soils.

These systems most commonly classify soils using soil color, texture, hardness, moisture, organic matter, and topography, as well as other soil properties (Figure 3.2). Most of these properties are those observable in the surface horizon, the part of the soil with which farmers come into daily contact. In this respect, the local classifications differ from most scientific classification schemes, which tend to focus on the subsurface horizons (as indicated in Figure 1.17). Rather than viewing this as a weakness, the two approaches can be seen as complementary.

History teaches us that people may rapidly degrade their new land and water resources when they move to an environment that is radically different from what they are used to. Whether these newcomers are Europeans settling the North American continent, lowland African tribes migrating into mountainous regions, or development specialists attempting to transfer technology from one country to another, they generally lack sufficient knowledge of the local soils to allow them to manage the resource in a sustainable way. Indigenous soil classification, reflecting knowledge gained over many generations of living in the local environment, can help and is too important to ignore when planning rural development projects.

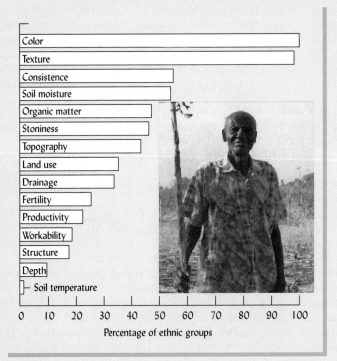

FIGURE 3.2 *Soil characteristics used in soil classification among 62 ethnic groups around the world. [Data from Barrera-Bassols et al. (2006); photo courtesy of R. Weil]*

[a] For a brief review of the study of local soil classifications, see Talawar and Rhoades (1998). For a summary of ancient Chinese soil classification, see Gong et al. (2003).

Bases of Soil Classification

Soil Taxonomy is based on the properties of soils as they are found today. This does not mean that the processes of soil genesis are ignored. In fact, one of the objectives of the system is to group soils that are similar in genesis. However, the specific criteria used to place soils in these groups are those of observable soil properties.

Most of the chemical, physical, and biological properties presented in this text are used as criteria for *Soil Taxonomy*. A few examples are moisture and temperature status throughout the year, as well as color, texture, and structure of the soil. Chemical and mineralogical properties, such as the contents of organic matter, clay, iron and aluminum oxides, silicate clays, salts, the pH, the percentage *base saturation,*[5] and soil depth are other important criteria for classification. While many of the properties used may be observed in the field, others require precise measurements on samples taken to a sophisticated laboratory. This precision makes the system more objective, but in some cases may make the proper classification of a soil quite expensive and time-consuming. Precise measurements are also used to define certain **diagnostic soil horizons**, the presence or absence of which help determine the place of a soil in the classification system.

Diagnostic Surface Horizons of Mineral Soils

The diagnostic horizons that occur at the soil surface are called **epipedons** (from the Greek *epi,* over, and *pedon,* soil). The epipedon includes the upper part of the soil darkened by organic matter, the upper eluvial horizons, or both. It may include part of the B horizon if the latter is significantly darkened by organic matter. Eight are recognized (Table 3.1), but only five occur naturally over wide areas (Figure 3.3). Two of the others, anthropic and plaggen, result from intensive human use. They are common in parts of Europe and Asia where soils have been utilized for many centuries.

Brief overview (with pictures) of diagnostic epipedons and subsurface horizons:

http://soils.umn.edu/academics/classes/soil2125/doc/s5chp1.htm

The **mollic epipedon** (Latin *mollis,* soft) is a mineral surface horizon noted for its dark color (see Plates 8 and 20, after page 112) associated with its accumulated organic matter (>0.6% organic C throughout), for its thickness (generally >25 cm), and for its softness even when dry. It has a high base saturation greater than 50%. Mollic epipedons are moist at least three months a year when the soil temperature is usually 5 °C or higher to a depth of 50 cm. These epipedons are characteristic of soils developed under grassland (Figure 3.4 and Plate 8).

The **umbric epipedon** (Latin *umbra,* shade; hence, dark) has the same general characteristics as the mollic epipedon except the percentage base saturation is lower. This mineral horizon commonly develops in areas with somewhat higher rainfall and where the parent material has lower content of calcium and magnesium.

The **ochric epipedon** (Greek *ochros,* pale) is a mineral horizon that is either too thin, too light in color, or too low in organic matter to be either a mollic or umbric horizon. It is usually not as deep as the mollic or umbric epipedons. As a consequence of its low organic matter content, it may be hard and massive when dry (see Plates 1, 4, 7, and 11).

The **melanic epipedon** (Greek *melas,* melan, black) is a mineral horizon that is very black in color due to its high organic matter content (organic carbon >6%). It is characteristic of soils high in such minerals as allophane, developed from volcanic ash. It is more than 30 cm thick and is extremely light in weight and fluffy for a mineral soil (see Plate 2).

The **histic epipedon** (Greek *histos,* tissue) is a 20 to 60 cm-thick layer of **organic soil materials**[6] overlying a mineral soil. Formed in wet areas, the histic epipedon is a layer of peat or muck with a black to dark brown color and a very low density.

---

[5]The percentage base saturation is the percentage of the soil's negatively charged sites (cation exchange capacity) that are satisfied by attracting nonacid (or *base*) cations (such as Ca^{2+}, Mg^{2+}, and K^+) (see Section 9.3).

[6]**Organic soil material**, such as peat and muck, may consist almost entirely of organic matter, but it is technically defined as containing more than a certain *minimum amount of organic matter* as follows: If the material is not water-saturated in its natural state, its organic matter content must be at least 35% (about 200 g/kg organic C). If it is water-saturated during part of the year in its natural state, then the minimum organic matter content varies with the amount of clay in the material, ranging from 20% (120 g/kg organic C) if no clay is present to 30% (180 g/kg organic C) if the clay content exceeds 600 g/kg. For a discussion of the relationship between organic matter and organic carbon, see Section 12.4 and the footnote in Table 12.1.

TABLE 3.1 Major Features of Diagnostic Horizons in Mineral Soils Used for Differentiation at the Higher Levels of *Soil Taxonomy*

Diagnostic horizon (and typical genetic horizon designation)	Major features
Surface horizons = epidedons	
Anthropic (A)	Human-modified mollic-like horizon, high in available P
Folistic (O)	Organic horizon saturated for less than 30 days per normal year
Histic (O)	Very high in organic content, wet during some part of year
Melanic (A)	Thick, black, high in organic matter (>6% organic C), common in volcanic ash soils
Mollic (A)	Thick, dark-colored, high base saturation, strong structure
Ochric (A)	Too light-colored, low organic content or thin to be Mollic; may be hard and massive when dry
Plaggen (A)	Human-made sodlike horizon created by years of manuring
Umbric (A)	Same as Mollic except low base saturation
Subsurface horizons	
Agric (A or B)	Organic and clay accumulation just below plow layer resulting from cultivation
Albic (E)	Light-colored, clay and Fe and Al oxides mostly removed
Argillic (Bt)	Silicate clay accumulation
Calcic (Bk)	Accumulation of $CaCO_3$ or $CaCO_3 \cdot MgCO_3$
Cambic (Bw, Bg)	Changed or altered by physical movement or by chemical reactions, generally nonilluvial
Duripan (Bqm)	Hard pan, strongly cemented by silica
Fragipan (Bx)	Brittle pan, usually loamy textured, dense, coarse prisms
Glossic (E)	Whitish eluvial horizon that tongues into a Bt horizon
Gypsic (By)	Accumulation of gypsum
Kandic (Bt)	Accumulation of low-activity clays
Natric (Btn)	Argillic, high in sodium, columnar or prismatic structure
Oxic (Bo)	Highly weathered, primarily mixture of Fe, Al oxides and nonsticky-type silicate clays
Petrocalcic (Ckm)	Cemented calcic horizon
Petrogypsic (Cym)	Cemented gypsic horizon
Placic (Csm)	Thin pan cemented with iron alone or with manganese or organic matter
Salic (Bz)	Accumulation of salts
Sombric (Bh)	Organic matter accumulation
Spodic (Bh, Bs)	Organic matter, Fe and Al oxide accumulation
Sulfuric (Cj)	Highly acid with Jarosite mottles

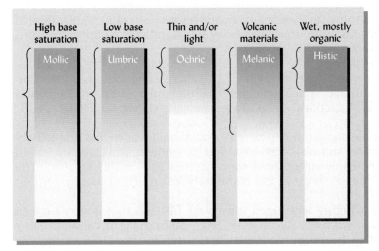

FIGURE 3.3 Representative profile characteristics of five surface diagnostic horizons (epidedons). The comparative organic matter levels and distribution are indicated by the darkness of colors. The mollic and umbric epidedons have similar organic matter distribution but the percentage base saturation is higher (greater than 50%) in the mollic epidedon and lower (less than 50%) in the umbric epidedon. The ochric epidedon is lower in organic matter content; consequently, it is light in color and sometimes hard when dry. Two other epidedons have very high organic matter contents and are very dark in color. The melanic epidedon is formed on recently deposited volcanic materials, usually in cool wet areas. The histic epidedon is formed from organic deposits laid down over mineral soils, usually in wet, boggy conditions. The relative depth of each epidedon is shown by the brackets.

Ap

A2 } Mollic epipedon

Bt1

} Argillic horizon

Bt2

Bk

FIGURE 3.4 The mollic epipedon (a diagnostic horizon) in this soil includes genetic horizons designated Ap, A2, and Bt1, all darkened by the accumulation of organic matter. A subsurface diagnostic horizon, the argillic horizon, overlaps the mollic epipedon. The argillic horizon is the zone of illuvial clay accumulation (Bt1 and Bt2 horizons in this profile). Scale marked every 10 cm. (Photo courtesy of R. Weil)

Diagnostic Subsurface Horizons

Many subsurface diagnostic horizons are used to characterize different soils in *Soil Taxonomy* (Table 3.1 and Figure 3.5). Each diagnostic horizon provides a characteristic that helps place a soil in its proper class in the system. We will briefly discuss a few of the more commonly encountered subsurface diagnostic horizons.

The **argillic horizon** is a subsurface accumulation of silicate clays that have moved downward from the upper horizons or have formed in place. Examples are shown in Figure 3.4 and in Plate 1 between 50 and 90 cm. The clays often are found as coatings on pore walls (as shown in Figure 4.2) and surfaces of the structural groupings. The coatings usually appear as shiny surfaces or as clay bridges between sand grains. Termed *argillans* or *clay skins,* they are concentrations of clay translocated from upper horizons (see Plates 18 and 24).

The **natric horizon** likewise has silicate clay accumulation (with clay skins), but the clays are accompanied by more than 15% exchangeable sodium on the colloidal complex and by columnar or prismatic soil structural units. The natric horizon is found mostly in arid and semiarid areas. Examples are shown in Figures 4.13*e* and 10.18.

The **kandic horizon** has an accumulation of Fe and Al oxides as well as low-activity silicate clays (e.g., kaolinite), but clay skins need not be evident. The clays are low in activity as shown by their low cation-holding capacities (<16 $cmol_c$/kg clay). The epipedon that overlies a kandic horizon has commonly lost much of its clay content (see Figure 3.6).

The **oxic horizon** is a highly weathered subsurface horizon that is very high in Fe and Al oxides and in low-activity silicate clays (e.g., kaolinite). The cation-holding capacity is <16 $cmol_c$/kg clay. The horizon is at least 30 cm deep and has <10% weatherable

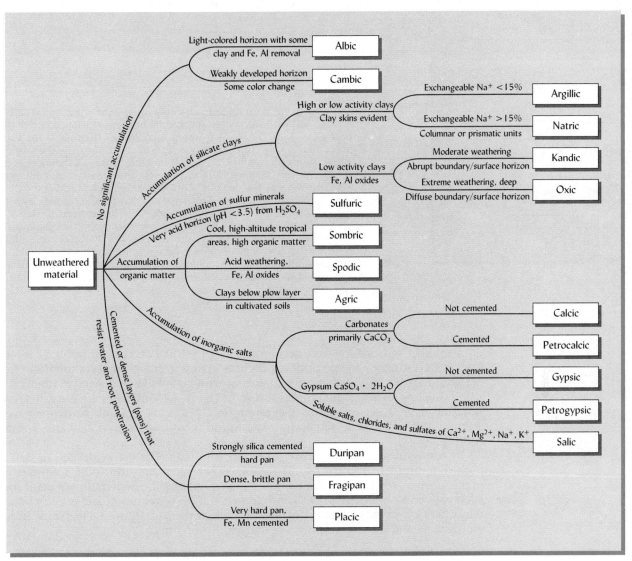

FIGURE 3.5 Names and major distinguishing characteristics of subsurface diagnostic horizons. Among the characteristics emphasized is the accumulation of silicate clays, organic matter, Fe and Al oxides, calcium compounds, and soluble salts, as well as materials that become cemented or highly acidified, thereby constraining root growth. The presence or absence of these horizons plays a major role in determining in which class a soil falls in *Soil Taxonomy*. See Chapter 8 for a discussion of low- and high-activity clays.

minerals in the fine fraction. It is generally physically stable, crumbly, and not very sticky, despite its high clay content. It is found mostly in humid tropical and subtropical regions (see Plate 9, between about 1 and 3 feet on the scale and Plate 25 at 70 to 170 cm).

The **spodic horizon** is an illuvial horizon that is characterized by the accumulation of colloidal organic matter and aluminum oxide (with or without iron oxide). It is commonly found in highly leached forest soils of cool humid climates, typically on sandy-textured parent materials (see Plate 10, reddish-brown and black layers below the whitish layer).

The **sombric horizon** is an illuvial horizon, dark in color because of high organic matter accumulation. It has a low degree of base saturation and is found mostly in the cool, moist soils of high plateaus and mountains in tropical and subtropical regions (Plate 23).

The **albic horizon** is a light-colored eluvial horizon that is low in clay and oxides of Fe and Al. These materials have largely been moved downward from this horizon (see Plate 10, starting at about 10 cm depth).

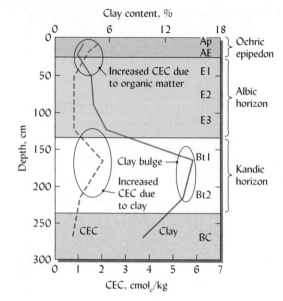

FIGURE 3.6 Vertical variation in clay content and cation exchange capacity (CEC) in a soil with thick albic and kandic horizons (E1-E2-E3 and Bt1-Bt2, respectively). Note the well-expressed "clay bulge" that marks the kandic horizon. Similar clay enrichment (plus clay skins or other visual evidence of clay illuviation) characterizes an argillic horizon. This is a kandic rather than an argillic horizon because there was no visible evidence of clay illuviation and because the accumulated clay is of *low-activity* types, meaning the CEC of the clay is less than 16 cmol$_c$/kg of clay. Note that the CEC of the clay = CEC of the soil × percent clay in the soil ÷ 100. The sharp increase in clay at the upper boundary of the kandic horizon, the considerable thickness (more than 100 cm) of the clay-rich layer, and the low CEC per kg of clay are all indications that this is a very old, highly mature soil. It formed under humid, subtropical conditions in sandy sediments in the upper coastal plain of Georgia. It is classified in the Kandiudults great group in *Soil Taxonomy*. (Data from Shaw et al., 2000)

A number of horizons have accumulations of saltlike chemicals that have leached from upper horizons in the profile. **Calcic horizons** contain an accumulation of carbonates (mostly $CaCO_3$) that often appear as white chalklike nodules (see the Bk horizon in the lower part of the profiles shown in Figure 3.4 and Plates 13 and 20). **Gypsic horizons** have an accumulation of gypsum ($CaSO_4 \cdot 2H_2O$), and **salic horizons** have an accumulation of soluble salts. These are found mostly in soils of arid and semiarid regions.

In some subsurface diagnostic horizons, the materials are cemented or densely packed, resulting in relatively impermeable layers called *pans* (**duripan, fragipan,** and **placic horizons**).[7] These can resist water movement and the penetration of plant roots. Such pans constrain plant growth and may encourage water runoff and erosion because rainwater cannot move readily downward through the soil. Figure 3.5 explains the genesis of these and the other subsurface diagnostic horizons.

Soil Moisture Regimes (SMR)

A soil moisture regime refers to the presence or absence of either water-saturated conditions (usually groundwater) or plant-available soil water during specified periods in the year in what is termed the *control section* of the soil. The upper boundary of the SMR control section is the depth that 2.5 cm of water will penetrate within 24 hours when added to a dry soil. The lower boundary is the depth that 7.5 cm of water will penetrate. The control section ranges from 10 to 30 cm for soils high in fine particles (clay) and from 30 to 90 cm for sandy soils. Several moisture regime classes are used to characterize soils.

Global soil moisture regimes map:
http://soils.usda.gov/use/worldsoils/mapindex/smr.html

Aquic. Soil is saturated with water and virtually free of gaseous oxygen for sufficient periods of time for evidence of poor aeration (gleying and mottling) to occur.

Udic. Soil moisture is sufficiently high year-round in most years to meet plant needs. This regime is common for soils in humid climatic regions and characterizes about one-third of the worldwide land area. An extremely wet moisture regime with excess moisture for leaching throughout the year is termed **perudic.**

---

[7] Well-developed argillic horizons may present such a great and abrupt increase in clay content that water and root movement are severely restricted; such a horizon is commonly referred to by the non-taxonomic term, *claypan*.

Ustic. Soil moisture is intermediate between Udic and Aridic regimes—generally there is some plant-available moisture during the growing season, although significant periods of drought may occur.

Aridic. The soil is dry for at least half of the growing season and moist for less than 90 consecutive days. This regime is characteristic of arid regions. The term *torric* is used to indicate the same moisture condition in certain soils that are both hot and dry in summer, though they may not be hot in winter.

Xeric. This soil moisture regime is found in typical Mediterranean-type climates, with cool, moist winters and warm, dry summers. Like the Ustic regime, it is characterized by having long periods of drought in the summer.

These terms are used to diagnose the soil moisture regime and are helpful not only in classifying soils but in suggesting the most sustainable long-term use of soils.

Soil Temperature Regimes

Soil temperature regimes, such as frigid, mesic, and thermic, are used to classify soils at some of the lower levels in *Soil Taxonomy*. The cryic (Greek *kryos*, very cold) temperature regime distinguishes some higher-level groups. These regimes are based on mean annual soil temperature, mean summer temperature, and the difference between mean summer and winter temperatures, all at 50 cm depth. The specific temperature regimes will be described in the discussion of soil families (Section 3.17).

3.3 CATEGORIES AND NOMENCLATURE OF SOIL TAXONOMY

There are six hierarchical categories of classification in *Soil Taxonomy:* (1) *order,* the highest (broadest) category, (2) *suborder,* (3) *great group,* (4) *subgroup,* (5) *family,* and (6) *series* (the most specific category). The lower categories fit within the higher categories (Figure 3.7). Thus, each order has several suborders, each suborder has several great groups, and so forth. This system may be compared with those used for the classification of plants or animals, as shown in Table 3.2. Just as *Trifolium repens* identifies a specific kind of plant, the Miami series identifies a specific kind of soil.

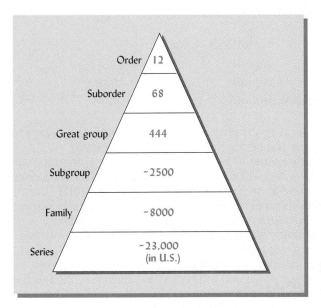

FIGURE 3.7 The categories of *Soil Taxonomy* and approximate number of units in each category.

TABLE 3.2 Comparison of the Classification of a Common Cultivated Plant, White Clover (*Trifolium repens*), and a Soil, Miami Series

Flowering Plant Classification			*Soil classification*	
Division	Magnoliophyta		Order	Alfisols
Class	Magnoliopsida		Suborder	Udalfs
Subclass	Rosidae	Increase specificity →	Great Group	Hapludalfs
Order	Fabales		Subgroup	Oxyaquic Hapludalfs
Family	Fabaceae		Family	Fine loamy, mixed, mesic, active
Genus	*Trifolium*			
Species	*repens*		Series	Miami
			Phase[a]	Miami silt loam

[a] Technically not a category in *Soil Taxonomy* but used in field surveying. *Silt loam* refers to the texture of the A horizon.

Nomenclature of Soil Taxonomy

Although unfamiliar at first sight, the nomenclature system has a logical construction and conveys a great deal of information about the nature of the soils named. The system is easy to learn after a bit of study. The nomenclature is used throughout this book, especially to identify the kinds of soils shown in illustrations. When reading, if you make a conscious effort to identify the parts of each soil class mentioned in the text and figure captions and recognize the level of category indicated, the system will become second nature.

The names of the classification units are combinations of syllables, most of which are derived from Latin or Greek, and are root words in several modern languages. Since each part of a soil name conveys a concept of soil character or genesis, the name automatically describes the general kind of soil being classified. For example, soils of the order **Aridisols** (from the Latin *aridus,* dry, and *solum,* soil) are characteristically dry soils in arid regions. Those of the order **Inceptisols** (from the Latin *inceptum,* beginning, and *solum,* soil) are soils with only the beginnings or inception of profile development. Thus, the names of orders are combinations of (1) formative elements, which generally define the characteristics of the soils, and (2) the ending *sols.*

The names of suborders automatically identify the order of which they are a part. For example, soils of the suborder **Aquolls** are the wetter soils (from the Latin *aqua,* water) of the Mollisols order. Likewise, the name of the great group identifies the suborder and order of which it is a part. **Argiaquolls** are Aquolls with clay or argillic (Latin *argilla,* white clay) horizons. In the following illustration, note that the three letters *oll* identify each of the lower categories as being in the Mollisols order.

Mollisols	Order
Aqu**oll**s	Suborder
Argiaqu**oll**s	Great group
Typic Argiaqu**oll**s	Subgroup

If one is given only the subgroup name, the great group, suborder, and order to which the soil belongs are automatically known.

Family names in general identify subsets of the subgroup that are similar in texture, mineral composition, and mean soil temperature at a depth of 50 cm. Thus the name **fine, mixed, mesic, active Typic Argiaquolls** identifies a family in the Typic Argiaquolls subgroup with a fine texture, mixed clay mineral content, mesic (8 to 15 °C) soil temperature, and clays active in cation exchange.

Soil series are named after a geographic feature (town, river, etc.) near where they were first recognized. Thus, names such as *Fort Collins, Cecil, Miami, Norfolk,* and *Ontario* identify soil series first described near the town or geographic feature named. Approximately 23,000 soil series have been classified in the United States alone.

In detailed field soil surveying, soil series are sometimes further differentiated on the basis of surface soil texture, degree of erosion, slope, or other characteristics. These

In the U.S., "official state soils" share the same level of distinction as official state flowers and birds: http://soils.usda.gov/gallery/state_soils/

practical subunits are called soil **phases**. Names such as *Fort Collins loam, Cecil clay,* or *Cecil clay loam, eroded phase* are used to identify such phases. Note, however, that soil phases, practical as they may be in local situations, are *not* a category in the *Soil Taxonomy* system.

With this brief explanation of the nomenclature of *Soil Taxonomy,* we will now consider the general nature of soils in each of the soil orders.

3.4 SOIL ORDERS

Maps and photos of each soil order: http://soils.ag.uidaho.edu/soilorders/index.htm

Each of the world's soils is assigned to one of 12 **orders**, largely on the basis of soil properties that reflect a major course of development, with considerable emphasis placed on the presence or absence of major diagnostic horizons (Table 3.3). As an example, many soils that developed under grassland vegetation have the same general sequence of horizons and are characterized by a mollic epipedon—a thick, dark, surface horizon that is high in non-acid cations. Soils with these properties are thought to have been formed by the same general genetic processes, but it is because of the properties they have in common that they are included in the same order: Mollisols. The names and major characteristics of each soil order are shown in Table 3.3. Note that all order names have a common ending, *sols* (from the Latin *solum,* soil).

The general conditions that enhance the formation of soils in the different orders are shown in Figure 3.8. From soil profile characteristics, soil scientists can ascertain the relative degree of soil development in the different orders, as shown in this figure. Note that soils with essentially no profile layering (Entisols) have the least development, while the deeply weathered soils of the humid tropics (Oxisols and Ultisols) show the greatest soil development. The effect of climate (temperature and moisture) and of vegetation (forests or grasslands) on the kinds of soils that develop is also indicated in Figure 3.8. Study Table 3.3 and Figure 3.8 to better understand the relationship between soil properties and the terminology used in *Soil Taxonomy.*

To some degree, most of the soil orders occur in climatic regions that can be described by moisture and temperature regimes. Figure 3.9 illustrates some of the relationships among the soil orders with regard to these climatic factors. While only the Gelisols and Aridisols orders are defined directly in relation to climate, Figure 3.9 indicates that orders with the most highly weathered soils tend to be associated with the warmer and wetter climates. Figure 3.10 is a simplified general soil map of North

TABLE 3.3 Names of Soil Orders in *Soil Taxonomy* with Their Derivation and Major Characteristics

The bold letters in the order names indicate the formative element used as the ending for suborders and lower taxa within that order.

Name	Formative element	Derivation	Pronunciation	Major characteristics
Alfisols	alf	Nonsense symbol	Ped**alf**er	Argillic, natric, or kandic horizon; high-to-medium base saturation
Andisols	and	Jap. *ando,* black soil	**And**esite	From volcanic ejecta, dominated by allophane or Al-humic complexes
Ar**id**isols	id	L. *aridus,* dry	Ar**id**	Dry soil, ochric epipedon, sometimes argillic or natric horizon
Entisols	ent	Nonsense symbol	Rec**ent**	Little profile development, ochric epipedon common
G**el**isols	el	Gk. *gelid,* very cold	J**el**ly	Permafrost, often with cryoturbation (frost churning)
H**ist**osols	ist	Gk. *histos,* tissue	H**ist**ology	Peat or bog; >20% organic matter
Inc**ept**isols	ept	L. *inceptum,* beginning	Inc**ept**ion	Embryonic soils with few diagnostic features, ochric or umbric epipedon, cambic horizon
M**oll**isols	oll	L. *mollis,* soft	M**oll**ify	Mollic epipedon, high base saturation, dark soils, some with argillic or natric horizons
Oxisols	ox	Fr. *oxide,* oxide	**Ox**ide	Oxic horizon, no argillic horizon, highly weathered
Sp**od**osols	od	Gk. *spodos,* wood ash	P**od**zol; odd	Spodic horizon commonly with Fe, Al oxides and humus accumulation
Ultisols	ult	L. *ultimus,* last	**Ult**imate	Argillic or kandic horizon, low base saturation
V**ert**isols	ert	L. *verto,* turn	Inv**ert**	High in swelling clays; deep cracks when soil is dry

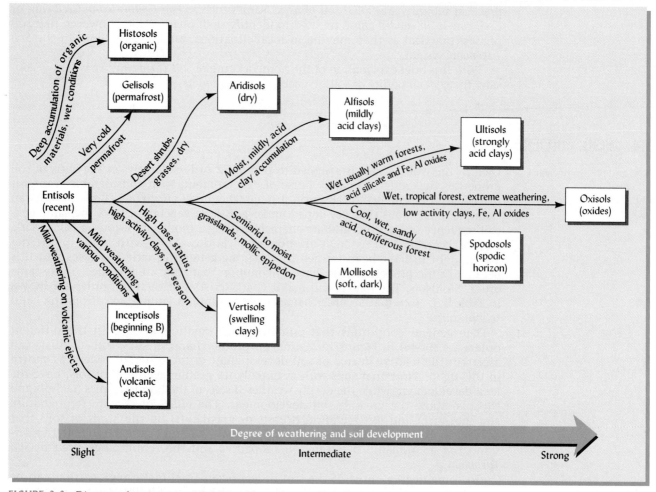

FIGURE 3.8 Diagram showing general degree of weathering and soil development in the different soil orders classified in *Soil Taxonomy.* Also shown are the general climatic and vegetative conditions under which soils in each order are formed.

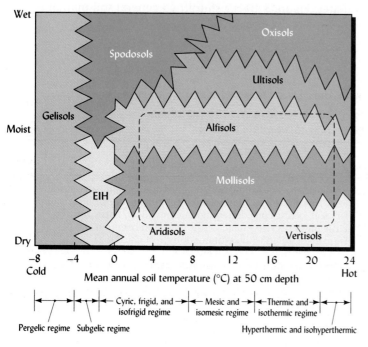

FIGURE 3.9 Diagram showing the general soil moisture and soil temperature regimes that characterize the most extensive soils in each of eight soil orders. Soils of the other four orders (Andisols, Entisols, Inceptisols, and Histosols) may be found under any of the soil moisture and temperature conditions (including the area marked EIH). Major areas of Vertisols are found only where clayey materials are in abundance and are most extensive where the soil moisture and temperature conditions approximate those shown inside the box with broken lines. Note that these relationships are only approximate and that less extensive areas of soils in each order may be found outside the indicated ranges. For example, some Ultisols (Ustults) and Oxisols (Ustox) have soil moisture levels for at least part of the year that are much lower than this graph would indicate. (The terms used at the bottom to describe the soil temperature regimes are those used in helping to identify soil families.)

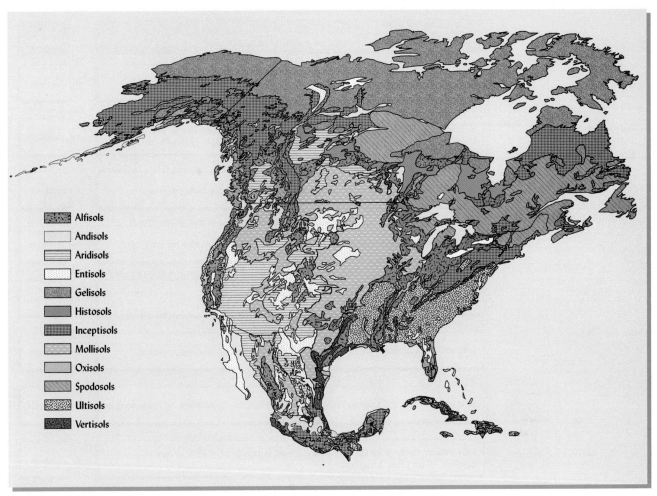

Legend:
- Alfisols
- Andisols
- Aridisols
- Entisols
- Gelisols
- Histosols
- Inceptisols
- Mollisols
- Oxisols
- Spodosols
- Ultisols
- Vertisols

FIGURE 3.10 Simplified soils map of North America showing the general distribution of the 12 soil orders defined by *Soil Taxonomy*. For a more detailed soils map of the United States showing the distribution of orders and suborders, see the back endpaper of this book. Map adapted from USDA/NRCS.

America showing the major areas dominated by each order. Profiles for each soil order are shown in color on Plates 1 through 12. A more detailed color-coded soil map for the United States can be found on the back endpaper of this book; a general world map of the 12 soil orders is printed in color on the front endpaper. In studying these maps, try to confirm that the distribution of the soil orders is in accordance with what you know about the climate in various regions of the world.

Although a detailed description of all the lower levels of soil categories is far beyond the scope of this (or any other) book, a general knowledge of the 12 soil orders is essential for understanding the nature and function of soils in different environments. The simplified key given in Figure 3.11 helps illustrate how *Soil Taxonomy* can be used to key out the order of any soil based on observable and measurable properties of the soil profile. Because certain diagnostic properties take precedence over others, the key must always be used starting at the top, and working down. It will be useful to review this key after reading about the general characteristics, nature, and occurrence of each soil order.

We will now consider each of the soil orders, beginning with those characterized by little profile development and progressing to those with the most highly weathered profiles (as represented from left to right in Figure 3.8).

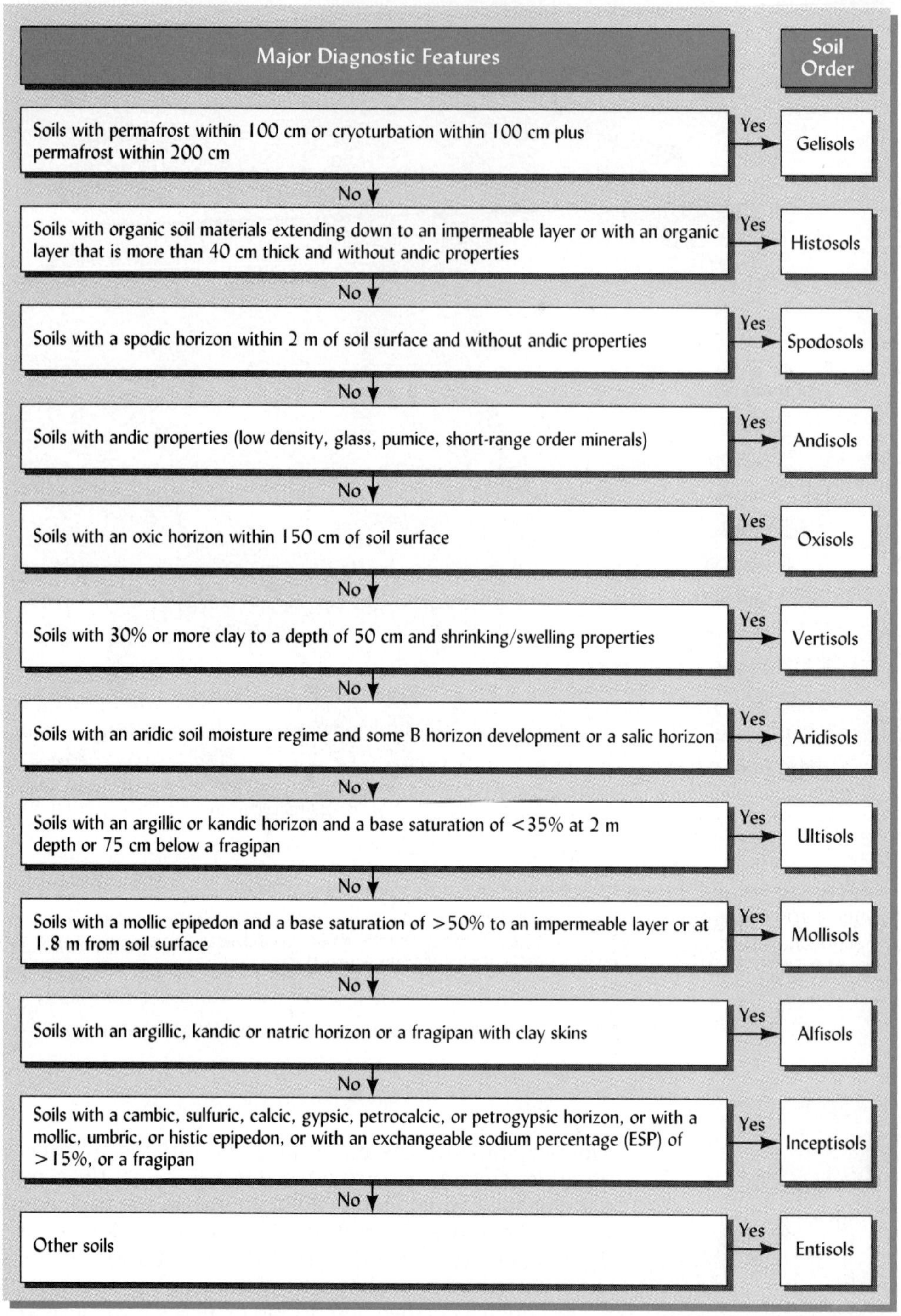
Major Diagnostic Features
Soil Order
Soils with permafrost within 100 cm or cryoturbation within 100 cm plus permafrost within 200 cm
Yes
Gelisols
No
Soils with organic soil materials extending down to an impermeable layer or with an organic layer that is more than 40 cm thick and without andic properties
Yes
Histosols
No
Soils with a spodic horizon within 2 m of soil surface and without andic properties
Yes
Spodosols
No
Soils with andic properties (low density, glass, pumice, short-range order minerals)
Yes
Andisols
No
Soils with an oxic horizon within 150 cm of soil surface
Yes
Oxisols
No
Soils with 30% or more clay to a depth of 50 cm and shrinking/swelling properties
Yes
Vertisols
No
Soils with an aridic soil moisture regime and some B horizon development or a salic horizon
Yes
Aridisols
No
Soils with an argillic or kandic horizon and a base saturation of <35% at 2 m depth or 75 cm below a fragipan
Yes
Ultisols
No
Soils with a mollic epipedon and a base saturation of >50% to an impermeable layer or at 1.8 m from soil surface
Yes
Mollisols
No
Soils with an argillic, kandic or natric horizon or a fragipan with clay skins
Yes
Alfisols
No
Soils with a cambic, sulfuric, calcic, gypsic, petrocalcic, or petrogypsic horizon, or with a mollic, umbric, or histic epipedon, or with an exchangeable sodium percentage (ESP) of >15%, or a fragipan
Yes
Inceptisols
No
Other soils
Yes
Entisols

FIGURE 3.11 A simplified key to the 12 soil orders in *Soil Taxonomy*. In using the key, always begin at the top. Note how diagnostic horizons and other profile features are used to distinguish each soil order from the remaining orders. Entisols, having no such special diagnostic features, key out last. Also note that the sequence of soil orders in this key bears no relationship to the degree of profile development and adjacent soil orders may not be more similar than nonadjacent ones. See Section 3.2 for explanations of the diagnostic horizons.

3.5 ENTISOLS (RECENT: LITTLE IF ANY PROFILE DEVELOPMENT)

Weakly developed mineral soils without natural genetic (subsurface) horizons or with only the beginnings of such horizons (see Plate 4, after page 112) belong to the Entisols order. Most have an ochric epipedon and a few have human-made anthropic or agric

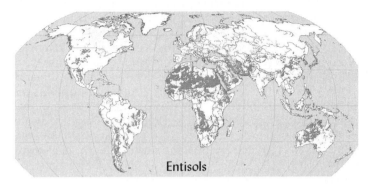

16.3% of global and 12.2% of U.S. ice-free land

Suborders are:

    Aquents (wet)
    Arents (mixed horizons)
    Fluvents (alluvial deposits)
    Orthents (typical)
    Psamments (sandy)

epipedons. Some have albic subsurface horizons. Soil productivity ranges from very high for certain Entisols formed in recent alluvium to very low for those forming in shifting sand or on steep rocky slopes.

This is an extremely diverse group of soils with little in common, other than the lack of evidence for all but the earliest stages of soil formation. Entisols are either young in years or their parent materials have not reacted to soil-forming factors. On parent materials, such as fresh lava flows or recent alluvium (Fluvents), there has been too little time for much soil formation. In extremely dry areas, scarcity of water and vegetation may inhibit soil formation. Likewise, frequent saturation with water (Aquents) may delay soil formation. Some Entisols occur on steep slopes, where the rates of erosion may exceed the rates of soil formation, preventing horizon development. Others occur on construction sites where bulldozers destroy or mix together the soil horizons, causing the existing soils to become Entisols (some have suggested that these be called *urbents,* or urban Entisols).

Distribution and Use

Globally, Entisols are found under a wide variety of environmental conditions (Figure 3.10 and endpapers). For example, in rocky and mountainous regions, shallow, medium-textured Entisols (Orthents) are common. These support mostly rangeland in

FIGURE 3.12 Profile of a Psamment formed on sandy alluvium in Virginia. Note the accumulation of organic matter in the A horizon but no other evidence of profile development. The A horizon is 30 cm thick. (Photo courtesy of R. Weil)

dry regions and forests in more humid areas. Sandy Entisols (Psamments; Figure 3.12) are found in parts of the Sahara desert, southern Africa, central Australia, northwest Nebraska, and the southeastern U.S. coastal plain. Psamments in the humid southern United States are successfully used for citrus, vegetable, and peanut production. Poorly drained and seasonally flooded Entisols (Aquents) occur in major river valleys. Fluvents on recent alluvium in Asia have produced rice crops for generations.

The agricultural productivity of the Entisols varies greatly depending on their location and properties. Entisols developed on alluvial floodplains are among the world's most productive soils. Such soils, with their level topography, proximity to water for irrigation, and periodic nutrient replenishment by floodwater sediments, have supported the development of many major civilizations. However, the productivity of most Entisols is restricted by inadequate soil depth, clay content, or water availability.

3.6 INCEPTISOLS (FEW DIAGNOSTIC FEATURES: INCEPTION OF B HORIZON)

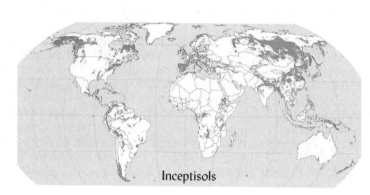

Inceptisols

9.9% of global and 9.1% of U.S. ice-free land

Suborders are:

Anthrepts (human-made, high phosphorus, dark surface)

Aquepts (wet)

Cryepts (very cold)

Gelepts (permafrost)

Udepts (humid climate)

Ustepts (semiarid)

Xerepts (dry summers, wet winters)

In Inceptisols the beginning or *inception* of profile development is evident, and some diagnostic features are present. However, the well-defined profile characteristics of soils thought to be more mature have not yet developed. For example, a cambic horizon showing some color or structural change is common in Inceptisols, but a more mature illuvial B horizon such as an argillic cannot be present. Other subsurface diagnostic horizons that may be present in Inceptisols include duripans, fragipans, and calcic, gypsic, and sulfuric horizons. The epipedon in most Inceptisols is an ochric, although a plaggen or weakly expressed mollic or umbric epipedon may be present. Inceptisols show more significant profile development than Entisols, but are defined to exclude soils with diagnostic horizons or properties that characterize certain other soil orders. Thus, soils with only slight profile development occurring in arid regions or containing permafrost or andic properties are excluded from the Inceptisols. They fall, instead, in the soil orders Aridisols, Gelisols, or Andisols, as discussed in later sections.

Distribution and Use

Inceptisols are widely distributed throughout the world. As with Entisols, Inceptisols are found in most climatic and physiographic conditions. They are often prominent in mountainous areas (Figure 3.13). They are also probably the most important soil order in the lowland rice-growing areas of Asia.

Inceptisols are found in each of the continents (see front papers). Inceptisols of humid regions, called *Udepts*, often have only thin, surface horizons (ochric epipedons). Udepts are common in the mountains from southern New York through the Carolinas. Udepts, along with Xerepts (Inceptisols in xeric climates), also dominate an area extending from southern Spain through central France to Germany and are present as well in Chile, North Africa, eastern China, and western Siberia. Wet Inceptisols or Aquepts are found in areas along the Amazon and Ganges rivers. The natural productivity of Inceptisols is highly variable.

FIGURE 3.13 Inceptisols, like the Humic Dystrudept shown here, are common in mountainous terrain where steep slopes retard profile development, forming a cambic diagnostic horizon (Bw), which features a change in color and structure but no illuvial accumulation of clay. Wayah series, Blue Ridge Mountains, North Carolina. Scale in cm. (Photos courtesy of R. Weil)

3.7 ANDISOLS (VOLCANIC ASH SOILS)

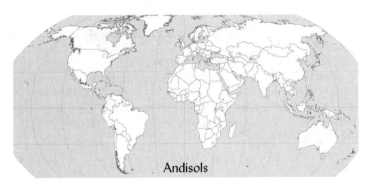

0.7% of global and 1.7% of U.S. ice-free land

Suborders are:

    Aquands (wet)
    Cryands (cold)
    Gelands (very cold)
    Torrands (hot, dry)
    Udands (humid)
    Ustands (moist/dry)
    Vitrands (volcanic glass)
    Xerands (dry summers, moist winters)

Andisols are usually formed on volcanic ash and cinders deposited in recent geological times. They are commonly found near the volcano source or in areas downwind from the volcano, where a sufficiently thick layer of ash has been deposited during eruptions. Andisols have not had time to become highly weathered. The principal soil-forming process has been the rapid weathering (transformation) of volcanic ash to produce amorphous or poorly crystallized silicate minerals such as **allophane** and **imogolite** and the iron oxy-hydroxide, **ferrihydrite**. Some Andisols have a melanic epipedon, a surface diagnostic horizon that has a high organic matter content and dark color (see Plate 2). The accumulation of organic matter is quite rapid due largely to its protection in aluminum–humus complexes. Little downward translocation of the colloids, or other profile development, has taken place. Like the

Melanic Epipedon

Pumice layer

Weathered layers of volcanic ash and pumice

Buried A horizon

Oldest layers of volcanic pumice

Underlying layer of expanding clay

FIGURE 3.14 An Andisol developed in layers of volcanic ash and pumice in central Africa. (Photo courtesy of R. Weil)

Entisols and Inceptisols, Andisols are young soils, usually having developed for only 5000 to 10,000 years.

Unlike the previous two orders of immature soils, Andisols have a unique set of **andic properties** in at least 35 cm of the upper 60 cm of soil due to common types of parent materials. Materials with andic properties are characterized by a high content of volcanic glass and/or a high content of amorphous or poorly crystalline iron and aluminum minerals. The combination of these minerals and the high organic matter results in light, fluffy soils that are easily tilled, yet have a high water-holding capacity and resist erosion by water. They are mostly found in regions where rainfall keeps them from being susceptible to erosion by wind. Andisols are usually of high natural fertility, except that phosphorus availability is severely limited by the extremely high phosphorus retention capacity of the andic materials (see Section 14.8). Fortunately, proper management of plant residues and fertilizers can usually overcome this difficulty.

Distribution and Use

Andisols are found in areas where significant depths of volcanic ash and other ejecta have accumulated (Figure 3.14). Globally, they make up less than 1% of the soil area. However, in the Pacific rim area they are important and productive soils that support intensive agriculture, especially in cool, high-elevation areas.

Andisols, having an udic (moist) soil moisture regime (Udands), are widely cultivated in Japan, producing enough food to support very high population densities. The somewhat drier Ustands are also used intensively for agriculture. Significant areas of Udands and Ustands occur along the Rift Valley of eastern Africa. Andisols are found to a minor extent in cold climates (Cryands) in Canada and Russia, and in hot, dry climates (Torrands) in Mexico and Syria. Very recent eruptions, such as that of Mount Saint Helens in the northwestern part of the United States and Mount Pinatubo in the Philippines, are giving rise to Vitrands that still have much volcanic glass and lower water-holding capacities.

In the United States, the area of Andisols is not extensive since recent volcanic action is not widespread. However, Andisols do occur in some very productive wheat- and timber-producing areas of Washington, Idaho, Montana, and Oregon. Likewise, this soil order represents some of the best farmland found in Chile, Ecuador, Colombia, and much of Central America.

3.8 GELISOLS (PERMAFROST AND FROST CHURNING)

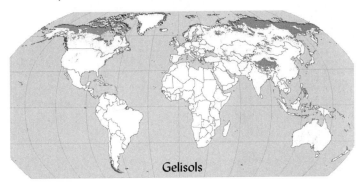

8.6% of global and 7.5% of U.S. ice-free land

Suborders are:

Histels (organic)
Orthels (no special features)
Turbels (cryoturbation)

Gelisols are young soils with little profile development. Cold temperatures and frozen conditions for much of the year slow the process of soil formation. The principal defining feature of these soils is the presence of a **permafrost** layer (see Plates 5 and 14). Permafrost is a layer of material that remains at temperatures below 0 °C for more than two consecutive years. It may be a hard, ice-cemented layer of soil material (e.g., designated Cfm in profile descriptions), or, if dry, it may be uncemented (e.g., designated Cff). In Gelisols, the permafrost layer lies within 100 cm of the soil surface, unless **cryoturbation** is evident within the upper 100 cm, in which case the permafrost may begin as deep as 200 cm from the soil surface.

Cryoturbation is the physical disturbance of soil materials caused by the formation of ice wedges and by the expansion and contraction of water as it freezes and thaws. This *frost churning* action moves the soil material so as to orient rock fragments along the lines of force and to form broken, convoluted horizons (e.g., designated Cjj) at the top of the permafrost (Figure 3.15). The frost

FIGURE 3.15 The broken and involuted patterns formed by cryoturbation. This example is actually a relic of cryoturbation during the Pleistocene ice age in Europe, when this soil would most likely have been in the Turbels suborder. However, it is located in Hungary where permafrost no longer occurs and today is found buried beneath a modern Mollisol formed in an overlying layer of windblown silt (loess). The dark round spots are filled-in animal burrows called *crotovinas* (see Section 2.5). Photograph is about 60 cm across. (Photo courtesy of Ericka Michéli, University of Agricultural Sciences, Gödöllö, Hungary)

Active layer

Permafrost

FIGURE 3.16 Gelisols in Alaska. (*Left*) The soil is in the suborder Histels and has a histic epipedon and permafrost. This soil was photographed in Alaska in July. Scale in cm. (*Right*) Melting of the permafrost under this section of the Alaska Highway caused the soil to lose all bearing strength and collapse. Scale in cm. (Left photo courtesy of James G. Bockheim, University of Wisconsin; right photo courtesy of John Moore, USDA/NRCS)

churning also may form patterns on the ground surface, such as hummocks and ice-rich polygons that may be several meters across. In some cases rocks forced to the surface form rings or netlike patterns.

Gelisols showing evidence of cryoturbation are called *Turbels*. Other Gelisols, often found in wet environments, have developed in accumulations of mainly organic materials, making them **Histels** (Greek *histos,* tissue; Figure 3.16). Most of the soil-forming processes that occur take place above the permafrost in the **active layer** that thaws every year or two. Various types of diagnostic horizons may have developed in different Gelisols, including mollic, histic, umbric, calcic, and, occasionally, argillic horizons.

Distribution and Use

Gelisols cover over 11 million km^2 or 8.6% of the Earth's land area—about 8 million km^2 in Northern Russia and another 4 million in Canada and Alaska. Blanketed under snow and ice for much of the year, most Gelisols support tundra vegetation of lichens, grasses, and low shrubs that grow during the brief summers. Large areas of Gelisols consist of bogs, some literally floating on layers of frozen or unfrozen water. Millions of caribou, reindeer, and musk ox survive on this vegetation during the summer, then migrate to the boreal forests during the coldest seasons. The many bogs and pools serve as nesting sites for migratory birds, which feed on the thick clouds of biting flies and mosquitoes. Human populations are very sparse in these inhospitable environments.

Very few areas of Gelisols are used for agriculture. Plant productivity is low because of the extremely short potential growing season in the far northern latitudes, the low levels of solar radiation (except during the fleeting summer), and the waterlogged condition of many Gelisols in which permafrost inhibits internal drainage during the summer thaw (Figure 3.16).

If the vegetation or surface peat layer on Gelisols is disturbed by cultivation or forest fires or if the soil is warmed by construction activities, the permafrost layer may melt. The permafrost in the southern part of the Gelisols region is only 1 or 2 °C below freezing, so even small changes can cause melting. Unless the soil is mainly gravel, the melting is likely to cause the soil to completely lose its bearing strength and collapse. This presents many serious engineering difficulties (see Figure 3.16, *right*). Houses built directly on Gelisols may sink into the ground as the heat from inside the building penetrates the soil and melts the permafrost. The trans-Alaska oil pipeline had to be constructed on stilts, rather than buried, where it crossed sensitive Gelisols as the heat from the flowing oil would have melted the permafrost, causing the pipes to rupture.

Scientists have observed regional permafrost melting in much of the Arctic. This response of Gelisols is thought to be an early symptom of global climate change caused by greenhouse gas emissions (see Section 12.9). Unfortunately, the melting of permafrost and deepening of the active layer in Gelisols are expected to accelerate this trend as the enormous pools of organic carbon once locked away in the permafrost become exposed to decay, thus releasing yet more greenhouse gases to the atmosphere.

3.9 HISTOSOLS (ORGANIC SOILS WITHOUT PERMAFROST)

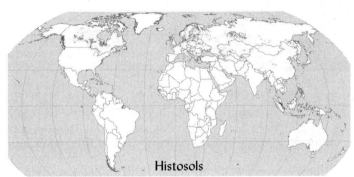

1.2% of global and 1.3% of U.S. ice-free land

Suborders are:

Fibrists (fibers of plants obvious)

Folists (leaf mat accumulations)

Hemists (fibers partly decomposed)

Saprists (fibers not recognizable)

Histosols are soils that have undergone little profile development because of the anaerobic environment in which they form. The main process of soil formation evident in Histosols is the accumulation of partially decomposed organic parent material without permafrost (which would cause the soil to be classified in the Histels suborder of Gelisols). Histosols consist of one or more thick layers of *organic soil material* (see footnote 6, page 80). Generally, Histosols have organic soil materials in more than half of the upper 80 cm of soil (Plate 6) or in two-thirds of the soil overlying shallow rock.

Organic deposits accumulate in marshes, bogs, and swamps, which are habitats for water-loving plants such as pond weeds, cattails, sedges, reeds, mosses, shrubs, and even some trees. Generation after generation, the residues of these plants sink into the water, which inhibits their oxidation by reducing oxygen availability and, consequently, acts as a partial preservative (see Figure 2.22).

The organic matter in Histosols ranges from peat to muck. *Peat* is comprised of the brownish, only partially decomposed, fibrous remains of plant tissues (see Figure 3.17, Plate 26). Some of these soils are mined and sold as peat, a material widely used in containerized plant production (see Box 12.4). *Muck,* on the other hand, is a black material in which decomposition is much more complete and the organic matter is highly humified (Figure 3.18). Muck is like a black ooze when wet and powdery when dry.

While not all wetlands contain Histosols, all Histosols (except Folists) occur in wetland environments. They can form in almost any moist climate in which plants can grow, from equatorial to arctic regions, but they are most prevalent in cold climates, up to the limit of permafrost. Horizons are differentiated by the type of vegetation contributing the residues, rather than by translocations and accumulations within the profile.

Whether artificially drained for cultivation or left in their natural water-saturated state, Histosols possess unique properties resulting from their high organic matter

FIGURE 3.17 A tidal marsh Histosol. The inset shows the *fibric* (peaty) organic material that contains recognizable roots and rhizomes of marsh grasses that died perhaps centuries ago, the anaerobic conditions having preserved the tissues from extensive decay. The soil core (held horizontally for the photograph) gives some idea of the soil profile, the surface layer being at the right and the deepest layer at the left. The water level is usually at or possibly above the soil surface. (Photos courtesy of R. Weil)

contents. Histosols are generally black to dark brown in color. They are extremely light-weight (0.15 to 0.4 Mg/m³) when dry, being only about 10 to 20% as dense as most mineral soils. Histosols also have high water-holding capacities on a mass basis. While a mineral soil will absorb and hold from 20 to 40% of its weight of water, a cultivated Histosol may hold a mass of water equal to 200 to 400% of its dry weight. These soils also possess very high cation exchange capacities (typically 150 to 300 cmol$_c$/kg) that increase with increasing soil pH. The water- and cation-holding capacities are much

Organic materials

Lacustrine mineral deposits

FIGURE 3.18 A drained Histosol on which onions are being produced in New York State. The organic soil rests on old lake-bottom (lacustrine) mineral sediment. The organic material is mucky (see insert), placing this soil in the Saprist suborder. The thickness of this Histosol has been much reduced by subsidence during the nearly 100 years of drainage and cultivation. The surface of this black soil appears light-colored because it was dry when photographed. (Photos courtesy of R. Weil)

higher than those of even clay-rich mineral soils on a weight basis, but similar to those of mineral soils rich in 2:1 silicate clays when considered on a volume basis (water or cations held per liter of soil).

Distribution and Use

Even though they cover only about 1% of the world's land area, Histosols, or **peat lands**, comprise significant areas in cold, wet regions of Alaska, Canada, Finland, Russia, Iceland, Ireland, and Scotland. Of approximately 2 million km^2 of Histosols worldwide, about 75,000 are found in the contiguous United States. Three-quarters of this peat land is in glaciated areas such as Wisconsin, Minnesota, New York, and Michigan. Other important areas of Histosols are found in the tule-reed beds of California and in low-lying parts of the Atlantic and Gulf coastal plains, especially in the Everglades of Florida, the bayous of Louisiana, and the tidal marshes of the mid-Atlantic states (Figure 3.17).

Because the ecological roles of natural wetland environments have not always been appreciated (or protected by law), more than 50% of the original wetland area in the lower 48 United States has been drained for agricultural or other uses, especially for vegetable and flower production. Some Histosols make very productive farmlands, but the organic nature of the materials requires liming, fertilization, tillage, and drainage practices quite different from those applied to soils in the other 11 orders.

If other than wetland plants are to be grown, the water table is usually lowered to provide an aerated zone for root growth. This practice, of course, alters the soil environment and causes the organic material to oxidize, resulting in the disappearance of as much as 5 cm of soil per year in warm climates. To slow the loss of valuable soil resources and avoid unnecessarily aggravating the global greenhouse effect (see Section 12.9), the water table in forested or agricultural Histosols should be kept no lower than is needed to assure adequate root aeration. A more sustainable approach would be to allow some Histosol areas to revert to their native wetland condition (see Section 7.7).

As the organic matter oxidizes in drained Histosols, the land surface is actually lowered as a result of compression and oxidation, a process termed **subsidence** (see Figure 3.19). To preserve these valuable soils and reduce land subsidence, the water table should be kept as near to the surface as possible. Except for the production of

FIGURE 3.19 Soil subsidence due to rapid organic matter decomposition after artificial drainage of Histosols in the Florida Everglades. The house was built at ground level, with the septic tank buried about 1 m below the soil surface. Over a period of about 60 years, more than 1.2 m of the organic soil has "disappeared." The loss has been especially rapid because of Florida's warm climate, but artificial drainage that lowers the water table and continually dries out the upper horizons is an unsustainable practice on any Histosol. (Photo courtesy of George H. Snyder, Everglades Research and Education Center, Belle Glade, FL)

such crops as rice and cranberries under flooded conditions, agriculture on Histosols is not sustainable in the long term.

In some places, Histosols are also mined for their peat, which is sold for use in potting media, as a mulch, and to make peat-fiber pots. Peat deposits are also used for fuel in some countries, especially in Russia, where several power stations are fueled by this material.

3.10 ARIDISOLS (DRY SOILS)

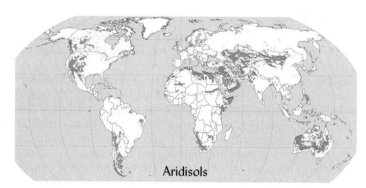

12.7% of global and 8.8% of U.S. ice-free land

Suborders are:

Argids (clay)

Calcids (carbonate)

Cambids (typical)

Cryids (cold)

Durids (duripan)

Gypsids (gypsum)

Salids (salty)

Aridisols occupy a larger area globally than any other soil order except Entisols. Water deficiency is a major characteristic of these soils. The soil moisture level is sufficiently high to support plant growth for no longer than 90 consecutive days. The natural vegetation consists mainly of scattered desert shrubs and short bunchgrasses. Soil properties, especially in the surface horizons, may differ substantially between interspersed bare and vegetated areas (see Section 2.5).

Aridisols are characterized by an ochric epipedon that is generally light in color and low in organic matter (see Plate 3). The processes of soil formation have brought about a redistribution of soluble materials, but there is generally not enough water to leach these materials completely out of the profile. Therefore, they often accumulate at a lower level in the profile. These soils may have a horizon of accumulation of calcium carbonate (calcic), gypsum (gypsic), soluble salts (salic), or exchangeable sodium (natric). Under certain circumstances, carbonates may cement together the soil particles and coarse fragments in the layer of accumulation, producing hard layers known as **petrocalcic** horizons (Figure 3.20). These hard layers act as impediments to plant root growth and also greatly increase the cost of excavations for buildings.

Some Aridisols (the *Argids*) have an argillic horizon, most probably formed under a wetter climate that long ago prevailed in many areas that are deserts today. With time

FIGURE 3.20 Two features characteristic of some Aridisols. (*Left*) Wind-rounded pebbles have given rise to a desert pavement. (*Right*) A petrocalcic horizon of cemented calcium carbonate. (Photo courtesy of R. Weil)

and the addition of carbonates from calcareous dust and other sources, many argillic horizons become engulfed by carbonates (*Calcids*). On steeper land surfaces subject to erosion, argillic horizons do not get a chance to form, and the dominant soils are often *Cambids* (Aridisols with only weakly differentiated cambic subsurface B horizons).

In stony or gravelly soils, erosion may remove all the fine particles from the surface layers, leaving behind a layer of wind-rounded pebbles that is called **desert pavement** (see Figure 3.20 and Plate 55). The surfaces of the pebbles in desert pavement often have a shiny coating called **desert varnish** (Plate 57). This coating is thought to be produced by algae that extract iron and manganese from the minerals and leave an oxide coating on the pebbles.

Except where there is groundwater or irrigation, the soil layers are only moist for short periods during the year. These short, moist periods may be sufficient for drought-adapted desert shrubs and annual plants, but not for conventional crop production. If groundwater is present near the soil surface, soluble salts may accumulate in the upper horizons to levels that most crop plants cannot tolerate.

Type "desert pavement" into Google and click on "Image results" to see a wide variety of photos.

Solving the mystery of desert varnish:
http://sciencenow.sciencemag.org/cgi/content/full/2006/707/1

Distribution and Use

In the United States, Aridisols occur mostly in the western region. A large area dominated by Argids (Aridisols with an argillic horizon) occupies much of the southern parts of California, Nevada, Arizona, and central New Mexico (endpapers). The Argids also extend down into northern Mexico. Smaller areas of Cambids (simple Aridisols without horizons of clay or salt accumulation) are found in several western states (see Plate 3).

Vast areas of Aridisols are present in the Sahara desert in Africa, the Gobi and Taklamakan deserts in China, and the Turkestan desert of the former Soviet Union. Most of the soils of southern and central Australia are Aridisols, as are those of southern Argentina, southwestern Africa, Pakistan, and much of the Middle East.

Without irrigation, Aridisols are not suitable for growing cultivated crops. Some areas are used for low-intensity grazing, especially with sheep or goats, but the production per unit area is low. The overgrazing of Aridisols leads to increased heterogeneity of both soils and vegetation. The animals graze the relatively even cover of palatable grasses, giving a competitive advantage to various shrubs not eaten by the grazing animals. The scattered shrubs compete against the struggling grasses for water and nutrients. The once-grassy areas become increasingly bare, and the soils between the scattered shrubs succumb to erosion by the desert winds and occasional thunderstorms. The desertification of areas of sub-Saharan Africa and the western United States is evidence of such degradation (Figure 3.21).

FIGURE 3.21 It is easy to overgraze the vegetation on an Aridisol, and an overgrazed Aridisol is subject to ready wind and water erosion. This gully was formed during infrequent but heavy rainstorms in an area in Arizona that at one time had been well covered with desert plants such as alkali sacaton and vine mesquite, plants that were destroyed by the overgrazing. (Photo courtesy of U.S. Natural Resource Conservation Service)

Some xerophytic plants, such as a jojoba, have been cultivated on Aridisols to produce various industrial feedstocks such as oil and rubber. Where irrigation water and fertilizers are available, some Aridisols can be made highly productive. Irrigated valleys in arid areas are among the most productive in the world. However, they must be carefully managed to prevent the accumulation of soluble salts (see Section 10.3).

3.11 VERTISOLS (DARK, SWELLING, AND CRACKING CLAYS)[8]

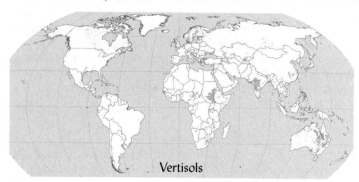

Vertisols

2.4% of global and
1.7% of U.S. ice-free land

Suborders are:

Aquerts (wet)
Cryerts (cold)
Torrerts (hot summer, very dry)
Uderts (humid)
Usterts (moist/dry)
Xererts (dry summers, moist winters)

The main soil-forming process affecting Vertisols is the shrinking and swelling of clay as these soils go through periods of drying and wetting. Vertisols have a high content (>30%) of sticky, swelling, and shrinking-type clays to a depth of 1 m or more. Most Vertisols are dark, even blackish in color, to a depth of 1 m or more (Plate 12). However, unlike for most other soils, the dark color of Vertisols is not necessarily indicative of a high organic matter content. The organic matter content of dark Vertisols typically ranges from as much as 5 or 6% to as little as 1%.

Vertisols typically develop from limestone, basalt, or other calcium- and magnesium-rich parent materials. In east Africa, they typically form in landscape depressions that collect the calcium and magnesium leached out of the surrounding upland soils. The presence of these cations encourages the formation of swelling-type clays (see Section 8.3).

Vertisols are found mostly in subhumid to semiarid environments in warm regions, but a few (Cryerts) occur where the average soil temperatures are as low as 0 °C (see Figure 3.9). The native vegetation is usually grassland. Vertisols generally occur where the climate features dry periods of several months. In dry seasons the clay shrinks, causing the soils to develop deep, wide cracks that are diagnostic for this order (Figure 3.22*a*). The surface soil generally forms granules, of which a significant number may slough off into the cracks, giving rise to a partial inversion of the soil (Figure 3.23*a*). This accounts for the association with the term *invert*, from which this order derives its name.

When the rains come, water entering the large cracks moistens the clay in the subsoils, causing it to swell. The repeated shrinking and swelling of the subsoil clay results in a kind of imperceptibly slow "rocking" movement of great masses of soil. As the subsoil swells, blocks of soil shear off from the mass and rub past each other under pressure, giving rise in the subsoil to shiny, grooved, tilted surfaces called **slickensides** (Figure 3.23*c*). Eventually, this back-and-forth motion may form bowl-shaped depressions with relatively deep profiles surrounded by slightly raised areas in which little soil development has occurred and in which the parent material remains close to the surface (see Figure 3.23*b*). The resulting pattern of micro-highs and micro-lows on the land surface, called **gilgai**, is usually discernable only where the soil is untilled (Figure 3.22*b*).

---

[8] See Coulombe, et al. (1996) for a detailed review of the properties and mode of formation of Vertisols.

(b)

(a)

FIGURE 3.22 (*a*) Wide cracks formed during the dry season in the surface layers of this Vertisol in India. Surface debris can slough off into these cracks and move to subsoil. When the rains come, water can move quickly to the lower horizons, but the cracks are soon sealed, making the soils relatively impervious to the water. (*b*) Once the cracks have sealed, water may collect in the "microlows," making the gilgai relief easily visible as in the Texas vertisol. (Photo (*a*) courtesy of N. C. Brady; (*b*) courtesy of K. N. Potter, USDA/ARS, Temple, Texas)

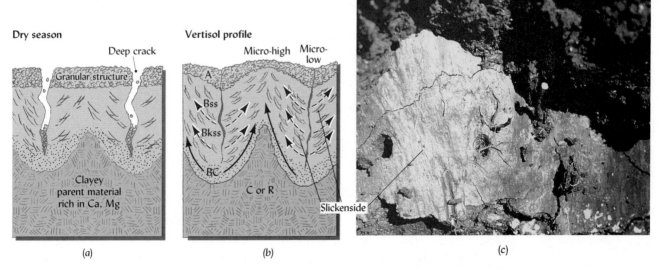

(a)   (b)   (c)

FIGURE 3.23 Vertisols are high in swelling-type clay and have developed wedgelike structures in the subsoil horizons. (*a*) During the dry season, large cracks appear as the clay shrinks upon drying. Some of the surface soil granules fall into cracks under the influence of wind and animals. This action causes a partial mixing, or *inversion*, of the horizons. (*b*) During the wet season, rainwater pours down the cracks, wetting the soil near the bottom of the cracks first, and then the entire profile. As the clay absorbs water, it swells the cracks shut, entrapping the collected granular soil. The increased soil volume causes lateral and upward movement of the soil mass. The soil is pushed up between the cracked areas. As the subsoil mass shears from the strain, smooth surfaces or *slickensides* form at oblique angles. These processes result in a Vertisol profile that typically exhibits *gilgai*, cracks more than 1 m deep and slickensides in a Bss horizon. (*c*) An example of a slickenside in a Vertisol. Note the grooved, shiny surface. The white spots in the lower right of the photo are calcium carbonate concretions that often accumulate in a Bkss horizon. (Diagrams and photo courtesy of R. Weil)

Globally, Vertisols comprise about 2.5% of the total land area. Large areas of Vertisols are found in India, Ethiopia, the Sudan, and northern and eastern Australia (see front papers). Smaller areas occur in sub-Saharan Africa and in Mexico, Venezuela, Bolivia, and Paraguay. These latter soils probably are of the Usterts or Xererts suborders, since dry conditions persist long enough for the wide cracks to stay open for periods of three months or longer.

There are several small but significant areas of Vertisols in the United States (see Figure 3.10 and endpapers). Two areas are located in humid areas, one in eastern Mississippi and western Alabama (the so-called *black belt*) and the other along the southeast coast of Texas. These soils are of the Uderts suborder, because their moist condition prevents cracks from persisting for more than three months of the year.

Two other Vertisol areas are found in east central and southern Texas, where the soils are drier. Since cracks persist for more than three months of the year in these areas, the soils belong to the Usterts suborder, characteristic of areas with hot, dry summers. An area of Cryerts is located in the Dakotas and Saskatchewan. Xererts (Greek *xeros,* dry) are also located in California.

The high shrink-swell potential of Vertisols makes them extremely problematic for any kind of highway or building construction (Figure 8.33 and Plate 43). This property also makes agricultural management very difficult. Because they are very sticky and plastic when wet and become very hard when dry, the timing of tillage operations is critical. Some farmers refer to Vertisols as *24-hour soils,* because they are said to be too wet to plow one day and too dry the next.

Even when the soil moisture is near optimal, the energy requirement for tillage is high. Therefore, except where heavy equipment is used for tillage, the operations are slow, and the amount of land a farmer can cultivate is much smaller than for other soil orders. In areas such as those in India and the Sudan, where slow-moving animals or human power are commonly used to till the soil, farmers cannot perform tillage operations on time and are limited to the use of very small tillage implements that their animals can pull through the heavy soil.

Recent research shows that the large areas of Vertisols in the tropics can produce greatly increased yields of food crops with improved soil management practices. Soils in this order are, however, very susceptible to physical degradation and erosion (despite their mainly gentle slopes), and conservation practices or reversion to rangeland are important management options to consider.

3.12 MOLLISOLS (DARK, SOFT SOILS OF GRASSLANDS)

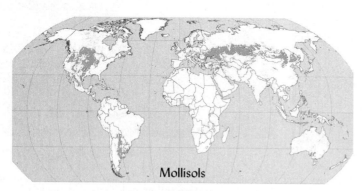

Mollisols

6.9% of global and
22.4% of U.S. ice-free land

Suborders are:

Albolls (albic horizon)

Aquolls (wet)

Cryolls (cold)

Gelolls (very cold)

Rendolls (calcareous)

Udolls (humid)

Ustolls (moist/dry)

Xerolls (dry summers, moist winters)

The principal process in the formation of Mollisols is the accumulation of calcium-rich organic matter, largely from the dense root systems of prairie grasses, to form the thick, soft Mollic epipedon that characterizes soils in this order (Plates 8, 13, and 20). This humus-rich surface horizon is often 60 to 80 cm in depth. Its cation exchange capacity

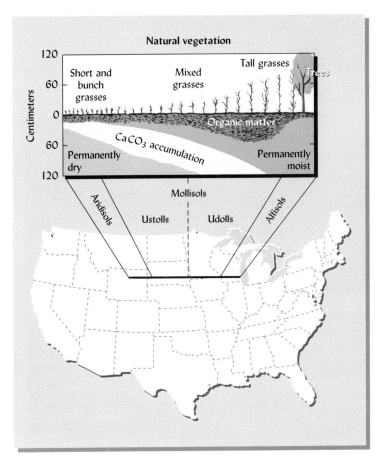

FIGURE 3.24 Correlation between natural grassland vegetation and certain soil orders is graphically shown for a transect across north central United States. The controlling factor, of course, is climate. Note the deeper organic matter and deeper zone of calcium accumulation, sometimes underlain by gypsum, as one proceeds from the drier areas in the west toward the more humid region where prairie soils are found. Alfisols may develop under grassland vegetation, but more commonly occur under forests and have lighter-colored surface horizons.

is more than 50% saturated with base cations (Ca^{2+}, Mg^{2+}, etc.). Mollisols in humid regions generally have higher organic matter and darker, thicker mollic epipedons than their lower-moisture-regime counterparts (see Section 12.8).

The surface horizon generally has granular or crumb structures, largely resulting from an abundance of organic matter and swelling-type clays. In many cases, the highly aggregated soil is not hard when dry, hence the name *Mollisol,* implying softness (Table 3.3). In addition to the mollic epipedon, Mollisols may have an argillic (clay), natric, albic, or cambic subsurface horizon, but not an oxic or spodic horizon.

Most Mollisols have developed under grass vegetation (Figure 3.24). Grassland soils of the central part of the United States, lying between Aridisols on the west and the Alfisols on the east, typify the central concept of this order. However, a few soils developed under forest vegetation (primarily in depressions) have a mollic epipedon and are included among the Mollisols.

Distribution and Use

Mollisols cover a larger land area in the United States than any other soil order. Mollisols are dominant in the Great Plains of North America, as well as in Illinois (see Figure 3.10). Where soil moisture is not limiting, Udolls are found. They are associated with nearby wet Mollisols termed Aquolls. A region characterized by Ustolls (intermittently dry during the summer) extends from Manitoba and Saskatchewan in Canada to southern Texas. Farther west are found sizable areas of Xerolls (with a Xeric moisture regime, which is very dry in summer but moist in winter). Landscapes common for Udolls and Ustolls can be seen in Figure 3.25. Conservation of soil water is a major consideration in the management of Ustolls, in particular. Two Mollisols profiles are included in Figure 3.26.

FIGURE 3.25 Typical landscapes dominated by Ustolls (Montana, *top*) and Udolls (Iowa, *bottom*). These productive soils produce much of the food and feed in the United States. (Photos courtesy of R. Weil)

The largest area of Mollisols in the world stretches from east to west across the heartlands of Kazakhstan, Ukraine, and Russia. Other sizable areas are found in Mongolia and northern China and in northern Argentina, Paraguay, and Uruguay. Mollisols occupy only about 7% of the world's total soil area, but because of their generally high fertility, they account for a much higher percentage of total crop production.

In the United States, efforts are underway to preserve the few remnants of the once vast and diverse prairie ecosystem. Because the high native fertility of Mollisols makes them among the world's most productive soils, few Mollisols have been left uncultivated in regions with sufficient rainfall for crop production. When they were first cleared and plowed, much of their native organic matter was oxidized, releasing nitrogen and other nutrients in sufficient quantities to produce high crop yields even without the use of fertilizers. Even after more than a century of cultivation, Mollisols are among the most productive soils, although some fertilization is generally required. However, continuous cultivation with row crops has led to serious deterioration of soil structure and to soil erosion where the land is sloping.

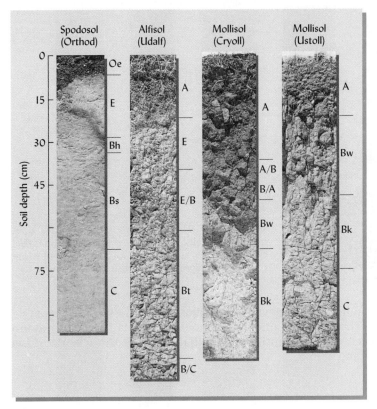

FIGURE 3.26 Monoliths of profiles representing three soil orders. The suborder names are in parentheses. Genetic (not diagnostic) horizon designations are also shown. Note the spodic horizons in the Spodosol characterized by humus (Bh) and iron (Bs) accumulation. In the Alfisol is found the illuvial clay horizon (Bt), and the structural B horizon (Bw) is indicated in the Mollisols. The thick dark surface horizon (mollic epipedon) characterizes both Mollisols. Note that the zone of calcium carbonate accumulation (Bk) is near the surface in the Ustoll, which has developed in a dry climate. The E/B horizon in the Alfisol has characteristics of both E and B horizons.

3.13   ALFISOLS (ARGILLIC OR NATRIC HORIZON, MODERATELY LEACHED)

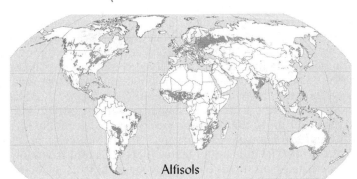

Alfisols

9.6% of global and
14.5% of U.S. ice-free land

Suborders are:

    Aqualfs (wet)

    Cryalfs (cold)

    Udalfs (humid)

    Ustalfs (moist/dry)

    Xeralfs (dry summers, moist winters)

The Alfisols are more strongly weathered than soils in the orders just discussed, but less so than Spodosols and Ultisols (see following). They are found in cool to hot humid areas (see Figure 3.9) as well as in the semiarid tropics and Mediterranean climates. Most often, Alfisols develop under native deciduous forests, although in some cases, as in California and parts of Africa, savanna (mixed trees and grass) is the native vegetation.

Alfisols are characterized by a subsurface diagnostic horizon in which silicate clay has accumulated by illuviation (see Plate 1). Clay skins or other signs of clay movement are present in such a B horizon (see Plates 18 and 24). In Alfisols, this clay-rich horizon is only moderately leached, and its cation exchange capacity is more than 35% base saturated (Ca^{2+}, Mg^{2+}, etc.). In most Alfisols this horizon is termed *argillic* because of its accumulation of silicate clays. The horizon is termed *natric* if, in addition to having an accumulation of clay, it is more than 15% saturated with sodium and has prismatic or columnar structure (see Figure 4.13). In some Alfisols in subhumid tropical regions, the accumulation is termed a *kandic* horizon (from the mineral kandite) because the clays have a low cation exchange capacity.

Alfisols very rarely have a mollic epipedon, for such soils would be classified in the Argiudolls or other suborder of Mollisols with an argillic horizon. Instead, Alfisols

FIGURE 3.27 A landscape in southern Portugal dominated by Xeralfs. Most of the original forest has been cleared and the soils plowed for production of winter cereals and grapes. (Photo courtesy of R. Weil)

typically have a relatively thin, gray to brown ochric epipedon (Plate 1 shows an example) or an umbric epipedon. Those formed under deciduous temperate forests commonly have a light-colored, leached *albic* E horizon immediately under the A horizon (see Plate 21 and Figures 1.15 and 3.26).

Distribution and Use

Udalfs (humid region Alfisols) dominate large areas in Ohio, Indiana, Michigan, Wisconsin, Minnesota, Pennsylvania, and New York in the United States (see Figure 3.10), as well as in central China, England, France, central Europe, and southeastern Australia. There are sizable areas of Xeralfs (Alfisols in regions of dry summers and moist winters) in central California, southwestern Australia, Italy, and central Spain and Portugal (Figure 3.27). Cryalfs (very cold) can be found in the Rocky Mountains; in south-central Canada; in Minnesota; in northern Europe, extending from the Baltic States through western Russia; and in Siberia. Where summers are hot and dry, including areas in Texas, New Mexico, sub-Saharan Africa, eastern Brazil, eastern India, and southeastern Asia, Ustalfs are prominent. Many Alfisols landscapes include wet depressions characterized by Aqualfs.

In general, Alfisols are productive soils. Good hardwood forest growth and crop yields are favored by their medium- to high-base-saturation status, generally favorable texture, and location (except for some Xeralfs) in regions with sufficient rainfall for plants for at least part of the year. In the United States these soils rank favorably with the Mollisols and Ultisols in their productive capacity. Many Alfisols, especially the sandier ones, are quite susceptible to erosion by heavy rains if deprived of their natural surface litter. Alfisols in Udic moisture regimes are sufficiently acidic in the A horizon to require amendment with limestone for many kinds of plants (see Chapter 9).

3.14 ULTISOLS (ARGILLIC HORIZON, HIGHLY LEACHED)

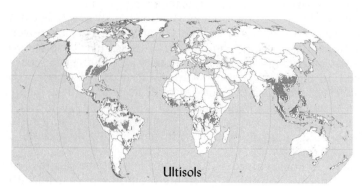

Ultisols

8.5% of global and
9.6% of U.S. ice-free land

Suborders are:

Aquults (wet)

Humults (high humus)

Udults (humid)

Ustults (moist/dry)

Xerults (dry summers, moist winters)

The principal processes involved in forming Ultisols are clay mineral weathering, translocation of clays to accumulate in an argillic or kandic horizon, and leaching of base-forming cations from the profile. Most Ultisols have developed under moist conditions in warm to tropical climates. Ultisols are formed on old land surfaces, usually under forest vegetation, although savanna or even swamp vegetation is also common. They often have an ochric or umbric epipedon, but are characterized by a relatively acidic B horizon that has less than 35% of the exchange capacity satisfied with base cations. The clay accumulation may be either an argillic horizon or, if the clay is of low activity, a kandic horizon. Ultisols commonly have both an epipedon and a subsoil that is quite acid and low in plant nutrients.

Ultisols are more highly weathered and acidic than Alfisols, but less acid than Spodosols and less highly weathered than the Oxisols. Except for the wetter members of the order, their subsurface horizons are commonly red or yellow in color, evidence of accumulations of oxides of iron (see Plate 11). Certain Ultisols that formed under fluctuating wetness conditions have horizons of iron-rich mottled material called *plinthite* (see Plates 15 and 37). This material is soft and can be easily dug from the profile so long as it remains moist. When dried in the air, however, plinthite hardens irreversibly into a kind of ironstone that is virtually useless for cultivation (Plates 29 and 33), but can be used to make durable bricks for building (Plate 48).

Distribution and Use

Most of the soils of the southeastern part of the United States fall in the suborder Udults (see Figure 3.10 and endpapers). Large areas of Udults are also located in southeastern Asia and in southern China. Extensive areas of Ultisols are found in the humid tropics in close association with some Oxisols. Important agricultural areas are found in southern Brazil and Paraguay.

Humults (high in organic matter) are found in the United States in Hawaii and in western California, Oregon, and Washington. Humults are also present in the highlands of some tropical countries. Xerults (Ultisols in Mediterranean-type climates) occur locally in southern Oregon and northern and eastern California. Ustults are found in semiarid areas with a marked dry season. Together with the Ustalfs, the Ustults

FIGURE 3.28   The soils in this high-elevation, tropical area of South Asia are Ultisols in the suborder Humults. These soils are being intensively used both for house construction and for market gardens. The combination of a favorable climate and soils that are high in organic matter (Humults have at least 9% down to the upper part of the B horizon) and respond well to fertilizer has encouraged local residents to use every bit of the land in producing vegetables to supplement their incomes. (Photo courtesy of R. Weil)

occupy large areas in Africa and India. Ultisols are prominent on the east and northeast coasts of Australia (see front papers).

Although Ultisols are not naturally as fertile as Alfisols or Mollisols, they respond well to good management. They are located mostly in regions of long growing seasons and of ample moisture for good crop production (Figure 3.28). The silicate clays of Ultisols are usually of the nonsticky type, which, along with the presence of iron oxides and aluminum, assures ready workability. Where adequate levels of fertilizers and lime are applied, Ultisols are quite productive. In the United States, well-managed Ultisols compete well with Mollisols and the Alfisols as first-class agricultural soils. They also support the most productive commercial softwood and hardwood forests in the country.

3.15   SPODOSOLS (ACID, SANDY, FOREST SOILS, HIGHLY LEACHED)

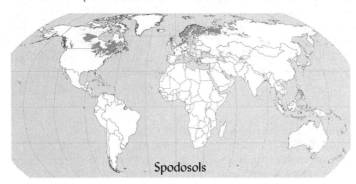

2.6% of global and 3.3% of U.S. ice-free land

Suborders are:

Aquods (wet)

Cryods (cold)

Gelods (very cold)

Humods (humus)

Orthods (typical)

Spodosols occur mostly on coarse-textured, acid parent materials subject to ready leaching. They occur only in moist to wet areas, commonly where it is cold or temperate (see Figure 3.9), but also in some tropical and subtropical areas. Intensive acid leaching is the principal soil-forming process. They are mineral soils with a *spodic* horizon, a subsurface accumulation of illuviated organic matter, and an accumulation of aluminum oxides with or without iron oxides (see Plates 10 and 31 and Figure 3.26). This usually thin, dark, illuvial horizon typically underlies a light, ash-colored, eluvial *albic* horizon.

Spodosols form under forest vegetation, especially under coniferous species whose needles are low in base-forming cations like calcium and high in acid resins. As this acid litter decomposes, strongly acid organic compounds are released and carried down into the permeable profile by percolating waters. Some of the leaching organic compounds may precipitate and form a black-colored Bh horizon. Leaching organic acids bind with iron and aluminum, removing these metals from the A and E horizons and carrying them downward. This iron and aluminum eventually precipitates in a reddish-brown-colored Bs horizon, usually just below the black-colored Bh horizon. Together, the Bh and Bs horizons constitute the spodic diagnostic horizon that defines the Spodosols. The depth at which the spodic horizon forms can vary from less than 20 cm to several meters. As iron oxides (and most other minerals except quartz) are stripped from the E horizon by the organic leaching process, this horizon may become a nearly white albic diagnostic horizon that consists mainly of clean quartz sand. The leaching and precipitation often occur along wavy wetting fronts, thus yielding the striking profiles seen in Spodosols (Figure 3.29).

Distribution and Use

Large areas of Spodosols are found in northern Europe and Russia and central and eastern Canada. Many of the soils in the northeastern United States, as well as those of northern Michigan and Wisconsin and southern Alaska, belong to this order (see Figures 3.10 and 3.29). Spodosols are found on about 3% of the land area both globally and in the United States. Small but important areas occur in the southern part of South America and in the cool mountainous areas of temperate regions.

Most Spodosols are Orthods, soils that typify the central concept of Spodosols described previously. Some, however, are Aquods because they are seasonally saturated

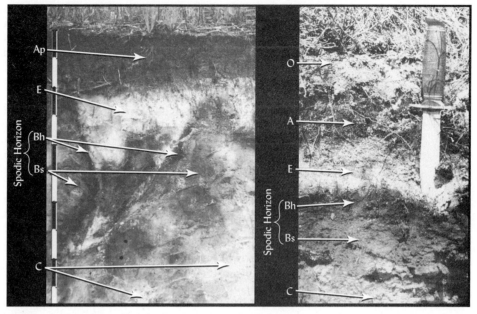

FIGURE 3.29 Two Spodosol profiles. (*Left*) A Spodosol in northern Michigan exhibits discontinuous, wavy Bh and Bs genetic horizons, which comprise the relatively deep spodic diagnostic horizon. The nearly white, discontinuous eluvial horizon (E) consists mainly of uncoated quartz sand particles and is an albic diagnostic horizon. The dark, organic-enriched surface horizon (an ochric diagnostic horizon) shows the smooth lower boundary and uniform thickness characteristic of an Ap horizon formed by plowing after the original coniferous forest was cleared. (*Right*) A much shallower Spodosol in Scotland also exhibits a spodic diagnostic horizon comprised of Bh and Bs genetic horizons. Above the spodic horizon is the light colored albic horizon (E) and above that, the darker ochric (A) horizon. These horizons, in combination with the relatively thick surface layer of pine needles in various states of decay (the O horizon), suggest an undisturbed forest floor. Both soils formed from the sandy parent material evident in their C horizons. Spodosols are typically sandy, acid, infertile, and best suited to supporting the coniferous forest vegetation under which they usually form. Scale at left marked every 10 cm, knife at right has a 12 cm long handle. (Photos courtesy of R. Weil)

with water and possess characteristics associated with this wetness. Important areas of Aquods occur in Florida and other areas with warm climates.

Spodosols are not naturally fertile. When properly fertilized, however, these soils can become quite productive. For example, most potato-producing soils of northern Maine are Spodosols, as are some of the vegetable- and fruit-producing soils of Florida, Michigan, and Wisconsin. Because of their sandy nature and occurrence in regions of high rainfall, groundwater contamination by leaching of soluble fertilizers and pesticides has proved to be a serious problem where these soils are used in crop production. They are now covered mostly with forests, the vegetation under which they originally developed. Most Spodosols should remain as forest habitats. Because they are already quite acid and poorly buffered, many Spodosols and the lakes in watersheds dominated by soils of this order are susceptible to damage from acid rain (see Section 9.6).

3.16 OXISOLS (OXIC HORIZON, HIGHLY WEATHERED)

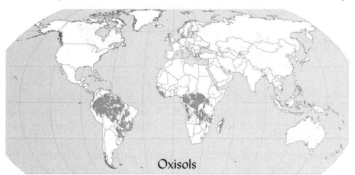

Oxisols

7.6% of global and <0.01% of U.S. ice-free land

Suborders are:

Aquox (wet)
Perox (very humid)
Torrox (hot, dry)
Udox (humid)
Ustox (moist/dry)

The Oxisols are the most highly weathered soils in the classification system (see Figure 3.8). They form in hot climates with nearly year-round moist conditions; hence, the native vegetation is generally thought to be tropical rain forest. However, some Oxisols (Ustox) are found in areas that are today much drier than was the case when the soils were forming their oxic characteristics. Their most important diagnostic feature is a deep oxic subsurface horizon. This horizon is generally very high in clay-size particles dominated by hydrous oxides of iron and aluminum. Weathering and intense leaching have removed a large part of the silica from the silicate materials in this horizon. Some quartz and 1:1-type silicate clay minerals remain, but the hydrous oxides are dominant (see Chapter 8 for information on the various clay minerals). The epipedon in most Oxisols is either ochric or umbric. Usually the boundaries between subsurface horizons are indistinct, giving the subsoil a relatively uniform appearance with depth.

The clay content of Oxisols is generally high, but the clays are of the low-activity, nonsticky type. Consequently, when the clay dries out it is not hard and cloddy, but is easily worked. Also, Oxisols are resistant to compaction, so water moves freely through the profile. The depth of weathering in Oxisols is typically much greater than for most of the other soils, 20 m or more having been observed. The low-activity clays have a very limited capacity to hold nutrient cations such as Ca^{2+}, Mg^{2+}, and K^+, so they are typically of low natural fertility and moderately acid. The high concentration of iron and aluminum oxides also gives these soils a capacity to bind so tightly with what little phosphorus is present that phosphorus deficiency often limits plant growth once the natural vegetation is disturbed.

Road and building construction is relatively easily accomplished on most Oxisols because these soils are easily excavated, do not shrink and swell, and are physically very stable on slopes. The very stable aggregation of the clays, stimulated largely by iron compounds, makes these soils quite resistant to erosion.

Distribution and Use

Oxisols occupy old land surfaces that have not been disturbed by glaciation or erosion. Although nearly all Oxisols occur in the tropics, most tropical soils are *not* Oxisols. Large areas of Oxisols occur in South America and Africa (see front papers). New data from Brazilian soil scientists suggests that some of the areas of the Amazon basin currently mapped as Oxisols are in reality dominated by Ultisols and other soils. Udox (Oxisols having a short dry season or none) occur in northern Brazil and neighboring countries as well as in the Caribbean area (see Plates 9 and 25). Important areas of Ustox (hot, dry summers) occur in Brazil to the south of the Udox. In the humid areas of central Africa, Oxisols are prominent and in some cases dominant.

Relatively less is known about Oxisols than about most other soil orders. They occur in large geographic areas, often associated with Ultisols. Millions of people in the tropics depend on them for food and fiber production. However, because of their low natural fertility, most Oxisols have been left under forest vegetation or are farmed by shifting cultivation methods. Nutrient cycling by deep-rooted trees is especially important to the productivity of these soils. Probably the best use of Oxisols, other than supporting rain forests, is the culture of mixed-canopy perennial crops, especially tree crops. Such cultures can restore the nutrient cycling system that characterized the soil–plant relationships before the rain forest was removed.

3.17 LOWER-LEVEL CATEGORIES IN *SOIL TAXONOMY*

Suborders

As indicated next to the global distribution maps in previous sections, soils within each order are grouped into suborders on the basis of soil properties that reflect major environmental controls on current soil-forming processes. Many suborders are indicative of the moisture regime or, less frequently, the temperature regime under which the soils are found. Thus, soils formed under wet conditions generally are identified under separate suborders (e.g., Aquents, Aquerts, and Aquepts), as being wet soils.

PLATE 1   Alfisols—a Glossic Hapludalf in New York. Scale in cm.

PLATE 2   Andisols—a Typic Melanudand from western Tanzania. Scale in 10 cm.

PLATE 3   Aridisols—a skeletal Ustollic Calcicambid from Nevada. Shovel handle is 60 cm long.

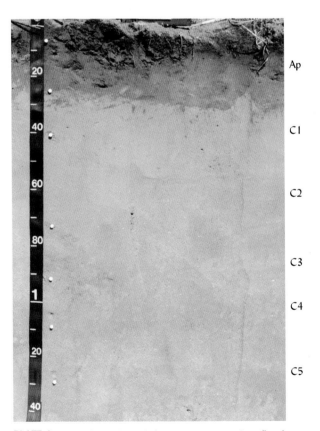

PLATE 4   Entisols—a Typic Udipsamment on a river flood plain in North Carolina.

O

Bw

Bf

Bf2

20

40

60

80

PLATE 5   Gelisols—a Typic Aquaturbel from Alaska. Permafrost below 32 cm on scale.

O1

O2

O3

Cgb

PLATE 6   Histosols—a Limnic Haplosaprist from southern Michigan. Buried mineral soil at bottom of scale. Scale in feet.

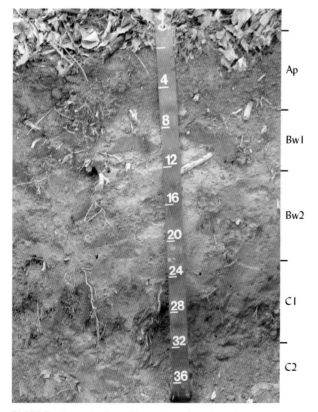

Ap

Bw1

Bw2

C1

C2

4

8

12

16

20

24

28

32

36

PLATE 7   Inceptisols—a Typic Eutrudept from Vermont.

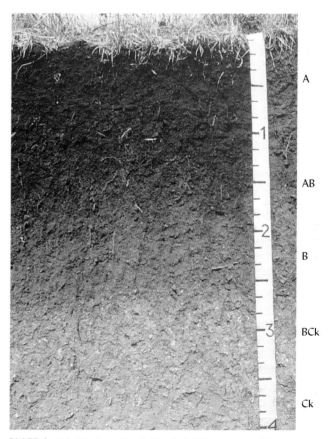

A

AB

B

BCk

Ck

1

2

3

4

PLATE 8   Mollisols—a Typic Hapludoll from central Iowa. Mollic epipedon to 1.8 ft. Scale in feet.

PLATE 9 Oxisols—a Udeptic Hapludox from central Puerto Rico. Scale in feet and inches.

PLATE 10 Spodosols—a Typic Haplorthod in New Jersey. Scale in 10 cm increments.

PLATE 11 Ultisols—a Typic Hapludult from central Virginia showing metamorphic rock structure in the saprolite below the 60-cm-long shovel.

PLATE 12 Vertisols—a Typic Haplustert from Queensland, Australia, during wet season. Scale in meters.

PLATE 13 Typic Argiustolls in eastern Montana with a chalky white calcic horizon (Bk and Ck) overlain by a Mollic epipedon (Ap, A2, and Bt).

PLATE 14 Ice wedge and permafrost underlying Gelisols in the Seward Peninsula of Alaska. Shovel is 1 m long.

PLATE 15 A Typic Plinthudult in central Sri Lanka. Mottled zone is plinthite, in which ferric iron concentrations will harden irreversibly if allowed to dry.

PLATE 16 A soil catena or toposequence in central Zimbabwe. Redder colors indicate better internal drainage. Inset: B-horizon clods from each soil in the catena.

PLATE 17 Redox concentrations (red) and depletions (gray) in Btg horizon from an Aquic Paleudalf.

PLATE 18 Clay skins (argillans) appear as dark, shiny coatings in this Ultisol Bt horizon. Bar = 1 cm.

PLATE 19 The boundary between the Oe and the E horizons of a forested Ultisol.

PLATE 20 The effect of moisture on soil color. Right side of this Mollisol profile was sprayed with water.

PLATE 21 Effect of poor drainage on soil color. Gray matrix colors and red redox concentrations in the B horizons of a Plinthaquic Paleudalf.

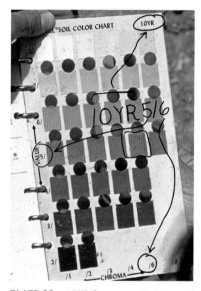

PLATE 22 10YR hue page in Munsell color book with standard notation for color of hue 10YR, value 5 and chroma 6.

PLATE 23 Roadcut in southern Brazil exposing the profile of an Udalf with a sombric horizon. This dark, humus rich subsurface horizon typically forms in humid, high altitude tropical and subtropical mountains.

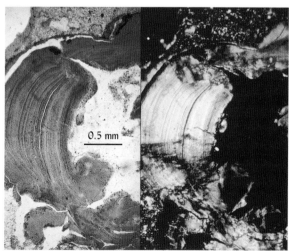

PLATE 24 Thick clay skins (argillans) in an argillic (Bt) horizon. Image made from a very thin, polished slice of soil, magnified with a petrographic microscope using plain polarized (*left*) and cross polarized light (*right*). Note the thin layers of illuvial clay.

PLATE 25 Oxisol profile in Central Brazil. Scale in 10 cm.

PLATE 26 Histosols—a Fibrist in central Scotland. Knife handle 11.5 cm.

Ap

A2

C

2Ab

2Bb

PLATE 27 Sloping Entisol formed in shaley colluvium deposited over buried soil in Pennsylvania.

PLATE 28 Green slickenside in glauconitic Marlton soil.

PLATE 29 Hard plinthite concretions in tropical Alfisol Bv horizon.

5 cm

PLATE 30 Yellow Jarosite formed by sulfidization.

PLATE 31 Spodosol formed in glacial outwash in Michigan.

1 cm

PLATE 32 Dark organic matter coatings (*arrows*) on blocky peds in a Kansas Mollisol.

PLATE 33 Shallow laterite (hard plinthite) layer (*knife point*) in tropical Alfisol. Scale marked in 10 cm.

PLATE 35   Gleyed colors along root channel in ped from C horizon.

PLATE 34   Traditional humid region farmers use slash and burn systems. They chop down patches of forest patches and burn the dead vegetation, returning many nutrients in the ash. Note fire in the background and burned logs in the foreground of this Sri Lankan woman's new clearing. See Section 20.7.

PLATE 36   Erosion of convex sites by tillage and water has exposed red B horizon material.

PLATE 37   Iron depletions and concentrations in C horizon of Ultisol in Alabama.

PLATE 38   Oxidized (red) root zones in the A and E horizons indicate a hydric soil. They result from oxygen diffusion out from roots of wetland plants having aerenchyma tissues (air passages).

PLATE 39   Dark (black) humic accumulation and gray humus depletion spots in the A horizon are indicators of a hydric soil. Water table is 30 cm below the soil surface.

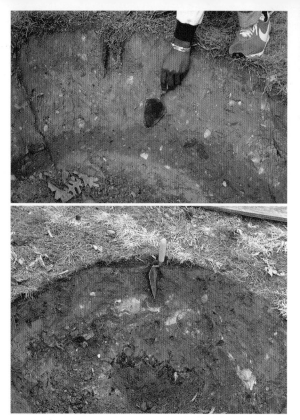

PLATE 40   Urban soils (referred by some as *Urbents*) often hold surprises in their profiles. Here a tree planting hole reveals a buried A horizon (*top*) and buried asphalt (*lower*).

PLATE 42   Mass wasting of clayey soils on steep slopes may occur when saturated with water as in this rotational block slide in East Africa. Note man in center for scale.

PLATE 44   Two large soil stockpiles on a construction site, the brown A horizon was set aside for landscaping topsoil, while the redder B horizon (*back*) was stockpiled for use as fill and road base.

PLATE 41   Soil saturated beyond its liquid limit by torrential rains caused this landslide and mudflow that pushed huge tropical cloud forest trees downslope and demolished a village at the foot of this mountain in Honduras.

PLATE 43   The red, kaolinitic soil in the foreground was hauled in to build up a stable roadbed across this low-lying landscape. The black soils are rich in expansive clays, which would break up the pavement if used for the sub-base. South Central Tanzania.

PLATE 45   An uneven layer of silt blankets over a glacial deposit of coarse sand and gravel in this Rhode Island Inceptisol. The profile water holding capacity varies with the thickness of the silt, causing an irregular pattern of drought stricken turf grass (*inset*).

PLATE 46 Bent trees indicate soil creep.

PLATE 47 Reduced interior of clay inclusions in sandy sediment.

PLATE 49 Tillage research plots with rails for implements to avoid wheel effects. Auburn, Alabama.

PLATE 48 Bricks made from hardened plinthite.

PLATE 50 Adult 17-year cicada and emergence holes.

PLATE 51 Dark green clover in N–deficient lawn.

PLATE 52 Morrow plots at Univ. of Illinois, Urbana.

PLATE 53 Grape vines growing in Jory soil (Ultisol with low P availability) are dramatically stimulated (*right*), but are little affected if growing in Chelalis soil (Mollisol with high P availability).

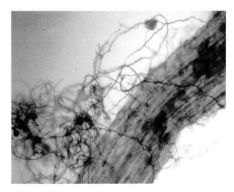

PLATE 54 Sorghum root colonized by AM fungi showing hyphae growing outside the root (extra radicular).

PLATE 55 Desert pavement with a pebble removed to reveal vesicular pores in soil.

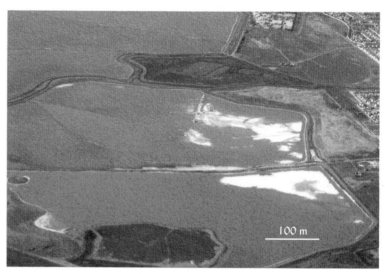

PLATE 56 Evaporating basins for sea salt. Bloom of *Halobacterium*, an archaean with red photosynthesizing pigment.

PLATE 57 Ancient carving in desert varnish, a Mn oxide rock coating deposited by bacteria. Nevada.

PLATE 58 Alluvium deposited 1 m thick in 8 years since riprap installed on stream bank in an urban watershed.

PLATE 59 Iron concentrations (orange lepidocrocite) on ped faces in Btg horizon of Alfisol in Chad.

PLATE 60 Connecticut River valley in western Massachusetts. Note variable alluvial soils and presence of riparian forest buffer along the river bank.

TABLE 3.4 Formative Elements in Names of Suborders in *Soil Taxonomy*

Formative element	Derivation	Connotation of formative element
alb	L. *albus*, white	Presence of albic horizon (a bleached eluvial horizon)
anthr	Gk. *anthropos*, human	Presence of anthropic or plaggen epipedon
aqu	L. *aqua*, water	Characteristics associated with wetness
ar	L. *arare*, to plow	Mixed horizons
arg	L. *argilla*, white clay	Presence of argillic horizon (a horizon with illuvial clay)
calc	L. *calcis*, lime	Presence of calcic horizon
camb	L. *cambriare*, to change	Presence of cambic horizon
cry	Gk. *kryos*, icy cold	Cold
dur	L. *durus*, hard	Presence of a duripan
fibr	L. *fibra*, fiber	Least decomposed stage
fluv	L. *fluvius*, river	Floodplains
fol	L. *folia*, leaf	Mass of leaves
gel	Gk. *gelid*, cold	Cold
gyps	L. *gypsum*, gypsum	Presence of gypsic horizon
hem	Gk. *hemi*, half	Intermediate stage of decomposition
hist	Gk. *histos*, tissue	Presence of histic epipedon
hum	L. *humus*, earth	Presence of organic matter
orth	Gk. *orthos*, true	The common ones
per	L. *per*, throughout time	Of year-round humid climates, perudic moisture regime
psamm	Gk. *psammos*, sand	Sand textures
rend	Modified from Rendzina	Rendzina-like—high in carbonates
sal	L. *sal*, salt	Presence of salic (saline) horizon
sapr	Gk. *sapros*, rotten	Most decomposed stage
torr	L. *torridus*, hot and dry	Usually dry
turb	L. *turbidus*, disturbed	Cryoturbation
ud	L. *udus*, humid	Of humid climates
ust	L. *ustus*, burnt	Of dry climates, usually hot in summer
vitr	L. *vitreus*, glass	Resembling glass
xer	Gk. *xeros*, dry	Dry summers, moist winters

To determine the relationship between suborder names and soil characteristics, refer to Table 3.4. Here the formative elements for suborder names are identified and their connotations given. Thus, the Ustolls are dry Mollisols. Likewise, soils in the Udults suborder (from the Latin *udus*, humid) are moist Ultisols.

Great Groups

The great groups are subdivisions of suborders. More than 400 great groups are recognized. They are defined largely by the presence or absence of diagnostic horizons and the arrangements of those horizons. These horizon designations are included in the list of formative elements for the names of great groups shown in Table 3.5. Note that these formative elements refer to epipedons such as umbric and ochric (see Table 3.1 and Figure 3.3), to subsurface horizons such as argillic and natric, and to certain diagnostic impervious layers such as duripans and fragipans (see Figure 3.30).

Remember that the great group names are made up of these formative elements attached as prefixes to the names of suborders in which the great groups occur. Thus, Ustolls with a natric horizon (high in sodium) belong to the Natrustolls great group. As can be seen in the example discussed in Box 3.2, soil descriptions at the great group level can provide important information not indicated at the higher, more general levels of classification.

The names of selected great groups from two orders are given in Table 3.6. This list illustrates again the usefulness of *Soil Taxonomy*, especially the nomenclature it employs. The names identify the suborder and order in which the great groups are found. Thus, Argiudolls are Mollisols of the Udolls suborder characterized by an argillic horizon. Cross-reference to Table 3.5 identifies the specific characteristics separating the great group classes from each other.

Note from Table 3.6 that not all possible combinations of great group prefixes and suborders are used. In some cases a particular combination does not exist. For example, Aquolls occur in lowland areas but not on very old landscapes. Hence, there are no "Paleaquolls." Also, since *all* Ultisols contain an argillic horizon, the use of terms such as "Argiudults" would be redundant.

TABLE 3.5 Formative Elements for Names of Great Groups and Their Connotation

These formative elements combined with the appropriate suborder names give the great group names.

Formative element	Connotation	Formative element	Connotation	Formative element	Connotation
acr	Extreme weathering	fol	Mass of leaves	petr	Cemented horizon
agr	Agric horizon	fragi	Fragipan	plac	Thin pan
al	High aluminum, low iron	fragloss	Combination of *fragi* and *gloss*	plagg	Plaggen horizon
alb	Albic horizon	fulv	Light-colored melanic horizon	plinth	Plinthite
and	Ando-like	gyps	Gypsic horizon	psamm	Sand texture
anhy	Anhydrous	gloss	Tongued	quartz	High quartz
aqu	Water saturated	hal	Salty	rhod	Dark red colors
argi	Argillic horizon	hapl	Minimum horizon	sal	Salic horizon
calc, calci	Calcic horizon	hem	Intermediate decomposition	sapr	Most decomposed
camb	Cambic horizon	hist	Presence of organic materials	somb	Dark horizon
chrom	High chroma	hum	Humus	sphagn	Sphagnum moss
cry	Cold	hydr	Water	sulf	Sulfuric
dur	Duripan	kand	Low-activity 1:1 silicate clay	torr	Usually dry and hot
dystr, dys	Low base saturation	lithic	Near stone	ud	Humid climates
endo	Fully water saturated	luv, lu	Illuvial	umbr	Umbric epipedon
epi	Perched water table	melan	Melanic epipedon	ust	Dry climate, usually hot in summer
eutr	High base saturation	molli	With a mollic epipedon	verm	Wormy, or mixed by animals
ferr	Iron	natr	Presence of a natric horizon	vitr	Glass
fibr	Least decomposed	pale	Old development	xer	Dry summers, moist winters
fluv	Floodplain				

Subgroups

Subgroups are subdivisions of the great groups. More than 2500 subgroups are recognized. The central concept of a great group makes up one subgroup, termed *Typic*. Thus, the Typic Hapludolls subgroup typifies the Hapludolls great group. Other subgroups may have characteristics that intergrade between those of the central concept and soils of other orders, suborders, or great groups. A Hapludoll with restricted drainage would be classified as an Aquic Hapludoll. One with evidence of intense earthworm activity would fall in the Vermic Hapludolls subgroup. Some intergrades may have properties in common with other orders or with other great groups. Thus, soils in the Entic Hapludolls subgroup are very weakly developed Mollisols, close to being in the Entisols order. The subgroup concept illustrates very well the flexibility of this classification system.

FIGURE 3.30 A forested fragiudalf in Missouri containing a typical well-developed fragipan with coarse prismatic structure (outlined by gray, iron-depleted coatings). Fragipans (usually Bx or Cx horizons) are extremely dense and brittle. They consist mainly of silt, often with considerable sand, but not very much clay. One sign of encountering a fragipan in the field is the ringing noise that your shovel will make when you attempt to excavate it. Digging through a fragipan is almost like digging concrete. Plant roots cannot penetrate this layer. Yet, once a piece of a fragipan is broken loose, it fairly easily crushes with hand pressure. It does not squash or act in a plastic manner as a claypan would; instead, it bursts in a *brittle* manner. (Photo courtesy of Fred Rhoton, Agricultural Research Service, U.S. Department of Agriculture)

BOX 3.2 GREAT GROUPS, FRAGIPANS, AND ARCHAEOLOGIC DIGS

Soil Taxonomy is a communications tool that helps scientists and land managers share information. In this box we will see how misclassification, even at a lower level in *Soil Taxonomy*, such as the great group, can have costly ramifications.

In order to preserve our historical and prehistorical heritage, laws require that an archaeological impact statement be prepared prior to starting major construction work on the land. The archaeological impact is usually assessed in three phases. Selected sites are then studied by archaeologists, with the hope that at least some of the artifacts can be preserved and interpreted before construction activities obliterate them forever. Only a few relatively small sites can be subjected to actual archaeological digs because of the expensive skilled hand labor involved (Figure 3.31).

Such an archaeological impact study was ordered as a precursor to construction of a new highway in a mid-Atlantic state. In the first phase, a consulting company gathered soils and other information from maps, aerial photographs, and field investigations to determine where neolithic people may have occupied sites. Then the consultants identified about 12 ha of land where artifacts indicated significant neolithic activities. The soils in one area were mapped mainly as Typic Dystrudepts. These soils formed in old colluvial and alluvial materials that, many thousands of years ago, had been along a river bank. Several representative soil profiles were examined by digging pits with a backhoe. The different horizons were described, and it was determined in which horizons artifacts were most likely to be found. What was not noted was the presence in these soils of a fragipan, a dense, brittle layer that is extremely difficult to excavate using hand tools.

A fragipan is a subsurface diagnostic horizon used to classify soils, usually at the great group or subgroup level (see Figure 3.30). Its presence would distinguish Fragiudepts from Dystrudepts.

When it came time for the actual hand excavation of sites to recover artifacts, a second consulting company was awarded the contract. Unfortunately, their bid on the contract was based on soil descriptions that did not specifically classify the soils as Fragiudepts—soils with very dense, brittle, hard fragipans in the layer that would need to be excavated by hand. So difficult was this layer to excavate and sift through by hand that it nearly doubled the cost of the excavation—an additional expense of about $1 million. Needless to say, there ensued a controversy as to whether this cost would be borne by the consulting firm that bid with faulty soils data, the original consulting firm that failed to adequately describe the presence of the fragipan, or the highway construction company that was paying for the survey.

This episode gives us an example of the practical importance of soil classification. The formative element *Fragi* in a soil great group name warns of the presence of a dense, impermeable layer that will be very difficult to excavate, will restrict root growth (often causing trees to topple in the wind or become severely stunted), may cause a perched water table (epiaquic conditions), and will interfere with proper percolation in a septic drain field.

FIGURE 3.31 An archaeological dig. (Photo courtesy of Antonio Segovia, University of Maryland)

TABLE 3.6 Examples of Great Group Names for Selected Suborders in the Mollisol and Ultisol Orders

	Dominant feature of great group			
	Argillic horizon	Central concept with no distinguishing features	Old land surfaces	Fragipan
Mollisols				
1. Aquolls (wet)	Argi*aquolls*	Hapl*aquolls*	—	—
2. Udolls (moist)	Argi*udolls*	Hapl*udolls*	Pale*udolls*	—
3. Ustolls (dry)	Argi*ustolls*	Hapl*ustolls*	Pale*ustolls*	—
4. Xerolls (Med.)[a]	Argi*xerolls*	Haplo*xerolls*	Pale*xerolls*	—
Ultisols				
1. Aquults (wet)	—	—	Pale*aquults*	Fragi*aquults*
2. Udults (moist)	—	Hapl*udults*	Pale*udults*	Fragi*udults*
3. Ustults (dry)	—	Hapl*ustults*	Pale*ustults*	—
4. Xerults (Med.)[a]	—	Haplo*xerults*	Pale*xerults*	—

[a] Med. = Mediterranean climate; distinct dry period in summer.

Families

Within a subgroup, soils fall into a particular family if, at a specified depth, they have similar physical and chemical properties affecting the growth of plant roots. About 8000 families have been identified. The criteria used include broad classes of particle size, mineralogy, cation exchange activity of the clay, temperature, and depth of the soil penetrable by roots. Table 3.7 gives examples of the classes used. Terms such as *loamy, sandy,* and *clayey* are used to identify the broad particle size classes. Terms used to describe the mineralogical classes include *smectitic, kaolinitic, siliceous, carbonatic,* and *mixed.* The clays are described as *superactive, active, semiactive,* or *subactive* with regard to their capacity to hold cations. For temperature classes, terms such as *cryic, mesic,* and *thermic* are used. The terms *shallow* and *micro* are sometimes used at the family level to indicate unusual soil depths.

Thus, a Typic Argiudoll from Iowa, loamy in texture, having a mixture of moderately active clay minerals and with annual soil temperatures (at 50 cm depth) between 8 and 15 °C, is classed in the *loamy, mixed, active, mesic Typic Argiudolls* family. In contrast, a sandy-textured Typic Haplorthod, high in quartz and located in a cold area in eastern Canada, is classed in the *sandy, siliceous, frigid Typic Haplorthods* family (note that clay activity classes are not used for soils in sandy textural classes).

TABLE 3.7 Some Commonly Used Particle-Size, Mineralogy, Cation Exchange Activity, and Temperature Classes Used to Differentiate Soil Families

The characteristics generally apply to the subsoil or 50 cm depth. Other criteria used to differentiate soil families (but not shown here) include the presence of calcareous or highly aluminum toxic (allic) properties, extremely shallow depth (shallow or micro), degree of cementation, coatings on sand grains, and the presence of permanent cracks.

Particle-size class	Mineralogy class	Cation exchange activity class[b] Term	CEC / % clay	Mean annual temperature, °C	Soil temperature regime class >6 °C difference between summer and winter	<6 °C difference between summer and winter
Ashy	Mixed	Superactive	0.60	<−10	Hypergelic[c]	—
Fragmental	Micaceous	Active	0.4 to 0.6	−4 to −10	Pergelic[c]	—
Sandy-skeletal[a]	Siliceous	Semiactive	0.24 to 0.4	+1 to −4	Subgelic[c]	—
Sandy	Kaolinitic	Subactive	<0.24	<+8	Cryic	—
Loamy	Smectitic			>+8	Frigid[d]	Isofrigid
Clayey	Gibbsitic			+8 to +15	Mesic	Isomesic
Fine-silty	Gypsic			+15 to +22	Thermic	Isothermic
Fine-loamy	Carbonic			>+22	Hyperthermic	Isohyperthermic
Etc.	Etc.					

[a] Skeletal refers to presence of up to 35% rock fragments by volume.
[b] Cation exchange activity class is not used for taxa already defined by low CEC (e.g., kandic or oxic groups).
[c] Permafrost present.
[d] Frigid is warmer in summer than cryic.

Series

The series category is the most specific unit of the classification system. It is a subdivision of the family, and each series is defined by a specific range of soil properties involving primarily the kind, thickness, and arrangement of horizons. Features such as a hard pan within a certain distance below the surface, a distinct zone of calcium carbonate accumulation at a certain depth, or striking color characteristics may aid in series identification.

In the United States, each series is given a name, usually from some town, river, or lake such as Fargo, Muscatine, Cecil, Mohave, or Ontario. There are about 23,000 soil series in the United States.

The complete classification of a Mollisol, the Kokomo series, is given in Figure 3.32. This figure illustrates how *Soil Taxonomy* can be used to show the relationship between *the soil*, a comprehensive term covering all soils, and a specific soil series. The figure deserves study because it reveals much about the structure and use of *Soil Taxonomy*. If a soil series name is known, the complete *Soil Taxonomy* classification of the soil may be found on the Internet at the URL in the margin. Box 3.3 illustrates how soil taxonomic information can assist in understanding the nature of a landscape such as shown in Figure 3.33.

Official Soil Taxonomy Soil Series Names and Descriptions. Click on "Soil Series Name Search." http://soils.usda.gov/technical /classification/scfile/index.html

3.18 CONCLUSION

Rationale for concepts in Soil Taxonomy: http://soils.usda.gov/ technical/classification/ taxonomy/rationale/index. html

The soil that covers the Earth is actually comprised of many individual soils, each with distinctive properties. Among the most important of these properties are those associated with the layers, or *horizons*, found in a soil profile. These horizons reflect the physical, chemical, and biological processes soils have undergone during their development. Horizon properties greatly influence how soils can and should be used.

Knowledge of the kinds and properties of soils around the world is critical to humanity's struggle for survival and well-being. A soil classification system based on

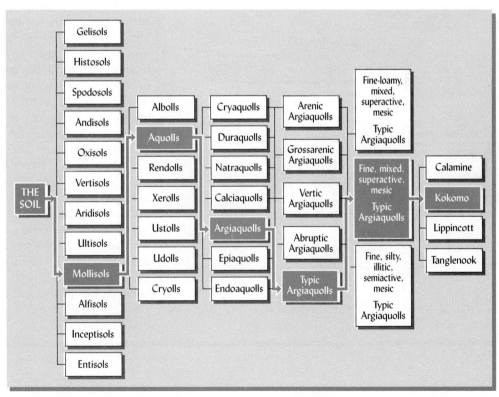

FIGURE 3.32 Diagram illustrating how one soil (Kokomo) keys out in the overall classification scheme. The shaded boxes show that this soil is in the Mollisols order, Aquolls suborder, Argiaquolls great group, and so on. In each category, other classification units are shown in the order in which they key out in *Soil Taxonomy*. Many more families exist than are shown.

BOX 3.3   USING SOIL TAXONOMY TO UNDERSTAND A LANDSCAPE

In real-world landscapes, different soils exist alongside each other, often in complex patterns. Adjacent soils on a tract of land may belong to different families, subgroups, great groups, or even different soil orders. Figure 3.33 depicts a landscape in a humid temperate region (Iowa) where 2 to 7 m of loess overlies leached glacial till and the native vegetation was principally tall grass prairie interspersed with small areas of trees. This landscape demonstrates how diagnostic horizons and other features of soil taxonomy are used to organize soils information. It also highlights the relationships among soils that allow us to make soil maps and interpret geographic soils information to help in planning projects on the land (see Chapter 19).

Seven soil map units are shown in the block landscape diagram, along with a profile diagram for the dominant soil series in each map unit. The soils include two Alfisols (Fayette and Downs) and five Mollisols (Tama, Wabash, Dinsdale, Muscatine, and Garwin). The particular set of soil horizons present in each profile relates to the (1) parent material, (2) vegetation, and (3) topography and drainage. Find the Dinsdale and Tama soils and notice where they occur in the landscape. The Dinsdale soil differs from the Tama because two parent materials (loess and glacial till) contributed to the Dinsdale profile, but the Tama soil is found where the loess layer by itself is thick enough to accommodate the entire profile. Both the Tama and Fayette soils exhibit argillic B horizons, but the Fayette has a thin ochric epipedon and a bleached albic horizon because it formed under forest vegetation, while the Tama has a thick mollic epipedon because it formed under grassland vegetation. The influence of topography can be seen by noting that soils on concave or level positions (the Garwin and Muscatine soils, which have slopes ranging from 0 to 1% and 1 to 3%, respectively) are wetter and less permeable than those on the steeper slopes (Tama and Dinsdale soils). In the less sloping, wetter soils, restricted drainage has retarded the development of an argillic horizon, so that only a gleyed (waterlogged) cambic B horizon (Bg) is present.

Are these relationships reflected in the soil taxonomy names? Note that the formative element aqu appears in the taxonomic name of the three wetter soils. Aqu appears at the suborder level (Endoaquolls) for the very wet, poorly drained soils, but only at the subgroup level for the less wet, somewhat poorly drained soil (Aquic Hapludoll). The formative element argi is used in the name of two soils (Argiudolls) to indicate that enough clay has accumulated in the B horizon of these Mollisols to develop into an argillic diagnostic horizon. Argi does not appear in the names of the two Alfisols, because an accumulation of clay (argillic or similar horizon) is a required feature of all Alfisols. Subgroup modifiers also provide important information about the interrelationships of these soils in the landscape. For example, the modifier Cumulic indicates that the Wabash soil has an unusually thick mollic epipedon because soil material washing off the uplands and carried by local streams has accumulated in the low-lying floodplains where this soil is found. The modifier Mollic used for the Downs soil indicates that this soil is transitional between the Alfisols and Mollisols, the A horizon in the Downs soils being slightly too thin to classify as a mollic epipedon.

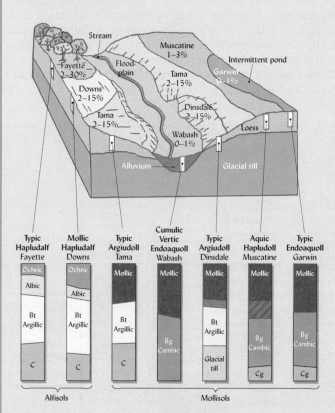

FIGURE 3.33 *Soil taxonomy reflects soil-landscape relationships.*

these properties is equally critical if we expect to use knowledge gained at one location to solve problems at other locations where similarly classed soils are found. *Soil Taxonomy*, a classification system based on measurable soil properties, helps fill this need in more than 50 countries. Scientists constantly update the system as they learn more about the nature and properties of the world's soils and the relationships among them. In the remaining chapters of this book we will use taxonomic names whenever appropriate to indicate the kinds of soils to which a concept or illustration may apply.

STUDY QUESTIONS

1. Diagnostic horizons are used to classify soils in *Soil Taxonomy*. Explain the difference between a diagnostic horizon (such as an argillic horizon) and a genetic horizon designation (such as a Bt1 horizon). Give a field example of a diagnostic horizon that contains several genetic horizon designations.

2. Explain the relationships among a *soil individual*, a *polypedon*, a *pedon*, and a *landscape*.

3. Rearrange the following soil orders from the *least* to the *most* highly weathered: Oxisols, Alfisols, Mollisols, Entisols, and Inceptisols.

4. What is the principal soil property by which Ultisols differ from Alfisols? Inceptisols from Entisols?

5. Use the key given in Figure 3.11 to determine the soil order of a soil with the following characteristics: a spodic horizon at 30 cm depth, permafrost at 80 cm depth. Explain your choice of soil order.

6. Of the five soil-forming factors discussed in Chapter 2 (parent material, climate, organisms, topography, and time), choose *two* that have had the dominant influence on developing soil properties characterizing each of the following soil orders: Vertisols, Mollisols, Spodosols, and Oxisols.

7. To which soil order does each of the following belong: Psamments, Udolls, Argids, Udepts, Fragiudalfs, Haplustox, and Calciusterts.

8. What's in a name? Write a hypothetical soil profile description and land-use suitability interpretation for a hypothetical soil that is classified in the Aquic Argixerolls subgroup.

9. Explain why *Soil Taxonomy* is said to be a hierarchical classification system.

10. Name the soil taxonomy category and discuss the engineering implications of these soil taxonomy classes: Aquic Paleudults, Fragiudults, Haplusterts, Saprists, and Turbels.

REFERENCES

Barrera-Bassols, N., J. Alfred Zinck, and E. Van Ranst. 2006. "Symbolism, knowledge and management of soil and land resources in indigenous communities: Ethnopedology at global, regional and local scales," *Catena* **65**:118–137.

Coulombe, C. E., L. P. Wilding, and J. B. Dixon. 1996. "Overview of Vertisols: Characteristics and impacts on society," *Advances in Agronomy* **17**:289–375.

Ditzler, C. A. 2005. "Has the polypedon's time come and gone?" *HPSSS Newsletter*, February 2005, pp. 8–11. Commission on History, Philosophy and Sociology of Soil Science, International Union of Soil Sciences. http://www.iuss.org/Newsletter12C4-5.pdf (verified 20 October 2005).

Eswaran, H. 1993. "Assessment of global resources: Current status and future needs," *Pedologie* **43**(1):19–39.

Eswaran, H., T. Rice, R. Ahrens, and B. A. Stewart (eds.). 2003. *Soil classification: A global desk reference.* CRC Press, Boca Raton, FL.

Gong, Z., X. Zhang, J. Chen, and G. Zhang. 2003. "Origin and development of soil science in ancient China." *Geoderma* **115**:3–13.

Riecken, F. F., and G. D. Smith. 1949. "Principal upland soils of Iowa, their occurrence and important properties," *Agron* **49** (revised). Iowa Agr. Exp. Sta.

Shaw, J. N., L. T. West, D. E. Radcliffe, and D. D. Bosch. 2000. "Preferential flow and pedotransfer functions for transport properties in sandy Kandiustults." *Soil Sci. Soc. Amer. J.* **64**:670–678.

Soil Survey Staff. 1975. *Soil taxonomy: A basic system of soil classification for making and interpreting soil surveys*. Natural Resources Conservation Service, Washington, DC.

Soil Survey Staff. 1999. *Soil taxonomy: A basic system of soil classification for making and interpreting soil surveys*, 2nd ed. Natural Resources Conservation Service, Washington, DC.

Soil Survey Staff. 2006. *Keys to soil taxonomy*. United States Department of Agriculture, Natural Resources Conservation Service. http://soils.usda.gov/technical/classification/tax_keys/keysweb.pdf.

SSSA. 1984. *Soil taxonomy, achievements and challenges*. SSSA Special Publication 14. Soil Sci. Soc. Amer., Madison, WI.

Talawar, S., and R. E. Rhoades. 1998. "Scientific and local classification and management of soils." *Agriculture and Human Values* **15**:3–14.

U.S. Department of Agriculture. 1938. *Soils and men*. USDA Yearbook. U.S. Government Printing Office, Washington, DC.

Structure and texture of a Mollisol. (Ray Weil)

4
SOIL ARCHITECTURE AND PHYSICAL PROPERTIES

And when that crop grew, and was harvested,
no man had crumbled a hot clod in his fingers
and let the earth sift past his fingertips.
—JOHN STEINBECK, THE GRAPES OF WRATH

Soil physical properties profoundly influence how soils function in an ecosystem and how they can best be managed. Success or failure of both agricultural and engineering projects often hinges on the physical properties of the soil used. The occurrence and growth of many plant species are closely related to soil physical properties, as is the movement over and through soils of water and its dissolved nutrients and chemical pollutants.

Soil scientists use the color, texture, and other physical properties of soil horizons in classifying soil profiles and in making determinations about soil suitability for agricultural and environmental projects. Knowledge of basic soil physical properties is not only of great practical value in itself, but will also help in understanding many aspects of soils considered in later chapters.

The physical properties discussed in this chapter relate to the solid particles of the soil and the manner in which they are aggregated. If we think of the soil as a house, the primary particles in soil are the building blocks from which the house is constructed. **Soil texture** describes the sizes of the soil particles. The larger mineral particles usually are embedded in, and coated with, clay and other colloidal size materials. Where the larger mineral particles predominate, the soil is gravelly or sandy; where the mineral colloids are dominant, the soil is claylike. All gradations between these extremes are found in nature.

In building a house, the manner in which the building blocks are put together determines the nature of the walls, rooms, and passageways. Organic matter and other substances act as cement between individual particles, encouraging the formation of clumps or aggregates of soil. **Soil structure** describes the manner in which soil particles are aggregated. This property, therefore, defines the nature of the system of pores and channels in a soil.

The physical properties considered in this chapter focus on soil solids and on the pore spaces between the solid particles. Together, soil texture and structure help determine the ability of the soil to hold and conduct the water and air necessary for sustaining life. These factors also determine how soils behave when used for highways and building foundations, or when manipulated by tillage. In fact, through their influence on the movement of water through and off soils, physical properties also exert considerable control over the destruction of the soil itself by erosion.

121

4.1 SOIL COLOR[1]

Color anarchy vs. system: http://www.urbanext.uiuc.edu/soil/less_pln/color/color.htm

Color is often the most obvious characteristic of a soil. Although color itself has little effect on the behavior and use of soils, it does provide clues about other soil properties and conditions. To obtain the precise, repeatable description of colors needed for soil classification and interpretation, soil scientists compare a small piece of soil to standard color chips in special Munsell[2] color charts. The Munsell charts use color chips arranged according to the three components of how people see color: the **hue** (in soils, usually redness or yellowness), the **value** (lightness or darkness, a value of 0 being black), and the **chroma** (intensity or brightness, a chroma of 0 being neutral gray). In a Munsell color book (carefully study Plate 22 after page 112), color chips are arranged on pages with values increasing from the bottom to the top, and the chromas increasing from left to right, while hues change from one page to another.

Soils display a wide range of reds, browns, yellows, and even greens (see Plates 16 and 35). Some soils are nearly black, others nearly white. Some soil colors are very bright, others are dull grays. Soil colors may vary from place to place in the landscape (e.g., Plates 16 and 36) as well as with depth through the various layers (horizons) within a soil profile, or even within a single horizon or clod of soil (Plates 10, 17, and 37). When making field soil descriptions, it is worth noting that horizons in a given profile that differ in chroma and value are often similar in hue.

Causes and Interpretation of Soil Colors

Three major factors influence soil colors: (1) organic matter content, (2) water content, and (3) the presence and oxidation states of iron and manganese oxides. Organic matter tends to coat mineral particles, darkening and masking the brighter colors of the minerals themselves (see Plates 8 and 32). Soils are generally darker (have low color value) when wet than when dry (Plate 20). Water content has a more profound indirect effect on soil colors. It influences the level of oxygen in the soil, and thereby the rate of organic matter accumulation which darkens the soil (Plate 39). Water also affects the oxidation state of iron and manganese (Plates 16, 17, and 21). In well-drained uplands, especially in warm climates, well-oxidized iron compounds impart bright (high chroma) reds and browns to the soil (Plates 9 and 11). Other minerals that influence soil color include manganese oxide (black) and glauconite (green—as in Plate 28). In dry regions, calcite and soluble salts impart a whitish color to many soils (Plates 13, 99, and 104). These colors are in contrast to the gray and bluish colors (low chroma) that reduced iron compounds impart to poorly drained soil profiles (Plates 35 and 38). Under prolonged anaerobic conditions, reduced iron (which is far more soluble than oxided iron) is removed from particle coatings, often exposing the light gray colors of the underlying silicate minerals. Soil exhibiting gray colors from reduced iron and iron depletion is said to be **gleyed** (Plates 21 and 38). If sulfur is present under anaerobic conditions, iron sulfides may color the soil black regardless of the organic matter level (Plate 109).

Color is used as a diagnostic criterion for classifying soils. For example, a mollic epipedon (see Section 3.2) is so dark that both its value and its chroma are 3 or less (Plate 20). In an other example, Rhodic subgroups of certain soil orders have B horizons that are very red, having hues between 2.5YR (the most red of the yellowish red pages) and 10R (the most red in the Munsell color book). The presence in upper horizons of gley (low-chroma colors), either alone or mixed in a mottled pattern with brighter colors (see Plate 38), is used in delineating **wetlands**, for it is indicative of waterlogged conditions during at least a major part of the plant growing season (see Section 7.7). The depth in the profile at which gley colors are found helps to define the **drainage class** of the soil (see Figure 19.3).

Finally, it is worth mentioning that soils, with their distinctive colors, are important aesthetic components of the landscape. For example, warm, reddish colors are characteristic of many tropical and subtropical landscapes, while dark grays and browns typify cooler, more temperate regions.

---

[1] For a collection of papers on causes and measurement of soil color, see Bigham and Ciolkosz (1993).

[2] Developed in 1905 by artist Albert Munsell to explain colors to his art students, the Munsell system is now also used to standardize colors in a wide range of scientific and commercial applications. To learn why American school children don't ask to borrow a "5YR 4/5" crayon, see Landa and Fairchild (2005).

4.2 SOIL TEXTURE (SIZE DISTRIBUTION OF SOIL PARTICLES)

Knowledge of the proportions of different-sized particles in a soil (i.e., the **soil texture**) is critical for understanding soil behavior and management. When investigating soils on a site, the texture of various soil horizons is often the first and most important property to determine, for a soil scientist can draw many conclusions from this information. Furthermore, the texture of a soil in the field is not readily subject to change, so it is considered a basic property of a soil.

Nature of Soil Separates

Baseball mud. By Lesley Bannatyne: http://www.csmonitor.com/ 2005/1018/p18s02-hfks. html?s=widep

Diameters of individual soil particles range over six orders of magnitude, from boulders (1 m) to submicroscopic clays ($<10^{-6}$ m). Scientists group these particles into soil **separates** according to several classification systems, as shown in Figure 4.1. The classification system established by the U.S. Department of Agriculture is used in this text. The size ranges for these separates are not purely arbitrary, but reflect major changes in how the particles behave and in the physical properties they impart to soils.

Gravels, cobbles, boulders, and other **coarse fragments** greater than 2 mm in diameter may affect the behavior of a soil, but they are not considered to be part of the **fine earth fraction** to which the term *soil texture* properly applies.

SAND. Particles smaller than 2 mm but larger than 0.05 mm are termed *sand*. Sand feels gritty between the fingers. The particles are generally visible to the naked eye and may be rounded or angular (Figure 4.2), depending on the degree of weathering and abrasion undergone. Coarse sand particles may be rock fragments containing several minerals, but most sand grains consist of a single mineral, usually quartz (SiO_2) or other primary silicate (Figure 4.3). The dominance of quartz means that the sand separate generally contains few plant nutrients. The large particle size means that what nutrients are present will not likely be released for plant uptake.

As sand particles are relatively large, so, too, the pores between them are relatively large. The large pores in sandy soils cannot hold water against the pull of gravity (see Section 5.2) and so drain rapidly and promote entry of air into the soil. The relationship between particle size and **specific surface area** (the surface area for a given mass of particles) is illustrated in Figure 4.4. The large particles of sand have low specific surface areas. Therefore, sand particles possess little capacity to hold water or nutrients and do not stick together into a coherent mass (see Section 4.9). Owing to the just described properties, most sandy soils are well aerated and loose, but also infertile and prone to drought.

SILT. Particles smaller than 0.05 mm but larger than 0.002 mm in diameter are classified as *silt*. Although similar to sand in shape and mineral composition, individual silt

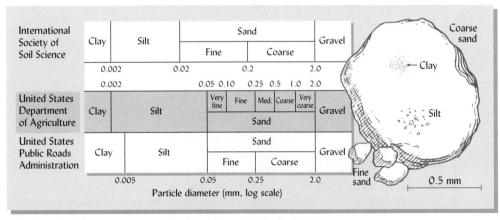

FIGURE 4.1 Classification of soil particles according to their size. The shaded scale in the center and the names on the drawings of particles follow the U.S. Department of Agriculture system, which is widely used throughout the world and in this book. The other two systems shown are also widely used by soil scientists and by highway construction engineers. The drawing illustrates the sizes of soil separates (note scale). (Diagram courtesy of R. Weil)

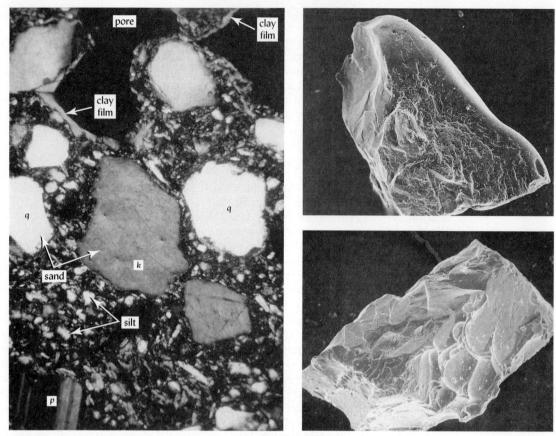

FIGURE 4.2 (*Left*) A thin section of a loamy soil as seen through a microscope using polarized light (empty pores appear black). The sand and silt particles shown are irregular in size and shape, the silt being only smaller. Although quartz (*q*) dominates the sand and silt fractions in this soil, several other silicate minerals can be seen (*p* = plagioclase, *k* = feldspar). Clay films coat the walls of the large pores (*arrows*). Scanning electron micrographs of sand grains show quartz sand (*bottom right*) and a feldspar grain (*upper right*) magnified about 40 times. (Left photo courtesy of Martin Rabenhorst, University of Maryland; right photos courtesy of J. Reed Glasmann, Union Oil Research)

particles are so small as to be invisible to the unaided eye (see Figure 4.2). Rather than feeling gritty when rubbed between the fingers, silt feels smooth or silky, like flour. Where silt is composed of weatherable minerals, the relatively small size (and large surface area) of the particles allows weathering rapid enough to release significant amounts of plant nutrients.

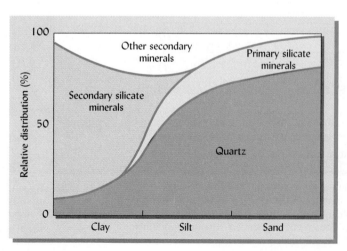

FIGURE 4.3 General relationship between particle size and kinds of minerals present. Quartz dominates the sand and coarse silt fractions. Primary silicates such as the feldspars, hornblende, and micas are present in the sands and, in decreasing amounts, in the silt fraction. Secondary silicates dominate the fine clay. Other secondary minerals, such as the oxides of iron and aluminum, are prominent in the fine silt and coarse clay fractions. (Diagram courtesy of N. Brady)

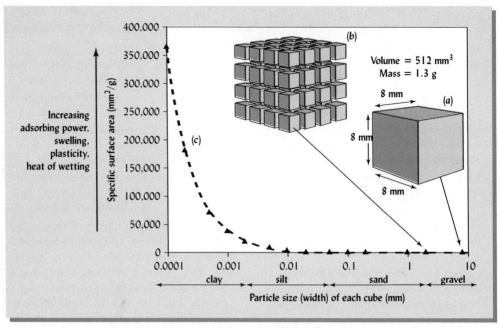

FIGURE 4.4 The relationship between the surface area of a given mass of material and the size of its particles. Consider a single, gravel-sized cube 8 mm on a side and weighing 1.3 g (*a*). Each face has 64 mm^2 of surface area. The cube has six faces with a total of 384 mm^2 surface area (6 faces · 64 mm^2 per face) or a specific surface of 295 cm^2/g (384/1.3). If this cube were cut into smaller cubes so that each cube was only 2 mm on each side (*b*), then the same mass of material would now be present as 64 (4 · 4 · 4) smaller, sand-sized cubes. Each face of each small cube would have 4 mm^2 (2 mm · 2 mm) of surface area, giving 24 mm^2 of surface area for each cube (6 faces · 4 mm^2 per face). The total surface area would therefore be 1536 mm^2 (24 mm^2 per cube · 64 cubes), or a specific surface of 1182 mm^2/g (1536/1.3). This is four times as much surface area as the single large cube. The curve (*c*) shows that clay particles, which are very, very small, have a surface area thousands of times greater than that of the same mass of silt particles and hundreds of thousands of times greater than the same mass of sand. The specific surface area curve explains why nearly all of the adsorbing power, swelling, plasticity, heat of wetting and other surface area related properties are associated with the clay fraction in mineral soils. (Diagram courtesy of R. Weil)

The pores between particles in silty material are much smaller (and much more numerous) than those in sand, so silt retains more water and lets less drain through. However, even when wet, silt itself does not exhibit much **stickiness** or **plasticity** (malleability). What little plasticity, cohesion, and adsorptive capacity some silt fractions exhibit is largely due to a film of adhering clay (see Figure 2.21). Because of their low stickiness and plasticity, soils high in silt and fine sand can be highly susceptible to erosion by both wind and water. Silty soil is easily washed away by flowing water in a process called **piping** (Box 4.1).

CLAY. Clay particles are smaller than 0.002 mm. They therefore have very large specific surface areas, giving them a tremendous capacity to adsorb water and other substances. A spoonful of clay may have a surface area the size of a football field (see Section 8.1). This large adsorptive surface causes clay particles to cohere in a hard mass after drying. When wet, clay is sticky and can be easily molded (exhibits high plasticity).

Fine clay–size particles are so small that they behave as **colloids**—if suspended in water they do not readily settle out. Unlike most sand and silt particles, clay particles tend to be shaped like tiny flakes or flat platelets. The pores between clay particles are very small and convoluted, so movement of both water and air is very slow. In clayey soil the pores between particles are tiny in size, but huge in number, allowing the soil to hold a great deal of water, however much of it may be unavailable to plants (see Section 5.9). Each unique clay mineral (see Chapter 8) imparts different properties to the soils in which it is prominent. Therefore, soil properties such as shrink-swell behavior, plasticity, water-holding capacity, soil strength, and chemical adsorption depend on the *kind* of clay present as well as the *amount*.

BOX 4.1   SILT AND THE FAILURE OF THE TETON DAM[a]

One of the most tragic and costly engineering failures in American history occurred in southern Idaho on 5 June 1977, less than a year after construction was completed on a large earth-fill dam across the Teton River. Eleven people were killed and 25,000 made homeless in the five hours it took to empty the 28 km long lake that had been held in place by the dam. Some $400 million (1977 dollars) worth of damages were caused as the massive wall of water surged through the collapsed dam and the valley below. The dam failed with little warning as small seepage leaks quickly turned into raging torrents that swept away a team of bulldozers sent to make repairs.

The Teton dam was built according to a standard, time-tested design for zoned earth-fill embankments (Figure 4.5). Essentially, after preparing a base in the rhyolite rock below the soil, a core (zone 1) of tightly compacted soil material was constructed and covered with a layer (zone 2) of coarser alluvial soil material to protect it from water and wind erosion.

The core is meant to be the watertight seal that prevents water from seeping through the dam. Normally, clayey material is chosen for the core, since the sticky, plastic qualities of moist clay allow it to be compacted into a malleable, watertight mass that holds together and does not crack so long as it is kept moist. Silt, on the other hand, though it may appear similar to clay in the field, has little or no stickiness or plasticity and therefore cannot be compacted into a coherent mass as clay can. In fact, a moist mass of compacted silt will crack as it settles because it lacks plasticity. If water seeps into these cracks, the silty material will rapidly wash away with the flowing water, enlarging the crack and inviting more water to flow through, which will wash away still more of the silt. This process of rapidly enlarging seepage channels is termed *piping*. Such piping was almost certainly a major cause of the Teton Dam failure, for the engineers built the zone 1 core of the dam using windblown silt deposits (termed *loess*—see Section 2.3) rather than clay.

This was a tragic but useful lesson about the role of texture in determining soil behavior and the importance of distinguishing clay from silt.

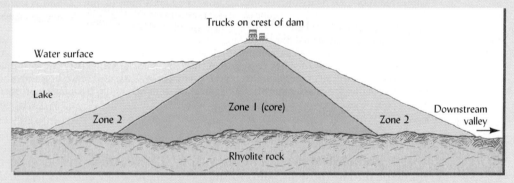

FIGURE 4.5   *Diagram of the Teton Dam.*

[a] Based on the report by the U.S. Department of Interior Teton Dam Failure Review Group (1977).

Influence of Surface Area on Other Soil Properties

When particle size decreases, specific surface area and related properties increase greatly, as shown graphically in Figure 4.4. Fine colloidal clay has about 10,000 times as much surface area as the same weight of medium-sized sand. Soil texture influences many other soil properties in far-reaching ways (see Table 4.1) as a result of five fundamental surface phenomena:

1. In addition to the tiny pools of water held in the smaller soil pores, water is also retained in soils as thin films on the surfaces of soil particles. The greater the surface area, the greater the soil's capacity for holding water films.

2. Both gases and dissolved chemicals are attracted to and adsorbed by mineral particle surfaces. The greater the surface area, the greater the soil's capacity to retain nutrients and other chemicals.

3. Weathering takes place at the surface of mineral particles, releasing constituent elements into the soil solution. The greater the surface area, the greater the rate of release of plant nutrients from weatherable minerals.

TABLE 4.1 Generalized Influence of Soil Separates on Some Properties and Behavior of Soils[a]

	Rating associated with soil separates		
Property/behavior	Sand	Silt	Clay
Water-holding capacity	Low	Medium to high	High
Aeration	Good	Medium	Poor
Drainage rate	High	Slow to medium	Very slow
Soil organic matter level	Low	Medium to high	High to medium
Decomposition of organic matter	Rapid	Medium	Slow
Warm-up in spring	Rapid	Moderate	Slow
Compactability	Low	Medium	High
Susceptibility to wind erosion	Moderate (high if fine sand)	High	Low
Susceptibility to water erosion	Low (unless fine sand)	High	Low if aggregated, high if not
Shrink-swell potential	Very Low	Low	Moderate to very high
Sealing of ponds, dams, and landfills	Poor	Poor	Good
Suitability for tillage after rain	Good	Medium	Poor
Pollutant leaching potential	High	Medium	Low (unless cracked)
Ability to store plant nutrients	Poor	Medium to high	High
Resistance to pH change	Low	Medium	High

[a] Exceptions to these generalizations do occur, especially as a result of soil structure and clay mineralogy.

4. The surfaces of mineral particles often carry both negative and some positive electromagnetic charges so that particle surfaces and the water films between them tend to attract each other (see Section 4.5). The greater the surface area, the greater the propensity for soil particles to stick together in a coherent mass, or as discrete aggregates.

5. Microorganisms tend to grow on and colonize particle surfaces. For this and other reasons, microbial reactions in soils are greatly affected by the specific surface area.

4.3 SOIL TEXTURAL CLASSES

Interactive textural triangle:
http://courses.soil.ncsu.edu/
resources/physics/texture/
soiltexture.swf

Within the three broad groups of *sandy soils, clayey soils*, and *loamy soils*, specific **textural class** names convey a more precise idea of the size distribution of particles and the general nature of soil physical properties. The 14 textural classes named in Table 4.2 form a graduated sequence from the sands, which are coarse in texture to the clays, which are fine. Sands and loamy sands are dominated by the properties of sand, for the sand separate comprises at least 70% of the material by weight and less than 15% of the material is clay (see boundaries on Figure 4.6). Clays, sandy clays, and silty clays are dominated by characteristics of clay. However, most soils are some type of **loam**.

Loams. The central concept of a **loam** may be defined as a mixture of sand, silt, and clay particles that exhibits the *properties* of those separates in about equal proportions. This definition does not mean that the three separates are present in equal *amounts* (that is why the loam class is not exactly in the middle of the triangle in Figure 4.6). This anomaly exists because a relatively small percentage of clay is required to engender

TABLE 4.2 General Terms Used to Describe Soil Texture in Relation to the Basic Soil Textural Class Names in the U.S. Department of Agriculture Classification System

Basic soil textural class names													
Sands	Loamy sands	Sandy loam	Fine sandy loam[a]	Very fine sandy loam[a]	Loam	Silt loam	Silt	Sandy clay loam	Silty clay loam	Clay loam	Sandy clay	Silty clay	Clay

General texture terms

------ Coarse ------ | ---- Moderately coarse ------- | ------------------ Medium ------------------ | ---------- Moderately fine ----------- | --------- Fine ----------

Sandy soils | --------------------------------------- Loamy soils -------------------------------------- | ----- Clayey soils -----

[a] Although not included as class names in Figure 4.6, these soils are usually treated separately because of their fine sand content.

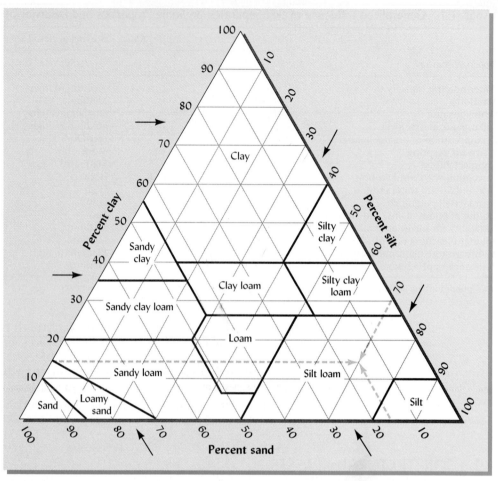

FIGURE 4.6 The major soil textural classes are defined by the percentages of sand, silt, and clay according to the heavy boundary lines shown on the textural triangle. If these percentages have been determined for a soil sample by particle size analysis, then the triangle can be used to determine the soil textural class name that applies to that soil sample. To use the graph, first find the appropriate clay percentage along the left side of the triangle, then draw a line from that location across the graph going parallel to the base of the triangle. Next find the sand percentage along the base of the triangle, then draw a line inward going parallel to the triangle side labeled "Percent silt." The small arrows indicate the proper direction in which to draw the lines. The name of the compartment in which these two lines intersect indicates the textural class of the soil sample. Percentages for any two of the three soil separates is all that is required. Because the percentages for sand, silt, and clay add up to 100%, the third percentage can easily be calculated if the other two are known. If all three percentages are used, the three lines will all intersect at the same point. Consider, as an example, a soil that has been determined to contain 15% sand, 15% clay, and 70% silt. This example is indicated by the light dashed lines that intersect in the compartment labeled "Silt loam." What is the textural class of another soil sample that has 33% sand, 33% silt, and 33% clay? The lines (not shown) for this second example would intersect in the center of the "Clay loam" compartment.

clayey properties in a soil, whereas small amounts of sand and silt have a lesser influence on how a soil behaves. A loam in which sand is dominant is classified as a *sandy loam*. In the same way, some soils are classed as *silt loams, silty clay loams, sandy clay loams*, and *clay loams*. Note from Figure 4.6 that a *clay loam* may have as little as 26% clay, but to qualify as *sandy loam* or *silt loam,* a soil must have at least 45% sand or 50% silt, respectively.

COARSE FRAGMENT MODIFIERS. If a soil contains a significant proportion of particles larger than sand (termed **coarse fragments**), a qualifying adjective may be used as part of the textural class name. Coarse fragments that range from 2 to 75 mm along their greatest diameter are termed *gravel* or *pebbles,* those ranging from 75 to 250 mm are called *cobbles* (if round) or *channers* (if flat), and those more than 250 mm across are called *stones* or *boulders.* A *gravelly, fine sandy loam* is an example of such a modified textural class.

FIGURE 4.7 Altering soil texture can be a big job. Here, large amounts of sand (darker material) and gravel (whitish) have been imported and are being spread to improve the all–weather playability of an athletic field. Insets show laser-guided bulldozers evenly spreading the sand and gravel materials. (Photos courtesy of R. Weil)

Alteration of Soil Textural Class

Over long periods of time, pedologic processes (see Chapter 2) such as illuviation and mineral weathering can alter the textures of certain soil horizons. Likewise, erosion and subsequent deposition downslope can selectively remove or deposit particles of certain sizes. However, management practices generally do not alter the textural class of a soil on a field scale. Changing the texture of a given soil would require mixing it with another soil material of a different textural class. For example, the incorporation of large quantities of sand to change the physical properties of a clayey soil for use in greenhouse pots or for turfgrass would be considered to change the soil texture. However, adding peat or compost to a soil while mixing a potting medium does not constitute a change in texture, since this property refers only to the mineral particles. In fact the term *soil texture* is not relevant to artificial media that contain mainly perlite, peat, styrofoam, or other nonsoil materials.

Great care must be exercised in attempting to ameliorate physical properties of fine-textured soils by adding sand. Where specifications (as for a landscape design) call for soil materials of a certain textural class, it is generally advisable to find a naturally occurring soil that meets the specification, rather than attempt to alter the textural class by mixing in sand or clay. If the sand is not of the proper size and not added in sufficient amounts, it may make matters worse, rather than better. While adjacent coarse sand grains form large pores between them, sand grains embedded in a silty or clayey matrix do not. Mixing in moderate amounts of fine sand or sand ranging widely in size may yield a product more akin to concrete than to a sandy soil. For some applications (such as golf putting greens and atheletic fields, see Figure 4.7), the need for rapid drainage and resistance to compaction even when wet may justify the construction of an artificial soil from carefully selected uniform sands. Similary, where a smooth, hard surface is required, such as for a tennis court, an artificial clay soil may be needed.

Determination of Textural Class by the "Feel" Method

Textural class determination is one of the first field skills a soil scientist should develop. Determining the textural class of a soil by its feel is of great practical value in soil survey, land classification, and any investigation in which soil texture may play a role. Accuracy depends largely on experience, so practice whenever you can, beginning with soils of known texture to calibrate your fingers.

The textural triangle (see Figure 4.6) should be kept in mind when determining the textural class by the feel method as explained in Box 4.2 and Figure 4.8.

BOX 4.2   A METHOD FOR DETERMINING TEXTURE BY FEEL

The first, and most critical, step in the texture-by-feel method is to knead a walnut-sized sample of moist soil into a uniform puttylike consistency, slowly adding water if necessary. This step may take a few minutes, but a premature determination is likely to be in error as hard clumps of clay and silt may feel like sand grains. The soil should be moist, but not quite glistening. Try to do this with only one hand so as to keep your other hand clean for writing in a field notebook (and shaking hands with your client).

While squeezing and kneading the sample, note its malleability, stickiness, and stiffness, all properties associated with the clay content. A high silt content makes a sample feel smooth and silky, with little stickiness or resistance to deformation. A soil with a significant content of sand feels rough and gritty, and makes a grinding noise when rubbed near one's ear.

Get a feel for the amount of clay by attempting to squeeze a ball of properly moistened soil between your thumb and the side of your forefinger, making a ribbon of soil. Make the ribbon as long as possible until it breaks from its own weight (see Figure 4.8).

Interpret your observations as follows:

1. Soil will not cohere into a ball, falls apart: **sand**

2. Soil forms a ball, but will not form a ribbon: **loamy sand**

3. Soil ribbon is dull and breaks off when less than 2.5 cm long and

   a. Grinding noise is audible; grittiness is prominent feel: **sandy loam**

   b. Smooth, floury feel prominent; no grinding audible: **silt loam**

   c. Only slight grittiness and smoothness; grinding not clearly audible: **loam**

4. Soil exhibits moderate stickiness and firmness, forms ribbons 2.5 to 5 cm long, and

   a. Grinding noise is audible; grittiness is prominent feel: **sandy clay loam**

   b. Smooth, floury feel prominent; no grinding audible: **silty clay loam**

   c. Only slight grittiness and smoothness; grinding not clearly audible: **clay loam**

5. Soil exhibits dominant stickiness and firmness, forms shiny ribbons longer than 5 cm, and

   a. Grinding noise is audible; grittiness is dominant feel: **sandy clay**

   b. Smooth, floury feel prominent; no grinding audible: **silty clay**

   c. Only slight grittiness and smoothness; grinding not clearly audible: **clay**

FIGURE 4.8 The "feel" method for determining soil textural class. A moist soil sample is rubbed between the thumb and forefingers and squeezed out to make a "ribbon." (Top) The gritty, noncohesive appearance and short ribbon of a sandy loam with about 15% clay. (Middle) The smooth, dull appearance and crumbly ribbon characteristic of a silt loam. (Bottom) The smooth, shiny appearance and long, flexible ribbon of a clay. (Photos courtesy of R. Weil)

A more precise estimate of sand content (and hence more accurate placement in the horizontal dimension of the textural class triangle) can be made by wetting a pea-sized clump of soil in the palm of your hand and smearing it around with your finger until your palm becomes coated with a souplike suspension of soil. The sand grains will stand out visibly and their volume as compared to the original "pea" can be estimated, as can their relative size (fine, medium, coarse, etc.).

It is best to learn the method using samples of known textural class. With practice, accurate textural class determinations can be made on the spot.

BOX 4.3   STOKES' LAW AND PARTICLE SIZE DETERMINED
BY SEDIMENTATION

The complete expression of Stokes' law tells us the velocity V of a particle falling through a fluid is directly proportional to the gravitational force g, the difference between the density of the particle and the density of the fluid ($D_s - D_f$) and the square of the effective particle diameter (d^2). The effective diameter is referred to because Stokes' law applies to smooth, round particles. Since most soil particles are neither smooth nor round, sedimentation techniques determine the *effective* diameters, not necessarily the actual diameter of the soil particle. The settling velocity is *inversely* proportional to the viscosity or "thickness" of the fluid η. Since velocity equals distance h divided by time t we can write Stokes' law as:

$$V = \frac{h}{t} = \frac{d^2 g(D_s - D_f)}{18\eta}$$

Where: g = gravitational force = 9.81 newtons per kilogram (9.81 N/kg)
η = viscosity of water at 20° C = 1/1000 newton-seconds per m^2 (10^3 Ns/m^2)
D_s = density of the solid particles, for most soils = $2.65 \cdot 10^3$ kg/m^3
D_f = density of the fluid (i.e., water) = $1.0 \cdot 10^3$ kg/m^3

Substituting these values into the equation, we can write:

$$V = \frac{h}{t} = \frac{d^2 * 9.81 \text{ N/kg} * (2.65 * 10^3 \text{ kg/m}^3 - 1.0 * 10^3 \text{ kg/m}^3)}{18 * 10^{-3} \text{ Ns/m}^2}$$

$$= \frac{9.81 \text{ N/kg} * 1.65 * 10^3 \text{ kg/m}^3}{18 * 10^{-3} \text{ Ns/m}^2} * d^2 = \frac{16.19 * 10^3 \text{ N/m}^3}{0.018 \text{ Ns/m}^2} * d^2 = \frac{9 \times 10^5}{\text{sm}} * d^2 = kd^2, \text{ where} \ldots k = \frac{9 \times 10^5}{\text{sm}}$$

Note that $V = kd^2$ is a highly simplified formula for Stoke's Law in which k represents a constant related to the acceleration of gravity and the nature of the liquid.

Let's choose to sample or measure a soil suspension at 0.1 m (10 cm) depth (Figure 4.9). We can calculate the seconds of settling time we must allow if we want the smallest silt particle to have just passed our sampling depth so our sample will contain only clay.

Chosen: $h = 0.1$ m   and $d = 2 \cdot 10^{-6}$ m (0.002 mm, smallest silt)

Solving for t we can write: $\dfrac{h}{t} = d^2 k \Rightarrow \dfrac{t}{h} = \dfrac{1}{d^2 k} \Rightarrow t = \dfrac{h}{d^2 k}$

Therefore: $t = \dfrac{0.1 \text{ m}}{(2*10^{-6} \text{ m})^2 * 9 * 10^5 \text{ s}^{-1}/\text{m}^{-1}} = 27{,}777$ sec = 463 min = 7.72 hours

By comparison, the smallest sand particle ($d = 0.05$ mm) would make the same journey in only 44 *seconds*.

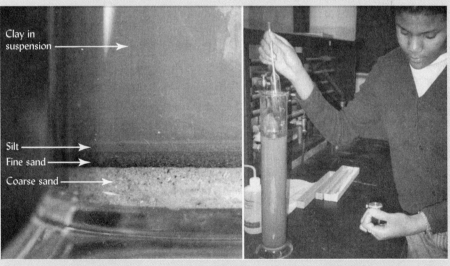

FIGURE 4.9 *In performing particle-size analysis, a sample of soil (with organic matter removed) is suspended in water, stirred vigorously, and then allowed to settle. A hydrometer (right) can indicate the mass of particles remaining in suspension after different settling times (it floats higher when more soil is in suspension). Stokes' Law is used to calculate the smallest effective diameter of the particles still in suspension at these times. (Left) The layers of sand and silt that have settled out after 7 hours. (Photos courtesy of R. Weil)*

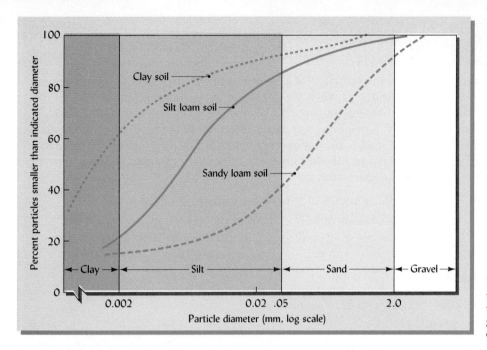

FIGURE 4.10 Particle-size distribution in three soils varying widely in their textures. Note that there is a gradual transition in the particle-size distribution in each of these soils.

Laboratory Particle-Size Analyses

The first and sometimes most difficult step in a particle-size analysis is the complete dispersion of a soil sample in water, so even the tiniest clumps are broken down into individual, primary particles. Dispersion is usually accomplished using chemical treatments along with a high-speed blender, shaker, or ultrasonic vibrator.

Separation into size groups can be accomplished by washing the suspended particles through standard sieves that are graded in size to separate the coarse fragments and sand separates, permitting the silt and clay fractions to pass through. A sedimentation procedure is usually used to determine the amounts of silt and clay (Figure 4.9). The principle involved is simple. Because soil particles are more dense than water, they tend to sink, settling at a velocity that is proportional to their size. In other words, "The bigger they are, the faster they fall." The equation that describes this relationship is referred to as *Stokes' law* (Box 4.3).

Figure 4.10 presents particle-size distribution curves for soils representative of three textural classes. The fact that these curves are smooth emphasizes that there is a continuum of particle sizes between coarse sand and fine clay.

It is important to note that soils are assigned to textural classes *solely* on the basis of the mineral particles of sand size and smaller; therefore, the percentages of sand, silt, and clay always add up to 100%. The amounts of stone and gravel are rated separately. Organic matter is not considered.

The relationship between such analyses and textural class names is commonly shown diagrammatically as a triangular graph (see Figure 4.6). This **textural triangle** also enables us to use laboratory particle-size analysis data to check the accuracy of field textural determinations by feel.

4.4 STRUCTURE OF MINERAL SOILS

The term **soil structure** relates to the *arrangement* of sand, silt, clay, and organic particles in soils. The particles become aggregated together due to various forces and at different scales to form distinct structural units called **peds** or **aggregates**. When a mass of soil is excavated and gently broken apart, it tends to break into peds along natural zones of weakness. These zones exhibit low tensile strength because particles within a ped or aggregate are more strongly attracted to one another than to the particles of the surrounding soil. Although *aggregate* and *ped* can be used synonymously, the term *ped* is most commonly used to describe the large-scale structure evident when observing soil profiles and involving structural units which range in size from a few mm to about 1 m. At this scale, the attraction of soil particles to one another in patterns that define structural units is influenced

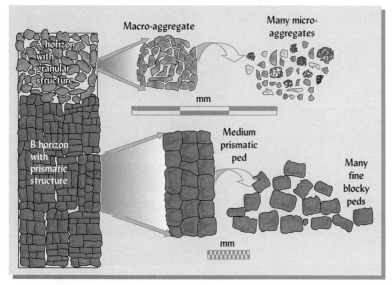

FIGURE 4.11 The hierarchical organization of soil structure. The larger structural units observable in a soil profile each contain many smaller units. The lower example shows how large prismatic peds typical of B horizons break down into smaller peds (and so on). The upper example illustrates how microaggregates smaller than 0.25 mm in diameter are contained within the granular macroaggregates of about 1 mm diameter that typify A horizons. The microaggregates often form around and occlude tiny particles of organic matter originally trapped in the macroaggregate. Note the two different scales for the prismatic and granular structures. (Diagram courtesy of R. Weil)

mainly by such physical processes as freeze-thaw, wet-dry, shrink-swell, the penetration and swelling of plant roots, the burrowing of soil animals, and the activities of people and machines. Structural peds should not be confused with **clods**—the compressed, cohesive chunks of soil that can form artificially when wet soil is plowed or excavated.

Most large peds are composed of, and can be broken into, smaller peds or aggregates (Figure 4.11). The networks of **pores** within and between the aggregates constitute a key aspect of soil structure (see Section 4.7). The pore network greatly influences the movement of air and water, the growth of plant roots, and the activities of soil organisms, including the accumulation and breakdown of organic matter. Such practices as timber harvest, grazing, tillage, trafficking, and manuring impact soils largely through their effects on soil structure, especially in the surface horizons.

The biological, chemical, and physical forces involved in the formation, stabilization, and management of soil aggregates of various sizes will be discussed in Sections 4.5 and 4.6. First we will examine the nature and types of structure observable in soil profiles.

Types of Soil Structure[3]

Many types or shapes of peds occur in soils, often within different horizons of a particular soil profile. Some soils may exhibit a **single grained** structural condition in which particles are not aggregated (Figure 4.12). The loose sand in wind-blown dunes or loose dust

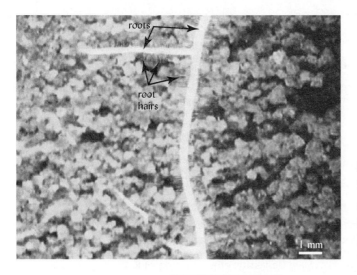

FIGURE 4.12 A plant root growing in a loamy sand soil (90% sand). The soil parent material was probably well sorted by wave action on an ancient beach resulting in uniform sand grain size. Very little aggregation is evident in this nearly structureless (or single grained structure) soil. The image was taken using a fiber-optic minirhizotron camera at about 50 cm depth in the Bw horizon of a psammentic Hapludult soil in the mid-Atlantic coastal plain of North America. (Photo courtesy of R. Weil and G. Chen,. University of Maryland)

---

[3] In the U.S. Soil Survey Handbook (USDA-NRCS, 2005), the single-grained and massive conditions are described as *structureless*.

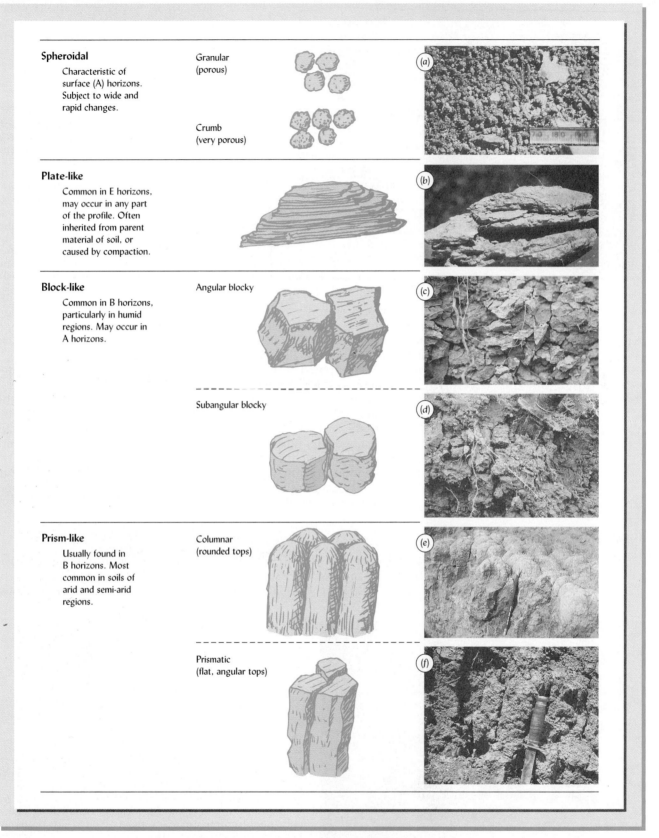

Spheroidal

Characteristic of surface (A) horizons. Subject to wide and rapid changes.

Granular (porous)

Crumb (very porous)

(a)

Plate-like

Common in E horizons, may occur in any part of the profile. Often inherited from parent material of soil, or caused by compaction.

(b)

Block-like

Common in B horizons, particularly in humid regions. May occur in A horizons.

Angular blocky

(c)

Subangular blocky

(d)

Prism-like

Usually found in B horizons. Most common in soils of arid and semi-arid regions.

Columnar (rounded tops)

(e)

Prismatic (flat, angular tops)

(f)

FIGURE 4.13 The various structure types (shapes) found in mineral soils. Their typical location is suggested. The drawings illustrate their essential features and the photos indicate how they look *in situ*. For scale, note the 15-cm-long pencil in (*e*) and the 3-cm-wide knife blade in (*d*) and (*f*). [Photo (*e*) courtesy of J. L. Arndt, North Dakota State University; others courtesy of R. Weil]

accumulations such as freshly deposited loess are examples of this single-grain structural condition. At the opposite extreme, some soils (such as certain clay sediments) occur as large, cohesive masses of material and are described as exhibiting a **massive** structural condition. However, most soils exhibit some type of aggregation and are composed of peds that can be characterized by their shape (or *type*), size, and distinctness (or *grade*). The four principal soil ped shapes are *spheroidal, platy, prismlike,* and *blocklike* (see below and Figure 4.13).

SPHEROIDAL.   **Granular** structure consists of spheroidal **aggregates** that may be separated from each other in a loosely packed arrangement (see Figure 4.13*a*). They typically range from less than 1 to as large as 10 mm in diameter. In reference to this type of structure, the term *aggregate* is used more commonly than *ped*. Granular structure characterizes many surface soils (usually A horizons), particularly those high in organic matter. Consequently, this is the principal type of soil structure affected by management. Granular aggregates are especially prominent in grassland soils and in soils that have been worked by earthworms.

PLATELIKE.   **Platy** structure, characterized by relatively thin horizontal peds or plates, may be found in both surface and subsurface horizons. In most instances, the plates have developed as a result of soil-forming processes. However, unlike other structure types, platy structure may also be inherited from soil parent materials, especially those laid down by water or ice. In some cases compaction of clayey soils by heavy machinery can create platy structure (see Figure 4.13*b*).

BLOCKLIKE.   **Blocky** peds are irregular, roughly cubelike (Figure 4.14), and range from about 5 to 50 mm across. The individual blocks are not shaped independently, but are molded by the shapes of the surrounding blocks. When the edges of the blocks are sharp and the rectangular faces distinct, the subtype is designated **angular blocky** (see Figure 4.13*c*). When some rounding has occurred, the peds are referred to as **subangular blocky** (see Figure 4.13*d*). These types are usually found in B horizons, where they promote drainage, aeration, and root penetration.

PRISMLIKE.   **Columnar** and **prismatic** structure are characterized by vertically oriented prisms or pillarlike peds that vary in height among different soils and may have a diameter of 150 mm or more. Columnar structure (see Figure 4.13*e*), which has pillars with distinct, rounded tops, is mainly found in subsoils high in sodium (e.g., natric

FIGURE 4.14   Strong, medium angular blocky peds in the B horizon of an Alfisol (Ustalf) in a semiarid region. The knifepoint is prying loose an individual blocky ped. Note the lighter colored surface coatings of illuvial clay that help define and bind the ped together. (Photo courtesy of R. Weil)

horizons; see Section 3.2). When the tops of the prisms are relatively angular and flat horizontally, the structure is designated as prismatic (see Figure 4.13*f*). Both prismlike structures are often associated with swelling types of clay. Prismatic structure commonly occurs in subsurface horizons in arid and semiarid regions and, when well developed, provides a very striking feature of the profile (see Figure 10.18). In humid regions, prismatic structure sometimes occurs in poorly drained soils and in fragipans (see Figure 3.30).

Description of Soil Structure in the Field

In describing soil structure (see Table 19.1), soil scientists note not only the *type* (shape) of the structural peds present, but also the relative *size* (fine, medium, coarse) and degree of development or distinctness of the peds (*grades* such as strong, moderate, or weak). For example, the soil shown in Figure 4.13*d* might be described as having "weak, fine, subangular blocky structure." Generally, the structure of a soil is easier to observe when the soil is relatively dry. When wet, structural peds may swell and press closer together, making the individual peds less well defined. We will now turn our attention to the formation and stabilization of soil structure, particularly the granular aggregates that characterize surface horizons.

4.5 FORMATION AND STABILIZATION OF SOIL AGGREGATES

The granular aggregation of surface soils is a highly dynamic soil property. Some aggregates disintegrate and others form anew as soil conditions change. Generally, smaller aggregates are more stable than larger ones, so maintaining the much-prized larger aggregates requires great care. We will discuss practical means of managing soil structure after we consider the factors responsible for aggregate formation and stabilization.

Hierarchical Organization of Soil Aggregates[4]

Surface horizons are usually characterized by roundish granular structure that exhibits a hierarchy in which relatively large **macroaggregates** (0.25 to 5 mm in diameter) are comprised of smaller **microaggregates** (2 to 250 μm). The latter, in turn, are composed of tiny packets of clay and organic matter only a few μm in size. You may easily demonstrate the existence of this *hierarchy of aggregation* by selecting a few of the largest aggregates in a soil and gently crushing or picking them apart to separate them into many smaller-sized pieces. Then try rubbing the tiniest of these soil crumbs between your thumb and forefinger. You will find that even the smallest specks of soil usually break down into a smear of still smaller particles of silt, clay, and humus. The hierarchical organization of aggregates seems to be characteristic of most soils, with the exception of certain Oxisols and some very young Entisols. Small particles of organic matter are often occluded inside the macro- and microaggregates. At each level in the hierarchy of aggregates, different factors are responsible for binding together the subunits (Figure 4.15).

Factors Influencing Aggregate Formation and Stability in Soils

Both biological and physical-chemical (abiotic) processes are involved in the formation of soil aggregates. Physical-chemical processes tend to be most important at the smaller end of the scale, biological processes at the larger end. The physical-chemical processes of aggregate formation are associated mainly with clays and, hence, tend to be of greater importance in finer-textured soils. In sandy soils that have little clay, aggregation is almost entirely dependent on biological processes.

Physical-Chemical Processes

Most important among the physical-chemical processes are (1) flocculation, the mutual attraction among clay and organic molecules; and (2) the swelling and shrinking of clay masses.

---

[4] The hierarchical organization of soil aggregates and the role of soil organic matter in its formation were first put forward by Tisdall and Oades (1982). For a review of advances in this area during the two decades following, see Six et al. (2004).

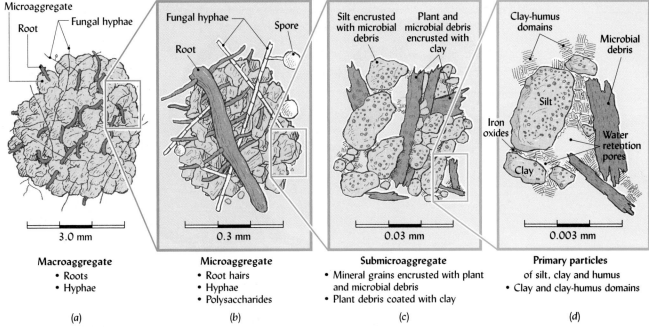

FIGURE 4.15 Larger aggregates are often composed of an agglomeration of smaller aggregates. This illustration shows four levels in this hierarchy of soil aggregates. The different factors important for aggregation at each level are indicated. (*a*) A *macroaggregate* composed of many microaggregates bound together mainly by a kind of sticky network formed from fungal hyphae and fine roots. (*b*) A *microaggregate* consisting mainly of fine sand grains and smaller clumps of silt grains, clay, and organic debris bound together by root hairs, fungal hyphae, and microbial gums. (*c*) A very small submicroaggregate consisting of fine silt particles encrusted with organic debris and tiny bits of plant and microbial debris (called *particulate organic matter*) encrusted with even smaller packets of clay, humus, and Fe or Al oxides. (*d*) Clusters of parallel and random clay platelets interacting with Fe or Al oxides and organic polymers at the smallest scale. These organoclay clusters or *domains* bind to the surfaces of humus particles and the smallest of mineral grains. (Diagram courtesy of R. Weil)

FLOCCULATION OF CLAYS AND THE ROLE OF ADSORBED CATIONS. Except in very sandy soils that are almost devoid of clay, aggregation begins with the **flocculation** of clay particles into microscopic clumps or *floccules* (Figure 4.16). If two clay platelets come close enough to each other, cations compressed in a layer between them will attract the negative charges on both platelets, thus serving as bridges to hold the platelets together. These processes lead to the formation of a small "stack" of parallel clay platelets, termed a *clay domain*. Other types of clay domains are more random in orientation, resembling a house of cards. These form when the positive charges on the edges of the clay platelets attract the negative charges on the planar surfaces (Figure 4.16). Multivalent cations (e.g., Ca^{2+}, Fe^{2+}, and Al^{3+}) also complex with hydrophobic humus molecules, allowing them to bind to clay surfaces. Clay/humus domains form bridges that bind to each other and to fine silt particles (mainly quartz), creating the smallest size groupings in the hierarchy of soil aggregates (Figure 4.15*d*). These domains, aided by the flocculating influence of polyvalent cations (e.g., Ca^{2+}, Fe^{2+}, and Al^{3+}) and humus, provide much of the long-term stability for the smaller (<0.25 mm) microaggregates. In certain highly weathered clayey soils (Ultisols and Oxisols) the cementing action of iron oxides and other inorganic compounds produces very stable small aggregates called **pseudosand**.

When monovalent cations, especially Na^+ (rather than polyvalent cations such as Ca^{2+} or Al^{3+}) are prominent, as in some soils of arid and semiarid areas, the attractive forces are not able to overcome the natural repulsion of one negatively charged clay platelet by another (Figure 4.16). The clay platelets cannot approach closely enough to flocculate, so remain in a dispersed, gel-like condition that causes the soil to become almost structureless, impervious to water and air, and very undesirable from the standpoint of plant growth (see Section 10.6).

VOLUME CHANGES IN CLAYEY MATERIALS. As a soil dries out and water is withdrawn, the platelets in clay domains move closer together, causing the domains and, hence, the soil mass to shrink in volume. As a soil mass shrinks, cracks will open up along zones of weakness. Over the course of many cycles (as occur between rain or irrigation events in the field) the network of cracks becomes better defined. Plant roots also have a distinct drying effect as they take up soil moisture in their immediate vicinity. Water

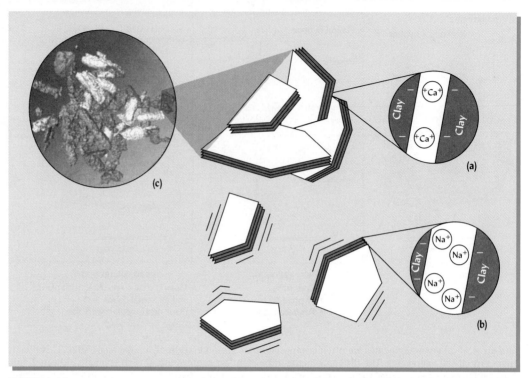

FIGURE 4.16   The role of cations and bacteria on the flocculation of clays. (*a*) Such di- and trivalent cations as Ca^{2+}, Fe^{3+} and Al^{3+} are tightly adsorbed and can effectively neutralize the negative surface charges on the clay particles. These cations can also form bridges that bring clay particles together. (*b*) Monovalent ions, especially Na^+, with relatively large hydrated radii can cause clay particles to repel each other and create a dispersed condition. Three things contribute to such dispersion: (1) the large hydrated Na^+ ion does not get close enough to the clay to effectively neutralize the negative charges, (2) the single charge on Na^+ is not effective in forming a bridge between clay particles, and (3) compared to di- or trivalent ions, 2 or 3 times as many monovalent ions must crowd between clay particles in order to neutralize the charges on the clay surfaces. (*c*) Due to charges on their cells walls and sticky exo-cellular compounds, bacteria (light colored) further flocculate clay particles (dark) as shown in this X-ray tomographic image. [Diagrams (*a*) and (*b*) courtesy of R. Weil; diagram (*c*) from Thieme et al., 2003]

uptake, especially by fibrous-rooted perennial grasses, accentuates the physical aggregation processes associated with wetting and drying. This effect is but one of many ways in which physical and biological soil processes interact.

Freezing and thawing cycles have a similar effect, since the formation of ice crystals is a drying process that also draws water out of clay domains. The swelling and shrinking actions that accompany freeze-thaw and wet-dry cycles in soils create fissures and pressures that alternately break apart large soil masses and compress soil particles into defined structural peds. The aggregating effects of these water and temperature cycles are most pronounced in soils with a high content of swelling-type clays (see Chapter 8), especially Vertisols, Mollisols, and some Alfisols (see Chapter 3).

Biological Processes

Glomalin: where the carbon is?
http://www.ars.usda.gov/is/AR/ archive/sep02/soil0902.htm

ACTIVITIES OF SOIL ORGANISMS.   Among the biological processes of aggregation, the most prominent are (1) the burrowing and molding activities of soil animals, (2) the enmeshment of particles by sticky networks of roots and fungal hyphae, and (3) the production of organic glues by microorganisms, especially bacteria and fungi. Earthworms (and termites) move soil particles about, often ingesting them and forming them into pellets or casts (see Chapter 11). In some forested soils, the surface horizon consists primarily of aggregates formed as earthworm castings (see, for example, Figure 4.13*a*). Plant roots also move particles about as they push their way through the soil. This movement forces soil particles to come into close contact with each other, encouraging aggregation. At the same time, the channels created by plant roots and soil animals serve as large pores, breaking up clods and helping to define larger soil structural units (see Plates 81 and 82, after page 656).

Plant roots (particularly root hairs) and fungal hyphae exude sugarlike polysaccharides and other organic compounds, forming sticky networks that bind together individual soil particles and tiny microaggregates into larger macroaggregates (see Figure 4.15*a* and Figure 4.19). The threadlike fungi that associate with plant roots (called *mycorrhizae;* see Section 11.9) produce a sticky sugar-protein called **glomalin**, which is thought to be an effective cementing agent (Box 4.4).

Bacteria also produce polysaccharides and other organic glues as they decompose plant residues. Bacterial polysaccharides are shown intermixed with clay at a very small scale in Figure 4.20. Many of these root and microbial organic glues resist dissolution by water and so not only enhance the formation of soil aggregates but also help ensure their

Adventures in the Rhizosphere: by Alex Blumberg: http://chicagowildernessmag.org/issues/spring1999/underground.html

BOX 4.4 GLOMALIN AND AGGREGATES TAKE SYMBIOSIS TO ANOTHER LEVEL[a]

Scientists continue to discover new ways in which soil microorganisms, especially fungi, help create and stabilize soil aggregates. Certain **mycorrhizal** fungi (see Section 11.9) live in a symbiotic state with the roots of many native and agricultural plants. The plants provide the fungi with sugars and the fungi help plants obtain nutrients from the soil. Recently, scientists discovered a mixture of glycoproteins (sugar-proteins), called glomalin, that are produced by the **hyphae** (thin strands) of these fungi. Although the function of glomalin in the fungi is still under investigation, substantial amounts of glomalin-related proteins have been identified in soils. Current thinking is that these iron-binding glycoproteins are quite resistant to decay and therefore accumulate to relatively high levels in soils.

Several types of observations point to a role for glomalin in soil aggregate stabilization. (1) The glomalin-related proteins are usually found in higher concentration in stable than in unstable aggregates in soils of both temperate and tropical regions. (2) Using immuno-fluorescence antibody techniques, soil scientists can detect glomalin-related proteins on the surfaces of soil aggregates (Figure 4.17). (3) An experiment using small and large glass beads to simulate unaggregated and aggregated soil suggested that fungi may produce glomalin partly to increase soil aggregation and create larger pores for better growth of their hyphae (Figure 4.18). Carrot roots and mycorrhizal fungi were grown in containers with barriers that allowed fungi but not roots to grow into chambers filled with large (~1 mm) or small (~0.1 mm) diameter glass beads. Although root growth and infection by mycorrhizal fungi were about the same for each bead size, the hyphae grew only 1/6 as much in the small as in the large beads, suggesting that lack of large pores can restrict fungal growth in soils. However, the hyphae that did grow in the small beads produced 7 times as much total glomalin and 44 times as much glomalin per unit of hyphal length as was produced in the large beads.

So it seems that the plant's investment of photosynthates (sugar) in the relationship with the fungi may indirectly help stimulate a better aggregated soil that is more suitable for the survival and growth of both plants *and* fungi. Land managers may be able to take advantage of this knowledge by using practices, such as reducing tillage and applying mulches, that favor the fungi that stimulate soil aggregation.

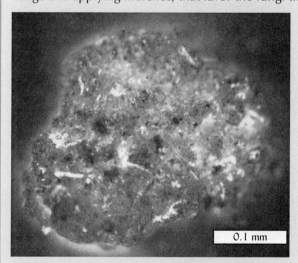

FIGURE 4.17 *Soil aggregate with fluorescing white areas indicating presence of glomalin by its reaction with a specific antibody. (Image courtesy of Kris Nichols, USDA/ARS, Mandan, ND)*

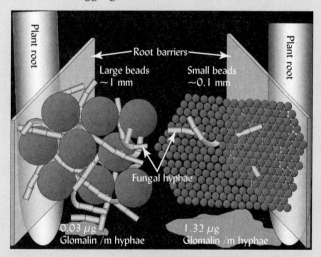

FIGURE 4.18 *Experiment suggesting fungi produce glomalin partly to aggregate soil and create large pores for better growth of their hyphae. [Diagram courtesy of R. Weil, from data in Rilling and Steinberg (2002)]*

[a] For correlations between glomalin proteins and soil aggregation, see Wright and Anderson (2000); for an update on the nature of the glomalin-related proteins see Rillig (2004).

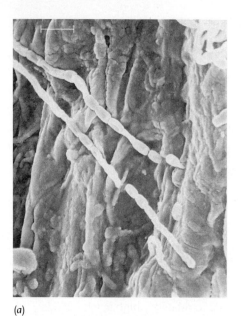
(a)

(b)

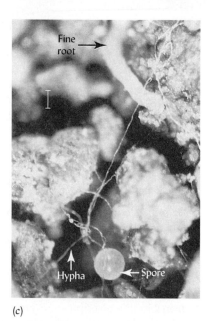
(c)

FIGURE 4.19 Fungal hyphae binding soil particles into aggregates. (*a*) Close-up of a hyphae growing over the surface of a mineral grain encrusted with microbial cells and debris. Bar = 10 μm. (*b*) An advanced stage of aggregation during the formation of soil from dune sands. Note the net of fungal hyphae and the encrustation of the mineral grains with organic debris. Bar = 50 μm. (*c*) Hyphae of the root-associated fungus from the genus *Gigaspora* interconnecting particles in a sandy loam from Oregon. Note also the fungal spore and the plant root. Bar = 320 μm. [Photos (*a*) and (*b*) courtesy of Sharon L. Rose, Willamette University; photo (*c*) courtesy of R. P. Schreiner, USDA-ARS, Corvallis, OR]

stability over a period of months to a few years. These processes are most notable in surface soils, where root and animal activities and organic matter accumulation are greatest.

INFLUENCE OF ORGANIC MATTER. In most temperate zone soils, the formation and stabilization of granular aggregates is primarily influenced by soil organic matter (see Figure 4.21). Organic matter provides the energy substrate that makes possible the previously mentioned biological activities. During the aggregation process, soil mineral particles (silts and fine sands) become coated and encrusted with bits of decomposed

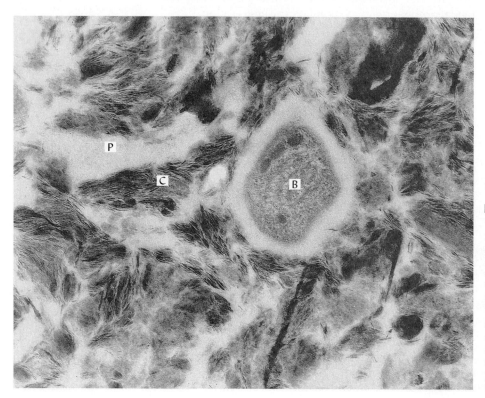

FIGURE 4.20 A transmission electron micrograph illustrating the interaction among organic materials and silicate clays in a water-stable aggregate. The dark-colored materials (C) are groups of clay particles that are interacting with organic polysaccharides (P). A bacterial cell (B) is also surrounded by polysaccharides. Note the generally parallel orientation of the clay particles, an orientation characteristic of clay domains. [From Emerson et al. 1986); photograph provided by R. C. Foster, CSIRO, Glen Osmond, Australia]

Before wetting   After wetting

High O.M.   Low O.M.   High O.M.   Low O.M.

FIGURE 4.21 The aggregates of soils high in organic matter are much more stable than are those low in this constituent. The low-organic-matter soil aggregates fall apart when they are wetted; those high in organic matter maintain their stability. (Photo courtesy of N. C. Brady)

BOX 4.5 PREPARING A GOOD SEEDBED

Early in the growing season, one of the main activities of a farmer or gardener is the preparation of a good seedbed to ensure that the sowing operation goes smoothly, and that the plants come up quickly, evenly, and well spaced.

A good seedbed consists of soil loose enough to allow easy root elongation and seedling emergence (Figure 4.22). At the same time, a seedbed should be packed firmly enough to ensure the seed can easily imbibe water to begin the germination process. The seedbed should also be relatively free of large clods. Seeds could fall between such clods and become lodged too deeply for proper emergence and lack sufficient soil contact.

Tillage may be needed to loosen compacted soil, help control weeds, and, in cool climates, help the soil dry out so it will warm more rapidly. On the other hand, the objectives of seedbed preparation may be achieved with little or no tillage if soil and climatic conditions are favorable and a mulch or herbicide is used to control early weeds.

FIGURE 4.22 A bean seed emerges from a seedbed. (Photo courtesy of R. Weil)

Mechanical planters can assist in maintaining a good seedbed. Most planters are equipped with coulters (sharp steel disks) designed to cut a path through plant residues on the soil surface. No-till planters usually follow the coulter with a pair of sharp cutting wheels called a *double disk opener* that opens a groove in the soil into which the seeds can be dropped (Figure 4.23). Most planters also have a press wheel that follows behind the seed dropper and packs the loosened soil just enough to ensure that the groove is closed and the seed is pressed into contact with moist soil.

Ideally, only a narrow strip in the seed row is packed down to create a seed germination zone, while the soil between the crop rows is left as loose as possible to provide a good rooting zone. The surface of the interrow rooting zone may be left in a rough condition to encourage water infiltration and discourage erosion. The above principles also apply to the home gardener who may be sowing seeds by hand.

Coulter

Double disk

Press wheel

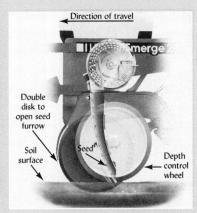

Direction of travel

Double disk to open seed furrow

Soil surface

Seed

Depth control wheel

FIGURE 4.23 No-till planter and a diagram showing how it works. (Photo and diagram courtesy of Deere & Company, Moline, IL)

FIGURE 4.24 Puddled soil (*left*) and well-granulated soil (*right*). Plant roots and especially humus play the major role in soil granulation. Thus a sod tends to encourage development of a granular structure in the surface horizon of cultivated land. (Courtesy USDA Natural Resources Conservation Service)

Compaction and its control:
http://www.extension.umn.edu/
distribution/cropsystems/
DC3115.html

plant residue and other organic materials. Complex organic polymers resulting from decay chemically interact with particles of silicate clays and iron and aluminum oxides. These compounds help orient the clays into packets (domains), which form bridges between individual soil particles, thereby binding them together in water-stable aggregates (see Figure 4.15*d*). In Figure 4.20 we can directly observe bacterial polymers and organomineral domains that bind soil particles.

INFLUENCE OF TILLAGE. Tillage can both promote and destroy aggregation. If the soil is not too wet or too dry, tillage can break large clods into natural aggregates, creating a temporarily loose, porous condition conducive to the easy growth of young roots and the emergence of tender seedlings (see Box 4.5). Tillage can also incorporate organic amendments into the soil and kill weeds.

Over longer periods, however, tillage greatly hastens the oxidative loss of soil organic matter, thus weakening soil aggregates. Tillage operations, especially if carried out when the soil is wet, also tend to crush or smear soil aggregates, resulting in loss of macroporosity and the creation of a *puddled* condition (see Figure 4.24).

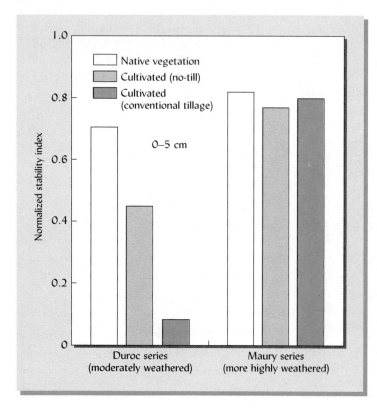

FIGURE 4.25 The interaction of soil organic matter with the clay fraction of moderately weathered soils such as the Duroc series accounts for most of the aggregate stability of these soils. Consequently, when these soils are cultivated the stability of soil aggregates declines since such cultivation decreases soil organic matter levels, especially where conventional tillage is followed. In contrast, in more highly weathered soils such as the Maury series, aggregate stability is less dependent on organic matter levels than on the interaction of iron oxide compounds with certain silicate clays such as kaolinite. The tillage system used also has less effect on aggregate stability in these more highly weathered soils. This figure suggests that cultivated soils of the highly weathered tropics have greater aggregate stability then their counterparts in temperate zones. [Redrawn from Six et al. (2000)]

Influence of Iron/Aluminum Oxides. Many of the highly weathered soils of the tropics (especially Oxisols) have large amounts of iron and aluminum sesquioxides (in largely amorphous forms) that coat soil particles and cement soil aggregates, thereby preventing their ready breakdown when the soil is tilled. Compared to soils of more temperate regions with similar amounts of organic matter, such tropical soils tend to have much greater aggregate stability, and their aggregation is less dependent on soil organic matter (Figure 4.25).

4.6 TILLAGE AND STRUCTURAL MANAGEMENT OF SOILS

Tillage implements and what they do:
http://www.WTAMU.EDU/~crobinson/TILLAGE/tillage.htm

When protected under dense vegetation and undisturbed by tillage, most soils (except perhaps some sparsely vegetated soils in arid regions) possess a surface structure sufficiently stable to allow rapid infiltration of water and to prevent crusting. However, for the manager of cultivated soils, the development and maintenance of stable surface soil structure is a major challenge. Many studies have shown that aggregation and associated desirable soil properties such as water infiltration rate decline under long periods of tilled row-crop cultivation (Table 4.3).

Tillage and Soil Tilth

Simply defined, **tilth** refers to the physical condition of the soil in relation to plant growth. Tilth depends not only on aggregate formation and stability, but also on such factors as bulk density (see Section 4.7), soil moisture content, degree of aeration, rate of water infiltration, drainage, and capillary water capacity. As might be expected, tilth often changes rapidly and markedly. For instance, the workability of fine-textured soils may be altered abruptly by a slight change in moisture.

A major aspect of tilth is soil **friability** (see also Section 4.9). Soils are said to be **friable** if their clods are not sticky or hard, but rather crumble easily, revealing their constituent aggregates. Generally, **soil friability** is enhanced when the **tensile strength** (i.e., the force required to pull apart) of individual aggregates is relatively high compared to the tensile strength of the clods. This condition allows tillage or excavation forces to easily break down the large clods, while the resulting aggregates remain stable. As might be expected, friability can be markedly affected by changes in soil water content, especially for fine-textured soils. Each soil typically has an optimum water content for greatest friability (Figure 4.26).

Clayey soils are especially prone to puddling and compaction because of their high plasticity and cohesion. When puddled clayey soils dry, they usually become dense and hard. Proper timing of trafficking is more difficult for clayey than for sandy soils, because the former take much longer to dry to a suitable moisture content and may also become too dry to work easily. Increased soil organic matter content usually

---

TABLE 4.3 **Effect of Period of Corn Cultivation on Soil Organic Matter, Aggregate Stability, and Water Infiltration[a] in Five Silt Loam Inceptisols from Southwest France**

In soils with depleted organic-matter levels, aggregates easily broke down under the influence of water, forming smaller aggregates and dispersed materials that sealed the soil surface and inhibited infiltration. A level of 3% soil organic matter seems to be sufficient for good structural stability in these temperate region silt loam soils.

| Period of cultivation, years | Organic matter, % | Aggregate stability, mm MWD[b] | Infiltration | |
			Of total rain, %	Prior to ponding, mm
100	0.7	0.35	25	6
47	1.6	0.61	34	9
32	2.6	0.76	38	15
27	3.1	1.38	47	25
15	4.2	1.52	44	23

[a] Infiltration is the amount of water that entered the soil when 64 mm of "rain" was applied during a 2-hour period.
[b] MWD = mean weighted diameter or average size of the aggregates that remained intact after sieving under water. [Data from Le Bissonnais and Arrouays (1997)]

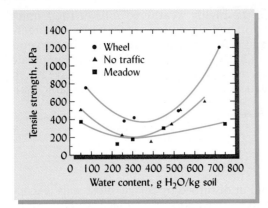

FIGURE 4.26 Influence of tillage and wheel traffic on soil strength at various water contents. Tensile strength is the force (measured in kilopascals, kPa) required to break apart soil clods. The lower this force, the easier it is to crumble the soil and the greater the soil friability and workability. The upper two curves represent soil from the wheel tracks (•) and the non-trafficked portion (▲) of a cultivated calcareous clay soil in England. The lowest curve represents the same soil in an adjacent long-term uncultivated meadow (■). There appears to be an optimum water content at which a soil has the lowest tensile strength and is therefore most friable (near 300 g/kg for this soil). Also, the less traffic and tillage, the more friable the soil (the lower the tensile strength), especially when very dry or very wet. [Redrawn from Watts and Dexter (1997)]

enhances soil friability and can partially alleviate the susceptibility of a clay soil to structural damage during tillage and traffic (Figure 4.27).

Some clayey soils of humid tropical regions are much more easily managed than those just described. The clay fraction of these soils is dominated by hydrous oxides of iron and aluminum, which are not as sticky, plastic, and difficult to work. These soils may have very favorable physical properties, since they hold large amounts of water but have such stable aggregates that they respond to tillage after rainfall much like sandy soils.

Farmers in temperate regions typically find their soils too wet for tillage just prior to planting time (early spring), while farmers in tropical regions may face the opposite problem of soils too dry for easy tillage just prior to planting (end of dry season). In tropical and subtropical regions with a long dry season, soil often must be tilled in a very dry state in order to prepare the land for planting with the onset of the first rains. Tillage under such dry conditions can be very difficult and can result in hard clods if the soils contain much sticky-type silicate clay.

Conventional Tillage and Crop Production

Since the Middle Ages, the moldboard plow has been the primary tillage implement most used in the Western world.[5] Its purpose is to lift, twist, and invert the soil while incorporating crop residues and animal wastes into the plow layer (Figure 4.28). The moldboard plow is often supplemented by the disk plow, which is used to cut up

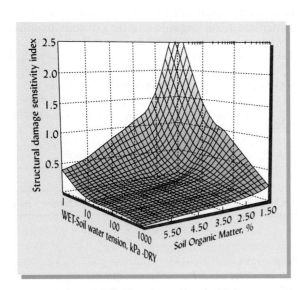

FIGURE 4.27 Sensitivity of fine-textured soils to mechanical damage by tillage is increased by wetness and decreased by soil organic matter. The sensitivity index on the vertical axis is based mainly on the tendency for clay dispersion that is associated with soil smearing and aggregate destruction. The soil is especially susceptible to this damage when excessively wet (wetter than the condition termed *field capacity*, as defined in Section 5.8). It is also evident that a high level of soil organic matter helps protect the aggregated clay during tillage. The relationships shown apply to clayey soils and would be much less extreme in sandy soils. [Graph based on data in Watts and Dexter (1997)]

---

[5] For an early but still valuable critique of the moldboard plow, see Faulkner (1943).

FIGURE 4.28 While the action of the moldboard plow lifts, turns, and loosens the upper 15 to 20 cm of soil (the furrow slice), the counterbalancing downward force compacts the next lower layer of soil. This compacted zone can develop into a *plowpan*. Compactive action can be understood by imagining that you are lifting a heavy weight—as you lift the weight your feet press down on the floor below. (Photo courtesy of R. Weil)

residues and partially incorporate them into the soil. In conventional practice, such primary tillage is followed by a number of secondary tillage operations, such as harrowing to kill weeds and to break up clods, thereby preparing a suitable seedbed.

After the crop is planted, the soil may receive further secondary tillage to control weeds and to break up crusting of the immediate soil surface. In mechanized agriculture, all conventional tillage operations are performed with tractors and other heavy equipment that may pass over the land several times before the crop is finally harvested. In many parts of the world, farmers use hand hoes or animal-drawn implements to stir the soil. Although humans and draft animals are not as heavy as tractors, their weight is applied to the soil in a relatively small area (foot- or hoofprint), and so can also cause considerable compaction.

Conservation Tillage and Soil Tilth

Conservation tillage in vegetable production, by Mary Peet:
http://www.cals.ncsu.edu/ sustainable/peet/tillage/c03tilla. html

In recent years, agricultural land-management systems have been developed that minimize the need for soil tillage. Since these systems also leave considerable plant residues on or near the soil surface, they protect the soil from erosion (see Section 17.6 for a detailed discussion). For this reason, the tillage practices followed in these systems are called *conservation tillage*. The U.S. Department of Agriculture defines *conservation tillage* as that which leaves at least 30% of the soil surface covered by residues. Figure 4.29 illustrates a no-till operation, where one crop is planted in the residue of another, with virtually no tillage. Other minimum-tillage systems such as chisel plowing permit some stirring of the soil, but still leave a high proportion of the crop residues on the surface. These organic residues protect the soil from the beating action of raindrops and the abrasive action of the wind, thereby reducing water and wind erosion and maintaining soil structure.

Soil Crusting

Falling drops of water during heavy rains or sprinkler irrigation beat apart the aggregates exposed at the soil surface. In some soils the dilution of salts by this water stimulates the dispersion of clays. Once the aggregates become dispersed, small particles and dispersed clay tend to wash into and clog the soil pores. Soon the soil surface is covered with a thin layer of fine, structureless material called a **surface seal**. The surface seal inhibits water infiltration and increases erosion losses.

As the surface seal dries, it forms a hard **crust**. Seedlings, if they emerge at all, can do so only through cracks in the crust. A crust-forming soil is compared to one with stable

FIGURE 4.29 An illustration of one conservation tillage system. Wheat is being harvested (background) and soybeans are planted (foreground) with no intervening tillage operation. This no-tillage system permits double-cropping, saves fuel costs and time, and helps conserve the soil. (Courtesy Allis-Chalmers Corporation)

aggregates in Figure 4.30. Formation of a crust soon after a crop is sown may allow so few seeds to emerge that the crop has to be replanted. In arid and semiarid regions, soil sealing and crusting can have disastrous consequences because high runoff losses leave little water available to support plant growth.

Crusting can be minimized by keeping some vegetative or mulch cover on the land to reduce the impact of raindrops. Once a crust has formed, it may be necessary to rescue a newly planted crop by breaking up the crust with light tillage (as with a rotary hoe), preferably while the soil is still moist.

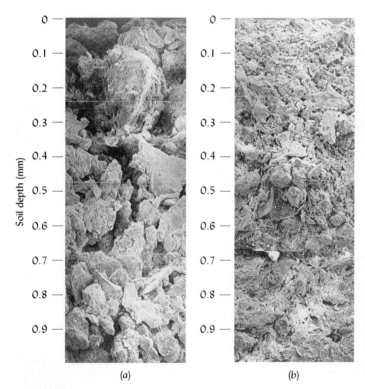

Soil depth (mm)

(a)   (b)

(c)

FIGURE 4.30 Scanning electron micrographs of the upper 1 mm of a soil with stable aggregation (a) compared to one with unstable aggregates (b). Note that the aggregates in the immediate surface have been destroyed and a surface crust has formed. The bean seedling (c) must break the soil crust as it emerges from the seedbed. [Photos (a) and (b) from O'Nofiok and Singer (1984), used with permission of Soil Science Society of America; photo (c) courtesy of R. Weil]

Improved management of soil organic matter and use of certain soil amendments can "condition" the soil and help prevent clay dispersion and crust formation (see also Section 10.10).

GYPSUM. Gypsum (calcium sulfate) is widely available in its relatively pure mined form, or as a major component of various industrial by-products. Gypsum has been shown effective in improving the physical condition of many types of soils, from some highly weathered acid soils to some low-salinity, high-sodium soils of semiarid regions (see Chapter 10). The more soluble gypsum products provide enough electrolytes (cations and anions) to promote flocculation and inhibit the dispersion of aggregates, thus preventing surface crusting. Field trials have shown that gypsum-treated soils permit greater water infiltration and are less subject to erosion than untreated soils. Similarly, gypsum can reduce the strength of hard subsurface layers, thereby allowing greater root penetration and subsequent plant uptake of water from the subsoil.

ORGANIC POLYMERS. Certain synthetic organic polymers can stabilize soil structure in much the same way as do natural organic polymers such as polysaccharides. While large applications of these polymers would be uneconomical, it has been shown that even very small amounts can effectively inhibit crust formation if applied properly. For example, polyacrylamide (PAM) is effective in stabilizing surface aggregates when applied at rates as low as 1 to 15 mg/L of irrigation water or sprayed on at rates as low as 1 to 4 kg/ha. Figure 4.31 shows the dramatic stabilizing effect of synthetic polyacrylamides used in irrigation water. A number of research reports indicate that the best results can be obtained by combining the use of PAM and gypsum products. These steps nearly eliminated irrigation-induced erosion.

OTHER SOIL CONDITIONERS. Several species of algae that live near the soil surface are known to produce quite effective aggregate-stabilizing compounds. Application of small quantities of commercial preparations containing such algae may bring about a significant improvement in surface soil structure. The amount of amendment required is very small because the algae, once established in the soil, can multiply.

Various humic materials are marketed for their soil conditioning effects when incorporated at low rates (<500 kg/ha). However, carefully conducted research at many universities has failed to show that these materials have significantly affected aggregate stability or crop yield, as claimed.

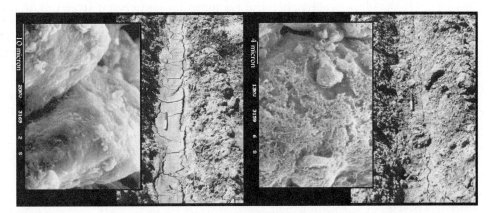

FIGURE 4.31 The dramatic stabilizing effect of synthetic polyacrylamide (PAM), which was used at rates of 1 to 2 kg/ha per irrigation in the furrow shown at right, but not in the furrow on the left (note the marker pens for scale). PAM increases infiltration by 15 to 50% on medium to fine-textured soils. Scanning electron micrographs (scale on left edge) reveal the structure of the respective soils. Note the meshlike network of PAM enveloping the treated soil particles. This PAM weighs about 12 to 15 Mg per mole of anionic, water soluble molecules. [Photos courtesy of C. Ross from Ross et al. (2003)]

Although each soil presents unique problems and opportunities, the following principles are generally relevant to managing soil tilth:

1. Minimizing tillage, especially moldboard plowing, disk harrowing, or rototilling, reduces the loss of aggregate-stabilizing organic matter.

2. Timing traffic activities to occur when the soil is as dry as possible and restricting tillage to periods of optimum soil moisture conditions will minimize destruction of soil structure.

3. Mulching the soil surface with crop residues or plant litter adds organic matter, encourages earthworm activity, and protects aggregates from beating rain and direct solar radiation.

4. Adding crop residues, composts, and animal manures to the soil is effective in stimulating microbial supply of the decomposition products that help stabilize soil aggregates.

5. Including sod crops in the rotation favors stable aggregation by helping to maintain soil organic matter, providing maximal aggregating influence of fine plant roots, and ensuring a period without tillage.

6. Using cover crops and green manure crops, where practical, provides another good source of root action and organic matter for structural management.

7. Applying gypsum (or calcareous limestone if the soil is acidic) by itself or in combination with synthetic polymers can be very useful in stabilizing surface aggregates, especially in irrigated soils.

The maintenance of a high degree of aggregation is one of the most important goals of soil management. This is so largely because the ability of soil to perform needed ecosystem functions is greatly influenced by soil porosity and density, properties which we shall now consider.

4.7 SOIL DENSITY

Particle Density

Soil **particle density** D_p is defined as the mass per unit volume of soil *solids* (in contrast to the volume of the *soil*, which would also include spaces between particles). Thus, if 1 cubic meter (m^3) of soil solids weighs 2.6 megagrams (Mg), the particle density is 2.6 Mg/m^3 (which can also be expressed as 2.6 grams per cubic centimeter).[6]

Particle density is essentially the same as the **specific gravity** of a solid substance. The chemical composition and crystal structure of a mineral determines its particle density. Particle density is *not* affected by pore space, and therefore is not related to particle size or to the arrangement of particles (soil structure).

Particle densities for most mineral soils vary between the narrow limits of 2.60 to 2.75 Mg/m^3 because quartz, feldspar, micas, and the colloidal silicates that usually make up the major portion of mineral soils all have densities within this range. For general calculations concerning arable mineral surface soils (1 to 5% organic matter), a particle density of about 2.65 Mg/m^3 may be assumed if the actual particle density is not known. This number would be adjusted upward to 3.0 Mg/m^3 or higher when large amounts of high-density minerals such as magnetite, garnet, epidote, zircon, tourmaline, or hornblende are present. Likewise, it would be reduced for soils known to be high in organic matter, which has a particle density of only 0.9 to 1.4 Mg/m^3.

Bulk Density

A second important mass measurement of soils is **bulk density** D_b, which is defined as the mass of a unit volume of dry soil. This volume includes both solids and pores. A careful study of Figure 4.32 should make clear the distinction between particle and

---

[6] Since 1 Mg = 1 million grams and 1 m^3 = 1 million cubic centimeters, 1 Mg/m^3 = 1 g/cm^3.

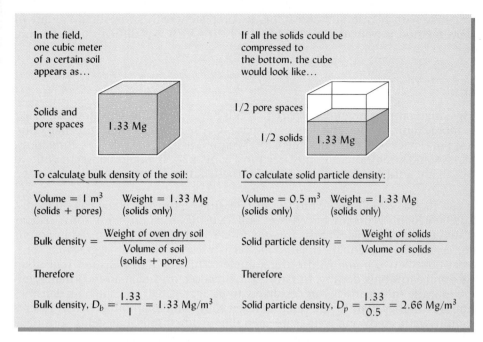

In the field, one cubic meter of a certain soil appears as...

Solids and pore spaces

1.33 Mg

If all the solids could be compressed to the bottom, the cube would look like...

1/2 pore spaces

1/2 solids | 1.33 Mg

To calculate bulk density of the soil:

Volume = 1 m^3 (solids + pores)   Weight = 1.33 Mg (solids only)

$$\text{Bulk density} = \frac{\text{Weight of oven dry soil}}{\text{Volume of soil (solids + pores)}}$$

Therefore

$$\text{Bulk density, } D_b = \frac{1.33}{1} = 1.33 \text{ Mg/m}^3$$

To calculate solid particle density:

Volume = 0.5 m^3 (solids only)   Weight = 1.33 Mg (solids only)

$$\text{Solid particle density} = \frac{\text{Weight of solids}}{\text{Volume of solids}}$$

Therefore

$$\text{Solid particle density, } D_p = \frac{1.33}{0.5} = 2.66 \text{ Mg/m}^3$$

FIGURE 4.32 Bulk density D_b and particle density D_p of soil. Bulk density is the weight of the solid particles in a standard volume of field soil (solids plus pore space occupied by air and water). Particle density is the weight of solid particles in a standard volume of those solid particles. Follow the calculations through carefully and the terminology should be clear. In this particular case the bulk density is one-half the particle density, and the percent pore space is 50.

Animation of particle vs. bulk density:
http://www.landfood.ubc.ca/soil200/components/mineral.htm#114

bulk density. Both expressions of density use only the mass of the solids in a soil; therefore, any water present is excluded from consideration.

There are several methods of determining soil bulk density by obtaining a known volume of soil, drying it to remove the water, and weighing the dry mass.[7] A special coring instrument (Figure 4.33) can obtain a sample of known volume without disturbing the natural soil structure. For surface soils, perhaps the simplest method is to dig a small hole, dry and weigh all the excavated soil, and then determine the soil volume by

(a)

(b)

FIGURE 4.33 A special sampler designed to remove a cylindrical core of soil without causing disturbance or compaction (a). The sampler head contains an inner cylinder and is driven into the soil with blows from a drop hammer. The inner cylinder (b) containing an undisturbed soil core is then removed and trimmed on the end with a knife to yield a core whose volume can easily be calculated from its length and diameter. The weight of this soil core is then determined after drying in an oven. (Photos courtesy of R. Weil)

[7] An instrumental method that measures the soil's resistance to the passage of gamma rays is used in soil research, but is not discussed here.

lining the hole with plastic film and filling it completely with a measured volume of water. This method is well adapted to stony soils in which it is difficult to use a core sampler.

Factors Affecting Bulk Density

Soils with a high proportion of pore space to solids have lower bulk densities than those that are more compact and have less pore space. Consequently, any factor that influences soil pore space will affect bulk density. Typical ranges of bulk density for various soil materials and conditions are illustrated in Figure 4.34. It would be worthwhile to study this figure until you have a good feel for these ranges of bulk density.

EFFECT OF SOIL TEXTURE.   As illustrated in Figure 4.34, fine-textured soils such as silt loams, clays, and clay loams generally have lower bulk densities than do sandy soils.[8] This is true because the solid particles of the fine-textured soils tend to be organized in porous granules, especially if adequate organic matter is present. In these aggregated soils, pores exist both between *and* within the granules. This condition ensures high total pore space and a low bulk density. In sandy soil, however, organic matter contents generally are low, the solid particles are less likely to be aggregated, and the bulk densities are commonly higher than in the finer-textured soils. Similar amounts of large pores are present in both sandy and well-aggregated fine-textured soils, but sandy soils have few of the fine, within-ped pores, and so have less total porosity (Figure 4.35).

While sandy soils generally have high bulk densities, the packing arrangement of the sand grains also affects their bulk density (see Figure 4.36). Loosely packed grains may fill as little as 52% of the bulk volume, while tightly packed grains may fill as much as 75% of the volume. If we assume that grains consist of quartz with a particle density of 2.65 Mg/m^3, then the corresponding range of bulk densities for loose to tightly packed sand would be 1.38 to 1.99 Mg/m^3 ($0.52 * 2.65 = 1.38$ and $0.75 * 2.65 = 1.99$), not too

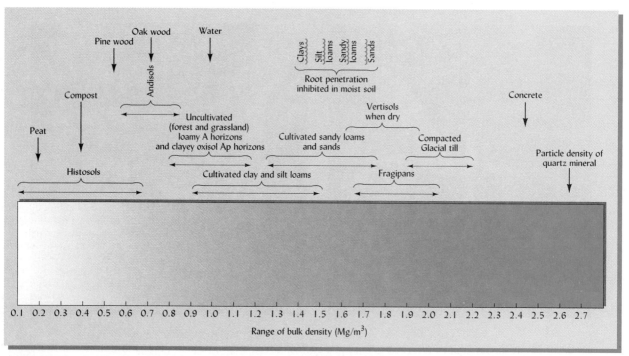

FIGURE 4.34   Bulk densities typical of a variety of soils and soil materials.

[8] This fact may seem counterintuitive at first because sandy soils are commonly referred to as "light" soils, while clays and clay loams are referred to as "heavy" soils. The terms *heavy* and *light* in this context refer not to the mass per unit volume of the soils, but to the amount of effort that must be exerted to manipulate these soils with tillage implements—the sticky clays being much more difficult to till.

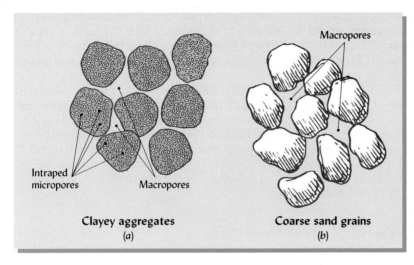

FIGURE 4.35 A schematic comparison of sandy and clayey soils showing the relative amounts of large (macro-) pores and small (micro-) pores in each. There is less total pore space in the sandy soils than in the clayey one because the clayey soil contains a large number of fine pores within each aggregate (a), but the sand particles (b), while similar in size to the clayey aggregates, are solid and contain no pore spaces within them. This is the reason why, among surface soils, those with coarse texture are usually more dense than those with finer textures. (Diagram courtesy of R. Weil)

different from the range actually encountered in very sandy soils. The bulk density is generally lower if the sand particles are mostly of one size class (i.e., *well-sorted* sand), while a mixture of different sizes (i.e., *well-graded* sand) is likely to have an especially high bulk density. In the latter case, the smaller particles partially fill in the spaces between the larger particles. The most dense materials are those characterized by both a mixture of sand sizes and a tight packing arrangement.

DEPTH IN SOIL PROFILE.   Deeper in the soil profile, bulk densities are generally higher, probably as a result of lower organic matter contents, less aggregation, fewer roots and other soil-dwelling organisms, and compaction caused by the weight of the overlying layers. Very compact subsoils may have bulk densities of 2.0 Mg/m^3 or even greater. Many soils formed from till (see Section 2.3) have extremely dense subsoils as a result of past compaction by the enormous mass of ice.

Useful Density Figures

Bulk densities - check "Earth" and "Sand":
http://www.asiinstr.com/ technical/Material_Bulk_ Density_Chart_A.htm

For engineers involved with moving soil during construction, or for landscapers bringing in topsoil by the truckload, a knowledge of the bulk density of various soils is useful in estimating the weight of soil to be moved. A typical medium-textured mineral soil might have a bulk density of 1.25 Mg/m^3, or 1250 kilograms in a cubic meter.[9] People are often surprised by how heavy soil is. Imagine driving your pickup truck to a nursery where natural topsoil is sold in bulk and filling your truck bed with a nice,

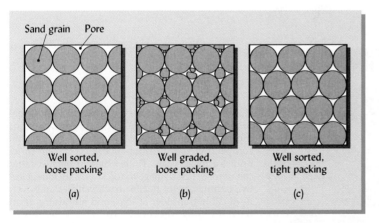

Well sorted, loose packing

(a)

Well graded, loose packing

(b)

Well sorted, tight packing

(c)

FIGURE 4.36 The uniformity of grain size and the type of packing arrangement significantly affect the bulk density of sandy materials. Materials consisting of all similar-sized grains are termed *well sorted* (or poorly graded). Those with a variety of grain sizes are *well graded* (or poorly sorted). In either case, compaction of the particles into a tight packing arrangement markedly increases the bulk density of the material and decreases its porosity. Note that the size distribution of sand and gravel particles can be described as either graded or sorted, the two terms having essentially opposite meanings. Geologists usually speak of rivers having sorted the sand grains by size as deposits were laid down. Engineers usually are concerned as to whether or not sand consists of a gradation of sizes (i.e., is well graded or not).

---

[9] Most commercial landscapers and engineers in the United States still use English units. To convert values of bulk density given in units of Mg/m^3 into values of lb/yd^3, multiply by 1686. Therefore, 1 yd^3 of a typical medium-textured mineral soil with a bulk density of 1.25 Mg/m^3 would weigh over a ton (1686 × 1.25 = 2108 lb/yd^3).

rounded load. Of course, you would not really want to do this, as the load might break your rear axle; you certainly would not be able to drive away with the load. A typical "half-ton" (1000 lb or 454 kg) load capacity pickup truck could carry less than 0.4 m^3 of this soil, even though the truck bed has room for about five times this volume of material.

The mass of soil in 1 ha to a depth of normal plowing (15 cm) can be calculated from soil bulk density. If we assume a bulk density of 1.3 Mg/m^3 for a typical arable surface soil, such a hectare-furrow slice 15 cm deep weighs about 2 million kg.[10] This estimate of the mass of surface soil in a hectare of land is very useful in calculating lime and fertilizer application rates and organic matter mineralization rates (see Boxes 8.4, 9.4, 10.4, and 13.1 for detailed examples). However, this estimated mass must be adjusted if the bulk density is other than 1.3 Mg/m^3 or the depth of the layer under consideration is more or less than 15 cm.

Management Practices Affecting Bulk Density

Changes in bulk density for a given soil are easily measured and can alert soil managers to changes in soil quality and ecosystem function. Increases in bulk density usually indicate a poorer environment for root growth, reduced aeration, and undesirable changes in hydrologic function, such as reduced water infiltration.

Trees on construction sites: http://www.treesaregood.com/treecare/avoiding_construction.aspx

FOREST LANDS.   The surface horizons of most forested soils have rather low bulk densities (see Figure 4.34). Tree growth and forest ecosystem function are particularly sensitive to increases in bulk density. Conventional timber harvest generally disturbs and compacts 20 to 40% of the forest floor (Figure 4.37) and is especially damaging along the skid trails where logs are dragged and at the landing decks—areas where logs are piled and loaded onto trucks (Table 4.4). An expensive, but effective, means of moving logs while minimizing compactive degradation of forest lands is the use of cables strung between towers or hung from large balloons.

Intensive recreational and transport use of soils in forests and other areas with natural vegetation can also lead to increased bulk densities. Such effects can be seen where access roads, trails, and campsites are found (Figure 4.38). An important consequence of increased bulk density is a diminished capacity of the soil to take in water, hence increased losses by surface runoff. Damage from hikers can be minimized by restricting foot traffic to well-designed, established trails that may include a thick layer of wood

FIGURE 4.37 Timber harvest with a conventional rubber-tired skidder in a boreal forest in western Alberta, Canada. Such practices cause significant soil compaction that can impair soil ecosystem functions for many years. Timber harvest practices that can reduce such damage to forest soils include selective cutting, use of flexible-track vehicles and overhead cable transport of logs, and abstaining from harvest during wet conditions. (Photo courtesy of Andrei Startsev, Alberta Environmental Center)

---

[10] 10,000 m^2/ha × 1.3 Mg/m^3 × 0.15 m = 1950 Mg/ha, or about 2 million kg per ha to a depth of 15 cm. A comparable figure in the English system is 2 million lb per acre–furrow slice 6 to 7 in. deep.

TABLE 4.4 **Effects of Timber Harvest on Bulk Density at Different Depths in Two Forested Ultisols in Georgia**

Rubber-wheeled skidders were used to harvest the logs. Note the generally higher bulk densities of the sandy loam soil compared to the clay loam, and the greater effect of timber harvest on the skidder trails.

Soil Depth, cm	Bulk density, Mg/m^3		
	Preharvest	Postharvest, off trails	Postharvest, skidder trails
Upper coastal plain, sandy loam			
0–8	1.25	1.50	1.47
8–15	1.40	1.55	1.71
15–23	1.54	1.61	1.81
23–30	1.58	1.62	1.77
Piedmont, clay loam			
0–8	1.16	1.36	1.52
8–15	1.39	1.49	1.67
15–23	1.51	1.51	1.66
23–30	1.49	1.46	1.61

Data from Gent, et al. (1984, 1986).

chips, or even a raised boardwalk in the case of heavily traveled paths over very fragile soils, such as in wetlands.

Cherry Trees, Loved to Death. By Adrian Higgins, Washington Post: http://www.washingtonpost.com/wp-dyn/articles/A30233-2005Apr6.html

URBAN SOILS.   In urban areas, trees planted for landscaping purposes must often contend with severely compacted soils. While it is usually not practical to modify the entire root zone of a tree, several practices can help (see also Section 7.6). First, making the planting hole as large as possible will provide a zone of loose soil for early root growth. Second, a thick layer of mulch spread out to the drip line (but not too near the trunk) will enhance root growth, at least near the surface. Third, the tree roots may be given paths for expansion by digging a series of narrow trenches radiating out from the planting hole and backfilled with loose, enriched soil.

In some urban settings it may be desirable to create an "artificial soil" that includes a skeleton of coarse angular gravel to provide strength and stability, and a mixture of

FIGURE 4.38 Impact of campers on the bulk density of forest soils, and the consequent effects on rainwater infiltration rates and runoff losses (see white arrows). At most campsites the high-impact area extends for about 10 m from the fire circle or tent pad. Managers of recreational land must carefully consider how to protect sensitive soils from compaction that may lead to death of vegetation and increased erosion. [Data from Vimmerstadt et al. (1982)]

loam-textured topsoil and organic matter to provide nutrient- and water-holding capacities. Also, large quantities of sand and organic materials are sometimes mixed into the upper few centimeters of a fine-textured soil on which putting green turf grass is to be grown.

GREEN ROOFS. Soil bulk density is critical in the design of rooftop gardens. The mass of soil involved must be minimized in order to design a cost-effective structure of sufficient strength to carry the soil load. One might choose to grow only such shallow-rooted plants as sedums or turfgrasses so that a relatively thin layer of soil (say, 15 cm) could be used, keeping the total mass of soil from being too great. It may also be possible to reduce the cost of construction by selecting a natural soil having a relatively low bulk density, such as some well aggregated loams or peat soils. Often an artificial growing medium is created from such light-weight materials as perlite and peat. However, if anchoring trees and other plants is an important function of the soil in such an installation, very low density materials would not be suitable. When very low density materials are used, they may require a surface netting system to prevent wind from blowing them off the roof.

AGRICULTURAL LAND. Although tillage may temporarily loosen the surface soil, in the long term intense tillage increases soil bulk density because it depletes soil organic matter and weakens soil structure (see Section 4.6). The data in Table 4.5 illustrate this trend. These data are from long-term studies in different locations where relatively undisturbed soils were compared to adjacent areas that had been cultivated for 12 to 80+ years. In all cases, cropping increased the bulk density of the topsoils. The effect of cultivation can be minimized by adding crop residues or farm manure in large amounts and rotating cultivated crops with a grass sod.

In modern agriculture, heavy machines used to pull implements, apply amendments, or harvest crops can create yield-limiting soil compaction. Certain tillage implements, such as the moldboard plow and the disk harrow, compact the soil below their working depth even as they lift and loosen the soil above. Use of these implements or repeated trips over the field by heavy machinery can form **plow pans** or **traffic pans**, dense zones immediately below the plowed layer (Figure 4.39). Other tillage implements, such as the chisel plow and the spring-tooth harrow, do not press down upon the soil beneath them, and so are useful in breaking up plow pans and stirring the soil with a minimum of compaction. Large chisel-type plows (Figure 4.40) can be used in **subsoiling** to break up dense subsoil layers, thereby permitting root penetration (Figure 4.41). However, in some soils the effects of subsoiling are quite temporary. Any tillage tends to reduce soil strength, thus making the soil less resistant to subsequent compaction.

Traffic is particularly damaging on wet soil. Generally, with heavier loads and on wetter soils, compactive effects are more pronounced and penetrate more deeply into the profile. To prevent compaction, which can result in yield reductions and loss of profitability, the number of tillage operations and heavy equipment trips over the field should be minimized and timed to avoid periods when the soil is wet. Unfortunately,

TABLE 4.5 **Bulk Density and Pore Space of Surface Soils from Cultivated and Nearby Uncultivated Areas**

With cultivation, the bulk density was increased, and the pore space proportionately decreased, in every case.

Soil	Texture	Years cropped	Bulk density, Mg/m³		Pore space, %	
			Cultivated soil	Uncultivated soil	Cultivated soil	Uncultivated soil
Mean of 6 Ustolls (S. Dakota)	Silt loam	80+	1.30	1.10	50.9	58.5
Mean of 2 Udults (Maryland)	Sandy loam	50+	1.59	0.84	40.0	66.4
Mean of 2 Udults (Maryland)	Silt loam	50+	1.18	0.78	55.5	68.8
Xeroll (Turkey)	Clay loam	12	1.34	1.25	49.6	52.7
Mean of 3 Ustalfs (Zimbabwe)	Clay	20–50	1.44	1.20	54.1	62.6
Mean of 3 Ustalfs (Zimbabwe)	Sandy loam	20–50	1.54	1.43	42.9	47.2

Data for South Dakota soils from Eynard et al. (2004), for Turkey from Celik (2005), for Maryland from Lucas and Weil (unpublished) and for Zimbabwe soils from Weil (unpublished).

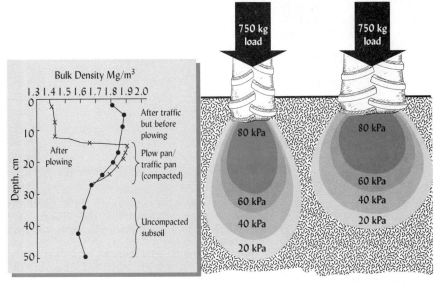

FIGURE 4.39 Vehicle tires compact soil to considerable depths. (*Left*) Representative bulk densities associated with traffic compaction on a sandy loam soil. Plowing can temporarily loosen the compacted surface soil (plow layer), but usually increases compaction just below the plow layer. (*Right*) Vehicle tires (750 kg load per tire) compact soil to about 50 cm. The more narrow the tire, the deeper it sinks and the deeper its compactive effect. The tire diagram shows the compactive pressure in kPa. For tire designs that reduce compaction, see Tijink and van der Linden (2000). (Diagrams courtesy of R. Weil)

traffic on wet agricultural soils is sometimes unavoidable in humid temperate regions in spring and fall.

Another approach for minimizing compaction is to carefully restrict all wheel traffic to specific lanes, leaving the rest of the field (usually 90% or more of the area) free from compaction. Such **controlled traffic** systems are widely used in Europe, especially on clayey soils. Gardeners can practice controlled traffic by establishing permanent foot

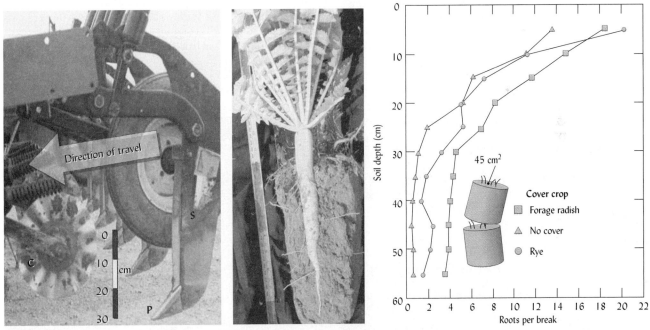

FIGURE 4.40 Two approaches to the alleviation of subsoil compaction. (*Left*) A heavy chisel plow, also known as a subsoiler or ripper. The heavy shanks (S) are pulled through the soil toward the left with the chisel point (P) about 40 cm deep. If the soil is relatively dry, the subsoiler will cause the mass of compacted soil to shatter leaving a network of cracks that enhance water, air, and root movement. However, pulling a subsoiler through dry, clayey soil is a slow, energy-intensive operation, the benefits usually last for only a year or two, and the operation usually disturbs the soil surface, leaving it more susceptible to erosion. Recently, researchers in humid regions have investigated an alternative approach that involves growing taprooted plants (such as the forage radish shown) in the fall and spring when the subsoil is relatively wet and easily penetrated by roots. The taproots then decay, leaving semipermanent channels in which roots of subsequent crops can grow to pass through the compacted subsoil zone, even during the summer when the soil is relatively dry and hard. (*Right*) Corn planted after forage radish had twice as many roots reaching the subsoil (below 30 cm) as corn planted after the fibrous rooted rye, and nearly 10 times as many as corn planted in soil that had no cover crop during the winter. Soil cores 7.5 cm in diameter (45 cm² area) by 60 cm long were broken at 5 cm intervals, allowing live roots, which stick out at the break (see inset diagram), to be counted. (Photos courtesy of R. Weil. Data from Chen and Weil, unpublished)

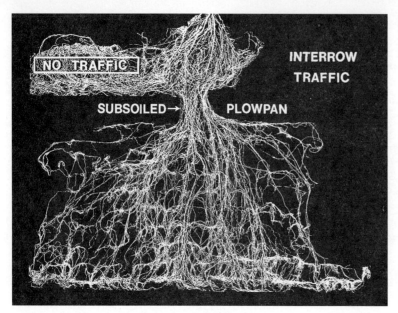

FIGURE 4.41 Root distribution of a cotton plant. On the right, interrow tractor traffic and plowing have caused a plowpan that restricts root growth. Roots are more prolific on the left where there had been no recent tractor traffic. The roots are seen to enter the subsoil through a loosened zone created by a subsoiling chisel-type implement. (Courtesy USDA National Tillage Machinery Laboratory)

paths between planting beds. The paths may be enhanced by covering with a thick mulch, planting to sod grass, or paving with flat stones.

Some managers attempt to reduce compaction using an opposite strategy in which special wide tires are fitted to heavy equipment so as to spread the weight over more soil surface, thus reducing the force applied per unit area (Figure 4.42a). Wider tires do lessen the compactive effect, but they also increase the percentage of the soil surface that is impacted. In a practice analogous to using wide wheels, home gardeners can avoid concentrating their body weight on just the few square centimeters of their footprints by standing on wooden boards when preparing seedbeds in relatively wet soil (Figure 4.42b).

(a)   (b)

FIGURE 4.42 One approach to reducing soil compaction is to spread the applied weight over a larger area of the soil surface. Examples are extra-wide wheels on heavy vehicles used to apply soil amendments (*left*) and standing on a wooden board while preparing a garden seedbed in early spring (*right*). (Photos courtesy of R. Weil)

High bulk density may occur as a natural soil profile feature (for example, a fragipan), or it may be an indication of human-induced soil compaction. In any case, root growth is inhibited by excessively dense soils for a number of reasons, including the soil's resistance to penetration, poor aeration, slow movement of nutrients and water, and the buildup of toxic gases and root exudates.

Roots penetrate the soil by pushing their way into pores. If a pore is too small to accommodate the root cap, the root must push the soil particles aside and enlarge the pore. To some degree, the density *per se* restricts root growth, as the roots encounter fewer and smaller pores. However, root penetration is also limited by **soil strength**, the property of the soil that causes it to resist deformation. One way to quantify soil strength is to measure the force needed to push a standard cone-tipped rod (a **penetrometer**) into the soil (see also Section 4.9). Compaction generally increases both bulk density and soil strength. At least two factors (both related to soil strength) must be considered to determine the effect of bulk density on the ability of roots to penetrate soil.

EFFECT OF SOIL WATER CONTENT. Soil water content and bulk density both affect *soil strength* (see Section 4.9 and Figure 4.43). Soil strength is increased when a soil is compacted to a higher bulk density, and also when finer-textured soils dry out and harden. Therefore, the effect of bulk density on root growth is most pronounced if those soils are dry, a higher bulk density being necessary to prevent root penetration when the soils are moist. For example, a traffic pan having a bulk density of 1.6 Mg/m^3 may completely prevent the penetration of roots when the soil is rather dry, yet roots may readily penetrate this same layer when it is in a moist condition.

EFFECT OF SOIL TEXTURE. The more clay present in a soil, the smaller the average pore size, and the greater the resistance to penetration at a given bulk density. Therefore, if the bulk density is the same, roots more easily penetrate a moist sandy soil than a moist clayey one. The growth of roots into moist soil is generally limited by bulk densities ranging from 1.45 Mg/m^3 in clays to 1.85 Mg/m^3 in loamy sands (see Figure 4.34). Viewed in this context, root growth was probably inhibited by the bulk density of the skidder trails in both soils illustrated in Table 4.4.

EFFECT OF LAND USE AND MANAGEMENT. As might be expected, such land uses as row-crop agriculture, pasture, rangeland, forestry, or off-road trafficking often markedly and simultaneously affect soil bulk density and strength in ways that restrict or enhance root growth and water movement. It is not always appreciated that tillage and traffic can compact soils quite deep into the subsoil. It may take many years of restorative

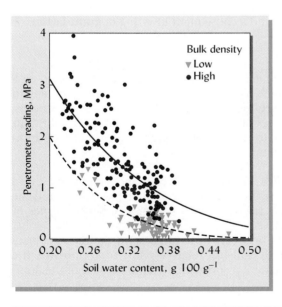

FIGURE 4.43 Both water content and bulk density affect soil strength as measured by penetrometer resistance. The data are for the clay textured Bt horizon of a Tatum soil in Virginia (Hapludults), which was either severely compacted (bulk density 1.7 Mg/m^3) or not compacted (bulk density 1.3 Mg/m^3). Note that soil strength decreases as water content increases and is very low regardless of bulk density when the soil is nearly saturated with water. [Graph based on study by Gilker et al. (2002)]

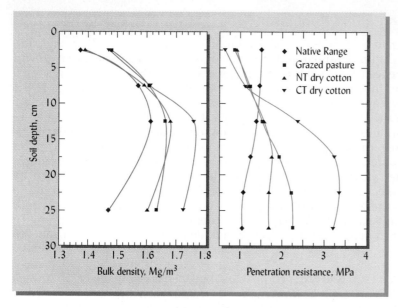

FIGURE 4.44 Bulk density and penetration resistance after about 30 years of no-till (NT) and conventional tillage (CT) dryland cotton, or 15 years of grazed pasture compared to native rangeland. Conventional tillage cotton resulted in the greatest bulk density and penetration resistance, especially below the depth of plowing (15 to 20 cm), suggesting the formation of a plow pan in the CT cotton soil. After 15 to 30 years without tillage, the pasture and no-till cotton systems had partially recovered the favorable physical properties of the native rangeland. The penetration resistance levels for all systems were quite high (2 MPa can restrict root growth) because water contents of these semi-arid region soils were quite low at the time of measurement. However, soil water was relatively uniform among the systems, at least in the lower layers where the major differences in penetration resistance occurred. Amarillo loamy fine sand (Aridic Paleustalfs) in the Southern High Plains of Texas. [Drawn from data in Halfmann (2005)]

management for the subsoil to recover its natural degree of porosity and friability. Figure 4.44 illustrates such a case.

4.8   PORE SPACE OF MINERAL SOILS

One of the main reasons for measuring soil bulk density is that this value can be used to calculate pore space. For soils with the same particle density, the lower the bulk density, the higher the percent pore space (**total porosity**). See Box 4.6 for derivation of the formula expressing this relationship.

Factors Influencing Total Pore Space

In Chapter 1 (Figure 1.18) we noted that for an "ideal" medium-textured, well-granulated surface soil in good condition for plant growth, approximately 50% of the soil volume would consist of pore space, and that the pore space would be about half-filled with air and half-filled with water. Actually, total porosity varies widely among soils for the same reasons that bulk density varies. Values range from as low as 25% in compacted subsoils to more than 60% in well-aggregated, high-organic-matter surface soils. As is the case for bulk density, management can exert a decided influence on the pore space of soils (see Table 4.5). Data from a wide range of soils show that cultivation tends to lower the total pore space compared to that of uncultivated soils. This reduction usually is associated with a decrease in organic matter content and a consequent lowering of **granulation**.

Size of Pores

Bulk density values help us predict only *total* porosity. However, soil pores occur in a wide variety of sizes and shapes that largely determine what role the pore can play in the soil (see Figure 4.45). Pores can be grouped by size into macropores, mesopores, micropores, and so on (Table 4.6 illustrates one such grouping). We will simplify our discussion at this point by referring only to **macropores** (larger than about 0.08 mm) and **micropores** (smaller than about 0.08 mm).

MACROPORES.   The macropores characteristically allow the ready movement of air and the drainage of water. They also are large enough to accommodate plant roots and the wide range of tiny animals that inhabit the soil (see Chapter 11). Several types of macropores are illustrated in Figure 4.46.

Macropores can occur as the spaces between individual sand grains in coarse-textured soils. Thus, even though a sandy soil has relatively low total porosity, the movement of air and water through such a soil is surprisingly rapid because of the dominance of the macropores.

BOX 4.6   CALCULATION OF PERCENT PORE SPACE IN SOILS

The bulk density of a soil can be easily measured and particle density can usually be assumed to be 2.65 Mg/m³ for most silicate-dominated mineral soils. Direct measurement of the pore space in soil requires the use of much more tedious and expensive techniques. Therefore, when information on the percent pore space is needed, it is often desirable to calculate the pore space from data on bulk and particle densities.

The derivation of the formula used to calculate the percentage of total pore space in soil follows:

Let D_b = bulk density, Mg/m³ V_s = volume of solids, m³

D_p = particle density, Mg/m³ V_p = volume of pores, m³

W_s = Weight of soil (solids), Mg $V_s + V_p$ = total soil volume V_t, m³

By definition,

$$\frac{W_s}{V_s} = D_p \qquad\qquad \text{and} \qquad\qquad \frac{W_s}{V_s + V_p} = D_b$$

Solving for W_s gives

$$W_s = D_p \times V_s \qquad\qquad \text{and} \qquad\qquad W_s = D_b(V_s + V_p)$$

Therefore

$$D_p \times V_s = D_b(V_s + V_p) \qquad\qquad \text{and} \qquad\qquad \frac{V_s}{V_s + V_p} = \frac{D_b}{D_p}$$

Since

$$\frac{V_s}{V_s + V_p} \times 100 = \text{\% solid space} \qquad \text{then} \qquad \text{\% solid space} = \frac{D_b}{D_p} \times 100$$

Since % pore space + % solid space = 100%, and % pore space = 100% − % solid space, then

$$\text{\% pore space} = 100\% - \left(\frac{D_b}{D_p} \times 100\right)$$

EXAMPLE

Consider a cultivated clay soil with a bulk density determined to be 1.28 Mg/m³. If we have no information on the particle density, we assume that the particle density is approximately that of the common silicate minerals (i.e., 2.65 Mg/m³). We calculate the percent pore space using the formula derived above:

$$\text{\% pore space} = 100\% - \left(\frac{1.28 \text{ Mg/m}^3}{2.65 \text{ Mg/m}^3} \times 100\right) = 100\% - 48.3 = 51.7$$

This value of pore space, 51.7%, is quite close to the typical percentage of air and water space described in Figure 1.18 for a well-granulated, medium- to fine-textured soil in good condition for plant growth. This simple calculation tells us nothing about the relative amounts of large and small pores, however, and so must be interpreted with caution.

For certain soils it is inaccurate to assume that the soil particle density is 2.65 Mg/m³. For instance, a soil with a high organic matter content can be expected to have a particle density somewhat lower than 2.65. Similarly, a soil rich in iron oxide minerals will have a particle density greater than 2.65 because these minerals have particle densities as high as 3.5. As an example of the latter type of soil, let us consider the uncultivated clay soils from Zimbabwe (Ustalfs) described in Table 4.5. These are red-colored clays, high in iron oxides. The particle density for these soils was determined to be 3.21 Mg/m³ (not shown in Table 4.5). Using this value and the bulk density value from Table 4.5, we calculate the pore space as follows:

$$\text{\% pore space} = 100\% - \left(\frac{1.20}{3.21} \times 100\%\right) = 100 - 37.4 = 62.6$$

Such a high percentage pore space is an indication that this soil is in an uncompacted, very well granulated condition typical for soils found under undisturbed natural vegetation.

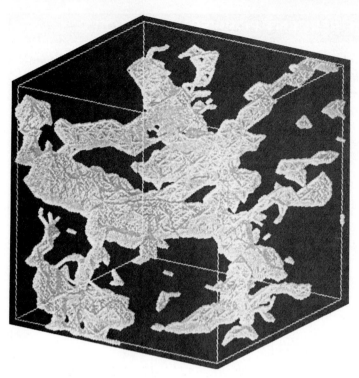

FIGURE 4.45 A three-dimensional representation of the network of pores in a small block of undisturbed soil in France (edges about 2 mm in length). The pores (light in color) exhibit great variability in size and cross-sectional area. Note that the pore channels are tortuous and that not all pores are connected to each other, some being isolated from channels that could transport water and air into and out of the soil. This suggests that tiny volumes of water and air may be trapped in localized pockets, thereby preventing their ready movement downward or upward in the soil. (Image courtesy of Dr. Isabelle Cousin, INRA Unité de Science du Sol—SESCPF Centre de Recherche d'Orléans Domaine de Limère, Ardon, France)

In well-structured soils, the macropores are generally found between peds. These **interped pores** may occur as spaces between loosely packed granules or as the planar cracks between tight-fitting blocky and prismatic peds (see Plate 82, after page 656).

Macropores created by roots, earthworms, and other organisms constitute a very important type of pores termed **biopores**. These are usually tubular in shape and may be continuous for lengths of a meter or more. (See Plate 81). In some clayey soils, biopores are the principal form of macropores, greatly facilitating the growth of plant roots (Table 4.7, Plates 81 and 83). Perennial plants, such as forest trees and certain forage crops, are particularly effective at creating channels that serve as conduits for roots, long after the death and decay of the roots that originally created them. Two such old root channels, each about 8 mm in diameter, can be seen perforating the clay slickenside shown in Figure 3.23c.

It is clear that both soil structure and texture influence the balance between macropores and micropores in a soil. Figure 4.47 shows that the decrease in organic matter and increase in clay that occur with depth in many profiles are associated with a shift from macropores to micropores.

TABLE 4.6 A Size Classification of Soil Pores and Some Functions of Each Size Class

Pore sizes are actually a continuum and the boundaries between classes given here are inexact and somewhat arbitrary. The term micropore *is often broadened to refer to all the pores smaller than macropores.*

Simplified class	Class[a]	Effective diameter range (mm)	Characteristics and functions
Macropores	Macropores	0.08–5+	Generally found between soil peds (interped); water drains by gravity; effectively transmit air; large enough to accommodate plant roots, habitat for certain soil animals.
Micropores	Mesopores	0.03–0.08	Retain water after drainage; transmit water by capillary action; accommodate fungi and root hairs.
	Micropores	0.005–0.03	Generally found within peds (intraped); retain water that plants can use; accommodate most bacteria.
	Ultramicropores	0.0001–0.005	Found largely within clay groupings; retain water that plants cannot use; exclude most microorganisms.
	Cryptopores	<0.0001	Exclude all microorganisms, too small for large molecules to enter.

[a] The pore size classes and boundary diameters are those cited in *Soil Sci. Soc. Amer.* (2001).

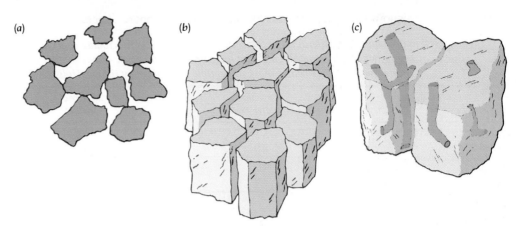

FIGURE 4.46   Various types of soil pores. (*a*) Many soil pores occur as *packing pores,* spaces left between primary soil particles. The size and shape of these spaces is largely dependent on the size and shape of the primary sand, silt, and clay particles and their packing arrangement. (*b*) In soils with structural peds, the spaces between the peds form *interped pores*. These may be rather planar in shape, as with the cracks between prismatic peds, or they may be more irregular, like those between loosely packed granular aggregates. (*c*) *Biopores* are formed by organisms such as earthworms, insects, and plant roots. Most of these are long, sometimes branched channels, but some are round cavities left by insect nests and the like. (Diagrams courtesy of R. Weil)

TABLE 4.7   Distribution of Different-Sized Loblolly Pine Roots in the Soil Matrix and in Old Root Channels in the Uppermost Meter of an Ultisol in South Carolina

The root channels were generally from 1 to 5 cm in diameter and filled with loose surface soil and decaying organic matter. Such channels are easy for roots to penetrate and have better fertility and aeration than the surrounding soil matrix.

	Numbers of roots counted per 1 m^2 of the soil profile		
Root size, diameter	Soil matrix	Old root channels	Comparative increase in root density in the old channels, %
Fine roots, <4 mm	211	3617	94
Medium roots, 4–20 mm	20	361	95
Coarse roots, >20 mm	3	155	98

Calculated from Parker and Van Lear (1996).

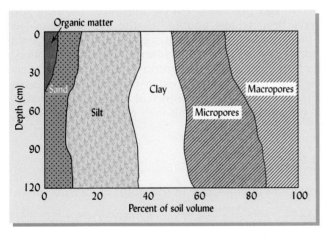

FIGURE 4.47   Volume distribution of organic matter, sand, silt, clay, and pores of macro- and microsizes in a representative medium–textured soil with good structure. Note that macropores are especially plentiful in the surface horizon (upper 30 cm). (Diagram courtesy of R. Weil)

MICROPORES.   In contrast to macropores, micropores are usually filled with water in field soils. Even when not water-filled, they are too small to permit much air movement. Water movement in micropores is slow, and much of the water retained in these pores is not available to plants (see Chapter 5). Fine-textured soils, especially those without a stable granular structure, may have a preponderance of micropores, thus allowing

TABLE 4.8   **Continuous Cropping Affects Macropore and Micropore Spaces**

Compared to undisturbed prairie soil, the cultivated soil has far less macropore space, but has gained some micropore space as aggregates were destroyed, changing large interped pores into much smaller micropores. Loss of organic matter made the aggregates more susceptible to damage by tillage. Houston clay (Hapludert).

Soil history	Soil depth, cm	Organic matter, %	Total pore space, %	Macropore space, %	Micropore space, %	Bulk density, Mg/m³
Prairie	0–15	5.6	58.3	32.7	25.6	1.11
Tilled 50 years	0–15	2.9	50.2	16.0	34.2	1.33
Prairie	15–30	4.2	56.1	27.0	29.1	1.16
Tilled 50 years	15–30	2.8	50.7	14.7	36.0	1.31

Data from Laws and Evans (1949).

relatively slow gas and water movement, despite the relatively large volume of total pore space. Aeration, especially in the subsoil, may be inadequate for satisfactory root development and desirable microbial activity. While the larger micropores accommodate plant root hairs and microorganisms, the smaller micropores (sometimes termed *ultramicropores* and *cryptopores*) are so tiny that their radii are measured in nanometers (10^{-9} meters), giving rise to the term *nanopores*. Such pores are too small to permit the entrance of even the smallest bacteria or some decay-stimulating enzymes produced by the bacteria. Thus, these pores can act as hiding places for some adsorbed organic compounds (both naturally occurring and pollutants), thereby protecting them from breakdown for long periods of time, perhaps for centuries (see Figure 18.18).

Clearly, the size, shape, and interconnection of soil pores, rather than their combined volume, are of greatest importance in determining soil drainage, aeration, and other such processes. Figure 4.46 illustrates the variability in size and shapes of the soil pores, which range from the micropores inside the aggregates where air and water movement is restricted, to the large macropores through which air and water will move freely.

Cultivation and Pore Size

Continuous cropping, particularly of soils originally high in organic matter, often results in a reduction of macropore spaces (see Table 4.8). When native prairie lands are plowed and planted to row crops such as corn or soybeans, soil organic matter contents and total pore space are reduced. But most striking is the effect of cropping on the size of the soil pores. Such cropping drastically reduces the amount of macropore space that is so critical for ready air movement.

In recent years, conservation tillage practices, which minimize plowing and associated soil manipulations, have been widely adopted in the United States (see Sections 4.6 and 17.6). Because of increased accumulation of organic matter near the soil surface and the development of a long-lived network of macropores (especially biopores), some conservation tillage systems lead to greater macroporosity of the surface layers. These benefits are particularly likely to accrue in soils with extensive production of earthworm burrows, which may remain undisturbed in the absence of tillage. Unfortunately, such improvements in porosity do not always occur in soils with poor internal drainage.

4.9   SOIL PROPERTIES RELEVANT TO ENGINEERING USES

Soil properties are obviously important to engineering. In the words of Richard Hand:

> *"Virtually every structure is supported by soil or rock. Those that aren't either fly, float, or fall over."*

Field Rating of Soil Consistence and Consistency

CONSISTENCE.   Soil **consistence** is a term used by soil scientists to describe the ease with which a soil can be reshaped or ruptured. As a clod of soil is squeezed between the thumb and forefinger (or crushed underfoot, if necessary), observations are made on

TABLE 4.9 Some Field Tests and Terms Used to Describe the Consistence and Consistency of Soils

The consistency of cohesive materials is closely related to, but not exactly the same as, their consistence. Conditions of least coherence are represented by terms at the top of each column, those of greater coherence near the bottom.

	Soil consistence[a]			Soil consistency[b]	
Dry soil	Moist to wet soil	Soil dried then submerged in water	Field rupture (crushing) test	Soil at in situ moisture	Field penetration test
Loose	Loose	Not applicable	Specimen not obtainable	Soft	Blunt end of pencil penetrates deeply with ease
Soft	Very friable	Noncemented	Crumbles under very slight force between thumb and forefinger	Medium firm	Blunt end of pencil can penetrate about 1.25 cm with moderate effort
Slightly hard	Friable	Extremely weakly cemented	Crumbles under slight force between thumb and forefinger	Firm	Blunt end of pencil can penetrate about 0.5 cm
Hard	Firm	Weakly cemented	Crushes with difficulty between thumb and forefinger	Very firm	Blunt end of pencil makes slight indentation; thumbnail easily penetrates
Very hard	Extremely firm	Moderately cemented	Cannot be crushed between thumb and forefinger, but can be crushed slowly underfoot	Hard	Blunt end of pencil makes no indentation; thumbnail barely penetrates
Extremely hard	Slightly rigid	Strongly cemented	Cannot be crushed by full body weight underfoot		

[a] Abstracted from USDA/NRCS (2005).
[b] Modified from McCarthy (1993).

the amount of force needed to crush the clod and on the manner in which the soil responds to the force. The degree of cementation of the soil by such materials as silica, calcite, or iron is also considered in identifying soil consistence.

Moisture content greatly influences how a soil responds to stress; hence, moist and dry soils are given separate consistence ratings (Table 4.9). A dry, clayey soil that cannot be crushed between the thumb and forefinger but can be crushed easily underfoot would be designated as *very hard*. The same soil, when wet, would exhibit much less resistance to deformation, and would be termed *plastic*. The degrees of *stickiness* and *plasticity* (malleability) of soil in the wet condition are often included in describing soil consistence (although not shown in Table 4.9). As described in Section 4.6, a moist clod that crumbles with only light pressure is said to be friable. Friable soils are easily excavated or tilled.

Consistency. Engineers use the term **consistency** to describe how a soil resists *penetration* by an object, while the soil scientist's consistence describes resistance to *rupture*. Instead of crushing a clod of soil, the engineer attempts to penetrate it with either the blunt end of a pencil (some use their thumbs) or a thumbnail. For example, if the blunt end of a pencil makes only a slight indentation, but the thumbnail penetrates easily, the soil is rated as *very firm* (Table 4.9). Consistency, then, is a kind of simple field estimation of soil strength or penetration resistance (see Section 4.7).

Field observations of both consistence and consistency provide valuable information to guide decisions about loading and manipulating soils. For construction purposes, however, more precise measurements are needed of a number of related soil properties that help predict how a soil will respond to applied stress.

Soil Strength and Sudden Failure

Excavation safety-animation of unstable soils:
http://physics.uwstout.edu/geo/exca_s.avi

Engineers define soil **bearing strength** as the capacity of a soil mass to withstand stresses without rupturing or becoming deformed. Failure of a soil to withstand stress can result in a building toppling over as its weight exceeds the soil's bearing strength. Similarly, an earthen dam or levee might give way under the pressure of impounded water, or pavements and structures might slide down unstable hillsides (Figure 1.10).

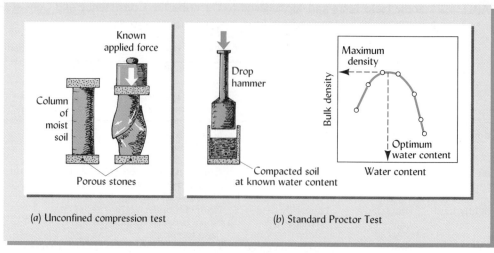

FIGURE 4.48 Two important tests to determine engineering properties of soil materials. (*a*) An unconfined compression test for soil strength. (*b*) The Proctor test for maximum density and optimum water content for compaction control.

Saturated soils lead to life in prison. By Adam Pitluk: http://www.riverfronttimes.com/issues/2000-01-26/news.html

COHESIVE SOILS. Two components of strength apply to **cohesive soils** (essentially soils with a clay content of more than about 15%): (1) the inherent electrostatic attractive forces between clay platelets and between clay surfaces and the water in very fine pores (see *clay flocculation* in Section 4.5), and (2) the frictional resistance to movement between soil particles of all sizes. While many different laboratory tests are used to estimate soil strength, perhaps the simplest to understand is the direct **unconfined compression test** using the apparatus illustrated in Figure 4.48*a*. A cylindrical specimen of cohesive soil is placed vertically between two flat, porous stones (which allow water to escape from the compressed soil pores) and a slowly increasing downward force is applied. The soil column will first bulge out a bit and then fail—that is, give way suddenly and collapse—when the force exceeds the soil strength.

The strength of cohesive soils declines dramatically if the material is very wet and the pores are filled with water. Then the particles are forced apart so that neither the cohesive nor the frictional component is very strong, making the soil prone to failure, often with catastrophic results (such as mudslides, Figure 4.49, or levee failures, Box 4.7).

FIGURE 4.49 Houses damaged by a mudslide that occurred when the soils of a steep hillside in Oregon became saturated with water after a period of heavy rains. The weight of the wet soil exceeded its shear strength, causing the slope to fail. Excavations for roads and houses near the foot of a slope can contribute to the lack of slope stability, as can removal of tree roots by large-scale clear-cutting on the slope itself. (Photo courtesy of John Griffith, Coos Bay, OR)

On the other hand, if cohesive soils become more compacted or dry down, their strength increases as particles are forced into closer contact with one another—a result that has implications for plant root growth as well as for engineering (see Section 4.7).

NONCOHESIVE SOILS.   The strength of dry, noncohesive soil materials such as loose sand depends entirely on frictional forces, including the interlocking of rough particle surfaces. One reflection of such interparticle friction is the **angle of repose**, the steepest angle to which a material can be piled without slumping. Smooth, rounded sand grains cannot be piled as steeply as can rough, interlocking sands. If a small amount of water bridges the gaps between particles, electrostatic attraction of the water for the mineral surfaces will increase the soil strength (as illustrated in Figure 4.50). Interparticle water bridges explain why cars can drive along the edge of the beach where the sand is moist, but their tires sink in and lose traction on loose, dry sand or in saturated quicksand (see Figure 5.27 for an example of the latter).

Soil liquefaction:
http://www.ce.washington.edu/
~liquefaction/html/content.html

COLLAPSIBLE SOILS.   Certain soils that exhibit considerable strength at low *in situ* water contents lose their strength suddenly if they become wet. Such soils may collapse without warning under a roadway or building foundation. A special case of soil collapse is **thixotropy**, the sudden liquification of a wet soil mass when subjected to vibrations, such as those accompanying earthquakes and blasting.

Most **collapsible soils** are noncohesive materials in which loosely packed sand grains are cemented at their contact points by small amounts of gypsum, clay, or water under tension. These soils usually occur in arid and semiarid regions, where such cementing agents are relatively stable. Many collapsible soils have derived their open

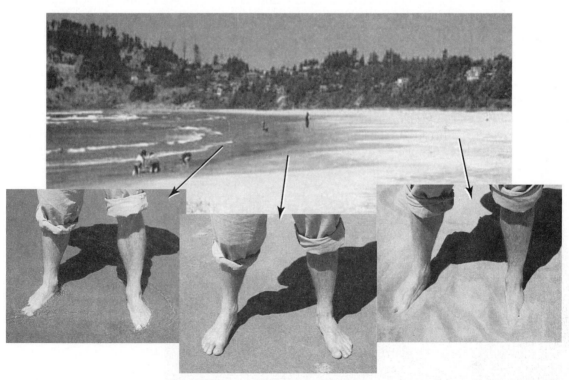

FIGURE 4.50   A walk along a beach such as this one in Oregon illustrates the concept of soil strength for sandy materials. The dry sand (*lower right*) has little strength and your feet easily mire into it as you walk along. There is nothing to hold the individual sand particles together to permit them to serve as a firm base. As you move toward the ocean where the soil has been thoroughly wetted by the incoming waves, but where there is no standing water (*lower center*), you find firm footing, indicating considerably higher soil strength. Thin water films act as bridges between sand particles, holding them together and thereby resisting penetration by the feet. If you stand in shallow water along the edge of the ocean (*lower left*), once again your feet penetrate the surface sand, indicating that soil strength has been reduced. Each sand particle is completely surrounded by water, which acts more as a lubricant than as a binding force. If you were to drive an automobile over the same areas, only the wetted but not submerged sand would provide a firm base. Soil strength is a property of great concern to engineers, but also must be reckoned with by plant roots as they try to penetrate soils with high bulk densities. (Photos courtesy of R. Weil)

particle arrangement from the process of sedimentation beneath past or present bodies of water. When these soils are wetted, excess water may dissolve cements such as gypsum or disperse clays that form bridges between particles, causing a sudden loss of strength. In some cases, similar behavior is exhibited by highly weathered Oxisols in humid tropical regions.

Settlement—Gradual Compression

While embankments and hill slopes often abruptly fail due to stresses that exceed the soil's strength, most foundation problems result from slow, often uneven, vertical subsidence or **settlement** of the soil.

COMPACTION CONTROL.   People attempting to grow plants should generally avoid any practice that might compact the soil. However, soils to be used for a foundation or roadbed are compacted on purpose using heavy rollers (Figure 4.51) or vibrators. Compaction occurring after construction would result in uneven settlement and cracked pavements or foundations.

Some soil particles, such as certain silicate clays and micas, can be compressed when a load is placed upon them. If that load is removed, these particles tend to regain their original shape, in effect reversing their compression. As a result, soils rich in these particles are not easily compacted into a stable base for roads and foundations.

The **Proctor test** is used to guide efforts at compacting soil materials before construction. A specimen of soil is mixed to a given water content and placed in a holder, where it is compacted by a drop hammer. The bulk density (usually referred to as the *dry density* by engineers) is then measured. The process is repeated with increasing water contents until a *Proctor curve* (Figure 4.48b) can be drawn from the data. The curve indicates the maximum bulk density achievable and the soil water content that maximizes compactability. On construction sites, tank trucks may spray water to bring the soil water content to the determined optimum level before heavy equipment (such as that shown in Figure 4.51) compacts the soil to the desired density.

COMPRESSIBILITY.   A **consolidation test** may be conducted on a soil specimen to determine its **compressibility**—how much its volume will be reduced by a given applied force. Because of the relatively low porosity and equidimensional shape of the individual mineral grains, very sandy soils resist compression once the particles have settled into a tight packing arrangement. They make excellent soils for foundations. The high porosity of clay floccules and the flakelike shape of clay particles give clayey soils much

FIGURE 4.51 Compaction of soils used as foundations and roadbeds is accomplished by heavy equipment such as this sheepsfoot roller. The knobs ("sheepsfeet") concentrate the mass on a small impact area, punching and kneading the loose, freshly graded soil to optimum density. (Photo courtesy of R. Weil)

greater compressibility. Soils consisting mainly of organic matter (peats) have the highest compressibilities and generally are unsuitable for foundations.

In the field, compression of wet, clayey soils may occur very slowly after a load (e.g., a building) is applied because compression can occur only as fast as water can escape from the soil pores—which for the fine pores in clayey materials is not very fast. Perhaps the most famous example of uneven settlement due to slow compression is the Leaning Tower of Pisa in Italy. Unfortunately, most cases of uneven settlement result in headaches, not tourist attractions.

Expansive Soils

Damage caused by expansive soils in the United States rarely makes the evening news programs, although the total cost annually exceeds that caused by tornados, floods, and earthquakes. Expansive clays occur on about 20% of the land area in the United States and cause upwards of $6 billion in damages annually to pavements, foundations, and utility lines. The damages can be severe in certain sites in all parts of the country, but are most extensive in regions that have long dry periods alternating with periods of rain (see distribution of Vertisols, endpapers).

Some clays, particularly the smectites, swell when wet and shrink when dry (see Section 8.14). Expansive soils are rich in these types of clay. The electrostatic charges on clay surfaces attract water molecules from larger pores into the micropores within clay domains. Also, the adsorbed cations associated with the clay surfaces tend to hydrate, drawing in additional water. The water pushes apart the layers of clay, causing the mass of soil to swell in volume. The reverse of these processes occurs when the soil dries and water is withdrawn from packets of clay platelets, causing shrinkage and cracking. After a prolonged dry spell, soils high in smectites can be recognized in the field by the criss-crossed pattern of wide, deep cracks (Figure 4.52). The swelling and shrinkage cause sufficient movement of the soil to crack building foundations, burst pipelines, and buckle pavements.

Atterberg Limits

Determine Atterberg Limits, U of TX at Arlington: http://geotech.uta.edu/lab/Main/atrbrg_lmts/

As a dry, clayey soil takes on increasing amounts of water, it undergoes dramatic and distinct changes in behavior and consistency. A hard, rigid solid in the dry state, it becomes a crumbly (friable) semisolid when a certain moisture content (termed the **shrinkage limit**) is reached. If it contains expansive clays, the soil also begins to swell in volume as this moisture content is exceeded. Increasing the water content beyond the **plastic limit** will transform the soil into a malleable, plastic mass and cause additional swelling. The soil will remain in this plastic state until its **liquid limit** is exceeded, causing it to transform into a viscous liquid that will flow when jarred. These critical water contents (measured in units of percent) are termed the **Atterberg limits**.

PLASTICITY INDEX. The plasticity index (PI) is the difference between the plastic limit (PL) and liquid limit (LL) and indicates the water-content range over which the soil has plastic properties:

$$PI = LL - PL$$

FIGURE 4.52 Certain types of clays, especially the smectites, undergo significant volume changes in conjunction with changes in water content. Here, an expansive soil rich in smectitic clay has shrunk during a dry period, causing a network of large cracks to open up in the soil surface. (Courtesy of USDA Natural Resources Conservation Service)

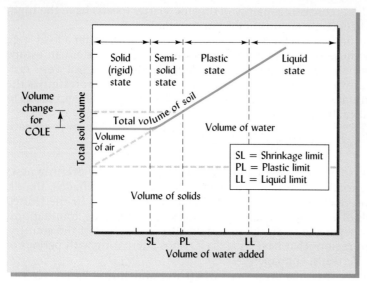

FIGURE 4.53 A common depiction of the Atterberg limits, which mark major shifts in the behavior of a cohesive soil as its water content changes (from left to right). As water is added to a certain volume of dry solids, first air is displaced; then, as more water is added, the total volume of the soil increases (if the soil has some expansive properties). When the shrinkage limit (SL) is reached, the once rigid, hard solid becomes a crumbly semisolid. With more water, the plastic limit (PL) is reached, after which the soil becomes plastic and can be molded. It remains in a plastic stage over a range of water contents until the liquid limit (LL) is exceeded, at which point the soil begins to behave as a viscous liquid that will flow when jarred. The volume change for calculating the coefficient of linear extensibility (COLE) is shown at the left.

Soils with a plasticity index greater than about 25 are usually expansive clays that make poor roadbeds (Plate 43) or foundations. Figure 4.53 shows the relationship among the Atterberg limits and the changes in soil volume associated with increasing water contents for a hypothetical soil.

Smectite clays (see Section 8.3) generally have high liquid limits and plasticity indices, especially if saturated with sodium. Kaolinite and other nonexpansive clays have low liquid limit values. The plastic and liquid limits of several soils, varying in clay content and type, and three clay samples are shown in Table 8.10. The tendency of expansive clay soils to literally flow down steep slopes when the liquid limit is exceeded, producing mass wasting and landslides, is illustrated in Plates 41 and 42, after page 112.

COEFFICIENT OF LINEAR EXTENSIBILITY. The expansiveness of a soil (and therefore the hazard of its destroying foundations and pavements) can be quantified as the *coefficient of linear extensibility* (COLE). Figure 4.53 indicates how the volume change used to calculate the COLE relates to the Atterberg limits. Suppose a sample of soil is moistened to its plastic limit and molded into the shape of a bar with length L_M. If the bar of soil is allowed to air dry, it will shrink to length L_D. The COLE is the percent reduction in length of the soil bar upon shrinking:

$$COLE = \frac{L_M - L_D}{L_M} \times 100$$

Unified Classification System for Soil Materials

The U.S. Army Corps of Engineers and the U.S. Bureau of Reclamation have established a widely used system of classifying soil materials in order to aid in predicting the engineering behavior of different soils (Table 4.10). The system first groups soils into coarse- and fine-grained soils. The coarse materials are further divided on the basis of grain size (gravels and sands), amount of fines present, and uniformity of grain size (well or poorly graded). The fine-grained soils are divided into silts, clays, and organic materials. These classes are further subdivided on the basis of their liquid limit (above or below 50) and their plasticity index. Each type of soil is then given a two-letter designation based primarily on its particle-size distribution (texture), Atterberg limits, and organic-matter content (e.g., GW for well-graded gravel, SP for poorly graded sands, CL for clay of low plasticity, and OH for organic-rich clays of high plasticity). This classification of soil materials helps engineers predict the soil strength, expansiveness, compressibility, and other properties so that appropriate engineering designs can be made for the soil at hand (Box 4.7).

BOX 4.7   TRAGEDY IN THE BIG EASY—A LEVEE DOOMED TO FAIL[a]

After Hurricane Katrina hit New Orleans in 2005, a 4 meter high storm surge breached the city's levees in several places, causing one of the worst natural disasters in American history with some 100,000 homes flooded and over 1,000 people killed. Some of the worst flooding occurred when the 17th Street levee failed (Figure 4.54). Investigations later showed that the 17th Street levee was doomed to fail by a faulty levee design that did not properly deal with underlying layers of organic soils and sands. Poor design, combined with poor levee maintenance, allowed water to seep under the levee and weaken the soil at its base.

The levee, essentially a gently sloping mound of compacted clay soils, was covered on the landward side with a thin veneer of topsoil to support a protective grass mantle. Hard armor of rock or concrete (see Section 17.10) was not used in most places. To hold back floodwater and storm surges, engineers had constructed a concrete seawall along the crest of the levee. The seawall was attached to long steel pilings driven deep into the soil of the levee. The pilings were meant to both anchor the seawall and to

FIGURE 4.54   A large helicopter attempts emergency repairs to the 17th Street levee breach several days after Hurricane Katrina hit New Orleans and toppled this section of the levee and seawall. Large chunks of the levee can be seen some 14 meters inland. (Photo courtesy of U.S. Army Corps of Engineers)

prevent water from seeping through or under the levee. Out of sight, under the layers of clayey and loamy materials from which the levee was constructed, several layers of peat (buried Histosols) and sand provided for the weak link in the levee design.

Organic soils are highly compressible and have very low bearing and shear strengths. Both peats and sands are also highly permeable and conduct water readily. To perform their intended functions, the steel pilings attached to the seawall had to be long enough to penetrate through the peat/sand layers and into the more cohesive, higher-strength soil below. Unfortunately, the pilings in the 17th Street levee were too short (less than 6 m) and failed to penetrate through the peat layer (Figure 4.55). Whenever storm surges—or even high flows—raised the water level in the canal, seepage would occur through the peat/sand layer under the levee. The seepage water would then rise to saturate the soil at the foot of levee. When saturated, the soil would lose most of its shear strength and resistance to compression. Apparently, the engineers designing the levee had data on the peat layers, but based their design on the *average* soil properties, rather than on the weakest soils present.

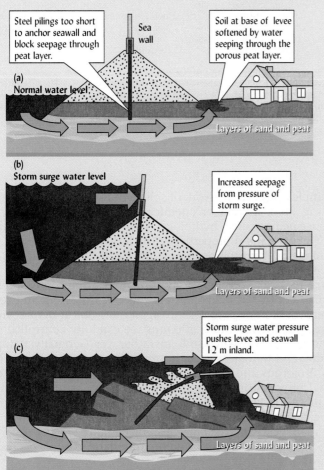

FIGURE 4.55   Illustration (not to scale) of how the buried layers of low-strength organic soil (a) allowed seepage of storm surge water (b) dooming the levee and its seawall to failure (c). (Diagram courtesy of R. Weil)

---

[a] Based on forensic engineering investigation information in Seed et al. (2005) and reporting by Marshall (2005) and Vartabedian and Braun (2006).

(continued)

BOX 4.7 (Cont.) TRAGEDY IN THE BIG EASY—A LEVEE DOOMED TO FAIL

On 21 August 2005, with seepage water saturating the soil at the levee base and turning the peat layer into little more than "soup," Katrina's storm surge "snapped the chain" at the weakest link, toppling a 140 m long section of the seawall, pushing both it and the levee some 14 m inland. The storm-churned waters poured through the breach, inundating the city of New Orleans.

TABLE 4.10 The Unified System of Classification Used to Classify Soil Materials (Not Natural Soil Bodies) for Engineering Uses[a]

Coarse-grained soils 50% or more retained on No. 200 (0.075 mm) sieve								Fine-grained soils 50% or more passes No. 200 (0.075 mm) sieve						
Gravels More than 50% of coarse fraction retained on No. 4 (2 mm) sieve				Sands More than 50% of coarse fraction passes No. 4 (2 mm) sieve				Silts and clays Liquid limit 50% or less			Silts and clays Liquid limit greater than 50%			Highly organic soils
Clean gravels		Gravels with fines		Clean sands		Sands with fines								
GW	GP	GM	GC	SW	SP	SM	SC	ML	CL	OL	MH	CH	OH	Pt
Well-graded gravels and gravel-sand mixtures, little to no fines	Poorly-graded gravels and gravel-sand mixtures, little to no fines	Silty gravels and gravel-sand mixtures	Clayey gravels and gravel-sand mixtures	Well-graded sands and gravelly sands, little to no fines	Poorly-graded sands and gravelly sands, little to no fines	Silty sands, sand-silt mixtures	Clayey sands, sand-clay mixtures	Inorganic silts, very fine sands, rock flour, silty or clayey fine sands	Inorganic clays of low to medium plasticity, gravelly, sandy or, silty clays, lean clays	Organic silts and organic silty clays of low plasticity	Inorganic silts, micaceous or diatomaceous fine sand or silts, elastic silts	Inorganic clays of high plasticity, fat clays	Organic clays of medium to high plasticity	Peat, muck, other highly organic soils

[a] The two-letter designations (SW, MH, etc.) help engineers predict the behavior of the soil material when used for construction purposes. The first letter is one of the following: G = gravel, S = sand, M = silts, C = clays, and O = organic-rich materials. The second letter indicates whether the sand or gravels are well graded (W) or poorly graded (P), and whether the silts, clays, and organic-rich materials have a high plasticity index (H) or a low plasticity index (L). Among the fine-grained materials, those closer to the right side of the table are the most troublesome materials for foundations and roadbeds.

4.10 CONCLUSION

Soils present an incredibly complex physical network of solid surfaces, pores, and interfaces that provides the setting for a myriad of chemical, biological, and physical processes. These in turn influence plant growth, hydrology, environmental management, and engineering uses of soil. The nature and properties of the individual particles, their size distribution, and their arrangement in soils determine the total volume of nonsolid pore space, as well as the pore sizes, thereby impacting on water and air relationships.

The properties of individual particles and their proportionate distribution (soil texture) are subject to little human control in field soils. However, it is possible to exert some control over the arrangement of these particles into aggregates (soil structure) and on the stability of these aggregates. Tillage and traffic must be carefully controlled to avoid undue damage to soil tilth, especially when soils are rather wet. Generally, nature takes good care of soil structure, and humans can learn much about soil management by studying natural systems. Vigorous and diverse plant growth, generous return of organic residues, and minimal physical disturbance are attributes of natural systems worthy of emulation. Proper plant species selection; crop rotation; and management of chemical, physical, and biological factors can help ensure maintenance of soil physical quality. In recent years, these management goals have been made more practical by the advent of conservation tillage systems that minimize soil manipulations while decreasing soil erosion and water runoff.

Particle size, moisture content, and plasticity of the colloidal fraction all help determine the stability of soil in response to loading forces from traffic, tillage, or building foundations. The physical properties presented in this chapter greatly influence nearly all other soil properties and uses, as discussed throughout this book.

STUDY QUESTIONS

1. If you were investigating a site for a proposed housing development, how could you use soil colors to help predict where problems might be encountered?

2. You are considering the purchase of some farmland in a region with variable soil textures. The soils on one farm are mostly sandy loams and loamy sands, while those on a second farm are mostly clay loams and clays. List the potential advantages and disadvantages of each farm as suggested by the texture of its soils.

3. Revisit your answer to question 2. Explain how soil structure in both the surface and subsurface horizons might modify your opinion of the merits of each farm.

4. Two different timber-harvest methods are being tested on adjacent forest plots with clay loam surface soils. Initially, the bulk density of the surface soil in both plots was 1.1 Mg/m^3. One year after the harvest operations, plot *A* soil had a bulk density of 1.48 Mg/m^3, while that in plot *B* was 1.29 Mg/m^3. Interpret these values with regard to the relative merits of systems *A* and *B*, and the likely effects on the soil's function in the forest ecosystem.

5. What are the textural classes of two soils, the first with 15% clay and 45% silt, and the second with 80% sand and 10% clay? (Hint: Use Figure 4.6.)

6. For the forest plot *B* in question 4, what was the change in percent pore space of the surface soil caused by timber harvest? Would you expect that most of this change was in the micropores or in the macropores? Explain.

7. Discuss the positive and negative impacts of tillage on soil structure. What is another physical consideration that you would have to take into account in deciding whether or not to change from a conventional to a conservation tillage system?

8. What would you, as a home gardener, consider to be the three best and three worst things that you could do with regard to managing the soil structure in your home garden?

9. What does the Proctor test tell an engineer about a soil, and why would this information be important?

10. In a humid region characterized by expansive soils, a homeowner experienced burst water pipes, doors that no longer closed properly, and large vertical cracks in the brick walls. The house had had no problems for over 20 years, and a consulting soil scientist blamed the problems on a large tree that was planted near the house some 10 years before the problems began to occur. Explain.

REFERENCES

Bigham, J. M., and E. J. Ciolkosz (eds.). 1993. *Soil Color.* SSSA Special Publication no. 31. Soil Science Society of America, Madison, Wis.

Celik, I. 2005. "Land-use effects on organic matter and physical properties of soil in a southern mediterranean highland of Turkey." *Soil Tillage Res.* **83**:270–277.

Emerson, W. W., R. C. Foster, and J. M. Oades. 1986. "Organomineral complexes in relation to soil aggregation and structure," in P. M. Huang and M. Schnitzer (eds.), *Interaction of Soil Minerals with Natural Organics and Microbes.* SSSA Special Publication no. 17. Soil Science Society of America, Madison, Wis.

Eynard, A., T. E. Schumacher, M. J. Lindstrom, and D. D. Malo. 2004. "Porosity and pore-size distribution in cultivated Ustolls and Usterts." *Soil Sci. Soc. Am. J.* **68**:1927–1934.

Faulkner, E. H. 1943. *Plowman's Folly* University of Oklahoma Press, Norman, Okla.

Gent, J. A., Jr., R. Ballard, A. E. Hassan, and D. K. Cassel. 1984. "Impact of harvesting and site preparation on physical properties of Piedmont forest soils," *Soil Sci. Soc. Amer. J.,* **48**:173–177.

Gent, J. A., Jr., and L. A. Morris. 1986. "Soil compaction from harvesting and site preparation in the Upper Gulf Coastal Plain," *Soil Sci. Soc. Amer. J.,* **50**:443–446.

Gilker, R. E., R. R. Weil, D. T. Krizek, and B. Momen. 2002. "Eastern gamagrass root penetration in adverse subsoil conditions." *Soil Sci. Soc. Amer. J.* **66**:931–938.

Halfmann, D. 2005. "Management system effects on water infiltration and soil physical properties." Master's Thesis, Texas Tech University, Lubbock, Texas.

Landa, E. R., and M. D. Fairchild. 2005. Charting color from the eye of the beholder. *American Scientist* 93:436–442.

Laws, W. D., and D. D. Evans. 1949. "The effects of long-time cultivation on some physical and chemical properties of two rendzina soils," *Soil Sci. Soc. Amer. Proc.,* **14**:15–19.

Le Bissonnais, Y., and D. Arrouays. 1997. "Aggregate stability and assessment of soil crustability and erodibility: II. Application to humic loamy soils with various organic carbon contents," *European J. Soil Sci.* **48**:39–48.

Marshall, B. 2005. "17th Street Canal levee was doomed—report blames Corps: Soil could never hold." *The Times-Picayune*, Wednesday, November 30, New Orleans.

McCarthy, D. F. 1993. *Essentials of Soil Mechanics and Foundations,* 4th ed. Prentice Hall, Englewood Cliffs, N.J.

Oades, J. M. 1993. "The role of biology in the formation, stabilization, and degradation of soil structure," *Geoderma* **56**:377–400.

O'Nofiok, O., and M. J. Singer. 1984. "Scanning electron microscope studies of surface crusts formed by simulated rainfall," *Soil Sci. Soc. Amer. J.* **48**:1137–1143.

Parker, M. M., and D. H. Van Lear. 1996. "Soil heterogeneity and root distribution of mature loblolly pine stands in Piedmont soils," *Soil Sci. Soc. Amer. J.* **60**:1920–1925.

Rillig, M. C. 2004. "Arbuscular mycorrhizae, glomalin, and soil aggregation." *Can. J. Soil Sci.* **84**:355–363.

Rillig, M. C., and P. D. Steinberg. 2002. "Glomalin production by an arbuscular mycorrhizal fungus: A mechanism of habitat modification?" *Soil Biol. Biochem.* **34**:1371–1374.

Ross, C., R. E. Sojka, and J. A. Foerster. 2003. "Scanning electron micrographs of polyacrylamide-treated soil in irrigation furrows." *J. Soil Water Conserv.* **58**:327–331.

Seed, R. B., P. G. Nicholson, R. A. Dalrymple, J. Battjes, R. G. Bea, G. Boutwell, J. D. Bray, B. D. Collins, L. F. Harder, J. R. Headland, M. Inamine, R. E. Kayen, R. Kuhr, J. M. Pestana, R. Sanders, F. Silva-Tulla, R. Storesund, S. Tanaka, J. Wartman, T. F. Wolff, L. Wooten, and T. Zimmie. 2005. "Preliminary report on the performance of the New Orleans levee systems in hurricane Katrina on August 29, 2005—Preliminary findings from field investigations and associated studies shortly after the hurricane." *Report UCB/CITRIS – 05/01.* University of California at Berkeley and the American Society of Civil Engineers, Berkeley, CA.

Six, J., H. Bossuyt, S. Degryze, and K. Denef. 2004. "A history of research on the link between (micro)aggregates, soil biota, and soil organic matter dynamics." *Soil Tillage Res.* **79**:7–31.

Six, J., E. T. Elliott, and K. Paustian. 2000. "Soil structure and soil organic matter: II. A normalized stability index and the effect of mineralogy," *Soil Sci. Soc. Amer. J.* **64**:1042–1049.

Soil Science Society of America. 2001. *Glossary of Soil Science Terms 1996* (: Soil Science Society of America, Madison, Wisc.

Thieme, J., G. Schneider, and C. Knochel. 2003. "X-ray tomography of a microhabitat of bacteria and other soil colloids with sub-100 nm resolution." *Micron* **34**:339–344.

Tijink, F. G. J., and J. P. Van Der Linden. 2000. "Engineering approaches to prevent compaction in cropping systems with sugar beet. In R. Horn et al., eds., *Subsoil compaction: Distribution, processes and consequences.* pp. 442–452. Catena Verlag, Reiskirchen, Germany.

Tisdall, J. M., and J. M. Oades. 1982. "Organic matter and water-stable aggregates in soils." *Soil Sci. Soc. Am. J.* **33**:141–163.

USDA-NRCS. 2005. *National soil survey handbook,* title 430-vi. U.S. Department of Agriculture, Natural Resources Conservation Service. http://soils.usda.gov/technical/handbook/ (posted September 2005; verified 23, 2005).

U.S. Department of Interior Teton Dam Failure Review Group. 1977. *Failure of Teton Dam.* Stock no. 024-003-00112-1. U.S. Government Printing Office, Washington, D.C.

Vartabedian, R., and S. Braun. 2006. "Fatal flaws: Why the walls tumbled in New Orleans". *Los Angeles Times,* Los Angeles. CA.

Vimmerstadt, J., F. Scoles, J. Brown, and M. Schmittgen. 1982. "Effects of use pattern, cover, soil drainage class, and overwinter changes on rain infiltration on campsites," *J. Environ. Qual.,* **11**:25–28.

Watts, C. W., and A. R. Dexter. 1997. "The influence of organic matter in reducing the destabilization of soil by simulated tillage." *Soil Tillage Res.* **42**:253–275.

Wright, S. F., and R. L. Anderson. 2000. "Aggregate stability and glomalin in alternative crop rotations for the central great plains." *Biol. Fertil. Soils* **31**:249–253.

5

SOIL WATER: CHARACTERISTICS AND BEHAVIOR

When the earth will . . . drink up the rain as fast as it falls.
—*H. D. THOREAU*, THE JOURNAL

One of nature's simplest chemical compounds, water is a vital component of every living cell. Its unique properties promote a wide variety of physical, chemical, and biological processes. These processes greatly influence almost every aspect of soil development and behavior, from the weathering of minerals to the decomposition of organic matter, from the growth of plants to the pollution of groundwater.

We are all familiar with water. We drink it, wash with it, and swim in it. But water in the soil is something quite different from water in a drinking glass. In the soil, the intimate association between water and soil particles changes the behavior of both. Water causes soil particles to swell and shrink, to adhere to each other, and to form structural aggregates. Water participates in innumerable chemical reactions that release or tie up nutrients, create acidity, and wear down minerals so that their constituent elements eventually contribute to the saltiness of the oceans.

Certain soil water phenomena seem to contradict our intuition about how water ought to behave. Attraction to solid surfaces restricts some of the free movement of water molecules, making it less liquid and more solidlike in its behavior. In the soil, water can flow up as well as down. Plants may wilt and die in a soil whose profile contains a million kilograms of water in a hectare. A layer of sand or gravel in a soil profile may actually inhibit drainage, rather than enhance it.

Soil–water interactions determine the rates of water loss by leaching, surface runoff and evapotranspiration, the balance between air and water in soil pores, the rate of change in soil temperature, the rate and kind of metabolism of soil organisms, and the capacity of soil to store and provide water for plant growth.

The characteristics and behavior of water in the soil comprise a common thread that interrelates nearly every chapter in this book. The principles contained in this chapter will help us understand why mudslides occur in water-saturated soils (Chapter 4), why earthworms may improve soil quality (Chapter 11), why wetlands contribute to global ozone depletion (Chapter 13), and why famine stalks humanity in certain regions of the world (Chapter 20). Mastery of the principles presented in this chapter is fundamental to your working knowledge of the soil system.

5.1 STRUCTURE AND RELATED PROPERTIES OF WATER[1]

Water properties:
http://www.biologylessons.
sdsu.edu/classes/lab1/semnet/
water.htm

The ability of water to influence so many soil processes is determined primarily by the structure of the water molecule. This structure also is responsible for the fact that water is mainly present as a liquid, not a gas, at temperatures found on Earth. Water is, with the exception of mercury, the *only* inorganic (not carbon-based) liquid found on Earth. Water is a simple compound, its individual molecules containing one oxygen atom and two much smaller hydrogen atoms. The elements are bonded together covalently, each hydrogen atom sharing its single electron with the oxygen.

Polarity

Instead of lining up symmetrically on either side of the oxygen atom (H-O-H), the hydrogen atoms are attached to the oxygen in a V-shaped arrangement at an angle of only 105°. Water is therefore an asymmetrical molecule with the shared electrons most of the time nearer to the oxygen than to the hydrogen (Figure 5.1). Consequently, water molecules exhibit *polarity*; that is, the charges are not evenly distributed. Rather, the side on which the hydrogen atoms are located tends to be electropositive and the opposite side electronegative.

Hydrogen Bonding

Through a phenomenon called **hydrogen bonding**, a hydrogen atom of one water molecule is attracted to the oxygen end of a neighboring water molecule, thereby forming a low-energy bond between the two molecules. This type of bonding accounts for the polymerization of water. Hydrogen bonding also accounts for the relatively high boiling point, specific heat, and viscosity of water compared to the same properties of other hydrogen-containing compounds, such as H_2S, which has a higher molecular weight but no hydrogen bonding.

Hydration

Polarity also explains why water molecules are attracted to electrostatically charged ions and to colloidal surfaces. Cations such as H^+, Na^+, K^+, and Ca^{2+} become hydrated through their attraction to the oxygen (negative) end of water molecules. Likewise, negatively charged clay surfaces attract water, this time through the hydrogen (positive) end of the molecule. Polarity of water molecules also encourages the dissolution of

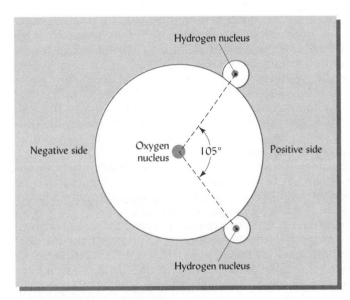

FIGURE 5.1 Two-dimensional representation of a water molecule showing a large oxygen atom and two much smaller hydrogen atoms. The H-O-H angle of 105° results in an asymmetrical arrangement. One side of the water molecule (that with the two hydrogens) is electropositive; the other is electronegative. This accounts for the polarity of water.

---

[1] For more in-depth discussions of water–soil interactions, see Hillel (1998) or Warrick (2001).

salts in water since the ionic components have greater attraction for water molecules than for each other.

When water molecules become attracted to electrostatically charged ions or clay surfaces, they are more closely packed than in pure water. In this packed state, their freedom of movement is restricted and their energy status is lower than in pure water. Thus, when ions or clay particles become hydrated, energy is released. That released energy is evidenced as *heat of solution* when ions hydrate or as *heat of wetting* when clay particles become wet. The latter phenomenon can be demonstrated by adding a few drops of water to dry, fine clay. A rise in temperature can be measured—or even felt if the clay is wetted in the palm of one's hand.

Cohesion Versus Adhesion

Hydrogen bonding accounts for two basic forces responsible for water retention and movement in soils: the attraction of water molecules for each other (**cohesion**) and the attraction of water molecules for solid surfaces (**adhesion**). By adhesion (also called *adsorption*), some water molecules are held rigidly at the surfaces of soil solids. In turn, these tightly bound water molecules hold, by cohesion, other water molecules farther removed from the solid surfaces (Figure 5.2). Together, the forces of adhesion and cohesion make it possible for the soil solids to retain water and control its movement and use. Adhesion and cohesion also make possible the property of plasticity possessed by clays (see Section 4.9).

Surface Tension

Another important property of water that markedly influences its behavior in soils is that of **surface tension**. This property is commonly evidenced at liquid–air interfaces and results from the greater attraction of water molecules for each other (cohesion) than for the air above. The net effect is an inward force at the surface that causes water to behave as if its surface were covered with a stretched elastic membrane (Figure 5.3). Because of the relatively high attraction of water molecules for each other, water has a high surface tension (72.8 millinewtons/m at 20 °C) compared to that of most other liquids (e.g., ethyl alcohol, 22.4 mN/m). As we shall see, surface tension is an important factor in the phenomenon of capillarity, which determines how water moves and is retained in soil.

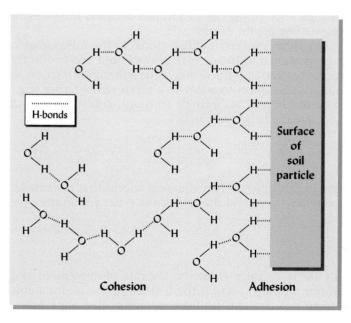

FIGURE 5.2 The forces of cohesion (between water molecules) and adhesion (between water and solid surface) in a soil–water system. The forces are largely a result of H-bonding shown as broken lines. The adhesive or adsorptive force diminishes rapidly with distance from the solid surface. The cohesion of one water molecule to another results in water molecules forming temporary clusters that are constantly changing in size and shape as individual water molecules break free or join up with others. The cohesion between water molecules also allows the solid to indirectly restrict the freedom of water for some distance beyond the solid–liquid interface.

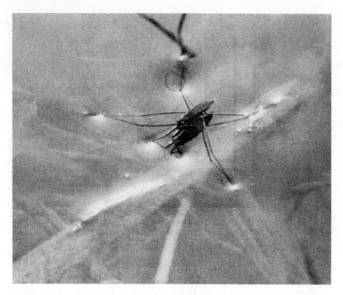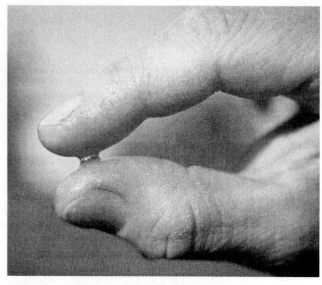

FIGURE 5.3 Everyday evidences of water's surface tension (*left*) as insects land on water and do not sink and of forces of cohesion and adhesion (*right*) as a drop of water is held between the fingers. (Photos courtesy of R. Weil)

5.2 CAPILLARY FUNDAMENTALS AND SOIL WATER

Capillary rise, click "start animation":
http://www.ito.ethz.ch:16080/filep/futuresite/inhalt_en/index.php?seite=1200#

The movement of water up a wick typifies the phenomenon of capillarity. Two forces cause capillarity: (1) the attraction of water for the solid (adhesion or adsorption), and (2) the surface tension of water, which is due largely to the attraction of water molecules for each other (cohesion).

Capillary Mechanism

Capillarity can be demonstrated by placing one end of a fine, clean glass tube in water. The water rises in the tube; the smaller the tube bore, the higher the water rises. The water molecules are attracted to the sides of the tube (adhesion) and start to spread out along the glass in response to this attraction. At the same time, the cohesive forces hold the water molecules together and create surface tension, causing a curved surface (called a *meniscus*) to form at the interface between water and air in the tube (Figure 5.4c). Lower pressure under the meniscus in the glass tube (P2) allows the higher pressure (P1) on the free water to push water up the tube. The process continues until the water in the tube has risen high enough that its weight just balances the pressure differential across the meniscus (see Box 5.1 for details).

The height of rise in a capillary tube is inversely proportional to the tube radius r. Capillary rise is also inversely proportional to the density of the liquid and is directly proportional to the liquid's surface tension and the degree of its adhesive attraction to the soil surface. If we limit our consideration to water at a given temperature (e.g., 20 °C), then these factors can be combined into a single constant, and we can use a simple capillary equation to calculate the height of rise h:

$$h = \frac{0.15}{r} \tag{5.1}$$

where both h and r are expressed in centimeters. This equation tells us that the smaller the tube bore, the greater the capillary force and the higher the water rise in the tube (Figure 5.5a).

Height of Rise in Soils

Capillary forces are at work in all moist soils. However, the rate of movement and the rise in height are less than one would expect on the basis of soil pore size alone. One reason is that soil pores are not straight, uniform openings like glass tubes.

BOX 5.1 THE MECHANISM OF CAPILLARITY

Capillary action is due to the combined forces of adhesion and cohesion, as seen when a drop of water is placed on a solid surface. Solid substances that have an electronegative surface (for example, due to the oxygens in the silica tetrahedra of quartz or glass) strongly attract the electropositive H-end of the water molecule. These substances are said to be *hydrophilic* (water-loving) because attraction of the water molecules for the solid surface (adhesion) is much greater than the attraction of the water molecules for each other (cohesion). Adhesion will cause a drop of water placed on a hydrophilic solid, such as clean glass, to spread out along the surface, thus forming an acute (<90°) angle between the water–air interface and the solid surface (see Figure 5.4a). This *contact angle* is characteristic for a particular liquid–solid pair (e.g., water on glass). The more strongly the water molecules are attracted to the solid, the closer to zero the contact angle.

In contrast, water molecules placed on a hydrophobic (water-hating) surface will pull themselves into a spherical mass. The resulting contact angle is obtuse (>90°), indicating that the adhesion is not as strong as the cohesion (see Figure 5.4b). This relationship explains why water beads up on a freshly waxed automobile.

Now instead of a flat surface and a drop of water, consider a small-diameter tube of clean glass dipped into a pool of water. Adhesion will again cause the water to spread out on the glass surface, forming the same contact angle α with the glass as was the case for the water drop. At the same time, cohesion among water molecules creates a surface tension that causes a curved surface (called a *meniscus*) to form at the interface between the water and the air in the tube (see Figure 5.4c). If the contact angle is nearly zero, the curvature of the meniscus will approximate a hemisphere.

The curved (rather than a flat) interface between water and air causes the pressure to be lower on the convex side (labeled P2 in Figure 5.4c) than on the concave side of the meniscus. Normal atmospheric pressure P1 is exerted both above the meniscus and on the pool of free water. Because the pressure under the meniscus P2 is less than the pressure on the free-water pool, water is pushed up the capillary tube. The water will rise in the tube until the meniscus reaches the height h at which the weight of the water in the tube just balances the pressure difference P2 − P1. In this condition, the forces pushing water up the tube will be balanced by the forces pulling it down.

The upward forces are determined by the product of surface tension T, the length of the contact between the tube and the meniscus (tube *circumference* $= 2\pi r$) and the upward component of this force ($\cos \alpha$).

The downward forces are determined by the product of the water density d, the water volume above the free-water surface $h\pi r^2$, and the acceleration of gravity g.

Thus, when capillary rise ceases we can equate:

Upward-acting force = Downward-acting force

$$T * 2\pi r * \cos \alpha = d * h * \pi r^2 * g \qquad (5.2)$$

Note that if the tube radius were made half as large (0.5r), the force acting upward would be cut in half, but the downward force would be ¼ as great [$(0.5r)^2 = 0.5r * 0.5r = 0.25r$]—hence, the height of rise would be twice as great when the forces come into balance again. Herein lies the reason why capillarity rise is greater in finer tubes. The equation balancing the upward- and downward-acting forces can be algebraically rearranged to give an equation describing the height of capillary rise:

$$h = \frac{2T \cos \alpha}{rdg} \qquad (5.3)$$

Most water–solid interactions in soils are of the hydrophilic type shown in Figure 5.4a and c. The attraction between water and soil particle surfaces is usually so strong that the angle of contact is very close to zero, making its cosine 1. The $\cos \alpha$ can therefore be ignored under these circumstances. Three of the other factors affecting capillary rise (T, d, and g) are constants at a given temperature and can therefore be combined into a single constant. Thus, we can rewrite the simplified capillary rise equation given on page 176.

$$h \, (\text{cm}) = \frac{0.15 \, (\text{cm}^2)}{r \, (\text{cm})} \qquad (5.4)$$

As one would expect, capillary rise will only occur if the tube is made of hydrophilic material. If a *hydrophobic* tube (such as one with a waxed surface) is dipped into a pool of water, the meniscus will be convex rather than concave to the air, so that the situation is reversed and capillary *depression* rather than capillary *rise* will occur (see Figure 5.4d). This is the case in certain water-repellent soil layers (see Figure 7.27 and Plate 71).

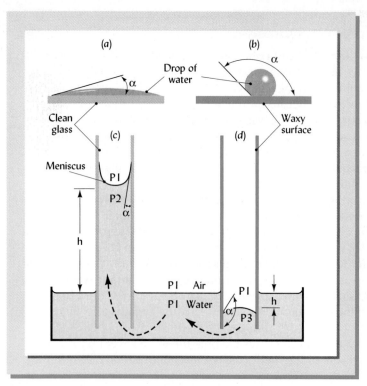

FIGURE 5.4 The interaction of water with a hydrophilic (a, c) or a hydrophobic (b, d) surface results in a characteristic *contact angle* (α). If the solid surface surrounds the water as in a tube, a curved water–air interface termed the *meniscus* forms because of adhesive and cohesive forces. When air and water meet in a curved meniscus, pressure on the convex side of the curve is lower than on the concave side. (c) Capillary rise occurs in a fine hydrophillic (e.g., glass) tube because pressure under the meniscus (P2) is less than pressure on the free water. (d) Capillary depression occurs if the tube is hydrophobic, and the meniscus is inverted. (Diagram courtesy of R. Weil)

Furthermore, some soil pores are filled with air, which may be entrapped, slowing down or preventing the movement of water by capillarity (see Figure 5.5b).

Since capillary movement is determined by pore size, it is the pore-size distribution discussed in Chapter 4 that largely determines the amount and rate of movement of capillary water in the soil. The abundance of medium- to large-sized capillary pores in sandy soils permits rapid initial capillary rise, but limits the ultimate height of

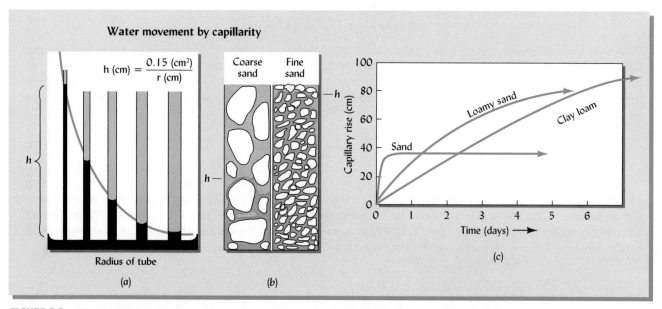

FIGURE 5.5 Upward capillary movement of water through tubes of different bore and soils with different pore sizes. (a) The capillary equation can be graphed to show that the height of rise *h* doubles when the tube inside radius is halved. The same relationship can be demonstrated using glass tubes of different bore size. (b) The same principle also relates pore sizes in a soil and height of capillary rise, but the rise of water in a soil is rather jerky and irregular because of the tortuous shape and variability in size of the soil pores (as well as because of pockets of trapped air). (c) The finer the soil texture, the greater the proportion of small-sized pores and, hence, the higher the ultimate rise of water above a free-water table. However, because of the much greater frictional forces in the smaller pores, the capillary rise is much slower in the finer-textured soil than in the sand. (Diagrams courtesy of R. Weil)

FIGURE 5.6 In this irrigated field in Arizona, water has moved up by capillarity from the irrigation furrow toward the top of the ridge (*left*), as well as horizontally to both sides and away from the irrigation water (*right*). (Photos courtesy of N. C. Brady)

rise[2] (Figure 5.5c). Clays have a high proportion of very fine capillary pores, but frictional forces slow down the rate at which water moves through them. Consequently, in clays the capillary rise is slow initially, but in time it generally exceeds that of sands. Loams exhibit capillary properties between those of sands and clays.

Capillarity is traditionally illustrated as an upward adjustment. But movement in any direction takes place, since the attractions between soil pores and water are as effective in forming a water meniscus in horizontal pores as in vertical ones (Figure 5.6). The significance of capillarity in controlling water movement in small pores will become evident as we turn to soil water energy concepts.

5.3 SOIL WATER ENERGY CONCEPTS

Soil water energy and dynamics:
faculty.washington.edu/slb/esc210/soils15.pdf

The retention and movement of water in soils, its uptake and translocation in plants, and its loss to the atmosphere are all energy-related phenomena. Different kinds of energy are involved, including *potential energy* and *kinetic energy*. Kinetic energy is certainly an important factor in the rapid, turbulent flow of water in a river, but the movement of water in soil is so slow that the kinetic energy component is usually negligible. Potential energy is most important in determining the status and movement of soil water. For the sake of simplicity, in this text we will use the term *energy* to refer to potential energy.

As we consider energy, we should keep in mind that all substances, including water, tend to move or change from a higher to a lower energy state. Therefore, if we know the pertinent energy levels at various points in a soil, we can predict the direction of water movement. It is the *differences* in energy levels from one contiguous site to another that influence this water movement.

Forces Affecting Potential Energy

The discussion of the structure and properties of water in the previous section suggests three important forces affecting the energy level of soil water. First, adhesion, or the attraction of water to the soil solids (matrix), provides a **matric** force (responsible for

---

[2] Note that if water rises by capillarity to a height of 37 cm above a free-water surface in a sand (as shown in the example in Figure 5.5c), then it can be estimated (by rearranging the capillary equation to $r = 0.15/h$) that the smallest continuous pores must have a radius of about 0.004 cm (0.15/37 = 0.004). This calculation gives an approximation of the minimum effective capillary pore radius in a soil.

adsorption and capillarity) that markedly reduces the energy state of water near particle surfaces. Second, the attraction of water to ions and other solutes, resulting in **osmotic** forces, tends to reduce the energy state of water in the soil solution. Osmotic movement of pure water across a semipermeable membrane into a solution (osmosis) is evidence of the lower energy state of water in the solution. The third major force acting on soil water is **gravity**, which always pulls the water downward. The energy level of soil water at a given elevation in the profile is thus higher than that of water at some lower elevation. This difference in energy level causes water to flow downward.

Soil Water Potential

The *difference* in energy level of water from one site or one condition to another (e.g., between wet soil and dry soil) determines the direction and rate of water movement in soils and in plants. In a wet soil, most of the water is retained in large pores or thick water films around particles. Therefore, most of the water molecules in a wet soil are not very close to a particle surface and so are not held very tightly by the soil solids (the matrix). In this condition, the water molecules have considerable freedom of movement, so their energy level is near that of water molecules in a pool of pure water outside the soil. In a drier soil, however, the water that remains is located in small pores and thin water films and is therefore held tightly by the soil solids. Thus the water molecules in a drier soil have little freedom of movement, and their energy level is much lower than that of the water in wet soil. If wet and dry soil samples are brought in touch with each other, water will move from the wet soil (higher energy state) to the drier soil (lower energy).

Determining the absolute energy level of soil water is a difficult and sometimes impossible task. Fortunately, it is not necessary to know the absolute energy level of water to be able to predict how it will move in soils and in the environment. Relative values of soil water energy are all that is needed. Usually the energy status of soil water in a particular location in the profile is compared to that of pure water at standard pressure and temperature, unaffected by the soil and located at some reference elevation. The *difference* in energy levels between this pure water in the reference state and that of the soil water is termed soil **water potential** (Figure 5.7), the term *potential*, like the term *pressure*, implying a difference in energy status.

If all water potential values under consideration have a common reference point (the energy state of pure water), differences in the water potential of two soil samples in fact reflect differences in their absolute energy levels. This means that water will move from a soil zone having a high soil water potential to one having a lower soil water

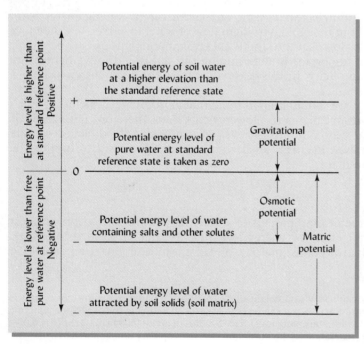

FIGURE 5.7   Relationship between the potential energy of pure water at a standard reference state (pressure, temperature, and elevation) and that of soil water. If the soil water contains salts and other solutes, the mutual attraction between water molecules and these chemicals reduces the potential energy of the water, the degree of the reduction being termed *osmotic potential*. Similarly, the mutual attraction between soil solids (soil matrix) and soil water molecules also reduces the water's potential energy. In this case the reduction is called *matric potential*. Since both of these interactions reduce the water's potential energy level compared to that of pure water, the changes in energy level (osmotic potential and matric potential) are both considered to be negative. In contrast, differences in energy due to gravity (*gravitational potential*) are always positive because the reference elevation of the pure water is purposely designated at a site in the soil profile below that of the soil water. A plant root attempting to remove water from a moist soil would have to overcome all three forces simultaneously.

potential. This fact should always be kept in mind when thinking about the behavior of water in soils.

Several forces are implicated in soil water potential, each of which is a component of the **total soil water potential** ψ_t. These components are due to differences in energy levels resulting from gravitational, matric, submerged hydrostatic, and osmotic forces and are termed **gravitational potential** ψ_g, **matric potential** ψ_m, **hydrostatic potential** ψ_h, and **osmotic potential** ψ_o, respectively. All of these components act simultaneously to influence water behavior in soils. The general relationship of soil water potential to potential energy levels is shown in Figure 5.7 and can be expressed as:

$$\psi_t = \psi_g + \psi_m + \psi_o + \psi_h + \cdots \tag{5.5}$$

where the ellipsis ($\cdots$) indicates the possible contribution of additional potentials not yet mentioned.

Gravitational Potential

Gravitational potential:
http://id.mind.net/~zona/
mstm/physics/mechanics/
energy/gravitationalPotential
Energy/gravitationalPotential
Energy.html

The force of gravity acts on soil water the same as it does on any other body (Figure 5.8), the attraction being toward the Earth's center. The gravitational potential ψ_g of soil water may be expressed mathematically as:

$$\psi_g = gh \tag{5.6}$$

where g is the acceleration due to gravity and h is the height of the soil water above a reference elevation. The reference elevation is usually chosen within the soil profile or at its lower boundary to ensure that the gravitational potential of soil water above the reference point will always be positive.

Following heavy precipitation or irrigation, gravity plays an important role in removing excess water from the upper horizons and in recharging groundwater below the soil profile. It will be given further attention when the movement of soil water is discussed (see Section 5.5).

Pressure Potential (Including Hydrostatic and Matric Potentials)

This component accounts for all other effects on soil water potential besides gravity and solute levels. Pressure potential most commonly includes (1) the positive hydrostatic pressure due to the weight of water in saturated soils and aquifers, and (2) the

FIGURE 5.8 Whether concerning matric potential, osmotic potential, or gravitational potential (as shown here), water always moves to where its energy state will be lower. In this case the energy lost by the water is used to turn the waterwheel and grind flour at historic Mabry Mills. (Photo courtesy of R. Weil)

negative pressure due to the attractive forces between the water and the soil solids or the soil matrix.[3]

The hydrostatic pressures give rise to what is often termed the **hydrostatic potential** ψ_h, a component that is operational only for water in saturated zones below the water table. Anyone who has dived to the bottom of a swimming pool has felt hydrostatic pressure on their eardrums.

The attraction of water to solid surfaces gives rise to the **matric potential** ψ_m, which is always negative because the water attracted by the soil matrix has an energy state lower than that of pure water. (These negative pressures are sometimes referred to as *suction* or *tension*. If these terms are used, their values are positive.) The matric potential is operational in unsaturated soil above the water table (Figure 5.9).

Matric potential ψ_m, which results from adhesive forces and capillarity, influences both the retention and movement of soil water. Differences between the ψ_m of two adjoining soil zones encourage the movement of water from wetter (high energy state) areas to drier (low energy state) areas or from large pores to small pores. Although this movement may be slow, it is extremely important in supplying water to plant roots and in engineering applications.

Osmotic Potential

Osmotic potential animation:

http://www.dlt.ncssm.edu/ TIGER/Flash/Colligative Properties/OsmoticPressure. html

The osmotic potential ψ_o is attributable to the presence of both inorganic and organic solutes in the soil solution. As water molecules cluster around solute ions or molecules, the freedom of movement (and therefore the potential energy) of the water is reduced. The greater the concentration of solutes, the more osmotic potential is lowered. As always, water will tend to move to where its energy level will be lower, in this case to the zone of higher solute concentration. However, liquid water will move in response to differences in osmotic potential (the process termed **osmosis**) only if a *semipermeable membrane* exists between the zones of high and low osmotic potential, allowing water through but *preventing the movement of the solute*. If no membrane is present, the solute, rather than the water, generally moves to equalize concentrations. The process of osmosis and the relationship between the matric and osmotic components of total soil water potential is shown in Figure 5.10.

Because soil zones are *not* generally separated by membranes, the osmotic potential ψ_o has little effect on the mass movement of water in soils. Its major effect is on the uptake of water by plant root cells that *are* isolated from the soil solution by their semipermeable cell membranes. In soils high in soluble salts, ψ_o may be lower (have a greater negative value) in the soil solution than in plant root cells. This leads to constraints in the uptake of water by plants. In very salty soil, the soil water osmotic

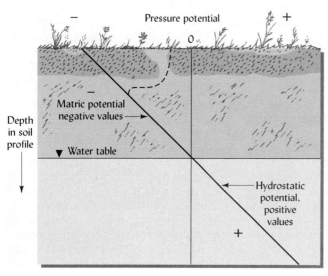

FIGURE 5.9 The matric potential and hydrostatic potential are both pressure potentials that may contribute to total water potential. The matric potential is always negative and the hydrostatic potential is positive. When water is in unsaturated soil above the water table (top of the saturated zone), it is subject to the influence of matric potentials. Water below the water table in saturated soil is subject to hydrostatic potentials. In the example shown here, the matric potential decreases linearly with elevation above the water table, signifying that water rising by capillary attraction up from the water table is the only source of water in this profile. Rainfall or irrigation (see dotted line) would alter or curve the straight line, but would not change the fundamental relationship described.

---

[3] In addition to matric and hydrostatic forces, in some situations the weight of the overburdened soil and the pressure of air in the soil also make a contribution to the total soil water potential.

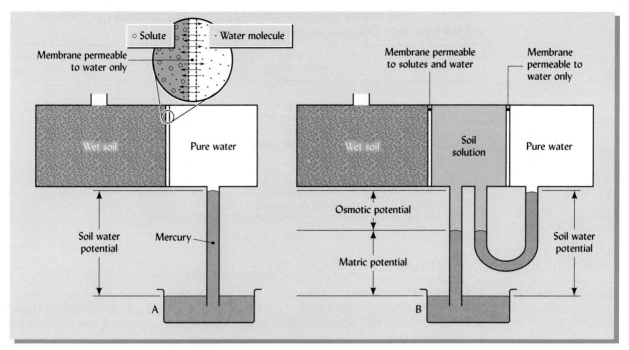

FIGURE 5.10 Relationships among osmotic, matric, and combined soil water potentials. (*Left*) A container of soil separated from pure water by a membrane permeable only to water (see inset showing osmosis across the membrane). The pure water is connected to a vessel of mercury through a tube. Water will move into the soil in response to the matric forces attracting water to soil solids and the osmotic forces attracting water to solutes. At equilibrium the height of the mercury column above vessel A is a measure of this combined soil water potential (matric plus osmotic). (*Right*) A second container is placed between the pure water and the soil, and this container is separated from the soil by a fine screen permeable to both solutes and water. Ions will move from the soil into this second container until the concentration of solutes in this water and in the soil water have equalized. Then the difference between the potential energies of this solution and of the pure water gives a measure of the *osmotic potential*. The *matric potential*, as measured by the column of mercury above vessel B, would then be the difference between the combined soil water potential and the osmotic component. The *gravitational potential* (not shown) is the same for all compartments and does not affect the outcome since the water movement is horizontal. [Modified from Richards (1965)]

potential may be low enough to cause cells in young seedlings to collapse (plasmolyze) as water moves from the cells to the lower osmotic potential zone in the soil.

The random movement of water molecules causes a few of them to escape a body of liquid water, enter the atmosphere, and become water vapor. Since the presence of solutes restricts the movement of water molecules, fewer water molecules escape into the air as the solute concentration of liquid water is increased. Therefore, water vapor pressure is lower in the air over salty water than in the air over pure water. By affecting water vapor pressure, ψ_o affects the movement of water vapor in soils (see Section 5.7).

Methods of Expressing Energy Levels

Several units can be used to express differences in energy levels of soil water. One is the *height of a water column* (usually in centimeters) whose weight just equals the potential under consideration. We have already encountered this means of expression since the *h* in the capillary equation (Section 5.2) tells us the matric potential of the water in a capillary pore. A second unit is the standard *atmosphere* pressure at sea level, which is 760 mm Hg or 1020 cm of water. Another unit termed *bar* approximates the pressure of a standard atmosphere. Energy may be expressed per unit of mass (**joules/kg**) or per unit of volume (**newtons/m²**). In the International System of Units (SI), 1 pascal (Pa) equals 1 newton (N) acting over an area of 1 m². In this text we will use Pa or kilopascals (kPa) to express soil water potential. Since other publications may use other units, Table 5.1 is provided to show the equivalency among common means of expressing soil water potential.

TABLE 5.1 **Approximate Equivalents Among Expressions of Soil Water Potential and the Equivalent Diameter of Pores Emptied of Water**

Height of unit column of water, cm	Soil water potential, bars	Soil water potential, kPa[a]	Equivalent diameter of pores emptied, μm[b]
0	0	0	—
10.2	−0.01	−1	300
102	−0.1	−10	30
306	−0.3	−30	10
1,020	−1.0	−100	3
15,300	−15	−1,500	0.2
31,700	−31	−3,100	0.97
102,000	−100	−10,000	0.03

[a] The SI unit kilopascal (kPa) is equivalent to 0.01 bars.
[b] Smallest pore that can be emptied by equivalent tension as calculated using Eq. 5.4.

5.4 SOIL WATER CONTENT AND SOIL WATER POTENTIAL

Types of soil water sensors: http://www.sowacs.com/sensors/index.html

The previous discussions suggest an inverse relationship between the water content of soils and the tenacity with which the water is held in soils. Many factors affect the relationship between soil water potential ψ and moisture content θ. A few examples will illustrate this point.

Soil Water Versus Energy Curves

The relationship between soil water potential ψ and moisture content θ of three soils of different textures is shown in Figure 5.11. Such curves are sometimes termed *water release characteristic curves*, or simply *water characteristic curves*. The absence of sharp breaks in the curves indicates a continuous range of pore sizes and therefore a gradual

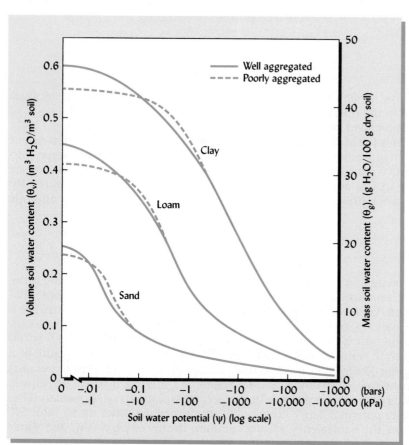

FIGURE 5.11 Soil water potential curves for three representative mineral soils. The curves show the relationship obtained by slowly drying completely saturated soils. The dashed lines show the effect of compaction or poor aggregation. The soil water potential ψ (which is negative) is expressed in terms of bars (*upper scale*) and kilopascals (kPa) (*lower scale*). Note that the soil water potential is plotted on a log scale.

change in the water potential with increased soil water content. The clay soil holds much more water at a given potential than does the loam or sand. Likewise, at a given moisture content, the water is held much more tenaciously in the clay than in the other two soils (note that soil water potential is plotted on a log scale). The amount of clay in a soil largely determines the proportion of very small micropores in that soil. As we shall see, about half of the water held by clay soils is held so tightly in these micropores that it cannot be removed by growing plants. Soil texture clearly exerts a major influence on soil moisture retention.

Soil structure also influences soil water content–energy relationships. A well-granulated soil has more total pore space and greater overall water-holding capacity than one with poor granulation or one that has been compacted. Soil aggregation especially increases the relatively large inter-aggregate pores (Section 4.5) in that water is held with little tenacity. In contrast, a compacted soil will hold less total water but is likely to have a higher proportion of small- and medium-sized pores, that hold water with greater tenacity than do larger pores. Therefore, soil structure predominantly influences the shape of the water characteristic curve in the portion where the potentials are between 0 and about 100 kPa. The shape of the remainder of the curve generally reflects the influence of soil texture.

The soil water characteristic curves in Figure 5.11 have great practical significance for various field measurements and processes. It will be useful to refer back to these curves as we consider the applied aspects of soil water behavior in the following sections.

Hysteresis

The relationship between soil water content and potential, determined as a soil dries out, will differ somewhat from the relationship measured as the same soil is rewetted. This phenomena, known as **hysteresis**, is illustrated in Figure 5.12. Hysteresis is caused by a number of factors, including the nonuniformity of soil pores. As soils are wetted, some of the smaller pores are bypassed, leaving entrapped air that prevents water penetration. Some of the macropores in a soil may be surrounded only by micropores, creating a bottleneck effect. In this case, the macropore will not lose its water until the

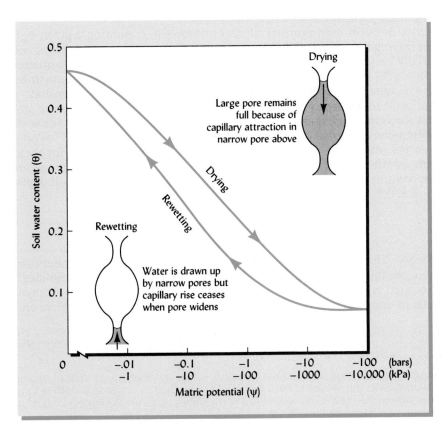

FIGURE 5.12 The relationship between soil water content and matric potential of a soil upon being dried and then rewetted. The phenomenon, known as *hysteresis*, is apparently due to factors such as the nonuniformity of individual soil pores, entrapped air, and the swelling and shrinking that might affect soil structure. The drawings show the effect of nonuniformity of pores.

matric potential is low enough to empty the water from the surrounding smaller pores (see Figure 5.12). Also, the swelling and shrinking of clays as the soil is dried and rewetted brings about changes in soil structure that affect the soil–water relationships. Because of hysteresis, it is important to know whether soils are being wetted or dried when properties of one soil are compared with those of another.

Measurement of Soil Water Status

The soil water characteristic curves just discussed highlight the importance of making two general kinds of soil water measurements: the *amount* of water present (water content) and the *energy status* of the water (soil water potential). In order to understand or manage water supply and movement in soils, it is essential to have information (directly measured or inferred) on *both* types of measurements. For example, a soil water potential measurement might tell us whether water will move toward the groundwater, but without a corresponding measurement of the soil water content, we would not know the possible significance of the contribution to groundwater.

Generally, the behavior of soil water is most closely related to the energy status of the water, not to the amount of water in a soil. Thus, a clay loam and a loamy sand will both feel moist and will easily supply water to plants when the ψ_m is, say, -10 kPa. However, the amount of water held by the clay loam, and thus the length of time it could supply water to plants, would be far greater at this potential than would be the case for the loamy sand.

We will briefly consider several methods for making each of these two types of soil water measurements. Researchers, land managers, and engineers may use a combination of several of these methods to study the storage and movement of water in soil, manage irrigation systems, and predict the physical behavior of soils.

Volumetric Water Content

The **volumetric water content** θ_v is defined as the volume of water associated with a given volume (usually 1 m³) of dry soil (see Figure 5.11). A comparable expression is the **mass water content** θ_m, or the mass of water associated with a given mass (usually 1 kg) of dry soil. Both of these expressions have advantages for different uses. In most cases we shall use the volumetric water content θ_v in this text.

As compaction reduces total porosity, it also increases θ_v (assuming a given θ_m), therefore often leaving too little air-filled pore space for optimal root activity. However, if a soil is initially very loose and highly aggregated (such as the forested A horizons described in Figure 5.13), moderate compaction may actually benefit plant growth by increasing the volume of pores that hold water between 10 and 1500 kPa of tension.

We think of plant root systems as exploring a certain depth of soil. We measure precipitation (and sometimes irrigation) as a depth of water (e.g., mm of rain). For such reasons, it is often convenient to express the volumetric water content as a *depth ratio* (depth of water per unit depth of soil). Conveniently, the numerical values for these two expressions are the same. For example, for a soil containing 0.1 m³ of water per m³ of soil (10% by volume) the depth ratio of water is 0.1 m of water per m of soil depth (see also Section 5.9).[4]

Measuring Water Content

GRAVIMETRIC METHOD. The gravimetric method is a direct measurement of soil water content and is therefore the standard method by which all indirect methods are calibrated. The water associated with a given mass (and, if the bulk density of the soil is known, a given volume) of dry soil solids is determined. A sample of moist soil is weighed and

---

[4] When measuring amounts of water added to soil by irrigation, it is customary to use units of volume such as m³ and hectare-meter (the volume of water that would cover a hectare of land to a depth of 1 m). Generally, farmers and ranchers in the irrigated regions of the United States use the English units ft³ or acre-foot (the volume of water needed to cover an acre of land to a depth of 1 ft).

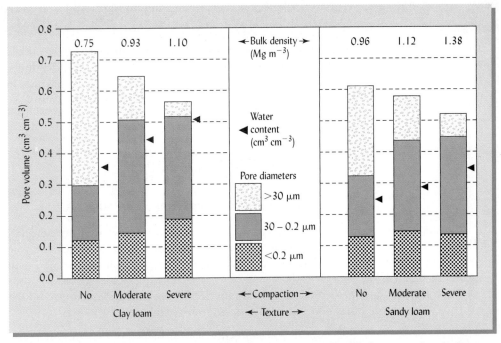

FIGURE 5.13 The compaction of two soils decreased total porosity mainly by converting the largest (usually air-filled) pores into smaller pores that hold water more tightly. These forested A horizon soils were initially so loose that the moderate compaction benefited plants by increasing the volume of water-holding 0.2 to 30 μm pores. On the other hand, the water originally in the uncompacted soil takes up a greater percentage (◀ indicates cm water/cm soil) of the pore volume when the soil is compacted, possibly leading to nearly water-saturated conditions. For example, here the clay loam with severe compaction contains 0.52 cm^3 water, but only 0.04 cm^3 air per cm^3 soil, less than the 0.10 cm^3 air per cm^3 soil (≈10% air porosity; see Section 7.2) thought to be required for good plant growth. [Adapted from Shestak and Busse (2005) with permission of The Soil Science Society of America]

then dried in an oven at a temperature of 105 °C for about 24 hours,[5] and finally weighed again. The weight loss represents the soil water. Box 5.2 provides examples of how θ_v and θ_m can be calculated. The gravimetric method is a *destructive* method (a soil sample must be removed for each measurement) and cannot be automated, thereby making it poorly suited to monitoring changes in soil moisture. Several indirect methods of measuring soil water content are nondestructive, easily automated, and very useful in the field (see Table 5.2).

NEUTRON SCATTERING. A **neutron scattering** probe, which is lowered into the soil via a previously installed access tube, contains a source of fast neutrons and a detector for slow neutrons. When fast neutrons collide with hydrogen atoms (most of which are part of water molecules), the neutrons slow down and scatter. The number of slow neutrons counted by a detector corresponds to the soil water content (see Table 5.2).

ELECTROMAGNETIC METHODS. A widely used electromagnetic method is **time-domain reflectometry** (TDR), which measures two parameters: (1) the time it takes for an electromagnetic impulse to travel down two or three parallel metal transmission rods (wave guides) buried in the soil, and (2) the degree of dissipation of the impulse as it impacts with the soil at the end of the lines. The transit time is related to the soil's apparent dielectric constant, which in turn is proportional to the amount of water in the soil. The dissipation of the signal is related to the level of salts in the soil solution. Thus, both soil *moisture content* and *salinity* can be measured using TDR.

The TDR wave guides may be portable (inserted into the soil for each reading) or may be installed in the soil at various depths and connected by wire to a meter or

---

[5] Enough drying time must be allowed so that the soil has stopped losing water and has reached a constant weight. To save time, a microwave oven may be used. A dozen small samples of soil (about 20 g each) in glass beakers may be dried on a turntable in a 1000-W microwave oven using three or more consecutive 3-minute periods, stirring the soil between periods.

BOX 5.2 GRAVIMETRIC DETERMINATION OF SOIL WATER CONTENT

The gravimetric procedures for determining mass soil water content θ_m are relatively simple. Assume that you want to determine the water content of a 100-g sample of moist soil. You dry the sample in an oven kept at 105 °C and then weigh it again. Assume that the dried soil now weighs 70 g, which indicates that 30 g of water has been removed from the moist soil. Expressed in kilograms, this is 30 kg water associated with 70 kg dry soil.

Since the mass soil water content θ_m is commonly expressed in terms of kg water associated with 1 kg dry soil (not 1 kg of wet soil), it can be calculated as follows:

$$\frac{30 \text{ kg water}}{70 \text{ kg dry soil}} = \frac{X \text{ kg water}}{1 \text{kg dry soil}}$$

$$X = \frac{30}{70} = 0.428 \text{ kg water/kg dry soil} = \theta_m$$

To calculate the volume soil water content θ_v, we need to know the bulk density of the dried soil, which in this case we shall assume to be 1.3 Mg/m³. In other words, a cubic meter of this soil (*when dry*) has a mass of 1300 kg. From the above calculations we know that the mass of water associated with this 1300 kg of dry soil is 0.428×1300 or 556 kg.

Since 1 m³ of water has a mass of 1000 kg, the 556 kg of water will occupy 556/1000 or 0.556 m³.

Thus, the volume water content is 0.556 m³/m³ of dry soil:

$$\frac{1300 \text{ kg soil}}{\text{m}^3 \text{ soil}} \times \frac{\text{m}^3 \text{ water}}{1000 \text{ kg water}} \times \frac{0.428 \text{ kg water}}{\text{kg soil}} = \frac{0.556 \text{ m}^3 \text{ water}}{\text{m}^3 \text{ soil}}$$

Assuming a soil that does not swell when wet, the relationship between the mass and volume water contents can be summarized as:

$$\theta_v = D_b \times \theta_m \tag{5.7}$$

TABLE 5.2 **Some Methods of Measuring Soil Water**

More than one method may be needed to cover the entire range of soil moisture conditions.

Method	Measures soil water Content	Potential	Useful range, kPa	Used mainly in Field	Lab	Comments
1. Gravimetric	x		0 to <−10,000		x	Destructive sampling; slow (1 to 2 days) unless microwave used. The standard for calibration.
2. Neutron scattering	x		0 to <−1,500	x		Radiation permit needed; expensive equipment; not good in high-organic-matter soils; requires access tube.
3. Time domain reflectometry (TDR)	x		0 to <−10,000	x	x	Can be automated; accurate to ±1 to 2% volumetric water content; very sandy or salty soils need calibration; requires wave guides; expensive instrument.
4. Capacitance sensors	x		0 to <−1,500	x		Can be automated; accurate to ±2 to 4% volumetric water content; sands or salty soils need calibration; simple, inexpensive sensors and recording instruments.
5. Resistance blocks		x	−90 to <−1,500	x		Can be automated; not sensitive near optimum plant water contents, may need calibration.
6. Tensiometer		x	0 to −85	x		Can be automated; accurate to ±0.1 to 1 kPa; limited range; inexpensive; needs periodic servicing to add water.
7. Thermocouple psychrometer		x	50 to <−10,000	x	x	Moderately expensive; wide range; accurate only to ±50 kPa.
8. Pressure membrane apparatus		x	50 to <−10,000		x	Used with gravimetric method to construct drier part of water characteristic curve.
9. Tension table		x	0 to −50		x	Used with gravimetric method to construct wetter part of water characteristic curve.

FIGURE 5.14 Instrumental measurement of soil water content using time domain reflectometry (TDR). The instrument sends a pulse of electromagnetic energy down the two parallel metal rods of a waveguide that the soil scientist is pushing into the soil (*inset photo*). The TDR instrument makes precise picosecond measurements of the speed at which the pulse travels down the rods, a speed influenced by the nature of the surrounding soil. Microprocessors in the instrument analyze the wave patterns generated and calculate the apparent dielectric constant of the soil. Since the dielectric constant of a soil is mainly influenced by its water content, the instrument can accurately convert its measurements into volumetric water content of the soil. (Photos courtesy of R. Weil)

data logger. The TDR instrument incorporates sophisticated electronics and computer software capable of measuring and interpreting minute voltage changes over precise picosecond time intervals (Figure 5.14). While quite expensive, the TDR instrument can be used (without repeated calibration) in most types of soils to obtain accurate readings for the entire range of soil water contents.

CAPACITANCE METHODS. By measuring the rate of change of voltage along a thin metal rod, a capacitance sensor determines the dielectric constant of the soil in which it is embedded. Because water has a far greater dielectric constant (~80) than does mineral soil (~4) or air (~1), variations in measured dielectric constant are mainly due to variations in volumetric water content of the soil. Capacitance sensors are less expensive than neutron or TDR probes and simpler to use. They also do not use hazardous radiation (as does the neutron probe). The sensors are normally accurate to within 3 to 5%, but changes in temperature and salinity, as well as air gaps that occur in very sandy or gravelly soils, can require special calibration of the probe.

Measuring Soil Water Potentials

TENSIOMETERS. The tenacity with which water is attracted to soil particles is an expression of matric water potential ψ_m. Field **tensiometers** (Figure 5.15) measure this attraction or *tension*. The tensiometer is basically a water-filled tube closed at the bottom with a porous ceramic cup and at the top with an airtight seal. Once placed in the soil, water in the tensiometer moves through the porous cup into the adjacent soil until the water potential in the tensiometer is the same as the matric water potential in the soil. As the water is drawn out, a vacuum develops under the top seal, which can be measured by a vacuum gauge or an electronic transducer. If rain or irrigation rewets the soil, water will enter the tensiometer through the ceramic tip, reducing the vacuum or tension recorded by the gauge.

Tensiometers are useful between 0 and −85 kPa potential, a range that includes half or more of the water stored in most soils. Laboratory tensiometers called **tension plates** operate over a similar range of potentials. As the soil dries beyond −80 to −85 kPa, tensiometers fail because air is drawn in through the pores of the ceramic, relieving the vacuum. A solenoid switch can be fitted to a field tensiometer in order to automatically turn an irrigation system on and off.

THERMOCOUPLE PSYCHROMETER. Since plant roots must overcome both matric and osmotic forces when they draw water from the soil, there is sometimes a need for an instrument that measures both. The relative humidity of soil air is affected by both matric and

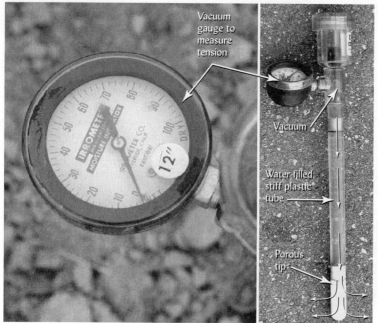

FIGURE 5.15 Tensiometer used to determine water potential in the field. The side view (*right*) shows the entire instrument. The tube is filled with water through the screw-off top. Once the instrument is tightly sealed, the white porous tip and the lower part of the plastic tube is inserted into a snug-fitting hole in the soil. The vacuum gauge (*close up, left*) will directly indicate the tension or negative potential generated as the soil draws the water out (curved arrows) through the porous tip. Note the scale goes up to only 100 centibars (= 100 kPa) tension at the driest. (Photos courtesy of R. Weil)

osmotic forces, for both constrain the escape of water molecules from liquid water. In a thermocouple psychrometer, a voltage generated by the evaporation of a water drop is converted into a readout of soil water potential ($\psi_o + \psi_m$). The thermocouple psychrometer is most useful in relatively dry soils in which imprecisions of ±50 kPa involve negligible quantities of water.

PRESSURE MEMBRANE APPARATUS.   A pressure membrane apparatus (Figure 5.16) is used to subject soils to matric potentials as low as −10,000 kPa. After application of a specific matric potential to a set of soil samples, their soil water contents are determined gravimetrically. This important laboratory tool makes possible accurate measurement of water content over a wide range of matric potentials in a relatively short time. It is used, along with the tension plate, to obtain data to construct soil water characteristic curves such as those shown in Figure 5.11.

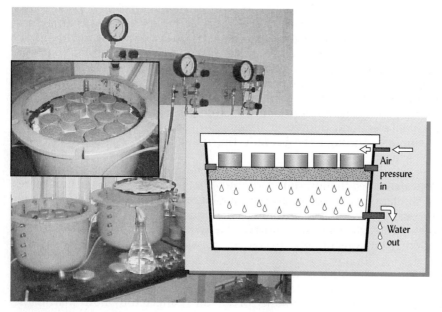

FIGURE 5.16 Pressure membrane apparatus used to determine relationships between water content and matric potential in soils. An outside source of gas creates a pressure inside the sealed chamber. Water is forced out of the soil through a porous plate (*see inset diagram*) into a cell at atmospheric pressure. The inset photo (*upper left*) shows a top view of soil samples contained in metal rings set on the porous plate before bolting on the cover. The pressure applied when the downward flow of water ceases reflects the water potential in the soil. This apparatus will measure much lower soil water potential values (drier soils) than will tensiometers or tension plates. (Photos and diagram courtesy of R. Weil)

FIGURE 5.17 A cutaway view of a commercial gypsum electrical resistance block installed about 45 cm below the soil surface. Thin wires lead from the block to the surface, where they can be connected to a special resistance meter. In the inset, another gypsum block has been broken open to reveal two concentric metal screen cylinders that serve as the electrodes between which moistened gypsum conducts a small electric current. The resistance to current flow is inversely proportional to the wetness of the gypsum block. (Photos courtesy of R. Weil)

ELECTRICAL RESISTANCE BLOCKS. Electrical resistance blocks are made of porous gypsum, nylon, or fiberglass, suitably embedded with electrodes. When placed in moist soil, the block absorbs water in proportion to the soil water content. The resistance to flow of electricity between the embedded electrodes decreases proportionately (Figure 5.17). These devices must be calibrated for each soil and the accuracy and range of soil moisture contents measured are limited (Table 5.2). However, they are inexpensive and can be used to measure approximate changes in soil moisture during one or more growing seasons. It is possible to connect them to data loggers or electronic switches so that irrigation systems can be turned on and off automatically at set soil moisture levels.

5.5 THE FLOW OF LIQUID WATER IN SOIL

All about groundwater flow:
http://environment.uwe.ac.uk/geocal/SoilMech/water/index.htm

Three types of water movement within the soil are recognized: (1) saturated flow, (2) unsaturated flow, and (3) vapor movement. In all cases water flows in response to energy gradients, with water moving from a zone of higher to one of lower water potential. *Saturated flow* takes place when the soil pores are completely filled (or saturated) with water. *Unsaturated flow* occurs when the larger pores in the soil are filled with air, leaving only the smaller pores to hold and transmit water. *Vapor movement* occurs as vapor pressure differences develop in relatively dry soils.

Saturated Flow Through Soils

Under some conditions, at least part of a soil profile may be completely saturated; that is, all pores, large and small, are filled with water. The lower horizons of poorly drained soils are often saturated, as are portions of well-drained soils above stratified layers of clay. During and immediately following a heavy rain or irrigation, pores in the upper soil zones are often filled entirely with water.

The quantity of water per unit of time Q/t that flows through a column of saturated soil (Figure 5.18) can be expressed by Darcy's law, as follows:

$$\frac{Q}{t} = AK_{\text{sat}}\frac{\Delta\psi}{L}$$

(5.8)

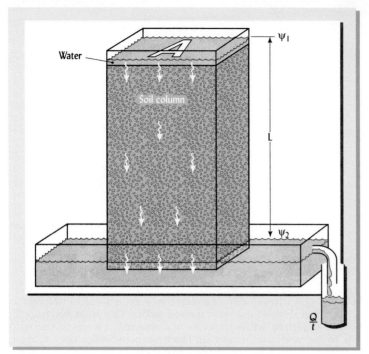

Water

Soil column

ψ_1

L

ψ_2

$\frac{Q}{t}$

FIGURE 5.18 Saturated flow (percolation) in a column of soil with cross-sectional area A, cm^2. All soil pores are filled with water. At lower right, water is shown running off into a container to indicate that water is actually moving down the column. The force driving the water through the soil is the water potential gradient, $\psi_1 - \psi_2/L$, where both water potentials and length are expressed in cm (see Table 5.1). If we measure the quantity of water flowing out Q/t as cm^3/s we can rearrange Darcy's law (from page 191) to calculate the saturated hydraulic conductivity of the soil K_{sat} in cm/s as:

$$K_{sat} = \frac{Q}{A \cdot t} \frac{L}{\psi_1 - \psi_2} \qquad (5.9)$$

Remember that the same principles apply where the water potential gradient moves the water in a horizontal direction.

where A is the cross-sectional area of the column through which the water flows, K_{sat} is the **saturated hydraulic conductivity**, $\Delta\psi$ is the change in water potential between the ends of the column (for example, $\psi_1 - \psi_2$), and L is the length of the column. For a given column, the rate of flow, is determined by the ease with which the soil transmits water (K_{sat}) and the amount of force driving the water, namely the **water potential gradient** $\Delta\psi/L$. For saturated flow, this force may also be called the **hydraulic gradient**. By analogy, think of pumping water through a garden hose, with K_{sat} representing the size of the hose (water flows more readily through a larger hose) and $\Delta\psi/L$ representing the size of the pump that drives the water through the hose.

The units in which K_{sat} is measured are length/time, typically cm/s or cm/h. The K_{sat} is an important property that helps determine how well a soil or soil material will perform in such uses as irrigated cropland, sanitary landfill cover material, wastewater storage lagoon lining, and septic tank drain field (Table 5.3).

It should not be inferred from Figure 5.18 that saturated flow occurs only down the profile. The hydraulic force can also cause horizontal and even upward flow, as occurs

TABLE 5.3 **Some Approximate Values of Saturated Hydraulic Conductivity (in Various Units) and Interpretations for Soil Uses**

K_{sat}, cm/s	K_{sat}, cm/h	K_{sat}, in/h	Comments
1×10^{-2}	36	14	Typical of beach sand.
5×10^{-3}	18	7	Typical of very sandy soil, too rapid to effectively filter pollutants in wastewater.
5×10^{-4}	1.8	0.7	Typical of moderately permeable soils, K_{sat} between 1.0 and 15 cm/h considered suitable for most agricultural, recreational, and urban uses calling for good drainage.
5×10^{-5}	0.18	0.07	Typical of fine-textured, compacted, or poorly structured soils. Too slow for proper operation of septic tank drain fields, most types of irrigation, and many recreational uses such as playgrounds.
$<1 \times 10^{-8}$	$<3.6 \times 10^{-5}$	$<1.4 \times 10^{-5}$	Extremely slow; typical of compacted clay. K_{sat} of 10^{-5} to 10^{-8} cm/h may be required where nearly impermeable material is needed, as for wastewater lagoon lining or landfill cover material.

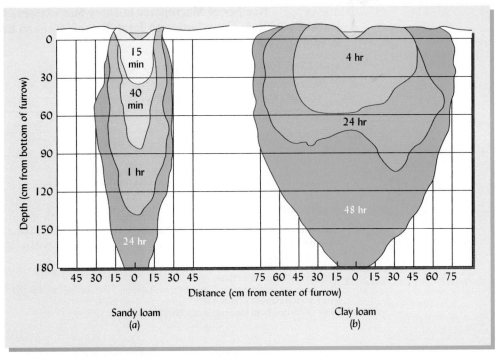

FIGURE 5.19 Comparative rates of irrigation water movement into a sandy loam and a clay loam. Note the much more rapid rate of movement in the sandy loam, especially in a downward direction. [Redrawn from Cooney and Peterson (1955)]

when groundwater wells up under a stream (see Section 6.6). The rate of such flow is usually not quite as rapid, however, since the force of gravity does not assist horizontal flow and hinders upward flow. Downward and horizontal flow is illustrated in Figure 5.19, which records the flow of water from an irrigation furrow into two soils, a sandy loam and a clay loam. The water moved down much more rapidly in the sandy loam than in the clay loam. On the other hand, horizontal movement (which would have been largely by unsaturated flow) was much more evident in the clay loam.

Factors Influencing the Hydraulic Conductivity of Saturated Soils

MACROPORES. Any factor affecting the size and configuration of soil pores will influence hydraulic conductivity. The total flow rate in soil pores is proportional to the fourth power of the radius. Thus, flow through a pore 1 mm in radius is equivalent to that in 10,000 pores with a radius of 0.1 mm (even though it takes only 100 pores of radius 0.1 mm to give the same cross-sectional area as a 1 mm radius pore). As a result, macropores (radius > 0.08 mm) account for nearly all water movement in saturated soils (see Table 5.4). However, air trapped in rapidly wetted soils can block pores and thereby reduce hydraulic conductivity. Similarly, the *interconnectedness* of pores is important as non-interconnected pores are like "dead-end streets" to flowing water. Vesicular pores in certain desert soils are examples (Plate 55).

The presence of biopores, such as root channels and earthworm burrows (typically >1 mm in radius), has a marked influence on the saturated hydraulic conductivity of different soil horizons (Table 5.5 and Plate 81). Because they usually have more macropore space, sandy soils generally have higher saturated conductivities than finer-textured soils. Likewise, soils with stable granular structure conduct water much more rapidly than do those with unstable structural units, which break down upon being wetted. Saturated conductivity of soils under perennial vegetation is commonly much higher than where annual plants have been grown (Figure 5.20).

PREFERENTIAL FLOW. Scientists have been surprised to find more extensive pollution of groundwater from pesticides and other toxicants than would be predicted from traditional hydraulic conductivity measurements that assume uniform soil porosity.

TABLE 5.4 Number of Macropores in Three Size Classes, Their Proportion of the Soil Porosity and Their Contribution to Total Water Flow in an Irrigated Maize Field in Southern Portugal[a]

Most of the flow took place through the largest class of pores even though the smaller pores were far greater in number and in percent of the total soil porosity. Note that only 5.5% of the total soil porosity is considered here, the other 94.5% being comprised of pores smaller than 0.1 mm.

	Effective pore radius (mm)[b]		
	>0.5	0.5 to 0.25	0.25 to 0.1
	Large macropores	Small macropores[c]	
Number of pores/m^2	235	167	2200
% of effective porosity	3.1	------ 2.4 ------	
% of flow	88	9	3

[a] The data pertain to the upper 30 cm of a Fluvent with a silt loam texture. Maize was planted on permanent ridges with furrow irrigation and minimum tillage.
[b] A disk infiltrometer was used to maintain tensions of 0.3, 0.6, and 1.5 kPa to exclude flow in pores with diameters >0.5, 0.25, and 0.1 mm, respectively.
[c] Although the original paper refers to these pores as "mesopores," we use the term "small macropores," to be consistent with the classes in Table 4.6.
[Data calculated from Cameira et al. (2003)]

Apparently solutes (dissolved substances) are carried downward rapidly by water that moves through large macropores such as cracks and biopores, often before the bulk of the soil is thoroughly wetted. Mounting evidence suggests that this type of nonuniform water movement, referred to as **preferential flow**, greatly increases the chances of groundwater pollution (see Figure 5.21).

Macropores with continuity from the soil surface down through the profile encourage preferential flow. Burrowing animals (e.g., worms, rodents, and insects) as well as decayed plant roots leave tubular channels through which water can flow rapidly. In very sandy soils, hydrophobic organic coatings on sand grains repel water, preventing it from soaking in uniformly. Where these coatings are absent or wear off, water rapidly enters and produces "fingers" of rapid wetting (see Plate 69). This "finger flow" probably is responsible for the finger-like shapes of the spodic horizon in some Spodosol profiles (e.g., Figure 3.29 and Plate 10). Fingering also occurs in stratified sandy layers that underlie finer-textured materials (see Section 5.6).

Preferential flow in fine-textured soils is enhanced by clay shrinkage, which leaves open cracks and fissures that can extend down into the lower subsoil horizons. In some clay soils, water from the first rain storm after a dry spell moves rapidly down the profile, carrying with it soluble pesticides or nutrients that may be on the soil surface (Table 5.6). Recognition that heterogeneity of field soils leads to preferential flow is stimulating more aggressive approaches to controlling groundwater contamination. When agri-chemicals and animal manures are applied to the soil surface, transport of

TABLE 5.5 The Saturated Hydraulic Conductivity K_{sat} and Related Properties of Various Horizons in a Typic Hapludult Profile

The upper horizons had many biopores (mainly earthworm burrows), which resulted in high values of K_{sat} as well as extreme variability from sample to sample. The presence of a clay-enriched argillic horizon resulted in reduced K_{sat} values. Apparently most large biopores in this soil did not extend below 30 cm.

Horizon	Depth, cm	Clay, %	Bulk density, Mg/m^3	Mean K_{sat}, cm/h	Range of K_{sat} values[a], cm/h
Ap	0–15	12.6	1.42	22.4	0.80–70
E	15–30	11.1	1.44	7.9	0.50–24
E/B	30–45	14.5	1.47	0.93	0.53–1.33
Bt	45–60	22.2	1.40	0.49	0.19–0.79
Bt	60–75	27.2	1.38	0.17	0.07–0.27
Bt	75–90	24.1	1.28	0.04	0.01–0.07

[a] For each soil layer K_{sat} was determined on 5 soil cores of 7.5 cm diameter.
Data from Waddell and Weil (1996).

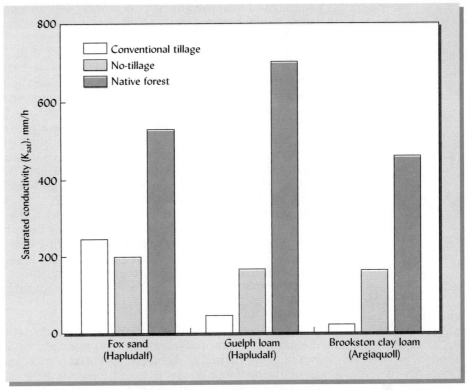

FIGURE 5.20 The effect of land management and soil texture on saturated conductivity (K_{sat}) of three soils in Canada. Soils under native woodlots had higher K_{sat} values, apparently due to higher organic matter contents and to preferential flow channels provided by decayed roots and burrowing animals. Tillage practices had little effect on conductivity in sand, but in loam and clay loam soils, conductivity was higher where no-tillage systems had been used, suggesting that no-till had increased the proportion of larger, water-conducting pores. [Drawn from averages of three methods of measuring K_{sat} in Reynolds et al. (2000)]

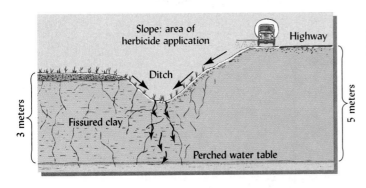

FIGURE 5.21 An illustration of preferential flow of water and pesticides downward to the water table. An herbicide (weed killer) was applied alongside a highway (*right*) with the expectation that downward movement into the water table would not be a serious problem since the surrounding soils were fine-textured and would not be expected to readily permit infiltration of the chemical. As the deep rooted vegetation dried the soil, wide cracks formed in the swelling clay. Because of these cracks, the first heavy rain after a dry spell carried the chemicals into the groundwater before the soil could swell and shut the cracks. Through the groundwater the herbicide could move into nearby streams. [From DeMartinis and Cooper (1994) with permission of Lewis Publishers]

TABLE 5.6   Leaching of Pesticides by Preferential Flow in a Slowly Permeable Alfisol

Most of the spring leaching of three widely used pesticides took place following the first major storm of the year.

	Leaching of pesticide applied, % (3-year average)		
Chemical	First storm	Spring season	First storm/spring season total, %
Carbofuran	0.22	0.25	88
Atrazine	0.037	0.053	68
Cyanazine	0.02	0.02	100

Calculated from Kladivco et al. (1999).

chemicals and fecal bacteria by preferential flow can threaten human health, as well as environmental quality.

Unsaturated Flow in Soils

Most of the time, water movement takes place when upland soils are *unsaturated*. Such movement occurs in a more complicated environment than that which characterizes saturated water flow. In saturated soils, essentially all the pores are filled with water, although the most rapid water movement is through the large and continuous pores. But in unsaturated soils, these macropores are filled with air, leaving only the finer pores to accommodate water movement. Also, in unsaturated soils the water content and, in turn, the tightness with which water is held (water potential) can be highly variable. This influences the rate and direction of water movement and also makes it more difficult to measure the flow of soil water.

As was the case for saturated water movement, the driving force for unsaturated water flow is differences in water potential. This time, however, the difference in the matric potential, not gravity, is the primary driving force. This **matric potential gradient** is the difference in the matric potential of the moist soil areas and nearby drier areas into which the water is moving. Movement will be from a zone of thick moisture films (high matric potential, e.g., −1 kPa) to one of thin films (lower matric potential, e.g., −100 kPa).

How soil texture affects hydraulic properties: http://www.pedosphere.com/resources/texture/triangle_us.cfm

INFLUENCE OF TEXTURE. Figure 5.22 shows the general relationship between matric potential ψ_m (and, in turn, water content) and hydraulic conductivity of a sandy loam and clay soil. Note that at or near zero potential (which characterizes the saturated flow region), the hydraulic conductivity is thousands of times greater than at potentials that characterize typical unsaturated flow (−10 kPa and below).

At high potential levels (high moisture contents), hydraulic conductivity is higher in the sand than in the clay. The opposite is true at low potential values (low moisture contents). This relationship is to be expected because the sandy soil contains many large pores that are water-filled when the soil water potential is high (and the soil is quite wet), but most of these have been emptied by the time the soil water potential becomes lower than about −10 kPa. The clay soil has many more micropores that are still water-filled at lower soil water potentials (drier soil conditions) and can participate in unsaturated flow.

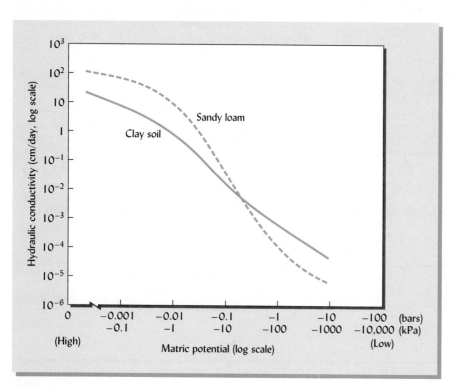

FIGURE 5.22 Generalized relationship between matric potential and hydraulic conductivity for a sandy soil and a clay soil (note log scales). Saturated flow takes place at or near zero potential, while much of the unsaturated flow occurs at a potential of −0.1 bar (−10 kPa) or below.

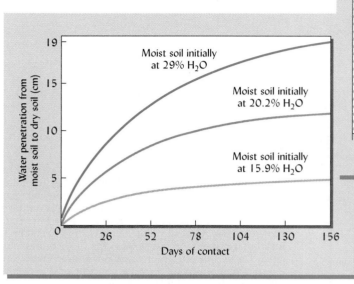

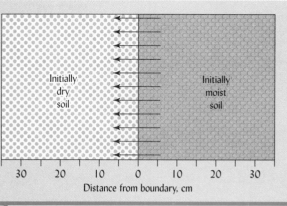

FIGURE 5.23 Water movement into a drier soil from a wetter one (*inset*). The greater the initial water content of the wetter soil, the greater the water potential gradient between the two soils and the more rapid the water movement between them (*graph*). Water adjustment between two slightly moist soils at about the same water content will be exceedingly slow. [Gardner and Widtsoe (1921)]

The influence of the magnitude of the potential gradient on water movement is illustrated by Figure 5.23. Laboratory measurements made on three moist soil samples adjacent to a dry soil show that the higher the water content in the moist soil, the greater the matric potential gradient between the moist and dry soil and, in turn, the more rapid the flow. The curves tend to level off over time as the water moves into the dry soil, because the distance *L* between the wettest soil and the as-yet-unwetted soil increases, reducing the potential gradient $\Delta\psi/L$.

5.6 INFILTRATION AND PERCOLATION

A special case of water movement is the entry of free water into the soil at the soil–atmosphere interface. As we shall explain in Chapter 6, this is a pivotal process in landscape hydrology that greatly influences the moisture regime for plants and the potential for soil degradation, chemical runoff, and down-valley flooding. The source of free water at the soil surface may be rainfall, snowmelt, or irrigation.

Infiltration

Infiltration and water budget:
http://www.scas.cit.cornell.edu/faculty/hmv1/watrshed.wtsd.htm

The process by which water enters the soil pore spaces and becomes soil water is termed *infiltration*, and the rate at which water can enter the soil is termed the *infiltrability i*:

$$i = \frac{Q}{A * t} \qquad (5.10)$$

where Q is the volume quantity of water (m³) infiltrating, A is the area of the soil surface (m²) exposed to infiltration, and t is time (s). Since m³ appears in the numerator and m² in the denominator, the units of infiltration can be simplified to m/s or, more commonly, cm/h. The infiltration rate is not constant over time, but generally decreases during an irrigation or rainfall episode. If the soil is quite dry when infiltration begins, all the macropores open to the surface will be available to conduct water into the soil. In soils with expanding types of clays, the initial infiltration rate may be particularly high as water pours into the network of shrinkage cracks. However, as infiltration proceeds, many macropores fill with water, and shrinkage cracks close up. The infiltration rate declines sharply at first and then tends to level off, remaining fairly constant thereafter (Figure 5.24).

MEASUREMENT. The infiltration capacity of a soil may be easily measured using a simple device known as a **double ring infiltrometer**. Two heavy metal cylinders, one smaller in diameter than the other, are pressed partially into the soil so that the smaller is

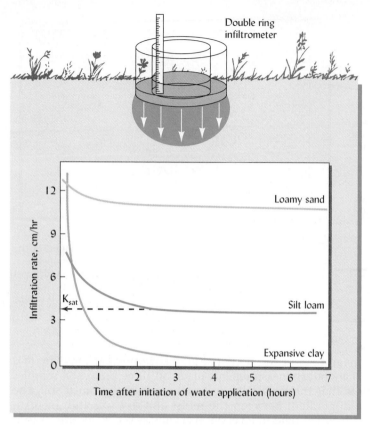

FIGURE 5.24 The potential rate of water entry into the soil, or infiltration capacity, can be measured by recording the drop in water level in a double ring infiltrometer (*top*). Changes in the infiltration rate of several soils during a period of water application by rainfall or irrigation are shown (*bottom*). Generally, water enters a dry soil rapidly at first, but its infiltration rate slows as the soil becomes saturated. The decline is least for very sandy soils with macropores that do not depend on stable structure or clay shrinkage. In contrast, a soil high in expansive clays may have a very high initial infiltration rate when large cracks are open, but a very low infiltration rate once the clays swell with water and close the cracks. Most soils fall between these extremes, exhibiting a pattern similar to that shown for the silt loam soil. The dashed arrow indicates the level of K_{sat} for the silt loam illustrated. (Diagram courtesy of R. Weil)

inside the larger (see Figure 5.24). A layer of cheesecloth is placed inside the rings to protect the soil surface from disturbance, and water is poured into both cylinders. The depth of water in the central cylinder is then recorded periodically as the water infiltrates the soil. The water infiltrating in the outer cylinder is not measured, but it ensures that the surrounding soil will be equally moist and that the movement of water from the central cylinder will be principally downward, not horizontal.

Percolation

Infiltration is a transitional phenomenon that takes place at the soil surface. Once the water has infiltrated the soil, the water moves downward into the profile by the process termed **percolation**. Both saturated and unsaturated flow are involved in percolation of water down the profile, and rate of percolation is related to the soil's hydraulic conductivity. In the case of water that has infiltrated a relatively dry soil, the progress of water movement can be observed by the darkened color of the soil as it becomes wet (Figure 5.25). There usually appears to be a sharp boundary, termed a **wetting front**, between the dry underlying soil and the soil already wetted. During an intense rain or heavy irrigation, water movement near the soil surface occurs mainly by saturated flow in response to gravity. At the wetting front, however, water is moving into the underlying drier soil in response to matric potential gradients as well as gravity. During a light rain, both infiltration and percolation may occur mainly by unsaturated flow as water is drawn by matric forces into the fine pores without accumulating at the soil surface or in the macropores.

Water Movement in Stratified Soils

The fact that, at the wetting front, water is moving by unsaturated flow has important ramifications for how percolating water behaves when it encounters an abrupt change in pore sizes due to such layers as fragipans or claypans, or sand and gravel lenses.

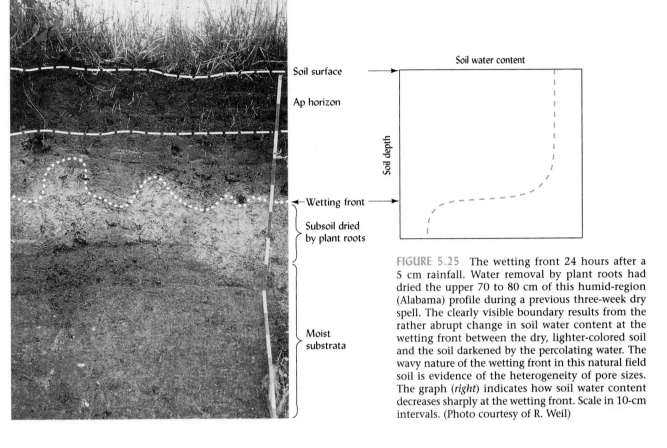

FIGURE 5.25 The wetting front 24 hours after a 5 cm rainfall. Water removal by plant roots had dried the upper 70 to 80 cm of this humid-region (Alabama) profile during a previous three-week dry spell. The clearly visible boundary results from the rather abrupt change in soil water content at the wetting front between the dry, lighter-colored soil and the soil darkened by the percolating water. The wavy nature of the wetting front in this natural field soil is evidence of the heterogeneity of pore sizes. The graph (*right*) indicates how soil water content decreases sharply at the wetting front. Scale in 10-cm intervals. (Photo courtesy of R. Weil)

In some cases, such pore-size stratification may be created by soil managers, as when coarse plant residues are plowed under in a layer or a layer of gravel is placed under finer soil in a planting container. In all cases, the effect on water percolation is similar—that is, the downward movement is impeded—even though the causal mechanism may vary. The contrasting layer acts as a barrier to water flow and results in much higher field-moisture levels above the barrier than what would normally be encountered in freely drained soils (Figure 5.26). It is not surprising that percolating water should slow down markedly when it reaches a layer with finer pores, which therefore

FIGURE 5.26 One result of soil layers with contrasting texture. This North Carolina soil has about 50 cm of loamy sand coastal plain material atop deeper layers of silty clay loam-textured material derived from the Piedmont. Rainwater rapidly infiltrates the sandy surface horizons, but its downward movement is arrested at the finer-textured layer, resulting in saturated conditions near the surface and a quicksand-like behavior. (Photo courtesy of R. Weil)

has a lower hydraulic conductivity. However, the fact that a layer of *coarser* pores will temporarily stop the movement of water may not be obvious.

In Figure 5.27, a layer of coarse sand underlies a fine-textured soil. Intuitively, one might expect the sand layer to speed the downward percolation of water. However, the coarser sand layer has just the opposite effect; it actually impedes the flow of water. The macropores of the sand offer less attraction for the water than do the finer pores of the overlying material. Since water always moves from higher to lower potential (to where it will be held more tightly), the wetting front cannot move readily into the sand. Eventually the downward-moving water will accumulate above the sand layer (if it cannot move laterally) and nearly saturate the pores at the soil–sand interface. The matric potential of the water at the wetting front will then fall to nearly zero or even become positive. Once this occurs, the water will be so

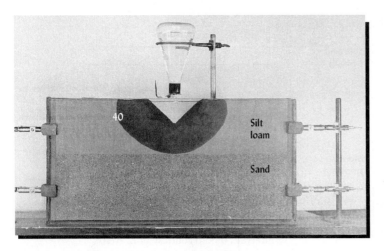

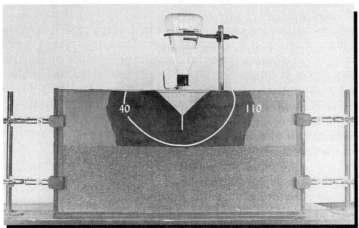

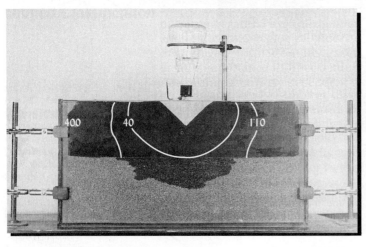

FIGURE 5.27 Downward water movement in soils with a stratified layer of coarse material. (*a*) Water is applied to the surface of a medium-textured topsoil. Note that after 40 min, downward movement is no greater than movement to the sides, indicating that in this case the gravitational force is insignificant compared to the matric potential gradient between dry and wet soil. (*b*) The *downward* movement stops when a coarse-textured layer is encountered. After 110 min, no movement into the sandy layer has occurred. The macropores of the sand provide less attraction for water than the finer-textured soil above. Only when the water content (and in turn the matric potential gradient) is raised sufficiently will the water move into the sand. (*c*) After 400 min, the water content of the overlying layer becomes sufficiently high to give a water potential of about −1 kPa or more, and downward movement into the coarse material takes place. (Courtesy W. H. Gardner, Washington State University)

BOX 5.3 PRACTICAL APPLICATIONS OF UNSATURATED WATER FLOW IN CONTRASTING LAYERS

Unsaturated flow is interrupted where soil texture abruptly changes from relatively fine to coarse. Capillary water held tightly by matric attraction in the smaller pores of the finer-textured layer cannot move into the larger pores in the underlying coarser-textured layer unless under positive pressure. That is, a larger pore cannot "pull" water from a smaller pore. In fact, unsaturated water flow always occurs in the reverse direction, from larger to smaller pores. A wetting front moving down a soil profile along the matric potential gradient therefore ceases downward movement when it encounters pores much larger than those through which it has been traversing. Instead of continuing downward, water will move laterally in the finer layer. If water is entering the system more rapidly than lateral capillarity can carry it away, a perched water table may develop above the interface between the two layers.

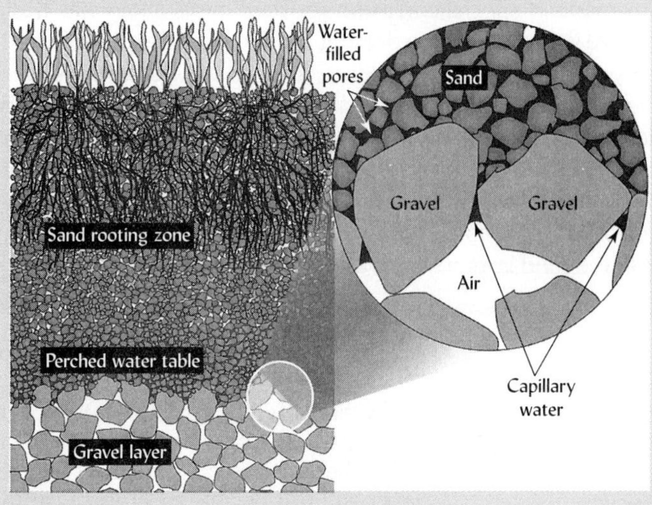

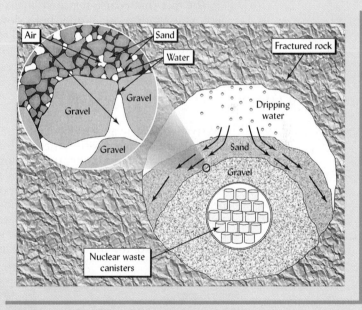

FIGURE 5.28 A gravel layer under a putting green is used to increase the water available to turfgrass roots in the sand rooting zone, while allowing rapid drainage if saturation occurs. (Diagram courtesy of R. Weil)

This phenomenon is applied in the design of soil profiles for golf course putting greens. The soil specified for the rooting zone consists almost entirely of sand in order to promote rapid infiltration of water and to resist compaction by foot traffic. However, water normally drains so fast through sand that too little is held to meet the needs of growing grass. This situation is remedied to some extent by constructing the putting green with a layer of gravel underneath the sand rooting zone. The large pores in the gravel temporarily stop the downward movement of water. The resulting perched water table (Figure 5.28) causes the sand layer to retain more water than it would otherwise, but still allows for rapid drainage of excess water—when a buildup of positive pressure allows gravitational forces to override matric forces.

FIGURE 5.29 Proposed use of capillary principles to prevent dripping water from corroding toxic nuclear waste containers inside Yucca Mountain in Nevada, thus protecting groundwater from contamination over hundreds of thousands of years. [Diagram based on Carter and Pigford (2005)]

The same principle is at the heart of a design proposed to keep nuclear wastes from contaminating groundwater during the many thousands of years required for the radionuclides to decay to harmless products. One plan is to store radioactive wastes in sealed containers kept in caverns carved from the rock deep inside Yucca Mountain, Nevada. Despite its desert location, the Yucca Mountain rock contains large amounts of water in pores and fractures, resulting in water dripping from the ceiling of the storage caverns. Although the containers for the heat-generating wastes are to be as corrosion-resistant as possible, no one believes they would remain intact if exposed to moisture and air for thousands of years. To fend off the dripping water, a huge

(continued)

canopy made of special corrosion-resistant metal alloys was designed. Construction of this canopy over the highly radioactive wastes would be extremely difficult and expensive, with no guarantee that the structure would not deteriorate over the millennia. Therefore, a far easier, less expensive, and more reliable alternative has been proposed. Burying the waste containers under mounds of first gravel and then sand (Figure 5.29) in a texture-stratified system would create a *capillary barrier* to protect the containers. Water dripping into the sand would be held by capillary forces in the relatively small pores between sand grains. As the water enters the sand, it moves by capillary flow along matric potential gradients. Its downward movement would be interrupted when it reached the much larger pores of the gravel layer. The containers would remain dry because both matric potential gradients and gravity would cause the water in the sand layer to move along the curved interface rather than into the gravel layer (arrows in Figure 5.29).

loosely held by the fine-textured soil that gravity or hydrostatic pressure will force the water into the sand layer.

Interestingly, a coarse sand layer in an otherwise fine-textured soil profile would also inhibit the *rise* of water from moist subsoil layers up to the surface soil, a situation that could be illustrated by turning Figure 5.27b upside down. The large pores in the coarse layer will not be able to support capillary movement up from the smaller pores in a finer layer. Consequently, water rises by capillarity up to the coarse-textured layer but cannot cross it to supply moisture to overlying layers. Thus, plants growing on some soils with buried gravel lenses are subject to drought since they are unable to exploit water in the lower soil layers. This principle also allows a layer of gravel to act as a capillary barrier under a concrete slab foundation to prevent water from soaking up from the soil and through the concrete floor of a home basement.

The fact that coarse-textured layers (e.g., gravel, sand, coarse organic materials, or geotextile fabrics) can hinder both downward and upward unsaturated flow of water must be considered—and may be used to advantage—when dealing with such materials in planting containers, landscape drainage schemes (see Section 6.7), or engineering works (see Box 5.3).

5.7   WATER VAPOR MOVEMENT IN SOILS

Two types of water vapor movement are associated with soils, *internal* and *external*. Internal movement takes place within the soil, that is, in the soil pores. External movement occurs at the land surface, and water vapor is lost by surface evaporation (see Section 6.4).

Water vapor moves from one point to another within the soil in response to differences in vapor pressure. Thus, water vapor will move from a moist soil where the soil air is nearly 100% saturated with water vapor (high vapor pressure) to a drier soil where the vapor pressure is somewhat lower. Also, water vapor will move from a zone of low salt content to one with a higher salt content (e.g., around a fertilizer granule). The salt lowers the vapor pressure of the water and encourages water movement from the surrounding soil.

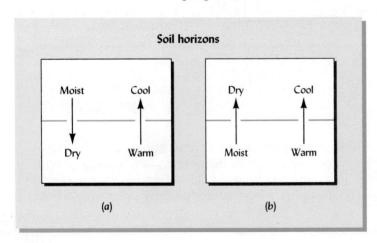

FIGURE 5.30 Vapor movement tendencies that may be expected between soil horizons differing in temperature and moisture. In (a) the tendencies more or less negate each other, but in (b) they are coordinated, and considerable vapor transfer might be possible if the liquid water in the soil capillaries does not interfere.

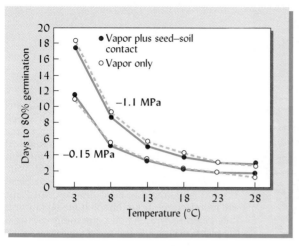

FIGURE 5.31 The rate of water vapor movement from nearby soil particles is sufficient to germinate the seeds of wheat as demonstrated in these graphs. In this experiment, either soil is firmly packed around the planted seed (vapor plus seed–soil contact), or the seed is kept from such contact by wrapping it in fiberglass cloth, permitting the movement of only water vapor to the seed (vapor only). Note that where water moves only in the vapor form, seed germination at different temperatures and at two soil water potentials is essentially the same as where there is soil–seed contact. [From Wuest, Albrecht, and Skirvin (1999)]

If the temperature of one part of a uniformly moist soil is lowered, the vapor pressure will decrease and water vapor will tend to move toward this cooler part. Heating will have the opposite effect in that heating will increase the vapor pressure, and the water vapor will move away from the heated area. Figure 5.30 illustrates these relationships.

The actual amount of water vapor in a soil at optimum moisture for plant growth is surprisingly small, being equivalent to perhaps no more than 10 L in the upper 15 cm of a hectare of a silt loam soil. This compares with some 600,000 L of liquid water in the same soil volume. Even though the amount of water vapor is small, its movement in soils can be of some practical significance. For example, seeds of some plants can absorb sufficient water vapor from the soil to stimulate germination (see Figure 5.31). Likewise, in dry soils, water vapor movement may be of considerable significance to drought-resistant desert plants (*xerophytes*), many of which can exist at extremely low soil water contents. For instance, at night the surface horizon of a desert soil may cool sufficiently to cause vapor movement up from deeper layers. If cooled enough, the vapor may then condense as dewdrops in the soil pores, supplying certain shallow-rooted xerophytes with water for survival.

5.8 QUALITATIVE DESCRIPTION OF SOIL WETNESS

The measurement of soil water potential and the observable behavior of soil water always depend on that portion of the soil water that is farthest from a particle surface and therefore has the highest potential. As an initially water-saturated soil dries down, both the soil as a whole and the soil water it contains undergo a series of gradual changes in physical behavior and in their relationships with plants. These changes are due mainly to the fact that the water remaining in the drying soil is found in smaller pores and thinner films where the water potential is lowered principally by the action of matric forces. Matric potential therefore accounts for an increasing proportion of the total soil water potential, while the proportion attributable to gravitational potential decreases.

To study these changes and introduce the terms commonly used to describe varying degrees of soil wetness, we shall follow the moisture and energy status of soil during and after wetting. The terms to be introduced describe various stages along a continuum of soil wetness and should not be interpreted to imply that soil water exists in different "forms." Because these terms are essentially qualitative and lack a precise scientific basis, some soil physicists object to their use. However, it would be a serious disservice to the reader to omit them from this text as they are widely used in practical soil management and help communicate important facts about soil water behavior.

Maximum Retentive Capacity

When all soil pores are filled with water, the soil is said to be *water-saturated* (Figure 5.32) and at its **maximum retentive capacity**. The matric potential is close to zero, nearly the same as that of free water. The volumetric water content is essentially the same as the total porosity. The soil will remain at maximum retentive capacity only so long as water continues to infiltrate, for the water in the largest pores (sometimes

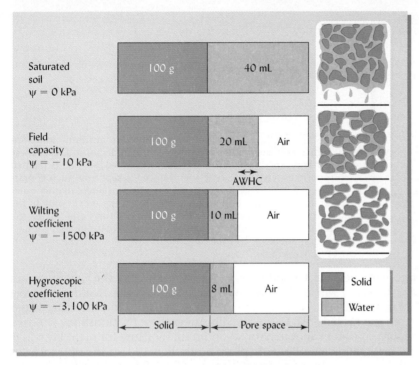

FIGURE 5.32 Volumes of water and air associated with 100 g of soil solids in a representative well-granulated silt loam. The top bar shows the situation when the soil is completely saturated with water. This situation will usually occur for short periods of time when water is being added. Water will soon drain out of the larger pores (*macropores*). The soil is then said to be at *field capacity*. Plants will remove water from the soil quite rapidly until they begin to wilt. When permanent wilting of the plants occurs, the soil water content is said to be at the *wilting coefficient*. There is still considerable water in the soil, but it is held too tightly to permit its absorption by plant roots. The water lost between field capacity and wilting coefficient is considered to be the soil's plant available water-holding capacity (AWHC). A further reduction in water content to the *hygroscopic coefficient* is illustrated in the bottom bar. At this point the water is held very tightly, mostly by the soil colloids.

termed **gravitational water**) will percolate downward, mainly under the influence of gravitational forces. Data on maximum retentive capacities and the average depth of soils in a watershed are useful in predicting how much rainwater can be stored in the soil temporarily, thus possibly avoiding downstream floods.

Field Capacity

Once the rain or irrigation has ceased, water in the largest soil pores will drain downward quite rapidly in response to the hydraulic gradient (mostly gravity). After one to three days, this rapid downward movement will become negligible as matric forces play a greater role in the movement of the remaining water (Figure 5.33). The soil then is said to be at its **field capacity**. In this condition, water has moved out of the macropores and air has moved in to take its place. The micropores or capillary pores are still filled with water and can supply plants with needed water. The matric potential will vary slightly from soil to soil but is generally in the range of −10 to −30 kPa, assuming drainage into a less-moist zone of similar porosity.[6] Water movement will continue to

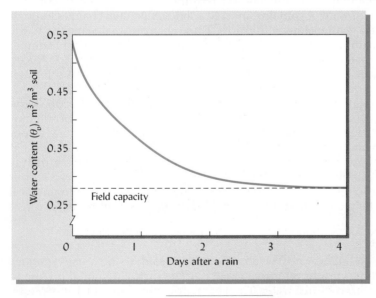

FIGURE 5.33 The water content of a soil drops quite rapidly by drainage following a period of saturation by rain or irrigation. After two or three days the rate of water drainage out of the soil is very slow and the soil is said to be at *field capacity*. (Diagram courtesy of R. Weil)

---

[6] Note that because of the relationships pertaining to water movement in stratified soils (see Section 5.6), soil in a flower pot will cease drainage while much wetter than field capacity.

take place by unsaturated flow, but the rate of movement is very slow since it now is due primarily to capillary forces, which are effective only in micropores (Figure 5.32). The water found in pores small enough to retain it against rapid gravitational drainage, but large enough to allow capillary flow in response to matric potential gradients, is sometimes termed **capillary water**.

While all soil water is affected by gravity, the term *gravitational water* refers to the portion of soil water that readily drains away between the states of maximum retentive capacity and field capacity. Most soil leaching occurs as gravitational water that drains from the larger pores before field capacity is reached. Gravitational water therefore includes much of the water that transports chemicals such as nutrient ions, pesticides, and organic contaminants into the groundwater and, ultimately, into streams and rivers.

Field capacity is a very useful term because it refers to an approximate degree of soil wetness at which several important soil properties are in transition:

1. At field capacity, a soil is holding the maximal amount of water useful to plants. Additional water, while held with low energy of retention, would be of limited use to plants because it would remain in the soil for only a short time before draining, and, while in the soil, it would occupy the larger pores, thereby reducing soil aeration. Drainage of gravitational water from the soil is generally a requisite for optimum plant growth (hydrophilic plants, such as rice or cattails, excepted).

2. At field capacity, the soil is near its lower plastic limit—that is, the soil behaves as a crumbly semisolid at water contents below field capacity, and as a plastic putty-like material that easily turns to mud at water contents above field capacity (see Section 4.9). Therefore, field capacity approximates the optimal wetness for ease of tillage or excavation.

3. At field capacity, sufficient pore space is filled with air to allow optimal aeration for most aerobic microbial activity and for the growth of most plants (see Section 7.6).

Permanent Wilting Percentage or Wilting Coefficient

Once an unvegetated soil has drained to its field capacity, further drying is quite slow, especially if the soil surface is covered to reduce evaporation. However, if plants are growing in the soil, they will remove water from their rooting zone, and the soil will continue to dry. The roots will remove water first from the largest water-filled pores where the water potential is relatively high. As these pores are emptied, roots will draw their water from the progressively smaller pores and thinner water films in which the matric water potential is lower and the forces attracting water to the solid surfaces are greater. Hence, it will become progressively more difficult for plants to remove water from the soil at a rate sufficient to meet their needs.

As the soil dries, the rate of plant water removal may fail to keep up with plant needs, and plants may begin to wilt during the daytime to conserve moisture. At first the plants will regain their turgor at night when water is not being lost through the leaves, and the roots can catch up with the plants' demand. Ultimately, however, the plants will remain wilted night and day when the roots cannot generate water potentials low enough to coax the remaining water from the soil. Although not yet dead, the plants are now in a permanently wilted condition and will die if water is not provided. For most plants this condition develops when the soil water potential ψ has a value of about -1500 kPa (-15 bars). A few plants, especially xerophytes (desert-type plants) can continue to remove water at even -1800 or -2000 kPa, but the amount of water available between -1500 kPa and -2000 kPa is very small (Figure 5.34).

The water content of the soil at this stage is called the **wilting coefficient** or **permanent wilting percentage** and by convention is taken to be that amount of water retained by the soil when the water potential is -1500 kPa. The soil will appear to be dusty dry, although some water remains in the smallest of the micropores and in very thin films (perhaps only 10 molecules thick) around individual soil particles (see Figure 5.32). As illustrated in Figure 5.34, plant **available water** is considered to be that water retained in soils between the states of field capacity and wilting coefficient (between -10 to -30 kPa and -1500 kPa). The amount of capillary water remaining in the soil that is unavailable to higher plants can be substantial, especially in fine-textured soils and those high in organic matter.

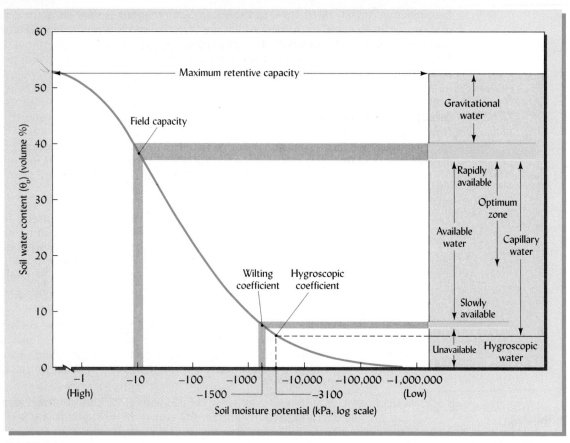

FIGURE 5.34 Water content–matric potential curve of a loam soil as related to different terms used to describe water in soils. The shaded bars in the diagram to the right suggest that measurements such as field capacity are only approximations. The gradual change in potential with soil moisture change discourages the concept of different "forms" of water in soils. At the same time, such terms as *gravitational* and *available* assist in the qualitative description of moisture utilization in soils.

Hygroscopic Coefficient

Although plant roots do not generally dry the soil beyond the permanent wilting percentage, if the soil is exposed to the air, water will continue to be lost by evaporation. When soil moisture is lowered below the wilting point, the water molecules that remain are very tightly held, mostly being adsorbed by colloidal soil surfaces. This state is approximated when the atmosphere above a soil sample is essentially saturated with water vapor (98% relative humidity) and equilibrium is established at a water potential of −3100 kPa. The water is thought to be in films only 4 or 5 molecules thick and is held so rigidly that much of it is considered nonliquid and can move only in the

TABLE 5.7 Volumetric Water Content θ_v at Field Capacity and Hygroscopic Coefficient for Three Representative Soils and the Calculated Capillary Water

Note that the clay soil retains most water at the field capacity, but much of that water is held tightly in the soil at −3100 kPa potential by soil colloids (hygroscopic coefficient).

	Volume % (θ_v)		
Soil	Field capacity −10 to −30 kPa	Hygroscopic coefficient, −3100 kPa	Capillary water, col. 1 − col. 2
Sandy loam	12	3	9
Silt loam	30	10	20
Clay	35	18	17

vapor phase. The moisture content of the soil at this point is termed the **hygroscopic coefficient**. Soils high in colloidal materials (clay and humus) will hold more water under these conditions than will sandy soils that are low in clay and humus (Table 5.7). Soil water considered to be *unavailable* to plants includes the hygroscopic water as well as that portion of capillary water retained at potentials below −1500 kPa (see Figure 5.34).

5.9 FACTORS AFFECTING AMOUNT OF PLANT-AVAILABLE SOIL WATER

The amount of soil water available for plant uptake is determined by a number of factors, including the water content–potential relationship for each soil horizon, soil strength and density effects on root growth, soil depth, rooting depth, and soil stratification or layering. Each will be discussed briefly.

Water Content–Potential Relationship

What is available water holding capacity? soils.usda.gov/sqi/publications/files/avwater.pdf

As illustrated in Figure 5.34, there is a relationship between the water potential of a given soil and the amount of water held at field capacity and at permanent wilting percentage, the two boundary properties determining the available water-holding capacity. This energy-controlling concept should be kept in mind as we consider the various soil properties that affect the amount of water a soil can hold for plant use.

The general influence of texture on field capacity, wilting coefficient, and **available water-holding capacity** is shown in Figure 5.35. Note that as fineness of texture increases, there is a general increase in available moisture storage from sands to loams and silt loams. Plants growing on sandy soils are more apt to suffer from drought than are those growing on a silt loam in the same area (see Plate 45, after page 112). However, clay soils frequently provide less available water than do well-granulated silt loams since the clays tend to have a high wilting coefficient.

The influence of organic matter deserves special attention. The available water-holding capacity of a well-drained mineral soil containing 5% organic matter is generally higher than that of a comparable soil with 3% organic matter. Evidence suggests that both direct and indirect factors contribute to the favorable effects of organic matter on soil water availability.

The direct effects are due to the very high water-holding capacity of organic matter, which, when the soil is at the field capacity, is much higher than that of an equal volume of mineral matter. Even though the water held by organic matter at the wilting

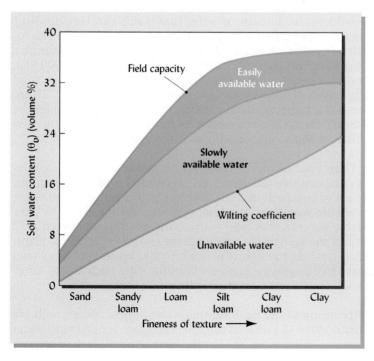

FIGURE 5.35 General relationship between soil water characteristics and soil texture. Note that the wilting coefficient increases as the texture becomes finer. The field capacity increases until we reach the silt loams, then levels off. Remember these are representative curves; individual soils would probably have values different from those shown.

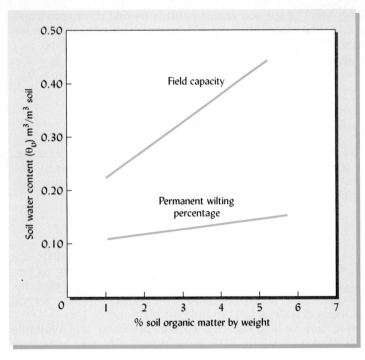

FIGURE 5.36 The effects of organic matter content on the field capacity and permanent wilting percentage of a number of silt loam soils. The differences between the two lines shown is the available soil moisture content, which was obviously greater in the soils with higher organic matter levels. [Redrawn from Hudson (1994); used with permission of the Soil & Water Conservation Society]

point is also somewhat higher than that held by mineral matter, the amount of water available for plant uptake is still greater from the organic fraction. Figure 5.36 provides data from a series of experiments to justify this conclusion.

Organic matter indirectly affects the amount of water available to plants through its influence on soil structure and total pore space. We learned in Section 4.5 that organic matter helps stabilize soil structure and that it increases the total volume as well as the size of the pores. This results in an increase in water infiltration and water-holding capacity with a simultaneous increase in the amount of water held at the wilting coefficient. Recognizing the beneficial effects of organic matter on plant-available water is essential for wise soil management.

Compaction Effects on Matric Potential, Aeration, and Root Growth

Soil compaction generally reduces the amount of water that plants can take up. Four factors account for this negative effect. First, compaction crushes many of the macropores and large micropores into smaller pores, and the bulk density increases. As the clay particles are forced closer together, soil strength may increase beyond about 2000 kPa, the level considered to limit root penetration. Second, compaction decreases the total pore space, which generally means that less water is retained at field capacity. Third, reduction in macropore size and numbers generally means less air pore space when the soil is near field capacity. Fourth, the creation of more very fine micropores will increase the permanent wilting coefficient and so decrease the available water content.

LEAST LIMITING WATER RANGE. These four factors associated with soil compaction are integrated in Figure 5.37, which compares two different concepts of water availability to roots. The figure illustrates the effect of compaction (bulk density) on the range of soil water contents defining **plant-available water** and the **least limiting water range**. We have already defined plant-available water as that held with matric potentials between field capacity (−10 to −30 kPa) and the permanent wilting point (−1500 kPa). Thus, plant-available water is that which is not held too tightly for roots to take up and yet is not held so loosely that it freely drains away by gravity. The least limiting water range is that range of water contents for which soil conditions do not severely restrict root growth.

OXYGEN SUPPLY IN WET SOIL. According to the least limiting water range concept, soils are too *wet* for normal root growth when so much of the soil pore space is filled with water that less than about 10% remains filled with air. At this water content, lack of oxygen

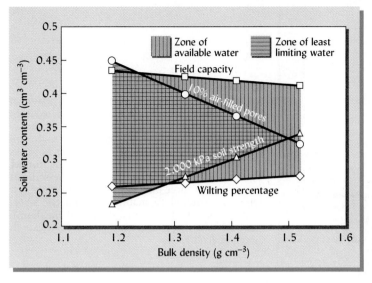

FIGURE 5.37 Influence of increased bulk density on the range of soil water contents available for plant uptake. Traditionally, plant-available water is defined as that retained between field capacity and wilting percentage (vertical hatching). If soils are compacted, however, plant use of water may be restricted by poor aeration (<10% air-filled pore space) at high water contents and by soil strength (>2000 kPa) that restricts root penetration at low water contents. The latter criteria define the *least limiting water range* shown by horizontal hatching. The two sets of boundary criteria give similar results when the soil is not compacted (bulk density is about 1.25 for the soil illustrated). [Adapted from Da Silva and Kay (1997)]

for respiration limits root growth. In loose, well-aggregated soils, this water content corresponds quite closely to field capacity. However, in a compacted soil with very few large pores, oxygen supply may become limiting at lower water contents (and potentials) because some of the smaller pores will be needed for air.

SOIL STRENGTH IN DRY SOIL. The least limiting water range concept tells us that soils are too *dry* for normal root growth when the soil strength (measured as the pressure required to push a pointed rod through the soil) exceeds about 2000 kPa. This level of soil strength occurs at water contents near the wilting point in loose, well-aggregated soils, but may occur at considerably higher water contents if the soil is compacted (see Figure 5.38). To summarize, the least limiting water range concept suggests that root growth is limited by lack of oxygen at the wet end of the range and by the inability of roots to physically push through the soil at the dry end. Thus, compaction effects on root growth are most pronounced in dry soils (Figure 5.39).

Osmotic Potential

The presence of soluble salts, either from applied fertilizers or as naturally occurring compounds, can influence plant uptake of soil water. For soils high in salts, the total moisture stress will include the osmotic potential ψ_o of the soil solution as well as the matric potential. The osmotic potential tends to reduce available moisture in such soils because more water is retained in the soil at the permanent wilting coefficient than

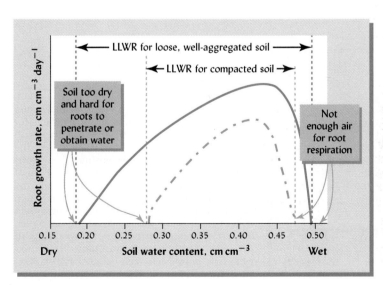

FIGURE 5.38 Compaction reduces the range of soil water contents suitable for plant growth (the *least limiting water range, LLWR*). Near the wet end of the soil water content scale (*right*), root growth (*curved line*) is limited by lack of air for root respiration. Once the soil dries a little, the largest pores drain and fill with air. Neither water nor air is then limiting and root growth rate becomes maximal. With further drying, low water potentials make it more difficult for roots to obtain moisture, and the soil increases its mechanical resistance to root penetration. Root growth declines until the soil is so dry that roots cannot grow at all (*left*). The lower (*dashed*) curve depicts the reduced rate of root growth that would pertain if the soil were compacted. Because compaction compresses the largest pores, it takes somewhat less water than before to create an oxygen-limited condition that reduces root growth. Toward the dry end of the scale, higher soil strength brings root growth to a halt at a water content that would still support considerable growth in an uncompacted soil. (Diagram courtesy of R. Weil)

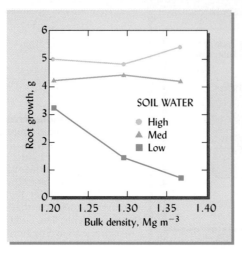

FIGURE 5.39 Root growth of Lodgepole pine seedlings in response to increased compaction at three soil water levels. Compaction affected root growth only when soil water was low, probably because root growth was limited by high soil strength. The tree seedlings were grown for 12 weeks in pots of mineral soil collected during timber harvest in British Columbia, Canada. The soil was compacted to three bulk density levels. Water was added as required to maintain volumetric water contents of 0.10–0.15 (low), 0.20–0.30 (medium), and 0.30–0.35 (high) cm³/cm³. [Drawn from data in Blouin et al. (2004)]

would be the case due to matric potential alone. In most humid region soils these osmotic potential effects are insignificant, but they become of considerable importance for certain soils in dry regions that may accumulate soluble salts through irrigation or natural processes.

Soil Depth and Layering

Our discussion thus far has referred to available water-holding capacity as the percentage of the soil volume consisting of pores that can retain water at potentials between field capacity and wilting percentage. The total volume of available water will depend on the total volume of soil explored by plant roots (see Plate 85, after page 656). This volume may be governed by the total depth of soil above root-restricting layers (see Figure 5.40), by the greatest rooting depth characteristic of a particular plant species, or even by the size of a flower pot chosen for containerized plants. The depth of soil available for root exploration is of particular significance for deep-rooted plants (see Figure 1.4), especially in subhumid to arid regions where perennial vegetation depends on water stored in the soils for survival during long periods without precipitation.

Soil stratification or layering can markedly influence the available water and its movement in the soil. Impervious layers drastically slow down the rate of water movement and also restrict the penetration of plant roots, thereby reducing the soil depth from which moisture is drawn. Sandy layers also act as barriers to soil moisture movement from the finer-textured layers above, as explained in Section 5.6.

The capacity of soils to store available water determines to a great extent their usefulness for plant growth. The productivity of forest sites is often related to soil

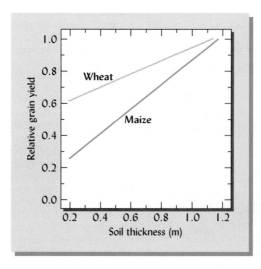

FIGURE 5.40 Relative grain yields for maize and wheat in relation to thickness of soil available for rooting. The crops were all grown with no-till management in Argiudolls and Paleudolls of the southeastern Pampas in Argentina. These soils held about 1.5 mm of available water per 1 cm of soil depth. Maize, which was grown during the hot, dry weather from spring to fall, was much more responsive to increased soil thickness than winter wheat, which was grown during the cooler, low water-demand period from fall to spring. In this case the soil thickness was limited by a root-restricting petrocalcic (cemented) horizon. [Redrawn from Sadras and Calvino (2001)]

BOX 5.4 TOTAL AVAILABLE WATER-HOLDING CAPACITY OF A SOIL PROFILE

The total amount of water available to a plant growing in a field soil can be estimated from the rooting depth of the plant and the amount of water held between field capacity and wilting percentage in each of the soil horizons explored by the roots. For each soil horizon, the mass available water-holding capacity *(AWHC)* is estimated as the difference between the mass water content at field capacity θ_{mFC} (g water per 100 g soil at field capacity) and that at permanent wilting percentage, θ_{mWP}. This value can be converted into a volume water content θ_v by multiplying by the ratio of the bulk density D_b of the soil to the density of water D_w. Finally, this volume ratio is multiplied by the thickness of the horizon to give the total centimeters of available water capacity *AWHC* in that horizon.

$$AWHC = (\theta_{mFC} - \theta_{mWP}) * D_b * D_w * L \tag{5.11}$$

For the first horizon described in Table 5.8 we can substitute values (with units) into eq. 5.11:

$$AWHC = \left(\frac{22\,g}{100\,g} - \frac{8\,g}{100\,g}\right) * \frac{1.2\,Mg}{m^3} * \frac{1\,m^3}{1\,Mg} * 20\,cm = 3.36\,cm$$

Note that all units cancel out except cm, resulting in the depth of available water (cm) held by the horizon. In Table 5.8, the *AWHC* of all horizons within the rooting zone are summed to give a total *AWHC* for the soil–plant system. Since no roots penetrated to the last horizon (1.0 to 1.25 m), this horizon was not included in the calculation. We can conclude that for the soil–plant system illustrated, 14.13 cm of water could be stored for plant use. At a typical summertime water-use rate of 0.5 cm of water per day, this soil could hold about a four-week supply.

TABLE 5.8 Calculation of Estimated Soil Profile Available Water-Holding Capacity

Soil depth, cm	Relative root length	Soil depth increment, cm	Soil bulk density, Mg/m^3	Field capacity (FC), g/100 g	Wilting percentage (WP), g/100 g	Available water-holding capacity (AWHC), cm
0–20	xxxxxxxxx	20	1.2	22	8	$20 * 1.2\left(\frac{22}{100} - \frac{8}{100}\right) = 3.36\,cm$
20–40	xxxx	20	1.4	16	7	$20 * 1.4\left(\frac{16}{100} - \frac{7}{100}\right) = 2.52\,cm$
40–75	xx	35	1.5	20	10	$35 * 1.5\left(\frac{20}{100} - \frac{10}{100}\right) = 5.25\,cm$
75–100	xx	25	1.5	18	10	$25 * 1.5\left(\frac{18}{100} - \frac{10}{100}\right) = 3.00\,cm$
100–125	—	25	1.6	15	11	No roots
Total						3.36 + 2.52 + 5.25 + 3.00 = 14.13 cm

water-holding capacity. This capacity provides a buffer between an adverse climate and plant production. In irrigated soils, it helps determine the frequency with which water must be applied. Soil water-holding capacity becomes more significant as the use of water for all purposes—industrial and domestic, as well as irrigation—begins to tax the supply of this all-important natural resource. To estimate the water-holding capacity of a soil, each soil horizon to which roots have access may be considered separately and then summed to give a total water-holding capacity for the profile (see Box 5.4).

5.10 MECHANISMS BY WHICH PLANTS ARE SUPPLIED WITH WATER

At any one time, only a small proportion of the soil water is adjacent to the absorptive plant-root surfaces. How then do the roots get access to the immense amount of water (see Section 6.3) used by vigorously growing plants? Two phenomena seem to account for this access: the capillary movement of the soil water to plant roots and the growth of the roots into moist soil.

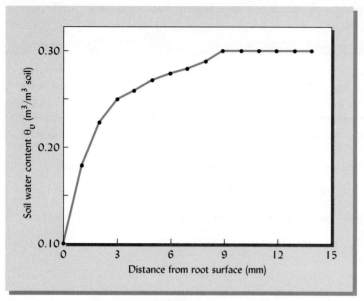

FIGURE 5.41 The drawdown of soil water levels surrounding a radish root after only two hours of transpiration. Water has moved by capillarity from a distance of at least 9 mm from the surface of the root. [Modified from Hamza and Aylmore (1992); used with permission of Kluwer Academic Publishers, The Netherlands]

Rate of Capillary Movement

When plant rootlets absorb water, they reduce the soil moisture content, thus reducing the water potential in the soil immediately surrounding them (see Figure 5.41). In response to this lower potential, water tends to move toward the plant roots. The rate of movement depends on the magnitude of the potential gradients developed and the conductivity of the soil pores. With some sandy soils, the adjustment may be comparatively rapid and the flow appreciable if the soil is near field capacity. In fine-textured and poorly granulated clays, the movement will be sluggish and only a meager amount of water will be delivered. However, in drier conditions with water held at lower potentials, the clay soils will be able to deliver more water by capillarity than the sand because the latter will have very few capillary pores still filled with water.

The total distance that water flows by capillarity on a day-to-day basis may be only a few centimeters. This might suggest that capillary movement is not a significant means of enhancing moisture uptake by plants. However, if the roots have penetrated much of the soil volume so that individual roots are rarely more than a few centimeters apart, movement over greater distances may not be necessary. Even during periods of hot, dry weather when evaporative demand is high, capillary movement can be an important means of providing water to plants. It is of special significance during periods of low moisture content when plant root extension is low.

Rate of Root Extension

Capillary movement of water is complemented by rapid rates of root extension, which ensure that new root–soil contacts are constantly being established. Such root penetration may be rapid enough to take care of most of the water needs of a plant growing in a soil at optimum moisture. The mats of roots, rootlets, and root hairs in a forest floor or a meadow sod exemplify successful adaptations of terrestrial plants for exploitation of soil water storage.

The primary limitation of root extension is the small proportion of the soil with which roots are in contact at any one time. Even though the root surface is considerable, root–soil contacts commonly account for less than 1% of the total soil surface area. This suggests that most of the water must move from the soil to the root even though the distance of movement may be no more than a few millimeters. It also suggests the complementarity of capillarity and root extension as means of providing soil water for plants.

Root Distribution

The distribution of roots in the soil profile determines to a considerable degree the plant's ability to absorb soil water. Most plants, both annuals and perennials, have the bulk of their roots in the upper 25 to 30 cm of the profile. For most plants in limited

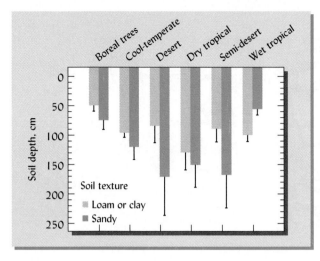

FIGURE 5.42 The soil depth above which 95% of all roots are located under different vegetation and soil types. The analysis used 475 root profiles reported from 209 geographic locations. The deepest rooting depths were found mainly in water-limited ecosystems. Within all but the wettest ecosystem, rooting was deeper in the sandier soils. Globally, 9 out of 10 profiles had at least 50% of all roots in the upper 30 cm of the soil profile and 95% of all roots in the upper 2 m (including any O horizons present). [Redrawn from Schenk and Jackson (2002)]

rainfall regions, roots explore relatively deep soil layers, but usually 95% of the entire root system is contained in the upper 2 m of soils. As Figure 5.42 shows, in more humid regions rooting tends to be somewhat shallower. Perennial plants, both woody and herbaceous, produce some roots that grow very deeply (>3 m) and are able to absorb a considerable proportion of their moisture from deep subsoil layers. Even in these cases, however, it is likely that much of the root absorption is from the upper layers of the soil, provided these layers are well supplied with water (Table 5.9). On the other hand, if the upper soil layers are moisture-deficient, even annual row crops such as sunflower, corn, and soybeans will absorb much of their water from the lower horizons (Figure 5.43), provided that adverse physical or chemical conditions do not inhibit their exploration of these horizons.

Root–Soil Contact

As roots grow into the soil, they move into pores of sufficient size to accommodate them. Contact between the outer cells of the root and the soil permits ready movement of water from the soil into the plant in response to differences in energy levels (Figure 5.44). When the plant is under moisture stress, however, the roots tend to shrink in size as their cortical cells lose water in response to this stress. Such conditions exist during hot, dry weather and are most severe during the daytime, when water loss through plant leaves is at a maximum. The diameter of roots under these conditions may shrink by 30%. This reduces the direct root–soil contact considerably, as well as the movement of liquid water and nutrients into the plants. While water vapor can still be

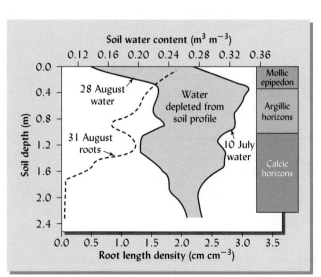

FIGURE 5.43 Distribution of roots and depletion of water in the profile of a soil growing sunflowers under dryland conditions in Texas. Notice the relationship between root density (cm of root length per cm^3 of soil) and amount of water used during the hot, dry weather between 10 July and 28 August. The water content on 28 August is near the wilting point in the upper 1 m of this Pullman clay loam (Torrertic Paleustolls). [Drawn from data in Moroke et al. (2005)]

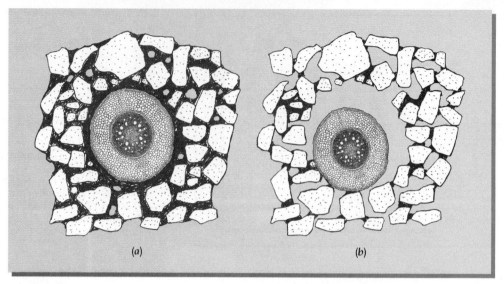

FIGURE 5.44 Cross-section of a root surrounded by soil. (*a*) During periods of adequate moisture and low plant moisture stress, the root completely fills the soil pores and is in close contact with the soil water films. (*b*) When the plant is under severe moisture stress, such as during hot, dry weather, the root shrinks (mainly in the cortical cells), significantly reducing root–soil contact. Such root shrinkage can occur on a hot day even if soil water content is high. [Diagram courtesy of R. Weil, based on concepts in Huck et al. (1970)].

absorbed by the plant, its rate of absorption is too low to keep any but the most drought-tolerant plants alive.

5.11 CONCLUSION

Water impacts all life. The interactions and movement of this simple compound in soils help determine whether these impacts are positive or negative. An understanding of the principles that govern the attraction of water for soil solids and for dissolved ions can help maximize the positive impacts while minimizing the less desirable ones.

The water molecule has a polar structure that results in electrostatic attraction of water to both soluble cations and soil solids. These attractive forces tend to reduce the potential energy level of soil water below that of pure water. The extent of this reduction, called soil water potential ψ, has a profound influence on a number of soil properties, but especially on the movement of soil water and its uptake by plants.

The water potential due to the attraction between soil solids and water (the matric potential ψ_m) combines with the force of gravity ψ_g to largely control water movement. This movement is relatively rapid in soils high in moisture and with an abundance of macropores. In drier soils, however, the adsorption of water on the soil solids is so

TABLE 5.9 Percentage of Root Mass of Three Crops and Two Trees Found in the Upper 30 cm Compared with Deeper Depths (30–180 cm)

	Percentage of roots	
Plant species	*Upper 30 cm*	*30–180 cm*
Soybeans	71	29
Corn	64	36
Sorghum	86	14
Pinus radiata	82	18
Eucalyptus marginata	86	14

Data for crops from [Mayaki et al. (1976)]; for trees, estimated from Bowen (1985).

strong that its movement in the soil and its uptake by plants are greatly reduced. As a consequence, plants die for lack of water—even though there are still significant quantities of water in the soil—because that water is unavailable to plants.

Water is supplied to plants by capillary movement toward the root surfaces and by growth of the roots into moist soil areas. In addition, vapor movement may be of significance in supplying water for drought-resistant desert species (xerophytes). The osmotic potential ψ_o becomes important in soils with high soluble salt levels that can impede plant uptake of water from the soil. Such conditions occur most often in soils with restricted drainage in areas of low rainfall and in potted indoor plants.

The characteristics and behavior of soil water are very complex. As we have gained more knowledge, however, it has become apparent that soil water is governed by relatively simple, basic physical principles. Furthermore, researchers are discovering the similarity between these principles and those governing the movement of groundwater and the uptake and use of soil moisture by plants—the subject of the next chapter.

STUDY QUESTIONS

1. What is the role of the *reference state of water* in defining soil water potential? Describe the properties of this reference state of water.

2. Imagine a root of a cotton plant growing in the upper horizon of an irrigated soil in California's Imperial Valley. As the root attempts to draw water molecules from this soil, what forces (potentials) must it overcome? If this soil were compacted by a heavy vehicle, which of these forces would be most affected? Explain.

3. Using the terms *adhesion, cohesion, meniscus, surface tension, atmospheric pressure,* and *hydrophilic surface,* write a brief essay to explain why water rises up from the water table in a mineral soil.

4. Suppose you were hired to design an automatic irrigating system for a wealthy homeowner's garden. You determine that the flower beds should be kept at a water potential above −60 kPa, but not wetter than −10 kPa, as the annual flowers here are sensitive to both drought and lack of good aeration. The rough turf areas, however, can do well if the soil dries to as low as −300 kPa. Your budget allows either tensiometers or electrical resistance blocks to be hooked up to electronic switching valves. Which instruments would you use and where? Explain.

5. Suppose the homeowner referred to in question 4 increased your budget and asked to use the TDR method to measure soil water contents. What additional information about the soils, not necessary for using the tensiometer, would you have to obtain to use the TDR instrument? Explain.

6. A greenhouse operator was growing ornamental woody plants in 15-cm-tall plastic containers filled with a loamy sand. He watered the containers daily with a sprinkler system. His first batch of 1000 plants yellowed and died from too much water and not enough air. As an employee of the greenhouse, you suggest that he use 30-cm-tall pots for the next batch of plants. Explain your reasoning.

7. Suppose you measured the following data for a soil:

Horizon	Bulk density, Mg/m^3	θ_m at different water tensions, kg water/kg dry soil		
		−10 kPa	−100 kPa	−1500 kPa
A (0–30 cm)	1.28	0.28	0.20	0.08
B_t (30–70 cm)	1.40	0.30	0.25	0.15
B_x (70–120 cm)	1.95	0.20	0.15	0.05

Estimate the total available water-holding capacity (AWHC) of this soil in centimeters of water.

8. A forester obtained a cylindrical core (L = 15 cm, r = 3.25 cm) of soil from a field site. She placed all the soil in a metal can with a tight-fitting lid. The empty metal can weighed 300 g and when filled with the field-moist soil

weighed 972 g. Back in the lab, she placed the can of soil, with lid removed, in an oven for several days until it ceased to lose weight. The weight of the dried can with soil (including the lid) was 870 g. Calculate both θ_m and θ_v.

9. Give four reasons why compacting a soil is likely to reduce the amount of water available to growing plants.

10. Since even rapidly growing, finely branched root systems rarely contact more than 1 or 2% of the soil particle surfaces, how is it that the roots can utilize much more than 1 or 2% of the water held on these surfaces?

11. For two soils subjected to "no," "moderate," or "severe" compaction, Figure 5.13 shows the volume fraction (cm^3 cm^{-3}) of pores in three size classes. The symbol ◀ indicates the volume fraction of water, θ_v (cm^3 cm^{-3}), in each soil. The figure indicates that $\theta_v \approx 0.35$ for the uncompacted clay loam (bulk density = 0.75 g cm^{-3}). Show a complete calculation (with all units) to demonstrate that ◀ in the figure correctly indicates that $\theta_v \approx 0.36$ for the severely compacted sandy loam (bulk density = 1.10 g/cm^3).

12. Fill in the shaded cells of this table to show the cm of available water capacity in the entire 90 cm profile. Show complete calculations for the first (upper left) and last (lower right) of the shaded cells.

Soil depth cm	Bulk density, D_b (Mg m^{-3})	Field capacity		Wilting point		Available water	
		θ_m, %[a]	θ_d, cm[b]	θ_m, %	θ_d, cm	θ_m, %	θ_d, cm
0–30	1.48	27.1		17.9		9.2	
30–60	1.51	27.5		18.1		9.4	
60–90	1.55	27.1		20.0		7.1	
0–90	—	—		—		—	

[a] θ_m is water content by mass and is reported here as percent (%), which is equivalent to g water/100 g dry soil.

[b] θ_d is water content by depth and is reported a mm of water held in the soil layer indicated.

REFERENCES

Blouin, V., M. Schmidt, C. Bulmer, and M. Krzic. 2004. "Soil compaction and water content effects on lodgepole pine seedling growth in British Columbia" B. Singh, ed., *Supersoil 2004*. Program and abstracts for the 3rd Australian-New Zealand soils conference. University of Sydney, Sydney, Australia, http://www.regional.org.au/au/asssi/supersoil2004/s14/oral/2036_blouinv.htm.

Bowen, G. D. 1985. "Roots as components of tree productivity," in M. G. R. Cannell and J. E. Jackson (eds.), *Attributes of Trees as Crop Plants*. Institute of Terrestial Ecology Midlothian, Scotland.

Cameira, M. R., R. M. Fernando, and L. S. Pereira. 2003. "Soil macropore dynamics affected by tillage and irrigation for a silty loam alluvial soil in southern Portugal," *Soil Tillage Res.* **70**:131–140.

Carter, L. J., and T. H. Pigford. 2005. "Proof of safety at Yucca Mountain," *Science* **310**: 447–448.

Cooney, J. J., and J. E. Peterson. 1955. *Avocado Irrigation*. Leaflet 50. California Agricultural Extension Service.

Da Silva, A. P., and B. D. Kay. 1997. "Estimating the least limiting water range of soil from properties and management," *Soil Sci. Soc. Amer. J.* **61**:877–883.

DeMartinis, J. M., and S. C. Cooper. 1994. "Natural and man-made modes of entry in agronomic areas," in R. Honeycutt and D. Schabacker (eds.), *Mechanisms of Pesticide Movement in Ground Water*. Lewis Publishers: Boca Raton, Fl: pp. 165–175.

Gardner, W., and J. A. Widtsoe. 1921. "The movement of soil moisture," *Soil Sci.* **11**:230.

Hamza, M., and L. A. G. Aylmore. 1992. "Soil solute concentrations and water uptake by single lupin and radish plant roots: 1. Water extraction and solute accumulation," *Plant Soil* **145**:187–196.

Hillel, D. 1998. *Environmental Soil Physics*. Academic Press: Orlando, Fl.

Huck, M. G., B. Klepper, and H. M. Taylor. 1970. "Diurnal variations in root diameter," *Plant Physiol.* **45**:529.

Hudson, B. D. 1994. "Soil organic matter and available water capacity," *J. Soil and Water Cons.* **49**:189–194.

Kladivko, E. J. et al. 1999. "Pesticide and nitrate transport into subsurface tile drains of different spacing," *J. Environ. Qual.* **28**:997–1004.

Mayaki, W. C., L. R. Stone, and I. D. Teare. 1976. "Irrigated and nonirrigated soybean, corn, and grain sorghum root systems," Agron. J. **68**: 532-534.

Moroke, T. S., R. C. Schwartz, K. W. Brown, and A. S. R. Juo. 2005. "Soil water depletion and root distribution of three dryland crops," *Soil Sci. Soc. Amer. J.* **69**:197–205.

Reynolds, W. D. et al. 2000. "Comparison of tension infiltrometer, pressure infiltrometer and soil core estimates of saturated conductivity," *Soil Sci. Soc. Amer. J.* **64**:478–484.

Richards, L. A. 1965. "Physical condition of water in soil," in *Agronomy 9: Methods of Soil Analysis, Part 1* American Society of Agronomy: Madison, WI.

Sadras, V. O., and O. A. Calvino. 2001. "Quantification of grain yield response to soil depth in soybean, maize, sunflower and wheat," *Agron J.* **93**:577–583.

Schenk, H., and R. Jackson. 2002. "The global biogeography of roots," *Ecological Monographs* **72**:311–328.

Shestak, C. J., and M. D. Busse. 2005. "Compaction alters physical but not biological indices of soil health," *Soil Sci. Soc. Amer. J.* **69**:236–246.

Waddell, J. T., and R. R. Weil. 1996. "Water distribution in soil under ridge-till and no-till corn," *Soil Sci. Soc. Amer. J.* **60**:230–237.

Warrick, A. W. 2001. *Soil Physics Companion.* (CRC Press, Boca Raton, Fl.) 400 pp.

Wuest, S. B., S. L. Albrecht, and K. W. Skirvin. 1999. "Vapor transport vs. seed–soil contact in wheat germination," *Agron. J.* **91**:783–787.

Water cycles in the Teton Mountains. (R. Weil)

6

SOIL AND THE HYDROLOGIC CYCLE

Both soil and water belong to the biosphere, to the order of nature,
and—as one species among many, as one generation
among many yet to come—we have no right to destroy them.
—*DANIEL HILLEL*, OUT OF EARTH

One of the most striking—and troubling—features of human society is the yawning gap in wealth between the world's rich and poor. The life experience of the one group is quite incomprehensible to the other. So it is with the distribution of the world's water resources. The rain forests of the Amazon and Congo basins are drenched by more than 2000 mm of rain each year, while the deserts of North Africa and central Asia get by with less than 200. South America and the Caribbean receive nearly one-third of the annual global precipitation, Australia only 1%. Nor is the supply of water distributed evenly throughout the year. Rather, periods of high rainfall and flooding alternate with dry spells or periods of drought.

Yet one could say that everywhere the supply of water is adequate to meet the needs of the plants and animals native to the natural communities of the area. Of course, this is so only because the plants and animals have adapted to the local availability of water. Early human populations, too, adapted to the local water supplies by settling where water was plentiful from rain or rivers, by learning which underground gourds and plant stems could quench one's thirst, by developing techniques to harvest water for agriculture and store it in underground cisterns, and by adopting nomadic lifestyles that allowed them and their herds to follow the rains and the grass supply.

But "civilized" humans have not been willing to adapt their needs and cultures to their environment. Rather, they have joined in organized efforts to adapt their environment to their desires. Hence, the ancients tamed the flows of the Tigris and Euphrates. We moderns dig wells in the Sahel, bottle up the mighty Nile at Aswan, channel the waters of the Colorado to the chaparral region of southern California, pump out the aquifers under farms and suburbs, and create sprawling cities (with swimming pools and bluegrass lawns!) in the deserts of the American Southwest or on the sands of Arabia. Truly, cities like Las Vegas are gambling in more ways than one.

There is plenty of room for improvement in managing water resources, and many of the improvements are likely to come as a result of better management of soils. The soil plays a central role in the cycling and use of water. For instance, by serving as a massive reservoir, soil helps moderate the adverse effects of excesses and deficiencies of water. It takes in water during times of surplus and then releases it in due time, either to satisfy the transpiration requirements of plants or to replenish groundwater. Stable structure at the surface of the soil ensures that a large fraction of the precipitation received will move slowly into the groundwater and from there to nearby streams or to deeper reservoirs under the earth. The soil can help us treat and reuse wastewaters from animal, domestic, and industrial sources. The flow of water through the soil in these and other circumstances connects the chemical pollution of soils to the possible contamination of groundwater.

In Chapter 5 we considered the nature and movement of water in soils. In this chapter we will see how those characteristics apply to the management of water as it cycles between the soil, the atmosphere, and vegetation. We will then examine the unique role of the soil in water-resource management.

6.1   THE GLOBAL HYDROLOGIC CYCLE

Global Stocks of Water

There are nearly 1400 million km³ of water on Earth—enough (if it were all above-ground at a uniform depth) to cover the Earth's surface to a depth of some 3 km. Most of this water, however, is relatively inaccessible and is not active in the annual cycling of water that supplies rivers, lakes, and living things. More than 97% is found in oceans (Figure 6.1) where the water is not only salty, but has an average **residence time** of several thousand years. Only the near-surface ocean layers take part in annual water cycling. An additional 2% of the water is in glaciers and ice caps of mountains with similarly long residence times (about 10,000 years). Some 0.7% is found in groundwater, most of which is more than 750 m underground. Except where it is pumped by humans, it, too, has a long average residence time measured in centuries.

The water with shorter residence times (that which cycles more actively) is in the surface layer of the oceans, in shallow groundwater, in lakes and rivers, in the atmosphere, and in the soil (see Figure 6.1). Although the combined volume is a tiny fraction of the water on Earth, these pools of water are accessible for movement in and out of the atmosphere and from one place on Earth's surface to another. The average residence time for water in the atmosphere is about 10 days, that for the longest rivers is 20 days or less, and that for soil moisture is about one month.

The Hydrologic Cycle

Primer on hydrologic cycle. Environment Canada: http://www.ec.gc.ca/water/en/nature/e_nature.htm

Solar energy drives the cycling of water from Earth's surface to the atmosphere and back again in what is termed the *hydrologic cycle* (Figure 6.2). About one-third of the solar energy that reaches Earth is absorbed by water, stimulating *evaporation*—the conversion of liquid water into water vapor. The water vapor moves up into the atmosphere, eventually forming clouds that can move from one region of the globe to another. Within an average of about 10 days, pressure and temperature differences in the atmosphere cause the water vapor to condense into liquid droplets or solid particles, which return to the Earth as rain or other precipitation.

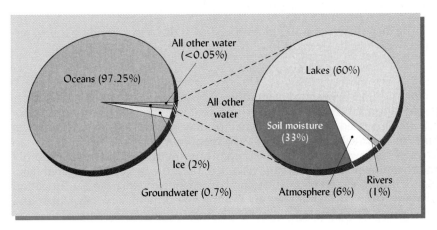

FIGURE 6.1 The sources of the Earth's water. The preponderance of water is found in the oceans, glaciers and ice caps, and deep groundwater (*left*) but most of these waters are inaccessible for rapid exchange with the atmosphere and the land. The sources on the right, though much smaller in quantity, are actively involved in water movement through the hydrologic cycle. (Data from several sources)

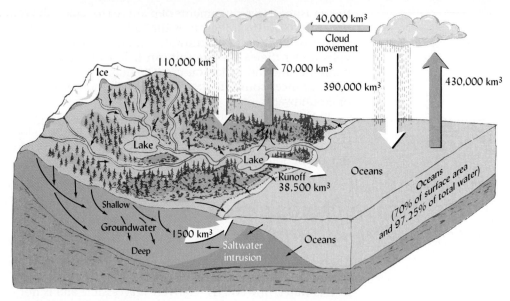

FIGURE 6.2 The hydrologic cycle upon which all life depends is very simple in principle. Water evaporates from the Earth's surface, both the oceans and continents, and returns in the form of rain or snowfall. The net movement of clouds brings some 40,000 km³ of water to the continents and an equal amount of water is returned through runoff and groundwater seepage that is channeled through rivers to the ocean. About 86% of the evaporation and 78% of the precipitation occurs in the ocean areas. However, the processes occurring on land areas where the soils are influential have impacts not only on humans but on all other forms of life, including those residing in the sea.

The water cycle:
http://micrometeorology.unl.
edu/et/flash/watercycle_f.html

About 500,000 km³ of water are evaporated from the Earth's surfaces and vegetation each year, some 110,000 km³ of which falls as rain or snowfall on the continents. Some of the water falling on land runs off the surface of the soil, and some infiltrates the soil and drains into the groundwater. Both the surface runoff and groundwater seepage enter streams and rivers that, in turn, flow into the oceans. The volume of water returned in this way is about 40,000 km³, which balances the same quantity of water that is transferred annually in clouds from the oceans to the continents.

Water Balance Equation

Earth's water budget:
http://ww2010.atmos.uiuc.
edu/(Gh)/guides/mtr/hyd/
bdgt.rxml

It is often useful to consider the components of the hydrologic cycle as they apply to a given **watershed**, an area of land drained by a single system of streams and bounded by ridges that separate it from adjacent watersheds. All the precipitation falling on a watershed is either stored in the soil, returned to the atmosphere (see Section 6.2), or discharged from the watershed as surface or subsurface flow (runoff). Water is returned to the atmosphere either by **evaporation** from the land surface (vaporization of soil water) or, after plant uptake and use, by vaporization from the stomata on the surfaces of leaves (a process termed **transpiration**). Together, these two pathways of evaporative loss to the atmosphere are called **evapotranspiration**.

WATER BALANCE.   The disposition of water in a watershed is often expressed by the water-balance equation, which in its simplest form is:

$$P = ET + SS + D \qquad (6.1)$$

where P = precipitation, ET = evapotranspiration, SS = soil storage, and D = discharge.

Tools to protect natural
waters:
http://www.cwp.org/
whywatersheds_files/
frame.htm

Rearranging equation 6.1 (D = P – ET – SS) shows that discharge can be increased only if ET and/or SS are decreased, changes that may or may not be desirable. For a forest watershed (sometimes termed a *catchment*), management may aim to maximize D so as to provide more water to downstream users. Clear-cutting the trees will almost certainly decrease ET and therefore increase discharge (Figure 6.3). In the case of an irrigated field, water applied in irrigation would be included on the left side of equation 6.1. Irrigation managers may want to save water applied by minimizing unnecessary losses in D and allowing negative values for SS (withdrawals of soil storage water) during parts of the year.

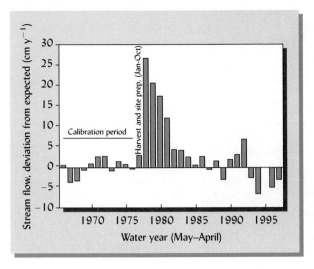

FIGURE 6.3 Discharge (stream flow) from a southern Appalachian catchment before and after commercial clear-cutting of mixed hardwoods. Flow is expressed as cm of water above or below the long-term average flow (as measured prior to the timber harvest). Note the very high values for several years after the trees were felled and the site prepared for planting. The high flows resulted mainly from reduced transpiration water use in the absence of large trees. Dominated by Typic Hapludults and Typic Dystrochrepts, this 59 ha catchment is part of the Coweeta Hydrologic Laboratory, North Carolina, U.S. [Data from Swank et al. (2001)]

6.2 FATE OF PRECIPITATION AND IRRIGATION WATER

Some precipitation is intercepted by plant foliage (Table 6.1) and returned to the atmosphere by evaporation without ever reaching the soil. In some forested areas, **interception** may prevent 30 to 50% of the precipitation from reaching the soil. Interception and subsequent sublimation (vaporization directly from the solid state) of snow is especially important in coniferous forests.

Water that does reach the ground may penetrate into the soil by the process of **infiltration**, especially if the soil surface structure is loose and open. If the rate of rainfall or snowmelt exceeds the infiltration capacity of the soil, some ponding may result (Figure 6.4) and considerable runoff and erosion may take place, thereby reducing the proportion of water that moves into the soil. In extreme cases, more than 50% of the precipitation may be lost as **surface runoff**, which usually carries with it dissolved chemicals and detached soil particles (**sediment**; see Chapter 17).

Once water penetrates the soil, some of it is subject to downward percolation and eventual loss from the root zone by **drainage**. In humid areas and on irrigated land, up to 50% of the water input may be lost as drainage below the root zone. However, during subsequent periods of low rainfall, some of this water may move back up into the plant-root zone by rise of **capillary water**. Such movement is important to plants in areas with deep soils, especially in dry climates.

The water retained by the soil is referred to as **soil storage** water, some of which eventually moves upward by capillarity and is lost by evaporation from the soil surface. Much of the remainder is absorbed by plants and then moves through the roots and stems to the leaves, where it is lost by transpiration. The water thus lost to the atmosphere by evapotranspiration may later return to the soil as precipitation or irrigation water, and the cycle starts again.

Timing of precipitation greatly influences the amount of water moving through each of the channels just discussed. Heavy rainfall, even if of short duration, can supply

TABLE 6.1 **Interception of Precipitation by Several Crop and Tree Species**

Species	U.S. state	Percent of precipitation intercepted by plant
Alfalfa	MO	22
Corn	MO	7
Soybeans	NJ	15
Ponderosa pine	ID	22
Douglas fir	WA	34
Maple, beech	NY	43

Crop data from Haynes (1954); forest data from Kittridge (1948).

FIGURE 6.4 A small depressional wetland in North Dakota called a *prairie pothole*. When rain falls faster than the soil infiltration capacity, water will begin to pond on the soil surface and then run off into the depressions. In some landscapes with relatively low infiltration capacities, water may stand in depressions for several months during the year. Farmers who cultivate these landscapes with large machinery may view these wetlands as nuisances in need of drainage. However, small and temporary though they are, wetlands like these scattered across the northern Great Plains provide nesting sites for about half of the ducks in North America. (Photo courtesy of R. Weil)

water faster than most soils can absorb it. This accounts for the fact that in some arid regions with only 200 mm of annual rainfall, a cloudburst that brings 20 to 50 mm of water in a few minutes can result in flash flooding and gully erosion. A larger amount of precipitation spread over several days of gentle rain could move slowly into the soil, thereby increasing the stored water available for plant absorption, as well as replenishing the underlying groundwater. As illustrated in Figure 6.5, the timing of snowfall in early winter can affect the partitioning of spring snowmelt water between surface runoff and infiltration. A blanket of snow is good insulation.

Effects of Vegetation and Soils on Infiltration

TYPE OF VEGETATION.   Plants help determine the proportion of water that runs off and that which penetrates the soil. The vegetation and surface residues of perennial grasslands and dense forests protect the porous soil structure from the beating action of raindrops. Therefore, they encourage water infiltration and reduce the likelihood that soil will be carried off by any runoff that does occur. In general, very little runoff occurs from land under undisturbed forests or well-managed turfgrass. However, as Figure 6.6 indicates, differences in plant species, even among grasses, can influence runoff.

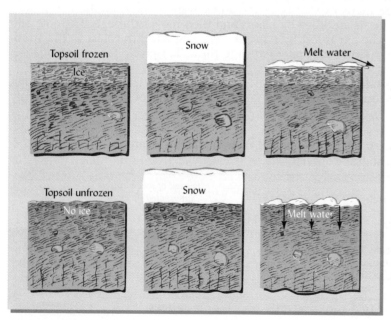

FIGURE 6.5 The relative timing of freezing temperatures and snowfall in the fall in some temperate regions drastically influences water runoff and infiltration into soils in the spring. The upper three diagrams illustrate what happens when the surface soil freezes before the first heavy snowfall. The snow insulates the soil so that it is still frozen and impermeable as the snow melts in the spring. The lower sequence of diagrams illustrates the situation when the soil is unfrozen in the fall when it is covered by the first deep snowfall.

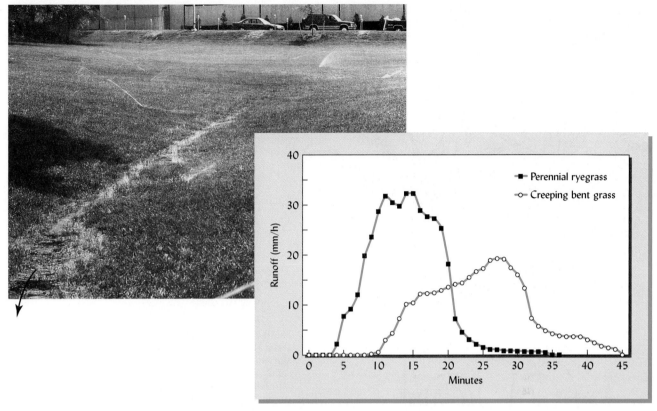

FIGURE 6.6 Little water generally runs off turfgrass, except during very intense rainfall, where the soils have been compacted or where irrigation water is applied unevenly or at too high a rate (as shown in the photo). The graph shows runoff from two golf fairway grasses following irrigation at the rate of 150 mm per hour. Note that the runoff peak was much lower on the creeping bent grass, which was characterized by a dense thatch of plant stems and many biopores near the soil surface. A lower rate of irrigation on either type of turfgrass could have eliminated the runoff, which represents a waste of water and a potential for soluble lawn chemicals to be carried to streams and rivers. [Data from Linde et al. (1995), used with permission of the American Society of Agronomy; photo courtesy of R. Weil]

STEM FLOW. Many plant canopies direct rainfall toward the plant stem, thus altering the spatial distribution of rain reaching the soil. Under a forest canopy, more than half of the rainfall may trickle down the leaves, twigs, and branches to the tree trunk, there to progress downward as **stem flow**. Likewise, certain crop canopies, such as that of corn, funnel a large proportion of the rainfall to the soil in the crop row (Figure 6.7). The concentration of water in limited zones around plant stems increases the opportunity for saturated flow to occur. Stem flow must be considered in studying the hydrology and nutrient cycling in many plant ecosystems.

SOIL MANAGEMENT. Encouraging infiltration rather than runoff is usually a major objective of soil and water management. One approach is to allow more time for infiltration to take place by enhancing soil surface storage (Figure 6.8, *left*). A second approach is to maintain dense vegetation during periods of high rainfall. For example, **cover crops**, plants established between the principal crop-growing seasons, can greatly enhance water infiltration by creating open root channels, encouraging earthworm activity, and protecting soil surface structure (Figure 6.8, *right*). However, remember that cover crops also transpire water. If the following crop will be dependent on soil storage water, care may be needed to kill the cover crop before it can dry out the soil profile.

A third approach is to maintain soil structure by minimizing compaction, whether by people in parks, heavy equipment on farm fields, or cattle on rangeland (see Section 10.1). For example, soil compaction by heavy forestry equipment used to skid logs or clear forestland for agricultural use can seriously impair soil infiltration capacity and watershed hydrology. While the ill effects of compaction can sometimes be partially overcome by deep tillage (see Section 4.7), environmental stewardship requires that caution should be exercised in the use of heavy equipment on any landscape and that disturbance of the natural soil and vegetation be completely avoided on as much of the land as possible.

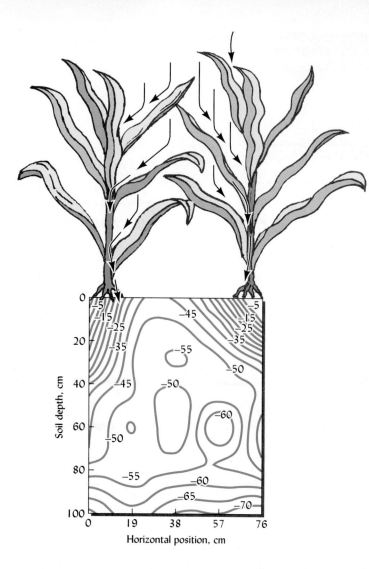

FIGURE 6.7 Vertical and horizontal distribution of soil water resulting from stem flow. The contours indicate the soil water potential in kPa between two corn rows in a sandy loam soil. During the previous two days, 26 mm of rain fell on this field. Many plant canopies, including that of the corn crop shown, direct a large proportion of rainfall toward the plant stem. Stem flow results in uneven spatial distribution of water. In cropland this may have ramifications for the leaching of chemicals such as fertilizers, depending on whether they are applied in or away from the zone of highest wetting near the plant stems. The concentration of water by stem flow may also increase the likelihood of macropore flow in soils after a moderate rainfall. [Data from Waddell and Weil (1996); used by permission of the Soil Science Society of America]

FIGURE 6.8 Managing soils to increase infiltration of rainwater. (*left*) Small furrow dikes on the right side of this field in Texas retain rainwater long enough for it to infiltrate rather than run off. (*right*) This saturated soil under a winter cover crop of hairy vetch is riddled with earthworm burrows that greatly increased the infiltration of water from a recent heavy rain. Scale in centimeters. [Photo (*left*) courtesy of O. R. Jones, USDA Agricultural Research Service, Bushland, Texas; photo (*right*) courtesy of R. Weil]

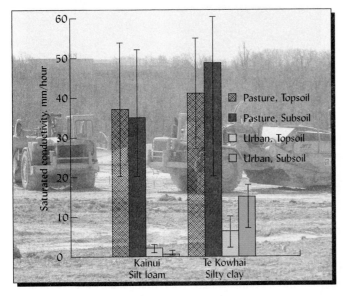

FIGURE 6.9 Grading and excavating land in preparation for urban/suburban development can greatly reduce soil permeability and hydraulic conductivity. The impairment results from compaction, damage to soil structure, and profile truncation (removal of A horizon). Both topsoil and subsoil horizons in these two New Zealand soils suffered three- to tenfold reductions in saturated hydraulic conductivity when the land was converted from a pasture to a housing development. The vertical lines indicate the variation among repeated measurements in the same soil, this variation being especially large in the pasture soils because of the random presence of earthworm channels. [Data from Zanders 2001), used with permission from Landcare Research NZ Ltd.; photo courtesy of R. Weil]

URBAN WATERSHEDS. The use of heavy equipment to prepare land for urban development can severely curtail soil infiltration capacity and saturated hydraulic conductivity in an urbanized watershed (Figure 6.9). Soil compaction activity by construction therefore results in drastically increased surface runoff during storms. The runoff burden carried by streams in urbanized watersheds is further increased because a large proportion of the land is covered with completely impermeable surfaces (rooftops, paved streets, and parking lots). Erosion of stream banks, toppling of trees, scouring of streambeds, and exposure of once-buried pipelines (Figure 6.10) are typical signs of the environmental degradation that ensues as streams become overwhelmed. Damage to streams in urban watersheds is accentuated because storm sewers and street gutters rush all this excess water off the land, requiring the stream to carry a huge volume of runoff concentrated in a short period of time. In recognition of the severe environmental problems caused by such excessive and concentrated runoff, urban planners and engineers are now working with soil scientists to reduce disruptions of the hydrologic cycle by urban development. Features such as permeable pavers that allow some infiltration even in parking lots (Figure 6.11, *left*) and rain gardens that catch runoff and release it slowly by infiltration (Figure 6.11, *right*) are part of what is termed *low impact* urban design.

Sprawl Aggravates Drought: http://www.smartgrowth america.org/waterandsprawl. html

Permeable paving vs runoff pollution. Audio report: http://www.npr.org/templates/ story/story.php?storyId= 6165654

FIGURE 6.10 Exposure of a once-buried storm sewer manhole along a highly degraded stream that collects runoff from part of Baltimore, Maryland. The drastically increased storm runoff volume from the largely impermeable urban watershed has caused the stream to erode its banks and deeply cut into the hillside. The round manhole cover indicates the former land surface. (Photo courtesy of R. Weil)

FIGURE 6.11 Two methods of increasing infiltration and slowing runoff in urbanized watersheds. Permeable pavers (*left*) allow grass to grow and water to infiltrate a parking lot while cars can still park without compacting soil or forming mud. Inlet to a small rain garden (*right*) that captures runoff from a suburban parking lot. The water is directed to a small depression. The pond that forms is ephemeral, holding the runoff water only temporarily. The permeable soil underlying the depression is designed to allow the water to infiltrate and seep away over a period of a few hours. The depression is planted with a variety of native plants that provide both beauty and wildlife habitat. (Photos courtesy of R. Weil)

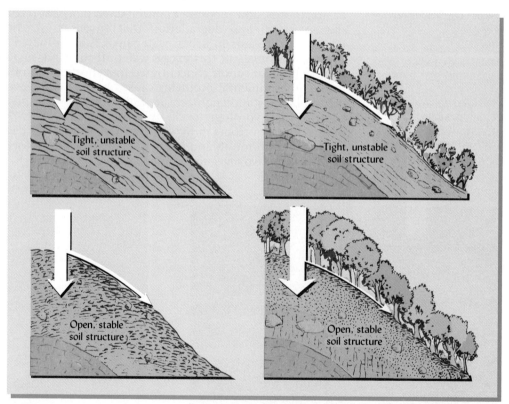

FIGURE 6.12 Influence of soil structure and vegetation on the partitioning of rainfall into infiltration and runoff. The upper two diagrams show soils with tight, unstable structure that resists infiltration and percolation. The bare soil is especially prone to surface sealing and resulting high losses by runoff. Even with forest cover, the low-permeability soils cannot accept all the rain in an intense storm. The two lower diagrams show much greater infiltration into soils that have open, stable structures with significant macropore space. The more open soil structure combined with the protective effects of the forest floor and canopy nearly eliminate surface runoff.

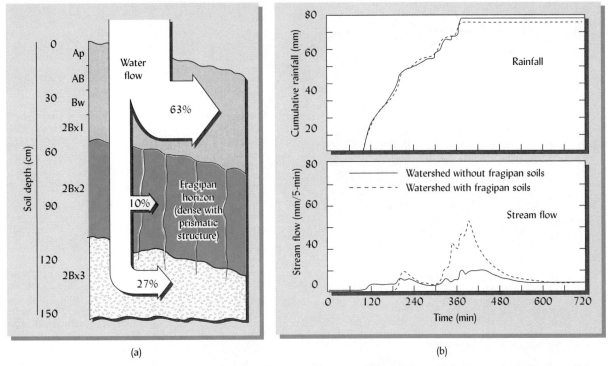

(a) (b)

FIGURE 6.13 Soil profile characteristics largely determine the vertical and lateral movement of water, including runoff from watersheds during rainstorms. In this example, a fragipan (*left,* also see Figure 3.30) provides a barrier to both root growth and downward percolation of water. The zone above the fragipan may become saturated during wet weather, giving rise to a perched water table. As more rain falls, little can enter the already saturated surface horizons. Most of the water flows laterally, either through the soil above the fragipan or as surface runoff. The graphs (*right*) show the cumulative rainfall (*upper*) and volume of stream flow (*lower*) during and after a storm in which nearly 80 mm of rain fell in a 300-minute period. Of the two small (13 to 20 ha) watersheds represented, one has a large area of fragipan-containing soils (Fragiochrepts) formed in colluvium near the stream channel; the other watershed has no such fragipans. Although the rainfall was almost identical for the two nearby watersheds, stream flow (both the maximum and the overall volume) was much greater for the watershed with fragipans than the one without these impermeable layers. [Soil profile based on data in Day et al. (1998); graphs from Gburek et al. (2006), with permission of Elsevier Science, Oxford, United Kingdom]

SOIL PROPERTIES. Inherent soil properties also affect the fate of precipitation. If the soil is loose and open (e.g., sands and well-granulated soils), a high proportion of the incoming water will infiltrate the soil, and relatively little will run off. In contrast, heavy clay soils with unstable soil structures resist infiltration and encourage runoff. These differences, attributable to soil properties as well as vegetation, are illustrated in Figure 6.12. Other factors that influence the balance between infiltration and runoff include the slope of the land (steep slopes favoring runoff over infiltration; see Box 2.1) and impermeable layers within the soil profile. Such impermeable layers as fragipans (see Figure 3.30) and clay pans (see Figure 3.6) can restrict infiltration and increase surface runoff once the upper horizons become saturated, even if the surface soil has intact structure and high porosity (Figure 6.13). All the soil and plant factors just discussed can result in some parts of a landscape contributing more runoff than others. Such spatial heterogeneity of infiltration and runoff is particularly important to ecosystem function in dry regions (see Box 10.1).

6.3 THE SOIL–PLANT–ATMOSPHERE CONTINUUM[1]

The flow of water through the soil–plant–atmosphere continuum (SPAC) is a major component of the overall hydrologic cycle. Figure 6.14 ties together many of the processes we have just discussed: *interception, surface runoff, percolation, drainage, evaporation, plant water uptake, ascent of water to plant leaves,* and *transpiration* of water from the leaves back into the atmosphere.

---

[1] The physics of water movement in plants has been quite controversial. For recent evidence that confirms that water moves up tall trees by the same capillary forces that control its movement in soil, see Tyree (2003).

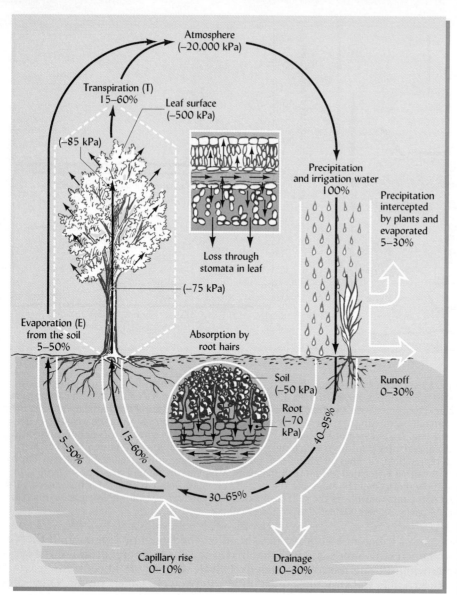

Atmosphere
(−20,000 kPa)

Transpiration (T)
15–60%

Leaf surface
(−500 kPa)

(−85 kPa)

Precipitation
and irrigation water
100%

Precipitation
intercepted
by plants and
evaporated
5–30%

Loss through
stomata in leaf

(−75 kPa)

Evaporation (E)
from the soil
5–50%

Absorption by
root hairs

Soil
(−50 kPa)

Root
(−70
kPa)

Runoff
0–30%

5–50%

15–60%

40–95%

30–65%

Capillary rise
0–10%

Drainage
10–30%

FIGURE 6.14 Soil–plant–atmosphere continuum (SPAC) showing water movement from soil to plants to the atmosphere and back to the soil in a humid to subhumid region. Water behavior through the continuum is subject to the same energy relations covering soil water that were discussed in Chapter 5. Note that the moisture potential in the soil is −50 kPa, dropping to −70 kPa in the root, declining still further as it moves upward in the stem and into the leaf, and is very low (−500 kPa) at the leaf–atmosphere interface, from whence it moves into the atmosphere, where the moisture potential is −20,000 kPa. Moisture moves from a higher to a lower moisture potential. Note the suggested ranges for partitioning of the precipitation and irrigation water as it moves through the continuum. Over 98% of the water absorbed by the roots of plants is transpired as water vapor over the course of the growing season.

WATER POTENTIALS.   In studying the SPAC, scientists have discovered that the same basic principles govern the retention and movement of water whether it is in soil, in plants, or in the atmosphere. In Chapter 5, we learned that water in soil moves to where its potential energy level will be lower. This principle applies to water movement between the soil and the plant root and between the plant and the atmosphere (see Figure 6.14). If a plant is to absorb water from the soil, the water potential must be lower (greater negative value) in the plant root than in the soil adjacent to the root. Likewise, movement up the stem to the leaf cells is in response to differences in water potential, as is the movement from leaf surfaces to the atmosphere. To illustrate the movement of water to sites of lower and lower water potential, Figure 6.14 shows that the water potential drops from −50 kPa in the soil, to −70 in the root, to −500 kPa at the leaf surfaces, and, finally, to −20,000 kPa in the atmosphere.

TWO POINTS OF RESISTANCE.   Water encounters major resistance to its movement as it crosses the root–soil water interface and again as it crosses the leaf cell–atmosphere interface. This means that two primary factors determine whether plants are well supplied with water: (1) the rate at which water is supplied by the soil to the absorbing roots, and (2) the rate at which water is transpired from the plant leaves. Since factors affecting the soil's ability to supply water were discussed in Section 5.9, we will now address the loss of water by evaporation from soil–plant systems.

Evapotranspiration

Animated water use by plants:
http://micrometeorology.unl.edu/et/flash/slide2.html

While it is relatively easy to measure the total change in soil water content due to vapor losses, it is quite difficult to determine just how much of that loss occurred directly from the soil (by **evaporation**, E) and how much occurred from the leaf surfaces after plant uptake (by **transpiration**, T). Therefore, information is most commonly available on **evapotranspiration** (ET), the combined loss resulting from these two processes. The evaporation component of ET may be viewed as a "waste" of water from the standpoint of plant productivity. However, at least some of the transpiration component is essential for plant growth, providing the water that plants need for cooling, nutrient transport, photosynthesis and turgor maintenance.

The **potential evapotranspiration** rate (PET) tells us how fast water vapor *would* be lost from a densely vegetated plant–soil system *if* soil water content were continuously maintained at an optimal level. The PET is largely determined by the *vapor pressure gradient* between a wet soil, leaf, or body of water and the atmosphere. This gradient is in turn influenced by solar radiation and such related climatic variables as temperature, relative humidity, cloud cover, and wind speed.

Evapotransipration and water use:
http://www.engineering.usu.edu/uwrl/atlas/ch3/index.html

A number of mathematical models have been devised to estimate PET from climatological data, but in practice, PET can be most easily estimated by applying a correction factor to the amount of water evaporated from an open pan of water of standard design (a class A evaporation pan; see Figure 6.15). Because of the resistances to water flow just mentioned, water loss by transpiration from well-watered, dense vegetation is typically only about 65% as rapid as loss by evaporation from an open pan; hence, the correction factor for dense vegetation such as a lawn is typically 0.65 (it is lower for less dense vegetation):

$$PET = 0.65 \times \text{pan evaporation} \tag{6.2}$$

Values of PET range from more than 1500 mm per year in hot, arid areas to less than 40 mm in very cold regions. During the winter in temperate regions, PET may be less than 1 mm per day. By contrast, hot, dry wind will continually sweep away water vapor from a wet surface, creating a particularly steep vapor pressure gradient and PET levels as high as 10 to 12 mm per day.

EFFECT OF SOIL MOISTURE. Evaporation from the soil surface at a given temperature is determined largely by soil surface wetness and by the ability of the soil to replenish this surface water as it evaporates. In most cases, the upper 15 to 25 cm of soil provides most of the water for surface evaporation. Unless a shallow water table exists, the upward capillary movement of water is very limited and the surface soil soon dries out, greatly reducing further evaporation loss.

However, because plant roots penetrate deep into the profile, a significant portion of the water lost by evapotranspiration can come from the subsoil layers. As Figure 5.42 shows, water stored deep in the profile is especially important to vegetation in regions

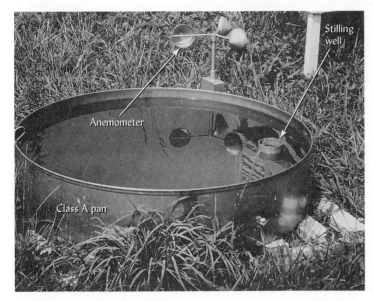

FIGURE 6.15 A class A evaporation pan used to help estimate potential evapotranspiration (PET). Once a day, the water level is determined in the stilling well (small cylinder) and a measured amount of water is added to bring the level back up to the original mark. Evaporation from the pan integrates the effects of relative humidity, temperature, wind speed, and other climatic variables related to the vapor pressure gradient. Also shown is an anemometer to measure wind speed. (Photo courtesy of R. Weil)

having alternating moist and dry seasons (such as Ustic or Xeric moisture regimes). Water stored in the subsoil during rainy periods is available for evapotranspiration during dry periods. Plate 85 following page 656 illustrates the death of a rooftop lawn because of the inability of the shallow soil to hold sufficient water during a prolonged summer drought.

PLANT WATER STRESS. For dense vegetation growing in soil well supplied with water, ET will nearly equal PET. When soil water content is less than optimal, the plant will not be able to withdraw water from the soil fast enough to satisfy PET. If water evaporates from the leaves faster than it enters the roots, the plant will lose turgor pressure and wilt. Under these conditions, actual evapotranspiration is less than potential evapotranspiration and the plant experiences *water stress*. The difference between PET and actual ET is termed the *water deficit*. A large deficit is indicative of high water stress and aridity.

Under water-stress conditions, plants first close the *stomata* (openings) on their leaf surfaces to reduce the vapor loss of water and prevent wilting. However, the closure of the stomata has two detrimental side effects: (1) plant growth is arrested because insufficient CO_2 for photosynthesis can pass through the closed stomata, and (2) the reduction in evaporative cooling results in detrimental *heating* of leaves as they continue to absorb solar radiation. The latter effect allows infrared-detecting instruments to estimate water stress in plants by sensing the increase in leaf temperature over air temperature. Stomatal closure in response to low leaf water potentials may also be "too little, too late" for plant survival (see Box 6.1).

BOX 6.1 "FEEDFORWARD"[a]

In response to the low water potentials transmitted up the xylem from the soil to the leaf cells, a plant may change its leaf characteristics in ways that provide feedback to reduce its rate of water use (Figure 6.16). But if a plant responds only after water potential is low enough to cause wilting, there may not be enough water left in the soil to see the plant through a drought period, even with reduced water use. Chances of survival may improve if a plant can restrict its water use while there is still a substantial supply of available water left in the soil; that is, *before* low water potentials actually are transmitted to the leaf and cause wilting. A review of Figure 5.34 tells us that most of the plant-available water in a soil is held at relatively high potentials (e.g., much more water is held between −10 and −60 kPa than between −200 and −1500 kPa). Can plants actually *anticipate* adverse soil water conditions and prepare for them ahead of time

FIGURE 6.16 *A plant wilts in bright sunlight as the supply of water from the shaley soil falls short of evaporative demand. (Photo courtesy of R. Weil)*

without having to suffer the drought stress first? Some plant scientists think that roots "may sense difficult conditions in the soil and then send inhibitory signals to the shoots which harden the plants against the consequences of . . . the soil becoming too dry or too hard. . . ."

The plant may reduce its leaf stomatal conductance, reduce the rate of leaf growth or limit the formation of new leaves. This kind of adaptive action taken by the plant *before* serious stress is encountered may be termed "feedforward" in contrast to "feedback" from lowered leaf water potential. Conservative "feedforward" behavior may confer a great drought survival advantage to wild plants. However, restriction of growth rate before drought stress is encountered might *not* be desirable in agricultural situations where high crop productivity is of more concern than mere plant survival.

[a] This concept of feedforward is based on the discussion in Passioura (2002).

PLANT CHARACTERISTICS. In vegetated areas, incoming solar radiation is either absorbed by plant leaves on its way through the plant canopy or it reaches the ground and is absorbed by the soil. Therefore, as the leaf area per unit land area (a ratio termed the **leaf area index, LAI**) increases, more radiation will be absorbed by the foliage to stimulate transpiration and less will reach the soil to promote evaporation. For a monoculture of annual plants, the LAI value typically varies from 0 at planting to a peak of perhaps 3 to 5 at flowering, then declines as the plant senesces, and finally drops back to 0 when the plant is removed at harvest (assuming no weeds are allowed to grow). In contrast, perennial vegetation, such as pastures and forests, has very high leaf area indices both early and late in the growing season. Where leaf litter has accumulated on the forest floor, very little direct sunlight strikes the soil, and evaporation is very low throughout the year (Figure 6.17). Other plant characteristics, including rooting depth, length of life cycle, and leaf morphology, can influence the amount of water lost by evapotranspiration over a growing season. Plates 62 and 84 illustrate the competitiveness of mature trees for soil water supplies.

Conference on drought resistant soils:
http://www.fao.org/landandwater/agll/soilmoisture/#the%20importance

WATER USE EFFICIENCY. Plant dry matter produced while using a given amount of water is an important measure of **water use efficiency**. This efficiency may be expressed in terms of dry matter per unit of water transpired (*T efficiency*) or the dry matter yield per unit of water lost by evapotranspiration (*ET efficiency*). There are a number of ways that researchers express water use efficiency, including kg of grain produced per m³ of water used in evapotranspiration (see Figure 6.18) and kg of plant dry matter per kg of water used in ET. For irrigation systems analysis, water use efficiency may consider additional aspects such as losses of water in storage reservoirs or leakage from canals (see Section 6.10).

Whatever the units of expression, data from studies around the world show that huge amounts of water are needed to produce the human food supply and that water use efficiency is largely driven by climatic factors. In arid regions, crops may use 1000 to more than 5000 kg of water to produce a single kg of grain. Ironically, in more humid regions where water and rain are plentiful, much less water is needed for each kg of grain produced because the evaporative demand is far lower. When we take into account the water used to grow grains, fruits, vegetables, and feed for cattle, almost 7000 L (1700 gal) of water are required to grow a *single day's* food supply for one adult in the United States!

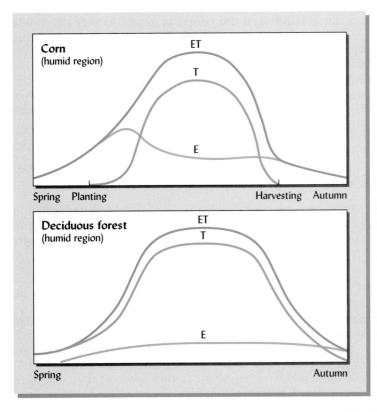

FIGURE 6.17 Relative rates of evaporation from the soil surface E, of transpiration from the plant leaves T, and the combined vapor loss ET for two field situations. (*Upper*) A field of corn in a humid region. Until the plants are well established, most of the vapor loss is from the soil surface E, but as the plants grow, T soon dominates. The soil surface is shaded and E may actually decrease somewhat since most of the moisture is moving through the plants. As the plants reach maturity, T declines, as does the combined ET. For a nearby deciduous forested area (*lower*) the same general trend is illustrated, except there is relatively less evaporation from the soil surface and a higher proportion of the vapor loss is by transpiration. The soil is shaded by the leaf canopy most of the growing season in the forested area. It should be noted that these figures relate to a representative situation and that the actual field losses would be influenced by rainfall distribution, temperature fluctuations, and soil properties. [Diagram courtesy of R. Weil and N. C. Brady]

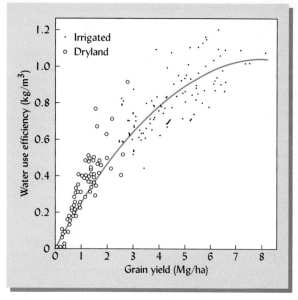

FIGURE 6.18 Evapotranspiration water-use efficiency, the dry matter produced for every unit of water used on the crop, generally increases as crop yield increases. This is true mainly because such yield-improving measures as fertilizer use, better cultivars, denser stands, and better pest control generally increase crop yield by a greater proportion than they do the use of water. This relationship generally applies where water is available to meet the demands of higher-yielding crops. Even so, the curve seems to level off at just over 1 kg grain per 1000 kg water (1 m^3 = 1000 kg water). The data shown are for irrigated and nonirrigated (dryland) wheat fields in the southern high plains region of the U.S. [From Howell (2001) with permission of The American Soc. of Agronomy]

Evaporation and evapotranspiration from Florida's wetlands: http://aquat1.ifas.ufl.edu/guide/evaptran.html

ET EFFICIENCY. Since evapotranspiration includes both transpiration from plants and evaporation from the soil surface, evapotranspiration efficiency is more subject to management than transpiration efficiency. Highest ET efficiency is attained where plant density and other growth factors minimize the proportion of ET attributable to evaporation from the soil. Figure 6.18 illustrates the relationship generally found between crop yields and water use efficiency. Higher yielding plants generally send their roots deeper into the soil profile and produce a denser canopy (higher LAI) that allows less solar radiation to pass through and cause evaporation from the soil surface. We can conclude that as long as the supply of water is not too limited, maintaining optimum conditions for plant growth (by closer plant spacing, fertilization, or selection of more vigorous varieties) increases the efficiency of water use by plants. If irrigation water is not available and the period of rainfall is very short, one should apply this principle with caution. The increase in evapotranspiration by more vigorously growing plants may deplete stored soil water, resulting in serious water stress or even plant death before any harvestable yield has been produced.

In summary, water losses from the soil surface and from transpiration are determined by (1) climatic conditions, (2) plant cover in relation to soil surface (LAI), (3) efficiency of water use by different plants and management regimes, and (4) length and season of the plant-growing period.

6.4 CONTROL OF EVAPOTRANSPIRATION (ET)

Because transpiration is closely related to the total leaf area exposed to solar radiation and evaporation is related to the amount of soil surface so exposed, measures taken to reduce these exposures can bring ET into closer balance with the available water supply and may, to some degree, lessen water stress on desirable vegetation.

Control of Transpiration

In various agricultural situations it may be necessary to limit transpiration by desirable plants (e.g., crops) and/or unwanted (weed) vegetation. Where rapid crop growth might prematurely deplete the water available, it may be wise to limit plant-growth factors such as nutrient supply to only moderate levels, thus keeping LAI in check. The LAI of desired plants can also be limited by spacing plants farther apart, as nature does in arid environments. It should be noted, however, that plants growing farther apart tend to individually produce a greater leaf area, compensating somewhat for the lower density. Also, wider spacing allows greater evaporation losses from the soil.

Integrated weed management. Agric. Canada: http://ssca.usask.ca/ conference/1997proceedings/ Odenovan.html

UNWANTED VEGETATION. It is largely through their heavy use of soil water that weeds interfere with the establishment and growth of desirable forest, range, and crop plants. Weeds in cropland have traditionally been controlled by cultivation (light soil tillage) designed to uproot young weeds and leave them to dry out and die. Disadvantages of cultivation include likely damage to roots of nearby desired plants and the exposure of bare soils, which in turn increases evaporative water loss and eventually encourages runoff and erosion.

HERBICIDES. In recent decades weeds have been widely controlled by spraying weed-killing chemicals called herbicides (see Section 18.2). Weed control with herbicides has several advantages over cultivation, among which are that it requires less labor and energy, and it allows the soil to be left undisturbed, with plant residues covering the soil surface (see below). Chemical weed control also has serious disadvantages in some situations, including high material costs, eventual evolution of weed resistance to specific compounds, damage to desired plants, and environmental toxicity. With regard to the latter, some herbicides are toxic to soil organisms, fish, or land animals, and have accumulated to undesirable levels in downstream waterways and underground drinking-water sources (see Section 18.3).

ALTERNATIVE WEED CONTROLS. Consequently, alternative weed control methods are being developed. For example, some weeds can be held in check by biological controls, that is, encouraging specific insects or diseases that attack only the weeds. Undesirable vegetation in rangeland and some forests can be suppressed by prescribed fires if the timing and temperature levels are carefully controlled to minimize damage to desirable vegetation. Well-timed mowing or grazing can also reduce weed problems. Furthermore, cultivation techniques to help control weeds are being improved, thereby minimizing the need for chemical herbicides. New implements capable of cultivating through high levels of plant residue hold promise for controlling weeds without herbicides while still maintaining some of the desirable surface residue cover.

FALLOW IN DRYLAND CROPPING. Farming systems that alternate bare **fallow** (unvegetated period) one year with traditional cropping the next have been used to conserve soil moisture in some low-rainfall environments. Transpiration water losses during the fallow year are minimized by controlling weeds with light tillage and/or with herbicides. Some of the water saved during the fallow year remains in the profile the following year when a crop is planted, resulting in higher yields of that second-year crop than if the soil had been cropped every year. Early research showed that yields in the alternate years were often sufficiently high as to make up for the harvest foregone during the fallow year. This cropping practice is responsible for the checkerboard of dark fallow soils and golden ripening wheat that can be seen when flying over such semiarid "breadbaskets" as the Great Plains region of the U.S. and Canada.

Research now shows that summer fallow is probably not in the best interests of farmers or natural resource conservation. The main disadvantage of this system is the soil degradation that occurs, largely because of a negative soil organic matter balance (see Section 12.7) and wind erosion (see Section 17.11) during the fallow years. The growing use of conservation tillage practices (see below and Section 17.6) in semiarid areas has led to sufficient savings in soil moisture to minimize the need for fallow cropping (Figure 6.19). Conservation tillage practices (including "no-till") leave most of the crop residues on the soil surface where they reduce evaporation and protect the soil. From research at many sites we can now conclude that long-term productivity, profitability, and soil quality are likely to be optimized with cropping systems that use conservation tillage and keep the soil continuously vegetated with a diversity of crops (Table 6.2).

Control of Surface Evaporation (E)

More than half the precipitation in semiarid and subhumid areas is usually returned to the atmosphere by evaporation (E) directly from the soil surface. In natural rangeland systems, E is a large part of ET because plant communities tend to self-regulate

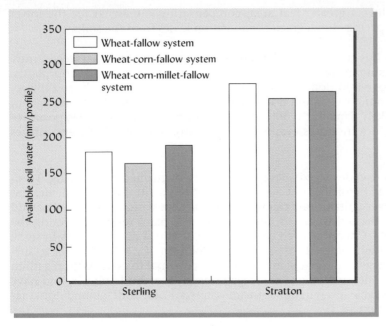

FIGURE 6.19 Available water (mm/profile) in the soil profile at wheat planting time in systems with fallow every second, third, or fourth year at two semiarid locations in eastern Colorado where conservation tillage practices had been adopted. Note that the available soil water level was about the same whether the soil was fallowed every other year or every third or fourth year. Apparently soil evaporation losses were so high during the fallow years that little additional water was conserved by leaving the land fallow. [Redrawn from Farahani et al. (1998)]

toward configurations that minimize the deficit between PET and ET—generally low plant densities with large unvegetated areas between scattered patches of shrubs or bunchgrass. In addition, plant residues on the soil surface are sparse. Evaporation losses are also high in arid-region irrigated soil, especially if inefficient practices are used (see Section 6.9). Even in humid-region rain-fed areas, E losses are significant during hot, rainless periods. Such moisture losses rob the plant community of much of its growth potential and reduce the water available for discharge to streams. Careful study of Figure 6.20 will clarify these relationships and the principles just discussed. Note that the gap between PET and ET represents the *soil water deficit*, which is a measure of how limiting water supply is for plant productivity (see also Plate 76).

For arable soils, the most effective practices aimed at controlling E are those that cover the soil. This cover can best be provided by mulches and by selected conservation tillage practices that leave plant residues on the soil surface, mimicking the soil cover of natural ecosystems.

TABLE 6.2 **Long-term Average Grain Yields in the Semiarid North American Great Plains Region from Cropping Systems With or Without Summer Fallow**

Note that in every case the system without fallow was most productive. Results like these are encouraging farmers to abandon summer fallow in favor of reduced tillage rotations that yield more and better maintain soil quality.

Location	Duration of experiment, years	Mean annual precipitation, cm	Rotation (and tillage)[a]	Long-term avg. yield, Mg ha^{-1} yr^{-1}
Akron, CO	10	41.8	WW–**Fallow** (CT)	1.1
			WW–C–M (NT)	1.5
Mandan, ND	12	42.7	SW–**Fallow** (CT)	1.1
			SW–WW–SF (NT)	1.7
Sidney, MT	13	34.5	SW–**Fallow** (CT)	1.2
			Cont. SW (NT)	1.9
Swift Current, SK	23	36.1	SW–**Fallow** (RT)	1.3
			Lentil–SW (RT)	1.6

[a] Abbreviations: C = corn, Cont. = continuous, CT = Conventional till, M = millet, NT = no-till, RT = reduced till with herbicides, SF = sunflower, SW = spring wheat, WW = winter wheat.
Data from Varvel et al. (2006).

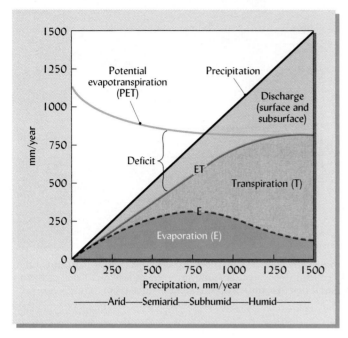

FIGURE 6.20 Partitioning of liquid water losses (discharge) and vapor losses (evaporation and transpiration) in regions varying from low (arid) to high (humid) levels of annual precipitation. The example shown assumes that temperatures are constant across the regions of differing rainfall. Potential evapotranspiration (PET) is somewhat higher in the low-rainfall zones because the lower relative humidity there increases the vapor pressure gradient at a given temperature. Evaporation (E) represents a much greater proportion of total vapor losses (ET) in the drier regions due to sparse plant cover caused by interplant competition for water. The greater the gap between PET and ET, the greater the deficit and the more serious the water stress to which plants are subject. (Diagram courtesy of R. Weil)

Organic vs plastic mulches:
http://jeq.scijournals.org/cgi/
content/full/30/5/1808

VEGETATIVE MULCHES. A *mulch* is a material used to cover the soil surface primarily for the purpose of controlling evaporation, soil erosion, temperature, and/or weeds. Examples of *organic* mulches include straw, leaves, and crop residues. Mulches can be highly effective in checking evaporation, but they may be expensive and labor-intensive to produce or purchase, transport to the field, and apply to the soil. Mulching is therefore most practical for small areas (gardens and landscaping beds) and for high-value crops such as cut flowers, berries, fruit trees, and certain vegetables. It is much less labor-intensive to produce a mulch "in place" by growing a cover crop or cash crop and letting the residues remain on the soil surface (see *Crop Residue* and *Conservation Tillage*, following).

In addition to reducing evaporation, vegetative mulches may provide these benefits: (1) reduce soilborne diseases spread by splashing water; (2) provide a clean path for foot traffic; (3) reduce weed growth (if applied thickly); (4) moderate soil temperatures, especially preventing overheating in summer months (see Section 7.11); (5) increase water infiltration; (6) provide organic matter and, possibly, plant nutrients to the soil; (7) encourage earthworm populations; and (8) reduce soil erosion. Most of these ancillary benefits do not accrue from the use of plastic mulches, discussed next.

PLASTIC MULCHES. Plastic films (and specially prepared paper) are also used as mulch to control evaporative water losses. In the case of black plastic and dark paper, they also effectively control weeds. The mulch is often applied by machine, and plants grow through holes made in the film (Figure 6.21). These mulches are widely used for vegetable and small fruit crops and in landscaping beds, where they are often covered with a layer of tree-bark mulch or gravel for longer life and a more pleasing appearance. As long as the ground is covered, evaporation is checked, and in some cases remarkable increases in plant growth have been reported. Rainwater can usually infiltrate reasonably well through the plant holes and in the uncovered inter-row spaces. A major problem with plastic mulches is the difficulty of completely removing the plastic at the end of the growing season; after a number of years scraps of plastic accumulate in the soil, causing an unsightly mess and interfering with water movement and cultivation. Some manufacterers now offer biodegradable and light-degradable plastic mulches (made from plant products) that remain intact for a month or so and then break down almost completely (see Figure 7.39).

CROP RESIDUE AND CONSERVATION TILLAGE. We have seen that plant residues on the soil surface conserve soil moisture by reducing evaporation and increasing infiltration. *Conservation tillage* practices leave a high percentage of the residues from the previous crop on or near the surface (Figure 6.22, *left*). A conservation tillage practice widely

FIGURE 6.21 For crops with high cash value, plastic mulches are commonly used. The plastic is installed by machine (*left*) and at the same time the plants are transplanted (*right*). Plastic mulches help control weeds, conserve moisture, encourage rapid early growth, and eliminate the need for cultivation. The high cost of plastic makes it practical only with the highest-value crops. (Photos courtesy K. Q. Stephenson, Pennsylvania State University)

used in subhumid and semiarid regions is **stubble mulch** tillage. With this method, residues such as wheat stubble or cornstalks from the previous crop are uniformly spread on the soil surface. The land is then tilled with special implements that permit much of the plant residue to remain on or near the surface. Conservation tillage planters are capable of planting through the stubble and allow much of it to remain on the surface during the establishment of the next crop (see Figure 6.22). In semiarid regions, summer fallow is commonly combined with stubble mulching to conserve soil moisture. Unfortunately, plant growth in dry regions is usually insufficient to produce the large amount of residue mulches needed to greatly reduce evaporation and maximize water conservation (Table 6.3).

Other conservation tillage systems that leave residues on the soil surface include no-tillage (see Figure 6.22, *right*), where the new crop is planted directly into the sod or residues of the previous crop, with almost no soil disturbance. The long-term soil water conserving effects of such tillage systems are shown in Box 6.2. Conservation tillage systems will receive further attention in Section 17.6.

FIGURE 6.22 Conservation tillage leaves plant residues on the soil surface, reducing both evaporation losses and erosion. (*Left*) In a semiarid region (South Dakota) the straw from the previous year's wheat crop was only partially buried to anchor it against the wind while still allowing it to cover much of the soil surface. In the next year the left half of the field, now shown growing wheat, will be **stubble mulched** and the right half sown to wheat. (*Right*) **No-till** planted corn in a more humid region grows up through the straw left on the surface by a previous wheat crop. Note that with no-till, almost no soil is directly exposed to solar radiation, rain, or wind. (Photos courtesy of R. Weil, *left*, and USDA Natural Resources Conservation Service, *right*)

TABLE 6.3   **Gains in Soil Water from Different Rates of Straw Mulch During Fallow Periods at Four Great Plains Locations**

Location	Mean annual precipitation, mm	Soil water gain, mm, at each mulch rate			
		0 Mg/ha	2.2 Mg/ha	4.4 Mg/ha	6.6 Mg/ha
Bushland, TX	508	71	99	99	107
Akron, CO	476	134	150	165	185
North Platte, NB	462	165	193	216	234
Sidney, MT	379	53	69	94	102
Average		107	127	145	157
Average gain by mulching			20	38	50

From Greb (1983).

BOX 6.2   WATER CONSERVATION PAYS OFF

Yields of grain crops have increased dramatically around the world during the past few decades. Three major beneficial factors have stimulated these increases: (1) improved crop varieties, (2) increased availability of chemical nutrients, and (3) increased amounts of available water (mostly from increased irrigation).

Figure 6.23 illustrates the contribution that conserved moisture has made to these yield increases, especially in semiarid areas of the United States. The sorghum grain yields from numerous plots at a USDA research facility in Lubbock, Texas increased about 300% over a 60-year period.

In 1938 the average yield was about 900 kg/ha. By 1997 this average had increased to 3830 kg/ha, an overall increase of some 325%. Further research suggests that about one-third of this yield increase could be attributed to improved varieties, the remaining two-thirds to other factors. Increased nutrient availability was ruled out as a yield-inducing factor since no fertilizers were applied to these plots.

This left increased water availability as the prime factor likely contributing to the higher yields. Analyses of rain and snowfall records showed no increase in precipitation that might have accounted for the increased yields. However, soil moisture measurements made at planting time (Figure 6.24) were found to be much higher during the period from 1972 to 1997 than in the earlier period (1956 to 1972).

Records showed that tillage and residue management were the primary factors that accounted for this soil moisture difference. During the earlier period (1956 to 1972) the plots were tilled between crops to control weeds, and little residue remained on the soil. During the later period (1972 to 1997) there was a major shift to no-tillage crop production, which maintained crop residues on the soil surface. Together with improved weed control using herbicides, the no-till systems reduced evaporative losses, leaving extra moisture in the soil to accommodate the increased sorghum yields.

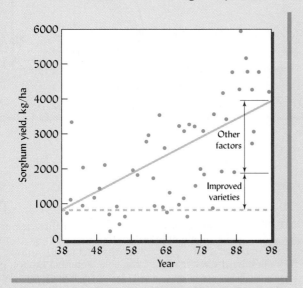

FIGURE 6.23 Sorghum yield increases 1938 to 1998. From Unger and Baumhardt (1999).

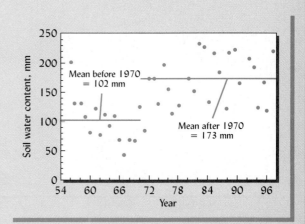

FIGURE 6.24 Changes in soil water storage 1954 to 1997. From Unger and Baumhardt (1999).

6.5 LIQUID LOSSES OF WATER FROM THE SOIL

In our discussion of the hydrologic cycle we noted two types of liquid losses of water from soils: (1) percolation or drainage water, and (2) runoff water (see Figure 6.2). *Percolation water* recharges the groundwater and moves chemicals out of the soil. *Runoff water* often carries appreciable amounts of soil (erosion) as well as dissolved chemicals.

Percolation and Leaching

When the amount of rainfall entering a soil exceeds the water-holding capacity of the soil, losses by percolation will occur. Percolation losses are influenced by (1) the amount of rainfall and its distribution, (2) runoff from the soil, (3) evaporation, (4) the character of the soil, and (5) the nature of the vegetation.

Percolation-Evaporation Balance

Figure 6.25 illustrates the relationships among precipitation, runoff, soil storage, soil water depletion, and percolation for representative humid and semiarid temperate regions and for an irrigated arid region. In the humid temperate region, the rate of water infiltration into the soil (precipitation minus runoff) is greater, at least during certain seasons, than the rate of evapotranspiration. As soon as the soil field capacity is reached, percolation into the substrata occurs.

In the example shown in Figure 6.25*a*, maximum percolation occurs during the winter and early spring, when evaporation is lowest. During the summer, little percolation occurs. In fact, evapotranspiration exceeds precipitation, resulting in a depletion of soil water. Normal plant growth is possible only because of water stored in the soil from the previous winter and early spring.

In the semiarid region, as in the humid region, water is stored in the soil during the winter months and is used to meet the moisture deficit in the summer. But because of the low rainfall, little runoff and essentially no percolation out of the profile occurs.

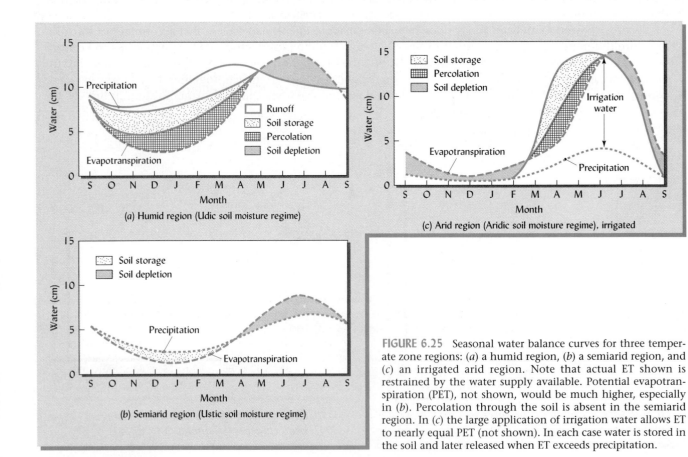

FIGURE 6.25 Seasonal water balance curves for three temperate zone regions: (*a*) a humid region, (*b*) a semiarid region, and (*c*) an irrigated arid region. Note that actual ET shown is restrained by the water supply available. Potential evapotranspiration (PET), not shown, would be much higher, especially in (*b*). Percolation through the soil is absent in the semiarid region. In (*c*) the large application of irrigation water allows ET to nearly equal PET (not shown). In each case water is stored in the soil and later released when ET exceeds precipitation.

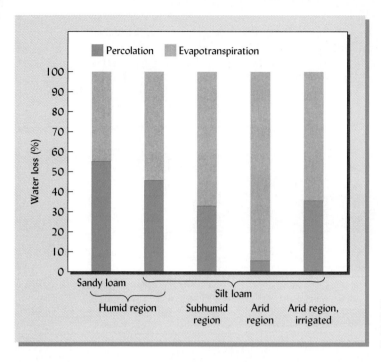

FIGURE 6.26 Percentage of the water entering the soil that is lost by downward percolation and by evapotranspiration. Representative figures are shown for different climatic regions. Surface runoff is assumed to be negligible.

Water may move to the lower horizons, but it is absorbed by plant roots and ultimately is lost by transpiration.

The irrigated soil in the arid region (Figure 6.25c) shows a unique pattern. Irrigation in the early spring, along with a little rainfall, provides more water than is being lost by evapotranspiration. The soil is charged with water, and some percolation may occur. As we shall see in Section 10.8, irrigation systems must provide enough water for some percolation, in order to remove excess soluble salts. During the summer, fall, and winter months, the stored water is depleted because the amount added is less than that removed by the very high evapotranspiration that takes place in response to large vapor pressure gradients.

The situations depicted in Figure 6.25 are typical of temperate zones where potential evapotranspiration (PET) varies seasonally with temperature. In the tropics, where temperatures are somewhat higher and less variable, PET is somewhat more uniform throughout the year, although it does vary with seasonal changes in humidity. In very high rainfall tropical areas (perudic moisture regimes), much more runoff and somewhat more percolation occurs than shown in Figure 6.25. In irrigated areas of the arid tropics, the relationships are similar to those shown for arid temperate regions.

The comparative losses of water by evapotranspiration and percolation through soils found in different climatic regions are shown in Figure 6.26. These differences should be kept in mind while reading the following section on percolation and groundwaters.

6.6 PERCOLATION AND GROUNDWATERS

Groundwater Basics:
http:www.groundwater.org/
gi/whatisgw.html

When drainage water moves downward through the soil and regolith it eventually encounters a zone in which the pores are all saturated with water. Often this saturated zone lies above an impervious soil horizon (Figure 6.27) or a layer of impermeable rock or clay. The upper surface of this zone of saturation is known as the **water table**, and the water within the saturated zone is termed **groundwater**. The water table (Figure 6.28) is commonly only 1 to 10 m below the soil surface in humid regions but may be several hundred or even thousands of meters deep in arid regions. In swamps it is essentially at the land surface.

The unsaturated zone above the water table is termed the **vadose zone** (see Figures 6.27 and 6.28). The vadose zone may include unsaturated materials underlying the soil

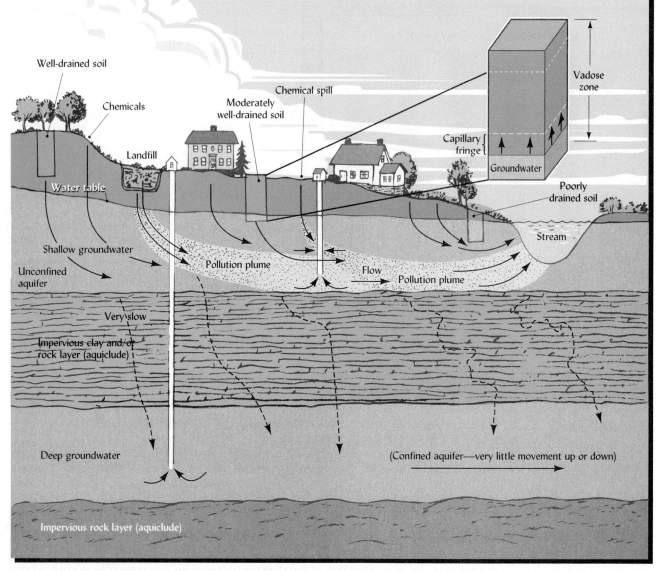

FIGURE 6.27 The water table and groundwater in relation to water movement into and out of the soil. Precipitation and irrigation water percolate down the soil profile, ultimately reaching the water table and underlying shallow groundwater. The unsaturated zone above the water table is known as the *vadose zone* (*upper right*). Groundwater moves up from the water table by capillarity into the *capillary fringe*. Groundwater also moves horizontally down the slope toward a stream, carrying with it chemicals that have leached through the soil, including plant nutrients (N, P, Ca, etc.) as well as pesticides and other pollutants. A shallow well pumps groundwater from the unconfined aquifer near the surface. A deeper well exploits groundwater from a deep, confined aquifer. Two plumes of pollution are shown, one originating from landfill leachate, the other from a chemical spill. The former appears to be contaminating the shallow well. (Diagram courtesy of R. Weil)

profile, and so may be considerably deeper than the soil itself. In some cases, however, the saturated zone may be sufficiently near the surface to include the lower soil horizons, with the vadose zone confined to the upper soil horizons.

Shallow groundwater receives downward-percolating drainage water. Most of the groundwater, in turn, seeps laterally through porous geological materials (termed **aquifers**) until it is discharged into springs and streams. Groundwater may also be removed by pumping for domestic and irrigation uses. The water table will move up or down in response to the balance between the amount of drainage water coming in through the soil and the amount lost through pumped wells and natural seepage to springs and streams. In humid temperate regions, the water table is usually highest in early spring following winter rains and snowmelt, but before evapotranspiration begins to withdraw the water stored in the soil above.

FIGURE 6.28 The water table, capillary fringe, zone of unsaturated material above the water table (vadose zone), and groundwater are illustrated in this photograph. The groundwater can provide significant quantities of water for plant uptake. (Photo courtesy of R. Weil)

Groundwater Resources

Colorado Farmers in Crisis. Audio report: http://www.npr.org/templates/story/story.php?storyId=5400947

Groundwater is a significant source of water for domestic, industrial, and agricultural use. For example, some 20% of all water used in the United States comes from groundwater sources, and about 50% of the people use groundwater to meet at least some of their needs. A shallow water-bearing layer that is not separated from the soil surface by any overlying impermeable layer is called an **unconfined aquifer**. Unconfined aquifers, which are replenished annually, commonly provide water for farm and rural dwellings (see Figure 6.27). The larger groundwater stores in deeper aquifers, which may take decades or centuries to recharge, are commonly pumped to meet municipal, industrial, and irrigation needs. In many cases deep wells are used to tap water from aquifers confined between impermeable strata (termed **aquicludes**). Water is replenished by slow, mainly horizontal seepage from **recharge areas** where the aquifer is exposed to the soil.

Water draining from soils is the main source of replenishment for most underground water resources. So long as water is not withdrawn faster than percolation allows it to be replenished, groundwater may be considered to be a renewable resource. However, in many areas of the world, people are pumping water out faster than it can be replenished, an activity that is lowering water tables and depleting the resource in a manner akin to mining an ore. When overpumping occurs in coastal areas, sea water may push its way into the aquifer, a process termed **saltwater intrusion** (as illustrated in Figure 6.2). Soon, deep municipal wells begin to pull in salty water instead of freshwater.

Shallow Groundwater

Groundwater that is near the surface can serve as a reciprocal water reservoir for the soil. As plants remove water from the soil, it may be replaced by upward capillary movement from a shallow water table. The zone of wetting by capillary movement is known as the **capillary fringe** (see Figure 6.28). Such movement can provide a steady and significant supply of water that enables plants to survive during periods of low rainfall.[2]

---

[2] Capillary rise from shallow groundwater may also bring a steady supply of salts to the surface if the groundwater is brackish (see Section 10.3 for details on this soil-degrading process).

Movement of Chemicals in the Drainage Water[3]

Quality of U.S. waters:
http://www.epa.gov/305b/
2000report/factsheet.pdf

Percolation of water through the soil to the water table not only replenishes the groundwater, it also dissolves and carries downward a variety of inorganic and organic chemicals found in the soil or on the land surface. Chemicals *leached* from the soil to the groundwater (and eventually to streams and rivers) in this manner include elements weathered from minerals, natural organic compounds resulting from the decay of plant residues, plant nutrients derived from natural and human sources, and various synthetic chemicals applied intentionally or inadvertently to soils.

In the case of plant nutrients, especially nitrogen, downward movement through the soil and into underlying groundwaters has two serious implications. First, the leaching of these chemicals represents a depletion of plant nutrients from the root zone (see Section 16.2). Second, accumulation of these chemical nutrients in ponds, lakes, reservoirs, and groundwater downstream may stimulate a process called *eutrophication*, which ultimately depletes the oxygen content of the water, with disastrous effects on fish and other aquatic life (see also Section 14.2). Also, in some areas underground sources of drinking water may become contaminated with excess nitrates to levels unsafe for human consumption (see Section 13.8).

Green roofs to counter
pollution. Audio report:
http://www.npr.org/templates/
story/story.php?storyId=
5454152

Of even more concern is the contamination of groundwater with human pathogens (Box 6.3) and various highly toxic synthetic compounds, such as pesticides and their

---

BOX 6.3 EXCESS LEACHING OF CONTAMINANTS CAN ENDANGER HUMAN HEALTH

Chemical contaminants readily leach into groundwaters from some soils but not others. Homogeneous, fine-textured soils that have no preferential flow channels (wide cracks, animal bore holes, old root channels, etc.) will hold contaminants within the top layers for periods long enough for most contaminants to be destroyed or removed from the water. In contrast, excessively well-drained coarse-textured soils allow these contaminants to move quickly through the soil horizons and down into the groundwater.

As we learned in Chapter 4, there are more and larger macropores in sandy than in clayey soils. Rapid leaching occurs through these large pores. In addition, a sandy soil that is already saturated will leach much more rapidly than one that is near the wilting point. The leaching waters carry nutrients as well as contaminants down to the water table, and then on to wetlands, streams, and underground aquifers, from which drinking water is drawn. An unfortunate example of well-water contamination is found in the following story pieced from news reports.

Over a thousand people were sickened, and two died, from the 0157:H7 strain of *Escherichia coli*, which apparently had contaminated a shallow well used for drinking water at an annual fair in Washington County, New York. Most *E. coli* bacteria are relatively harmless, but this particular strain can be deadly. An overflow crowd of 110,000 people attended the six-day fair. However, only those who consumed food and beverages from the western side of the 25-hectare fairgrounds on the final two days of the fair became ill. On this side, a shallow 7-m well was used as a source of water for the fire department and some 12 vendors. The water system passed all tests conducted by state officials in June and just before the fair in August. However, the DNA of the bacteria in the well-water samples after the fair matched that of the bacteria in the victims.

Findings by Department of Health and by fair officials suggest that four main factors likely led to this tragic situation. First, a summer-long drought preceding the fair had lowered water tables below the inlet level of some shallow wells. Second, within about 30 m of the well in question, some 600 cattle that were shown at the fair were temporarily housed. Third, on the fourth day of the fair (August 26) a 5-cm downpour of rainwater soaked the area and swept through the manure from the animal housing area and into the soil whose characteristics were the fourth factor. The soil was a Hoosic gravelly sandy loam with little if any slope and a high sand content from the surface to the bedrock. This type of soil allows very rapid water percolation and possesses little capacity to remove contaminants by filtering or adsorption of clays.

These four factors allowed the *E. coli*–contaminated water from the cattle barns to percolate downward and undiluted to the well's aquifer, raising its level so that shallow wells could pump out the contaminated water and sicken hundreds of people. Had the soil been more fine-textured, this sad event might well have been avoided. In any case, this story reminds us of the need to take soil properties into consideration in managing the movement of pollutants in the environment.

---

[3] For a review of chemical transport through field soils, see Jury and Fluhler (1992).

breakdown products or chemicals leached out of waste disposal sites (see Chapter 18 for a detailed discussion of these pollution hazards). Figure 6.27 illustrates how the groundwater can be charged with these pathogens or chemicals, and how a plume of contamination spreads to downstream wells and bodies of water.

Chemical Movement Through Macropores

Studies of the movement of chemicals in soils and from soils into the groundwater and downstream bodies of water have called attention to the critical role played by large macropores in determining field hydraulic conductivity (see Section 5.5). The pore configuration in most soils is nonuniform. Old root channels, earthworm burrows, and clay shrinkage cracks commonly contribute large macropores that may provide channels for rapid water flow from the soil surface to depths of 1 m or more.

Once chemicals are carried below the zone of greatest root and microbial activity, they are less likely to be removed or degraded before being carried further down to the groundwater. This means that chemicals that are normally broken down in the soil by microorganisms within a few weeks may move down through large macropores to the groundwater before their degradation can occur.

PREFERENTIAL OR BYPASS FLOW. In some cases, leaching of chemicals is most serious if the chemicals are merely applied on the soil surface. As shown in Figure 6.29, chemicals may be washed from the soil surface into large pores, through which they can quickly move downward. Research suggests that most of the water flowing through large macropores does not come into contact with the bulk of the soil. Such flow is sometimes termed **preferential** or **bypass flow**, as it tends to move rapidly around, rather than through, the soil matrix. As a result, if chemicals have been incorporated into the upper few centimeters of soil, their movement into the larger pores is reduced, and downward leaching is greatly curtailed.

INTENSITY OF RAIN OR IRRIGATION. During a high-intensity rainfall event following a few days of fair weather, water and its associated chemicals move downward rapidly through the channels and cracks (macropores), bypassing the bulk of the soil. In contrast, a gentle rain that in time may provide as much water as the more intense event would likely thoroughly wet the upper soil aggregates, thereby minimizing the rapid downward percolation of both water and the chemicals and pathogens it carries. Table 6.4 illustrates the effects of rainfall intensity on the leaching of pesticides applied to turfgrass.

Manipulations of soil macropores and irrigation intensity provide us with two examples of how environmental quality can be impacted by human activities that influence the hydrologic cycle. Likewise, we have considered steps that can be taken to influence the infiltration of water into soils and to maximize plant biomass production from the water stored in soils. The picture would not be complete if we did not consider briefly three other anthropogenic modifications to the hydrologic cycle: (1) the use of artificial drainage to enhance the downward percolation of excess water from some soils; conversely, (2) the application of additional water to soils as a means of wastewater disposal; and (3) the use of irrigation to supplement water available for plant growth. We shall consider artificial drainage first.

6.7 ENHANCING SOIL DRAINAGE[4]

Some soils tend to be water-saturated in the upper part of their profile for extended periods during part or all of the year. The prolonged saturation may be due to the low-lying landscape position in which the soil is found, such that the regional water table is at or near the soil surface for extended periods. In other soils, water may accumulate above an impermeable layer in the soil profile, creating a *perched water table* (see Figure 6.30). Soils with either type of saturation may be components of wetlands, transitional ecosystems between land and water that are characterized by anaerobic (no oxygen) conditions (see Section 7.7).

---

[4] For a review of all aspects of artificial drainage, see Skaggs and van Schilfgaarde (1999).

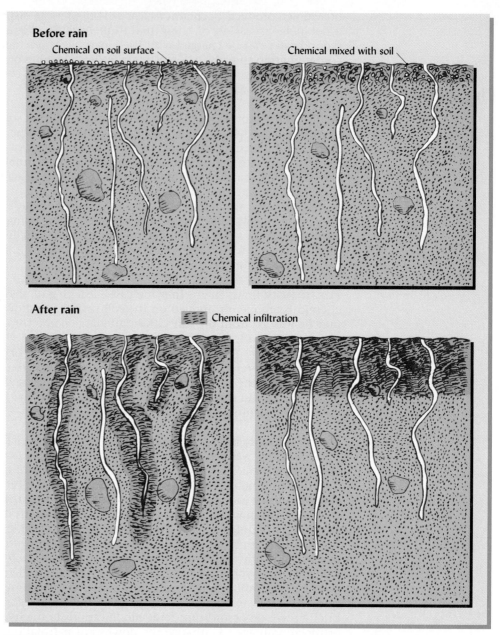

FIGURE 6.29 Preferential or bypass flow in macropores transports soluble chemicals downward through a soil profile. Where the chemical is on the soil surface (*left*) and can dissolve in surface-ponded water when it rains, it may be transported rapidly down cracks, earthworm channels, and other macropores. Where the chemical is dispersed within the soil matrix in the upper horizon (*right*), most of the water moving down through the macropores will bypass the chemical, and thus little of the chemical will be carried downward. Note that channels *not* open all the way to the surface do not conduct water or contaminants by preferential flow.

Reasons for Enhancing Soil Drainage

Water-saturated, poorly aerated soil conditions are essential to the normal functioning of wetland ecosystems and to the survival of many wetland plant species. However, for most other land uses, these conditions are a distinct detriment.

ENGINEERING PROBLEMS.   During construction, the muddy, low-bearing-strength conditions of saturated soils make it very difficult to operate machinery (see Section 4.9). By the same token, soils used for recreation can withstand traffic much better if they are well drained. Houses built on poorly drained soils may suffer from uneven settlement and flooded basements during wet periods. Similarly, a high water table will result in capillary rise of water into roadbeds and around foundations, lowering the soil strength and

TABLE 6.4 Influence of Water Application Intensity on the Leaching of Pesticides Through the Upper 50 cm of a Grass Sod-Covered Mollisol

Soil columns with the natural structure undisturbed received heavy rain (four doses of 2.54 cm each) or light rain (16 doses of 0.64 cm each). Metalaxyl is far more water soluble than Isazofos, but in each case the heavy rain stimulated much more pesticide leaching through the macropores.

	Pesticide leached as percentage of that applied to surface	
Pesticide	Heavy rains	Light rains
Isazofos	8.8	3.4
Metalaxyl	23.8	13.9

Data from Starrett et al. (1996).

leading to damage from frost-heaving (see Section 7.8) if the water freezes in winter. Heavy trucks traveling over a paved road underlaid by a high water table create pot-holes and eventually destroy the pavement.

PLANT PRODUCTION. Water-saturated soils make the production of most upland crops and forest species difficult, if not impossible. In wet soil, farm equipment used for planting, tillage, or harvest operations may bog down. Except for a few specially adapted plant species (bald cypress trees, rice, cattails, etc.), most crop and forest species grow best in well-drained soils since their roots require adequate oxygen for respiration (see Section 7.1). Furthermore, a high water table early in the growing season will confine the plant roots to a shallow layer of partially aerated soil; the resulting restricted root system can lead to water stress later in the year, when the weather turns dry and the water table drops rapidly (Figure 6.31).

For these and other reasons, artificial drainage systems have been widely used to remove excess (gravitational) water and lower the water table in poorly drained soils. Land drainage is practiced in select areas in almost every climatic region, but is most

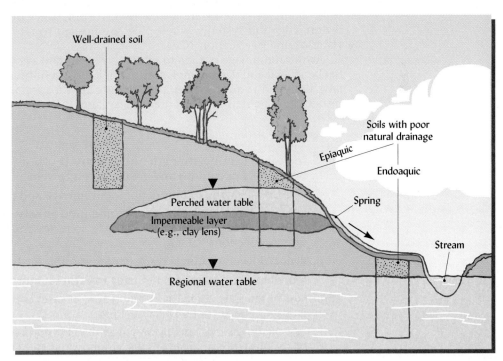

FIGURE 6.30 Cross-section of a landscape showing the regional and perched water tables in relation to three soils, one well-drained and two with poor internal drainage. By convention, a triangle (▼) is used to identify the level of the water table. The soil containing the perched water table is wet in the upper part, but unsaturated below the impermeable layer, and therefore is said to be *epiaquic* (Greek *epi*, upper), while the soil saturated by the regional water table is said to be *endoaquic* (Greek *endo*, within). Artificial drainage can help to lower both types of water tables. (Diagram courtesy of R. Weil)

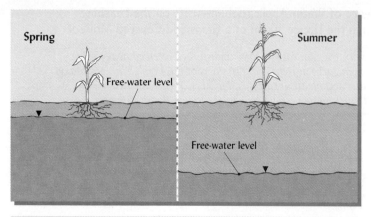

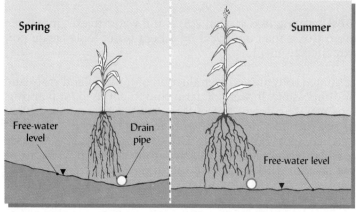

FIGURE 6.31 Illustration of water levels of undrained and tile-drained land in the spring and summer. Benefits of the drainage include more vigorous early growth, as well as a more extensive root system capable of exploiting a large volume of soil when the profile dries out in summer. [Redrawn from Hughes (1980); used with permission of Deere & Company, Moline, IL.]

widely used to enhance the agricultural productivity of clayey alluvial and lacustrine soils. Drainage systems are also a vital, if sometimes neglected, component of arid region irrigation systems, where they are needed to remove excess salts and prevent waterlogging (see Section 10.8).

Artificial drainage is a major alteration of the soil system, and the following potential beneficial and detrimental effects should be carefully considered. In many instances laws designed to protect wetlands require that a special permit be obtained for the installation of a new artificial drainage system.

Benefits of Artificial Drainage

1. Increased bearing strength and improved soil workability, which allow more timely field operations and greater access to vehicular or foot traffic.
2. Less frost-heaving of foundations, pavements, and plants (e.g., see Figure 7.25).
3. Enhanced rooting depth, growth, and productivity of most upland plants due to improved oxygen supply and, in acid soils, lessened toxicity of manganese and iron (see Sections 7.3 and 7.5).
4. Reduced levels of fungal disease infestation in seeds and on young plants.
5. More rapid soil warming, resulting in earlier maturing crops (see Section 7.11).
6. Less production of methane and nitrogen gases that cause global environmental damages (see Sections 12.9 and 13.9).
7. Removal of excess salts from irrigated soils and prevention of salt accumulation by capillary rise in areas of salty groundwater (see Section 10.3).

Detrimental Effects of Artificial Drainage

1. Loss of wildlife habitat, especially waterfowl breeding and overwintering sites.
2. Reduction in nutrient assimilation and other biochemical functions of wetlands (see Section 7.7).

3. Increased leaching of nitrates and other contaminants to groundwater.

4. Accelerated loss of soil organic matter, leading to subsidence of certain soils (see Sections 3.9 and 12.8).

5. Increased frequency and severity of flooding due to loss of runoff water retention capacity.

6. Greater cost of damages when flooding occurs on alluvial lands developed after drainage.

Artificial drainage systems are designed to promote two general types of drainage: (1) *surface drainage* and (2) internal or *subsurface drainage*. Each will be discussed briefly.

Surface Drainage Systems

This type of drainage is extensively used, especially where the landscape is nearly level and soils are fine-textured with slow internal drainage (percolation). Its purpose is to remove water from the land before it infiltrates the soil.

SURFACE DRAINAGE DITCHES. Most surface drainage systems hasten the surface runoff of water by construction of shallow ditches with gentle side slopes that do not interfere with equipment traffic. If there is some slope on the land, the shallow ditches are usually oriented across the slope and across the direction of planting and cultivating, thereby permitting the interception of water as it runs off down the slope. These ditches can be made at low cost with simple equipment. For removing surface water from landscaped lawns, this system of drainage can be modified by constructing gently sloping swales rather than ditches.

LAND SMOOTHING. Often, surface drainage ditches are combined with **land smoothing** to eliminate the ponding of water and facilitate its removal from the land. Small ridges are cut down and depressions are filled in using precision, laser-guided field-leveling equipment. The resulting land configuration permits excess water to move at a controlled rate over the soil surface to the outlet ditch and then on to a natural drainage channel. Land smoothing is also commonly used to prepare a field for flood irrigation (see Section 6.9).

Subsurface (Internal) Drainage

The purpose of subsurface drainage systems is to remove the groundwater from within the soil and to subsequently lower the water table. They require channels such as deep ditches, underground pipes, or "mole" tunnels into which excess water can flow. Internal drainage occurs only when the pathway for drainage is located below the level of the water table (Figure 6.32). The flow of water from a saturated soil into a drainage outlet is illustrated in Figure 6.33. Box 6.4 provides an example of how knowledge of basic soil properties and water-movement principles can be applied in designing a system to alleviate drainage problems in an ornamental garden.

The network of drainage channels may be laid out in several types of patterns, depending on the nature of the area to be drained (Figure 6.34). Where the landscape is too level or at too low an elevation to provide for sufficient fall for gravity removal, expensive pumping operations are sometimes used to remove the drainage water (e.g., in parts of Florida and the Netherlands).

DEEP OPEN-DITCH DRAINAGE. If a ditch is excavated to a depth below the water table (Figures 6.32 and 6.36), water will seep from the saturated soil, where it is under a positive pressure, into the ditch, where its potential will be essentially zero. Once in the ditch, the water can flow rapidly off the field as it no longer must overcome the frictional forces that would delay its twisting journey through tiny soil pores. However, the ditches, being 1 m or more deep, present barriers to equipment. Therefore, deep-ditch drainage is generally practical only for sandy soils in which the ditches may be spaced quite far apart. The high-saturated hydraulic conductivity on these soils

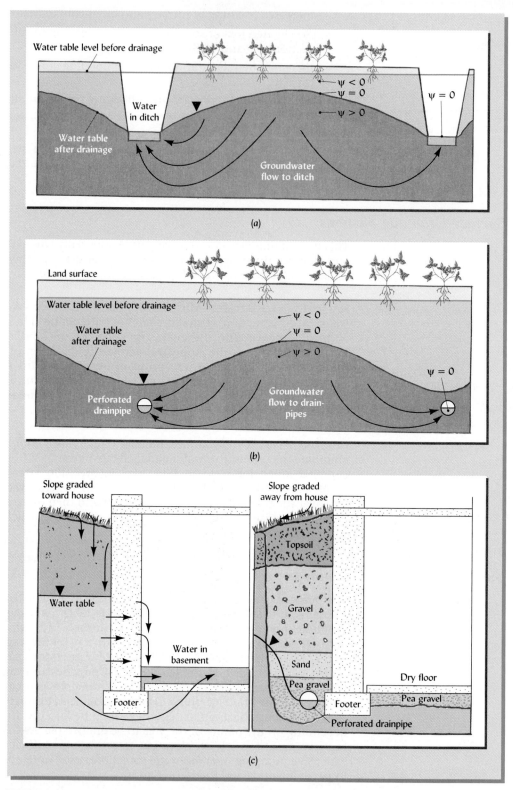

FIGURE 6.32 Three types of subsurface drainage systems. *(a)* Open ditches are used to lower the water table in a poorly drained soil. The wet season levels of the water table before and after ditch installation are shown. The water table is deepest next to the ditch, and the drainage effect diminishes with distance from the ditch. *(b)* Buried "tile lines" made of perforated plastic pipe act very much as the ditches in *(a)*, but have two advantages: they are not visible after installation and they do not present any obstacle for surface equipment. Note the flow lines indicating the paths taken by water moving to the drainage ditches or pipes in response to the water potential gradients between the submerged water ($\psi > 0$) and free water in the drainage ditch and pipes ($\psi = 0$). *(c)* The water table around a building foundation before *(left)* and after *(right)* installation of a *footer drain* and correction of surface grading. The principles of water movement in soils are applied to keep the basement dry. (Diagrams courtesy of R. Weil)

FIGURE 6.33 Demonstration of the saturated flow patterns of water toward a drainage tile. Water containing a colored dye was added to the surface of the saturated soil, and drainage was allowed through the simulation drainage tile shown on the extreme right. (Courtesy G. S. Taylor, Ohio State University)

ensures that the water table will be lowered for a considerable distance from each ditch.[5] Open ditches need regular maintenance to control vegetation and sediment buildup.

BURIED PERFORATED PIPES (DRAIN TILES).[6] A network of perforated plastic pipe can be laid underground using specialized equipment (Figure 6.37, *right*). Water moves into the pipe through the perforations. The pipe should be laid with the slotted or perforated side *down*. This allows water to flow up into the pipe (as shown in Figure 6.33) but protects against soil falling into and clogging the pipe (see Figure 6.37, *left*). Sediment buildup will also be avoided if the pipe has the proper slope (usually a 0.5 to 1% drop) so water flows rapidly to the outlet ditch or stream. The pipe outlet should be covered by a gate or wire mesh to prevent the entrance of rodents in dry weather but still allow the free flow of water.

BUILDING FOUNDATION DRAINS. Surplus water around building foundations can cause serious damage. The removal of this excess water is commonly accomplished using buried perforated pipe, placed alongside and slightly below the foundation or underneath the floor (Figure 6.32c). The perforated pipe must be sloped to allow water to move rapidly to an outlet ditch or sewer. If the drain successfully prevents the water table from rising above the floor level, water will not seep into the basement for the same reason that it will not seep into a drainpipe placed above the water table.

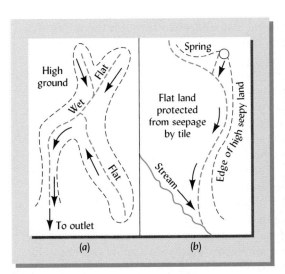

FIGURE 6.34 Three types of subsurface tile drainage systems. (*a*) *natural*, which merely follows the natural drainage pattern; (*b*) *interception*, which cuts off water seeping into lower ground from higher lands above; and (*c*) *gridiron*, which uniformly drains the entirety of an area. In the freshly installed drainage system in the photo, tile lines feed into main drains at the bottom, the direction of flow being shown by the arrows. (Photo courtesy of U.S. Department of Agriculture)

[5] The required spacing of ditches and buried drainpipes ranges from about 10 m apart for very low permeability soils to more than 100 m apart for high-permeability soils such as peats and sands.

[6] Before the adoption of plastic perforated pipe in the 1960s, short sections of ceramic pipe known as *tiles* were laid down end to end with small gaps between them through which water could enter. The terms *drain tile* or *tile line* are still often used in reference to the newer plastic pipe.

BOX 6.4   SUCCESS OR FAILURE IN LANDSCAPE DRAINAGE DESIGN

A hedge of hemlock trees, all carefully pruned into ornamental shapes as part of an intricate landscape design, were dying again because of poor drainage. A very expensive effort to improve the drainage under the hemlocks had proved to be a failure, and the replanted hemlocks were again causing an unsightly blemish in an otherwise picture-perfect garden.

Finally, the landscape architect for the world-famous ornamental garden called in a soil scientist to assist her in finding a solution to the dying hemlock problem. Records showed that in the previous failed attempt to correct the drainage problem, contractors had removed all the hemlock trees in the hedge and had dug a trench under the hedge some 3 m deep (a). They had then backfilled the trench with gravel up to about 1 m from the soil surface, completing the backfill with a high-organic-matter, silt loam topsoil. It was into this silt loam that the new hemlock trees had been planted. Finally a bark mulch had been applied to the surface.

The landscaper who had designed and installed the drainage system apparently had little understanding of the various soil horizons and their relation to the local hydrology. When the soil scientist examined the problem site, he found an impermeable claypan that was causing water from upslope areas to move laterally into the hemlock root zone.

Basic principles of soil water movement also told him that water would not drain from the fine pores in the silt loam topsoil into the large pores of the gravel, and therefore the gravel in the trench would do no good in draining the silt loam topsoil (compare the situation to that in Figures 5.26 and 5.28). In fact, the water moving laterally over the impermeable layer created a perched water table that poured water into the gravel-filled trench, soon saturating both the gravel and the silt loam topsoil.

To cure the problem, the previous "solution" had to be undone (b). The dead hemlocks were removed, the ditches were reexcavated, and the gravel was removed from the trenches. Then the trench, except for the upper $\frac{1}{2}$ m, was filled with a sandy loam subsoil to provide a suitable rooting medium for the replacement evergreen trees (a different species was chosen for reasons unrelated to drainage). The upper $\frac{1}{2}$ m of the trench was filled with a sandy loam topsoil, which was also acid but contained a higher level of organic matter. The interface between the subsoil and surface soil was mixed so that there would be no abrupt change in pore configuration.

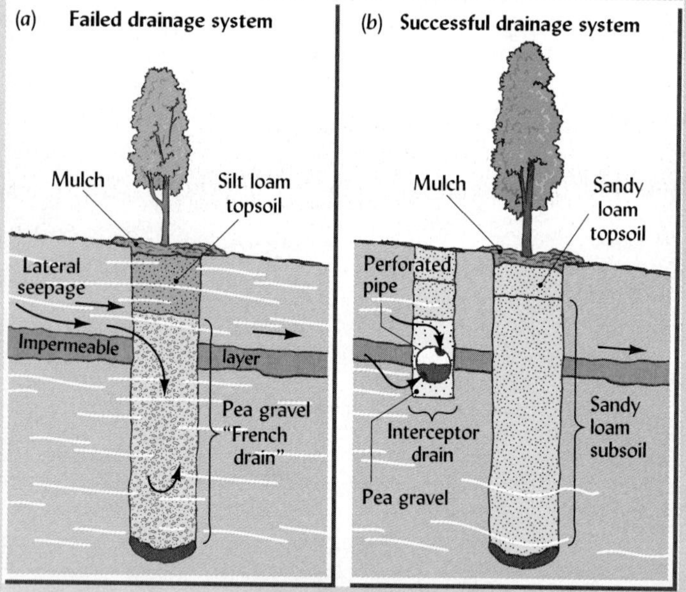

FIGURE 6.35   Failed and successful drainage designs.

This would allow an unsaturated wetting front to move down from the upper to the lower layers, drawing down any excess water.

About 1 m uphill from the trench an *interceptor drain* was installed (b). This involved digging a small trench through the impermeable clay layer that was guiding water to the area. A perforated drainage pipe surrounded by a layer of gravel was laid in the bottom of this trench with about a 1% slope to allow water to flow away from the area to a suitable outlet. The interceptor drain prevented the water moving laterally over the impermeable soil layer from reaching the evergreen hedge root zone.

Even though the replanting of the hedge was followed by an exceptionally rainy year, the new drainage system kept the soil well aerated, and the trees thrived. The principles of water movement explained in Sections 5.6 and 6.7 of this text were applied successfully in the field.

MOLE DRAINAGE.   A **mole drain** system can be created by pulling a pointed shank followed by an attached bullet-shaped steel plug about 7 to 10 cm in diameter through the soil at the desired depth. The compressed wall channel thus formed provides a pathway for the removal of excess water, similar to a buried pipe. Mole drainage is quite inexpensive to install, but is efficient only in fine-textured soils in which the channel is likely to remain open for a number of years.

FIGURE 6.36 An open-ditch drainage system designed to lower the water table during the wet season. The water flowing in the ditch has seeped in from the saturated soil. Such ditches also speed the removal of surface runoff water. The capillary fringe above the water table can be seen as a dark band of soil. In order to protect water quality from chemicals and sediments that might reach the ditch via surface runoff, buffer strips of grass or forest vegetation should be established on both sides. Note in the background the piles of *spoil*, subsoil excavated to make the ditches. This material is usually spread thinly and mixed by tillage with the surface soil in the field, but it should first be tested for extreme acidity or other detrimental properties that could impair the quality of the surface soil in which it is to be mixed. (Photo courtesy of R. Weil)

FIGURE 6.37 (*Left*) A perforated plastic drainage pipe or "tile line" clogged with sediment. The sediment entered the pipe because of improper installation with the perforations facing upward instead of downward. (*Right*) Specialized, laser-guided equipment laying a drain line made of corrugated plastic pipe. The pipe has perforations on the underside that allow water to seep in from a saturated soil. The trench will be backfilled with the soil shown piled on the left. The drain line must be laid deeper than the seasonal high water table if it is to remove any water. [Photo courtesy of Ray Weil (*left*) and USDA/NRCS (*right*)]

6.8 SEPTIC TANK DRAIN FIELDS

Septic tank drainage fields:
http://www.soil.ncsu.edu/
publications/Soilfacts/
AG-439-13/

Thousands of ordinary suburbanites get their first exposure to the importance of water movement through soils when they apply for the building permit for their new dream house. The local authorities will usually not allow a home to be built until arrangements are made for wastewater treatment. Typically, a soil scientist will come out to inspect the soils at the homesite and judge their suitability for use as a septic tank **drain field**. If the soils are found unsuitable, the landowner may be denied the permit to build.

Thus, in many regions, it is through their wastewater treatment function that soil properties influence the value of land and the spread of residential development. For consulting soil scientists in industrialized nations, determination of soil suitability for septic tank drain fields is probably the single most commonly rendered service. Furthermore, in rapidly suburbanizing areas improperly sited septic tank drain fields may contribute significantly to pollution of groundwater and streams. For all these reasons, our discussion of practical water management in soil systems would be incomplete without consideration of the role of water movement through soil in treating wastewater in areas not served by a centralized sewage treatment system.

Operation of a Septic System

Septic systems and their maintenance:
http://www.agnr.umd.edu/users/wye/personel/Miller/septic.html

The most common type of on-site wastewater treatment for homes not connected to municipal sewage systems is the *septic tank* and associated *drain field* (sometimes called *filter field* or *absorption field*). Over 15 million homes in the United States use septic tank drain fields to treat sewage and wastewater. This method of sewage and wastewater treatment depends on several soil processes, of which the most fundamental is the movement of water through the soil.

In essence, a septic drain field operates like artificial soil drainage in reverse. A network of perforated underground pipes is laid in trenches, very much like the network of drainage pipes used to lower the water table in a poorly drained soil. But instead of draining water away from the soil, the pipes in a septic drain field carry wastewater *to* the soil, the water entering the soil via slits or perforations in the pipes. In a properly functioning septic drain field, the wastewater will enter the soil and percolate downward, undergoing several purifying processes before it reaches the groundwater. One of the advantages of this method of sewage treatment is that it has the potential to replenish local groundwater supplies for other uses.

THE SEPTIC TANK.   Water carrying wastes from toilets, sinks, and bathtubs flows by gravity through sealed pipes to a large underground concrete box, called the *septic tank* Figure 6.38a. Baffles in the tank cause the inflowing wastewater to slow down and drop most (70%) of its load of suspended solid materials, which subsequently settle to

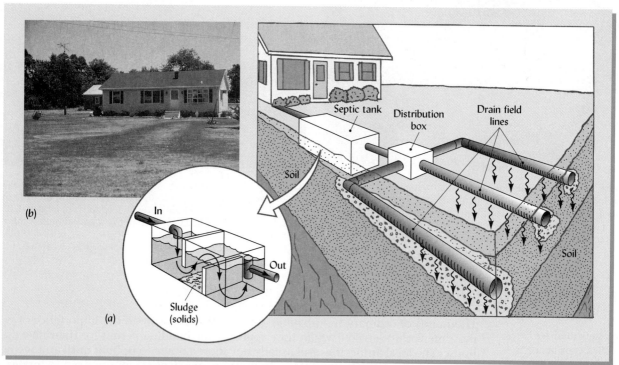

FIGURE 6.38   (a) A septic tank and drain field constituting a standard system for on-site wastewater treatment. Most of the solids suspended in the household wastewater settle out in the concrete septic tank. Effluent from the tank flows to the drain field, where it seeps out of the perforated pipes and into the soil. In the soil, the effluent is purified by microbial, chemical, and physical processes as it percolates toward the groundwater. (b) Photo shows the telltale dark strips of lawn where poorly functioning septic tank drain lines are stimulating the grass with wastewater and nitrogen. (Photo courtesy of R. Weil)

the bottom of the septic tank. As these organic solids partially decompose by microbial action in the septic tank, their volume is reduced so that many years usually pass before the septic tank becomes too full and the accumulated **sludge** (also called *septage*) has to be pumped out.

THE DRAIN FIELD. The water exiting the septic tank via a pipe near the top is termed the septic tank **effluent**. Although its load of suspended solids has been much reduced, it still carries organic particles, dissolved chemicals (including nitrogen), and microorganisms (including pathogens). The flow is directed to one or more buried pipes that constitute the **drain field**. Blanketed in gravel and buried in trenches about 0.6 to 2 m under the soil surface, these pipes are perforated on the bottom to allow the wastewater to seep out and enter the soil. It is at this point that soil properties play a crucial role. Septic systems depend on the soil in the drain field to (1) keep the effluent out of sight and out of contact with people, (2) treat or purify the effluent, and (3) conduct the purified effluent to the groundwater.

As the wastewater percolates, soil microbial action removes the organic materials and pathogens, but releases nitrates that are subject to leaching into the groundwater. Some 4 to 5 kg of nitrogen per year per person using the system may be released in this manner. This suggests that septic tanks from a group of family homes in near proximity to each other may release to the groundwater sufficient nitrates to be of environmental concern.

Soil Properties Influencing Suitability for a Septic Drain Field

Drainage tips for hillside homeowners:
http://www.wy.nrcs.usda.gov/technical/ewpfactsheets/homedrain.html

The soil should have a *saturated hydraulic conductivity* (see Section 5.5) that will allow the wastewater to enter and pass through the soil profile rapidly enough to avoid backups that might saturate the surface soil with effluent, but slowly enough to allow the soil to purify the effluent before it reaches the groundwater. The soil should be sufficiently *well aerated* to encourage *microbial breakdown* of the wastes and *destruction of pathogens*. The soil should have some fine pores and clay or organic matter to adsorb and filter contaminants from the wastewater.

Soil properties that may disqualify a site for use as a septic drain field include impermeable layers such as a fragipan or a heavy claypan, gleying in the upper horizons, too steep a slope, or excessively drained sand and gravel.

Septic tank drain fields installed where soil properties are not appropriate may result in extensive pollution of groundwater and in health hazards caused by seepage of untreated wastewater. In Figure 6.38b the dark lines in the lawn show where the grass has responded to the water and nitrogen in the waste stream. Such signs indicate that the soil has too slow a percolation rate or too high a water table and that the wastewater has moved upward rather than downward (see Plate 67).

SUITABILITY RATING. The suitability of a site for septic drain field installation depends largely on soil properties that affect water movement and the ease of installation (see Table 6.5). For example, too steep a slope may interfere both with the ease of installation and with the operation of a septic drain field. A septic drain field laid out on a slope greater than 15% may allow considerable lateral movement of the percolating water such that at some point downslope, the wastewater will seep to the surface and present a potential health hazard (Plate 66).

The soil properties ideal for a septic drain field are nearly the opposite of those associated with the need for tile drainage. For example, instead of a high water table that requires lowering by drainage, septic drain field sites should have a low water table so that there is plenty of well-aerated soil to purify the wastewater before it reaches the groundwater. Application of large quantities of wastewater through septic drain fields will actually raise the water table somewhat under the drain field.

PERC TEST. This is a test that determines the *percolation rate* (which is related to the saturated hydraulic conductivity described in Section 5.5) expressed in millimeters (or other unit of depth) of water entering the soil per hour. The percolation rate indicates whether or not the soil can accept wastewater rapidly enough to provide a practical disposal medium (Table 6.5). The test is simple to conduct (Figure 6.39) and should be carried out during the wettest season of the year.

TABLE 6.5 Soil Properties Influencing Suitability for a Septic Tank Drain Field

Note that most of these soil properties pertain to the movement of water through the soil profile.

Soil property[a]	Limitations		
	Slight	Moderate	Severe
Flooding	—	—	Floods frequent to occasional
Depth to bedrock or impermeable pan, cm	>183	102–183	<102
Ponding of water	No	No	Yes
Depth to seasonal high water table, cm	>183	122–183	<122
Permeability (perc test) at 60 to 152 cm soil depth, mm/h	50–150	15–50	<15 or >150[b]
Slope of land, %	<8	8–15	>15
Stones >7.6 cm, % of dry soil by weight	<25	25–50	>50

[a] Assumes soil does not contain permafrost and has not subsided more than 60 cm.
[b] Soil permeability (as determined by a perc test) greater than 150 mm/h is considered too fast to allow for sufficient filtering and treatment of wastes.
Adapted from Soil Survey Staff (1993), Table 620–17.

To some degree, a low percolation rate can be compensated for by increasing the total length of drain field pipes, and hence increasing the area of land devoted to the drain field. The size of the septic drain field is also influenced by the amount of waste-water that is likely to be generated (which may be estimated by the number of bedrooms in the house being served).

Alternative Systems

In certain low-lying regions, it is virtually impossible to find sites where the water table is deep enough to be suitable for installation of a standard septic tank drain field. When this is the case, alternatives to the traditional underground septic drain field must be found.

One of these alternatives is the *mound drain field system*, which involves the construction of a septic drain field above ground (Figure 6.40). The perforated drain field pipes are laid above ground on a bed of sand and are covered by a mound of sandy soil material. A final covering of loamy soil is used to support grass vegetation. This type of system must rely on pumps to deliver the wastewater up to the perforated pipes in the mound. Vegetation established on the mound will use some of the water for evapotranspiration, but most of the wastewater will seep through the porous mound material into the underlying soil.

A few communities are experimenting with artificial wetlands constructed to treat septic tank effluent where soils are too impermeable or low-lying to be suitable for a

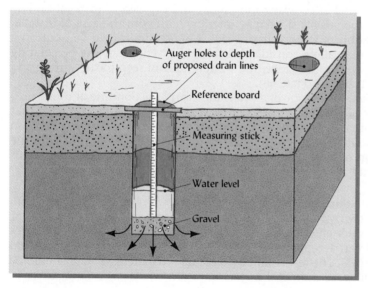

FIGURE 6.39 The perc test used to help determine soil suitability for a septic tank drain field. On the site proposed for the drain field, a number of holes are drilled to the depth where the perforated pipes are to be laid. The bottom of each hole is lined with a few centimeters of gravel, and the holes are filled with water as a pretreatment to ensure that the soil is wet when the test is conducted. After the water has drained, the hole is refilled with water and a measuring rod is used to determine how rapidly the water level drops during a period of several hours or a day. This measurement is related to the saturated hydraulic conductivity. (Diagram courtesy of R. Weil)

FIGURE 6.40 A mound-type septic tank effluent treatment system installed for a newly built house in a flat, poorly drained region. Note the air vent on the top of the mound and the ponded rainwater in the foreground. (Photo courtesy of R. Weil)

standard drain field. The effluent is made to slowly flow through a series of shallow, vegetated ponds that purify the wastewater by physically filtering some of the solids and by removing or destroying nutrients and organic compounds by plant uptake and biochemical reactions.

Such alternatives to septic drain fields require a special permit and are not allowed in all localities. Still more uncommon, but more environmentally friendly, are composting toilets that convert human wastes directly into a humuslike soil amendment rather than using water to flush wastes away.

6.9 IRRIGATION PRINCIPLES AND PRACTICES[7]

FAO's map/data of irrigated areas:
http://www.fao.org/ag/agl/aglw/aquastat/irrigationmap/index.stm

In most regions of the world, insufficient water is the prime limitation to agricultural productivity. In semiarid and arid regions intensive crop production is all but impossible without supplementing the meager rainfall provided by nature. However, if given supplemental water through irrigation, the sunny skies and fertile soils of some arid regions stimulate extremely high crop yields. It is no wonder, then, that many of the earliest civilizations depended on irrigated agriculture (and vice versa). The history of irrigation is nearly as old as the history of agriculture itself. Rice producers in Asia, wheat and barley producers in the Middle East, and corn producers in Central and South America were irrigating their crops well over 2000 years ago. The ancient Mesopotamian civilizations of the Tigris and Euphrates river valleys created complex networks of canals and ditches to divert water from the great rivers to extensive areas of cultivated land. In other places, farmers filled simple buckets with water from a stream or open well and carried it to the thirsty plants in their gardens. Today these methods of water conveyance can still be seen, although in many agricultural and landscaping systems they have been replaced by modern pumps and pipes.

Importance of Irrigation Today

FOOD PRODUCTION. During the twentieth century, the area of irrigated cropland expanded greatly in many parts of the world, including the 17 semiarid states of the western United States (see Plate 111, after page 656). The total area under irrigation

---

[7] For a fascinating account of water resources and irrigation management in the Middle East, see Hillel (1995). For a practical manual on small-scale irrigation with simple but efficient microtechnology, see Hillel (1997). For extensive information on irrigation of crops, see Lascano and Sojka (2007).

seems to be leveling off at about 210 million hectares worldwide, about 15% of the world's agricultural land (see Figure 20.11). The high productivity induced by irrigation is evident in the fact that this irrigated land is responsible for some 40% of global crop production.

Expanded and improved irrigation, especially in Asia, has been a major factor in helping global food supplies keep up with, and even surpass, the global growth in population. As a result, irrigated agriculture remains the largest *consumptive* user[8] of water resources, accounting for about 80% of all water consumed worldwide, in both developing and developed countries (Figure 6.41).

Guide to desert landscaping: http://www.vvwater.org/guide/index.htm

LANDSCAPING.   Irrigation is an integral part of such landscaping installations as golf courses, home lawns, and flower beds (see Figure 6.6). Irrigation serves to keep the grass green during summer dry spells in humid regions, but is necessary almost year-round to keep certain species thriving in drier regions. In arid regions, irrigation can alter a glaring, brown landscape into one of cooling shade trees and colorful flowers. In most cases, the use of irrigation for landscaping is predicated on the desire to maintain vegetation that conforms to an ideal notion of perpetual green lushness. For example, turfgrass species such as Kentucky bluegrass, which is naturally verdant in Kentucky only during a few cool, wet months, are made to stay green year-round in some desert gardens.

In contrast, lawns, landscapes, and golf courses in many parts of the world utilize only adapted vegetation that is capable of surviving the periods of dry weather and other adverse conditions characteristic of the local climate without irrigation. Increasing environmental awareness has engendered a growing trend toward more xerophytic landscaping (utilizing desert plants and rocks) in arid regions and generally toward more use of locally native vegetation that requires little or no irrigation.

FUTURE PROSPECTS.   The water for expanded irrigation has come largely from reservoirs formed by the construction of dams and from the pumping of groundwater out of deep aquifers. Among the problems facing irrigated agriculture in the future is the slowly dwindling availability of irrigation water from both of these sources. The reasons for reduction include (1) increased competition for water from a growing population of urban water users, (2) the overpumping of aquifers that has led to falling water tables, (3) reduction of storage capacity of existing reservoirs by siltation with

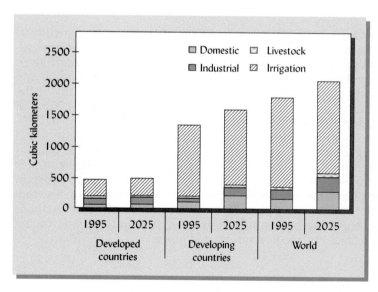

FIGURE 6.41 Consumptive use of water for domestic, industrial, livestock, and irrigation purposes in developing and developed countries and for the world. Consumptive use refers to water that is withdrawn by pumping from rivers or aquifers but not returned to the source after use. Most of it eventually evaporates. The values for 2025 are predicted. Note that irrigation is the dominant consumer of water in all cases. [From Rosegrant et al. (2002) Reproduced and adapted with permission from the International Food Policy Research Institute and the International Water Management Institute]

---

[8] A distinction is made between *withdrawals* and *consumptive use*. The latter term refers to water withdrawn from surface or groundwater sources and used in a way that does not return the water directly to those sources for reuse. Thus, water withdrawn for irrigation and lost to the atmosphere by evapotranspiration is a consumptive use, but water withdrawn to cool an electric generation turbine and then returned to the river from which it came is not consumptive use. On average, about 30 to 35% of the water withdrawals for irrigation are returned to water sources for reuse and are not part of the consumptive use, while 90 to 95% of the withdrawals for industrial use are returned and so are not considered consumptive use.

eroded sediments (see Section 17.2), and (4) increased recognition of the need to allow a portion of river flows to go unused by irrigation in order to maintain fish habitats downstream.

Water resources for irrigation are most rapidly becoming scarce and/or expensive in arid regions where they are most needed. Reducing waste and achieving greater efficiency of water use in irrigation are increasingly important aspects that will be emphasized in this section. One of the other major problems associated with irrigation, the salinization of soils and drainage waters, is considered in Chapter 10.

Water-Use Efficiency

Various measures of water-use efficiency are used to compare the relative benefits of different irrigation practices and systems. The most meaningful overall measure of efficiency would compare the output of a system (crop biomass or value of marketable product) to the amount of water allocated as an input into the system. There are many factors to consider (such as type of plants grown, reuse of "wasted" water by others downstream, etc.), so such comparisons must be made with caution.

APPLICATION EFFICIENCY.   A simpler measure of water-use efficiency, sometimes termed the **water application efficiency**, compares the amount of water available or allocated to irrigate a field to the amount of water actually used in transpiration by the irrigated plants. In this regard, most irrigation systems are quite inefficient, with only about 10 to 30% of the water that is taken from the source transpired by the desired plants. Table 6.6 gives estimates of the various water losses that typically occur between allocation and transpiration. The table compares average losses in semiarid regions under both rainfed and irrigated agriculture.

Much of the water loss occurs by evaporation and leakage in the reservoirs, canals, and ditches used to store and deliver water to irrigated fields. To reduce these losses during water distribution, ditches can be lined with concrete or plastic (see Figure 6.42). Evaporative losses can be all but eliminated during distribution by the use of pipelines instead of open canals, but this is very much more expensive.

FIELD WATER EFFICIENCY.   Water-use efficiency *in the field* may be expressed as

$$\text{Field water efficiency, \%} = \frac{\text{Water transpired by the crop}}{\text{Water applied to the field}} \times 100 \qquad (6.3)$$

Values are usually less than 50%. As shown in Table 6.6, field water efficiency for traditional irrigation systems used in semiarid regions is often as low as 20 to 25%. For example, if transpiration uses 18% out of the total allocated water of the 70% that reaches the field, the field water efficiency = 100 (18/70) = 25.7%. The water delivered to the field that is not transpired by the crop is lost as surface runoff, deep percolation below the root zone, and/or evaporation from the soil surface. Achieving a high level of field water efficiency (or a low level of water wastage) is very dependent on the skill of the irrigation manager and on the methods of irrigation used. As shown in

TABLE 6.6   Estimates of Water Losses in Traditional Irrigated and Rainfed Agriculture in Semiarid Areas of the World

	Irrigated agriculture	Rainfed agriculture
	Percent of available water (%)[a]	
Storage and conveyance losses	30	0
Surface runoff and drainage losses[b]	44	40–50
Evaporation (from soil or water)	8–13	30–35
Transpiration by the crop	13–18	15–30

[a] Available water for irrigated agriculture = water stored in reservoirs or pumped from groundwater, available water for rainfed agriculture = rainfall.
[b] Some of the water lost by drainage and runoff may be reused by downstream irrigators.
From Wallace (2000).

FIGURE 6.42 Concrete-lined irrigation ditches (*center*) and standard-sized siphon pipes (*right*) can increase the efficiency of water delivery to the field. Unlined ditches (*left*) lose much of the water to adjacent soil areas or to the groundwater. Note the evidence of capillary movement above the water level in the unlined ditch. (Left and right photos courtesy of N. C. Brady, center photo courtesy of R. Weil)

Table 6.7, different irrigation systems vary greatly in their field water efficiencies. We will now turn our attention to a brief description of the principal methods of irrigation in use today.

Surface Irrigation

In these systems water is applied to the upper end of a field and allowed to distribute itself by gravity flow. Usually the land must be leveled and shaped so that the water will flow uniformly across the field. The water may be distributed in **furrows** graded to a slight slope so that water applied to the upper end of the field will flow down the furrows at a controlled rate (Figure 6.43). In **border irrigation** systems, the land is shaped into broad strips 10 to 30 m wide, bordered by low dikes.

WATER CONTROL. Water is usually brought to surface-irrigated fields in supply ditches or gated pipes (such as are shown in Figures 6.42–6.44). The amount of water that enters the soil is determined by the permeability of the soil and by the length of time a given spot in the field is inundated with water. Achieving a uniform infiltration of exactly the required amount of water is very difficult and depends on controlling the slope and length of the irrigation runs across the field. If the soil is highly permeable (e.g., some loamy sands), too much water may infiltrate near the upper end of the field and too little may reach the lower end (Figure 6.44). On the other hand, for a fine-textured soil, infiltration may be so slow that water flows across the field and ponds up or runs off the lower end without a sufficient amount soaking into the soil. Therefore, in the very sandy soils, leaching loss of water and chemicals is a problem at the upper end of the field. In clayey soils, erosion, runoff, and waterlogging may be problems at the lower end.

The **level basin** technique of surface irrigation, as is used for paddy rice and certain tree crops, alleviates these problems because each basin has no slope and is completely surrounded by dikes that allow water to stand on the area until infiltration is complete.

TABLE 6.7 Some Characteristics of the Three Principal Methods of Irrigation

Methods and specific examples	Direct costs of installation, 2006 dollars/ha[a]	Labor requirements	Field water efficiency, %[b]	Suitable soils
Surface: basin, flood, furrow	600–900	High to low, depending on system	20–50	Nearly level land; not too sandy or rocky
Sprinkler: center pivot, movable pipe, solid set	900–1800	Medium to low	60–70	Level to moderately sloping; not too clayey
Microirrigation: drip, porous pipe, spitter, bubbler	1000–2000	Low	80–90	Steep to level slopes; any texture, including rocky or gravelly soils

[a] Average ranges from many sources. Costs of required drainage systems not included.
[b] Field water efficiency = $100 \times$ (water transpired by crop/water applied to field).

FIGURE 6.43 Typical irrigation scene in early spring in the western United States. The water is delivered to the field in concrete ditches, and then (*left*) siphoned from the ditch into miniponds that can supply water to every row or alternate rows as shown (*right*). These small ponds also reduce furrow head erosion and encourage any suspended soil particles to settle out before the slow-moving water starts flowing down the furrows. Note the upward capillary movement of the water along the sides of the rows. (Photos courtesy of N. Brady)

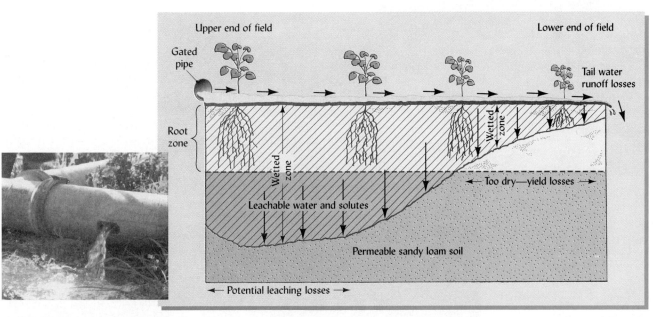

FIGURE 6.44 Penetration of water into a coarse-textured soil under surface irrigation. The high infiltration rate causes most of the water to soak in near the gated pipe at the upper end of the field. The uneven penetration of water results in the potential for leaching losses of water and dissolved chemicals at the upper end of the field, while plants at the lower end may receive insufficient water to moisten the entire potential root zone. On a less permeable soil, or on a field with a steep slope, the tail water runoff losses would likely be greater at the lower end and leaching potential less at the upper end. (Photo and diagram courtesy of R. Weil)

This method is not practical for highly permeable soils into which water would infiltrate so fast that the basin would never fill. On sloping land, terraces can be built in a modification of the level-basin method (Figure 6.45).

Variants of the surface systems have been in use for 5000 years. They require little equipment and are relatively inexpensive to operate. The principal capital cost is usually the initial land shaping, which may be quite expensive if the land is not nearly level to begin with. However, control over leaching and runoff losses is difficult, and the entire soil surface is wetted so that much water is lost by evaporation from the soil and by weed transpiration.

Sprinkler Systems

In sprinkler irrigation, water is sprayed through the air onto a field, simulating rainfall. Thus the entire soil surface, as well as plant foliage (if present), is wetted. This leads to evaporative losses similar to those described for surface systems. Furthermore, an additional 5 to 20% of the applied water may be lost by evaporation or windblown mist as the drops fly through the air. One advantage is that plants often respond positively to the cooler, better-aerated sprinkler water. A disadvantage is that wet leaves may increase the incidence of fungal diseases in some plants, such as grapes, fruit trees, and roses, so sprinkler systems are not often used for these plants.

WATER CONTROL.   A sprinkler system should be designed to deliver water at a rate that is less than the infiltration capacity of the soil, so that runoff or excessive percolation will not occur. In practice, runoff and erosion may be problems if the soil infiltration capacity is low, the land is relatively steep, or too much water is applied in one place. Water is sometimes lost by deep percolation because more water falls near the sprinkler than farther from it. Overlapping of spray circles can help achieve a more even distribution of water. Because of better control over application rates, the field water-use efficiency is generally higher for sprinkler systems than for surface systems, especially on coarse-textured soils.

FIGURE 6.45   The level-basin type of surface irrigation modified for sloping land in South Asia by construction of terraces. Paddy rice is growing in the flooded terrace basins. This type of irrigation is practical only on soils of low permeability. (Upper photo courtesy of R. Weil)

SUITABLE SOILS.   Sprinkler irrigation is practical on a wider range of soil conditions than is the case for the surface systems. Various types of sprinkler systems are adapted to moderately sloping as well as level land. They can be used on soils with a wide range of textures, even those too sandy for surface irrigation systems.

EQUIPMENT.   The equipment costs for sprinkler systems are higher than those for surface-flow systems. Large pressure pumps and specialized pipes and nozzles are required. Some types of sprinkler systems are set in place, others are moved by hand, and still others are self-propelled, either moving in large circles around a central pivot (Figure 6.46) or rolling slowly across a rectangular field. Most systems can be automated and adapted to deliver doses of pesticides or soluble fertilizers to plants.

Microirrigation

The most efficient irrigation systems in use today are those using microirrigation, whereby only a small portion of the soil is wetted (Figure 6.47, *left*) in contrast to the complete wetting accomplished by most surface and sprinkler systems. Microirrigation may also refer to the tiny amounts of water applied at any one time and the miniature size of the equipment involved.

Perhaps the best-established microirrigation system is *drip* (or *trickle*) *irrigation*, in which tiny emitters attached to plastic tubing apply water to the soil surface alongside individual plants (Plate 74). In some cases the tubing and emitters are buried 20 to 50 cm deep so the water soaks directly into the root zone. In either case, water is applied at a low rate (sometimes drop by drop) but at a high frequency, with the objective of maintaining optimal soil water availability in the immediate root zone while leaving most of the soil volume dry (Figure 6.47). Table 6.8 illustrates the high field-water efficiency of drip irrigation.

Other forms of microirrigation that are especially well adapted for irrigating individual trees include *spitters* (microsprayers) and *bubblers* (small vertical standpipes); (Figure 6.47). The bubblers (and usually the spitters) require that a small level basin be formed in the soil under each tree.

WATER CONTROL.   Water is normally carried to the field in pipes, run through special filters to remove any grit or chemicals that might clog the tiny holes in the emitters (filtering is not necessary for bubblers), and then distributed throughout the field by means of a network of plastic pipes. Soluble fertilizers may be added to the water as needed.

FIGURE 6.46   Center pivot irrigation systems. The system at right is making a low-energy, precision application of water to a soybean crop and is rotating slowly towards the left. The photo at left shows another center pivot system with a heavy duty motor used to pump the water up from the groundwater. Also shown are a large tank of liquid fertilizer and the computer controls that allow nutrients to be injected into the water at precise rates during irrigation events. [Photos courtesy of R. Weil]

FIGURE 6.47 Two examples of microirrigation. (*Left*) *Drip* or *trickle* irrigation with a single emitter for each seedling in a cabbage field in Africa. (*Right*) A *microsprayer* or *spitter* irrigating an individual tree in a home garden in Arizona. In both cases, irrigation wets only the small portion of the soil in the immediate root zone. Small quantities of water applied at high frequency (such as once or twice a day) ensures that the root zone is kept almost continuously at an optimal moisture content. (Photos courtesy of R. Weil, *left,* and N. C. Brady, *right.*)

If properly maintained and managed, microirrigation allows much more control over water application rates and spatial distribution than do either surface or sprinkler systems. Losses by supply-ditch seepage, sprinkler-drop evaporation, runoff, drainage (in excess of that needed to remove salts), soil evaporation, and weed transpiration can be greatly reduced or eliminated. Once in place, the labor required for operation is modest.

Microirrigation often produces healthier plants and higher crop yields because the plant is never stressed by low water potentials or low aeration conditions that are associated with the feast-or-famine regime of infrequent, heavy water applications made by all surface irrigation systems and most sprinkler systems. A disadvantage or risk is that there is very little water stored in the soil at any time, so even a brief breakdown of the system could be disastrous in hot, dry weather.

EQUIPMENT. The capital costs for microirrigation tend to be higher than for other systems (see Table 6.7), but the differences are not so great if the cost of drainage systems for control of salinity and waterlogging in surface systems, the costs of high-pressure pumping in sprinkler systems, and the real value of wasted water is taken into account. Because of its high water-use efficiency, microirrigation is most profitable where water supplies are scarce and expensive and where high-valued plants such as fruit trees are being grown.

TABLE 6.8 **Efficiencies of Selected Irrigation Methods in the High Plains of Texas**

Note that improved irrigation methods show 20–35% higher efficiency and considerably lower water requirements.

Irrigation method	Typical efficiency (percent)	Water application needed to add 100 mm to root zone (millimeters)	Water savings over conventional furrow (percent)
Conventional furrow	60	167	—
Furrow with surge valve	80	125	25
Low-pressure sprinkler	80	125	25
LEPA sprinkler[a]	90–95	105	37
Drip	95	105	37

[a] LEPA refers to low-energy precision application that delivers the water close to the plants.
From Postel (1999) based on data from the High Plains Underground Water Conservation District (Lubbock, Texas).

6.10 CONCLUSION

The hydrologic cycle encompasses all movements of water on or near the Earth's surface. It is driven by solar energy, which evaporates water from the ocean, the soil, and vegetation. The water cycles into the atmosphere, returning elsewhere to the soil and the oceans in rain and snow.

The soil is an essential component of the hydrologic cycle. It receives precipitation from the atmosphere, rejecting some of it, which is then forced to run off into streams and rivers, and absorbing the remainder, which then moves downward to be either transmitted to the groundwater, taken up and later transpired by plants, or evaporated directly from soil surfaces and returned to the atmosphere.

The behavior and movement of water in soils and plants are governed by the same set of principles: water moves in response to differences in energy levels, moving from higher to lower water potential. These principles can be used to manage water more effectively and to increase the efficiency of its use.

Management practices should encourage movement of water into well-drained soils while minimizing evaporative (E) losses from the soil surface. These two objectives will provide as much water as possible for plant uptake and groundwater recharge. Water from the soil must satisfy the transpiration (T) requirements of healthy leaf surfaces; otherwise, plant growth will be limited by water stress. Practices that leave plant residues on the soil surface and that maximize plant shading of this surface will help achieve high efficiency of water use.

Extreme soil wetness, characterized by surface ponding and saturated conditions, is a natural and necessary condition for wetland ecosystems. However, for most other land uses, extreme wetness is detrimental. Drainage systems have therefore been developed to hasten the removal of excess water from soil and lower the water table so that upland plants can grow without aeration stress, and so the soil can better bear the weight of vehicular and foot traffic.

A septic tank drain field operates as a drainage system in reverse. Septic wastewaters can be disposed of and treated by soils if the soils are freely draining. Soils with low permeability or high water tables may indicate good conditions for wetland creation or appropriate sites for installation of artificial drainage for agricultural use, but they are not generally suited for septic tank drain fields.

Irrigation waters from streams or wells greatly enhance plant growth, especially in regions with scarce precipitation. With increasing competition for limited water resources, it is essential that irrigators manage water with maximal efficiency so that the greatest production can be achieved with the least waste of water resources. Such efficiency is encouraged by practices that favor transpiration over evaporation, such as mulching and the use of microirrigation.

As the operation of the hydrologic cycle causes constant changes in soil water content, other soil properties are also affected, most notably soil aeration and temperature, the subjects of the next chapter.

STUDY QUESTIONS

1. You know that the forest vegetation that covers a 120 km^2 wildland watershed uses an average of 4 mm of water per day during the summer. You also know that the soil averages 150 cm in depth and at field capacity can store 0.2 mm of water per mm of soil depth. However, at the beginning of the season the soil was quite dry, holding an average of only 0.1 mm/mm. As the watershed manager, you are asked to predict how much water will be carried by the streams draining the watershed during the 90-day summer period when 450 mm of precipitation falls on the area. Use the water balance equation to make a rough prediction of the stream discharge as a percentage of the precipitation and in cubic meters of water.

2. Draw a simple diagram of the hydrologic cycle using a separate arrow to represent these processes: *evaporation, transpiration, infiltration, interception, percolation, surface runoff,* and *soil storage*.

3. Describe and give an example of the *indirect* effects of plants on the hydrologic balance through their effects on the soil.

4. State the basic principle that governs how water moves through the SPAC. Give two examples, one at the soil–root interface and one at the leaf–atmosphere interface.

5. Define *potential evapotranspiration* and explain its significance to water management.

6. What is the role of evaporation from the soil (E) in determining water-use efficiency, and how does it affect ET? List three practices that can be used to control losses by E.

7. Weed control should reduce water losses by what process?

8. Comment on the relative advantages and disadvantages of organic versus plastic mulches.

9. What does conservation tillage conserve? How does it do it?

10. The small irrigation project you manage collects 2,000,000 m³ of water annually in a reservoir. Of this, 20% evaporates from the reservoir surface during the year. Of the remaining water, 25% is lost by evaporation and percolation into the soil during distribution via unlined canals before the water reaches the fields. The water is then applied by furrow irrigation, with averages 20% of the water applied percolating below the crop root zone, 20% running off into collection canals at the low end of the field, and 30% evaporating from the soil surface. By the time of crop harvest, ET has dried the soil to about the same water content it had before irrigation began. Average ET is 7 mm/day for a 180-day irrigation season and crop water-use efficiency averages 1.1 kg of dry matter/m³ water transpired. Show calculations to estimate:

    (a) the overall water-use efficiency of the project (kg output/m³ water allocated),
    (b) the application water efficiency for the project,
    (c) the field water efficiency for the project,
    (d) the number of hectares that can be irrigated in this project.

11. Explain under what circumstances earthworm channels might increase downward saturated water flow, but not have much effect on the leaching of soluble chemicals applied to the soil.

12. What will be the effect of placing a perforated drainage pipe in the capillary fringe zone just above the water table in a wet soil? Explain in terms of water potentials.

13. What soil features may limit the use of a site for a septic tank drain field?

14. Which irrigation systems are likely to be used where: (a) water is expensive and the market value of crops produced per hectare is high, and (b) the cost of irrigation water is subsidized and the value of crop products that can be produced per hectare is low? Explain.

REFERENCES

Day, R. L. et al. 1998. "Water balance and flow patterns in a fragipan using in situ soil block." *Soil Sci.* **163**:517–528.

Farahani, H. J. et al. 1998. "Soil water storage in dryland cropping systems: The significance of cropping intensification." *Soil Sci. Soc. Am. J.* **62**:984–991.

Gburek, W. J., B. A. Needelman, and M. S. Srinivasan. 2006. Fragipan controls on runoff generation: Hydropedological implications at landscape and watershed scales. *Geoderma* **131**:330–344.

Greb, B. W. 1983. "Water conservation: Central Great Plains." In H. E. Dregue and W. O. Willis, eds., *Dryland Agriculture*. Agronomy Series no. 23. (Madison, WI.: Amer. Soc. of Agron.).

Haynes, J. L. 1954. "Ground rainfall under vegetative canopy of crops." *J. Amer. Soc. Agron.* **46**:67–94.

Hillel, D. 1995. *The Rivers of Eden* (New York: Oxford University Press).

Hillel, D. 1997. *Small-Scale Irrigation for Arid Zones*. FAO Development Series 2. (Rome: U.N. Food and Agriculture Organization).

Howell, T. A. 2001. Enhancing water use efficiency in irrigated agriculture. *Agron. J.* **93**:281–289.

Hughes, H. S. 1980. *Conservation Farming.* (Moline, Ill.: John Deere and Company).

Jury, W. A., and H. Fluhler. 1992. "Transport of chemicals through soil: Mechanisms, models, and field applications." *Advances in Agronomy* 47:141–201.

Kittridge, J. 1948. *Forest Influences: The Effects of Woody Vegetation on Climate, Water and Soil* (New York: McGraw-Hill).

Lascano, R. and R. Sojka. 2007. *Irrigation of Agricultural Crops.* Agron. Monograph No. 30. 2nd Ed. (Madison: Amer. Soc. Agronomy).

Linde, D. T. et al. 1995. "Surface runoff assessment from creeping bent grass and perennial ryegrass turf." *Agron. J.* **87**:176–182.

Passioura, J. B. 2002. Soil conditions and plant growth. *Plant, Cell and Environment* **25**:311–318.

Postel, S. 1999. *Pillar of Sand: Can the Irrigation Miracle Last?* (Washington, DC: Worldwatch Institute).

Rosegrant, M. W., X. Cai, and S. A. Cline. 2002. Global water outlook to 2025: Averting in impending crisis. International Food Policy Research Institute and the International Water Management Institute, Washington, DC, http://www.ifpri.org/pubs/fpr/fprwater2025.pdf

Skaggs, R. W., and J. van Schilfgaarde (eds.). 1999. *Agricultural Drainage.* Agronomy Series no. 38. (Madison, WI: Amer. Soc. Agron., Crop Sci. Soc. Amer., Soil Sci. Soc. Amer.).

Soil Survey Staff. 1993. *National Soil Survey Handbook.* Title 430-VI. (Washington, DC: USDA Natural Resources Conservation Service).

Starrett, S. K., N. E. Christians, and T. Al Austin. 1996. "Movement of pesticides under two irrigation regimes applied to turfgrass." *J. Environ. Qual.* **25**:566–571.

Swank, W. T., J. M. Vose, and K. J. Elliot. 2001. Long-term hydrologic and water quality responses following commercial clear cutting of mixed hardwoods on a southern appalachian catchment. *Forest Ecology & Management* **143**:163–178.

Tyree, M. T. 2003. The ascent of water. *Nature* **423**:923.

Unger, P. W., and R. L. Baumhardt. 1999. "Factors related to dryland grain sorghum yield increases: 1939 through 1997." *Agron J.* **91**:870–875.

Varvel, G., W. Riedell, E. Deibert, B. Mcconkey, D. Tanaka, M. Vigil, and R. Schwartz. 2006. Great Plains cropping system studies for soil quality assessment. *Renewable Agriculture and Food Systems* **21**:3–14.

Waddell, J., and R. Weil. 1996. "Water distribution in soil under ridge-till and no-till corn." *Soil Sci. Soc. Amer. J.* **60**:230–237.

Wallace, J. S. 2000. Increasing agricultural water use efficiency to meet future food production. *Agric. Ecosyst. Environ.* **82**:105–119.

Zanders, J. 2001. Urban development and soils. Soil Horizons—A Newsletter of Landcare Research New Zealand, Ltd. 6(Nov.):6.

Plant-animal community on poorly aerated wetland soils (R. Weil)

7

SOIL AERATION AND TEMPERATURE

The naked earth is warm with Spring. . . .
—JULIAN GRENFELL,
INTO BATTLE

It is a central maxim of ecology that "everything is connected to everything else." This interconnectedness is one reason why soils are such fascinating (and challenging) objects of study. In this chapter we shall explore two aspects of the soil environment, aeration and temperature; they are not only closely connected to each other but are both also intimately influenced by many of the soil properties discussed in other chapters.

Since air and water share the pore space of soils, it is not surprising that much of what we learned about the texture, structure, and porosity of soils (Chapter 4) and the retention and movement of water in soils (Chapters 5 and 6) will have direct bearing on soil aeration. These are some of the physical parameters affecting aeration status, but chemical and biological processes also affect, and are affected by, soil aeration.

For the growth of plants and the activity of microorganisms, soil aeration status can be just as important as soil moisture status and can sometimes be even more difficult to manage. In most forest, range, agricultural, and ornamental applications, a major management objective is to maintain a high level of oxygen in the soil for root respiration. Yet it is also vital that we understand the chemical and biological changes that take place when the oxygen supply in the soil is depleted.

Soil temperatures affect plant and microorganism growth and also influence soil drying by evaporation. The movement and retention of heat energy in soils are often ignored, but they hold the key to understanding many important soil phenomena, from frost-damaged pipelines and pavements to the Spring awakening of biological activity in soils. The unusually high soil temperatures that result from fires on forest-, range-, or croplands can markedly change critical physical and chemical soil properties.

We will see that increasing soil temperatures influence soil aeration largely through their stimulating effects on the growth of plants and soil organisms and on the rates of biochemical reactions. Nowhere are these interrelationships more critical than in the water-saturated soils of wetlands, ecosystems that will therefore receive special attention in this chapter.

7.1 SOIL AERATION—THE PROCESS

In order for plant roots and other soil organisms to readily carry on respiration, the soil must be well ventilated. Good ventilation allows the exchange of gases between the soil and the atmosphere to supply enough oxygen (O_2) while preventing the potentially toxic accumulation of gases such as carbon dioxide (CO_2), methane (CH_4) and ethylene (C_2H_6). Soil aeration status involves the rate of such ventilation, as well as the proportion of pore spaces filled with air, the composition of that soil air, and the resulting chemical oxidation or reduction potential in the soil environment.

Soil Aeration in the Field

Soil aeration and plant growth:

www.uoguelph.ca/~mgoss/five/410_N06.html

Oxygen availability in field soils is regulated by three principal factors: (1) *soil macroporosity* (as affected by texture and structure), (2) *soil water content* (as it affects the proportion of porosity that is filled with air), and (3) O_2 *consumption* by respiring organisms (including plant roots and microorganisms). The term *poor soil aeration* refers to a condition in which the availability of O_2 in the root zone is insufficient to support optimal growth of upland plants and aerobic microorganisms. Typically, poor aeration becomes a serious impediment to plant growth when O_2 concentration drops below 0.1 L/L. This often occurs when more than 80 to 90% of the soil pore space is filled with water (leaving less than 10 to 20% of the pore space filled with air). The high soil water content not only leaves little pore space for air storage, but, more important, the water blocks the pathways by which gases could exchange with the atmosphere. Compaction can also cut off gas exchange, even if the soil is not very wet and has a large percentage of air-filled pores.

Excess Moisture

The extreme case of excess water occurs when all or nearly all of the soil pores are filled with water. The soil is then said to be **water saturated** or **waterlogged**. Waterlogged soil conditions are typical of wetlands and may also occur for short periods of time on upland sites. In well-drained soils, saturated conditions may occur temporarily during a heavy rainstorm, when excess irrigation water is applied, or if wet soil has been compacted by plowing or by heavy machinery.

Plants adapted to life in waterlogged soils are termed **hydrophytes**. For example, a number of grass species, including rice, eastern gamagrass, and spartina marsh grasses transport oxygen for respiration down to their roots via hollow structures in their stems and roots known as **aerenchyma** tissues. Mangroves (Figure 7.1) and other

FIGURE 7.1   Relatively few plants, such as these mangrove trees along the coast of the Indian Ocean, are able to grow in saturated soils that are virtually devoid of oxygen. The mangroves are partially submerged during high tide, but have developed aerial "knees" on their roots that enable them to access oxygen directly from the atmosphere. To survive in this environment, the mangroves have also evolved means of excluding the salts in the ocean water. (Photo courtesy of R. Weil)

hydrophytic trees produce aerial roots and other structures that allow their roots to obtain O_2 while growing in water-saturated soils.

Most plants, however, are dependent on a supply of oxygen from the soil and suffer dramatically if good soil aeration is not maintained by drainage or other means (Figure 7.2). Some plants succumb to O_2 deficiency or toxicity of other gases within hours after the soil becomes saturated.

Gaseous Interchange

The more rapidly roots and microbes use up oxygen and release carbon dioxide, the greater is the need for the exchange of gases between the soil and the atmosphere. This exchange is facilitated by two mechanisms, ***mass flow*** and ***diffusion***. Mass flow of air is much less important than diffusion in determining the total exchange that occurs. It is enhanced, however, by fluctuations in soil moisture content that force air in or out of the soil or by wind and changes in barometric pressure.

The great bulk of the gaseous interchange in soils occurs by *diffusion*. Through this process, each gas moves in a direction determined by its own partial pressure. The *partial pressure* of a gas in a mixture is simply the pressure this gas would exert if it alone were present in the volume occupied by the mixture. Thus, if the pressure of air is 1 atmosphere (~100 kPa), the partial pressure of oxygen, which makes up about 21% (0.21 L/L) of the air by volume, is approximately 21 kPa.

Diffusion allows extensive gas movement from one area to another even though there is no overall pressure gradient for the total mixture of gases. There is, however, a concentration gradient for each individual gas, which may be expressed as a *partial pressure gradient*. As a consequence, the higher concentration of oxygen in the atmosphere will result in a net movement of this particular gas into the soil. Carbon dioxide and water vapor normally move in the opposite direction, since the partial pressures of these two gases are generally higher in the soil air than in the atmosphere. A representation of the principles involved in diffusion is given in Figure 7.3.

FIGURE 7.2 Most plants depend on the soil to supply oxygen for root respiration and therefore are disastrously affected by even relatively brief periods of soil saturation during which oxygen becomes depleted. (*Left*) Sugar beets on a clay loam soil dying where the soil has become water-saturated in a compacted area. (*Right*) Pine trees dying in a sandy soil area that has become saturated as a result of flooding by beavers. A new community of plants better adapted to poorly aerated soil conditions is taking over the site. (Photos courtesy of R. Weil)

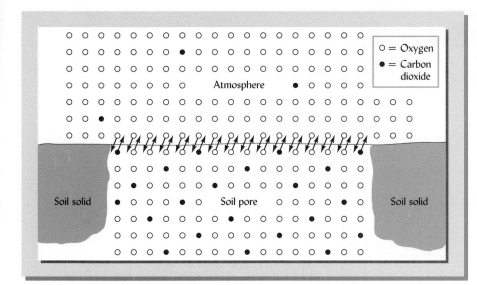

FIGURE 7.3 The process of diffusion between gases in a soil pore and in the atmosphere. The total gas pressure is the same on both sides of the boundary. The partial pressure of oxygen is greater, however, in the atmosphere. Therefore, oxygen tends to diffuse into the soil pore where fewer oxygen molecules per unit volume are found. The carbon dioxide molecules, on the other hand, move in the opposite direction owing to the higher partial pressure of this gas in the soil pore. This diffusion of O_2 into the soil pore and of CO_2 into the atmosphere will continue as long as the respiration by roots and microorganisms in the soil consumes O_2 and releases CO_2.

7.2 MEANS OF CHARACTERIZING SOIL AERATION

The aeration status of a soil can be characterized in several ways, including (1) the content of oxygen and other gases in the soil atmosphere, (2) the air-filled soil porosity, and (3) the chemical oxidation-reduction (redox) potential.

Gaseous Composition of the Soil Air

OXYGEN. The atmosphere above the soil contains nearly 21% O_2, 0.035% CO_2, and more than 78% N_2. In comparison, soil air has about the same level of N_2 but is consistently lower in O_2 and higher in CO_2. The O_2 content may be only slightly below 20% in the upper layers of a soil with an abundance of macropores. It may drop to less than 5% or even to near zero in the lower horizons of a poorly drained soil with few macropores. Once the supply of O_2 is virtually exhausted, the soil environment is said to be **anaerobic**.

Low O_2 contents are typical of wet soils. Even in well-drained soils, marked reductions in the O_2 content of soil air may follow a heavy rain, especially if oxygen is being rapidly consumed by actively growing plant roots or by microbes decomposing readily available supplies of organic materials (Figure 7.4). Oxygen depletion in this manner occurs most rapidly when the soil is warm.

CARBON DIOXIDE. Since the N_2 content of soil air is relatively constant, there is a general inverse relationship between the contents of the other two major components of soil air—O_2 and CO_2—with O_2 decreasing as CO_2 increases. Although the actual differences in CO_2 amounts may not be impressive, they are significant, comparatively speaking. Thus, when the soil air contains only 0.35% CO_2, this gas is about 10 times as concentrated as it is in the atmosphere. In cases where the CO_2 content becomes as high as 10%, it may be toxic to some plant processes.

OTHER GASES. Soil air usually is much higher in water vapor than is the atmosphere, being essentially saturated except at or very near the surface of the soil (see Section 5.7). Also, under waterlogged conditions, the concentrations of gases such as methane (CH_4) and hydrogen sulfide (H_2S), which are formed as organic matter decomposes, are notably higher in soil air. Another gas produced by anaerobic microbial metabolism is ethylene (C_2H_4). This gas is particularly toxic to plant roots, even in concentrations lower than 1 μL/L (0.0001%). Root growth of a number of plants has been shown to be inhibited by ethylene that accumulates when gas exchange rates between the atmosphere and the soil are too slow.

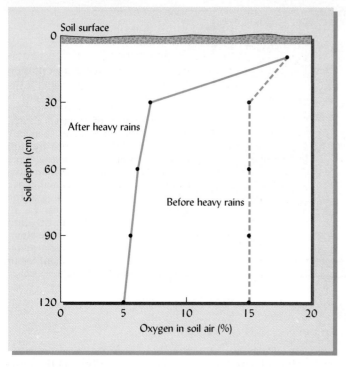

FIGURE 7.4 Oxygen content of soil air before and after heavy rains in a soil on which cotton was being grown. The rainwater replaced most of the soil air. The small amount of oxygen remaining was consumed in the respiration of roots and soil organisms. The carbon dioxide content (not reported) probably increased accordingly. [Redrawn from Patrick (1977); used with permission of the Soil Science Society of America]

Air-Filled Porosity

Oxidation-reduction reactions:
www.shodor.org/UNChem/advanced/redox/index.html

In Chapter 1 (Figure 1.18) we noted that the ideal soil composition for plant growth would include close to a 50:50 mix of air and water in the soil pore space, or about 25% air in the soil, by volume (assuming a total porosity of 50%). Many researchers believe that microbiological activity and plant growth become severely inhibited in most soils when air-filled porosity falls below 20% of the pore space or 10% of the total soil volume (with correspondingly high water contents).

One of the principal reasons that high water contents cause oxygen deficiencies for roots is that water-filled pores block the diffusion of oxygen into the soil to replace that used by respiration. In fact, oxygen diffuses 10,000 times faster through a pore filled with air than through a similar pore filled with water.

7.3 OXIDATION–REDUCTION (REDOX) POTENTIAL[1]

Important chemical characteristics related to soil aeration are the reduction and oxidation states of the chemical elements.

Redox Reactions

The reaction that takes place as the reduced state of an element is changed to the oxidized state may be illustrated by the oxidation of two-valent iron [Fe^{2+} or Fe(II)] in FeO to the trivalent form [Fe^{3+} or Fe(III)] in FeOOH.

$$\underset{\text{Fe(II)}}{\overset{(2+)}{2FeO}} + 2H_2O \rightleftharpoons \underset{\text{Fe(III)}}{\overset{(3+)}{2FeOOH}} + 2H^+ + 2e^- \tag{7.1}$$

As reaction 7.1 proceeds to the *right*, each Fe(II) loses an electron (e^-) to become Fe(III) and forms H^+ ions by hydrolyzing H_2O. These H^+ ions lower the pH. When the reaction proceeds to the *left*, FeOOH acts as an **electron acceptor** and the pH rises as H^+ ions are consumed. The tendency or potential for electrons to be transferred from one

---

[1] For reviews of redox reactions in soils, see Bartlett and James (1993) and Bartlett and Ross (2005).

substance to another in such reactions can be measured using a platinum electrode and is termed the **redox potential** (E_h).

The redox potential is usually measured in volts or millivolts. As is the case for water potential (Section 5.3), redox potential is related to a reference state, in this case the hydrogen couple $\frac{1}{2}H_2 \rightleftharpoons H^+ + e^-$, whose redox potential is arbitrarily taken as zero. If a substance will accept electrons easily, it is known as an *oxidizing agent;* if a substance supplies electrons easily, it is a *reducing agent.*

Role of Oxygen Gas

Oxygen gas (O_2) is an important example of a strong oxidizing agent, since it rapidly accepts electrons from many other elements. All aerobic respiration requires O_2 to serve as the electron acceptor as living organisms oxidize organic carbon to release energy for life.

Oxygen can oxidize both organic and inorganic substances. Keep in mind, however, that as it oxidizes another substance, O_2 is in turn reduced. This reduction process can be seen in the following reaction.

$$\overset{(0)}{\frac{1}{2}O_2} + 2H^+ + 2e- \rightleftharpoons \overset{(2-)}{H_2O} \tag{7.2}$$

Note that the oxygen atom which has zero charge in elemental O_2 accepts two electrons, taking on a charge of -2 when it becomes part of the water molecule. These electrons could have been donated by two molecules of FeO undergoing oxidation, as shown in reaction 7.1. If we combine the equations 7.1 and 7.2, we can see the overall effect of oxidation and reduction.

$$\begin{aligned} 2FeO + 2H_2O &\rightleftharpoons 2FeOOH + 2H^+ + 2e^- \\ \frac{1}{2}O_2 + 2H^+ + 2e^- &\rightleftharpoons H_2O \\ \hline 2FeO + \frac{1}{2}O_2 + H_2O &\rightleftharpoons 2FeOOH \end{aligned} \tag{7.3}$$

The donation and acceptance of electrons (e^-) and H^+ ions on each side of the equations have balanced each other and therefore do not appear in the combined reaction. However, for the specific reduction and oxidation reactions, they are both very important.

In a well-aerated soil with plenty of gaseous O_2, the E_h is in the range of 0.4 to 0.7 volt (V). As aeration is reduced and gaseous O_2 is depleted, the E_h declines to about 0.32 to 0.38 V. If organic-matter-rich soils are flooded under warm conditions, E_h values as low as -0.3 V can be found.

The relationship between changes in O_2 content and E_h of a wet soil is shown in Figure 7.5. Within a day or two after the warm soil became water saturated, aerobic and

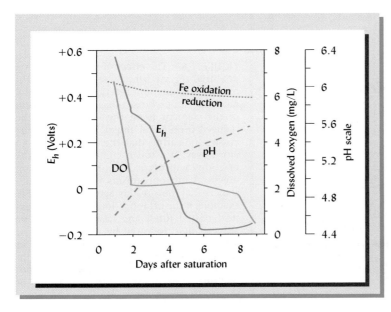

FIGURE 7.5 Changes in soil chemistry following water-saturation of a silt loam A horizon. In the first two days, aerobic and facultative microorganisms respired most of the dissolved O_2 (DO), thereby lowering the E_h of the soil solution. As DO was depleted and E_h dropped, conditions became suitable for anaerobic microorganisms. Substances such as the iron (Fe) in soil minerals are reduced by anaerobic microbes, which use them as the terminal electron acceptor for their metabolism. This process lowers the E_h further. The dotted line indicates that the E_h at which Fe^{3+} tends to be reduced to Fe^{2+} declines somewhat as the pH rises. The rise in pH depicted by the dashed line in the graph is in turn partly caused by the consumption of H^+ ions in reactions that reduce Fe^{3+} (see reactions 7.1 and 7.4). Such reactions also change the color of the soil by dissolving certain minerals. [Adapted from Jenkinson and Franzmeier (2006) with permission of the Soil Sci. Soc. of America]

facultative microorganisms oxidizing organic carbon in the soil (see Section 11.2) respired most of the O_2 initially present, thereby lowering the E_h of the soil solution.

Other Electron Acceptors

Redoximorphic features in soils:

http://nesoil.com/images/redox.htm

As O_2 becomes depleted and E_h drops, reducing conditions become established. With no O_2 available, only **anaerobic** microorganisms can survive. They must use substances other than O_2 as the terminal electron acceptors for their metabolism. For example, they may use iron in soil minerals. As they reduce iron, the E_h drops still further because electrons are consumed. The pH simultaneously rises because the reaction consumes H^+ ions:

$$\underset{\text{Fe(III)}}{Fe(OH)_3} + e^- + 3H^+ \longrightarrow \underset{\text{Fe(II)}}{Fe^{2+}} + 3H_2O \tag{7.4}$$

When these reduction reactions proceed, soil colors change from the reds of iron oxides to the greys of reduced iron (see Plate 63 and Sections 4.1 and 7.7). Similar reactions involve the reduction or oxidation of C, N, Mn, S, and other elements from the solution, organic matter, and mineral components of the soil.

The E_h value at which a particular oxidation–reduction (**redox**) reaction can occur depends on the chemical to be oxidized or reduced and the pH at which the reaction takes place, as illustrated by the sloping lines in Figure 7.6. At E_h values above and below the line, a particular chemical element would be found in oxidized and reduced forms, respectively. The fact that these lines are sloping downward suggests that the E_h for a reaction is lower when the pH of the soil solution is higher. Since both pH and E_h are easily measured, it is not too difficult to ascertain whether a certain reaction could occur in a given soil under specific conditions.

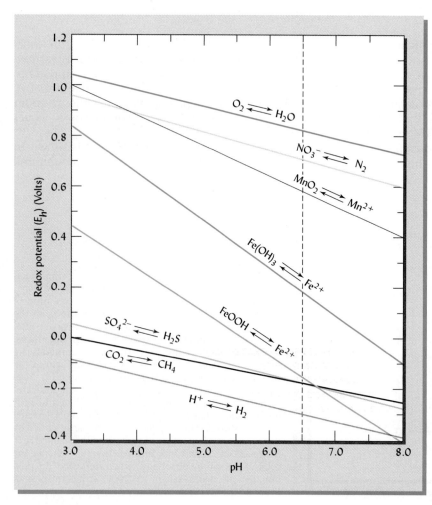

FIGURE 7.6 The effect of pH on the redox potential E_h at which several important reduction–oxidation reactions theoretically take place. In each reaction the oxidized form of the element is on the left side and the reduced form is on the right side with the arrows indicating that the reactions can go either way. If a soil becomes saturated, the reactions will occur in order from the top down, as microorganisms first use the substances that most readily accept electrons. For example, if the pH is 6.5 (indicated by the vertical dotted line), any gaseous O_2 will be reduced first, followed by nitrogen (NO_3^-), then manganese (MnO_2), then iron in $Fe(OH)_3$, iron in FeOOH, and finally sulfur (in SO_4^{2-}) and carbon (in CO_2). [Modified from McBride (1994); used with permission of Oxford University Press]

TABLE 7.1 **Oxidized to Reduced Forms and Charge for Several Elements and Redox Potentials E_h at Which the Redox Reactions Occur in a Soil at pH 6.5**

The relative order of elements reduced is the same as in Figure 7.6, but E_h values measured were generally lower than the theoretical values indicated in the figure. At E_h levels lower than about 0.38 to 0.32 V, microorganisms utilize elements other than oxygen as their electron acceptor.

Element	Oxidized form	Charge on oxidized element	Reduced form	Charge on reduced element	E_h at which change of form occurs, V
Oxygen	O_2	0	H_2O	-2	0.38 to 0.32
Nitrogen	NO_3^-	$+5$	N_2	0	0.28 to 0.22
Manganese	Mn^{4+}	$+4$	Mn^{2+}	$+2$	0.22 to 0.18
Iron	Fe^{3+}	$+3$	Fe^{2+}	$+2$	0.11 to 0.08
Sulfur	SO_4^{2-}	$+6$	H_2S	-2	-0.14 to -0.17
Carbon	CO_2	$+4$	CH_4	-4	-0.20 to -0.28

E_h values from Patrick and Jugsujinda (1992).

The order of E_h lines from the top to the bottom of Figure 7.6 illustrates the *sequence* in which the various reduction reactions are likely to occur when a well-aerated soil becomes saturated with water. For example, at pH of 6.5 (indicated by the dotted vertical line), the reduction of nitrate (NO_3^-) to nitrogen gas (N_2) occurs before manganese Mn^{4+} (in MnO_2) is reduced to Mn^{2+}. It is important to note, however, that E_h lines are theoretical and based on the assumptions that all the *reactions are at equilibrium* and that the electrons and protons from one reaction are perfectly free to participate in any of the other reactions. In real soils, these assumptions are hardly ever met.

Table 7.1 lists oxidized and reduced forms of several elements, along with the approximate redox potentials at which the oxidation–reduction reactions were observed to actually occur in a soil at pH 6.5. Because in actual soils the various reactions may be spatially separated from each other and some reaction rates are slower than others, equilibrium is rarely, if ever, achieved. Therefore, the measured E_h values for the reaction (Table 7.1) tend to be quite a bit lower than the theoretical values indicated by the lines in Figure 7.6.

When the soil becomes essentially devoid of O_2 the soil redox potential falls below levels of about 0.38 to 0.32 V (at pH 6.5). After O_2 is used up, the next most easily reduced substance present is usually the N^{5+} in nitrate (NO_3^-). If the soil contains much NO_3^- the E_h will remain near 0.28 to 0.22 V as the nitrate is reduced:

$$\underset{N(V)}{\overset{(5+)}{NO_3^-}} + 2e^- + 2H^+ \longleftrightarrow \underset{N(III)}{\overset{(3+)}{NO_2^-}} + H_2O \tag{7.5}$$

Once all the N^{5+} in nitrate has been transformed into NO_2^-, N_2, and other N species, the E_h will drop further. At this point, organisms capable of reducing Mn^{4+} will become active. Thus, as E_h values fall, the elements N, Mn, Fe, and S (in SO_4^{2-}) and C (in CO_2) accept electrons and become reduced, predominantly in the order shown in Figure 7.6 and Table 7.1.

In other words, transformation of different elements requires different degrees of reducing conditions. The soil E_h must be lowered to -0.2 V before methane is produced, but reduction of NO_3^- to N_2 gas takes place when the E_h is as high as $+0.28$ V. We can conclude that soil aeration helps determine the specific chemical species present in soils and, in turn, the availability, mobility, and possible toxicity of many chemical elements.

7.4 FACTORS AFFECTING SOIL AERATION AND E_h

Drainage of Excess Water

Drainage of gravitational water out of the profile and concomitant diffusion of air into the soil takes place most readily in macropores. The most important factors influencing the aeration of well-drained soils are therefore those that determine the volume of the

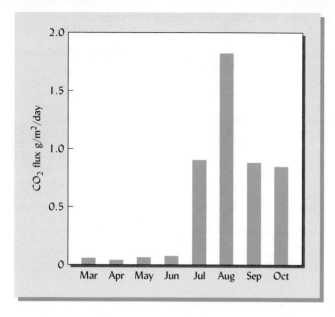

FIGURE 7.7 The rate of CO_2 movement (flux) from the surface of a soil within a northern hardwood forest ecosystem in New York State. Note the high rates from July to October when forest growth and microbical action were highest. [From Yavitt et al. (1995); used with permission of Soil Science Society of America]

soil macropores. Soil texture, bulk density, aggregate stability, organic matter content, and biopore formation are among the soil properties that help determine macropore content and, in turn, soil aeration (see Section 4.8).

Rates of Respiration in the Soil

The concentrations of both O_2 and CO_2 are largely dependent on microbial activity, which in turn depends on the availability of organic carbon compounds as food. Incorporation of large quantities of manure, crop residues, or sewage sludge may alter the soil air composition appreciably. Likewise, the cycling of plant residues by leaf fall, root mass decay, and root excretion in natural ecosystems provides the substrate for microbial activity. Respiration by plant roots and enhanced respiration by soil organisms near the roots are also significant processes (Figure 7.7). All these processes are very much enhanced as soil temperature increases (see Section 7.8).

The Soil Profile

Subsoils are usually more deficient in oxygen than are topsoils. Not only is the water content usually higher (in humid climates), but the total pore space, as well as the macropore space, is generally much lower in the deeper horizons. In addition, the pathway for diffusion of gases into and out of the soil is longer for deeper horizons. However, if organic substrates are in low supply, the subsoil may still be aerobic because O_2 can diffuse fast enough to replace that used by respiration. For this reason, certain recently flooded soils are anaerobic in the upper 50 to 100 cm and are aerobic below.

Tree roots may extend into layers that are very low in oxygen and high in carbon dioxide, especially if the subsoil is high in clay. Carbon dioxide levels of nearly 15% (150 mL/L) have been observed in some such subsoils. Studies in deep, highly weathered soils under tropical rain forests indicate that respiration is carried out, and the concentration of CO_2 continues to rise, down into the deepest subsoil layers (Figure 7.8).

Soil Heterogeneity

TILLAGE. One cause of soil heterogeneity is tillage, which has both short-term and long-term effects on soil aeration. In the short term, stirring the soil often allows it to dry out faster and also mixes in large quantities of air. These effects are especially evident on somewhat compacted, fine-textured soils, on which plant growth often responds immediately after a cultivation to control weeds or "knife in" fertilizer. In the long term, however, tillage may reduce macroporosity (see Section 4.6).

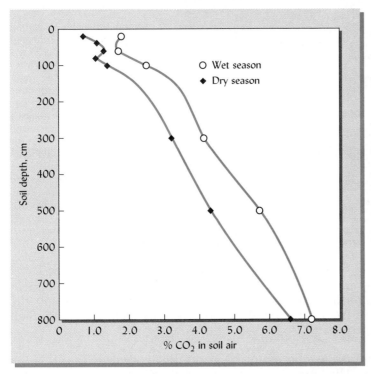

FIGURE 7.8 Changes in the concentration of CO_2 in soil air with depth into the profile of a Haplustox soil under a tropical rain forest in the Brazilian Amazon region. The source of the CO_2 was likely a combination of root and microbial respiration. Although by far the highest rates of CO_2 production were in the upper soil layers, gas produced there had little distance to travel to reach the atmosphere and many large pores to travel through. The concentration of CO_2 increased with depth because the increasing travel distance to the surface and the much lower macroporosity at depth greatly slowed the movement of gases, causing CO_2 to accumulate. [Data from Davidson and Trumbore (1995)]

PORE SIZES. Very small pores in a clayey or compacted soil layer may also slow water percolation so that the overlying soil becomes waterlogged and poorly aerated. Such layers may also slow the diffusion of O_2 and thereby lower redox potentials within that layer, whether or not the soil is saturated. Similarly, O_2 will diffuse much more slowly through the small, largely water-filled pores within a soil aggregate than through the largely air-filled pores between aggregates. Therefore, anaerobic conditions may occur in the center of an aggregate, only a few mm from well-aerated conditions near the surface of the same aggregate (Figure 7.9).

In some upland soils, large subsoil pores such as cracks between peds and old root channels may periodically fill with water, causing localized zones of poor aeration. This

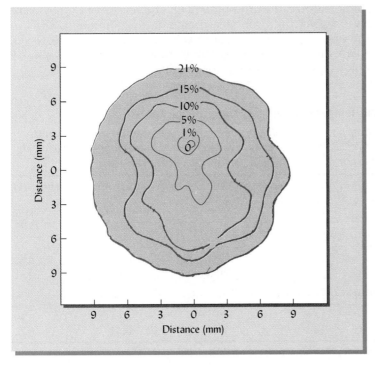

FIGURE 7.9 The oxygen content of soil air in a wet aggregate from an Aquic Hapludoll (Muscatine silty clay loam) from Iowa. The measurements were made with a unique microelectrode. Note that the oxygen content near the aggregate center was zero, while that near the edge of the aggregate was 21%. Thus pockets of oxygen deficiency can be found in a soil whose overall oxygen content may not be low. [From Sexstone et al. (1985)]

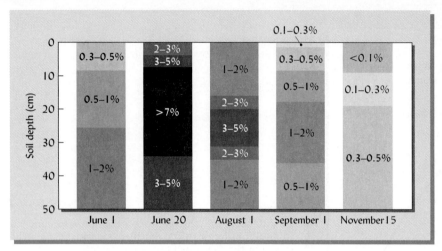

FIGURE 7.10 Seasonal changes in carbon dioxide content in the upper 50 centimeters of an Alfisol in Missouri on which corn was grown from early May until about September 1. By June 20, the corn was growing vigorously, the soil moisture level was still quite high, and the CO_2 level increased to more than 7% in the 10 to 30 cm zone. By August 1 gaseous exchange had probably increased and the CO_2 level had declined to 3 to 5% in the 20 to 30 cm layer, but even this is 100 times the level in the atmosphere. By November 15 activities of plants and microorganisms had declined because of lower temperatures and the CO_2 below 20 cm was only 10 times that in the atmosphere. [Data from Buyanovsky and Wagner (1983); used with permission of the Soil Science Society of America]

condition is expressed by gray, reduced surfaces on peds with reddish, oxidized interiors (see Plate 82). In a normally saturated soil, such large pores may cause the opposite effect (peds with oxidized faces but reduced interiors), as they facilitate O_2 diffusion into the soil during dry periods.

PLANT ROOTS. Respiration by the roots of upland plants usually depletes the O_2 in the soil just outside the root. The opposite can occur in wet soils growing hydrophytic plants. Aerenchyma tissues in these plants may transport surplus O_2 into the roots, allowing some to diffuse into the soil and produce an oxidized zone in an otherwise anaerobic soil (see for example, Plate 38, after page 112).

For all these reasons, aerobic and anaerobic processes may proceed simultaneously and in very close proximity to each other in the same soil. This heterogeneity of soil aeration should be kept in mind when attempting to understand the role that soil plays in elemental cycling and ecosystem function.

Seasonal Differences

In many soils, aeration status varies markedly with the seasons. In temperate humid regions, soils are often wet early in spring and opportunities for ready gas exchange are poor. But due to low soil temperatures, the utilization of O_2 and release of CO_2 by plant roots and microorganisms is restricted. In summer, these soils are commonly lower in moisture content, and the opportunity for gaseous exchange is increased. However, plant roots and microorganisms, stimulated by the warmer temperatures, quickly deplete O_2 and release copious quantities of CO_2 (for example, see Figure 7.10).

TABLE 7.2 **Tree Removal Effect on Soil Aeration and Temperature in a Subtropical Pine Forest**

Removal of the 55-year-old loblolly pine trees reduced transpiration and shading, resulting in a higher water table, warmer spring temperatures, and lower redox potentials in this Vertic Ochraqualf. Because warmer temperatures stimulated microbial activity, redox potentials were lower in spring, even though the soil was wetter in winter.

Site treatment	Time soil is saturated, %	Soil temperature (°C)		Soil redox potentials E_h (V)	
		Winter	Spring	Winter	Spring
Measured at 50-cm depth					
Undisturbed pine stand	39	11.8	18.3	0.83	0.65
Trees cut, soil not compacted	71	11.7	20.5	0.51	0.11
Measured at 100-cm depth					
Undisturbed pine stand	57	13.3	17.3	0.83	0.49
Trees cut, soil not compacted	69	13.2	18.7	0.54	0.22

Data from Tiarks et al. (1996).

Effects of Vegetation

In addition to the root respiration effects mentioned previously, vegetation may affect soil aeration by removing large quantities of water via transpiration, enough to lower the water table in some poorly drained soils. The data in Table 7.2 illustrate the effects of vegetation removal, soil depth, and season on E_h in a forested soil.

7.5 ECOLOGICAL EFFECTS OF SOIL AERATION

Effects on Organic Residue Degradation

Soil aeration influences many soil reactions and, in turn, many soil properties. The most obvious of these reactions are associated with microbial activity, especially the breakdown of organic residues and other microbial reactions. Poor aeration slows down the rate of decay, as evidenced by the relatively high levels of organic matter that accumulate in poorly drained soils.

The nature as well as the rate of microbial activity is determined by the O_2 content of the soil. Where O_2 is present, aerobic organisms are active, see Section 12.2. In the absence of gaseous oxygen, anaerobic organisms take over. Much slower breakdown occurs through reactions such as the following:

$$C_6H_{12}O_6 \longrightarrow 2CO_2 + 2CH_3CH_2OH \qquad (7.6)$$

$$\text{Sugar} \qquad\qquad\qquad \text{Ethanol}$$

Poorly aerated soils therefore tend to contain a wide variety of only partially oxidized products such as ethylene gas (C_2H_4), alcohols, and organic acids, many of which can be toxic to higher plants and to many decomposer organisms. The latter effect helps account for the formation of Histosols in wet areas where inhibition of decomposition allows thick layers of organic matter to accumulate. In summary, the presence or absence of oxygen gas completely modifies the nature of the decay process and its effect on plant growth.

Oxidation–Reduction of Elements

NUTRIENTS. Through its effects on the redox potential, the level of soil oxygen largely determines the forms of several inorganic elements, as shown in Table 7.3. The oxidized states of the nitrogen and sulfur are readily utilizable by higher plants. In general, the oxidized conditions are desirable for iron and manganese nutrition of most plants in acid soils of humid regions because, in these soils, the reduced forms of these elements are so soluble that toxicities may occur. However, some reduction of iron may be beneficial as it will release phosphorus from insoluble iron-phosphate compounds. Such phosphorus release has implications for eutrophication (see Section 14.2) when it occurs in saturated soils or in underwater sediments.

In drier areas, the opposite is generally true, and reduced forms of elements such as iron and manganese are preferred. In the neutral-to-alkaline soils of these drier areas, oxidized forms of iron and manganese are tied up in highly insoluble compounds, resulting in deficiencies of these elements. Such differences illustrate the interaction of aeration and soil pH in supplying available nutrients to plants (see Chapter 15).

TABLE 7.3 Oxidized and Reduced Forms of Several Important Elements

Element	Normal form in well-oxidized soils	Reduced form found in waterlogged soils
Carbon	CO_2, $C_6H_{12}O_6$	CH_4, C_2H_4, CH_3CH_2OH
Nitrogen	NO_3^-	N_2, NH_4^+
Sulfur	SO_4^{2-}	H_2S, S^{2-}
Iron	Fe^{3+} [Fe(III) oxides]	Fe^{2+} [Fe(II) oxides]
Manganese	Mn^{4+} [Mn(IV) oxides]	Mn^{2+} [Mn(II) oxides]

TOXIC ELEMENTS.   Redox potential determines the species of such potentially toxic elements as chromium, arsenic, and selenium, markedly affecting their impact on the environment and food chain (see Section 18.8). Reduced forms of arsenic are most mobile and toxic, giving rise to toxic levels of this element in drinking water, a serious human health problem in many places around the world, but especially in Bangladesh. In contrast, it is the oxidized hexavalent form of chromium (Cr^{6+}) that is mobile and very toxic to humans. In neutral to acid soils, easily decomposed organic materials can be used to reduce the chromium to the Cr^{3+} form, which is not subject to ready reoxidation (Figure 7.11).

SOIL COLORS.   As was discussed in Section 4.1, soil color is influenced markedly by the oxidation status of iron and manganese. Colors such as red, yellow, and reddish brown are characteristic of well-oxidized conditions. More subdued shades such as grays and blues predominate if insufficient oxygen is present. Soil color can be used in field methods for determining the status of soil drainage. Imperfectly drained soils are characterized by contrasting streaks of oxidized and reduced materials (see Plates 15 and 37, after page 112). Such a mottled condition indicates a zone of alternating good and poor aeration, a condition not conducive to the optimum growth of most plants.

GREENHOUSE GASES.   The production of nitrous oxide (N_2O) and methane (CH_4) in wet soils is of universal significance. These two gases, along with carbon dioxide (CO_2), are responsible for about 80% of the anthropogenic global warming caused by the "greenhouse effect" (see Section 12.9). The atmospheric concentrations of these gases have been increasing at alarming rates each year for the past half century or more.

Methane gas is produced by the reduction of CO_2. Its formation occurs when the E_h is −0.2 V, a condition common in natural wetlands and in rice paddies (see Figure 12.31). It is estimated that wetlands in the United States emit about 100 million metric tons of methane annually. Nitrous oxide is also produced in large quantities by wetland soils, as well as sporadically by upland soils. Because of the biological productivity and

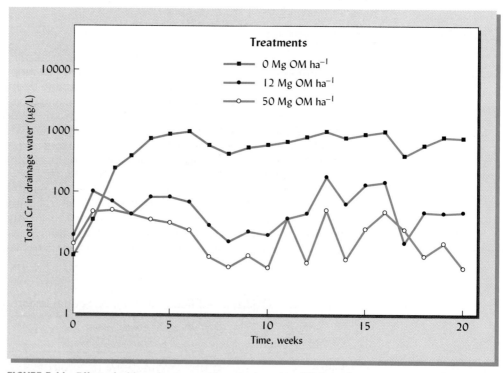

FIGURE 7.11   Effect of adding decomposable organic matter (OM) on the concentration of chromium in water draining from a chromium-contaminated soil. Here dried cattle manure was added as the decomposable OM. As the manure oxidized, it caused the reduction of the toxic, mobile Cr^{6+} to the relatively immobile, nontoxic Cr^{3+}. Note the log scale for the Cr in the water, indicating that the 50 Mg manure ha^{-1} addition caused the Cr level to be lowered approximately 100-fold. The coarse-textured soil was a Typic Torripsamment in California. [Data from Losi et al. (1994)]

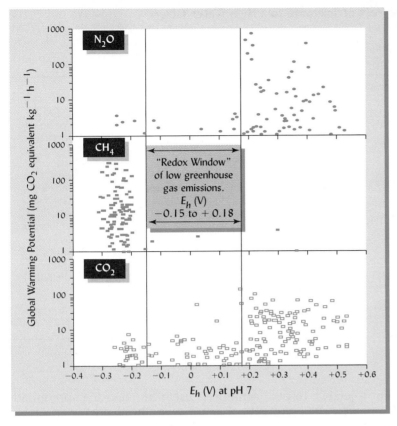

FIGURE 7.12 The relationship between soil redox potentials (E_h) and emissions of three "greenhouse" gases from soils. Because the three gases differ widely in their potential for global warming per mole of gas, their emissions are expressed here as CO_2 equivalents. Note the low global warming potential from all three gases when E_h is between −0.15 and +0.18 V. Although this study used small containers of flooded soil in the lab, it may be desirable to manage the aeration status of flooded soils in rice paddies and wetlands with these results in mind. Manipulating organic matter additions, water table levels, water flow rates, and the duration of flooding might allow managers to maintain the soil within the "window" where E_h is too high to stimulate microbial methanogenesis (CH_4 production) but also too low to stimulate production of much N_2O or CO_2. Such management, if feasible, could potentially reduce the contribution of wet soils to global warming. [From Yu and Patrick (2004) with permission of The Soil Sci. Soc. of America.]

diversity of wetlands (Section 7.7), soil scientists are seeking means of managing greenhouse gases from these wetland soils without resorting to draining (i.e., destroying) them. Fortunately, it may be possible to minimize production of all three major greenhouse gases by maintaining soil E_h in a moderately low range that is feasible for many rice paddies and natural wetlands (Figure 7.12).

Effects on Activities of Higher Plants

Hemoglobin helps plants survive flooding:

http://www. umanitoba.ca/afs/fiw/04072

It is the lack of oxygen in the root environment rather than the excessive wetness itself that impairs plant growth in flooded or overly wet soils. This fact explains why flooding a soil with stagnant water is generally much more damaging to plants—even for some hydrophytes—than flooding with flowing water. The O_2 content of the soil is continually replenished by flowing water, but roots and microbes will completely deplete the O_2 supply under stagnant water.

PLANT GROWTH. The lack of oxygen in the soil alters root function and changes the metabolism of the entire plant. Often poor soil aeration will reduce shoot growth even more than root growth. Among the first plant responses to low soil oxygen is the closure of leaf stomata, followed by reductions in photosynthesis and sugar translocation within the plant. The ability of the root to take up water and nutrients is inhibited, and as a result of impaired root metabolism, plant hormones are thrown out of balance.

Plant species vary in their ability to tolerate poor aeration (Table 7.4). Among crop plants, sugar beet is an example of a species that is very sensitive to poor soil aeration (see Figure 7.2, *left*). At the opposite extreme, paddy rice (Figure 13.15*a*) is an example of a species that can grow with its roots completely submerged in water. Furthermore, for a given species of plant, the young seedlings may be more tolerant of low soil aeration porosity than are older plants. A case in point is the tolerance of red pine to restricted drainage during its early development and its poor growth or even death on the same site at later stages (see Figure 7.2, *right*).

TABLE 7.4 Examples of Plants with Varying Degrees of Tolerance to a High Water Table and Accompanying Restricted Aeration

The plants in the leftmost column commonly thrive in wetlands. Those in the rightmost column are very sensitive to poor aeration.

Plants adapted to grow well with a water table at the stated depth				
<10 cm	*15 to 30 cm*	*40 to 60 cm*	*75 to 90 cm*	*>100 cm*
Bald cypress	Alsike clover	Birdsfoot trefoil	Beech	Arborvitae
Black spruce	Bermuda grass	Black locust	Birch	Barley
Common cattail	Black willow	Bluegrass	Cabbage	Beans
Cranberries	Cottonwood	Linden	Corn	Cherry
Duckgrass	Creeping bentgrass	Mulberry	Hairy vetch	Hemlock
Phragmites grass	Deer tongue	Mustard	Millet	Oats
Maiden cane	Eastern gamagrass	Red maple	Peas	Peach
Mangrove	Ladino clover	Sorghum	Red oak	Sand lovegrass
Pitcher plant	Loblolly pine	Sycamore		Sugar beets
Reed canary grass	Orchard grass	Weeping love grass		Walnut
Rice	Redtop grass	Willow oak		Wheat
Skunk cabbage	Tall fescue			White pine
Spartina grass				
Swamp white oak				
Swamp rose mallow				
Water tupelo				

Knowledge of plant tolerance to poor aeration is useful in choosing appropriate species to revegetate wet sites. The occurrence of plants specially adapted to anaerobic conditions is useful in identifying wetland sites (see Section 7.7).

NUTRIENTS AND WATER UPTAKE. Low O_2 levels constrain root respiration and impair root function. The root cell membrane may become less permeable to water, so that plants may actually have difficulty taking up water and some species will wilt and desiccate in a waterlogged soil. Likewise, plants may exhibit nutrient deficiency symptoms on poorly drained soils even though the nutrients may be in good supply. Furthermore, toxic substances (e.g., ethylene gas) produced by anaerobic microorganisms may harm plant roots and adversely affect plant growth.

SOIL COMPACTION. Soil compaction does decrease the exchange of gases; however, the negative effects of soil compaction are not all owing to poor aeration. Soil density and strength itself can impede the root growth even if adequate oxygen is available (see Sections 4.7 and 5.9).

7.6 AERATION IN RELATION TO SOIL AND PLANT MANAGEMENT

For crops in the field, aeration may be enhanced by implementing the principles outlined in Sections 4.5–4.7 regarding the maintenance of soil aggregation and tilth. Equally important are systems to increase both surface and subsurface drainage (see Section 6.7) and encourage the production of vertically oriented biopores (e.g., earthworm and root channels) that are open to the surface (Sections 4.5 and 6.7). Here we will briefly consider steps that can be taken to avoid aeration problems for container-grown plants, landscape trees, and lawns.

Container-Grown Plants

Potted plants frequently suffer from waterlogging and poor aeration even though potting mixes are engineered to meet the requirements of the containerized plants and minimize waterlogging. Mineral soil generally makes up no more than one-third of the volume of most potting mixes, the remainder being composed of inert, lightweight and coarse-grained materials such as perlite (expanded volcanic glass), vermiculite (expanded mica—not soil vermiculite), or pumice (porous volcanic rock). In order to achieve maximum aeration and minimum weight, some potting mixes contain no mineral soil at all. Most modern mixes also contain some peat, shredded bark, wood chips, compost, or other stable organic material that adds macroporosity and holds water.

Despite the fact that most planting containers have holes to allow drainage of excess water, the bottom of the container still creates a perched water table. As is the case with stratified soils in the field (Section 5.6), water drains out of the holes at the bottom of the pot *only* when the soil at the bottom is saturated with water and the water potential is positive. The finer pores in the medium remain filled with water, leaving no room for air, and anaerobic conditions soon prevail. The situation is aggravated if the potting medium contains much mineral soil. In any case, the use of as tall a pot as possible will allow for better aeration in the upper part of the medium. Watering should be deferred until the soil *near the bottom of the container* has begun to dry.

Tree and Lawn Management

In transplanting a tree seedling or any woody species, special caution must be taken to prevent poor aeration or waterlogging immediately around the young root systems. Figure 7.13 illustrates the right and the wrong way to transplant trees into a compacted soil.

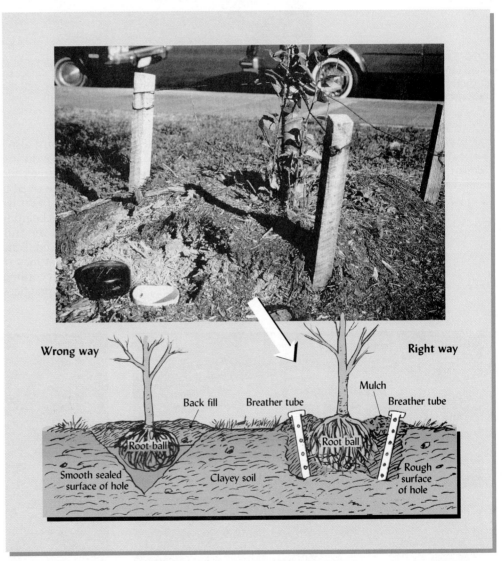

FIGURE 7.13 Providing a good supply of air to tree roots can be a problem, especially when trees are planted in fine-textured, compacted soils of urban areas. A machine-dug hole with smooth sides will act as a "tea cup" and fill with water, suffocating tree roots. Breather tubes, a larger rough-surfaced hole, and a layer of surface mulch in which some fine tree roots can grow are all measures that can improve the aeration status of the root zone. (Photo courtesy of R. Weil)

FIGURE 7.14 Protection of valuable trees during landscape grading operations. Even a thin layer of soil spread over a large tree's root system can suffocate the roots and kill the tree. (*Inset*) In order to preserve the original ground surface so that tree feeder roots can obtain sufficient oxygen, a dry well may be constructed of brick or any decorative material. The dry well may be incorporated into the final landscape design or filled in at a rate of a few centimeters per year. (Photos courtesy of R. Weil)

Healthy city trees:

www.forestry.iastate.edu/ext/ roadside_tree_management/ Growth_Requirements.pdf

The aeration of well-established, mature trees must also be safeguarded. If operators push surplus excavated soil around the base of a tree during landscape grading, serious consequences are soon noticed (Figure 7.14). The tree's feeder roots near the original soil surface become deficient in oxygen even if the overburden is no more than 5 to 10 cm in depth. One should build a protective wall (a *dry well*) or install a fence around the base of a valuable tree before grading operations begin in order to preserve the original soil surface for a radius of several meters from the trunk. This measure will allow the tree's roots to continue to access the O_2 they need. Failure to observe these precautions can easily kill a large, valuable tree, although it may take a year or two to do so.

Management systems for heavily trafficked lawns commonly include the installation of perforated drainage pipes (tiles) as well as other practices that enhance soil aeration

FIGURE 7.15 One way of increasing the aeration of compacted soil is by core cultivation. The machine removes small cores of soil, leaving holes about 2 cm in diameter and 5 to 8 cm deep. This method is commonly used on high-traffic turf areas. Note that the machine *removes* the cores and does not simply punch holes in the soil, a process that would increase compaction around the hole and impede air diffusion into the soil. (Photos courtesy of R. Weil)

(Section 6.7). For example, *core cultivation* can be used to increase the aeration in compacted lawn areas. This procedure removes thousands of small cores of soil from the surface horizon, thereby permitting gas exchange to take place more easily (Figure 7.15). Spikes that merely punch holes in the soil are much less effective than corers, since compaction is increased in the soil surrounding a spike.

7.7 WETLANDS AND THEIR POORLY AERATED SOILS[2]

Wetland definition and function:
www.stemnet.nf.ca/CITE/ecowetlands.htm

Poorly aerated areas called **wetlands** cover approximately 14% of the world's ice-free land, with the greatest areas occurring in the cold regions of Canada, Alaska, and Russia (Table 7.5). In the continental United States, about 400,000 km[2] exist today, less than half of the area that is estimated to have existed when European settlement of the nation began. Most wetland losses occurred as farmers used artificial drainage (see Section 6.7) to convert them into cropland. This conversion process was assisted by some of the same government agencies (e.g., U.S. Army Corps of Engineers and USDA) that are now working to protect wetlands from further damage. In recent decades, filling and drainage for urban development has also taken its toll of wetland areas (see Box 7.1). Since environmental consciousness has become a force in modern societies, wetland preservation has become a major issue, and the loss of wetlands has been slowed from about 114,000 ha/yr during the 1970s and 1980s to 23,000 ha/yr during the 1990s.

TABLE 7.5 Major Types of Wetlands and Their Global Areas

All together, wetlands constitute perhaps 14% of the world's land area.

Wetland type	Global area, 1000s km[2]	Percent of ice-free land area	Percent of all wetland areas
Inland (swamps, bogs, etc.)	5415	3.9	28.8
Riparian or ephemeral	3102	2.3	16.5
Organic (Histosols)	1366	1.0	7.3
Salt-affected, including coastal	2230	1.6	11.9
Permafrost-affected (Histels)	6697	4.9	35.6

Data from Eswaren et al. (1996).

BOX 7.1 IT'S THE LAW

Not only is it a bad idea ecologically to drain or fill wetlands, it's also against the law! In the United States and many other countries, knowingly destroying a wetland can bring severe penalties. The case reported here (Figure 7.16) reflects the change from a generation ago when most wetland destruction was caused by farmers installing agricultural drainage, to today when wetlands in industrial countries are most threatened by urban/suburban development. The newspaper reports that the developer had been informed (even warned) about the wetland area on his 1,000-hectare development tract. He nonetheless filled the wetland areas so he could build hundreds of homes on these sites. To add further environmental insult, he installed septic drain fields (See Section 6.8) on these seasonally saturated soils. The judge sentenced the man to nine years in prison followed by three years of supervised release. His business partners were also imprisoned and fined. The severity of this punishment should "send a message" to other would-be violators.

Developers Sentenced in Wetlands Case

A federal judge sentenced three Mississippi real estate developers to prison yesterday for filling in wetlands and selling the property to low- and fixed-income families, marking the end of the nation's largest wetlands criminal prosecution.

FIGURE 7.16 *Clipping of Washington Post story by Eilperin (2005). (Photo courtesy of R. Weil)*

---

[2] Two well-illustrated, nontechnical, yet informative publications on wetlands are Welsh et al. (1995) and CAST (1994). For a compilation of technical papers on hydric soils and wetlands, see Richardson and Vepraskas (2001).

Wetland is a scientific term for ecosystems that are transitional between land and water. These systems are neither strictly terrestrial (land-based) nor aquatic (water-based). While there are many different types of wetlands, they all share a key feature, namely *soils that are water-saturated near the surface for prolonged periods when soil temperatures and other conditions are such that plants and microbes can grow and remove the soil oxygen, thereby assuring anaerobic conditions.* It is largely the prevalence of anaerobic conditions that determines the kinds of plants, animals, and soils found in these areas. There is widespread agreement that the wetter end of a wetland occurs where the water is too deep for rooted, emergent vegetation to take hold. The difficulty is in precisely defining the so-called *drier end* of the wetland, the boundary beyond which exist nonwetland, upland systems in which the plant–soil–animal community is no longer predominantly influenced by the presence of anaerobic soils.

The controversy is probably more political than scientific, however. Since uses and management of wetlands are regulated by governments in the United States and in many other countries, billions of dollars are at stake in determining what is and what is not protected as a wetland. (Consider, for example, a developer who wants to buy a 100-ha tract of land on which to build a shopping center. If 20 versus 60 of those hectares are declared off-limits to development, how will that affect the developer's willingness to pay?)

Because so much money is at stake, thousands of environmental professionals are employed in the process of **wetland delineation**—finding the exact drier-end boundaries of wetlands on the ground. Wetland delineation is *not* done in front of a computer screen, but is a sweaty, muddy, tick- and mosquito-ridden business that those trained in soil science are uniquely qualified to carry out.

What are the characteristics these scientists look for to indicate the existence of a wetland system? Most authorities agree that three characteristics can be found in any wetland:

1. A wetland hydrology or water regime
2. Hydric soils
3. Hydrophytic plants

We shall briefly examine each.

Wetland Hydrology

Wetlands characterized:
http://psybergate.com/wetfix/shareNet/Sharenet2/Share.htm

WATER BALANCE.   Water flows into wetlands from surface runoff (e.g., bogs and marshes), from groundwater seepage, and from direct precipitation. It flows out by surface and subsurface flows, as well as by evaporation and transpiration (see Section 6.3 and Table 7.2). The balance between inflows and outflows, as well as the water storage capacity of the wetland itself, determines how wet it will be and for how long.

HYDROPERIOD.   The temporal pattern of these water table changes is termed the **hydroperiod**. For a coastal marsh, the hydroperiod may be daily, as the tides rise and fall (Figure 7.17). For inland swamps, bogs, or marshes, the hydroperiod is more likely to be seasonal. Some wetlands may be flooded for only a month or so each year, while some may never be flooded, although they are saturated within the upper soil horizons.

Also, if the period of saturation occurs when the soil is too cold for microbial or plant-root activity to take place, oxygen may be dissolved in the water or entrapped in aggregates within the soil. Consequently, true anaerobic conditions may not develop, even in flooded soils. Remember, it is the anaerobic condition, not just saturation, that makes a wetland a wetland.

---

[3] In 1987, the U.S. Army Corps of Engineers and the Environmental Protection Agency agreed on the following definition to be used in enforcing the Clean Water Act: "The term wetlands means those areas that are inundated or saturated by surface or ground water at a frequency and duration sufficient to support, and that under normal circumstances do support, a prevalence of vegetation typically adapted for life in saturated soil conditions."

FIGURE 7.17 An example of a wetland with a daily hydroperiod that follows the rise and fall of the slightly brackish estuary tides. This tidal marsh with a beaver house is seen at low tide when some saturated soils and emergent plants are exposed. The drier-end boundary of the wetland is probably just beyond the treeline in the background. (Photo courtesy of R. Weil).

RESIDENCE TIME. The more slowly water moves through a wetland, the longer the *residence time* and the more likely that wetland functions and reactions will be carried out. For this reason, actions that speed water flow, such as creating ditches or straightening stream meanders, are generally considered degrading to wetlands and are to be avoided.

INDICATORS. All wetlands are water-saturated some of the time, but many are not saturated all of the time. Systematic field observations, assisted by instruments to monitor the changing level of the water table, may be required to ascertain the frequency and duration of flooding or saturated conditions.

In the field, even during dry periods, there are many signs one can look for to indicate where saturated conditions frequently occur. Past periods of flooding will leave water stains on trees and rocks and a coating of sediment on the plant leaves and litter. Drift lines of once-floating branches, twigs, and other debris also suggest previous flooding. Trees with extensive root masses above ground indicate an adaptation to saturated conditions. But perhaps the best indicator of saturated conditions is the presence of **hydric soils**.

Hydric Soils[4]

In order to assist in delineating wetlands, soil scientists developed the concept of hydric soils. In *Soil Taxonomy* (see Chapter 3) these soils are mostly (but not exclusively) classified in the order Histosols, in Aquic suborders such as Aquents, Aquepts, and Aqualfs, or in Aquic subgroups. These soils generally have an aquic or peraquic moisture regime (see Section 3.2).

DEFINED. Three properties help define hydric soils. First, they are subject to *periods of saturation* that inhibit the diffusion of O_2 into the soil. Second, for substantial periods of time they undergo *reduced conditions* (see Section 7.3); that is, electron acceptors other than O_2 are reduced. Third, they exhibit certain features termed *hydric soil indicators*. Such indicators are discussed in Box 7.2.

---

[4] The U.S. Department of Agriculture Natural Resources Conservation Service defines a hydric soil as one "that formed under conditions of saturation, flooding or ponding long enough during the growing season to develop anaerobic conditions in the upper part." For an illustrated field guide to features that indicate hydric soils, see Hurt et al. (1996). For a current list of soil series considered to be hydric soils, see http://soils.usda.gov/use/hydric/.

BOX 7.2   HYDRIC SOIL INDICATORS

Hydric soil indicators are features associated (sometimes only in specific geographic regions) with the occurrence of saturation and reduction. Most of the indicators can be observed in the field by digging a small pit to a depth of about 50 cm. They principally involve the loss or accumulation of various forms of Fe, Mn, S, or C. The carbon (organic matter) accumulations are most evident in Histosols, but thick, dark-surface layers in other soils can also be indicators of hydric conditions in which organic matter decomposition has been inhibited (see, for example, Plates 6 and 39, after page 112).

Iron, when reduced to Fe(II), becomes sufficiently soluble that it migrates away from reduced zones and may precipitate as Fe(III) compounds in more aerobic zones. Zones where reduction has removed or depleted the iron coatings from mineral grains are termed *redox depletions*. They commonly exhibit the gray, low-chroma colors of the bare, underlying minerals (see Section 4.1 for an explanation of chroma). Also, iron itself turns gray to blue-green when reduced. The contrasting colors of redox depletions or reduced iron and zones of reddish oxidized iron result in unique mottled **redoximorphic features** (see, for example, Plates 15 and 17). Other redoximorphic features involve reduced Mn. These include the presence of hard black *nodules* that sometimes resemble shotgun pellets. Under severely reduced conditions the entire soil matrix may exhibit low-chroma colors, termed gley. Colors with a chroma of 1 or less quite reliably indicate reduced conditions (Figure 7.18).

Always keep in mind that redoximorphic features are indicative of hydric soils only when they occur in the upper horizons. Many soils of upland areas exhibit redoximorphic features only in their deeper horizons, due to the presence of a fluctuating water table at depth. Upland soils that are saturated or even flooded for short periods, especially if during cold weather, are not wetland (hydric) soils.

A unique redoximorphic feature associated with certain wetland plants is the presence, in an otherwise gray matrix, of reddish oxidized iron around root channels where O_2 diffused out from the aerenchyma-fed roots of a hydrophyte (see Plate 38). These *oxidized root zones* exemplify the close relationship between hydric soils and hydrophytic vegetation.

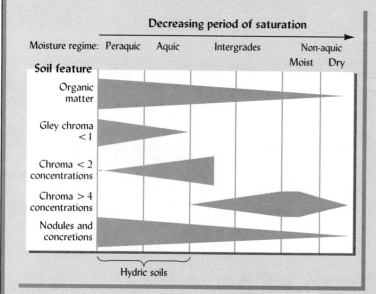

FIGURE 7.18 *The relationship between the occurrence of some soil features and the annual duration of water-saturated conditions. The absence of iron concentrations (mottles) with colors of chroma >4 and the presence of strong expressions of the other features are indications that a soil may be hydric. Peraquic refers to a moisture regime in which soils are saturated with water throughout the year. For other moisture regimes, see Section 3.2. [Adapted from Veneman (1999)]*

Hydrophytic Vegetation

Photos of hydrophytic vegetation:
http://www.bixby.org/parkside/multimedia/vegetation

Although the vast majority of plant species cannot survive the conditions characteristic of wetlands, there do exist varied and diverse communities of plants that have evolved special mechanisms to adapt to life in saturated, anaerobic soils. These plants comprise the **hydrophytic vegetation** that distinguishes wetlands from other systems.

Typical adaptive features include hollow aerenchyma tissues that allow plants like spartina grass to transport O_2 down to their roots. Certain trees (such as bald cypress) produce adventitious roots, buttress roots, or knees (see Figure 7.1). Other species spread their roots in a shallow mass on or just under the soil surface, where some O_2 can diffuse even under a layer of ponded water. The leftmost column in Table 7.4 lists a few common hydrophytes. Not all the plants in a wetland are likely to be hydrophytes, but the majority usually are.

Wetland Chemistry

The central characteristic of wetland chemistry is the low redox potential (see Section 7.3) that pertains. Furthermore, many wetland functions depend on *variations* in the redox potential; that is, in certain zones or for certain periods of time oxidizing conditions alternate with reducing ones.

LOW OXYGEN.   For example, even in a flooded wetland, O_2 will be able to diffuse from the atmosphere or from oxygenated water into the upper 1 or 2 cm of soil, creating a thin *oxidizing zone* (see Figure 7.19 and Plate 63). The diffusion of O_2 within the saturated soil is extremely limited, so that a few centimeters deeper into the profile, O_2 is eliminated and the redox potential becomes low enough for reactions such as nitrate reduction to take place. The close proximity of the oxidized and anaerobic zones allows water passing through wetlands to be stripped of N by the sequential oxidation of ammonium N to nitrate N, and then the reduction of the nitrate to various nitrogen gases that escape into the atmosphere (see Section 13.9).

REDOX.   To be considered a wetland, redox potentials should become low enough for iron reduction to produce redoximorphic features. Still lower E_h will allow sulfate reduction to produce rotten-egg-smelling hydrogen sulfide (H_2S) gas. The anaerobic zone may extend downward or in some cases may be limited to the upper horizons where microbial activity is high. The anaerobic carbon reactions discussed in Section 7.5 are characteristic of this zone, including methane (swamp gas) production. These and other chemical reactions involving the cycling of C, N, and S are explained in Sections 12.2, 13.9, and 13.20. Toxic elements such as chromium and selenium undergo redox reactions that may help remove them from the water before it leaves the wetland. Acids from industry or mine drainage may also be neutralized by reactions in hydric soils. This array of unique chemical reactions contributes greatly to the benefits that society and the environment gain from wetlands (see Box 7.3).

Constructed Wetlands

Realizing all the beneficial functions of wetlands, scientists and engineers have begun not only to find ways to preserve natural wetlands, but to construct artificial ones for specific purposes, such as wastewater treatment (see, for example, Box 14.2).

Another reason for attempting to construct wetlands is the provision in several regulations that allows for the destruction of certain natural wetland areas, provided that

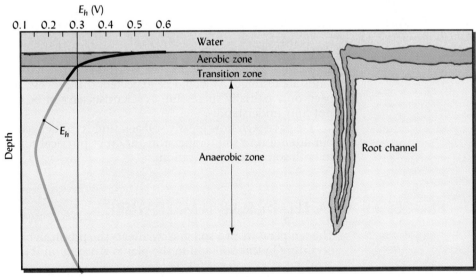

FIGURE 7.19   Representative redox potentials within the profile of an inundated hydric soil. Many of the biological and chemical functions of wetlands depend on the close proximity of reduced and oxidized zones in the soil. The changes in redox potential at the lower depths depend largely on the vertical distribution of organic matter. In some cases, low subsoil organic matter results in a second oxidized zone beneath the reduced zone. [Diagram courtesy of R. Weil]

BOX 7.3 WETLAND FUNCTIONS OF VALUE TO ECOSYSTEMS AND SOCIETY

Once considered to be nothing but disease-breeding wastelands, wetlands are now widely recognized as performing many extremely valuable functions. In fact, a group of scientists and economists (Costanza et al., 1997) has estimated that on average, one hectare of wetlands provides nonmarket services worth about 15 times more than those provided by one hectare of forestland. Globally, wetlands annually perform needed services that would cost some $10 trillion (in 2007 dollars) to replace. We can summarize at least six types of benefits derived from wetlands, the recognition of which has motivated the intensification of wetland study and protective regulation.

1. *Species habitat.* Wetlands provide special environmental conditions required by a wide array of wild species. Plants like bald cypress trees and pitcher plants, and animals like salamanders and muskrats, use wetlands as their primary habitat (that is, they live in wetlands). Furthermore, about 40% of all endangered and threatened species depend on wetlands in some way (e.g., for food or shelter). About one-third of all bird species in the United States also depend on wetlands.

2. *Water filtration.* Water is filtered and purified as it passes through wetlands on its way from the land to rivers, bays, lakes, estuaries, and groundwater aquifers. Wetlands physically filter out most sediment, remove a high proportion of plant nutrients (especially N and P) that could otherwise cause eutrophication of aquatic systems (see Section 14.2), and break down many organic substances that could cause toxic effects or deplete O_2 supplies in aquatic systems and drinking water.

3. *Flooding reduction.* Wetlands act as giant reservoirs to hold back large volumes of stormwater runoff from the land. They then release the water slowly, either to surface flow or to groundwater, thereby avoiding high peak flows that lead to floods and damage to homes and developed land along rivers. Studies have shown that floods are far less severe and less frequent where river systems have been allowed to retain their undisturbed wetlands.

4. *Shoreline protection.* Coastal wetlands serve as a buffer between the high-energy ocean waves and the shore, preventing the rapid shoreline erosion that is observed when the wetlands are drained or filled in.

5. *Commercial and recreational activities.* In the United States alone, people spend many millions of dollars annually in order to catch, hunt, view, or photograph birds, animals, and fish that either live in wetlands or depend on wetlands for their food supply or for nesting sites. Two examples of this relationship stand out. First, the populations of ducks have declined in parallel to the loss of prairie pothole wetlands (see Figure 6.4) since the 1940s. Second, the catch of striped bass (rockfish) in the Chesapeake Bay declined precipitously following the drainage and degradation of the coastal wetlands that provide the base of their food chain.

6. *Natural products.* Valuable products such as timber, blueberries, cranberries, fish, and wild rice can be sustainably harvested from certain wetlands under careful management. Scenic beauty could also be considered a natural product of wetlands that people can enjoy without damaging the system.

an equal or larger area of new wetlands is constructed or that previously degraded wetlands are restored. As would be expected, this process, termed **wetland mitigation,** has been only partially successful, as scientists cannot be expected to create what they do not fully understand.

We have seen how greatly soil aeration is influenced by soil water. We will now turn our attention to soil temperature, another physical soil property that is closely related to both soil water and aeration.

7.8 PROCESSES AFFECTED BY SOIL TEMPERATURE

The temperature of a soil greatly affects the physical, biological, and chemical processes occurring in that soil and in the plants growing on it (Figure 7.20).

Plant Processes

Most plants are actually more sensitive to *soil* temperature than to aboveground *air* temperature, but this is not often appreciated since air temperature is more commonly measured. Research also shows that, contrary to what one might expect, adverse soil

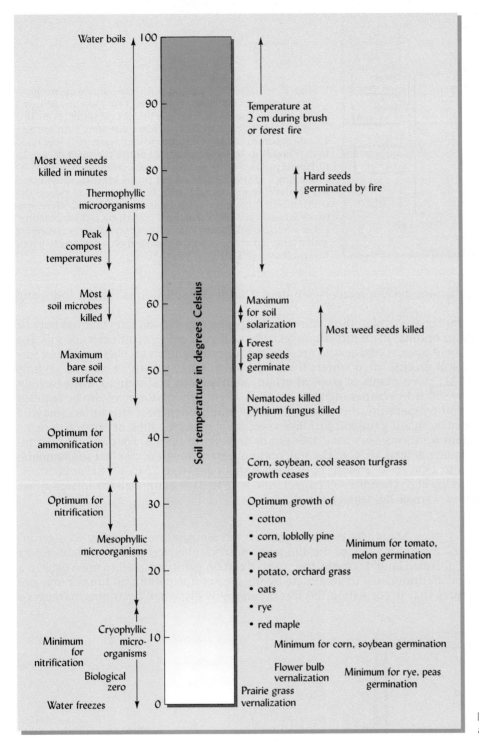

Water boils — 100
Temperature at
2 cm during brush
or forest fire — 90

Most weed seeds
killed in minutes — 80
Thermophyllic
microorganisms

Hard seeds
germinated by fire

Peak
compost — 70
temperatures

Most
soil microbes — 60
killed
Maximum
for soil
solarization
Forest
gap seeds
germinate

Most weed seeds killed

Maximum
bare soil — 50
surface

Nematodes killed
Pythium fungus killed

Optimum for — 40
ammonification

Corn, soybean, cool season turfgrass
growth ceases

Optimum for — 30
nitrification

Optimum growth of

Mesophyllic
microorganisms — 20

• cotton
• corn, loblolly pine
• peas
• potato, orchard grass
• oats
• rye
• red maple

Minimum for tomato,
melon germination

Cryophyllic — 10
micro-
organisms

Minimum for corn, soybean germination

Minimum
for
nitrification
Biological
zero

Flower bulb
vernalization

Minimum for rye, peas
germination

Water freezes — 0

Prairie grass
vernalization

FIGURE 7.20 Soil temperature ranges associated with a variety of soil processes.

temperature generally influences shoot growth and photosynthesis more than root growth (as was seen in the study described in Figure 7.21). Most plants have a rather narrow range of soil temperatures for optimal growth. For example, two species that evolved in warm regions, corn and loblolly pine, grow best when the soil temperature is about 25 to 30 °C. In contrast, the optimal soil temperature for cereal rye and red maple, two species that evolved in cool regions, is in the range of 12 to 18 °C.

In temperate regions, cold soil temperature often limits the productivity of crops and natural vegetation. Although artificial soil warming is cost-effective for only the most high-value crops, higher soil temperatures can markedly increase crop yields (Figure 7.22). The life cycles of plants are also greatly influenced by soil temperature. For example, tulip bulbs require chilling in early winter to develop flower buds,

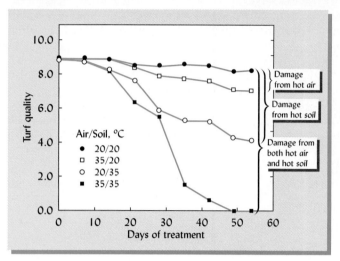

FIGURE 7.21 Effects of air and soil temperatures on turfgrass quality. Heat stress is a major problem for bent grass (*Agrostis spp.*) grown on golf greens in warmer climates. Researchers in this study grew bent grass for 60 days while controlling both soil and air temperatures. Turfgrass quality (color, vigor, etc.) was rated from 0 (dead) to 10 (best quality). Compare the small decline in turf quality caused by increasing air temperature from 20 to 35 °C to the much greater decline caused by the same increase in soil temperature. High temperature in both air and soil caused the worst effects. Among the plant parameters measured, photosynthesis was affected more than root growth by soil temperature. Other research has shown that a fine spray of water combined with a large fan can cool both air and soil on a golf green. [From Xu and Huang (2000)]

although flower development is suppressed until the soil warms up the following spring.

In warm regions, and in the summer in temperate regions, soil temperatures may be too high for optimal plant growth, especially in the upper few centimeters of soil. For example, bent grass, a cool-season grass prized for providing an excellent "playing surface" on golf greens, often suffers from heat stress when grown in warmer regions (Figure 7.21). Even plants of tropical origin, such as corn and tomato, are adversely affected by soil temperatures higher than 35 °C. Seed germination may also be reduced by high soil temperatures. In hot conditions, root growth near the surface may be encouraged by shading the soil with live vegetation, plant residues, or organic mulch.

Different plant processes have different optimal temperatures. For corn, seed germination requires at least 10 °C in the soil, optimum root growth occurs at a soil temperature of about 23 to 25 °C, and the optimum for shoot growth is 25 to 30 °C. For potatoes, tubers develop best when the soil temperature is 16 to 21 °C, although the foliage grows quite well at warmer soil temperatures.

SEED GERMINATION. Many plants require specific soil temperatures to trigger seed germination, accounting for much of the difference in species between early- and late-season weeds in cultivated land. Likewise, the seeds of certain plants adapted to open gaps in a forest stand are stimulated to germinate by the greater fluctuations and maximum soil temperatures that occur where the forest canopy is disturbed by timber harvest or

Measuring soil temperature video:

http://videogoogle.com/videoplay?docid=8773072375890666921&pr=goog-sl

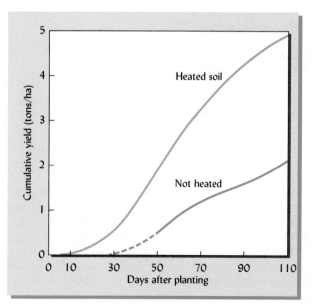

FIGURE 7.22 The influence of heating the soil in an experiment in Oregon on the yield of broccoli. Heating cables were buried about 92 cm deep. They increased the average temperature of the 0 to 100 cm layer by about 10 °C, although the upper 10 cm of soil was warmed by only about 3 °C. [From Rykbost et al. (1975); used with permission of American Society of Agronomy]

wind-thrown trees. The seeds of certain prairie grasses and grain crops require a period of cold soil temperatures (2 to 4 °C) to enable them to germinate the following spring, a process termed *vernalization*.

ROOT FUNCTIONS. Root functions such as nutrient uptake and water uptake are sluggish in cool soils with temperatures below the optimum for the particular species. One result is that nutrient deficiencies, especially of phosphorus, often occur in young plants in early spring, only to disappear when the soil warms later in the season. The phenomenon of *winter burn* of plant foliage is another consequence of low soil temperature that particularly affects evergreen shrubs. On bright, sunny days in winter and early spring when the soil is still cold, evergreen plants may become desiccated and even die because the slow water uptake by roots in the cold soils cannot keep up with the high evaporative demand of bright sun on the foliage. The problem can be prevented by covering the shrubs with a shade cloth.

Microbial Processes

Microbial processes are influenced markedly by soil temperature changes (Figure 7.23). Although it is commonly assumed that microbial activity ceases at temperatures that freeze water (<0 °C), low rates of soil microbial activity and organic matter decomposition have been measured in the permafrost layers of Gelisols at temperatures as low as −20 °C. In fact, given that some 80% of the Earth's biosphere is colder than 5 °C, it should not be surprising that microbes have widely adapted to life at cold temperatures.

Nonetheless, microbial activity is far greater at warm temperatures; the rates of microbial processes such as respiration typically more than double for every 10 °C rise in temperature (see Figure 7.24). The optimum temperature for microbial decomposition processes may be 35 to 40 °C, considerably higher than the optimum for plant growth. The dependence of microbial respiration on warm soil temperatures has important implications for soil aeration (see Section 7.7) and for the decomposition of plant residues and, hence, the cycling of the nutrients they contain. The productivity of northern (boreal) forests is probably limited by the inhibiting effect of low soil

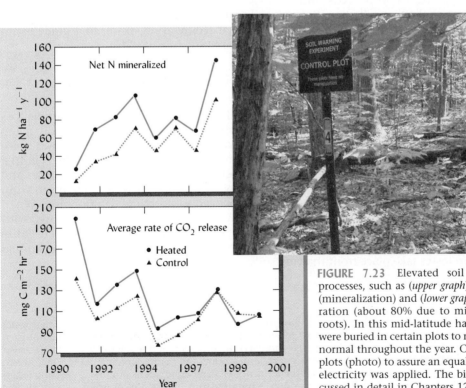

FIGURE 7.23 Elevated soil temperature accelerates biological processes, such as (*upper graph*) nitrogen release from organic matter (mineralization) and (*lower graph*) carbon dioxide release by soil respiration (about 80% due to microorganisms and 20% due to plant roots). In this mid-latitude hardwood forest, electric heating cables were buried in certain plots to maintain soil temperature at 5 °C above normal throughout the year. Cables were also installed in the control plots (photo) to assure an equal degree of physical disturbance, but no electricity was applied. The biological processes represented are discussed in detail in Chapters 12 and 13. [Graphs from data in Melillo et al. (2002); photo courtesy of R. Weil]

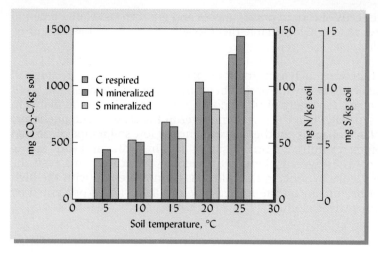

FIGURE 7.24 Effect of soil temperature on cumulative microbial respiration (CO_2 release) and net nitrogen and sulfur mineralization in surface soils from hardwood forest in Michigan. The soil water content was adequate for microbial growth throughout the 32-week study period. Note the near doubling of the microbial activity with a 10-degree change in soil temperature (compare findings for 15 vs. 25 °C). Means from four sites are shown. [From MacDonald et al. (1995)].

temperatures on microbial recycling and release of nitrogen from the tree litter and soil organic matter.

The microbial oxidation of ammonium ions to nitrate ions, which occurs most readily at temperatures near 30 °C, is also negligible when the soil temperature is low, below about 8 to 10 °C. Farmers in cool regions can take advantage of this fact by injecting ammonia fertilizers into cold soils in the early spring, expecting that ammonium ions will not be readily oxidized to leachable nitrate ions until the soil temperature rises. Unfortunately, in some years a warm spell allows the production of nitrates earlier than expected, with the result that much nitrate is lost by leaching, to the detriment of both the farmer and the downstream water quality.

In environments with hot, sunny summers (maximum daily air temperatures >35 °C) a controlled heating process called **soil solarization** can be used to control pests and diseases in some high-value crops. In this process, the ground is covered with a clear plastic film that traps enough heat to raise the temperature of the upper few centimeters of soil to as high as 50 to 60 °C. Such high temperatures can effectively suppress certain fungal pathogens, weed seeds, and insect pests.

As we shall see in Chapter 18, warm soil temperatures are also critical for new pollution remediation technologies that utilize specialized microorganisms to degrade petroleum products, pesticides, and other organic contaminants in soils.

Freezing and Thawing

Ice lens formation:
http://www.aip.org/png/html/frost.htm

When soil temperatures fluctuate above and below 0 °C, the water in the soil undergoes cycles of freezing and thawing. Alternate freezing and thawing subject the soil aggregates to pressures as zones of pure ice, called *ice lenses,* form within the soil and as ice crystals form and expand.[5] These pressures alter the physical structure of the soil. In a saturated soil with a puddled structure, the frost action breaks up the large masses and greatly improves granulation. In contrast, for soils with good aggregation to begin with, freeze-thaw action when the soil is very wet can lead to structural deterioration. When the surface of a frozen soil thaws, the upper few mm may be completely saturated with water, making it easy for spring rains to detach soil particles that are then easily removed by erosion. In some soils of temperate regions more than 50% of the annual erosion losses take place from thawing soils.

---

[5] The pressure is due mainly to ice lens growth rather than to the 9% increase in volume that water undergoes when it freezes. Ice lenses grow as water is drawn to the freezing zone from adjacent unfrozen areas. The flow to the growing ice lenses is encouraged by the fact that fine soil particles remain coated with a film of liquid (unfrozen) water at temperatures below the normal freezing point. The lowered freezing point occurs for two reasons: (1) the influence of water–solid interactions near the particle surface, and (2) the presence of dissolved and exchangeable ions in this film of water. For details, see Torrance and Schellekens (2006).

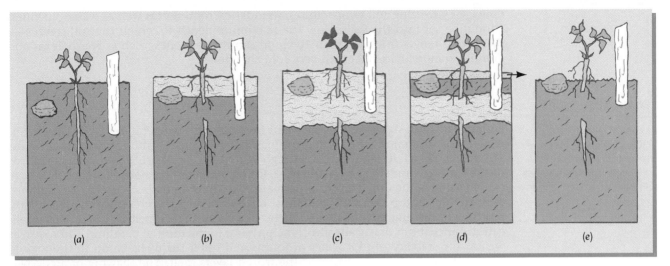

FIGURE 7.25 How frost heaving moves objects upward. (*a*) Position of the object (stone, plant, or fence post) before the soil freezes. (*b*) As lenses of pure ice form in the freezing soil by attraction of water from the unfrozen soil below, the frozen soil tightens around the upper part of the object, lifting it somewhat—enough to break the root in the case of the plant. (*c*) The objects are lifted upwards as ice-lens formation continues with deeper penetration of the freezing front. (*d*) As for freezing, thawing commences from the surface downward. Water from thawing ice lenses escapes to the surface because it cannot drain downward through the frozen soil. The soil surface subsides while the heaved objects are held in the "jacked-up" position by the still-frozen soil around their lower parts. (*e*) After complete thaw, the stone is closer to the surface than previously (although rarely at the surface unless erosion of the thawed soil has occurred), and the upper part of the broken plant's root is exposed, so that is likely to die. (*f*) Alfalfa plants lifted out of the ground by frost action. (*g*) Fence posts encased in concrete that have been progressively "jacked out" of the ground by frost action over several years. [Photo (*f*) courtesy of R. Weil; photo (*g*) courtesy of R. L. Berg, Corps of Engineers, Cold Regions Research and Engineering Laboratory, Hanover, N.H.]

Alternate freezing and thawing can force objects upward in the soil, a process termed *frost heaving*. Objects subject to heaving include stones, fence posts, and perennial tap-rooted plants. This action, which is most severe where the soil is silty in texture, wet, and lacking a covering of snow or dense vegetation, can drastically reduce stands of alfalfa (Figure 7.25), some clovers, and trefoil.

Freezing can also heave shallow foundations, roads, and runways that have fine material as a base. Gravels and pure sands are normally resistant to frost damage, but silts and sandy soils with modest amounts of finer particles are particularly susceptible. Very clay-rich soils do not usually exhibit much frost heave, but ice-lens segregation can still occur and can lead to severe loss of strength when thawing occurs. To avoid damage by freezing soil temperatures, foundation footings (as well as water pipelines) should be set into the soil below the maximum depth to which the soil freezes—a depth that ranges from less than 10 cm in subtropical zones, such as South Texas and Florida, to more than 200 cm in very cold climates.

Permafrost[6]

Images of permafrost and more:

http://www.earthscience world.org/images/search/result s.html?Keyword=permafrost

Perhaps the most significant global event involving soil temperatures is the thawing in recent years of some of the permafrost (permanently frozen ground) in northern areas of Canada, Russia, Iceland, China, Mongolia, and Alaska. Nearly 25% of the land areas of the Earth are underlain by permafrost. Rising temperatures since the late 1980s have caused some of the upper layers of permafrost to thaw. In parts of Alaska, for example, temperatures in top layers of permafrost have risen about 3.5 °C since the late 1980s, resulting in melting rates of about a meter in a decade. Such melting has serious implications since it can drastically affect the physical foundation of buildings and roads, as well as the stability of root zones of forests and other such vegetation in the region. Trees fall and buildings collapse as the frozen layers melt. The thawing of arctic permafrost is expected to further accelerate global warming, as decomposition of organic materials long trapped in the frozen layers of Histels releases vast quantities of carbon dioxide into the atmosphere (see Figure 3.16).

Soil with ice lenses may contain much more water than would be needed to saturate the soil in the unfrozen state. When the ice lenses thaw, the soil becomes supersaturated because the excess water cannot drain away through the underlying still-frozen soil. Soil in this condition readily turns into noncohesive mud that is very susceptible to erosion and movement by mudslides.

Soil Heating by Fire

Fire is one of the most far-reaching ecosystem disturbances in nature. In addition to the obvious aboveground effects of forest, range, or crop-residue fires, the brief but sometimes dramatic changes in soil temperature also may have lasting impacts below ground. Unless the fire is artificially stoked with added fuel, the temperature rise itself is usually very brief and is limited to the upper few centimeters of soil. But the temperatures resulting from "slash and burn" practices in the tropics may be sufficiently high in the upper few mm of soil to cause the breakdown of minerals such as gibbsite and kaolinite (Figure 7.26).

The heat may also affect the breakdown and movement of organic compounds. Figure 7.27 shows the results of a wildfire on a lodgepole pine stand in Oregon. The high temperatures (>125 °C being common) essentially distill various fractions of the organic matter, with some of the volatilized hydrocarbon compounds moving quickly through the soil pores to deeper, cooler areas. As these compounds reach cooler soil particles deeper in the soil, they condense (solidify) on the surface of the soil particles and fill some of the surrounding pore spaces. Some of the condensed compounds are water-repellent (hydrophobic) hydrocarbons. Consequently, when rain comes, water infiltration in even a sandy soil is greatly reduced in comparison to unburned areas. This effect of soil temperature is quite common on chaparral lands in semiarid regions and may be responsible for the disastrous mudslides that occur when the layer of soil above the hydrophobic zone becomes saturated with rainwater (see Plate 71).

Fires also affect the germination of certain seeds, which have hard coatings that prevent them from germinating until they are heated above 70 to 80 °C. On the other hand, burning of straw in wheat fields generates similar soil temperatures, but with the effect of killing most of the weed seeds near the surface and thus greatly reducing subsequent weed infestation. The heat and ash may also hasten the cycling of plant nutrients. Fires set to clear land of timber slash may burn long and hot enough to seriously

---

[6] For a discussion of permafrost in Alaska, see Wuethrich (2000).

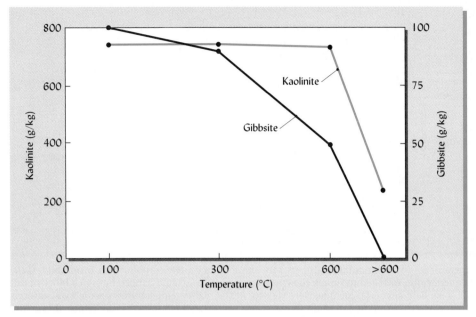

FIGURE 7.26 Some farmers in the tropics "slash and burn" forested areas to clear land for cultivation. The burning raises the temperature of the upper few millimeters of soil to levels sufficiently high to cause the breakdown of some soil minerals. Such breakdown is illustrated for two minerals (kaolinite and gibbsite) in a soil from Indonesia. [Drawn from data in Ketterings et al. (2000)]

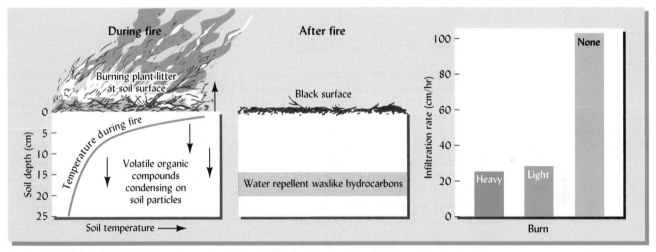

FIGURE 7.27 (*Left*) Wildfires of a lodgepole pine stand heat up the surface layers of this sandy soil (an Inceptisol) in Oregon. (*Center*) Note that the soil temperature is increased sufficiently near the surface to volatilize organic compounds, some of which then move down into the soil and condense (solidify) on the surface of cooler soil particles. These condensed compounds are waxlike hydrocarbons that are water repellent. As a consequence (*right*) the infiltration of water into the soil is drastically reduced and remains so for a period of at least six years. [From Dryness (1976)]

deplete soil organic matter and kill so many soil organisms that forest regrowth is inhibited.

Contaminant Removal

The removal of certain organic pollutants from contaminated soils can be accomplished by raising the soil temperature. But the process may be prohibitively expensive if the soil has to be excavated and hauled to and from an extraction bin. Techniques are under development to warm the soil in place in the field using electromagnetic radiation. The resulting temperatures are sufficiently high to vaporize some contaminants which can then be flushed from the soil by air. Figure 7.28 shows the results of one such operation set up to remove diesel fuel from soil under an air force base.

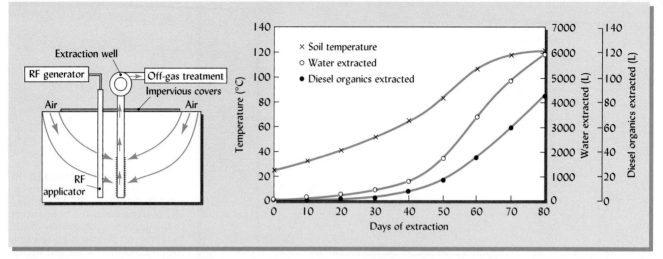

FIGURE 7.28 By increasing soil temperatures, environmental engineers can extract some organic pollutants from soils. (*Left*) Electromagnetic radiation from radio frequency (RF) was used to supply the energy to gradually increase temperatures in a block of soil containing diesel fuel at the Kirkland Air Force Base near Albuquerque, New Mexico. At the higher temperatures organic hydrocarbons were vaporized (along with water) and then extracted from the soil using an extraction well. They were subsequently removed from the air by off-gas treatment. (*Right*) Soil temperatures increase with time near the RF applicator, along with the quantity of organic compounds and water extracted from the soil. While this procedure is rather expensive, it permits remediation of the soil without having to remove it from its natural setting and does not result in extremely high temperatures as the soil is heated. [Modified from figures in Lowe et al. (2000)]

7.9 ABSORPTION AND LOSS OF SOLAR ENERGY[7]

Green roofs vs urban heat islands:

http://www.arctic.edu/webspaces/greeninitiatives/greenroofs/

The temperature of soils in the field is directly or indirectly dependent on at least three factors: (1) the net amount of heat energy the soil absorbs; (2) the heat energy required to bring about a given change in the temperature of a soil; and (3) the energy required for processes such as evaporation, which are constantly occurring at or near the surface of soils.

Solar radiation is the primary source of energy to heat soils. But clouds and dust particles intercept the sun's rays and absorb, scatter, or reflect most of the energy (Figure 7.29). Only about 35 to 40% of the solar radiation actually reaches the Earth in cloudy humid regions and 75% in cloud-free arid areas. The global average is about 50%.

Little of the solar energy reaching the Earth actually results in soil warming. The energy is used primarily to evaporate water from the soil or leaf surfaces or is radiated or reflected back to the sky. Only about 10% is absorbed by the soil and can be used to warm it. Even so, this energy is of critical importance to soil processes and to plants growing on the soils.

ALBEDO. The fraction of incident radiation that is reflected by the land surface is termed the **albedo** and ranges from as low as 0.1 to 0.2 for dark-colored, rough soil surfaces to as high as 0.5 or more for smooth, light-colored surfaces. Vegetation may affect the surface albedo either way, depending on whether it is dark green and growing or yellow and dormant.

The fact that dark-colored soils absorb more energy than lighter-colored ones does not necessarily imply, however, that dark soils are always warmer. In fact, the opposite is often true. In most landscapes, the darkest soils are those found in the low spots where excessive wetness has caused organic matter to accumulate. Therefore, the darkest soils are also usually the wettest. The water in these soils requires much energy to be warmed, and it also cools the soil when it evaporates.

ASPECT. The angle at which the sun's rays strike the soil also influences soil temperature. If the sun is directly overhead, the incoming path of the rays is perpendicular to the soil surface, and energy absorption (and soil temperature increase) is greatest (Figure 7.30).

---

[7] For application of these principles to the role of soil moisture in models of global warming, see Lin et al. (2003).

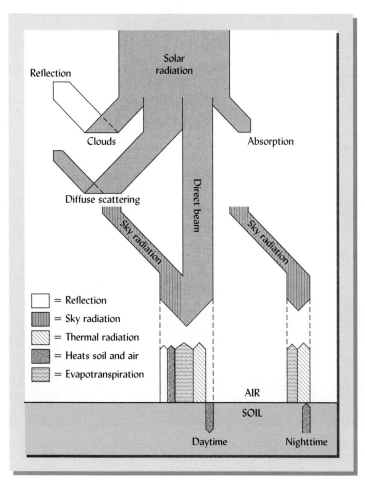

FIGURE 7.29 Schematic representation of the radiation balance in daytime and nighttime in the spring or early summer in a temperate region. About half the solar radiation reaches the Earth, either directly or indirectly, from sky radiation. Most radiation that strikes the Earth in the daytime is used as energy for evapotranspiration or is radiated back to the atmosphere. Only a small portion, perhaps 10%, actually heats the soil. At night the soil loses some heat, and some evaporation and thermal radiation occur.

The effect of the direction of slope, or **aspect**, on forest species is illustrated in the photo in Figure 7.30.

Planting crops on soil ridges is one method of controlling the soil aspect on a microscale. This is most effectively done at high latitudes by planting crops on the south- or southwest-facing sides of ridges. The ridges need to be only about 25 cm tall to have a major effect. In Fairbanks, Alaska, midafternoon soil temperatures (at 1 cm depth) in early May can be about 15 °C warmer on the south side of such a ridge than on the north side and about 8 °C warmer than on level ground.

RAIN. Mention should also be made of the effect of rain or irrigation water on soil temperature. For example, in temperate zones, spring rains definitely warm the surface soil as the water moves into it. Conversely, in the summer, rainfall cools the soil, since it is often cooler than the soil it penetrates. However, spring rain, by increasing the amount of solar energy used in evaporating water from the soil, often accentuates low temperatures.

SOIL COVER. Whether the soil is bare or is covered with vegetation, mulch, or snow is another factor markedly influencing the amount of solar radiation reaching the soil. Bare soils warm up more quickly and cool off more rapidly than those covered with vegetation, with snow, or with plastic mulches. Frost penetration during the winter is considerably greater in bare, noninsulated land.

Even low-growing vegetation such as turfgrass has a very noticeable influence on soil temperature and on the temperature of the surroundings (Table 7.6). Much of the cooling effect is due to heat dissipated by transpiration of water. To experience this effect, on a blistering hot day, try having a picnic on an asphalt parking lot instead of on a growing green lawn!

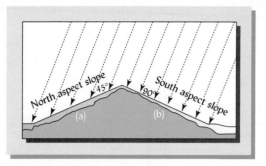

FIGURE 7.30 (*Upper*) Effect of slope aspect on solar radiation received per unit land area. Slope (*a*) is north facing and receives solar radiation at an angle of 45° to the ground surface so only 5 units of solar radiation (arrows) hit the unit of land area. The same land area on the south-facing slope (*b*) receives 7 units of radiation at a 90° angle to the ground. In other words, if a given amount of radiation from the sun strikes the soil at right angles, the radiation is concentrated in a relatively small area, and the soil warms quite rapidly. This is one of the reasons why north slopes tend to have cooler soils than south slopes. It also accounts for the colder soils in winter than in summer. (*Lower*) A view looking eastward toward a forested mountain in Virginia illustrates the temperature effect. The main ridge (left to right) is running north and south and the smaller side ridges east and west (up and down). The dark patches are pine trees in this predominantly hardwood deciduous forest. The pines dominate the southern slopes on each east-west ridge. The soils on the southern slopes are warmer and therefore drier, less deeply weathered, and lower in organic matter. (Photo and diagram courtesy of R. Weil)

The effect of a dense forest is universally recognized. Timber-harvest practices that leave sufficient canopy to provide about 50% shade will likely prevent undue soil warming that could hasten the loss of soil organic matter or the onset of anaerobic conditions in wet soils. The effect of timber harvest on soil temperature as deep as 50 cm is seen in the case presented in Table 7.2, where tree removal warmed the soil in spring, even though it also raised the soil water content. However, as we shall see in the next section, a higher water content normally slows the warming of soils in spring.

TABLE 7.6 **Maximum Surface Temperatures for Four Types of Surfaces on a Sunny August Day in College Station, Texas**

	Maximum temperature, °C	
Type of surface	*Day*	*Night*
Green, growing turfgrass	31	24
Dry, bare soil	39	26
Brown, summer-dormant grass	52	27
Dry synthetic sports turf	70	29

Data from Beard and Green (1994).

7.10 THERMAL PROPERTIES OF SOILS

Specific Heat of Soils

Heat energy concepts:

http://hyperphysics.phy-astr. gsu.edu/hbase/thermo/ heat.html

A dry soil is more easily heated than a wet one. This is because the amount of energy required to raise the temperature of water by 1 °C (its heat capacity) is much higher than that required to warm soil solids by 1 °C. When heat capacity is expressed per unit mass—for example, in calories per gram (cal/g)—it is called **specific heat** or heat capacity c. The specific heat of pure water is about 1.00 cal/g (or 4.18 joules per gram, J/g); that of dry soil is about 0.2 cal/g (0.8 J/g).

The specific heat largely controls the degree to which soils warm up in the spring, wetter soils warming more slowly than drier ones (see Box 7.4). Furthermore, if the

---

BOX 7.4 CALCULATING THE SPECIFIC HEAT OR HEAT CAPACITY OF MOIST SOILS

Soil water content markedly impacts soil temperature changes through its effect on the specific heat or heat capacity c of a soil. For example, consider two soils with comparable characteristics, soil A, a relatively dry soil with 10 g water/100 g soil solids, and soil B, a wetter soil with 30 g water/100 g soil solids.

We can assume the following values for specific heat:

Water = 1.0 cal/g and dry mineral soil = 0.2 cal/g

For soil A with 10 g water/100 g dry soil, or 0.1 g water/g dry soil, the number of calories required to raise the temperature of 0.1 g of water by 1 °C is

$$0.1 \text{ g} \times 1.0 \text{ cal/g} = 0.1 \text{ cal}$$

The corresponding figure for the 1.0 g of soil solids is

$$1 \text{ g} \times 0.2 \text{ cal/g} = 0.2 \text{ cal}$$

Thus, a total of 0.3 cal (0.1 + 0.2) is required to raise the temperature of 1.1 g (1.0 + 0.1) of the moist soil by 1 °C. Since the specific heat is the number of calories required to raise the temperature of 1 g of moist soil by 1 °C, we can calculate the specific heat of soil A as follows:

$$c_{\text{soil A}} = \frac{0.3}{1.1} = 0.273 \text{ cal/g}$$

These calculations can be expressed as a simple equation to calculate the weighted average specific heat of a mixture of substances:

$$c_{\text{moist soil}} = \frac{c_1 m_1 + c_2 m_2}{m_1 + m_2} \tag{7.7}$$

where c_1 and m_1 are the specific heat and mass of substance 1 (the dry mineral soil, in this case), and c_2 and m_2 are the specific heat of substance 2 (the water, in this case).

Applying this equation to soil A, we again calculate that $c_{\text{soil A}}$ is 0.273 cal/g, as follows:

$$c_{\text{soil A}} = \frac{0.2 \text{ cal/g} * 1.0 \text{ g} + 1.0 \text{ cal/g} * 0.10 \text{ g}}{1.0 \text{ g} + 0.10 \text{ g}} = \frac{0.30 \text{ cal}}{1.1 \text{ g}} = 0.273 \text{ cal/g}$$

In the same manner, we calculate the specific heat of the wetter soil B:

$$c_{\text{soil B}} = \frac{0.2 \text{ cal/g} * 1.0 \text{ g} + 1.0 \text{ cal/g} * 0.30 \text{ g}}{1.0 \text{ g} + 0.30 \text{ g}} = \frac{0.50 \text{ cal}}{1.3 \text{ g}} = 0.385 \text{ cal/g}$$

The wetter soil B has a specific heat c_B of 0.385 cal/g, whereas the drier soil A has a specific heat c_A of 0.273 cal/g. Because it must absorb an additional 0.112 cal (0.385 − 0.273) of solar radiation for every degree of temperature rise, the wetter soil will warm up much more slowly than the drier soil.

---

water does not drain freely from the wet soil, it must be evaporated, a process that is very energy consuming, as the next section will show.

The high specific heat of soils is also used in the design of energy-efficient geothermic temperature control systems that both warm and cool buildings. To maximize heat-exchange contact with the soil, a network of pipes is laid underground near the building to be heated and cooled. Advantage is taken of the fact that subsoils are generally warmer than the atmosphere in the winter and cooler than the atmosphere in the summer. Water circulating through the network of pipes absorbs heat from the soil during the winter and releases it to the soil in the summer. The high specific heat of soils permits a large exchange of energy to take place without greatly modifying the soil temperature.

Heat of Vaporization

The evaporation of water from soil surfaces requires a large amount of energy, 540 kilocalories (kcal) or 2.257 megajoules (mJ) for every kilogram of water vaporized. This energy must be provided by solar radiation or it must come from the surrounding soil. In either case, evaporation has the potential of cooling the soil, much the way it chills a person who comes out from the water after swimming on a windy day.

For example, if the amount of water associated with 100 g of dry soil was reduced by evaporation from 25 g to 24 g (only about a 1% decrease) and if all the thermal energy needed to evaporate the water came from the moist soil, the soil would be cooled by about 12 °C. Such a figure is hypothetical because only a part of the heat of vaporization comes from the soil itself. Nevertheless, it indicates the tremendous cooling influence of evaporation.

The low temperature of a wet soil is due partially to evaporation and partially to high specific heat. The temperature of the upper few centimeters of wet soil is commonly 3 to 6 °C lower than that of a moist or dry soil. This is a significant factor in the spring in a temperate zone, when a few degrees will make the difference between the germination or lack of germination of seeds, or the microbial release or lack of release of nutrients from organic matter.

Thermal Conductivity of Soils

Soil heat capacity saves energy in buildings:

http://www.geoexchange.org/about/movie.htm

As shown in Section 7.9, some of the solar radiation that reaches the Earth slowly penetrates the profile largely by conduction; this is the same process by which heat moves to the handle of a cast-iron frying pan. The movement of heat in soil is analogous to the movement of water (see Section 5.5), the rate of flow being determined by a driving force and by the ease with which heat flows through the soil. This can be expressed as Fourier's Law:

$$Q_h = K * \frac{\Delta T}{x} \tag{7.8}$$

where Q_h is the *thermal flux,* the quantity of heat transferred across a unit cross-sectional area in a unit time; K is the **thermal conductivity** of the soil; and $\Delta T/x$ is the temperature gradient over distance x that serves as the driving force for the conduction of heat.

The thermal conductivity K of soil is influenced by a number of factors, the most important being the moisture content of the soil and the degree of compaction (see Figure 7.31). Heat passes through water many times faster than through air. As the water content increases in a soil, the air content decreases, and the transfer resistance is decidedly lowered. When sufficient water is present to form a bridge between most of the soil particles, further additions will have little effect on heat conduction. Heat moves through mineral particles even faster than through water, so when particle-to-particle contact is increased by soil compaction, heat-transfer rates are also increased. Therefore, a wet, compacted soil would be the poorest insulator or the best conductor of heat. Here again the interconnectedness of soil properties is demonstrated.

Relatively dry soil makes a good insulating material. Buildings built mostly underground can take advantage of both the low thermal conductivity and relatively high heat capacity of large volumes of soil (see Figure 7.32).

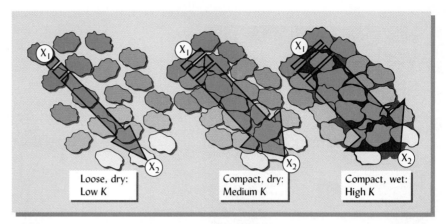

FIGURE 7.31   Bulk density and water content affect the rate at which heat is transferred through soils from a warm zone (X_1) to a cooler zone (X_2). The arrows represent heat transfer through the soil, the rate of transfer being in proportion to the arrow thickness. Higher bulk density from soil compaction increases particle-to-particle contact, which in turn hastens heat transfer because the thermal conductivity of mineral particles is much higher than that of air. If the remaining gaps between particles become filled with water instead of air, thermal conductivity increases still more because water also conducts heat better than air. Therefore wet, compacted soils transfer heat most rapidly. (Diagram courtesy of R. Weil)

The significance of heat conduction with respect to field soil temperature is not difficult to comprehend. It provides a means of temperature adjustment, but, because it is slow, changes in subsoil temperature lag behind those of the surface layers. Moreover, temperature changes are always less in the subsoil. In temperate regions, surface soils in general are expected to be warmer in summer and cooler in winter than the subsoil, especially the lower horizons of the subsoil. Soil thermal conductivity can also affect air temperature above the soil, as shown in Figure 7.33.

FIGURE 7.32   This high school in Oklahoma has been built into the ground with only one side exposed. This design takes advantage of the high specific heat and low thermal conductivity of the overlying soils, keeping the school warm in winter and cool in summer with a minimum of energy used for heating or air-conditioning. (Photo courtesy of R. Weil)

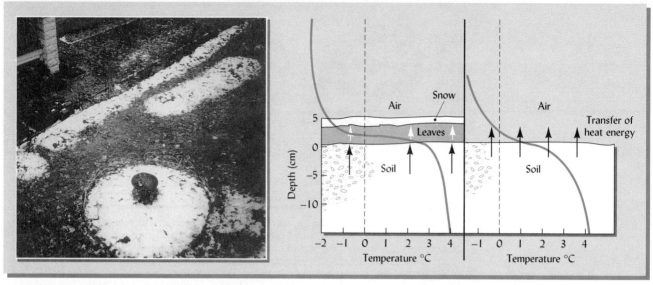

FIGURE 7.33 Transfer of heat energy from soil to air. The scene, looking down on a garden after an early fall snowstorm, shows snow on the leaf-mulched flower beds, but not on areas where the soil is bare or covered with thin turf. The reason for this uneven accumulation of snow can be seen in the temperature profiles. Having stored heat from the sun, the soil layers are often warmer than the air as temperatures drop in fall (this is also true at night during other seasons). On bare soil, heat energy is transferred rapidly from the deeper layers to the surface, the rate of transfer being enhanced by high moisture content or compaction, which increases the *thermal conductivity* of the soil. As a result, the soil surface and the air above it are warmed to above freezing, so snow melts and does not accumulate. The leaf mulch, which has a low thermal conductivity, acts as an insulating blanket that slows the transfer of stored heat energy from the soil to the air. The upper surface of the mulch is therefore hardly warmed by the soil, and the snow remains frozen and accumulates. A heavy covering of snow can itself act as an insulating blanket. (Photo and diagram courtesy of R. Weil)

Variation with Time and Depth

The temperature of the soil at any time depends on the ratio of the energy absorbed to that being lost. The constant change in this relationship is reflected in seasonal, monthly, and daily temperature fluctuations. Figure 7.34 illustrates the considerable seasonal variations of soil temperature that occur in temperate region soils. The surface

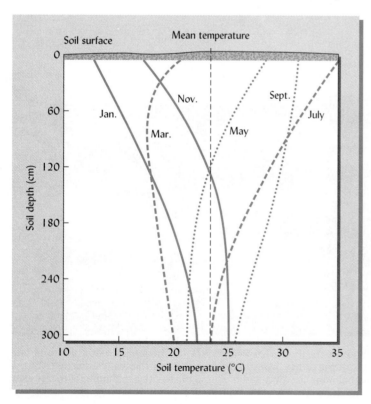

FIGURE 7.34 Average monthly soil temperatures for 6 of the 12 months of the year at different soil depths at College Station, TX (1951–1955). Note the lag in soil temperature change at the lower depths. [From Fluker (1958)]

layer temperatures vary more or less according to the temperature of the air, although these layers are generally warmer than the air throughout the year.

In the subsoil, the seasonal temperature increases and decreases lag behind changes registered in the surface soil and in the air. Accordingly, Figure 7.34 shows surface soil temperatures in March responding to the warming of spring, while temperatures of the deep subsoil still reflect the cold of winter. Subsoil temperatures are less variable than air and surface soil temperatures, although there is some temperature fluctuation even at 300 cm deep (Figure 7.34).

Compared to the surface soil and the air, deep soil layers are generally warmer in the late fall and winter and cooler in the spring and summer. Soil reaches its maximum daily temperature later in the afternoon or evening than does the air, the lag time being greater and the fluctuation less pronounced for greater depths. Deeper than 4 to 5 m, the temperature changes little and approximates the mean annual air temperature (a fact experienced by people visiting deep caverns).

7.11 SOIL TEMPERATURE CONTROL

The temperature of field soils is not subject to radical human regulation. However, two kinds of management practice have significant effects on soil temperature: those that affect the cover or mulch on the soil and those that reduce excess soil moisture. These effects have meaningful biological implications.

Organic Mulches and Plant-Residue Management

Mulching effects:
http://www.ianrpubs.unl.edu/epublic/pages/publicationD.jsp?publicationId=187

Soil temperatures are influenced by soil cover and especially by organic residues or other types of mulch on the soil surface. Figure 7.35 shows that mulches effectively buffer extremes in soil temperatures. In periods of hot weather, they keep the surface

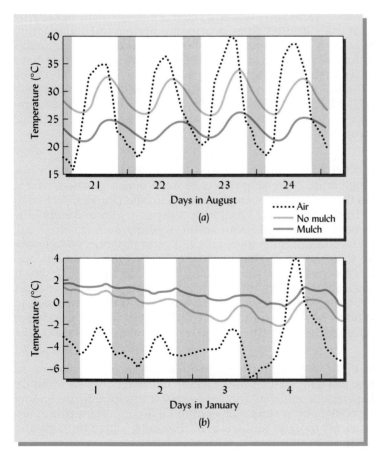

FIGURE 7.35 (a) Influence of straw mulch (8 tons/ha) on air temperature at a depth of 10 cm during an August hot spell in Bushland, TX. Note that the soil temperatures in the mulched area are consistently lower than where no mulch was applied. (b) During a cold period in January, the soil temperature was higher in the mulched than in the unmulched area. The shaded bars represent nighttime. [Redrawn from Unger (1978); used with permission of American Society of Agronomy]

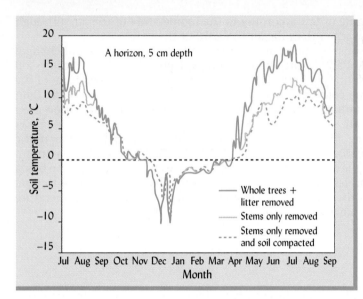

FIGURE 7.36 Soil temperature in an aspen-spruce boreal forest after two levels of harvest and soil compaction. One harvest procedure removed only the stems (tree trunks) with branches and foliage left on the soil, while a second procedure removed whole trees and stripped woody materials and litter to expose the mineral soil (this was done to simulate the kind of damage often inflicted by poorly managed harvest equipment). The soils were either left undisturbed during harvest, or they were severely compacted. The compacted treatment is shown only for the whole-tree removal procedure as compaction did not affect soil temperature where harvest removed only tree trunks. Exposure of the mineral soil A horizon resulted in much warmer temperatures in summer and somewhat colder soil in winter. Compaction of this soil mainly slowed warming in summer, partly because of a higher water content (and therefore a higher heat capacity). The Aquepts (Luvic Gleysols in the Canadian soil classification) at this site in British Columbia included about 20–30 cm of silt loam material over a clay loam. [From Tan et al. (2005)]

soil cooler than where no cover is used; in contrast, during cold weather they keep the soil warmer than it would be if bare.

The forest floor is a prime example of a natural temperature-modifying mulch. It is not surprising, therefore, that timber harvest practices can markedly affect forest soil temperature regimes (Figure 7.36). Disturbance of the leaf mulch, changes in water content due to reduced evapotranspiration, and compaction by machinery are all factors that influence soil temperatures through thermal conductivity. Reduced shading after tree removal also lets in more solar radiation.

MULCH FROM CONSERVATION TILLAGE. Until fairly recently, the labor and expense of carrying and spreading mulch materials limited their use in modifying soil temperature extremes mostly to home gardens and flower beds. Although these uses are still important, the use of mulches has been extended to field-crop culture in areas that have adopted conservation tillage practices. Conservation tillage leaves most or all of the crop residues at or near the soil surface, thereby growing the mulch in place, rather than transporting it to the field. Partly due to the influence of surface residues, soil temperatures to depths as great as 70 cm are consistently lower during spring in no-tillage systems, which leave all crop residues as a mulch.

CONCERNS IN COOL CLIMATES. While the mulch provides great control over erosion (see Section 17.6), the soil-temperature-depressing effects of some mulch practices have a serious negative impact on the production of crops like corn in cold climates, such as in Canada and the northern United States. The lower temperatures in May and early June resulting from these practices inhibit seed germination, seedling performance, and, often, the yields of corn. The effect of the residue mulch is most pronounced in lowering the midday maximum temperature and has much less effect on the minimum temperature reached at night. This effect is well illustrated by the data in Figure 7.37, which also features an innovative way to alleviate this problem by pushing aside the residues in just a narrow band over the seed row in the no-tillage system. Another approach to solving this problem is to ridge the soil, permit water to drain out of the ridge, and then plant on the drier, warmer ridgetop (or on the south side of the ridge—see Section 7.9).

ADVANTAGES IN WARM CLIMATES. In warm regions, delayed planting is not a problem. In fact, the cooler near-surface soil temperatures under a mulch may reduce heat stress on roots during summer. Plant-residue mulches also conserve soil moisture by decreasing evaporation. The resulting cooler, moist surface layer of soil is an important part of no-tillage systems because it allows roots to proliferate in this zone, where nutrient and aeration conditions are optimal.

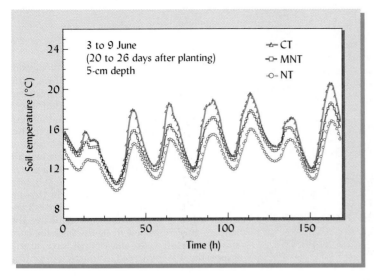

FIGURE 7.37 Tillage effects on hourly temperature changes near the surface of a cold Alfisol in northern British Columbia. The soil had been managed to grow barley under no-tillage (NT) and conventional (clean surface without residues) tillage (CT) systems for the previous 14 years. In the clean-tilled soil, midafternoon temperatures peaked at 4 °C higher than those in the residue-mulch-covered no-tillage soil. A modification of the no-tillage system (MNT) that pushed aside the residues in a narrow (7.5-cm wide) band over the seeding row eliminated much of the temperature depression while keeping most of the surface covered by the soil- and water-conserving mulch. Note the daily temperature changes and the general warming trend during the seven days shown. [From Arshad and Azooz (1996)]

Plastic Mulches

One of the reasons for the popularity of plastic mulches for gardens and high-value specialty crops is their effect on soil temperature (see Section 6.4 for their effect on soil moisture). In contrast to organic mulches, plastic mulches generally increase soil temperature, clear plastic having a greater heating effect than black plastic. In temperate regions, this effect can be used to extend the growing season or to hasten production to take advantage of the higher prices offered by early-season markets. Figure 7.38 shows the use of clear plastic mulch for winter-grown strawberries in Southern California.

Major disadvantages of both clear and black plastic mulch are the nonrenewable fossil fuels used in their manufacture, the difficulty of removing the material from the field at the end of the season, and the problem of properly disposing of all that shredded plastic waste. One solution may be found in newer biodegradable plastic films that have been manufactured from such natural renewable raw materials as corn starch. Figure 7.39 illustrates the use of biodegradable plastic film on a residue-covered soil in order to allow earlier than normal planting of a sweet corn crop in a cool climate.

In warmer climates, and during the summer months, the soil-heating effect of plastic mulches may be quite detrimental, inhibiting root growth in the upper soil layers and sometimes seriously decreasing crop yields. Table 7.7 provides an example in

FIGURE 7.38 These winter-grown Southern California strawberries will come to market when prices are still high because of the effect of the clear plastic mulch on soil temperature. (Photo courtesy of R. Weil)

FIGURE 7.39 Laying down a special biodegradable clear plastic mulch to hasten soil warming for a no-till sweet corn crop on hilly cropland in Pennsylvania. The farmer's no-till system keeps the soil covered with plant residues that prevent erosion but also slow soil warming in spring. Each strip of plastic mulch covers two rows of corn seeds that have already been planted with a no-till planter. The plastic film acts like a greenhouse to trap solar energy, warm the soil, and hasten the corn germination and early growth. When the corn seedlings are about 20 cm tall the farmer will slit the plastic film, allowing the plants to grow unimpeded. By the time the corn canopy has closed, the plastic will have largely disappeared, having served its function of getting the corn off to an early start. (Photo courtesy of R. Weil)

TABLE 7.7 **Soil Temperature and Tomato Yield with Straw or Black Plastic Mulch[a]**

The data are averages for two years of tomato production on a sandy loam Ultisol near Griffin, Georgia.
The straw kept the surface soil from rising to detrimentally high temperatures, while it also increased infiltration
of rainwater and reduced soil compaction. Daily drip irrigation supplied plenty of water, but could not
overcome the temperature effects of the black plastic mulch.

	Not irrigated		Irrigated daily	
	Straw mulch	*Plastic mulch*	*Straw mulch*	*Plastic mulch*
Average soil temperature, °C	24	37	24	35
Tomato yield, Mg/ha	68	30	70	24

[a] Soil temperature measured at 5 cm below the soil surface, average of weeks 2–10 of the growing season.
Data calculated from Tindall et al. (1991).

which the benefits of weed control and moisture conservation by plastic mulch were outweighed by the detrimental effects of excessive soil heating.

Moisture Control

Another means of exercising some control over soil temperature is by controlling soil moisture. Poorly drained soils in temperate regions that are wet in the spring have temperatures 3 to 6 °C lower than comparable well-drained soils. By removing excess water temperature depression can be alleviated. Water can be removed by installing drainage systems using ditches and underground pipes (see Section 6.7). Where this is not feasible, the ridging systems of tillage just referred to can be used.

As is the case with soil air, the controlling influence of soil water on soil temperature is apparent everywhere. Whether a problem concerns capture of solar energy, loss of energy to the atmosphere, or the movement of heat within the soil, the amount of water present is always important. Water regulation seems to be a key to what little practical temperature control is possible for field soils.

7.12 CONCLUSION

Soil aeration and soil temperature critically affect the quality of soils as habitats for plants and other organisms. Most plants have definite requirements for soil oxygen along with limited tolerance for carbon dioxide, methane, and other such gases found in poorly aerated soils. Some microbes, such as the nitrifiers and general-purpose decay organisms, are also constrained by low levels of soil oxygen. Through its effect on soil redox potential (E_h) and acidity (pH), soil aeration status helps determine the forms present, availability, mobility, and possible toxicity of such elements as nitrogen, sulfur, carbon, iron, manganese, chromium, and many others.

Soils with extremely wet moisture regimes are unique with respect to their morphology and chemistry and to the plant communities they support. Such hydric soils are characteristic of wetlands and help these ecosystems perform a myriad of valuable functions.

Plants as well as microbes are also quite sensitive to differences in soil temperature, particularly in temperate climates where low temperatures can limit essential biological processes. Soil temperature also impacts the use of soils for engineering purposes, again primarily in the cooler climates. Frost action, which can move perennial plants such as alfalfa out of the ground, can also cause damage to building foundations, fence posts, sidewalks, and highways.

Soil water exerts a major influence over both soil aeration and soil temperature. It competes with soil air for the occupancy of soil pores and interferes with the diffusion of gases into and out of the soil. Soil water also resists changes in soil temperature by virtue of its high specific heat and its high energy requirement for evaporation.

STUDY QUESTIONS

1. What are the two principal gases involved with soil aeration, and how do their relative amounts change as one samples deeper into a soil profile?

2. What is aerenchyma tissue, and how does it affect plant–soil relationships?

3. If the redox potential for a soil at pH 6 is near zero, write two reactions that you would expect to take place. How would the presence of a great deal of nitrate compounds affect the occurrence of these reactions?

4. It is sometimes said that organisms in anaerobic environments will use the combined oxygen in nitrate or sulfate instead of the oxygen in O_2. Why is this statement incorrect? What actually happens when organisms reduce sulfate or nitrate?

5. If an alluvial forest soil were flooded for 10 days and you sampled the gases evolving from the wet soil, what gases would you expect to find (other than oxygen and carbon dioxide)? In what order of appearance? Explain.

6. Explain why warm weather during periods of saturation is required in order to form a hydric soil.

7. If you were in the field trying to delineate the so-called drier end of a wetland area, what are three soil properties and three other indicators that you might look for?

8. For each of these gases, write a sentence to explain its relationship to wetland conditions: *ethylene, methane, nitrous oxide, oxygen,* and *hydrogen sulfide.*

9. What are the three major components that define a wetland?

10. Discuss four plant processes that are influenced by soil temperature.

11. Explain how a brush fire might lead to subsequent mudslides, as often occurs in California.

12. If you were to build a house below ground in order to save heating and cooling costs, would you firmly compact the soil around the house? Explain your answer.

13. If you measured a daily maximum air temperature of 28 °C at 1 P.M., what might you expect the daily maximum temperature to be at a 15-cm depth in the soil? At about what time of day would the maximum temperature occur at this depth? Explain.

14. In relation to soil temperature, explain why conservation tillage has been more popular in Missouri than in Minnesota.

REFERENCES

Arshad, A., and R. H. Azooz. 1996. "Tillage effects on soil thermal properties in a semi-arid cold region," *Soil Sci. Soc. Amer. J.* **60**:561–567.

Bartlett, R. J., and B. R. James. 1993. "Redox chemistry of soils," *Advances in Agronomy* **50**:151–208.

Bartlett, R. J., and D. S. Ross. 2005. "Chemistry of redox process in soils," pp. 461–487, in A. Tabatabai and D. Sparks," (eds.), *Chemical Processes In Soils*. SSSA Book Series N 8. (Madison, WI: Soil Science Society of America).

Beard, J. B., and R. L. Green. 1994. "The role of turfgrasses in environmental protection and their benefits to humans," *J. Environ. Qual.* **23**:452–460.

Buyanovsky, G. A., and G. H. Wagner. 1983. "Annual cycles of carbon dioxide level in soil air," *Soil Sci. Soc. Amer. J.* **47**:1139–1145.

CAST. 1994. *Wetland Policy Issues*. Publication No. CC1994–1. (Ames, IA: Council for Agricultural Science and Technology).

Costanza, R., R. D'arge, R. De Groot, S. Farber, M. Grasso, B. Hannon, K. Limburg, S. Naeem, R. V. O'Neill, J. Paruelo, R. G. Raskin, P. Sutton, and M. Van Den Belt. 1997. "The value of the world's ecosystem services and natural capital," *Nature* **387**:253–260.

Davidson, E. A., and S. E. Trumbore. 1995. "Gas diffusivity and production of CO_2 in deep soils of the eastern Amazon," *Tellus* **47B**:550–565.

Dryness, C. T. 1976. "Effects of wildfire on soil wetability in the high cascades of Oregon," USDA Forest Service Research Paper PNW-202. (Washington, DC: USDA).

Eilperin, J. 2005. "Developers sentenced in wetlands case," p. A-14, *The Washington Post,* December 07, 2005, Washington, DC.

Eswaren, H., P. Reich, P. Zdruli, and T. Levermann. 1996. "Global distribution of wetlands," *Amer. Soc. Agron. Abstracts* 328.

Fluker, B. J. 1958. "Soil temperature," *Soil Sci.* **86**:35–46.

Hurt, G. W., P. M. Whited, and R. F. Pringle (eds.). 1996. *Field Indicators of Hydric Soils in the United States.* (Fort Worth, TX: USDA Natural Resources Conservation Service).

Jenkinson, B. J., and D. P. Franzmeier. 2006. "Development and evaluation of iron-coated tubes that indicate reduction in soils," *Soil Sci. Soc. Amer. J.* **70**:183–191.

Ketterings, Q. M., J. M. Bigham, and V. Laperche. 2000. "Changes in soil mineralogy and texture caused by slash and burn fires in Sumatra, Indonesia," *Soil Sci. Soc. Amer. J.* **64**:1108–1117.

Lin, X., J. E. Smerdon, A. W. England, and H. N. Pollack. 2003. "A model study of the effects of climatic precipitation changes on ground temperatures," *J. Geophys. Res.* **108**(D7):4230, doi:10.1029/2002JD002878.

Losi, M. E., C. Amrheim, and W. T. Frankenberger, Jr. 1994. "Bioremediation of chromatic contaminated groundwater by reduction and precipitation in surface soils," *J. Environ. Qual.* **23**:1141–1150.

Lowe, D. F., C. L. Oubre, and C. H. Ward (eds.). 2000. *Soil Vapor Extraction using Radio Frequency Heating: Resource Manual and Technology Demonstration* (New York: Lewis).

MacDonald, N. W., D. R. Zac, and K. S. Pregitzer. 1995. "Temperature effects on kinetics of microbial respiration and net nitrogen and sulfur mineralization," *Soil Sci. Soc. Amer. J.* **59**:233–240.

McBride, M. B. 1994. *Environmental Chemistry of Soils.* (New York: Oxford University Press).

Melillo, J. M., P. A. Steudler, J. D. Aber, K. Newkirk, H. Lux, F. P. Bowles, C. Catricala, A. Magill, T. Ahrens, and S. Morrisseau. 2002. "Soil warming and carbon-cycle feedbacks to the climate system," *Science* **298**:2173–2176.

Patrick, W. H., Jr. 1977. "Oxygen content of soil air by a field method," *Soil Sci. Soc. Amer. J.* **41**:651–652.

Patrick, W. H., Jr., and A. Jugsujinda. 1992. "Sequential reduction and oxidation of inorganic nitrogen, manganese, and iron in flooded soil," *Soil Sci. Soc. Amer. J.* **56**: 1071–1073.

Richardson, J. L., and M. J. Vepraskas. 2001. *Wetland Soils—Genesis, Hydrology, Landscapes, and Classification.* (Boca Raton, FL: Lewis).

Rykbost, K. A., L. Boersma, J. J. Mack, and W. E. Schmisseur. 1975. "Yield response to soil warming: Vegetable crops," *Agron. J.,* **67**:738–743.

Sexstone, A. J., N. P. Revsbech, T. B. Parkin, and J. M. Tiedje. 1985. "Direct measurement of oxygen profiles and denitrification rates in soil aggregates," *Soil Sci. Soc. Amer. J.* **49**:645–651.

Tan, X., S. X. Chang, and R. Kabzems. 2005. "Effects of soil compaction and forest floor removal on soil microbial properties and N transformations in a boreal forest long-term soil productivity study," *Forest Ecology and Management* **217**:158–170.

Tiarks, A. E., W. H. Hudnall, J. F. Ragus, and W. B. Patterson. 1996. "Effect of pine plantation harvesting and soil compaction on soil water and temperature regimes in a semi-tropical environment," in A. Schulte and D. Ruhiyat, (eds.), *Proceedings of International Congress on Soils of Tropical Forest Ecosystems 3rd Conference on Forest Soils: Vol. 3, Soil and Water Relationships* (Samarinda, Indonesia: Mulawarmon University Press).

Tindall, J. A., R. B. Beverly, and D. E. Radcliff. 1991. "Mulch effect on soil properties and tomato growth using micro-irrigation," *Agron. J.* **83**:1028–1034.

Torrance, J. K., and F. J. Schellekens. 2006. "Chemical factors in soil freezing and frost heave," *Polar Record* **42**:33–42.

Unger, P. W. 1978. "Straw mulch effects on soil temperatures and sorghum germination and growth," *Agron. J.* **70**:858–864.

Veneman, P.L.M., D. L. Lindbo, and L. A. Spokas. 1999. "Soil moisture and redoximorphic features: A historical perspective," in M. J. Rabenhorst, J. C. Bell, and P. A. McDaniel (eds.), *Quantifying Soil Hydromorphology.* Special Publication No. 54. Soil Science Society of America. (Madison, WI).

Welsh, D., D. Smart, J. Boyer, P. Minkin, H. Smith, and T. McCandless (eds.). 1995. *Forested Wetlands: Functions, Benefits, and Use of Best Management Practices.* (Radnor, Pa.: USDA Forest Service).

Wuethrich, B. 2000. "When permafrost isn't," *Smithsonian,* Comments and Notes, February 2000.

Xu, Q., and B. Huang. 2000. "Growth and physiological responses of creeping bentgrass to changes in air and soil temperatures," *Crop Sci* **40**:1363–1368.

Yavitt, J. B., T. J. Fahley, and J. A. Simmons. 1995. "Methane and carbon dioxide dynamics in a northern hardwood ecosystem," *Soil. Sci. Soc. Amer. J.* **59**:796–804.

Yu, K., and W. H. Patrick, Jr. 2004. "Redox window with minimum global warming potential contribution from rice soils," *Soil Sci. Soc. Amer. J.* **68**:2086–2091.

Mica weathers to clay. (Serge Jolicoeur, Université de Moncton)

8

THE COLLOIDAL FRACTION: SEAT OF SOIL CHEMICAL AND PHYSICAL ACTIVITY

The landscape of the clays is like—the intricate folds of the womb—whose activity is to receive, contain, enfold, and give birth.
—WILLIAM BRYANT LOGAN

How can using sewage effluent for irrigation contribute to the safe recharge of groundwater aquifers? Why is it more difficult to restore productivity after logging a tropical rain forest on Oxisols than a temperate forest on Alfisols? Why would a nuclear power plant accident seriously contaminate food grown on some downwind soils, but not on others? The answers to these and other environmental mysteries lie in the nature of the smallest of soil particles, the clay and humus **colloids**. These particles are not just extra-small fragments of rock and organic matter. They are highly reactive materials with electrically charged surfaces. Because of their size and shape, they give the soil an enormous amount of reactive **surface area**. It is the colloids, then, that allow the soil to serve as nature's great electrostatic chemical reactor.

Each tiny colloid particle carries a swarm of positively and negatively charged ions (cations and anions) that is attracted to electrostatic charges on its surface. The ions are held tightly enough by the **soil colloids** to greatly reduce their loss in drainage waters, but loosely enough to allow plant roots access to the nutrients among them. Other modes of adsorption bind ions more tightly so that they are no longer available for plant uptake, reaction with the soil solution, or leaching loss to the environment. In addition to plant nutrient ions, soil colloids also bind with water molecules, biomolecules (e.g., DNA or antibiotics), viruses, toxic metals, pesticides, and a host of other mineral and organic substances. Hence, soil colloids greatly impact nearly all ecosystem functions.

We shall see that different soils are endowed with different types of clays that, along with humus, elicit very different types of physical and chemical behaviors. Certain clay minerals are much more reactive than others. Some are more dramatically influenced than others by the acidity of the soil and other environmental factors. Studying the soil colloids in some detail will deepen your understanding of soil architecture (Chapter 4) and soil water (Chapters 5 and 6). Knowledge of the structure, origin, and behavior of the different types of soil colloids will also help you understand soil chemical and biological processes so you can make better decisions regarding the use of soil resources.

310

8.1 GENERAL PROPERTIES AND TYPES OF SOIL COLLOIDS

Size

The clay and humus particles in soils are referred to collectively as the **colloidal fraction** because of their extremely small size and colloid-like behavior. Too small to be seen with an ordinary light microscope, they can be made visible only with an electron microscope. Particles behave as colloids if they are less than about 1 μm (0.000001 meter) in diameter, although some soil scientists consider 2 μm to mark the upper boundary of the colloidal fraction to coincide with the definition of the clay particle size fraction.

Surface Area

As discussed in Section 4.2, the smaller the size of the particles in a given mass of soil, the greater the surface area exposed for adsorption, catalysis, precipitation, microbial colonization, and other surface phenomena. Because of their small size, all soil colloids expose a large **external surface** area per unit mass, more than 1000 times the surface area of the same mass of sand particles. Some silicate clays also possess extensive **internal surface** area between the layers of their platelike crystal units. To grasp the relative magnitude of the internal surface area, remember that these clays are structured much like this book. If you were to paint the external surfaces of this book (the covers and edges), a single brush of paint would do. However, to cover the internal surfaces (both sides of each page in the book) you might need a very large can of paint.

The total surface area of soil colloids ranges from 10 m^2/g for clays with only external surfaces, to more than 800 m^2/g for clays with extensive internal surfaces. To put this in perspective, we can calculate that the surface area exposed within 1 ha (about the size of a football field) of a 1.5-m-deep fine-textured soil (45% clay) might be as great as 8,700,000 km^2 (the land area of the entire United States).

Surface Charges

The internal and external surfaces of soil colloids carry positive and/or negative electrostatic charges. For most soil colloids, electronegative charges predominate, although some mineral colloids in very acid soils have a net electropositive charge. As we shall see in Sections 8.3 to 8.7, the amount and origin of surface charge differs greatly among the different types of soil colloids and, in some cases, is influenced by changes in chemical conditions, such as soil pH. The charges on the colloid surfaces attract or repulse substances in the soil solution as well as neighboring colloid particles. These reactions, in turn, greatly influence soil chemical and physical behavior.

Adsorption of Cations and Anions

Of particular significance is the attraction of positively charged ions (**cations**) to the surfaces of negatively charged soil colloids. Each colloid particle attracts thousands of Al^{3+}, Ca^{2+}, Mg^{2+}, K^+, H^+, and Na^+ ions and lesser numbers of other cations. In moist soils the cations exist in the hydrated state (surrounded by a shell of water molecules), but for simplicity in this text, we will show just the cations (e.g., Ca^{2+} or H^+) rather than the hydrated forms (e.g., $Ca(H_2O)_6^{2+}$ or the hydronium ion, H_3O^+). These hydrated cations constantly vibrate about in a swarm near the colloid surface, held there by electrostatic attraction to the colloid's negative charges. Frequently, an individual cation will break away from the swarm and move out to the soil solution. When this happens, another cation of equal charge will simultaneously move in from the soil solution and take its place. This process of **cation exchange** will be discussed in detail (Section 8.8) because of its fundamental importance in nutrient cycling and other environmental processes. The cations swarming about near the colloidal surface are said to be **adsorbed** (loosely held) on the colloid surface. Because these cations can *exchange places* with those moving freely about in the soil solution, the term **exchangeable ions** is also used to refer to the ions in this adsorbed state.

The colloid with its adsorbed cations is sometimes described as an **ionic double layer** in which the negatively charged colloid acts as a huge anion constituting the inner ionic layer, and the swarm of adsorbed cations constitutes the outer ionic layer

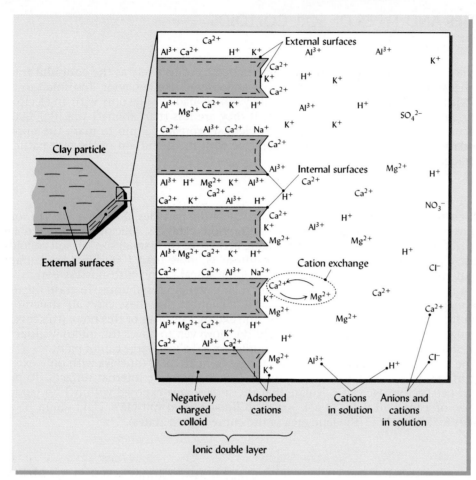

Clay particle

External surfaces

External surfaces

Internal surfaces

Cation exchange

Ca^{2+} ⟷ Mg^{2+}

Negatively charged colloid

Adsorbed cations

Cations in solution

Anions and cations in solution

Ionic double layer

FIGURE 8.1 Simplified representation of a silicate clay crystal, its complement of adsorbed cations, and ions in the surrounding soil solution. The enlarged view (*right*) shows that the clay comprises sheet-like layers with both external and internal negatively charged surfaces. The negatively charged particle acts as a huge anion and a swarm of positively charged cations is adsorbed to it because of attraction between charges of opposite sign. Cation concentration decreases with distance from the clay. Anions (such as Cl^-, NO_3^-, and SO_4^{2-}), which are repulsed by the negative charges, can be found in the bulk soil solution farthest from the clay (*far right*). Some clays (not shown) also exhibit positive charges that can attract anions.

(Figure 8.1). Because cations from the soil solution are constantly trading places with those that are adsorbed to the colloid, the ionic composition of the soil solution reflects that of the adsorbed swarm. For example, if Ca^{2+} and Mg^{2+} dominate the exchangeable ions, they will also dominate the soil solution. Under natural conditions, the proportions of specific cations present are largely influenced by the soil parent material and the degree to which the climate has promoted the loss of cations by leaching (see Section 8.9).

Anions such as Cl^-, NO_3^-, and SO_4^{2-} (also surrounded by water molecules, though, again, we do not show these water shells) may also be attracted to certain soil colloids that have *positive* charges on their surfaces. While adsorption of **exchangeable anions** is not as extensive as that for exchangeable cations, we shall see (Section 8.11) that it is an important mechanism for holding negatively charged constituents, especially in acid subsoils. When thinking about the colloids in soil, we should always keep in mind that they carry with them a complement of exchangeable cations and anions, along with certain other more tightly bound ions and molecules.

Adsorption of Water

In addition to adsorbing cations and anions, soil colloids attract and hold a large number of water molecules. Generally, the greater the external surface area of the soil colloids, the greater the amount of water held when the soil is air-dry (Figure 8.2). While this water may not be available for plant uptake (see Section 5.8), it does play a role in the survival of soil microorganisms, especially bacteria. The charges on the internal and external colloid surfaces attract the oppositely charged end of the polar water molecule. Some water molecules are attracted to the exchangeable cations, each of which is hydrated with a shell of water molecules. Water adsorbed between the clay layers can cause the layers to move apart, making the clay more plastic and swelling its volume. Colloids that adsorb a great deal of water may make soil unsuitable for construction

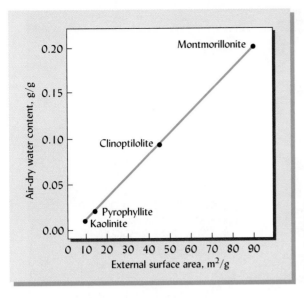

FIGURE 8.2 The amount of water held in air-dry clay as influenced by the external surface area of the clay. The clays were dried in low-humidity air for 48 hours at 20 °C. The names refer to four silicate clay minerals that are characterized by differing amounts of external surface area per unit mass. Of the four, kaolinite and montmorillonite are by far the most common in soils and are discussed in detail in Section 8.2. [Drawn from data in Morra et al. (1998)]

purposes (see Sections 4.9 and 8.14). As a soil colloid dries, any water between the layers is removed, and the layers are brought closer together.

Types of Soil Colloids[1]

Soils contain numerous types of colloids, each with its particular composition, structure, and properties (Table 8.1). The colloids most important in soils can be grouped in four major types:

CRYSTALLINE SILICATE CLAYS. These clays are the dominant type in most soils (except in Andisols, Oxisols, and Histosols—see Chapter 3). Their crystalline structure is layered much like pages in a book (clearly visible in Figure 8.3a). Each layer (page) consists of two to four sheets of closely packed and tightly bonded oxygen, silicon, and aluminum atoms. Although all are predominately negatively charged, silicate clay minerals differ widely with regard to their particle shapes (**kaolinite, a fine-grained mica,** and a **smectite** are shown in Figure 8.3a–c), intensity of charge, stickiness, plasticity, and swelling behavior.

TABLE 8.1 **Major Properties of Selected Soil Colloids**

Colloid	Type	Size, μm	Shape	Surface area, m^2/g External	Surface area, m^2/g Internal	Interlayer spacing[a], nm	Net charge[b], $cmol_c/kg$
Smectite	2:1 silicate	0.01–1.0	Flakes	80–150	550–650	1.0–2.0	−80 to −150
Vermiculite	2:1 silicate	0.1–0.5	Plates, flakes	70–120	600–700	1.0–1.5	−100 to −200
Fine mica	2:1 silicate	0.2–2.0	Flakes	70–175	—	1.0	−10 to −40
Chlorite	2:1 silicate	0.1–2.0	Variable	70–100	—	1.41	−10 to −40
Kaolinite	1:1 silicate	0.1–5.0	Hexagonal crystals	5–30	—	0.72	−1 to −15
Gibbsite	Al-oxide	<0.1	Hexagonal crystals	80–200	—	0.48	+10 to −5
Goethite	Fe-oxide	<0.1	Variable	100–300	—	0.42	+20 to −5
Allophane & Imogolite	Noncrystalline silicates	<0.1	Hollow spheres or tubes	100–1000	—	—	+20 to −150
Humus	Organic	0.1–1.0	Amorphous	Variable[c]	—	—	−100 to −500

[a] From the top of one layer to the next similar layer, 1 nm = 10^{-9} m = 10 Å.
[b] Centimoles of unbalanced or net charge per kilogram of colloid ($cmol_c/kg$), a measure of ion exchange capacity (see Section 8.9).
[c] It is very difficult to determine the surface area of organic matter. Different procedures give values ranging from 20 to 800 m^2/g.

---

[1] For a review of the structures and properties of the clays, see Meunier (2005), and for properties of clay and humus in soils, see Dixon and Weed (1989).

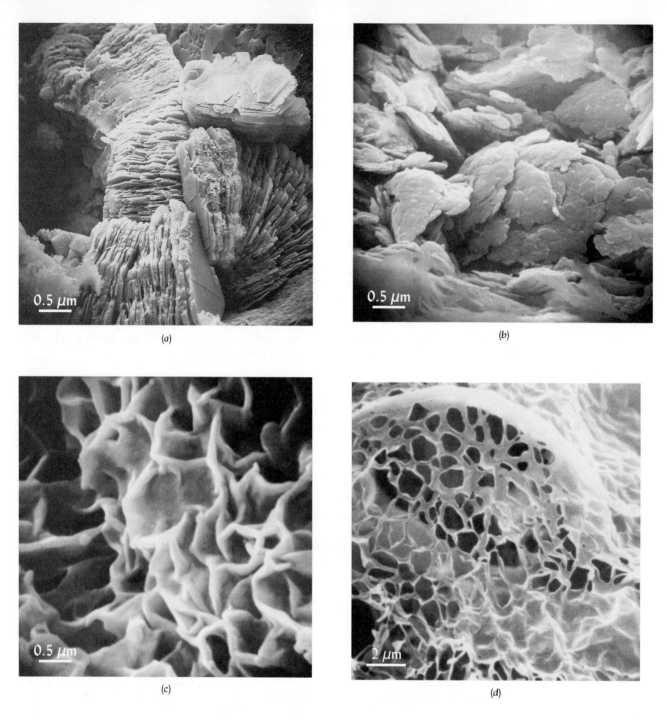

FIGURE 8.3 Crystals of three silicate clay minerals and a photomicrograph of humic acid found in soils. (*a*) Kaolinite from Illinois (note hexagonal crystal at upper right). (*b*) A fine-grained mica from Wisconsin. (*c*) Montmorillonite (a smectite group mineral) from Wyoming. (*d*) Fulvic acid (a humic acid) from Georgia. [(*a*)–(*c*) Courtesy of Dr. Bruce F. Bohor, Illinois State Geological Survey; (*d*) from Dr. Kim H. Tan, University of Georgia; used with permission of Soil Science Society of America]

NONCRYSTALLINE SILICATE CLAYS. These clays also consist mainly of tightly bonded silicon, aluminum, and oxygen atoms, but they do not exhibit ordered, crystalline sheets. The two principal clays of this type, **allophane** and **imogolite**, usually form from volcanic ash and are characteristic of Andisols (Section 3.7). They have high amounts of both positive and negative charge, and high water-holding capacities. Although malleable (plastic) when wet, they exhibit a very low degree of stickiness. Allophane and imogolite are also known for their extremely high capacities to strongly adsorb phosphate and other anions, especially under acid conditions.

IRON AND ALUMINUM OXIDES. These are found in many soils, but are especially important in the more highly weathered soils of warm, humid regions (e.g., Ultisols and Oxisols). They consist mainly of either iron or aluminum atoms coordinated with oxygen atoms (the latter are often associated with hydrogen ions to make hydroxyl groups). Some, like **gibbsite** (an Al-oxide) and **goethite** (an Fe-oxide) consist of crystalline sheets. Other oxide minerals are noncrystalline, often occurring as **amorphous** coatings on soil particles. The oxide colloids are relatively low in plasticity and stickiness. Their net charge ranges from slightly negative to moderately positive. Although for simplicity we will use term *Fe, Al oxides* for this group, many are actually hydroxides or oxyhydroxides because of the presence of hydrogen ions.

ORGANIC (HUMUS). Organic colloids are important in nearly all soils, especially in the upper parts of the soil profile. Humus colloids are not minerals, nor are they crystalline (Figure 8.3*d*). Instead, they consist of convoluted chains and rings of carbon atoms bonded to hydrogen, oxygen, and nitrogen. Humus particles are often among the smallest of soil colloids and exhibit very high capacities to adsorb water, but almost no plasticity or stickiness. Because humus is noncohesive, soils composed mainly of humus (Histosols) have very little bearing strength and are unsuitable for making building or road foundations. Humus has high amounts of both negative and positive charge per unit mass, but the net charge is always negative and varies with soil pH. The negative charge on humus is extremely high in neutral to alkaline soils.

8.2 FUNDAMENTALS OF LAYER SILICATE CLAY STRUCTURE[2]

To see why soils rich in one silicate clay mineral, say kaolinite, behave so very differently from soils dominated by another silicate clay, say montmorillonite, it is necessary to understand the main structural features of the silicate clay minerals. We will begin by examining the main building blocks from which the layer silicates are constructed, then consider the particular arrangements that give rise to the critically important surface charges.

Silicon Tetrahedral and Aluminum-Magnesium Octahedral Sheets

The most important silicate clays are known as **phyllosilicates** (Greek *phyllon,* leaf) because of their leaflike or planar structure. As shown in Figure 8.4, they are composed of two kinds of horizontal **sheets**.

TETRAHEDRAL SHEETS. This kind of sheet consists of two **planes** of oxygens with mainly silicon in the spaces between the oxygens. The basic building block for the tetrahedral sheet is a unit composed of one silicon atom surrounded by four oxygen atoms. It is called a **tetrahedron** because (as shown in Figure 8.4, top left) the oxygens define the apices of a *four*-sided geometric solid that resembles a pyramid (having three "sides" and a bottom). An interlocking array of such tetrahedra, each sharing its basal oxygens with its neighbor, give rise to a **tetrahedral sheet**.

Three-dimensional rotatable models of silicate clay minerals and their building blocks:
http://www.soils1.cses.vt.edu/MJE/VR_exports/intro.shtml

OCTAHEDRAL SHEETS. Six oxygen atoms coordinating with a central aluminum or magnesium atom form the shape of an *eight*-sided geometric solid, or **octahedron**. Numerous octahedra linked together horizontally constitute the **octahedral sheet**. If *three* Mg^{2+} atoms are coordinated with (and balance the charges on) the six oxygens/hydroxyls, then the sheet is called a *tri*octahedral sheet. If, instead, the six oxygens/hydroxyls are coordinated with *two* Al^{3+} atoms, then the sheet is called **dioctahedral**. Note that the distinction is based on the number of metal atoms required to satisfy the six negative charges from the oxygen/hydroxyls (see Figure 8.5, left and middle). As we will see later, numerous intergrades are possible in which both 2+ and 3+ cations are present.

The tetrahedral and octahedral sheets are the fundamental structural units of silicate clays. Two to four of these sheets may be stacked together in sandwich-like

---

[2] The authors are indebted to Dr. Darrel G. Schultze of Purdue University for kindly providing the structural models for the silicate clay minerals.

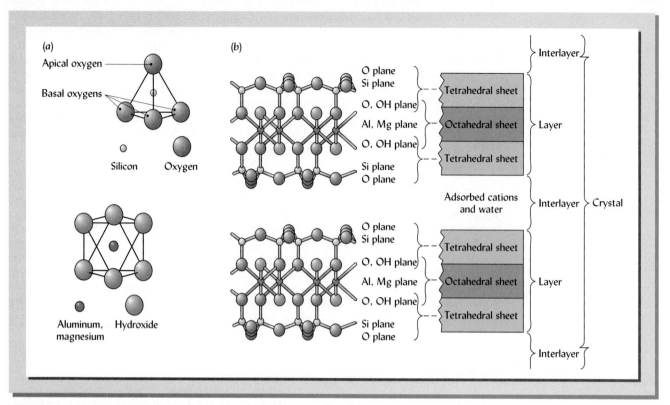

FIGURE 8.4 The basic molecular and structural components of silicate clays. (*a*) A single *tetrahedron*, a four-sided building block composed of a silicon ion surrounded by four oxygen atoms; and a single eight-sided *octahedron*, in which an aluminum (or magnesium) ion is surrounded by six hydroxy groups or oxygen atoms. (*b*) In clay crystals thousands of these tetrahedral and octahedral building blocks are connected to give planes of silicon and aluminum (or magnesium) ions. These planes alternate with planes of oxygen atoms and hydroxy groups. Note that apical oxygen atoms are common to adjoining tetrahedral and octahedral sheets. The silicon plane and associated oxygen-hydroxy planes make up a *tetrahedral sheet*. Similarly, the aluminum-magnesium plane and associated oxygen-hydroxy planes constitute the *octahedral sheet*. Different combinations of tetrahedral and octahedral sheets are termed *layers*. In some silicate clays these layers are separated by *interlayers* in which water and adsorbed cations are found. Many layers are found in each *crystal*.

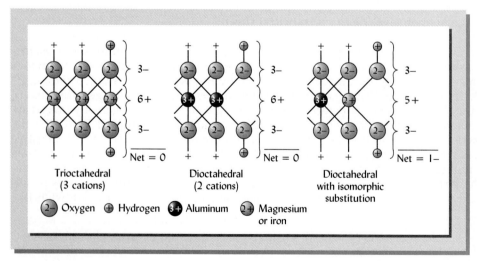

FIGURE 8.5 Simplified diagrams of octahedral sheets in silicate minerals illustrating *tri-* and *di*octahedral structures and isomorphous substitution. For each oxygen atom, one of the two − charges is balanced by a + charge from either a H^+ (making a hydroxyl group) or from a Si atom in the tetrahedral sheet (not shown, but represented by a +). In a *tri*octahedral sheet (*left*) *three* out of three octahedral positions are occupied by a metal cation with a 2+ charge (typically either Fe^{2+} or Mg^{2+}). The + and − charges are balanced so there is no net charge. Biotite is an example of trioctahedral clay. In a *di*octahedral sheet (*middle*) only two out of three octahedral positions are occupied, but by a metal cation with a 3+ charge (Al^{3+} is most common). Again, the + and − charges are balanced so there is no net charge on the octahedral sheet. Muscovite is an example of dioctahedral clay. In the dioctahedral sheet shown on the right, a Mg^{2+} atom occupies one of the positions normally occupied by an Al^{3+} atom, thus leaving a 1− net charge on the sheet.

TABLE 8.2 Ionic Radii and Location of Elements Found in Silicate Clays

Ion	Radius, nm (10^{-9} m)	Found in
Si^{4+}	0.042	Tetrahedral sheet
Al^{3+}	0.051	
Fe^{3+}	0.064	
Mg^{2+}	0.066	Octahedral sheet
Zn^{2+}	0.074	Exchange or interlayer sites
Fe^{2+}	0.076	
Na^+	0.095	
Ca^{2+}	0.099	
K^+	0.133	
O^{2-}	0.140	Both sheets
OH^-	0.155	

arrangements, with adjacent sheets strongly bound together by sharing some of the same oxygen atoms (see Figure 8.4). The specific nature and combination of sheets in these layers vary from one type of clay to another and largely control the physical and chemical properties exhibited. The relationship between *planes, sheets,* and *layers* shown in Figure 8.4 should be carefully studied.

Isomorphous Substitution

The structural arrangements just described suggest a very simple relationship among the elements making up silicate clays. In nature, however, clays have formulas that are more complex. During the weathering of rocks and minerals, many different elements are present in the weathering solution. As clay minerals or their precursors crystallize, cations of comparable size (see Table 8.2) may substitute for silicon, aluminum, and magnesium ions in the respective tetrahedral and octahedral sheets.

Note from Table 8.2 that aluminum is only slightly larger than silicon. Consequently, aluminum can fit into the center of the tetrahedron in the place of the silicon without much change in the basic structure of the crystal. This process by which one element fills a position usually filled by another of similar size is called **isomorphous substitution**. This phenomenon is responsible for much of the variability in the nature of silicate clays.

Isomorphous substitution can also occur in the octahedral sheets. For example, iron and zinc ions are not much different in size from aluminum and magnesium ions (Table 8.2). Any of these ions can fit in the central position of an octahedra. In some layer silicates, isomorphous substitution occurs in both tetrahedral and in octahedral sheets.

Source of Charges

Isomorphous substitution is of vital importance because it is the primary source of both negative and positive charges of silicate clays. For example, the Mg^{2+} ion is only slightly larger than the Al^{3+} ion, but it has one less positive charge. If a Mg^{2+} ion substitutes for an Al^{3+} ion in a dioctahedral sheet, there will be insufficient positive charges to balance the negative charges from the oxygens; hence, the lattice is left with a 1− net charge (see Figure 8.5, right). Similarly, every Al^{3+} that substitutes for a Si^{4+} in a tetrahedral sheet creates a net negative charge at that site because the negative charges from the four oxygens will be only partially balanced. In a trioctahedral sheet, if an Al^{3+} substitutes for the usual Mg^{2+} or Fe^{2+}, then a net *positive* charge is created. The net charge associated with a clay crystal is the sum of the positive and negative charges. In most silicate clays, the negative charges predominate (as will be discussed in Section 8.8). As we shall see (Sections 8.3 and 8.6), additional, more temporary charges can also develop on the edges of the tetrahedral and octahedral surfaces.

8.3 MINERALOGICAL ORGANIZATION OF SILICATE CLAYS

Based on the number and arrangement of tetrahedral (Si) and octahedral (Al, Mg, Fe) sheets contained in the crystal units or layers, crystalline clays may be classed into two main groups: **1:1 silicate clays,** in which each layer contains *one* tetrahedral and *one*

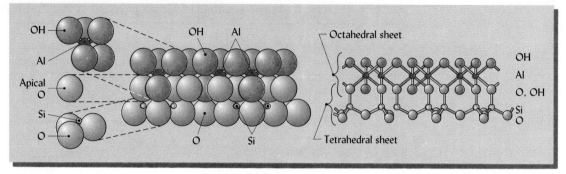

FIGURE 8.6 Models of the 1:1-type clay kaolinite. The primary elements of the octahedral (*upper left*) and tetrahedral (*lower left*) sheets are depicted as they might appear separately. In the crystal structure, however, these sheets are held together by common *apical* oxygen atoms. Note that each layer consists of alternating octahedral and tetrahedral sheets—hence, the designation 1:1. The octahedral and tetrahedral sheets are bound together (*center*) by mutually shared (apical) oxygen atoms. The result is a layer with hydroxyls on one surface and oxygens on the other. To permit us to view the front silicon atoms, we have not shown some basal oxygen atoms that are normally present. The diagram at right shows the bonds between atoms. The kaolinite mineral is comprised of a series of these flat layers tightly held together with no interlayer spaces.

octahedral sheet, and **2:1 silicate clays**, in which each layer has *one* octahedral sheet sandwiched between *two* tetrahedral sheets.

1:1-Type Silicate Clays

Interactive 3-D model of Kaolinite structure:

http://www.soils.wise.edu/ virtual_museum/kaolinite/ index.html

The 1:1 silicate clays include **kaolinite,** *halloysite, nacrite,* and *dickite.* To illustrate the properties of 1:1 silicate clays we will focus on kaolinite, which is by far the most common in soils.

As implied by the term *1:1 silicate clay,* each kaolinite layer consists of one silicon tetrahedral sheet and one aluminum octahedral sheet. The two types of sheets are tightly held together because the apical oxygen atom (the oxygen atom that forms the apex or tip of the "pyramid") in each tetrahedron also forms a bottom corner of one or more of the octahedra in the adjoining sheet (Figure 8.6). Note that because a kaolinite crystal layer consists of these two sheets, it exposes a plane of oxygen atoms on the bottom surface, but a plane of hydroxyls on the upper surface.

This arrangement has two very important consequences. First, as will be discussed in Section 8.6, where the hydroxyl plane is exposed on the clay particle surface, removal or addition of hydrogen ions can produce either positive or negative charges, depending on the pH of the soil. The exposed **hydroxylated surface** can also react with and strongly bind specific anions. Second, when the layers consisting of alternating tetrahedral and octahedral sheets are stacked on top of one another, the hydroxyls of the octahedral sheet in one layer are adjacent to the basal oxygens of the tetrahedral sheet of the next layer. Therefore, adjacent layers are bound together by **hydrogen bonding** (see Section 5.1).

Kaolin clay used for pest control:

http://www.nysaes.cornell. edu/pp/resourceguide/mfs/ 07kaolin.php

Because of the interlayer hydrogen bonding, the structure of kaolinite is fixed, and no expansion can occur between the layers when the clay is wetted. Cations and water generally do not enter between the structural layers of a 1:1 mineral particle. The effective surface of kaolinite is thus restricted to its outer faces or external surface area. This fact and the lack of significant isomorphous substitution in this mineral account for the relatively small capacity of kaolinite to adsorb exchangeable cations (see Table 8.1).

Kaolinite crystals are usually hexagonal in shape (see Figure 8.3*a*) and larger than most other clays (Table 8.1). In contrast to some 2:1 silicate clays, 1:1 clays like kaolinite exhibit less plasticity, stickiness, cohesion, shrinkage, and swelling and can also hold less water than other clays (Figure 8.2). Because of these properties, soils dominated by 1:1 clays are relatively easy to cultivate for agriculture and, with proper nutrient management, can be quite productive. Kaolinite-containing soils are well suited for use in roadbeds and building foundations (Plate 43). The nonexpanding 1:1 structure also makes kaolinite clays useful for making bricks and ceramics (see Box 8.1).

BOX 8.1   KAOLINITE CLAY—THE STORY OF WHITE GOLD[a]

FIGURE 8.7 *Kaolinitic clay soil is dug, molded, dried, stacked to form a kiln, and fired to make bricks. (Courtesy of R. Weil)*

Kaolinite, the most common of the 1:1 clay minerals, has been used for thousands of years to make pottery, roofing tiles, and bricks. The basic processes have changed little to this day. The clayey material is saturated with water, kneaded and molded or thrown on a potter's wheel to obtain the desired shape, and then hardened by drying or firing (Figures 8.7, 8.8). The mass of cohering clay platelets hardens irreversibly when fired and the nonexpanding nature of kaolinite allows it to be fired without cracking from shrinkage. The heat also changes the typical gray color of the soil material to "brick red" because of the irreversible oxidation and crystallization of the iron-oxyhydroxides that often coat soil kaolinite particles. In contrast, kaolinite mined from pure deposits fires to a light, creamy color. Kaolinite is not as plastic (moldable) as some other clays, however, and so is usually mixed with more plastic types of clays for making pottery.

It was in seventh-century China that pure kaolinite deposits were first used to make objects of a translucent, lightweight, and strong ceramic called porcelain. The name *kaolinite* derives from the Chinese words *kan* and *ling*, meaning "high ridge," as the material was first mined from a hillside in Kiangsi Province. The Chinese held a monopoly on porcelain-making technology (hence the term *china* for porcelain dishes) until the early 1700s. English colonists, in what is now Georgia in the United States, noted outcrops of white kaolinite clay in areas of rather unproductive soil. The colonialists soon were exporting this kaolinite as the main ingredient for making porcelain in England, where the now-famous pottery was first manufactured from the Georgia kaolinite clay.

The market for pure, white kaolinite clay greatly expanded when paper manufacturers started using kaolinite clay to make sizing, the coating that makes high-quality papers smoother, whiter, and more printable. Other industrial uses now include paint pigments, fillers in plastic manufacture, and ceramic materials used for electrical insulation and heat shielding (as on the belly of the space shuttle). The kaolinite in kaopectin-type medications lines the stomach walls and inactivates diarrhea-causing bacteria by adsorbing them on the clay particle surfaces. A development that is likely to increase the demand for kaolinite is its use as a spray-on coating for fruit tree leaves that provides nontoxic protection from insect pests and fungal diseases. All these uses have made industrial kaolinite clay mining a big business in Georgia. Unfortunately, surface mining of kaolinite to meet these needs can cause both environmental and social disruptions.

FIGURE 8.8 *Kaolinite clay in African pottery and early 19th-century English china (inset). (Courtesy of R. Weil)*

[a] For more on social aspects, see Seabrook, 1995.

Expanding 2:1-Type Silicate Clays

The four general groups of 2:1 silicate clays are characterized by *one* octahedral sheet sandwiched between *two* tetrahedral sheets. Two of these groups, **smectite** and **vermiculite**, include expanding-type minerals; the other two, **fine-grained micas (illite)** and **chlorite**, are relatively nonexpanding.

SMECTITE GROUP.   The flakelike crystals of smectites (see Figure 8.3c) have a high amount of mostly negative charge resulting from isomorphous substitution. Most of the charge derives from Mg^{2+} ions substituted in the Al^{3+} positions of the octahedral sheet, but some also derives from substitution of Al^{3+} ions for Si^{4+} in the tetrahedral sheets (Figure 8.9).

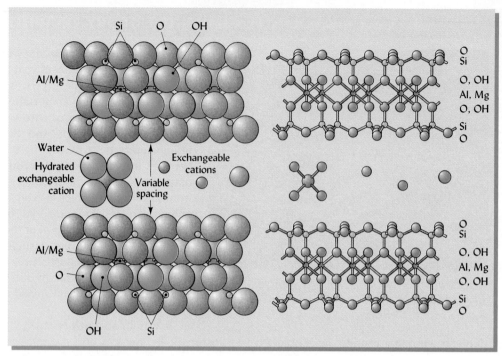

FIGURE 8.9 Model of two crystal layers and an interlayer characteristic of montmorillonite, a smectite expanding-lattice 2:1-type clay mineral. Each layer is made up of an octahedral sheet sandwiched between two tetrahedral sheets with shared apical oxygen atoms. There is little attraction between oxygen atoms in the bottom tetrahedral sheet of one unit and those in the top tetrahedral sheet of another. This permits a variable space between layers, which is occupied by water and exchangeable cations. The internal surface area thus exposed far exceeds the surface around the outside of the crystal. Note that magnesium has replaced aluminum in some sites of the octahedral sheet. Likewise, some silicon atoms in the tetrahedral sheet may be replaced by aluminum (not shown). These substitutions give rise to a negative charge, which accounts for the high cation exchange capacity of this clay mineral. A ball-and-stick model of the atoms and chemical bonds is at the right.

Interactive 3-D model of Smectite structure:

http://virtual-museum.soils.wisc.edu/soil_smectite/index.html

Because of these substitutions, the capacity to adsorb cations is very high—about 20 to 40 times that of kaolinite.

In contrast to kaolinite, smectites have a 2:1 structure that exposes a layer of oxygen atoms at both the top and bottom planes. Therefore, adjacent layers are only loosely bound to each other by very weak oxygen-to-oxygen and cation-to-oxygen linkages and the space between is variable (Figure 8.9). The internal surface area exposed between the layers by far exceeds the external surface area of these minerals and contributes to the very high total **specific surface area** (Table 8.1). Exchangeable cations and associated water molecules are attracted to the spaces between the interlayer spaces.

Flakelike smectite crystals tend to pile upon one another, forming wavy stacks that contain many extremely small *ultramicropores* (see Table 4.6). When soils high in smectite are wetted, adsorption of water in these ultramicropores leads to severe swelling; when they are dried, the soils shrink in volume (see Section 8.14). The expansion upon wetting contributes to the high degree of plasticity, stickiness, and cohesion that make smectitic soils very difficult to cultivate or excavate. Wide cracks commonly appear during the drying of smectite-dominated soils (such as Vertisols, Figure 3.22). The shrink/swell behavior makes smectitic soils quite undesirable for most construction activities, but they are well suited for a number of applications that require a high adsorptive capacity and the ability to form seals of very low permeability (see Section 8.14). **Montmorillonite** is the most prominent of the smectites in soils, although others are also found.

Safe storage for nuclear wastes: the Swedish approach using Bentonite clay:

http://www.skb.se/templates/SKBPage____8776.aspx

VERMICULITE GROUP. The most common **vermiculites** are 2:1-type minerals in which the octahedral sheet is aluminum dominated (dioctahedral), but some magnesium-dominated (trioctahedral) vermiculites also exist. The tetrahedral sheets of most vermiculites have considerable substitution of aluminum in the silicon positions, giving rise to a cation exchange capacity that usually exceeds that of all other silicate clays, including smectites (Table 8.1).

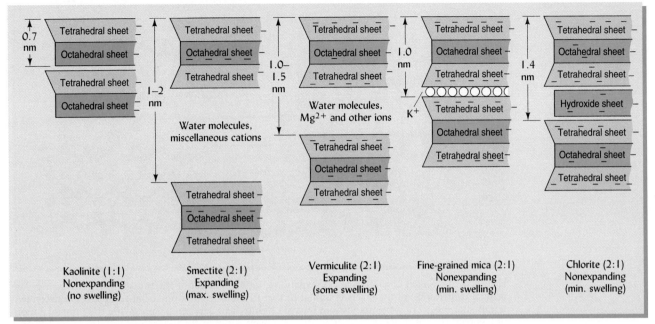

FIGURE 8.10 Schematic drawing illustrating the organization of tetrahedral and octahedral sheets in one 1:1-type mineral (kaolinite) and four 2:1-type minerals. The octahedral sheets in each of the 2:1-type clays can be either aluminum dominated (dioctahedral) or magnesium dominated (trioctahedral). However, in most chlorites the trioctahedral sheets are dominant while the dioctahedral sheets are generally most prominent in the other three 2:1 types. Note that kaolinite is nonexpanding, the layers being held together by hydrogen bonds. Maximum interlayer expansion is found in smectite, with somewhat less expansion in vermiculite because of the moderate binding power of numerous Mg^{2+} ions. Fine-grained mica and chlorite do not expand because K^+ ions (fine-grained mica) or an octahedral-like sheet of hydroxides of Al, Mg, Fe, and so forth (chlorite) tightly bind the 2:1 layers together. The interlayer spacings are shown in nanometers (1 nm = 10^{-9} m).

The interlayer spaces of vermiculites usually contain strongly adsorbed water molecules, Al-hydroxy ions, and cations such as magnesium (Figure 8.10). However, these interlayer constituents act primarily as bridges to hold the units together, rather than wedges driving them apart. The degree of swelling and shrinkage is, therefore, considerably less for vermiculites than for smectites. For this reason, vermiculites are considered limited-expansion clays, expanding more than kaolinite, but much less than the smectites.

Nonexpanding 2:1 Silicate Minerals

The main nonexpanding 2:1 minerals are the **fine-grained micas** and the **chlorites**. We will discuss the fine-grained micas first.

MICA GROUP. Biotite and muscovite are examples of unweathered micas typically found in the sand and silt fractions. The more weathered **fine-grained micas**, such as **illite** and **glauconite**, are found in the clay fraction of soils. Their 2:1-type structures are quite similar to those of their unweathered cousins. Unlike in smectites, the main source of charge in fine-grained micas is the substitution of Al^{3+} in about 20% of the Si^{4+} sites in the tetrahedral sheets. This results in a high net negative charge in the tetrahedral sheet, even higher than that found in vermiculites. The negative charge attracts cations, among which potassium (K^+) is just the right size to fit snugly into certain hexagonal "holes" between the tetrahedral oxygen groups (Figures 8.10 and 8.11) and thereby get very close to the negatively charged sites. By their mutual attraction for the K^+ ions in between, adjacent layers in fine-grained micas are strongly bound together. Hence, the fine-grained micas are quite **nonexpansive**. Because of their nonexpansive character, the fine-grained micas are more like kaolinite than smectites with regard to their capacity to adsorb water and their degree of plasticity and stickiness.

CHLORITES. In most soil **chlorites**, iron or magnesium, rather than aluminum, occupy many of the octahedral sites. Commonly, a magnesium-dominated trioctahedral hydroxide sheet is sandwiched in between adjacent 2:1 layers (Figure 8.10). Thus, chlorite is sometimes said to have a 2:1:1 structure. Chlorites are nonexpansive because the

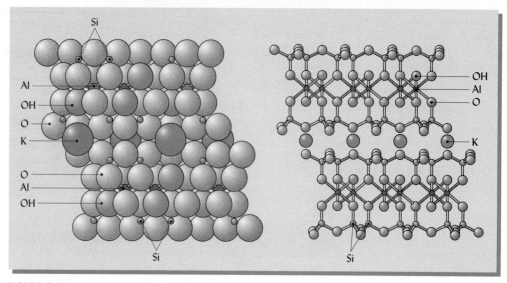

FIGURE 8.11 Model of a 2:1-type nonexpanding lattice mineral of the fine-grained mica group. The general constitution of the layers is similar to that in the smectites, one octahedral sheet between two tetrahedral sheets. However, potassium ions are tightly held between layers, giving the mineral a more or less rigid type of structure that prevents the movement of water and cations into the space between layers. The internal surface and cation exchange capacity of fine-grained micas are thus far below those of the smectites.

hydroxylated surfaces of an intervening Mg-octahedral sheet are hydrogen-bonded to the oxygen atoms of the two adjacent tetrahedral sheets, binding the layers tightly together. The colloidal properties of the chlorites are therefore quite similar to those of the fine-grained micas (Table 8.1).

X-Ray Diffraction Analysis

Use of X-ray diffraction in art conservation and detecting art fraud: http://www.cci-icc.gc.ca/about-cci/cci-in-action/view-document_e.aspx?Type_ID=6&Document_ID=101

Until the invention of suitable means of investigating mineral structure at the atomic level, it was thought that clays consisted of inert mineral fragments coated in amorphous gels of iron oxides. Now, the relative amounts of various types of clay minerals present in a soil can be determined by a procedure called **X-ray diffraction analysis**, which measures the distance between layers (the **d-spacing**) in the mineral structure (see Box 8.2).

8.4 STRUCTURAL CHARACTERISTICS OF NONSILICATE COLLOIDS

Iron and Aluminum Oxides

These clays consist of modified octahedral sheets with either iron (Fe^{3+}) or aluminum (Al^{3+}) in the cation positions. They have neither tetrahedral sheets nor silicon in their structures. Isomorphous substitution by ions of varying charge rarely occurs, so these clays do not have a large negative charge. The small amount of net charge these clays possess (positive and negative) is caused by the removal or addition of hydrogen ions at the surface oxy-hydroxyl groups. The presence of these bound oxygen and hydroxyl groups enables the surfaces of these clays to strongly adsorb and combine with anions such as phosphate or arsenate. The oxide clays are nonexpansive and generally exhibit relatively little stickiness, plasticity, and cation adsorption. They make quite stable materials for construction purposes.

Gibbsite [$Al(OH)_3$], the most common soil aluminum oxide, is a prominent constituent of highly weathered soils (e.g., Oxisols and Ultisols). Figure 8.13 shows gibbsite to consist of a series of aluminum octahedral sheets linked to one another by hydrogen-bonding between their hydroxyls. Note that a plane of hydroxyls is exposed at the upper and lower surfaces of gibbsite crystals. These hydroxylated surfaces can strongly adsorb certain anions.

Other oxide-type clays have iron instead of aluminum in the central cation positions, and their octahedral structures are somewhat distorted and less regular than that of gibbsite. **Goethite** ($FeOOH$) and **ferrihydrite** ($Fe_2O_3 \cdot nH_2O$) are common iron oxide

BOX 8.2   X-RAYS UNLOCK THE MYSTERIES
OF CRYSTALLINE CLAY STRUCTURE

When a thin smear of clay powder or paste is slowly rotated in an X-ray beam of a particular wavelength, the X-ray waves reflecting off parallel planes of atoms in the clay crystal create a diffraction pattern that can be detected by X-ray diffraction analysis. The detected X-ray energy is low when the waves bouncing off different layers are out of phase and therefore cancel each other out. The detected energy is magnified when the waves reflected from two layers are in phase (synchronous) and therefore reinforce each other (Figure 8.12, *left*).

These reinforced waves create peaks of energy, which are recorded on a graph called a **diffractogram**.

When incoming waves strike two (or more) parallel surfaces, the angle of incidence theta (θ) will determine how much farther the waves striking the second layer must travel compared to those striking the first layer. **Bragg's law** tells us there is a specific angle that will cause the waves striking the second layer to travel an additional distance exactly equal to their wavelength. Therefore, the waves reflected off the second layer will be in phase with those reflected off the first layer. The distance (*d*, in nm) between layers is calculated from Bragg's law:

$$n\lambda = 2d \sin \theta \qquad (8.1)$$

where *n* is an integer, λ is the known wavelength (in nm) of the X-rays, and θ is the glancing angle that causes the waves reflected from different layers to reinforce each other.

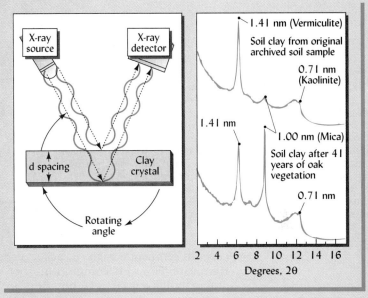

FIGURE 8.12   *A simplified diagram of X-ray diffraction used to identify clay minerals (left) and two X-ray **diffractograms** (right) showing peaks indicative of vermiculite, fine-grained mica, and kaolinite clay minerals. [Diffractograms redrawn from Tice et al. (1996)]*

Layers at specific distances apart characterize the crystal structure for each clay mineral (note the nanometer d-spacings shown in Figure 8.10), so one can determine which minerals are present by examining the specific angles that cause reinforced X-ray energy peaks. The angle at which a peak forms is indicative of the layer spacing. The sizes of the X-ray peaks are semiquantitatively related to the relative amount of a specific mineral present. X-ray diffraction can best identify clay structures if the clay is given special pretreatments by saturating it with specific cations, washing it with various solvents, or heating it to remove interlayer impurities.

Figure 8.12 (*right*) shows two diffractograms that suggest the presence of mainly fine-grained mica, vermiculite, and kaolinite clays. The diffractograms shown are from a long-term experiment in the San Demas Mountains of southern California. Scientists began by digging special pits (called lysimeters) fit with drains that allowed leaching water to be collected. Each pit was then uniformly packed full with soil material of known composition. Oak trees (*Quercus dumosa*) were planted in some of the lysimeters. After 41 years, the surface layer of soil was sampled and compared to archived samples of the original soil material. Both X-ray diffractograms showed a distinct peak for the 1.41-nm interlayer spacing characteristic of vermiculite. However, a strong peak for 1.0-nm spacing (characteristic of mica) appeared only in the soil under oak forest. Chemical analyses showed that the oak trees had increased the amount of potassium in the A horizon by cycling this element up from the deeper layers (see Section 2.5). Together, these two diffractograms illustrate (1) the utility of X-ray diffraction as a tool for identifying clay minerals and (2) the chemical alteration processes by which one clay mineral may be transformed into another. The latter topic will be further explored in Section 8.6.

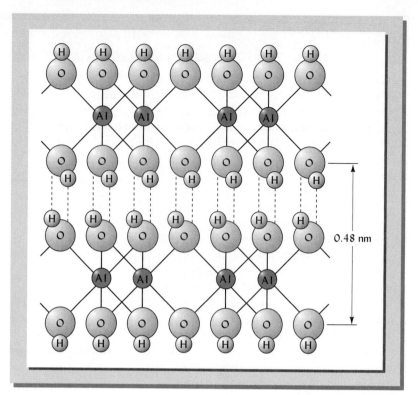

FIGURE 8.13 A simplified diagram showing the structure of gibbsite, an aluminum oxide clay common in highly weathered soils. This clay consists of dioctahedral sheets (two are shown) that are hydrogen-bonded together. Other oxide-type clays have iron instead of aluminum in the octahedral positions, and their structures are somewhat less regular and crystalline than that shown for gibbsite. The surface plane of covalently bonded hydroxyls gives this, and similar clays, the capacity to strongly adsorb certain anions (see Section 8.8).

0.48 nm

clays in temperate regions, accounting for the yellow-brown colors of many soils. **Hematite** (Fe_2O_3) is common in drier environments and gives redder colors to well-drained soils, especially in hot, dry climates.

In many soils, iron and aluminum oxide minerals are mixed with silicate clays. The oxides may form coatings on the external surfaces of the silicate clays, or they may occur as "islands" in the interlayer spaces of such 2:1 clays as vermiculites and smectites. In either case, the presence of iron and aluminum oxides can substantially alter

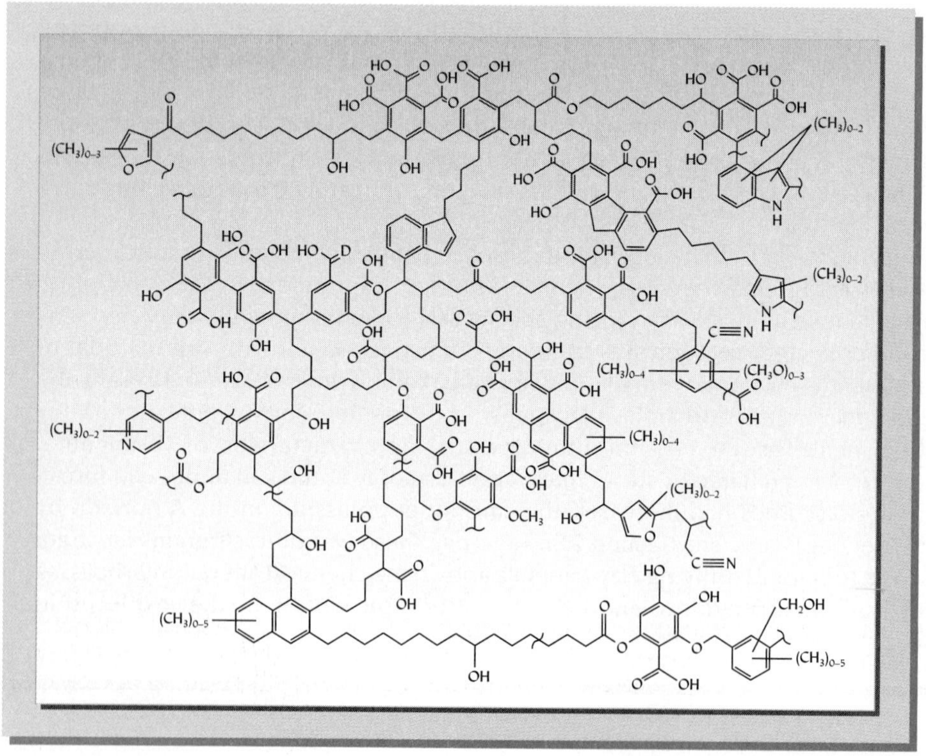

FIGURE 8.14 A possible structure for humic acid, a primary constituent of colloidal humus in soils. Careful inspection will reveal the presence of many of the active –OH groups illustrated in Figure 8.15, as well as certain nitrogen- and sulfur-containing groups. [From Schulten and Schnitzer (1993) with kind permission of Springer-Verlag Publishers]

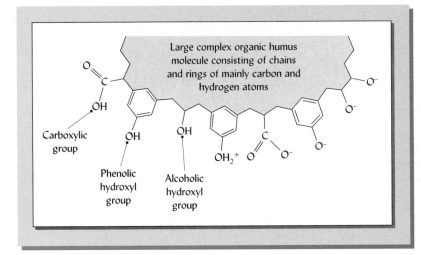

FIGURE 8.15 A simplified diagram showing the principal chemical groups responsible for the high amount of negative charge on humus colloids. The three groups highlighted all include –OH that can lose its hydrogen ion by dissociation and thus become negatively charged. Note that the **carboxylic, phenolic,** and **alcoholic** groups on the right side of the diagram are shown in their disassociated state, while those on the left side still have their associated hydrogen ions. Note also that association with a second hydrogen ion causes a site to exhibit a net positive charge. (Diagram courtesy of R. Weil)

the colloidal behavior of the associated silicate clays by masking charge sites, interfering with shrinkage and swelling, and providing anion-retentive surfaces.

Humus

All about humic substances:
http://www.ar.wroc.pl/
~weber/humic.htm#start

As mentioned in Section 8.1, humus is a noncrystalline organic substance. It consists of very large organic molecules whose chemical composition varies considerably, but generally contains 40 to 60% C, 30 to 50% O, 3 to 7% H, and 1 to 5% N. The molecular weights of humic acids, a major type of colloidal humus, range from 10,000 to 100,000 g/mol. Identification of the actual structure of humus colloids is very difficult. A proposed structure typical of humic acid is shown in Figure 8.14. Note that it contains a very complex series of carbon chains and ring structures, with numerous chemically active functional groups throughout. Figure 8.15 provides a simplified diagram to illustrate the three main types of —OH groups thought to be responsible for the high amount of charge associated with these colloids. Negative or positive charges on the humus colloid develop as H^+ ions are either lost or gained by these groups. Both cations and anions are therefore attracted to and adsorbed by the humus colloid. The negative sites always outnumber the positive ones, and a very large *net* negative charge is associated with humus (Table 8.1). Because of its great surface area and many hydrophilic (water-loving) groups, humus can adsorb very large amounts of water per unit mass. However, humus also contains many hydrophobic sites and therefore can strongly adsorb a wide range of hydrophobic, nonpolar organic compounds (see Section 8.12). Because of its extraordinary influence on soil properties and behavior, we will delve much more deeply into the nature and function of soil humus in Chapter 12.

8.5 GENESIS AND GEOGRAPHIC DISTRIBUTION OF SOIL COLLOIDS

Genesis of Colloids

The silicate clays are developed from the weathering of a wide variety of minerals by at least two distinct processes: (1) a slight physical and chemical **alteration** of certain primary minerals, and (2) a **decomposition** of primary minerals with the subsequent **recrystallization** of certain of their products into the silicate clays. These processes will each be given brief consideration.

ALTERATION. The changes that occur as muscovite mica is altered to fine-grained mica represent a good example of alteration. Muscovite is a dioctahedral 2:1-type primary mineral with a nonexpanding crystal structure and a formula of $KAl_2(Si_3Al)O_{10}(OH)_2$. As weathering occurs, the mineral is broken down in size to the colloidal range. Part of the interlayer potassium is lost, and some silicon along with such cations as Ca^{2+} or Mg^{2+} are added from weathering solutions. The net result is a less rigid crystal structure and the availability of free electronegative charges at sites formerly occupied by the

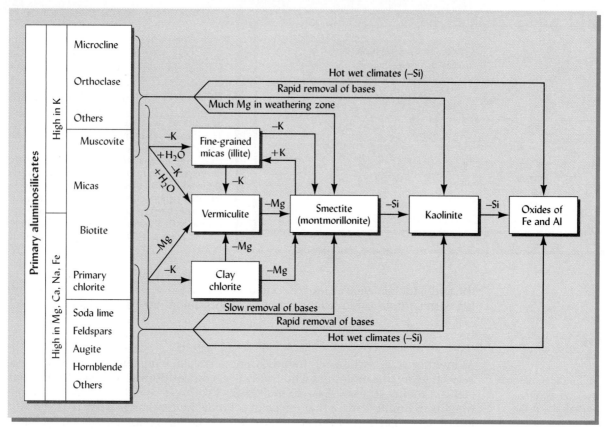

FIGURE 8.16 General conditions for the formation of the various layer silicate clays and oxides of iron and aluminum. Fine-grained micas, chlorite, and vermiculite are formed through rather mild weathering of primary aluminosilicate minerals, whereas kaolinite and oxides of iron and aluminum are products of much more intense weathering. Conditions of intermediate weathering intensity encourage the formation of smectite. In each case silicate clay genesis is accompanied by the removal in solution of such elements as K, Na, Ca, and Mg. Several members of this weathering series may be present in a single soil profile, with the less weathered clay in the C horizon and the more weathered clay minerals in the B or A horizons.

fixed interlayer potassium. The fine mica colloid that emerges still has a 2:1-type structure, having only been *altered* in the process.

RECRYSTALLIZATION. This process involves the complete breakdown of the crystal structure and recrystallization of clay minerals from products of this breakdown. It is the result of much more intense weathering than that required for the alteration process just described.

An example of recrystallization is the formation of kaolinite (a 1:1-type clay mineral) from solutions containing soluble aluminum and silicon that came from the breakdown of primary minerals having a 2:1-type structure. Such recrystallization makes possible the formation of more than one kind of clay from a given primary mineral. The specific clay mineral that forms depends on weathering conditions and the specific ions present in the weathering solution as crystallization occurs.

RELATIVE STAGES OF WEATHERING. Specific conditions conducive to the formation of important clay types are shown in Figure 8.16. Note that fine-grained micas and magnesium-rich chlorites represent earlier weathering stages of the silicates, and kaolinite and (ultimately) iron and aluminum oxides the most advanced stages. The smectites (e.g., montmorillonite) represent intermediate stages. Different weathering stages may occur across climatic zones or across horizons within a single profile. As noted in Section 2.1, silicon tends to be lost as weathering progresses, leaving a lower Si:Al ratio in more highly weathered soil horizons.

MIXED AND INTERSTRATIFIED LAYERS. In a given soil, it is common to find several silicate clay minerals in an intimate mixture. In fact, the properties and compositions of some mineral colloids are intermediate between those of the well-defined minerals described in Section 8.3. For example, a **mixed layer** or **interstratified** clay mineral in which some layers are more like mica and some more like vermiculite might be called *fine-grained mica-vermiculite*.

IRON AND ALUMINUM OXIDES.   Iron oxides often are produced by the weathering of iron-containing primary silicate minerals or by precipitation of iron from soil solutions. Under aerated weathering conditions, divalent iron (Fe^{2+}) oxidizes rapidly to trivalent iron (Fe^{3+}), either while still within the structure of primary minerals or after its release into the soil solution. The Fe^{3+} forms stable oxides and hydroxides by reacting with oxygen atoms and water. The yellow-brown colored **goethite** (FeOOH) is the most stable iron oxide under most conditions and therefore tends to be the dominant iron oxide in most soils, especially in temperate, humid regions. The red-colored **hematite** (Fe_2O_3) tends to form under drier, warmer, more oxidized conditions. It can also be inherited from such soil parent materials as red shales. Other iron oxides precipitate under wet, poorly oxygenated conditions. Interestingly, certain highly weathered soils contain significant amounts of *maghematite*, a magnetic iron oxide that forms in surface horizons under the influence of the heat from brush fires. If a water suspension of such a soil is stirred with a magnetic stir-bar, this iron oxide can be readily observed as tiny particles clinging to the magnet.

Aluminum oxides, mainly **gibbsite** [$Al(OH)_3$], are produced by strong weathering environments in which acid leaching rapidly removes the Si released from the breakdown of primary and secondary silicate minerals. During weathering, hydrogen ions replace the K^+, Mg^{2+}, and other such ions in the crystal, causing the framework to break down and releasing the silicon and aluminum. The aluminum released by the breakdown of dark-colored, iron-rich rocks such as gabbro and basalt often forms gibbsite directly. The weathering of light-colored rocks such as granite and gneiss may first produce 1:1 silicate minerals such as kaolinite or halloysite, which yield gibbsite upon further weathering. Gibbsite is extremely stable in soils and typically represents the most advanced stage of weathering in soils.

ALLOPHANE AND IMOGOLITE.   Relatively little is known of factors influencing the formation of allophane and imogolite. While they are commonly associated with materials of volcanic origin, they are also formed from igneous rocks and are found in some Spodosols. Apparently, volcanic ashes release significant quantities of $Si(OH)_x$ and $Al(OH)_x$ materials that precipitate as gels in a relatively short period of time. These minerals are generally poorly crystalline in nature, imogolite being the product of a more advanced state of weathering than that which produces allophane. Both types of minerals have a pronounced capacity to strongly retain anions as well as to bind with humus, protecting it from decomposition.

HUMUS.   The breakdown and alteration of plant residues by microorganisms and the concurrent synthesis of new, more stable, organic compounds results in the formation of the dark-colored colloidal organic material called *humus* (see Section 12.4 for details). The various organic structural units associated with the decay and synthesis provide charged sites for the attraction of both cations and anions.

Distribution of Clays by Geography and Soil Order

The clay of any particular soil is generally made up of a mixture of different colloidal minerals. In a given soil, the mixture may vary from horizon to horizon, because the kind of clay that develops depends not only on climatic influences and profile conditions but also on the nature of the parent material. The situation may be further complicated by the presence in the parent material itself of clays that were formed under a preceding and perhaps an entirely different type of climatic regime. Nevertheless, some broad generalizations are possible.

Table 8.3 shows the dominant clay minerals in different soil orders, descriptions of which were given in Chapter 3. The well-drained and highly-weathered Oxisols and Ultisols of warm humid and subhumid tropics tend to be dominated by kaolinite, along with oxides of iron and aluminum. The smectite, vermiculite, and fine-grained mica groups are more prominent in Alfisols, Mollisols, and Vertisols, where weathering is less intense. Where the parent material is high in micas, fine-grained micas such as illite are apt to be formed. Parent materials that are high in metallic cations (particularly magnesium) or are subject to restricted drainage, which discourages the leaching of these cations, encourage smectite formation.

Soil order[a]	General weathering intensity	Typical location in U.S.	Fe, Al oxides	Kaolinite	Smectite	Fine-grained mica	Vermiculite	Chlorite	Intergrades
Aridisols	Low	Dry areas			XX	XX		X	X
Vertisols[b]	↑	Alabama, Texas			XXX				X
Mollisols		Kansas, Iowa		X	XX	X	X	X	X
Alfisols		Ohio, New York		X	X	X	X	X	X
Spodosols	↓	New England	X	X					
Ultisols		Southeast	XX	XXX			X	X	X
Oxisols	High	Hawaii, Puerto Rico	XX	XXX					

[a] See Chapter 3 for soil descriptions.
[b] By definition these soils have swelling-type clays, which account for the dominance of smectites.

8.6 SOURCES OF CHARGES ON SOIL COLLOIDS

There are two major sources of charges on soil colloids: (1) hydroxyls and other functional groups on the surfaces of the colloidal particles that by releasing or accepting H$^+$ ions can provide either negative or positive charges, and (2) the charge imbalance brought about by the isomorphous substitution in some clay crystal structures of one cation by another of similar size but differing in charge.

All colloids, organic or inorganic, exhibit the surface charges associated with OH$^-$ groups, charges that are largely **pH dependent**. Most of the charges associated with humus, 1:1-type clays, the oxides of iron and aluminum, and allophane are of this type. In the case of the 2:1-type clays, however, these surface charges are complemented by a much larger number of charges emanating from the isomorphous substitution of one cation for another in the octahedral and/or tetrahedral sheets. Since these charges are not dependent on the pH, they are termed **permanent** or **constant charges**. We will consider these constant charges first.

Constant Charges on Silicate Clays

We noted in Section 8.2 that isomorphous substitution could be the source of both negative and positive charges. Examples of specific substitutions will now be considered.

NEGATIVE CHARGES.   A net negative charge is found in minerals where there has been an isomorphous substitution of a lower-charged ion (e.g., Mg^{2+}) for a higher-charged ion (e.g., Al^{3+}). Such substitution commonly occurs in some aluminum-dominated dioctahedral sheets. As shown in Figure 8.5 (*right*), this leaves an unsatisfied negative charge. The substitution of Mg^{2+} for Al^{3+} is an important source of the negative charge on the smectite, vermiculite, and chlorite clay micelles.

A second example is the substitution of an Al^{3+} for an Si^{4+} in the tetrahedral sheet, which also leaves one unsatisfied negative charge from the tetrahedral oxygen atoms. Such a substitution is common in several of the important soil silicate clay minerals, such as the fine-grained micas, vermiculites, and even some smectites.

POSITIVE CHARGES.   Isomorphous substitution can also be a source of positive charges if the substituting cation has a higher charge than the ion for which it substitutes. In a trioctahedral sheet, there are three magnesium ions surrounded by oxygen and hydroxy groups, and the sheet has no charge (review Figure 8.5). However, if an Al^{3+} ion substitutes for one of the Mg^{2+} ions, a positive charge results.

Such positive charges are characteristic of the trioctahedral hydroxide sheet in the interlayer of clay minerals such as chlorites, a charge that is overbalanced by negative charges in the tetrahedral sheet. Indeed, in several 2:1-type silicate clays, including chlorites and smectites, substitutions in both the tetrahedral and octahedral sheets can occur. The net charge in these clays is the balance between the negative and positive charges. In all 2:1-type silicate clays, however, the **net charge** is negative since those

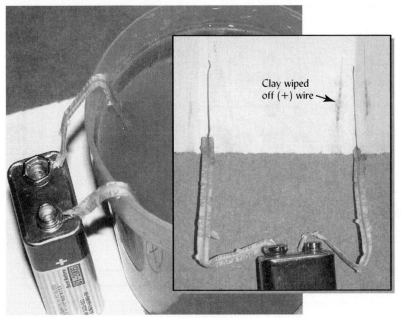

FIGURE 8.17 Simple demonstration of the negatively charged nature of clay. Wires connected to the (−) and (+) terminals of a 9-volt battery are dipped for a few minutes in a suspension of clayey soil in water. The wires are then wiped on a piece of paper (*inset*), showing that the wire on the (+) terminal has attracted the clay while the (−) wire has not. [Adapted from Weil (2009)]

substitutions leading to negative charges far outweigh those producing positive charges (see Figure 8.17).

pH-Dependent Charges

The second source of charges noted on some layer silicate clays (e.g., kaolinite) and on humus, allophane, and Fe, Al oxides, is dependent on the soil pH and consequently is termed **variable** or **pH-dependent**. Both negative and positive charges come from this source.

NEGATIVE CHARGES.   The pH-dependent charges are associated primarily with hydroxyl (OH) groups on the surfaces of the inorganic and organic colloids. Broken edges of mineral colloids also generate pH-dependent charges (see Figure 8.18). The OH groups or oxygen atoms are attached to iron and/or aluminum in the inorganic colloids (e.g., $>$ Al—OH) and to the carbon in humus (e.g., —C—OH). Under moderately acid conditions, there is little or no charge on these particles, but as the pH increases, the hydrogen dissociates from the colloid OH group, and negative charges result.

$$\boxed{\text{Colloid}}\begin{matrix}H^+\\H^+\end{matrix} + Ca(OH)_2 \rightleftharpoons \boxed{\text{Colloid}}\ Ca^{2+} \tag{8.2}$$

As indicated by the $\rightleftharpoons$ arrows, such reactions are reversible. If the pH increases, more OH^- ions are available to force the reactions to the right, and the negative charge on the particle surfaces increases. If the pH is lowered, OH^- ion concentrations are reduced, the reaction goes back to the left, and the negative charge is reduced.

Another source of increased negative charges as the pH is increased is the removal of positively charged complex aluminum hydroxy ions [e.g., $Al(OH)_2{}^+$]. At low pH levels, these ions block negative sites on the silicate clays (e.g., vermiculite) and make them unavailable for cation exchange. As the pH is raised, the $Al(OH)_2{}^+$ ions react with the OH^- ion in the soil solution to form insoluble $Al(OH)_3$, thereby freeing the negatively charged sites.

$$>\!\!Al-(OH)_2{}^- Al(OH)_2{}^+ + OH^- \longrightarrow >\!\!Al-(OH)_2{}^- + Al(OH)_3 \tag{8.3}$$

Negative charged   Negative charge   No charge
site is blocked   site is freed

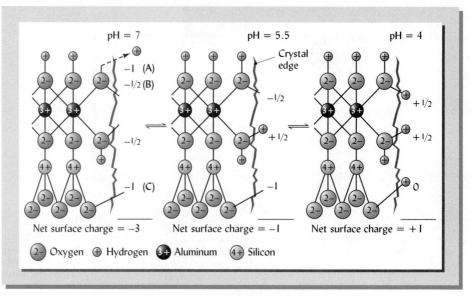

FIGURE 8.18 How pH-dependent charges develop at the broken edge of a kaolinite crystal. Three sources of net negative surface charge at a high pH are illustrated (*left*): (A) One (–1) charge from octahedral oxygen that has lost its H⁺ ion by dissociation (the H broke away from the surface hydroxyl group and escaped into the soil solution). *Note that such dissociation can generate negative charges all along the surface hydroxyl plane, not just at a broken edge.* (B) One half (–½) charge from each octahedral oxygen that would normally be sharing its electrons with a second aluminum. (C) One (–1) charge from a tetrahedral oxygen atom that would normally be balanced by bonding to another silicon if it were not at the broken edge. The middle and right diagrams show the effect of acidification (lowering the pH), which increases the activity of H⁺ ions in the soil solution. At the lowest pH shown (*right*), all of the edge oxygens have an associated H⁺ ion, giving rise to a net positive charge on the crystal. These mechanisms of charge generation are similar to those illustrated for humus in Figure 8.15.

POSITIVE CHARGES. Under moderate to extreme acid soil conditions, some silicate clays and Fe, Al oxides may develop positive charges by **protonation**—the attachment of H⁺ ions to the surface OH groups (Figure 8.18, right).

Since a mixture of humus and several inorganic colloids is usually found in soil, it is not surprising that positive and negative charges may be exhibited at the same time. In most soils of temperate regions, the negative charges far exceed the positive ones (Table 8.4). However, in some acid soils high in Fe, Al oxides or allophane, the overall net charge may be positive. The effect of soil pH on positive and negative charges on such soils is illustrated in Figure 8.19.

The charge characteristics of selected soil colloids are shown in Table 8.4. Note the high percentage of constant negative charges in some 2:1-type clays (e.g., smectites

TABLE 8.4 **Charge Characteristics of Representative Colloids Showing Comparative Levels of Permanent (Constant) and pH-Dependent Negative Charges as Well as pH-Dependent Positive Charges**

	Negative charge			
Colloid type	*Total at pH 7, $cmol_c/kg$*	*Constant, %*	*pH dependent, %*	*Positive charge, $cmol_c/kg$*
Organic	200	10	90	0
Smectite	100	95	5	0
Vermiculite	150	95	5	0
Fine-grained micas	30	80	20	0
Chlorite	30	80	20	0
Kaolinite	8	5	95	2
Gibbsite (Al)	4	0	100	5
Goethite (Fe)	4	0	100	5
Allophane	30	10	90	15

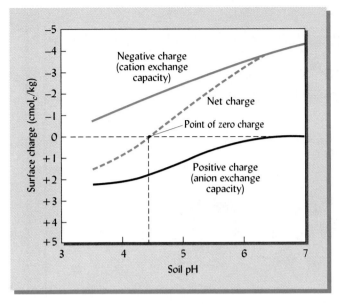

FIGURE 8.19 Relationship between soil pH and positive and negative charges on an Oxisol surface horizon in Malaysia. The negative charges (cation exchange capacity) increase and the positive charges (anion exchange capacity) decrease with increasing soil pH. The point of zero charge is about pH 4.4. [Redrawn from Shamshuddin and Ismail (1995)]

and vermiculites). Humus, kaolinite, allophane and Fe, Al oxides have mostly variable (pH-dependent) negative charges and exhibit modest positive charges at low pH values. The negative and positive charges on soil colloids are of vital importance to the behavior of soils in nature, especially with regard to the adsorption of oppositely charged ions from the soil solution. This subject will be taken up next.

8.7 ADSORPTION OF CATIONS AND ANIONS

In soil, the negative and positive surface charges on the colloids attract and hold a complex swarm of cations and anions. Table 8.5 lists some important cations and anions. The adsorption of these ions by soil colloids greatly affects their biological availability and mobility, thereby influencing both soil fertility and environmental quality. Note that the soil solution and colloidal surfaces in most soils are dominated mainly by just a few of the cations and anions, the others being found in much smaller amounts or

TABLE 8.5   Selected Cations and Anions Commonly Adsorbed to Soil Colloids and Important in Plant Nutrition and Environmental Quality

The listed ions form inner- and/or outer-sphere complexes with soil colloids. Ions marked by an asterisk (*) are among those that predominate in most soil solutions. Many other ions may be important in certain situations.

Cation	Formula	Comments	Anion	Formula	Comments
Ammonium	NH_4^+	Plant nutrient	Arsenate	AsO_4^{3-}	Toxic to animals
Aluminum	Al^{3+} etc.[a]	Toxic to many plants	Borate	$B(OH)_4^-$	Plant nutrient, can be toxic
Calcium*	Ca^{2+}	Plant nutrient	Bicarbonate	HCO_3^-	Toxic in high-pH soils
Cadmium	Cd^{2+}	Toxic pollutant	Carbonate*	CO_3^{2-}	Forms weak acid
Cesium	Cs^+	Radioactive contaminant	Chromate	CrO_4^{2-}	Toxic pollutant
Copper	Cu^{2+}	Plant nutrient, toxic pollutant	Chloride*	Cl^-	Plant nutrient, toxic in large amounts
Hydrogen*	H^+	Causes acidity	Fluoride	Fl^-	Toxic, natural and pollutant
Iron	Fe^{2+}	Plant nutrient	Hydroxyl*	OH^-	Alkalinity factor
Lead	Pb^{2+}	Toxic to animals, plants	Nitrate*	NO_3^-	Plant nutrient, pollutant in water
Magnesium*	Mg^{2+}	Plant nutrient	Molybdate	MoO_4^{2-}	Plant nutrient, can be toxic
Manganese	Mn^{2+}	Plant nutrient	Phosphate	HPO_4^{2-}	Plant nutrient, water pollutant
Nickel	Ni^{2+}	Plant nutrient, toxic pollutant	Selenate	SeO_4^{2-}	Animal nutrient and toxic pollutant
Potassium*	K^+	Plant nutrient	Selenite	SeO_3^{2-}	Animal nutrient and toxic pollutant
Sodium*	Na^+	Used by animals, some plants, can damage soil	Silicate*	SiO_4^{4-}	Mineral weathering product, used by plants
Strontium	Sr^{2+}	Radioactive contaminant	Sulfate*	SO_4^{2-}	Plant nutrient
Zinc	Zn^{2+}	Plant nutrient, toxic pollutant	Sulfide	S^{2-}	In anaerobic soils, forms acid on oxidation

[a] Important aluminum cations include Al^{3+}, $AlOH^{2+}$, and $Al(OH)_2^+$.

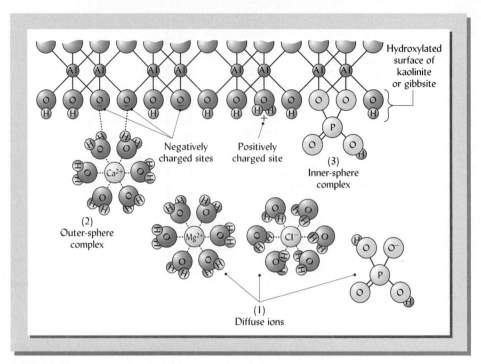

FIGURE 8.20 A diagrammatic representation of the adsorption of ions on a colloid by the formation of outer-sphere and inner-sphere complexes. (1) Water molecules surround diffuse cations and anions (such as the Mg^{2+}, Cl^-, and HPO_4^- shown) in the soil solution. (2) In an **outer-sphere complex** (such as the adsorbed Ca^{2+} ion shown), water molecules form a bridge between the adsorbed cation and the charged colloid surface. (3) In the case of an **inner-sphere complex** (such as the adsorbed $H_2PO_4^-$ anion shown), no water molecules intervene, and the cation or anion binds directly with the metal atom (aluminum in this case) in the colloid structure. Outer-sphere complexes typify easily exchangeable ions that satisfy, in a general way, the net charge on the colloid surface. Inner-sphere complexes, on the other hand, are not easily replaced from the colloid surface, as they represent strong bonding of specific ions to specific sites on the colloid. Adsorption on an exposed hydroxylated surface octahedral sheet, such as that in kaolinite or gibbsite, is shown. In this example, all the charges originate with the dissociation of H^+ ions from surface hydroxyl groups. Although not shown in this example, permanent charges from isomorphous substitution in the interior structure of a colloid could also cause adsorption of outer-sphere complexes. In other colloids (not shown), charged silica tetrahedral surfaces form inner- and outer-sphere complexes by similar mechanisms. (Diagram courtesy of R. Weil)

only in special situations such as contaminated soils. In Figure 8.1 ion adsorption was illustrated in a simplified manner, showing positive cations held on the negatively charged surfaces of a soil colloid. Actually, both cations and anions are usually attracted to the same colloid. In temperate-region soils, anions are commonly adsorbed in much smaller quantities than cations because these soils generally contain predominately 2:1-type silicate clays on which negative charges predominate. In the tropics, where soils are more highly weathered, acid, and rich in 1:1 clays and Fe, Al oxides, the amount of negative charge on the colloids is not so high, and positive charges are more abundant. Therefore, the adsorption of anions is more prominent in these soils.

Figure 8.20 shows how both cations and anions may be attracted to the same colloid if it has both positively and negatively charged sites. This figure also illustrates that adsorption of ions on colloidal surfaces occurs by the formation of two quite different general types of colloid-ion complexes referred to as *outer-sphere* and *inner-sphere* complexes.

Outer- and Inner-Sphere Complexes

Remembering that water molecules surround (hydrate) the cations and anions in the soil solution, we can visualize that in an **outer-sphere complex** water molecules form a bridge between the adsorbed ion and the charged colloid surface. Sometimes several layers of water molecules are involved. Thus, the ion itself never comes close enough to the colloid surface to form a bond with a specific charged site. Instead, the ion is only weakly held by electrostatic attraction, the charge on the oscillating hydrated ion balancing, in a general way, an excess charge of opposite sign on the colloid surface. Ions in an outer-sphere complex are therefore easily replaced by other similarly charged ions.

In contrast, adsorption via formation of an **inner-sphere complex** does *not* involve any intervening water molecules. Therefore, one or more direct bonds are formed between the adsorbed ion and the atoms in the colloid surface. One example already discussed is the case of the K^+ ions that fit so snugly into the spaces between silicon tetrahedra in a mica crystal (see Figure 8.11). Since there are no intervening water molecules, the K^+ ions are directly bonded by sharing electrons with the negatively charged tetrahedral oxygen atoms. Similarly, strong inner-sphere complexes may be formed by reactions of Cu^{2+} or Ni^{2+} with the oxygen atoms in silica tetrahedra.

Another important example, this time involving an anion, occurs when a $H_2PO_4^+$ ion is directly bonded by shared electrons with the octahedral aluminum in the colloid structure (Figure 8.20). Other ions cannot easily replace an ion held in an inner-sphere complex because this type of adsorption involves relatively strong bonds that are dependent on the compatible nature of specific ions and specific sites on the colloid.

Figure 8.20 illustrates only two examples of adsorption complexes on one type of colloid. In other colloids, charged silica tetrahedral surfaces form inner- and outer-sphere complexes by mechanisms similar to those shown in the figure. Permanent charges from isomorphous substitution in the interior structure of a colloid (not shown in Figure 8.20) can also cause adsorption of outer-sphere complexes.

8.8 CATION EXCHANGE REACTIONS

Animation explaining cation exchange:

http://hintze-online.com/ sos/1997/Articles/Art5/ animat2.dcr

Let us consider the case of an outer-sphere complex between a negatively charged colloid surface and a hydrated cation (such as the Ca^{2+} ion shown in Figure 8.20). Such an outer-sphere complex is only loosely held together by electrostatic attraction, and the adsorbed ion remains in constant motion near the colloid surface. There are moments (microseconds) when the adsorbed cation is located a bit farther than average from the colloid surface. This moment provides an opportunity for another hydrated cation from the soil solution (say the Mg^{2+} ion shown in Figure 8.20) to diffuse into a position a bit closer to the negative site on the colloid. The instant this occurs, the second ion would replace the first ion, freeing the formerly adsorbed ion to diffuse out into the soil solution. In this manner, an exchange of cations between the adsorbed and diffuse state takes place. As mentioned in Section 8.1, this process is referred to as **cation exchange**. If a hydrated anion similarly replaces another hydrated anion at a positively charged colloid site, the process is called **anion exchange**. The ions held in outer-sphere complexes from which they can be replaced by exchange reactions are said to be **exchangeable** cations or anions. As a group, all the colloids in a soil, inorganic and organic, capable of holding exchangeable cations or anions are termed the cation or anion **exchangeable complex**.

Principles Governing Cation Exchange Reactions

REVERSIBILITY.   We can illustrate the process of cation exchange using a simple reaction in which a hydrogen ion (perhaps generated by organic matter decomposition—see Section 9.1) displaces a sodium ion from its adsorbed state on a colloid surface:

$$\boxed{\text{Colloid}} \; Na^+ + H^+ \; \rightleftharpoons \; \boxed{\text{Colloid}} \; H^+ + Na^+ \qquad (8.4)$$
$$\qquad\quad \text{(soil} \qquad\qquad\qquad\qquad\qquad \text{(soil}$$
$$\qquad\quad \text{solution)} \qquad\qquad\qquad\qquad\quad \text{solution)}$$

The reaction takes place rapidly and, as shown by the double arrows, the reaction is reversible. It will go to the left if sodium is added to the system. This reversibility is a fundamental principle of cation exchange.

CHARGE EQUIVALENCE.   Another basic principle of cation exchange reactions is that the exchange is chemically equivalent; that is, it takes place on a *charge-for-charge* basis. Therefore, although one H^+ ion exchanged with *one* Na^+ ion in the reaction just shown, it would require *two* singly charged H^+ ions to exchange with or replace *one* divalent

Ca^{2+} ion. If the reaction is reversed, one Ca^{2+} ion will displace two H^+ ions. In other words, two charges from one cation species replace two charges from the other:

$$\boxed{Colloid}\,Ca^{2+} + 2\,H^+ \rightleftharpoons \boxed{Colloid}\begin{matrix}H^+\\H^+\end{matrix} + Ca^{2+} \qquad (8.5)$$

$$\text{(soil solution)} \qquad\qquad\qquad\qquad \text{(soil solution)}$$

Note that by this principle, it would require three Na^+ ions to replace a single Al^{3+} ion, and so on.

RATIO LAW. Consider an exchange reaction between two similar cations, say Ca^{2+} and Mg^{2+}. If there are a large number of Ca^{2+} ions adsorbed on a colloid and some Mg^{2+} is added to the soil solution, the added Mg^{2+} ions will begin displacing the Ca^{2+} from the colloid. This will bring more Ca^{2+} into the soil solution and these Ca^{2+} ions will, in turn, displace some of the Mg^{2+} back off the colloid. Theoretically, these exchanges will continue back and forth until equilibrium is reached. At this point, there will be no further *net* change in the number of adsorbed Ca^{2+} and Mg^{2+} ions (although the exchanges will continue each balancing the other). The **ratio law** tells us that, at equilibrium, the ratio of Ca^{2+} to Mg^{2+} on the colloid will be the same as the ratio of Ca^{2+} to Mg^{2+} in the solution and both will be the same as the ratio in the overall system. To illustrate this concept, assume that 20 Ca^{2+} ions are initially adsorbed on a soil colloid and either 5 or 80 Mg^{2+} ions are added to the system:

Detailed tutorial on CEC, M.J. Eick, VA Tech:
http://www.soils1.cses.vt.edu/MJE/shockwave/cec_demo/version1.1/cec.shtml

$$\boxed{Colloid}\,20\,Ca^{2+} + 5\,Mg^{2+} \rightleftharpoons \boxed{Colloid}\begin{matrix}16\,Ca^{2+}\\4\,Mg^{2+}\end{matrix} + 1\,Mg^{2+} + 4\,Ca^{2+} \quad \text{Ratio: 4 Ca:1 Mg} \quad (8.6)$$

$$\text{(soil solution)} \qquad\qquad\qquad\qquad \text{(soil solution)}$$

$$\boxed{Colloid}\,20\,Ca^{2+} + 80\,Mg^{2+} \rightleftharpoons \boxed{Colloid}\begin{matrix}4\,Ca^{2+}\\16\,Mg^{2+}\end{matrix} + 64\,Mg^{2+} + 16\,Ca^{2+} \quad \text{Ratio: 1 Ca:4 Mg} \quad (8.7)$$

$$\text{(soil solution)} \qquad\qquad\qquad\qquad \text{(soil solution)}$$

If the two exchanging ions are not of the same charge (for example, K^+ exchanging with Mg^{2+}), the reaction becomes somewhat more complicated and a modified version of the ratio law would apply. We will not go into any more detail on this point, except to note that the reactions just given suggest that in order to completely replace one element with a second element by cation exchange, an overwhelming amount of the second ion must be added. This fact must be considered when using displacement to measure the amounts of exchangeable ions or exchange sites in soils (see Section 8.9).

Up to this point, our discussion of exchange reactions has assumed that both ionic species (elements) exchanging places take part in the exchange reaction in exactly the same way. This assumption must be modified to take into account three additional factors if we are to understand how exchange reactions actually proceed in nature.

ANION EFFECTS ON MASS ACTION. In the reactions just discussed, we have not mentioned the anions that always accompany cations in solution. We also showed the exchange reactions as being completely reversible, with an equal chance of proceeding to the right or the left. In reality, the laws of **mass action** tell us that an exchange reaction will be more likely to proceed to the right if the released ion is prevented from reacting in the reverse direction. This may be accomplished if the released cation on the right side of the reaction either *precipitates, volatilizes,* or *strongly associates* with an anion. In each case, most of the displaced cations will be removed from solution and so will not be able to reverse the exchange. To illustrate this concept, consider the displacement of H^+ ions on an acid colloid by Ca^{2+} ions added to the soil solution, either as calcium chloride or as calcium carbonate:

$$\boxed{Colloid}\begin{matrix}H^+\\H^+\end{matrix} + CaCl_2 \rightleftharpoons \boxed{Colloid}\,Ca^{2+} + 2\,H^+ + 2\,Cl^- \qquad (8.8)$$

$$\text{(added)} \qquad\qquad\qquad \text{(dissociated ions)}$$

$$\boxed{Colloid}\begin{matrix}H^+\\H^+\end{matrix} + CaCO_3 \longrightarrow \rightleftharpoons \boxed{Colloid}\,Ca^{2+} + H_2O + CO_2\uparrow \qquad (8.9)$$

$$\text{(added)} \qquad\qquad\qquad\quad \text{Water} \qquad \text{(gas)}$$

TABLE 8.6   Some Exchangeable Ions, Their Hydrated Size, and Their Expected Replacement by NH_4^+ Ions

Among ions of a given charge, the larger the hydrated radius, the more easily it is replaced.

Element	Ion	Hydrated ionic radius, (nm)[a]	Likely replacement of ion initially saturating a kaolinite clay if $cmol_c$ NH_4^+ added = CEC of the soil, (%)[b]
Lithium	Li^+	1.00	80
Sodium	Na^+	0.79	67
Ammonium	NH_4^+	0.54	50
Potassium	K^+	0.53	49
Rubidium	Rb^+	0.51	48
Cesium	Cs^+	0.50	47
Magnesium	Mg^{2+}	1.08	31
Calcium	Ca^{2+}	0.96	29
Strontium	Sr^{2+}	0.96	29
Barium	Ba^{2+}	0.88	26

[a] Not to be confused with nonhydrated radii (Table 8.2), hydrated radii are from Evangelou and Phillips (2005).
[b] Based on empirical data from various sources and assumes no special affinity by kaolinite for any of the listed ions.

In the first reaction, $CaCl_2$ is added, but relatively little calcium will end up on the colloid because the displaced hydrogen ions remain active in the solution and can reverse the reaction, thus displacing calcium back off the colloid. In the second reaction, where $CaCO_3$ is added, when a hydrogen ion is displaced off the colloid, it combines with an oxygen atom from the $CaCO_3$ to form water. Furthermore, the CO_2 produced is a gas, which can volatilize out of the solution and leave the system. The removal of these products pulls the reaction to the right. Therefore, much more Ca will be adsorbed on the colloid and much more H displaced if $CaCO_3$, rather than an equivalent amount of $CaCl_2$, is added to the hydrogen-dominated soil. This principle explains why $CaCO_3$ (in the form of limestone) is effective in neutralizing an acid soil, while calcium chloride is not (see Section 9.11).

CATION SELECTIVITY.   Up to now, we have assumed that both cation species taking part in the exchange reaction are held with equal tenacity by the colloid and therefore have an equal chance of displacing each other. In reality, some cations are held much more tightly than others and so are less likely to be displaced from the colloid. In general, the higher the charge and the smaller the hydrated radius of the cation, the more strongly it will adsorb to the colloid. Because exchangeable adsorption involves creation of an outer-sphere complex, it is the ion's hydrated radius, not the ionic radius, that affects the strength of adsorption. For example, Na^+ is very weakly held because, while it has a relatively small ionic radius (Table 8.2), it carries a large shell of water, giving it a relatively large hydrated radius of 79 nm (Table 8.6). The order of strength of adsorption for selected cations is:

$$Al^{3+} > Sr^{2+} > Ca^{2+} > Mg^{2+} > Cs^+ > K^+ = NH_4^+ > Na^+ > Li^+$$

The less tightly held cations oscillate farther from the colloid surface and therefore are the most likely to be displaced into the soil solution and carried away by leaching. This series therefore explains why the soil colloids are dominated by Al^{3+} (and other aluminum ions) and Ca^{2+} in humid regions and by Ca^{2+} in drier regions, even though the weathering of minerals in many parent materials provides relatively larger amounts of K^+, Mg^{2+}, and Na^+ (see Section 8.10). The strength of adsorption of the H^+ ion is difficult to determine because hydrogen-dominated mineral colloids break down to form aluminum-saturated colloids.

The relative strengths of adsorption order may be altered on certain colloids whose properties favor adsorption of particular cations. An important example of such colloidal "preference" for specific cations is the very high affinity for K^+ ions (and the similarly sized NH_4^+ and Cs^+ ions) exhibited by vermiculite and fine-grained micas (Section 8.3), which attract these ions to inter-tetrahedral spaces exposed at weathered crystal edges. The influence of different colloids on the adsorption of specific cations impacts the availability of cations for leaching or plant uptake (see Section 14.15 and Box 8.3). Certain metals such as copper, mercury, and lead have very high selective affinities for sites on humus and iron oxide colloids, making most soils quite efficient at removing these potential pollutants from water leaching through the profile.

BOX 8.3 CATION EXCHANGE AND FOOD CONTAMINATION BY NUCLEAR FALLOUT

A nuclear accident can contaminate the environment with radioactive isotopes such as cesium (^{137}Cs) and strontium (^{90}Sr) that readily move into the food chain because of their chemical similarity to K and Ca, respectively (see periodic table in Appendix B). If taken into the body, the radiation released by these radioisotopes causes a high incidence of cancer. Contamination of soil with radioisotopes can lead to their uptake by agricultural crops and hence the contamination of human food supplies (see Section 18.11). The 1986 meltdown of a nuclear power plant in Chernobyl spewed about 7 tons of radioactive material, contaminating millions of hectares of land, including some 4 million hectares in Europe outside the immediate area in Belarus, Ukraine, and Russia. World health experts estimate that this radiation may cause between 4,000 and 60,000 more people to die of cancer than would have died if the accident had not occurred. In the event of such widespread soil contamination, safeguarding the food supply requires the ability to predict which soils are likely to produce contaminated vegetables, cattle fodder, and dairy products. One obvious factor to consider is the amount of radioactive contamination in a soil. However, the ratio of ^{137}Cs in the soil to ^{137}Cs in plants varies by four orders of magnitude from one soil to another. In other words, even if two soils have the same ^{137}Cs concentration, the food produced on one soil may be 10,000 times more ^{137}Cs-contaminated than that produced on the other soil. This is because ^{137}Cs availability for plant uptake is largely governed by cation exchange reactions in the soil. If the soil has little capacity to adsorb cations, most of the ^{137}Cs will remain in the soil solution, where roots will easily take it up. Soils with more clay and humus will adsorb much of the ^{137}Cs, releasing into the soil solution such previously adsorbed cations as K^+ and Ca^{2+}.

A study in Belgium found that the risk of ^{137}Cs uptake by plants was closely related to soil clay content in northern Belgium, but only weakly related in southern Belgium (see Figure 8.21). Differences in the types of clays present in soils from each region explain this finding. Apparently, in northern Belgium (but not in southern Belgium), soils have formed mainly in eolian parent materials rich in dioctahedral mica. The clay colloids in the northern Belgium soils therefore are dominated by dioctahedral fine-grained mica and vermiculite, both minerals that exhibit a high selectivity for the K^+ (as well as for the similarly sized NH_4^+ and Cs^+) ions that fit so well into their inter-layer spaces. The ^{137}Cs$^+$ ions in the soil solution preferentially displace Ca^{2+} and other ions from the colloids. These soils are therefore capable of "intercepting" most of the ^{137}Cs$^+$ before plants can take it up. However, because K^+ ions are equally attracted to these clays, repeated additions of potassium fertilizer to these soils would release much of the adsorbed ^{137}Cs$^+$ by cation exchange, making it available again for plant uptake.

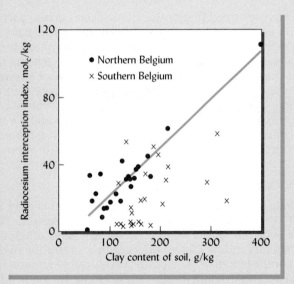

FIGURE 8.21 *Influence of clay content on the ability of soils to reduce plant uptake of radioactive cesium (Radiocesium Interception Index). The soils were sampled in pastures used to feed dairy cows and produce milk in several regions of southern (×) and northern (•) Belgium. In southern Belgium, little relationship existed between the index and the soil clay contents. However, a strong relationship existed in the northern regions, where micaceous parent materials are common and soils are dominated by dioctahedral fine-grained mica and vermiculite clays. [Redrawn from data in Waegeneers et al. (1999)]*

COMPLEMENTARY CATIONS. In soils, colloids are always surrounded by many different adsorbed cation species. The likelihood that a given adsorbed cation will be displaced from a colloid is influenced by how strongly its neighboring cations are adsorbed to the colloid surface. For example, consider an adsorbed Mg^{2+} ion. An ion diffusing in from the soil solution is more likely to displace one of the neighboring ions rather than the Mg^{2+} ion, if the neighboring adsorbed ions (sometimes called **complementary ions**) are loosely held. If they are tightly held, then the chances are greater that the Mg^{2+} ion will be displaced. In Section 8.10, we shall discuss the influence of complementary ions on the availability of nutrient cations for plant uptake.

In soils, cation exchange reactions follow the basic principles just illustrated. However, because there are many different cations, both adsorbed on the cation exchange complex and free in the soil solution, the overall reactions are much more complicated than the simple two-species interactions we have discussed so far. One additional example will illustrate the combined influence of charge equivalence, ion selectivity, and complementary ions in cation exchange.

$$\boxed{\text{Colloid}}\begin{array}{l} 20\ Ca^{2+} \\ 5\ K^{+} \\ 10\ Mg^{2+} \\ 3\ Na^{+} \\ 10\ Al^{3+} \end{array} + 20\ NH_4^{+} + 20\ Cl^{-} \rightleftharpoons \boxed{\text{Colloid}}\begin{array}{l} 19\ Ca^{2+} \\ 1\ K^{+} \\ 8\ Mg^{2+} \\ 13\ NH_4^{+} \\ 10\ Al^{3+} \end{array} + 4\ K^{+} + 3\ Na^{+} + 2\ Mg^{2+} + Ca^{2+} + 7\ NH_4^{+} + 20\ Cl^{-} \qquad (8.10)$$

Note that 13 charges from NH_4^{+} displaced a total of 13 charges from the initially adsorbed cations. Also note that a much larger proportion of loosely held Na^{+}, K^{+}, and Mg^{2+} were exchanged rather than the more tightly held Al^{3+} and Ca^{2+}. *If* more of the complementary ions had been Al^{3+}, then more of the Ca^{2+} would have been displaced and released to the soil solution, where it would have been easily leached or taken up by plants. Although the example merely approximates the ratio law, it does show that only a portion of the added NH_4^{+} ions were adsorbed. Leaching loss of the adsorbed NH_4^{+} ions would be retarded, but they could still serve as a source of nitrogen for plant roots by further exchange reactions.

8.9 CATION EXCHANGE CAPACITY

Previous sections have dealt qualitatively with exchange reactions. We now turn to a consideration of the quantitative **cation exchange capacity** (CEC). This property is defined simply as the sum total of the exchangeable cations that a soil can absorb.

Means of Expression

The cation exchange capacity (CEC) is expressed as the number of moles of positive charge adsorbed per unit mass. In order to be able to deal with whole numbers of a convenient size, many publications, including this textbook, report CEC values in centimoles of charge per kilogram ($cmol_c/kg$). Some publications still use the older unit, milliequivalents per 100 grams (me/100 g), which gives the same value as $cmol_c/kg$ (1 me/100 g = 1 $cmol_c/kg$). A particular soil may have a CEC of 15 $cmol_c/kg$, indicating that 1 kg of the soil can hold 15 $cmol_c$ of H^{+} ions, for example, and can exchange this number of charges from H^{+} ions for the same number of charges from any other cation. This means of expression emphasizes that exchange reactions take place on a charge-for-charge (not an ion-for-ion) basis. The concept of a mole of charges and its use in CEC calculations are reviewed in Box 8.4.

Methods of Determining CEC

The CEC is an important soil chemical property that is used for classifying soils in *Soil Taxonomy* (e.g., in defining an Oxic, Mollic, or Kandic diagnostic horizon, Section 3.2) and for assessing their fertility and environmental behavior. Several different standard methods can be used to determine the CEC of a soil. In general, a concentrated solution of a particular exchanger cation (for example, Ba^{2+}, NH_4^{+}, or Sr^{2+}) is used to leach the soil sample. This provides an overwhelming number of the exchanger cations that can completely replace all the exchangeable cations initially in the soil. Then, the CEC can be determined by measuring either the number of exchanger cations adsorbed (Figure 8.22d) or the amounts of each of the displaced elements originally held on the exchange complex (usually Ca^{2+}, Al^{3+}, Mg^{2+}, K^{+}, and Na^{+}, Figure 8.22b). See Box 8.5 for detailed calculations.

BUFFER CEC METHODS. The CEC procedure often calls for use of a solution buffered to maintain a certain pH (usually either pH 7.0 using ammonium as the exchanger cation or pH 8.2 using barium as the exchanger cation). If the native soil pH is less than the pH of the buffered solution, then these methods measure not only the cation exchange sites active at the pH of the particular soil, but also any pH-dependent exchange sites (see Section 8.6) that would become negatively charged at pH 7.0 or 8.2.

BOX 8.4   CHEMICAL EXPRESSION OF CATION EXCHANGE

One mole of any atom, molecule, or charge is defined as 6.02×10^{23} (Avogadro's number) of atoms, molecules, or charges. Thus, 6.02×10^{23} negative charges associated with the soil colloidal complex would attract 1 mole of positive charge from adsorbed cations such as Ca^{2+}, Mg^{2+}, and H^+. The number of moles of the positive charge provided by the adsorbed cations in any soil gives us a measure of the cation exchange capacity (CEC) of that soil.

The CEC of soils commonly varies from 0.03 to 0.5 mole of positive charge per kilogram. To express the CEC in whole numbers, the charge is usually indicated in centimoles per kilogram ($cmol_c/kg$) of soil. Since there are 100 centimoles in 1 mole, the preceding range of CEC of soils is 3 to 50 $cmol_c/kg$.

CALCULATING MASS FROM MOLES

Using the mole concept, it is easy to relate the mole charges to the mass of ions or compounds involved in cation or anion exchange. Consider, for example, the exchange that takes place when adsorbed sodium ions in an alkaline arid-region soil are replaced by hydrogen ions:

$$\boxed{\text{Colloid}}\ Na^+ + H^+ \rightleftharpoons \boxed{\text{Colloid}}\ H^+ + Na^+$$

If 1 $cmol_c$ of adsorbed Na^+ ions per kilogram of soil were replaced by H^+ ions in this reaction, how many grams of Na^+ ions would be replaced?

Since the Na^+ ion is singly charged, the mass of Na^+ needed to provide 1 mole of charge (1 mol_c) is the gram atomic weight of sodium, or 23 g (see periodic table in Appendix B). The mass providing 1 *centimole* of charge ($cmol_c$) is 1/100 of this amount; thus, the mass of the 1 $cmol_c$ Na^+ replaced is 0.23 g Na^+/kg soil. The 0.23 g Na^+ would be replaced by only 0.01 g H, which is the mass of 1 $cmol_c$ of this much lighter element.

Another example is the replacement of H^+ ions when hydrated lime [$Ca(OH)_2$] is added to an acid soil. This time assume that 2 $cmol_c$ H^+/kg soil is replaced by the $Ca(OH)_2$, which reacts with the acid soil as follows:

$$\boxed{\text{Colloid}}\ {}^{H^+}_{H^+} + Ca(OH)_2 \rightleftharpoons \boxed{\text{Colloid}}\ Ca^{2+} + 2H_2O$$

Since the Ca^{2+} ion in each molecule of $Ca(OH)_2$ has two positive charges, the mass of $Ca(OH)_2$ needed to replace 1 mole of charge from the H^+ ions is only one-half of the gram molecular weight of this compound, or $74/2 = 37$ g. A comparable figure for 1 *centimole* is 37/100, or 0.37 grams. The mass of $Ca(OH)_2$ needed to replace 2 $cmol_c$ H^+/kg soil is:

$$2\ cmol_c\ Ca(OH)_2/kg \times 0.37\ g\ Ca(OH)_2/cmol_c = 0.74\ g\ Ca(OH)_2/kg\ soil$$

The 0.74 g $Ca(OH)_2$/kg soil can be converted to the amount of $Ca(OH)_2$ needed to replace 2 $cmol_c$ H^+/kg from the surface 15 cm of 1 ha of field soil, remembering from Chapter 4 (Section 4.7, footnote 11) that this depth of soil typically weighs 2 million kg/ha.

$$0.74\ g/kg \times 2 \times 10^6\ kg = 1.48 \times 10^6\ g;\quad 1.48 \times 10^3\ kg;\quad \text{or } 1.48\ Mg$$

CHARGE AND CHEMICAL EQUIVALENCY

In each preceding example, the number of charges provided by the replacing ion is equivalent to the number associated with the ion being replaced. Thus, 1 mole of negative charges attracts 1 mole of positive charges whether the charges come from H^+, K^+, Na^+, NH_4^+, Ca^{2+}, Mg^{2+}, Al^{3+}, or any other cation. Keep in mind, however, that only one-half the atomic weights of divalent cations, such as Ca^{2+} or Mg^{2+}, and only one-third the atomic weight of trivalent Al^{3+} are needed to provide 1 mole of charge. This *chemical equivalency* principle applies to both cation and anion exchange.

EFFECTIVE CEC (UNBUFFERED). Alternatively, the CEC procedure may use unbuffered solutions to allow the exchange to take place at the actual pH of the soil. The buffered methods (NH_4^+ at pH 7.0 or Ba^{2+} at pH 8.2) measure the *potential* or *maximum* cation exchange capacity of a soil. The unbuffered method measures only the **effective cation**

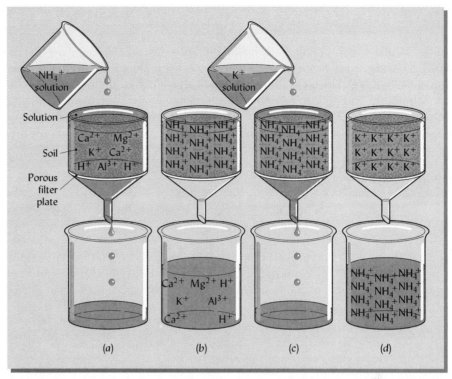

FIGURE 8.22 Illustration of a method for determining the cation exchange capacity of soils. (*a*) A given mass of soil containing a variety of exchangeable cations is leached with an ammonium (NH_4^+) salt solution. (*b*) The NH_4^+ ions replace the other adsorbed cations, which are leached into the container below. (*c*) After the excess NH_4^+ salt solution is removed with an organic solvent, such as alcohol, a K^+ salt solution is used to replace and leach the adsorbed NH_4^+ ions. (*d*) The amount of NH_4^+ released and washed into the lower container can be determined, thereby measuring the chemical equivalent of the cation exchange capacity (i.e., the negative charge on the soil colloids). (Diagram courtesy of R. Weil)

BOX 8.5   CALCULATING SOIL CEC FROM LAB DATA

Most procedures for measuring soil CEC use cation exchange reactions similar to those illustrated in Figure 8.22. The various cations initially adsorbed by the soil colloids are replaced by exposing the soil sample to a salt solution containing an overwhelming number of cations of a single element. The element chosen is usually one not found in large quantities on the soil exchange complex (e.g., NH_4^+, Ba^{2+}, or Sr^{2+}). Examine Figure 8.22 and assume that 100 g (0.1 kg) of dry soil was placed in the funnel and that 400 mL (0.4 L) of solution was added at each step. There are methods by which one can estimate the CEC of the soil.

METHOD 1: MEASURE ALL THE CATIONS ORIGINALLY HELD ON THE EXCHANGE COMPLEX

After leaching the soil with 0.4 L of NH_4^+ solution, all the exchangeable cations shown in the soil sample in Figure 8.22*a* were displaced off the colloids and washed into the beaker (Figure 8.22*b*) along with the excess NH_4^+ ions. The solution in this beaker (*b*) was analyzed for Ca, Mg, K, Al, and H with the following results: 200 mg/L Ca^{2+}, 60 mg/L Mg^{2+}, 97.5 mg/L K^+, 5 mg/L H^+, and 67.5 mg/L Al^{3+}. Because only 0.4 L of solution was collected from the soil sample and the soil sample weighed only 0.1 kg, these results can be multiplied by 0.4 and 10 to give the amounts of each ion collected in mg/kg soil. As an example we can show the calculation for Ca^{2+} as:

$$\frac{200 \text{ mg } Ca^{2+}}{L} \times \frac{0.4 \text{ L}}{\text{sample}} \times \frac{10 \text{ samples}}{\text{kg soil}} = \frac{800 \text{ mg } Ca^{2+}}{\text{kg soil}}$$

(continued)

BOX 8.5 (*Cont.*)   CALCULATING SOIL CEC FROM LAB DATA

Similar calculations show that the concentration of exchangeable cations in the soil were: 800 mg Ca^{2+}/kg, 240 mg Mg^{2+}/kg, 390 mg K^+/kg, 20 mg H^+/kg, and 270 mg Al^{3+}/kg. The CEC is normally expressed as the cmol of charge per kg of soil ($cmol_c$/kg). Since the atomic weight of an element is defined as the grams per mole of that element, we now turn to the periodic table in Appendix B for the atomic weights (g/mol) needed to convert our mg/kg concentrations into mol/kg for each cation species measured. We divide the mg/kg soil (from above) by 1000 to give g/kg and then divide by the atomic weight to give mol/kg soil. We then multiply the mol by 100 to give the $cmol_c$/kg soil (see Box 8.4).

For example, for Ca^{2+} the atomic weight is approximately 40 g/mol, so we calculate the cmol of exchangeable Ca^{2+} in 1 kg of our soil as follows:

$$\frac{800 \text{ mg Ca}^{2+}}{\text{kg soil}} \times \frac{1 \text{ g}}{1000 \text{ mg}} \times \frac{1 \text{ mol Ca}^{2+}}{40 \text{ g}} \times \frac{100 \text{ cmol}}{\text{mol}} = \frac{2 \text{ cmol Ca}^{2+}}{\text{kg soil}}$$

Repeating this calculation provides the following results: 2 cmol Ca^{2+}/kg, 1 cmol Mg^{2+}/kg, 1 cmol K^+/kg, 2 cmol H^+/kg, and 1 cmol Al^{3+}/kg. We now must multiply the cmol/kg for each element by the valence of the ion to convert to the cmol of *charge* ($cmol_c$/ kg) from that element. Using Ca^{2+} again as an example:

$$\frac{2 \text{ cmol Ca}^{2+}}{\text{kg soil}} \times \frac{2 \text{ cmol}_c \text{ from Ca}^{2+}}{\text{cmol Ca}^{2+}} = \frac{4 \text{ cmol}_c \text{ from Ca}^{2+}}{\text{kg soil}}$$

Repeating this calculation provides the following results: 4 $cmol_c$ Ca^{2+}/kg, 2 $cmol_c$ Mg^{2+}/kg, 1 $cmol_c$ K^+/kg, 2 $cmol_c$ H^+/kg, and 3 $cmol_c$ Al^{3+}/kg. Assuming the five elements measured account for nearly all the exchangeable cations, the sum of their charges ($4 + 2 + 1 + 2 + 3 = 12$) equals the CEC of our soil: 12 $cmol_c$/kg.

METHOD 2: MEASURE ALL THE NH_4^+ IONS IN THE FINAL LEACHATE

Assume that after washing out and discarding all excess (nonexchangeable) NH_4^+ ions, the remaining exchangeable NH_4^+ ions were replaced by K^+ ions and washed into beaker *d* (Figure 8.22). Therefore the number of charges from NH_4^+ ions in beaker *d* is equal to the CEC of the soil. Note that by this method, the lab needs to determine only one element. Assume the NH_4^+ concentration in beaker *d* to be 540 mg NH_4^+/L. As in method 1 (above), because only 0.4 L of solution was collected from the soil sample and the soil sample weighed only 0.1 kg, these results can be calculated as follows to give the amount of NH_4^+ ions collected in mg/kg soil:

$$\frac{540 \text{ mg NH}_4^+}{\text{L}} \times \frac{0.4 \text{ L}}{\text{sample}} \times \frac{10 \text{ samples}}{\text{kg soil}} = \frac{2160 \text{ mg NH}_4^+}{\text{kg soil}}$$

The periodic table in Appendix B provides the atomic weights needed to convert our mg NH_4^+/kg concentration into mol NH_4^+/kg. We divide the mg/kg (from above) by 1000 to give g/kg. Using atomic weights from the periodic table (Appendix B) we calculate that 1 mol of NH_4^+ = 18 g [14g/mol N + 4(1 g/mol H) = 18 g/mol NH_4^+]. We then divide the g/kg by the 18 g/mol to give the moles of NH_4^+. We then multiply the moles NH_4^+ by 100 to give the $cmol_c$/ kg (see Box 8.4). We can therefore calculate the cmol of exchangeable NH_4^+ in 1 kg of our soil as follows:

$$\frac{2160 \text{ mg NH}_4^+}{\text{kg soil}} \times \frac{1 \text{ mol NH}_4^+}{18 \text{ g}} \times \frac{1 \text{ g}}{1000 \text{ mg}} \times \frac{100 \text{ cmol}}{\text{mol}} = \frac{12 \text{ cmol NH}_4^+}{\text{kg soil}}$$

Since each NH_4^+ ion carries only 1 charge, the cmol of *charge* from NH_4^+ ($cmol_c$/kg) is the same as the cmol NH_4^+/kg:

$$\frac{12 \text{ cmol NH}_4^+}{\text{kg soil}} \times \frac{1 \text{ cmol}_c \text{ from NH}_4^+}{\text{cmol NH}_4^+} = \frac{12 \text{ cmol}_c \text{ from NH}_4^+}{\text{kg soil}}$$

Therefore the CEC of our soil = 12 $cmol_c$/kg. [If you happened to count the ions in Figure 8.22, you may have noticed that the concentrations chosen here are those that would apply if each ion shown actually represented 1 cmol of ions.]

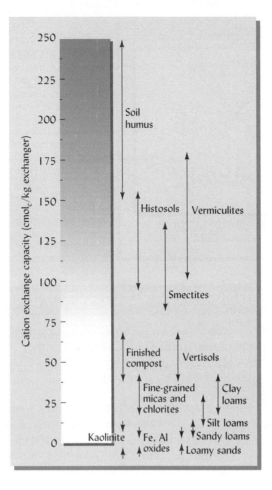

FIGURE 8.23 Ranges in the cation exchange capacities (at pH 7) that are typical of a variety of soils and soil materials. The high CEC of humus shows why this colloid plays such a prominent role in most soils and especially those high in kaolinite and Fe, Al oxides, and clays that have low CECs. (Diagram courtesy of R. Weil)

exchange capacity (ECEC), which can hold exchangeable cations at the pH of the soil as sampled. As the different methods may yield significantly different values of CEC, it is important that the method used be known when comparing soils based on their CEC. This is especially significant if the soil pH is much below the buffer pH used.

Cation Exchange Capacities of Soils

The cation exchange capacity (CEC) of a given soil horizon is determined by the relative amounts of different colloids in that soil and by the CEC of each of these colloids. Figure 8.23 illustrates the common range in CEC among different soils and other organic and inorganic exchange materials. Note that sandy soils, which are generally low in all colloidal material, have low CECs compared to those exhibited by silt loams and clay loams. Also note the very high CECs associated with humus compared to those exhibited by the inorganic clays, especially kaolinite and Fe, Al oxides. The CEC coming from humus generally plays a very prominent role, and sometimes a dominant one, in cation exchange reactions in A horizons. For example, in a clayey Ultisol (pH = 5.5) containing 2.5% humus and 30% kaolinite, about 75% of the CEC is associated with humus. Figure 8.24 illustrates the contribution of organic matter to the CEC of various forested soils and how that contribution increases at higher soil pH levels.

Using the CEC range from Figure 8.23 it is possible to estimate the CEC of a soil if the quantities of the different soil colloids in the soil are known (see Box 8.6).

Data in Table 8.7 show the average CEC values for seven different soil orders. Note the very high CEC for the Histosols, verifying the high CEC of the organic colloids. The Vertisols, which are very high in swelling-type clays (mostly smectite), had the highest average CEC of the mineral soils. Next came the Aridisols and Mollisols, which are also commonly high in 2:1-type clays. The Ultisols, whose clays are dominantly kaolinite and hydrous oxides of iron and aluminum, had relatively low CEC values. Despite large variations in soil organic matter and texture, these data appear to reflect the quantities and kinds of soil colloids found in the soils.

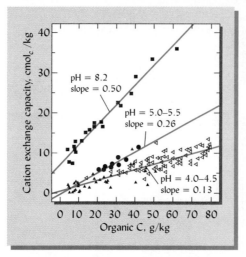

FIGURE 8.24 Soil organic carbon (SOC) contributes markedly to the cation exchange capacity (CEC) of soils and more so at higher pH levels. In this graph, the sloping lines indicate the relationship between increasing SOC and CEC. The greater slopes for the higher pH soils indicate that the CEC of humus increases with increased pH. At pH 4.0 to 4.5, every 1 g increase in organic C contributed to the CEC an additional 0.13 $cmol_c$/kg soil. At the pH 5.0 to 5.5 level, this contribution was twice as great (0.26 $cmol_c$) and at pH 8.2 the contribution was about four times as great (0.50 $cmol_c$). At each pH level the data represent soils with similar clay contents, but with SOC contents that varied because of differences in soil depth or land use. [Data sources: squares and filled triangles, Krishnaswamy and Richter (2002); circles, Bayer et al. (2001); open triangles, Sullivan et al. (2006)]

BOX 8.6 ESTIMATING CEC AND CLAY MINERALOGY

Data on clay mineralogy and cation exchange capacity are rather time-consuming to obtain and not always available. Fortunately, one can often estimate one of these types of data from the other, assuming that data on soil pH, clay content, and organic matter (OM) level are available.

1. Example of Estimating CEC from Data on Mineralogy

Assume you know that a cultivated Mollisol in Iowa contains 20% clay and 4% organic matter (OM) and its pH = 7.0. The dominant clays in Mollisols are likely 2:1 types such as vermiculite and smectite (see Table 8.3). We estimate the average CEC of the clays of these types to be about 100 $cmol_c$/kg clay (Tables 8.1 and 8.4). At pH 7.0 the CEC of OM is about 200 $cmol_c$/kg (Table 8.4). Since 1 kg of this soil has 0.20 kg (20%) of clay and 0.04 kg (4%) of OM, we can calculate the CEC associated with each of these sources.

From the clays in this Mollisol:	0.2 kg × 100 $cmol_c$/kg = 20 $cmol_c$
From the OM in this Mollisol:	0.04 kg × 200 $cmol_c$/kg = 8 $cmol_c$
The total CEC of this Mollisol:	20 + 8 = 28 $cmol_c$/kg soil

2. Example of Estimating the Clay Mineralogy from Information on CEC

Assume you know that a soil contains 60% clay and 4% organic matter and the pH = 4.2. You also know the CEC is 5.8 $cmol_c$/kg. You want to estimate the types of clays present. At pH 4.2 the CEC of the organic matter would be comparatively low, about 100 $cmol_c$/kg (Figure 8.21). Therefore we estimate:

CEC from OM in 1 kg soil = 0.04 kg OM × 100 $cmol_c$/kg OM = 4.0 $cmol_c$

The remaining portion of the CEC contributed by the clay can be estimated as:

CEC from the clay in 1 kg soil = 5.8 $cmol_c$ − 4.0 $cmol_c$ = 1.8 $cmol_c$

Since this 1.8 $cmol_c$/kg soil is provided by 0.60 kg of clay (60% of 1 kg soil), we can estimate:

CEC of the pure clay = 1.8 $cmol_c$/kg soil × 1 kg soil/0.60 kg clay = 3 $cmol_c$/kg clay

From Tables 8.1 and 8.4 we see that the Fe and Al oxides at pH 7 have a CEC of about 4 $cmol_c$/kg clay. We know (from Section 8.6) their CEC would be lower at pH 4. Likewise, kaolinite has a CEC of about 8 $cmol_c$/kg clay at pH 7 and perhaps only about 4 $cmol_c$/kg clay at pH 4. The CEC values for the other types of clay listed in Table 8.4 are far higher. *Therefore, it is reasonable to conclude that the clays in this Oxisol consist mainly of Fe and Al oxides and kaolinite.*

pH and Cation Exchange Capacity

In previous sections it was pointed out that the cation exchange capacity of most soils increases with pH. At very low pH values, the cation exchange capacity is also generally low (Figure 8.25). Under these conditions, only the permanent charges of the 2:1-type

TABLE 8.7 The Average Cation Exchange Capacities (CEC) and pH Values of More Than 3000 Surface Soil Samples Representing Seven Different Soil Orders

Organic colloids give Histosols a very high CEC. Compare to data in Tables 8.3 and 8.4 to see the relationship between the average CEC and the main types of colloids in the other soil orders.

Soil order	CEC, cmol$_c$/kg	pH
Ultisols	3.5	5.60
Alfisols	9.0	6.00
Spodosols	9.3	4.93
Aridisols	15.2	7.26
Mollisols	18.7	6.51
Vertisols	35.6	6.72
Histosols	128.0	5.50

From Holmgren et al. (1993).

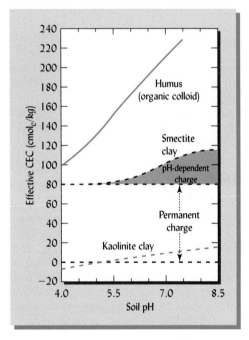

FIGURE 8.25 Influence of pH on representative cation exchange capacities of two clay minerals and humus. Below pH 6.0 the charge for smectite has a fairly *constant* charge that is due mainly to isomorphic substitution and is considered *permanent*. Above pH 6.0 the charge on smectite increases somewhat with pH (shaded area) because of the ionization of hydrogen from the exposed hydroxyl groups at crystal edges. In contrast, the charges on kaolinite and humus are all *variable*, increasing with increasing pH. Humus carries a far greater number of charges than kaolinite. At low pH kaolinite carries a net negative CEC because its positive charges outnumber the negative ones.

clays (see Section 8.8) and a small portion of the pH-dependent charges of organic colloids, allophane, and some 1:1-type clays hold exchangeable ions. As the pH is raised, the negative charges on some 1:1-type silicate clays, allophane, humus, and even Fe, Al oxides increases, thereby increasing the cation exchange capacity. As noted earlier, to obtain a measure of this maximum retentive capacity, the CEC is commonly determined at a pH of 7.0 or 8.2. At neutral or slightly alkaline pH, the CEC reflects most of those pH-dependent charges as well as the permanent ones.

8.10 EXCHANGEABLE CATIONS IN FIELD SOILS

The specific exchangeable cations associated with soil colloids differ from one climatic region to another—Ca^{2+}, Al^{3+}, complex aluminum hydroxy ions, and H^+ being most prominent in humid regions, and Ca^{2+}, Mg^{2+}, and Na^+ dominating in low-rainfall areas (Table 8.8). The cations that dominate the exchange complex have a marked influence on soil properties.

In a given soil, the proportion of the cation exchange capacity satisfied by a particular cation is termed the **saturation percentage** for that cation. Thus, if 50% of the CEC

TABLE 8.8   Cation Exchange Properties Typical for Unamended Clay Loam Surface Soils in Different Climatic Regions

Note that soils with coarser textures would have less clay and organic matter and therefore lower amounts of exchangeable cations and lower CEC values.

Property	Warm, humid region (Ultisols)[a]	Cool, humid region (Alfisols)	Semiarid region (Ustolls)	Arid region (Natrargids)[b]
Exchangeable H^+ and Al^{3+}, $cmol_c/kg$ (% of CEC)	7.5 (75%)	5 (28%)	0 (0%)	0 (0%)
Exchangeable Ca^{2+}, $cmol_c/kg$ (% of CEC)	2.0 (20%)	9 (50%)	17 (65%)	13 (50%)
Exchangeable Mg^{2+}, $cmol_c/kg$ (% of CEC)	0.4 (4%)	3 (17%)	6 (23%)	5 (19%)
Exchangeable K^+, $cmol_c/kg$ (% of CEC)	0.1 (1%)	1 (5%)	2 (8%)	3 (12%)
Exchangeable Na^+, $cmol_c/kg$ (% of CEC)	Tr	0.02 (0.1%)	1 (4%)	5 (19%)
Cation exchange capacity (CEC)[c], $cmol_c/kg$	10	18	26	26
Probable pH	4.5–5.0	5.0–5.5	7.0–8.0	8–10
Nonacid cations (% of CEC)[d]	25%	68%	100%	100%

[a] See Chapter 3 for explanation of soil group names.
[b] Natrargids are Aridisols with natric horizons. They are sodic soils, high in exchangeable sodium, as explained in Section 10.5.
[c] The sum of all the exchangeable cations measured at the pH of the soil. This is termed the effective CEC or ECEC (see Section 8.9).
[d] Traditionally referred to as "base" saturation.

is satisfied by Ca^{2+} ions, the exchange complex is said to have a ***calcium saturation percentage*** of 50.

This terminology is especially useful in identifying the relative proportions of sources of acidity and alkalinity in the soil solution. Thus, the percentage saturation with Al^{3+} and H^+ ions gives an indication of the acid conditions, while increases in the percentage **nonacid cation saturation** (sometimes referred to as the ***base saturation percentage***[3]) indicate the tendency toward neutrality and alkalinity. These relationships will be discussed further in Chapter 9.

Cation Saturation and Nutrient Availability

Exchangeable cations generally are available to both higher plants and microorganisms. By cation exchange, hydrogen ions from the root hairs and microorganisms replace nutrient cations from the exchange complex. The nutrient cations are forced into the soil solution, where they can be assimilated by the adsorptive surfaces of roots and soil organisms, or they may be removed by drainage water. Cation exchange reactions affecting the mobility of organic and inorganic pollutants in soils will be discussed in Section 8.12. Here we focus on the plant nutrition aspects.

The percentage saturation of essential nutrient cations such as calcium and potassium greatly influences the uptake of these elements by growing plants. For example, if the percentage calcium saturation of a soil is high, the displacement of this cation is comparatively easy and rapid. Thus, 6 cmol/kg of exchangeable calcium in a soil whose exchange capacity is 8 cmol/kg (75% calcium saturation) probably would mean ready availability, but 6 cmol/kg when the total exchange capacity of a soil is 30 cmol/kg (20% calcium saturation) would produce lower availability. This is one reason that, for calcium-loving plants such as alfalfa, the calcium saturation of at least part of the soil should approach 80 to 85%.

Influence of Complementary Cations

A second factor influencing plant uptake of a given cation is the effect of the complementary ions held on the colloids. As was discussed in Section 8.8, the strength of adsorption of common cations is in the following order for most colloids:

$$Al^{3+} > Sr^{2+} > Ca^{2+} > Mg^{2+} > Cs^+ > K^+ = NH_4^+ > Na^+ > Li^+$$

---

[3]Technically speaking, nonacid cations such as Ca^{2+}, Mg^{2+}, K^+, and Na^+ are not bases. When adsorbed by soil colloids in the place of H^+ ions, however, they reduce acidity and increase the soil pH. For that reason, they are traditionally referred to as *bases* and the portion of the CEC that they satisfy is often termed *base saturation percentage*.

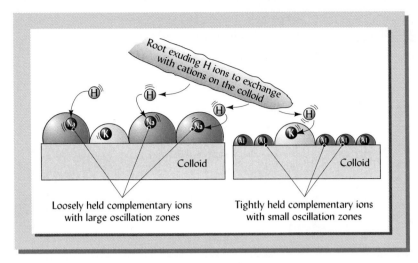

FIGURE 8.26 Effect of complementary ions on the availability of a particular exchangeable nutrient cation. The half spheres represent the zones in which the ion oscillates, the more loosely held ions moving within larger zones of oscillation. For simplicity, the water molecules that hydrate each ion are not shown. (*Left*) H⁺ ions from the root are more likely to encounter and exchange with loosely held Na⁺ ions rather than the more tightly held K⁺ ion. (*Right*) The likelihood that H⁺ ions from the root will encounter and exchange with a K⁺ ion is increased by the inaccessibility of the neighboring tightly held Al³⁺ ions. The K⁺ ion on the right colloid is comparatively more vulnerable to being replaced and sent into the soil solution and is therefore more available for plant uptake or leaching than the K⁺ ion on the left colloid. (Diagram courtesy of R. Weil)

Consider, for example, the relatively loosely held ion, K^+. If the complementary ions surrounding a K^+ ion are held tightly (that is, they oscillate very close to the colloid surface), then a H^+ ion from a root is less likely to "find" a complementary ion and more likely, instead, to "bump into" and replace a K^+ ion (see Figure 8.26). If, on the other hand, the complementary ions are loosely held (oscillating quite far from the colloid surface), then the H^+ is more likely to "bump into" and replace one of the complementary ions and less likely to "find" the K^+. Consequently, K^+ is more likely to be replaced off the colloid if the complementary ions are mainly tightly held Al^{3+} or H^+ (as in acid soils) than if they are mainly Mg^{2+} and Na^+ (as in neutral to alkaline soils). This is why, at a given level of K^+ saturation, K^+ is more readily available for both plant uptake and for leaching in acid soils than in neutral to alkaline soils (see also Section 14.13).

There are also some nutrient antagonisms that in certain soils cause inhibition of uptake of some cations by plants. For instance, potassium uptake by plants is limited by high levels of calcium in some soils. Likewise, high potassium levels are known to limit the uptake of magnesium even when significant quantities of magnesium are present in the soil.

Effect of Type of Colloid

Differences exist in the tenacity with which several types of colloidal micelles hold specific cations and in the ease with which they exchange cations. At a given percentage base saturation, smectites—which have a high charge density per unit of colloid surface—hold calcium much more strongly than does kaolinite (low charge density). As a result, smectite clays must be raised to about 70% base saturation before calcium will exchange easily and rapidly enough to satisfy most plants. In contrast, a kaolinite clay exchanges calcium much more readily, serving as a satisfactory source of this constituent at a much lower percentage base saturation. The need to add limestone to the two soils will be somewhat different, partly because of this factor.

8.11 ANION EXCHANGE

Anions are held by soil colloids in two major ways. First, they are held by anion adsorption mechanisms similar to those responsible for cation adsorption. Second, they may actually react with surface oxides or hydroxides, forming more definitive **inner-sphere complexes**. We shall consider anion adsorption first.

The basic principles of **anion exchange** are similar to those of cation exchange, except that the charges on the colloids are positive and the exchange is among negatively charged anions. The positive charges associated with the surfaces of kaolinite,

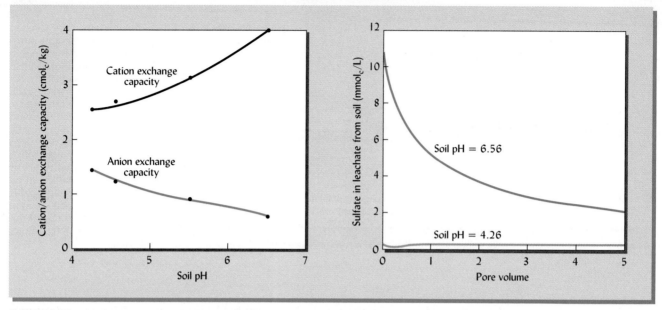

FIGURE 8.27 (*Left*) Effect of increasing the pH of subsoil material from an Ultisol from Georgia on the cation and anion exchange capacities. Note the significant decrease in anion exchange capacity associated with the increased soil pH. When a column of the low-pH material (pH = 4.26) was leached with $Ca(NO_3)_2$ (*right*), little sulfate was removed from the soil. In contrast, similar leaching of a column of the soil with the highest pH (6.56), where the anion exchange capacity had been reduced by half, resulted in anion exchange of NO_3^- ions for SO_4^{2-} ions and significant leaching of sulfate from the soil. The importance of anion adsorption in retarding movement of specific anions or other negatively charged substances is illustrated. [Data from Bellini et al. (1996)]

iron and aluminum oxides, and allophane attract anions such as SO_4^{2-} and NO_3^-. A simple example of an anion exchange reaction is as follows:

$$\boxed{\text{Colloid}}\ NO_3^- + Cl^- \rightleftharpoons \boxed{\text{Colloid}}\ Cl^- + NO_3^-$$

(positively charged
soil solid)   (soil solution)   (positively charged
soil solid)   (soil solution)

(8.11)

Just as in cation exchange, *equivalent* quantities of NO_3^- and Cl^- are exchanged; the reaction can be reversed; and plant nutrients so released can be absorbed by plants.

In contrast to cation exchange capacities, anion exchange capacities of soils generally *decrease* with increasing pH. Figure 8.27 illustrates this fact for an Ultisol in Georgia. In some very acid tropical soils that are high in kaolinite and iron and aluminum oxides, the anion exchange capacity may actually exceed the cation exchange capacity.

Anion exchange is very important in making anions available for plant growth while at the same time retarding the leaching of such anions from the soil. For example, anion exchange restricts the loss of sulfates from subsoils in the southern United States (see Figure 8.27 and Section 13.21). Even the leaching of nitrate may be retarded by anion exchange in the subsoil of certain highly weathered soils of the humid tropics. Similarly, the downward movement into groundwater of some charged organic pollutants found in organic wastes can be retarded by such anion and/or cation exchange reactions.

Inner-Sphere Complexes

Some anions, such as phosphates, arsenates, molybdates, and sulfates, can react with particle surfaces, forming **inner-sphere complexes** (see Figure 8.20). For example, the $H_2PO_4^-$ ion may react with the protonated hydroxyl group rather than remain as an easily exchanged anion.

$$\diagdown Al-OH_2^+ + H_2PO_4^- \longrightarrow \diagdown Al-H_2PO_4 + H_2O$$

(soil solid)   (soil solution)   (soil solid)   (soil solution)

(8.12)

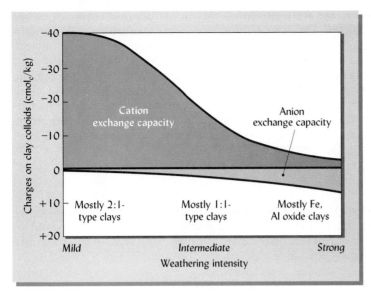

FIGURE 8.28 The effect of weathering intensity on the charges on clay minerals and, in turn, on their cation and anion exchange capacities (CECs and AECs). Note the high CEC and very low AEC associated with mild weathering, which has encouraged the formation of 2:1-type clays such as fine-grained micas, vermiculites, and smectites. More intense weathering destroys the 2:1-type clays and leads to the formation of first kaolinite and then oxides of Fe and Al. These have much lower CECs and considerably higher AECs. Such changes in clay dominance account for the curves shown.

This reaction actually reduces the net positive charge on the soil colloid. Also, the $H_2PO_4^-$ is held very tightly by the soil solids and is not readily available for plant uptake.

Anion adsorption and exchange reactions regulate the mobility and availability of many important ions. Together with cation exchange they largely determine the ability of soils to hold nutrients in a form that is accessible to plants and to retard movement of pollutants in the environment.

Weathering and CEC/AEC Levels

The range of CEC levels of different clay minerals shown in Figure 8.28 shows that clays developed under mild weathering conditions (e.g., smectites and vermiculites) have much higher CEC levels than those developed under more extreme weathering pressures. Also shown are the AEC levels that, in turn, tend to be much higher in clays developed under strong weathering conditions (e.g., kaolinite) than in those found under more mild weathering. This generalized figure is helpful in obtaining a first approximation of CEC and AEC levels in soils of different climatic regions. It must be used with caution, however, since some soils high in 2:1-type clays are found in areas currently undergoing intensive weathering. The nature of the parent material and the time allowed for the weathering to occur also influence the clay types present and the CEC/AEC relations.

8.12 SORPTION OF PESTICIDES AND GROUNDWATER CONTAMINATION

Soil colloids help control the movement of pesticides and other organic compounds into groundwater. The retention of these chemicals by soil colloids can prevent their downward movement through the soil or can delay that movement until the compounds are broken down by soil microbes.

By accepting or releasing protons (H^+ ions), groups such as —OH, —NH$_2$, and —COOH in the chemical structure of some organic compounds provide positive or negative charges that stimulate anion or cation exchange reactions. Other organic compounds participate in inner-sphere complexation and adsorption reactions just as do the inorganic ions we have discussed. However, it is more common for organic colloids to be **absorbed** within the soil organic colloids by a process termed **partitioning**. The soil organic colloids tend to act as a solvent for the applied chemicals, thereby partitioning their concentrations between those held on the soil colloids and those left in the soil solution.

Since we seldom know for certain the exact involvement of the adsorption, complexation, or partitioning processes, we use the general term **sorption** to describe the retention by soils of these organic compounds. Nonionic organic compounds are **hydrophobic**, being repelled by water. As a result, moist clays contribute little to partitioning since their adsorbed water molecules prevent the movement of the nonionic organic chemicals into or around the clay particles. The hydrated metal cations (e.g., Ca^{2+}) that are adsorbed on the surface of smectites can be replaced with large organic cations, giving rise to what are termed **organoclays**. Such clay surfaces are more friendly toward applied organic compounds, making it possible for the clay to participate in partitioning. Environmental scientists use this phenomenon by making smectite organoclays that can remove organic contaminants from wastewaters and from contaminated groundwaters (see Section 18.5).

Distribution Coefficients

The tendency of a pesticide or other organic compound to leach into the groundwater is determined by the solubility of the compound and by the ratio of the amount of chemical sorbed by the soil to that remaining in solution. This ratio is known as the **soil distribution coefficient K_d**.

$$K_d = \frac{\text{mg chemical sorbed/kg soil}}{\text{mg chemical/L solution}} \tag{8.13}$$

The K_d therefore is typically expressed in units of L/kg. Researchers have found that the K_d for a given compound may vary widely depending on the nature of the soil in which the compound is distributed. The variation is related mainly to the amount of organic matter (organic carbon) in the soils. Therefore, most scientists prefer to use a similar ratio that focuses on sorption by organic matter. This ratio is termed the organic carbon distribution coefficient K_{oc}:

$$K_{oc} = \frac{\text{mg chemical sorbed/kg organic carbon}}{\text{mg chemical/L solution}} = \frac{K_d}{\text{g org. C/g soil}} \tag{8.14}$$

The K_{oc} can be calculated by dividing the K_d by the fraction of organic C (g/g) in the soil. This relationship can be seen in Table 8.9, which shows both K_d and K_{oc} for several commonly used herbicides and metabolites. Higher K_d or K_{oc} values indicate the chemical is more tightly sorbed by the soil and therefore less susceptible to leaching and movement to the groundwater. On the other hand, if the management objective is to wash the chemical out of a soil, this will be more easily accomplished for chemicals with lower coefficients. Equations 8.13 and 8.14 emphasize the importance of the sorbing power of the soil colloidal complex, and especially of humus, in the management of organic compounds added to soils.

TABLE 8.9 **Partitioning Coefficients for Soil (K_d) and for Organic Carbon (K_{oc}) for Several Widely Used Herbicides**

Three of the listed compounds are metabolites that form when microorganisms decompose Atrazine. Higher K_d or K_{oc} values indicate stronger attraction to the soil solids and lower susceptibility to leaching loss. The values were measured for a particular soil (an Ultisol in Virginia). Using the relationship between K_d and K_{oc}, it can be ascertained that this soil contained 0.013 g C/g soil (1.3%).

Herbicide	K_d	K_{oc}
Atrazine	1.82	140
Diethyl atrazine	0.99	80
Diisopropyl atrazine	1.66	128
Hydroxy atrazine	7.92	609
Metolachlor	2.47	190

Data from Seybold and Mersie (1996).

The enormous surface area and charged sites on the clay and humus in soils attract and bind many types of organic molecules. These molecules include such biologically active substances as DNA (genetic code material), enzymes, antibiotics, toxins, and even viruses. The initial attraction may be between charged colloidal surfaces and positively or negatively charged functional groups on the biomolecule, similar to the ion exchange reactions just discussed. The data in Figure 8.29 show that adsorption of biomolecules takes place rapidly (in a matter of minutes) and the amount adsorbed is related to the type of clay mineral. The bond between the biomolecule and the colloid is often quite strong so that the biomolecule cannot be easily removed by washing or exchange reactions. In most cases, biomolecules bound to clays do not enter the interlayer spaces, but are attached to the outer planar surfaces and edges of the clay crystals.

The binding of biomolecules to soil colloids in this manner has important environmental implications for two reasons. First, such binding usually protects the biomolecules from enzymatic attack, meaning that the molecules will persist in the soil much longer than studies of unbound biomolecules might suggest. Apparently, interaction with the charged colloid surface changes the three-dimensional shape of the biomolecule and its electron distribution so that enzymes, which would normally cleave (cut) the biomolecule, are unable to recognize and react with their target sites. Second, it has been shown that many biomolecules retain their biological activity in the bound state. Toxins remain toxic to susceptible organisms, enzymes continue to catalyze reactions, viruses can lyse (break open) cells or transfer genetic information to host cells, and DNA strands retain the ability to transform the genetic code of living cells, even while bound to colloidal surfaces and protected from decay.

Figure 8.30 reveals the nature of one type of organoclay complex involving DNA, a long-chain organic biomolecule. Researchers believe the DNA is adsorbed by hydrogen-bonding to negatively charged sites on clay crystal planar surfaces and by electrostatic attraction of negatively charged DNA to positively charged sites on clay crystal edges. Strands of DNA bound to clay or humus may survive intact in soils for long periods because the adsorption on the colloid surfaces hinders microbial enzyme attack. One

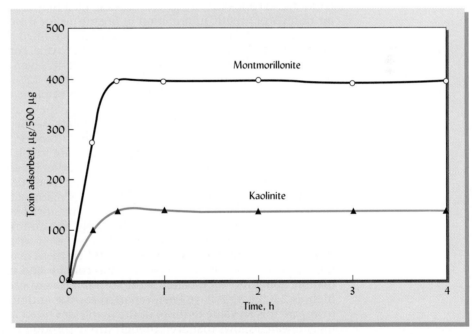

FIGURE 8.29   Adsorption of a toxin called *Bt*, the insecticidal protein produced by the soil bacteria, *Bacillus thuringiensis*, and used to protect some crop plants. Highly active clay minerals such as montmorillonite (a member of the smectite group of 2:1 clays) adsorb and bind much larger amounts of these biomolecules than do low-activity clays such as kaolinite (a 1:1 mineral). In both cases, the adsorption reaction was completed in 30 minutes or less. Since only 500 µg of either clay mineral was used in the experiment, it appears that the clays adsorbed an amount of the toxin equal to 30 to 80% of their mass. [Redrawn from Stotzky (2000)]

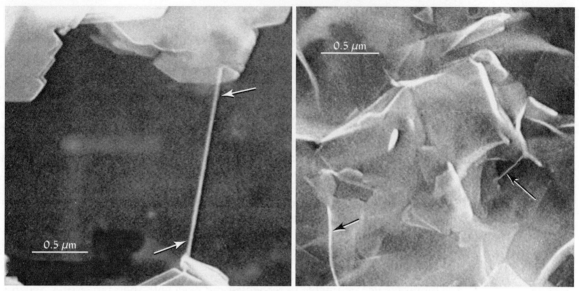

FIGURE 8.30 Scanning electron micrographs (SEMs) of DNA from *Bacillus subtilis* bound on kaolinite clay (*left*) and on montmorillonite clay (*right*). Note the hexagonal crystal shape of the kaolinite and the flake-like shape of the montmorillonite (compare to Figure 8.3a and c). The arrows point to strands of bound DNA. As indicated by the 0.5-μm bars, the clays are magnified about 40,000 times. Research shows that, in soil, DNA bound to clay or humus is protected from decomposition but retains the capability of transferring genetic information to living cells. [Images courtesy of Dr. Guenther Stotzky, Dept. of Biology, New York University]

end of each DNA strand seems to be bound to the colloid, with the other end extended into the soil solution, where it can interact with living cells. Therefore, such bound DNA has the potential to be taken up by microbial cells in the soil, which could then express the genes contained in the DNA. The presence of such DNA is not detectable by the usual chemical tests since it is not in a living cell and therefore not expressing its genes.

When genetically modified organisms (GMOs) are introduced to the soil environment, the cryptic (hidden) genes just described may represent an undetected potential for transfer of genetic information to organisms for which it was not intended. A similar concern exists regarding plants genetically modified to produce such compounds as human drugs ("pharma crops") or insecticidal toxins. For example, millions of hectares are planted each year with corn and cotton plants given a bacterial gene that codes for production of the insecticidal toxin Bt (see Figure 8.29). The Bt toxin is released into the soil by root excretion and decomposition of plant residues containing the toxin. We know little about what effect the toxins may have on soil ecology if it accumulates in soils in the colloid-bound, but still active state (see Sections 11.2 to 11.5).

Antibiotics comprise another class of important organic compounds that sorb to soil colloids. These unique natural chemicals are irreplaceable life-saving compounds. Think of the last time you were given a prescription for an antibiotic drug; had the drug not ended your bacterial infection, the infection may have ended your life! Yet, in most industrial countries, about 80% of antibiotics manufactured are *not* used to cure diseases (in animals or humans) but are used in livestock feed to stimulate faster growth of cattle, swine, and chickens. Not surprisingly, the animal manures produced on most industrial-style farms have been found to be laden with antibiotics that have passed through the animals' digestive systems. When these manures are applied to farmland, the antibiotics become sorbed on the soil colloids and may accumulate with repeated manure applications. Apparently the sorption is very strong. For example, K_d values as high as 2300 L kg^{-1} have been reported for the antibiotic tetracycline in some soils (compare this K_d value to those of the herbicides listed in Table 8.9). However, increasingly research shows that even though strong sorption to soil colloids may reduce their efficacy somewhat, the soil-bound antibiotics still work against bacteria (Figure 8.31). This revelation raises concerns that the huge amounts of antibiotics exposed in the environment will select for resistant strains of "super bacteria" (including human pathogens), which would then no longer be controllable by these (once) life-saving drugs. It is clear that soil colloids and soil science have important roles to play with respect to environmental health.

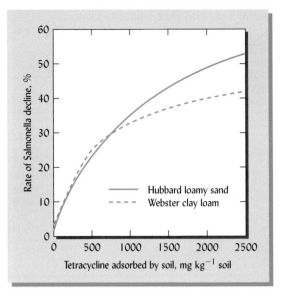

FIGURE 8.31 Antibiotics fed to livestock are known to appear in animal manure. When such manure is applied to farmland, the antibiotics become adsorbed to the soil colloids and therefore may accumulate in soils. The graph shows that the number of *Salmonella* bacteria declined where more of the tetracycline was present, even though it was sorbed by the soil colloids. Adsorption to the Webster clay loam soil was apparently stronger than to the Hubbard loamy sand, resulting in a slightly diminished level of antibiotic activity in the clay loam. The researchers concluded that "even though antibiotics are tightly adsorbed by clay particles, they are still biologically active and may influence the selection of antibiotic resistant bacteria in the terrestrial environment." [Redrawn from Chander et al. (2005)]

8.14 PHYSICAL IMPLICATIONS OF SWELLING-TYPE CLAYS

Engineering Hazards

Soil colloids differ widely in their physical properties, including plasticity, cohesion, swelling, shrinkage, dispersion, and flocculation. These properties greatly influence the usefulness of soils for both engineering and agricultural purposes. As discussed in Sections 4.9 and 8.3, the tendency of certain clays to swell in volume when wetted is a major concern for the construction of roads and foundations. The worst clays in this regard are the smectites, which form wavy stacks or microscopic clay domains containing extremely small ultramicropores. These ultramicropores attract and hold large quantities of water (Figure 8.32), accounting for much of the swelling and plasticity of these clays. The coefficient of linear extensibility [COLE] and the plasticity index (PI) are two measures of soil expansiveness. Both are much higher for smectitic soils than kaolinitic ones (Table 8.10). Relatively pure, mined smectite clay (often sold as

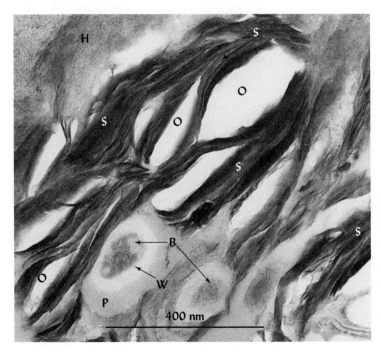

FIGURE 8.32 Transmission electron micrograph showing smectite clay (S) along with humus (H), bacterial cells (B), cell walls (W), and polysaccharides (P). Stacks of parallel crystals of smectite clay (S) form an open wavy clay domain structure (dark) in which ultramicropores (O) are visible as white areas. Water drawn into these ultramicropores in the clay domains accounts for most of the swelling of smectite clay upon wetting. It is less likely, as was once thought, that water causes swelling by entering the interlayers between smectite clay crystal units. Note that the entire image is about 1 micron across. [Image courtesy of M. Thompson, T. Pepper, and A. Carmo, Iowa State University]

TABLE 8.10 Plastic and Liquid Limits of the Clayey B Horizons of Several Soils and Pure Smectite Clay

All soils listed are high in clay, but those dominated by smectite and other high-activity clays tend to have the highest plasticity indices and coefficients of linear extensibility (COLE). Also, note the effect of saturating cation on the liquid limit of pure smectite clay.

Soil	Clay content, %	Clay mineralogy	Plastic limit, %	Liquid limit, %	Plasticity index	COLE[a]
Bashaw (Aquerts)	65	Smectitic	18	71	53	—
Jackland (Udalfs)	68	Smectitic	42	90	48	0.16
Waxpool (Aqualfs)	68	Smectitic	36	76	40	0.18
Kelly (Udalfs)	59	Vermiculitic	12	45	33	0.10
Creedmoor (Udults)	54	Mixed, semiactive	36	67	31	0.09
Cecil (Udults)	68	Kaolinitic	43	61	18	0.03
Davidson (Udults)	68	Kaolinitic	40	56	16	0.04
Slickrock (Udands)	45	Iron oxides	46	59	13	—
Brazil Oxisol (Aquox)	60	Kaolinitic	25	45	20	0.03
Na-saturated smectite	100	—	—	950	—	—
Ca-saturated smectite	100	—	—	360	—	—
Na-saturated kaolinite	100	—	—	36	—	—

[a]Coefficients of linear extensibility, see Section 4.9.

Alfisols and Ultisols data from Thomas et al. (2000); Oxisol data from USDA/NRCS; Udands and Aquerts data from McNabb (1979); pure clay data from Warkentin (1961).

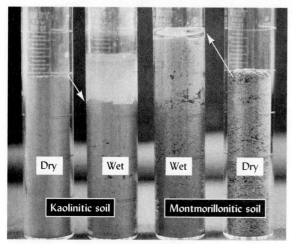

FIGURE 8.33 The different swelling tendencies of two types of clay are illustrated in the lower left. All four cylinders initially contained dry, sieved clay soil, the two on the left from the B horizon of a soil high in kaolinite, the two on the right from one of a soil high in montmorillonite. An equal amount of water was added to the two center cylinders. The kaolinitic soil settled a bit and was not able to absorb all the water. The montmorillonitic soil swelled about 25% in volume and absorbed nearly all the added water. The scenes to the right and above show a practical application of knowledge about these clay properties. Soils containing large quantities of smectite undergo pronounced volume changes as the clay swells and shrinks with wetting and drying. Such soils (e.g., the California Vertisol shown here) make very poor building sites. The normal-appearing homes (*upper*) are actually built on deep, reinforced-concrete pilings (*lower right*) that rest on nonexpansive substrata. Construction of the 15 to 25 such pilings needed for each home more than doubles the cost of construction. (Photos courtesy of R. Weil)

BOX 8.7   ENVIRONMENTAL USES OF SWELLING-TYPE CLAYS[a]

The physical and chemical properties of swelling-type clays make them extremely useful in certain environmental engineering applications. A common use of swelling clays—especially a mined mixture of smectite clays called bentonite—is as a sealant layer placed on the bottom and sides of ponds, waste lagoons, and landfill cells (see Section 18.10). The clay material expands when wetted and forms a highly impermeable barrier to the movement of water as well as organic and inorganic contaminants contained in the water. The contaminants are thus held in the containment structure and prevented from polluting the groundwater.

A more exotic use for swelling clays is proposed in Sweden for the final repository of that country's highly radioactive and toxic nuclear power plant wastes. The plan is to place the wastes in large (about 5 m × 1 m) copper canisters and bury them deep underground in chambers carved from solid rock. As a final defense against leakage of the highly toxic material to the groundwater, the canisters will be surrounded by a thick buffer layer of bentonite clay. The clay is packed dry around the canisters and is expected to absorb water to saturation during the first century of storage, thus gradually swelling into a sticky, malleable mass that will fill any cavities or cracks in the rock. The clay buffer will serve three protective functions: (1) cushion the canister against small (10 cm) movements in the rock formation, (2) form a seal of extremely low permeability to keep corrosive substances in the groundwater away from the canister, and (3) act as a highly efficient electrostatic filter to adsorb and trap cationic radionuclides that might leak from the canister in some far future time.

Figure 8.34 shows how bentonite is used as a plug or sealant to prevent leakage around an environmental groundwater monitoring well. For most of the well depth, the gap between the bore hole wall and the well tube is back-filled with sand to support the tube and allow vertical movement of the groundwater to be sampled. About 30 cm below the soil surface, the space around the well casing is filled instead with air–dry granulated bentonite (white substance being poured from bucket in the photograph). As the bentonite absorbs water, it swells markedly, taking on an almost rubbery consistency and forming an impermeable seal that fits tightly against both the well casing and the soil bore hole wall. This seal prevents contaminants from the soil surface from leaking down the outside of the well casing. In the case of groundwater contaminated with volatile organics like gasoline, the bentonite also prevents vapors from escaping without being properly sampled.

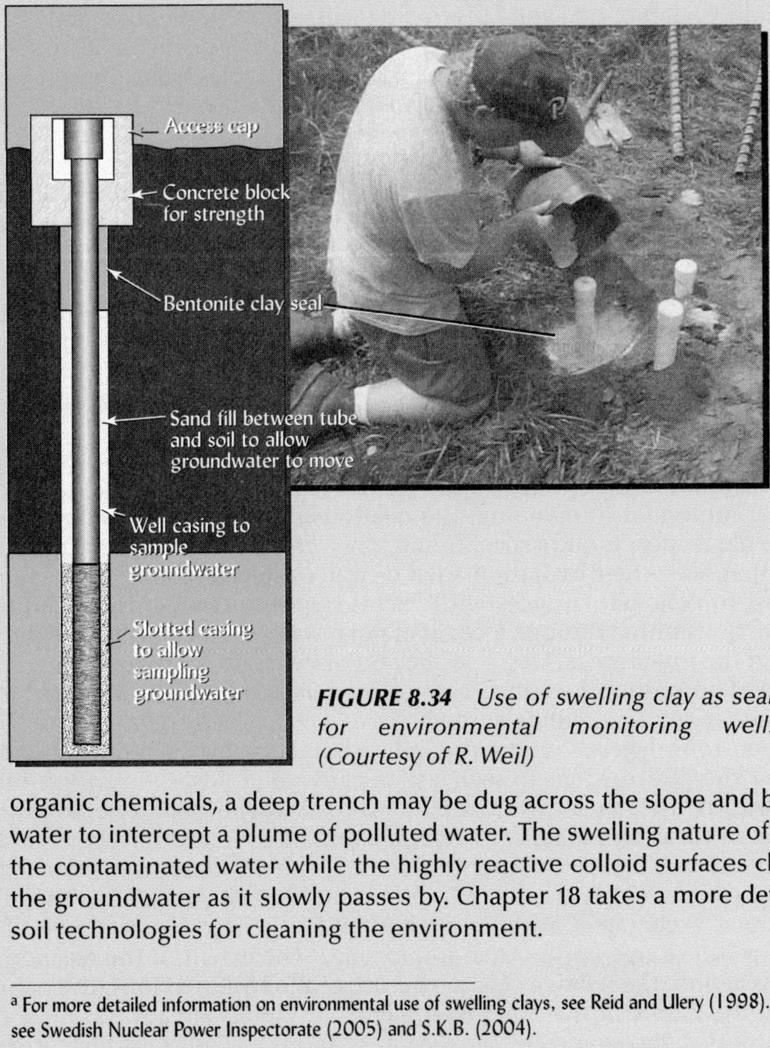

Increasingly, environmental scientists are using swelling-type clays for the removal of organic chemicals from water by partitioning. For example, where there has been a spill of toxic organic chemicals, a deep trench may be dug across the slope and back-filled with a slurry of swelling clay and water to intercept a plume of polluted water. The swelling nature of the smectites prevents the rapid escape of the contaminated water while the highly reactive colloid surfaces chemically sorb the contaminants, purifying the groundwater as it slowly passes by. Chapter 18 takes a more detailed look at such "slurry walls" and other soil technologies for cleaning the environment.

FIGURE 8.34   Use of swelling clay as seal for environmental monitoring well. (Courtesy of R. Weil)

Labels in figure:
- Access cap
- Concrete block for strength
- Bentonite clay seal
- Sand fill between tube and soil to allow groundwater to move
- Well casing to sample groundwater
- Slotted casing to allow sampling groundwater

[a] For more detailed information on environmental use of swelling clays, see Reid and Ulery (1998). For details of the Swedish nuclear repository use of bentonite, see Swedish Nuclear Power Inspectorate (2005) and S.K.B. (2004).

"bentonite"), especially when saturated with Na^+ ions, can have far greater potential for swelling and plasticity than the impure clays in soil.

Figure 8.33 gives an example of special steps needed to safely build homes on soils dominated by smectitic clay. The cost of building homes on smectitic soils may be double that of building on soils dominated by nonswelling clays, for which conventional foundation designs can be safely used. If preventative design measures are not taken during the construction of houses on smectite clays, homeowners will pay dearly in the future. The building foundation is likely to move with the swelling and shrinking of the soil, misaligning doors and windows and eventually cracking foundations, walls, and pipes. The same swelling properties that make smectitic soils so problematic for construction activities make them attractive for certain environmental applications (see Box 8.7) and well-suited for siting ponds and lagoons or creating wetlands. This is just one example of the critical role soil colloids play in determining the usefulness of our soils.

8.15 CONCLUSION

The complex structures, enormous surface area (both internal and external), and tremendous numbers of charges associated with soil colloids combine to make these tiniest of soil particles the seats of chemical and physical activity in soils. The physical activity of the colloids, their adsorption of water, swelling, shrinking, and cohesion are discussed in detail in Chapters 4, 5, and 6. Here we focused on the chemical activity of the colloids, activity that results largely from charged sites on or near colloid surfaces. These charged sites attract oppositely charged ions and molecules from the soil solution. The negative sites attract positive ions (cations) such as Ca^{2+}, Cu^{2+}, K^+, or Al^{3+} and the positive sites attract negative ion (anions) such as Cl^-, SO_4^{2-}, NO_3^-, or HPO_4^{2-}.

Although both positive and negative charges occur on colloids, in most soils the negative charges far outnumber the positive. Most elements dissolved from rocks by weathering or added to soils in lime or fertilizer will eventually end up in the oceans, but it is very fortunate for land plants and animals that attraction to soil colloids greatly slows the journey. Colloidal attraction is a major mechanism by which soils accumulate the stocks of nutrients necessary to support forests, crops, and, ultimately, civilizations. This role is especially critical for forests when nutrient storage in plant biomass is disrupted by fire or timber harvest. The colloidal attraction also enables soils to act as effective filters, sinks, and exchangers, protecting groundwater and food chains from excessive exposure to many pollutants.

When ions are attracted to a colloid, they may enter into two general types of relationships with the colloid surface. If the ion bonds directly to atoms of colloidal structure with no water molecules intervening, the relationship is termed an *inner-sphere complex*. This type of reaction is quite specific and, once created, is not easily reversed. In contrast, ions that keep their hydration shell of water molecules around them are generally attracted to colloidal surfaces with excess opposite charge. However, the attractive forces are transmitted through a chain of polar water molecules and are therefore weakened, and the interaction is less specific and quite easily reversed.

The latter type of adsorption is termed *outer-sphere complexation*. The adsorbed ion and its shell of water molecules oscillate or move about within a zone of attraction. The size of the oscillation zone depends on the strength of attraction between the particular ion and the type of charged site. Ions in such a state of dynamic adsorption are termed *exchangeable ions* because they break away from the colloid whenever another ion from the solution moves in closer and takes over, neutralizing the colloid's charges.

The replacement of one ion for another in the outer-sphere complex is termed *ion exchange*. Except in certain highly weathered, subsurface horizons, cation exchange is far greater than anion exchange. Cation and anion exchange reactions are reversible and balanced charge-for-charge (rather than ion-for-ion). The extent of the reaction is influenced by mass action, the relative charge and size of the hydrated ions, the nature of the colloid, and the nature of the other (complementary) ions already adsorbed on the colloid. Plant roots can exchange H^+ for nutrient cations or OH^- ions for nutrient anions.

The colloids in soils are both organic (humus) and mineral (clays) in nature. In most surface soils, half or more of the charges are contributed by organic matter colloids, while in most subsurface horizons, clays provide the majority of charges. The total number of negative colloid charges per unit mass is termed the *cation exchange capacity* (CEC). The CEC of different colloids varies from about 1 to over 200 $cmol_c/kg$ and that of whole mineral soils commonly varies from about 1 to 40 $cmol_c/kg$. The CEC of a soil, as well as its capacity to strongly adsorb particular ions (such as K^+ or HPO_4^{2-}), depends on the amount of humus in the soil and on both the amount and type of clays present. Low-activity clays (iron and aluminum oxides and 1:1-type silicate clays like kaolinite) tend to dominate highly weathered soils of warm, humid regions. High-activity clays (expanding 2:1 silicates like smectite and vermiculite, and nonexpanding 2:1 silicates like fine-grained mica and chlorite) tend to dominate soils in cooler or drier regions where weathering is less advanced. Most of the charge on humus and low-activity clays is pH-dependent (becomes more negative as pH rises), while most of the charge on high-activity clays is permanent.

The differing ability of soil colloids to adsorb ions and molecules is key to managing soils, both for plant production and to understand how CEC may regulate movement of both nutrients and toxins in the environment. Among the important properties influenced by colloids is the acidity or alkalinity of the soil, the topic of the next chapter.

STUDY QUESTIONS

1. Describe the *soil colloidal complex*, indicate its various components, and explain how it tends to serve as a "bank" for plant nutrients.

2. How do you account for the difference in surface area associated with a grain of kaolinite clay compared to that of montmorillonite, a smectite?

3. Contrast the difference in crystalline structure among *kaolinite, smectites, fine-grained micas, vermiculites,* and *chlorites*.

4. There are two basic processes by which silicate clays are formed by weathering of primary minerals. Which of these would likely be responsible for the formation of (1) fine-grained mica, and (2) kaolinite from muscovite mica? Explain.

5. If you wanted to find a soil high in kaolinite, where would you go? The same for (1) smectite and (2) vermiculite?

6. Which of the silicate clay minerals would be *most* and *least* desired if one were interested in (1) a good foundation for a building, (2) a high cation exchange capacity, (3) an adequate source of potassium, and (4) a soil on which hard clods form after plowing?

7. Which of the following would you expect to be *most* and *least* sticky and plastic when wet: (1) a soil with significant sodium saturation in a semiarid area, (2) a soil high in exchangeable calcium in a subhumid temperate area, or (3) a well-weathered acid soil in the tropics? Explain your answer.

8. A soil contains 4% humus, 10% montmorillonite, 10% vermiculite, and 10% Fe, Al oxides. What is its approximate cation exchange capacity?

9. Calculate the number of grams of Al^{3+} ions needed to replace 10 $cmol_c$ of Ca^{2+} ion from the exchange complex of 1 kg of soil.

10. A soil has been determined to contain the exchangeable cations in these amounts: Ca^{2+} = 9 $cmol_c$, Mg^{2+} = 3 $cmol_c$, K^+ = 1 $cmol_c$, Al^{3+} = 3 $cmol_c$. (a) What is the CEC of this soil? (b) What is the aluminum saturation of this soil?

11. A 100 g sample of a soil has been determined to contain the exchangeable cations in these amounts: Ca^{2+} = 90 mg, Mg^{2+} = 35 mg, K^+ = 28 mg, Al^{3+} = 60 mg. (a) What is the CEC of this soil? (b) What is the aluminum saturation of this soil?

12. A 100 g sample of a soil was shaken with a strong solution of $BaCl_2$ buffered at pH 8.2. The soil suspension was then filtered, the filtrate was discarded, and the soil was thoroughly leached with distilled water to remove any nonexchangeable Ba^{2+}. Then the sample was shaken with a strong solution of $MgCl_2$ and again filtered. The last filtrate was found to contain 10,520 mg of Mg^{2+} and 258 mg of Ba^{2+}. What is the CEC of the soil?

13. Explain the importance of K_d and K_{oc} in assessing the potential pollution of drainage water. Which of these expressions is likely to be most consistently characteristic of the organic compounds in question regardless of the type of soil involved? Explain.

14. An accident at a nuclear power plant has contaminated soil with strontium-90 (Sr^{2+}), a dangerous radionuclide. Health officials order forages growing in the area to be cut, baled, and destroyed. However, there is concern that as the forage plants regrow, they will take up the strontium from the soil and cows eating this contaminated forage will excrete the strontium into their milk. You are the only soil scientist assigned to a risk assessment team consisting mainly of distinguished physicians and statisticians. Write a brief memo to your colleagues explaining how the properties of the soil in the area, especially those related to cation exchange, could affect the risk of contaminating the milk supply.

15. Explain why there is environmental concern about the adsorption by soil colloids of such normally beneficial substances as antibiotic drugs and natural insecticides.

REFERENCES

Bayer, C., L. Martin-Neto, J. Mielniczuk, C. N. Pillon, and L. Sangoi. 2001. "Changes in soil organic matter fractions under subtropical no-till cropping systems," *Soil Sci. Soc. Amer. J.,* **65**:1473–1478.

Bellini, G., M. E. Sumner, D. E. Radcliffe, and N. P. Qafoku. 1996. "Anion transport through columns of highly weathered acid soil: Adsorption and retardation," *Soil Sci. Soc. Amer. J.,* **60**:132–137.

Chander, Y., K. Kumar, S. M. Goyal, and S. C. Gupta. 2005. "Antibacterial activity of soil-bound antibiotics," *J. Environ. Qual.,* **34**:1952–1957.

Dixon, J. B., and S. B. Weed. 1989. *Minerals in Soil Environments,* 2nd ed. (Madison, Wis.: Soil Science Society of America).

Evangelou, V. P., and R. E. Phillips. 2005. "Cation exchange in soils," pp. 343–410, in A. Tabatabai and D. Sparks (eds.), *Chemical Processes in Soils*. SSSA Book Series No. 8. (Madison, Wis: Soil Science Society of America).

Holmgren, G.G.S., M. W. Meyer, R. L. Chaney, and R. B. Daniels. 1993. "Cadmium, lead, zinc, copper, and nickel in agricultural soils in the United States of America," *J. Environ. Qual.,* **22**:335–348.

Krishnaswamy, J., and D. D. Richter. 2002. "Properties of advanced weathering-stage soils in tropical forests and pastures," *Soil Sci. Soc. Amer. J.,* **66**:244–253.

McNabb, D. H. 1979. "Correlation of soil plasticity with amorphous clay constituents," *Soil Sci. Soc. Amer. J.,* **43**:613–616.

Meunier, A. 2005. *Clays.* (Berlin: Springer-Verlag). 470 pp.

Morra, M. J., M. H. Chaverra, L. M. Dandurand, and C. S. Orser. 1998. "Survival of *Pseudomonas flourescens* 2-79RN$_{10}$ in clay powders undergoing drying," *Soil Sci. Soc. Amer. J.,* **62**:663–670.

Reid, D. A., and A. L. Ulery. 1998. "Environmental applications of smectites," in J. Dixon, D. Schultze, W. Bleam, and J. Amonette (eds.), *Environmental Soil Mineralogy* (Madison, Wis.: Soil Science Society of America).

Schulten, H. R., and M. Schnitzer. 1993. "A state of the art structural concept for humic substances," *Naturwissenschaften,* **80**:29–30.

Seabrook, Charles. 1995. *Red Clay, Pink Cadillacs and White Gold: The Kaolin Chalk War* (Atlanta: Longstreet Press).

Seybold, C. A., and W. Mersie. 1996. "Adsorption and desorption of atrazine, deethylatrazine, deisopropylatrazine, hydroxyatrazine, and metolachlor in two soils in Virginia," *J. Environ. Qual.,* **25**:1179–1185.

Shamshuddin, J., and H. Ismail. 1995. "Reactions of ground magnesium limestone and gypsum in soils with variable-charged minerals," *Soil Sci. Soc. Amer. J.,* **59**:106–112.

S.K.B. 2004. Final repository—properties of the buffer. Svensk Kärnbränslehantering AB (Swedish Nuclear Fuel and Waste Management Company) http://www.skb.se/templates/SKBPage_8776.aspx (posted 27 April 2004; verified 24 April 2006).

Stotzky, G. 2000. "Persistence and biological activity in soil of insecticidal proteins from *Bacillus thuringiensis* and of bacterial DNA bound on clays and humic acids," *J. Environ. Quality,* **29:**691–705.

Sullivan, T. J., I. J. Fernandez, A. T. Herlihy, C. T. Driscoll, T. C. Mcdonnell, N. A. Nowicki, K. U. Snyder, and J. W. Sutherland. 2006. "Acid-base characteristics of soils in the Adirondack Mountains, New York," *Soil Sci. Soc. Amer. J.,* **70:**141–152.

Swedish Nuclear Power Inspectorate. 2005. Engineered barrier system-long-term stability of buffer and backfill. Report from a workshop in Lund, Sweden, November 15–17, 2004, synthesis and extended abstracts Swedish Nuclear Power Inspectorate. *SKI Research Report,* 48, 120.

Thomas, P. J., J. C. Baker, and L. W. Zelazny. 2000. "An expansive soil index for predicting shrink-swell potential," *Soil Sci. Soc. Amer. J.,* **64:**268–274.

Tice, K. R., R. C. Graham, and H. B. Wood. 1996. "Transformations of 2:1 phyllosilicates in 41-year-old soils under oak and pine," *Geoderma,* **70:**49–62.

Waegeneers, N., E. Smolders, and R. Merckx. 1999. "A statistical approach for estimating radiocesium interception potential of soils," *J. Environ. Quality,* **28:**1005–1011.

Warkentin, B. P. 1961. "Interpretation of the upper plastic limits of clays," *Nature,* **190:**287–288.

Weil, R. R. 2009. *Laboratory Manual for Introductory Soils,* 8th ed. (Dubuque, Ia.: Kendall/Hunt).

An acid soil community. (R. Weil)

9

SOIL ACIDITY

What have they done to the rain?
—*SONG LYRICS BY*
MALVINA REYNOLDS

The degree of soil acidity or alkalinity, expressed as soil pH, is a *master variable* that affects a wide range of soil chemical and biological properties. This chemical variable greatly influences the root uptake availability of many elements, including both nutrients and toxins. The activity of soil microorganisms is also affected. The mix of plant species that dominate a landscape under natural conditions often reflects the pH of the soil. For people attempting to produce crops or ornamental plants, soil pH is a major determinant of which species will grow well or even grow at all in a given site.

Soil pH affects the *mobility* of many pollutants in soil by influencing the rate of their biochemical breakdown, their solubility, and their adsorption to colloids. Thus, soil pH is a critical factor in predicting the likelihood that a given pollutant will contaminate groundwater, surface water, and food chains. Furthermore, there are certain situations in which so much acidity is generated that the acid itself becomes a significant environmental pollutant. For example, soils on certain types of disturbed land generate extremely acid drainage water that can cause massive fish kills when it reaches a lake or stream.

Many complex factors affect soil pH, but none more than two simple balances: the balance between acid and nonacid cations on colloid surfaces and the balance between H^+ and OH^- ions in the soil solution. These balances, in turn, are largely controlled by the nature of the soil colloids. Therefore, to understand and manage soil acidity, it is essential to have a good grasp of the concepts of charged colloidal surfaces and cation exchange as presented in Chapter 8. Also inextricably tied to soil acidity is the toxicity of the nonnutrient element, **aluminum**.

Acidification is a natural process involved in soil formation. It reaches its greatest expression in humid regions where rainfall is sufficient to thoroughly leach the soil profile. Not surprisingly, acidity is one of the main constraints to crop production in these areas. Forests, which comprise the natural vegetation in most humid regions, are generally better adapted to acid soils than are most crops. Nonetheless, forest health has declined where soils are incapable of resisting the effects of human-induced acidification.

By contrast, in drier regions leaching is much less extensive, allowing soils to retain enough Ca^{2+}, Mg^{2+}, K^+, and Na^+ to prevent a buildup of acid cations. Soils in semiarid and arid regions, therefore, tend to have **alkaline** pH levels (that is, pH > 7), sometimes accompanied by high levels of soluble salts and exchangeable sodium, two related soil conditions that will be discussed in Chapter 10. In this chapter we will focus on acid soils, those with pH levels well below 7.

9.1 THE PROCESS OF SOIL ACIDIFICATION

Acidity and alkalinity are usually quantified using the pH scale, which expresses the activity or concentration of H^+ ions present in a solution. Therefore, before we begin our discussion of the chemical nature and causes of acidity in soils, it will be useful to review the fundamentals about pH in Box 9.1. It will also be of value to develop a feel for the range of pH values found in soils and to compare these to the pH values of other common substances (see Figure 9.2).

In order to understand soil acidification and its management, we need to know where the H^+ ions come from, how they enter the soil system, and how they may be lost. We will begin by describing a variety of processes that add H^+ ions to the soil system and therefore contribute to acidification. Although most of these acid-producing processes occur naturally, human activities have a major impact on some of them, as we shall see in Section 9.6.

Sources of Hydrogen Ions

CARBONIC AND OTHER ORGANIC ACIDS.   Perhaps the most ubiquitous contributor to soil acidity is the formation and subsequent dissociation of H^+ ions from **carbonic acid**. This weak acid is formed when carbon dioxide gas from soil air dissolves in water. Root respiration and the decomposition of soil organic matter by microorganisms produce high levels of CO_2 in soil air, pushing the following reaction to the right.

$$CO_2 + H_2O \longrightarrow H_2CO_3 \rightleftharpoons HCO_3^- + H^+ \quad pK_a = 6.35 \tag{9.1}$$

Because H_2CO_3 is a weak acid (has a relatively high pK_a),[1] its contribution of H^+ ions is negligible when the pH is much below 5.0.

Many other organic acids are also generated as microbes break down soil organic matter. Some of these are low molecular weight organic acids, such as citric or malic acids, that only weakly dissociate. Others are more complex and stronger acids, such as the carboxylic and phenolic acid groups in humic substances produced by litter breakdown. A generalized reaction showing a carboxylic group is given:

$$\underset{\substack{\text{Organic matter} \\ \text{(generalized)}}}{[RCH_2OH...]} + O_2 + H_2O \rightleftharpoons \underset{\substack{\text{Strong organic} \\ \text{acids}}}{RCOOH} \rightleftharpoons RCOO^- + H^+ \quad pK_a = 3 \text{ to } 5 \tag{9.2}$$

The pK_a of carboxylic groups ranges from 3 to 5, depending on the nature of the associated organic structures (R). Other organic groups, such as phenolic (R-OH), dissociate at higher pH levels.

ACCUMULATION OF ORGANIC MATTER.   The accumulation of organic matter tends to acidify the soil for two reasons. First, organic matter forms soluble complexes with nonacid nutrient cations such as Ca^{2+} and Mg^{2+}, thus facilitating the loss of these cations by leaching. Second, organic matter is a source of H^+ ions because it contains numerous acid **functional groups** from which these ions can dissociate. Different groups dissociate at different pH levels. If pH is increased, more functional groups undergo dissociation

---

[1] The pK_a is the negative logarithm (analogous to the pH) of the equilibrium constant for a dissociation reaction—for example, $[H^+] \times [HCO_3^-]/[H_2CO_3] = K_a$. It indicates the pH at which one-half of the acid will be dissociated. Dissociation of the acid can usually be considered negligible when the pH is more than 1.0 to 2.0 units lower than the pK_a.

BOX 9.1   SOIL pH, SOIL ACIDITY, AND ALKALINITY

Whether a soil is acid, neutral, or alkaline is determined by the comparative concentrations of H^+ and OH^- ions.[a] Pure water provides these ions in equal concentrations:

$$H_2O \rightleftharpoons H^+ + OH^-$$

The equilibrium for this reaction is far to the left; only about 1 out of every 10 million water molecules is dissociated into H^+ and OH^- ions. The ion product of the concentrations of the H^+ and OH^- ions is a constant (K_w), which at 25 °C is known to be 1×10^{-14}:

$$[H^+] \times [OH^-] = K_w = 10^{-14}$$

Since in pure water the concentration of H ions $[H^+]$ must be equal to that of OH^- ions $[OH^-]$, this equation shows that the concentration of each is 10^{-7} ($10^{-7} \times 10^{-7} = 10^{-14}$). It also shows the inverse relationship between the concentrations of these two

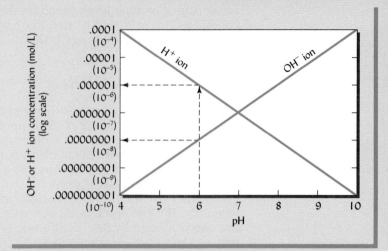

FIGURE 9.1   *The relationship between pH, pOH, and the concentrations of hydrogen and hydroxyl ions in water solution.*

ions (Figure 9.1). As one increases, the other must decrease proportionately. Thus, if we were to increase the H^+ ion concentration $[H^+]$ by 10 times (from 10^{-7} to 10^{-6}), the $[OH^-]$ would be decreased by 10 times (from 10^{-7} to 10^{-8}) since the product of these two concentrations must equal 10^{-14}:

$$10^{-6} \times 10^{-8} = 10^{-14}$$

Scientists have simplified the means of expressing the very small concentrations of H^+ and OH^- ions by using the *negative logarithm of the H^+ ion concentration,* termed the *pH*. Thus, if the H^+ concentration in an acid medium is 10^{-5}, the pH is 5; if it is 10^{-9} in an alkaline medium, the pH is 9.

Note that the pH also gives us an indirect measure of the OH^- ion concentration since the product of $[H^+] \times [OH^-]$ must always equal 10^{-14}. Thus at pH 5 the $[OH^-]$ is 10^{-9} ($10^{-5} \times 10^{-9} = 10^{-14}$); at pH 8 it is 10^{-6} ($10^{-8} \times 10^{-6} = 10^{-14}$).

Figure 9.1 above shows the relationship between pH and the concentrations of H^+ and OH^- ions. Note that as one goes down, the other goes up, and vice versa. The dotted line illustrates the concentrations of these two ions at pH 6.0: $[H^+] = 10^{-6}$; $[OH^-] = 10^{-8}$. This reciprocal relationship between H^+ and OH^- ions should always be kept in mind in studying soil acidity and alkalinity.

---

[a] Technically speaking, chemical reactions are influenced by the activity of an ion rather than by its concentration. This is because of the electrostatic effect of one ion on the activity of its nearby neighbors. Ionic activities are thus essentially effective concentrations. Since the difference between ionic activities and concentrations in soils are not so great, we will use the term *concentration* in this text.

of H^+ ions, leaving behind an increasing number of negatively charged sites on the molecule. It is these *pH-dependent* negative charges that give rise to the large cation exchange capacity for which humus is well known (see Sections 8.4 and 8.6).

OXIDATION OF NITROGEN (NITRIFICATION). Oxidation reactions generally produce H^+ ions as one of their products. Reduction reactions, on the other hand, tend to consume H^+ ions and raise soil pH. Ammonium ions (NH_4^+) from organic matter or from most fertilizers are subject to microbial oxidation that converts the N to nitrate ions (NO_3^-). The reaction with oxygen, termed nitrification, releases two H^+ ions for each NH_4^+ ion oxidized. Because the NO_3^- produced is the anion of a **strong acid** (nitric acid, HNO_3), it does not tend to recombine with the H^+ ion to make the reaction go to the left.

$$NH_4^+ + 2O_2 \rightleftharpoons H_2O + H^+ + \underbrace{H^+ + NO_3^-}_{\substack{\text{Dissociated} \\ \text{nitric acid}}} \tag{9.3}$$

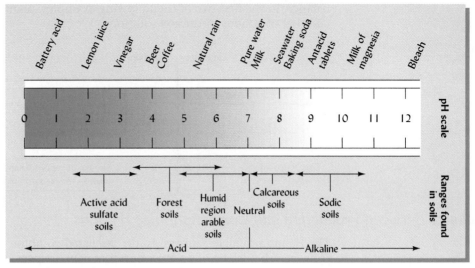

FIGURE 9.2 Some pH values for familiar substances (*above*) compared to ranges of pH typical for various types of soils (*below*).

OXIDATION OF SULFUR. The decomposition of plant residues commonly involves the oxidation of organic −SH groups to yield sulfuric acid (H_2SO_4). Another important source of this strong acid is the oxidation of reduced sulfur in minerals such as pyrite. This and related reactions are responsible for producing large amounts of acidity in certain soils in which reduced sulfur is plentiful and oxygen levels are increased by drainage or excavation (see Section 9.6).

$$FeS_2 + 3\frac{1}{2}O_2 + H_2O \rightleftharpoons FeSO_4 + 2H^+ + SO_4^{2-}$$

Pyrite — Ferrous sulfate — Dissociated sulfuric acid

(9.4)

ACIDS IN PRECIPITATION. Precipitation (rain, snow, fog, and dust) contains a variety of acids that contribute H^+ ions to the soil receiving the precipitation. As raindrops fall through unpolluted air, they dissolve CO_2 and form enough carbonic acid to lower the pH of the water from 7.0 (the pH of pure water) to about 5.6. Varying amounts of sulfuric and nitric acids form in precipitation from certain sulfur and nitrogen gases produced by lightning, volcanic eruptions, forest fires, and the combustion of fossil fuels. Unlike carbonic acid, these strong acids completely dissociate to form H^+ ions and sulfate or nitrate anions:

$$H_2SO_4 \rightleftharpoons SO_4^{2-} + 2H^+$$

(9.5)

$$HNO_3 \rightleftharpoons NO_3^- + H^+$$

(9.6)

In recent decades, combustion of coal and petroleum products has significantly increased the amounts of these strong acids found in precipitation (see Section 9.6).

PLANT UPTAKE OF CATIONS. Plants must maintain a balance between the positive and negative charges on the ions they take up from the soil solution. For every positive charge taken in on a cation, a root can maintain charge balance either by taking up a negative charge as an anion or by exuding a positive charge as a different cation. When they take up far more of certain cations (e.g., K^+, NH_4^+, and Ca^{2+}) than they do of anions (e.g., NO_3^-, SO_4^{2-}), plants usually exude H^+ ions into the soil solution to maintain charge balance. In the first two of the following examples, plant nutrient uptake results in the addition of H+ ions to the soil solution:

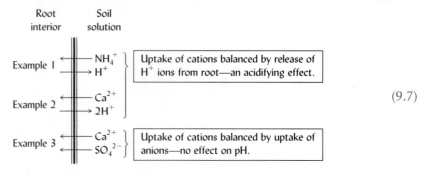

(9.7)

Balance Between Production and Consumption of H+ Ions

The H+ ion-producing processes just discussed combine to stimulate soil acidification. However, the degree of acidification that actually occurs in a given soil is determined by the balance between those processes that produce H+ ions and other processes that *consume* them (Table 9.1). An example of the latter is the case in which plant uptake of an anion such as NO_3^- exceeds the uptake of associated cations. In this case, the root exudes the anion, bicarbonate (HCO_3^-), to maintain charge balance. The increased concentration of bicarbonate ions tends to reverse the dissociation of carbonic acid (see Equation 9.1), thereby *consuming* H+ ions and raising the pH of the soil solution.

(9.8)

Another H+ ion-consuming process involving nitrogen is the reduction of nitrate to nitrogen gases under anaerobic conditions (see denitrification, Sections 7.3 and 13.9).

The weathering of nonacid cations from minerals (Section 2.1) is a slow but very important H+ ion-consuming process that counteracts acidification. An example is the weathering of calcium from a silicate mineral:

$$\text{Ca-silicate} + 2H^+ \longrightarrow H_4SiO_4 + Ca^{2+}$$

(9.9)

Some of the nonacid cations (Ca^{2+}, Mg^{2+}, K^+, and Na^+) so released become exchangeable cations on the soil colloids. Hydrogen ions added to the soil solution from acids in rain (and other sources just discussed) may replace these cations on the cation exchange sites of humus and clay. The displaced nonacid cations are then subject to loss by leaching along with the anions of the added acids (Figure 9.3). The soil

TABLE 9.1 The Main Processes that Produce or Consume Hydrogen Ions (H+) in Soil Systems

Production of H+ ions increases soil acidity, while consumption of H+ ions delays acidification and leads to alkalinity. The pH level of a soil reflects the long-term balance between these two types of processes.

Acidifying (H+ ion-producing) processes	Alkalinizing (H+ ion-consuming) processes
Formation of carbonic acid from CO_2	Input of bicarbonates or carbonates

Acid dissociation such as:
  $RCOOH \rightarrow RCOO^- + H^+$
	Anion protonation such as:
	  $RCOO^- + H^+ \rightarrow RCOOH$

Oxidation of N, S, and Fe compounds	Reduction of N, S, and Fe compounds
Atmospheric H_2SO_4 and HNO_3 deposition	Atmospheric Ca, Mg deposition
Cation uptake by plants	Anion uptake by plants

Accumulation of acidic organic matter
  (e.g., fulvic acids)
	Specific (inner sphere) adsorption of anions
	  (especially SO_4^{2-})

Cation precipitation such as:
  $Al^{3+} + 3H_2O \rightarrow 3H^+ + Al(OH)_3^0$
  $SiO_2 + 2Al(OH)_3 + Ca^{2+} \rightarrow CaAl_2SiO_6 + 2H_2O + 2H^+$
	Cation weathering from minerals such as:
	  $3H^+ + Al(OH)_3^0 \rightarrow Al^{3+} + 3H_2O$
	  $CaAl_2SiO_6 + 2H_2O + 2H^+ \rightarrow SiO_2 + 2Al(OH)_3 + Ca^{2+}$

Deprotonation of pH-dependent charges	Protonation of pH-dependent charges

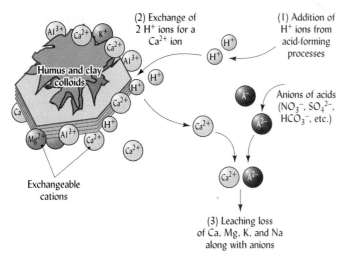

(2) Exchange of
2 H⁺ ions for a
Ca²⁺ ion

(1) Addition of
H⁺ ions from
acid-forming
processes

Humus and clay
colloids

Exchangeable
cations

Anions of acids
(NO_3^-, SO_4^{2-},
HCO_3^-, etc.)

(3) Leaching loss
of Ca, Mg, K, and Na
along with anions

FIGURE 9.3 Soils become acid when H^+ ions added to the soil solution exchange with nonacid Ca^{2+}, Mg^{2+}, K^+, and Na^+ ions held on humus and clay colloids. The nonacid cations can then be exported in leaching water along with accompanying anions. As a result, the exchange complex (and therefore also the soil solution) becomes increasingly dominated by acid cations (H^+ and Al^{3+}). Because of this sequence of events, H^+ ion-producing processes acidify soils in humid regions where leaching is extensive, but cause little long-term soil acidification in arid regions where the Ca^{2+}, Mg^{2+}, K^+, and Na^+ are mostly *not* removed by leaching. In the latter case, the Ca^{2+}, Mg^{2+}, K^+, and Na^+ remain in the soil and reexchange with the acid cations, preventing a drop in pH level. (Diagram courtesy of R. Weil)

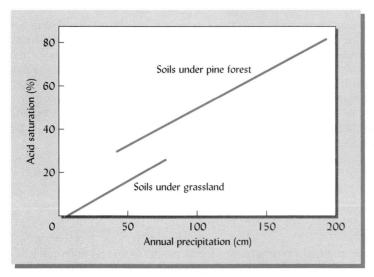

Soils under pine forest

Soils under grassland

FIGURE 9.4 Effect of annual precipitation on the percent of acid saturation of untilled California soils under grassland and pine forest vegetation. Note that the degree of acidity goes up as the annual precipitation increases. Also note that the forest produced a higher degree of acidity than did the grassland at the same annual precipitation. [From Jenny et al. (1968)]

slowly becomes more acid if the leaching of Ca^{2+}, Mg^{2+}, K^+, and Na^+ proceeds faster than the release of these cations from weathering minerals. Thus, the formation of acid soils is favored by high rainfall, parent materials low in Ca, Mg, K, and Na, and a high degree of biological activity (favoring formation of H_2CO_3). The positive relationship between annual rainfall and soil acidity is well documented (Figure 9.4).

9.2 ROLE OF ALUMINUM IN SOIL ACIDITY

Although low pH is defined as a high concentration of H^+ ions, **aluminum** also plays a central role in soil acidity. Aluminum is a major constituent of most soil minerals (aluminosilicates and aluminum oxides), including clays. When H^+ ions are adsorbed on a clay surface, they usually do not remain as exchangeable cations for long, but instead they attack the structure of the minerals, releasing Al^{3+} ions in the process. The Al^{3+} ions then become adsorbed on the colloid's cation-exchange sites. These exchangeable Al^{3+} ions, in turn, are in equilibrium with dissolved Al^{3+} in the soil solution:

$$
\begin{array}{cc}
\text{Structural} & \text{Exchangeable} \\
\text{Al atom} & H^+ \text{ ions}
\end{array}
\quad
\begin{array}{c}
\text{Added} \\
H^+ \text{ ions}
\end{array}
\qquad\qquad
\begin{array}{cc}
\text{Exchangeable} & Al^{3+} \text{ ion in} \\
Al^{3+} \text{ ion} & \text{soultion}
\end{array}
$$

$$
\begin{array}{c}
-O \\
-O-Al-O^-H^+ \\
-O \quad O^-H^+ \quad O^-H^+
\end{array}
+ \; 3H^+ \;
\xrightarrow{3H_2O}
\begin{array}{c}
-O^- \\
-O^- \\
-O^-
\end{array}
Al^{3+} \; \rightleftharpoons \; Al^{3+} \qquad (9.10)
$$

Edge of mineral → | Edge of mineral → |

The exchangeable and soluble Al^{3+} ions play two critical roles in the soil acidity story. First, aluminum is *highly toxic* to most organisms and is responsible for much of the deleterious impact of soil acidity on plants and aquatic animals. We will discuss this role in Section 9.7.

Second, Al^{3+} ions have a strong tendency to hydrolyze, splitting water molecules into H^+ and OH^- ions.[2] The aluminum combines with the OH^- ions, leaving the H^+ to lower the pH of the soil solution. For this reason, Al^{3+} and H^+ together are considered **acid cations**. A single Al^{3+} ion can thus release up to three H^+ ions as the following reversible reaction series proceeds to the right in stepwise fashion:

$$Al^{3+} \rightleftharpoons AlOH^{2+} \rightleftharpoons Al(OH)_2^+ \rightleftharpoons Al(OH)_3^0$$
Gibbsite or amorphous (solid)

$$pK_a = 5.0 \qquad pK_a = 5.1 \qquad pK_a = 6.7$$

(9.11)

Most of the hydroxy aluminum ions $[Al(OH)_x^{y+}]$ formed as the pH increases are strongly adsorbed to clay surfaces or complexed with organic matter. Often the hydroxy aluminum ions join together, forming large polymers with many positive charges. When tightly bound to the colloid's negative charge sites, these polymers are not exchangeable and so mask much of the colloid's potential cation exchange capacity. As the pH is raised and more of the hydroxyl aluminum ions precipitate as uncharged $Al(OH)_3^0$, the negative sites on the colloids become available for cation exchange. This is one reason for the increase in soil cation exchange capacity (CEC) as the pH is raised from 4.5 to 7.0 (at which pH virtually all the aluminum cations have precipitated as $Al(OH)_3^0$).

9.3 POOLS OF SOIL ACIDITY

Principal Pools of Soil Acidity

Fundamentals of acid-base chemistry:
http://www.shodor.org/unchem/basic/ab/index.html

Research suggests that three major pools of acidity are common in soils: (1) **active acidity** due to the H^+ ions in the soil solution; (2) **salt-replaceable (exchangeable) acidity**, involving the aluminum and hydrogen that are *easily exchangeable* by other cations in a simple unbuffered salt solution, such as KCl; and (3) **residual acidity**, which can be neutralized by limestone or other alkaline materials but cannot be detected by the salt-replaceable technique. These types of acidity all add up to the **total acidity** of a soil. In addition, a much less common, but sometimes very important fourth pool, namely **potential acidity**, can arise upon the oxidation of sulfur compounds in certain acid sulfate soils (see Section 9.6).

ACTIVE ACIDITY. The active acidity pool is defined by the H^+ ion activity in the soil solution. This pool is very small compared to the acidity in the exchangeable and residual pools. For example, only about 2 kg of calcium carbonate are needed to neutralize the active acidity in the upper 15 cm of a hectare of an average mineral soil at pH 4 and 20% moisture. Even so, the active acidity is extremely important, since it determines the solubility of many substances and provides the soil solution environment to which plant roots and microbes are exposed. Very acid soils contain aluminum ions in solution, which can add to the active acidity as they hydrolyze.

EXCHANGEABLE (SALT-REPLACEABLE) ACIDITY. Salt-replaceable acidity is primarily associated with exchangeable aluminum and hydrogen ions that are present in large quantities in very acid soils (see Figure 9.5). These ions can be released into the soil solution by cation exchange with an unbuffered salt, such as KCl.

---

[2] Iron (Fe^{3+}) has properties similar to those of aluminum, undergoes similar reactions, and can also produce H^+ ions by hydrolysis. However, because Fe^{3+} hydrolyzes at pH levels (<3.0) that are below those commonly encountered in soils, we will confine our discussion to aluminum.

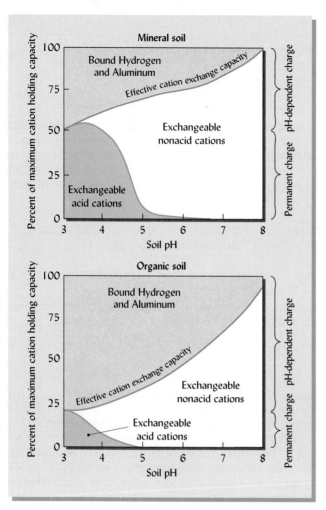

FIGURE 9.5 General relationship between soil pH and cations held in exchangeable form or tightly bound to colloids in two representative soils. Note that any particular soil would give somewhat different distributions. (*Upper*) A mineral soil with mixed mineralogy and a moderate organic matter level exhibits a moderate decrease in effective cation exchange capacity as pH is lowered, suggesting that *pH-dependent charges* and *permanent charges* (see Section 8.6 for explanation of these terms) each account for about half of the maximum CEC. At pH values above 5.5, the concentrations of exchangeable aluminum and H^+ ions are too low to show in the diagram, and the effective CEC is essentially 100% saturated with exchangeable nonacid cations (Ca^{2+}, Mg^{2+}, K^+, and Na^+, the so-called base cations). As pH drops from 7.0 to about 5.5, the effective CEC is reduced because H^+ ions and $Al(OH)_x^{y+}$ ions (which may include $AlOH^{2+}$, $Al(OH)_2^+$, etc.) are tightly bound to some of the pH-dependent charge sites. As pH is further reduced from 5.5 to 4.0, aluminum ions (especially Al^{3+}), along with some H^+ ions, occupy an increasing portion of the remaining exchange sites. Exchangeable H^+ ions occupy a major portion of the exchange complex only at pH levels below 4.0. (*Lower*) The CEC of an organic soil is dominated by pH-dependent (variable) charges with only a small amount of permanent charge. Therefore, as pH is lowered, the effective CEC of the organic soil declines more dramatically than the effective CEC of the mineral soil. At low pH levels, exchangeable H^+ ions are more prominent and Al^{3+} less prominent on the organic soil than on the mineral soil. (Diagram courtesy of R. Weil)

$$\text{Clay}\begin{vmatrix} - & Al^{3+} \\ - & \\ - & H^+ \end{vmatrix} + \underset{\text{(soil solution)}}{4KCl} \rightleftharpoons \text{Clay}\begin{vmatrix} - & K^+ \\ - & K^+ \\ - & K^+ \\ - & K^+ \end{vmatrix} + \underset{\text{(soil solution)}}{AlCl_3 + HCl} \qquad (9.12)$$

(soil solid)   (soil solid)

Once released to the soil solution, the aluminum hydrolyzes to form additional H^+, as explained in Section 9.2. The chemical equivalent of salt-replaceable acidity in strongly acid soils is commonly thousands of times that of active acidity in the soil solution. Even in moderately acid soils, the limestone needed to neutralize this type of acidity is commonly more than 100 times that needed to neutralize the soil solution (active acidity). At a given pH value, exchangeable acidity is generally highest for smectites, intermediate for vermiculites, and lowest for kaolinite.

RESIDUAL ACIDITY. Together, exchangeable (salt-replaceable) and active acidity account for only a fraction of the total soil acidity. The remaining **residual acidity** is generally associated with hydrogen and aluminum ions (including the aluminum hydroxy ions) that are bound in nonexchangeable forms by organic matter and clays (see Figure 9.5). As the pH increases, the bound hydrogen dissociates and the bound aluminum ions are released and precipitate as amorphous $Al(OH)_3^0$. These changes free up negative cation exchange sites and increase the cation exchange capacity. The reaction with a liming material [e.g., $Ca(OH)_2$] shows how the bound hydrogen and aluminum can be released.

$$\text{Clay}\begin{array}{c}\text{Al}\\ \\ \text{H}\end{array}\ \underset{\text{(soil solid)}}{} + \underset{\text{(soil solution)}}{2Ca(OH)_2} \rightleftharpoons \text{Clay}\begin{array}{c}^-\ Ca^{2+}\\ ^-\\ ^-\ Ca^{2+}\end{array}\underset{\text{(soil solid)}}{} + Al(OH)_3 + H_2O \qquad (9.13)$$

The residual acidity is commonly far greater than either the active or salt-replaceable acidity. It may be 1000 times greater than the soil solution or active acidity in a sandy soil and 50,000 or even 100,000 times greater in a clayey soil high in organic matter. The amount of ground limestone recommended to at least partly neutralize residual acidity in the upper 15 cm of soil is commonly 5 to 10 metric tons (Mg) per hectare (2.25 to 4.5 tons per acre).

TOTAL ACIDITY. For most soils (not potential acid-sulfate soils) the total acidity that must be overcome to raise the soil pH to a desired value can be defined as:

$$\text{Total acidity} = \text{active acidity} + \text{salt-replaceable acidity} + \text{residual acidity} \qquad (9.14)$$

We can conclude that the pH of the soil solution is only the tip of the iceberg in determining how much lime may be needed to overcome the ill effects of soil acidity.

Soil pH and Cation Associations

EXCHANGEABLE AND BOUND CATIONS. Figure 9.5 illustrates the relationship between soil pH and the prevalence of various exchangeable and tightly bound cations in a mineral and an organic soil. Two forms of hydrogen and aluminum are shown in Figure 9.5: (1) that tightly held by the pH-dependent sites (*bound*), and (2) that associated with negative charges on the colloids (*exchangeable*). The bound forms contribute to the residual acidity pool, but only the exchangeable ions have an immediate effect on soil pH. As we shall see later, both forms are very much involved in determining how much lime or sulfur is needed to change soil pH (see Section 9.8).

EFFECTIVE CEC AND pH. Note that in both soils illustrated in Figure 9.5, the effective CEC increases as the pH level rises. This change in effective CEC results mainly from two factors: (1) the binding and release of H^+ ions on pH-dependent charge sites (as explained in Section 8.6), and (2) the hydrolysis reactions of aluminum species (as explained in Section 9.2). The change in effective CEC will be most dramatic for organic soils (Figure 9.5, *lower*) and highly weathered mineral soils dominated by iron and aluminum oxide clays. However, effective CEC changes with pH even in surface soils dominated by 2:1 clays, which carry mainly permanent charges. This is because a substantial amount of variable charge is usually supplied by the organic matter and the weathered edges of clay minerals.

Cation Saturation Percentages

The proportion of the CEC occupied by a given ion is termed its **saturation percentage**. Consider a soil with a CEC of 20 $cmol_c/kg$ holding these amounts of exchangeable cations (in $cmol_c/kg$): 10 of Ca^{2+}, 3 of Mg^{2+}, 1 of K^+, 1 of Na^+, 1 of H^+, and 4 of Al^{3+}. This soil with 10 $cmol_c\ Ca^{2+}/kg$ and a CEC of 20 $cmol_c/kg$ is said to be 50% calcium saturated. Likewise, the aluminum saturation of this soil is 20% (4/20 = 0.20 or 20%). Together, the 4 $cmol_c/kg$ of exchangeable Al^{3+} and 1 $cmol_c/kg$ of exchangeable H^+ ions give this soil an **acid saturation** of 25% [(4 + 1)/20 = 0.25]. Similarly, the term **nonacid saturation** can be used to refer to the proportion of Ca^{2+}, Mg^{2+}, K^+, and Na^+, etc. on the CEC. Thus, the soil in our example has a nonacid saturation of 75% [(10 + 3 + 1 + 1)/20 = 0.75].

Traditionally, the nonacid cations have been referred to as "base" cations and their proportion on the CEC the **percent "base" saturation**. Cations such as Ca^{2+}, Mg^{2+}, K^+, and Na^+ do not hydrolyze as Al^{3+} and Fe^{3+} do, and therefore are not acid-forming cations. However, they are also *not* bases and do not necessarily form bases in the chemical

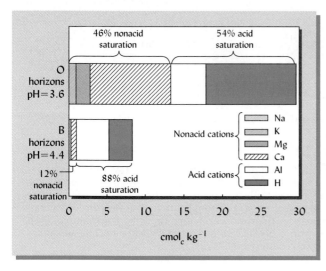

FIGURE 9.6 Saturation of the exchange capacity with acid and nonacid cations helps characterize the acidification of soils in the Adirondack Mountains of New York. The data represent the averages for O horizons and B horizons from more than 150 pedons in 144 watersheds. From the graph we can see that the effective cation exchange capacity (ECEC), the sum of all the exchangeable cations, was almost 30 cmol$_c$ kg^{-1} in the O horizons compared to only about 8 cmol$_c$ kg^{-1} in the B horizons. As is typical of temperate forested soils, the O horizons (which were about 90% organic) exhibited an extremely acid pH but a relatively low acid saturation, and the acid cations were mainly H$^+$. By contrast, the B horizons (which were about 90% mineral) had a more moderate pH but were 88% acid-saturated, and most of the acid cations were aluminum. [Modified from Sullivan et al. (2006)]

sense of the word.[3] Because of this ambiguity, it is more straightforward to refer to *acid saturation* when describing the degree of acidity on the soil cation exchange complex. Figure 9.6 uses these concepts to characterize acidified soil in the Adirondack Mountains of New York. The relationships among these terms can be summarized as follows:

$$\text{Percent acid saturation} = \frac{\text{cmol}_c \text{ of exchangeable Al}^{3+} + \text{H}^+}{\text{cmol}_c \text{ of CEC}} \tag{9.15}$$

$$\begin{matrix} \text{Percent} \\ \text{nonacid} \\ \text{saturation} \end{matrix} = \begin{matrix} \text{percent} \\ \text{"base"} \\ \text{saturation} \end{matrix} = \frac{\text{cmol}_c \text{ of exchangeable Ca}^{2+} + \text{Mg}^{2+} + \text{K}^+ + \text{Na}^+}{\text{cmol}_c \text{ of CEC}}$$

$$= 100 - \begin{matrix} \text{percent acid} \\ \text{saturation} \end{matrix} \tag{9.16}$$

Acid (or Nonacid) Cation Saturation and pH

The percentage saturation of a particular cation (e.g., Al^{3+} or Ca^{2+}) or class of cations (e.g., nonacid cations or acid cations) is often more closely related to the nature of the soil solution than is the absolute amount of these cations present. Generally, as illustrated in Figure 9.7, when the acid cation percentage increases, the pH of the soil solution decreases. However, a number of factors can modify this relationship.

EFFECT OF TYPE OF COLLOID. The type of clay minerals or organic matter present influences the pH of different soils at the same percent acid saturation. This is due to differences in the ability of various colloids to furnish H$^+$ ions to the soil solution. For example, the dissociation of adsorbed H$^+$ ions from smectites is much higher than that from Fe and Al oxide clays. Consequently, the pH of soils dominated by smectites is appreciably lower than that of the oxides at the same percent acid saturation. The dissociation of adsorbed hydrogen from 1:1-type silicate clays and organic matter is intermediate between that from smectites and from the hydrous oxides.

EFFECT OF KIND OF ADSORBED NONACID CATIONS. The relative amounts of each of the nonacid cations present on the colloidal complex is another factor influencing soil pH. For example, soils with high sodium saturation (usually highly alkaline soils of arid and

---

[3] A base is a substance that combines with H$^+$ ions, while an acid is a substance that releases H$^+$ ions. The anions OH$^-$ and HCO$_3^-$ are strong bases because they react with H$^+$ to form the weak acids, H$_2$O and H$_2$CO$_3$, respectively.

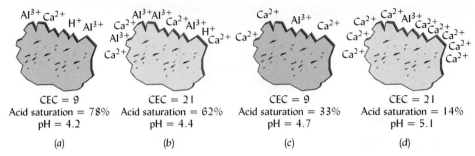

CEC = 9	CEC = 21	CEC = 9	CEC = 21
Acid saturation = 78%	Acid saturation = 62%	Acid saturation = 33%	Acid saturation = 14%
pH = 4.2	pH = 4.4	pH = 4.7	pH = 5.1
(a)	(b)	(c)	(d)

FIGURE 9.7 Acid cation saturation and soil pH in four hypothetical soils, two with a high CEC and two with a low CEC. The *relative proportion* of the effective CEC occupied by Al^{3+} and H^+ (the *percent acid saturation*) is more closely related to the soil pH than is the *absolute amount* of these acid cations present. Thus, the pH can be seen to *increase* from left to right as the percent acid saturation *decreases*, regardless of changes in the cation exchange capacity. For example, even though soil (b) has about twice as many mol_c of acid cations, soil (a) has a lower pH because it has a lower percent acid saturation. (Diagram courtesy of R. Weil)

semiarid regions) have much higher pH values than those dominated by calcium and magnesium (as explained in Section 10.1).

EFFECT OF METHOD OF MEASURING CEC. An unfortunate ambiguity in the cation saturation percentage concept is that the actual percentage calculated depends on whether the effective CEC (which itself changes with pH) or the maximum potential CEC (which is a constant for a given soil) is used in the denominator. The different methods of measuring CEC are explained in Section 8.12.

When the concept of cation saturation was first developed, the percent nonacid saturation (then termed "base saturation") was calculated by dividing the level of these exchangeable cations by the *potential* cation exchange capacity that is measured at high pH values (7.0 or 8.2). Thus, if a representative mineral soil such as shown in Figure 9.5 has a potential CEC of 20 $cmol_c$/kg, and at pH 6 has a nonacid exchangeable cation level of 15 $cmol_c$/kg, the percent nonacid cation saturation would be calculated as 15 $cmol_c$/20 $cmol_c$ × 100 = 75%. The percent "base" saturation determined by this method is still used as a soil classification criterion in *Soil Taxonomy*.

A second method relates the exchangeable cation levels to the *effective* CEC at the pH of the soil. As Figure 9.5 shows, the effective CEC of the representative soil at pH 6 would be only about 15 $cmol_c$/kg. At this pH level, essentially all the exchangeable sites are occupied by nonacid cations (15 $cmol_c$/kg). Using the effective CEC as our base, we find that the nonacid cation saturation is 15 $cmol_c$/15 $cmol_c$ × 100 = 100%. Thus, this soil at pH 6 is either 75% or 100% saturated with nonacid cations, depending on whether we use the potential CEC or the effective CEC in our calculations.

USES OF CATION SATURATION PERCENTAGES. Which nonacid saturation percentage just described is the correct one? It depends on the purpose at hand. The first percentage (75% of the potential CEC) indicates that significant acidification has occurred and is used in soil classification (for example, by definition Ultisols must have a nonacid or "base" saturation of less than 35%). The second percentage (100% of the effective CEC) is more relevant to soil fertility and the availability of nutrients. It indicates what proportion of the total exchangeable cations at a given soil pH is accounted for by nonacid cations. For example, when the effective CEC of a mineral soil is less than 80% saturated with nonacid cations (that is, more than 20% acid saturated), aluminum toxicity is likely to be a problem in many soils. Figure 9.7 illustrates that soil pH is more closely related to the acid cation saturation *percentage* than to the *absolute amount* of exchangeable acid cations in a soil.

While the factors responsible for soil acidity are far from simple, the described relationships indicate that two dominant groups of elements are in control. The different aluminum-containing ions and H^+ ions generate acidity, and most of the other cations do not. This simple statement is worth remembering.

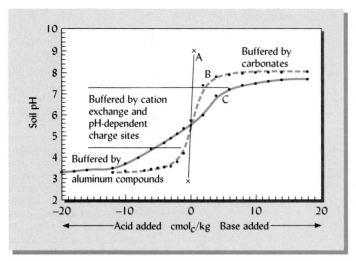

FIGURE 9.8 Buffering of soils against changes in pH when acid (H_2SO_4) or base ($CaCO_3$) is added. A well-buffered soil (C) and a moderately buffered soil (B) are compared to unbuffered water (A). Most soils are strongly buffered at low pH by the **hydrolysis and precipitation of aluminum** compounds and at high pH by the precipitation and dissolution of **calcium carbonate**. Most of the buffering at intermediate pH levels (pH 4.5 to 7.5) is provided by **cation exchange** and **protonation or deprotonation** (gain or loss of H^+ ions) of pH-dependent exchange sites on clay and humus colloids. The well-buffered soil (C) would have a higher amount of organic matter and/or highly charged clay than the moderately buffered soil (B). [Curves based on data from Magdoff and Bartlett (1985) and Lumbanraja and Evangelou (1991)]

9.4 BUFFERING OF pH IN SOILS[4]

Soils tend to resist change in the pH of the soil solution when either acid or base is added. This resistance to change is called **buffering** and can be demonstrated by comparing the *titration curves* for pure water with those for various soils (Figure 9.8).

Titration Curves

A titration curve is obtained by monitoring the pH of a solution as an acid or base is added in small increments. For example, consider the addition of 0.1 $cmol_c$ of a strong acid like HCl to a liter of water initially at pH 6 (in which the H^+ ion concentration = 10^{-6} = 0.000001 mol/L). The acid supplies 0.001 mol (= 0.1 cmol) of H^+ ions. Because the water is unbuffered, the new H^+ concentration is 0.001001, which is approximately 10^{-3} mol/L or pH 3. The pH has dropped about three units (from pH 6 to about pH 3) in response to this tiny addition of acid (Curve A in Figure 9.8). If the same amount of acid were added to soil, the change in pH would be almost too small to measure (Curves B and C in Figure 9.8). By comparing the slopes of these titration curves, we can conclude that the better buffered the soil, the smaller the change in pH caused by a given addition of acid (or base).

The titration curves shown in Figure 9.8 suggest that the soils are most highly buffered when aluminum compounds (low pH) and carbonates (high pH) are controlling the buffer reactions. The soil is least well buffered at intermediate pH levels where H^+ ion dissociation and cation exchange are the primary buffer mechanisms. However, considerable variability exists in the titration curves for various soils. This may be due to differences among soils with regard to the amounts and types of dominant colloids and contents of bound Al–hydroxy complexes that can absorb OH^- ions as the pH rises.

Mechanisms of Buffering

Compare pH changes in "water" and "buffer" by adding an acid or base:
http://michele.usc.edu/java/acidbase/acidbase.html

For soils with intermediate pH levels (5 to 7), buffering can be explained in terms of the equilibrium that exists among the three principal pools of soil acidity: active, salt-replaceable, and residual (Figure 9.9). If just enough base (lime, for example) is applied to neutralize the H^+ ions in the soil solution, they are largely replenished as the reactions move to the right, thereby minimizing the change in soil solution pH (Figure 9.9). Likewise, if the H^+ ion concentration of the soil solution is increased (for example, by organic decay or fertilizer applications) the reactions in Figure 9.9 are forced to the left, consuming H^+ and again minimizing changes in soil solution pH. Because of the involvement of residual and exchangeable acidity, we can see that soils with higher clay and organic matter contents are likely to be better buffered in this pH range.

Throughout the entire pH range, reactions that either consume or produce H^+ ions provide mechanisms to buffer the soil solution and prevent rapid changes in soil pH. We will now briefly consider some of the specific mechanisms involved.

---

[4] For a detailed discussion of the chemical principles behind this and related topics, see Bloom et al. (2005).

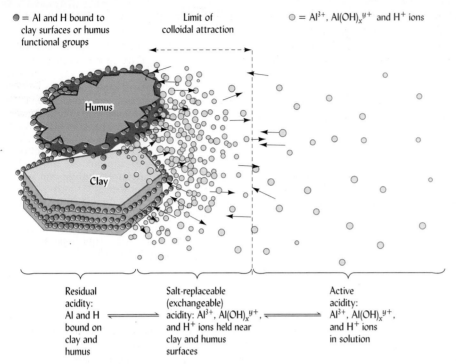

= Al and H bound to clay surfaces or humus functional groups

Limit of colloidal attraction

= Al^{3+}, $Al(OH)_x^{y+}$ and H^+ ions

| Residual acidity: Al and H bound on clay and humus | Salt-replaceable (exchangeable) acidity: Al^{3+}, $Al(OH)_x^{y+}$, and H^+ ions held near clay and humus surfaces | Active acidity: Al^{3+}, $Al(OH)_x^{y+}$, and H^+ ions in solution |

FIGURE 9.9 Equilibrium relationship among residual, salt-replaceable (exchangeable), and soil solution (active) acidity in a soil with organic and mineral colloids. Note that the adsorbed (exchangeable) and residual (bound) ions are much more numerous than those in the soil solution, even when only a small portion of the ions associated with the colloids is shown. Most of the bound aluminum is in the form of $Al(OH)_x^{y+}$ ions that are held tightly on the surfaces of the clay or complexed with the humus; relatively few $Al(OH)_x^{y+}$ ions are exchangeable. Remember that the aluminum ions, by hydrolysis, also supply H^+ ions in the soil solution. It is obvious that neutralizing only the hydrogen and aluminum ions in the soil solution will be of little consequence. They will be quickly replaced by ions associated with the colloid. The soil, therefore, demonstrates high buffering capacity. (Diagram courtesy of R. Weil)

ALUMINUM HYDROLYSIS. In very acid soils (pH 4 to 5.5), hydrolysis, dissolution, or precipitation of gibbsite [$Al(OH)_3$] and other aluminum and iron hydroxyoxide clay minerals provide a major mechanism for pH buffering. These reactions were discussed in Sections 9.1 and 9.2. They include the associated hydrolysis reactions by which the released Al and Fe ions react with water. For example:

$$Al(OH)_2^+ + H_2O \rightleftharpoons Al(OH)_3 + H^+ \qquad (9.17)$$

The additional H^+ ions would drive these reactions to the left, so relatively few of the H^+ ions would accumulate in the soil solution and the pH would be reduced only modestly. Likewise, if OH^- ions were added, they would remove H^+ from the solution by forming water and therefore force the reaction to the right. In both cases, changes in soil solution pH would be buffered.

ORGANIC MATTER REACTIONS. The protonation or deprotonation of organic matter functional (R-OH) groups provides much of the pH buffering in soils. As discussed in Section 9.1, the tenacity with which H^+ ions are held by organic matter is different for different functional groups and is also influenced by the nature of the nearby C-to-C bonds. Therefore, soil organic matter, with its complex structure, has many types of sites to which H^+ ions can bond with various degrees of strength. At low pH, H^+ ions will dissociate from the most acid sites (that is, from sites with a low pK_a). If a base is added to such an acid soil, these sites will dissociate, providing H^+ ions to replace those neutralized by the base. The sites from which the H^+ ions dissociated then become negatively charged, adding to the effective CEC. The reverse occurs if an acid is added to the soil. Organic matter buffers the soil over a wide range of pH levels as different sites on organic matter dissociate or associate.

Organic matter also buffers soil pH by the release of aluminum ions from organic complexes in which they are held with varying degrees of strength. If a base is added to raise the pH of the solution, more aluminum escapes from these complexes and subsequently replaces the neutralized H^+ ions by hydrolysis reactions. Again, the pH is stabilized.

PH-DEPENDENT CHARGE SITES ON CLAY. As discussed in Section 8.6, aluminum and silicon atoms at the surface of clay minerals are bonded to hydroxyl groups and oxygen atoms that can associate with or dissociate with H^+ ions. Adding a base to raise the solution

OH$^-$ concentration would induce the dissociation of H$^+$ ions from these sites, especially on the weathered edges of clays. This release of H$^+$ ions would buffer the pH by neutralizing most of the added OH$^-$ ions. At the same time, the sites from which the H$^+$ ions dissociated would become negatively charged and would increase the soil's CEC.

CATION EXCHANGE. Exchangeable ions on clay and humus are in equilibrium with the ions in solution. Therefore, if additional H$^+$ is added to the soil, most of these ions will be attracted to the cation exchange complex where they will exchange with other ions such as Ca^{2+}. The H$^+$ concentration in the solution will therefore not change very much. If H$^+$ ions in solution are neutralized by addition of a base, they will be immediately replaced by H$^+$ ions and various aluminum ions released from exchange sites on clay and humus. Because of the relatively small amount of exchangeable ions present in soils, these reactions contribute most significantly to pH buffering when the amount of H$^+$ ions in solution is relatively low (that is, when the pH is above 6).

CARBONATE DISSOLUTION AND PRECIPITATION. Buffer reactions involving carbonates, bicarbonates, carbonic acid, and water are most important in alkaline soils (see also Section 10.2). Acid dissolution of the carbonate mineral releases bicarbonate ions which then hydrolyze to produce carbonic acid and hydroxyl. The combined reaction is:

$$CaCO_3 + H_2O + H^+ \rightleftharpoons Ca^{2+} + H_2CO_3 + OH^-$$

(9.18)

If acid were added, the H$^+$ ions would be consumed as the reactions shifted to the right. If a base were added, the OH$^-$ ions would be consumed and H$^+$ ions produced as the reactions shifted to the left. In both cases, the pH change would be buffered.

Importance of Soil Buffering Capacity

Soil buffering is important for two primary reasons. First, buffering tends to ensure some stability in the soil pH, preventing drastic fluctuations that might be detrimental to plants, soil microorganisms, and aquatic ecosystems. For example, well-buffered soils resist the acidifying effect of acid rain, preventing the acidification of both the soil and the drainage water. Second, buffering influences the amount of amendments, such as lime or sulfur, required to bring about a desired change in soil pH.

Soils vary greatly in their buffering capacity. Other things being equal, the higher the cation exchange capacity (CEC) of a soil, the greater its buffering capacity. This relationship exists because in a soil with a high CEC, more reserve and exchangeable acidity must be neutralized or increased to affect a given change in soil pH. Thus, a clay loam soil containing 6% organic matter and 20% of a 2:1-type clay would be more highly buffered than a sandy loam with 2% organic matter and 10% kaolinite (Figure 9.10).

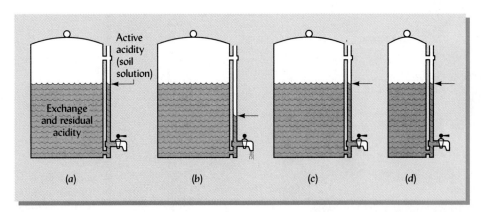

FIGURE 9.10 The buffering capacity of soils can be described by using the analogy of a coffee dispenser. (*a*) The active acidity, which is represented by the coffee in the indicator tube on the outside of the urn, is small in quantity. (*b*) When H$^+$ ions are removed, this active acidity falls rapidly. (*c*) The active acidity is quickly restored to near the original level by movement from the exchange and residual acidity. By this process, the active acidity resists change. (*d*) A second soil with the same active acidity (pH) level but much less exchange and residual acidity would have a lower buffering capacity. Much less coffee would have to be added to raise the indicator level in the last dispenser. So too, much less liming material must be added to a soil with a small buffering capacity in order to achieve a given increase in the soil pH.

9.5 DETERMINATION OF SOIL pH

Several methods can be used
to measure soil pH:
http://soils.usda.gov/technical/
technotes/note8.html

More may be inferred regarding the chemical and biological conditions in a soil from the pH value than from any other single measurement. Soil pH can be easily and rapidly measured in the field or in the laboratory. Simple field kits use certain organic dyes that change color as the pH is increased or decreased. A few drops of the dye solutions are placed in contact with the soil, usually on a white spot plate (see Plate 91 after page 656), and the color of the dye is compared to a color chart that indicates the pH to within about 0.2 to 0.5 pH unit.

Potentiometric Methods

The most accurate method of determining soil pH is with a pH electrode (Figure 9.11). In this method, a pH-sensitive *glass* electrode and a standard reference electrode (or a probe that combines both electrodes in one) are inserted into a soil:water suspension (usually at a ratio of 1:1 or 1:2.5). The difference between the H^+ ion activities in the soil suspension and in the glass electrode gives rise to an electrometric potential that is related to the soil solution pH. A special meter (called a *pH meter*) is used to measure the electrometric potential in millivolts and convert them into pH readings. The glass electrode is what makes this system sensitive to H^+ ion activities. Be advised that certain metallic soil probes on the market that do *not* contain a glass electrode cannot measure pH as claimed and may give highly misleading readings.

Most soil-testing laboratories (see Section 16.11) in the United States measure the pH of a suspension of soil in water. This is designated the pH_{water}. Other labs may report lower soil pH readings because they suspend the soil in a salt solution instead of in pure water (see Box 9.2). In this textbook (as in most U.S. publications), we will report values for pH_{water}, unless specified otherwise.

Variability in the Field

SPATIAL VARIATION.   Soil pH may vary dramatically over very small distances (millimeter or smaller). For example, plant roots may raise or lower the pH in their immediate vicinity, making the rhizosphere soil pH quite different from that in the bulk soil just a few mm away (Figure 9.13 and Plate 93). Thus, the root may experience a very different chemical environment from that indicated by lab measurements of bulk soil samples. Such mm–scale variability may account in part for the great diversity in microbial

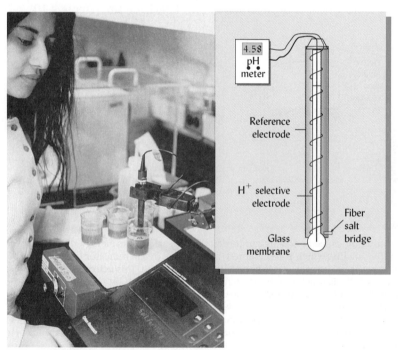

FIGURE 9.11   Soil pH is measured accurately and inexpensively in the laboratory using a pH meter (a sensitive millivolt meter that converts readings to pH) connected to a combination pH electrode that consists of a reference electrode constructed around a H^+-sensing glass membrane electrode (*diagram*). A mixture of soil and water (in beakers) is stirred and the electrodes immersed in the suspension. The H^+ ions create an electrical potential once the salt bridge completes the circuit. One technician can make hundreds of determinations in a day. (Photo and diagram courtesy of R. Weil)

BOX 9.2 TAKE YOUR SOIL pH WITH A PINCH OF SALT

In North America, most labs use **pure water** to make the soil suspension used in measuring pH, giving results reported as pH_{water}. Two important drawbacks to this method are that it is sensitive to (1) the soil:water ratio used and (2) small variations in the soluble salt content of soil. For example, fertilizer additions or evaporative salt accumulation can cause pH_{water} readings to differ by as much as 0.5 units even though soil acidity remains constant.

These problems can usually be overcome by a second method using a weak, unbuffered salt solution instead of pure water to make the soil suspension. Most commonly, a **0.01 M CaCl$_2$ solution** is used to provide a background electrolyte concentration sufficient to minimize variations caused by most salt accumulations or chemical fertilizer applications. The Ca^{2+} ions added in the solution force a portion of the exchange acidity to move into the active pool, giving pH_{CaCl} readings that are typically 0.2 to 0.5 units lower than pH_{water} readings for the same soil. Many soil test labs in Europe and Australia routinely report pH_{CaCl} rather than pH_{water}.

A third method in common use involves mixing the soil with a **solution of 1 M KCl**. Enough K^+ ions are supplied to completely exchange with cations on the soil's CEC, thus

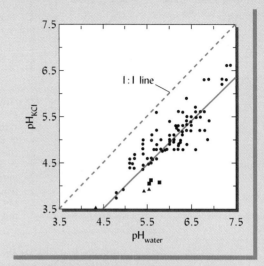

FIGURE 9.12 *Relationship between soil pH_{KCl} and pH_{water} for 151 diverse surface soils. [Data from Weil (2000), Lumbanraja and Evangelou (1991), Falkengren-Grerup et al. (2006), and Stine and Weil (2002).]*

forcing the exchangeable pool of acidity into the soil solution where the pH is measured. The pH_{KCl} is unaffected by salinity variations in the soil and may give an indication of the active plus exchangeable (salt-replaceable) pools of acidity. Figure 9.12 shows that values of pH_{KCl} average about 1 unit lower than for pH_{water}. The data points in Figure 9.12 are scattered rather widely because for any individual soil, the effect of KCl depends on such soil properties as ion exchange capacity, type of colloids, and initial soluble salt content. In certain highly weathered, acid soils, pH_{KCl} may actually be *higher* than pH_{water}. Such would be the case if *anion* exchange capacity exceeds *cation* exchange capacity so that the dominant effect of using KCl would be that Cl^- ions force more OH^- ions from anion exchange sites into the bulk solution.

If subsamples of a soil were sent to three labs, the labs might report that the soil pH was 6.5, 6.0 or 5.5 (if the labs used methods for pH_{water}, pH_{CaCl} and pH_{KCl}, respectively). All three pH values indicate the same level of acidity—and a suitable pH for most crops. Therefore, to interpret soil pH readings or compare reports from different laboratories, it is essential to know the method used.

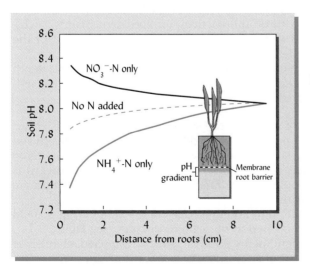

FIGURE 9.13 Soil pH at different distances from the roots of wheat plants receiving either ammonium (NH_4^+) or nitrate (NO_3^-) or no nitrogen fertilizer. Uptake of NH_4^+ cations causes the roots to release equivalent positive charges in the form of H^+ cations, which lower the pH (see rxn. 9.7). When a NO_3^- anion is taken up, the roots release a bicarbonate anion (HCO_3^-), which raises the pH (see rxn. 9.8). The soil used was a calcareous sandy clay loam in the Aridisols order with pH = 8.1. In this experiment, the lowered pH near the roots using NH_4^+ markedly enhanced the plant's uptake of phosphorus by increasing the solubility of calcium phosphate minerals near the root. In more acid soils, the reduced pH might increase the toxicity of aluminum. A barrier membrane allowed soil solution to pass through, but prevented root growth into the lower soil where pH was measured. Plants were watered from the bottom by capillary rise. [Redrawn from Zhang et al. (2004) with permission of the Soil Science Soc. of America.]

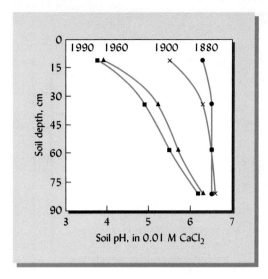

FIGURE 9.14 The change in soil pH (measured in 0.01 M $CaCl_2$) during a 110-year period in which a former agricultural field was allowed to revert to natural vegetation (eventually a mature oak forest). The fine-textured (clay loam to clay) Alfisol at Rothamstead in England was untilled, unfertilized, and unlimed. Note that in the first 20 years acidification was most pronounced near the soil surface. In the ensuing years acidity continued to increase most dramatically at the surface, but eventually increased throughout the profile. By 1960, the surface horizon had reached the pH range in which strong buffering by aluminum compounds probably slowed acidification. [Drawn from data in Blake et al. (1999); used with permission of Blackwell Science, Ltd.]

species present in normal soils (see Chapter 11). For example, organisms unfavorably influenced by a given H^+ ion concentration may find, at an infinitesimally short distance away, a different environment that is more satisfactory.

Concentrations of fertilizers or ashes from forest fires may cause sizeable pH variations within the space of a few centimeters to a few meters. Other factors, such as erosion or drainage, may cause pH to vary considerably over larger distances (hundreds of m), often ranging over two or more pH units within a few hectares. A carefully planned sampling procedure may minimize errors due to such variability (see Section 16.11). In addition to differences from place to place, a sampling scheme should also recognize variations with depth and time.

WITH SOIL DEPTH. Different horizons, or even parts of horizons, within the same soil may exhibit substantial differences in pH. In many instances, the pH in the upper horizons is lower than in the deeper horizons (see Figure 9.6), but many patterns of variability exist. Acidifying processes usually proceed initially near the soil surface and slowly work their way down the profile. Examples include the acid input from rainfall, the oxidation of nitrogen applied as fertilizer to the soil surface, and the decomposition of plant litter falling on the soil surface. Reinforcing this vertical pH trend, many natural alkalizing processes such as mineral weathering are typically most active in the lower soil horizons where the more weatherable minerals from the parent material are still present.

Human application of liming materials is one of the more obvious exceptions to the already described trends since this practice raises the pH mainly in the upper horizons into which the lime is incorporated. The incorporation of liming materials into the Ap horizon by tillage usually results in relatively uniform pH readings within the top 15 to 20 cm of cultivated soil. However, severe acidity may occur in the subsoil beyond the depth of lime incorporation but within the reach of most plant roots. Untilled soils—including croplands managed with no-till practices, unplowed grasslands, lawns, and forest land—often show marked vertical variation in soil pH, with most of the changes in pH occurring in the upper few cm (see Figure 9.14).

For all these reasons, it is often advisable to obtain soil samples from various depth increments within the root zone and determine the pH level for each. Otherwise, serious acidity problems may be overlooked.

EFFECTS OF SEASON AND TIME. A buildup of salts near the surface during dry periods versus the leaching of these salts during wet periods often produce seasonal variations in soil pH_{water}. Other causes of seasonal variation include periods of intense organic decay with the onset of warm temperatures or first rains. Because of such variations, it is advisable to obtain successive soil samples at the same time of year if changes in pH are to be monitored over a number of years.

Most acidification processes are quite slow and must overcome soil buffering, so field soil pH generally changes slowly over a period of years or decades (see Figure 9.14). However, if finely ground, reactive amendments (lime or sulfur) are applied, changes of 1 or more pH units may occur in as little as a few months (see Section 9.8).

9.6 HUMAN-INFLUENCED SOIL ACIDIFICATION

In certain situations, the natural processes of soil acidification are greatly (and usually inadvertently) accelerated by human activities. We will consider three major types of human-influenced soil acidification: (1) nitrogen amendments, (2) acid precipitation, and (3) acid sulfate soils.

Nitrogen Fertilization

During the past 100 years, agricultural activities have greatly increased the amount of nitrogen cycling through the world's soils. This intensification of the nitrogen cycling, largely by the use of chemical fertilizers, has helped stimulate remarkable increases in global food supplies, but has also brought about significant acceleration of soil acidification.

CHEMICAL FERTILIZERS. Widely used ammonium-based fertilizers, such as ammonium sulfate [$(NH_4)_2 SO_4$] and urea [$CO(NH_2)_2$] are oxidized in the soil by microbes to produce strong inorganic acids by reactions such as the following:

$$(NH_4)_2 SO_4 + 4O_2 \rightleftharpoons 2HNO_3 + H_2SO_4 + 2H_2O \tag{9.19}$$

These strong acids provide H^+ ions that lower pH (Figure 9.15). However, since H^+ ions are consumed by the bicarbonate released when plants take up anions (Reaction 9.8), soil acidification results largely from that portion of applied nitrogen that is not actually used. Excessive nitrogen fertilization rates popularized since the 1960s (Sections 13.15, 16.13) have ensured that soil acidification from this cause is not trivial—the amount of limestone that would be required for its neutralization each year in the United States alone is some 20 million Mg.

ACID-FORMING ORGANIC MATERIALS. Deposition of organic materials such as sewage sludge or animal manures on agricultural and forested lands can decrease soil pH, both by oxidation of the ammonium nitrogen released and by organic and inorganic acids formed during decomposition. Therefore, a program of regular organic matter additions should also include regular additions of liming materials to counteract this acidification. In many countries, control of soil pH after amendment with sewage sludge is regulated, with the objective of minimizing the mobility of toxic metals found in this type of waste material (see Section 18.8). It should be noted that certain types of sewage sludge have been made with large quantities of lime to control pathogens and odors. Rather than acidifying the soil, application of such lime-stabilized sewage sludge may result in **overliming** if the lime content is not taken into account when determining application rates.

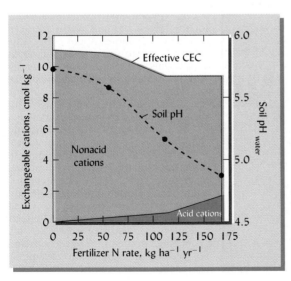

FIGURE 9.15 Soil pH can be significantly lowered by fertilization with ammonium forms of nitrogen. Excess H^+ ions are generated during the bacterial conversion of NH_4^+ to NO_3^-. The acidification is especially severe if more NO_3^- is created than plants can take up and if most of the nonacid cations taken up by the plants are removed by harvest. As a result of the declining pH, the effective cation exchange capacity (CEC) of the soil also declines (see also Section 8.9). In the case illustrated, a Mollisol in Wisconsin was fertilized with N (urea or ammonium nitrate) for 30 years at the rates indicated. Conventional plow tillage was used to grow corn, soybean, and tobacco crops, and all the aboveground residues were removed. [Redrawn from data in Barak et al. (1997)]

Acid Deposition from the Atmosphere

Animated maps showing acid deposition trends:

http://nadp.sws.uiuc.edu/amaps2/

ORIGINS OF ACID RAIN. Industrial activities such as the combustion of coal and oil in electric power generation, the combustion of fuel in vehicles, and the smelting of sulfur-containing metal ores emit enormous quantities of nitrogen and sulfur-containing gases into the atmosphere (Figure 9.16). Other sources of these gases include forest fires and the burning of crop residues. The gases (mainly sulfur dioxide and oxides of nitrogen) react with water and other substances in the atmosphere to form HNO_3 and H_2SO_4. These strong acids are then returned to the Earth in rain (as well as in snow, fog, and dry deposition). This precipitation is called **acid rain**. Normal rainwater that is in equilibrium with atmospheric carbon dioxide has a pH of about 5.5. The pH of acid rain commonly is between 4.0 and 4.5, but may be as low as 2.0. The greatest amount of acidity currently falls on regions downwind from major industrial centers, including the eastern part of North America, much of Europe, central Russia, and eastern China.

EFFECTS OF ACID RAIN. Acid rain causes expensive damage to buildings and car finishes, but the principal environmental reasons for concern about acid rain are its effects on (1) fish and (2) forests. Since the 1970s, scientists have documented the loss of normal fish populations in thousands of lakes and streams. More recently, studies have suggested that the health of certain forest ecosystems is suffering because of acid rain. Furthermore, scientists have learned that the health of both the lakes and the forests is not usually affected directly by the rain, but rather by the interaction of the acid rain with the soils in the watershed (Figure 9.16).

SOIL ACIDIFICATION. The incoming strong acids mobilize aluminum in the soil minerals, and the aluminum displaces Ca^{2+} and other nonacid cations from the exchange complex. The presence of the strong acid anions (SO_4^{2-} and NO_3^-) facilitates the leaching of the displaced Ca^{2+} ions (as explained in Figure 9.3). Soon Al^{3+} and H^+ ions, rather than Ca^{2+} ions, become dominant on the exchange complex, as well as in the soil solution and drainage waters. Figure 9.17 illustrates the increase in acid saturation of the cation exchange complex over a 28-year period. However, in this and other studies, it is not

FIGURE 9.16 Simplified diagram showing the formation of acid rain in urban areas and its impact on distant watersheds. Combustion of fossil fuels in electric power plants and in vehicles accounts for the largest portions of the nitrogen and sulfur emissions. About 60% of the acidity is due to sulfur gases and about 40% is due to nitrogen gases. The gases are carried hundreds of kilometers by the wind and are oxidized to form sulfuric and nitric acid in the clouds. These acids then return to Earth in precipitation and in dry deposition. The H^+ cations and NO_3^- and SO_4^{2-} anions cause acidification to occur in soils, soil aluminum to mobilize, and the loss of calcium and magnesium to accelerate. The mobilized aluminum percolates through the soil mantle, eventually reaching lakes and streams. The principal ecological effects of concern in sensitive watersheds are (1) possible decline in forest health and (2) decline or even death of aquatic ecosystems.

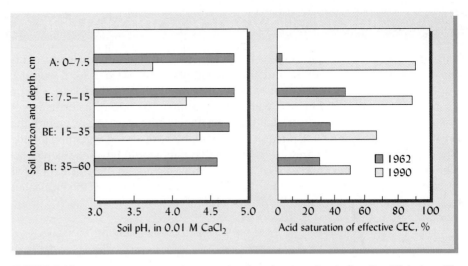

FIGURE 9.17 Changes in two indicators of soil acidity in a South Carolina Ultisol profile during 28 years under a loblolly pine forest. Note that the greatest acidification took place in the uppermost mineral horizon (data for the O horizon were not reported). The soils at this site were limed regularly for agricultural crops prior to 1952 and therefore were less acid at the beginning of the study than they would be in their natural state. The acid saturation percentage = $(cmol_c/Al + H)/(cmol_c$ effective CEC). The effective CEC for these soil horizons ranged from 0.8 to 3.3 $cmol_c/kg$ (data not shown). [Calculated and drawn from data in Markewitz et al. (1998); with permission of the Soil Science Society of America].

easy to sort out how much acidification is due to natural processes internal to the soil ecosystem (see left side of Table 9.1) and how much is due to acid rain. Different studies estimate that acid deposition has caused between 30 and 80% of the observed acidification in humid regions receiving highly acid rain.

EFFECTS ON FORESTS. Some scientists are concerned that trees, which have a high requirement for calcium to synthesize wood, may eventually suffer from insufficient supplies of this and other nutrient cations in acidified soils. Research has shown reductions in Ca^{2+}, Mg^{2+}, and K^+ and declines in pH levels over several decades in forested watersheds subject to acid rain deposits. The leaching of calcium and the mobilization of aluminum may result in Ca/Al ratios (mol_c/mol_c) of less than 1.0 in both the soil solution and on the exchange complex. A Ca/Al ratio of 1.0 is widely considered a threshold for aluminum toxicity, reduced calcium uptake, and reduced survival for forest vegetation. While there is little doubt that aluminum in acidified soils is toxic to many forest species (see Section 9.7), the scientific evidence for forest calcium deficiencies is less clear. The calcium supply in most forested soils in the humid eastern United States is being depleted as the rate of calcium loss by leaching, tree uptake, and harvest exceeds the rate of calcium deposition. However, it seems that even in very acid soils low in exchangeable Ca^{2+}, the weathering of soil minerals often releases sufficient calcium for good tree growth—at least in the short term.

EFFECTS ON AQUATIC ECOSYSTEMS. The acid soil water, containing elevated levels of aluminum, and often of sulfate and nitrate, eventually drains into streams and lakes. The water in the lakes and streams becomes lower in calcium, less well-buffered, more acid, and higher in aluminum. The aluminum is directly toxic to fish, partly because it damages the gill tissues. As the lake water pH drops to about 6.0, acid-sensitive organisms in the aquatic food web die off, and reproductive performance of such fish as trout and salmon declines. With a further drop in water pH to about 5.0, virtually all fish are killed. Although the acidified water may be crystal clear (in part due to the flocculating influence of aluminum), the lake or stream is considered to be "dead" except for a few algae, mosses, and other acid-tolerant organisms.

The quantity of HNO_3 and H_2SO_4 deposited globally in acid rain is enormous, but the amount falling on a given hectare in a year is usually not enough to significantly change soil pH in the short term (see Box 9.3). In time, however, the cumulative acid deposition negatively impacts soils, the plants growing in them, and the aquatic ecosystems receiving their drainage waters.

BOX 9.3 HOW MUCH ACIDITY FALLS IN ACID RAIN?

To use an example typical of the area most impacted by acid rain in North America, consider 1 m^2 of land in a humid region with 1000 mm of annual rainfall at pH 4. This area of land would annually receive 1000 L of rain (1 m^3 of water) carrying 0.0001 mol of H^+ ion/L for a total of $1000 \times 0.0001 = 0.1$ mol of H^+ ions per m^2 of land. Assume that this rain reacts mainly with the top 20 cm of soil (0.2 m^3), consisting of 10 cm of organic O_i, O_e horizons underlain by 10 cm of sandy E horizon. The O horizons might weigh about 50 kg, assuming the bulk density = 0.5 Mg/m^3, and the E horizon might weigh about 140 kg assuming a bulk density = 1.4 Mg/m^3. The soil, a Spodosol typical of many areas sensitive to acid rain, would be already quite acid, say, pH = 4.0. It might have an effective CEC in the O horizons of about 50 $cmol_c/kg$ and in the E horizon of about 4 $cmol_c/kg$. Therefore, in this case, the acidity in the annual rainfall (10 cmol H^+/m^2) would be the equivalent of about 0.4% of the O horizons' CEC or about 1.8% of the E horizon's CEC:

Acid Input as Percentage of O horizon CEC

Acid input: 0.1 mol H^+/50 kg soil = 2×10^{-3} mol H^+/kg = 2 mmol H^+/kg = 0.2 $cmol_c$ H^+/kg

Percentage of CEC: (0.2 $cmol_c$ H^+/kg)/(50 $cmol_c$ CEC/kg) × 100 = 0.4%

Acid Input as Percentage of E horizon CEC

Acid input: 0.1 mol H^+/140 kg soil = 7×10^{-4} mol H^+/kg = 0.7 mmol H^+/kg = 0.07 $cmol_c$ H^+/kg

Percentage of CEC: (0.07 $cmol_c$ H^+/kg)/(4 $cmol_c$ CEC/kg) × 100 = 1.8%

The degree to which the incoming H^+ ions actually would replace exchangeable nonacid cations would depend on the conditions of cation exchange (see Section 8.8) and on such factors as how much of this acidity could be consumed by weathering Ca and other cations from the soil minerals, etc. Remember that acidification will occur only to the degree that H^+ ion generation exceeds H^+ ion consumption (by the processes listed in Table 9.1).

In our example, a H^+ ion input of the magnitude just considered would cause only a very small change in residual acidity and almost no measurable change in pH from year to year. However, in a more poorly buffered soil (for example, if some disturbance had caused our Spodosol to lose its O horizons), residual acidity would first increase, soon to be followed by a significant decline in pH of the soil solution and drainage water.

Overview of environmental damages from acid rain:

http://www.epa.gov/airmarkets/acidrain/effects/index.html

SENSITIVE SOILS. Ecological damage from acid rain is most likely to occur where the rain is most acid and the soils are most susceptible to acidification. Figure 9.18a delineates areas of the world where soils belong to five classes of acidification susceptibility that are based mainly on the soils' CEC and acid saturation percentage. Inclusion of other factors such as the soils' content of weatherable minerals would make this classification more precise, but the general trends are clear. Ecosystems in eastern China and Brazil are among those most likely to be damaged in the future (Figure 9.18b).

SLOW PROGRESS.[5] Since the passage of the U.S. Clean Air Act in 1970, the atmospheric deposition of sulfates in North America has declined significantly. Unfortunately, the alkalinity of many streams and lakes in the Northeast has not increased as rapidly as would be expected to accompany the decline in sulfates. One explanation for the delayed recovery is that air pollution controls did an even better job of removing particulate pollutants from smokestack emissions. These tiny soot particles happen to be rich in calcium, so their removal in the interest of clean air resulted in less calcium deposition on acidified watersheds. Another important source of Ca deposition—dust from traffic on unpaved roads—has also declined in recent decades. Applying liming materials to huge tracts of forested land is not the answer, as this is both impractical (even by helicopter) and, in the long term, not sustainable.

Nonetheless, tightening air quality standards in industrialized countries should continue to reduce acid inputs into sensitive ecosystems, eventually restoring a suitable chemical balance in the soils of these areas (and therefore in the lakes as well).

---

[5] For an assessment of the recovery of lakes and streams from acidification in North America and Europe, see Stoddard et al. (1999). For evidence of a slow reduction in aluminum mobilization in the eastern United States, see Palmer and Driscoll (2002).

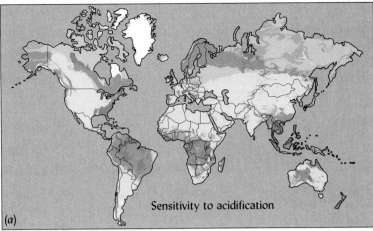

(a)

Sensitivity to acidification

No data available for areas shaded in white

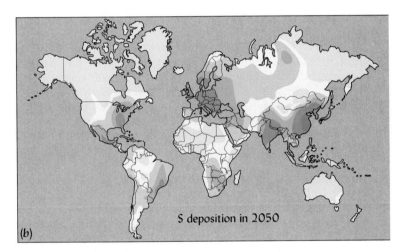

(b)

S deposition in 2050

FIGURE 9.18 Sensitivity of global ecosystems to damage by acid deposition (*a*) and global distribution of sulfur deposition predicted for the year 2050 (*b*). The darkest areas in (*b*) are those most sensitive to damage by acid deposition. These areas have poorly buffered soils with low CEC values and high acid saturation percentages. A comparison to the global soil orders map (front endpaper) will reveal that the most sensitive areas are dominated by soils in the orders Spodosols, Ultisols, and Oxisols. The darkest areas in (*b*) are those with the highest predicted sulfur deposition rates (50 to 100 kg S $ha^{-1}y^{-1}$). Regions most likely to suffer severe acidification damage in the future are those areas that are dark on both maps, indicating that they are both sensitive and likely to receive high amounts of acid deposition. [Map (*a*) from Kuylenstierna et al. (2001) and map (*b*) from Rodhe et al. (1995)].

Exposure of Potential Acid Sulfate Materials[6]

POTENTIAL ACIDITY FROM REDUCED SULFUR.   A large pool of potential soil acidity may occur in soils or sediments that contain reduced sulfur. If drainage, excavation, or other disturbance introduces oxygen into these normally anaerobic soils, oxidation of the sulfur may produce large amounts of acidity. The adjective **sulfidic** is used to describe such materials with enough reduced sulfur to markedly lower the pH within two months of becoming aerated. The term **potential acidity** refers to the acidity that could be produced by such reactions.

DRAINAGE OF CERTAIN COASTAL WETLANDS.   Due to the microbial reduction of sulfates originally in seawater, certain coastal sediments contain significant quantities of pyrite (FeS_2), iron monosulfides (FeS), and elemental sulfur (S). Coastal wetland areas in the

[6] Many of these sulfide-rich soils are clayey in texture and termed cat clays. Fanning et al. (2002) describe some of the properties of these soils and discuss environmental problems that arise from their misuse. For a report on the iron and sulfur oxidizing roles of acidophile microorganisms discovered growing at pH 0.5 in acid mine drainage, see Edwards et al. (2000).

southeastern United States, Southeast Asia, coastal Australia, and West Africa commonly contain soils formed in such sediments. So long as waterlogged conditions prevail, the *potential* acid sulfate soils retain the sulfur and iron in their reduced forms. However, if these soils are drained for agriculture, forestry, or other development, air enters the soil pores and both the sulfur (S^0, S^- or S^{2-}) and the iron (Fe^{2-}) are oxidized, changing the potential acid sulfate soils into *active* acid sulfate soils. Ultimately, such soils earn their name by producing prodigious quantities of sulfuric acid, resulting in soil pH values below 3.5 and in some cases as low as 2.0. The principal reactions involved are:

$$Fe^{II}S^{-I}_2 + 3\tfrac{1}{2}O_2 + H_2O \rightleftharpoons Fe^{II}S^{VI}O_4 + H_2S^{VI}O_4$$

Pyrite $\qquad\qquad\qquad$ Ferrous sulfate $\quad$ Sulfuric acid

$$Fe^{II}SO_4 + \tfrac{1}{4}O_2 + 1\tfrac{1}{2}H_2O \rightleftharpoons Fe^{III}OOH + H_2SO_4$$

Ferrous sulfate $\qquad\qquad\qquad$ Iron oxyhydroxides $\quad$ Sulfuric acid

$$(9.20)$$

$$S^0 + 1\tfrac{1}{2}O_2 + H_2O \longrightarrow H_2SO_4$$

Elemental S $\qquad\qquad\qquad$ Sulfuric acid

$$(9.21)$$

Note that both oxidation of the S in pyrite and oxidation and hydrolysis of the $FeSO_4$ produce acidity. The sulfur is oxidized from a -1 to a $+6$ valence state and the iron is oxidized from $+2$ to $+3$. Inspection of Reactions 9.20 and 9.21 will reveal that two moles of acidity (one mole of H_2SO_4) are ultimately produced for each mole of S that reacts, whether from FeS or elemental S. The oxidation reactions can occur by purely chemical means, but generally they are facilitated by microorganisms (such as the Bacterium *Thiobacillus ferrooxidans* and the Archaeon *Ferroplasma acidipilum*) that make them proceed thousands of times faster, especially when warm, moist conditions favor microbial activity.

The iron sulfide compounds in the potential acid sulfate soils often give these soils a black color (Plates 47 and 109). The black color has sometimes led, with disastrous results, to the use of such soils by those seeking black, organic-matter-rich "topsoil" material for landscaping installations. The pH of the potential acid sulfate soils is in the neutral range (typically near 7.0) while they are still reduced, but drops precipitously within days or weeks of the soil being exposed to air. When in doubt, the pH of the soil should be monitored for several weeks while a sample is incubated in a moist, well-aerated, warm condition. See Plate 64 for an example of inappropriate, albeit inadvertent, engineering use of a potential acid sulfate clay.

EXCAVATION OF PYRITE-CONTAINING MATERIALS. Coastal marsh soils are not the only places where problems with acid sulfate soils occur. The sediments dredged to deepen shipping lanes in coastal harbors may also contain high concentrations of reduced sulfur compounds (see Plate 109). Furthermore, many sulfide-rich types of geological sediment long ago became sedimentary rock or parent materials for upland soils. Since coal seams usually occur between layers of sedimentary rock, it is not surprising that coal-mining operations often uncover large quantities of pyrite-containing shale and similar rocks. When these once deeply buried materials are exposed to air and water, the reduced sulfur and iron compounds in them begin to oxidize and hydrolyze by the types of reactions already shown. Again, the result is the production of sulfuric acid in large quantities.

As water percolates through such oxidizing materials, it becomes an extremely acid and toxic brew known as **acid mine drainage**. Typical acid mine drainage has a pH in the range of 0.5 to 2.0, but pH values *below zero* have been measured! When this drainage water reaches a stream (as in Plate 108), iron sulfates dissolved in the drainage water continue to produce acid by oxidation and hydrolysis. The aquatic community can be devastated by the pH shock and the iron and aluminum that is mobilized. Similar problems occur when road cuts or building excavations expose buried sulfide-containing layers. An indication of the tremendous scale of this environmental problem is the fact that acid drainage from mining and other excavations is thought to account for nearly a quarter of the global sulfate input into the ocean. If sulfide-rich excavated material (often black or dark gray in color) is used to cover a site after mining has been

Research on acid mine drainage with pH<0:

http://www.pnas.org/cgi/content/full/96/7/3455

completed, acid sulfate soils are likely to form, making revegetation of the site all but impossible. In many countries, environmental regulations are designed to prevent this from happening.

AVOIDANCE AS THE BEST SOLUTION. The amount of sulfur present in such soil material is considered an indication of its **potential acidity**. If calcite is either naturally occurring in the soil or is added as a neutralizing agent, sulfuric acid from S oxidation may react with it to form gypsum:

$$\underset{\text{Sulphuric acid}}{H_2SO_4} + \underset{\text{Calcite}}{CaCO_3} + H_2O \longrightarrow \underset{\text{Gypsum}}{CaSO_4 \cdot 2H_2O} + \underset{\text{Carbon dioxide}}{CO_2} \qquad (9.22)$$

Int'l Soil Reference and Info. Center tutorial on Acid Sulfate soils:

http://www.isric.org/isric/webdocs/tutorial/WHStart.htm

Equation 9.22 indicates that 1 mole of liming material ($CaCO_3$) would be required to eventually neutralize the sulfuric acid produced by the oxidation of 1 mole of reduced S. Since 1 mole of $CaCO_3$ equals 100 g and 1 mole of S equals 32 g, about 3 kg of limestone would be needed to neutralize the acidity from the oxidation of 1 kg of S. The enormous amounts of $CaCO_3$ required often make neutralization impractical in the field.

Usually the best approach to solving this environmental challenge is to *prevent* the S oxidation in the first place. This means that sulfide-bearing wetland soils are best left undisturbed or returned to their undrained, wetland conditions. In this manner, both damage to agricultural crops and the considerable expense of attempting to neutralize the acidity can be avoided. Preserving the wetland condition of the soils will also avoid contamination of water with acid drainage and will maintain the natural habitat for a diversity of wild plants and animals.

In the case of mining or other excavation, any sulfide-bearing materials exposed must be identified and eventually deeply reburied to prevent their oxidation. If some acid drainage is unavoidable (as from abandoned, poorly designed mines), an effective treatment is to route the acid water through a wetland, either natural or constructed for the purpose (see Section 7.7). The anaerobic wetland conditions will re-reduce the iron and sulfur, causing iron sulfide to precipitate, simultaneously raising the pH of the water and reducing the iron content.

9.7 BIOLOGICAL EFFECTS OF SOIL pH

The pH of the soil solution is a critical environmental factor for the growth of all organisms that live in the soil, including plants, animals, and microbes. Although we have already mentioned some ways that soil acidity and alkalinity affects plant growth, we will now take a more detailed look at the impacts that various pH conditions have on soil organisms.

To nonadapted plants, strongly acid soil presents a host of problems. These include toxicities of aluminum, manganese, and hydrogen, as well as deficiencies of calcium, magnesium, molybdenum, and phosphorus. As it is difficult to separate one problem from another, the situation is sometimes simply referred to as the acid soil "headache."

Aluminum Toxicity[7]

Aluminum toxicity stands out as the most common and severe problem associated with acid soils. Not only plants are affected; many bacteria, such as those that carry out transformations in the nitrogen cycle, are also adversely impacted by the high levels of Al^{3+} and $AlOH^{2+}$ that come into solution at low soil pH. As can be deduced from Figure 9.5, aluminum toxicity is rarely a problem when the soil pH is above about 5.2 (above pH_{CaCl} 4.8) because little aluminum exists in the solution or exchangeable pools above this pH level. Figure 9.19 shows the exponential increase in Al^{3+} concentration of the soil solution as pH drops from 5 to 4. Other toxic Al species, namely $AlOH^{2+}$ and $Al(OH)_2^+$, also increase in solubility below pH 5, but they are not shown in Figure 9.19 as their concentrations near pH 4 are 10 to 100 times smaller than that of Al^{3+}. At comparable pH levels in most organic soils (or in organic soil horizons), aluminum toxicity is much less of a problem because there is far less total aluminum in these soils—and

---

[7] For a review of aluminum toxicity and the development of plant tolerance to aluminum, see de la Fuente-Martinez and Herrera-Estrella (2000).

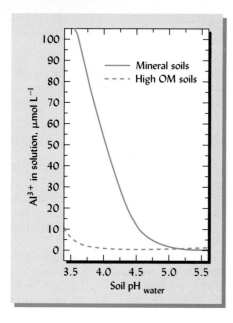

FIGURE 9.19 Relationship between toxic dissolved aluminum and soil pH in mineral and high organic matter (OM) soils. The amount of Al^{3+} in solution increases exponentially as the soil pH declines, but much less so for organic soils. Even for mineral soils high in total aluminum, little Al^{3+} exists in solution (or in the exchangeable form, not shown here) when the pH is above 5. The Al^{3+} in solution in mineral soils is thought to be largely controlled by the dissolution or precipitation of $Al[OH]_3^0$ coatings on particle surfaces. Below pH 4.0, both iron and H ions (not shown) may also be toxic to roots. Data for the graph came from 89 mineral soils (various horizons) and 30 organic soils (surface horizons with >10% organic matter, OM). All the soils were under forest or grass vegetation. [From equations and data compiled by Tipping (2005)]

because aluminum ions are strongly attracted and bound to the carboxylic (R-COO$^-$) and phenolic (R-CO$^-$) sites on soil organic matter (review Figure 8.15), leaving much less Al^{3+} in solution (Figure 9.19, *lower curve*).

EFFECTS ON PLANTS. When aluminum, which is not a plant nutrient, is taken into the root, most remains there and little is translocated to the shoot (except in aluminum accumulator plants such as tea, which may contain as much as 5000 ppm in the dry leaves). Therefore, analysis of leaf tissue is rarely a good diagnostic technique for aluminum toxicity. In the root, aluminum damages membrane sites where calcium is normally taken in and restricts cell wall expansion so roots cannot grow properly (Figure 9.20, *left*). Aluminum also interferes with the metabolism of phosphorus-containing compounds essential for energy transfers (ATP) and genetic coding (DNA).

SYMPTOMS. The most common symptom of aluminum toxicity is a stunted root system with short, thick, stubby roots that show little branching or growth of laterals. The root tips and lateral roots often turn brown. In some plants, the leaves may show chlorotic (yellowish) spots. Because of the restricted root system, plants suffering from aluminum toxicity often show symptoms of drought stress and phosphorus deficiency (stunted growth, dark green foliage, and purplish stems).

TOLERANCE. Among and within plant species there exists a great deal of genetic variability in sensitivity to aluminum toxicity. Generally, plant species that originated in areas dominated by acid soils (such as most humid regions) tend to be less sensitive than

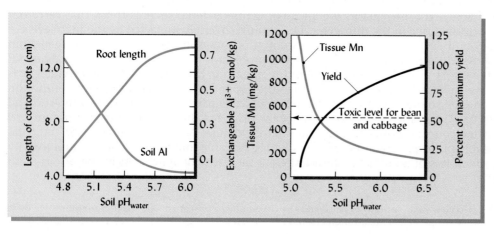

FIGURE 9.20 Plant responses to toxicity of aluminum (*left*) and manganese (*right*) at low soil pH. (*Left*) As soil pH$_{water}$ drops below 5.2, exchangeable Al increases and cotton root length is severely restricted in an Ultisol. (*Right*) Plant shoot growth (the average of bean and cabbage) declines and Mn content of foliage increases at low pH levels in manganese-rich soils from East Africa (average data for an Andisol and an Alfisol). [(*left*) From Adams and Lund (1966); (*right*) Redrawn from data in Weil (2000)]

Tolerant   Sensitive   Tolerant   Sensitive

pH 4.4   pH 5.7

FIGURE 9.21 Influence of pH on the growth of shoots (*a*) and roots (*b*) from two wheat varieties, one sensitive and one tolerant to aluminum. Note the stunted shoot growth and extremely stunted, stubby root system of the sensitive variety in the low-pH treatment. [Photos courtesy of C. D. Foy, USDA/ARS, Beltsville, MD]

species originating in areas of neutral to alkaline soils (such as the Mediterranean region). Fortunately, plant breeders have been able to find genes that confer tolerance to aluminum even in species that are typically sensitive to this toxicity (Figure 9.21). Most Al-tolerant plants avoid the problem by excluding aluminum from their roots. To do this, some species raise the pH of the soil just outside the root, causing the aluminum to precipitate. Others excrete organic mucilage that complexes with the aluminum, preventing its uptake into the root. Still others produce certain organic acids that combine with the aluminum to form nontoxic compounds.

Manganese, Hydrogen, and Iron Toxicity to Plants

MANGANESE TOXICITY. Although not as widespread as aluminum toxicity, **manganese toxicity** is a serious problem for plants in acid soils derived from manganese-rich parent minerals. Unlike Al, Mn is an essential plant nutrient (see Section 15.5) that is toxic only when taken up in excessive quantities. Like aluminum, manganese becomes increasingly soluble as pH drops, but in the case of Mn, toxicity is common at pH_{water} levels as high as 5.6 (about 0.5 units higher than for aluminum).

Plant species and genotypes within species vary widely with regard to their susceptibility to manganese toxicity. Symptoms of Mn toxicity vary among plant species, but often include crinkling or cupping of leaves and interveinal patches of chlorotic tissue. Unlike for Al, the leaf tissue content of Mn usually correlates with toxicity symptoms, toxicity beginning at levels that range from 200 mg/kg in sensitive plants to over 5000

TABLE 9.2 Relationship between Soil Properties Associated with Acidity and Survival of Sugar Maple Seedlings

Data are averages for 18 forested sites dominated by overstory sugar maples.
Al was most abundant in the low-organic-matter B horizons and Mn most abundant
in the high-organic-matter O horizons. The ratio of Ca/Al was <1.0 only in the B horizons,
while a ratio of Ca/Mn <30 in all horizons was associated with tree mortality.

Seedlings present?	Exchangeable cations, mg/kg			Ratio, mol_c/mol_c		Soil pH_{water}
	Mn	Ca	Al	Ca/Al	Ca/Mn	
			O Horizons			
No	188	2738	53	23.1	20	4.02
Yes	89	6371	38	74.1	98	4.45
			A Horizons			
No	59	1252	143	3.9	28	4.34
Yes	33	2755	142	8.5	114	4.58
			B Horizons			
No	15	305	279	0.5	28	4.62
Yes	8	1061	202	2.3	180	4.90

[Data from Demchik et al. (1999b)]. Ca/Al and Ca/Mn ratios calculated here to give units shown.

mg/kg in tolerant plants. Figure 9.20 (*right*) illustrates a case in which low soil pH induced plant uptake of Mn to toxic levels.

Since the reduced form [Mn(II)] is far more soluble than the oxidized form [Mn(IV)], toxicity is greatly increased by low oxygen conditions associated with a combination of oxygen-demanding, decomposable organic matter and wet conditions. Manganese toxicity is also common in certain high organic matter surface horizons of volcanic soils (e.g., Melanudands). Unlike that of Al, the solubility and toxicity of Mn is commonly accentuated, rather than restricted, by higher soil organic matter (Table 9.2).

HYDROGEN ION TOXICITY. At pH levels below 4.0 to 4.5, the H^+ ions themselves are of sufficient concentration to be toxic to some plants, mainly by damaging the root membranes. Likewise, such low pH, even in the absence of high Al or Mn, has been found to kill certain beneficial soil bacteria such as the *Rhizobium* bacteria that help supply legume plants with nitrogen (see Section 13.11).

IRON TOXICITY. Iron can become toxic to plants in the oxidized form at very low pH levels (usually less than 4.0) or if anaerobic conditions cause the iron to become reduced to the Fe(II) form, which is far more soluble than the oxidized form. Toxicity of reduced iron can be a problem in acid rice paddies. Again, plant species and strains vary widely with regard to their susceptibility to iron toxicity.

Nutrient Availability to Plants

Figure 9.22 shows in general terms the relationship between the pH of mineral soils and the availability of plant nutrients. Note that in strongly acid soils the availability of the macronutrients (Ca, Mg, K, P, N, and S) as well as the two micronutrients, Mo and B, is curtailed. In contrast, availability of the micronutrient cations (Fe, Mn, Zn, Cu, and Co) is increased by low soil pH, even to the extent of toxicity.

In slightly to moderately alkaline soils, molybdenum and all of the macronutrients (except phosphorus) are amply available, but levels of available Fe, Mn, Zn, Cu, and Co are so low that plant growth is constrained. Phosphorus and boron availability is likewise reduced in alkaline soil—commonly to a deficiency level.

It appears from Figure 9.22 that the pH range of 5.5 to 7.0 may provide the most satisfactory plant nutrient levels overall. However, this generalization may not be valid for all soil and plant combinations. For example, certain micronutrient deficiencies are common in some plants when soils are limed to pH values of only 6.5 to 7.0.

Microbial Effects

Fungi are particularly versatile, flourishing satisfactorily over a wide pH range. Fungal activity tends to predominate in low pH soils because bacteria are strong competitors

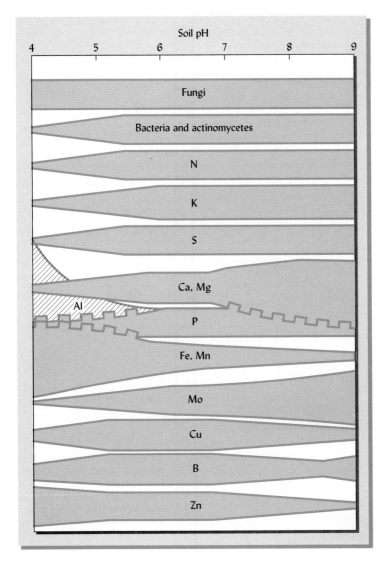

Soil pH

FIGURE 9.22 Relationships existing in mineral soils between pH and the availability of plant nutrients. The relationship with activity of certain microorganisms is also indicated. The width of the bands indicates the relative microbial activity or nutrient availability. The jagged lines between the P band and the bands for Ca, Al, and Fe represent the effect of these metals in restraining the availability of P. When the correlations are considered as a whole, a pH range of about 5.5 to perhaps 7.0 seems to be best to promote the availability of plant nutrients. In short, if the soil pH is suitably adjusted for phosphorus, the other plant nutrients, if present in adequate amounts, will be satisfactorily available in most cases.

and tend to dominate the microbial activity at intermediate and higher pH. Individual microbial species exhibit pH optima that may differ from this generality. Manipulation of soil pH can help control certain soilborne plant diseases (see Section 11.13).

Optimal pH Conditions for Plant Growth

Plants vary considerably in their tolerance to acid and/or alkaline conditions (Figure 9.23). For example, certain legume crops such as alfalfa and sweet clover grow best in near-neutral or alkaline soils, and most humid-region mineral soils must be limed to grow these crops satisfactorily. Asparagus and cantaloupe are two food crops that have a high calcium requirement and grow best at pH levels near or above 7.0.

Because forests exist mainly in humid regions where acid soils predominate, many forest plants are at the opposite end of the scale. Forest species such as rhododendrons, azaleas, blueberries, larch, some oaks, and most pines are inefficient in taking up the iron they need. Since high soil pH and high calcium saturation reduce the availability of iron, these plants will show **chlorosis** (yellowing of the leaves) and other symptoms indicative of iron deficiency under these soil conditions (see Plates 86, 88, and 107). Even forest trees differ in their tolerance of soil acidity and high aluminum (Table 9.3). For example, elm, poplar, honey locust, and the tropical legume tree *Leucaena* are known to be less tolerant to acid soils than are most other forest species.

Most cultivated crop plants (except those such as sweet potato, cassava, and others that originated in the humid tropics) grow well on soils that are just slightly acid to near neutral. Therefore, for most grain and vegetable crops a soil pH in the range of perhaps 5.5 to 7.0 is most suitable.

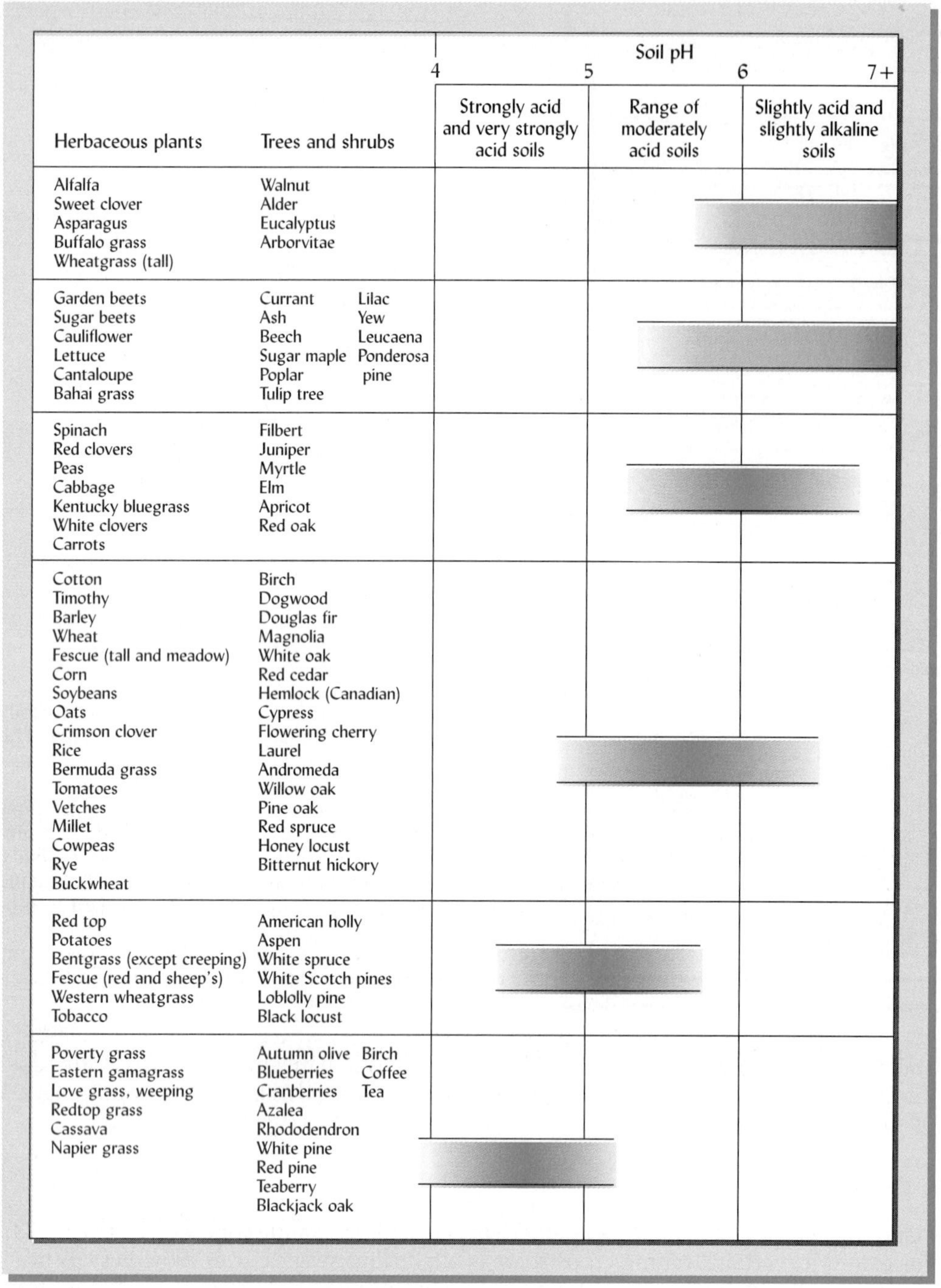

		Soil pH		
		4 5 6 7+		
Herbaceous plants	Trees and shrubs	Strongly acid and very strongly acid soils	Range of moderately acid soils	Slightly acid and slightly alkaline soils
Alfalfa Sweet clover Asparagus Buffalo grass Wheatgrass (tall)	Walnut Alder Eucalyptus Arborvitae			▓▓▓
Garden beets Sugar beets Cauliflower Lettuce Cantaloupe Bahai grass	Currant Lilac Ash Yew Beech Leucaena Sugar maple Ponderosa Poplar pine Tulip tree		▓▓▓	
Spinach Red clovers Peas Cabbage Kentucky bluegrass White clovers Carrots	Filbert Juniper Myrtle Elm Apricot Red oak		▓▓▓	
Cotton Timothy Barley Wheat Fescue (tall and meadow) Corn Soybeans Oats Crimson clover Rice Bermuda grass Tomatoes Vetches Millet Cowpeas Rye Buckwheat	Birch Dogwood Douglas fir Magnolia White oak Red cedar Hemlock (Canadian) Cypress Flowering cherry Laurel Andromeda Willow oak Pine oak Red spruce Honey locust Bitternut hickory	▓▓▓		
Red top Potatoes Bentgrass (except creeping) Fescue (red and sheep's) Western wheatgrass Tobacco	American holly Aspen White spruce White Scotch pines Loblolly pine Black locust	▓▓▓		
Poverty grass Eastern gamagrass Love grass, weeping Redtop grass Cassava Napier grass	Autumn olive Birch Blueberries Coffee Cranberries Tea Azalea Rhododendron White pine Red pine Teaberry Blackjack oak	▓▓▓		

FIGURE 9.23 Ranges of pH in mineral soils that present appropriate conditions for optimal growth of various plants. Note that the pH ranges are quite broad, but that plant requirement for calcium and sensitivity to aluminum toxicity generally decreases from the top group to the bottom group. Other factors, such as fertility level, will influence the actual relationships between plant growth and soil pH in specific cases. However, such a chart is of great value when choosing which species to plant or when deciding whether to change the pH with soil amendments.

Soil pH and Organic Molecules

Soil pH influences environmental quality in many ways, but we will discuss only one example here—the influence of pH on the mobility of ionic organic molecules in soils. The molecular structure of certain ionic herbicides includes such chemical groups as $-NH_2$ and $-COO^-$. If soil pH is low, the excess H^+ ions (protons) present in solution are attracted to and bond with these chemical groups, neutralizing negative charges and creating positively charged sites on the molecule. This process is called *protonation*.

Atrazine, a chemical that is widely used to control weeds in corn, is an example of a chemical whose mobility is greatly influenced by soil pH. In a low pH environment, the atrazine molecule protonates, developing a positive charge. The positively charged molecule is then adsorbed on the negatively charged soil colloids, where it is held until it can be decomposed by soil organisms. The adsorbed pesticide is less likely to move downward and into the groundwater. At pH values above 5.7, however, the adsorption is greatly

TABLE 9.3 **Forest Species Response to Calcium Oxide on an Extremely Acid Soil**

Adding 1 g of CaO to 180 g of this soil (a Dystrudept in Pennsylvania) raised the pH from 3.8 to 6.8 and the Ca/Al ratio from 0.24 to 15.5. The treated soil was overlimed with respect to such acid-loving species as Teaberry and White Pine.

Positive response to CaO (Al-sensitive)		Negative or no response to CaO	
Plant species	Root response to CaO addition, %	Plant species	Root response to CaO addition, %
Sugar maple	+150	Black locust	0
Pin oak	+104	Mountain laurel	−15
Quaking aspen	+83	Blueberry	−21
Bitternut hickory	+52	Norway spruce	−31
Red cedar	+46	Black birch	−47
Grey dogwood	+41	Chestnut oak	−52
White spruce	+40	White pine	−53
Honey locust	+17	Teaberry	−68

[Data selected from Demchik et al. (1999a)].

reduced, and the tendency for the herbicide to move downward in the soil is increased. Of course, the adsorption in acidic soils also reduces the availability of atrazine to weed roots, thus reducing its effectiveness as a weed killer. Farmers using such chemicals may notice excessive weediness in their fields as the first sign that their soils need liming.

9.8 RAISING SOIL pH BY LIMING

Agricultural Liming Materials

To decrease soil acidity (raise the pH), the soil is usually amended with alkaline materials that provide conjugate bases of weak acids. Examples of such conjugate bases include carbonate (CO_3^{2-}), hydroxide (OH^-), and silicate (SiO_3^{2-}). These conjugate bases are anions that are capable of consuming (reacting with) H^+ ions to form weak acids (such as water). For example:

$$CO_3^{2-} + 2H^+ \longrightarrow CO_2 + H_2O \tag{9.23}$$

Most commonly, these bases are supplied in their calcium or magnesium forms ($CaCO_3$, etc.) and are referred to as **agricultural limes**. Some liming materials contain oxides or hydroxides of alkaline earth metals (e.g., CaO or MgO), which form hydroxide ions in water.

$$CaO + H_2O \longrightarrow Ca(OH)_2 \longrightarrow Ca^{2+} + 2OH^- \tag{9.24}$$

Unlike fertilizers, which are used to supply plant nutrients in relatively small amounts for plant nutrition, *liming materials are used to change the chemical makeup of a substantial part of the root zone.* Therefore, lime must be added in large enough quantities to chemically react with a large volume of soil. This requirement dictates that inexpensive, plentiful materials are normally used for liming soils—most commonly finely ground limestone or materials derived from it (see Table 9.4). Wood ashes (which contain oxides of Ca, Mg, and K) are also effective. As a practical matter, the choice of which liming material to use is often based mainly on the relative costs of transporting large amounts of these materials from their source to the site to be limed.

Dolomitic limestone products should be used if magnesium levels are low. In some highly weathered soils, small amounts of lime may improve plant growth, more because of the enhanced calcium or magnesium nutrition than from a change in pH.

How Liming Materials React to Raise Soil pH

CHEMICAL REACTIONS. Most liming materials—whether they be oxide, hydroxide, or carbonate—react with carbon dioxide and water to yield bicarbonate when applied to an acid soil. The carbon dioxide partial pressure in the soil, usually several hundred

TABLE 9.4 Common Liming Materials: Their Composition and Use

The two limestones are by far the most commonly used. Use of the other materials is largely dependent on the need for fast reaction, cost, and local availability. The relative amounts needed of the different materials can be judged by comparing the respective $CaCO_3$ equivalent values.

Common name of liming material	Chemical formula (of pure materials)	% $CaCO_3$ equivalent	Comments on manufacture and use
Calcitic limestone	$CaCO_3$	100	Natural rock ground to a fine powder. Low solubility; may be stored outdoors uncovered. Noncaustic, slow to react.
Dolomitic limestone	$CaMg(CO_3)_2$	95–108	Natural rock ground to a fine powder; somewhat slower reacting than calcitic limestone. Supplies Mg to plants.
Burned lime (oxide of lime)	CaO (+ MgO)[a]	178	Caustic, difficult to handle, fast-acting, can burn foliage, expensive. Made by heating limestone. Protect from moisture.
Hydrated lime (hydroxide of lime)	$Ca(OH)_2$ (+ $Mg(OH)_2$)[a]	134	Caustic and difficult to handle. Fast-acting, can burn foliage, expensive. Made by slaking hot CaO with water. Must be protected from moisture.
Basic slag	$CaSiO_3$	70	By-product of pig-iron industry. Must be finely ground. Also contains 1–7% P.
Marl	$CaCO_3$	40–70	Usually mined from shallow coastal beds, dried, and ground before use. May be mixed with soil or peat.
Wood ashes	CaO, MgO, K_2O, $K(OH)$, etc.	40	Caustic, largely water-soluble, must be protected from water during storage.
Misc. lime-containing by-products	Usually $CaCO_3$ with various impurities	20–100	Variable composition; test for toxic impurities.

[a] If made from dolomitic limestone.

times greater than that in atmospheric air, is generally high enough to drive such reactions to the right. For example:

$$CaMg(CO_3)_2 + 2H_2O + 2CO_2 \rightleftharpoons Ca + 2HCO_3^- + Mg + 2HCO_3^- \qquad (9.25)$$

Dolomitic limestone   Bicarbonate   Bicarbonate

The Ca and Mg bicarbonates are much more soluble than are the carbonates, so the bicarbonate formed is quite reactive with the exchangeable and residual soil acidity. The Ca^{2+} and Mg^{2+} replace H^+ and Al^{3+} on the colloidal complex:

$$\boxed{\begin{array}{l}\text{Clay or}\\ \text{humus}\end{array}} \begin{array}{l} H^+ \\ Al^{3+}\end{array} + 2Ca^{2+} + 4HCO_3^- \rightleftharpoons \boxed{\begin{array}{l}\text{Clay or}\\ \text{humus}\end{array}} \begin{array}{l} Ca^{2+}\\ Ca^{2+}\end{array} + Al(OH)_3 + H_2O + 4CO_2\uparrow \qquad (9.26)$$

Bicarbonate   (solid)

$$\boxed{\begin{array}{l}\text{Clay or}\\ \text{humus}\end{array}} \begin{array}{l} H^+ \\ Al^{3+}\end{array} + 2CaCO_3 + H_2O \rightleftharpoons \boxed{\begin{array}{l}\text{Clay or}\\ \text{humus}\end{array}} \begin{array}{l} Ca^{2+}\\ Ca^{2+}\end{array} + Al(OH)_3 + 2CO_2\uparrow \qquad (9.27)$$

Limestone   (solid)
(calcite)

The insolubility of $Al(OH)_3$, the weak dissociation of water, and the release of CO_2 gas to the atmosphere all pull these reactions to the right. In addition, the adsorption of the calcium and magnesium ions lowers the percentage acid saturation of the colloidal complex, and the pH of the soil solution increases correspondingly.

Lime Requirement: How Much Lime is Needed to Do the Job?

The amount of liming material required to ameliorate acid soil conditions is determined by several factors, including (1) the change required in the pH or exchangeable Al saturation, (2) the buffer capacity of the soil, (3) the amount or depth of soil to

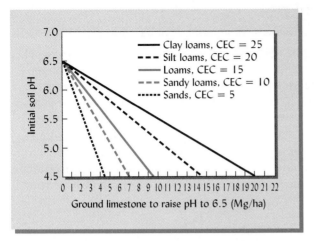

FIGURE 9.24 Effect of soil textural class on the amount of limestone required to raise the pH of soils from their initial level to pH 6.5. Note the very high amounts of lime needed for fine-textured soils that are strongly buffered by their high levels of clay, organic matter, and CEC. The chart is most applicable to soils in cool, humid regions where 2:1 clays predominate. In warmer regions where organic matter levels are lower and clays provide less CEC, the target pH would likely be closer to 5.8 and the amounts of lime required would be one-half to one-third of those indicated here. In any case, however, it is unwise to apply more than 7 to 9 Mg/ha (3 to 4 tons/acre) of liming materials in a single application. If more is needed, subsequent applications can be made at 2- to 3-year intervals until the desired pH is achieved.

Soil Acidity & Liming Recommendations in Ontario:

http://www.omafra.gov.on.ca/english/crops/pub811/2limeph.htm#changes

ameliorate, (4) the chemical composition of the liming materials to be used, and (5) the fineness of the liming material.

The limestone requirements of soils with several different textures (and therefore likely to have different buffering capacities) are estimated in Figure 9.24. Because of greater buffering capacity, the lime requirement for a clay loam is much higher than that of a sandy loam with the same pH value (see Section 9.4).

Within the pH range of 4.5 to 7.0, the degree of change in pH brought about by additions of base to an acid soil is determined by the buffering capacity of the particular soil (Figure 9.8). In turn, the capacity to buffer pH is closely related to organic matter and clay contents of a soil as reflected in the CEC. Given the titration curve for a given type of soil (such as the two examples shown in Figure 9.8), the pH of the unlimed soil, and the desired pH level, it is possible to calculate the approximate amount of liming material needed (see Box 9.4).

BUFFER pH METHODS FOR LIME REQUIREMENT. A rapid and inexpensive lab approach to estimating lime requirements is to equilibrate a sample of soil with a special salt solution that has a known initial pH value and is buffered to resist change in pH. Several different buffering solutions are in common use by soil-testing laboratories. The most widely used of these is the SMP (Shoemaker, McLean, and Pratt, 1961) method, which uses a solution buffered at pH = 7.5. It is well suited for moderately weathered soils with high CEC values, such as most soils in the northern and midwestern United States. Also popular is the Adams-Evans (Adams and Evans, 1962) solution, which is buffered at pH = 8 and was designed for use with lower CEC Ultisols, such as those common in the southeastern United States.

For all such methods, the greater the total acidity of the soil, the more the solution buffering is overcome and the solution pH reduced. Therefore, the pH drop in the buffer solution equilibrated with a soil is proportional to the amount of lime that would be needed to raise the pH of that soil. The amount of lime required to make a desired change in pH is determined from equations or tables that describe this proportionality as it pertains to the type of soils in question. The important thing to remember is that *the buffer pH indicates how much the soil acidity was able to change the buffer solution pH, but is not a measure of the soil pH itself.* For example, in the soil test report depicted in Figure 16.37, the soil pH_{water} = 5.2, in the very acid range, but the Adams-Evans buffer pH = 7.63, only slightly lowered from the initial pH of 8.0; therefore a relatively small application of lime was recommended (1 ton/acre = 2.2 Mg/ha).

EXCHANGEABLE ALUMINUM. Liming to eliminate exchangeable aluminum, rather than to achieve a certain soil pH, has been found appropriate for highly weathered soils such as Ultisols and Oxisols. By this approach, the required amount of lime can be calculated using values for the initial CEC and the percent Al saturation. For example, if a soil

BOX 9.4   CALCULATING LIME NEEDS BASED ON pH BUFFERING

Your client wants to grow a high-value crop of asparagus in a 2-ha field. The soil is a sandy loam with a current pH of 5.0. Asparagus is a calcium-loving crop that requires a high pH for best production (see Figure 9.23), so you recommend that the soil pH be raised to 6.8. Since the soil texture is a sandy loam, we will assume that its buffer curve is similar to that of the moderately buffered soil B in Figure 9.8. In actual practice, a soil test laboratory using this method to calculate the lime requirement should have buffer curves for the major types of soils in its service area.

1. Extrapolating from curve B in Figure 9.8, we estimate that it will require about 2.5 $cmol_c$ of lime/kg of soil to change the soil pH from 5.0 to 6.8. (Draw a horizontal line from 5.0 on the y-axis of Figure 9.8 to curve B, then draw a vertical line from this intersection down to the x-axis. Repeat this procedure beginning at 6.8 on the y-axis. Then use the x-axis scale to compare the distance between where your two vertical lines intersect the x-axis.)

2. Each molecule of $CaCO_3$ neutralizes 2 H^+ ions.

$$CaCO_3 + 2H^+ \rightarrow \rightarrow Ca^{2+} + CO_2 + H_2O$$

3. The mass of 2.5 $cmol_c$ of pure $CaCO_3$ can be calculated using the molecular weight of $CaCO_3$ = 100 g/mol.

$$(2.5 \text{ } cmol_c \text{ } CaCO_3/\text{kg soil}) \times (100 \text{ g/mol } CaCO_3) \times (1 \text{ mol } CaCO_3/2 \text{ } mol_c) \times (0.01 \text{ } mol_c/cmol_c)$$

$$= 1.25 \text{ g } CaCO_3/\text{kg soil}$$

4. Using the conversion factor of 2×10^6 kg/ha of surface soil (see footnote, page 152), we calculate the amount of pure $CaCO_3$ needed per hectare:

$$(1.25 \text{ g } CaCO_3/\text{kg soil}) \times (2 \times 10^6 \text{ kg soil/ha}) = 2,500,000 \text{ g } CaCO_3/\text{ha}$$

$$(2,500,000 \text{ g } CaCO_3/\text{ha}) \times (1 \text{ kg } CaCO_3/1000 \text{ g } CaCO_3) = 2500 \text{ kg } CaCO_3/\text{ha}$$

$$2500 \text{ kg } CaCO_3/\text{ha} = 2.5 \text{ Mg/ha (or about 1.1 tons/acre)}$$

5. Since our limestone has a $CaCO_3$ equivalence of 90%, 100 kg of our limestone would be the equivalent of 90 kg of pure $CaCO_3$. Consequently, we must adjust the amount of our limestone needed by a factor of 100/90:

$$2.5 \text{ Mg pure } CaCO_3 \times 100/90 = 2.8 \text{ Mg limestone/ha}$$

6. Finally, because not all the $CaCO_3$ in the limestone will completely react with the soil, the amount calculated from the laboratory buffer curve is usually increased by a factor of 2.

$$(2.8 \text{ Mg limestone/ha}) \times 2 = 5.6 \text{ Mg limestone/ha}$$

(using Appendix B, this value can be converted to about 2.5 tons/acre)

Note that this result is very similar to the amount of lime indicated by the chart in Figure 9.24 for this degree of pH change in a sandy loam.

has a CEC of 10 $cmol_c$/kg and is 50% Al-saturated, then 5 $cmol_c$/kg of Al^{3+} ions must be displaced (and their acidity from Al hydrolysis neutralized). This would require 5 $cmol_c$/kg of $CaCO_3$:

$$5 \text{ } cmol_c/\text{kg} \times (100 \text{ g/mol } CaCO_3) \times (1 \text{ mol } CaCO_3/2 \text{ } mol_c) \times (0.01 \text{ } mol_c/cmol_c)$$
$$= 2.5 \text{ g } CaCO_3/\text{kg soil.}$$

This amount is equivalent to 5000 kg/ha (2.5 g/kg $\times$ 2×10^6 kg/ha). Experience suggests that to assure a complete reaction in the field, the amount of limestone so calculated must be multiplied by a factor of 1.5 or 2.0 to give the actual amount of lime to apply.

INFLUENCE OF LIME COMPOSITION, FINENESS, AND DEPTH OF INCORPORATION. Liming materials are usually sold and rates for their application recommended on the basis of pure Ca, CaO, or $CaCO_3$. Since limestones vary, the recommended rates must be adjusted to reflect the actual composition of a particular liming material to be used. The general values in Table 9.4 suggest, for example, that a recommendation for 1.0 Mg of $CaCO_3$ equivalent could be met by using either 0.56 Mg of burned lime (1.0/1.78 = 0.56) or 1.43 Mg of

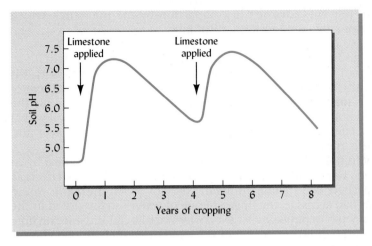

FIGURE 9.25 Gradual changes in soil pH in response to liming illustrate why repeated applications of limestone are needed to maintain the appropriate chemical balance in the soil. Soil pH increases during the months after an application of limestone and reaches a peak after about one year. However, the pH gradually declines thereafter until a new application of lime is required.

basic slag (1.0/0.7 = 1.43). In addition, for limestone, consideration must be given to the fineness of grind. If not finely ground, limestone reacts so slowly with the soil as to have little ameliorating effect on soil acidity. Materials with at least 50% of the particles small enough to pass through a 60-mesh screen (smaller than 0.25 mm in diameter) are quite satisfactory for most liming purposes. Finally, the recommendation must be adjusted if the depth of incorporation is other than that assumed or specified. In the example just given, if the recommendation was for incorporation to 18 cm, but you plan to incorporate to only 9 cm, then half as much liming material should be used.

How Lime is Applied

Practical tutorial on agricultural liming practices: http://hubcap.clemson.edu/~blpprt/acidity.html

FREQUENCY. Liming materials slowly react with soil acidity, gradually raising the pH to the desired level over a period ranging from a few weeks in the case of hydrated lime to a year or so with finely ground limestone. As Ca and Mg are removed from the soil by plants or by leaching, they are replaced by acid cations and the pH of the soil gradually drops. In humid regions the forces of acidification proceed relentlessly and application of lime to arable soils is not a one-time proposition, but must be repeated every 3 to 5 years (Figure 9.25).

Because of its gradual effects, lime should be spread about 6 to 12 months ahead of the crop that has the highest pH and calcium requirements. Thus, in a rotation with corn, wheat, and two years of alfalfa, the lime may be applied after the corn harvest to favorably influence the growth of the alfalfa crop that follows. However, since most lime is bulk-spread using heavy trucks (Figure 9.26), applications are most feasible on

FIGURE 9.26 Bulk application by specially equipped trucks is the most widespread method of applying ground limestone. The scene pictured occurred on a windy day and the dispersion by wind illustrates the finely ground nature of the agricultural limestone applied. To avoid problems with heavy trucks bogging down in soft, recently tilled soils, it is often preferred to make lime applications to land that is in sod, under no-till management, or frozen hard. (Photo courtesy of R. Weil)

the sod or hay crop. This prevents adverse soil compaction by the truck tires, which might occur if the lime were applied on recently tilled land.

DEPTH OF INCORPORATION.   Liming will be most beneficial to acid-sensitive plants if as much as possible of the root environment is altered. However, in most instances it is economically and physically feasible to mix lime into only the upper 15 to 20 cm of soil. The Ca^{2+} and Mg^{2+} ions provided by limestone replace acid cations on the exchange complex and do not move readily down the profile. Therefore, for soil with a high CEC, the short-term effects of limestone are mainly limited to the soil layer into which the material was incorporated.

Since subsoils are low in organic matter, subsoil acidity is often accompanied by Al toxicity causing plant roots to be restricted to the upper, limed soil layer. The effects of subsoil acidity will be especially detrimental during dry periods, when the plants would most benefit if their roots had access to the huge volume of water usually stored in the deeper soil horizons. Heavy plows pulled by large crawler tractors can incorporate lime into subsoil layers down to 50 or 60 cm deep, but the expense of such an operation is usually prohibitive in terms of both money and energy. Somewhat less expensive are special tillage machines that mix lime into narrow slots, creating low acidity zones that roots can use to explore the subsoil. A more sustainable alternative may be to create conditions that encourage the activity of certain species of deep-burrowing earthworms that have been found effective in moving lime down the soil profile. Other alternatives for ameliorating subsoil acidity are discussed in Section 9.9.

OVERLIMING.   Another practical consideration is the danger of **overliming**—the application of so much lime that the resultant pH values are too high for optimal plant growth. Overliming is not very common on fine-textured soils with high buffer capacities, but it can occur easily on coarse-textured soils that are low in organic matter. The detrimental results of excess lime include deficiencies of iron, manganese, copper, and zinc; reduced availability of phosphate; and constraints on the plant absorption of boron from the soil solution. It is an easy matter to add a little more lime later, but quite difficult to counteract the results of applying too much. Therefore, liming materials should be added conservatively to poorly buffered soils. For some Ultisols and Oxisols, overliming may occur if the pH is raised even to 6.0.

Special Liming Situations

LIMING FORESTS.   Spreading of limestone on forested watersheds is rarely practical (see Section 9.6). These areas are not accessible to ground-based spreaders, and manual application is too tedious and expensive. In some cases, landowners have turned to the use of helicopters for aerial spreading of small quantities of liming material. For very acid sandy soils, such small applications can ameliorate the ill effects of soil acidity and provide sufficient calcium for the trees.

UNTILLED SOILS.   Some soil–plant systems, such as no-till farming, make it difficult to incorporate and mix limestone with the soil. Fortunately, the undisturbed residue mulches in these systems tend to encourage earthworm activity, which, as we have just mentioned, can help distribute lime down the profile (Plate 81). Although apparently not critical on low CEC soils (Figure 9.27), it is considered advisable to incorporate a generous application of limestone as deeply as possible to correct any root-limiting acidity before commencing a no-till system on an acid high-CEC soil. Subsequent surface applications of limestone should adequately combat the acidity, which forms near the surface due to the decomposition of residues and the application of nitrogen fertilizers.

In lawns, golf greens, and other turfgrass areas it is also not possible to till the needed limestone into the soil. Again, the time to correct any problems with soil acidity is during soil preparation for the initial establishment of the grass. By proper timing of future liming applications with annual aeration tillage operations that leave openings down into the soil (see Figure 7.15), some downward movement of the lime can be achieved. In untilled systems, which require surface application of lime, frequent application of small quantities is most effective. In orchards, incorporation of lime into the planting hole, followed by periodic surface applications, is recommended for acid soils.

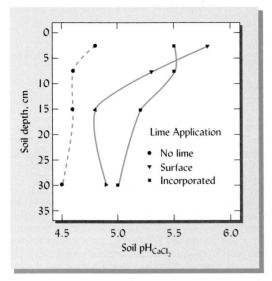

FIGURE 9.27 Soil pH 5 years after lime was applied to the soil surface or incorporated in preparation for converting grass pasture to no-till cropland. Dolomitic limestone was applied at 4.5 Mg/ha, enough to eliminate exchangeable Al^{3+} in the upper 20 cm of this clayey Oxisol in Brazil. Four months before the first crop was sown, three treatments were established: (1) no lime, (2) lime applied on the soil surface, and (3) lime incorporated into the soil, half plowed down to 20 cm and the remainder harrowed into the upper 10 cm. No other tillage was used during the 5 years of cropping. The incorporated lime raised the pH significantly down to 40 cm, but most markedly in the 20 cm zone of incorporation. The surface-applied lime also raised the pH to a considerable depth, thanks to the earthworm activity and biopore leaching characteristic of pasture soils. Exchangeable Al was negligible at $pH_{CaCl_2} > 4.8$. The cumulative crop yield over 5 years was higher where lime was applied, but was not affected by the method of application. Profitability was greatest for the surface-applied lime because no expensive tillage was required (the cost of tillage to incorporate the lime was nearly equal to the cost of the lime itself). [Graphed from data in Caires et al. (2006)]

9.9 ALTERNATIVE WAYS TO AMELIORATE THE ILL EFFECTS OF SOIL ACIDITY

Where the principal acidity problem is in the surface horizon and where sources of limestone are readily accessible, traditional liming procedures, as just described, are quite effective and economical. However, where subsoil acidity is a problem or where liming materials are not accessible or affordable, several other approaches to combating the negative effects of soil acidity may be appropriate for use with or without traditional liming. Particularly deserving of attention are the use of gypsum and organic materials to reduce aluminum toxicity and the use of plant species or genotypes that tolerate acid conditions.

Gypsum Applications

Gypsum ($CaSO_4 \cdot 2H_2O$) is a widely available material found in natural deposits or as an industrial by-product from the manufacture of high-analysis fertilizers, from flue-gas desulfurization by air-pollution "scrubbers" and in dry-wallboard wastes from construction or demolition of buildings. Researchers, mainly in the southeastern United States, Brazil, and South Africa, have found that gypsum can ameliorate aluminum toxicity despite the fact that it does not increase soil pH. In fact, gypsum has been found more effective than lime in reducing exchangeable aluminum in subsoils and thereby in improving root growth and crop yields (see Figure 9.28).

One reason for the effectiveness of gypsum is that the calcium from surface-applied gypsum moves down the soil profile more readily than that from lime. As lime dissolves, its reactions raise the pH, thus increasing the pH-dependent charges on the soil colloids, which in turn retain the released Ca^{2+}, preventing its downward leaching. In addition, the anion released by lime is either OH^- or CO_3^{2-}, both of which are largely removed by the lime reactions (either by forming water or carbon dioxide gas), thus depriving the Ca^{2+} cations of surplus anions that could accompany them in the leaching process. By contrast, gypsum as a neutral salt does not raise the soil pH and so does not increase the CEC. Furthermore, the SO_4^{2-} anion released by the dissolution of gypsum is available to accompany Ca^{2+} cations in leaching.

Once the Ca^{2+} and SO_4^{2-} ions move down to the subsoil, the mechanisms for amelioration of the effects of acidity may vary from soil to soil. However, gypsum is known to increase the level of calcium and reduce the level of aluminum both in the soil solution and on the exchange complex. The Ca^{2+} ions replace Al^{3+} ions from the exchange site, and the released aluminum is thought to react directly, or indirectly, with the sulfate ion. For example, the SO_4^{2-} can replace certain OH^- ions associated with Fe, Al, oxyhydroxides through reactions such as the following:

$$\boxed{\begin{array}{c}\text{Fe, Al} \\ \text{oxyhydroxides}\end{array}} {\overset{\displaystyle -OH}{\underset{\displaystyle -OH}{}}} + Ca^{2+} + SO_4^{2-} \longrightarrow \boxed{\begin{array}{c}\text{Fe, Al} \\ \text{oxyhydroxides}\end{array}} = SO_4 + Ca^{2+} + 2OH^- \quad (9.28)$$

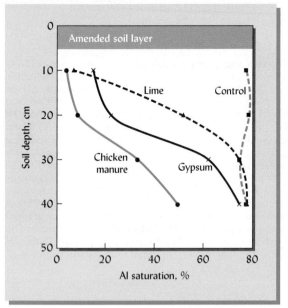

FIGURE 9.28 Aluminum saturation percentage in the subsoil of a fine-textured Hawaiian Ultisol after treatment of the surface soil with chicken manure, limestone, or gypsum. The soil was slowly leached with the 380 mm of water. The lime raised the pH and thereby effectively reduced Al saturation, though only in the upper 10 to 20 cm. The effect of gypsum extended somewhat more deeply. Gypsum does not raise the pH or increase the CEC, but is more soluble than lime and provides the SO_4^{2-} anions to accompany Ca^{2+} cations as they leach downward in solution. The greatest and deepest reduction in Al saturation resulted from the chicken manure. Another Ca-rich organic amendment, sewage sludge, gave similar results (not shown). The manure probably formed soluble organic complexes with Ca^{2+} ions, which then moved down the profile where Ca exchanged with Al to form nontoxic organic Al complexes. [Redrawn from Hue and Licudine (1999); used with permission of the American Society of Agronomy]

The $Ca(OH)_2$ can then react with Al^{3+} ions in the soil solution to form insoluble $Al(OH)_3$, thereby reducing the concentration of the Al^{3+} while increasing that of Ca^{2+}:

$$3\ Ca(OH)_2 + 2Al^{3+} \longrightarrow 2Al(OH)_3 + 3Ca^{2+} \tag{9.29}$$
$$\text{(insoluble)}$$

A second probable mechanism of aluminum detoxification by gypsum is the formation of $AlSO_4^+$ ions which are nonphytotoxic.

Using Organic Matter

Practices such as the application of organic wastes, production of cover crops (see Section 16.2), and mulching or return of crop residues all increase the organic matter in soil. In so doing they can ameliorate the effects of soil acidity in at least three ways:

1. Humified organic matter can bind tightly with aluminum ions and prevent them from reaching toxic concentrations in the soil solution (see Figure 9.19).
2. Low-molecular-weight organic acids produced by microbial decomposition or root exudation can form soluble complexes with aluminum ions that are non-toxic to plants and microbes. Table 9.5 provides an example of the ameliorating effect of adding decomposable organic materials.

TABLE 9.5 **Effects of Organic Residues on Aluminum Ions in Solution and Crop Yield in an Aluminum-Toxic Oxisol**

Initially the soil pH was 4.4 and the Al saturation of the effective CEC was 74%. The two tree residues and the manure were incorporated into the upper 5 cm of soil at the rate of 6 Mg/ha. Adequate fertilizer was applied to all plots to minimize any crop response to the nutrients added in the residues.

Organic residue	Inorganic Al in solution mg/L	Average yield of maize and beans Mg/ha
Crop residues alone (CR)	2.91	1.13
Grevillea leaves and twigs + CR	1.24	1.97
Leucaena leaves and twigs + CR	1.24	2.65
Farmyard manure + CR	0.93	2.67
None	2.92	0.79

Compiled from Wong et al. (1995)

3. Many organic amendments contain substantial amounts of calcium held in organic complexes that leach quite readily down the soil profile. Therefore, if such amendments as legume residues, animal manure, or sewage sludge are high in Ca, they can effectively combat aluminum toxicity and raise Ca and pH levels, not only in the surface soil where they are incorporated, but also quite deep into the subsoil (see Figure 9.28).

Enhancing these organic matter reactions may be more practical than standard liming practices for resource-poor farmers or those in areas far from limestone deposits. Green manure crops (vegetation grown specifically for the purpose of adding organic matter to the soil) and mulches can provide the organic matter needed to stimulate such interactions and thereby reduce the level of Al^{3+} ions in the soil solution. Aluminum-sensitive crops can then be grown following the green manure crop. One caution regarding use of organic amendments to ameliorate soil acidity is that the amounts of these materials effective for this purpose may exceed amounts suggested by nutrient management guidelines designed to prevent polluting water from excessive leaching and runoff losses of nitrogen and phosphorus (see Sections 16.4 and 16.6).

Selecting Adapted Plants

SELECTION OF TOLERANT SPECIES. It is often more judicious to solve soil acidity problems by changing the plant to be grown rather than by trying to change the soil pH. Table 9.3 and Figure 9.23 give some idea of the variation in plant tolerance to soil acidity. The choice of grass and tree species should consider soil pH adaptation, whether revegetating a former coal mine site on acidic mining debris or landscaping a suburban front yard on alkaline desert soils.

GENETIC IMPROVEMENT OF ACID TOLERANCE. Plant breeders and biotechnologists have developed cultivars that are quite tolerant of very acid conditions (Figure 9.29). For example, high-yielding varieties of wheat, soybeans, and corn that can tolerate soil acidity and high aluminum levels have been developed (see Figure 9.21). These varieties are especially valuable in some areas of the tropics where even modest liming applications are economically impractical. These advances show how collaboration between plant and soil scientists can enhance the ability of low-income families to grow their traditional food crops in acid, degraded soils.

9.10 LOWERING SOIL pH

It is often desirable to reduce the pH of highly alkaline soils. Furthermore, some acid-loving plants cannot even tolerate near-neutral pH values. For example, rhododendrons and azaleas, favorites of gardeners around the world, grow best on soils having

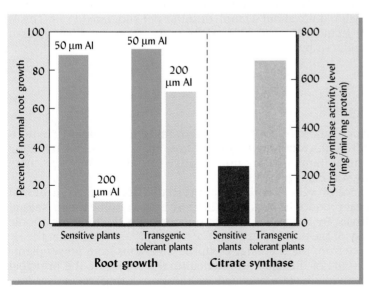

FIGURE 9.29 (*Left*) The effect of two levels of aluminum on the root growth of Al-sensitive plants compared to that of tolerant plants created by transgenic engineering. (*Right*) The comparative production by sensitive and tolerant plants of citrate synthase, an enzyme that stimulates the production of citric acid. Citric acid is produced by roots of tolerant plants to complex with (chelate) Al ions, thus preventing their absorption into the root. High citric acid producing, Al-tolerant strains have been developed by genetic engineering for cassava, maize, and other crops grown in acid tropical soils. [Drawn from data of de la Fuente et al. (1997)]

pH values of 5.0 and below. Blueberry is an acid-loving crop that requires similar low pH levels for good production. To accommodate these plants, it is sometimes desirable to increase the acidity of even acid soils. This is done by adding acid-forming organic and inorganic materials.

Acid Organic Matter

As organic residues decompose, organic and inorganic acids are formed. These can reduce the soil pH if the organic material is low in calcium and other nonacid cations. Leaf mold from coniferous trees, pine needles, tanbark, pine sawdust, and acid peat moss are quite satisfactory organic materials to add around ornamental plants (but see Section 12.3 for nitrogen considerations with these materials). However, farm manures (particularly poultry manures) and leaf mold of such high-base trees as beech and maple may be alkaline and may increase the soil pH, as discussed in Section 9.9.

Inorganic Chemicals

When the addition of acid organic matter is not feasible, inorganic chemicals such as aluminum sulfate ("alum") or ferrous sulfate ($Fe^{II}SO_4$) may be used. The latter chemical provides available iron (Fe^{2+} ions) for the plant and, upon hydrolysis, enhances acidity by reactions such as the following:

$$FeSO_4 + \tfrac{1}{4}O_2 + 1\tfrac{1}{2}H_2O \longrightarrow FeOOH + H_2SO_4 \tag{9.30}$$
$$\text{Sulfuric acid}$$

Ferrous sulfate thus serves a double purpose for iron-loving plants by supplying available iron directly and by reducing the soil pH—a process that may cause a release of fixed iron present in the soil (see Plates 86 and 88). Ferrous sulfate should be worked into the soil around ornamental plants, while taking care to avoid overly disturbing the root system. Contact with ferrous sulfate (but not alum) may cause black discoloration of foliage or mulch from the formation of iron sulfides.

Another material often used to increase soil acidity is elemental sulfur (Box 9.5). As the sulfur undergoes microbial oxidation in the soil (see Section 13.20), 2 moles of acidity (as sulfuric acid) are produced for every mole of S oxidized:

$$2S + 3O_2 + 2H_2O \longrightarrow 2H_2SO_4 \tag{9.31}$$

Under favorable conditions, sulfur is 4 to 5 times more effective, kilogram for kilogram, in developing acidity than is ferrous sulfate. Although ferrous sulfate brings about more rapid plant response, sulfur is less expensive, is easy to obtain, and is often used for other purposes. The quantities of ferrous sulfate or sulfur that should be applied will depend upon the buffering capacity of the soil and its original pH level. Figure 9.8 suggests that for each unit drop in pH desired, a well-buffered soil (e.g., a silty clay loam with 4% organic matter) will require about 4 $cmol_c$ of sulfur per kilogram of soil. This is about 1200 kg S/ha (since 1 $cmol_c$ of S = 0.32/2 = 0.16 g, the 2 mol_c/mol being based on the 2 mol of H^+ ion produced by each mole of S).

9.11   CONCLUSION

No other soil characteristic is more important in determining the chemical environment of higher plants and soil microbes than the pH. There are few reactions involving the soil or its biological inhabitants that are not sensitive to soil pH. This sensitivity must be recognized in any soil-management system.

Acidification is a natural process in soil formation that is accentuated in humid regions where processes that produce H^+ ions outpace those that consume them. Natural acidification is largely driven by the production of organic acids (including carbonic acid) and the leaching away of the nonacid cations (Ca^{2+}, Mg^{2+}, K^+, and Na^+) that the H^+ ions from the acids displace from the exchange complex. Emissions from power plants and vehicles, as well as inputs of nitrogen into agricultural systems are the principal means by which human activities have accelerated acidification during recent decades.

BOX 9.5 COSTLY AND EMBARRASSING SOIL pH MYSTERY

Work was nearing completion on an elaborate, 1-ha public garden in the heart of Washington, D.C. A ribbon-cutting ceremony planned for the following week would include the wealthy donors and powerful politicians who had supported the creation of the multimillion-dollar garden. The horticulturalist in charge now paced nervously in the light June rain as half a dozen soil scientists worked feverishly to collect soil and plant samples. Something had gone terribly wrong. Large ugly areas of brown, dying turfgrass were scattered throughout the garden.

Now, after so many rare and exotic trees, shrubs, and flowers had been planted, the horticulturalist feared the worst—the soil might be toxic, need to be removed, and the whole garden started

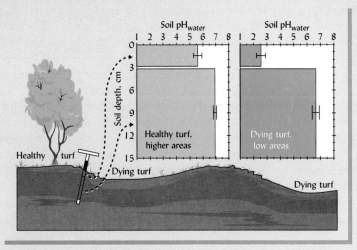

FIGURE 9.30 *Dying turf and soil acidity in a rooftop garden. [Courtesy of R. Weil]*

over. The garden was built partly over an underground museum. Beneath the pleasing, undulating surface topography, the "topsoil" lay over a meter thick in places, covering a tangle of pipes, conduits, and wires on the museum roof. If the soil needed to be removed, it would have to be done the slow and expensive way—by hand. The horticulturalist suspected some toxic factor in the soil was killing the grass and would soon start damaging other plants as well. He knew that the growing media installed was not exactly "topsoil" in the usual sense of the word. The landscape design specifications had called for "a natural friable soil . . . with 2% organic content . . . USDA textural class of loam and . . . pH 5.5 to 7.0." The lowest bidder had offered to make a "topsoil" using sediments dredged from a nearby tidal river, modified with enough lime and sand to meet the pH and texture requirements. The consulting engineers had run their lab tests and determined that the material met the specifications.

In late April, as the grand opening date approached, the grass began to turn brown and die in small patches. Although the shrubs, trees, and flowers in the cultivated flower beds were still looking good, the dead patches of turfgrass grew larger with every passing day. Turf specialists were called in, but could find no diseases or pests to account for the dead patches. Now, in desperation, the horticulturalist had hired several soil scientists to check the soil.

Some of the soil scientists knew where the "topsoil" had come from. Those trained as pedologists augered deep, looking in vain for telltale signs of acid sulfate weathering (Section 9.6) they suspected might solubilize toxic aluminum and heavy metals from the river bottom. Others, noting that it was the shallow-rooted turfgrass that seemed to suffer first, went shallow instead of deep (see Box 1.2), obtaining samples from several depths, including a separate set of samples from the topmost 3 cm. Since the turf was most damaged in the low spots between the little hillocks, they collected pairs of soil samples from several areas of dead turf and from adjacent areas where the turf was relatively healthy. Back at the lab, they stirred each soil sample in water and measured the pH. The results were completely normal until they got to the samples of that thin surface layer. Then they couldn't believe their eyes—all the samples of the upper 3 cm from the dead turf areas gave pH readings below pH 3, one as low as pH 1.9 (see bar graph in Figure 9.30). Looking closely at the 0-3 cm cores, they noticed small yellow flecks that smelled like sulfur. The pieces to the puzzle began to fall into place.

What was going on? The previous summer, shortly after the sod had been installed, the horticulturalist had pulled "normal" 20 cm deep soil cores for soil testing (see Section 16.11). The results indicated the soil pH was 7.2, a bit higher than the initial tests and considerably above the pH 6.0 to 6.5 recommended for the fine fescue turfgrass. Therefore, the horticulturalist had applied about 1000 kg/ha of elemental sulfur powder, as recommended to lower the pH by about 1 unit. He pulled another set of 20 cm deep soil samples about two months later and found the pH was still about 7.0. Therefore, he repeated the sulfur application. The lawn looked healthy during the cool, rainy winter, while the landscapers installed the valuable trees and shrubs. What he had failed to consider

(continued)

BOX 9.5 *(Cont.)*   COSTLY AND EMBARRASSING SOIL pH MYSTERY

was that sufficient time and warm weather would be needed for the soil microorganisms to oxidize the sulfur and produce sulfuric acid to acidify the soil. So the second sulfur application had been an over-response to the normal delay and had doubled the amount of S available to oxidize. Sulfur powder is quite water repellent and buoyant, so rain easily washed much of it off the high areas into the low spots, thus doubling or tripling the already doubled application—giving 5 or 6 times the recommended S concentration in those areas. When warm, wet weather the following spring stimulated the S-oxidizing bacteria to go into high gear, extreme acidity was produced in the thin surface layer of soil where the S was located and most of the turfgrass roots proliferated. Thankfully, the remedy would be simple and inexpensive: remove the sod along with about 5 cm of soil and lay down new sod. This they did and everyone at the opening ceremony was impressed by the beautiful lawn and garden.

The lessons learned? (1) Many soil processes are biological in nature—be patient and realize they will respond to environmental conditions with time. (2) Taking deep soil samples may "dilute out" evidence of extreme conditions near the soil surface. Therefore, be sure to sample the upper few cm separately for untilled soils, especially if amendments have been applied to the surface.

Aluminum is the other principal acid cation besides hydrogen. Its hydrolysis reactions produce H^+ ions and its toxicity comprises one of the main detrimental effects of soil acidity. The total acidity in soils is the sum of the active, exchangeable (salt replaceable), and residual pools of soil acidity. Changes in the soil solution pH (the active acidity) are buffered by the presence of the other two pools. In certain anaerobic soils and sediments, the presence of reduced sulfur provides the potential for enormous acid production if the material is exposed to air by drainage or excavation.

Soil pH is largely controlled by the humus and clay fractions and their associated exchangeable cations. The maintenance of satisfactory soil fertility levels in humid regions depends considerably on the judicious use of lime to balance the losses of calcium and magnesium from the soil. Liming not only maintains the levels of exchangeable calcium and magnesium but in so doing also provides a chemical and physical environment that encourages the growth of most common plants. Gypsum and organic matter (either applied or grown) represent other tools that can be used to ameliorate soil acidity instead of, or in addition to, liming. On the other hand, it is sometimes most judicious to use acid-tolerant plants, rather than attempt to change the chemistry of the soil.

Knowing how pH is controlled, how it influences the supply and availability of essential plant nutrients as well as toxic elements, how it affects higher plants and human beings, and how it can be ameliorated, is essential for the conservation and sustainable management of soils throughout the world. Understanding soil processes that occur at high pH levels is especially important in arid regions, as the next chapter will explain.

STUDY QUESTIONS

1. Soil pH gives a measure of the concentration of H^+ ions in the soil solution. What, if anything, does it tell you about the concentration of OH^- ions? Explain.

2. Describe the role of aluminum and its associated ions in soil acidity. Identify the ionic species involved and the effect of these species on the CEC of soils.

3. If you could somehow extract the soil solution from the upper 16 cm of 1 ha of moist acid soil (pH = 5), how many kg of pure $CaCO_3$ would be needed to neutralize the soil solution (bring its pH to 7.0)? Under field conditions, up to 6 Mg of limestone may be required to bring the pH of this soil layer to a pH of 6.5. How do you explain the difference in the amounts of $CaCO_3$ involved?

4. What is meant by *buffering?* Why is it so important in soils, and what are the mechanisms by which it occurs?

5. What is acid rain, and why does it seem to have greater impact on forests than on commercial agriculture?

6. Calculate the amount of pure $CaCO_3$ that could theoretically neutralize the H^+ ions in a year's worth of acid rain if a 1-ha site received 500 mm of rain per year and the average pH of the rain was 4.0.

7. Discuss the significance of soil pH in determining specific nutrient availabilities and toxicities, as well as species composition of natural vegetation in an area.

8. How much limestone with a $CaCO_3$ equivalent of 90% would you need to apply to eliminate exchangeable aluminum in an Ultisol with CEC = 8 $cmol_c$/kg and an aluminum saturation of 60%?

9. Based on the buffer pH of your soil sample, a lab recommends that you apply 2 Mg of $CaCO_3$ equivalent to your field and plow it in 18 cm deep to achieve your target pH of 6.5. You actually plan to use the lime to prepare a large lawn and till it in only 12 cm deep. The lime you purchase has a carbonate equivalent = 85%. How much of this lime do you need for 2.5 ha?

10. A landscape contractor purchased 10 dump-truck loads of "topsoil" excavated from a black, rich-appearing soil in a coastal wetland. Samples of the soil were immediately sent to a lab to be sure they met the specified properties (silt loam texture, pH 6 to 6.5) for the topsoil to be used in a landscaping job. The lab reported that the texture and pH were in the specified range, so the topsoil was installed and an expensive landscape of beautiful plants established. Unfortunately, within a few months all the plants began to die. Replacement plants also died. The topsoil was again tested. It was still a silt loam, but now its pH was 3.5. Explain why this pH change likely occurred and suggest appropriate management solutions.

11. A neighbor complained when his azaleas were adversely affected by a generous application of limestone to the lawn immediately surrounding the azaleas. To what do you ascribe this difficulty? How would you remedy it?

12. The ill effects of acidity in subsoils can be ameliorated by adding gypsum ($CaSO_4 \cdot 2H_2O$) to the soil surface. What are the mechanisms responsible for this effect of the gypsum?

REFERENCES

Adams, F., and C. E. Evans. 1962. "A rapid method of measuring lime requirement of red-yellow podzolic soils," *Soil Sci. Soc. Am. Proc.*, **26**:355–357.

Adams, F., and Z. F. Lund. 1966. "Effect of chemical activity of soil solution aluminum on cotton root penetration of acid subsoils," *Soil Sci.*, **101**:193–198.

Barak, P., B. O. Jobe, A. R. Krueger, L. A. Peterson, and D. A. Laird. 1997. "Effects of long-term soil acidification due to nitrogen fertilizer inputs in Wisconsin," *Plant Soil*, **197**:61–69.

Beyrouty, C. A., J. K. Keino, E. E. Gbur, and M. G. Hanson. 1999. "Phytotoxic concentrations of subsoil aluminum as influenced by soils and landscape position," *Soil Science*, **165**:135–143.

Blake, L., W. T. Goulding, C.J.B. Mott, and A. E. Johnston. 1999. "Changes in soil chemistry accompanying acidification over more than 100 years under woodland and grass at Rothamsted Experiment Station, UK," *European J. Soil Sci.*, **50**:401–412.

Bloom, P. R., U. L. Skyllberg, and M. E. Sumner. 2005. "Soil acidity," pp. 411–459, in A. Tabatabai and D. Sparks (eds.), *Chemical Processes in Soils*. SSSA Book Series No. 8. (Madison, Wis.: Soil Science Society of America).

Caires, E. F., G. Barth, and F. J. Garbuio. 2006. "Lime application in the establishment of a no-till system for grain crop production in southern Brazil," *Soil Tillage Res.*, **89**:3–12.

de la Fuente et al. 1997. "Aluminum tolerance in transgenic plants by alteration of citrate synthesis," *Science*, **276**:1566–1568.

de la Fuente-Martinez, J. M., and L. Herrera-Estralla. 2000. "Advances in the understanding of aluminum toxicity and the development of aluminum-tolerant transgenic plants," *Advances in Agronomy*, **66**:103–120.

Demchik, M. C., W. E. Sharpe, T. Yangkey, B. R. Swistock, and S. Bubalo. 1999a. "The effect of calcium/aluminum ratio on root elongation of twenty-six Pennsylvania plants,"pp. 211–217, in W. E. Sharpe and J. R. Drohan (eds.), *The Effects of Acidic Deposition on Pennsylvania's Forests*. Proceedings of the Sept. 14–16, 1998 PA Acidic Deposition Conference. (University Park: Environmental Resources Research Institute, Pennsylvania State University).

Demchik, M. C., W. E. Sharpe, T. Yangkey, B. R. Swistock, and S. Bubalo. 1999b. "The relationship of soil Ca/Al ratio to seedling sugar maple population, root characteristics, mycorrhizal infection rate, and growth and survival," (pp. 201–210), in W. E. Sharpe and J. R. Drohan (eds.), *The Effects of Acidic Deposition on Pennsylvania's Forests.* Proceedings of the Sept. 14–16, 1998 PA Acidic Deposition Conference. (University Park, Environmental Resources Research Institute, Pennsylvania State University).

Edwards, K. J., P. L. Bond, T. M. Gihring, and J. F. Banfield. 2000. "An archaeal iron-oxidizing extreme acidophile important in acid mine drainage," *Science,* **287:** 1796–1799.

Falkengren-Grerup, U., D.-J. T. Brink, and J. Brunet. 2006. "Land use effects on soil N, P, C and pH persist over 40–80 years of forest growth on agricultural soils," *Forest Ecology and Management,* **225:**74–81.

Fanning, D. S., M. Rabenhorst, S. N. Burch, K. R. Islam, and S. A. Tangen. 2002. "Sulfides and sulfates," pp. 229–260, in J. B. Dixon and D. G. Schultz (eds.), *Soil Mineralogy with Environmental Applications.* SSSA Book Series No. 7. (Madison, Wis.: Soil Science Society of America).

Hue, N. V., and D. L. Licudine. 1999. "Amelioration of subsoil acidity through surface applications of organic manures," *J. Environ. Qual.,* **28:**623–632.

Jenny, H. et al. 1968. "Interplay of soil organic matter and soil fertility with state factors and soil properties," in *Organic Matter and Soil Fertility.* Pontificiae Academia Scientiarum Scripta Varia 32 (New York: Wiley).

Kuylenstierna, J.C.I., H. Rodhe, S. Cinderby, and K. Hicks. 2001. "Acidification in developing countries: Ecosystem sensitivity and the critical load approach on a global scale," *Ambio,* **30:**20–28.

Likens, G. E., C. T. Driscoll, and D. C. Buso. 1996. "Longterm effects of acid rain: Response and recovery of a forest ecosystem," *Science,* **272:**244–246.

Lumbanraja, J., and V. P. Evangelou. 1991. "Acidification and liming influence on surface charge behavior of Kentucky subsoils," *Soil Sci. Soc. Amer. J.,* **54:**26–34.

Magdoff, F. R., and R. J. Barlett. 1985. "Soil pH buffering revisited," *Soil Sci. Soc. Amer. J.,* **49:**145–148.

Markewitz, D., D. D. Richter, H. L. Allen, and J. B. Urrego. 1998. "Three decades of observed soil acidification in the Calhoun experimental forest: Has acid rain made a difference?" *Soil Sci. Soc. Amer. J.,* **62:**1428–1439.

Okusmi, T. A., H. Rust, and A.S.R. Juo. 1987. "Reactive characteristics of certain soils from south Nigeria," *Soil Sci. Soc. Amer. J.,* **51:**1256–1262.

Palmer, S. M., and C. T. Driscoll. 2002. "Acidic deposition: Decline in mobilization of toxic aluminium," *Nature,* **417:**242–243.

Rodhe, H., J. Langner, L. Gallardo, and B. Kjellström. 1995. "Global scale transport of acidifying pollutants," *Water, Air and Soil Pollution,* **85:**37–50.

Shoemaker, H. E., E. O. McLean, and P. F. Pratt. 1961. "Buffer methods for determining lime requirements of soils with appreciable amounts of extractable aluminum," *Soil Sci. Soc. Amer. Proc.,* **25:**274–277.

Stine, M. A., and R. R. Weil. 2002. "The relationship between soil quality and crop productivity across three tillage systems in south central Honduras," *Am. J. Alternative Agric.,* **17:**2–8.

Stoddard, J. L., D. S. Jeffries, A. Lükewille et al. 1999. "Regional trends in recovery from acidification in North America and Europe," *Nature,* **401:**575–578.

Sullivan, T. J., I. J. Fernandez, A. T. Herlihy, C. T. Driscoll, T. C. McDonnell, N. A. Nowicki, K. U. Snyder, and J. W. Sutherland. 2006. "Acid-base characteristics of soils in the Adirondack mountains, New York," *Soil Sci. Soc. Amer. J.,* **70:**141–152.

Sumner, M. E. 1993. "Gypsum and acid soils: The world scene," *Advances in Agronomy,* **51:**1–32.

Tipping, E. 2005. "Modelling Al competition for heavy metal binding by dissolved organic matter in soil and surface waters of acid and neutral pH," *Geoderma,* **127:**293–304.

Weil, R. R. 2000. "Soil and plant influences on crop response to two African phosphate rocks," *Agron. J.,* **92:**1167–1175.

Wong, M.T.F., E. Akyeampong, S. Nortcliff, M. R. Rao, and R. S. Swift. 1995. "Initial responses of maize and beans to decreased concentrations of monomeric inorganic aluminum with application of manure or tree prunings to an Oxisol in Burundi," *Plant and Soil,* **171:**275–282.

Zhang, F., S. Kang, J. Zhang, R. Zhang, and F. Li. 2004. "Nitrogen fertilization on uptake of soil inorganic phosphorus fractions in the wheat root zone," *Soil Sci. Soc. Amer. J.,* **68:**1890–1895.

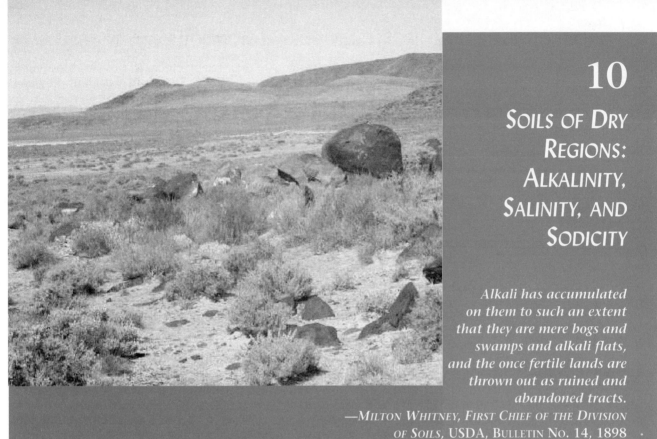

An alkaline soil landscape in Nevada. (R. Weil)

10

SOILS OF DRY REGIONS: ALKALINITY, SALINITY, AND SODICITY

Alkali has accumulated on them to such an extent that they are mere bogs and swamps and alkali flats, and the once fertile lands are thrown out as ruined and abandoned tracts.
—MILTON WHITNEY, FIRST CHIEF OF THE DIVISION OF SOILS, USDA, BULLETIN NO. 14, 1898 ·

More than 2100 years ago, the Roman armies finally emerged victorious in a long and bitter series of wars against their great rival, Carthage, a city-state located on the semi-arid north coast of Africa near modern day Tunis. To seal their victory and ensure that the Carthaginians would not arise once more to challenge the supremacy of Rome, the victors not only razed the city itself, but are said to have also *plowed salt into the land* around it. Common table salt (sodium chloride) was the ultimate weapon because it made the lands around Carthage useless for growing the food that would be required by the population (and armies) of a resurgent city-state.

In this chapter, we will learn why soluble salts can have such detrimental effects on soils and how for centuries people in many lands have acted as their own worst enemies by unwittingly bringing salt to their soils. The accumulation of soluble salts is perhaps the most vexing problem for long-term human land use in dry regions, but it is only one of several characteristics of dry-region soils that requires special attention.

In Chapter 9, we focused on humid regions where soil acidification is unrelenting. In this chapter, we will focus on arid and semiarid regions where most soils are alkaline (have a pH greater than 7). Many of these soils also accumulate detrimental levels of soluble salts (saline soils) or sodium ions (sodic soils), or both. Conditions associated with aridity, alkalinity, salinity, and sodicity can lead to a number of problems in the physical condition and fertility of soils in these regions. Some of these problems will be discussed in this chapter, while others can be found throughout the book. The principles outlined here will help improve your understanding of many concepts dealt with elsewhere in the book: mineral weathering and soil formation (Chapter 2); soil dispersion and crusting (Section 4.6); plant water stress (Section 6.3); infiltration and irrigation (Sections 6.1 and 6.9); plant deficiencies of phosphorus (Section 14.1), zinc, and iron (Section 15.5); and the toxicity to plants and animals of selenium and molybdenum.

For eons, plants and animals native to our deserts and rangelands have adapted to all but the most severe levels of soil alkalinity, salinity, and sodicity. People who wish to inhabit these dry and sunny environments must likewise adapt their actions to these soil conditions. This chapter will provide some of the tools needed to lead the way.

10.1 CHARACTERISTICS AND PROBLEMS OF DRY REGION SOILS

The principal soils that characterize low-rainfall regions include most Aridisols, many types of Entisols, and certain suborders (mainly Ustic, Xeric, and Natric) in the orders Mollisols, Inceptisols, Alfisols, and Vertisols (see Sections 3.6 to 3.13 for a description of these soils). The water-limited, high pH, carbonate-rich nature of these dry-region soils results in many characteristics and problems that are not generally found in the acid soils of more humid regions. We will begin by focusing on the nature of alkaline soils that do *not* have excessive levels of salts or sodium, conditions that will be taken up in Sections 10.5 to 10.7.

Soil Water Supply

Unlike their counterparts in humid regions, subsoil layers in low-rainfall areas are generally dry, except where the groundwater table is near the surface. Limited moisture in the upper soil horizons and competition for water among plants are principal factors determining the nature and productivity of the native vegetation (see Section 2.4), as well as the capacity of the land to support wild and domesticated animals. In turn, trampling by the hooves of grazing animals impacts the ability of the soil to absorb what limited rain does fall (Figure 10.1).

For agricultural cultivation of arid- and semiarid-region soils, water management must receive high priority. Irrigation, although extremely important, is possible on only a small fraction of the soils. For nonirrigated (dry land) cultivation, tillage practices that increase water infiltration during periods of rainfall or snowmelt should be encouraged (see Figure 6.8). Likewise, it is often wise to use tillage practices such as stubble mulch or no-till (see Section 6.4) that help reduce evaporative losses and soil erosion by keeping some vegetative residue cover at or near the soil surface. Although some 40% of the world's cultivated land is located in dry regions, most land there is uncultivated, occupied by rangelands and deserts.

Heterogeneity of Soils, Vegetation and Hydrology of Noncultivated Soils

Seen from the highway, many semiarid landscapes appear to be quite densely covered with vegetation. However, if one walks out into the landscape it is immediately evident that the plants are widely spaced, with much of the surface being quite bare of

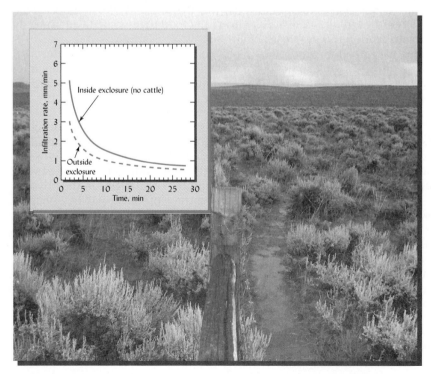

FIGURE 10.1 Cattle grazing may reduce water infiltration into semiarid rangeland soils. The effect of cattle is illustrated here by a photo taken immediately after a thundershower and by the mean infiltration measurements presented in the graph. The fence excludes cattle from the area on the left. Note the surface ponding and water running off between the shrubs on the right side where cattle have grazed. Among their other effects, the scattered shrubs partially protect the soil under them from trampling by animal hooves. (Photo and data courtesy of Sally Madden and Larry Munn, University of Wyoming)

vegetation (Chapter 10 opener photo and Figure 2.28). Such patchy vegetation is characteristic of environments in which water is too sparse to support a complete vegetative cover (see Section 6.4). Less obvious to the casual observer is that the soil under the plants is quite different from that in the bare spaces and that this difference is both a result of—and a cause of—the scattered plant distribution.

ISLANDS OF FERTILITY. Soil scientists studying arid lands have found that plants generally enhance the soil beneath them in several ways. Plants add litter, host macro- and microorganisms and trap windblown particles. Soils under clumps of vegetation therefore become higher in organic matter, silt, and clay, as well as richer in nitrogen and other nutrients.

Increased soil organic matter from plant litter leads to greater aggregation, while generations of plant roots create a network of biopores open to the surface, and the plant canopy partially protects the surface soil structure when it rains (see Section 17.5). All these effects lead to a notably enhanced infiltration of water into the soils in the vegetated patches (Figure 10.2).

The increased availability of soil moisture and nutrients enhances plant growth, which in turn further promotes the processes just mentioned. Over time, this *positive feedback loop* promotes a level of soil productivity in the vegetated patches distinctly more favorable than that in the unvegetated bare spaces. These so-called "islands of fertility" are a common feature of many arid land soils. Vegetation can also encourage "islands of fertility" by protecting soil from erosive desert winds (see Figure 10.3 for an extreme example). In certain tropical arid lands with long, gentle slopes, these processes lead to 2- to 4-m wide bands of shrub vegetation alternating with bands of bare soil, giving the terrain an appearance termed "tiger stripes" when seen from a distance.

Hydrologic heterogeneity (e.g., differences in infiltration capacity from place to place) may enhance ecosystem productivity and resilience in water-scarce environments. On a scale of centimeters to meters, an uneven soil surface with many small depressions that collect water increases the time for infiltration and reduces runoff from that part of the landscape. This natural process has been mimicked to improve rangeland productivity by use of heavy rollers with raised knobs that imprint a pattern of small depressions into the soil surface. In some desert areas, a surface layer of rock fragments plays a critical role in infiltration and runoff processes.

DESERT PAVEMENT. A large percentage of desert land (especially in North America) is naturally covered by a thin (usually single) layer of rock fragments ranging in size from coarse gravel to cobbles (Plate 55). When this layer covers at least 65% of the soil surface it is termed *desert pavement*, as it has the appearance of a street carefully paved with cobblestones (see also Section 3.10). The rock fragment layer protects the soil underneath

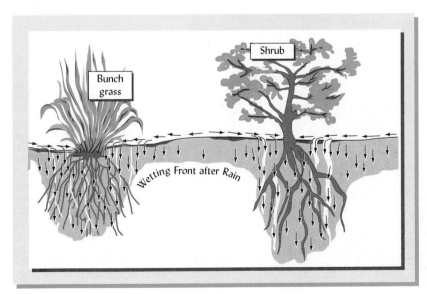

FIGURE 10.2 Differential runoff and infiltration rates may give rise to "islands" of enhanced soil water availability in arid and semiarid rangelands. Slight depressions and other features initially concentrate rainwater in small areas. Once plants establish in a pocket of relatively high moisture, they tend to amplify the soil heterogeneity because their litter and roots further enhance water infiltration in their vicinity. (Diagram courtesy of R. Weil)

FIGURE 10.3 "Islands" of relatively fertile soils commonly form under the influence of perennial vegetation in arid environments. This rather extreme example of the "island of fertility" phenomenon is in the Black Desert of the Sahara in western Egypt. The "islands" of soil preserved under the shrubs are 2 to 3 m high and 8 to 10 m in diameter. The inter-shrub areas are covered by virtually barren, shifting sands. (Photo courtesy of R. Weil)

from wind erosion, while trapping windblown particles that add more soil. The more complete the rock fragment cover, the less water enters the soil and the more it runs off. In fact, the density of desert pavement can profoundly affect the amount of water available for plant uptake and for such soil formation processes as the leaching of salts. Soils covered with desert pavement may exhibit much higher salt contents in the upper horizons than do adjacent soils without desert paving (Figure 10.4). These effects on water and salt in the soil, in turn, influence the distribution of desert shrubs in the landscape; more shrubs tend to grow where the desert pavement is less complete or absent.

Concentrating water in localized areas of enhanced infiltration assures that at least some parts of the landscape receive enough water for plant survival and productivity. Those parts of the landscape that experience low infiltration and high runoff rates serve as *water sources*. In other parts of the landscape, soils exhibit high infiltration rates and

FIGURE 10.4 Soils with only scattered rock fragments (generally less than two-thirds cover) have relatively high rates of water infiltration and therefore are largely leached of salts in the upper layers (*right graph*). Where the rock fragment cover is more complete, it is termed *desert pavement*. Like the pavement on a parking lot, desert pavement greatly reduces water infiltration; hence salts are not leached very deeply in the soil (*left graph*). In the photograph rock fragment cover varies from nearly 90% on the left to less than 50% on the right. [Data for sites in the Mojave Desert from Wood et al. (2005); photo courtesy of R. Weil]

BOX 10.1   RUNOFF FARMING: PEOPLE USING DESERT SOILS[a]

FIGURE 10.5 *Rows of fruit trees grow in the middle of an extremely arid landscape. The trees grow only because the small area devoted to their cultivation receives much of the rainwater that falls over the large, barren runoff catchment area. This scientific experiment in the Negev Desert of southern Israel reconstructs the runoff farming techniques of the ancient Nabetean people. (Photo courtesy of R. Weil)*

Several thousand years ago, two completely independent civilizations learned to manipulate landscape hydrology to allow farming in their respective desert habitats, the Anasazi Indians in the deserts of the American Southwest and the Nabateans in the Negev Desert. These people constructed diversions to collect runoff water from large areas of low-infiltration, sloping land (typically 20 to 100 ha) and concentrate this water on small, cultivated areas in the valley bottoms (typically about 1 or 2 ha). To enhance the performance of their *runoff farms*, they terraced and built berms around their valley bottom fields to maximize the infiltration and storage of water there. On the sloping catchment land their aim was to maximize the runoff. This they did by removing scattered surface rocks and debris that normally slow down runoff and increase infiltration. These practices may have supplied their cultivated fields with 5 to 6 times the amount of water that would normally have fallen directly on these fields as rain. The ruins of their farms and water collection systems can still be seen in parts of Arizona and southern Israel (Figure 10.5).

[a] Norton et al. (2007) discuss hydrologic and fertility aspects of present day runoff farming in New Mexico.

low runoff and serve as *water sinks*. On a landscape scale, topography plays an especially important role in the hydrology of arid lands. Steep slopes, especially when barren of vegetation, tend to absorb little water during rainstorms and generate large volumes of runoff that moisten downslope soils. Ancient peoples learned to harness these landscape–scale processes to allow agricultural production in regions where rainfall is far too low to sustain crops or orchards (Box 10.1).

MICROBIOTIC CRUSTS.   Another special feature that influences the hydrology of arid regions' soils in their natural state is a biological crust, which often appears as a thick, dark-colored, sometimes jagged, coating on the soil surface. We shall discuss these fragile living structures in Section 11.14, but they deserve mention here because of their role in the hydrology of many desert areas. Formed by algae, fungi, and cyanobacteria, microbiotic crusts originally covered much of the inter-shrub surface on undisturbed, fine-textured alkaline soils not covered by rock fragments. However, in many places, traffic by humans and cattle has degraded much of the original crust (Figure 10.6). While all forms of microbiotic crusts dramatically reduce soil erosion by wind and water (Section 17.11), the rough or pinnacled types also increase water infiltration rates. The smooth-surfaced types may have the opposite effect.

Calcium-Rich Layers

Soils of low-rainfall areas commonly accumulate calcium carbonate that forms a *calcic horizon* at some depth in the soil profile (Figure 10.7 and Plate 13). Calcareous soil materials (with free calcium carbonates) can be distinguished in the field by the

FIGURE 10.6 The sparse vegetation and exposed soil surface of a degraded rangeland in Arizona. Only a few scattered fragments remain of the biological crust (see Section 11.14) that once covered the soil between clumps of vegetation. Much of the soil is now bare and its degraded structure seals the surface, reducing its capacity to absorb water from the infrequent but high-intensity rainstorms that occur in this region. Ecosystems in arid and semiarid regions are often quite sensitive to such disturbances as overgrazing and off-road vehicle traffic. Partly because of the precarious water balance in these systems, recovery of rangeland vegetation may be very slow. Knife handle is 12 cm long. (Photo courtesy of R. Weil)

Illustrated tutorial on carbonates in soils:

http://edafologia.ugr.es/carbonat/indexw.htm

effervescence (fizzing) that occurs if a drop of acid (10% HCl or strong vinegar) is applied. The high carbonate concentrations in these calcic horizons can inhibit root growth for some plants. In eroded spots, or in regions of very low rainfall (<25 cm/yr), carbonate concentrations may be found at or near the soil surface (Plate 99 after page 656). In these cases, serious micronutrient and phosphorus deficiencies can be induced in plants that are not adapted to calcareous conditions (see following).

In other alkaline soils, one or more subsoil layers may be cemented into hard, concretelike horizons such as petrocalcic layers or duripans (Figure 3.20). Many alkaline soils also contain layers rich in calcium sulfate (gypsum), a mineral much more soluble than calcium carbonate. The depth of calcic horizons and gypsic horizons is largely determined by the age of the soil and by the amount of rainfall available to leach these minerals downward.

FIGURE 10.7 Calcium carbonate accumulation in the lower part of the B horizon (Btk) and the upper part of the C horizon (the knife is inserted at the upper boundary of the Ck horizon) characterizes this Ustoll and many other soils of arid and semiarid regions. Such calcareous layers maintain high pH and calcium levels, constraining the availability of nutrients such as phosphorus, iron, manganese, and zinc. (Photo courtesy of R. Weil)

Colloidal Properties

CATION EXCHANGE CAPACITY.   The cation exchange capacities (CECs) of alkaline soils are commonly higher than those of acid soils with comparable soil textures. This is true for two reasons. First, the 2:1-type clays that are most common in alkaline soils possess high amounts of permanent charge. Second, the high pH levels of alkaline soils stimulate high levels of pH-dependent charges on the soil colloids, especially humus (Section 8.9).

CLAY DISPERSION.   Clays in alkaline soils are particularly subject to deflocculation or dispersion because (1) the iron and aluminum coatings that act as strong flocculating and cementing agents in acid soils are largely lacking in alkaline soils, (2) the types of clays dominant in alkaline soils are especially susceptible to dispersion, and (3) monovalent ions (Na^+ and K^+) that are easily leached from acid soils (see Section 8.10) are still largely unleached in the soils of dry regions. Dispersion of soil clay leads to drastically reduced macroporosity, aeration, water percolation, and to the sealing of the soil surface (see Sections 10.5–10.6).

Nutrient Deficiencies

As discussed in Chapter 9 (see Figure 9.22), the availability of most nutrient elements is markedly influenced by soil pH, so alkaline soils can be expected to exhibit special problems regarding the solubility of plant nutrients and other elements. In addition, unlike in humid regions, the minimal weathering allows many weatherable and relatively soluble minerals to remain in the soil, in some cases contributing high levels of certain elements to the soil–plant–animal system.

MICRONUTRIENT METALS.   The micronutrients zinc, copper, iron, and manganese are readily available in acid soils, but are much less soluble at pH levels above 7. Therefore, in alkaline soils plant growth is commonly limited by deficiencies of these elements, deficiencies of iron and zinc being especially common in irrigated tree crops, as well as in dryland wheat and sorghum (see Plates 97–99). The low organic-matter levels of most dry-region soils further reduce the availability of these metals. In fact, the solubility of iron is so low at alkaline pH levels that both plants and microorganisms, in order to survive, have had to evolve special mechanisms for obtaining this essential nutrient (see Section 15.7).

Problems with these deficiencies can be especially acute when alkaline soils are brought under irrigation. Because of adverse reactions in alkaline soils, correction of these problems often requires special protective organic complexes called **chelates** (see Section 15.8) and the direct application of these micronutrients to plant foliage via foliar sprays or irrigation water.

Micronutrient deficiencies in alkaline soils can be especially troublesome for ornamental plantings (Figure 10.8). Unfortunately, some of the most popular ornamental plants are native to humid regions and not well adapted to alkaline soils, especially if the soils are calcareous in the upper horizons. Ornamental landscape managers should select plants that are adapted to alkaline soil—or they must change the soil pH to suit the plant by adding acid-forming chemicals, such as sulfur (see Section 9.10).

BORON.   Boron is thought to form difficult-to-reverse inner-sphere complexes (see Sections 8.7 and 8.11) on surfaces of iron and aluminum oxides and silicate clays. The strength of this adsorption increases as soil pH increases up to about 9 (Figure 10.9). Fine-textured alkaline soils can therefore tightly adsorb much more boron than can sandy soils. Boron deficiency is common at high pH levels in both sandy soils (because of low boron content) and clayey soils (because the boron is tightly held by the clay). In addition, plants tend to have a higher requirement for boron if calcium is abundant. For all these reasons, boron deficiencies are quite common in alkaline soils (see Section 15.9).

MOLYBDENUM.   In contrast to boron, molybdenum availability is high under alkaline conditions—so high that in some areas molybdenum toxicity is a problem. For ruminant animals (e.g., cattle and sheep) grazing on high molybdenum soils, the

FIGURE 10.8 Ornamental oak trees that thrive in acid soils of humid regions commonly suffer from deficiencies of iron and other micronutrients when grown on alkaline soils, as in the Denver, Colorado, suburb shown. The pale-colored foliage (chlorosis) and dying branches of the tree in the foreground are typical symptoms of iron or manganese deficiency. The close-up shows that the chlorosis is interveinal (the veins remain green) and that the pH is near 8. The situation could be remedied by either (1) selecting a species better adapted to alkaline soils, (2) acidifying the soil around the tree, or (3) supplying the needed micronutrients by applying them to the soil in chelated form or spraying them directly onto the foliage. (Photos courtesy of R. Weil)

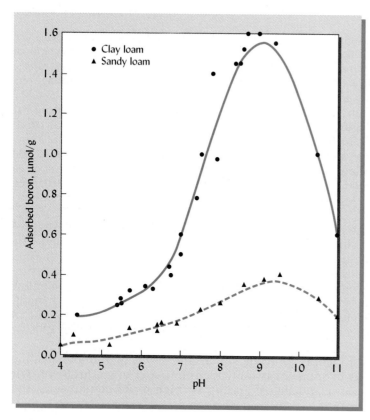

FIGURE 10.9 Adsorption of boron increases as soil pH increases up to about 9. Boron is thought to form difficult-to-reverse (inner-sphere) complexes with hydroxyls on the surfaces of iron and aluminum oxides and silicate clays. Because of their greater surface area, fine-textured soils can tightly adsorb much more boron than can sandy soils, which may lose most of their boron by leaching. Therefore, boron deficiency is common at high pH in both sandy soils (because of low boron content) and in fine-textured clayey soils (because the boron is tightly held by the clay). [Redrawn from Goldberg et al. (2000) and used with permission of the Soil Science Society of America]

high solubility of molybdenum and low solubility of copper at high pH can combine to cause a disease known as **molybdenosis**. Compounds containing sulfur and molybdenum (thiomolybdates) apparently block the proper absorption of copper by the animals. Molybdenosis is most common in areas of alkaline soils formed from granites and shales and is usually associated with wet floodplains and alluvial fans in these areas.

PHOSPHORUS. Deficiency of the macronutrient phosphorus (P) is a widespread problem in arid regions because the abundant Ca^{2+} and Mg^{2+} ions dissolved in alkaline soils act as common ions to constrain the dissolution of phosphorus-carrying minerals. Even P added as soluble fertilizer soon reacts with the abundant Ca^{2+} in solution to form calcium-phosphate compounds that become increasingly insoluble over time (see Section 14.5). To help them survive in the low-phosphorus environment of alkaline soils, certain fungi, bacteria, and higher plants (including some crops such as those in the Brassica family) have evolved the ability to excrete organic acids that can dissolve these calcium phosphates (see Sections 14.7 to 14.9 for more detail on P availability).

In light of the many impacts of high pH just discussed, we now turn our attention to why the pH in dry region soils is almost always alkaline.

10.2   CAUSES OF ALKALINITY: HIGH SOIL pH

In regions where precipitation is less than potential evapotranspiration (see Section 6.3 and Plate 76), the cations released by mineral weathering accumulate because there is not enough rain to thoroughly leach them away. The pH of soils in these arid and semi-arid environments is generally in the alkaline range—that is, 7 or above.

There seems to be considerable confusion about the terms **alkaline** and **alkalinity**, and people often use these terms *wrongly* to describe soils characterized by detrimental levels of soluble salts or sodium. Alkaline soils are simply those with a pH above 7.0. *Alkalinity* refers to the concentration of OH^- ions, much as *acidity* refers to that of H^+ ions. *Alkaline* soils should not be confused with *alkali* soils.[1] The latter name is an obsolete term for what are now called **sodic** or **saline-sodic** soils, those with levels of sodium high enough to be detrimental to plant growth (see Section 10.5). Current classification of salt-affected soils includes consideration of both pH and salinity measures.

Sources of Alkalinity

Minimal leaching in dry environments means that soil acidification is minimized. The cations in the soil solution and on the exchange complex are mainly Ca^{2+}, Mg^{2+}, K^+, and Na^+. As explained in Section 9.1, these cations are nonhydrolyzing and so do not produce acid (H^+) upon reacting with water as Al^{3+} or Fe^{3+} do. However, they generally do not produce OH^- ions either. Rather, their effect in water is neutral,[2] and soils dominated by them have a pH no higher than 7 unless certain *anions* are present in the soil solution. The basic hydroxyl (OH^-)–generating anions are principally **carbonate** (CO_3^{2-}) and **bicarbonate** (HCO_3^-). These anions originate from the

---

[1] The terms *alkali* and *alkaline* are derived from the chemical grouping of elements into the monovalent alkali earth metals (lithium, sodium, potassium, rubidium, and cesium) and the divalent alkaline earth metals (beryllium, magnesium, calcium, strontium, and barium). See the periodic table in Appendix B.

[2] The cations Ca^{2+}, Mg^{2+}, K^+, Na^+, and NH_4^+ have been traditionally called *base* or *base-forming* cations as a convenient way of distinguishing them from the acid cation, H^+, and the H^+-forming cation Al^{3+}. However, since Ca^{2+} and associated cations do not accept protons, they are technically not bases. Consequently, it is less misleading to refer simply to acid cations (H^+, Al^{3+}) and nonacid cations (most other cations). Likewise, the term *nonacid saturation* should be used rather than *base saturation* to refer to the percentage of the exchange capacity satisfied by nonacid cations (usually Ca^{2+}, Mg^{2+}, K^+, and Na^+) on the exchange complex (see also Section 9.3).

dissolution of such minerals as calcite ($CaCO_3$) or from the dissociation of carbonic acid (H_2CO_3).

$$CaCO_3 \rightleftharpoons Ca^{2+} + CO_3^{2-} \qquad (10.1)$$

Calcite (solid)　　　(dissolved　(dissolved
　　　　　　　　　　in water)　in water)

$$CO_3^{2-} + H_2O \rightleftharpoons HCO_3^- + OH^- \qquad (10.2)$$

$$HCO_3^- + H_2O \rightleftharpoons H_2CO_3 + OH^- \qquad (10.3)$$

$$H_2CO_3 \rightleftharpoons H_2O + CO_2 \uparrow \qquad (10.4)$$

Carbonic　　　　　　(gas)
acid

In this series of linked equilibrium reactions, carbonate and bicarbonate act as bases because they react with water to form hydroxyl ions and thus raise the pH. The relationships among the concentrations of these ions and that of the H^+ ion (expressed as pH) are illustrated in Figure 10.10. The importance of these reactions in **soil buffering** (resistance to pH change) is discussed in Section 9.4.

Influence of Carbon Dioxide and Carbonates

The direction of the overall reaction (Reactions 10.1 to 10.4) determines whether OH^- ions are consumed (proceeding to the left) or produced (proceeding to the right). The reaction is controlled mainly by the precipitation or dissolution of calcite on the one end and by the production (by respiration) or loss (by volatilization to the atmosphere) of carbon dioxide at the other end. The concentration of CO_2 in the atmosphere is about 0.0035%, but may be as high as 0.5% in soil air due to respiration by roots and microorganisms (see Section 7.2). Therefore, biological activity in soils tends to lower the pH by driving the reaction series to the left.

The other process that limits the rise in pH is the precipitation of $CaCO_3$ that occurs when the soil solution becomes saturated with respect to Ca^{2+} ions. Such precipitation removes Ca from the solution, again driving the reaction series to the left (lowering pH). Because of the limited solubility of $CaCO_3$, the pH of the solution cannot rise above 8.4 when the CO_2 in solution is in equilibrium with that in the atmosphere. The pH at

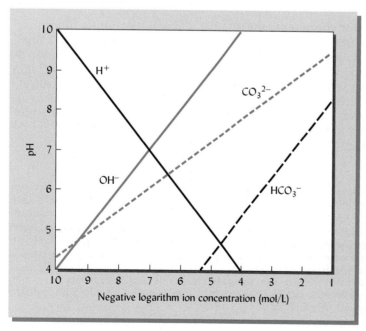

FIGURE 10.10 Effect of carbonates (CO_3^{2-}) and bicarbonates (HCO_3^-) on the concentrations of H^+ and OH^- ions in aqueous solution, expressed as negative logarithms. The H^+ line represents an increase in pH ($-\log H^+$ ions) from 4 to 10. Note that as the concentrations of CO_3^{2-} and HCO_3^- go up (to the right), so does the concentration of OH^-. Since the H^+ ion concentration is inversely related to the OH^- concentration, the pH also rises.

which $CaCO_3$ precipitates in soil is typically only about 7.0 to 8.0, depending on how much the CO_2 concentration is enhanced by biological activity. This is an important point to remember because it suggests that if other carbonate minerals more soluble than $CaCO_3$ (for example, Na_2CO_3) were present, the pH would rise considerably higher [(see Section 10.4).] Indeed this is the case; **calcareous** (calcite-laden) soil horizons range in pH from 7 to 8.4 (tolerable by most plants), while sodic (sodium carbonate-laden) horizons may range in pH from 8.5 to as high as 10.5 (levels toxic to many plants).

Role of the Cations (Na^+ versus Ca^{2+})

As just suggested, the particular cation associated with carbonate and bicarbonate anions influences the pH level attained. Calcium and sodium are the principal cations involved (although magnesium or others can play a role, as well). If Na^+ is prominent on the exchange complex and in the soil solution, the set of reactions just described (Reactions 10.1 to 10.4) will still apply, except that the first step (Reaction 10.1) will be replaced by the following:

$$Na_2CO_3 \rightleftharpoons 2Na^+ + CO_3^{2-}$$ (10.5)

     (solid) (dissolved (dissolved
         in water) in water)

Because sodium carbonate (and sodium bicarbonate) is much more water-soluble than calcium carbonate, the set of reactions will proceed more readily to the right, producing more hydroxyl ions and thus a higher pH. With high concentrations of CO_3^{2-} ions in solution, the pH can rise to 10 or higher (see Figure 10.10). It is fortunate for plants that Ca^{2+}, not Na^+, ions are dominant in most soils.

Influence of Soluble Salt Level

The presence in the soil solution of high levels of neutral salts (from sources other than carbonates, such as $CaSO_4$, Na_2SO_4, $NaCl$, and $CaCl_2$) tends to lower the pH by moderating the alkalinizing reactions just discussed. Increasing the concentration of Ca^{2+} or Na^+ ions on the right side of Reaction 10.1 or 10.5 drives the reaction to the left by *the common ion effect*. The *common ion effect* is a shift in equilibrium that occurs because of the addition of an ion already involved in an equilibrium reaction. In this example, Ca^{2+} or Na^+ added from sources other than $CaCO_3$ or Na_2CO_3 will reduce the dissolution of these carbonates. With less $CaCO_3$ or Na_2CO_3 going into solution, fewer CO_3^{2-} and HCO_3^- ions are formed, and the pH does not rise as high as it would have if less salt were present.

Nonetheless, salt accumulation is one of the most troubling problems in certain alkaline soils. Hence, we will now turn our attention to the characteristics and management of salt-affected soils.

10.3 DEVELOPMENT OF SALT-AFFECTED SOILS[3]

Extent and causes of salt-affected soils around the world:

http://www.fao.org/AG/AGL/agll/spush/topic2.htm

Salt-affected soils are widely distributed throughout the world (Table 10.1), the largest areas being found in Australia, Africa, Latin America, and the Near and Middle East. They typically occur in areas with precipitation-to-evaporation ratios of 0.75 or less and in low, flat areas with high water tables that may be subject to seepage from higher elevations (Plate 104). Figure 10.11 illustrates the distribution of these soils in the contiguous United States. Nearly 50 million ha of cropland and pasture are currently affected by salinity, and in some regions the area of land so affected is growing by about 10% annually.

In most cases, the soluble salts in soils originate from the weathering of primary minerals in rocks and parent materials. In extremely dry regions (<25 cm annual precipitation), calcium sulfate accumulations (gypsic horizons) may form near the soil

---

[3] For a discussion of these soils, which are also referred to as *halomorphic soils*, see Abrol et al. (1988) and Szabolcs (1989).

TABLE 10.1 Area of Salt-Affected Soils in Different Regions

Region	Area, million ha
Africa	69.5
Near and Middle East	53.1
Asia and Far East	19.5
Latin America	59.4
Australia	84.7
North America	16.0
Europe	20.7
World total	322.9

From Beek et al. (1980).

surface where this relatively soluble mineral may create a saline condition. However, in most cases, salts are transported to a developing salt-affected soil as ions dissolved in water. The salt-containing water moves through a landscape from areas of higher to lower elevations and from soil zones that are wetter to those that are drier. The water is eventually lost by evaporation. However, the dissolved salts cannot evaporate, and therefore they are left behind to accumulate in the soil. This is true in both irrigated and unirrigated landscapes.

Many salt-affected soils develop because changes in the local water balance, usually brought about by human activities, increase the input of salt-bearing water more than they increase the output of drainage water. Increased evaporation, waterlogging, and rising water tables usually result. It is worth remembering the irony that *salts usually become a problem when too much water is supplied, not too little.*

Accumulation of Salts in Nonirrigated Soils

In the United States, about one-third of the soils in arid and semiarid regions are affected by some degree of salinity. The salts are primarily chlorides and sulfates of calcium, magnesium, sodium, and potassium. These salts accumulate naturally in some

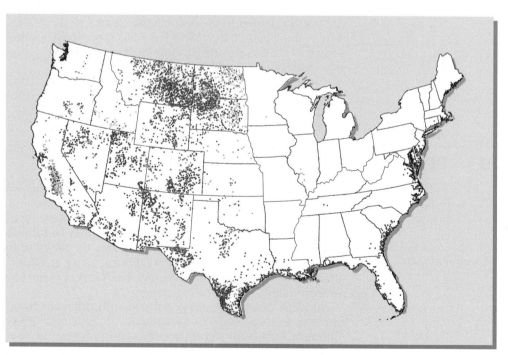

FIGURE 10.11 Distribution of salt-affected soils in the contiguous United States. Some small areas near the coasts are influenced by seawater, but the great majority of salt-affected soils lie in the vast dry regions of the West. Each dot represents about 4050 ha. [From USDA Natural Resources Conservation Service (1996)]

surface soils because there is insufficient rainfall to flush them from the upper soil layers. In coastal areas, sea spray (Plate 105) and inundation with seawater can be locally important sources of salt in soils.

Other localized but important sources are fossil deposits of salts laid down during geological time in the bottom of now-extinct lakes or oceans or in underground saline water pools. These fossil salts can be dissolved in underground waters that move horizontally over underlying impervious geological layers and ultimately rise to the surface of the soil in the low-lying parts of the landscape. The salts then concentrate near or on the surface of the soil in these low-lying areas, creating a saline soil. The low-lying areas where the saline groundwater emerges are termed **saline seeps**.

Saline seeps occur naturally in some locations, but their formation is often greatly increased when the water balance in a semiarid landscape is disturbed by bringing land under cultivation (Figure 10.12). Replacement of native, deep-rooted perennial vegetation with annual crop species greatly reduces the annual evapotranspiration, especially if the cropping system includes periods of fallow during which the soil is unvegetated. The decreased evapotranspiration allows more rainwater to percolate through the soil, thus raising the water table and increasing the flow of groundwater to lower elevations. In dry regions, soils and substrata may contain substantial amounts of soluble salts that can be picked up by the percolating water. If a shallow layer restricts percolation, it will further encourage the flow of the salt-laden groundwater across the landscape toward the lowest elevations. Eventually, the water table may rise to within 1 m or less of the soil surface, and capillary rise will begin to contribute a continuous stream of salt-laden water to replace the water lost at the surface by evaporation. The evaporating water will

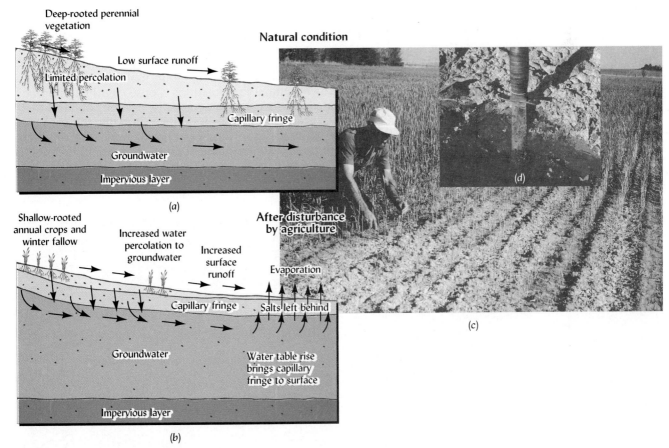

FIGURE 10.12 Saline seep formation in a semiarid area where the salt-rich substrata are underlain by an impermeable layer. (*a*) Under deep-rooted perennial vegetation, transpiration is high and the water table is kept low. (*b*) After conversion to agriculture, shallow-rooted annual crops take up much less water, especially if fallow is practiced, allowing more water to percolate through the salt-bearing substrata. Consequently, in lower-elevation landscape positions, the wet-season water table rises close to the soil surface. This allows the salt-laden groundwater to rise by capillary flow to the surface, from which it evaporates, leaving behind an increasing accumulation of salts. Note that the diagrams greatly exaggerate vertical distances. (*c*) A spreading saline seep area in eastern Montana where wheat fallow cropping has replaced natural deep-rooted prairie vegetation. (*d*) Close-up of salt crust over moist soil. (Photos and drawings courtesy of R. Weil)

leave behind the salts, which soon will accumulate to levels that inhibit plant growth. Year by year the evaporation zone will creep up the slope, and the barren, salinized area will become larger and more saline. Millions of hectares of land in North America, Australia, and other semiarid regions have been degraded in this fashion.

The exchange complex of most salt-affected soils is dominated by Ca^{2+} and Mg^{2+} ions, with little exchangeable Na^+. However, where the level of Na^+ is significant, soil aggregation deteriorates and permeability is low. If the Na^+ saturation exceeds 15%, the term *sodic soil* applies, signaling extremely adverse conditions (see Sections 10.5 and 10.6).

Irrigation-Induced Salinity and Alkalinity

Australia's mirage of green pastures evaporates: http://www.csmonitor.com/ 1996/0508/050896.intl.intl .6.html

Irrigation not only alters the water balance by bringing in more water, it also brings in more salts. Whether taken from a river or pumped from the groundwater, even the best quality freshwater contains some dissolved salts (see Section 10.8). The amount of salt brought in with the water may seem negligible, but the amounts of water applied over the course of time are huge. Again, pure water is lost by evaporation, but the salt stays and accumulates (Figure 10.13). The effect is accentuated in arid regions for two reasons: (1) the water available from rivers or from underground is relatively high in salts because it has flowed through dry-region soils that typically contain large amounts of easily weatherable minerals, and (2) the dry climate creates a relatively high evaporative demand, so large amounts of water are needed for irrigation. An arid-region farmer may need to apply 90 cm of water to grow an annual crop. Even if this is good-quality water relatively low in salts, it will likely dump more than 6 Mg/ha (3 tons/acre) of salt on the land every year (see Section 10.8).

If irrigation water carries a significant proportion of Na^+ ions compared to Ca^{2+} and Mg^{2+} ions, and especially if the HCO_3^- ion is present, sodium ions may come to saturate a major part of the colloidal exchange sites, creating an unproductive *sodic* soil (Section 10.6).

During the past three decades, low-income countries in the dry regions of the world have greatly expanded the area of their land under irrigation in order to produce the food needed by their rapidly growing human populations. Consequently, the proportion of arable land that is irrigated has increased dramatically, reaching about 40% in China, 35% in India, 94% in Pakistan, and 21% in Indonesia. Initially, the expanded irrigation stimulated phenomenal increases in food-crop production. Unfortunately, many irrigation projects failed to provide for adequate drainage. As a result, the process of *salinization* has accelerated, and salts have accumulated to levels that are already adversely affecting crop production. In some areas, sodic soils have been created.

FIGURE 10.13   Salinization, the accumulation of soluble salts in soils, can be observed in the potting medium of houseplants (*left*) and in irrigated fields (*right*). The salt accumulates because of evaporative water loss from soil that is repeatedly supplied with water that contains dissolved salts, even if in low concentrations. Only pure water evaporates; the salts dissolved in the water do not. Note that the salt tends to concentrate at the highest points of the soil surface, from which evaporation loss is greatest and to which the soil solution flows by capillary action. (Photos courtesy of R. Weil)

These events remind the world of the flat wastelands of southeastern Iraq that have been barren since the 12th century. In biblical times, these areas were fertile and productive. With irrigation water supplied by the Euphrates and Tigris, crop productivity was so high that the overall region was called the Fertile Crescent. Unfortunately, salinization eventually set in because many of the soils were not naturally well drained and the societies sometimes failed to maintain what artificial drains were created. Salts eventually accumulated to such a degree that crop production declined and the area had to be abandoned. Today, societies around the world are repeating the mistakes of the past (see, for example, Figure 19.15). Some observers believe that each year, the area of previously irrigated land degraded by severe salinization is greater than the area of land newly brought under irrigation. Truly, the world needs to give serious attention to the large-scale problems associated with salt-affected soils.

10.4 MEASURING SALINITY AND SODICITY

Salt-affected soils adversely affect plants because of the total concentration of salts (*salinity*) in the soil solution and because of concentrations of specific ions, especially sodium (*sodicity*). In order to assist in characterizing and managing salt-affected soils, techniques have been developed to measure and quantify the degree of soil salinity and sodicity. Salinity is measured primarily as the **total dissolved solids** (TDS) or more conveniently, as **electrical conductivity** (EC). Sodicity is characterized primarily by **exchangeable sodium percentage** (ESP) and the **sodium adsorption ratio** (SAR). Each measure will be discussed briefly.

Salinity[4]

TOTAL DISSOLVED SOLIDS.   In concept, the simplest way to determine the total amount of dissolved salt in a sample of water is to heat the solution in a container until all of the water has evaporated and only a dry residue remains. A temperature of 180 °C is used to ensure that water of hydration is removed from the salt residue. The residue can then be weighed and the **total dissolved solids** (TDS) expressed as milligrams of solid residue per liter of water (mg/L). In water to be used for irrigation, TDS typically ranges from about 5 to 1000 mg/L, while in the solution extracted from a soil sample, TDS may range from about 500 to 12,000 mg/L.

U.S. Soil Salinity Laboratory
"News and Events":
http:www.ars.usda.gov/pwa/
riverside/gebjsl

ELECTRICAL CONDUCTIVITY.   Pure water is a poor conductor of electricity, but conductivity increases as more and more salt is dissolved in the water. Thus, the **electrical conductivity** (EC) of the soil solution gives us an indirect measurement of the salt content. The EC can be measured both on samples of soil or on the bulk soil *in situ* (Table 10.2). It is expressed in terms of deciSiemens per meter (dS/m).[5]

TABLE 10.2   Different Measurements for Estimating Soil Salinity

The methods are well correlated, so each can be converted to any other. The EC_e is the most common standard for comparison.

Measured on soil sample	
EC_e	Conductivity of the solution extracted from a water-saturated soil paste
EC_p	Conductivity of the water-saturated soil paste itself
EC_w	Conductivity of the solution extracted from a 1:2 soil–water mixture
TDS	Total dissolved solids in water or the solution extracted from a water-saturated soil paste[a]
Measured on bulk soil in place	
EC_a	Apparent conductivity of bulk soil sensed by metal electrodes in soil
EC_a^*	Electromagnetic induction of an electric current using surface transmitter and receiving coils

[a] Note that TDS (mg/L) can be converted to EC_w using these relationships: for Na salts, TDS = 640 × EC_w; for Ca salts, TDS = 800 × EC_w.

---

[4] For an informative discussion of these methods, see Rhoades et al. (1999) or the Web site of the U.S. Salinity Laboratory at www.ars.usda.gov/main/site_main.htm?modecode=53102000
[5] Formerly expressed as millimhos per centimeter (mmho/cm). Since 1 S = 1 mho, 1 dS/m = 1 mmho/cm.

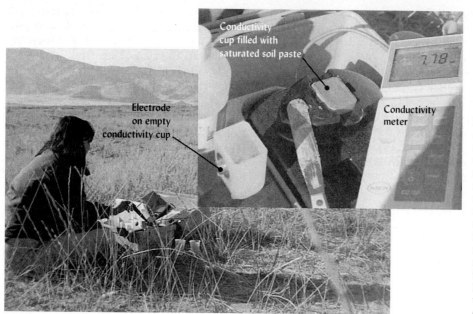

Saturated paste and dilute extracts in soil testing for salt-affected turfgrass. http://gcsaa.org/gcm/2003/sept03/PDFs/09Clarify.pdf

FIGURE 10.14 Measuring the electrical conductivity (EC) of a soil sample in a field of wheatgrass to determine the level of salinity. A sample of the soil is stirred with pure water until a saturated paste is made. The paste is then transferred into a special conductivity cup that has a flat, circular electrode on either side (*inset*). This is then inserted into a stand that connects the electrodes to a conductivity meter. Note readout of 7.78 dS/m on the conductivity meter. This level of EC_p indicates a highly saline soil that would inhibit the growth of many crops. (Photos courtesy of R. Weil)

The **saturation paste extract** method is the most commonly used procedure and the standard to which the others are usually compared. A soil sample is saturated with distilled water and mixed to the consistency of a paste that glistens with water and flows slightly if jarred. After standing overnight to thoroughly dissolve the salts, the solution is extracted by suction filtration and its electrical conductivity is measured (EC_e). A variant of this method involves measuring the EC of the solution extracted from a 1:2 soil–water mixture after 0.5 h of shaking (EC_w). The latter method takes less time but often is not as well related to the soil solution as is the saturation paste extract method. An even more rapid method that is well correlated to the salinity of the soil solution is the conductivity of the soil paste itself (EC_p). Because the tedious extraction step is eliminated, EC_p can be determined on samples in the field (Figure 10.14). Values for EC_w can be converted to total dissolved solids if the type of salt is known (see footnote, Table 10.2).

MAPPING EC IN THE FIELD. Advances in instrumentation now allow rapid, continuous field measurement of bulk soil conductivity, which, in turn, is directly related to soil salinity (see Table 10.2). One such method involves inserting four carefully spaced electrodes into moist soil to make direct measurements of apparent EC in the field (EC_a). The depth to which the electrodes sense the EC is related to the spacing between the electrodes. Such a **four-electrode conductivity apparatus** can be constructed with tillage shanks or rolling colter blades that move through the soil as a tractor or other vehicle pulls them across a field. This technique is rapid, simple, and practical and gives values that can be correlated with EC_e. This type of apparatus can generate a continuous readout of EC measurements as it is driven across a field (Figure 10.15). If a geopositioning system receiver (see Section 19.3) is integrated with the apparatus, the resulting data can be transformed into a map showing the spatial variation of soil salinity across a parcel of land.

ELECTROMAGNETIC INDUCTION. A second rapid field method employs **electromagnetic induction** (EM) of electrical current in the body of the soil, the level of which is related to electrical conductivity and, in turn, to soil salinity. A small transmitter coil located in one end of the battery-powered EM instrument generates a magnetic field within the soil. This magnetic field, in turn, induces small electric currents within the soil whose values are related to the soil's conductivity. These small currents generate their own secondary magnetic fields, which can be measured by a small receiving cell in the opposite end of the EM instrument. The EM instrument thus can measure ground EC (designated E_a^*) to considerable depths in the soil profile without mechanically probing the soil. A handheld model of such an EM conductivity sensor is shown in Figure 10.16.

 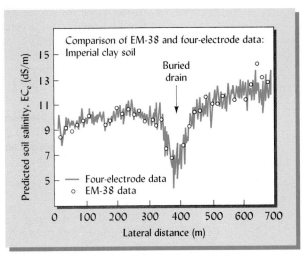

FIGURE 10.15 Surveying electrical conductivity in the field. (*Left*) A mobile four-electrode apparatus with integrated geopositioning system receiver generates a continuous readout of soil conductivity across a field. (*Right*) Soil electrical conductivity data from a transect across an irrigated field show that the EC (measured in the upper meter of soil) tends to increase down the length of an irrigation run across the field, indicating that the lower end of the field is more saline. This end of the field probably gets less irrigation water, and the salts in it are therefore less thoroughly leached out. The dip in the middle corresponds to the location of a buried drain that has reduced the salinity by encouraging drainage. The data points (small circles) in the graph were measured using an EM conductivity sensor such as that illustrated in Figure 10.16. (Data and photo courtesy of J. D. Rhoades and S. Lesch, U.S. Salinity Lab)

The same type of instrument can be vehicle-mounted and used to rapidly map the soil salinity levels across a field.

It should be noted that each of the methods discussed estimates soil salinity indirectly by measuring electrical conductivity. The values of EC_e, EC_w, EC_p, EC_a, or EC_a^* obtained by these procedures will not be identical, however. For example, in highly saline soils EC_a values measured by the four-electrode-sensor method are about one-fifth of those measured using the standard saturated paste extract procedure (EC_e). Fortunately, however, these values are all sufficiently well correlated with each other that the results from any of the methods can easily be converted to the standard EC_e.

FIGURE 10.16 A portable electromagnetic (EM) soil conductivity sensor used to estimate the electrical conductivity in the soil profile. When placed on the soil surface in the horizontal position (*lower left*), this instrument senses electrical conductivity of the soil down to about 1 m depth. When placed in the vertical position (*as in the inset photo*), the effective depth is about 2 m. This type of EM sensor (model EM-38, made by Geonics, Ltd., Ontario, Canada) was mounted on a special vehicle for the mobile soil salinity mapping that produced the data points in Figure 10.15. (Photos courtesy of R. Weil)

Advances in mobile salinity sensors have made it possible to produce detailed maps of the variation in salinity within a given field. The information from these maps can then be used in the techniques of **precision agriculture** (see Sections 19.3 and 19.10), which are capable of applying corrective measures tailored to match the degree of salinity in each small part of a large field.

Sodium Status

Two expressions are commonly used to characterize the sodium status of soils. The **exchangeable sodium percentage** (ESP) identifies the degree to which the exchange complex is saturated with sodium:

$$\text{ESP} = \frac{\text{Exchangeable sodium, cmol}_c/\text{kg}}{\text{Cation exchange capacity, cmol}_c/\text{kg}} \times 100 \tag{10.6}$$

ESP levels greater than 15 are associated with severely deteriorated soil physical properties and pH values of 8.5 and above.

The **sodium adsorption ratio** (SAR) is a second, more easily measured property that is becoming even more widely used than ESP. The SAR gives information on the comparative concentrations of Na^+, Ca^{2+}, and Mg^{2+} in soil solutions. It is calculated as follows:

$$\text{SAR} = \frac{[Na^+]}{(0.5[Ca^{2+}] + 0.5[Mg^{2+}])^{1/2}} \tag{10.7}$$

where $[Na^+]$, $[Ca^{2+}]$, and $[Mg^{2+}]$ are the concentrations (in mmol of charge per liter) of the sodium, calcium, and magnesium ions in the soil solution. An SAR value of 13 for the solution extracted from a saturated soil paste is approximately equivalent to an ESP value of 15. The SAR of a soil extract takes into consideration that the adverse effect of sodium is moderated by the presence of calcium and magnesium ions. The SAR also is used to characterize irrigation water applied to soils (see Section 10.8).

High amounts of other monovalent ions such as potassium (K^+) can also promote soil structure degradation, though less so than sodium. Therefore, some soil scientists suggest that the SAR should be modified to include the sum of (Na^+) + (K^+) in the numerator of Equation 10.7. Excessive K^+ may originate from soil minerals or irrigation water as typically is the case for Na^+, but it may also come from overapplication of potassium fertilizer or from manure generated by animals fed a high K diet, such as the alfalfa-rich diets used by many dairy farms.

10.5 CLASSES OF SALT-AFFECTED SOILS

Using EC, ESP (or SAR), and soil pH, salt-affected soils are classified as **saline**, **saline-sodic**, and **sodic** (Figure 10.17). Soils that are not greatly salt-affected are classed as **normal**.

Saline Soils

The processes that result in the accumulation of neutral soluble salts are referred to as **salinization**. The salts are mainly chlorides and sulfates of calcium, magnesium, potassium, and sodium. The concentration of these salts sufficient to interfere with plant growth (see Section 10.7) is generally defined as that which produces an electrical conductivity in the saturation extract (EC_e) greater than 4 dS/m. However, some sensitive plants are adversely affected when the EC_e is only about 2 dS/m.

Saline soils are those soils that contain sufficient salinity to give EC_e values greater than 4 dS/m, but have an ESP less than 15 (or an SAR less than 13) in the saturation extract. Thus, the exchange complex of saline soils is dominated by calcium and magnesium, not sodium. The pH of saline soils is usually below 8.5. Because soluble salts help prevent dispersion of soil colloids, plant growth on saline soils is not generally constrained by poor infiltration, aggregate stability, or aeration. In many cases, the evaporation of water creates a white salt crust on the soil surface (see Figures 10.12 and 10.13 and Plate 104), which accounts for the name *white alkali* that was previously used to designate saline soils.

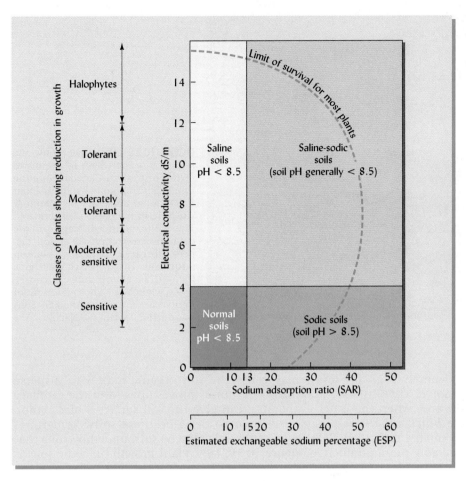

FIGURE 10.17 Diagram illustrating the classification of normal, saline, saline-sodic, and sodic soils in relation to soil pH, electrical conductivity, sodium adsorption ratio (SAR), and exchangeable sodium percentage (ESP). Also shown are the ranges for different degrees of sensitivity of plants to salinity.

Saline-Sodic Soils

Soils that have both detrimental levels of neutral soluble salts (EC_e greater than 4 dS/m) *and* a high proportion of sodium ions (ESP greater than 15 or SAR greater than 13) are classified as **saline-sodic soils** (see Figure 10.17). Plant growth in these soils can be adversely affected by both excess salts and excess sodium levels.

Saline-sodic soils exhibit physical conditions intermediate between those of saline soils and those of sodic soils. The high concentration of neutral salts moderates the dispersing influence of the sodium. The salts provide excess cations that move in close to the negatively charged colloidal particles, thereby reducing their tendency to repel each other, or to disperse. The salts, therefore, help keep the colloidal particles associated with each other in aggregates.

Unfortunately, this situation is subject to rather rapid change if the soluble salts are leached from the soil, especially if the SAR of the leaching waters is high. In such a case, salinity will drop, but the exchangeable sodium percentage will increase, and the saline-sodic soil will become a sodic soil.

Sodic Soils

DEFINITION. Sodic soils are, perhaps, the most troublesome of the salt-affected soils. While their levels of neutral soluble salts are low (EC_e less than 4.0 dS/m), they have relatively high levels of sodium on the exchange complex (ESP and SAR values are above 15 and 13, respectively). Some sodic soils in the order Alfisols (Natrustalfs) have a very thin A horizon overlying a clayey layer with columnar structure, a profile feature closely associated with high sodium levels (Figure 10.18). The pH values of sodic soils exceed 8.5, rising to 10 or higher in some cases. As explained in Section 10.2, these extreme pH levels are largely due to the fact that sodium carbonate is much more soluble than calcium or magnesium carbonate and so maintains high concentrations of CO_3^{2-} and HCO_3^- in the soil solution.

FIGURE 10.18 The upper profile of a sodic soil (a Natrustalf) in a semiarid region of western Canada. Note the thin A horizon (knife handle is about 12 cm long) underlain by columnar structure in the natric (Btn) horizon. The white, rounded "caps" of the columns are comprised of soil dispersed because of the high sodium saturation. The dispersed clays give the soil an almost rubbery consistency when wet. [Photo courtesy of Agriculture Canada, Canadian Soils Information System (CANSIS)]

The extremely high pH levels may cause the soil organic matter to disperse and/or dissolve. The dispersed and dissolved humus moves upward in the capillary water flow and, when the water evaporates, can give the soil surface a black color. The name **black alkali** was previously used to describe these soils. Sometimes located in small areas called **slick spots**, sodic soils may be surrounded by soils that are considerably more productive (Figure 10.19, *left*). Plant growth on sodic soils is often constrained by specific toxicities of Na^+, OH^-, and HCO_3^- ions. However, the main reason for the poor plant growth—often to the point of complete barrenness—is that few plants can tolerate the extremely poor soil physical conditions and slow permeability to water and air characteristic of sodic soils. We will next take a closer look at this chemically-induced physical degradation that is such a widespread phenomenon of great importance, even in many soils that do not fall within the formal definition of sodic.

10.6 PHYSICAL DEGRADATION OF SOIL BY SODIC CHEMICAL CONDITIONS

The high sodium and low salt levels in sodic soils (and, to a lesser degree, in some "normal" soils) can cause serious degradation of aggregate structure and loss of macroporosity such that the movement of water and air into and through the soil is severely restricted. This structural degradation is most commonly measured in terms of the readiness of water movement—the saturated hydraulic conductivity of the soil (Figure 10.20 and Section 5.5). Often sodic soils exhibit such low K_{sat} values that the infiltration rate is reduced almost to zero, causing water to form puddles rather than soak into the soil. The soil is therefore said to be *puddled,* a condition characteristic of sodic soils. Physically, the puddled condition of a sodic soil is much like that of a rice paddy soil in which a farmer has mechanically destroyed the soil structure in order to be able to keep the paddy inundated with water.

Slaking, Swelling, and Dispersion

In most soils, the low permeability related to sodic conditions has three underlying causes. First, exchangeable sodium increases the tendency of aggregates to break up or *slake* upon becoming wet. The clay and silt particles released by slaking aggregates clog soil pores as they are washed down the profile. Second, when expanding-type clays (e.g., montmorillonite) become highly Na^+-saturated, their degree of swelling is increased. As these clays expand, the larger pores responsible for water drainage in the soil are

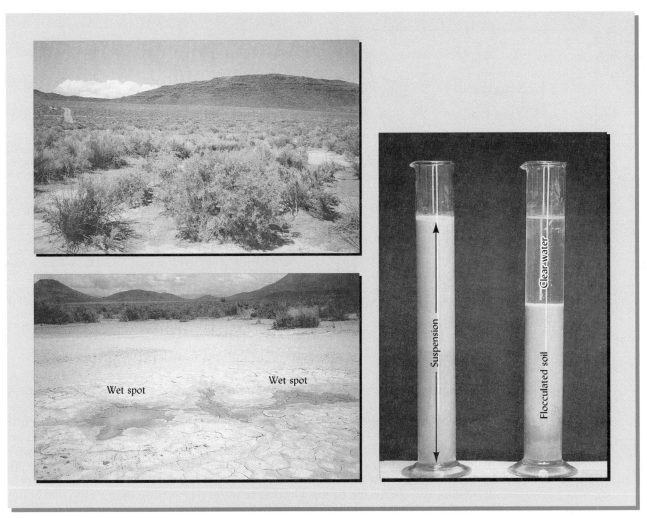

FIGURE 10.19 Illustration of the dispersion of colloids in a sodic soil from southern Colorado. The sodic soil (*lower left*) is bare of vegetation compared to the adjacent alkaline but nonsodic soil (*upper left*). Because of the complete dispersion of the soil colloids, water moves through the sodic soil very slowly, as the wet spots in the lower left photo suggest. To verify the dispersion, samples of the sodic soil were placed in cylinders and shaken thoroughly with water. The clay was highly dispersed, no observable settling occurring over a period of three weeks (left cylinder). To demonstrate how the salts in a saline-sodic soil can prevent the undesirable dispersion of the soil colloids, a small quantity of table salt was added to the other cylinder, and the mixture was again thoroughly shaken with water. Flocculation (clumping together) of the colloids began to occur, and within 24 hours, settling was quite noticeable (right cylinder). The reasons for this action are explained on pp. 422–424. In a sodic soil, salt-stimulated flocculation could be the first step toward aggregation, improved drainage, and good aeration. (Photos courtesy of N. C. Brady)

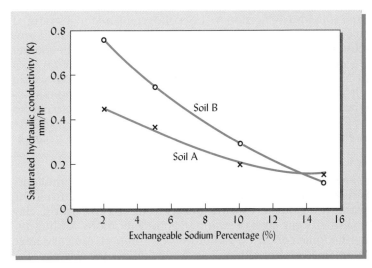

FIGURE 10.20 Effect of increasing exchangeable sodium percentage (ESP) on the saturated hydraulic conductivity of two Vertisols in Italy. Note the steady decline in conductivity with increasing ESP and the very low conductivity at ESP = 15%. Such low conductivity will slow infiltration to the point that water will form puddles on the surface rather than soak into the soil. Hence the term *puddled condition*. [Redrawn from (Crescimanno et al., 1995)]

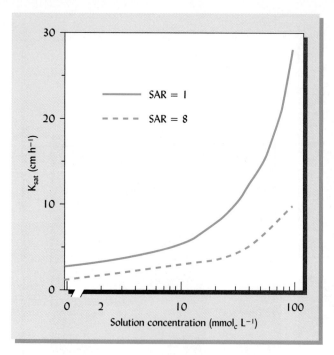

FIGURE 10.21 Opposing influences of sodium absorption ratio (SAR) and solution electrolyte (salt) concentration on the saturated hydraulic conductivity (K_{sat}) of a clay loam soil (an Aridisol) with smectitic clay mineralogy. Increasing solution electrolyte concentration stimulated clay flocculation and reduced clay swelling, both effects that promote more rapid flow of water through the soil (higher K_{sat}). Note that the solution concentration ranged from 0 (distilled water) to 100 mmol$_c$ L^{-1}, the highest concentration being quite saline—roughly equivalent to 5000 mg/L TDS or $EC_w \approx 7$ dS/m. In nearly pure water, K_{sat} was very low, regardless of the SAR level. When the solution concentration was stronger, a moderately high SAR (recall that SAR = 13 defines a sodic soil) reduced the K_{sat} from nearly 30 to less than 10 cm h^{-1}. These effects are explained conceptually by Figures 10.13 and 10.14 in terms of dispersion and flocculation. [Redrawn from Mace and Amrhein (2001) with permission from The Soil Sci. Soc. of Amer.]

squeezed shut. Third, and perhaps most important, sodic conditions—the combination of high sodium and low dissolved salt concentrations—lead to soil dispersion. Dispersion is the opposite of flocculation. In normal soils, clay particles flocculate together, giving rise to tiny clumps (floccules) that create pores between them and promote the formation of larger aggregates (see Section 4.5). In dispersed soils, the clay particles separate from one another, creating an almost gel-like condition. While these phenomena are most pronounced in sodic soils, they can occur to some degree in other dry-region soils subjected to low electrolyte water (e.g., rain!) and/or moderately high levels of monovalent cations (Figure 10.21).

Two Causes of Soil Dispersion

Two chemical conditions promote dispersion. One is a high proportion of Na$^+$ ions on the exchange complex. The second is a low concentration of electrolytes (salt ions) in the soil water.

High Sodium. Exchangeable Na$^+$ ions promote dispersion for two reasons. First, because of their single charge and large hydrated size, they are attracted only weakly to soil colloids, and so they spread out to form a relatively broad swarm of ions held in very loose outer-sphere complexes around the colloids (see Section 8.7). Second, compared to a swarm of divalent cations (which have two positive charges each), twice as many monovalent ions (with only one charge each) are needed to provide enough positive charges to counter the negative charges on a clay surface. As illustrated in Figure 10.22, the layer of exchangeable monovalent Na$^+$ ions is therefore much thicker than that which would form with the more strongly attracted divalent ions such as Ca^{2+}. The highly sodium-saturated colloids are kept so far apart that the forces of cohesion cannot come into play to attract one colloid surface to another. Instead, the poorly balanced electronegativity of each colloidal surface repels other electronegative colloids, and the soil becomes dispersed.

Low Salt Concentration. A low ionic concentration in the bulk soil solution simultaneously increases the gradient causing exchangeable cations to diffuse away from the clay surface while it decreases the gradient causing anions to diffuse toward the clay (Figure 10.23). The result is a thick ionic layer or swarm of absorbed cations. Adding *any* soluble salt would increase the ionic concentration of the soil solution and encourage the opposite effects—resulting in a compressed ionic layer that allows the clay particles

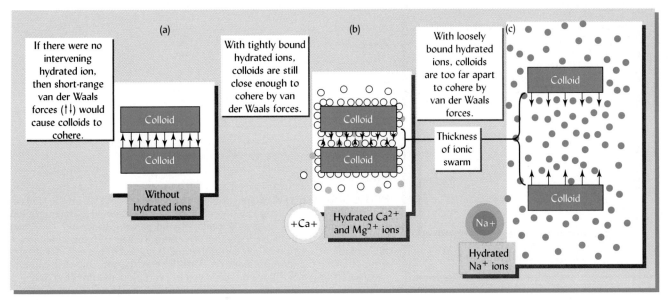

FIGURE 10.22   Conceptual diagrams showing how the *type* of cations present on the exchange complex influences clay dispersion. If colloids could approach closely (*a*), they would be held together (cohere) by short-range van der Waals forces. In soil, the colloids are surrounded by a swarm of hydrated exchangeable ions, which prevent the colloids from approaching so closely. If these are strongly attracted calcium and magnesium ions (*b*), they do not keep the colloids very far apart, so cohesive forces still have some effect. However, if they are sodium ions (*c*), the more spread-out ionic swarm keeps the colloids too far apart for cohesive forces to come into play. Sodium ions cause a spread-out ion swarm for two reasons: (1) their large hydration shell of water allows them to be only loosely attracted to the colloids, and (2) twice as many monovalent Na^+ ions as divalent (Ca^{2+} or Mg^{2+}) ions are attracted to a given colloid charge. When the colloid particles are separated from each other, the soil is in a dispersed condition. (Diagrams courtesy of R. Weil)

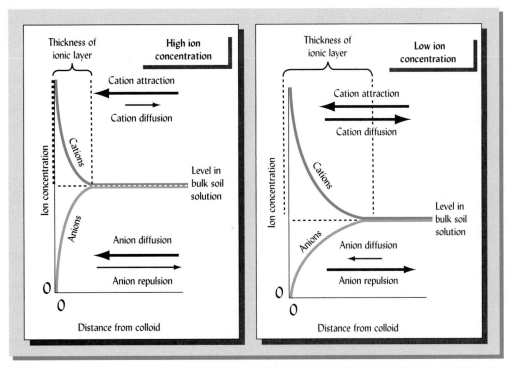

FIGURE 10.23   Ion (electrolyte) concentration in the soil solution influences soil dispersion. When the ion concentration is high (*left*), anions diffuse more strongly toward the colloid surface because their concentration gradient is steeper—that is, the difference is large between the anion concentration at the clay surface (nearly zero) and in the bulk soil solution. Since cation concentration is very high at the clay surface, an increased ion concentration in the bulk solution diminishes the cation concentration gradient. Therefore, the tendency is weaker for cations to diffuse away from the clay surface. To summarize: with higher salt concentration, cations stay closer and anions come closer to the clay; hence the ion swarm is compressed and clay particles can get close enough to one another to flocculate. When the ionic concentration is low (*right*), the opposite trends pertain: the ion swarm becomes more diffuse, and the colloids become dispersed. (Diagrams courtesy of R. Weil)

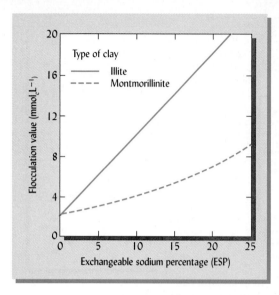

FIGURE 10.24 Effect of exchangeable sodium percentage (ESP) on the "floccu-lation value" for two clays. The flocculation value was defined as the minimum ionic strength needed to cause flocculation and prevent dispersion. Note that when sodium accounts for a greater percentage of the exchangeable ions, a higher ionic strength (i.e., salt concentration) is needed to prevent clay disper-sion. The effect is more pronounced for illite clay than for montmorillonite clay (see Section 8.3). [Based on data cited in Evangelou and Phillips (2005)]

to come close enough together to form floccules (Figure 10.19, *right*). The dissolved salt concentration that is just high enough to cause flocculation and prevent dispersion is termed the *flocculation value*. This value increases as the proportion of exchangeable Na[+] increases, illustrating that the effect of sodium can be counteracted by increasing the dissolved salt concentration and that the damaging effects of sodium are greatest when salt concentrations are lowest. Figure 10.24 shows that the flocculation value for a given level of sodium depends on the type of clay present, the micaceous clays (e.g., illite) being more susceptible to dispersion (i.e., requiring a higher ionic concentration to prevent dispersion) than the smectites (e.g., montmorillonite).

It is worth remembering that *low salt (ion) concentrations and weakly attracted ions (e.g., sodium) encourage soil dispersion and puddling, while high salt concentrations and strongly attracted ions (e.g., calcium) promote clay flocculation and soil permeability.*

10.7 GROWTH OF PLANTS ON SALT-AFFECTED SOILS[6]

How Salts Affect Plants

Plants respond to the various types of salt-affected soils in different ways. In addition to the nutrient-deficiency problems associated with high pH (see Section 10.1), high lev-els of soluble salts affect plants by two primary mechanisms: **osmotic effects** and **specific ion effects**.

OSMOTIC EFFECTS.   Soluble salts lower the osmotic potential of the soil water (see Section 5.3), making it more difficult for roots to remove water from the soil. For established plants, this rarely results in wilting or even reduced water uptake, but it does require that plants expend more energy making osmotic adjustments—accumulating organic and inorganic solutes to lower the osmotic potential *inside* their cells to counteract the low osmotic potential of the soil solution outside. The lost energy results in reduced growth.

Plants are most susceptible to salt damage in the early stages of growth. Salinity may delay, or even prevent, the germination of seeds (Figure 10.25). Young seedlings may be killed by saline conditions that older plants of the same species could tolerate. The radi-cle (root precursor) of a germinating seed appears to be particularly sensitive to salinity. As young root cells encounter a soil solution high in salts, they may lose water by

---

[6] For a review of plant response to water stress and salinity, see Munns (2002). For a description of nat-ural halophytes, see NAS (1990). For a perspective on the potential of biotechnology to deal with salt-affected soils, see Frommer et al. (1999).

FIGURE 10.25 Foliar symptoms on oldest leaves (*inset*), reduced germination, and stunted growth of soybean plants with increasing levels of soil salinity due to additions of NaCl to a sandy soil. The large numbers written on the pots indicate the electrical conductivity (EC_e) of the soil. Note that serious growth reductions occurred for this sensitive cultivar even at EC_e levels considered normal. The soybean cultivar used (Jackson) is more sensitive to salinity than most other cultivars of soybean. (Photos courtesy of R. Weil)

osmosis to the more concentrated soil solution. The cells then collapse. The same can happen to the tender young stems of certain seedlings.

SPECIFIC ION EFFECTS. The kind of salt can make a big difference in how plants respond to salinity. Certain ions, including Na^+, Cl^-, $H_3BO_4^-$, and HCO_3^- are quite toxic to many plants. However, as we will see, plant species, and even strains within a species, differ widely in their sensitivity to these ions. In addition to specific toxic effects, high levels of Na^+ can cause imbalances in the uptake and utilization of other cations. For example, Na^+ competes with the essential nutrient ion K^+ in the process of transport across the cell membrane during uptake, making it difficult for plants to obtain the K^+ they need from saline-sodic or sodic soils. The presence of adequate Ca^{2+} helps the plant to discriminate against Na^+ and for K^+. This is but one example of how the balance of specific ions in saline soils can be as important to plant health as the total salt concentrations.

PHYSICAL EFFECTS OF SODICITY. Deterioration of physical properties may also be a factor in determining which plants can grow in sodic soils. The colloidal dispersion caused by sodicity may harm plants in at least two ways: (1) oxygen becomes deficient due to the breakdown of soil structure and the very limited air movement that results, and (2) water relations are poor due largely to the very slow infiltration and percolation rates.

PLANT SYMPTOMS. In response to excessive soil salinity, many plants become severely stunted and exhibit small dark-bluish green leaves with dull surfaces. High levels of sodium or chloride typically produce scorching or necrosis of the leaf margins and tips (see Figure 10.25, *inset*). These symptoms appear first and most severely on the oldest leaves as these have been transpiring water and accumulating salts for the longest period. Salt-stressed plants may also lose their leaves prematurely.

Selective Tolerance of Higher Plants to Saline and Sodic Soils

Satisfactory plant growth on salty soils depends on a number of interrelated factors, including the physiological constitution of the plant, its stage of growth, and its rooting habits. For example, old alfalfa plants are more tolerant to salt-affected soils than young ones, and deep-rooted legumes show a greater resistance to such soils than those that are shallow-rooted.

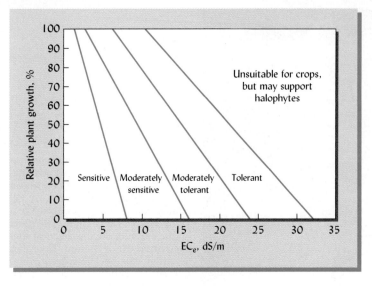

FIGURE 10.26 Relative productivity of five groups of plants classified by their sensitivity to salinity as measured by electrical conductivity of the soil. See Table 10.3 for examples of plants in each sensitivity class. [From Maas and Grattan (1999); with permission of the American Society of Agronomy]

Identifying a high-value crop that will grow on salt-affected soils: http://www.ars.usda.gov/is/AR/archive/aug04/salt0804.htm

PLANT SENSITIVITY. While it is difficult to forecast precisely the tolerance of plant species to salty soils, numerous tests have made it possible to classify many plants into four general salt-tolerance groups. Figure 10.26 shows how soil salinity, as measured by electrical conductivity (EC_e), affects the relative productivity of these general salinity-tolerance groups. Table 10.3 groups selected plants according to this classification. Note that trees, shrubs, grasses, fruits, vegetables, and field crops are included in the different categories.

Two particular groups of plant species contribute several plants that can be grown on salty soils: (1) wild *halophytes* (salt-loving plants), and (2) salt-tolerant varieties developed by plant breeders. A number of wild halophytes have been found that are quite tolerant to salts and that possess qualities that could make them useful for human and/or animal consumption. They are also valuable for restoring disturbed or degraded land under saline conditions.

GENETIC IMPROVEMENTS. Plant breeders have been able to develop new plant strains with salt tolerance greater than that possessed by conventional varieties. A major advance was made by the discovery of a single gene that enables halophytes to sequester high amounts of Na^+ in their cellular vacuoles (large, membrane-enclosed storage structures inside individual cells). In the vacuole, the Na^+ ions can act to the plant's advantage by contributing to low internal osmotic potential while remaining isolated from cellular systems that are susceptible to Na^+ toxicity. Work is underway to use genetic engineering techniques to transfer this gene to economically important plants, with the aim of producing crops that can tolerate saline soils and the use of salty water for irrigation. Plant selection and improvement will almost certainly make important contributions to food production on salt-affected soils in the future. However, improved plant tolerance must not be viewed as a substitute for proper salinity control, as discussed in Section 10.9.

Salt Problems Not Related to Arid Climates

DEICING SALTS. In areas where deicing salts are used to keep roads and sidewalks free of snow and ice during winter months, these salts may impact roadside soils and plants. Repeated application of deicing salts can result in salinity levels sufficiently high as to adversely affect plants and soil organisms living alongside highways or sidewalks. Figure 10.27 illustrates tree damage from such treatment. In humid regions, such salt contamination is usually temporary, as the abundant rainfall leaches out the salts in a matter of weeks or months. To avoid the specific chemical and physical problems associated with sodium salts, many municipalities have switched from NaCl to KCl for deicing purposes. Sand can be used to improve traction, thereby reducing the need for deicing salt.

CONTAINERIZED PLANTS. Salinity can also be a serious problem for potted plants, particularly perennials that remain in the same pot for long periods (Figure 10.13, *left*). Greenhouse operators producing containerized plants must carefully monitor the quality of the

TABLE 10.3 Relative Salt Tolerance of Selected Plants

Approximate EC$_e$ resulting in a 10% reduction in plant growth for the most sensitive species in each column.

Tolerant, 12 dS/m	Moderately tolerant, 8 dS/m	Moderately sensitive, 4 dS/m	Sensitive, 2 dS/m
Alkali grass, Nutall	Ash (white)	Alfalfa	Alders
Alkali sacaton	Asparagus	Arborvitae	Almond
Barley (grain)	Aspen	Boxwood	Apple
Bent grass	Barley (forage)	Broad bean	Apricot
Bermuda grass	Beet (garden)	Cabbage	Azalea
Bougainvillea	Birch (black)	Cauliflower	Beech
Boxwood, Japanese	Black cherry	Celery	Bean
Canola (rapeseed)	Broccoli	Clover (alsike,	Birch
Cotton	Bromegrass	ladino, red,	Blackberry
Date	Cedar (red)	strawberry,	Burford holly
Guayule	Cowpea	and berseem)	Carrot
Hawthorn Indian	Elm	Corn	Dogwood
Jojoba	Fescue (tall)	Cucumber	Elm (American)
Kallar grass	Fig	Dallas grass	Grapefruit
Kenaf	Honeysuckle	Grape	Hemlock
Natal plum	Hydrangea	Hickory (shagbark)	Hibiscus
Oak (red and white)	Juniper	Juniper	Larch
Oleander	Kale	Lettuce	Lemon
Olive	Locust (honey)	Locust (black)	Linden
Prostrate kochia	Oak (red and white)	Maple (red)	Maple (sugar and red)
Redwort	Oats	Pea	Onion
Rescue grass	Orchard grass	Peanut	Orange
Rosemary	Pomegranate	Radish	Peach
Rugosa	Privet	Rice (paddy)	Pear
Rye (grain)	Ryegrass (perennial)	Soybean (sens. var.)	Pine (red and white)
Salt grass (desert)	Safflower	Squash	Pineapple
Sugar beet	Sorghum	Sugar cane	Plum (prune)
Tamarix	Soybean (tol. var.)	Sweet clover	Potato
Wheat grass (crested)	Squash (zucchini)	Sweet potato	Raspberry
Wheat grass (fairway)	Sudan grass	Timothy	Rose
Wheat grass (tall)	Trefoil (birdsfoot)	Tomato	Silk tree
Wild rye (altai)	Wheat	Turnip	Star jasmine
Wild rye (Russian)	Wheat grass (western)	Vetch	Strawberry
Willow	Winged bean	Viburnum	Tomato

FIGURE 10.27 Salts spread on roadways to melt winter ice and snow in cold regions can cause salt injury to vegetation growing in roadside soils. (*Left*) Localized salt concentrations are visible on the soil surface some 3 months after the last salt application. Dieback of the maple trees is also evident. (*Right*) Close-up view of chlorotic and dead leaf-margin tissue caused by the salts. (Photos courtesy of R. Weil)

water used for irrigation. Salts in the water, as well as those applied in fertilizers, can build up if care is not taken to flush them out occasionally with excess water. Chlorinated urban tap water used for indoor plants should be left overnight in an open container to allow some of the dissolved chlorine to escape, thus reducing the load of Cl^- ions added to the potting soil.

10.8 WATER-QUALITY CONSIDERATIONS FOR IRRIGATION[7]

Whether in a single field or in a large regional watershed, understanding the **salt balance** is a basic prerequisite for wise management of salt-affected soils. To achieve salt balance, the amount of salt coming in must be matched by the amount being removed. Meeting this condition is a fundamental challenge to the long-term sustainability of irrigated agriculture. In irrigated areas, this principally means managing the quality and amount of the irrigation water brought in and the quality and amount of soil drainage water removed.

IRRIGATION WATER QUALITY. Table 10.4 provides some guidelines on water quality for irrigation. Monitoring the chemical quality of water added to salt-affected soils is a prime management strategy. For example, when dealing with a saline-sodic soil, water that is too low in salts, such as rainwater, can hasten the change from saline-sodic to sodic soil conditions. Rain not only leaches soluble salts from the upper few centimeters of soil, but the impact of the raindrops encourages dispersion of the soil colloids, an initial step in the process of sodic soil development.

If the salt content of irrigation water is high, salt balance will be difficult to achieve. However, even very salty water can be used successfully if soils are sufficiently well drained to allow careful management of salt inputs and outputs. Where irrigation water is low in salts but has a high SAR, the formation of sodic soils is likely to accelerate. In

TABLE 10.4 Water-Quality Guidelines for Irrigation

Note that with regard to effects on physical structure of soils, higher total salinity (EC_w) in the irrigation water compensates, somewhat, for increasing sodium hazard (SAR). In addition, note that while water low in salts (low EC_w) avoids problems of restricted water availability to plants, it may worsen soil physical properties, especially if the SAR is high.

		Degree of restriction on use		
Water Property	*Units*	*None*	*Slight to moderate*	*Severe*
Salinity (affects crop water availability)				
EC_w	dS/m	<0.7	0.7–3.0	>3.0
TDS	mg/L	<450	450–2000	>2000
Physical structure and water infiltration (Evaluate using EC_w and SAR together)				
SAR = 0–3 and EC_w =	dS/m	>0.7	0.7–0.2	<0.2
SAR = 3–6 and EC_w =	dS/m	>1.2	1.2–0.3	<0.3
SAR = 6–12 and EC_w =	dS/m	>1.9	1.9–0.5	<0.5
SAR = 12–20 and EC_w =	dS/m	>2.9	2.9–1.3	<1.3
SAR = 20–40 and EC_w =	dS/m	>5.0	5.0–2.9	<2.9
Sodium (Na) specific ion toxicity (affects sensitive crops)				
Surface irrigation	mmol/L	<3	3–9	>9
Sprinkler irrigation	mmol/L	<3	>3	
Chloride (Cl) specific ion toxicity (affects sensitive crops)				
Surface irrigation	mmol/L	<4	4–10	>10
Sprinkler irrigation	mmol/L	<3	>3	
Boron (B) specific ion toxicity (affects sensitive crops)				
	mg/L	<0.7	0.7–3.0	>3.0

Modified from Abrol et al. (1988) with permission of the Food and Agriculture Organization of the United Nations.

---

[7] For an overview of water-quality problems facing irrigated agriculture in California, see Letey (2000).

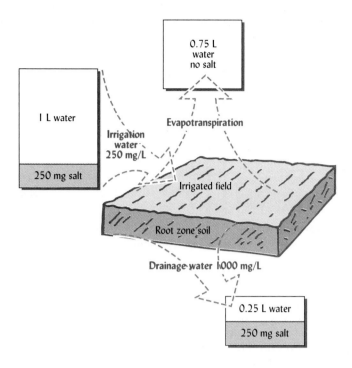

FIGURE 10.28 Evapotranspiration and salt balance together ensure that the drainage water from irrigated fields is much saltier than the irrigation water applied. In this example, the irrigation water contains 250 mg salts per liter. Some 75% of the applied water is lost to the atmosphere by evapotranspiration. About 25% of the water applied is used for drainage, which is necessary to maintain the salt balance (prevent the buildup of salts) in the field. The added salts are leached away with the drainage water, which then contains the same amount of salt as was added, but in only 25% of the added amount of water. The concentration of salt in the drainage water is thereby four times as great (1000 mg/L) as in the irrigation water. Disposal and/or reuse of the highly saline drainage water present challenges for any irrigation project. (Diagram courtesy of R. Weil)

addition, irrigation water high in carbonates or bicarbonates can reduce Ca^{2+} and Mg^{2+} concentrations in the soil solution by precipitating these ions as insoluble carbonates. This leaves a higher proportion of Na^+ in the soil solution and can increase its SAR, moving the soil toward the sodic class.

DRAINAGE WATER SALINITY. Since some portion of added water must be drained away to combat salt buildup, the quality and disposition of *waste irrigation waters* must also be carefully monitored and controlled to minimize potential harm to downstream users and habitats. In any irrigation system, the drainage water leaving a field will be considerably more concentrated in salts than the irrigation water applied to the same field (Figure 10.28 explains why). What to do with the increasingly saline drainage water presents a major challenge to the sustainability of irrigated agriculture.

Different irrigation projects take different approaches, but rarely is the problem solved without some downstream environmental damage. Perhaps the most efficient approach is to collect the drainage water, keep it isolated from the relatively high-quality canal water, and reuse it to irrigate a more salt-tolerant crop in a lower field. This approach provides both high-quality canal water and lower-quality drainage water for use on appropriate crops. Generally, the salt-tolerant crops are less valuable than the more salt-sensitive ones (e.g., the yield of salt-tolerant cotton from 1 ha is worth much less than the yield of salt-sensitive tomato); still, recycling drainage water saves money as well as water. Often some fresh water must be mixed with the recycled drainage water to bring its salinity level down to what can be tolerated by even the salt-tolerant crop (see Figure 10.29). After several cycles of reuse, the drainage water must be disposed of, as it will have become too saline for irrigating even the most salt-tolerant species.

A more common (though less water-efficient) approach is to route the drainage water back into the canal. This mixes the poor-quality drainage water with the high-quality canal water, improving the one but degrading the other. Again, after several cycles, the downstream water becomes too saline for use and must be disposed of (or treated to remove the salts—a very expensive process). In many cases, the irrigation wastewater is eventually channeled into shallow ponds that allow the water to evaporate and the salts to collect.

The Colorado River provides one example of the regional impacts of irrigation systems. By the time this river reaches the United States/Mexico border, it is so loaded with salts from upstream irrigation systems and from domestic and industrial uses that a huge desalinization plant has been constructed to enable the United States to meet its treaty obligations with Mexico. Up to now it has rarely been necessary to use the plant, since drainage water from the lowest U.S. irrigation system (Welton-Mohawk Valley) is being diverted to the Sea of Cortez through a canal that runs parallel to the Colorado River.

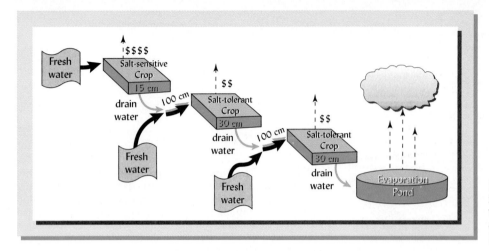

FIGURE 10.29 Generalized scheme for recycling irrigation drainage water by mixing it with fresh water and irrigating salt-tolerant crops. [Redrawn from Letey et al. (2003)]

TOXIC ELEMENTS IN DRAINAGE WATER. If either the irrigation water or the soil of the irrigated fields contains significant quantities of certain toxic trace elements, these too will become increasingly concentrated in the drainage water. The elements of concern include molybdenum (Mo), arsenic (As), boron (B), and selenium (Se). Molybdenum and selenium are necessary nutrients for animals and humans in trace amounts, but all four elements can be toxic to cattle, wildlife, or people if they become concentrated in water or food. At some locations in the western United States, such trace elements have accumulated to toxic levels in downstream wetlands or evaporative ponds. Table 10.5 shows that plants growing on these affected areas can accumulate levels of these elements that are unsafe for livestock and/or wildlife.

Indeed, just two years after irrigation wastewater began to flow into holding ponds established in California's Kesterson National Wildlife Refuge, scientists there began to observe widespread death and deformities among wildfowl, mammals, and amphibians that lived in or near the ponds. The wildlife damages were traced to poisoning by the selenium that had accumulated in the irrigation waters and been **biomagnified** up the wildlife food chain, as small fish ate water plants and birds ate the fish, etc. Box 10.2 illustrates how such concentrations can occur in an irrigation system such as that in California's San Joaquin Valley, where the Kesterson Refuge is located. The chemistry and cycling of selenium and other trace elements is discussed further in Chapter 15.

10.9 RECLAMATION OF SALINE SOILS

Salinity stress and its mitigation:
http://www.plantstress.com/Articles/salinity_m/salinity_m.htm

The restoration of soil chemical and physical properties conducive to high productivity is referred to as soil *reclamation*. Reclamation of saline soils is largely dependent on the provision of effective drainage and the availability of good-quality irrigation water (see Table 10.4) so that salts can be leached from the soil. In areas where irrigation water is not available, such as in saline seeps in the Northern Plains states, the leaching of salts is not practical. In these areas, deep-rooted vegetation may be used to lower the water table and reduce the upward movement of salts.

TABLE 10.5 **Levels of Molybdenum and Selenium in Three Salinity-Tolerant Forage Species Grown on a Highly Salinized Soil That Had Received Large Quantities of Waste Irrigation Water in the San Joaquin Valley of California**

The upper "safe" limit for animal consumption (potential toxicity level) is also shown.

	Potential toxicity level,	Level in the forages, mg/kg		
Trace elements	mg/kg	Tall wheatgrass	Alkali sacaton	Astragalus racemosus
Molybdenum	5	26	10	18
Selenium	5	12	9	670

Data selected from Retana et al. (1993).

BOX 10.2   SELENIUM IN IRRIGATED SOILS

The use of selenium-laden irrigation waters can result in the buildup of selenium (Se) to toxic levels in soils and in irrigation wastewaters coming from these soils. Selenium accumulates in plants growing on these soils and in nearby wetland areas where irrigation wastewaters are disposed.

High-selenium marine sediments and shales underlie some areas of the western United States. Under alkaline and well-drained soil conditions, the selenate (SeO_4^{2-}) form of selenium prevails. Much like sulfates, selenates are relatively water-soluble and are readily available for plant uptake. They may accumulate in rangeland plants to levels toxic for grazing animals. Or they may be leached from the soil and substrata into the streams draining the watershed. Downstream, the high-Se water in these streams may be used for irrigation.

Figure 10.30 illustrates how return of Se-laden irrigation wastewater to streams for reuse can cause serious environmental problems. Streams from high-Se watersheds provide water for irrigated fields in district A. The Se level in the water (shown by the darkness of the shading in the drainage) is significant, but not sufficient to cause environmental problems.

As with other salts, the Se accumulates in the soil and concentrates in the waste drainage water coming from the soil. This drainage water moves (or is pumped) back into a stream that later supplies water for irrigated field B downstream. The process is repeated, Se in the soil is increased, and the Se level in the drainage water is increased (darker shading). After several such cycles, the concentration of Se (and possibly of other salts) is high enough to make the water unusable for further irrigation.

The water is then diverted or pumped into nearby wetlands or shallow holding ponds, such as those established in the Kesterson National Wildlife Refuge in California. Birds and other wildlife find these wetlands and areas of open water inviting places to live and breed. As the water evaporates, the Se concentrates in the remaining water and moves into the soils of the holding area. Plants living in the wetland or pond area absorb the selenium, accumulating it to levels that are toxic to farm animals and wildlife. Selenium poisoning was determined to be responsible for the deformed and dead chicks that were found in some 40% of monitored bird nests at Kesterson.

Steps are being taken to reduce or eliminate the release of high-Se waters into wetlands. Researchers are also seeking plants that are very high Se accumulators so that Se can be removed from holding areas by harvesting the plant biomass.[a] Other research has shown that certain bacteria form volatile organic Se compounds, which may be slowly released into the atmosphere. The impacts of irrigation on wildlife remind us of the interconnectedness of all parts of the environment and the need to manage soil and water resources with a holistic view of their roles in the larger ecosystem.

---

[a] For example, see Banuelos et al. (1997); for additional information on selenium in irrigation, see Nolan and Clark (1997).

FIGURE 10.30   *Concentrating Se in an irrigation project. (Diagram courtesy of R. Weil)*

If the natural soil drainage is inadequate to accommodate the leaching water, an artificial drainage network must be installed. Intermittent applications of excess irrigation water may be required to effectively reduce the salt content to a desired level. The process can be monitored by measuring the soil's EC, using either the saturation extract procedure or one of the field instruments described in Section 10.4.

Leaching Requirement

The amount of water needed to remove the excess salts from saline soils, called the **leaching requirement (LR)**, is determined by the characteristics of the crop to be grown, the irrigation water, and the soil. As demonstrated in Box 10.3, an approximation of the LR is given for relatively uniform salinity conditions by the ratio of the salinity of the irrigation water (expressed as its EC_{iw}) to the maximum acceptable salinity of the soil solution for the crop to be grown (expressed as EC_{dw}, the EC of the drainage water).

$$LR = \frac{EC_{iw}}{EC_{dw}} \qquad (10.8)$$

The LR indicates water added in excess of that needed to thoroughly wet the soil and meet the crop's evapotranspiration needs. Note that if EC_{iw} is high and a salt-sensitive crop is chosen (dictating a low EC_{dw}), a very large leaching requirement LR will result. As mentioned in Section 10.8, disposal of the drainage water that has leached through the soil can present a major problem. Therefore, it is generally desirable to use management techniques that minimize the LR and the amount of drainage water that requires disposal.

Management of Soil Salinity

Management of irrigated soils should aim to simultaneously minimize drainage water and protect the root zone (usually the upper meter of soil) from damaging levels of salt accumulation. These two goals are obviously in conflict. The irrigator can attempt to find the best compromise between the two and can use certain management techniques that allow plants to tolerate the presence of higher salt levels in the soil profile. One option is to plant salt-tolerant species or choose the most salt-tolerant varieties within a crop species (as discussed in Section 10.7).

IRRIGATION TIMING. The timing of irrigation is extremely important on saline soils, particularly early in the growing season. Germinating seeds and young seedlings are especially sensitive to salts. Therefore, irrigation should precede or immediately follow planting to move the salts downward and away from the seedling roots. The irrigator can use high-quality water to keep root-zone salinity low during the sensitive early growth stages and then switch to lower-quality water as the maturing plants become more salt-tolerant.

LOCATION OF SALTS IN THE ROOT ZONE. Tillage and planting practices can influence the location and accumulation of salts in arid-region soils. Tillage or surface-residue management practices (such as mulches or conservation tillage) that reduce evaporation from the soil surface should also reduce the upward transport of soluble salts. Likewise, specific techniques for applying irrigation water that direct salt concentrations away from young plant roots can allow higher levels of salt to accumulate without damage to the crop. Applying water in every other furrow and asymmetrically planting only on the wet side of the furrows can provide significant protection to young plants (Figure 10.31).

Frequent application of water, as with sprinkler- or drip-irrigation systems (see Section 6.9), can help move salts away from plant roots. Proper placement of water emitters in drip-irrigation systems is crucial in establishing a low-salt zone around sensitive young plants. Buried drip-emitter lines, in particular, can cause problems by moving salts to the soil surface where seeds are germinating (Figure 10.31). The water emitted below the surface moves toward the soil surface by capillary rise, carrying

BOX 10.3   LEACHING REQUIREMENT FOR SALINE SOILS

The salt balance in a field can be described by equating the salt inputs and outputs:

$$S_{iw} + S_p + S_f + S_m \;=\; S_{dw} + S_c + S_{ppt}$$

$$\underbrace{\phantom{S_{iw} + S_p + S_f + S_m}}_{\text{Salt inputs}} \qquad \underbrace{\phantom{S_{dw} + S_c + S_{ppt}}}_{\text{Salt outputs}} \tag{10.9}$$

The salt inputs include those from irrigation water (S_{iw}), atmospheric deposition (S_p), fertilizers (S_f), and the weathering or dissolution of existing soil minerals (S_m). The salt outputs are due to drainage water (S_{dw}), crop removal (S_c), and chemical precipitation (S_{ppt}) of carbonates and sulfates. Usually S_{iw} and S_{dw} are far larger than the other terms in Equation 10.9, so the main concern is to balance the salt coming in with the irrigation water and that leaving with the drainage water:

$$\underset{\text{Salt in with irrigation water}}{S_{iw}} \;=\; \underset{\text{Salt out with drainage water}}{S_{dw}} \tag{10.10}$$

We can estimate the quantity of salt S carried in drainage or irrigation as the product of the volume of the water (expressed as cm depth applied to an area of land) and the concentration of salt in that water (as approximated by its electrical conductivity, EC). Therefore, we can rewrite Equation 10.10 as follows (using the same subscripts as before):

$$D_{iw} \times EC_{iw} = D_{dw} \times EC_{dw} \tag{10.11}$$

Rearranging the terms, we obtain the following expression that defines the leaching requirement (LR) or the ratio of drainage water depth to irrigation water depth (D_{dw}/D_{iw}) needed to maintain salt balance:

$$\frac{EC_{iw}}{EC_{dw}} = \frac{D_{dw}}{D_{iw}} = LR \tag{10.12}$$

The *LR* tells farmers how much irrigation water (in excess of that required to wet the soil) they should apply for sufficient leaching. The goal is usually to assure that the upper two-thirds of the root zone does not accumulate salts beyond the level acceptable for a particular crop.

From Equation 10.12, we see that *LR* also equals the ratio of the irrigation water EC to the drainage water EC:

$$LR = \frac{EC_{iw}}{EC_{dw}} \tag{10.8}$$

where EC_{dw} is an acceptable level for the crop being grown. What is considered an acceptable EC_{dw} is open to interpretation. An acceptable EC_{dw} might be interpreted to mean the EC_e that allows 90% of maximum crop yield. A more conservative interpretation of acceptable EC_{dw} is the threshold EC_e at which the growth of the particular crop just begins to decline (usually 1 to 2 dS/m lower than that which gives the 90% yield level).

As an example, consider the situation where the irrigation water has an EC_{iw} of 2.5 dS/m and a moderately tolerant crop (e.g., broccoli) is to be grown. If information that is more specific is unavailable, the acceptable EC_{dw} for the crop can be (roughly) estimated from the column heading in Table 10.3. For a moderately tolerant crop, we can use 8 dS/m as the acceptable EC_{dw} to produce 90% of the maximum yield. Then,

$$LR = \frac{2.5 \; dS/m}{8 \; dS/m} = 0.31 \tag{10.13}$$

If this *LR* (0.31) is multiplied by the amount of water needed to wet the root zone—let us suppose it is 12 cm of water—the amount of water to be leached would be 3.7 cm (12 cm × 0.31). This is the minimum amount of water that must be leached through a water-saturated soil to maintain the root zone salinity at the acceptable level. A more sensitive crop could be grown in this soil, but it would have a lower acceptable EC_{dw} and therefore would require the application of a greater amount of water for leaching.

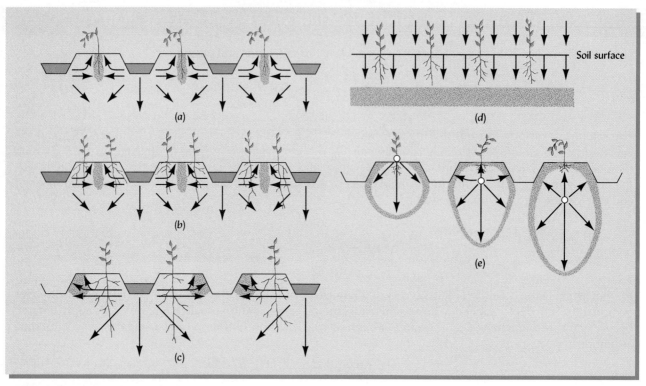

FIGURE 10.31 Effect of irrigation techniques on salt movement and plant growth in saline soils. (*a*) With irrigation water applied to furrows on both sides of the row, salts move to the center of the ridge and damage young plants. (*b*) Placing plants on the edges of the bed rather than in the center helps them avoid the most concentrated salts. (*c*) Application of water to every other furrow and placement of plants on the side of the bed nearest the water helps plants avoid the highest salt concentrations. (*d*) Sprinkler irrigation or uniform flooding temporarily moves salts downward out of the root zone, but the salts will return afterward as the soil surface dries out and water moves up by capillary flow. (*e*) Drip irrigation at low rates provides a nearly continuous flow of water, creating a low-salt soil zone with the salts concentrated at the wetting front. The placement of the drip emitters largely determines whether the salts are moved toward or away from the plant roots. (Diagram courtesy of Wesley M. Jarrell)

dissolved salts with it. When the water evaporates or is taken up by plant roots, the salts are left to concentrate at the soil surface. The size and shape of the low-salinity zone created by drip irrigation is also dependent on the rate of water application and the texture of the soil (Figure 10.32).

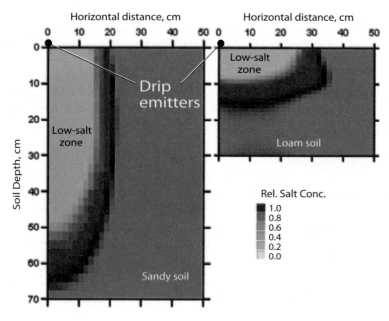

FIGURE 10.32 Effect of soil texture on distribution of salts after irrigation from surface drip emitters. The salts were originally distributed evenly within each soil, but the slowly applied irrigation water (4 L/h for 3 h) dissolved the salts and moved them away from the emitters toward the border between the wet and dry soil (the wetting front). Note that the low-salt zone in the sandy soil is much deeper, but more narrow, than in the loam. Repeated irrigation at short intervals may keep the salts away from the root zone. However, if the soil is allowed to dry, the dissolved salts will begin to return to the root zone as the matric potential gradient draws water toward the drier soil. [Redrawn from data in Bresler (1975)]

SPATIAL VARIABILITY. The actual leaching fraction applied to a field may exceed the theoretical amount calculated from the *LR* if soil salinity levels vary from spot to spot in a field. In order to avoid yield losses from salinity, a farmer may irrigate a field with the amount of water determined by the *LR* of the *most saline* parts of the field. This amount will assure that the most saline parts of the field are properly leached, but it will waste valuable water and create more than the necessary amount of drainage water in the less saline parts of the field.

It may be feasible to avoid such inefficiencies by using *precision agriculture* technology. For example, irrigation systems (such as sprinkler or drip) could be designed with variable rate controls and used in conjunction with soil-salinity sensing devices (such as the four-electrode probe or the electromagnetic induction sensor described in Section 10.4) so that they apply just the right amount of leaching water for each part of a field.

Some Limitations of the Leaching Requirement Approach[8]

The leaching requirement approach to managing irrigated soils is only an approximation and has several inherent weaknesses. First, additional leaching may be needed, in some cases, to reduce the excess concentration of specific elements, such as boron. Second, the *LR* by itself does not take into account the rise in the water table that is likely to result from increased leaching, and so it may lead to waterlogging and, eventually, increased salinization. Third, irrigation using a simple *LR* approach usually overapplies water because an entire field is treated to avoid salt damage in its most saline spots. Fourth, the *LR* method does not consider salts that may be picked up from fossil salt deposits already in the soil and substrata. Fifth, it assumes that the EC of the drainage water is known, but in fact this may be largely unknown, since it may take years or even decades for the water applied in irrigation to reach the main drains where it can be easily sampled. In other words, the drainage water sampled today may represent the leaching conditions of several months or years ago.

An alternative approach would be to closely monitor the salinity in the soil profile by taking repeated measurements across the field, using the EM sensor or four-electrode methods discussed in Section 10.4. The sufficiency of leaching and the dominant direction of water movement could then be judged from the type of salinity profile observed (Figure 10.33). This more complex approach, combined with site-specific management techniques, seems to hold promise for future management and reclamation of salt-affected soils under irrigation.

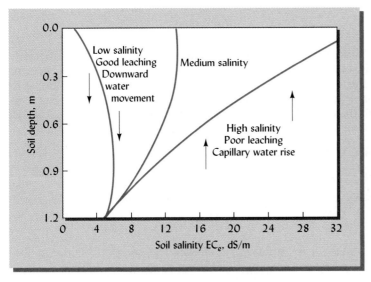

FIGURE 10.33 Soil salinity profile curves show the levels of salts throughout the root zone. The shapes of the curves also indicate whether leaching has been sufficient and whether saline water is rising from a shallow water table. The arrows indicate the direction of water flow for the low- and high-salinity soils. The curve on the right shows increasing salinity near the soil surface, a pattern that is typical of saline seeps and waterlogged irrigated fields. Note that the leftmost curve is similar in shape to the salinity profile of a dry, well-drained soil under desert pavement in which there is no influence of groundwater (compare to Figure 10.4, *left*). [Redrawn from Rhoades et al. (1999); used with permission of the Food and Agriculture Organization of the United Nations]

---

[8] For a discussion of how detailed spatial assessment of salinity can be used as an alternative to the *LR* approach, see Rhoades et al. (1999).

Saline-sodic soils have some of the adverse properties of both saline and sodic soils. If attempts are made to leach out the soluble salts in saline-sodic soils, as was discussed for saline soils, the exchangeable Na^+ level as well as the pH would likely increase, and the soil would take on adverse characteristics of sodic soils. Consequently, for both saline-sodic and sodic soils, attention must first be given to reducing the level of exchangeable Na^+ ions and then to the problem of excess soluble salts.

Gypsum

Removing Na^+ ions from the exchange complex is most effectively accomplished by replacing them with either the Ca^{2+} or the H^+ ion. Providing Ca^{2+} in the form of gypsum ($CaSO_4 \cdot 2H_2O$) is the most practical way to bring about this exchange. When gypsum is added, reactions such as the following take place:

$$2NaHCO_3 + CaSO_4 \longrightarrow CaCO_3 + Na_2SO_4 + CO_2 \uparrow + H_2O$$
$$\text{(leachable)} \tag{10.14}$$

$$Na_2CO_3 + CaSO_4 \rightleftharpoons CaCO_3 + Na_2SO_4$$
$$\text{(insoluble)} \quad \text{(leachable)} \tag{10.15}$$

$$\begin{matrix} Na^+ \\ Na^+ \end{matrix} \boxed{Colloid} + CaSO_4 \rightleftharpoons Ca^{2+} \boxed{Colloid} + Na_2SO_4 \tag{10.16}$$

Note that in each case the soluble salt Na_2SO_4 is formed, which can be easily leached from the soil as was done in the case of the saline soils.

Several tons of gypsum per hectare are usually necessary to achieve reclamation. In Box 10.4, calculations are made to approximate the amount of gypsum that is theoretically needed to remove an acceptable portion of the Na^+ ion from the exchange complex. The soil must be kept moist to hasten the reaction, and the gypsum should be thoroughly mixed into the surface by cultivation—not simply plowed under. The treatment must be supplemented later by a thorough leaching of the soil with irrigation water to leach out most of the sodium sulfate.

Gypsum is inexpensive, widely available in both natural and in industrial by-product forms, and easily handled. Care must be taken, however, to be certain that the gypsum is finely ground and that it is well mixed with the upper soil horizons so that its solubility and rate of reaction are maximized.

Sulfur and Sulfuric Acid

Elemental sulfur and sulfuric acid can be used to advantage on sodic soils, especially where sodium bicarbonate abounds. The sulfur, upon biological oxidation (see Sections 9.6 and 13.19), yields sulfuric acid, which not only changes the sodium bicarbonate to the less harmful and more leachable sodium sulfate but also decreases the pH. The reactions of sulfuric acid with the compounds containing sodium may be shown as follows:

$$2NaHCO_3 + H_2SO_4 \longrightarrow 2CO_2 \uparrow + 2H_2O + Na_2SO_4$$
$$\text{(leachable)} \tag{10.17}$$

$$Na_2CO_3 + H_2SO_4 \longrightarrow CO_2 \uparrow + H_2O + Na_2SO_4$$
$$\text{(leachable)} \tag{10.18}$$

$$\begin{matrix} Na^+ \\ Na^+ \end{matrix} \boxed{Colloid} + H_2SO_4 \rightleftharpoons \begin{matrix} H^+ \\ H^+ \end{matrix} \boxed{Colloid} + Na_2SO_4$$
$$\text{(leachable)} \tag{10.19}$$

Not only are the sodium carbonate and bicarbonate changed to sodium sulfate, a mild neutral salt, but the carbonate anion is removed from the system. When gypsum is used, however, a portion of the carbonate may remain as a calcium compound ($CaCO_3$).

BOX 10.4   CALCULATING THE THEORETICAL GYPSUM REQUIREMENT

PROBLEM

How much gypsum is needed to reclaim a sodic soil with an exchangeable sodium percentage (ESP) of 25% and a cation exchange capacity of 18 $cmol_c$/kg? Assume that you want to reduce the ESP of the upper 30 cm of soil to about 5% so that a crop like alfalfa could be grown.

SOLUTION

First, determine the amount of Na^+ ions to be replaced by multiplying the CEC (18) by the change in Na^+ saturation desired (25 − 5 = 20%).

$$18 \ cmol_c/kg \times 0.20 = 3.6 \ cmol_c/kg$$

From the reaction that occurs when the gypsum ($CaSO_4 \cdot 2H_2O$) is applied,

$$\boxed{Colloid} \begin{array}{c} Na^+ \\ Na^+ \end{array} + CaSO_4 \cdot 2H_2O \rightleftharpoons \boxed{Colloid} \ Ca^{2+} + Na_2SO_4 + 2H_2O \qquad (10.20)$$

We know that the Na^+ is replaced by a chemically equivalent amount of Ca^{2+} in the gypsum ($CaSO_4 \cdot 2H_2O$). In other words, 3.6 $cmol_c$ of $CaSO_4 \cdot 2H_2O$ will be needed to replace 3.6 $cmol_c$ of Na^+.

Second, calculate the weight in grams of gypsum needed to provide the 3.6 $cmol_c$/kg soil. This can be done by first dividing the molecular weight of $CaSO_4 \cdot 2H_2O$ (172) by 2 (since Ca^{2+} has two charges and Na^+ only one) and then by 100 since we are dealing with $centimole_c$ rather than $mole_c$.

$$\frac{172}{2} = 86 \ g \ CaSO_4 \cdot 2H_2O/mol_c$$

and $\frac{86}{100} = 0.86 \ g \ CaSO_4 \cdot 2H_2O/cmol_c$ required to replace 1 $cmol_c$ Na^+

The 3.6 $cmol_c Na^+$/kg would require

$$3.6 \ cmol_c/kg \times 0.86 \ g/cmol_c = 3.1 \ g \ CaSO_4 \cdot 2H_2O/kg \ of \ soil$$

Last, to express this in terms of the amount of gypsum needed to treat 1 ha of soil to a depth of 30 cm, multiply by 4×10^6, which is twice the weight in kg of a 15-cm-deep hectare–furrow slice (see Section 4.7, footnote 11).

$$3.1 \ g/kg \times 4 \times 10^6 \ kg/ha = 12,400,000 \ g \ gypsum/ha$$

This is 12,400 kg/ha, 12.4 Mg/ha, or about 5.5 tons/acre.

Because of impurities in the gypsum and the inefficiency of the overall process, these amounts would likely be adjusted upward by 20 to 30% in actual field practice to account for less than complete reactivity. However, if the soil in question naturally contains some gypsum, that amount should be subtracted from the total just calculated. Finally, it is advisable to apply only half of the recommended amount at one time, with the remaining half applied about 6 months later, if necessary.

In research trials, sulfur and even sulfuric acid have proven to be very effective in the reclamation of sodic soils, especially if large amounts of $CaCO_3$ are present. In practice, however, gypsum is much more widely used than the acid-forming materials.

Physical Condition

The effects of gypsum and sulfur on the physical condition of sodic soils is perhaps more spectacular than are the chemical effects. Sodic soils are almost impermeable to rainwater and irrigation water, since the soil colloids are largely dispersed and the soil is essentially void of stable aggregates (see Figure 10.19, *right* and Table 10.6). When the exchangeable Na^+ ions are replaced by Ca^{2+} or H^+, soil aggregation and improved water infiltration results (Figure 10.34). The neutral sodium salts (e.g., Na_2SO_4) formed when the exchange takes place can then be leached from the soil, thereby reducing both

TABLE 10.6 **The Degree of Aggregation of Two Salt-Affected Alfisols (Xeralfs) after Repeated Wetting (8 times) With Solutions Differing in Sodium and Calcium Contents**

The Na treatment simulates the effect of applying irrigation water with high SAR. The effect of adding gypsum is simulated by the Ca treatment. Note that in each case the Na treatment reduced aggregation and the Ca treatment increased it.

Soil	Aggregate size, μm	Original, %	Na-treated, %	Ca-treated, %
Farrell	>50	5.1	3.1	9.9
	20–50	12.8	2.1	18.0
Tarlee	>50	15.6	0.0	34.1
	20–50	11.2	5.4	42.0

From Barzegar et al. (1996).

salinity and sodicity. Some research suggests that aggregate-stabilizing synthetic polymers may be helpful in at least temporarily increasing the water infiltration capacity of gypsum-treated sodic soils. Data in Table 10.7 from one experiment show the possible potential of these soil conditioners, especially when used in combination with gypsum.

DEEP-ROOTED VEGETATION.   The reclamation effects of gypsum or sulfur are greatly accelerated by plants growing on the soil. Crops that have some degree of tolerance to saline and sodic soils, such as sugar beets, cotton, barley, sorghum, berseem clover, or rye, can be grown initially. Their roots help provide channels through which gypsum can move downward into the soil. Deep-rooted crops, such as alfalfa, are especially effective in improving the water conductivity of gypsum-treated sodic soils. Figure 10.34 illustrates the ameliorating effects of the combination of gypsum and deep-rooted crops.

AIR INJECTION.   In addition to reduced diffusion of air because of soil dispersion, irrigated soils may provide less than optimal oxygen availability to roots because irrigation water—especially applied by drip systems—contains much less dissolved oxygen than rainwater. One approach to improving the aeration of the root zone in heavy-textured, irrigated soils is to mechanically add air. This can be easily accomplished by injecting air into drip irrigation lines (Figure 10.35) and may be practical for high-value crops and ornamental landscape containers.

Ammonia Effects

The practice in recent years of adding nitrogen as anhydrous ammonia (NH_3) to irrigation water as it is applied to a field has created some soil problems. The NH_3 reacts with the irrigation water to form NH_4OH.

$$NH_3 + H_2O \rightarrow NH_4OH \qquad (10.21)$$

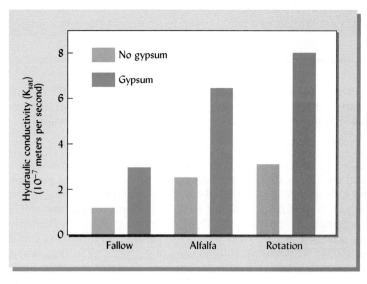

FIGURE 10.34 The influence of gypsum and growing crops on the hydraulic conductivity (K_{sat}) of the upper 20 cm of a saline-sodic soil (Natrustalf) in Pakistan. The use of gypsum to increase water conductivity in all plots was more effective when deep-rooted alfalfa and rotation of sesbania-wheat were grown. [Drawn from selected data from Ilyas et al. (1993); used with permission of the Soil Science Society of America]

TABLE 10.7 **Use of Gypsum and Synthetic Polymers on Reclamation of a Sodic Soil**

Adding gypsum to samples of a fine-textured (clay) saline-sodic soil, a Mollisol from California, increased both the hydraulic conductivity and the salts leached in the experiment while decreasing the exchangeable sodium percentage (ESP). Adding two experimental synthetic polymers (T4141 and 21J) with the gypsum gave even greater increases in hydraulic conductivity and leached salts.

	Characteristic measured					
	Hydraulic conductivity, mm/h		Total salts leached, mg/kg		ESP, %	
Gypsum added→	No	Yes	No	Yes	No	Yes
Polymer treatment						
No polymer	0.0	0.06	0.0	4.7	22.9	9.6
T4141	0.0	0.28	0.0	10.1	25.4	9.6
21J	0.0	0.28	0.0	9.7	25.5	9.6

Data from Zahow and Amrhein (1992).

The high pH brought about by this reaction causes the precipitation of calcium and magnesium carbonates, as indicated by downward arrows.

$$Ca(HCO_3)_2 + 2NH_4OH \rightarrow CaCO_3\downarrow + (NH_4)_2CO_3 + 2H_2O \qquad (10.22)$$

$$Mg(HCO_3)_2 + 2NH_4OH \rightarrow MgCO_3\downarrow + (NH_4)_2CO_3 + 2H_2O \qquad (10.23)$$

This removal of the Ca^{2+} and Mg^{2+} ions from the irrigation water raises the sodium adsorption ratio (SAR) and the hazard of increased exchangeable sodium percentage (ESP). To counteract these difficulties, sulfuric acid is sometimes added to the irrigation water to reduce its pH as well as that of the soil. This practice may well spread where there are economical sources of sulfuric acid and where the personnel applying it have been trained and alerted to the serious hazards of using this strong acid.

10.11 MANAGEMENT OF RECLAIMED SOILS

Once salt-affected soils have been reclaimed, prudent management steps must be taken to be certain that the soils remain productive. For example, surveillance of the EC and SAR and trace element composition of the irrigation water is essential. Management

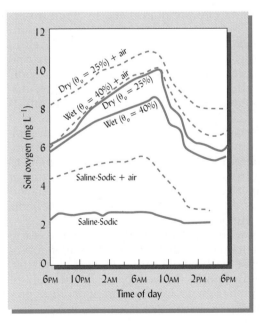

FIGURE 10.35 Soil oxygen content in normal and saline-sodic soils as affected by adding air to water supplied by buried drip irrigation emitters. Mature tomato plants were grown in large pots filled with heavy clay soil (a Vertisol) and either kept wet (θ_v = 40% water, near field capacity) or dry (θ_v = 25% water). In the normal soil, adding air (dotted lines) to wet soil achieved the same O_2 level as in the drier soil. Saline-sodic conditions were created by applying enough NaCl to raise the EC to 8.8 dS/m. Under these conditions, soil dispersion impeded air movement causing low O_2 levels even when air was added. Note that for all treatments, the O_2 levels declined during the day when plant roots are most actively respiring and taking up water and nutrients. Air was added into the drip lines by a venturi inlet and O_2 was measured by fiber-optic mini-sensors located at 15 cm depth from the soil surface. [Drawn from data in Bhattarai et al. (2006)]

adjustment is needed to accommodate any change in water quality that could affect the soil. The number and timing of irrigation episodes helps determine the balance of salts entering and leaving the soil. Likewise, the maintenance of good internal drainage is essential for the removal of excess salts.

Steps should also be taken to monitor appropriate chemical characteristics of the soils, such as pH, EC, and SAR, as well as specific levels of such elements as boron, chlorine, molybdenum, and selenium that could lead to chemical toxicities. These measurements will help determine the need for subsequent remedial practices and/or chemicals.

Crop and soil fertility management for satisfactory yield levels is essential to maintain the overall quality of salt-affected soils. The crop residues (roots and aboveground stalks) will help maintain organic matter levels and good physical condition of the soil. To maintain high yields, micronutrient and phosphorus deficiencies characteristic of other high-pH soils will need to be overcome by adding appropriate organic and inorganic sources.

10.12 CONCLUSION

Arid and semiarid regions predominantly feature alkaline and salt-affected soils. These soils typically exhibit above-neutral pH values throughout their profiles and often have calcic (calcium carbonate-rich) or gypsic (gypsum-rich) horizons at some depth. Such soils cover vast areas and support important—though water-limited—desert and grassland ecosystems. Soils of dry regions interact with scattered vegetation to influence the hydrology and fertility of the landscapes that are characterized by small "islands of fertility." Among the unique features that play a role in shaping these ecosystems are desert pavement and microbiologic crusts.

People use large areas of these soils for rangeland and dryland farming. In fact, about half of the world's arable soils are salt-affected or alkaline in nature. Many of these soils are quite well endowed with plant nutrients and, if irrigated, can be among the most productive soils in the world. Irrigation of alkaline soils of arid regions almost inevitably leads to the accumulation of salts, which must be carefully managed.

Since their pH levels are high, management practices used on these soils are quite different from those applied to acid-soil regions. The high-pH, calcium-rich conditions of alkaline soils commonly lead to deficiencies of certain essential micronutrients (especially iron and zinc) and macronutrients (especially phosphorus). In localized areas boron, molybdenum, and selenium may be so readily available as to accumulate in plants to levels that can harm grazing animals.

Scientists have grouped salt-affected soils into three classes based on their total salt content (indicated by electrical conductivity [EC]) and the proportion of sodium among the cations (indicated by either the sodium adsorption ration [SAR] or the exchangeable sodium percentage [ESP]). *Saline soils* are dominated by neutral salts (EC > 4 ds/m) and by pH values less than 8.5. *Saline-sodic* soils have similar salt and pH levels, but with EPS > 15 and SAR > 13. *Sodic soils* also have the same high levels of sodium, but their soluble salt concentrations are relatively low (EC < 4 dS/m), and their pH values are higher than 8.5. The physical conditions of saline and saline-sodic soils are satisfactory for plant growth, but the colloids in sodic soils are largely dispersed, the soil is puddled and poorly aerated, and the water infiltration rate is extremely slow.

If irrigators in arid regions apply only enough water to meet plant evapotranspiration needs, salts will build up relentlessly in the soil until the land becomes too salinized to grow crops. Therefore, just as the humid-region farmer must be ever on guard to counteract acidification, so the arid-region irrigator must be ever vigilant to combat salinization. Whereas the humid-region farmer periodically uses lime to restore a favorable balance between H^+ ion consumption and production, the irrigator uses periodic leaching to restore the balance between the import and export of salts.

Furthermore, there can be no effective leaching if drainage is insufficient to carry away the leaching water. Thus, leaching and drainage are both essential components of any successful irrigation scheme. However, the leaching of salts is not a simple matter, and it inevitably leads to additional problems both in the field being leached and in sites further downstream. At best, salinity management in irrigation agriculture provides a compromise between unavoidable evils.

The reclamation of saline-sodic and sodic soils requires an additional process before leaching of excess salinity can be achieved. In order to make these soils permeable enough for leaching to take place, the excess exchangeable sodium ions must first be removed from the exchange complex. This is accomplished by replacing the Na^+ ions with either Ca^{2+} or H^+ ions. The Ca^{2+} or H^+ ions then stimulate flocculation and increased permeability to the point that the replaced sodium and other salts can be leached downward and out of the profile. Gypsum ($CaSO_4 \cdot 2H_2O$) and elemental sulfur (S) are two amendments that can supply the Ca^{2+} and H^+ ions needed. Monitoring the chemical content of both the irrigation water and the soil is essential to achieve this goal of removing the Na^+ ions from the exchange complex and, ultimately, the soil.

About 20% of the world's farmland is irrigated, up from only 10% in 1960. Impressively, this land produces some 45% of our food supply. The world increasingly depends on irrigated agriculture in dry regions for the production of food for its growing population. Unfortunately, irrigated agriculture is in inherent disharmony with the nature of arid-region ecology. We have seen that it is therefore fraught with difficulties that require constant and careful management if human and environmental tragedies are to be avoided.

STUDY QUESTIONS

1. What are the primary sources of alkalinity in soils? Explain.

2. Compare the availability of the following essential elements in alkaline soils with that in acid soils: (1) iron, (2) nitrogen, (3) molybdenum, and (4) phosphorus.

3. The iron analysis of an arid-region soil showed an abundance of this element, yet a peach crop growing on the soil showed serious iron deficiency symptoms. What is a likely explanation?

4. A soil with an abundance of $CaCO_3$ may have a pH no higher than about 8.3, while a nearby soil with high Na_2CO_3 content has a pH of 10.5. What is the primary reason for this difference?

5. An arid-region soil, when it was first cleared for cropping, had a pH of about 8.0. After several years of irrigation, the crop yield began to decline, the soil aggregation tended to break down, and the pH had risen to 10. What is the likely explanation for this situation?

6. What physical and chemical treatments would you suggest to bring the soil described in question 5 back to its original state of productivity?

7. What are some of the adverse consequences of using wetlands as recipients of irrigation wastewater?

8. Using the information in Table 10.2 and assuming only Na^+ and Ca^{2+} are present and the SAR = 1, calculate the TDS (mg/L) and EC_w of the solution represented by 10 $mmol_c\ L^{-1}$ on the x-axis scale in Figure 10.21.

9. Calculate the leaching requirement to prevent the buildup of salts in the upper 45 cm of a soil if the EC_{dw} of the drainage water is 6 dS/m and the EC_{iw} of the irrigation water is 1.2 dS/m.

10. What are the advantages of using gypsum ($CaSO_4 \cdot 2H_2O$) in the reclamation of a sodic soil? Show the chemical reactions that take place.

11. Calculate the quantity of gypsum needed to reclaim a sodic soil (EPS = 30%) when CEC = 25 $cmol_c$/kg and pH is 10.2. Assume you want an ESP no higher than 4%.

REFERENCES

Abrol, I. P., J.S.P. Yadov, and F. I. Massoud. 1988. "Salt-affected soils and their management," *FAO Soils Bulletin* **39**. Rome: Food and Agriculture Organization of the United Nations. Complete text available at www.fao.org/docrep/x5871e/x5871e00.htm.

Banuelos, G. S., H. A. Ajwa, B. Mackey, L. Wu, C. Cook, S. Akohoue, and S. Zambruzuski. 1997. "Evaluation of different plant species used for phytoremediation of high selenium," *J. Envir. Qual.*, **26**:639–648.

Barzegar, A. R., J. M. Oades, and P. Rengasamy. 1996. "Soil structure degradation and mellowing of compacted soils by saline-sodic solutions," *Soil Sci. Soc. Amer. J.*, **60**:583–588.

Beek, K. L., W. A. Blokhuis, P. M. Driessen, N. Van Breeman, N. Brinkman, and L. J. Pons. 1980. "Problem soils: Their reclamation and management," in ILRI Publication No. 27 (Wageningen, Netherlands: ILRI), pp. 47–72.

Bhattarai, S. P., L. Pendergast, and D. J. Midmore. 2006. "Root aeration improves yield and water use efficiency of tomato in heavy clay and saline soils," *Scientia Horticulturae*, **108**:278–288.

Bresler, E. 1975. "Two-dimensional transport of solutes during non-steady infiltration from a trickle source," *Soil Sci. Soc. Amer. Proc.*, **39**:604–613.

Crescimanno, G., M. Iovino, and G. Provenzano. 1995. "Influence of salinity and sodicity on soil structural and hydraulic characteristics," *Soil Sci. Soc. Amer. J.*, **59**: 1701–1708.

Evangelou, V. P., and R. E. Phillips. 2005. "Cation exchange in soils," pp. 343–410, in A. Tabatabai and D. Sparks (eds.). *Chemical processes in soils*. SSSA Book Series No. 8. (Madison, WI.: Soil Science Society of America).

Frommer, W. B., U. Ludewig, and D. Rentsch. 1999. "Taking transgenic plants with a pinch of salt," *Science*, **285**:1222–1223.

Goldberg, S., S. M. Lesch, and D. L. Suarez. 2000. "Predicting boron adsorption by soils using soil chemical parameters in the constant capacitance model," *Soil Sci. Soc. Amer. J.*, **64**:1356–1363.

Ilyas, M., R. W. Miller, and R. H. Qureski. 1993. "Hydraulic conductivity of saline-sodic soil after gypsum application," *Soil Sci. Soc. Amer. J.*, **57**:1580–1585.

Jacobsen, T., and R. M. Adams. 1958. "Salt and silt in ancient Mesopotamian agriculture," *Science*, **128**:1251–1258.

Letey, J. 2000. "Soil salinity poses challenges for sustainable agriculture and wildlife," *Calif. Agric.*, **54**(2):43–48.

Letey, J., D. E. Birkle, W. A. Jury, and I. Kan. 2003. "Model describes sustainable long-term recycling of saline agricultural drainage water," *Calif. Agric.*, **57**:24–27.

Maas, E. V., and S. R. Grattan. 1999. "Crop yields as affected by salinity," Chap. 3 in R. W. Skaggs and J. van Schilfgaarde (eds.), *Agricultural Drainage*, Agronomy Monograph No. 38 (Madison, Wis.: ASA, CSSA, SSSA).

Mace, J. E., and C. Amrhein. 2001. "Leaching and reclamation of a soil irrigated with moderate sar waters," *Soil Sci. Soc. Amer. J.*, **65**:199–204.

Munns, R. 2002. "Comparative physiology of salt and water stress," *Plant, Cell and Environment*, **25**:239–250.

National Academy of Sciences. 1990. *Saline Agriculture: Salt Tolerant Plants for Developing Countries.* (Washington, D.C.: National Academy Press).

Nolan, B. T., and M. L. Clark. 1997. "Selenium in irrigated agricultural areas of the western United States," *J. Envir. Qual.*, **26**:849–857.

Norton, J. B., J. A. Sandor and C. S. White. 2007. "Runoff and sediments from hillslope soils within a Native American agroecosystem." Soil Sci. Soc. Amer. J. **71**:476–483.

Retana, J., D. R. Parker, C. Amrhein, and A. L. Page. 1993. "Growth and trace element concentrations of 5 plant species grown on a highly saline soil," *J. Environ. Qual.*, **22**:805–811.

Rhoades, J. D., F. Chanduvi, and S. Lesch. 1999. *Soil Salinity Assessment Methods and Interpretation of Electrical Conductivity Measurements.* FAO Irrigation and Drainage Paper No. 57. (Rome: Food and Agriculture Organization of the United Nations).

Rhoades, J. D., A. Kandiah, and A. M. Mashali. 1992. *The Use of Saline Waters for Crop Production.* FAO Irrigation and Drainage Paper No. 48. (Rome: Food and Agriculture Organization of the United Nations).

Szabolcs, I. 1989. *Salt-Affected Soils.* (Boca Raton, Fla.: CRC Press).

USDA. 1996. *America's Private Land: A Geography of Hope.* (Washington, D.C.: USDA Natural Resources Conservation Service).

Wood, Y. A., R. C. Graham, and S. G. Wells. 2005. "Surface control of desert pavement pedologic process and landscape function, cima volcanic field, Mojave Desert, California," *CATENA*, **59**:205–230.

Zahow, M. F., and C. Amrhein. 1992. "Reclamation of a saline sodic soil using synthetic polymers and gypsum," *Soil Sci. Soc. Amer. J.*, **56**:1257–1260.

Head of a bacteria-feeding nematode.
(Sven Boström, Swedish Museum of Natural History)

11

ORGANISMS AND ECOLOGY OF THE SOIL[1]

*Under the silent, relentless chemical jaws of the fungi,
the debris of the forest floor quickly disappears. . . .*
—A. FORSYTH AND K. MIYATA, TROPICAL NATURE

The terms *ecosystem* and *ecology* usually call to mind scenes of lions stalking vast herds of wildebeest on the grassy savannas of East Africa or the interplay of phytoplankton, fish, and fishermen in some great estuary. Like a savanna or an estuary, a soil is an ecosystem in which thousands of different creatures interact and contribute to the global cycles that make all life possible. This chapter will introduce some of the actors in the living drama staged largely unseen in the soil beneath our feet. If our bodies were small enough to enter the tiny passages in the soil, we would discover a world populated by a wild array of creatures all fiercely competing for every leaf, root, fecal pellet, and dead body that reaches the soil. We would also find predators of all kinds lurking in the dark, some with fearsome jaws to snatch unwary victims, others whose jellylike bodies simply engulf and digest their hapless prey.

Most of the work of the soil community is carried out by creatures whose "jaws" are chemical enzymes that eat away at organic substances left in the soil by their coinhabitants. Some of these enzymes are located inside microbial cells, others may be found in the guts of tiny animals, but many—perhaps most—are found adsorbed to soil particles, having been excreted by microorganisms into their immediate environment. The diversity of substrates and environmental conditions found in every handful of soil spawns a diversity of adapted organisms that staggers the imagination. The collective vitality, diversity, and balance among these organisms make possible the functions of a high-quality soil.

We will learn how these organisms, both flora and fauna, interact with one another, what they eat, how they affect the soil, and how soil conditions affect them. The central theme will be how this community of organisms assimilates plant and animal materials, creating soil humus, recycling carbon and mineral nutrients, and supporting

---

[1] For stories about how life underground affects everything on Earth, see Baskin (2005); for an introduction to the actors in the soil drama, see Nardi (2003); for soil ecology with emphasis on the meso- and microfauna, see Coleman et al. (2004); for reviews of soil microbiology, see Sylvia et al. (2005), Tate (2001), or Paul (2006).

plant growth. A subtheme will be how people can manage soils to encourage a healthy, diverse soil community that efficiently makes nutrients available to higher plants, protects plant roots from pests and disease, and helps protect the global environment from some of the excesses of the human species.

11.1 THE DIVERSITY OF ORGANISMS IN THE SOIL

The Soil Biodiversity Program, Scotland: http://soilbio.nerc.ac.uk/

Soil organisms are creatures that spend all or part of their lives in the soil environment (Plate 50). Every handful of soil is likely to contain billions of organisms, with representatives of nearly every phylum of living things. A simplified, general classification of soil organisms is shown in Table 11.1. In this book we will emphasize activities rather than the scientific classification; consequently, we will consider only very broad, simple taxonomic categories.

The term *fauna* is used in a very general way to distinguish animals (including single-celled protista) from *flora*, a term which is used in an equally general manner to refer to the true plants (including single-celled algae) as well as to all the nonanimal microorganisms. Based on similarities in genetic material, biologists classify all living organisms into three primary domains: *Eukarya* (which includes all plants, animals, and fungi), *Bacteria*, and *Archaea*. Organisms can also be grouped by what they "eat." Some organisms subsist on living plants (**herbivores**), others on dead plant debris (**detritivores**). Some consume animals (**predators**), some devour fungi (**fungivores**) or bacteria (**bacterivores**), and some live off of, but do not consume, other organisms (**parasites**). **Heterotrophs** rely on organic compounds for their carbon and energy needs while **autotrophs** obtain their carbon mainly from carbon dioxide and their energy from photosynthesis or oxidation of various elements.

TABLE 11.1 General Classification by Size of Some Important Groups of Soil Organisms

Generalized grouping (body width in mm)	Major taxonomic groups	Examples
Macrofauna (>2 mm)		
All heterotrophs, largely herbivores and detritivores	Vertebrates	Gophers, mice, moles
	Arthropods	Ants, beetles and their larvae, centipedes, grubs, maggots, millipedes, spiders, termites, woodlice
	Annelids	Earthworms
	Mollusks	Snails, slugs
Macroflora		
Largely autotrophs	Vascular plants	Feeder roots
	Bryophytes	Mosses
Mesofauna (0.1–2 mm)		
All heterotrophs, largely detritivores	Arthropods	Mites, collembola (springtails)
All heterotrophs, largely predators	Annelids	Enchytraeid (pot) worms
	Arthropods	Mites, protura
Microfauna (<0.1 mm)		
Detritivores, predators, fungivores, bacterivores	Nematodes	Nematodes
	Rotifera	Rotifers
	Protozoa	Amoebae, ciliates, flagellates
	Tardigrades	Water bears, *Macrobiotus sp.*
Microflora (<0.1 mm)		
Largely autotrophs	Vascular plants	Root hairs
	Algae	Greens, yellow-greens, diatoms
Largely heterotrophs	Fungi	Yeasts, mildews, molds, rusts, mushrooms
Heterotrophs and autotrophs	Bacteria	Aerobes, anaerobes
	Cyanobacteria	Blue-green algae, autotrophs
	Actinomycetes	Many kinds of actinomycetes, heterotrophs
	Archaea	Methanotrophs, *Thermoplasma sp.*, halophiles

More on Soil Biodiversity from FAO: http://www.fao.org/ag/AGL/agll/soilbiod/soilbtxt.stm

SIZES OF ORGANISMS. The animals (*fauna*) of the soil range in size from **macrofauna** (such as moles, prairie dogs, earthworms, and millipedes) through **mesofauna** (such as tiny springtails and mites) to **microfauna** (such as nematodes and single-celled protozoans). Plants (*flora*) include the roots of higher plants, as well as microscopic algae and diatoms. Other microorganisms (too small to be seen without the aid of a microscope) include fungi, bacteria, and actinomycetes, which tend to predominate in terms of numbers, mass, and metabolic capacity.

Note that we have classified the organisms according to their size (*macro* being larger than 2 mm, *meso* being between 0.1 and 2 mm, and *micro* being less than 0.1 mm), as well as by their ecological functions (what they eat). A typical, healthy soil might contain several species of vertebrate animals (mice, gophers, snakes, etc.), a half dozen species of earthworms, 20 to 30 species of mites, 50 to 100 species of insects (collembola, beetles, ants, etc.), dozens of species of nematodes, hundreds of species of fungi, and perhaps thousands of species of bacteria and actinomycetes.

DIVERSITY AND ISOLATION. Tremendous diversity is possible because of the nearly limitless variety of foods and the wide range of habitat conditions found in soils. Within a handful of soil there may be areas of good and poor aeration, high and low acidity, cool and warm temperatures, moist and dry conditions, and localized concentrations of dissolved nutrients, organic substrates, and competing organisms. The populations of soil organisms tend to be concentrated in zones of favorable conditions, rather than evenly distributed throughout the soil. The soil aggregate (Section 4.5) can be considered a fundamental unit of habitat for meso- and microorganisms, providing a complex range of hiding places, food sources, environmental gradients, and genetic isolation on a micro-scale. We know that on islands isolated by the ocean or in valleys isolated by mountains, local populations of birds or insects are genetically isolated and tend to evolve over time into separate species. In the same way, microorganisms living in different aggregates may go down their separate evolutionary paths. The soil provides both spatial diversity and a high degree of genetic isolation, helping to explain why soils contain so many more species of organisms than do aquatic environments (in which the moving fluid media is far more homogenous and provides few physical barriers to genetic commingling).

Critter clips from Iowa State U:http://www.agron.iastate.edu/%7Eloynachan/mov/

TYPES OF DIVERSITY. Aquatic ecologists have long considered a highly diverse community of organisms to indicate good water quality in a lake or stream. In the same way, soil scientists are now using the concept of biological diversity as an indicator of soil quality. A high **species diversity** indicates that the organisms present are fairly evenly distributed among a large number of species. Most ecologists believe that such complexity and species diversity are usually paralleled by a high degree of **functional diversity**—the capacity to utilize a wide variety of substrates and carry out a wide array of processes. Ecologists often turn to the soil as a model ecosystem with which to test such fundamental tenets of ecology. Table 11.2 provides an example in which researchers manipulated the level of biodiversity and measured the effects on ecological functions.

Arid soil biological communities: http://www.blm.gov/nstc/soil/communities/index.html

ECOSYSTEM DYNAMICS. In most healthy soil ecosystems there are several—and in some cases many—different species capable of carrying out each of the thousands of different enzymatic or physical processes that proceed every day. This **functional redundancy**—the presence of several organisms to carry out each task—leads both to ecosystem **stability** and **resilience**. *Stability* describes the ability of soils, even in the face of wide variations in environmental conditions and inputs, to continue to perform such functions as the cycling of nutrients, assimilation of organic wastes, and maintainence of soil structure. *Resilience* describes the ability of the soil to "bounce back" to functional health after a severe disturbance has disrupted normal processes.

Given a high degree of diversity, no single organism is likely to become completely dominant. By the same token, the loss of any one species is unlikely to cripple the entire system. Nonetheless, for certain soil processes, such as ammonium oxidation (see Section 13.7), methane oxidation (see Section 12.9), or the creation of aeration macropores (see Section 4.8), primary responsibility may fall to only one or two species. The activity and abundance of these **keystone species** (for example, certain nitrifying

TABLE 11.2   The Relationship Between Biodiversity and Ecosystem Function in a Soil Under Grazed Grassland Vegetation

Species diversity was adjusted by fumigating soil samples with chloroform for 0, 0.5, 2, or 24 hours. The longer the fumigation time, the fewer the number of species and individuals that survived. The soils were then incubated for 5 months. The total abundance and biomass of organisms rebounded to pre-fumigation levels, but consisted only of those species that had survived the fumigation treatments. Soil functions were then measured. Specialized ecological functions (that only a few species can perform) declined consistently with the decrease in biodiversity. In contrast, the more general types of functions (that most species can perform) fluctuated somewhat, but were not significantly affected by the differences in species diversity.

	Hours of initial fumigation			
	0	*0.5*	*2*	*24*
Examples of Biodiversity Measures				
Bacterial-feeding nematode groups present	8	7	1	0
Flagellate protozoa groups present	23	18	10	5
Examples of General Decomposition Functions				
Microbial respiration rate (μg CO_2/g per h)	0.87	0.37	0.29	0.83
Assimilation of added amino acid, pmol/g per h	90	114	113	159
Examples of Specialized Ecosystem Functions				
Methane oxidation pg/g per h	77	16	4	−14[a]
Nitrate formation, μg N/g	45	42	24	6

[a] Negative value indicates net methane production rather than oxidation (see Section 12.9).
Data compiled from Griffiths et al. (2000).

bacteria or burrowing earthworms) merit special attention, for their populations may indicate the health of the entire soil ecosystem.

GENETIC RESOURCES.   The diversity of organisms in a soil is important for reasons in addition to the safeguarding of ecological functions—soils also make an enormous contribution to **global biodiversity.** Many scientists believe that there are more species in existence below the surface of the Earth than above it. The soil is therefore a major storehouse of the genetic innovations that nature has written into the DNA code over hundreds of millions of years. Humans have always found ways to make use of some of the genetic material in soil organisms (beer, yogurt, and antibiotics are examples). However, with recent developments in biological engineering that allow the transfer of genetic material from one type of organism to another, the soil DNA bank has taken on a much larger practical importance for human welfare. The genes from soil organisms may now be used to produce plants and animals of superior utility to the human community.

11.2   ORGANISMS IN ACTION[2]

Virtual soil tour. Click on each number:
http://www.fieldmuseum.org/undergroundadventure/flash/VirtualTour.swf

The activities of soil flora and fauna are intimately related in what ecologists call a *food chain* or, more accurately, a *food web.* Some of these relationships are shown in Figure 11.1, which illustrates how various soil organisms participate in the degradation of residues from higher plants. As one organism eats another, nutrients and energy are said to be passed from one **trophic level** to a higher one. The first trophic level is that of the **primary producers.** The second trophic level consists of the **primary consumers** that eat those producers. The third trophic level would be those **predators** that eat the primary consumers, the fourth level would be predators that eat predators, and so on.

---

[2] For a succinct and well-illustrated summary of soil communities and ecological dynamics, see Tugel and Lewandowski (1999).

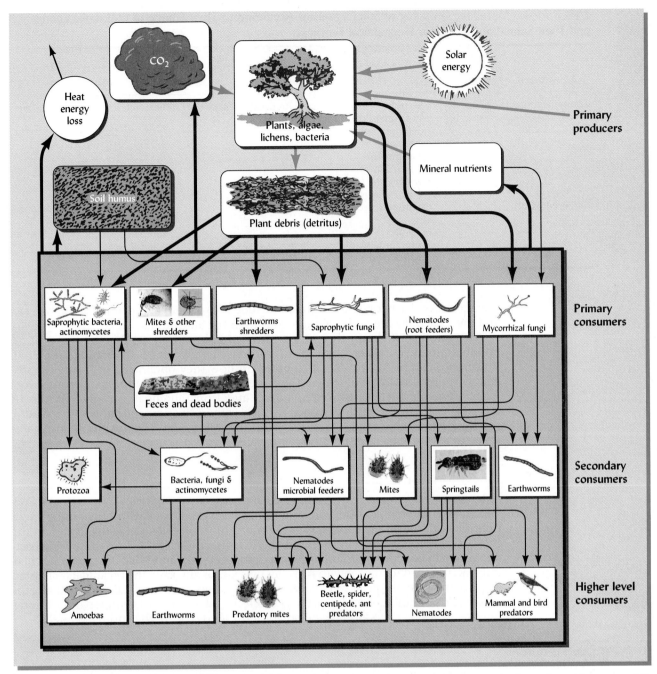

FIGURE 11.1 Generalized diagram of the soil food web involved in the breakdown of plant tissue, the formation of humus, and the cycling of carbon and nutrients. The large shaded compartment represents the community of soil organisms. The rectangular boxes represent various groups of organisms; the arrows represent the transfer of carbon from one group to the next as predator eats prey. The thick arrows entering the top of the shaded compartment represent primary consumption of carbon originating from the tissue of the producers—higher plants, algae, and cyanobacteria. The rounded boxes represent other inputs that support, and outputs that result from, the soil food web. Although all groups shown play important roles in the process, some 80 to 90% of the total metabolic activity in the food web can be ascribed to the fungi and the bacteria (including actinomycetes). As a result of this metabolism, soil humus is synthesized, and carbon dioxide, heat energy, and mineral nutrients are released into the soil environment. (Diagram courtesy of R. Weil)

Source of Energy and Carbon

Soil organisms may be classified as either **autotrophic** or **heterotrophic** based on where they obtain the *carbon* needed to build their cell constituents (Table 11.3). The heterotrophic soil organisms obtain their carbon from the breakdown of organic materials previously produced by other organisms. Nearly all heterotrophs also obtain their *energy* from the oxidation of the carbon in organic compounds. They are responsible for

TABLE 11.3 Metabolic Grouping of Soil Organisms According to Their Source of Metabolic Energy and Their Source of Carbon for Biochemical Synthesis

Source of carbon	Source of energy	
	Biochemical oxidation	Solar radiation
Combined organic carbon	**Chemoheterotrophs:** Many Archaea and Bacteria All animals, plant roots, fungi, actinomycetes, and most bacteria Examples: Earthworms *Aspergillus sp.* *Azotobacter sp.* *Pseudomonas sp.*	**Photoheterotrophs:** A few algae
Carbon dioxide	**Chemoautotrophs** Examples: Ammonia oxidizers—*Nitrosomonas sp.* Sulfur oxidizers—*Thiobacillus denitrificans*	**Photoautotrophs:** Plant shoots, algae and cyanobacteria Examples: *Chorella sp.* *Nostoc sp.*

organic decay. These organisms, which include the soil fauna, the fungi, actinomycetes, and most bacteria, are far more numerous than the autotrophs.

The autotrophs obtain their carbon from simple carbon dioxide gas (CO_2) or carbonate minerals, rather than from carbon already fixed in organic materials. Autotrophs can be further classified based on how they obtain energy. Some use solar energy (photoautotrophs), while others use energy released by the oxidation of inorganic elements such as nitrogen, sulfur, and iron (chemoautotrophs). While the autotrophs are in the distinct minority, their carbon fixation and inorganic oxidation reactions allow them to play crucial roles in the soil system. The autotrophs are mainly algae, cyanobacteria, and certain other bacteria.

Primary Producers

As in most aboveground ecosystems, vascular plants play the principal role as primary producers. By combining carbon from atmospheric carbon dioxide with water, using energy from the sun, the producer organisms make organic molecules and living tissues. In this, they are commonly joined by mosses, algae, lichens, and certain photosynthesizing bacteria. In certain environments, other primary producers are found that use inorganic chemical reactions to obtain energy to make living cells. In any case, the organic materials created by the primary producers contain both carbon and chemical energy that other organisms can utilize, either directly or indirectly, after having been passed on through intermediaries. The producers therefore form the food base for the entire food web.

Primary Consumers

As soon as a leaf, a stalk, or a piece of bark drops to the ground, it is subject to coordinated attack by microflora and by macro- and mesofauna (see Figure 11.1). The animals, which include mites, springtail insects, woodlice, and earthworms, chew or tear holes in the tissue, opening it up to more rapid attack by the microflora. The animals and microflora that use the energy stored in the plant residues are termed *primary consumers*.

HERBIVORES.   Certain soil organisms that eat live plants are called **herbivores**. Examples are parasitic nematodes and insect larvae that attack plant roots, as well as termites, ants, beetle larvae, woodchucks, and mice that devour aboveground plant parts. Because they attack living plants that may be of value to humans, many of these soil herbivores are considered pests. On the other hand, certain herbivores, such as soil-dwelling immature cicadas, commonly do more good than harm for higher plants (see Plate 83 after page 656).

DETRITIVORES.   For the vast majority of soil organisms, however, the principal source of food is the debris of dead tissues left by plants on the soil surface and within the soil pores. This debris is called **detritus**, and the animals that directly feed on it are called

detritivores. Both herbivores and detritivores that eat the tissues of primary producers are considered primary consumers. However, many of the animals that chew up plant detritus actually get most of their nutrition from the microorganisms that live in the detritus, not from the dead plant tissues themselves. Such animals are not really primary consumers (second trophic level), but belong to a higher trophic level as they eat detritus from other animals (fecal pellets, dead bodies) (see "Secondary Consumers," following).

SAPROPHYTIC MICROORGANISMS.　　On balance, most actual decomposition of dead plant and animal debris is carried out by **saprophytic** (feeding on dead tissues) microflora, both fungi and bacteria. With the assistance of the animals that physically shred and chew the plant debris (see following), the saprophytes break down all kinds of dead plant and animal compounds, from simple sugars to woody materials (Figure 11.1). Saprophytes feed on detritus, dead animals (corpses), and animal feces. These microorganisms commonly grow to form large colonies of microbial cells on the decaying material. These microbial colonies, in turn, soon provide nutrition to a myriad of other soil organisms—the secondary consumers.

Secondary Consumers

A virtual introduction to mites:
http://www.sel.barc.usda.gov/acari/index.html

PREDATORS AND MICROBIAL FEEDERS.　　The bodies (cells) of primary consumers become food sources for an array of predators and parasites in the soil. These *secondary consumers* include microflora, such as bacteria, fungi, and actinomycetes, as well as **carnivores**, which consume other animals. Examples of carnivores include centipedes (Figure 1.8) and mites that attack small insects or nematodes (see Figure 11.2), spiders, predatory nematodes, and snails. **Microbivorous feeders**, organisms that use microflora as their source of food, include certain collembola (springtail insects), mites (Figure 11.3), termites, certain nematodes (see chapter opening photo), and protozoa. These microphytic feeders exert considerable influence over the activity and growth of fungal and bacterial populations. The grazing by these meso- and microfauna on microbial colonies may stimulate faster growth and activity among the microbes, in much the same way that grazing animals can stimulate the growth of pasture grasses. In other cases, the attack of the microphytic feeders may kill off so much of a microbial colony as to inhibit the work of the microorganisms.

STIMULATION OF DECOMPOSITION.　　While the actions of the microflora are mostly biochemical, those of the fauna are both physical and chemical. The mesofauna and macrofauna chew the plant residues into small pieces and move them from one place to another on the soil surface and even into the soil.

The actions of these animals enhance the activity of the microflora in several ways. First, the chewing action fragments the litter, cutting through the resistant waxy coatings

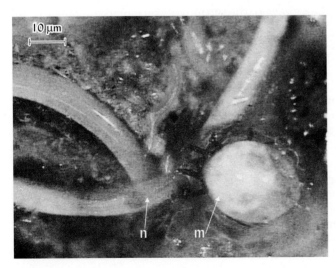

FIGURE 11.2　A predatory mite (an Astigmatid, *m*) dining on its prey, a microscopic roundworm (a nematode, *n*). Predation of this type keeps the populations of various groups of organisms in balance and releases nutrients previously tied up in the bodies of the prey. (Photo courtesy of Marie Newman, North Carolina State University)

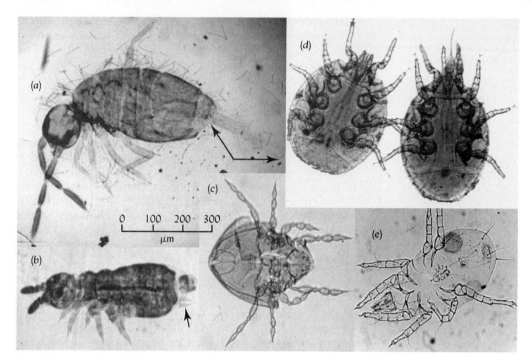

FIGURE 11.3 Collembola (*a, b*) and mites (*c, d, e*) are prominent among the mesofauna that play important roles in soil food webs. Collembola are also called springtails because of their springlike furcula or "tail" (arrows). The larger springtail (*a*) lives in and on the O horizons where there is light to see (hence the operational eyes) and room to hop (hence the well-developed furcula). The smaller springtail (*b*) lives in the mineral soil horizons where there is no light and no room to hop (hence it has only tiny vestigial "eyes" and furcula). Both springtails shown probably eat detritus and the fungi growing in it. The Orabatid mite (*c*) also feeds on detritus. The Mesostigmatic mites (*d, e*) are predators of smaller arthropods, nematodes, and other denizens of the soil. Note that mites (arachnids) have eight legs, while springtails (insects) have six. (Photos courtesy of R. Weil)

on many leaves to expose the more easily decomposed cell contents for microbial digestion. Second, the chewed plant tissues are thoroughly mixed with microorganisms in the animal gut, where conditions are ideal for microbial action. Third, the mobile animals carry microorganisms with them and help the latter to disperse and find new food sources to decompose. The ability of different size groups of soil fauna to enhance plant residue decomposition is illustrated in Figure 11.4.

Tertiary Consumers

PREDATORS.   In the next level of the food web, the secondary consumers are prey for still other carnivores, called *tertiary consumers*. For example, ants consume centipedes, spiders, mites, and scorpions—all of which can themselves prey on primary or secondary consumers. Many species of birds specialize in eating soil animals such as beetles and earthworms. Robins are widely known to pull earthworms from their burrows, while other birds such as gulls and blackbirds commonly follow behind a farmer's plow to feed on the exposed earthworms and other soil macrofauna. Such mammals as moles can be effective predators of macrofauna. Whether the prey is a nematode or an earthworm, predation serves important ecological functions, including the release of nutrients tied up in the living cells.

MICROBIAL DECOMPOSERS.   The microflora are intimately involved in every level of decomposition. In addition to their direct attack on plant tissue (as primary consumers), microflora are active within the digestive tracts of many soil animals, helping these animals digest more resistant organic materials. The microflora also attack the finely shredded organic material in animal feces, and they decompose the bodies of dead animals. For this reason they are referred to as the ultimate decomposers.

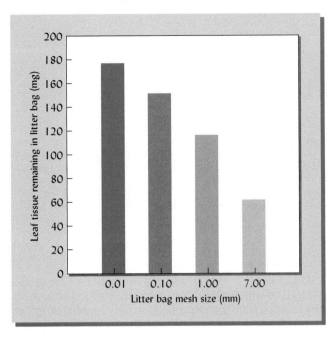

FIGURE 11.4 Influence of various sizes of soil organisms on the decomposition of corn leaf tissue buried in soil. Small bags made of nylon material with four different-size openings (mesh size) were filled with 558 mg (dry weight) of corn leaf tissue and buried in the soil for 10 weeks. The amount of corn leaf tissue remaining in the bags was considerably greater (less decomposition had taken place) when the meso- and macrofauna were excluded by the smaller mesh sizes. [Data from Weil and Kroontje (1979)]

DIVERSIFICATION WITHIN BROAD GROUPS. As indicated in Figure 11.1, broad groups of organisms, such as fungi or mites, include some species that are primary consumers, some that are secondary consumers, and others that are tertiary consumers. For example, some mites attack detritus directly, while others eat mainly fungi or bacteria that grow on the detritus. Still others attack and devour the mites that eat fungi. Two-way interactions exist between many groups of organisms. For example, some nematodes eat fungi, and some fungi attack nematodes.

Ecosystem Engineers[3]

Certain organisms make major alterations to their physical environment that influence the habitats of many other organisms in the ecosystem. These organisms are sometimes referred to as *ecosystem engineers*. For example, some of these "engineer" species are microorganisms that create an impermeable surface crust that spatially concentrates scarce water supplies in certain desert soils.

BURROWING ANIMALS. Others are burrowing animals that create opportunities and challenges for other organisms by digging channels that greatly alter air and water movement in soils. Termites (see Section 11.5) and ants, for example, may literally invert the soil profile in local areas (bringing subsoil material to the surface) and denude the soil surface in wide areas. Earthworms literally eat their way through the soil as they incorporate plant residues into the mineral soil by passing them through their bodies along with mineral soil particles (see Section 11.4). Larger animals, such as gophers, moles, prairie dogs, and rats, also burrow into the soil and bring about considerable soil mixing and granulation. Not only do such burrows encourage more water and air to enter the soil, they also provide passages that plant roots can easily follow to penetrate dense subsurface soil layers. Furthermore, the underground chambers and burrows made by the engineers provide new kinds of habitat in which other organisms (from frogs to fungi) soon take up residence.

DUNG BEETLES. Certain beetles of the *Scarabaeidae* family greatly enhance nutrient cycling by burying animal dung in the upper soil horizons. Many of these **dung beetles** cut round balls from large mammal feces, enabling them to roll the dung balls to a new

---

[3] For a discussion of the ecosystem engineer concept in soils, see Jouquet et al. (2006) and Lavelle et al. (1997). For applications of the concept to coral reef and desert soils, see Alper (1998).

FIGURE 11.5 Two dung beetles (*Scarabaeidae*), one on top and the other underneath, roll a ball that they have fashioned out of buffalo dung over the leaf-strewn surface of a sandy soil. The female will lay her eggs in the ball of dung and bury it in the soil (some coarse sand grains are sticking to the surface of the dung ball). Burying the dung is very important for making the nutrients therein available to the soil food web. Dung burial also prevents the reproduction of carnivorous flies and other pests of dung-producing mammals. Different dung beetles have evolved to specialize in the burial of dung from particular species of animals. (Photo courtesy of R. Weil)

location (Figure 11.5). The female dung beetle then lays her eggs in the ball of dung and buries it in the soil. Dispersal and burial of the dung not only provides a food source for the beetle larvae, it also protects the nutrients in the manure from easy loss by runoff or volatilization—fates to which the nutrients would most likely succumb if left on the soil surface. Dung beetles therefore play important roles in nutrient cycling and conservation in many grazed ecosystems.

Several thousand different dung beetle species are known worldwide, many having evolved to specialize in the burial of dung from particular mammal species (elephant, cattle, buffalo, etc.). Rapid burial of dung by beetles also prevents the reproduction of carnivorous flies and other pests of large dung-producing mammals. If native dung-burying "engineer" species are lacking in a grazed ecosystem (such as a savanna, a prairie, or a pasture), scientists have found that the introduction of appropriate species of dung beetles or earthworms can greatly increase the amount of vegetation produced and the number of grazing animals supported.

11.3 ORGANISM ABUNDANCE, BIOMASS, AND METABOLIC ACTIVITY

Soil organism numbers are influenced primarily by the amount and quality of food available. Other factors affecting their numbers include physical factors (e.g., moisture and temperature), biotic factors (e.g., predation and competition), and chemical characteristics of the soil (e.g., acidity, dissolved nutrients, and salinity). The species that inhabit the soil in a desert will certainly be different from those in a humid forest, which, in turn, will be quite different from those in a cultivated field. Acid soils are populated by species different from those in alkaline soils. Likewise, species diversification and abundance in a tropical rain forest are different from those in a cool temperate area.

Despite these variations, a few generalizations can be made. For example, forested areas usually support a more diverse soil fauna than do grasslands, although the total faunal mass per hectare and level of faunal activity are generally higher in grasslands (see Table 11.4). Cultivated fields are generally lower than undisturbed native lands in numbers and biomass of soil organisms, especially the fauna, partly because tillage destroys much of the soil habitat.

Total **soil biomass**, the living fraction of the soil, is generally related to the amount of organic matter present. On a dry-weight basis, the living portion is usually between 1 and 5% of the total soil organic matter. In addition, scientists commonly observe that the ratios of soil organic matter to detritus to microbial biomass to faunal biomass are approximately 1000:100:10:1.

Comparative Organism Activity

The importance of specific groups of soil organisms is commonly identified by (1) the numbers of individuals in the soil, (2) their weight (biomass) per unit volume or area of soil, and (3) their metabolic activity (often measured as the amount of carbon

TABLE 11.4   Biomass of Groups of Soil Animals Under Grassland and Forest Cover

The mass and, in turn, the metabolism are greatest under grasslands. The spruce, with low-calcium-containing leaves, encourages acid conditions and slow organic matter decomposition.

| | Biomass,[a] g/m^2 | | |
| | | Forest | |
Group of organisms	Grassland meadow	Oak	Spruce
Herbivores	17.4	11.2	11.3
Detritivores			
Large	137.5	66.0	1.0
Small	25.0	1.8	1.6
Predators	9.6	0.9	1.2
Total	189.5	79.9	15.1

[a] In upper 15 cm.
Data from Macfadyen (1963).

dioxide given off in respiration). The numbers and biomass of groups of organisms that commonly occur in soils are shown in Table 11.5. Although the relative metabolic activities are not shown, they are generally related to the biomass of the organisms. Representative measurements of soil biological activity are not easily made because the soil is such a heterogeneous medium and the organisms living in it are very unevenly distributed. Concentrations of microbial activity (hotspots) occur in the immediate vicinity of living plant roots and decaying bits of detritus, in the organic material lining earthworm burrows, in fecal pellets of soil fauna, and in other favored soil environments.

As might be expected, the microorganisms are the most numerous and have the highest biomass. Together with earthworms (or termites, in the case of some tropical soils), the microflora dominate the biological activity in most soils. It is estimated that about 80% of the total soil metabolism is due to the microflora, although, as previously mentioned, their activity is enhanced by the actions of soil fauna. Despite their relatively small total biomass, such microfauna as nematodes and protozoa play important roles in nutrient cycling by preying on bacteria and fungi. For these reasons, major attention will be given to the microflora and microfauna, along with earthworms, termites, and certain other fauna.

TABLE 11.5   Relative Numbers and Biomass of Fauna and Flora Commonly Found in Surface Soil Horizons

Microflora and earthworms dominate the life of most soils.

| | Number[a] | | Biomass[b] | |
Organisms	Per m^2	Per gram	kg/ha	g/m^2
Microflora				
Bacteria and Archaea	10^{14}–10^{15}	10^9–10^{10}	400–5000	40–500
Actinomycetes	10^{12}–10^{13}	10^7–10^8	400–5000	40–500
Fungi	10^6–10^8 m	10–10^3 m	1000–15,000	100–1500
Algae	10^9–10^{10}	10^4–10^5	10–500	1–50
Fauna				
Protozoa	10^7–10^{11}	10^2–10^6	20–300	2–30
Nematodes	10^5–10^7	1–10^2	10–300	1–30
Mites	10^3–10^6	1–10	2–500	0.2–5
Collembola	10^3–10^6	1–10	2–500	0.2–5
Earthworms	10–10^3		100–4000	10–400
Other fauna	10^2–10^4		10–100	1–10

[a] For fungi the individual is hard to discern, so meters of hyphal length is given as a measure of abundance.
[b] Biomass values are on a liveweight basis. Dry weights are about 20 to 25% of these values.
[c] Estimated numbers of Bacteria and Archaea from Torsvik et al. (2002); others from many sources.

Even those animals that account for only very small fractions of the total metabolism in the soil can play important roles in soil formation (see Chapter 2) and management. Rodents pulverize, mix, and granulate soil, as well as incorporate surface organic residues into lower horizons. They provide large channels through which water and air can move freely. Ants are important in localized areas, especially in warm-region grasslands and in boreal forests, for their exceptional ability to break down woody materials and turn over soil materials as they build their nests. Mesofauna detritivores (mostly mites and collembola; see Figure 11.3) translocate and partially digest organic residues and leave their excrement for microfloral degradation (see Plate 80 noting in the center of the left image, where mites feeding on soft inner root tissue have left a pile of fecal pellets). By their movements in the soil, many animals rearrange soil particles to form biopores, thus favorably affecting the soil's physical condition (also visible in Plate 80). Other animals, especially certain insect larvae, feed directly on plant roots, causing considerable damage.

11.4 EARTHWORMS[4]

Earthworm information:
http://www.sarep.ucdavis.edu/worms/

Earthworms are probably the most important macroanimals in soils. They (along with their much smaller mesofauna cousins, the **enchytraeid worms**) are egg-laying *hermaphrodites* (organisms without separate male and female genders) that eat detritus, soil organic matter, and microorganisms found on these materials (Figure 11.6). They do not eat living plants or their roots, and so do not act as pests to crops (and so should not be confused with root-feeding insect larvae with common names such as "army worm" or "wireworm").

EPIGEIC, ENDOGEIC, AND ANECIC EARTHWORMS.   The 7000 or so species of earthworms reported worldwide can be grouped according to their burrowing habits and habitat. The relatively small **epigeic** earthworms live in the litter layer or in the organic-rich soil very near the surface. Epigeic earthworms, which include the common compost

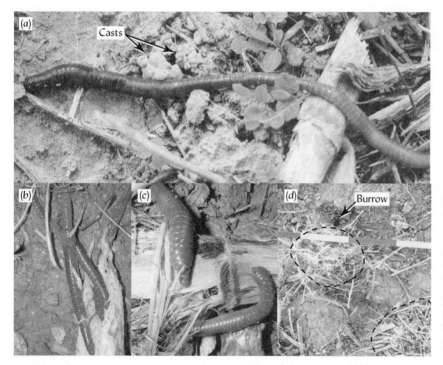

FIGURE 11.6 Anecic earthworm species such as these *Lumbricus terrestris* (*a*) come to the surface to feed on litter (*c*), excrete soil casts (*a*, arrows) and reproduce (*b*). They incorporate large amounts of plant litter into the soil and gather plant debris into piles called *middens* to cover their burrow entrances (*d*, 10 cm scale markings). Earthworms are perhaps the most significant macroorganism in soils of humid temperate regions, particularly in relation to their effects on the physical conditions of soils. (Photos *a* and *d* courtesy of R. Weil; photos *b* and *c* courtesy of Steve Groff)

---

[4] For extensive information on earthworm ecology, biology, and distribution, see Edwards (2004). For a brief discussion of earthworms in relation to agriculture, soil organic matter, and soil microbiology, see Edwards and Arancon (2004).

worm, *Eisenia foetida*, hasten decomposition of litter but do not mix it into the mineral soil. **Endogeic** earthworms, such as the pale, pink *Allolobophora caliginosa* (known as "red worm"), live mainly in the upper 10 to 30 cm of mineral soil where they make shallow, largely horizontal burrows. Finally, the relatively large **anecic** earthworms make vertical, relatively permanent burrows as much as several meters deep. They emerge in wet weather or at night to forage on the surface for pieces of litter that they drag back into their burrows, often covering the entrance to their burrow with a **midden** of leaves (Figure 11.6*d*). The well-known "night crawler" (*Lumbricus terrestris*) was accidentally introduced to North America from Europe (perhaps in the root balls of settlers' fruit trees) and is now the most common anecic earthworm on both continents.

Influence on Soil Fertility, Productivity, and Environmental Quality

BURROWS.   Earthworms literally eat their way through the soil, sometimes ingesting soil equal to 2 to 30 times their own weight in a single day. In a year, the earthworms living in 1 ha of land may ingest between 50 and 1000 Mg of soil, the higher figure occurring in moist, tropical climates. In so doing, they create extensive systems of burrows. The burrows may be mainly vertical or horizontal, depending on the species, some providing continuous macropores to depths exceeding 1 m. Earthworm channels, whether empty or filled with casts, offer important pathways by which plant roots can penetrate dense soil layers, especially in compacted soils (see Plate 81). Their extensive physical activity, which is particularly important in untilled soils (including grasslands and no-till croplands) has earned earthworms the title of "nature's tillers" (see Figure 11.7). The middens of plant litter and the circular burrow entrances (often hidden by middens) are important signs used to assess earthworm activity (Figure 11.6*d*).

Winogradsky column-perpetual life in a tube:
http://helios.bto.ed.ac.uk/bto/microbes/winograd.htm

CASTS.   After passing through the earthworm gut, ingested soil is expelled as globules of soil called **casts** (Figure 11.6*a*). During the passage through the earthworm's gut, organic materials are thoroughly shredded and mixed with mineral soil materials.

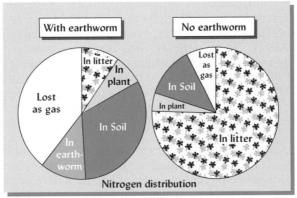

FIGURE 11.7 Burrowing activities of earthworms and their effect on the distribution of nitrogen from plant litter. Soil materials from A and B horizons were packed into a 35-cm-tall glass cylinder (*photo*, scale in cm), and the cylinder was wrapped to exclude light. Then the soil was moistened, a few grass seeds planted, a layer of leaves placed on the soil surface, and two anecic earthworms were added. The photo, taken 1 week later, shows the soil riddled with burrows (some occupied by grass roots), illustrating the aerating and mixing action of these invertebrate creatures. The pie charts show data from a study using similar soil columns (called mesocosms). Corn residue litter labeled with ^{15}N (isotope of nitrogen) was incubated for 3 weeks with or without one earthworm (*Lumbricus terrestris*) per column. After the incubation, the earthworms and remaining litter were removed, and corn plants were grown for 30 days. Without earthworms, most of the N remained in the litter. In contrast, the earthworms made nearly all the N available for biological processes, including plant uptake (which nearly doubled) and gaseous losses (which increased 5-fold) (see Sections 13.6 and 13.9). Although the distribution among specific processes likely differs in the field, these studies demonstrate that earthworms dramatically hasten the cycling of nutrients in the soil–plant system. [Photo courtesy of Ray Weil; charts based on data in Amador and Gorres, (2005)]

TABLE 11.6 Comparative Characteristics of Earthworm Casts and Soils

Average of six Nigerian soils.

Characteristic	Earthworm casts	Soils
Silt and clay, %	38.8	22.2
Bulk density, Mg/m^3	1.11	1.28
Structural stability[a]	849	65
Cation exchange capacity, cmol/kg	13.8	3.5
Exchangeable Ca^{2+}, cmol/kg	8.9	2.0
Exchangeable K$^+$, cmol/kg	0.6	0.2
Soluble P, ppm	17.8	6.1
Total N, %	0.33	0.12

[a] Numbers of raindrops required to destroy structural aggregates.
From de Vleeschauwer and Lal (1981).

Probably because of enhanced bacterial activity, earthworm casts are usually high in polysaccharides, which are credited with stabilizing the casts into granular structure (see Section 4.5). The casting behavior of earthworms, therefore, generally enhances the aggregate stability of the soil (Table 11.6). The casts are deposited within the soil profile or on the soil surface, depending on the species of earthworm, and are another sign used to assess earthworm activity in soil.

Groff family farm no-till worm walk: http://www.newfarm.org/ depts/notill/features/worms .shtml

NUTRIENTS. The activities of earthworms also enhance soil fertility and productivity by altering chemical conditions in the soil, especially in the upper 15 to 35 cm. Earthworms hasten the cycling and increase the availability of mineral nutrients to plants in three ways. First, as soil and organic materials pass through an earthworm, they are ground up physically as well as attacked chemically by the digestive enzymes of the earthworm and its gut microflora. Compared to the bulk soil, the casts are significantly higher in bacteria, organic matter and available plant nutrients (see Table 11.6). Roots growing down earthworm burrows also find rich sources of nutrients in the casts and burrow lining material.

Second, although earthworms may feed on detritus and soil organic matter of relatively low nitrogen, phosphorus, and sulphur concentrations, their own body tissues have high concentrations of these nutrients. When the earthworms die and decay, the nutrients in their bodies are readily released into plant-available form. Studies have shown that where earthworm populations are large, a major proportion of the N taken up by plants (50 to 90 kg N/ha) can be made available by this mechanism.

Third, physical incorporation of surface residues into the soil reduces the loss of nutrients, especially of nitrogen, by erosion and volatilization. However, the same action may increase losses by other pathways (Figure 11.7, *pie charts*).

BENEFICIAL PHYSICAL EFFECTS. Earthworms are important in other ways. The holes left in the soil (see, for example, Figure 6.8) serve to increase aeration and drainage, an important consideration in plant productivity and soil development. In turfgrass, the mixing activity of earthworms can alleviate or reduce compaction problems and nearly eliminate the formation of an undesirable thatch layer. Under conditions of heavy rainfall, earthworm burrows may greatly increase the infiltration of water into the soils (Figure 11.8), thus playing an important role in water conservation and prevention of soil erosion. Moreover, earthworm casts may form stable macroaggregates that cover as much as 50 to 60% of the surface soil.

Deleterious Effects of Earthworms

EXPOSURE OF SOIL ON THE SURFACE. Not all effects of earthworms are beneficial. For example, in the process of building its middens, *Lumbricus terrestris* has been observed to leave some 60% of the soil surface bare of residues (see Figure 11.6*d*). To some extent this action leaves the exposed soil susceptible to the impact of raindrops, crust formation, and increased erosion. In other situations, the middens themselves may be considered a nuisance—for example, on closely cropped golf greens. Even the burrowing action may not always be welcome—as in forests with thick litter layers (see Box 11.1).

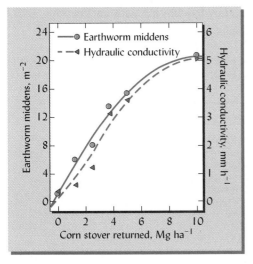

Figure 11.8 graph

FIGURE 11.8 Influence of crop residue return to the soil on earthworm activity and soil hydraulic conductivity. The total amount of residues produced by the corn crop was about 10 Mg ha^{-1}. The lower rates of residue return represent the situations when some or all of the corn stover is collected and removed for the production of ethanol biofuel. The graph shows the effect of just 1 year of differential stover returns on numbers of earthworm middens present and the soil hydraulic conductivity (rate of water infiltration). The curves follow one another very closely, most likely because reductions in residue return negatively affect earthworm burrowing activity and fewer burrows result in less ready infiltration of rainwater. The data suggests two conclusions: (1) there exists a close positive relationship between earthworm activity and soil infiltration rate, and (2) if more than about one-third of the corn stover is removed, significant damage may be done to soil quality needed to sustain future production of crops and biofuels (see Section 20.6). Unfortunately, many models of potential biofuel production fail to take this relationship into account. The data are averages for three medium- to fine-textured soils under long-term no-till management in Ohio. [Graphed from data in Blanco-Canqui et al. (2007)]

INFLUENCE ON CHEMICAL LEACHING. Another aspect of concern is that water rapidly percolating down vertical earthworm burrows may carry potential pollutants toward the groundwater (see Figure 6.29). In artificially drained fields, earthworm channels may provide direct paths from the soil surface to the drain tiles, leading to increased leaching losses of soluble agri-chemicals and nutrients. However, the organic-matter-enriched material lining

BOX 11.1   GARDENERS' FRIEND NOT ALWAYS SO FRIENDLY[a]

Earthworms are legendary for the many ways they enhance the productivity of gardens, pastures, and croplands. However, in other ecological settings, "nature's tillers" can be quite damaging. In particular, these active immigrants are starting to wreak havoc in some of North America's forest ecosystems. Since the Pleistocene glaciers wiped out the native earthworms some 10,000 years ago, the plant–soil communities in these boreal forests have evolved without substantial earthworm populations. Without the soil-mixing action of earthworms, these forests have developed a thick, stratified forest floor, usually consisting of several distinct O horizons. This loose, thick litter layer is essential habitat for certain native millipedes, ground-nesting birds, and salamanders (Figure 11.9). Native plants such as mayflowers, wood anemone, and trillium also depend on a thick layer of forest litter.

Recently, scientists have observed that the forest floor has all but disappeared from certain boreal forest stands. They think they know the culprit—invading populations of *Lumbricus terrestris* and similar large exotic *anecic*-type earthworms. This type of earthworm is particularly damaging to the forest ecosystem for precisely the same actions that make it so beneficial in gardens and pastures—the incorporation of surface litter into the soil and production of deep vertical burrows. These actions rapidly destroy the forest floor O horizons and greatly accelerate the normally conservative cycling of nutrients.

Some of the worst impacts are being felt in Minnesota, where European earthworms (including *Aporrectodea* sp., *Lumbricus rubellus*, and *L. terrestris*) have invaded the lake-dotted boreal forests. One approach to slowing the invasion of these uninvited soil engineers might be to surround the forest with buffer zones of habitat unsuitable for earthworms. Researchers believe the advance of these earthworms has resulted largely from their use as fishing bait. The take-home message: bring those bait worms home and dump them in your compost pile, not on the shore of your favorite fishing lake!

FIGURE 11.9 *The marbled salamander (Ambystoma opacum), one of the creatures whose habitat is threatened by invading earthworms. (Photo courtesy of Stan Trauth, Arkansas State University)*

[a]For more on invasive earthworms, see Hendrix and Bohlen (2002) and Minnesota Worm Watch at www.nrri.umn.edu/worms.

TABLE 11.7 Improvements in Soil Quality of Restored Coal Mine Land by Inoculating with Earthworms

Earthworms were initially absent from the soils being restored. The mine site in Wales (U.K.) was covered with clay loam surface horizon material that had been pushed aside and stored for four months during the coal mining operation. Earthworms of several indigenous species were added ($70/m^2$) to certain experimental plots, but not to the control plots. Without inoculation, normal populations of earthworms may take 20 to 30 years to recolonize such soils.

	Earthworm population 6 years after inoculation	Soil properties 9 years after earthworm inoculation		
	number/m^2	Carbohydrate[a] content in soil, g/100 g soil	Clay dispersed upon wetting (susceptibility to erosion), g/100 g soil	Respiration per g of microbial biomass (metabolic quotient, of CO_2[b]), mg CO_2–C g^{-1} h^{-1}
No earthworms added	8	0.90	23.7	15.3
Earthworms added	106	1.28	16.9	10.9
Percent difference	+1200	+42	−29	−29

[a] Mainly polysaccharides produced by earthworms themselves or by bacteria stimulated by the earthworms.
[b] A higher value indicates a stressed ecosystem in which microorganisms must devote increased energy to survival rather than growth. See also Tables 20.1 and 20.8.
[Data from Scullen and Malik (2000)]

earthworm burrows has two to five times as great a capacity to adsorb certain herbicides as the bulk soil. Therefore, transport of such pollutants through earthworm burrows may be much less than is suggested by the mass flow of water through these large biopores. On balance, earthworms are very beneficial, and soil managers would do well to encourage their activity in most situations. Table 11.7 illustrates the value of using earthworms to assist in the restoration of soil after a strip-mining operation.

Factors Affecting Earthworm Activity

Earthworms prefer cool, moist, but well-aerated soils well supplied with decomposable organic materials, preferably supplied as surface mulch. In temperate regions they are most active in the spring and fall, often curling into a tight ball (aestivating) to ride out hot, dry periods in summer. They do not live under anaerobic conditions, nor do they thrive in coarse sands. A few species are reasonably tolerant to low pH, but most earthworms thrive best where the soil is not too acid (pH 5.5 to 8.5) and has an abundant supply of calcium (which is an important component of their mucilage excretions). Enchytraeid worms are much more tolerant of acid conditions and are more active than earthworms in some forested Spodosols. Most earthworms are quite sensitive to excess salinity.

Other factors that depress earthworm populations include predators (moles, mice, and certain mites and millipedes), very sandy soils (partly because of the abrasive effect of sharp sand grains), direct contact with ammonia fertilizer, application of certain insecticides (especially carbamates), and tillage. The last factor is often the overriding deterrent to earthworm populations in agricultural soils. Minimum tillage, with plenty of crop residues left as a mulch on the soil surface, is ideal for encouraging earthworms. The numbers commonly found in agricultural soils range from 0 to 300+ per square meter, the highest populations occurring in meadows, pastures, and under no-till rotations with cover crops. A simple method of assessing earthworm populations in grassland or arable soils is to dig into the surface soil to a depth of about 25 to 30 cm. During relatively cool, moist conditions, an average of 5 to 10 earthworms in every shovelful of upturned soil would indicate a reasonably good-sized population of these ecosystem engineers.

11.5 ANTS AND TERMITES[5]

Some mention has already been made (Section 11.2) of the varied food web roles of mesofauna arthropods (animals with a hard exoskeleton), namely the mites (eight-legged arachnids, which account for about 30,000 soil-dwelling species) and the springtails

---

[5] For an excellent discussion of termites and ants in the tropics, see Lal (1987).

(wingless insects in the order *Collembola*, of which some 6500 soil species are known). We will now turn our attention to ants and termites, two groups of insects that are considered to be important ecosystem engineers.

Ants

Ants and termites, the real kings of the jungle (video): http://www.nhm.ac.uk/ nature-online/life/insects-spiders/webcast-antsandtermitesvid/ants-and-termites-the-real-kings-of-the-jungle.html

Nearly 9000 species of soil-inhabiting ants have been identified. They are most diverse in the humid tropics, but are perhaps most functionally prominent in temperate semi-arid grasslands (Figure 11.10). Ants play important roles in forests from the tropics to the taiga. Some ant species act as detritivores, others as herbivores, and still others as predators. A rather notorious example of the latter is the imported red fire ant (*Solenopsis invicta*), which arrived in the southeastern United States from South America in the 1940s and by the late 1990s had spread as far as California. Like many ants, the fire ant lives in large underground nests topped by mounds of soil about 30 cm tall. This ant is known for its painful, venomous bite (actually a combination of a nonvenomous bite followed by a venomous sting). As the fire ant destroys insect pests and weed seeds, it can actually benefit farmers' crops in some situations, but it is also known to physically damage many crops and to feed on beneficial native insects, soft-shelled eggs of native birds and reptiles, and even newborn mammals.

Some ants feed on the sugary fluid (*honeydew*) produced by aphids as these sucking insects attack higher plants. The ants may "raise" the aphids in their nest or "herd" them while they feed on plant sap. Ants belonging to the genus *Formica* have been shown to offer forest trees significant relief from canopy-feeding aphids, a single nest of these ants collecting and devouring some 10 kg of aphids in a single summer. Ant nest-building activity can improve soil aeration, increase water infiltration, and modify soil pH. The nests, and the microbial populations they encourage, also stimulate the cycling of soil nitrogen (Figure 11.10). Although ants occur widely and play vital and varied ecological roles, their cousins, the termites, have been much more intensively studied and will be considered next.

Termites

DIVERSITY AND DIET.   Termites are sometimes called *white ants*, although they are quite distinct from ants (one can distinguish an ant by its narrow "waist" between the abdomen and thorax; compare Figures 11.10 and 11.11). There are about 2000 species of termites,

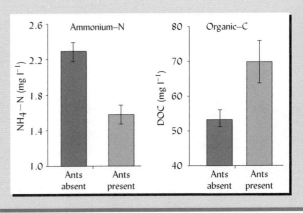

FIGURE 11.10 Ants exert important direct and indirect influences on the soil ecosystem. (*Left*) The herbivorous harvester ant (*Messor andrei*) forages for plant material to bring back to its nests in semiarid grasslands. Studies have shown that the soil in its nests is enriched in mineral nutrients and supports higher and more diverse populations of bacteria, fungi, nematodes, and microarthropods compared to soil away from the ant nests. (*Right*) The predatory ant (*Formica polyctena*) influences the concentrations of ammonium-N and organic carbon in water leaching through O horizons in a Norway spruce forest. These measurements were made within a few cm of the tree trunks, where the ants tend to build their nests. The effects were not seen 30 cm away from the tree trunks. Ants belonging to this genus also protect both coniferous and deciduous trees by preying on leaf-eating insects in the tree canopies. [Photo courtesy of Alex Wild, University of California, Davis; graphs modified from Stadler et al. (2006)]

most of which use cellulose in the form of plant fiber as their primary food. Yet most termites cannot themselves digest cellulose. Instead, a termite depends on a mutualistic relationship with protozoa and bacteria that live in its gut. These gut microorganisms produce the enzymes that degrade cellulose and allow the termite to derive energy from it. Termites (and their gut microorganisms) are major contributors to the breakdown of organic materials near or on the soil surface. The bacterial metabolism that takes place under anaerobic conditions in termite guts also accounts for a substantial fraction of the global production of methane (CH_4), an important greenhouse (global warming) gas (see Section 12.9).

Termites are found in about two-thirds of the world's land area but are most prominent in the grasslands (savannas) and forests of tropical and subtropical areas (both humid and semiarid). Their activity is on a scale comparable to that of earthworms. In the drier tropics (less than 800 mm annual rainfall) termites surpass earthworms in dominating the soil fauna.

How to tell ants and termites apart:
http://drdons.net/whatr.htm

Most termite species eat decaying logs, grasses, or fallen tree leaves, but some attack the sound wood in standing trees. These groups have become quite infamous because of their habit of invading (and subsequently destroying) the houses people build of wood. The most damaging of the wood-eating species is the highly invasive Formosan subterranean termite (*Coptotermes formosanus*), which now infests wooden structures around the world in the tropical and subtropical zones. Most tree- and house-invading termites build protective tubes made of compacted soil and termite feces, which enable them to return to the soil for their daily water supply. Termite inspectors look for these earthen tunnels running up foundation walls under a house to indicate an infestation.

MOUND-BUILDING ACTIVITIES. Termites are social animals that live in very complex labyrinths of nests, passages, and chambers that they build both below and above the soil surface. Termite mounds built from soil particles and feces cemented with saliva are characteristic features of many landscapes in Africa, Latin America, Australia, and Asia (Figure 11.11). These mounds are essentially termite "cities" with a network of underground passages and aboveground covered runways that typically spread 20 to 30 m beyond the mound. Several species, such as *Macrotermes spp.* in Africa, use plant residues to "cultivate" fungi in their mounds as a source of food.

In building their mounds, termites transport soil from lower layers to the surface, thereby extensively mixing the soil and incorporating into it the plant residues they use as food. Scavenging a large area around each mound, these insects remove up to 4000 kg/ha of leaf and woody material annually, a substantial portion of the plant litter produced in many tropical ecosystems. They also annually move 300 to 1200 kg/ha of

FIGURE 11.11   Termites in southern Africa. (*Left*) A termite mound in a cultivated field constructed of soil cemented hard with termite saliva. (*Right*) Individual worker termites (striped) drag cut pieces of leaves into their underground nest as soldier termites (large heads and mandibles) stand guard. (Photos courtesy of R. Weil)

Soil casings

FIGURE 11.12 Termites are quickly making these sorghum residues disappear, depriving the sandy, erosion-prone soil of the benefits of a protective mulch. However, the termites are also creating numerous stable macrochannels, which will aid in capturing water when the rains finally come to this field in the Sahelian region of West Africa. Note the soil casings that the termites have built around parts of the plant residues. Under these casings the termites are protected from the drying sun and wind. The knife blade in the picture is 7 cm long. (Photo courtesy of R. Weil)

soil in their mound-building activities. These activities have significant impacts on soil formation, as well as on current soil fertility and productivity.

The quantity of soil materials incorporated into termite mounds can be enormous; up to 2.4 million kg/ha has been recorded (recall, from Chapter 4, that a 15-cm-deep layer of soil over 1 ha weighs approximately 2 million kg). Depending on the species and environmental conditions, termites may build mounds 6 m or more in height and may extend them to an even greater depth into the soil in search of water or clay layers. Each mound provides a home for 1 million or more termites. The mounds are abandoned after 10 to 20 years and can then be broken down to level the land for crop production. Attempts to level an occupied mound are usually frustrating, as the termites rebuild very rapidly unless the queen (egg-laying) termite is destroyed.

EFFECT OF TERMITES ON SOIL PRODUCTIVITY.   Unlike earthworms, termites do not generally have a beneficial effect on soil productivity. This is because the digestive processes of termites, aided by microorganisms in their gut, are generally more efficient than those of earthworms. Also, earthworms incorporate organic matter into the soil in a relatively uniform manner over a hectare of land. In contrast, termites mix plant residues into the soil only in the localized areas of their nests, while denuding the remaining land area of surface residues. This termite behavior can make it extremely difficult to provide cropland with the protective benefits of a crop residue mulch (Figure 11.12).

PLANT GROWTH ON TERMITE MOUND MATERIAL.   Termite mound material often has a lower organic matter and nutrient content than the surrounding undisturbed topsoil. This is because termites build their mounds mainly with subsoil, which is typically lower in organic matter content than topsoil. Plant growth in soil derived from these mounds is often poor, not only because of low nutrient content in the surface soil, but also because of the greater density of some of the mound material.

However, where the subsoil is richer in mineral nutrients than the topsoil or is rich in clay compared to a very sandy surface soil, the material from abandoned mounds may provide islands of relatively high plant production, due to greater availability of phosphorus, potassium, calcium, and moisture. On soils with a high water table, termite mounds may provide islands of better drainage and aeration that produce much better plant growth. In certain semiarid and savanna regions, the stable macrochannels constructed by termites greatly increase water infiltration into soils that otherwise tend to form impermeable surface crusts.

11.6   SOIL MICROANIMALS

From the viewpoint of microscopic animals, soils present many habitats that are essentially aquatic, at least intermittently so. For this reason the soil microfauna are closely related to the microfauna found in lakes and streams. The two groups exerting the greatest influence on soil processes are the nematodes and protozoa.

Nematodes

Links to movies of the nematode, C. elegans, in action:
http://www.bio.unc.edu/faculty/goldstein/lab/movies.html

Nematode Caenorhabditis elegans on the move!
http://video.google.com/videoplay?docid=20195700875678727668q=soil

Nematodes—commonly called *threadworms* or *eelworms*—are found in almost all soils, often in surprisingly great numbers (see Table 11.5) and diversity. Some 20,000 species have been identified of the 100,000 nematode species that are thought to exist. These unsegmented roundworms are highly mobile creatures about 4 to 100 μm in cross section and up to several millimeters in length. They wriggle their way through the labyrinth of soil pores, sometimes swimming in water-filled pores (like their aquatic cousins), but more often pushing off the moist particle surfaces of partially air-filled pores. The latter mode of locomotion helps explain why some nematodes are active and reproduce when the soil water potential is so low (see Section 5.4) that only pores smaller than 1 μm (far too small for nematodes to enter) are filled with water. Nematodes are very sensitive to soil water content and porosity. Moist, well-aggregated, or sandy soils typically have especially high nematode populations, as these soils contain abundant pores large enough to accommodate their movements. When the soil becomes too dry, nematodes survive by coiling up into a **cryptobiotic** or resting state, in which they seem to be nearly impervious to environmental conditions and use no detectable oxygen for respiration. In semiarid rangelands, nematode activity has been shown to be largely restricted to the first few days after each rainfall that awakens the nematodes from their cryptobiotic state.

FEEDING HABITS.   Most nematodes feed on fungi, bacteria, and algae or are predatory on other nematodes, protozoa, or insect larvae. The different trophic groups of nematodes can often be distinguished by the type of mouth parts present. Figure 11.13 illustrates a predator and plant parasite. The head of a bacteria-eating nematode is detailed in the opening photo for this chapter. Grazing by nematodes can have a marked effect on the growth and activities of fungal and bacterial populations. Since bacterial cells contain more nitrogen than the nematodes can use, nematode activity often stimulates the cycling and release of plant-available nitrogen in the soil, accounting for 30 to 40% of the nitrogen released in some ecosystems. Certain predatory nematodes that efficiently attack insect larvae in the soil are sold for use as biological control agents, an environmentally beneficial alternative to the use of toxic pesticides. Once established, they can

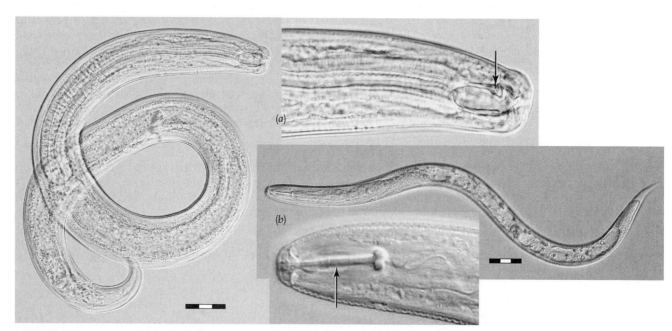

FIGURE 11.13 Two soil nematodes, one (*a*) a predator, and the other (*b*) a plant root parasite (soybean cyst nematode). The head and mouth parts of a nematode often reflect its trophic role, as the inset enlargements of these nematodes illustrate. Predators (such as this one in the Mononchidae family) usually have hard teeth (*a*, arrow) and a large mouth for capturing and swallowing prey. Nematodes that feed on plant roots or fungal hyphae are characterized by a retractable spearlike mouth part (*b*, arrow) that pierces the targeted cell to feed on the liquid contents. Mouth parts of a bacteria-eating nematode can be seen in the opening photo for Chapter 11. Scale bars marked in 10 μm units. (Photos courtesy of Lisa Stocking Gruver, University of Maryland)

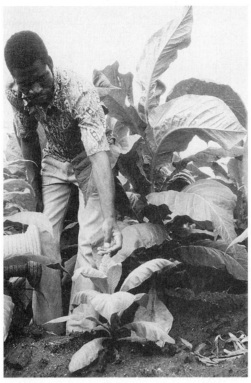

FIGURE 11.14 (*Left*) The stunted (foreground) and large (background) tobacco plants pictured both received identical management. (*Right*) Examination of the root systems of the stunted plant showed that it was heavily infested with root-knot nematodes, which stunted the roots and produced knotlike deformities. (Photos courtesy of R. Weil)

provide effective, long-term control of such soilborne insect pests as the corn rootworm or the grubs (Japanese beetle larvae) that destroy homeowners' lawns.

PLANT PARASITES. Some nematodes, especially those of the genus *Heterodera*, can infest the roots of practically all plant species by piercing the plant cells with a sharp, spearlike mouth part. These wounds often allow infection by secondary pathogens and may cause the formation of knotlike growths on the roots. Minor nematode infestations are nearly ubiquitous and often have little observable effect on the host plant. However, infestations beyond a certain threshold level result in serious stunting of the plant. Cyst- (egg-sac-) forming nematodes are major pests of soybeans, while root-knot-forming nematodes cause widespread damage to fruit trees and solanaceous crops (Figure 11.14).

Until recently, the principal methods of controlling plant-parasitic nematodes were long rotations with nonhost crops (often 5 years is required for the parasitic nematode populations to sufficiently dwindle), use of genetically resistant crop varieties, and soil fumigation with highly toxic chemicals (**nematicides**). The use of soil nematicides, such as methyl bromide, has been sharply restricted because of undesirable environmental effects.

New, less dangerous approaches to nematode control include the use of hardwood bark for containerized plants and interplanting or rotating susceptible crops with plants such as marigolds that produce root exudates with nematicidal properties (Figure 11.15). Certain plants in the Brassicaceae family (such as mustards, radishes, and rapeseed) contain sulfurous compounds called glucosinolates. When the residues of these plants break down in the soil, their glucosinolates produce volatile biofumigants that can kill certain nematodes. The use of these plants as green manures has successfully substituted for the use of more broadly toxic synthetic chemicals to control nematodes in potato, strawberry, and other nematode-sensitive crops. Progress has also been made in the development of nematode-resistant varieties of such plants as soybeans. Often a series of different varieties will be planted in succeeding years to prevent the buildup of parasitic nematode populations.

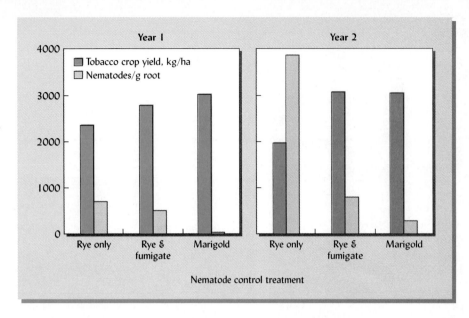

FIGURE 11.15 Marigolds control plant-parasitic nematodes. A susceptible host plant (tobacco) was grown in the summers of year 1 and year 2 on a sandy soil in Ontario, Canada. Rye was grown as a cover crop each winter in the untreated plots (rye only) and fumigated plots (rye and fumigate). The fumigated plots were injected in year 1 *and* year 2 with a chemical fumigant consisting mainly of 1,3-dichloropropene. The remaining plots (marigold) were planted to marigolds (*Tagetes patula* cv. "Creole") *only* in the summer of the year before the susceptible crop was first planted. The nematode infestation grew worse in the second year that a susceptible crop was grown (especially in rye only plots) and without nematode control the crop yield declined substantially. However, growing marigolds in one year controlled nematodes in susceptible crops for the following two years. [Based on data from Reynolds et al. (2000); used with permission of the American Society of Agronomy]

Protozoa

See an image gallery of protozoa at http://www.pirx.com/droplet/gallery.html

Protozoa are mobile, single-celled creatures that capture and engulf their food. With some 50,000 species in existence, they are the most varied and numerous of the soil microfauna (Table 11.5). Most are considerably larger than bacteria (see Figure 11.16), having a diameter range of 4 to 250 μm. Their cells do not have true cell walls and have a distinctly more complex organization than bacterial cells. Soil protozoa include amoebas (which move by extending and contracting pseudopodia), ciliates (which move by waving hairlike structures; see Figure 11.17), and flagellates (which move by waving a whiplike appendage called a *flagellum*). They swim about in the water-filled pores and water films in the soil and can form resistant resting stages (called *cysts*) when the soil dries out or food becomes scarce.

Sometimes as many as 40 or 50 different protozoa species may occur in a single sample of soil. The liveweight of protozoa in surface soil ranges from 20 to 200 kg/ha (see Table 11.5). A considerable number of serious animal and human diseases are attributed to infection by protozoa, but mainly by those that are waterborne, rather than soil-borne. Most soil-inhabiting protozoa prey upon soil bacteria, exerting a significant influence on the populations of these microflora in soils.

Protozoa generally thrive best in moist, well-drained soils and are most numerous in surface horizons. Protozoa are especially active in the area immediately around plant roots. Their main influence on organic matter decay and nutrient release is through their effects on bacterial populations. In pursuit of their bacterial prey, some soil-dwelling protozoa are adapted to squeezing into soil pores with openings as small as 10 μm. Still, soil aggregates often provide even smaller pores where bacterial cells can hide to escape predation. This protection from protozoa predation is a characteristic of the soil environment that helps explain the greater diversity of bacteria in soils than in aquatic habitats where such hiding places do not exist.

Other Fascinating Soil Micro-Creatures

Space does not allow us to describe in detail the many other invertebrate animals of the soil, such as the hundreds of species of **rotifers** (somewhat like zooplankton of the soil water) and **tardigrades** (sometimes called water bears for their teddybear-like appearance), most of which are smaller than 1 mm and either prey on protozoa or feed

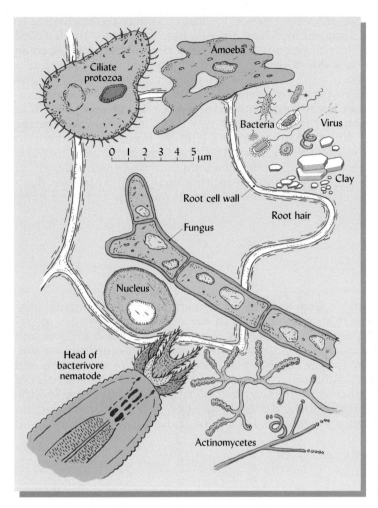

FIGURE 11.16 A depiction of representative groups of soil microorganisms, showing their approximate relative sizes. The large white-outlined structure in the center background is a plant root cell. (Drawing courtesy of R. Weil)

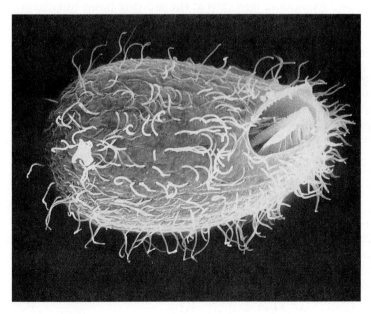

FIGURE 11.17 A ciliated protozoan (*Glaucoma scintillans*), one of the types of microanimals found in soils. Predation by protozoa exerts a major influence on the abundance and diversity of soil bacteria. [Scanning electron micrograph courtesy of J. O. Corless, University of Maryland]

on fungi, algae, and bacteria. However, one soil life-form is simply too fascinating to omit: the ***Dictyostelium* slime mold.** These amoeba-like, eukaryotic single-celled microorganisms live in the litter layers and the upper few centimeters of soil, feeding on bacteria, much as protozoa do. However, when the local food supply runs out, a truly amazing phenomenon occurs. The individual amoeboid cells begin to congregate. In response to a complex pattern of chemical signals, some 50,000 of these individual organisms stream together to form a mound about 0.1 mm in size. The cells in this mound then differentiate themselves into two types, most conglomerating together to form a "foot," while the others form a kind of "head." The 50,000 cells now behave as a single, multicellular organism! This tiny "creature" crawls, slug-like, up to the soil surface, where the cells undergo yet another rearrangement. The cells of the slug's "foot" transform into the stalk of a fruiting body, atop which the slug's "head" transforms into spores—which can now disperse through the air to new hunting grounds!

11.7   PLANT ROOTS

Higher plants store the sun's energy and are the primary producers of organic matter (see Figure 11.1). Their roots grow and die in the soil and are classified as soil organisms in this text. They typically occupy about 1% of the soil volume and may be responsible for a quarter to a third of the respiration occurring in a soil. Roots usually compete for oxygen, but they also supply much of the carbon and energy needed by the soil community of fauna and microflora. The activities of plant roots greatly influence soil chemical and physical properties, the specific effects depending on the type of soil and plant in question (see, for example, Section 7.8). As we shall see, plant roots interact with other soil organisms in varied and complex ways.

Morphology of Roots

Depending on their size, roots may be considered to be either meso- or microorganisms. Fine feeder roots range in diameter from 100 to 400 µm, while root hairs are only 10 to 50 µm in diameter—similar in size to the strands of microscopic fungi (see Figure 11.16). Root hairs are elongated protuberances of single cells of the outer (epidermal) layer (Figure 11.18). One function of root hairs is to anchor the root as it pushes its way through the soil. Another function is to increase the amount of root surface area available to absorb water and nutrients from the soil solution.

Roots grow by forming and expanding new cells at the growing point (meristem), which is located just behind the root tip. The root tip itself is shielded by a protective cap of expendable cells that slough off as the root pushes through the soil. Root morphology is affected by both type of plant and soil conditions. For example, fine roots may proliferate in localized areas of high nutrient concentrations. Root-hair formation is stimulated by contact with soil particles and by low nutrient supply. When soil water is scarce, plants typically put more energy into root growth than into shoot growth, decreasing the shoot-to-root ratio (thus increasing the uptake of water and minimizing its loss by transpiration). Many roots become thick and stubby in response to high soil bulk density or high aluminum concentrations in soil solution (see Chapters 4 and 9).

Living roots physically modify the soil as they push through existing cracks and make new openings of their own. Roots move through the soil following paths of least resistance, growing between soil peds and into existing cracks and channels. Once extended into a pore, the root matures and expands, exerting lateral forces that enlarge the pore. By removing moisture from the soil, plant roots stabilize organic–mineral bonds and encourage soil shrinkage and cracking, which, in turn, increase stable soil aggregation. Root exudates also support a myriad of microorganisms, which help to further stabilize soil aggregates. In addition, when roots die and decompose, they provide building materials for humus, not only in the top few centimeters, but also to greater soil depths. If soil conditions permit, the roots of annual herbaceous plants

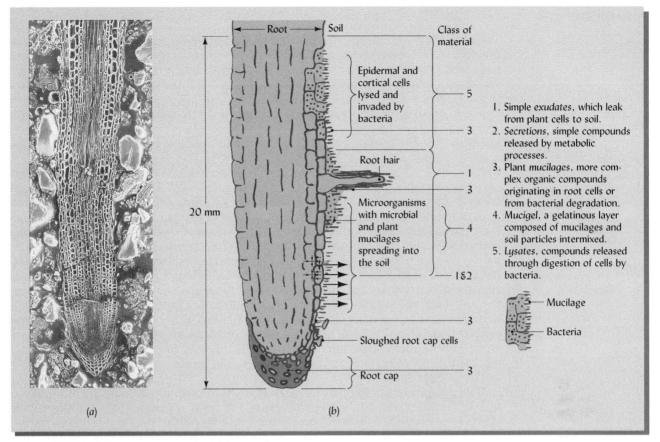

FIGURE 11.18 (*a*) Photograph of a root tip illustrating how roots penetrate soil and emphasizing the root cells through which nutrients and water move into and up the plant. (*b*) Diagram of a root showing the origins of organic materials in the rhizosphere. [(*a*) From Chino (1976), used with permission of Japanese Society of Soil Science and Plant Nutrition, Tokyo; (*b*) redrawn from Rovira et al. (1979), used with permission of Academic Press, London]

(such as most crops) send their roots down 1 to 2 m. Perennial plants, especially woody species, may send some roots more than 5 m deep if restricting layers are not encountered. Generally, rooting is deepest in hot, dry climates and most shallow in boreal or wet tropical environments (see Figure 5.42).

Amount of Organic Tissue Added

The importance of root residues in helping to maintain soil organic matter is often overlooked. In grasslands, about 50 to 60% of the net primary production (total plant biomass) is commonly in the form of roots. In addition, prairie fires may remove most of the aboveground biomass, so that the deep, dense root systems are the main source of organic matter added to these soils. In plantation and natural forests, 40 to 70% of the total biomass production may be in the form of tree roots. In arable soils, the mass of roots remaining in the soil after crop harvest is commonly 15 to 40% that of the aboveground crop. If an average figure of 25% is used, good crops of oats, corn, and sugar cane would be expected to leave about 2500, 4500, and 8500 kg/ha of root residues, respectively. When considering the contribution of a relatively young cover crop plants (Section 16.2), it should be remembered that in the early stages of development, plants give priority to roots, so the biomass below ground may be 10 times the biomass above ground. The mass of organic compounds contributed by roots is seen to be even greater when the rhizosphere effects discussed in the following subsections are considered.

Rhizosphere[6]

The zone of soil significantly influenced by living roots is termed the **rhizosphere** and usually extends about 2 mm out from the root surface (Figure 1.26). The chemical and biological characteristics of this zone can be very different from those of the bulk soil. Soil acidity may be 10 times higher (or lower) in the rhizosphere than in the bulk soil (e.g., see Figure 9.13). Roots greatly affect the nutrient supply in this zone by withdrawing dissolved nutrients on one hand and by solubilizing nutrients from soil minerals on the other. By these and other means, roots affect the mineral nutrition of soil microbes, just as the microbes affect the nutrients available to the plant roots.

RHIZODEPOSITION. Significant quantities of at least three broad types of organic compounds are released at the surface of young roots (see Figure 11.18). First, low-molecular-weight organic compounds are exuded by root cells, including organic acids, sugars, amino acids, and phenolic compounds. Some of these root exudates, especially the phenolics, exert growth-regulating influences on other plants and soil microorganisms in a phenomenon called **allelopathy** (see Section 12.5). Second, high-molecular-weight mucilages secreted by root-cap cells and epidermal cells near apical zones form a substance called **mucigel** when mixed with microbial cells and clay particles. This mucigel appears to have several beneficial functions: It lubricates the root's movement through the soil; it improves root–soil contact, especially in dry soils when roots may shrink in size and lose direct contact with the soil (see Figure 5.44); it may protect the root from certain toxic chemicals in the soils; and it provides an ideal environment for the growth of the rhizosphere microorganisms. Third, cells from the root cap and epidermis continually slough off as the root grows and enrich the rhizosphere with a wide variety of cell contents.

Taken together, these types of **rhizodeposition** typically account for 2 to 30% of total dry-matter production in young plants. The roots of common grain and vegetable plants have been observed to rhizodeposit 5 to 40% of the organic substances translocated to them from the plant shoot. Sometimes when a plant is carefully uprooted from a loose soil, these root exudates cause a thin layer of soil, approximating the rhizosphere, to form a sheath around the actively growing root tips (Figure 11.19). The amount of organic material lost to the rhizosphere during the growing season of annual plants may be more than twice the amount remaining in the root system at the end of the growing season. Rhizodeposition decreases with plant age but increases with soil stresses, such as compaction and low nutrient supply.

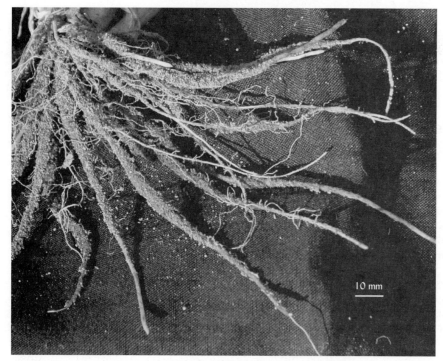

FIGURE 11.19 The zone of soil within 1 to 2 mm of living plant roots is termed the *rhizosphere*. This zone is greatly enriched in organic compounds excreted by the roots. These exudates and the microorganisms they support, as well as root hairs, have caused the rhizosphere soil to adhere to these grass roots as a sheath, except near the root tips where the bare white root is visible. (Photo courtesy of R. Weil)

10 mm

[6] For a detailed review of the rhizosphere and its management, see Bowen and Rovira (1999).

Because of the rhizodeposition of carbon substrates and specific growth factors (such as vitamins and amino acids), microbial numbers in the rhizosphere are typically 2 to 10 times as great as in the bulk soil (sometimes expressed as a *R/S ratio*, typically ranging from 2 to 10).

The processes just described explain why plant roots are among the most important organisms in the soil ecosystem.

11.8 SOIL ALGAE

Directory of algal images including those that can be found in soil: http://vis-pc.plantbio.ohiou .edu/algaeimage/imageindex .htm

Like higher plants, algae consist of eukaryotic cells, those with nuclei organized inside a nuclear membrane. (Organisms formerly called *blue-green algae* are prokaryotes and therefore will be considered with the bacteria.) Also like higher plants, algae are equipped with chlorophyll, enabling them to carry out photosynthesis. As photoautotrophs, algae need light and are therefore mostly found very near the surface of the soil. Some species can also function as heterotrophs in the dark. A few species are photoheterotrophs that use sunlight for energy but cannot synthesize all of the organic molecules they require (see Table 11.3).

Most soil algae range in size from 2 to 20 μm. Many algal species are motile and swim about in soil pore water, some by means of flagella (whiplike "tails"). Most grow best under moist to wet conditions, but some are also very important in hot or cold desert environments. Sometimes the growth of algae may be so great that the soil surface is covered with a green or orange algal mat. Some algae (as well as certain cyanobacteria) form *lichens*, symbiotic associations with fungi. These are important in colonizing bare rock and other low-organic-matter environments (see Figure 2.8). In unvegetated patches in deserts, algae commonly contribute to the formation of **microbiotic crusts** (see Section 11.14).

Several hundred species of algae have been isolated from soils, but a small number of species are the most prominent in soils throughout the world. The mass of these live algae may range from 10 to 500 kg/ha (see Table 11.5). In addition to producing a substantial amount of organic matter in some fertile soils, certain algae excrete polysaccharides that have very favorable effects on soil aggregation (see Section 4.5).

11.9 SOIL FUNGI

See pictures of fungal mycelia: http://ic.ucsc.edu/~wxcheng/ wewu/soilfungi.htm

Soil fungi comprise an extremely diverse group of microorganisms. Tens of thousands of species have been identified in soils, representing some 170 genera. As many as 2500 species have been reported to occur in a single location. Scientists, using DNA and fatty acids extracted from soils, estimate that there are at least 1 million fungal species in the soil still awaiting discovery. Because of the extensive filamentous morphology of many fungi, it is difficult to define the numbers of fungi in soil. For example, scientists using DNA molecular analysis techniques determined that the fungal strands permeating the soil and tree roots of an entire 20-hectare forest stand belonged to *a single organism* that weighed more than 10,000 kg and was over 1500 years old! Instead of counting numbers, scientists use the biomass or hyphal length per m^2 as more meaningful measures of fungal presence. Total fungal biomass typically ranges from 1000 to 15,000 kg/ha in the upper 15 cm (Table 11.5). The fungi dominate the biomass in many soils, exceeding even that of the bacteria.

Fungi are eukaryotes with a nuclear membrane and cell walls. As heterotrophs, they depend on living or dead organic materials for both their carbon and their energy. Fungi are aerobic organisms, although some can tolerate the rather low oxygen concentrations and high levels of carbon dioxide found in wet or compacted soils. Strictly speaking, fungi are not entirely microscopic, since some of these organisms, such as mushrooms, form macroscopic structures that can easily be seen without magnification.

For convenience of discussion, fungi may be divided into three groups: (1) yeasts, (2) molds, and (3) mushroom fungi. *Yeasts*, which are single-celled organisms, live principally in waterlogged, anaerobic soils. *Molds* and *mushrooms* are both considered to be filamentous fungi, because they are characterized by long, threadlike, branching chains of cells. Individual fungal filaments, called **hyphae** (Figure 11.27 on page 476), are often twisted together to form **mycelia** that appear somewhat like woven ropes. Fungal

FIGURE 11.20 Fungal mycelia, consisting of bundles of microscopic hyphae, grow up from the soil into the leaves and woody debris of the forest floor. The ability of fungi to "reach out" from the soil in this manner helps explain why they dominate the decay of surface litter and mulch, while bacteria are more prominent in decaying organic material incorporated into the soil. Scale marked in cm. (Photo courtesy of R. Weil)

mycelia are often visible as thin, white or colored strands running through decaying plant litter (Figure 11.20). Filamentous fungi reproduce by means of spores, often formed on fruiting bodies, which may be microscopic (e.g., molds) or macroscopic (such as that shown in Figures 11.21 and 11.22).

Molds

Hunting slime molds, by Adele Conover, Smithsonian: http://www.smithsonianmag.com/issues/2001/march/phenom_mar01.php

The molds are distinctly filamentous, microscopic, or semimacroscopic fungi that play a much more important role in soil organic matter breakdown than the mushroom fungi. Molds develop vigorously in acid, neutral, or alkaline soils. Some are favored, rather than harmed, by lowered pH. Consequently, they may dominate the microflora in acid surface soils, where bacteria and actinomycetes offer only mild competition. The ability of molds to tolerate low pH is especially important in decomposing organic residues in acid forest soils.

Many genera of molds are found in soils. Four of the most common are *Penicillium*, *Mucor*, *Fusarium*, and *Aspergillus*. The complexity of the organic compounds being attacked seems to determine which particular mold (or molds) prevail. Their biomass fluctuates greatly with soil conditions. In some cases much of the hyphal length measured may no longer be living.

FIGURE 11.21 Fruiting bodies of a Bird's Nest Fungus (*Cyathus olla*) contain disc-shaped spores ("eggs") that are scattered forcefully when the cone is hit by a raindrop. Like other members of the Basidiomycota, this fungus produces enzymes that break down cellulose, hemicellulose, and lignin in materials such as woody forest litter or corncobs left on agricultural fields after harvest. See tip of ballpoint pen for scale. (Photo courtesy of R. Weil)

FIGURE 11.22 A "fairy ring" of fungal growth and the fungi's fruiting bodies (mushrooms). As the fairy ring fungi (most commonly *Marasmius spp.*) metabolize accumulated grass thatch and residues, they release excess nitrogen that stimulates the lush green growth of the grass. Later, bacteria decompose the aging and dead fungi, producing a second release of nitrogen. The fungus produces a chemical (thought to be hydrogen cyanide) that is toxic to itself. Therefore, each generation must grow into uncolonized soil, producing an ever-expanding circle of fungi and decay marked by an ever-larger ring of dark green grass. The grass in the center of the ring is often brown, stunted, and water-stressed, most likely because the fungi render the upper soil layers somewhat hydrophobic. (Photos courtesy of R. Weil)

Mushroom Fungi

These fungi are associated with forest and grass vegetation where moisture and organic residues are ample. Although the mushrooms of many species are extremely poisonous to humans, some are edible—and a few have been domesticated.

The aboveground fruiting body of most mushrooms is only a small part of the total organism. An extensive network of hyphae permeates the underlying soil or organic residue. While mushrooms are not as widely distributed as the molds, these fungi are very important, especially in the breakdown of woody tissue, and because some species form a symbiotic relationship with plant roots (see "Mycorrhizae," following).

Activities of Fungi

As decomposers of organic materials in soil, fungi are the most versatile and persistent of any group. Cellulose, starch, gums, and lignin, as well as the more easily metabolized proteins and sugars, succumb to their attack. Fungi play major roles in the processes of humus formation (see Section 12.4) and aggregate stabilization (see Section 4.5). They usually dominate in the upper horizons of forested soils, as well as in very acid or sandy soils. They carry out the largest share of the decomposition in many cultivated soils, as well.

Fungi are quite efficient in using the organic materials they metabolize. Up to 50% of the substances decomposed by fungi may become fungal tissue, compared to about 20% for bacteria. Soil *fertility* depends in no small degree on nutrient cycling by fungi, since they continue to decompose complex organic materials after bacteria and actinomycetes have essentially ceased to function. Soil *tilth* also benefits from fungi as their hyphae stabilize soil structure (see Box 4.4 and Figure 4.15). Some of the nutrient cycling and ecological activities of soil fungi are made easily visible in the case of "fairy rings," commonly seen on lawns in early spring (Figure 11.22).

In addition to the breakdown of organic residues and the formation of humus, numerous other fungal activities have significant impact on soil ecology. Certain species even trap nematodes (see Figure 11.23). Soil fungi can synthesize a wide range of complex organic compounds in addition to those associated with soil humus. Certain fungi produce compounds that kill other fungi or bacteria and provide a competitive edge over rival microorganisms in the soil. These fungi have proved to be highly beneficial to humankind (see Sections 11.10 and 11.14).

Unfortunately, not all the compounds produced by soil fungi benefit humans or higher plants. A few fungi produce chemicals (**mycotoxins**) that are highly toxic to

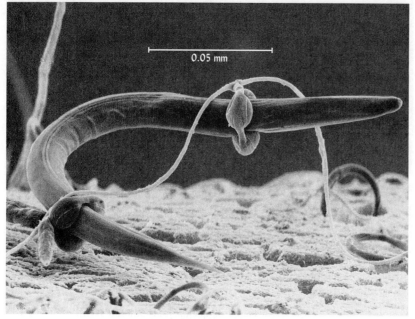

FIGURE 11.23 Several species of fungi prey on soil nematodes—often on those nematodes that parasitize higher plants. Some species of nematode-killing fungi attach themselves to, and slowly digest, the nematodes. Others, like this *Arthrobotrys anchonia*, make loops with their hyphae and wait for a nematode to swim through these lasso-like structures. The loop is then constricted, and the nematode is trapped. The nematode shown here is being crushed by two such fungal loops. Additional loops can be seen in their nonconstricted configuration. (Photo courtesy of George L. Barron, University of Guelph)

plants or animals (including humans). An important example of the latter is the production of highly carcinogenic aflatoxin by the fungus *Aspergillus flavus* growing on grains such as corn or peanuts, especially when grain is exposed to soil and moisture. Other fungi produce compounds that allow them to invade the tissues of higher plants (see Section 11.13), causing such serious plant diseases as wilts (e.g., *Verticillium*) and root rots (e.g., *Rhizoctonia*).

On the other hand, efforts are now under way to develop the potential of certain fungi (such as *Beauveria*) as biological control agents against some insects and mites that damage higher plants. These examples merely hint at the impact of the complex array of fungal activities in the soil.

Mycorrhizae[7]

Overview of mycorrhizal symbioses:
http://cropsoil.psu.edu/sylvia/mycorrhiza.htm

One of the most ecologically and economically important activities of soil fungi is the mutually beneficial association (**symbiosis**) between certain fungi and the roots of higher plants. This association is called **mycorrhizae**, a term meaning "fungus root." In natural ecosystems many plants are quite dependent on mycorrhizal relationships and cannot survive without them. Mycorrhizae are the rule, not the exception, for most plant species, including the majority of economically important plants. Mycorrhizal structures have been found in fossils of plants that lived some 400 million years ago, indicating that mycorrhizal infection may have played a role in the evolutionary adaptation of plants to the land environment.

Mycorrhizal fungi derive an enormous survival advantage from teaming up with plants. Instead of having to compete with all the other soil heterotrophs for decaying organic matter, the mycorrhizal fungi obtain sugars directly from the plant's root cells. This represents an energy cost to the plant, which may lose as much as 5 to 30% of its total photosynthate production to its mycorrhizal fungal symbiont.

In return, plants receive some extremely valuable benefits from the fungi. The fungal hyphae grow out into the soil some 5 to 15 cm from the infected root, reaching farther and into smaller pores than could the plant's own root hairs. This extension of the plant root system increases its efficiency, providing perhaps 10 times as much absorptive surface as the root system of an uninfected plant.

Mycorrhizae greatly enhance the ability of plants to take up phosphorus and other nutrients that are relatively immobile and present in low concentrations in the soil

---

[7] For an excellent book on mycorrhizal associations and their ecological effects, see Smith and Read (1997).

TABLE 11.8   Effect of Seedling Inoculation with Arbuscular Mycorrhizae (AM) on Root Colonization, Fruit Yield, and Shoot Nutrient Contents for Tomatoes Irrigated with Nonsaline or Saline Water

Very small amounts of inoculum were added to the potting mix in seedling trays. Mycorrhizal inoculation increased all parameters, but the greatest benefits of AM accrued under saline conditions. AM-inoculated plants under saline conditions yielded 5.3 kg fruit m^{-2}, not statistically different from the 5.8 kg fruit m^{-2} yield of noninoculated plants under nonsaline conditions.

Irrigation water treatment	AM root colonization	Fruit yield	Nutrient content of plant shoot					
			Increase from AM inoculation, %					
			P	K	Na	Cu	Fe	Zn
Nonsaline ($EC_w = 0.5$)	166	29	44	33	21	93	33	51
Saline ($EC_w = 2.4$)	293	60	192	138	7	193	165	120

Data selected from Al-Karaki (2006)

solution. Water uptake may also be improved by mycorrhizae, making plants more resistant to drought and salinity stress (Table 11.8). In soils contaminated with high levels of metals, mycorrhizae protect the plants from excessive uptake of these potential toxins (see Section 18.7). There is evidence that mycorrhizae also protect plants from certain soilborne diseases and parasitic nematodes by producing antibiotics, altering the root epidermis, and competing with fungal pathogens for infection sites. For all these reasons, the use of mycorrhizae can be a powerful tool in land restoration projects as well as in some agricultural situations.

ECTOMYCORRHIZA.   Two types of mycorrhizal associations are of considerable practical importance: **ectomycorrhiza** and **endomycorrhiza**. The ectomycorrhiza group includes hundreds of different fungal species associated primarily with temperate- or semiarid-region trees and shrubs, such as pine, birch, hemlock, beech, oak, spruce, and fir. These fungi, stimulated by root exudates, cover the surface of feeder roots with a fungal mantle. Their hyphae penetrate the roots and develop in the free space around the cells of the cortex but do not penetrate the cortex cell walls (hence the term *ecto*, meaning outside). Ectomycorrhizae cause the infected root system to consist primarily of stubby, white rootlets with a characteristic Y shape (Figure 11.24). These Y-shaped rootlets provide visible evidence of mycorrhizal infection.

ENDOMYCORRHIZA.   The most important members of the endomycorrhiza group are called **arbuscular mycorrhizae** (AM). When forming AM, fungal hyphae actually penetrate the cortical root cell walls and, once inside the plant cell, form small, highly branched structures known as **arbuscules**. These structures serve to transfer mineral nutrients from the fungi to the host plants and sugars from the plant to the fungus. Other structures, called **vesicles**, are usually also formed and serve as storage organs for the mycorrhizae (see Figure 11.24 and Plate 54).

Nearly 100 identified species of fungi form these endomycorrhizal associations in soils from the tropics to the arctic. Most native plants and agricultural crops can form AM associations and do not grow well on unfertilized soil in their absence. Many plants in the Legume family are especially dependent on mycorrhizae, not only to obtain sufficient phosphorus, but also to enhance their nitrogen-fixing symbiosis with rhizobia bacteria (Section 13.10). Two important groups of plants that do *not* form mycorrhizae are the *Cruciferae* (mustards, cabbage, rapeseed) and the *Chenopodiaceae* (beet and spinach). AM fungi are agriculturally most important where soils are low in nutrients, especially phosphorus (Figure 11.25 and Plate 53).

Research on AM continues to reveal the ecological and practical significance of this symbiosis. The importance of mycorrhizal hyphae in stabilizing soil aggregate structure is becoming increasingly clear (see Box 4.4). Also, AM fungi have been observed to form hyphal interconnections among nearby plants in forest, grassland, and pasture ecosystems. These hyphal connections can transfer nutrients from one plant to another, sometimes resulting in a complex, four-way symbiotic relationship (Figure 11.26). The ecological significance of these AM-mediated nutrient transfers is not yet well understood.

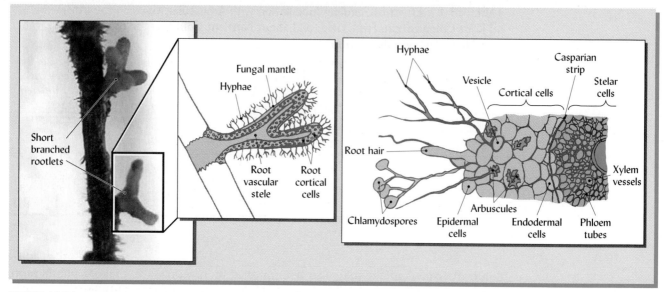

FIGURE 11.24 Diagram of ectomycorrhiza and arbuscular mycorrhiza (AM) associations with plant roots. (*Left*) The ectomycorrhiza association produces short branched rootlets that are covered with a fungal mantle, the hyphae of which extend out into the soil and between the plant cells, but do not penetrate the cells. (*Right*) In contrast, the AM fungi penetrate not only between cells but into certain cells as well. Within these cells, the fungi form structures known as *arbuscules* and *vesicles*. The former transfer nutrients to the plant, and the latter store these nutrients. In both types of association, the host plant provides sugars and other food for the fungi and receives in return essential mineral nutrients that the fungi absorb from the soil. [Redrawn from Menge (1981); photo courtesy of R. Weil]

Studying Mycorrhizas in Forestry and Agriculture: http://www.ffp.csiro.au/ research/mycorrhiza/index .html

MANAGING MYCORRHIZAE. Because of the near ubiquitous distribution of native mycorrhizal fungi, adding mycorrhizal inoculum rarely makes a difference in normal, biologically active soils. However, there are steps that can ensure good mycorrhizal infection. Soil tillage destroys hyphal networks; therefore physical soil disruption is likely to decrease the effectiveness of native mycorrhizae. It is also best to avoid too-frequent use of nonhost species, long periods of soil bareness, or heavy fertilization with phosphorus. In addition, the buildup of effective mycorrhizae in soils is favored by growing a diversity of host plant species as continuously as possible. The importance of maintaining host plants is exemplified by the case of Douglas fir forest regrowth. Clear-cutting of Douglas fir stands is commonly followed by suppression of hardwood trees, which are viewed as weed species that compete with the regenerated Douglas fir trees. However, it has been found that the young Douglas firs survive and grow much better

FIGURE 11.25 The effect of mycorrhizae on availability of phosphorus to a tropical legume, *Pueraria phaseoloides:* (Nil) no treatment, (PR) phosphate rock, and (PR + M) phosphate rock plus mycorrhizae. Mycorrhizal hyphae can dissolve phosphate rock and other low solubility forms of P that plant roots cannot otherwise utilize. (Courtesy of Dr. Fritz Kramer, CIAT, Cali, Colombia)

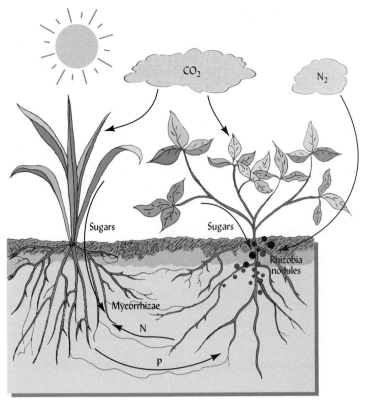

FIGURE 11.26 Mycorrhizal fungi, rhizobia bacteria, legumes, and nonlegume plants can all interact in a four-way, mutually beneficial relationship. Both the fungi and the bacteria obtain their energy from sugars supplied through photosynthesis by the plants. The rhizobia form nodules on the legume roots and enzymatically capture atmospheric nitrogen, providing the legume with nitrogen to make amino acids and proteins. The mycorrhizal fungi infect both types of plants and form hyphal interconnections between them. The mycorrhizae then not only assist in the uptake of phosphorus from the soil, but can also directly transfer nutrients from one plant to the other. Isotope tracer studies have shown that, by this mechanism, nitrogen is transferred from the nitrogen-fixing legume to the nonlegume (e.g., grass) plant, and phosphorus is mostly transferred to the legume from the nonlegume. The nonlegume grass plant has a fibrous root system and an extensive mycorrhizal network, which is relatively more efficient in extracting P from soils than the root system of the legume. Research indicates that some direct transfer of nutrients via mycorrhizal connections occurs in many mixed plant communities, such as in forest understories, grass–legume pastures, and mixed cropping systems. (Diagram courtesy of R. Weil)

if some hardwood saplings are allowed to remain to provide host continuity and serve as an inoculum source for AM fungi.

APPLYING MYCORRHIZAL INOCULUM. There may be a need to inoculate soils with mycorrhizal fungi where native populations are very low or conditions for infection are unusually adverse. Examples include soils that have been subjected to broad spectrum fumigation; extreme soil heating, drying, or salinization; drastic disturbance such that subsoil layers are brought to the surface; or long periods without vegetative cover (such as surface soil stockpiled during mining or construction activities as shown in Figure 1.16). Successful restoration of healthy vegetation to such denuded soils often requires inoculation with effective mycorrhizal fungi.

COMMERCIAL INOCULANTS Many ectomycorrhizal fungi are facultative symbionts (they can also live independently in the soil) and therefore can be cultured in large quantities on artificial media. Effective and pathogen-free ectomycorrhizae inoculants can, therefore, be easily produced for use on such trees as conifers, oaks, hickory, birch, willow, poplar, pecan, and eucalyptus. Commercial ectomycorrhizal inoculants for use in planting such trees in warm climates usually contain the fungus *Pisolithus tinctorius*, while those used to inoculate soils in cold climates typically contain *Rhizopogon* species. On infertile soils, use of these ectomycorrhizal inoculants is likely to increase tree survival and growth by 50 to 500%.

Inoculation with AM fungi may be more difficult, although sometimes soil from a mature ecosystem containing the desired plant species can be brought in as inoculum to supply the necessary mycorrhizal fungi. Although AM fungi cannot be grown in pure culture, concentrated, disease-free inoculants can be made. An appropriate host plant (often a grass species) is grown in a potting medium inoculated with a small amount of roots infected with AM fungi, usually of the *Glomus* genus. After growing the host for 3 or 4 months, the soil is allowed to dry out and the plants to slowly die, stimulating AM fungi sporulation. The dry potting medium, which now contains infected root fragments, hyphal fragments, and spores, is ground up and mixed with sand or other media to create a concentrated inoculum. Again, little response to inoculation is likely on healthy, biologically active soils, but dramatically positive effects are often obtained with such inoculum on highly disturbed or infertile sites.

11.10 SOIL PROKARYOTES: BACTERIA AND ARCHAEA

Into the Archaea:
http://www.ucmp.berkeley
.edu/archaea/archaea.html

The organisms described in the previous sections—from mammals to molds—all belong to the Eukarya domain. The organisms in the other two domains of life, the Bacteria and the Archaea, are prokaryotes—their cells lack a nucleus surrounded by a membrane. However, despite their similar appearance under the microscope, archaea are evolutionarily quite distinct from bacteria. For example, archaean cell membranes exhibit major chemical differences from bacterial membranes, such as the use of isoprene derivatives instead of fatty acids in their construction. In some respects, genetic analysis suggests that archaeans may be as closely related to plants or people as they are to bacteria! Until very recently, the archaea were thought of as rare and primitive creatures that live in only the most extreme and unusual environments on Earth—salt-saturated waters (see Plate 56), extremely acid or alkaline soils, deeply frozen ice, boiling hot water, anaerobic sediments, and the like. However, molecular identification techniques now suggest that archaeans are also common in more "normal" environments and probably represent about 10% of the microbial biomass in typical upland soils.

Another recent insight provided by molecular identification techniques is that although we previously had no idea how little we knew, now we do! Traditionally, scientists enumerated and identified soil microbes by culturing them on agar of various kinds. We now know that the thousands of species so identified represent less than 0.1% of the species present in soils—most prokaryotes simply cannot be cultured in the lab. While the usual concept of a "species" is difficult to apply to single-celled microorganisms that reproduce asexually, "kinds" of organisms are now estimated by "genome equivalents" of DNA information. We will consider the archaea together with the bacteria in this section, calling them prokaryotes when the discussion applies to members of both domains.

Characteristics

Prokaryotes range in size from 0.5 to 5 µm, considerably smaller in diameter than most fungal hyphae (Figure 11.27). The smaller ones approach the size of the average clay particle (see Figure 11.16). Prokaryotes are found in various shapes: nearly round (coccus), rodlike (bacillus), or spiral (spirillum). In the soil, the rod-shaped prokaryotes seem to predominate. Many prokaryotes are motile, swimming about in the soil water films by means of hairlike cilia or whiplike flagella. Others heavily colonize the nutrient-rich surface of plant roots (Figure 11.28).

Prokaryotes Populations in Soils

The numbers of prokaryotes are extremely variable but high, ranging from a few billion to more than a trillion in each gram of soil. A biomass of a 400 to 5000 kg/ha liveweight is commonly found in the upper 15 cm of fertile soils (see Table 11.5).

Their small size and ability to form extremely resistant resting stages that survive dispersal by winds, sediments, ocean currents, and animal digestive tracts have allowed prokaryotes to spread to almost all soil environments. The prokaryote diversity in a

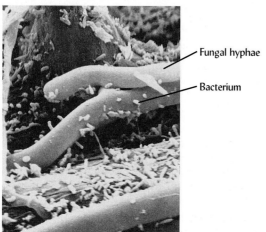

FIGURE 11.27 Fungal hyphae associated with much smaller rod-shaped bacteria. (Scanning electron micrograph courtesy of R. Campbell, University of Bristol, used with permission of American Phytopathological Society)

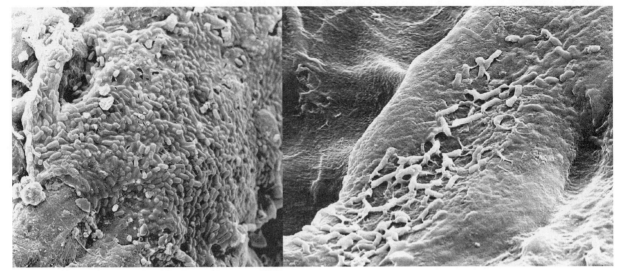

FIGURE 11.28 Bacteria colonies on the surface of a wheat root (*left*) and a corn root (*right*). The bacteria on the wheat root are embedded in their own secreted mucilage. The rhizosphere soil and the root surface itself are usually crowded with bacteria that are integral to the plant–soil–microbe system. The roots were growing in a sandy loam soil (Haplorthods) near Ottawa, Canada. [Cryo scanning electron micrographs courtesy of Margaret McCully, CSIRO Plant Industry, Canberra, Australia]

handful of soil is said to be comparable to the diversity of insects, birds, and mammals in the Amazon Basin! Their extremely rapid reproduction (generation times of a few hours in the lab to a few days in favorable soil) enables prokaryotes to increase their populations quickly in response to favorable changes in soil environment and food availability. As the early soil microbiologist Martinus Beijerinck expressed it, when it comes to microorganisms, "everything is everywhere and the milieu selects." That is, the microbial community reflects the soil environment. For example, the prokaryote community in a well-drained soil with a neutral pH is likely to be quite similar to a well-drained neutral soil on another continent, but quite different from the community in a geographically nearby soil that is very acid or poorly drained. By the same token, if a new substrate is added—even an industrial waste—populations of prokaryotes capable of feeding on it will likely soon emerge.

Source of Energy

Microorganisms that grow under extreme conditions studied at Oak Ridge National Lab: http://www.ornl.gov/info/ornlreview/rev32_3/amazing.htm

Soil prokaryotes are either autotrophic or heterotrophic (see Section 11.2). The autotrophs obtain their energy from sunlight (photoautotrophs) or from the oxidation of inorganic constituents such as ammonium, sulfur, and iron (chemoautotrophs) and obtain their carbon from carbon dioxide or dissolved carbonates. Autotrophic bacteria are not as diverse as heterotrophs in soils, but they play vital roles in controlling nutrient availability to higher plants and other organisms.

Most soil bacteria are heterotrophic—both their energy and their carbon come from organic matter. Heterotrophic bacteria, along with fungi, account for the general breakdown of organic matter in soil. The bacteria often predominate on easily decomposed substrates, such as animal wastes, starches, and proteins. Where oxygen supplies are depleted, as in wetlands, nearly all decomposition is mediated by prokaryotes. Certain gaseous products of anaerobic metabolism, such as methane and nitrous oxide, have major effects on the global environment (see Sections 12.9 and 13.9).

Importance of Prokaryotes

Prokaryotes participate vigorously in virtually all of the organic transactions that characterize a healthy soil. Scientists are working to harness, even improve, the prokaryotes' broad range of enzymatic capabilities to help with the remediation of soils polluted by crude oil, pesticides, and various other organic toxins (see Section 18.6). The archaea are the most important group in the breakdown of hydrocarbon compounds, such as petroleum products.

Prokaryotes hold near monopolies in the oxidation or reduction of certain chemical elements in soils (see Sections 7.4 and 7.6). Some autotrophic prokaryotes obtain their energy from such inorganic oxidations, while anaerobic and facultative bacteria reduce a number of substances other than oxygen gas. Many of these biochemical oxidation and reduction reactions have significant implications for environmental quality as well as for plant nutrition. For example, through nitrogen oxidation (nitrification), selected bacteria oxidize relatively stable ammonium nitrogen to the much more mobile nitrate form of nitrogen. Likewise, certain archaeans oxidize sulfur, yielding plant-available sulfate ions, but also potentially damaging sulfuric acid (see Sections 9.6 and 13.20). Prokaryote oxidation and reduction of inorganic ions such as iron and manganese not only influence the availability of these elements to other organisms (see Section 15.5), but also help determine soil colors (see Section 4.1). A critical process in which bacteria are prominent is nitrogen fixation—the biochemical combining of atmospheric nitrogen with hydrogen to form organic nitrogen compounds usable by plants (see Section 13.10).

Cyanobacteria

Previously classified as blue-green algae, **cyanobacteria** contain chlorophyll, which allows them to photosynthesize like plants. Cyanobacteria are especially numerous in rice paddies and other wetland soils and fix appreciable amounts of atmospheric nitrogen when such lands are flooded (see Section 13.12). These organisms also exhibit considerable tolerance to saline environments and are important in forming microbiotic crusts on desert soils (Section 11.14).

Soil Actinomycetes

Actinomycetes are filamentous and often profusely branched (see Figure 11.29), appearing somewhat like tiny fungi. However, their genetic makeup and cellular properties clearly place them in the Bacteria domain—they have no nuclear membrane, are about the same diameter as other bacteria, and often break up into spores that closely resemble cocci bacterial cells.

DECOMPOSITION ACTIVITIES. Generally aerobic heterotrophs, the actinomycetes live on decaying organic matter in the soil or on compounds supplied by plants with which certain species form parasitic or symbiotic relationships. Actinomycetes undoubtedly are of great importance in the decomposition of soil organic matter and the liberation of its nutrients. They are capable of breaking even resistant compounds, such as cellulose, chitin, and phospholipids, into simpler forms. They often become dominant in the later stages of decay when the easily metabolized substrates have been used up. They are very important in the final (curing) stages of composting (see Section 12.10).

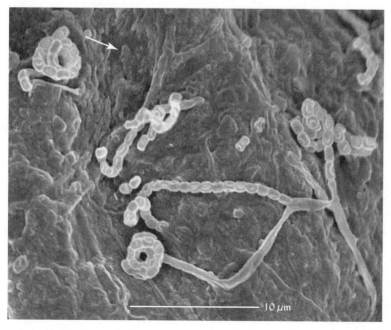

FIGURE 11.29 Strands of an actinomycete, a type of filamentous bacteria, growing on the surface of a soil biopore (an old root channel). The filaments, some breaking into the beadlike spores by which this organism reproduces, are about 0.8 μm in diameter. Some of the actinomycete filaments are embedded in mucilage of the soil pore (e.g., at arrow). The image is from 1.5 m deep in a clayey soil (poorly structured Alfisols) of a wheat field in eastern New South Wales, Australia. Almost all the wheat roots observed in this dense, hard subsoil were clustered into biopores made by roots of preceding alfalfa crops. The biopore surfaces are generally smooth, coated with illuvial clay and residues of old alfalfa roots. Although fungi commonly occupy old root channels, few were found in this soil, perhaps because of the antibiotic and chitinase secretions of the actinomycetes. This actinomycete image is quite unique in that the cryo scanning electron micrograph (SEM) was made directly from field material (frozen in liquid nitrogen), not from lab cultures. [SEM image courtesy of Margaret McCully, CSIRO Plant Industry, Canberra, Australia]

NUMBERS OF ACTINOMYCETES. Actinomycetes are often among the most abundant of the prokaryote groups (see Table 11.5), and their biomass often exceeds that of the other bacteria. Where the acidity is not too great, actinomycetes are especially numerous in soils high in humus, such as old meadows or pastures. The earthy aroma of organic-rich soils and freshly plowed land is mainly due to actinomycete-produced *geosmins*, volatile derivatives of terpene.

Penicillin and other antibiotics: http://www.biology.ed.ac.uk/research/groups/jdeacon/microbes/penicill.htm

SPECIAL ATTRIBUTES. Actinomycetes develop best in moist, warm, well-aerated soil. However, they tolerate low osmotic potential and are active in arid-region, salt-affected soils, and during periods of drought. They are generally rather sensitive to acid soil conditions, with optimum development occurring at pH values between 6.0 and 7.5. Some actinomycete species tolerate relatively high temperatures. In forest ecosystems, much of the nitrogen supply depends on actinomycetes that fix atmospheric nitrogen gas into ammonium nitrogen that is then available to plants (see Section 13.12). Many actinomycete species, especially in the genus *Streptomyces*, produce compounds that kill other microorganisms, and these "antibiotics" have become extremely important in human medicine (see Box 11.2).

11.11 CONDITIONS AFFECTING THE GROWTH OF SOIL MICROORGANISMS

Organic Matter Requirements

The addition of almost any energy-rich organic substance, including the compounds excreted by plant roots, stimulates microbial growth and activity. Certain bacteria and fungi are stimulated by specific amino acids and other growth factors found in the rhizosphere or produced by other organisms.

Generalist bacteria tend to respond most rapidly to additions of simple compounds such as starch and sugars, while fungi and actinomycetes overshadow the other microbes if added organic materials are rich in cellulose and other resistant compounds. In addition, if organic materials are left on the soil surface (as in forest litter), fungi dominate the microbial activity. Bacteria commonly play a larger role if the substrates are mixed into the soil, as by earthworms, root distribution, or tillage.

Oxygen Requirements

While most soil microorganisms are *aerobic* and use O_2 as the electron acceptor in their metabolism, some prokaryotes are *anaerobic* and use substances other than O_2 (e.g., NO_3^-, SO_4^{2-}, or other electron acceptors). *Facultative* bacteria can use either aerobic or anaerobic forms of metabolism. All three of these types of metabolism are usually carried out simultaneously in different habitats within a soil. The zone of greatest microbial activity usually occurs just a few cm below the soil surface where oxygen is high and the soil is not too dry (Figure 11.31).

Moisture and Temperature

Optimum moisture potential for higher plants (−10 to −70 kPa) is also usually best for aerobic microbes. Too high a water content will limit the oxygen supply. Microbial activity is generally greatest when temperatures are 20 to 40 °C. The warmer end of this range tends to favor actinomycetes. Ordinary soil temperature extremes seldom kill bacteria and commonly only temporarily suppress their activity. However, except for certain **psychrophilic** species, most microorganisms cease metabolic activity below 3 to 5 °C, a temperature sometimes referred to as *biological zero* (see Section 7.8).

Exchangeable Calcium and pH

Levels of exchangeable calcium and pH help determine which specific organisms thrive in a particular soil. Although in any chemical condition found in soils some bacterial species will thrive, high calcium and near-neutral pH generally result in the largest, most diverse bacterial populations. Regardless of other soil properties, bacterial diversity has been found to increase dramatically with soil pH, from very acid soils to

BOX 11.2   A POST-ANTIBIOTIC AGE ON THE HORIZON?[a]

Antimicrobial agents—commonly called antibiotics—are chemical compounds that inhibit or kill specific microorganisms. Most antibiotics kill bacteria by interfering with a few physiological reactions specific to bacteria, for instance, the generation of bacterial cell walls; therefore they are not toxic to unrelated organisms, such as humans and other animals. Certain soil bacteria and fungi have evolved the capability of producing these compounds to provide a competitive edge in their struggle for survival in the soil. Being bathed in their own chemical warfare agents, most of these spore-forming microbes have also evolved immunities to many types of antibiotics. With the advent of antibiotic "miracle drugs" in the middle of the 20th century, these same compounds were harnessed to enable humans to all but conquer infectious bacterial diseases, which up until that time were the most common cause of human deaths. The first antibiotic compound discovered (and eventually put to use as the human medicine, penicillin) was produced by a soil fungus (*Penicillium* spp.) that contaminated some laboratory petri dishes in 1928. In 1943, streptomycin was discovered, leading to the first of many antibiotic drugs synthesized by soil bacteria belonging to the genus *Streptomyces*. It is likely that you yourself are alive reading this book today because when you came down with a bacterial infection (perhaps pneumonia or a dirty wound), an antibiotic produced by a soil actinomycete (chloramphenicol, erythromycin, tetracycline, and vancomycin, to name a few) was available to save your life.

Unfortunately, the efficacy of these drugs is being rapidly eroded by the global spread of resistant strains of pathogenic bacteria. For example, enterococci and staphylococci that cause potentially fatal human diseases have now developed resistance to virtually every antibiotic drug available in the medical arsenal. What is causing this resistance that threatens to return humankind to the bad old days of the pre-antibiotic era? The answer is largely that gross overexposure to the various antibiotic drugs has exerted tremendous selection pressure for resistance in the pathogen populations. Antibiotic drugs now permeate the environment—some 18 million kg are used annually in the United States alone. Part of the problem stems from overuse and misuse of human drugs by doctors, patients (did you take *all* the pills prescribed—or did you stop when you felt better, letting the most resistant bacteria live on?), hospitals, and consumers (does bath soap really need to contain an antibiotic compound when the soap itself kills bacteria by lysing their cell walls?).

But the principal source of antibiotics released into the environment is the enormous amount of these compounds used for nonmedical purposes (Figure 11.30). In fact, in the United States, some 87% of the antimicrobial drugs produced (nearly 15 million kg/y) is devoted to nonhuman uses, most of this as a growth-promoting additive for poultry, hog, and cattle feed in industrial-style farms. Much of the antibiotic ingested by the livestock passes through the digestive tract unchanged and accumulates in the manure that is eventually spread on farm fields. Once in the soil, the compounds are known to retain their antibiotic activity even if adsorbed for long periods to clay surfaces (see Figure 8.31). In 2006, it was reported that crops (corn, onion, and cabbage) growing on soils fertilized with such manure can take up small amounts of the antibiotic chlortetracycline, presenting the

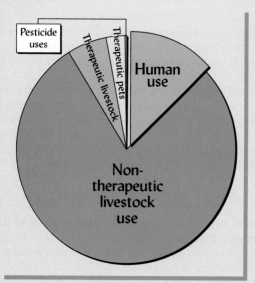

FIGURE 11.30 *Estimated antibiotic uses in the United States. Rather than curing infections in people, the greatest use of antibiotics is in livestock feed to stimulate faster meat production. Smaller proportions are used to cure sick animals (including pets). Even the 17% or so devoted to humans is thought to represent a dangerous level of over-use. [Based on data in Mellon et al. (2001)].*

possibility that the antibiotics added to livestock feed may end up causing human allergic reactions and selecting for resistance in the human digestive tract. In any case, the large and continuous antibiotic presence in industrial animal facilities and in manured soils almost certainly hastens the evolution of antibiotic resistance in bacteria, including in human pathogens. While the problem has been known by scientists since the mid-1980s, policy makers in industrial countries have been slow to realize the need to eliminate such careless use of these life-saving substances.

[a] For more information on this topic, see also Kumar et al. (2005), D'Costa et al. (2006), and Chander et al. (2007).

FIGURE 11.31 A wood fence post pulled up after several years in the soil. Lines on the post indicate the soil surface when it was set in place. Decay of wood was greatest a few cm below the soil surface, where oxygen and moisture are in plentiful supply and create a zone of maximum biological activity. Keeping this phenomenon in mind (commonly referred to as the "fence post principle"), farmers often avoid deep incorporation of organic amendments. (Photo courtesy of R. Weil)

slightly alkaline ones. Low pH allows fungi to become dominant. The effect of pH and calcium helps explain why fungi tend to dominate in forested soils, while bacterial biomass generally exceeds fungal biomass in most subhumid to semiarid prairie and rangeland soils.

11.12 BENEFICIAL EFFECTS OF SOIL ORGANISMS ON PLANT COMMUNITIES

The soil fauna and flora are indispensable to plant productivity and the ecological functioning of soils. Of their many beneficial effects, only the most important can be emphasized here.

Organic Material Decomposition

Perhaps the most significant contribution of the soil fauna and flora to higher plants is the decomposition of dead leaves, roots, and other plant tissues. Soil organisms also assimilate wastes from animals (including human sewage) and other organic materials added to soils. As a by-product of their metabolism, microbes synthesize new compounds, some of which help to stabilize soil structure and others of which contribute to humus formation. The bacteria, archaeans, and fungi assimilate some of the N, P, and S in the organic materials they digest. Excess amounts of these nutrients may be excreted into the soil solution in inorganic form either by the microflora themselves or by the nematodes and protozoa that feed on them. In this manner, the soil food web converts organically bound forms of nitrogen, phosphorus, and sulfur into mineral forms that can be taken up once again by higher plants.

Breakdown of Toxic Compounds

Many organic compounds toxic to plants or animals find their way into the soil. Some of these toxins are produced by soil organisms as metabolic by-products, some are applied purposefully by humans as agri-chemicals to kill pests, and some are deposited in the soil because of unintentional environmental contamination. If these compounds accumulated unchanged, they would do enormous ecological damage. Fortunately, most biologically produced toxins do not remain long in the soil, for soil ecosystems include organisms that not only are unharmed by these compounds but can produce enzymes that allow them to use these toxins as food.

Some toxins are **xenobiotic** (artificial) compounds foreign to biological systems, and these may resist attack by commonly occurring microbial enzymes. Soil prokaryotes and fungi are especially important in helping maintain a nontoxic soil environment by breaking down toxic compounds (see Section 18.5). The detoxifying activity of these microorganisms is by far the greatest in the surface layers of soil, where microbial numbers are concentrated in response to the greater availability of organic matter and oxygen (Figure 11.32).

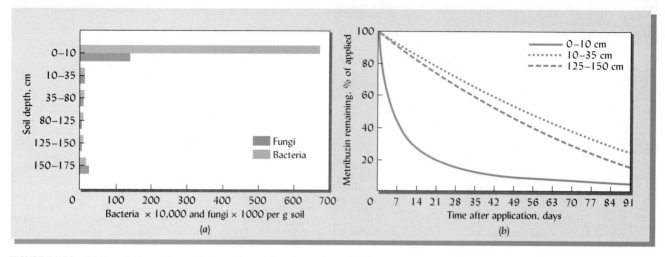

FIGURE 11.32   (a) Populations of aerobic bacteria and fungi at various depths in a Dundee soil (Aqualfs). (b) Breakdown of the herbicide metribuzin in the same soil at various depths. Soil was sampled from each depth and incubated with the herbicide. Note that the fungi and bacteria were concentrated in the upper 10 cm and that the breakdown of the herbicide was far more rapid in this layer than in the deeper layers. [From Moorman and Harper (1989)]

Inorganic Transformations

Nitrates, sulfates, and, to a lesser degree, phosphate ions are present in soils primarily due to inorganic transformations, such as the oxidation of sulfide to sulfate or ammonium to nitrate stimulated by microorganisms. Likewise, the availabilities of other essential elements, such as iron and manganese, are determined largely by microbial action. In well-drained soils, these elements are oxidized by autotrophic organisms to their higher valence states, in which forms they are quite insoluble. This keeps iron and manganese mostly in low solubility and nontoxic forms, even under fairly acid conditions (Section 15.7). If such oxidation did not occur, plant growth would be jeopardized because of toxic quantities of these elements in solution. Microbial oxidation also controls the potential for toxicity in soil contaminated with selenium or chromium (see Sections 15.9 and 18.8).

Nitrogen Fixation

The fixation of elemental nitrogen gas, which cannot be used directly by higher plants, into compounds usable by plants is one of the most important microbial processes in soils (see Section 13.10). Actinomycetes in the genus *Frankia* fix major amounts of nitrogen in forest ecosystems; cyanobacteria are important in flooded rice paddies, wetlands, and deserts; and rhizobia bacteria are the most important group for the capture of gaseous nitrogen in agricultural soils. By far the greatest amount of nitrogen fixation by these organisms occurs in root nodules or in other associations with plants.

Rhizobacteria

As pointed out in Section 11.7 and in Figure 11.28, the zone immediately around plant roots (the rhizosphere soil and the root surface itself, or **rhizoplane**) supports a dense population of microorganisms. Bacteria especially adapted to living in this zone are termed **rhizobacteria**, many of which are beneficial to higher plants (the so-called **plant growth-promoting rhizobacteria**). In nature, root surfaces are almost completely encrusted with bacterial cells, so little interaction between the soil and root can take place without some intervening microbial influence. The world of rhizobacteria is still largely uncharted, but research is beginning to uncover useful ways to take advantage of interactions that can benefit higher plants. In addition to those that ward off plant diseases (see Section 11.13), certain rhizobacteria promote plant growth in other ways, such as enhanced nutrient uptake or hormonal stimulation (for an example, see Table 11.9).

TABLE 11.9 Rice Plants Respond to Inoculation with Growth-Promoting Rhizobacteria

Rice plants were grown in pots with clay soil that was puddled and flooded with water. All pots were fertilized with adequate N fertilizer, but only some were inoculated with various strains of rhizobia or bradyrhizobia bacteria. The bacteria colonized the rice rhizosphere (hence the term rhizobacteria) and changed the physiology of the rice plant, partly by producing the plant growth hormone IAA. Inoculated roots were more efficient at nutrient uptake. Analysis of N isotope tracers showed that the rhizobacteria did not cause significant amounts of N fixation.

| Treatment | Grain yield, g/pot | Uptake of nutrients by rice plants, mg/pot | | | | IAA[a] in the rhizosphere, mg/L |
		N	P	K	Fe	
Control—no inoculation	36.7	488	111	902	18.9	1.0
Inoculated with rhizobacteria	44.3	612	134	1020	23.6	2.1
Percent change	+21	+25	+21	+13	+25	+110

[a] IAA = indol-3-acetic acid, a plant growth hormone.
Data calculated from Biswas et al. (2000)

Plant Protection

Certain soil organisms attack higher plants, but others act to protect plant roots from invasion by soil parasites and pathogens. Plant diseases and the protective action of the soil microflora will be discussed in the next section.

11.13   SOIL ORGANISMS AND DAMAGE TO HIGHER PLANTS

Although most of the activities of soil organisms are vital to a healthy soil ecosystem and economic plant production, some soil organisms affect plants in detrimental ways that cannot be overlooked. For example, soil organisms successfully compete with plants for soluble nutrients (especially for nitrogen), as well as for oxygen in poorly aerated soils. Here we focus on the soil organisms that act as herbivores, parasites, or pathogens.

Plant Pests and Parasites

Forage brassicas for control of nematodes:
http://www.abc.net.au/gardening/stories/s124457.htm

SOIL FAUNA.   The herbivorous soil fauna are by definition injurious to higher plants. Some rodents may severely damage young trees and farm crops. Snails and slugs in some climates are dreaded pests, especially of vegetables. Some ants are herbivorous (e.g., leaf-cutting ants), while others transfer aphids onto plants and so contribute to plant damage. Undoubtedly, the greatest damage to plants by soil fauna is caused by the feeding of nematodes and insect larvae. In nature, plants commonly sustain a low level of nematode and insect infestation, but under certain circumstances, infestation may be so great that the plant is killed or severely stunted. In agriculture, such infestations are often associated with a lack of proper crop rotation and insufficient organic matter addition. To prevent or diminish such infestations, large amounts of nematicide and insecticide chemicals are used in agriculture, often with unintended ecological results (see Section 11.14).

MICROFLORA AND PLANT DISEASE.   Disease infestations occur in great variety and are induced by many different organisms. Although bacterial blights and wilts are common, the fungi are responsible for the majority of soilborne plant diseases. Fungi of the genera *Pythium, Fusarium, Phytophthora,* and *Rhizoctonia* are especially prominent as soilborne agents of plant diseases described by such symptoms as *damping-off, root rots, leaf blights,* and *wilts.* Soils are easily infested with disease organisms, which are transferred from soil to soil by many means, including tillage or planting implements, transplant material, manure from animals that were fed infected plants, soil erosion, and windborne fungal spores and bacteria. Once a soil is infested, it is apt to remain so for a long time. Splash during rainstorms from an unmulched soil surface is a major cause of plant disease initiation and spread.

DELETERIOUS RHIZOBACTERIA. Some bacteria that live in the rhizosphere or on the rhizoplane inhibit root growth and function by various noninvasive chemical interactions. These nonparasitic **deleterious rhizobacteria** can cause stunting, wilting, foliar discoloration, nutrient deficiency, and even death of affected plants, but often the effects are subtle and difficult to detect. Their buildup may contribute to yield declines during long-term monoculture and aggravate problems in planting new trees in old orchards. On the other hand, by management that favors the deleterious rhizobacteria associated with certain weeds, scientists hope to be able to reduce weed seed germination and seedling growth and thereby reduce the use of herbicide (weed-killer) sprays on cropland and rangeland.

Plant Disease Control by Soil Management[8]

Prevention is the best defense against soilborne diseases. Strict quarantine systems will restrict the transfer of soilborne pathogens from one area to another. Crop rotation can be very important in controlling a disease by growing nonsusceptible plants for several years between susceptible crops. Tillage may help by burying plant residues on which fungal spores might overwinter. However, disease problems are often lessened in no-tillage systems in which the soil surface remains mulched with plant residues that maintain a diverse soil community and prevent the splashing of soil onto foliage by rain or irrigation. Residues from certain green manure crops have been shown to chemically inhibit specific plant diseases. Direct management of soil physical and chemical properties can also be useful in disease control.

SOIL pH. Regulation of soil pH is effective in controlling some diseases. For example, keeping the pH low (<5.2) can control both the actinomycete-caused *potato scab* and the fungal disease of turfgrass known as *spring dead spot*. Raising soil pH to about 7.0 can control *clubroot* disease in the cabbage family, because the spores of the fungal pathogen germinate poorly, if at all, under neutral to alkaline conditions.

SOIL NUTRIENTS. Healthy, vigorous plants usually can resist or outgrow diseases better than weaker plants, so provision of an optimal level of balanced nutrition is an important step in disease management. High levels of nitrogen fertilization tend to increase plants' susceptibility to fungal diseases; high levels of ammonium (as compared to nitrate) nitrogen especially increase wilt diseases caused by *Fusarium* fungi. However, potassium fertilizers often reduce fungal disease severity, as do relatively high levels of calcium and manganese. Nutrient imbalances and micronutrient deficiencies can make plants especially susceptible to attack.

ORGANIC TOXINS. As an alternative to the use of synthetic, broad-spectrum fungicides or fumigants, certain natural organic anti-fungals can be introduced to the soil via microbial breakdown of organic amendments. An example is the rotation of cauliflower with broccoli that has been shown, even in fields infested with the disease-causing fungus, to provide practical control of *verticillium wilt*, a serious disease of cauliflower. The broccoli leaf residues left after harvest are tilled into the soil. There they break down to release volatile compounds specifically toxic to the *Verticillium dahliae* fungus, providing a level of disease control in the following cauliflower crop equal to that achieved by synthetic fumigants.

SOIL PHYSICAL PROPERTIES. Soil compaction often aggravates fungal root diseases by slowing root growth, inducing more root excretions that attract the pathogens, and by promoting wet, poorly aerated conditions. Wet, cold soils favor some seed rots and seedling diseases such as *damping-off*. Good drainage and planting on ridges can help control these diseases. High soil temperature can be used to control a number of pathogens. **Solarization,** the use of sunlight to heat soil under clear plastic sheeting, is a practical way to partially sterilize the upper few cm of soil in some field situations. Steam or chemical sterilization is a practical method of treating greenhouse potting media.

---

[8] For a general overview of disease control by management of the soil ecosystem, see Stone et al. (2004); for a review of the scientific literature on disease and pest management using organic amendments, see Litterick and Harrier (2004).

It should be remembered, however, that sterilization kills beneficial microorganisms, such as mycorrhizal fungi, as well as pathogens, and so may do more harm than good!

Disease-Suppressive Soils

Research on plant diseases ranging from *Phytophthora root rot* in eucalyptus forests to *Fusarium wilts* in banana plantations has documented the existence of **disease-suppressive soils,** in which a disease fails to develop *even though both the virulent pathogen and a susceptible host are present.* The reason that certain soils become disease-suppressive is not entirely understood, but much evidence suggests that the pathogenic organisms are inhibited by **antagonism** from beneficial bacteria and fungi. Two broad types of disease suppression are recognized: general and specific.

Rhizobacteria, underground biocontrol allies?
http://www.ars.usda.gov/is/AR/archive/oct98/rhizo1098.htm

GENERAL SUPPRESSION.   General disease suppression is caused by high levels of overall microbial activity in a soil, especially at times critical in the development of a disease, such as when the pathogenic fungus is generating propagules or preparing to penetrate the plant cells. The presence of particular organisms is less important than the total level of activity. The mechanisms responsible for general suppression are thought to include (1) competition by beneficial microorganisms in the rhizosphere for carbon (energy) sources, (2) competition for mineral nutrients (such as nitrogen and iron), (3) colonization and decomposition of pathogen propagules (e.g., spores), (4) antibiotic production by varied actinomycete and fungal populations (see Section 11.10), and (4) lack of suitable root infection sites due to surface colonization by beneficial bacteria (see Figure 11.28) or previous infection by beneficial mycorrhizal fungi. In natural systems, the highly organic litter layer often provides an environment of such high microbial activity that most pathogens cannot compete. In agricultural systems, general suppression can often be encouraged by the addition of large amounts of decomposable organic matter from composts, manures, and cover crop residues, and by developing a "litter layer" through no-till and mulching techniques.

SPECIFIC SUPPRESSION.   Specific suppression is attributable to the actions of a single species or a narrow group of microorganisms that inhibit or kill a particular pathogen. The effective presence of the specific suppressing organism may result from the same types of organic matter management just described or from introduction of an inoculum containing high numbers of the desired organism.

In some cases, specific disease suppressiveness has developed through long-term crop monoculture in which the buildup of the pathogen during the first few years is eventually overshadowed by a subsequent buildup of specific organisms antagonistic to the pathogen (Figure 11.33). Specific organisms known to be antagonistic to pathogens

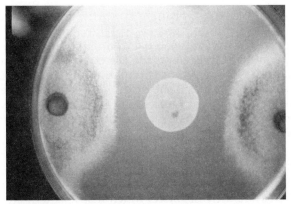 

FIGURE 11.33   The biological basis of a disease-suppressive soil. (*Left*) A colony of certain *Pseudomonas* bacteria (*center of plate*) produces an antibiotic toxic to *Gaeumannomyces graminis* (the fungal pathogen that causes take-all disease), preventing the pathogen colonies from growing to the center of the plate. (*Right*) Plots in eastern Washington that grew monoculture wheat for 15 years and developed high populations of organisms antagonistic to the take-all pathogen. In the 15th year of the study, the entire field was inoculated with *G. graminis* for experimental purposes, but the disease developed (seen as light-colored, prematurely ripened plots) only where the soil was fumigated prior to the inoculation. The fumigation killed most of the antagonistic organisms, leaving the pathogenic fungi free to infect the wheat plants. (Photos by R. J. Cook; courtesy of the American Phytopathological Society)

BOX 11.3 CHOOSING SIDES IN THE MICROBE WARS

Recent advances in our understanding of microbial genetics, combined with new techniques for transferring genetic material among different species of microorganisms, have opened many interesting and promising opportunities to harness the antagonisms long observed among microorganisms.

Seed treatments are now commercially available that use antagonistic soil bacteria instead of chemical fungicide to protect seed from rot-causing soil pathogens. Strains of *Pseudomonas fluorescens* have been specially selected and enhanced to inhabit rhizosphere soil and help protect plant roots from soilborne diseases. When a preparation of these bacteria is applied as seeds are sown, the seeds germinate safely, even in soil known to be infested with seed-rotting fungi. Seeds sown without the treatment decay before they can germinate (Figure 11.34). Widespread use of such biologically based seed protection in the place of conventional fungicidal chemicals could reduce the loading of toxic compounds into the environment.

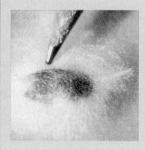 

FIGURE 11.34 *Seeds protected with antagonistic bacteria remain healthy (right), while pathogenic fungi infect unprotected seed (left). (Photo courtesy of Ecogen, Inc.)*

include *Trichoderma viride* fungi and certain fluorescent *Pseudomonas* bacteria, which produce antibiotics specific against pathogens or produce compounds that bind so tightly with iron that the pathogen spores cannot get enough of this nutrient to germinate. Treating plant seeds with specific microbes that fend off seed rot pathogens may be one of the most practical applications of the inoculation approach (Box 11.3). Despite the existence of these and many other commercial products containing beneficial microorganisms, we should remember the quote from Beijerinck (Section 11.10) and not place too much faith in the specific inoculation approach, because successful suppression is usually limited by appropriate conditions in the soil rather than lack of a particular organism.

USE OF COMPOSTS.   Horticulturalists have been able to control *Fusarium* diseases in containerized plants by replacing traditional potting mixes with growing media made mainly from certain well-aged **composts**. Apparently, large numbers of beneficial antagonistic organisms colonize the organic material during the final stages of composting (see Section 12.10), and the stabilized organic substrate stimulates the activity of indigenous beneficial organisms without stimulating the pathogens. Similar success in practical disease suppression has been experienced with the use of composted materials on turfgrass, especially for replacing peat (which is relatively inert and does not stimulate disease suppression) in topdressing golf course greens (see Figure 11.35).

INDUCED SYSTEMIC RESISTANCE.   The role of soil ecology in protecting plants from disease is not limited to belowground infections. Beneficial rhizobacteria have an intriguing mode of action called **induced systemic resistance**, which helps plants ward off infection by diseases or insect pests both above and below ground. The process begins when a plant root system is colonized by beneficial rhizobacteria that cause the accumulation of a signaling chemical. The chemical signal is translocated up to the shoot, where it induces leaf cells to mount a chemical defense against a specific pathogen, even before the pathogen has arrived on the scene. When the pathogen (perhaps a fungal spore) does arrive on the leaf, its infection process is aborted almost before it can begin. In many cases studied so far on crop plants, the resistance-inducing organism

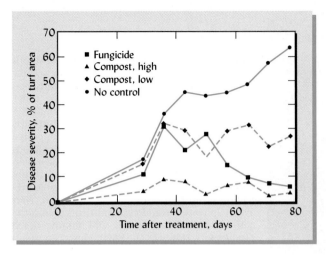

FIGURE 11.35 Topdressing with compost may be a practical, non-toxic means of suppressing dollar spot disease on bentgrass putting greens. The disease, caused by the fungus *Sclerotinia homoeocarpa*, was controlled as well or better by the high rate of compost (4900 kg/ha topdressed every 3 weeks) as by the synthetic fungicide (Chlorthalonil, sprayed on every 2 weeks). Even the lower rate (1200 kg/ha) of compost provided some control of the disease. All the turfgrass plots were inoculated with the disease organism. The researchers suggest that topdressing with compost provided a general type of disease suppression since they found little difference among composts made from many different materials. The means from 2 years' data are shown. [Based on data in Boulter et al. (2002)]

has been a *Pseudomonas* or *Serratia* bacteria. This mechanism has been shown to effectively reduce damages by numerous fungal, bacterial, and viral pathogens and several leaf-eating insect pests.

These examples merely hint at the potential that exists for controlling plant diseases and pests through ecological management rather than applications of toxic chemicals.

11.14 ECOLOGICAL RELATIONSHIPS AMONG SOIL ORGANISMS

Mutualistic Associations

We have already mentioned a number of mutually beneficial associations between plant roots and other soil organisms (e.g., mycorrhizae and nitrogen-fixing nodules) and between several microorganisms (e.g., lichens). Other examples of such associations abound in soils. For example, photosynthetic algae reside within the cells of certain protozoans. Several types of associations, among them algal-fungal associations on or in soils and rocks, are very important cyclers of nutrients and producers of biomass in desert ecosystems. Next, we will briefly consider the nature of such associations.

Microbiotic Crusts[9]

Soil biological crusts:
http://www.soilcrust.org/

In relatively undisturbed arid- and semiarid-region ecosystems, where the vegetation cover is quite patchy, it is common to find an irregular, usually dark-colored crust covering the soil in the areas between clumps of grasses and shrubs. In many cases this forms a sort of miniature landscape of tiny, jagged pinnacles only a few centimeters tall (see Figure 11.36). This crust is not at all like the physical-chemical crusts associated with degraded soils that have a hard, smooth surface seal (see Section 4.6). Rather, the **microbiotic crusts** of arid rangelands are intricate living systems that greatly benefit the associated natural vegetation communities. They consist of mutualistic associations that usually include algae or cyanobacteria along with fungi, mosses, bacteria, and/or liverworts. An intact microbiotic crust is considered a sign of a healthy ecosystem.

Desert ecology- From biological crust to dust (audio report):

http://www.npr.org/templates/story/story.php?storyId=5415315

Such crusts provide considerable protection against erosion by wind and water by binding soil particles together, protecting the soil from raindrop impact, and increasing surface roughness, which reduces wind velocity. They also improve arid-region ecosystem productivity by (1) helping to conserve and cycle nutrients, (2) increasing nitrogen supplies via the nitrogen-fixing activities of the cyanobacteria, (3) enhancing water supplies in some cases by increasing infiltration and reducing evaporation, and (4) contributing

---

[9] Other names used to refer to these microbiotic crusts include *biological crusts*, *cryptograms*, *cryptobiotic crusts*, *microfloral crusts*, and *microphytic crusts*. All refer to the same thing. See Belnap (2003) for a brief overview and further reference citations on the ecology of microbial crusts. For a technical review of scientific knowledge and management options, see Belnap et al. (2001).

FIGURE 11.36 Tiny pinnacles of a microbiotic crust in Arches National Monument, Utah, seem to reflect the larger pinnacles of an arid landscape. These crusts consist of algae, cyanobacteria, fungi, and other organisms living together in a mutualistic relationship. The inset shows a scanning electron micrograph of cyanobacteria filaments that make up the backbone of many crusts. Microbiotic crusts typically cover the soil surface in the unvegetated patches between clumps of desert shrubs and grasses. The crusts provide considerable protection against erosion by wind and water. They also help conserve and cycle nutrients, add nitrogen, enhance water supplies, and improve desert productivity. However, the fragile crusts can be easily destroyed by wheels, feet, and hooves. [Large photo courtesy of Ben Waterman; inset courtesy of Jayne Belnap (U.S. Geologic Service, Moab, Utah) and John Gardner (Brigham Young University, Provo, Utah)]

Biological Soil Crusts: Ecology and Management: http://www.id.blm.gov/publications/crust/part1.pdf

to net organic matter production by crust photosynthesis, which may continue during environmental conditions that inhibit photosynthesis by higher plants in the ecosystem. The filamentous cyanobacteria make a particularly important contribution to these functions, as they not only photosynthesize but also fix from 2 to 40 kg/ha of nitrogen annually and form sticky polysaccharide coatings or sheaths that catch nutrient-rich dust and bind soil particles. Unfortunately, the crusts can be easily destroyed by trampling or burial under windblown soil and are very slow to reestablish. Off-road vehicles, off-trail hiking, and overgrazing are especially damaging to these important ecosystem components.

Effects of Management Practices on Soil Organisms

Changes in environment affect both the number and kinds of soil organisms. Clearing forests or grasslands for cultivation drastically changes the soil environment. Monocultures or even common crop rotations greatly reduce the number of plant species and so provide a much narrower range of plant materials and rhizosphere environments than nature provides in forests or grasslands.

While agricultural practices have different effects on different organisms, a few generalizations can be made (Table 11.10). For example, some agricultural practices

Case Studies and Practices for Improved Soil Biological Management: http://www.fao.org/landwater/agll/soilbiod/cases.stm

TABLE 11.10 Soil-Management Practices and the Diversity and Abundance of Soil Organisms

Note that the practices that tend to enhance biological diversity and activity in soils are also those associated with efforts to make agricultural systems more sustainable.

Decreases biodiversity and populations	Increases biodiversity and populations
Fumigants	Balanced fertilizer use
Nematicides	Lime on acid soils
Some insecticides	Proper irrigation
Compaction	Improved drainage and aeration
Soil erosion	Animal manures and composts
Industrial wastes and heavy metals	Domestic (clean) sewage sludge
Moldboard plow–harrow tillage	Reduced or zero tillage
Monocropping	Crop rotations
Row crops	Grass–legume pastures
Bare fallows	Cover crops or mulch fallows
Residue burning or removal	Residue return to soil surface
Plastic mulches	Organic mulches

TABLE 11.11 **Effect of Tillage Systems on Biomass Carbon of Microbial and Faunal Groups in Soil**

Researchers in Georgia grew grain sorghum in summer and a rye cover crop each winter on a sandy loam (Kanhapludults). With plow tillage, plant residues were mixed into the soil, but with no-till residues were left as surface mulch. Decomposer, microphytic feeder, and detritivore functions were dominated by larger organisms (fungi, microarthropods, and earthworms) in the no-till system, while smaller organisms (bacteria, protozoa, and nematodes) were more prominent in the plowed system.

| | | | | Carbon, kg/ha[a] | | | | |
| | | | | Nematodes | | | | |
Tillage	Fungi	Bacteria	Protozoa	Fungivores	Bacterivores	Microarthropods	Enchytraeids	Earthworms
No-till	360	260	24	0.14	0.82	1.31	5.55	60
Plowed	240	270	39	0.47	1.27	0.49	4.79	21

[a] Depth of sampling was 0–5 cm for microflora and arthropods, 0–21 cm for nematodes, 0–15 cm for worms.
Calculated from data of various sources presented by Beare (1997).

(e.g., extensive tillage and monoculture) generally reduce the diversity of soil organisms as well as the abundance (number) of individuals. However, monoculture may *increase* the population of a few species.

Adding lime and fertilizers (either organic or inorganic) to an infertile soil generally will increase microbial and fauna activity, largely due to the increase in the plant biomass that is likely to be returned to the soil as roots, root exudates, and shoot residues. Tillage, on the other hand, is a drastic disturbance of the soil ecosystem, disrupting fungal hyphae networks and earthworm burrows, as well as speeding the loss of organic matter. Reduced tillage therefore tends to increase the role of fungi at the expense of the bacteria and usually increases overall organism numbers as organic matter accumulates (Table 11.11). Addition of animal manure or compost stimulates even higher microbial and faunal (especially earthworm) activity.

Pesticides are highly variable in their effects on soil ecology (see Section 18.4). Soil fumigants and nematicides can sharply reduce organism numbers, especially for fauna, at least on a temporary basis. On the other hand, application of a particular pesticide often stimulates the population of a specific microorganism, either because the organism can use the pesticide as food or, more likely, because the predators of that organism have been killed. Figure 11.37 illustrates how a practice that affects one group of organisms will likely affect other groups as well and will eventually impact the productivity and functioning of the whole soil ecosystem. It is wise to remember that the interrelationships among soil organisms are intricate, and the effects of any perturbation of the system are difficult to predict.

Links Between Communities Above- and Belowground[10]

Structure and function of soil biota on anthropogenic landscapes, Baltimore Ecosystem Study: http://www.beslter.org/frame4-page_3a_02.html

The communities of plants and animals we see aboveground greatly influence—and are profoundly influenced by—the communities belowground that we rarely see. The connections and interactions between them are both direct and indirect. Direct effects include the damage done to plants by pathogenic soil fungi, root-feeding nematodes, and the like. Direct effects also include the positive influences of mycorrhizae and beneficial rhizosphere bacteria, including those that cause the induced systemic resistance just described. Indirect effects include the complex feeding activities within the soil food web that eventually release nutrients that plants can use. In these and many other ways, the soil food web alters the types and productivity of plants in the aboveground world and, in so doing, affects the food supply and habitat for aboveground animals as well.

---

[10] For an introduction to how soil ecology influences plant invasiveness, see Wolfe and Klironomos (2005). For an authoritative treatise on the aboveground–belowground interactions, see Wardle (2002).

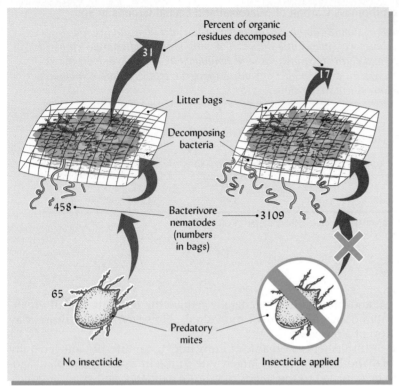

FIGURE 11.37 The indirect effects of insecticide treatment on the decomposition of creosote bush litter in desert ecosystems. Litter bags filled with creosote bush leaves and twigs were buried in desert soils in Arizona, Nevada, and California, either with or without an insecticide (chlordane) treatment. The insecticide killed virtually all the insects and mites. Without predatory mites to hold them in check, bacterivore nematodes multiplied rapidly and devoured a large portion of the bacterial colonies responsible for litter decomposition and nutrient cycling. Thus the insecticide reduced the rate of litter decomposition nearly in half, not by any direct effect on the bacteria, but by the indirect effect of killing the predators of their predators. [Data calculated from Whitford et al. (1982)]

In turn, inputs from the aboveground communities influence the communities in the soil. From the beginning of this chapter (Figure 11.1) we have seen that the energy that drives the soil food web comes from aboveground photosynthesis via the organic carbon in plant litter, rhizodeposition, and herbivore excretions. Therefore, the numbers and activities of the belowground organisms are highly responsive to the amount of such inputs, especially plant litter. However, it should also be noted that the *type* and *quality* of plant litter produced by the aboveground community (see also Section 12.3) has an enormous impact on the abundances of the various creatures living in the soil (Table 11.12).

It is no wonder that ecologists are coming to recognize that ecosystems on land can be understood only when sufficient attention is paid to the world beneath the land's surface. For example, conservation biologists urgently need to learn what makes some exotic plants so invasive that they destroy native plant communities. It turns out that at least part of the answer may be found belowground in the community of organisms that colonize the plant rhizospheres (see Figure 11.38).

TABLE 11.12 **Type of Forest Litter Influences the Abundances of Selected Groups of Fauna**

Leaves fallen from two species of forest plants were placed in mesh litter bags (100 g dry matter per bag) and pinned to the soil surface in a temperate forest in New Zealand. After 279 days the more nutrient-rich litter supported greater numbers of most faunal groups, except predatory nematodes. The activities of these and other soil organisms will in turn influence the types and productivity of plants in the aboveground forest community.

Source of litter plant species	Nutrients in litter (%)		Numbers of organisms/litter bag			
	N	P	Microbial feeding nematodes	Predatory nematodes	Tardigrades	Coleoptera beetles
Metrosideno umbellata	0.35	0.03	6,600	210	35	4
Aristotelia serrata	2.94	0.24	12,800	73	670	364

[Calculated from selected data in Wardle et al. (2006)]

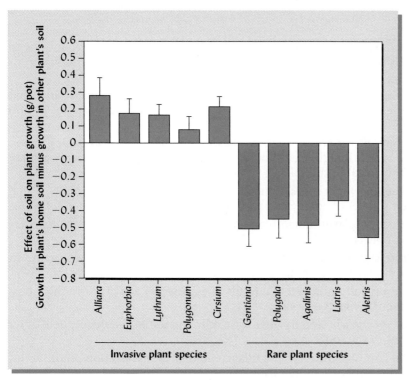

FIGURE 11.38 Soil feedback responses help explain why some plants become invasive and others rare. Invasive plants (from Canadian grasslands) did well with soil in which they had already grown (their "home soil"). Rare plants did not. Research shows that certain plants quickly accumulate species-specific pathogenic fungi and bacteria in their rhizosphere, making these plants such weak competitors that they eventually become rare species. Other plants are strong competitors and become invasive because they avoid a rapid buildup of specific soil pathogens in their rhizosphere. The invasive plants may also efficiently exploit mycorrhizal fungi, even competing for nutrients by connecting to mycorrhizal hyphae already associated with neighboring plants' roots. In this manner, the ecology belowground can influence the diversity and productivity of plant species we see aboveground. [Graph modified from Klironomos (2002)]

11.15   CONCLUSION

The soil is a complex ecosystem with a highly diverse community of organisms that are vital to the cycle of life on Earth. Soil organisms incorporate plant and animal residues into the soil and digest them, returning carbon dioxide to the atmosphere, where it can be recycled through higher plants. Simultaneously, they create humus, the organic constituent so important to good physical and chemical soil conditions. During digestion of organic substrates, they release essential plant nutrients in inorganic forms that can be absorbed by plant roots or be leached from the soil. They also mediate the redox reactions that influence soil colors, nutrient cycling, and the production of gases that contribute to global warming.

Animals, particularly earthworms, ants, and termites, mechanically incorporate residues into the soil and leave open channels through which water and air can flow. As such, they are examples of soil ecosystem engineers that change the soil environment for all its inhabitants and create niches in which other organisms can live. Microorganisms such as fungi, archaeans, and bacteria are responsible for most organic decay, although their activity is greatly influenced by the soil fauna. Certain of the microorganisms form symbiotic associations with higher plants, playing special roles in plant nutrition and nutrient cycling. Competition for mineral nutrients among soil microbes, and between these organisms and higher plants, can result in plant nutrient deficiencies. Microbial requirements are factors in determining the success of most soil-management systems. A high level of general microbial activity fed by organic inputs can help suppress plant pathogens. Several specific fungi and bacteria produce antibiotic compounds that help them compete and also can inhibit plant pathogens, as well as form the basis for life-saving human drugs. Scientific understanding is just beginning to scratch the surface of the complex communities beneath our feet.

The soil community must have energy and nutrients if it is to function efficiently. To obtain these, soil organisms break down organic matter, aid in the production of humus, and leave behind compounds that are useful to higher plants. Organic matter, its effects on soil behavior, and its decomposition are topics of the next chapter.

STUDY QUESTIONS

1. What is *functional redundancy*, and how does it help soil ecosystems continue to function in the face of environmental shocks such as fire, clear-cutting, or tillage?

2. In the example illustrated in Table 11.11, identify the organisms, if any, that play the roles of *primary producers, primary consumers, secondary consumers*, and *tertiary consumers*.

3. Describe some of the ways in which mesofauna play significant roles in soil metabolism even though their biomass and respiratory activity is only a small fraction of the total in the soil.

4. What are the four main types of metabolism carried out by soil organisms relative to their sources of energy and carbon?

5. What role does O_2 play in aerobic metabolism? What elements take its place under anaerobic conditions?

6. A *mycorrhiza* is said to be a symbiotic association. What are the two parties in this symbiosis, and what are the benefits derived by each party?

7. In what ways is soil improved as a result of earthworm activity? Are there possible detrimental effects as well?

8. What is the *rhizosphere*, and in what ways does the soil in the rhizosphere differ from the rest of the soil?

9. Explain and compare the effects of tillage and manure application on the abundance and diversity of soil organisms.

10. What is *induced systemic resistance*, and how does it work?

11. What is a disease-suppressive soil? Explain the difference between general and specific forms of suppression.

12. Discuss the value and limitations of using specific inoculants for (a) mycorrhizae and (b) disease suppression. For each type of inoculation, describe a situation for which the chances would be very good for improving plant growth.

13. What are the main food web roles played by nematodes, and how can we visually (with a microscope) distinguish among nematodes that play these roles?

14. In what ways are actinomycetes like other groups of bacteria, and in what ways are they special?

15. Through appropriate extractions and counting you determine that there are 58 nematodes in a 1-gram sample of soil. How many nematodes would occur in 1.0 m^2 area of this soil? In 1 hectare? Assume the samples came from the upper 10 cm of soil with a bulk density = 1.3 Mg/m^3.

16. Explain with two soil examples the concept of an *ecosystem engineer*.

REFERENCES

Al-Karaki, G. N. 2006. "Nursery inoculation of tomato with arbuscular mycorrhizal fungi and subsequent performance under irrigation with saline water," *Scientia Horticulturae*, **109**:1–7.

Alper, J. 1998. "Ecosystem 'engineers' shape habitats for other species," *Science*, **280**: 1195–1196.

Amador, J. A., and J. H. Gorres. 2005. "Role of the anecic earthworm lumbricus terrestris L. in the distribution of plant residue nitrogen in a corn (zea mays)-soil system," *Appl. Soil Ecol.*, 30:203–214.

Baskin, Y. 2005. *Under Ground: How Creatures of Mud and Dirt Shape Our World* (Washington, D.C.: Island Press), 237 pp.

Beare, M. H. 1997. "Fungal and bacterial pathways of organic matter decomposition and nitrogen mineralization in arable soils," in L. Brussaard and R. Ferrera-Cerrato (eds.), *Soil Ecology in Sustainable Agricultural Systems* (Boca Raton, Fla.: Lewis Publishers).

Belnap, J. 2003. "The world at your feet: Desert biological soil crusts," *Frontiers of Ecology and the Environment*, 1:181–189.

Belnap, J., J. H. Kaltenecker, R. Rosentreter, J. Williams, S. Leonard, and D. Eldridge. 2001. "Biological soil crusts: Ecology and management." Technical Reference 1730-2. United States Department of the Interior, Bureau of Land Management, Denver, CO.110. http://www.id.blm.gov/publications/crust/part1.pdf.

Biswas, J. C., J. K. Ladha, and F. B. Dazzo. 2000. "Rhizobia inoculation improves nutrient uptake and growth of lowland rice," *Soil Sci. Soc. Amer. J.*, **64**:1644–1650.

Blanco-Canqui, H., R. Lal, W. M. Post, R. C. Izaurralde, and M. J. Shipitalo. 2007. "Soil hydraulic properties influenced by corn stover removal from no-till corn in Ohio," *Soil Tillage Res.*, **92**:144–155.

Boulter, J. I., G. J. Boland, and J. T. Trevors. 2002. "Evaluation of composts for supression of dollar spot (*sclerotinia homoeocarpa*) of turfgrass," *Plant Dis.*, **86**:405–410.

Bowen, G. D., and A. D. Rovira. 1999. "The rhizosphere and its management to improve plant growth," *Advances in Agronomy*, **66**:1–102.

Brown, E. D. 2006. "Microbiology: Antibiotic stops 'ping-pong' match," *Nature*, **441**:293–294.

Chander, Y., S. C. Gupta, S. M. Goyal, and K. Kumar. 2007. "Antibiotics: Has the magic gone?" *Journal of the Science of Food and Agriculture*, **87**:739–742.

Chino, M. 1976. "Electron microprobe analysis of zinc and other elements within and around rice root growth in flooded soils," *Soil Sci. and Plant Nut. J.*, **22**:449.

Coleman, D. C., D.A.J. Crossley, and P. F. Hendrix. 2004. *Fundamentals of Soil Ecology* (London: Elsevier Academic Press). 386 pp.

de Vleeschauwer, D., and R. Lal. 1981. "Properties of worm casts under secondary tropical forest regrowth," *Soil Sci.*, **132**:175–181.

D'Costa, V. M., K. M. McGrann, D. W. Hughes, and G. D. Wright. 2006. "Sampling the antibiotic resistome," *Science*, **311**:374–377.

Edwards, C. A. (ed.). 2004. *Earthworm Ecology* (Boca Raton, Fla.: CRC Press), 448 pp.

Edwards, C. A., and N. Q. Arancon. 2004. "Interactions among organic matter, earthworms and microorganisms in promoting plant growth," in F. Magdoff and R. R. Weil (eds.), *Soil Organic Matter in Sustainable Agriculture* (Boca Raton, Fla.: CRC Press).

Griffiths, B. S. et al. 2000. "Ecosystem response of pasture soil community to fumigation-induced microbial diversity reductions: An examination of the biodiversity–ecosystem function relationship," *Oikos*, **90**:279–294.

Hendrix, P. F., and P. J. Bohlen. 2002. "Exotic earthworm invasions in North America: Ecological and policy implications," *BioScience*, **52**:801–811.

Ingham, E. R., and W. G. Thies. 1996. "Responses of soil food web organisms in the first year following clearcutting and application of chloropicrin to control laminated root rot," *Applied Soil Ecology*, **3**:35–47.

Jouquet, P., J. Dauber, J. Lagerlöfe, P. Lavelle, and M. Lepage. 2006. "Soil invertebrates as ecosystem engineers: Intended and accidental effects on soil and feedback loops," *Appl. Soil Ecol.*, **32**:153–164.

Klironomos, O. N. 2002. "Feedback with soil biota contributes to plant rarity and invasiveness in communities," *Nature*, **417**:67–70.

Kumar, K., S. C. Gupta, S. K. Baidoo, Y. Chander, and C. J. Rosen. 2005. "Antibiotic uptake by plants from soil fertilized with animal manure," *J. Environ Qual*, **34**:2082–2085.

Lal, R. 1987. *Tropical Ecology and Physical Edaphology* (New York: Wiley).

Lavelle, P. 1997. "Faunal activities and soil processes: Adaptive strategies that determine ecosystem function," *Advances in Ecological Research*, **27**:93–132.

Lavelle, P., D. Bignell, M. Lepage, V. Wolters, P. Roger, P. Ineson, O. W. Heal, and S. Dhillion. 1997. "Soil function in a changing world: The role of invertebrate ecosystem engineers," *European Journal of Soil Biology*, **33**:159–193.

Litterick, A. M., and L. Harrier. 2004. "The role of uncomposted materials, composts, manures, and compost extracts in reducing pest and disease incidence and severity in sustainable temperate agricultural and horticultural crop production: A review," *Critical Reviews in Plant Sciences*, **23**:453–479.

Macfadyen, A. 1963. in J. Doeksen and J. van der Drift (eds.), *Soil Organisms* (Amsterdam: North-Holland).

Mellon, M., C. Benbrook, and K. L. Benbrook. 2001. *Hogging It: Estimates of Antimicrobial Abuse in* Livestock. (Cambridge, Mass.: Union of Concerned Scientists). www.ucsusa.org/publications.

Menge, J. A. 1981. "Mycorrhizae agriculture technologies," in *Background Papers for Innovative Biological Technologies for Lesser Developed Countries*, Paper No. 9., Office of

Technology Assessment Workshop, Nov. 24–25, 1980. (Washington, D.C.: U.S. Government Printing Office), pp. 383–424.

Mooreman, T. B., and S. S. Harper. 1989. "Transformation and mineralization of Metribuzin in surface and subsurface horizons of a Mississippi Delta soil," *J. Environ. Qual.*, **18**:302–306.

Nardi, J. B. 2003. *The World Beneath Our Feet: A Guide to Life in the Soil* (New York: Oxford University Press), 224 pp.

Paul, E. A. (ed.) 2006. *Soil microbiology, ecology and biochemistry.* (San Diego: Academic Press), 552 pp.

Reynolds, L. B., J. W. Potter, and B. R. Ball-Coelho. 2000. "Crop rotation with *Tagetes* sp. is an alternative to chemical fumigation for control of root-lesion nematodes," *Agronomy J.*, **92**:957–966.

Rovira, A. D., R. C. Foster, and J. K. Martin. 1979. "Origin, nature and nomenclature of the organic materials in the rhizosphere," in J. L. Harley and R. S. Russell (eds.), *The Soil–Root Interface* (New York: Academic Press).

Scullen, J., and A. Malik. 2000. "Earthworm activity affecting organic matter, aggregation and microbial activity in soils restored after opencast mining for coal," *Soil Biology and Biochemistry*, **32**:119–126.

Smith, S. E., and D. J. Read. 1997. *Mycorrhizal Symbiosis*, 2nd ed. (San Diego, Calif.: Academic Press), 605 pp.

Stadler, B., A. Schramm, and K. Kalbitz. 2006. "Ant-mediated effects on spruce litter decomposition, solution chemistry, and microbial activity," *Soil Biol. Biochem.*, **38**:561–572.

Stone, A. G., S. J. Scheuerell, and H. M. Darby. 2004. "Suppression of soil-borne fungal diseases in field agricultural systems: Organic matter management, cover cropping, and cultural practices," in F. Magdoff and R. R. Weil (eds.), *Soil Organic Matter in Sustainable Agriculture* (Boca Raton, Fla.: CRC Press).

Sylvia, D. M., J. J. Fuhrmann, P. G. Hartel, and D. A. Zuberer. 2005. *Principles and Applications of Soil Microbiology* (Upper Saddle River, N.J.: Prentice Hall), 640 pp.

Tate, R. L., III. 2001. *Soil microbiology,* 2nd ed., (New York: John Wiley), 536 pp.

Torsvik, V., L. Ovreas, and T. F. Thingstad. 2002. "Prokaryotic diversity—Magnitude, dynamics, and controlling factors," *Science*, **296**:1064–1066.

Tugel, A. J., and A. M. Lewandowski, eds. 1999. *Soil Biology Primer.* (Ames, Iowa: Natural Resource Conservation Service Soil Quality Institute). www.statlab.iastate.edu/survey/SQI/SoilBiologyPrimer/index.htm.

Wardle, D. A. 2002. *Communities and Ecosystems: Linking the Aboveground and Belowground Components* (Princeton, N.J.: Princeton University Press).

Wardle, D. A., G. W. Yeates, G. M. Barker, and K. I. Bonner. 2006. "The influence of plant litter diversity on decomposer abundance and diversity," *Soil Biol. Biochem.*, **38**:1052–1062.

Weil, R. R., and W. Kroontje. 1979. "Organic matter decomposition in a soil heavily amended with poultry manure," *J. Environ. Qual.*, **8**:584–588.

Whitford, W. G., D. W. Freckman, P. F. Santos, N. Z. Elkins, and L. W. Parker. 1982. "The role of nematodes in decomposition in desert ecosystems," in Diana Freckman (ed.), *Nematodes in Soil Ecosystems* (Austin, Tex.: University of Texas Press), pp. 98–116.

Wolfe, B. E., and J. N. Klironomos. 2005. "Breaking new ground: Soil communities and exotic plant invasion," *BioScience*, **55**:477–487.

Carbon cycles in a mountain meadow. (R. Weil)

12
SOIL ORGANIC MATTER

*I bequeath myself to the dirt
to grow from the grass I love,
If you want me again look for
me under your boot-soles.*
—WALT WHITMAN,
SONG OF MYSELF

In most soils, the percentage of soil organic matter[1] is small, but its effects on soil function are profound. This ever-changing soil component exerts a dominant influence on many soil physical, chemical, and biological properties, especially in the surface horizons. Soil organic matter provides much of the soil's cation exchange capacity (discussed in Chapter 8) and water-holding capacity (Chapter 5). Certain components of soil organic matter are largely responsible for the formation and stabilization of soil aggregates (Chapter 4). Soil organic matter also contains large quantities of plant nutrients and acts as a slow-release nutrient storehouse, especially for nitrogen (Chapter 13). Furthermore, organic matter supplies energy and body-building constituents for most of the microorganisms whose general activities were discussed in Chapter 11. In addition to enhancing plant growth through the just-mentioned effects, certain organic compounds found in soils have direct growth-stimulating effects on plants. For all these reasons, the quantity and quality of soil organic matter are central in determining **soil quality** (Chapter 20).

Soil organic matter is a complex and varied mixture of organic substances. All organic substances, by definition, contain the element **carbon**, and, on average, carbon comprises about half of the mass of soil organic matter. Organic matter in the world's soils contains two to three times as much carbon as is found in all the world's vegetation. Soil organic matter, therefore, plays a critical role in the global carbon balance that is thought to be the major factor affecting global warming, or the **greenhouse effect.**

We will first examine the role of soil organic matter in the **global carbon cycle** and the process of **decomposition** of organic residues. Next, we will focus on inputs and losses with regard to soil carbon in specific ecosystems. Finally, we will study the processes and consequences involved in soil organic matter management.

---

[1]For an explanation of the chemical nature of soil organic matter, see Clapp et al. (2005). For a broad review of the nature, function, and management of organic matter in agricultural soils, see Magdoff and Weil (2004).

12.1 THE GLOBAL CARBON CYCLE

The element *carbon* is the foundation of all life. From cellulose to chlorophyll, the compounds that comprise living tissues are made of carbon atoms arranged in chains or rings and associated with many other elements. The cycle of carbon on Earth is the story of life on this planet. The carbon cycle is all-inclusive because it involves the soil, higher plants of every description, and all animal life, including humans. Disruption of the carbon cycle would mean disaster for all living organisms (Box 12.1).

Pathways

The basic processes involved in the global carbon cycle are shown in Figure 12.3. Plants take in carbon dioxide from the atmosphere. Then, through the process of photosynthesis, the energy of sunlight is trapped in the carbon-to-carbon bonds of

BOX 12.1   CARBON CYCLING—UP CLOSE AND PERSONAL

FIGURE 12.1   *The Biosphere 2 structure, a huge, sealed, ecological laboratory. (Photo by C. Allen Morgan. © 1995 by Decisions Investments Corp. Reprinted with permission.)*

FIGURE 12.2   *Biospherians at work growing their own food supply in the intensive agriculture biome with compost-amended soils rich in organic matter. (Photo by Pascale Maslin. © 1995 by Decisions Investments Corp. Reprinted with permission.)*

Imagine that you were one of the eight biospherians living a scientific game of survival in Biosphere 2, a giant 1.3-ha sealed glass building in the Arizona desert (Figure 12.1). Biosphere 2 contained a miniature ocean, coral reef, marsh, forest, and farm in a self-contained, self-supporting ecosystem in which the biospherians could live as part of the ecosystem they were studying. Instruments throughout the structure constantly monitored environmental parameters. What a great physical model (rather than a computer model) to study how a balanced ecosystem *really* works!

But it didn't take long for trouble to develop. First, the biospherians found it was no easy task to grow all the food they needed (Figure 12.2). Dependent on their meager harvests, they began to lose weight. Then, as if slowly starving were not bad enough, they soon began feeling short of breath. Instruments showed the oxygen level of the air was falling from its normal 21% to levels typical of high mountaintops (it eventually fell to as low as 14.2%). But unlike in "thin" air at high elevations, the carbon dioxide content of the air was rising. This wasn't supposed to be happening. Weren't all the green plants supposed to *use up* the carbon dioxide and *replenish* the oxygen supply? Unusual changes began talking place in the biosphereans' blood chemistry and metabolism—changes eerily like those of a bear in hibernation (during which both oxygen and food are limited)! Eventually, the atmosphere became so low in oxygen and high in carbon dioxide that engineers had to give up on the "fully self-contained" aspect of the project and pump in oxygen and remove carbon dioxide from the air.

What had they overlooked? It turned out that the ecosystem was thrown out of kilter by the organic-matter-rich soil hauled in for the Biosphere farm. The soil, made from a mixture of pond sediment (1.8% C), compost (22% C), and peat moss (40% C), was installed uniformly about 1 m deep. This artificial soil contained about 2.5% organic C at all depths—far more than the 0.5% C or less expected in a typical desert soil. Had the designers read this book they would have realized that peat might be stable in a boreal wetland (cool and anaerobic), but that aerobic soil microorganisms would rapidly use up oxygen and give off carbon dioxide as they metabolized organic matter in warm, moist garden soil aerated by tillage (see Section 12.2). This tale reminds us of the importance of soils in cycling C within the real biosphere! [For more on Biosphere 2, see: http://www.biospheres.com/, Torbert and Johnson (2001), and Walford (2002).]

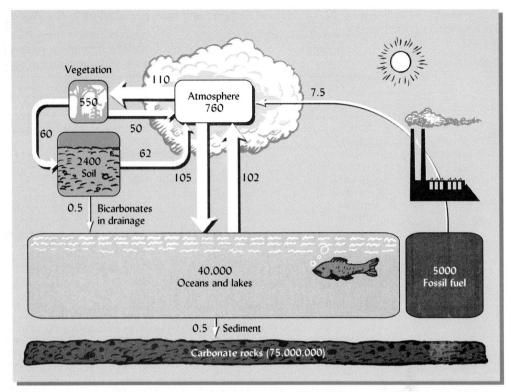

FIGURE 12.3 A simplified representation of the global carbon cycle emphasizing those pools of carbon which interact with the atmosphere. The numbers in the boxes indicate the petagrams (Pg = 10^{15} g) of carbon stored in the major pools. The numbers by the arrows show the amount of carbon annually flowing (Pg/yr) by various processes between the pools. Note that the soil contains almost twice as much carbon as the vegetation and the atmosphere combined. Imbalances caused by human activities can be seen in the flow of carbon to the atmosphere from fossil fuel burning (7.5) and in the fact that more carbon is leaving (62 + 0.5) than entering (60) the soil. These imbalances are only partially offset by increased absorption of carbon by the oceans. The end result is that a total of 221.5 Pg/yr enters the atmosphere while only 215 Pg/yr of carbon is removed. It is easy to see why carbon dioxide levels in the atmosphere are rising. [Data from IPCC (2007); soil carbon estimate from Batjes (1996)]

Satellite imagery showing C moving from the atmosphere to the biosphere:

http://www.gsfc.nasa.gov/topstory/20010327colors_of_life.html

organic molecules (such as those described in Section 12.2). Some of these organic molecules are used as a source of energy (via respiration) by the plants themselves (especially by the plant roots), with the carbon being returned to the atmosphere as carbon dioxide. The remaining organic materials are stored temporarily as constituents of the standing vegetation, most of which is eventually added to the soil as plant litter (including crop residues) or root deposition (see Section 11.7). Some plant material may be eaten by animals (including humans), in which case about half of the carbon eaten is exhaled into the atmosphere as carbon dioxide. The carbon not returned to the atmosphere is eventually returned to the soil as bodily wastes or body tissues. Once deposited on or in the soil, these plant or animal tissues are metabolized (digested) by soil organisms, which gradually return this carbon to the atmosphere as carbon dioxide.

Carbon dioxide also reacts in the soil to produce carbonic acid (H_2CO_3) and the carbonates and bicarbonates of calcium, potassium, magnesium, and sodium. The bicarbonates are readily soluble and may be removed in drainage. The carbonates, such as calcite ($CaCO_3$), are much less soluble and tend to accumulate in soils under alkaline conditions. Although this chapter focuses on the organic C in soils, the inorganic C content of soils (mainly as carbonates) may be substantial, especially in arid regions (Table 12.1). Eventually, as with the C in soil organic matter, most of the bicarbonate C and some of the carbonate C in soils is returned to the atmosphere as CO_2.

Microbial metabolism in the soil produces some organic compounds of such stability that decades or even centuries may pass before the carbon in them is returned to the atmosphere as carbon dioxide. Such resistance to decay allows organic matter to accumulate in soils.

TABLE 12.1   Mass of Organic and Inorganic Carbon in the World's Soils

Values for the upper 1 m represent 75 to 90% of the carbon in most soil profiles. Inorganic carbon is present mainly as calcium carbonates in soils of dry regions. Wetland soils as a group contain 468 Pg of organic C, some 30.3% of the total organic C in global soils.

| Soil order | Global area, 10^3 km^2 | Global carbon[a] in upper 100 cm | | | Total % |
| | | Organic | Inorganic | Total | |
		Pg			
Entisols	21,137	90	263	353	14.2
Inceptisols	12,863	190	34	224	9.0
Histosols	1,526	179	0	180	7.2
Andisols	912	20	0	20	0.8
Gelisols	11,260	316	7	323	12.9
Vertisols	3,160	42	21	64	2.6
Aridisols	15,699	59	456	515	20.6
Mollisols	9,005	121	116	237	9.5
Spodosols	3,353	64	0	64	2.6
Alfisols	12,620	158	43	201	8.0
Ultisols	11,052	137	0	137	5.5
Oxisols	9,810	126	0	126	5.1
Misc. land	18,398	24	0	24	1.0
Total	130,795	1,526	940	2,468	100.0

[a] Organic matter may be roughly estimated as 2.0 times this value, although the multiplier traditionally used is 1.72. Organic nitrogen may also be estimated from organic carbon values by dividing by 12 for most soils, but see Section 12.3. Pg = Petagram = 10^{15} g.
Data selected from Eswaran et al. (2000).

Carbon Sources

The original source of soil organic matter is plant tissue. Animals are secondary sources of organic matter. As they eat the original plant tissues, they contribute waste products, and they leave their own bodies when they die (review Figure 11.1). Certain forms of animal life, especially earthworms, termites, ants, and dung beetles, also play an important role in the incorporation and translocation of organic residues.

Globally, at any one time, approximately 2400 petagrams (Pg or 10^{15}g) of carbon are stored in soil profiles as soil organic matter (excluding surface litter), about one-third of that at depths below 1 m. An additional 940 Pg are stored as soil carbonates that can release CO_2 upon weathering. Altogether, nearly twice as much carbon is stored in the soil than in the world's vegetation and atmosphere combined (see Figure 12.1). Of course, this carbon is not equally distributed among all types of soils (Table 12.1). About 45% of the total organic carbon is contained in soils of just three orders, Histosols, Inceptisols, and Gelisols. Histosols (and Histels in the order Gelisols) are of limited extent but contain very large amounts of organic matter per unit land area. Inceptisols (and nonhistic Gelisols) contain only moderate concentrations of carbon, but cover vast areas of the globe. The reasons for the varying amounts of organic carbon in different soils will be detailed in Section 12.8.

In a mature natural ecosystem or a stable agroecosystem, the release of carbon as carbon dioxide by oxidation of soil organic matter (mostly by microbial respiration) is balanced by the input of carbon into the soil as plant residues (and, to a far smaller degree, animal residues). However, as discussed in Section 12.8, certain perturbations of the system, such as deforestation, some types of fires, tillage, and artificial drainage, result in a net loss of carbon from the soil system.

C emissions data, forecasts and analyses:

http://www.eia.doe.gov/environment.html

Figure 12.3 shows that, globally, the release of carbon from soils into the atmosphere is about 62 Pg/yr, while only about 60 Pg/yr enter the soils from the atmosphere via plant residues. This imbalance of about 2 Pg/yr, along with about 7.5 Pg/yr of carbon released by the burning of fossil fuels (in which carbon was sequestered from the atmosphere millions of years ago) is only partially offset by increased absorption of atmospheric carbon dioxide by the ocean. Fossil fuel burning and degrading land-use practices have increased the concentration of carbon dioxide in the atmosphere at an accelerating rate since the beginning of the industrial revolution, some 400 years ago. The levels have increased from 290 to 390 ppm during the past century alone. The implications of carbon dioxide imbalances and of other gaseous emissions on the greenhouse effect will be discussed in Section 12.9, after we consider the processes involved in the carbon cycle.

12.2 THE PROCESS OF DECOMPOSITION IN SOILS

Because plant residues are the principal material undergoing decomposition in soils and, hence, are the primary source of soil organic matter, we will begin by considering the makeup of these materials.

Composition of Plant Residues

Green plant tissues contain from 60 to 90% water by weight (Figure 12.4). If plant tissues are dried to remove all water, analysis of the *dry matter* remaining shows that, on a weight basis, the dry matter consists mostly (at least 90 to 95%) of carbon, oxygen, and hydrogen.

During photosynthesis, plants obtain these elements from carbon dioxide and water. If plant dry matter is burned (oxidized), these elements become carbon dioxide and water once more. Of course, some ash and smoke will also be formed upon burning, accounting for the remaining 5 to 10% of the dry matter. In the ash and smoke can be found the many nutrient elements originally taken up by the plants from the soil. Even though these elements are present in relatively small quantities, they play a vital role in plant and animal nutrition and in meeting the requirements of microorganisms. The essential nutrient elements in the ash, such as nitrogen, sulfur, phosphorus, potassium, and micronutrients, will be given detailed consideration in later chapters.

ORGANIC COMPOUNDS IN PLANT RESIDUES. The organic compounds in plant tissue can be grouped into broad classes. Although representative percentages of these classes are shown in Figure 12.4, tissues from different plant species, as well as from different parts (leaves, roots, stems, etc.) of a given plant differ considerably in their makeup.

Genetic diversity associated with chitin degradation in soil:
http://soilbio.nerc.ac.uk/Download/newsletter5.PDF

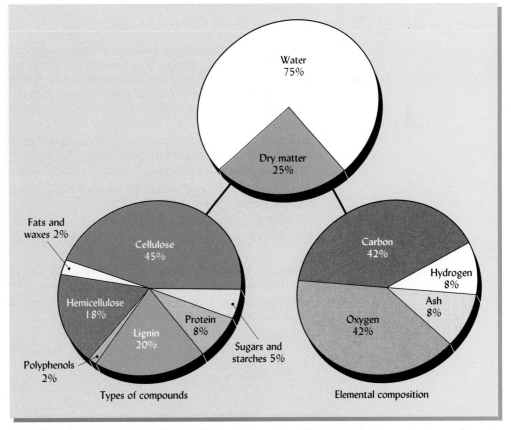

FIGURE 12.4 Typical composition of representative green-plant materials. The major types of organic compounds are indicated at left and the elemental composition at right. The *ash* is considered to include all the constituent elements other than carbon, oxygen, and hydrogen (nitrogen, sulfur, calcium, etc.).

Carbohydrates, which range in complexity from simple sugars and starches to cellulose, are usually the most plentiful of plant organic compounds.

Lignins, complex compounds with multiple ring-type or *phenol* structures, are components of plant cell walls. The content of lignin increases as plants mature and is especially high in woody tissues. Other **polyphenols**, such as tannins, may comprise as much as 6 or 7% of the leaves and bark of certain plants (for example, the brown color of steeped tea is due to tannins as are the brown leaf prints left by wet oak leaves on concrete sidewalks). Lignins and polyphenols are notoriously resistant to decomposition. Certain plant parts, especially seed and leaf coatings, contain significant amounts of fats, waxes, and oils, which are more complex than carbohydrates but less so than lignins.

Proteins contain about 16% nitrogen and smaller amounts of other essential elements, such as sulfur, manganese, copper, and iron. Simple proteins decompose and release their nitrogen easily, while complex crude proteins are more resistant to breakdown.

RATE OF DECOMPOSITION. Organic compounds may be listed in terms of ease of decomposition as follows:

1. Sugars, starches, and simple proteins   Rapid decomposition
2. Crude proteins
3. Hemicellulose
4. Cellulose
5. Fats and waxes
6. Lignins and phenolic compounds   Very slow decomposition

Decomposition of Organic Compounds in Aerobic Soils

Decomposition involves the breakdown of large organic molecules into smaller, simpler components. When organic tissue is added to an aerobic soil, three general reactions take place. Although in most cases mechanical shredding by soil fauna or physical processes must occur before these reactions can efficiently take place, the reactions themselves result from microbial activity.

1. Enzymatic oxidation of carbon compounds to produce carbon dioxide, water, energy, and decomposer biomass.
2. Release and/or immobilization of the essential nutrient elements, such as nitrogen, phosphorus, and sulfur, by a series of specific reactions that are relatively unique for each element.
3. Formation of compounds very resistant to microbial action, either through modification of compounds in the original tissue or by microbial synthesis.

DECOMPOSITION: AN OXIDATION PROCESS. In a well-aerated soil, all of the organic compounds found in plant residues are subject to oxidation. Since the organic fraction of plant materials is composed largely of carbon and hydrogen, the oxidation of the organic compounds in soil can be represented as:

$$\underset{\substack{\text{Carbon- and}\\\text{hydrogen-containing}\\\text{compounds}}}{R\!-\!(C, 4H)} + 2O_2 \xrightarrow[\text{oxidation}]{\text{Enzymatic}} CO_2\uparrow + 2H_2O + \text{energy (478 kJ mol}^{-1}\text{ C)} \qquad (12.1)$$

Many intermediate steps are involved in this overall reaction, and it is accompanied by important side reactions that involve elements other than carbon and hydrogen. Even so, this basic reaction accounts for most of the organic matter decomposition in the soil, as well as for the oxygen consumption and CO_2 release.

BREAKDOWN OF CELLULOSE AND STARCH.   Cellulose and starch are polysaccharides—long chains (polymers) of sugar molecules. Enzymatic degradation proceeds in steps: first the long chains are broken down by rather specialized organisms into short chains, then into individual sugar (glucose) molecules, which many different organisms can metabolize as in equation 12.1. Cellulose is the most abundant polysaccharide on the Earth. Because the C–O–C chemical bonds linking the sugar molecules of cellulose together are much more difficult to break than those in starch, the initial chain-breaking step in cellulose decomposition requires the activity of specialized organisms that produce the enzyme *cellulase*.

BREAKDOWN OF PROTEINS.   The plant proteins also succumb to microbial decay, yielding not only carbon dioxide and water, but amino acids such as glycine (CH_2NH_2COOH) and cysteine ($CH_2HSCHNH_2COOH$). In turn, these nitrogen and sulfur compounds are further broken down, eventually yielding such simple inorganic ions as ammonium (NH_4^+), nitrate (NO_3^-), and sulfate (SO_4^{2-}), forms available for plant nutrition.

BREAKDOWN OF LIGNIN.   Lignin molecules are very large and complex, consisting of hundreds of interlinked phenolic ring subunits, most of which are phenylpropene-like structures with various methoxyl (—OCH_3) groups attached (shown here as R or R′):

$$
\begin{array}{c}
R \\
|\\
CH{-}CH \\
HO{-}CH \qquad CH{=}CH{-}CH{-}CH_2O \\
CH{=}CH \\
|\\
R'
\end{array}
$$

Because the linkages among these structures are so varied and strong, only a few microorganisms (mainly *white rot fungi*) can break them down. Decomposition proceeds very slowly at first and is generally assisted by the physical activities of soil fauna. Once the lignin subunits are separated, many types of microorganisms participate in their breakdown. It is thought that microorganisms use some of the ring structures from lignin in the synthesis of stable soil organic matter.

Example of Organic Decay

The process of organic decay in time sequence is illustrated in Figure 12.5. Assume the soil has not been disturbed or amended with plant residues for some time. Initially, little or no readily decomposable materials are present. Competition for food is severe and microbial activity is relatively low, as reflected in the low **soil respiration** rate or level of CO_2 emission from the soil. The supply of soil carbon is steadily being depleted. Small populations of microorganisms survive by slowly digesting the very resistant, stable soil organic matter. Soil ecologists consider these microorganisms to exhibit a K-strategy for survival, so named because they have developed enzymes with high affinity constants (k) for specific types of resistant substrates. The **k-strategists** have a competitive advantage when the soil is poor in easily digested organic materials. These organisms maintain a low but fairly constant population by carrying out specialized functions.

Now suppose that deciduous trees in a forest begin to lose their leaves in fall or a farmer plows in residues of a harvested crop. The appearance of easily decomposable and often water-soluble compounds, such as sugars, starches, and amino acids, stimulates an almost immediate increase in metabolic activity among the soil microbes. Soon the slower-acting k-strategists are overtaken by rapidly multiplying populations of *opportunist* or *colonizing* organisms that have been awakened from their dormant state by the presence of new food supplies. These organisms are known as **r-strategists**, so named for their rapid rate (r) of growth and reproduction that allows them to take advantage of a sudden influx of food.

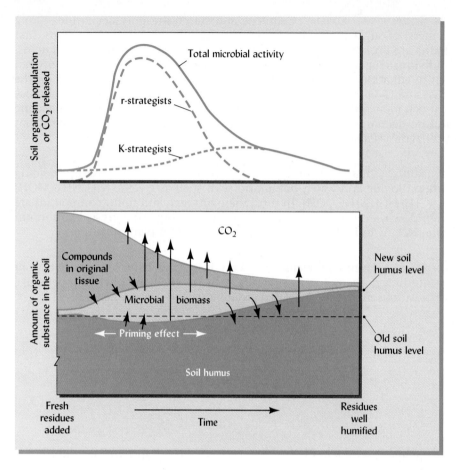

FIGURE 12.5 Schematic of the general changes occurring when fresh plant residues are added to a soil. The arrows indicate transfers of carbon among compartments. The upper panel shows the relative growth or activity of r-strategist (opportunist), K-strategist (more specialized) microorganisms, and the sum of these two groups. The time required for the process will depend on the nature of the residues and the soil. Most of the carbon released during the initial rapid breakdown of the residues is converted to carbon dioxide, but smaller amounts of carbon are converted into microbial biomass (and synthesis products) and, eventually, into soil humus. The peak level of microbial activity appears to accelerate the decay of the original humus, a phenomenon known as the *priming effect*. However, the humus level is increased by the end of the process. Where vegetation, environment, and management remain stable for a long time, the soil humus content will reach an equilibrium level at which the carbon added to the humus pool through the decomposition of plant residues is balanced by carbon lost through the decomposition of existing soil humus.

Microbial numbers and carbon dioxide evolution from microbial respiration both increase exponentially in response to the new food resource (upper panel in Figure 12.5). Soon microbial activity is at peak intensity, energy is being rapidly liberated, and carbon dioxide is being formed in large quantities. As organisms multiply, they increase the **microbial biomass** and also synthesize new exocellular organic compounds. The microbial biomass at this point may account for as much as one-sixth of the organic matter in the soil. The intense microbial activity may even stimulate the breakdown of some resistant soil organic matter, a phenomenon known as the **priming effect.**

With all this frenetic microbial activity, the easily decomposed compounds are soon exhausted. While the specialized K-strategists continue their slow work, degrading cellulose and lignin, r-strategists begin to die of starvation. As microbial populations plummet, the dead cells provide a readily digestible food source for the survivors, which continue to evolve carbon dioxide and water. The decomposition of the dead microbial cells is also associated with the **mineralization** or release of simple inorganic products, such as nitrates and sulfates. As food supplies are further reduced, microbial activity continues to decline, and the general-purpose r-strategists again sink back into comparative quiescence. A little of the original residue material persists, mainly as tiny particles that have become **physically protected** from decay by lodging inside soil pores too tight to allow access by most organisms. Some of the remaining carbon has also been **chemically protected** by conversion into **soil humus**, which is a dark-colored, heterogeneous, mostly colloidal mixture of modified lignin and newly synthesized organic compounds. Humus is highly resistant to attack and may be further protected by binding strongly to clay particles. Thus, a small percentage of the carbon in the added residues has been retained, increasing slightly the pool of stable soil organic matter. In a mature ecosystem, this increase will likely be offset during each annual cycle by slow, steady K-strategy decomposition, resulting in little net change in the level of soil organic matter from year to year.

Decomposition in Anaerobic Soils

Microbial decomposition proceeds most rapidly in the presence of plentiful supplies of O_2, which acts as the electron acceptor during aerobic oxidation of organic compounds. Oxygen supplies may become depleted when soil pores filled with water prevent the diffusion of O_2 into the soil from the atmosphere. Without sufficient oxygen present, aerobic organisms cannot function, so anaerobic or facultative organisms become dominant. Under low-oxygen or anaerobic conditions, decomposition takes place much more slowly than when oxygen is plentiful. Hence, wet, anaerobic soils tend to accumulate large amounts of organic matter in a partially decomposed condition.

The products of anaerobic decomposition include a wide variety of partially oxidized organic compounds, such as organic acids, alcohols, and methane gas. Anaerobic decomposition releases relatively little energy for the organisms involved; therefore, the end products still contain much energy. (For this reason, alcohol and methane can serve as fuel.) Some of the products of anaerobic decomposition are of concern because they produce foul odors or inhibit plant growth. The methane gas produced in wet soils is a major contributor to the greenhouse effect (Section 12.9). The following reactions are typical of those carried out in wet soils by various **methanogenic bacteria** and **archaea**:

$$4C_2H_5COOH + 2H_2O \xrightarrow{\text{Bacteria}} 4CH_3COOH + CO_2 \uparrow + 3CH_4 \uparrow \qquad (12.2)$$

Propionate   Acetate   Carbon   Methane
dioxide

$$CH_3COOH \xrightarrow{\text{Bacteria}} CO_2 \uparrow + CH_4 \uparrow \qquad (12.3)$$

$$CO_2 + 4H_2 \xrightarrow{\text{Bacteria}} 2H_2O + CH_4 \uparrow \qquad (12.4)$$

Production of Simple Inorganic Products

As proteins are attacked by microbes, the long chains of amino acids are broken, and individual amino acids appear in the soil solution along with dissolved CO_2. The amide (—R—NH_2) and sulfide (—R—S) groups of the amino acids, in turn, are broken off to produce, first, ammonium (NH_4^+) and sulfide (S^{2-}) compounds and, finally, nitrates (NO_3^-) and sulfates (SO_4^{2-}). Similar decomposition of other organic compounds releases these and other inorganic nutrient ions. The process that releases elements from organic compounds to produce inorganic (mineral) forms is known as **mineralization**. This is usually the last step in the overall decomposition process. Most of the inorganic ions released by mineralization are readily available to higher plants and to microorganisms. The decay of organic tissues is an important source of nitrogen, sulfur, phosphorus, and other essential elements for plants.

12.3 FACTORS CONTROLLING RATES OF DECOMPOSITION AND MINERALIZATION[2]

The time needed to complete the processes of decomposition and mineralization may range from days to years, depending mainly on two broad factors: (1) the environmental conditions in the soil, and (2) the quality of the added residues as a food source for soil organisms.

The environmental conditions conducive to rapid decomposition and mineralization (see also Section 11.11 and 12.8) include a near-neutral pH, sufficient soil moisture, good aeration (about 60% of the soil pore space filled with water), and warm temperatures (25 to 35 °C). Ironically, periodic stresses such as episodes of severe drying actually accelerate overall mineralization due to the dramatic burst of microbial activity that occurs each time the soil re-wets (e.g., Figure 13.7). These conditions

---

[2] For an excellent collection of papers dealing with litter decomposition in soils, see Cadisch and Giller (1997). Many of our current ideas about decomposition were first put forward in a classic book by Swift et al. (1979).

were discussed in Section 11.11 in relation to microbial activity and will be considered again in Section 12.8 as they affect the levels of organic matter accumulating in soils. Here we will focus on factors that determine the quality of the residues as a food resource for microbes, including the physical condition of the residues, their C/N ratio, and their content of lignins and polyphenols.

Physical Factors Influencing Residue Quality

The location of residues in or on the soil is a physical factor that has a critical impact on decomposition rates. Surface placement of plant residues, as in forest litter or conservation tillage mulch, usually results in slower, more variable rates of decomposition than where similar residues are incorporated into the soil by root deposition, faunal action, or tillage. Surface residues are subject to drying, as well as extremes of temperature. Nutrient elements mineralized from surface-applied residues are also more susceptible to loss in runoff or by volatilization than are those from incorporated residues. Surface residues are physically out of reach for most soil organisms, save the larger fauna such as earthworms and fungal mycelia. The latter have been shown to grow up from the soil to invade surface litter (see Figure 11.20). If the surface litter is low in nitrogen, fungi may even transfer nitrogen from the soil through their hyphae to lower the C/N ratio of the litter (see following). Compared to surface residue, incorporated residues are in intimate contact with soil moisture and soil organisms, decompose more quickly, and may lose nutrients more easily by leaching.

Residue particle size is another important physical factor—the smaller the particles, the more rapid the decomposition. Small particle size may result from the nature of the residues (e.g., twigs versus branches), from mechanical treatment (grinding, chopping, tillage, etc.), or from the chewing action of soil fauna. Diminution of residues into smaller particles physically exposes more surface area to decomposition and also breaks up lignacious cell walls and waxy outer coatings on leaves so as to expose the more readily decomposed tissues and cell contents. In addition, some organic materials, including tiny bits of tissues lodged in aggregates, exhibit hydrophobicity (water repellency), making them slow to wet and difficult to attack by water-soluble microbial enzymes.

Carbon/Nitrogen Ratio of Organic Materials and Soils

The carbon content of typical plant dry matter is about 42% (see Figure 12.4). The nitrogen content of plant residues is much lower and varies widely (from <1 to >6%). The ratio of carbon to nitrogen (C/N) in organic residues applied to soils is important for two reasons: (1) intense competition among microorganisms for available soil nitrogen occurs when residues having a high C/N ratio are added to soils, and (2) the C/N ratio in residues helps determine their rate of decay and the rate at which nitrogen is made available to plants.

C/N RATIO IN PLANTS AND MICROBES.   The C/N ratio in plant residues ranges from between 10:1 to 30:1 in legumes and young green leaves to as high as 600:1 in some kinds of sawdust (Table 12.2). Generally, as plants mature, the proportion of protein in their tissues declines, while the proportion of lignin and cellulose, and the C/N ratio, increase. As can be seen from the decay curves in Figure 12.6, these differences in composition have pronounced effects on the rate of decay when plant residues are added to the soil.

In the bodies and cells of microorganisms, the C/N ratio is not only less variable than in plant tissues, but also much lower, ordinarily falling between 5:1 and 10:1. Among microorganisms, bacteria are generally richer in protein than fungi and, consequently, have a lower C/N ratio.

C/N RATIO IN SOILS.   The C/N ratio in the organic matter of arable (cultivated) surface (Ap) horizons commonly ranges from 8:1 to 15:1, the median being near 12:1. The ratio is generally lower for subsoils than for surface layers in a soil profile. In a given climatic region, little variation occurs in the C/N ratio for similarly managed soils. For instance, in calcium-rich soils of semiarid grasslands (e.g., Mollisols and tropical Alfisols), the C/N ratio is relatively narrow. In more severely leached and acidic A horizons in humid

TABLE 12.2 Typical Carbon and Nitrogen Contents and C/N Ratios of Some Organic Materials Commonly Associated with Soils

Organic material	% C	% N	C/N
Spruce sawdust	50	0.05	600
Hardwood sawdust	46	0.1	400
Newspaper	39	0.3	120
Wheat straw	38	0.5	80
Corn stover	40	0.7	57
Sugar cane trash	40	0.8	50
Rye cover crop, anthesis	40	1.1	37
Maple leaf litter	48	1.4	34
Rye cover crop, vegetative stage	40	1.5	26
Mature alfalfa hay	40	1.8	25
Rotted barnyard manure	41	2.1	20
Bluegrass from fertilized lawn	42	2.2	20
Broccoli residues	35	1.9	18
Finished household compost	30	2.0	15
Young alfalfa hay	40	3.0	13
Hairy vetch cover crop	40	3.5	11
Digested municipal sewage sludge	31	4.5	7
Soil microorganisms			
Bacteria	50	10.0	5
Actinomycetes, nematodes	50	8.5	6
Fungi	50	5.0	10
Soil organic matter			
Spodosol O horizon	50	0.5	90
Average forest O horizons	50	1.3	45
Average forest A horizons	50	2.8	20
Tropical evergreen litter	50	2.0	25
Mollisol Ap horizon	56	4.9	11
Average B horizon	46	5.1	9

Data calculated from many sources.

regions, the C/N is relatively wide; C/N ratios as high as 30:1 are not uncommon. Forest O horizons commonly have C/N ratios of 30 to 40. When such soils are brought under cultivation and limed to increase their pH and calcium content, the enhanced decomposition tends to lower the C/N ratio to near 12:1.

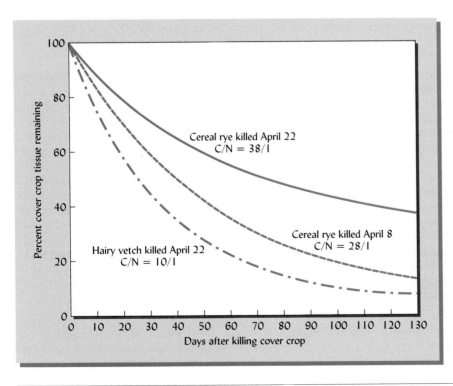

FIGURE 12.6 Rates of decomposition of various cover crop residues. The cover crops were grown over the winter and early spring, then killed with a herbicide and the residues left as a mulch on the soil surface. Corn was planted into this mulch without tillage. The lower the initial C/N ratio of the residues, the more rapid the decomposition process. Note that the legume (hairy vetch) had a much lower C/N ratio than the grass (cereal rye). Also note that a two-week delay in killing the rye cover crop resulted in more mature plants with a considerably higher C/N ratio and slower rate of decomposition. (Redrawn from S. Davis and R. Weil, unpublished)

Influence of Carbon/Nitrogen Ratio on Decomposition

Soil microbes, like other organisms, require a balance of nutrients from which to build their cells and extract energy. The majority of soil organisms metabolize carbonaceous materials both in order to obtain carbon for building essential organic compounds and to obtain energy for life processes. However, no creature can multiply and grow on carbon alone. Organisms must also obtain sufficient nitrogen to synthesize nitrogen-containing cellular components, such as amino acids, enzymes, and DNA.

On the average, soil microbes must incorporate into their cells about eight parts of carbon for every one part of nitrogen (i.e., assuming the microbes have an average C/N ratio of 8:1). Because only about one-third of the carbon metabolized by microbes is incorporated into their cells (the remainder is respired and lost as CO_2), the microbes need to find about 1 g of N for every 24 g of C in their "food".

This requirement results in two extremely important practical consequences. First, if the C/N ratio of organic material added to soil exceeds about 25:1, the soil microbes will have to scavenge the soil solution to obtain enough nitrogen. Thus, the incorporation of high C/N residues will deplete the soil's supply of soluble nitrogen, causing plants to suffer from nitrogen deficiency. Second, the decay of organic materials can be delayed if sufficient nitrogen to support microbial growth is neither present in the material undergoing decomposition nor available in the soil solution. These concepts are illustrated by the example in Figure 12.7.

Examples of Inorganic Nitrogen Release During Decay

The practical significance of the C/N ratio becomes apparent if we compare the changes that take place in the soil when residues of either high or low C/N ratio are added (Figure 12.8). Consider a soil with a moderate level of soluble nitrogen (mostly nitrates). General-purpose decay organisms are at a low level of activity in this soil, as evidenced by low carbon dioxide production. If no nitrogen were lost or taken up by plants, the level of nitrates would very slowly increase as the native soil organic matter decays.

LOW NITROGEN MATERIAL.   Now consider what happens when a large quantity of readily decomposable organic material is added to this soil. If this material has a C/N ratio greater than 25, changes will occur according to the pattern shown in Figure 12.8a. In the example shown, the initial C/N ratio of the residues is about 55, typical for cornstalks or many kinds of leaf litter. As soon as the residues contact the soil, the microbial community responds to the new food supply (see Section 12.2). Heterotrophic k-strategist microbes become active, multiply rapidly, and yield carbon dioxide in large quantities. Because of the microbial demand for nitrogen, little or no mineral nitrogen (NH_4^+ or NO_3^-) is available to higher plants during this period.

NITRATE DEPRESSION.   This condition, often called the **nitrate depression period**, persists until the activities of the decay organisms gradually subside due to lack of easily oxidizable carbon. As their numbers decrease, carbon dioxide formation drops off, and nitrogen demand by microbes becomes less acute. As decay proceeds, the C/N ratio of the remaining plant material decreases because carbon is being lost (by respiration) and nitrogen is being conserved (by incorporation into microbial cells). Generally, one can expect mineral nitrogen to begin to be released when the C/N ratio of the remaining material drops below about 20. Then, nitrates appear again in quantity, and the original conditions prevail, except that the soil is somewhat richer in both nitrogen and humus.

The nitrate depression period may last for a few days, a few weeks, or even several months. A longer, more severe period of nitrate depression is typical when added residues are easily decomposed and have a higher C/N ratio and when a larger quantity of residues is added. To avoid producing seedlings that are stunted, chlorotic, and nitrogen-starved, planting should be delayed until after the nitrate depression period or additional sources of nitrogen can be applied to satisfy the nutritional requirements of both the microbes and the plants.

HIGH NITROGEN MATERIAL.   The effects on soil nitrate level will be quite different if the residues added have a C/N ratio lower than 20, as in the case represented by Figure 12.8b. With organic materials of low C/N ratio, more than enough nitrogen is present

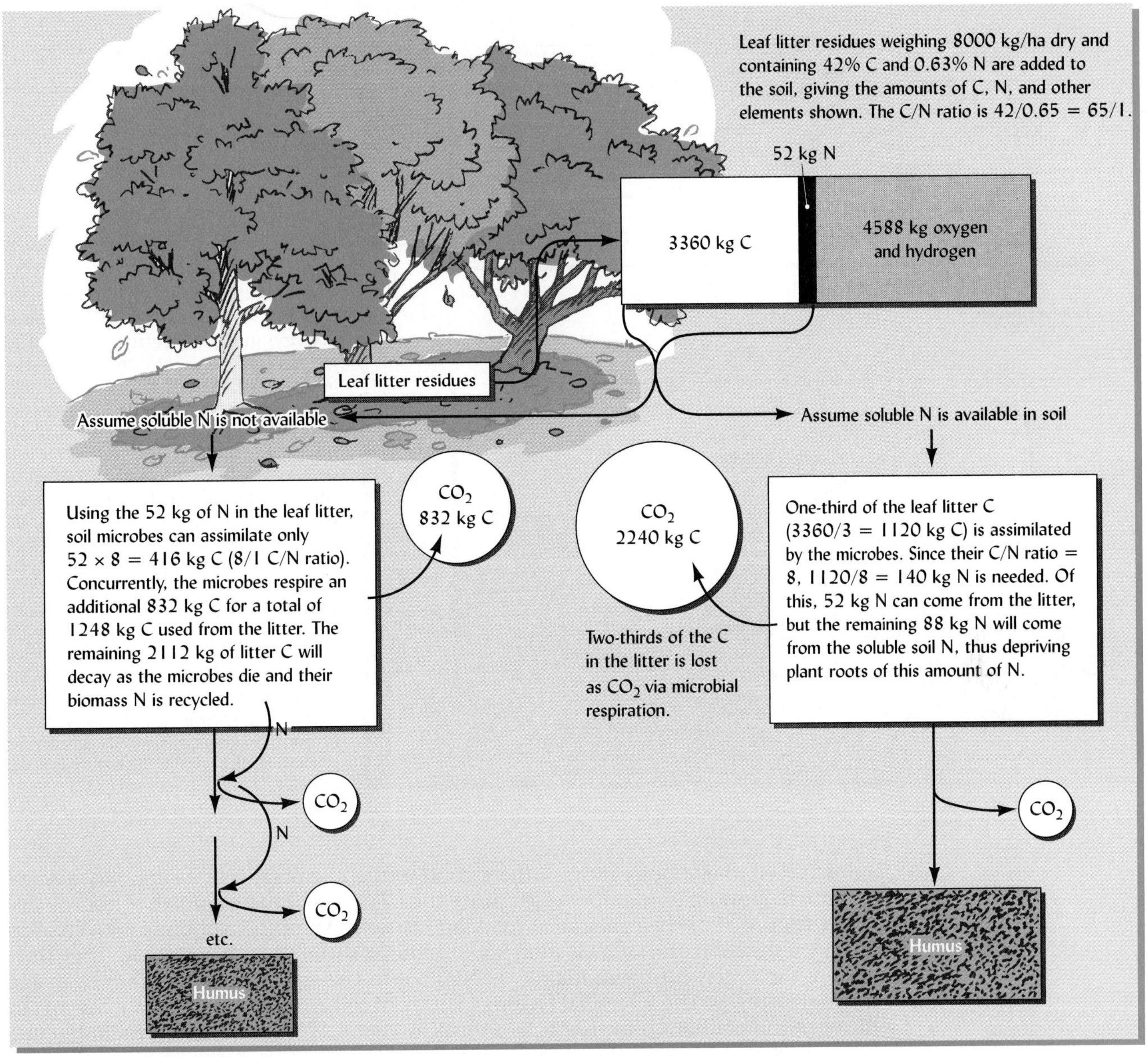
Leaf litter residues weighing 8000 kg/ha dry and containing 42% C and 0.63% N are added to the soil, giving the amounts of C, N, and other elements shown. The C/N ratio is 42/0.65 = 65/1.
52 kg N
3360 kg C
4588 kg oxygen and hydrogen
Leaf litter residues
Assume soluble N is not available
Assume soluble N is available in soil
Using the 52 kg of N in the leaf litter, soil microbes can assimilate only 52 × 8 = 416 kg C (8/1 C/N ratio). Concurrently, the microbes respire an additional 832 kg C for a total of 1248 kg C used from the litter. The remaining 2112 kg of litter C will decay as the microbes die and their biomass N is recycled.
CO2 832 kg C
CO2 2240 kg C
Two-thirds of the C in the litter is lost as CO2 via microbial respiration.
One-third of the leaf litter C (3360/3 = 1120 kg C) is assimilated by the microbes. Since their C/N ratio = 8, 1120/8 = 140 kg N is needed. Of this, 52 kg N can come from the litter, but the remaining 88 kg N will come from the soluble soil N, thus depriving plant roots of this amount of N.
N
CO2
N
CO2
etc.
Humus
CO2
Humus

FIGURE 12.7 A simplified, quantitative example of plant residue decay illustrating the fates of carbon and nitrogen and the consequences for decomposition and soil nitrogen availability. Note that if an adequate supply of nitrogen is available, the potential for humus creation is increased.

to meet the needs of the decomposing organisms. Therefore, soon after decomposition begins, some of the nitrogen from organic compounds is released into the soil solution, augmenting the level of soluble nitrogen available for plant uptake. Generally, nitrogen-rich materials decompose quite rapidly, resulting in a period of intense microbial growth and activity, but no nitrate depression period.

Influence of Soil Ecology

In nature, the process of nitrogen mineralization involves the entire food web (see Section 11.2), not just the saprophytic bacteria and fungi. For example, when organic residues are added to soil, bacteria and fungi grow rapidly on this food source, producing a large biomass of bacterial and fungal cells that contain much of the nitrogen originally in the residues. Until the microbial biomass begins to die off, this nitrogen is immobilized and not available to plants. However, a healthy soil ecosystem is likely to contain certain nematodes, protozoa, and earthworms that feed on bacteria and fungi. As these

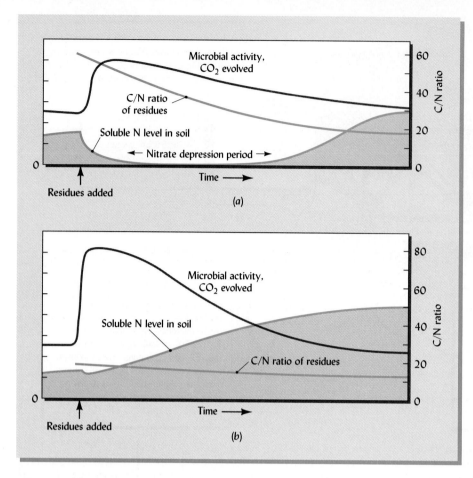

(a)

(b)

FIGURE 12.8 Changes in microbial activity, in soluble nitrogen level, and in residual C/N ratio following the addition of either high (*a*) or low (*b*) C/N ratio organic materials. Where the C/N ratio of added residues is above 25, microbes digesting the residues must supplement the nitrogen contained in the residues with soluble nitrogen from the soil. During the resulting nitrate depression period, competition between higher plants and microbes would be severe enough to cause nitrogen deficiency in the plants. Note that in both cases soluble N in the soil ultimately increases from its original level once the decomposition process has run its course. The trends shown are for soils without growing plants, which, if present, would continually remove a portion of the soluble nitrogen as soon as it is released.

animals feed, they respire most of the carbon in the microbial cells, using only a small fraction to grow on (or produce eggs). Since the C/N ratio of these animals is not too different from that of their microbial food, and since most of the carbon is converted to CO_2 by respiration, the animals soon ingest more nitrogen than they can use. They then excrete the excess nitrogen, mainly as NH_4^+, into the soil solution as plant-available mineral nitrogen. The microbial feeding activity of soil animals may increase the rate of nitrogen mineralization by 100%, as shown in Figure 12.9 for bacterial-feeding nematodes. Bacterial-feeding nematodes seem to have less effect on nitrogen mineralization when the C/N ratio of residues is very wide, because nitrogen excreted by the nematodes under these conditions would be quickly reimmobilized by bacteria that have plenty of carbon, but little nitrogen, to grow on. In any case, soil management that favors a complex food web (Section 11.14) with many trophic levels can be expected to enhance the cycling and efficient use of nitrogen (and of other nutrients).

Influence of Lignin and Polyphenol Content of Organic Materials

Complexities of decay and nutrient release from logs in Northwest forests:

http://oregonstate.edu/dept/ncs/newsarch/2005/Aug05/decay.htm

The lignin contents of plant litter range from less than 2% to more than 50%. Those materials with high lignin content decompose very slowly. Polyphenol compounds found in plant litter may also inhibit decomposition. These phenolics are often water-soluble and may be present in concentrations as high as 5 to 10% of the dry weight. By forming highly resistant complexes with proteins during residue decomposition, these phenolics can dramatically slow the rates of both nitrogen mineralization and carbon oxidation.

LITTER QUALITY. Because they support only low levels of microbial activity and biomass, residues high in phenols and/or lignin are considered to be *poor quality resources* for the soil organisms that cycle carbon and nutrients. The production of such slow-to-decompose residues by certain forest plants may help explain the accumulation of extremely high levels of humified nitrogen and carbon in the soils of mature boreal forests.

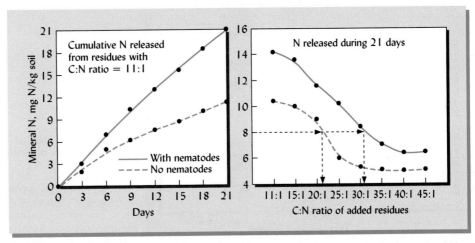

FIGURE 12.9 Bacteria-feeding animals, such as certain nematodes, enhance the release of plant-available mineral nitrogen from organic residues. The study illustrated here used columns of sandy soil with well-established bacterial communities. The researchers amended the soil with ground alfalfa tissue (low C/N) and cellulose (high C/N) in varying proportions to give the indicated C/N ratios. Some of the soil columns were inoculated with bacteria-feeding nematodes (*Cephalobus persegnis*) and some were free of nematodes. (*Left*) By feeding on the bacteria, the nematodes nearly doubled the amount of mineral N released from the added residues over a 21-day period. Without nematodes, much of the N was tied up in the bacterial biomass, but when the nematodes ate the bacteria, they excreted the excess N into the soil solution as NH_4^+. (*Right*) Nematodes influenced the release of mineral N from residues with different C/N ratios. Enhanced mineralization in the presence of bacteria-feeding nematodes suggests that a level of mineral N satisfactory for plant growth (say 8 mg/kg) could be maintained with residues of relatively high C/N ratio (about 32:1), but that without these nematodes, a more N-rich type of residue (C/N ratio about 22:1) would be required. [Redrawn from Ferris et al. (1998)]

The lignin and phenol contents also influence the decomposition and release of nitrogen from **green manures**—plant residues used to enrich agricultural soils (Table 12.3). For example, in the leaves of certain legume trees, the C/N ratio is quite narrow, but the phenol content is quite high, so that when these leaves are added to

TABLE 12.3 Litter Quality in Relation to the Lignin Content, Polyphenol Content, and C/N Ratio and Rate of Decay of Several Types of Materials as Measured at Tropical and Temperate Sites

Low values of C/N, lignin, and polyphenols all contribute to high litter quality and speed of decomposition. Crop residues tend to be low in lignin and polyphenol. For similar materials, decay is much faster (k values are larger) in the tropical climate. The inhibitory effect of polyphenol content can be seen by comparing Gliricidia to Leucaena. The effect of climate and C/N ratio (related to N fertilization) can be seen by comparing the two maize residues.

Plant species	Plant parts	Lignin, %	Polyphenols,%	Total N,%	C/N	Decay constant,[a] k, week^{-1}	Litter quality
Measured under humid tropical conditions in Nigeria							
Gliricidia sepium	Prunings	12	1.6	3.5	13	0.255	High
Leucaena leucocephala	Prunings	13	5.0	3.5	13	0.166	Medium-high
Oryza sativa (Rice)	Residues	5	0.6	1.0	42	0.124	Medium
Zea mays (Corn)	Residues	7	0.6	1.1	43	0.119	Medium
Dactyladenia barteri	Prunings	47	4.1	1.6	28	0.011	Low
Measured under temperate subhumid conditions in Missouri, U.S.							
Poa trivialis (Bluegrass)	Hay	8	0.2	2.2	19	0.027	Medium-high
Glycine max (Soybean)	Residues	9	0.2	2.2	20	0.020	Medium-high
Zea mays (Corn)	Residues	7	0.3	1.5	28	0.014	Medium
Acer saccharinum (Silver maple)	Fallen leaves	11	4.8	1.4	34	0.013	Medium
Carya illinoinensis (Pecan)	Fallen leaves	25	2.1	1.1	42	0.012	Medium

[a]As each type of residue decomposed, researchers periodically determined the proportion Y of the original residue dry matter remaining. The long-term decomposition rate k was determined from the equation $Y = e^{-kt}$, in which e is the base of natural logarithms (2.72), and t is time in weeks. Therefore, the larger the decomposition constant k, the faster the decomposition.

Data selected from Mungai and Motavalli (2006), and Tian et al. (1992 and 1995).

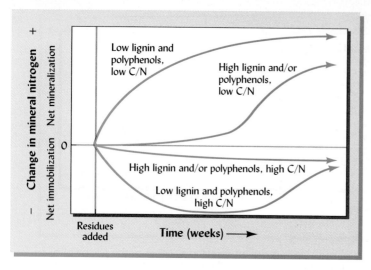

FIGURE 12.10 Temporal patterns of nitrogen release from organic residues differing in quality based on their C/N ratios and contents of lignin and polyphenols. Lignin contents greater than 20%, polyphenol contents greater than 3%, and C/N ratios greater than 30 would all be considered high in the context of this diagram, the combination of these properties characterizing litter of poor quality—that is, litter that has a limited potential for microbial decomposition and mineralization of plant nutrients. (Diagram courtesy of R. Weil)

soil, nitrogen is released only slowly—often too slowly to keep up with the needs of a growing crop. Similarly, residues with a lignin content of more than 20 to 25% will decompose too slowly to be effective as green manure for rapidly growing annual crops. However, for perennial crops, or forests, the slow release of nitrogen from such residues may be advantageous in the long run, as the nitrogen may be less subject to losses. By the same token, the slow decomposition of phenol- or lignin-rich materials means that even if their C/N ratio is very high, the nitrate depression will not be pronounced.

Figure 12.10 illustrates the combined effects of C/N ratio and lignin or phenol content on the balance between immobilization and mineralization of nitrogen during plant residue decomposition.

12.4   GENESIS AND NATURE OF SOIL ORGANIC MATTER AND HUMUS[3]

Everything about humus—a view from Poland:

http://www.ar.wroc.pl/~weber/humic.htm

The organic portion of soil is a complex mixture of substances, and there are many terms used to describe it and its components. In this textbook, we use the general term **soil organic matter** (SOM) to encompass all the organic components of a soil: (1) living **biomass** (intact plant and animal tissues and microorganisms); (2) dead roots and other recognizable plant residues or litter (although in practice, residue particles that do not pass 2-mm sieve openings are often excluded from consideration); and (3) a largely amorphous and colloidal mixture of complex organic substances no longer identifiable as tissues. Only the third category of organic material is properly referred to as **soil humus** (Figure 12.11). Since the element carbon (C) plays a prominent role in the chemical structure of all organic substances, it is not surprising that the term **soil organic carbon** (SOC) is often used to refer to the C component of soil organic matter. This term is particularly appropriate for quantitative discussions of soil organic matter because most methods of determining soil organic matter actually measure the C in the material and then use a conversion factor to estimate the organic matter. Since soil organic matter commonly contains about half carbon by weight (50% C), it is usually appropriate to estimate soil organic matter as two times the organic C (SOM = 2 × SOC).[4] However, the C content of soil organic matter does vary, so caution must be used when comparing values reported as soil organic matter and soil organic carbon!

---

[3] For a discussion of humus formation, composition, and reactions, see Stevenson (1994). Modern spectroscopic methods for studying the structure of humic substances are reviewed by Hatcher et al. (2001).

[4] Traditionally SOM has been estimated as 1.72 × SOC, a conversion that assumes 58% C in the soil organic matter. However, mainly highly stabilized humic acids contain that much C, and there is little basis for the general use of this value, so rounding to 2 × SOC is recommended.

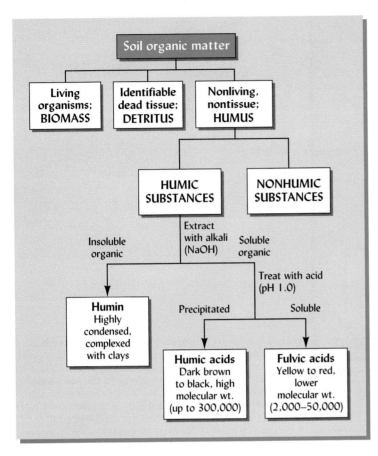

FIGURE 12.11 Classification of soil organic matter components separable by chemical and physical criteria. Although surface residues (litter) are not universally considered to be part of the soil organic matter, we include them because they are the principal components of the O horizons in soil profiles. Solubility in alkali and acid is a widely used criterion for grouping different fractions of soil humus. The classical scheme dividing soil humic substances into humin, fulvic acids, and humic acids fractions (shown in the lower part of the flowchart) is based on their insolubility in NaOH (humin) and their subsequent solubility (fulvic acids) and insolubility (humic acids) in acid solutions (pH = 1).

Microbial Transformations

As decomposition of plant residues proceeds, microbes slowly break down complex components into simpler compounds. In this process some of the lignin is broken down into its phenolic subunits. The soil microbes then metabolize the resulting simpler compounds. Using some of the carbon not lost as carbon dioxide in respiration, along with most of the nitrogen, sulfur, and oxygen from these compounds, the microorganisms synthesize new cellular components and biomolecules. Some of the original lignin is not completely broken down, but only modified to form complex residual molecules that retain many of the characteristics of lignin. The microbes polymerize (link together) some of the simpler new compounds with each other and with the complex residual products into long, complex chains that resist further decomposition. These high-molecular-weight compounds interact with nitrogen-containing amino compounds, giving rise to a significant component of resistant humus. The presence of colloidal clays stimulates the complex polymerization. These ill-defined, complex, resistant, polymeric compounds are called **humic substances**. The term **nonhumic substances** refers to the group of identifiable biomolecules that are mainly produced by microbial action and are generally less resistant to breakdown.

One year after plant residues are added to the soil, most of the carbon has returned to the atmosphere as CO_2, but one-fifth to one-third is likely to remain in the soil either as live biomass (~5%) or as the humic (~20%) and nonhumic (~5%) fractions of soil humus (Figure 12.12). The proportion remaining from root residues tends to be somewhat higher than that remaining from incorporated leaf litter.

Humic Substances

Humic substances comprise about 60 to 80% of the soil organic matter. They are comprised of huge molecules with variable, rather than specific, structures and composition (see Figure 8.14). Humic substances are characterized by aromatic, ring-type structures

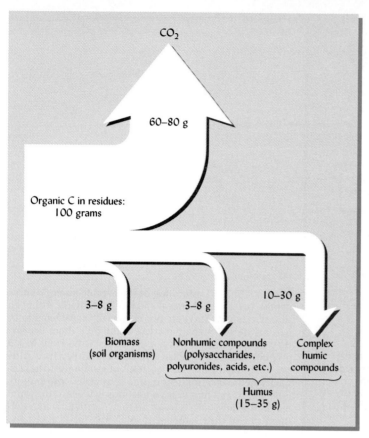

FIGURE 12.12 Disposition of 100 g of organic carbon in residues one year after they were incorporated into the soil. More than two-thirds of the carbon has been oxidized to CO_2, and less than one-third remains in the soil—some in the cells of soil organisms, but a larger component as soil humus. The amount converted to CO_2 is generally greater for aboveground residues than for belowground (root) residues. (Estimates from many sources)

that include polyphenols (numerous phenolic compounds linked together) and comparable polyquinones, which are even more complex. Humic substances generally are dark-colored, amorphous substances with molecular weights varying from 2000 to 300,000 g/mol. Because of their complexity, they are the organic materials most resistant to microbial attack.

SOLUBILITY GROUPINGS. Historically, humic substances have been classified into three chemical groupings based on solubility (see Figure 12.11): (1) *fulvic acid*, lowest in molecular weight and lightest in color, soluble in both acid and alkali, and most susceptible to microbial attack; (2) *humic acid*, medium in molecular weight and color, soluble in alkali but insoluble in acid, and intermediate in resistance to degradation; and (3) *humin*, highest in molecular weight, darkest in color, insoluble in both acid and alkali, and most resistant to microbial attack.

All three groups of humic substances are relatively stable in soils. Even fulvic acid, the most easily degraded, is more resistant to microbial attack than most freshly applied plant residues. Depending on the environment, the half-life (the time required to destroy half the amount of a substance) of fulvic acid may be 10 to 50 years, while the half-life of humic acid is generally measured in centuries.

Nonhumic Substances

3-D models of biomolecules—check out the DNA-protein complex and water:
http://www.umass.edu/microbio/chime/

About 20 to 30% of the humus in soils consists of nonhumic substances. These substances are less complex and less resistant to microbial attack than those of the humic group. Unlike humic substances, they are comprised of specific biomolecules with definite physical and chemical properties. Some of these nonhumic substances are microbially modified plant compounds, while others are compounds synthesized by soil microbes as by-products of decomposition.

Included among the nonhumic substances are polysaccharides, polymers that have sugarlike structures and a general formula of $C_n(H_2O)_m$, where *n* and *m* are variable.

Polysaccharides are especially important in enhancing soil aggregate stability (see Section 4.5). Also included are polyuronides, which are not found in plants, but are synthesized by soil microbes.

Some even simpler compounds (such as low-molecular-weight organic acids and some proteinlike materials) are part of the nonhumic group. Although none of these simpler materials are present in large quantities, they may influence the availability of plant nutrients, such as nitrogen and iron, and may also directly affect plant growth.

Colloid Characteristics of Humus

The colloidal nature and chemistry of humus was described in Section 8.4. Humus colloids exhibit very high levels of surface area and negative charge—similar per unit volume to those of high-activity clay, but much greater than clay per unit mass. Depending on the pH, the cation exchange capacity of humus may range from about 150 to as high as 500 $cmol_c/kg$ (about 40 to 120 $cmol_c/L$). The water-holding capacity of humus on a mass basis (but not on a volume basis) is four to five times that of the silicate clays. Humus promotes aggregate formation and stability. The highly complex humus molecules contain chemical structures that absorb nearly all wavelengths of visible light, giving the substance its characteristic black color (see the humus-rich A horizons in Plates 2 and 8, following page 112).

Stability of Humus

Studies using radioactive isotopes have shown that some organic carbon incorporated into humus thousands of years ago is still present in soils, evidence that humic materials can be extremely resistant to microbial attack. This resistance of humic substances to oxidation is important in maintaining soil organic matter levels and in protecting associated nitrogen and other essential nutrients against rapid mineralization and loss from the soil. For example, the formation of polyphenol–protein complexes can protect the protein nitrogen from microbial attack. Yet, despite its relative resistance, humus is subject to continual slow decomposition. Without annual additions of sufficient plant residues, microbial oxidation of humus will result in reduced soil organic matter levels.

CLAY–HUMUS COMBINATIONS.[5]   Interaction with clay minerals provides another means of stabilizing soil organic matter and the nitrogen it contains. Organic matter that is entrapped in the ultra-micropores (<1 μm) formed by clay particles is physically inaccessible to decomposing organisms (see Figure 8.32). High-activity clays attract and hold such substances as amino acids, peptides, and proteins, forming complexes that protect these nitrogen-containing compounds from microbial degradation. Research suggests that organic matter associated with smectite clay is highly aromatic (containing many ring structures) and on average remains in the soil for 1000 or more years. Organic matter associated with kaolinite clay seems to contain more polysaccharides and remains in the soil for an average of 300 to 400 years. Iron and aluminum oxide coatings in highly weathered soils and allophane in volcanic soils also can bind with nitrogenous organic molecules and protect them from decay. Although the extent and mechanisms are not yet fully understood, clay–humus interactions undoubtedly contribute to the high organic matter content of clay soils (see Section 12.8).

12.5   INFLUENCES OF ORGANIC MATTER ON PLANT GROWTH AND SOILS

Long ago, the observation that plants generally grow better on organic-matter-rich soils led people to think that plants derive much of their nutrition by absorbing humus from the soil. We now know that higher plants derive their carbon from carbon dioxide and that most of their nutrients come from inorganic ions dissolved in the soil solution. In fact, plants can complete their life cycles growing totally without

---

[5] For a detailed account of clay–humus complexes, see Huang and Schnitzer (1986) and Loll and Bollag (1983).

humus, or even without soil (as in soilless or **hydroponic** production systems using only aerated nutrient solutions). This is not to say that soil organic matter is less important to plants than was once supposed, but rather that most of the benefits accrue to plants indirectly through the many influences of organic matter on soil properties. These will be discussed later in this section, after we consider two types of direct organic matter effects on plants.

Direct Influence of Humus on Plant Growth

It is well established that certain organic compounds are absorbed by higher plants. For example, plants can absorb a varying proportion of their nitrogen and phosphorus needs as soluble organic compounds. In addition various growth-promoting compounds such as vitamins, amino acids, auxins, and gibberellins, are formed as organic matter decays. These substances may at times stimulate growth in both higher plants and microorganisms.

Small quantities of both fulvic and humic acids in the soil solution are known to enhance certain aspects of plant growth (Table 12.4). Some scientists have suggested that the humic substances may act as hormone-like regulators of specific plant-growth functions such as cell elongation or lateral root initiation. However, little evidence is available that supports this mechanism. Other scientists suggest that the humic substances stimulate plant growth by improving the availability of micronutrients, especially iron and zinc.

The concentrations of humic substances commonly present in the soil solution in humid regions (50 to 100 mg/L or ppm) are effective in stimulating plant growth. Commercial humate products have been marketed with claims that small amounts enhance plant growth, but scientific tests of many of these products have failed to show any benefit from their use. Perhaps this is because effective levels of humic substances are naturally present in most soils.

Allelochemical Effects[6]

Allelopathy is the process by which one plant infuses the soil with a chemical that affects the growth of other plants. The plant may do this by directly exuding **allelochemicals,** or the compounds may be leached out of the plant foliage by throughfall rainwater. In other cases, microbial metabolism of dead plant tissues (residues) forms the allelochemicals (Figure 12.13). Occasionally, the term *allelochemical* is also applied to plant chemicals that inhibit microorganisms. In principle, the interactions are much like the antagonistic relationships among certain microorganisms discussed in Section 11.14.

A review of allelopathy:
http://www.colostate.edu/Depts/Entomology/courses/en570/papers_2002/mccollum.htm

Allelochemicals present in the soil are apparently responsible for many of the effects observed when various plants grow in association with one another. Because they produce such chemicals, certain weeds (e.g., johnsongrass and giant foxtail) damage crops far out of proportion to the size and number of weeds present. Crop residues left on the soil surface may inhibit the germination and growth of the next crop planted (e.g., wheat residues often inhibit sorghum plants).

Other allelopathic interactions influence the succession of species in natural ecosystems. Allelopathy may be partially responsible for the invasiveness of certain exotic

TABLE 12.4 **Some Direct Effects of Humic Substances on Plant Growth**

Effect on plant growth	Humic substance	Concentration range, mg/L
Accelerated water uptake and enhanced germination of seeds	Humic acid	1–100
Stimulated root initiation and elongation	Humic and fulvic acids	50–300
Enhanced root cell elongation	Humic acid	5–25
Enhanced growth of plant shoots and roots	Humic and fulvic acids	50–300

From Chen and Aviad (1990).

[6] For a comprehensive review of allelopathy, see Inderjit et al. (1999). For the role of allelopathy in plant species invasiveness, see Hierro and Callaway (2003).

FIGURE 12.13 Positive and negative allelopathic effects of winged beans on grain amaranth plants. In the pot on the left (T4) amaranth is growing in fresh soil (no association with winged beans). In the center pot (T20) amaranth is growing in soil previously used to grow winged beans (positive effect). In the pot on the right (T28) the amaranth is growing in fresh soil, but the plant was watered three times with a water extract of winged bean tissue (negative effect). The average dry weight of the amaranth plants for each treatment is shown. All pots were watered with a complete nutrient solution. [From Weil and Belmont (1987)]

plant species that rapidly dominate a new ecosystem to which they have been recently introduced. The invaders' allelochemicals may be more effective in the new ecosystem than they were in their territory of origin where neighboring plant species had time to evolve tolerance.

The walnut tree—allelopathic effects and tolerant plants:

http://www.ext.vt.edu/pubs/ nursery/430-021/430-021 .html

Allelopathic interactions are usually very specific, involving only certain species, or even varieties, on both the producing and receiving ends. The effects of allelopathic chemicals are many and varied. Although the term *allelopathy* most commonly refers to negative effects, allelochemical effects can also be positive (as in certain **companion plantings**). While they vary in chemical composition, most allelochemicals are relatively simple phenolic or organic acid compounds that could be included among the nonhumic substances found in soils. Because most of these compounds can be rapidly destroyed by soil microorganisms or easily leached out of the root zone, effects are usually relatively short-lived once the source is removed.

Influence of Organic Matter on Soil Properties and Indirectly on Plants

Soil organic matter affects so many soil properties and processes that a complete discussion of the topic is beyond the scope of this chapter. Indeed, in almost every chapter in this book there is mention of the roles of soil organic matter. Figure 12.14 summarizes some of the more important effects of organic matter on soil properties and on soil–environment interactions. Often one effect leads to another, so that a complex chain of multiple benefits results from the addition of organic matter to soils. For example (beginning at the upper left in Figure 12.14), adding organic mulch to the soil surface encourages earthworm activity, which in turn leads to the production of burrows and other biopores, which in turn increases the infiltration of water and decreases its loss as runoff, a result that finally leads to less pollution of streams and lakes.

INFLUENCE ON SOIL PHYSICAL PROPERTIES. Humus tends to give surface horizons dark brown to black colors. Granulation and aggregate stability are encouraged, especially by the nonhumic substances produced during decomposition (see Section 4.5). The humic fractions help reduce the plasticity, cohesion, and stickiness of clayey soils, making these soils easier to manipulate. Soil water retention is also improved, since organic matter increases both infiltration rate and water-holding capacity. Organic matter has an especially pronounced effect on the water-holding capacity of very sandy soils, which can often be improved by adding stable organic amendments (Figure 12.15).

INFLUENCE ON SOIL CHEMICAL PROPERTIES. Humus generally accounts for 50 to 90% of the cation-adsorbing power of mineral surface soils. Like clays, humus colloids hold nutrient cations (potassium, calcium, magnesium, etc.) in easily exchangeable form, wherein they can be used by plants but are not too readily leached out of the profile by percolating waters. Through its cation exchange capacity and acid and base functional groups,

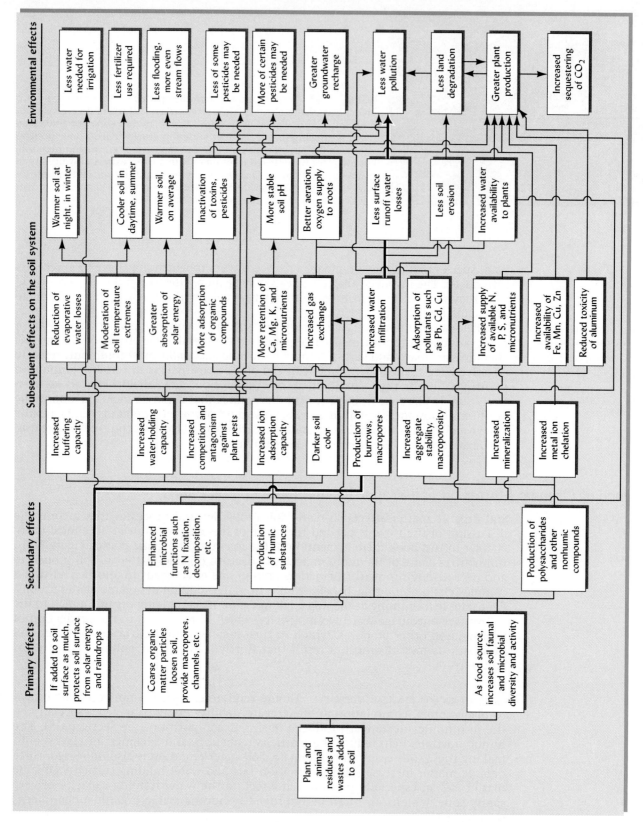

FIGURE 12.14 Some of the ways in which soil organic matter influences soil properties, plant productivity, and environmental quality. Many of the effects are indirect, the arrows indicating the cause-and-effect relationships. It can readily be seen that the influences of soil organic matter are far out of proportion to the relatively small amounts present in most soils. Many of these influences are discussed in this and other chapters in this book. The thicker line shows the sequence of effects referred to in the text in this section. (Diagram courtesy of R. Weil)

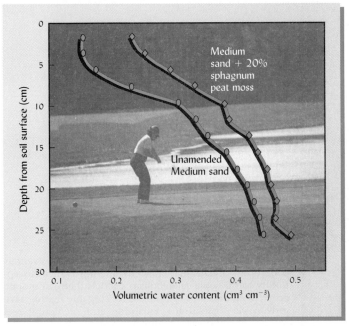

FIGURE 12.15 Effect of organic amendment on soil water retention. Columns 30 cm tall were filled with medium-sized sand of the type used for the root zone in golf course greens. The sand in some columns was mixed with peat (20% by volume). The columns were saturated with water and allowed to drain freely for 24 hours before the volumetric water was measured. While both columns were near 100% saturation at the lowest depth (where the water potential was near zero, see Section 5.3), the peat more than doubled the water held in the upper part of the profile. Improved water retention is a major reason why amending and topdressing sand-based golf greens are popular practices. However, the use of large amounts of peat mined from sphagnum wetlands cannot be considered environmentally sustainable (see Box 12.3). Compost made from various organic wastes (Sections 11.13 and 12.10) is a practical and environmentally beneficial substitute for peat in this kind of application. [Redrawn from data in Bigelow et al. (2004)].

organic matter also provides much of the pH buffering capacity in soils (see Section 9.4). In addition, nitrogen, phosphorus, sulfur, and micronutrients are stored as constituents of soil organic matter, from which they are slowly released by mineralization.

Humic acids also attack soil minerals and accelerate their decomposition, thereby releasing essential nutrients as exchangeable cations. Organic acids, polysaccharides, and fulvic acids all can attract such cations as Fe^{3+}, Cu^{2+}, Zn^{2+}, and Mn^{2+} from the edges of mineral structures and **chelate** or bind them in stable organomineral complexes. Some of these metals are made more available to plants as micronutrients because they are kept in soluble, chelated form (see Chapter 15). In very acid soils, organic matter alleviates aluminum toxicity by binding the aluminum ions in nontoxic complexes (see Sections 9.2 and 9.9).

BIOLOGICAL EFFECTS.   Soil organic matter—especially the detritus fraction—provides most of the food for the community of heterotrophic soil organisms described in Chapter 11. In Section 12.3 it was shown that the quality of plant litter and soil organic matter markedly affects decomposition rates and, therefore, the amount of organic matter accumulating in soils. The type and diversity of organic residues added to a soil can influence the type and diversity of organisms that make up the soil community.

12.6 AMOUNTS AND QUALITY OF SOIL ORGANIC MATTER

Perhaps the most useful approach to defining soil organic matter quality is to recognize different portions or **pools** of organic carbon that vary in their susceptibility to microbial metabolism. A model identifying five such pools of carbon in plant residues and soil organic matter is illustrated in Figure 12.16.

As discussed in Section 12.2, plant residues contain some components such as sugars, proteins, and starches, which are quite readily metabolized by soil microbes. Plant residues also contain carbon compounds that resist decomposition. The latter are found largely in the structure of the plant cell wall and include lignin, polyphenols, cellulose, and waxes. The model in Figure 12.16 denotes these groups as the *metabolic C pool* and *structural C pool* within the plant residues. The total organic matter in a soil also contains several pools, namely, the **active**, **slow**, and **passive** pools.

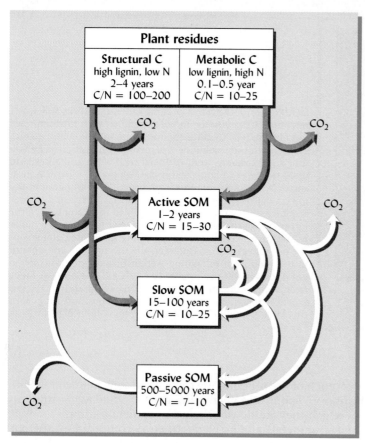

Plant residues	
Structural C high lignin, low N 2–4 years C/N = 100–200	**Metabolic C** low lignin, high N 0.1–0.5 year C/N = 10–25

CO_2 CO_2

CO_2

Active SOM
1–2 years
C/N = 15–30

CO_2

CO_2

Slow SOM
15–100 years
C/N = 10–25

Passive SOM
500–5000 years
C/N = 7–10

CO_2

FIGURE 12.16 A conceptual model that recognizes various pools of soil organic matter (SOM) differing by their susceptibility to microbial metabolism. Models that incorporate *active, slow,* and *passive* pools of soil organic matter have proven very useful in explaining and predicting real changes in soil organic matter levels and in associated soil properties. Note that microbial action can transfer organic carbon from one pool to another. For example, when the nonhumic substances and other components of the active fraction are rapidly broken down, some resistant, complex by-products may be formed, adding to the slow and passive pools. Note that all these metabolic changes result in some loss of carbon from the soil as CO_2. [Adapted from Paustian et al. (1992)]

Active Organic Matter

The **active pool** of soil organic matter consists of labile (easily decomposed) materials with half-lives (the time it takes for half of a mass of material to decay) of only a few days to a few years. Organic matter in the active pool has a relatively high average C/N ratio (about 15 to 30) and includes such organic matter fractions as the living biomass, tiny pieces of detritus (termed **particulate organic matter**, or POM), most of the polysaccharides, and other nonhumic substances described in Section 12.4, as well as some of the more labile fulvic acids. This active pool provides most of the readily accessible food for soil organisms and most of the readily mineralizable nitrogen. It is responsible for most of the beneficial effects on structural stability that lead to enhanced infiltration of water, resistance to erosion, and ease of tillage. The active pool can be readily increased by the addition of fresh plant and animal residues, but it is also very readily lost when such additions are reduced or tillage is intensified. This pool rarely comprises more than 10 to 20% of the total soil organic matter.

Passive and Slow Organic Matter

The **passive pool** of soil organic matter consists of very stable materials remaining in the soil for hundreds or even thousands of years. This pool includes most of the humus physically protected in clay–humus complexes, most of the humin, and much of the humic acid. The passive pool accounts for 60 to 90% of the organic matter in most soils, and its quantity is increased or diminished only slowly. The passive pool is most closely associated with the colloidal properties of soil humus, and it is responsible for most of the cation- and water-holding capacities contributed to the soil by organic matter.

Intermediate in properties between the active and passive pools is the **slow pool** of soil organic matter. This pool probably includes the finest fractions of particulate organic matter that are high in lignin and other slowly decomposable and chemically resistant components. The half-lives of these materials are typically measured in decades. The slow pool is an important source of mineralizable nitrogen and other plant nutrients, and it provides much of the underlying food source for the steady metabolism of the K-strategist soil microbes (see Section 12.2). The slow pool also probably makes some contribution to the effects associated primarily with the active and passive pools.

Changes in Active and Passive Pools with Soil Management

Changes in Soil Organic Matter in Canada:

http://www.agr.gc.ca/nlwis-snite/index_e.cfm?s1=pub&s2=hs_ss&page=11

Although scientists cannot yet accurately isolate and measure the active, slow, and passive pools of soil organic matter, analytical methods can evaluate chemical or physical fractions of soil organic matter (e.g., sugars, particulate organic matter, humin, etc.) that contribute to these functionally defined pools. Soil scientists have consistently observed that productive soils managed with conservation-oriented practices contain relatively high amounts of organic matter fractions associated with the active pool, including microbial biomass, particulate organic matter, and oxidizable sugars (see, for example, Table 20.8).

Despite the analytical difficulties, models that assume the existence of active, slow, and passive pools have proven very useful in explaining and predicting real changes in soil organic matter levels and in attendant soil properties. Studies on the dynamics of soil organic matter have established that the different pools of soil organic matter play quite different roles in the soil system and in the carbon cycle. The presence of a resistant (structural) pool of carbon in plant residues, as well as an easily decomposed (metabolic) pool, explains the initially rapid but decelerating rate of decay that occurs when plant tissues are added to a soil (see Figures 12.5 and 12.6). Similarly, the existence of a pool of complex, chemically and physically protected soil organic matter (passive pool), as well as a pool of easily metabolized soil organic matter (active pool), explains why conversion of native forests or grassland into cultivated cropland results in a very rapid decline in soil organic matter during the first few years, followed by a much slower decline thereafter (see, for example, Figure 12.17).

Soil-management practices that cause only very small changes in total soil organic matter often cause rather pronounced alterations in aggregate stability, nitrogen mineralization rate, or other soil properties attributed to organic matter. This occurs because the relatively small pool of active organic matter may undergo a large percentage increase or decrease without having a major effect on the much larger pool of total organic matter. Figure 12.17 shows how the different pools of soil organic matter are affected by changing management (in this case, cultivating a previously undisturbed soil). The figure also indicates how each pool contributes to the total soil organic matter level.

Accumulated plant residues and active organic matter are the first to be affected by changes in land management, accounting for most of the early losses in soil organic matter when cultivation of virgin soil begins. In contrast, losses from the passive pool are very gradual. As a result, the soil organic matter remaining after some years is far less effective in promoting structural stability and nutrient cycling than the original organic matter in the virgin soil. If a favorable change in environmental conditions or management regime occurs, the plant litter and active pools of soil organic matter are also the first to positively respond (see right-hand portion of Figure 12.17).

12.7 CARBON BALANCE IN THE SOIL–PLANT–ATMOSPHERE SYSTEM

Whether the goal is to reduce greenhouse gas emissions or to enhance soil quality and plant production, proper management of soil organic matter requires an understanding of the factors and processes influencing the cycling and balance of carbon in an ecosystem. Although each type of ecosystem, whether a deciduous forest, a prairie, or a wheat field, will emphasize particular compartments and pathways in the carbon cycle, consideration of a specific example, such as that described in Box 12.2, can help us develop a general model that can be applied to many different situations.

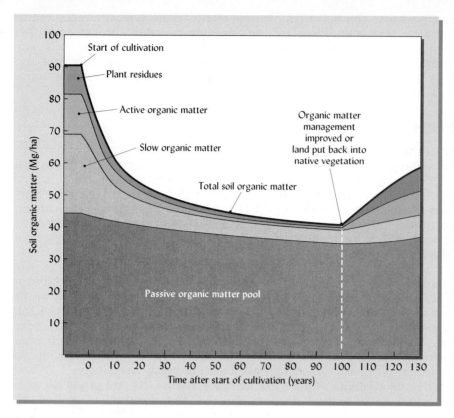

FIGURE 12.17 Changes in various fractions of organic matter in the upper 25 cm of a representative soil after bringing virgin land under cultivation. Initially, under natural vegetation, this soil contained about 91 Mg/ha of total organic matter. The resistant *passive pool* accounted for about 44 Mg/ha, or about half of the total soil organic matter. The rapidly decomposing *active pool* accounted for about 14 Mg, or about 16% of the total soil organic matter. After about 40 years of cultivation, the passive pool had declined by about 11% to about 39 Mg/ha, while the active pool had lost 90% of its mass, declining to only 1.4 Mg/ha. Note that much of the organic loss due to the change in land management came at the expense of the active pool. This was also the pool that most quickly increased when improved organic matter management was adopted after the 100th year. The susceptibility of the active pool to rapid change explains why even relatively small changes in total soil organic matter can produce dramatic changes in important soil properties, such as aggregate stability and nitrogen mineralization. (Diagram courtesy of R. Weil)

The rate at which soil organic matter either increases or decreases is determined by the balance between *gains* and *losses* of carbon. The gains come primarily from plant residues grown in place and from applied organic materials. The losses are due mainly to respiration (CO_2 losses), plant removals, and erosion (Table 12.5).

Agroecosystems

CONSERVATION OF SOIL CARBON.   In order to halt or reverse the net carbon loss shown in Figure 12.18, management practices would have to be implemented that would either *increase the additions* of carbon to the soil or *decrease the losses* of carbon from the soil. Since all crop residues and animal manures in the example are already being returned to the soil, additional carbon inputs could most practically be achieved by growing more plant material (i.e., increasing crop production or growing cover crops during the winter).

TABLE 12.5   **Factors Affecting the Balance between Gains and Losses of Organic Matter in Soils**

Factors promoting gains	*Factors promoting losses*
Green manures or cover crops	Erosion
Conservation tillage	Intensive tillage
Return of plant residues	Whole plant removal
Low temperatures and shading	High temperatures and exposure to sun
Controlled grazing	Overgrazing
High soil moisture	Low soil moisture
Surface mulches	Fire
Application of compost and manures	Application of only inorganic materials
Appropriate nitrogen levels	Excessive mineral nitrogen
High plant productivity	Low plant productivity
High plant root:shoot ratio	Low plant root:shoot ratio

BOX 12.2   CARBON BALANCE—AN AGROECOSYSTEM EXAMPLE

The principal carbon pools and annual flows in a terrestrial ecosystem are illustrated in Figure 12.18 using a hypothetical cornfield in a warm temperate region. During a growing season the corn plants produce (by photosynthesis) 17,500 kg/ha of dry matter containing 7500 kg/ha of carbon (C). This C is equally distributed (2500 kg/ha each) among the roots, grain, and unharvested aboveground residues. In this example, the harvested grain is fed to cattle, which oxidize and release as CO_2 about 50% of this C (1250 kg/ha), assimilate a small portion as weight gain, and void the remainder (1100 kg/ha) as manure. The corn stover and roots are left in the field and, along with the manure from the cattle, are incorporated into the soil by tillage or by earthworms.

The soil microbes decompose the crop residues (including the roots) and manure, releasing as CO_2 some 75% of the manure C, 67% of the root C, and 85% of the C in the surface residues. The remaining C in these pools is assimilated into the soil as humus. Thus, during the course of one year, some 1475 kg/ha of C enters the humus pool (825 kg from roots, plus 375 from stover, plus 275 from manure). These values are in general agreement with Figure 12.12, but they will vary widely among different soil conditions and ecosystems.

At the beginning of the year, the upper 30 cm of soil in our example contained 65,000 kg/ha organic C in humus. Such a soil cultivated for row crops would typically lose about 2.5% of its organic C by soil respiration each year. In our example this loss amounts to some 1625 kg/ha of C. Smaller losses of soil organic C occur by soil erosion (160 kg/ha), leaching (10 kg/ha), and formation of carbonates and bicarbonates (10 kg/ha).

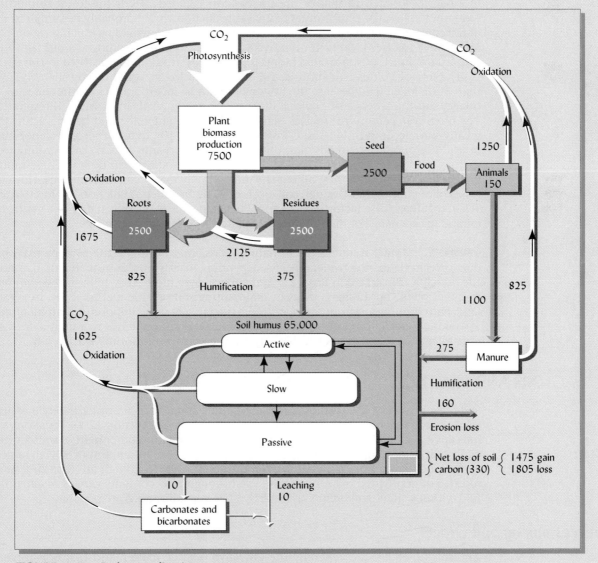

FIGURE 12.18   *Carbon cycling in an agroecosystem.*

(continued)

BOX 12.2 *(Cont.)* CARBON BALANCE—AN AGROECOSYSTEM EXAMPLE

Comparing total losses (1805 kg/ha) with the total gains (1475 kg/ha) for the pool of soil humus, we see that the soil in our example suffered a net annual loss of 330 kg/ha of C, or 0.5% of the total C stored in the soil humus. If this rate of loss were to continue, degradation of soil quality and productivity would surely result.

Specific practices to reduce carbon losses would include better control of soil erosion and the use of conservation tillage. Using a no-till production system would leave crop residues as mulch on the soil surface where they would decompose much more slowly. Refraining from tillage might also reduce the annual respiration losses from the original 2.5% to perhaps 1.5%. A combination of these changes in management would convert the system in our example from one in which soil organic matter is degrading (declining) to one in which it is aggrading (increasing).

Natural Ecosystems

FORESTS. Those interested in natural ecosystems may want to compare the carbon cycle of a natural forest with that of the cornfield in Figure 12.18. If the forest soil fertility were not too low, the total annual biomass production would probably be similar to that of the cornfield. The standing biomass, on the other hand, would be much greater in the forest since the tree crop is not removed each year. While some litter would fall to the soil surface, much of the annual biomass production would remain stored in the trees.

The rate of humus oxidation in the undisturbed forest would be considerably lower than in the tilled field because the litter would not be incorporated into the soil through tillage, and the absence of physical disturbance would result in slower soil respiration. The litter from certain tree species may also be rich in phenolics and lignin, factors that greatly slow decomposition and C losses (see Figure 12.10). In forest soils, decomposition of leaf litter produces copious quantities of dissolved organic carbon (DOC) compounds such as fulvic acids, and 5 to 40% of the total C losses may occur by leaching—a much greater proportion than from all but the most heavily manured cropland soils. However, losses of organic matter through soil erosion would be much smaller on the forested site. Taken together, these factors allow annual net gains in soil organic matter in a young forest and maintenance of high soil organic matter levels in mature forests.

GRASSLANDS. Similar trends occur in natural grasslands, although the total biomass production is likely to be considerably less, depending mainly on the annual rainfall. Among the principles illustrated in Box 12.2, and applicable to most ecosystems, is the dominant role that plant root biomass plays in maintaining soil organic matter levels. In a grassland, the contribution from the plant roots is relatively more important than in a forest. Therefore, a greater proportion of the total biomass produced tends to accumulate as soil organic matter, and this soil organic C is distributed more uniformly with depth.

12.8 FACTORS AND PRACTICES INFLUENCING SOIL ORGANIC LEVELS

Soil organic *carbon* (SOC) is more precisely defined and usually more accurately measured than the soil organic matter (SOM) of which it is a part (see Section 12.4). Therefore, various environmental and management effects on the organic component of soils are generally best compared, as following, in terms of SOC levels. The amount of SOC in mineral surface soils varies from a mere trace (sandy, desert soils) to as high as 10 or 20% (some forested or poorly drained A horizons). Even though SOC contents vary as much as tenfold within a single soil order (Table 12.1), a few generalizations are possible.

Differences among Soil Orders

Aridisols (dry soils) are generally the lowest in organic matter, and Histosols (organic soils) are definitely the highest (compare Plates 3 and 6). Contrary to popular myth, forested soils in humid tropical regions (e.g., Oxisols and some Ultisols) contain similar

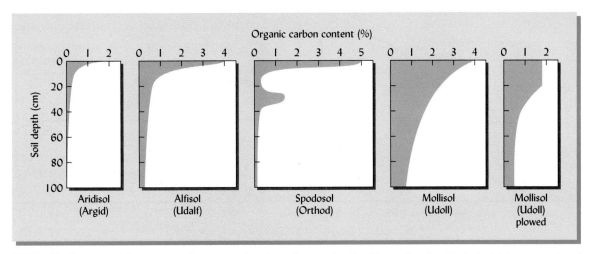

FIGURE 12.19 Vertical distribution of organic carbon in well-drained soils of four soil orders. Note the higher content and deeper distribution of organic carbon in the soils formed under grassland (Mollisols) compared to the Alfisol and Spodosol, which formed under forests. Also note the bulge of organic carbon in the Spodosol subsoil due to illuvial humus in the spodic horizon (see Chapter 3). The Aridisol has very little organic carbon in the profile, as is typical of dry-region soils.

amounts of organic carbon to those in humid temperate regions (e.g., Alfisols and Spodosols). Andisols (volcanic ash soils) generally have some of the highest organic carbon contents of any mineral soils, probably because association with allophane clay in these soils protects the organic carbon from oxidation (see Plate 2). Among cultivated soils in humid and subhumid regions, Mollisols (prairie soils) are known for their dark, organic, carbon-rich surface layers (see Plate 8).

Most organic residues in both cultivated and undisturbed soils are incorporated in, or deposited on, the soil surface. Therefore, organic matter tends to accumulate in the upper layers with much lower organic carbon contents in the subsurface horizons (Figure 12.19). Note that the organic carbon content decreases less abruptly with depth in grassland soils than in forested ones, because much of the annual carbon addition in grasslands comes in the form of fibrous roots extending deep into the profile (see Section 2.6).

Balance between Gains and Losses of C

As was indicated in Section 12.7, the level to which organic matter accumulates in soils is determined by the balance of gains and losses of organic carbon. In this regard, the level of SOM is analogous to the level of water in a bathtub, which is determined by the rates of inflow through the faucet and outflow through the drain. If the drain is fully opened, the water level will be low—even if the faucet is turned on full. In soils, organic carbon gains are principally governed by the amounts and types of residues added to the soil each year, while the losses result from oxidation, as well as from erosion. We will now consider the numerous factors that influence the rates of gain (how fully the faucet is turned on) and loss (how far the drain is opened). Table 12.5 lists some of the management-oriented factors that promote gains and losses of organic carbon.

Influence of Climate

TEMPERATURE.   The processes of organic matter production (plant growth) and destruction (microbial decomposition) respond differently to increases in temperature. Figure 12.20 shows that at low temperatures plant growth outstrips decomposition and organic matter accumulates. However, the opposite is true where mean annual temperature exceeds approximately 25 to 35 °C. At high temperatures, decomposition surpasses plant growth, so nutrient release is rapid, but organic matter accumulation is lower than in cooler soils. Some of the most rapid rates of organic matter decomposition occur in irrigated soils of hot desert regions.

Within zones of comparable moisture and vegetation, the average soil organic carbon and nitrogen contents increase from two to three times for each 10 °C decline in mean annual temperature. This temperature effect can be readily observed by noting

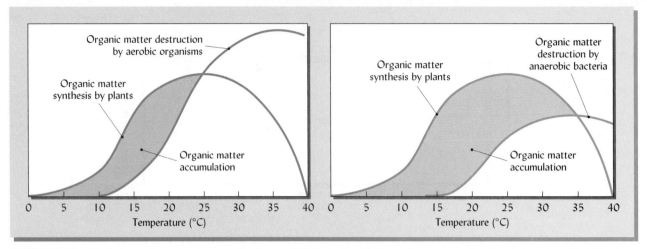

FIGURE 12.20 The balance between plant production and biological oxidation of organic matter determines the effect that temperature has upon organic matter accumulation in soils. The shaded areas indicate organic matter accumulation under aerobic (*left*) and anaerobic (*right*) conditions. Soil organic matter will accumulate to higher levels in cool climates, especially in waterlogged, anaerobic soils. Note that anaerobic accumulation is greater at most temperatures and continues at higher temperatures than under aerobic conditions. This explains why subtropical areas in Florida can contain both organic soils (e.g., the Everglades) and soils containing very little organic matter (e.g., in better drained parts of the state). [Adapted from Mohr and van Baren (1954)]

the greater organic carbon content of well-drained surface soils as one travels from south (Louisiana) to north (Minnesota) in the humid grasslands of the North American Great Plains region (Figure 12.21). Similar changes in soil organic carbon are evident as one climbs from warm lowlands to cooler highlands in mountainous regions.

MOISTURE. Under comparable conditions, soil organic carbon and nitrogen increase as the effective moisture becomes greater. The C/N ratio also tends to be higher in the more thoroughly leached soils of the higher rainfall areas. These relationships are evidenced by the darker and thicker A horizons encountered as one travels across the North American Great Plains region from the drier zones in the West (Colorado) to the more humid East (Illinois). The explanation lies mainly in the sparser vegetation of the drier regions. The lowest levels of soil organic matter and the greatest difficulty in maintaining those levels are found where annual mean temperature is high and rainfall is low. These relationships are extremely important to the relative difficulty of sustainable natural resource management. It must be remembered, as is shown following, that organic matter levels are influenced not only by temperature and precipitation, but also by vegetation, drainage, aspect, texture, and soil management.

Influence of Natural Vegetation

Climate and vegetation usually act together to influence the soil content of organic carbon. The greater plant productivity engendered by a well-watered environment leads to greater additions to the pool of soil organic matter. Grasslands generally dominate the subhumid and semiarid areas, while trees are dominant in humid regions. In climatic zones where the natural vegetation includes both forests and grasslands, the total organic matter is higher in soils developed under grasslands than under forests (see Figure 12.19). With grassland vegetation, a relatively high proportion of the plant residues consist of root matter, which decomposes more slowly and contributes more efficiently to soil humus formation than does forest leaf litter.

Effects of Texture and Drainage

While climate and natural vegetation affect soil organic matter over broad geographic areas, soil texture and drainage are often responsible for marked differences in soil organic matter within a local landscape. Under aerobic conditions, soils high in clay and silt are generally richer in organic matter than are nearby sandy soils (Figure 12.22). The finer-textured soils accumulate more organic matter for several reasons: (1) they

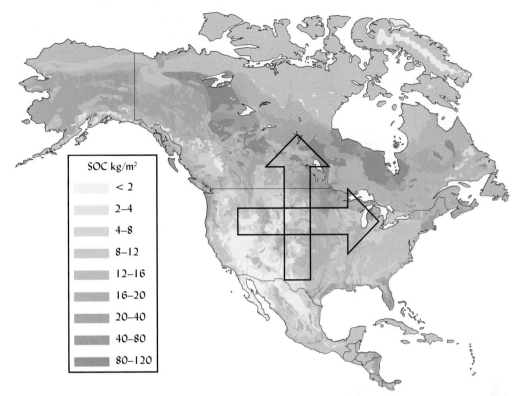

FIGURE 12.21 Distribution of soil organic carbon (SOC) in the upper 1 m of the soils of North America. The influences of precipitation and temperature are illustrated by the arrows indicating increasing soil organic C from west (drier) to east (wetter) and from south (warmer) to north (cooler) across the Great Plains region in the central part of the continent. The importance of Canadian Histosols and Gelisols as major repositories of organic C is also well illustrated. [Adapted from map by USDA/NRCS World Soils Resources Division]

produce more plant biomass, (2) they lose less organic matter because they are less well aerated, and (3) more of the organic material is protected from decomposition by being bound in clay–humus complexes (see Section 12.4) or sequestered inside soil aggregates. A given amount and type of clay can be expected to have a finite capacity to stabilize organic matter in organomineral complexes. Once this capacity is saturated, further additions of organic matter are likely to add little to humus accumulation, as they will remain readily accessible to microbial decomposition.

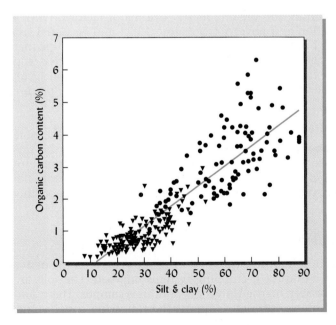

FIGURE 12.22 Soils high in silt and clay tend to contain high levels of organic carbon. The data shown are for surface soils in 279 tilled maize fields in subhumid regions of Malawi (▼) and Honduras (•). All soils were moderately well to well-drained and tilled. Variability (scatter of data points) among soils with the same silt + clay content is probably due to differences in (1) the type of clay minerals present (2:1 silicates tend to stabilize more organic carbon), (2) site elevation (cooler, high elevation locations being conducive to greater organic carbon accumulation), and (3) years since cultivation began (longer history of cultivation leading to lower organic carbon levels). (Data courtesy of R. Weil and M. A. Stine, University of Maryland, and S. K. Mughogho, University of Malawi)

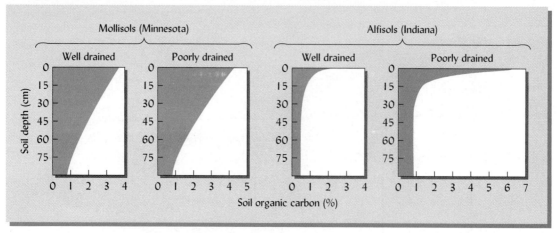

FIGURE 12.23 Distribution of organic carbon in four soil profiles, two well drained and two poorly drained. Poor drainage results in higher organic carbon content, particularly in the surface horizon.

DRAINAGE EFFECTS. In poorly drained soils, the high moisture supply promotes plant dry-matter production and relatively poor aeration inhibits organic-matter decomposition (see Figure 12.20b). Poorly drained soils therefore generally accumulate much higher levels of organic matter and nitrogen than similar but better-aerated soils (Figure 12.23).

Influence of Agricultural Management and Tillage

Except where barren deserts are brought under irrigation, it is safe to generalize that cultivated land contains much lower levels of organic matter than do comparable areas under natural vegetation. This is not surprising; under natural conditions all the organic matter produced by the vegetation is returned to the soil, and the soil is not disturbed by tillage. By contrast, in cultivated areas much of the plant material is removed for human or animal food and relatively less finds its way back to the land. Also, soil tillage aerates the soil and breaks up the organic residues, making them more accessible to microbial decomposition.

CONVERSION TO CROPLAND. A very rapid decline in soil organic matter occurs when a virgin soil is brought under cultivation. Eventually, the gains and losses of organic carbon reach a new equilibrium, and the soil organic matter content stabilizes at a much lower value (Figure 12.17). Similar declines in soil organic matter are seen when tropical rain forests are cleared; however, the losses may be even more rapid because of the higher soil temperatures involved. Declining soil organic matter is a major factor driving a downward spiral of soil degradation and poverty (see Figure 12.24 and Section 17.1) that plagues many of the world's 1 billion subsistence farmers. There is some hope that efforts to introduce reduced tillage, increased plant nutrient inputs, and better crop residue management will be able to reverse this trend (see following and Section 17.6). Organic matter losses are not so dramatic if forests or prairies are converted to pasture or hay production.

No-till farming—carbon as a new cash crop:

http://www.washingtonpost.com/ac2/wp-dyn?pagename=article&node=contentId=A55389-2002Aug23

CONSERVATION TILLAGE. Compared to conventional tillage, practices such as stubble mulching and no-till leave a higher proportion of the residues on or near the soil surface. These *conservation tillage* (see Section 17.6) techniques protect the soil from erosion and also discourage the rapid decomposition of crop residues that conventional tillage stimulates. These practices can help maintain or restore high surface soil organic carbon levels (Figure 12.25).

Influence of Rotations, Residues, and Plant Nutrients

Figure 12.26a illustrates changes in soil organic carbon during 120 years after the native prairie was first plowed and several different cropping systems imposed. The soils are Mollisols, and the plots can still be seen on the University of Illinois campus. These, and similar long-term experimental plots elsewhere, support the following conclusions.

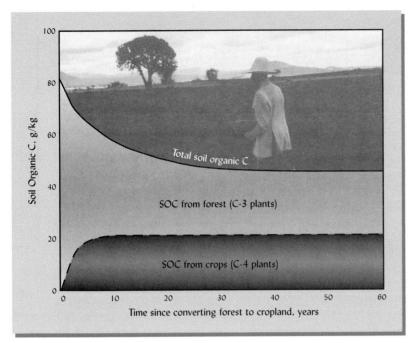

FIGURE 12.24 Amounts and sources of soil organic carbon (SOC) following conversion of forest to cropland. Researchers in Ethiopia sampled soils from neighboring farm fields cleared at various times in the past from native forest. The rapid initial decline of total SOC followed by a leveling off is typical for cultivated soils previously under grass or forest vegetation (compare to Figures 12.17 and 12.26a). The source of the SOC in these Humic Haplustands was determined by the relative amounts of the naturally occurring isotopes, ^{13}C and ^{12}C. The forest trees used the C-3 photosynthesis pathway (which favors use of $^{13}CO_2$ over the more abundant $^{12}CO_2$). Corn and sorghum use C-4 photosynthesis, which assimilates about half as much $^{13}CO_2$. Initially the forest-derived SOC declined sharply as the active pool decomposed, but after about 30 years a steady state was approached. The SOC with the $^{13}C/^{12}C$ signature of corn and sorghum increased rapidly at first, but reached a near steady state (no net change) plateau in only 10 years, as most of this C was in active fractions that turned over rapidly. Accumulation of SOC was limited by the small amounts of crop residues returned annually (240 kg C/ha) and by their rapid rate of decay. Rapid decay and low rate of residue return are typical where farmers plow their soil excessively, produce low yields, and burn crop residues. [Redrawn from Lemenih et al. (2005); photo courtesy of R. Weil].

A complex rotation (corn, oats, and clovers) achieved higher soil organic carbon levels than monocropping (continuous corn), regardless of fertility inputs, probably because the rotation used tillage less frequently and produced more root residues. Systems that maintain soil fertility with manure, lime, and phosphorus stimulate much higher organic carbon levels, especially where a complex rotation was also followed. This result was likely related to the greater additions of organic matter in the manure and in the residues from higher-yielding crops. Application of lime and fertilizers (N, P, and K) to previously unfertilized and unmanured plots (dashed lines starting in 1955) noticeably increased soil organic matter levels, probably due to the production and return of larger amounts of crop residues and the addition of sufficient nitrogen to compliment the carbon in humus formation.

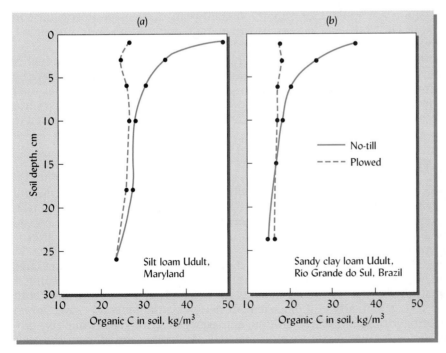

FIGURE 12.25 Less tillage means more soil organic carbon. In each case, the no-till system had been used on the experimental plots for 8 to 10 years when the data were collected. In the plowed plots, the soil was disturbed annually by tillage to about 20 cm deep. The soils in Maryland (a) and Brazil (b) were well-drained Ultisols and the climate was temperate (Maryland) to subtropical (Brazil). In Maryland, corn was grown every year with a rye cover crop. In Brazil, oats were rotated with corn using legume cover crops in between. In both cases, no-till encouraged the accumulation of organic C, but only in the upper 5 to 10 cm of the soil. [Data from Weil et al. (1988) and Bayer et al. (2000)]

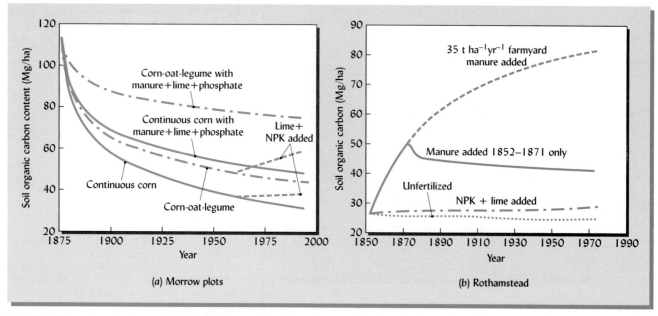

FIGURE 12.26 Soil organic carbon contents of selected treatments of (*a*) the Morrow plots at the University of Illinois and (*b*) the classical experiments at Rothamstead Experiment Station in England. The Morrow plots were begun on virgin grassland soil in 1876 and so suffered rapid loss of organic carbon in the early years of the experiment. The Rothamstead plots were established on soils with a long history of previous cultivation. As a result, the soil at Rothamstead had reached an equilibrium level of organic carbon characteristic of the unfertilized small-grains (barley and wheat) cropping system traditionally practiced in the area. [Data recalculated from Darmody and Peck (1997) and Jenkinson and Johnson (1977); used with permission of the Rothamstead Experiment Station, Harpenden, England]

Another famous long-term experiment was begun in 1855 at Rothamsted, England, on land that had been farmed to wheat for centuries (Figure 12.26*b*). Because the previous history of land use differed from that of the Morrow plots, the Rothamsted experiment provides lessons that are applicable to the old agricultural systems of Europe and Asia. Because this soil was already in equilibrium with the carbon and nitrogen gains and losses characteristic of unfertilized, small-grain cropping, continued production (and harvest) of barley and wheat resulted in little change, or a slow decline, in the already low level of soil organic carbon. Annual applications of animal manure at rates sufficient to supply all needed nitrogen resulted in a dramatic initial rise in soil organic carbon, until a new equilibrium state was approached at a much higher level. In the plots where manure was applied for only the first 20 years of the experiment, the soil organic carbon began to decline as soon as the manure applications were suspended, but the positive effect of the manure was still evident some 100 years later!

Results from such long-term experimental plots demonstrate that soils kept highly productive by supplemental applications of nutrients, lime, and manure and by the choice of high-yielding cropping systems are likely to have more organic matter than comparable, less productive soils. High productivity is sustainable if it involves not only greater economic harvests, but also larger amounts of roots and shoots returned to the soil.

The Conundrum of Soil Organic Matter Management

Land managers are faced with an inherent conflict between the need to *use* SOM and the need to *conserve* it. While the total carbon stabilized in the soil is important in relation to the global greenhouse effect (see Section 12.9), many beneficial effects on soil productivity and ecosystem function are realized only when the organic matter is destroyed by microbial metabolism.

In this regard, SOM is a lot like money—while it may be nice to accumulate large amounts, the benefits of having money are realized by spending it, not by keeping it "under the mattress." What is most desirable is an income large enough to allow both spending and saving. Similarly, soil management needs to simultaneously save carbon losses (by reducing losses from tillage, erosion, and aggregate disruption) while also providing a large income of high-quality organic materials. Although contradictory, the goals of using and conserving must be pursued simultaneously. The *decomposition* of SOM is necessary for its use as a source of nutrients for plant growth and organic compounds that promote biological diversity, disease suppression, aggregate stability, and metal chelation. In contrast, the *accumulation* of SOM is necessary for these functions in the long term, as well as for the sequestering of C, the enhancement of soil water-holding, the adsorption of exchangeable cations, the immobilization of pesticides, and the detoxification of metals.

General Guidelines for Managing Soil Organic Matter

Can Farmers Make Money With C Sequestration? (Texas A & M. Univ.):

http://www.oznet.ksu.edu/ctec/CASMGSnewsletter/Jan04-3.htm

A continuous supply of plant residues (roots and tops), animal manures, composts, and other materials must be added to the soil to maintain an appropriate level of soil organic matter, especially in the active pool. It is almost always preferable to keep the soil vegetated than to keep it in bare fallow. Even if some plant parts are removed in harvest, vigorously growing plants provide below- and aboveground residues as major sources of organic matter for the soil. Moderate applications of lime and nutrients may be needed to help free plant growth from the constraints imposed by chemical toxicities and nutrient deficiencies. Where climate permits, cover crops often present a tremendous opportunity to provide protective cover and additional organic material for the soil.

There is no "ideal" amount of soil organic matter. It is generally not practical to try to maintain higher soil organic matter levels than the soil–plant–climate control mechanisms dictate. For example, 1.5% organic matter might be an excellent level for a sandy soil in a warm climate, but would be indicative of a very poor condition for a finer-textured soil in a cool climate. It would be foolhardy to try to achieve as high a level of organic matter in a well-drained Texas silt loam soil as might be desirable for a similar soil found in Canada.

Adequate nitrogen is requisite for adequate organic matter because of the relationship between nitrogen and carbon in humus and because of the positive effect of nitrogen on plant productivity. Accordingly, the inclusion of leguminous plants and the judicious use of nitrogen-containing fertilizers to enhance high plant productivity are two desirable practices. At the same time, steps must be taken to minimize the loss of nitrogen by leaching, erosion, or volatilization (see Chapter 13).

Storing Carbon in Soil: Why and How?

http://www.agiweb.org/geotimes/jan02/feature_carbon.html

Tillage should be eliminated or limited to that needed to control weeds and to maintain adequate soil aeration. The more tillage that is performed, the faster organic matter is lost from the surface horizons.

Perennial vegetation, especially natural ecosystems, should be encouraged and maintained wherever feasible. Improved agricultural production on existing farmlands should be pursued to allow land currently supporting natural ecosystems to be left relatively undisturbed. In addition, there should be no hesitation about taking land out of cultivation and encouraging its return to natural vegetation where such a move is appropriate. In the United States, the Conservation Reserve Program provides incentives for such action (see Section 17.14). The fact is that large areas of land under cultivation today in every continent never should have been cleared.

12.9   THE GREENHOUSE EFFECT: SOILS AND CLIMATE CHANGE[7]

Free-air CO_2 enrichment (FACE) research:

http://cdiac.ornl.gov/programs/FACE/face.html

Soil is a major component of the Earth's system of self-regulation that has created (and, we hope, will continue to maintain) the environmental conditions necessary for life on this planet. Biological processes occurring in soils have major long-term effects on the composition of the Earth's atmosphere, which in turn influences all living things, including those in the soil.

Global Climate Change

Intergovernmental Panel on Climate Change: www.ipcc.ch/

Of particular concern today are increases in the levels of certain gases in the Earth's atmosphere. Known as **greenhouse gases**, they cause the Earth to be much warmer than it would otherwise be. Like the glass panes of a greenhouse, these gases allow short-wavelength solar radiation in but trap much of the outgoing long-wavelength radiation. This heat-trapping **greenhouse effect** of the atmosphere is a major determinant of global temperature and, hence, global climates. Gases produced by biological

---

[7] For a review of the potential for soil management to mitigate climate changes, see Paustian and Babcock (2004).

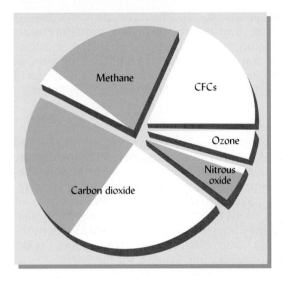

FIGURE 12.27 Relative contribution of different gases to the global greenhouse effect. The shaded portions indicate the emissions related to biological systems (in which soils play a role), the white portions being industrial contributions. [Modified from Dale et al. (1993)]

processes, such as those occurring in the soil, account for approximately half of the rising greenhouse effect (Figure 12.27). Of the five primary greenhouse gases, only chlorofluorocarbons (CFCs) are exclusively of industrial origin.

While it is certain that the concentrations of most greenhouse gases are increasing, there is less certainty about how rapidly global temperatures are actually rising and about how these increases are likely to affect climate in different regions of the world. Predicting changes in global temperature is complicated by numerous factors, such as cloud cover and volcanic dust, which can counteract the heat-trapping effects of the greenhouse gases. However, most scientists believe that the average global temperature has increased by 0.5 to 1.0 °C during the past century, and predict that it is likely to increase by another 1 to 2 °C in this century. If this increase in fact takes place, major changes in the Earth's climate are sure to result, including changes in rainfall distribution and growing season length, increases in sea level, and greater frequency and severity of storms. The rise in sea level alone, as predicted by some climate models, would threaten the homes of hundreds of millions of people living in coastal areas, mainly in Asia and North America. Through national programs and international treaties (such as those growing out of the Kyoto Protocols), much effort and expense are currently being directed at reducing the anthropogenic (human-caused) contributions to climate change. Soil science has the potential to contribute greatly to our ability to deal with global warming and the increasing levels of greenhouse gases.

Carbon Dioxide

Climate change will affect agriculture—the view from Australia:

http://www.greenhouse.gov .au/impacts/agriculture.html

Photo-documentation of global climate change. See links under "references":

http://www.worldviewofglobal warming.org/

Global climate change and agricultural production:

http://www.fao.org/docrep/ W5183E/W5183E00.htm

In 2007, the atmosphere contained about 390 ppm CO_2, as compared to about 280 ppm before the Industrial Revolution. Levels are increasing at about 0.5% per year. Although the burning of fossil fuels is a major contributor, much of the increase in atmospheric CO_2 levels comes from a net loss of organic matter from the world's soils. Through aerobic decomposition, the carbon in plant biomass and soil organic matter—carbon that originated from CO_2 in the atmosphere—is eventually converted back into CO_2 and returned to the atmosphere. Box 12.1 illustrated the importance of soil organic matter in regulating atmospheric CO_2 levels. Research (Figure 12.28) indicates that the feedback between the soil and atmosphere works both ways—changes in the levels of gases beneficial or harmful to plants influence the rate at which carbon accumulates in soil organic matter.

We have already discussed many ways that land managers can increase levels of soil organic matter (Section 12.8) by changing the balance between gains and losses. Gains in soil organic matter occur first in the active fractions, but eventually some of the carbon moves into the stable passive fraction, where it may be *sequestered* for hundreds or thousands of years. The opportunities for sequestering carbon are greatest for degraded soils that currently contain only a small portion of the organic matter levels they contained originally under natural conditions. Reforestation of denuded areas is one such opportunity. Others include switching cropland from conventional tillage to no-tillage,

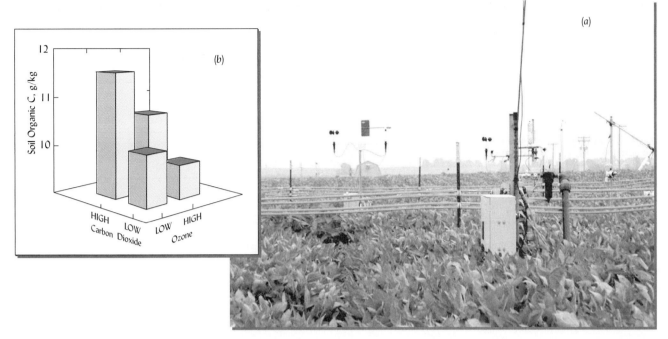

FIGURE 12.28 Changing atmospheric composition affects the C cycle. (*a*) Within a circular FACE (free air carbon enrichment) experimental plot on Mollisols at the University of Illinois, soybeans experience air altered to simulate the atmospheric composition expected in 50 years. Horizontal tubing and vertical poles supply computer-controlled concentrations of carbon dioxide and/or ozone, which are continuously monitored by various sensors. Increasing atmospheric CO_2 from low (350 mg/L, the ambient level) to high (500 mg/L, the level expected by the year 2050) enhanced photosynthesis and plant growth, thus increasing the amount of fixed carbon available for translocation to the roots and eventually to the soil. (*b*) Data from a different carbon-ozone enrichment experiment on Ultisols in Maryland show measurably increased soil organic carbon after 5 years of soybean and wheat crops grown in high CO_2 air. Increased root:shoot ratio, rhizodeposition of carbon compounds, and overall greater litter inputs all played a role. Ozone, a pollutant at ground level, injures plants, reduces photosynthesis, and therefore impacts the soil in a manner opposite to that of CO_2. The data suggest the full effect of CO_2 is seen only when ozone remains at low levels. [Data from Weil et al. (2000); photo courtesy of R. Weil]

which commonly sequesters 0.2 to 0.5 Mg/ha of carbon during the first decade. Conversion of cultivated land to perennial vegetation may sequester carbon at twice these rates.

By slowly increasing soil organic matter to near precultivation levels, such management changes could significantly enhance society's efforts to stem the rise in atmospheric CO_2 and at the same time improve soil quality and plant productivity. Some estimates suggest that during a 50-year period, improved management of agricultural lands could provide about 15% of the CO_2 emission reductions that the United States will need to make. It should be noted, however, that in accordance with the factors discussed in Section 12.8, soils have only a finite capacity to assimilate carbon into stable soil organic matter. Therefore carbon sequestration in soils can only buy time before other kinds of actions (shifts to renewable energy sources, increased fuel efficiency, etc.) are fully implemented to reduce carbon emissions to levels that will not threaten climate stability.

Calculate your greenhouse gases:

http://www.greentagsusa.org/ GreenTags/calculator_intro.cfm

Soil-related concerns in biofuel production:

http://www.culturechange.org/ cms/index.php?option=com_ content&task=view&id=107& Itemid=1

BIOFUELS PRODUCTION. The production of *biofuels* as substitutes for gasoline and diesel fuel is attracting much attention and money. One such fuel, biodiesel, consists of fuel-grade oil generally extracted from the same crops (e.g., soybeans, rapeseed, and sunflower) used to provide cooking oil. Even more attention is turning toward the production of ethanol, not from grain as has been done for decades, but from cellulosic residues such as corn stover (the stalks and leaves) and switchgrass, which can be grown much more cheaply and abundantly than grain. Much controversy still surrounds the question as to whether such fuels will actually produce more energy than is consumed in the fossil fuel used to grow and process the crop. However, one aspect that should *not* be controversial is the critical importance of leaving enough crop residues on the soil to maintain the levels of soil organic matter consistent with high quality and continued productivity of

TABLE 12.6 Deterioration of Soil Organic Carbon and Structural Properties in the Upper 5 to 10 cm of Three Ohio Soils after One Year of Corn Stover Removal for Biofuel Production[a]

Soil series, Great Group	Cropping history	Stover returned, Mg/ha[b]	Soil organic C, g/kg	Soil bulk density, Mg/m^3	Tensile strength of soil aggregates, kPa	Water stable aggregates < 4.75 mm diameter, %
Rayne silt loam (Hapludults)	33-yr. continuous no-till corn	5	29	1.46	140	18
		0	19	1.50	50	13
Hoytville clay loam (Epiaqualfs)	8-yr. corn/soybean minimum till	5	26	1.31	380	20
		0	22	1.49	120	8
Celina silt loam (Hapludalfs)	15-yr. corn/soybean no-till	5	28	1.25	225	36
		0	21	1.42	80	13

[a] Data selected from Blanco-Canqui et al. (2006).

[b] 5 Mg/ha corn stover represented the normal practice of returning all aboveground residues after corn grain harvest. Return of 0 Mg/ha represented removal of all stover for biofuel production.

the soil resource. Recent research suggests that soil organic matter levels, aggregate stability, and other soil properties critical to sustaining productivity may rapidly suffer if all, or even half, of the corn stover is removed to make biofuels (Table 12.6). Unfortunately, the need to "share" the plant residues with the soil has been overlooked by many energy engineers and planners advising policy makers on biofuel production strategies. Some have advised not only that all aboveground residues be harvested to make the biofuels, but that the crops be genetically altered to reduce the ratio of roots to shoots—just the opposite of what sustainable soil management requires!

Environmentalist assessment of climate change impacts in U.S. Gulf Coast:

http://www.ucsusa.org/gulf/

WETLAND SOILS. Discussion up to this point has focused on organic matter in mineral soils. Under waterlogged conditions in bogs and marshes, the oxidation of plant residues is so retarded that the soils in these environments consist mainly of organic matter and are classified as Histosols in *Soil Taxonomy* (see Chapter 3). Although they cover only about 2% of the world's land area, Histosols (and Histels—permafrost soils with organic surface horizons) are important in the global carbon cycle because they hold about 20% of global soil carbon. Drainage of these soils for high-value horticultural production speeds the oxidation of organic matter, which over time destroys the soil itself. Draining Histosols, or mining them for peat (Box 12.3), can make major contributions to the rise in atmospheric CO_2. On the other hand, Histosol formation may buffer climate change because increased global temperatures will cause the sea level to rise as seawater expands and ice caps melt. The rising sea level will flood more coastal land area, creating conditions conducive to the formation of more Histosols in tidal marshes. The organic matter accumulation in these new Histosols would represent a significant sequestering of CO_2 that might help to buffer the global warming trend (as would the increased absorption of CO_2 by the greater volume of ocean waters). However, such predictions are complicated by the involvement of another greenhouse gas, methane, which is emitted by some Histosols and other wetland soils.

METHANE. Methane (CH_4) occurs in the atmosphere in far smaller amounts than CO_2. However, methane's contribution to the greenhouse effect is nearly half as great as that from CO_2 because each molecule of CH_4 is about 25 times as effective as CO_2 in trapping outgoing radiation. The level of CH_4 is rising at about 0.6% per year. In 2000, there were about 1.8 ppm CH_4 in the atmosphere, more than double the preindustrial level. Soils serve as both a source and a sink for CH_4—that is, they both add CH_4 and remove it from the atmosphere.

Biological soil processes account for much of the methane emitted into the atmosphere. When soils are strongly anaerobic, as in wetlands and rice paddies, bacteria produce CH_4, rather than CO_2, as they decompose organic matter (see Sections 7.4 and 12.2). Among the factors influencing the amount of CH_4 released to the atmosphere from wet soils are: (1) the maintenance of a redox potential (Eh) near 0 mV, (2) the availability of easily oxidizable carbon, either in the soil organic matter or in plant residues returned to the soil, and (3) the nature and management of the plants growing

BOX 12.3 PEAT FOR POTS AND POWER: *UNSUSTAINABLE SOIL MINING*

The peat you may have purchased in large bags at the garden center is produced by literally mining wetland soils. More than half of the world's wetlands are considered to be *peatlands*, deep Histosols that occur mainly in cool climates (especially in Canada) and contain layers of *woody*, *sedimentary*, and *fibrous peat*. The fibrous peat is most desirable for horticultural uses such as mulching flower beds and topdressing or amending turfgrass, especially in sand-based golf greens (see Section 12.7). Most containerized plants are grown in soilless media consisting largely of peat. Peat and similar organic materials are lightweight, have great water-holding capacity (10 to 20 times their dry weight), contribute considerable cation exchange capacity for holding available plant nutrients, and provide large pores to promote drainage and aeration.

With the phenomenal growth in demand by horticultural "green" industries, mining of Canadian Histosols for peat is increasing. This activity increases greenhouse gas emissions (Figure 12.29) for two reasons. First, mining destroys the wetlands and eliminates their C sequestration function. Second, peat used for horticultural applications eventually decomposes, and its carbon is returned to the atmosphere. Even if peatland restoration efforts are successful and the cutover

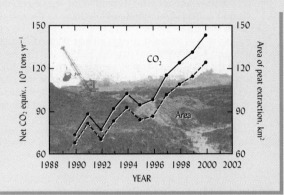

FIGURE 12.29 *Area of peat mining in Canada and the resulting net greenhouse gas emissions. The CO_2 equivalents take into account the fact that methane production is reduced when the peat bogs are mined. [Graphed from data in Cleary et al. (2005)].*

peatlands are restored to their role as net carbon sinks, research estimates it would take 2000 years to sequester the C lost during the past 50 years.

An even more direct conversion of sequestered C in peat to CO_2 is the extensive use of peat for fuel. In Scotland and Ireland, rural people still cut peat blocks from bog soils and dry them for use in home heating (Figure 12.30). In Russia, peat is mined on a much larger scale to provide fuel for electricity generation. While such peat utilization may be profitable in the short term, it is an inherently destructive and unsustainable use of the Histosol resource. It is also incompatible with global efforts to slow climate change by reducing CO_2 emissions.

Alternatives do exist. Organic wastes stabilized by composting can be used in place of peat for potting mixes and in turfgrass amendments. Composted materials provide most of the same ben-

FIGURE 12.30 *Peat blocks cut from the Histosol profile (see trench) are stacked to dry so they can be burned to heat rural homes in western Scotland. (Photo courtesy of Josh Weil).*

efits as peat, but, in addition, they may also suppress plant diseases (Section 11.13) and release significant levels of nutrients as they slowly decompose. Use of compost made from sewage sludge, farm manure, or municipal garbage not only provides a low-cost, renewable, greenhouse-neutral organic material for horticulture, it also solves some vexing waste disposal problems. With the mining of large quantities of peat continuing, the shift to such alternative materials needs to be encouraged.

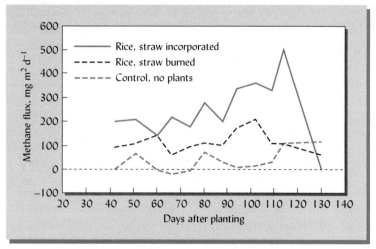

FIGURE 12.31 Factors affecting the emission of methane, a very active greenhouse gas, from a flooded rice paddy soil in California. The methane is generated by microbial metabolism in the soil and transported to the atmosphere through the rice plant. Emissions were greatest during the period of most active rice growth in mid-season. The sharp increase in methane near the end of the season was due to a rapid release of accumulated methane as the soil dried down and cracks opened up in the swelling clay. Very little methane was released if no rice was planted. Moderate amounts were released if rice was planted but rice residues from the previous crop had been burned off. The highest amounts of methane were released where rice was planted and the straw from the previous crop had been incorporated into the soil. Once the soil was drained at the end of the season, no more methane was produced. [Redrawn from Redeker et al. (2000), with permission of the American Association for the Advancement of Science]

on these soils (70 to 80% of the CH_4 released from flooded soils escapes to the atmosphere through the hollow stems of wetland plants). Figure 12.31 shows how these factors can influence CH_4 emissions from flooded rice paddies. Although not part of the study illustrated here, it has been demonstrated that periodically draining rice paddies prevents the development of extremely anaerobic conditions and therefore can substantially decrease CH_4 emissions. Such management practices should be given serious consideration, as rice paddies are thought to be responsible for up to 25% of global CH_4 production.

Wetland soils are not the only ones that contribute to atmospheric CH_4. Significant quantities of methane are also produced by the anaerobic decomposition of cellulose in the guts of termites living in well-aerated soils (see Section 11.5), and of garbage buried deep in landfills (see Section 18.10).

In well-aerated soils, certain **methanotrophic bacteria** produce the enzyme *methane monooxygenase*, which allows them to oxidize methane as an energy source:

$$CH_4 + 1/2O_2 \longrightarrow CH_3OH$$

Methane   Methanol

This reaction, which is largely carried out in soils, reduces the global greenhouse gas burden by about 1 billion Mg of methane annually. Unfortunately, the long-term use of inorganic (especially ammonium) nitrogen fertilizer on cropland, pastures, and forests has been shown to reduce the capacity of the soil to oxidize methane (see also Section 13.14). The evidence suggests that the rapid availability of ammonium from fertilizer stimulates ammonium-oxidizing bacteria at the expense of the methane-oxidizing bacteria. Long-term experiments in Germany and England indicate that supplying nitrogen in organic form (as manure) actually enhances the soil's capacity for methane oxidation (Table 12.7).

TABLE 12.7   Effect of Nitrogen Fertility Management Systems on Methane Oxidation by an Arable Soil (Mollisol) in Germany

The four nitrogen treatments were applied annually for 92 years to a rotation of sugar beets, spring barley, potatoes, and winter wheat. The measurements were made on soil sampled in spring, just before the annual nitrogen applications were made. Note that farmyard manure increased methane oxidation, while inorganic nitrogen fertilizer (NH_4NO_3) reduced methane oxidation from the levels in the control and the manure-only treatment.

Soil treatment	Soil pH	Soil NO_3^--N, kg/ha	Soil NH_4^+-N, kg/ha	Methane oxidation rate, nL CH_4 L^{-1}/hr^{-1}
1. Control—no N added in any form	6.8	0.83	0.20	4.60
2. Fertilizer N—40 to 130 kg/ha N as NH_4NO_3 to meet crop needs	6.9	15.36	3.1	1.34
3. Farmyard manure applied at 20 Mg/ha	7.0	1.98	0.22	11.2
4. Farmyard manure plus N fertilizer as in # 2.	7.2	5.01	0.71	3.76

Data from the Static Fertilization Experiment begun in 1902 at Bad Lauchstädt, Germany and reported in Willison et al. (1996).

Nitrous oxide (N_2O) is another greenhouse gas produced by microorganisms in poorly aerated soils, but since it is not directly involved in the carbon cycle, it will be discussed in the next chapter (see Section 13.9) on the nitrogen cycle.

Because the soil can act as a major source or sink for carbon dioxide, methane, and nitrous oxide, it is clear that, together with steps to modify industrial outputs, soil management has a major role to play in controlling the atmospheric levels of greenhouse gases.

12.10 COMPOSTS AND COMPOSTING

Making compost, step-by-step:

http://www.klickitatcounty. org/SolidWaste/default.asp? fCategoryIDSelected= 965105457

Composting is the practice of creating humuslike organic materials outside of the soil by mixing, piling, or otherwise storing organic materials under conditions conducive to aerobic decomposition and nutrient conservation (Figure 12.32). The decomposition processes and organisms involved are similar to those already described for the formation of humus in soils. Important differences include the fact that with composting, decay occurs outside of the soil and commonly undergoes periods of high temperatures not found in soils. The finished product, **compost**, is popular as a mulching material, as an ingredient for potting mixes, as an organic soil conditioner, and as a slow-release fertilizer.

High-quality compost can be made at ambient temperatures by a slow decomposition process, or more quickly by a process called *vermicomposting,* in which certain litter-dwelling (epigeic) earthworms are added to help transform the material. Vermicompost essentially consists of the casts made by earthworms eating the raw organic materials in moist, aerated piles. The piles are kept shallow to avoid heat buildup that could kill the worms.

We will focus, however, on the most commonly practiced type of composting, in which intense decomposition activity occurs within large, well-aerated piles. This approach is called *thermophilic composting* because the large mass of rapidly decomposing material, combined with the insulating properties of the pile, results in a considerable buildup of heat.

COMPOSTING PROCESS.   Thermophilic compost typically undergoes a three-stage process of decomposition (Figure 12.33): (1) During a brief, initial *mesophilic* stage, sugars and readily available microbial food sources are rapidly metabolized, causing the temperature in the compost pile to gradually rise from ambient levels to over 40 °C. (2) A *thermophilic* stage occurs during the next few weeks or months, during which temperatures rise to 50 to 75 °C while oxygen-using thermophilic organisms decom-

FIGURE 12.32   A special machine turns large-scale compost windrows (direction of travel is away from the reader) to mix the material and maintain well-aerated conditions at a facility in North Carolina where university dining hall food scraps are processed into compost for campus landscaping. (Photo courtesy of R. Weil)

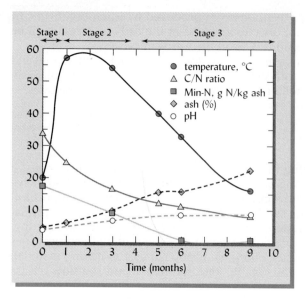

FIGURE 12.33 Physical and chemical changes during the composting process. In this case, easily decomposable plant materials were composted in a covered structure and turned monthly. This resulted in a very rapid rise in temperature in Stage 1, temperatures maintained above 40 °C for a long thermophilic Stage 2, until the readily decomposed compounds were depleted and temperatures dropped to the mesophilic range again during curing (Stage 3). Because 83% of the initial C but only 28% of the initial N was lost as gases, the C/N ratio declined from about 35 in the initial material to 9 in the finished compost. Most of the elements other than C and O were conserved, resulting in an increasing ash content (shown as a percentage of the dry weight). Mineral-N ($NH_4^+ + NO_3^-$), shown here as proportion of the ash, was high in the initial material, but declined to near zero as composting proceeded and the N was assimilated into microbial biomass and synthesized organic compounds. The pH of the material increased from 4 to 8 as organic acids were decomposed. [Adapted from data in Lynch et al. (2006)]

Controlled microbial composting—the Luebke method:

http://www.ibiblio.org/steved/Luebke/Luebke-compost2.html

pose cellulose and other more resistant materials. Frequent mixing during this stage is essential to maintain oxygen supplies and assure even heating of all the material. The easily decomposed compounds are used up, and humuslike compounds are formed during this stage. (3) A second mesophilic or *curing stage* follows for several weeks to months, during which the temperature falls back to near ambient, and the material is recolonized by mesophilic organisms, including certain beneficial microorganisms that produce plant-growth-stimulating compounds or are antagonistic to plant pathogenic fungi (see Section 11.14).

NATURE OF THE COMPOST PRODUCED.   As raw organic materials are humified in a compost pile, the content of nonhumic substances declines, and the content of humic acid increases markedly. During the composting process, the C/N ratio of organic materials in the pile decreases until a fairly stable ratio, in the range of 10:1 to 20:1, is achieved (Figure 12.33). The CEC of the organic matter may increase to about 50 to 70 cmol$_c$/kg of compost.

Although 50 to 75% of the carbon in the initial material is typically lost during composting, mineral nutrients are mostly conserved. The mineral content (referred to as *ash* content) therefore increases over time, making finished compost more concentrated in nutrients than the initial combination of raw materials used. Properly prepared finished compost should be free of viable weed seeds and pathogenic organisms, as these are generally destroyed during the thermophilic phase. However, inorganic contaminants such as heavy metals are *not* destroyed by composting. Proper management of compost is essential if the finished product is to be desirable for use as a potting media or soil amendment (see Box 12.4).

To compost or not to compost?

www.organicaginfo.org/upload/Compost.MarkMeasures.pdf

BENEFITS OF COMPOSTING.   Although making compost may involve more work and expense than applying uncomposted organic materials directly to the soil, the process offers several distinct advantages: (1) Composting provides a means of safely storing organic materials with a minimum of odor release until it is convenient to apply them to soils. (2) Compost is easier to handle than the raw materials as a result of the 30 to 60% smaller volume and greater uniformity of the resulting material. (3) For residues with a high initial C/N ratio, proper composting ensures that any nitrate depression period (see Figure 12.8) will occur in the compost pile, not in the soil, thereby avoiding induced plant nitrogen deficiency. (4) When applied to the soil, composted materials generally decompose and mineralize much more slowly than uncomposted organic materials. **Cocomposting** low-C/N-ratio materials (such as livestock manure and sewage sludge) with high-C/N-ratio materials (such as sawdust, wood chips, senescent tree leaves, or municipal solid waste) provides sufficient carbon for microbes to immobilize the excess nitrogen and minimize any nitrate leaching hazard (see Section 13.8) from the low-C/N materials. It also provides sufficient nitrogen to speed the decomposition of the high-C/N materials. (5) High

BOX 12.4   MANAGEMENT OF A COMPOST PILE

MATERIALS TO USE. Materials good for home composting include tree leaves (best if shredded), grass clippings, weeds (best before going to seed), kitchen scraps, wood shavings, gutter cleanings, pine needles, spoiled hay, straw, and even vacuum cleaner dust. Large-scale commercial compost is often made from materials such as municipal garbage, sewage sludge, wood chips, animal manures, municipal leaves, and food-processing wastes.

MATERIALS TO AVOID. Some materials to avoid include meat scraps (odors, rodents), cat droppings (carry microbes harmful to infants and pregnant women), sawdust from pressure-treated lumber and plywood (heavy metals and arsenic), and plastics and glass (nonbiodegradable; nuisance or dangerous in final product).

BALANCING NUTRIENTS. Although highly carbonaceous materials can be made to compost satisfactorily if they are turned frequently and kept moist, best results are obtained if high-C/N-ratio materials (e.g., brown leaves, straw, or paper) are mixed with low-C/N-ratio materials (e.g., green grass clippings, legume hay, blood meal, sewage sludge, or livestock manure) so as to achieve an overall C/N ratio between 20 and 30. Nitrogen fertilizer can also be added to lower the C/N ratio.

Other materials commonly added to improve nutrient balance and content include mixed fertilizers (N, P, and K), wood ashes (K, Ca, and Mg), bone meal or phosphate rock powder (P and Ca), and seaweed (K, Mg, Ca, and micronutrients). Some of these materials contain enough soluble salts to necessitate leaching of the finished compost before using it on salt-sensitive plants.

COMPOSTING METHODS. Provide good aeration throughout the pile, but build the pile large enough to provide mass sufficient to generate heat and prevent excessive drying. Backyard compost piles should be at least 1 m square and 1 m high. Compost bins are available to make turning the compost easier. Large-scale composting is usually carried out in windrows, about 2 to 3 m wide, 1 to 2 m tall, and many meters long (Figure 12.34). In dry climates, compost may be made in pits dug about 1 m deep to protect the material from drying. The various materials to be composted can be mixed together or applied in thin layers. Often, a small amount of garden soil or finished compost is added to

FIGURE 12.34 *(a) An efficient and easily managed method of composting suitable for homeowners is the three-bin method in which materials are turned with a pitchfork from one bin to the next. The perforated white plastic pipes enhance aeration. The leftmost bin contains relatively fresh materials, while the one at the right contains finished compost. (b) Sewage sludge mixed with wood chips is composted by a static (unturned), aerated pile method in which air is drawn through the large compost windrow via perforated pipes and small blowers. The air is expelled through smaller piles of finished compost, which absorb any odors. (Photos courtesy of R. Weil)*

ensure that plenty of decomposer organisms will be immediately available. Compost activators containing microbial inoculum or herbal extracts are available, but while some may speed the initial heating of the pile, scientific tests rarely show any other advantage to using these preparations.

OXYGEN AND MOISTURE CONTROL. Low oxygen levels, usually due to inadequate turning combined with excessive moisture, can produce putrid odors as anaerobic decomposition takes over. Monitoring temperature and oxygen levels in the pile can help avoid this situation. To promote good aeration it is best to mix in a bulking agent such as wood chips, avoid excessive packing, and either turn the pile or pull a stream of air through it (Figure 12.34). If turning is used, this should be done during the thermophilic stage whenever the temperature begins to drop. The compost water content should be maintained at 50 to 70%. Properly moist compost will feel damp—but not dripping wet—when squeezed. Turning the pile during dry weather can help reduce excess moisture, while turning it during rain can help moisten a too-dry pile.

The Don't Bag It composting plan to save landfill space:

http://aggie-horticulture.tamu.edu/earthknd/compost/compost.html

temperatures during the thermophilic stage in well-managed compost piles kill most weed seeds and pathogenic organisms in a matter of a few days. Under less ideal conditions, temperatures in parts of the pile may not exceed 40 to 50 °C, so weeks or months may be required to achieve the same results. (6) Most toxic compounds that may be in organic wastes (pesticides, natural phytotoxic chemicals, etc.) are destroyed by the time the compost is mature and ready to use. Composting is even used as a method of biological treatment of polluted soils and wastes (see Section 18.5). (7) Some composts can effectively suppress soilborne plant diseases by encouraging microbial antagonisms (see Section 11.13). Most success in disease suppression has occurred when well-cured compost is used as a main component of potting mixes for greenhouse-grown plants. (8) Because compost is made from organic waste materials that recently used up CO_2 in the process of their production (plant photosynthesis), compost is considered carbon-neutral, making it a much more environmentally sustainable choice than peat (see Box 12.3).

DISADVANTAGES OF COMPOST. Compost often has low nutrient contents and very low availability of the nutrients present. It also usually has a relatively high P to N ratio in comparison to plant needs. Because of this ratio, attempts to use compost as the principal source of plant nutrients can easily result in the application of potentially polluting levels of phosphorus (see Section 14.2). In addition, because the more labile organic substances are decomposed during the composting process, compost usually provides less benefit to soil aggregation than would the fresh residues from which the compost was made.

12.11   CONCLUSION

Organic matter is a complex and dynamic soil component that exerts a major influence on soil behavior, properties, and functions in the ecosystem. Because of the enormous amount of carbon stored in soil organic matter and the dynamic nature of this soil component, soil management may be an important tool for moderating the global greenhouse effect.

Organic residue decay, nutrient release, and humus formation are controlled by environmental factors and by the quality of the organic materials. High contents of lignin and polyphenols, along with high C/N ratios, markedly slow the decomposition process, causing organic matter to accumulate while reducing the availability of nutrients.

Soil organic matter comprises three major pools of organic compounds. The *active pool* consists of microbial biomass and relatively easily decomposed compounds, such as polysaccharides and other nonhumic substances. Although only a small percentage of the total carbon, the active pool plays a major role in nutrient cycling, micronutrient chelation, maintenance of structural stability and soil tilth, and as a food source underpinning biological diversity and activity in soils.

Most of the organic matter is in the *passive pool,* which contains very stable materials that resist microbial attack and may persist in the soil for centuries. This pool provides cation exchange and water-holding capacities, but is relatively inert biologically. The so-called *slow pool* is intermediate in stability and resistant to decomposition. It provides sources of food and energy for the steady metabolism of the authochthonous soil organisms that exist in the soil between times of residue additions. When soil is cleared of natural vegetation and brought under cultivation, the initial decline in soil organic matter is principally at the expense of the active pool. The passive pool is depleted very slowly and over very long periods of time.

The carbon-to-nitrogen (C/N) ratio of most soils is relatively constant, generally near 12:1. This means that the level of organic matter will be partially determined by the level of nitrogen available for assimilation into humus. Soil management for enhancing organic matter levels must, therefore, include some means of supplying nitrogen, for example, by the inclusion of leguminous plants.

The level of soil organic matter is influenced by climate (being higher in cool, moist regions), drainage (being higher in poorly drained soils), and by vegetation type (being generally higher where root biomass is greatest, as under grasses).

The maintenance of soil organic matter, especially the active pools, in mineral soils is one of the great challenges in natural resource management around the world. By encouraging vigorous growth of crops or other vegetation, abundant residues (which contain both carbon and nitrogen) can be returned to the soil directly or through feed-consuming animals. Also, the rate of destruction of soil organic matter can be minimized by restricting soil tillage, controlling erosion, and keeping most of the plant residues at or near the soil surface.

For some purposes it is advantageous to manage the decomposition of organic matter outside of the soil in a process known as *composting*. Composting transforms various organic waste materials into a humuslike product that can be used as a soil amendment or a component of potting mixes. The aerobic decomposition in a compost pile can conserve nutrients while avoiding certain problems, such as noxious odors and the presence of either excessive or deficient quantities of soluble nitrogen, which can occur if fresh organic wastes are applied directly to soils.

The decay and mineralization of soil organic matter is one of the main processes governing the economy of nitrogen and sulfur in soils—the subject of the next chapter.

STUDY QUESTIONS

1. Compare the amounts of carbon in Earth's standing vegetation, soils, and atmosphere.

2. If you wanted to apply an organic material that would make a long-lasting mulch on the soil surface, you would choose an organic material with what chemical and physical characteristics?

3. Describe how the addition of certain types of organic materials to soil can cause a nitrate depression period. What are the ramifications of this phenomenon for plant growth?

4. In addition to humic substances, what other categories of organic materials are found in soils?

5. Some scientists include plant litter (surface residues) in their definition of soil organic matter, while others do not. Write two brief paragraphs, one justifying the inclusion of litter as soil organic matter and one justifying its exclusion.

6. What soil properties are mainly influenced by the active and passive pools, respectively, of organic matter?

7. In this book and elsewhere, the terms *soil organic carbon* and *soil organic matter* are used to mean almost the same thing. How are these terms related, conceptually and quantitatively? Why is the term *organic carbon* generally more appropriate for quantitative scientific discussions?

8. Explain, in terms of the balance between gains and losses, why cultivated soils generally contain much lower levels of organic carbon than similar soils under natural vegetation.

9. In what ways are soils involved in the greenhouse effect that is warming up the Earth? What are some common soil-management practices that could be changed to reduce the negative effects and increase the beneficial effects of soils on the greenhouse effect?

10. Explain why compost is more environmentally sustainable than peat for use in potting media and as an amendment for golf course greens.

REFERENCES

Batjes, N. H. 1996. "Total carbon and nitrogen in the soils of the world," *European J. Soil Sci.,* **47**:151–163.

Bayer, C., J. Mielniczuk, T. Amado, L. Martin-Neto, and S. Fernandes. 2000. "Organic matter storage in a sandy clay loam Acrisol affected by tillage and cropping system in southern Brazil," *Soil and Tillage Research,* **54**:101–109.

Bigelow, C. A., D. C. Bowman, and D. K. Cassel. 2004. "Physical properties of sand amended with inorganic materials or sphagnum peat moss." USGA Turfgrass and Environmental Research Online, **3**(6):1–14. http://usgatero.msu.edu/v03/n06.pdf (posted 15 March 2004; verified 20 July 2006).

Blanco-Canqui, H., R. Lal, W. M. Post, R. C. Izaurralde, and L. B. Owens. 2006. "Rapid changes in soil carbon and structural properties due to stover removal from no-till corn plots," *Soil Sci.*, **171**:468–482.

Cadisch, G., and K. E. Giller (eds.). 1997. *Driven by Nature—Plant Litter Quality and Decomposition.* (Walingford, U.K.: CAB International).

Chen, Y., and T. Aviad. 1990. "Effects of humic substances in plant growth," pp. 161–186, in P. MacCarthy, C. E. Clapp, R. L. Malcolm, and P. R. Bloom (eds.), *Humic Substances in Soil and Crop Sciences: Selected Readings* (Madison, Wis.: ASA Special Publications).

Clapp, C. E., M.H.B. Hayes, A. J. Simpson, and W. L. Kingery. 2005. "Chemistry of soil organic matter", pp. 1–150, in M. A. Tabatabai and D. L. Sparks (eds.), *Chemical Processes in Soils*. SSSA Book Series No. 8. (Madison, Wis.: Soil Science Society of America).

Cleary, J., N. T. Roulet, and T. R. Moore. 2005. "Greenhouse gas emissions from Canadian peat extraction, 1990–2000: A life-cycle analysis," *AMBIO: A Journal of the Human Environment,* **34**:456–461.

Dale, V. H., R. A. Houghton, A. Grainger, A. E. Lugo, and S. Brown. 1993. "Emissions of greenhouse gases from tropical deforestation and subsequent uses of the land," pp. 215–260, in National Research Council, *Sustainable Agriculture and the Environment in the Humid Tropics* (Washington, D.C.: National Academy Press).

Darmody, R. G., and T. R. Peck. 1997. "Soil organic matter changes through time at the University of Illinois Morrow plots," pp. 161–169, in E. A. Paul, K. Paustian, E. T. Elliott, and C. V. Cole (eds.), *Soil Organic Matter in Temperate Agroecosystems: Long Term Experiments in North America* (Boca Raton, Fla.: CRC Press).

Eswaran, H., P. F. Reich, J. Kimble, F. H. Beinroth, E. Padmanabhan, and P. Moncharoen. 2000. "Global carbon stocks," pp. 15–26, in R. Lal et al. (eds). *Global Climate Change and Pedogenic Carbonates* (Boca Raton, Fla: Lewis Publishers).

Ferris, H., R. Venette, H. van der Meulen, and S. Lau. 1998. "Nitrogen mineralization by bacterial-feeding nematodes: Verification and measurement," *Plant and Soil,* **203**:159–171.

Hatcher, P. G., K. J. Dria, S. Kim, and S. W. Frazier. 2001. "Modern analytical studies of humic substances," *Soil Sci.*, **166**:770–794.

Hierro, J. L., and R. M. Callaway. 2003. "Allelopathy and exotic plant invasion," *Plant Soil*, **256**:29–39.

Huang, P. M., and M. Schnitzer (eds.). 1986. *Interactions of Soil Minerals with Natural Organics and Microbes*. SSSA Special Publication No. 17 (Madison, Wis.: Soil Science Society of America).

Inderjit, K. M., M. Dakshini, and C. L. Foy (eds.). 1999. *Principles and Practices in Plant Ecology: Allelochemical Interactions* (Boca Raton, Fla.: CRC Press).

IPCC. 2007. Climate change 2007: The physical science basis. Summary for policymakers. [Online]. Available by Intergovernmental Panel on Climate Change, United Nations. http://www.ipcc.ch/SPM2feb07.pdf (posted 2 February 2007; verified 3 February, 2007).

Jenkinson, D. S., and A. E. Johnson. 1977. "Soil organic matter in the Hoosfield barley experiment," *Rep. Rothamstead Exp. Stn. for 1976,* **2**:87–102.

Lemenih, M., E. Karltun, and M. Olsson. 2005. "Soil organic matter dynamics after deforestation along a farm field chronosequence in southern highlands of Ethiopia," *Agric. Ecosyst. Environ.,* **109**:9–19.

Loll, M. J., and J. M. Bollag. 1983. "Protein transformation in soil," *Advances in Agronomy,* **36**:352–382.

Lynch, D. H., R. P. Voroney, and P. R. Warman. 2006. "Use of 13c and 15n natural abundance techniques to characterize carbon and nitrogen dynamics in composting and in compost-amended soils," *Soil Biol. Biochem.,* **38**:1037–1114.

Magdoff, F., and R. R. Weil (eds.) 2004. *Soil Organic Matter in Sustainable Agriculture* (Boca Raton, Fla.: CRC Press), 398 pp.

Mohr, E.C.J., and P. A. van Baren. 1954. *Tropical Soils* (The Hague: N. V. Uitgeverij, W. Van Hoeve).

Mungai, N. W., and P. P. Motavalli. 2006. "Litter quality effects on soil carbon and nitrogen dynamics in temperate alley cropping systems," *Appl. Soil Ecol.* **31**:32–42.

Paustian, K., and B. Babcock. 2004. "Climate change and greenhouse gas mitigation: Challenges and opportunities," Task Force Report 141. Council on Agricultural Science and Technology, Ames, Iowa,120 pp.

Paustian, K., W. J. Parton, and J. Persson. 1992. "Modeling soil organic matter-amended and nitrogen-fertilized long-term plots," *Soil Sci. Soc. Amer. J.,* **56**:476–488.

Redeker, K., N. Wang, J. Low, A. McMillan, S. Tyler, and R. Cicerone. 2000. "Emissions of methyl halides and methane from rice paddies," *Science*, **290**:966–969.

Stevenson, F. J. 1994. *Humus Chemistry—Genesis, Composition Reactions*, 2nd ed. (New York: Wiley).

Swift, M. J., O. W. Heal, and J. M. Anderson. 1979. *Decomposition in Terrestrial Ecosystems.* Studies in Ecology, vol. 5 (Berkeley: University of California Press).

Tian, G. B., L. Brussaard, and B. T. Kang. 1995. "An index for assessing the quality of plant residues and evaluating their effects on soil and crop in the (sub-) humid tropics," *Applied Soil Ecol.*, **2**:25–32.

Tian, G., B. T. Kang, and L. Brussaard. 1992. "Biological effects of plant residues with contrasting chemical compositions under humid, tropical conditions—Decomposition and nutrient release," *Soil Biol. and Biochem.*, **24**:1051–1060.

Torbert, H., and H. Johnson. 2001. "Soil of the intensive agriculture biome of Biosphere 2. *J. Soil Water Conservation*, **56**:4–11.

Walford, R. L. 2002. "Biosphere 2 as voyage of discovery: The serendipity from inside," *BioScience*, **52**:259–263.

Weil, R. R., and G. S. Belmont. 1987. "Interactions between winged bean and grain amaranth," *Amaranth Newsletter*, **3**(1):3–6.

Weil, R. R., P. W. Benedetto, L. J. Sikora, and V. A. Bandel. 1988. "Influence of tillage practices on phosphorus distribution and forms in three Ultisols," *Agron. J.*, **80**:503–509.

Weil, R. R., K. R. Islam, and C. L. Mulchi. 2000. "Impact of elevated CO_2 and ozone on C cycling processes in soil," p. 47, in *Agronomy Abstracts* (Madison, Wis.: American Society of Agronomy).

Willison, T., R. Cook, A. Müller, and D. Powlson. 1996. "CH_4 oxidation in soils fertilized with organic and inorganic N: Differential effects," *Soil Biol. and Biochem.*, **28**:135–136.

Legumes enrich soils with nitrogen. (R. Weil)

13

NITROGEN AND SULFUR ECONOMY OF SOILS

The pulse and body of the soil . . .
—*D. H. LAWRENCE,*
THE RAINBOW

The biogeochemistries of nitrogen and sulfur have much in common. Both are essential nutrients found primarily in organic forms in soils. Both move in soils and plants mostly in the form of anions. Both undergo oxidation and reduction to form gases, and both are responsible for serious environmental problems. This chapter explores the stories of these two elements, starting with nitrogen.[1]

More money and effort are spent on the management of nitrogen than on any other nutrient element. And for good reason. Deficiencies or excesses of nitrogen have major impacts on the health and productivity of the world's ecosystems. Were it not for the biological fixation of nitrogen from the atmosphere by certain soil microbes, and for the recycling back to the soil of much of the nitrogen taken up by the vegetation, deficiencies of nitrogen would be ubiquitous, and most ecosystems would grind to a halt. Where humans harvest food from the land, these cycles have been disrupted. Yellowish leaves of nitrogen-starved crops forebode hunger or financial ruin for people in all corners of the globe. Nitrogen fertilizer is an expensive but often necessary input in agricultural systems. Its manufacture accounts for a large part of the fossil fuel energy used by the agricultural sector. The quest for protein—which contains about one-sixth nitrogen—by billions of humans is fueling demand for nitrogen fertilizer and transforming both global agriculture and the global N cycle.

Excesses of nitrogen are now nearly as common as deficiencies and can adversely affect both human and ecosystem health. Nitrogen leaking from overly enriched soils can lead to nitrate pollution of groundwater, making it unfit for human or animal consumption. In addition, the movement of soluble nitrogen compounds from soils to aquatic systems can disrupt the balance of those systems, leading to algae blooms, declining levels of dissolved oxygen, and subsequent death of fish and other aquatic species. Yet another way in which nitrogen links soils to the wider environment is the ozone-destroying and climate-forcing action of nitrous oxide gas generated in soils.

---

[1] For a review of nitrogen in agriculture, with special emphasis on environmental and human health impacts, see Addiscott (2005). For a comprehensive assessment of natural and anthropogenic reactive nitrogen in global ecosystems, see Galloway et al. (2003).

542

Clearly, soil processes are central to the global nitrogen cycle. The ecological, financial, and environmental stakes could not be higher.

13.1 INFLUENCE OF NITROGEN ON PLANT GROWTH AND DEVELOPMENT

Rate nitrogen sufficiency for corn:

http://www.ipm.iastate.edu/ipm/icm/2006/9-18/ntool.html

ROLES IN THE PLANT. Nitrogen is an integral component of many essential plant compounds. It is a major part of all amino acids, which are the building blocks of all proteins—including the enzymes, which control virtually all biological processes. Other critical nitrogenous plant components include the nucleic acids, in which hereditary control is vested, and chlorophyll, which is at the heart of photosynthesis. Nitrogen is also essential for carbohydrate use within plants. A good supply of nitrogen stimulates root growth and development, as well as the uptake of other nutrients.

Plants respond quickly to increased availability of nitrogen, their leaves turning deep green in color. Nitrogen increases the plumpness of cereal grains, the protein content of both seeds and foliage, and the succulence of such crops as lettuce and radishes. It can dramatically stimulate plant productivity, whether measured in tons of grain, volume of lumber, carrying capacity of pasture, or thickness of lawn. Healthy plant foliage generally contains 2.5 to 4.0% nitrogen, depending on the age of the leaves and whether the plant is a legume.

DEFICIENCY. Plants deficient in nitrogen tend to exhibit **chlorosis** (yellowish or pale green leaf colors), a stunted appearance, and thin, spindly stems (see Plates 87 and 94 after page 656). In nitrogen-deficient plants, the protein content is low and the sugar content is high, because carbon compounds normally destined to build proteins cannot be used to do so without sufficient nitrogen. Nitrogen is quite mobile (easily translocated) within the plant. When plant uptake is inadequate, supplies are transferred to the newest foliage, causing the older leaves to show pronounced chlorosis. The older leaves of nitrogen-starved plants are therefore the first to turn yellowish, possibly becoming prematurely senescent and dropping off (Figures 13.1a and b). Nitrogen-deficient plants often have a low shoot-to-root ratio, and they mature more quickly than healthy plants. The negative effects of nitrogen deficiency on plant size and vigor are often dramatic (Figure 13.1c).

Global Nitrogen: Cycling out of Control:

http://www.ehponline.org/members/2004/112-10/focus.html

OVERSUPPLY. When too much nitrogen is applied, excessive vegetative growth occurs; the cells of the plant stems become enlarged but relatively weak, and the top-heavy plants are prone to falling over (lodging) with heavy rain or wind (Figure 13.1d). High nitrogen applications may delay plant maturity and cause the plants to be more susceptible to disease (especially fungal disease) and to insect pests. These problems are especially noticeable if other nutrients, such as potassium, are in relatively low supply.

An oversupply of nitrogen degrades crop quality, resulting in undesirable color and flavor of fruits and low sugar and vitamin levels of certain vegetables and root crops. Flower production in ornamentals is reduced in favor of abundant foliage. An oversupply can also cause the buildup of nitrates that are harmful to livestock in the case of foliage and to babies in the case of leafy vegetables. Growth-inhibiting environmental conditions (e.g., drought; dark, cloudy days; or cool temperatures) can exacerbate the accumulation of nitrate in tissues because plant conversion of nitrate into protein is slowed. The leaching of excess nitrates from the soil can lead to environmental degradation of downstream water basins (see Sections 13.8 and 16.2).

FORMS OF NITROGEN TAKEN UP BY PLANTS. Plant roots take up nitrogen from the soil principally as dissolved nitrate (NO_3^-) and ammonium (NH_4^+) ions. Although certain plants grow best when provided mainly by one or the other of these forms, a relatively equal mixture of the two ions gives the best results with most plants. As explained in Chapter 9 (see Sections 9.1 and 9.5 and Figure 9.13), uptake of ammonium markedly lowers the pH of the rhizosphere soil while uptake of nitrate tends to raise it. These pH changes, in turn, influence the uptake of other ions such as phosphates and micronutrients. In addition to NO_3^- and NH_4^+ ions, nitrite ions (NO_2^-) can also be taken up, but fortunately only trace quantities of this toxic ion usually occur in field soils.

Although NO_3^- and NH_4^+ ions are the dominant forms of N taken up by plants in many systems, numerous plants have been shown to also take up low molecular weight dissolved organic compounds (mainly soluble proteins and amino acids). Generally,

FIGURE 13.1 Symptoms of too little and too much nitrogen. (*a*) The nitrogen-starved bean plant (*right*) shows the typical chlorosis of lower leaves and markedly stunted growth compared to the normal plant on the left. (*b*) The oldest leaves at right near the base of this potted cucurbit vine are chlorotic because N has been transferred to the newest leaves on the left which are dark green. (*c*) The oldest bottom corn leaves yellowed, beginning at the tip and continuing down the midrib—a pattern typical of nitrogen deficiency in corn. (*d*) Asian rice heavily fertilized with nitrogen. The traditional tall cultivar in the left field has lodged, but the modern rice cultivar at right yields well with high nitrogen inputs because of its short stature and stiff straw. (Photos courtesy of R. Weil)

when soluble organic and inorganic N sources are both present in equal amounts, the plant uptake of organic N is an order of magnitude smaller than the uptake of mineral N. However, there is much variation, even within a family of plants. For example, among the Gramineae, sorghum and rice seem to efficiently take up soluble protein N, while corn and millet do not. Also, some evidence suggests that if grown with soil solutions relatively high in organic N and low in inorganic N, some plant species may become conditioned to utilize organic N more effectively. The direct uptake of soluble organic N is of particular significance in natural grasslands and forests.

13.2 DISTRIBUTION OF NITROGEN AND THE NITROGEN CYCLE

Nitrogen sources and transformations: http://www.ext.colostate.edu/pubs/crops/00550.html

The air above each hectare of land contains some 75,000 Mg of nitrogen. The atmosphere, which is 78% gaseous nitrogen (N_2), appears to be a virtually limitless reservoir of this element.[2] However, the very strong triple bond between two nitrogen atoms

---

[2] We will not consider here the vast pool of nitrogen contained in igneous rocks deep under the planet's crust, where it is effectively out of contact with the biosphere.

Nitrogen species found in
soil:
http://www.sws.uiuc.edu/
nitro/nspecies.asp

(N≡N) makes this gas extremely inert and not directly usable by plants or animals. Little nitrogen would be found in soils and little vegetation would grow in terrestrial ecosystems around the world were it not for certain natural processes (principally microbial nitrogen fixation and lightning) that can break this triple bond and form **reactive nitrogen** (see Section 13.10). In reactive nitrogen, which includes any form of nitrogen that is readily available to living organisms, the N is usually bonded to hydrogen, oxygen, or carbon (e.g., NH_4^+, NO_3^-, and amino acids, R—C—NH_2).

Most of the nitrogen in terrestrial systems is found in the soil, with A horizons normally ranging from 0.02 to 0.5% N and a value of about 0.15% N being representative for cultivated soils. A hectare of such a soil would contain about 3.5 Mg nitrogen in the A horizon and perhaps an additional 3.5 Mg in the deeper layers. In forest soils, the litter layer (O horizons) might contain another 1 to 2 Mg of nitrogen. The soil contains 10 to 20 times as much nitrogen as does the standing vegetation (including roots) of either forested or cultivated areas. Most soil nitrogen occurs as part of organic molecules. Soil organic matter typically contains about 5% nitrogen; therefore, the distribution of soil nitrogen closely parallels that of soil organic matter (see Sections 12.3 and 12.10).

Except where large amounts of chemical fertilizers have been applied, inorganic (i.e., mineral) nitrogen seldom accounts for more than 1 to 2% of the total nitrogen in the soil. Unlike the bulk of organic nitrogen, most mineral forms of nitrogen are quite soluble in water and may be easily lost from soils through leaching and volatilization.

The Nitrogen Cycle

Nitrogen cycle links:
http://users.rcn.com/jkimball
.ma.ultranet/Biologypages/N/
NitrogenCycle.html

As it moves through the **nitrogen cycle**, an atom of nitrogen may appear in many different chemical forms, each with its own properties, behaviors, and consequences for the ecosystem. This cycle explains why vegetation (and, indirectly, animals) can continue to remove nitrogen from a soil for centuries without depleting the soil of this essential nutrient. The **biosphere** does not run out of nitrogen, because it uses the same nitrogen over and over again. This cycling of N is highly visible in Figure 11.22, which shows a dark green circle of grass plants that have responded to the fungal release of N from the residues of previous grass growth.

The nitrogen cycle has long been the subject of intense scientific investigation, since understanding the translocations and transformations of this element is fundamental to solving many environmental, agricultural, and natural-resource-related problems. The principal pools and forms of nitrogen, and the processes by which they interact in the cycle, are illustrated in Figure 13.2. This figure deserves careful study; we will refer to it frequently as we discuss each of the major divisions of the nitrogen cycle.

Ammonium and nitrate are two critical forms of inorganic nitrogen in the cycle. In addition to its possible loss by erosion and runoff, the N in ammonium is subject to five fates: (1) *immobilization* by microorganisms; (2) removal by *plant uptake;* (3) *fixation* in the interlayers of certain 2:1 clay minerals; (4) *volatilization* after being transformed into ammonia gas; and (5) oxidation to nitrite and subsequently to nitrate by a microbial process called *nitrification*. Similarly, the N in nitrate is subject to four possible fates: (1) *immobilization* by microorganisms; (2) removal by *plant uptake;* (3) loss to groundwater by *leaching* in drainage water; or (4) *volatilization* to the atmosphere as several nitrogen-containing gases formed by *denitrification*.

13.3 IMMOBILIZATION AND MINERALIZATION

Effects of field conditions:
http://muextension.missouri
.edu/explore/envqual/wq0260
.htm

The great bulk (95 to 99%) of the soil nitrogen is in organic compounds that protect it from loss but leave it largely unavailable to higher plants. Much of this nitrogen is present as amine groups (R—NH_2), largely in proteins or as part of humic compounds. When soil microbes attack these compounds, simple amino compounds[3] (R—NH_2) are formed. Then the amine groups are hydrolyzed, and the nitrogen is released as ammonium ions (NH_4^+), which can be oxidized to the nitrate form. The decomposition process involves the breakdown of large, insoluble N-containing organic molecules into

---

[3] Amino acids such as lysine (CH_2NH_2COOH) and alanine (CH_3CHNH_2COOH) are examples of these simpler compounds. The R in the generalized formula represents the part of the organic molecule with which the amino group (NH_2) is associated. For example, for lysine, the R is CH_2COOH.

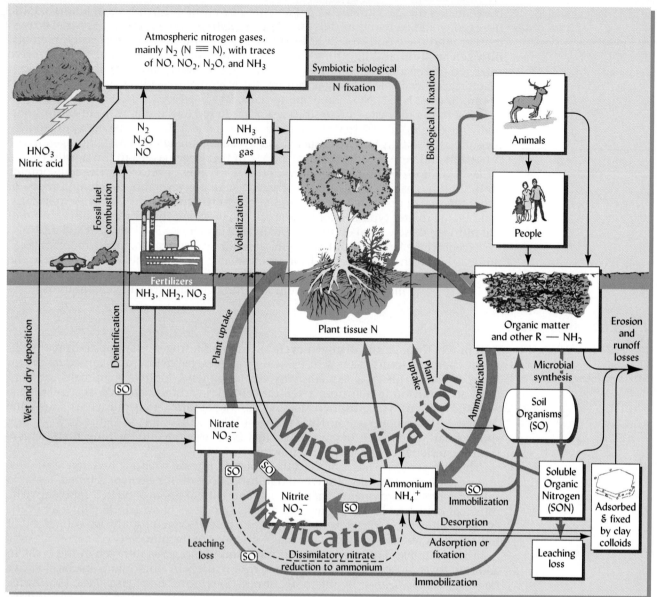

FIGURE 13.2 The nitrogen cycle, emphasizing the primary cycle (thick, gray arrows) in which organic nitrogen is mineralized, plants take up the mineral nitrogen, and eventually organic nitrogen is returned to the soil as plant residues. Note the processes by which soil nitrogen is lost and replenished. The boxes represent various forms of nitrogen; the arrows represent processes by which one form is transformed into another. Soil organisms, whose enzymes drive most of the reactions in the cycle, are represented as rounded boxes labeled "SO". [Diagram courtesy of R. Weil]

smaller and smaller units with the eventual release of the nitrogen as NH_4^+. The enzymes that bring about this process are produced mainly by microorganisms (but some are produced by plant roots and soil animals) and include hydrolases and deaminases that break C—H and $C—NH_2$ bonds. The enzymes may carry out the reactions inside microbial cells, but most often they are excreted by the microbes and work extracellularly in the soil solution or while adsorbed to colloidal surfaces. This enzymatic process termed **mineralization** (Figure 13.2) may be indicated as follows, using an amino compound ($R—NH_2$) as an example of the organic nitrogen source:

$$\xrightarrow{\hspace{3cm}} \text{Mineralization} \xrightarrow{\hspace{3cm}}$$

$$R—NH_2 \underset{-2H_2O}{\overset{+2H_2O}{\rightleftharpoons}} OH^- + R—OH + NH_4^+ \underset{-O_2}{\overset{+O_2}{\rightleftharpoons}} 4H^+ + energy + NO_2^- \underset{-^{1/2}O_2}{\overset{+^{1/2}O_2}{\rightleftharpoons}} energy + NO_3^-$$

$$\xleftarrow{\hspace{3cm}} \text{Immobilization} \xleftarrow{\hspace{3cm}}$$

(13.1)

Many studies have shown that only about 1.5 to 3.5% of the organic nitrogen of a soil mineralizes annually.[4] Even so, this rate of mineralization provides sufficient mineral nitrogen for normal growth of natural vegetation in most soils excepting those with low organic matter, such as the soils of deserts and sandy areas. Furthermore, *isotope tracer studies* of farm soils that have been amended with synthetic nitrogen fertilizers show that mineralized soil nitrogen constitutes a major part of the nitrogen taken up by crops. If the organic matter content of a soil is known, one can make a rough estimate of the amount of nitrogen likely to be released by mineralization during a typical growing season (Box 13.1).

The opposite of mineralization is **immobilization**, the conversion of inorganic nitrogen ions (NO_3^- and NH_4^+) into organic forms (see equation 13.1 and Figure 13.2).

BOX 13.1   CALCULATION OF NITROGEN MINERALIZATION

If the organic matter content of a soil, soil management practices, climate, and soil texture are known, it is possible to make a rough estimate of the amount of N likely to be mineralized each year. The following equation may be used:

$$\frac{\text{kg N mineralized}}{\text{ha 15 cm deep}} = \left(\frac{A \text{ kg SOM}}{100 \text{ kg soil}}\right)\left(\frac{B \text{ kg soil}}{\text{ha 15 cm deep}}\right)\left(\frac{C \text{ kg N}}{100 \text{ kg SOM}}\right)\left(\frac{D \text{ kg SOM mineralized}}{100 \text{ kg SOM}}\right) \quad (13.2)$$

where A = The amount of soil organic matter (SOM) in the upper 15 cm of soil, given in kg SOM per 100 kg soil. This value may range from close to zero to over 75% (in a Histosol) (see Section 3.9). Values between 0.5 and 5% are most common. ☛ Use a value of 2.5% (2.5 kg SOM/100 kg soil) for the example shown below.

B = The weight of soil per hectare to the depth of 15 cm. Most nitrogen used by plants is likely to come from this upper horizon. If it is 15 cm deep, 2×10^6 kg/ha is a reasonable estimate of its weight per hectare. See Section 4.7 to calculate the weight of this horizon if bulk density of a soil is known. ☛ Use 2×10^6 kg soil/ha 15 cm deep in the example shown below.

C = The amount of nitrogen in the SOM (see Section 12.3). ☛ Use the typical figure of 5 kg N/100 kg SOM in the example shown below.

D = The amount of SOM likely to be mineralized in one year for a given soil. This figure depends upon the soil texture, climate, and management practices. Values of around 2% are typical for a fine-textured soil, while values of around 3.5% are typical for coarse-textured soils. Slightly higher values are typical in warm climates; slightly lower values are typical in cool climates. ☛ Assume a value of 2.5 kg SOM mineralized/100 kg SOM for the example shown below.

The amount of nitrogen likely to be released by mineralization during a typical growing season may be calculated by substituting the example values indicated above into the equation:

$$\frac{\text{kg N mineralized}}{\text{ha 15 cm deep}} = \left(\frac{2.5 \text{ kg SOM}}{100 \text{ kg soil}}\right)\left(\frac{2 \times 10^6 \text{ kg soil}}{\text{ha 15 cm deep}}\right)\left(\frac{5 \text{ kg N}}{100 \text{ kg SOM}}\right)\left(\frac{2.5 \text{ kg SOM mineralized}}{100 \text{ kg SOM}}\right)$$

$$\frac{\text{kg N mineralized}}{\text{ha}} = \left(\frac{2.5}{100}\right)\left(\frac{2 \times 10^6}{1}\right)\left(\frac{5}{100}\right)\left(\frac{2.5}{100}\right) = 62.5 \text{ kg N/ha}$$

Most nitrogen mineralization occurs during the growing season when the soil is relatively moist and warm. Contributions from the deeper layers of this soil might be expected to bring total nitrogen mineralized in the root zone of this soil during a growing season to over 120 kg N/ha.

These calculations estimate the nitrogen mineralized annually from a soil that has not had large amounts of organic residues added to it. Animal manures, legume residues, or other nitrogen-rich organic soil amendments would mineralize much more rapidly than the native soil organic matter and thus would substantially increase the amount of nitrogen available in the soil.

---

[4] Studies using ^{15}N tracers suggest that the rate of mineralization of fertilizer nitrogen that had been immobilized the previous year can be as much as seven times these modest rates.

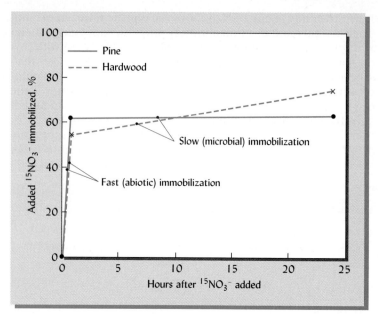

FIGURE 13.3 The rate of immobilization of tracer NO_3^- added in solution to samples of forest soils where either pine or hardwood stands prevailed. Very rapid immobilization took place in both soils during the first hour, apparently by nonbiological (abiotic) processes. The nitrates probably reacted chemically with the soil organic matter. This fast reaction was followed by the more commonly observed slower immobilization in response to biological reactions stimulated by soil microorganisms. This graph suggests that both biological and abiotic reactions must be taken into consideration in studying the nitrogen-retentive capacity of forest soils. [Redrawn from Bernston and Aber (2000); used with permission of Elsevier Science, Oxford, United Kingdom]

Immobilization can take place by both biological and nonbiological (abiotic) processes. The latter probably involves chemical reactions with high C/N ratio soil organic matter and can be quite important in forested soils (Figure 13.3). Biological immobilization occurs when microorganisms decomposing organic residues require more N than they can obtain from the residues they are metabolizing. The microorganisms then scavenge NO_3^- and NH_4^+ ions from the soil solution to incorporate into such cellular components as proteins, leaving the soil solution essentially devoid of mineral N (see also Section 12.3). When the organisms die, some of the organic nitrogen in their cells may be converted into forms that make up the humus complex, and some may be released as NO_3^- and NH_4^+ ions. Mineralization and immobilization occur simultaneously in the soil; whether the *net* effect is an increase or a decrease in the mineral nitrogen available depends primarily on the ratio of carbon to nitrogen in the organic residues undergoing decomposition (see Section 12.3).

13.4 SOLUBLE ORGANIC NITROGEN (SON)[5]

Until recently, most studies of nitrogen uptake and leaching focused exclusively on mineral N, especially nitrate. However, modern analytical tools have now shown conclusively that soluble organic nitrogen (SON) compounds are subject to plant uptake and leaching losses in both natural and agroecosystems. They account for about 0.3 to 1.5% of the total organic nitrogen in soils, a pool size similar to that of mineral nitrogen (NH_4^+ and NO_3^-). In fact, where organic manures have been applied to arable soils or where permanent grassland has been grown for many years, SON contents are often considerably higher than those of mineral nitrogen. Also the nitrogen mobilized from the litter of some forest species may have SON:inorganic N ratios of 10:1 or higher.

PLANT ABSORPTION OF SON. In nitrogen-limited ecosystems, such as those in strongly acidic and infertile soils (including some organic soils), SON may be the primary source of absorbed nitrogen. This helps explain the fact that plant growth, particularly of some forest species, is considerably greater than one would expect based on the limited supply of inorganic nitrogen at any one time. SON may be taken up directly by plant roots, or it may be assimilated through mycorrhizal associations. Most root uptake of SON occurs, however, after the readily decomposable organic N compounds are mineralized to NH_4^+ and NO_3^- ions.

---

[5] For reviews of the nature, significance, and analyses of soluble organic nitrogen, see Murphy et al. (2000). For agricultural soils and a focus on natural ecosystems, see Neff et al. (2003).

MICROBIAL UTILIZATION. Low molecular weight SON compounds can be transported into microbial cells for direct assimilation. Once inside the microbial cell, enzymes decompose these SON compounds, and the N is used to make proteins and other components needed by the microbe. If the N from the SON exceeds the immediate need in the cell, the excess will be released back into the soil solution (mineralization). Microbial uptake of both SON and mineral N take place concurrently in soils, giving rise in some soils to direct competition between plants and microbes for both forms of nitrogen.

LEACHING POTENTIAL OF SON. Soluble organic nitrogen is also a significant component of the N lost by leaching. For example, SON may comprise nearly all the N leached from some pristine forests and typically 30 to 60% of that leached from dairy farms and beef feedlots. In fact, SON comprises about 25% of the N carried by the Mississippi River into the Gulf of Mexico. Thus, the SON likely contributes to the environmental problems downstream and should be studied along with nitrate N to understand and solve nitrogen-pollution problems.

CHEMICAL MAKEUP OF SON. The chemical constituents of SON have not been fully identified. However, we know that some of the compounds are hydrophilic and that others are hydrophobic. This suggests some may be able to interact with inorganic colloids, but others would react primarily with the soil organic matter. About a third of the SON is in the form of amino-compounds such as amino sugars and amino acids. However, more research is needed to further ascertain the compounds that make up SON and the role they play in soil and related ecosystems.

13.5 AMMONIUM FIXATION BY CLAY MINERALS[6]

Like other positively charged ions, ammonium ions are attracted to the negatively charged surfaces of clay and humus, where they are held in exchangeable form, available for plant uptake, but partially protected from leaching. However, because of the particular size of the ammonium ion (and potassium also), it can become entrapped within cavities in the crystal structure of certain clays (see Figure 13.2). Several 2:1-type clay minerals, especially vermiculites, have the capacity to *fix* both ammonium and potassium ions in this manner (see Figures 8.11 and 14.33). Vermiculite has the greatest capacity, followed by fine-grained micas and some smectites. Ammonium and potassium ions fixed in the rigid part of a crystal structure are held in a nonexchangeable form, from which they are released only slowly.

Ammonium fixation by clay minerals is generally greater in subsoil than in topsoil, due to the higher clay content of subsoils (Table 13.1). In soils with considerable 2:1 clay content, interlayer-fixed NH_4^+ typically accounts for 5 to 10% of the total nitrogen in the surface soil and up to 20 to 40% of the nitrogen in the subsoil. In highly weathered soils, on the other hand, ammonium fixation is minor because little 2:1 clay is

TABLE 13.1   Total Nitrogen Levels of A and B Horizons of Four Cultivated Virginia Soils and the Percentage of the Nitrogen Present as Nonexchangeable or Fixed NH_4^+

Note the higher percentage of fixation in the B horizon.

Soil great group (series)	Total N, mg/kg		Nitrogen fixed as NH_4^+, %	
	A Horizon	B Horizon	A Horizon	B Horizon
Hapludults (Bojac)	812	516	5	18
Paleudults (Dothan)	503	336	5	14
Hapludults (Groseclose)	1792	458	3	17
Hapludults (Elioak)	1110	383	6	26

From Baethgen and Alley (1987).

---

[6] This chemical fixation of ammonia is caused by the entrapment or other strong binding of NH_4^+ ions by certain silicate clays. This type of fixation (by which K ions are similarly bound) is not to be confused with the very beneficial biological fixation of atmospheric nitrogen gas into compounds usable by plants (see Section 13.10).

present. In some forest soils, about half the nitrogen in the O and A horizons is immobilized by either ammonium fixation or chemical reactions with humus. While ammonium fixation may be considered an advantage because it provides a means of conserving nitrogen, the rate of release of the fixed ammonium is often too slow to be of much practical value in fulfilling the needs of fast-growing annual plants.

13.6 AMMONIA VOLATILIZATION

Ammonia gas (NH_3) can be produced from the breakdown of organic materials and farm manures and from such fertilizers as anhydrous ammonia and urea. The ammonia gas is in equilibrium with ammonium ions according to the following reversible reaction:

$$\underset{\text{Dissolved ions}}{NH_4^+ + OH^-} \rightleftharpoons H_2O + \underset{\text{Gas}}{NH_3 \uparrow}$$

(13.3)

From reaction 13.3 we can draw two conclusions. First, ammonia volatilization will be more pronounced at high pH levels (i.e., OH^- ions drive the reaction to the right); second, ammonia-gas-producing amendments will drive the reaction to the left, raising the pH of the solution in which they are dissolved.

Soil colloids, both clay and humus, adsorb ammonia gas, so ammonia losses are greatest where low quantities of these colloids are present or where the ammonia is not in close contact with the soil. For these reasons, ammonia losses can be quite large from sandy soils and from alkaline or calcareous soils, especially when the ammonia-producing materials are left at or near the soil surface and when the soil is drying out. High temperatures, as often occur on the surface of the soil, also favor the volatilization of ammonia (Figure 13.4).

Incorporation of manure and fertilizers into the top few centimeters of soil can reduce ammonia losses by 25 to 75% from those that occur when the materials are left on the soil surface. In natural grasslands and pastures, incorporation of animal wastes by earthworms and dung beetles is critical in maintaining a favorable nitrogen balance and a high animal-carrying capacity in these ecosystems (see Section 11.2).

VOLATILIZATION FROM WETLANDS. Gaseous ammonia loss from nitrogen fertilizers applied to the surface of fishponds and flooded rice paddies can also be appreciable, even on slightly acid soils. The applied fertilizer stimulates algae growing in the paddy water. As the algae photosynthesize, they extract CO_2 from the water and reduce the amount of carbonic acid formed. As a result, the pH of the paddy water increases markedly, especially during daylight hours, to levels commonly above 9.0. At these pH levels, ammonia is released from ammonium compounds and goes directly into the atmosphere. As with upland soils, this loss can be reduced significantly if the fertilizer is placed below the soil surface. Natural wetlands lose ammonium by a similar daily cycle.

AMMONIA ABSORPTION. By the reverse of the ammonium loss mechanism just described, both soils and plants can absorb ammonia from the atmosphere. Thus the soil–plant system can help cleanse ammonia from the air, while deriving usable nitrogen for

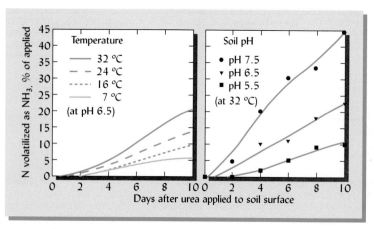

FIGURE 13.4 Ammonia volatilization is markedly affected by temperature and pH. Here, urea fertilizer (NH_2—CO—NH_2) was applied to a silt loam soil surface. Urea absorbs moisture from the air or soil and then hydrolyzes to form ammonia. The loss of ammonia gas is especially rapid when pH exceeds 7 and temperature exceeds 16 °C. Ammonia loss can be even faster from animal feces (manure) than from urea. Ammonia-forming amendments should not be left on the surface of warm, high pH soil for more than a day. [Redrawn from Glibert et al. (2006) using data in Franzen (2004)]

plants and soil microbes. Forests may receive a significant proportion of their nitrogen requirements as ammonia carried by wind from fertilized cropland and cattle feedlots located many kilometers away.

13.7 NITRIFICATION

Ammonium ions in the soil may be enzymatically oxidized by certain soil bacteria, yielding first nitrites and then nitrates. These bacteria are classed as **autotrophs** because they obtain their energy from oxidizing the ammonium ions rather than organic matter. The process termed **nitrification** (see Figure 13.2) consists of two main sequential steps. The first step results in the conversion of ammonium to nitrite by a specific group of autotrophic bacteria (**Nitrosomonas**). The nitrite so formed is then immediately acted upon by a second group of autotrophs, **Nitrobacter**. Therefore, when NH_4^+ is released into the soil it is usually converted rapidly into NO_3^- (see Figure 13.5). The enzymatic oxidation releases energy and may be represented very simply as follows:

Step 1

$$NH_4^+ + 1\tfrac{1}{2}O_2 \xrightarrow[\text{bacteria}]{\textit{Nitrosomonas}} NO_2^- + 2H^+ + H_2O + 275 \text{ kJ energy}$$ (13.4)

Ammonium   Nitrite

Step 2

$$NO_2^- + \tfrac{1}{2}O_2 \xrightarrow[\text{bacteria}]{\textit{Nitrobacter}} NO_3^- + 76 \text{ kJ energy}$$

Nitrite   Nitrate (13.5)

So long as conditions are favorable for both reactions, the second transformation is thought to follow the first closely enough to prevent accumulation of much nitrite. This is fortunate, because even at concentrations of just a few mg/kg, nitrite is quite toxic to most plants. When oxygen supplies are marginal, the nitrifying bacteria may also produce some NO and N_2O, which are potent greenhouse gases (see Section 12.9).

Regardless of the source of ammonium (i.e., ammonia-forming fertilizer, sewage sludge, animal excreta, or any other organic nitrogen source), nitrification will significantly increase soil acidity by producing H^+ ions, as shown in reaction 13.4. See also Sections 9.6 and 13.14.

Nitrification can be "reversed" by several bacterial processes, the best known of which is **denitrification**, an anaerobic process by which heterotrophic bacteria reduce nitrate to such gases as NO, N_2O and N_2 (Section 13.9). **Dissimilatory nitrate reduction to Ammonium (DNRA)** is another anaerobic bacterial process that in effect reverses nitrification; it reduces NO_3^- to NO_2^- and then to NH_4^+. Although denitrification is considered to be an anaerobic bacterial process, fungi have been discovered that produce N_2O gas from nitrate in quite dry arid rangeland soils.

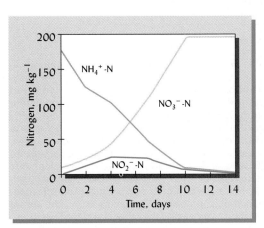

FIGURE 13.5 Transformation of ammonium into nitrite and nitrate by nitrification. On day zero, the silt loam soil was amended with enough $(NH_4)_2SO_4$ to supply 170 mg of N/kg soil. It then underwent a warm, well-aerated incubation for 14 days. Every second day, soil samples were extracted and analyzed for various forms of nitrogen. Note that the increase in nitrate-N (NO_3^--N) almost mirrored the decline in ammonium-N (NH_4^+-N), except for the small amount of nitrite-N (NO_2^--N) that accumulated temporarily between day 2 and 10. This pattern is consistent with the two-step process depicted by eqs. 13.4 and 13.5. No plants were grown during the study. [Data selected from Khalil et al. (2004)]

The nitrifying bacteria are much more sensitive to environmental conditions than are the broad groups of heterotrophic organisms responsible for the release of ammonium from organic nitrogen compounds (**ammonification**). Nitrification requires a supply of ammonium ions, but excess NH_4^+ can be toxic to *Nitrobacter*. The nitrifying organisms, being aerobic, require oxygen to make NO_2^- and NO_3^- ions and are therefore favored in well-drained soils. The optimum moisture for these organisms is about the same as that for most plants (about 60% of the pore space filled with water, Figure 13.6). Since they are autotrophs, their carbon sources are bicarbonates and CO_2. They perform best if the temperature is kept between 20 and 30 °C and perform very slowly if the soil is cold (below 5 °C).

Nitrification proceeds most rapidly where there is an abundance of exchangeable Ca^{2+} and Mg^{2+} and nutrient levels are optimum for the growth of higher plants. Nitrification is often constrained in soils high in smectite or allophane clays. These clays hold certain nitrogen-containing organic compounds in their intercolloid pores, thereby protecting them from microbial attack, including nitrification.

Nitrification inhibitors:
www.extension.iastate.edu/
Publications/NCH55.pdf

Nitrifying organisms are quite sensitive to some pesticides applied at high rates, but most studies suggest that at ordinary field rates, the majority of the pesticides have only a minimal effect on nitrification.

In recent years, chemicals have been found that can inhibit or slow down the nitrification process, thereby reducing the nitrate leaching potential. Such compounds, as well as others that retard the dissolution of urea, are discussed in Section 13.15.

Provided that all the preceding conditions are favorable, nitrification is such a rapid process that nitrate is generally the predominant mineral form of nitrogen in most soils. Irrigation of an initially dry arid-region soil, the first rains after a long dry season, the thawing and rapid warming of frozen soils in spring, and sudden aeration by tillage are examples of environmental fluctuations that typically cause a flush of soil nitrate production (Figure 13.7). The growth patterns of natural vegetation and the optimum planting dates for crops are greatly influenced by such seasonal changes in nitrate levels.

13.8   THE NITRATE LEACHING PROBLEM

Nitrates and water quality:
http://www.soil.ncsu.edu/
publications/Soilfacts/
AG-439-02/

In contrast to positively charged ammonium ions, negatively charged nitrate ions are not adsorbed by the negatively charged colloids that dominate most soils. Therefore, nitrate ions move downward freely with drainage water and are readily leached from the soil. The loss of nitrogen in this manner is of concern for three basic reasons: (1) the loss of this valuable nutrient is a waste that impoverishes the ecosystem (as discussed in Section 13.1), (2) leaching of nitrate anions stimulates the acidification of the soils and the co-leaching of such cations as Ca^{2+}, Mg^{2+}, and K^+ (as described in Section 9.6), and (3) the movement of nitrate to groundwater causes several serious water-quality problems downstream. We will now examine the nature of these environmental impacts.

WATER-QUALITY IMPACTS.   Water-quality problems caused by nitrogen are mainly associated with the movement of nitrate with drainage waters to the groundwater. The nitrate may contaminate drinking water causing health hazards (see Box 13.2) for people as well as

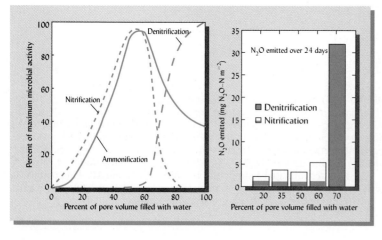

FIGURE 13.6 (*Left*) Rates of nitrification, ammonification, and denitrification are closely related to the availability of oxygen and water as depicted by percentage of water-filled pore space. Both nitrification and ammonification proceed at their maximal rates near 55 to 60% water-filled pore space; however, ammonification proceeds in soils too waterlogged for active nitrification. Only a small overlap exists in the conditions suitable for nitrification and denitrification. (*Right*) The greenhouse gas, nitrous oxide (N_2O), is mainly produced by denitrification, but it is also a minor by-product of nitrification. An experiment that used ^{15}N tracers shows the abrupt shift from one process to the other at water-filled porosities between 60 and 70%. [Bar graph from Bateman and Baggs (2005)]

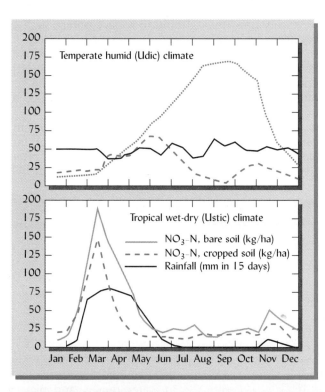

FIGURE 13.7 Seasonal patterns of nitrate-N concentration in representative surface soils with and without growing plants. (*Upper*) In a representative humid temperate region with cool winters and rainfall rather uniformly distributed throughout the year, NO_3^--N accumulates as the soil warms up in May and June. The nitrates are lost by leaching in the fall. (*Lower*) In a representative tropical region with four rainy months followed by eight months of dry, hot weather, a large flush of NO_3^--N appears when the rains first moisten the dry soil. This nitrate flush is caused by the rapid decomposition and mineralization of the dead cells of microorganisms previously killed by the dry, hot conditions. Note that soil nitrate is lower in both climates when plants are grown, because much of the nitrate formed is removed by plant uptake. (Diagrams courtesy of R. Weil)

BOX 13.2   SOIL, NITRATE, AND YOUR HEALTH[a]

Mismanagement of soil nitrogen can result in levels of nitrates in drinking water (usually in groundwater) and food (mainly leafy vegetables) that may seriously threaten human health. While nitrate itself is not directly toxic, once ingested, a portion of the nitrate is reduced by bacterial enzymes to nitrite, which *is* considered toxic.

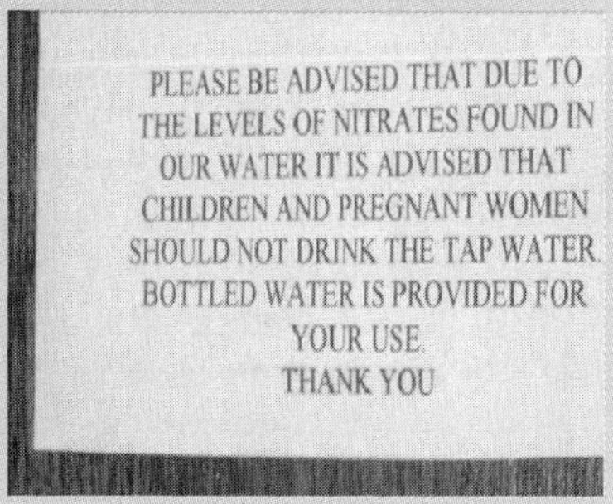

> PLEASE BE ADVISED THAT DUE TO THE LEVELS OF NITRATES FOUND IN OUR WATER IT IS ADVISED THAT CHILDREN AND PREGNANT WOMEN SHOULD NOT DRINK THE TAP WATER. BOTTLED WATER IS PROVIDED FOR YOUR USE.
> THANK YOU

FIGURE 13.8 *Hotel warning sign in a heavily agricultural watershed.*

The most widely known (though actually quite rare) malady caused by nitrite is **methemoglobinemia,** in which the nitrites decrease the ability of hemoglobin in the blood to carry oxygen to the body cells. Since inadequately oxygenated blood is blue rather than red, people with this condition take on a bluish skin color. This symptom, and the fact that infants under three months of age are much more susceptible to this illness than older individuals, accounts for the condition being commonly referred to as "blue baby syndrome." Most known deaths from this disease have been caused by infant formula made with high-nitrate water. With the aim of protecting infants from methemoglobinemia, governments have set standards to limit the nitrate concentrations allowed in drinking water (Figure 13.8). In the United States this limit is 10 mg/L NO_3^--N (=45 mg/L nitrate) and in the European Union it is 50 mg/L nitrate (=11 mg/L NO_3^--N).

Of greater potential concern is the tendency of nitrate to form N-nitroso compounds in the stomach by binding with such organic precursors as amines derived from proteins. Certain N-nitroso compounds are known to be highly toxic, causing cancer in some 40 species of test animals, so the threat to humans must be given serious consideration. Nitrates (or nitrites formed therefrom) have also been reported to promote certain types of diabetes, stomach cancers, interference with iodine uptake by the thyroid gland, and certain birth defects. While documenting cause and effect in chronic diseases is always uncertain, many of these effects appear to be associated with nitrate concentrations much lower than the drinking water limits just mentioned.

On the other hand, several research studies suggest that ingestion of nitrate does no harm and may actually provide protection against bacterial infections and some forms of cardiovascular diseases and stomach cancers. Therefore, while a precautionary approach is probably wise, we must conclude that the "jury is still out" on the health risks of nitrates in drinking water and vegetables.

[a] For reviews of nitrate effects on health, see Santamaria (2006) and L'hirondel and L'hirondel (2002). For a contrary view, see Addiscott (2006).

Nitrate pollution threatens amphibians:
http://www.on.ec.gc.ca/wildlife/factsheets/nitrate-e.html

livestock. The nitrates may also eventually flow underground to surface waters, such as streams, lakes, and estuaries. The key factor for health hazards is *concentration* of nitrate in the drinking water and the level of exposure (amount of water ingested, especially over long periods). Even more widespread are the damages to water quality and to the health of aquatic ecosystems, especially those with salty or brackish water (Box 13.3). The key factor for this kind of damage is often the *total load* (mass flux) of nitrogen delivered to the sensitive ecosystem.

The total nitrogen load may be comprised partly of organic and ammonium forms of N transferred from the land in surface runoff or on eroded soil material, but N leached through the soil profile as nitrate (along with soluble organic N) is often the main contributor. The quantity of nitrate lost in drainage water depends on two

BOX 13.3   NITROGEN POLLUTION: DEAD ZONE IN THE GULF OF MEXICO

Under natural conditions, low levels of nitrogen limit aquatic algae growth—especially in salty and brackish water, which inhibits N-fixing algae. Increased human input of nitrogen can remove this constraint. The resulting degradation of the aquatic ecosystem—particularly estuaries and coastal waters—is undoubtedly the most widespread water-quality problem induced by nitrogen pollution. Severe drops in water quality and biodiversity in the lower Chesapeake Bay on the Atlantic Coast and at the mouth of the Mississippi River in the Gulf of Mexico provide two major examples.

In the Gulf of Mexico off the coast of Louisiana, an enormous "dead zone" of water some 4 to 60 m deep reaches from the mouth of the Mississippi River westward nearly 500 km to Texas. Nutrient-rich freshwater carried by the river glides over the cooler, saltier (and therefore heavier) Gulf water. The nutrients (mainly nitrogen, but also P and Si) stimulate explosive growth of algae, which sink to the bottom

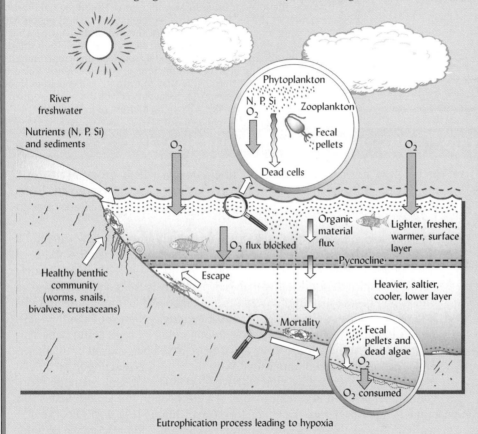

Eutrophication process leading to hypoxia

FIGURE 13.9 *Eutrophication process leading to hypoxia. From CAST (1999). For worldwide examples of hypoxia, see Diaz (2001).*

when they die (Figure 13.9). In decomposing this dead tissue, microorganisms deplete the oxygen dissolved in the water to levels unable to sustain animal life. Fish, shrimp, and other aquatic species either migrate out of the zone or die. This state of low oxygen in the water (less than 2 to 3 mg O_2/L) is known as **hypoxia,** and the process that brings it about is called **eutrophication** (see also Box 14.1 for eutrophication caused by phosphorus mainly in fresh waters).

Concentrations of N in the Mississippi River have tripled in the past 30 years, mainly due to human activities, especially those in agriculture. Critical assessments suggest that only about 11% of the N delivered by the river comes from sewage treatment plants and other point sources; nearly 50% comes from fertilizers and the rest mainly from farmland runoff and manure. Major efforts are required to help farmers and others improve their N use efficiency and reduce the transformation of this valuable nutrient into a pollutant.

Nitrates and water quality:
http://www.ext.colostate
.edu/PUBS/crops/00517.html

factors: (1) the volume of water leaching through the soil and (2) the concentration of nitrates in that drainage water.

VOLUME OF LEACHING WATER. The volume of leaching water is influenced by rates of precipitation, irrigation, and evapotranspiration (Section 6.1), as well as by soil texture and structure. Sandy soils in humid regions are therefore highly susceptible to leaching. By contrast, unirrigated soils in arid regions are expected to undergo very little leaching (but read Box 13.4!). Conservation tillage increases water infiltration at the expense of surface runoff, thereby also increasing the volume of water available for leaching.

CONCENTRATON OF NITROGEN. The concentration of nitrate in the leaching water is largely dependent on the size of the soil nitrate pool during periods of leaching. The nitrate present, in turn, reflects the balance between removal of nitrogen from this pool by plant uptake or immobilization and the input of nitrogen into the nitrate pool by mineralization, fertilization, and atmospheric deposition. For example, pristine mature forests usually maintain a very close balance between nitrogen taken up by trees and nitrogen returned as litter. Leaching water contains only about 0.1 mg nitrate-N/L, and only 1 or 2 kg of N/ha is lost annually to groundwater. However, deposition from the atmosphere can increase inputs to the nitrate pool, while disturbances such as timber harvest can reduce plant uptake out of the pool, leading to leachate concentrations of 2 to 3 mg nitrate-N/L and annual leaching losses greater than 25 kg N/ha.

Groundwater and streams draining watersheds with intense agricultural land use are commonly much higher in nitrate than those draining forested watersheds. Heavy nitrogen fertilization of crops (especially vegetables and some grain crops) can be a major source of excessive nitrate because the crops usually take up only a portion of the nitrogen applied (Figure 13.11). Watersheds with many concentrated animal-feeding

BOX 13.4   SURPRISE: NITRATE LEACHING IN THE DESERT!

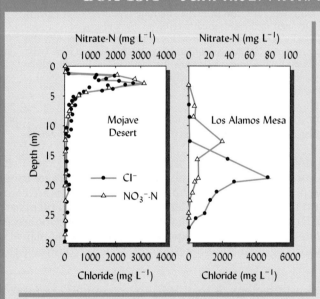

FIGURE 13.10 *Large accumulations of nitrate-N deep in desert soils. Shown are two examples of this recently discovered phenomenon. Note that the depth scale is in meters and that the NO_3^--N (but not Cl^-) concentration scales differ widely between sites. [From Walvoord et al. (2003), reprinted with permission from AAAS]*

Leaching of N is usually associated with high rainfall and is not expected in very dry environments. However, scientists were recently surprised to discover huge amounts of NO_3^--N accumulated deep within desert soils, evidence of leaching over thousands of years. This nitrogen was found at depths ranging from 3 to 30 meters—much deeper than was previously studied (Figure 13.10). That the nitrate got there by leaching is evidenced by the accumulation of another common anion, chloride (Cl^-), at similar depths (but not necessarily at similar concentrations). Apparently, desert vegetation does not use all the nitrate produced by nitrification, leaving some to leach beyond the rooting depth during rare heavy rain events. The leached nitrate then concentrates when water escapes upward as vapor (Section 5.5). The nitrate remains because deep in the profile the desert soil is well-aerated and very low in organic matter, water, and microbes—conditions not conducive to denitrification. The accumulations of nitrate, though variable, typically exceed 1000 kg N/ha. Their discovery suggests that previous models of the global N cycle may have significantly underestimated the pool of vadose-zone nitrogen in the world's warm deserts. The phenomenon may also have serious impacts on water quality in some desert

areas. For example, if desert soils are brought under irrigation, the necessary drainage water (see Section 10.8) may mobilize the accumulated nitrate, contaminating groundwater and streams. Such contamination could also result from changes in the local water balance caused by dam construction or increased regional rainfall.

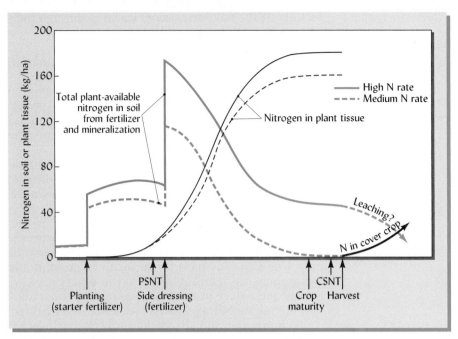

FIGURE 13.11 Two nitrogen fertilizer systems for corn production in the Midwest. The system that has been dominant in the past (solid lines) involves very high nitrogen applications and continuous corn culture. At least 150 kg N/ha is applied—part as starter fertilizer at or before planting time, the remainder as a side dressing just before the most rapid growth stage. Unfortunately, when the excess crop matures and is harvested, much soluble nitrogen remains in the soil, probably in the form of nitrates. If not captured by a cover crop planted in the corn at harvesttime, the excess nitrogen is subject to leaching during the fall and winter months. This leads to contamination of groundwater and, eventually, surface water. More environmentally sound systems involve crop rotation that include legumes, or if corn is grown continuously, the rate of nitrogen fertilizer is greatly reduced (broken lines). A presidedress nitrate soil test (PSNT) is used to determine the amounts of nitrogen to apply. At the end of the season, a cornstalk nitrate test (CSNT) is used to assess plant nitrogen status. With reduced N application, crop yields and economic returns are about the same, but the nitrogen remaining in the soil at crop harvest is low and nitrate contamination is minimized.

operations (CAFOs) are often the most polluted with nitrate because of the enormous quantities of nitrogen-rich animal manure produced (see Sections 13.15 and 16.4).

TIMING OF NITROGEN INPUT. In humid-temperate (Udic) and Mediterranean (Xeric) climates, the potential for nitrate leaching is lowest in midsummer when plants are using large amounts of both water and N. Leaching water volumes and nitrate levels are commonly highest in early spring and late fall or early winter (before frost) when plants are not intensively using water or N, leaving both available to leach. The synchrony between the production of nitrate by mineralization and the uptake of N by plants is greatest where perennial vegetation is grown and poorest where annual crops leave the soil bare in early spring and late fall.

MANAGEMENT TO REDUCE LOSSES. Even in regions of high leaching potential, careful soil management can prevent excessive nitrate losses. Applications of fertilizer and manure should be modest in amount and timed to provide nitrogen when the plants need it, not much before or after the period of active plant uptake. Other crops, including nitrogen-demanding winter cover crops (Section 16.2), should be planted immediately following the summer annual crop to take up the unused nitrates before they can leach away (see Figure 13.11). If such guidelines are followed, nitrogen leaching may be kept to less than 5 or 10% of the nitrogen applied.

REVERSING NITRATE LEACHING IN THE HUMID TROPICS. Much of the nitrate mineralized in certain highly weathered, tropical Oxisols and Ultisols leaches below the root zone before annual crops such as corn can take it up. Soil scientists in Africa recently discovered that some of this leached nitrate is not lost to groundwater. Instead, the highly weathered and acid clays deep in the subsoil adsorb the nitrate on their anion exchange sites

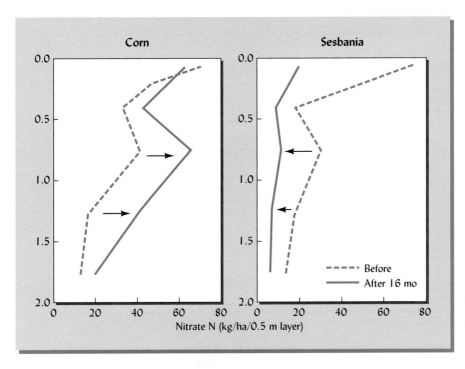

FIGURE 13.12 Depth distribution of nitrates in an Oxisol in western Kenya before planting either corn or a fast-growing tree, *Sesbania*, and after 16 months of growing fertilized corn (3 crops) and continuous *Sesbania*. Note that the *Sesbania* markedly decreased soil nitrates to a depth of about 2 meters. Separate root studies (Mekonnen et al. (1997) showed that 31 percent of the *Sesbania* roots were located between 2.5 and 4 m in depth. [From Sanchez et al. (1997)]

(see Section 8.11). Deep-rooted trees such as *Sesbania* are capable of taking up this sub-soil nitrate. If grown in rotation with annual food crops, the trees subsequently enrich the surface soil when they shed their leaves, making this pool of once-leached nitrogen available again for food production (Figure 13.12). Such **agroforestry** practices as this have the potential to improve both crop production and environmental quality in the humid tropics (see also Section 20.10).

13.9 GASEOUS LOSSES BY DENITRIFICATION

Nitrogen may be lost to the atmosphere when nitrate ions are converted to gaseous forms of nitrogen by a series of widely occurring biochemical reduction reactions termed **denitrification**.[7] The organisms that carry out this process are commonly present in large numbers and are mostly facultative anaerobic bacteria in genera, such as *Pseudomonas*, *Bacillus*, *Micrococcus*, and *Achromobacter*. These organisms are *heterotrophs*, which obtain their energy and carbon from the oxidation of organic compounds. Other denitrifying bacteria are *autotrophs*, such as *Thiobacillus denitrificans*, which obtain their energy from the oxidation of sulfide. The exact mechanisms vary depending on the conditions and organisms involved. In the reaction, NO_3^- [N(V)] is reduced in a series of steps to NO_2^- [N(III)], and then to nitrogen gases that include NO [N(II)], N_2O [N(I)], and eventually N_2 [N(0)]:

$$2NO_3^- \xrightarrow{-2O} 2NO_2^- \xrightarrow{-2O} 2NO\uparrow \xrightarrow{-O} N_2O\uparrow \xrightarrow{-O} N_2\uparrow \qquad (13.6)$$

| Nitrate ions | Nitrite ions | Nitric oxide gas | Nitrous oxide gas | Dinitrogen gas |
| (+5) | (+3) | (+2) | (+1) | (0) ← Valence state of nitrogen |

Although not shown in the simplified reaction given here, the oxygen released at each step would be used to form CO_2 from organic carbon (or SO_4^{2-} from sulfides if *Thiobacillus* is the nitrifying organism).

For these reactions to take place, sources of organic residues should be available to provide the energy the denitrifiers need. The soil air in the microsites where denitrification occurs should contain no more than 10% oxygen, and lower levels of oxygen are preferred. Optimum temperatures for denitrification are from 25 to 35 °C, but the

---

[7] Nitrate can also be reduced to nitrite and to nitrous oxide gas by nonbiological chemical reaction and under some circumstances by nitrifier bacteria, but these reactions are quite minor in comparison with biological denitrification. A recently discovered bacterial process, the anaerobic oxidation of ammonium (**anammox**), converts ammonium and nitrate to N_2 gas. It is widespread in oceans and may be important in hydric soils as well.

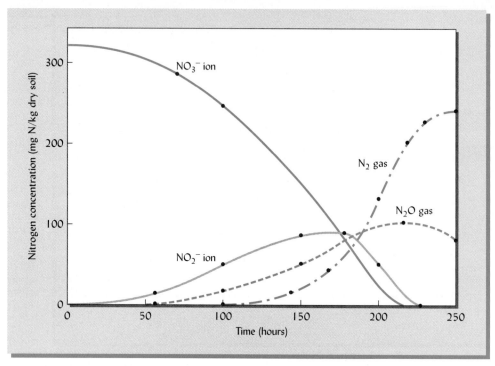

FIGURE 13.13 Changes in various forms of nitrogen during the process of denitrification in a moist soil incubated in the absence of atmospheric oxygen. [From Leffelaar and Wessel (1988)]

process will occur between 2 and 50 °C. Very strong acidity (pH < 5.0) inhibits rapid denitrification and favors the formation of N_2O.

Generally, when oxygen levels are very low, the end product released from the overall denitrification process is dinitrogen gas (N_2). It should be noted, however, that NO and N_2O are commonly also released during denitrification under the fluctuating aeration conditions that often occur in the field (Figure 13.13). The proportion of the three main gaseous products seems to be dependent on the prevalent pH, temperature, degree of oxygen depletion, and concentration of nitrate and nitrite ions available. For example, the release of nitrous oxide (N_2O) is favored if the concentrations of nitrite and nitrate are high and the supply of oxygen is not too low (see Figures 13.6 and 7.12). Under very acid conditions, almost all of the loss occurs in the form of N_2O. Nitric oxide (NO) formation is generally small and apparently occurs most readily under acid conditions.

Atmospheric Pollution

N and greenhouse gases from row crops:

http://www.oznet.ksu.edu/ctec/CASMGSnewsletter/Jan04-1.htm

The question of how much of each nitrogen-containing gas is produced is not merely of academic interest. Dinitrogen gas is quite inert and environmentally harmless, but the oxides of nitrogen are very reactive gases and have the potential to do serious environmental damage in at least four ways. First, NO and N_2O released into the atmosphere by denitrification can contribute to the formation of nitric acid, one of the principal components of acid rain. Second, the nitrogen oxide gases can react with volatile organic pollutants to form ground-level ozone, a major air pollutant in the photochemical smog that plagues many urban areas. Third, when NO rises into the upper atmosphere, it contributes to the greenhouse effect (as much as 300 times that of an equal amount of CO_2) by absorbing infrared radiation that would otherwise escape into space (see Section 12.9).

Finally, and perhaps most significantly, as N_2O moves up into the stratosphere, it may participate in reactions that result in the destruction of ozone (O_3), a gas that helps shield the Earth from harmful ultraviolet solar radiation. In recent decades this protective ozone layer has been measurably depleted by reaction with industrial CFCs as well as with N_2O and other gases. As this protective layer is further degraded, thousands of additional cases of skin cancer are likely to occur annually. While there are other important sources of N_2O, such as automobile exhaust fumes, a major contribution to the problem is being made by denitrification in soils, especially in rice paddies, wetlands, and heavily fertilized or manured agricultural soils.

Quantity of Nitrogen Lost Through Denitrification

As might be expected, the exact magnitude of denitrification loss is difficult to predict and will depend on management practices and soil conditions. Studies of forest ecosystems have shown that during periods of adequate soil moisture, denitrification results in a slow but relatively steady loss of nitrogen from these undisturbed natural systems. In contrast, most field measurements of gaseous nitrogen loss from agricultural soils reveal that the losses are highly variable in both time and space. The greater part of the annual nitrogen loss often occurs during just a few days in summer, when heavy rain has temporarily caused the warm soils to become poorly aerated (Figure 13.14).

Low-lying, organic-rich areas and other hot spots may lose nitrogen 10 times as fast as the average rate for a typical field. Although as much as 10 kg/ha of nitrogen may be lost in a single day from the sudden saturation of a well-drained, humid-region soil, such soils rarely lose more than 5 to 15 kg N/ha annually by denitrification. But where drainage is restricted and where large amounts of nitrogen fertilizer are applied, substantial losses might be expected. Losses of 30 to 60 kg N/ha/yr of nitrogen have been observed in such agricultural systems (Plate 94).

Much of the nitrogen that moves from watersheds into shallow streams and slow-moving rivers is lost by denitrification. As the water moves through these shallow transport channels, there is opportunity for the nitrogen to come in contact with the river bottoms where the denitrification occurs. Research suggests that 5 to 20% of the nitrogen in some streams or rivers may be lost by denitrification.

Denitrification in Flooded Soils

In flooded soils, such as those found in natural wetlands or rice paddies (Figure 13.15), losses by denitrification may be very high. The soils of rice paddies are commonly subject to alternate periods of wetting and drying. Nitrates that are produced by nitrification during the dry periods are often subject to denitrification when the soils are submerged. Even when submerged, the soil permits both reactions to take place at once—nitrification at the soil–water interface where some oxygen derived from the water is present and

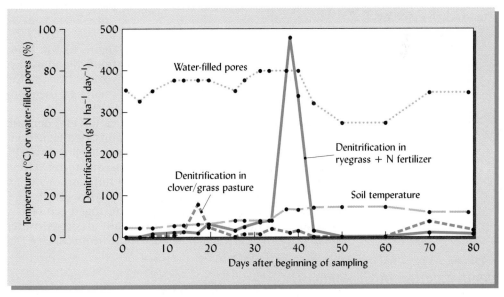

FIGURE 13.14 Changes in denitrification and soil conditions in spring on a well-drained, but somewhat fine-textured soil (East Keswick silty clay loam, Udalfs) in Wales, U.K. Two pasture systems were studied: ryegrass receiving 150 kg N/ha annually and a clover–grass mixture. The fertilizer nitrogen applied to the ryegrass system was intended to approximately equal the nitrogen fixed from the atmosphere by the clover in the mixed system (see Section 13.10). Note the sporadic nature of the denitrification process, with most of the nitrogen loss occurring during a brief period when the soil was warm, wet, and high in nitrate. This episodic pattern is typical of disturbed agroecosystems and stands in contrast to the slow, steady pattern of denitrification usually found in undisturbed natural forests. [Data from Colbourn (1993)]

FIGURE 13.15 Denitrification can be very efficient in removing nitrogen in flooded systems that combine aerobic and anaerobic zones and have high concentrations of available organic carbon. Some examples of such systems are (*a*) a flooded rice paddy, (*b*) a tidal wetland, (*c*) a site for treating sewage effluent by overland flow (effluent applied by sprinklers, see arrows), and (*d*) a manure storage lagoon. (Photos courtesy of R. Weil)

denitrification at lower soil depths (see Figure 13.16). However, nitrogen losses can be dramatically reduced by keeping the soil flooded and by deep placement of the fertilizer into the reduced zone of the soil. In this zone, because there is insufficient oxygen to allow nitrification to proceed, nitrogen remains in the ammonium form and is not susceptible to loss by denitrification.

The sequential combination of nitrification and denitrification also operates in natural and artificial wetlands. Tidal wetlands (see Figure 13.15*b*), which become alternately anaerobic and aerated as the water level rises and falls, have particularly high potentials for converting nitrogen to gaseous forms. Often the resulting rapid loss of nitrogen is considered to be a beneficial function of wetlands, in that the process protects estuaries and lakes from the eutrophying effects of too much nitrogen. In fact, wastewater high in organic carbon and nitrogen can be cleaned up quite efficiently by allowing it to flow slowly over a specially designed water-saturated soil system in a process known as *overland flow wastewater treatment* (see Figure 13.15*c*).

Denitrification in Groundwater

Recent studies on the movement of nitrate in groundwater have documented the significance of denitrification taking place in the poorly drained soils under **riparian** vegetation (mainly woodlands adjacent to streams). In most cases studied in humid temperate regions, contaminated groundwater lost most of its nitrate load as it flowed through the riparian zone on its way to the stream. The apparent removal of nitrate may be quite dramatic, whether the nitrate source is septic drainfields or fertilized cropland (Figure 13.17). Most of nitrate is believed to be lost by denitrification, stimulated

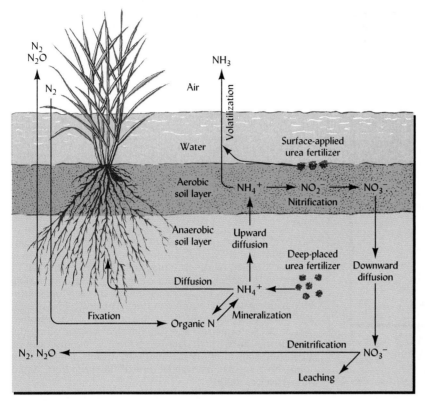

FIGURE 13.16 Nitrification–denitrification reactions and kinetics of the related processes controlling nitrogen loss from the aerobic–anaerobic layers of a flooded soil system. Nitrates, which form in the thin aerobic soil layer just below the soil–water interface, diffuse into the anaerobic (reduced) soil layer below and are denitrified to the N_2 and N_2O gaseous forms, which are lost to the atmosphere. Placing the urea or ammonium-containing fertilizers deep in the anaerobic layer prevents N oxidation of ammonium ions to nitrates, thereby greatly reducing N loss. [Modified from Patrick (1982)]

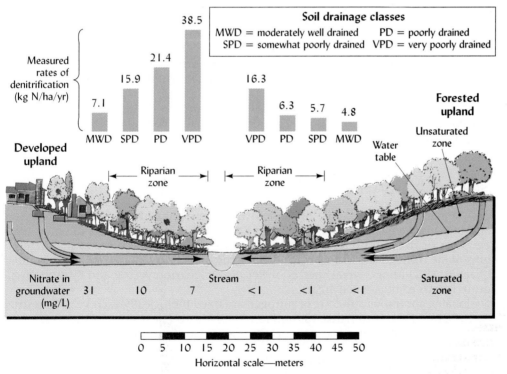

FIGURE 13.17 Denitrification in riparian wetlands receiving groundwater with high- and low-nitrate contents. The high-nitrate groundwater (*left*) came from sites with heavy development of houses using septic drainfields for sewage disposal (see Section 6.8). The low-nitrate groundwater (*right*) came from an area of undeveloped forest. The riparian zones on both sides of the stream were covered with red maple-dominated forest. Within a few meters after entering the riparian wetland, the nitrate content of the contaminated groundwater was reduced by 75%, from 31 mg/L nitrate to less than 7 mg/L nitrate. The soils in this study site are very sandy Inceptisols and Entisols. In other regions, riparian zones with finer-textured soils have shown even more complete removal of nitrate from groundwater. [Data shown are from Hanson et al. (1994)]

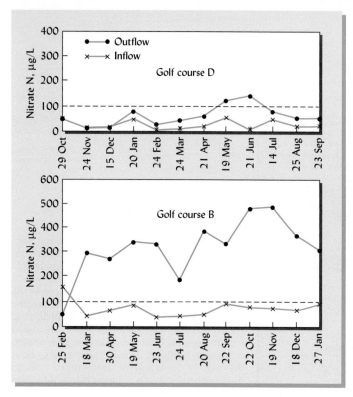

FIGURE 13.18 Monthly levels of nitrate –N in streams entering (inflow) and leaving (outflow) two golf courses in New Hanover County, North Carolina. Note the very high levels of nitrate in the outflow from Course B, far in excess of the 100-μg level that is known to encourage undesirable phytoplankton blooms in coastal waters (dotted line). The much lower levels in Course D are thought to be due to two ponds, through which the water passes, and to a forested wetland along the edge of the course. Such data suggest that golf courses can be significant sources of nitrate input into streams, but that such inputs are subject to managerial control. [Redrawn from Mallin and Wheeler (2000)]

by organic compounds leached from the decomposing forest litter and by the anaerobic conditions that prevail in the wet riparian zone soils.

Constructed wetlands can be used to reduce the nitrate content of surface waters moving toward streams. When coupled with buffer strips, such wetlands can remove half or more of the nitrates of the surface water before it enters the stream channel (see Figure 13.18).

We have just discussed a number of biological processes that lead to losses of nitrogen from the soil system. We turn next to the principal biological process by which soil nitrogen is replenished.

13.10 BIOLOGICAL NITROGEN FIXATION

Nitrogen fixation links:
http://academic.reed.edu/
biology/Nitrogen/Nfix1.html

Next to plant photosynthesis, **biological nitrogen fixation** is probably the most important biochemical reaction for life on Earth. This process converts the inert dinitrogen gas of the atmosphere (N_2) to reactive nitrogen that becomes available to all forms of life through the nitrogen cycle. The process is carried out by a limited number of bacteria, including several species of *Rhizobium*, actinomycetes, and cyanobacteria (formerly termed blue-green algae).

Globally, enormous amounts of nitrogen are fixed biologically each year. Terrestrial systems alone fix an estimated 139 million Mg. However, the amount that is fixed in the manufacture of fertilizers is now nearly as great (see Figure 13.19).

THE MECHANISM.   Regardless of the organisms involved, the key to biological nitrogen fixation is the enzyme *nitrogenase*, which catalyzes the reduction of dinitrogen gas to ammonia.

$$N_2 + 8H^+ + 6e^- \xrightarrow[\text{(Fe,Mo)}]{\text{(Nitrogenase)}} 2NH_3 + H_2 \qquad (13.7)$$

Lesson on N fixation:
http://www.soils.umn.edu/
academics/classes/soil2125/
doc/s9chap2.htm

The ammonia, in turn, is combined with organic acids to form amino acids and, ultimately, proteins.

$$NH_3 + \text{organic acids} \rightarrow \text{amino acids} \rightarrow \text{proteins} \qquad (13.8)$$

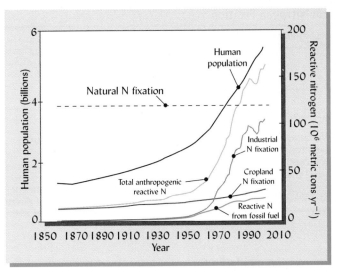

FIGURE 13.19 Changes in the human population and its contribution to global reactive nitrogen. Note that by the early 1980s human-caused N fixation (industrial fertilizer production, agricultural crop legumes, and combustion of fossil fuels) had surpassed natural N fixation (by legumes, cyanobacteria, and actinomycetes in natural terrestrial ecosystems as well as by lightning). [Modified from Lambert and Driscoll (2003) based on data in Galloway and Cowling (2002) with permission from Hubbard Brook Research Fndn.]

The site of N_2 reduction is the enzyme **nitrogenase**, a complex consisting of two proteins, the smaller of which contains iron while the larger contains molybdenum and iron (Figure 13.20). Several salient facts about this enzyme and its function are worth noting, for nitrogenase is unique and its role in the nitrogen cycle is of great importance to humankind.

1. Breaking the N≡N triple bond in N_2 gas requires a great deal of energy. Therefore, the process is greatly enhanced by association with higher plants, which can supply this energy from photosynthesis.

2. Nitrogenase is destroyed by free O_2, so organisms that fix nitrogen must protect the enzyme from exposure to oxygen. When nitrogen fixation takes place in root nodules (see Section 13.11), one means of protecting the enzyme from free oxygen is the formation of *leghemoglobin*.[8] This compound, which gives active nodules a red interior color, binds oxygen in such a way as to protect the

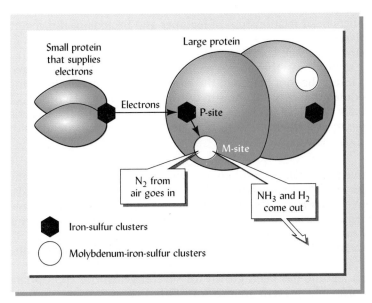

FIGURE 13.20 The nitrogenase complex consists of two proteins. The larger protein converts atmospheric N_2 to NH_3 using electrons provided by the smaller protein. The M-sites on the large protein capture nitrogen (N_2) from the air, while the P-sites receive the electrons provided by the small protein so that N_2 can be reduced to NH_3. [From Emsley (1991); reprinted with permission of *New Scientist*]

[8] Leghemoglobin is virtually the same molecule as the hemoglobin that gives human blood its red color when oxygenated. The use of hemoglobin to perform essentially similar functions in both legume root nodules and mammalian blood is a striking example of nature's conservative tendency and the unity of all life.

N-fixing systems	Organisms involved	Plants involved	Site of fixation
Symbiotic			
Obligatory			
Legumes	Bacteria *Rhizobia* and *Bradyrhizobia*	Legumes	Root nodules
Nonlegumes (angiosperms)	Actinomycetes (*Frankia*)	Nonlegumes (angiosperms)	Root nodules
Associative			
Morphological involvement	Cyanobacteria, bacteria	Various higher plants and microorganisms	Leaf and root nodules, lichens
Nonmorphological involvement	Cyanobacteria, bacteria	Various higher plants and microorganisms	Rhizosphere (root environment) Phyllosphere (leaf environment)
Nonsymbiotic	Cyanobacteria, bacteria	Not involved with plants	Soil, water independent of plants

nitrogenase while making oxygen available for respiration in other parts of the nodule tissue.

3. The reduction reaction is end-product inhibited—for example, an accumulation of ammonia will inhibit nitrogen fixation. Also, too much nitrate in the soil will inhibit the formation of nodules (see Section 13.11).

4. Nitrogen-fixing organisms have a relatively high requirement for molybdenum, iron, phosphorus, and sulfur, because these nutrients are either part of the nitrogenase molecule or are needed for its synthesis and use.

FIXATION SYSTEMS.   Biological nitrogen fixation occurs through a number of microbial systems that may or may not be directly or indirectly associated with higher plants (Table 13.2). Although the legume–bacteria symbiotic systems have received the most attention, recent findings suggest that the other systems involve many more families of plants worldwide and may even rival the legume-associated systems as suppliers of biological nitrogen to the soil. Each major system will be discussed briefly.

13.11   SYMBIOTIC FIXATION WITH LEGUMES

The process with legumes: http://academic.reed.edu/ biology/Nitrogen/Nfix1 (legumes).html

The **symbiosis** (mutually beneficial relationship) of legumes and bacteria of the genera *Rhizobium* and *Bradyrhizobium* provide the major biological source of fixed nitrogen in agricultural soils. The genus *Rhizobium* contains fast-growing, acid-producing bacteria, while the *Bradyrhizobia* are slow growers that do not produce acid. Both will be considered together. These organisms infect the root hairs and the cortical cells, ultimately inducing the formation of **root nodules** that serve as the site of nitrogen fixation (Figure 13.21 and Plate 103). In a mutually beneficial association, the host plant supplies the bacteria with carbohydrates for energy, and the bacteria reciprocate by supplying the plant with reactive-nitrogen compounds (Plate 51).

ORGANISMS INVOLVED.   A given *Rhizobium* or *Bradyrhizobium* species will infect some legumes but not others. For example, *Rhizobium trifolii* inoculates *Trifolium* species (most clovers), but not sweet clover, which is in the genus *Melilotus*. Likewise, *Rhizobium phaseoli* inoculates *Phaseolus vulgaris* (beans), but not soybeans, which are in the genus *Glycine*. This specificity of interaction is one basis for classifying rhizobia (see Table 13.3). Legumes that can be inoculated by a given *Rhizobium* species are included in the same cross-inoculation group.

In areas where a given legume has been grown for several years, the appropriate species of *Rhizobium* is probably present in the soil. Often, however, the natural *Rhizobium* population in the soil is too low or the strain of the *Rhizobium* species present is not effective (Figure 13.22). In such circumstances, special mixtures of the appropriate *Rhizobium* and *Bradyrhizobium* inoculant may be applied, either by coating the legume seeds or by applying the inoculant directly to the soil. Effective and competitive strains of *Rhizobium,* which are available commercially, often give significant yield increases,

(a)   (b)   (c)

FIGURE 13.21 Photos illustrating soybean nodules. In (a) the nodules are seen on the roots of the soybean plant, and a closeup (b) shows a few of the nodules associated with the roots. A scanning electron micrograph (c) shows a single plant cell within the nodule stuffed with the bacterium *Bradyrhizobium japonicum*. (Courtesy of W. J. Brill, University of Wisconsin)

but only if used on the proper crops. You may want to refer to Table 13.3 when planting legume rotations or purchasing commercial inoculant.

QUANTITY OF NITROGEN FIXED.   The rate of biological fixation is greatly dependent on soil and climatic conditions. The legume–*Rhizobium* associations generally function best on soils that are not too acid (although *Bradyrhizobium* associations generally can tolerate considerable acidity) and that are well supplied with essential nutrients. However, high levels of available nitrogen, whether from the soil or added in fertilizers, tend to depress biological nitrogen fixation (Figure 13.23). Apparently, plants make the heavy energy investment required for symbiotic nitrogen fixation only when short supplies of mineral nitrogen make nitrogen fixation necessary.

TABLE 13.3   Classification of Rhizobia Bacteria and Associated Legume Cross-Inoculation Groups

The genus **Rhizobium** *contains fast-growing, acid-producing bacteria, while those of* **Bradyrhizobium** *are slow growers that do not produce acid. A third genus,* **Azorhizobium**, *which is not shown, produces stem nodules on* **Sesbania rostrata**.

Genus	Bacteria	Host legume
	Species/subgroup	
Rhizobium	R. leguminosarum	
	bv. *viceae*	*Vicia* (vetch), *Pisum* (peas), *Lens* (lentils), *Lathyrus* (sweet pea)
	bv. *trifolii*	*Trifolium* spp. (most clovers)
	bv. *phaseoli*	*Phaseolus* spp. (dry bean, runner bean, etc.)
	R. Meliloti	*Melilotus* (sweet clover, etc.), *Medicago* (alfalfa), *Trigonella* (fenugreek)
	R. loti	*Lotus* (trefoils), *Lupinus* (lupins), *Cicer* (chickpea), *Anthyllis, Leucaena,* and many other tropical trees
	R. Fredii	*Glycine* spp. (e.g., soybean)
Bradyrhizobium	B. japonicum	*Glycine* spp. (e.g., soybean)
	B. sp.	*Vigna* (cowpeas), *Arachis* (peanut), *Cajanus* (pigeon pea), *Pueraria* (kudzu), *Crotolaria* (crotolaria), and many other tropical legumes

FIGURE 13.22 This soybean crop in East Africa was a total failure. The soybean seeds were not inoculated with the proper bacteria prior to planting in the newly cleared field that had been cleared from forest vegetation and had never grown soybeans before. (Photo courtesy of R. Weil)

Although quite variable from site to site, the amount of nitrogen biologically fixed can be quite high, especially for those systems involving nodules, which supply energy from photosynthates and protect the nitrogenase enzyme system (Table 13.4). Nonnodulating or nonsymbiotic systems generally fix relatively small amounts of nitrogen. Nonetheless, many natural plant communities and agricultural systems (generally involving legumes) derive the bulk of their nitrogen needs from biological fixation (Plate 101).

EFFECT ON SOIL NITROGEN LEVEL. Over time, the presence of nitrogen-fixing species can significantly increase the nitrogen content of the soil and benefit nonfixing species grown in association with fixing species (see Figure 13.24).

Although some direct transfer may take place via mycorrhizal hyphae connecting two plants, most of the transfer results from mineralization of nitrogen-rich compounds in root exudates and in sloughed-off root and nodule tissues. Ammonium and nitrate thus released into the soil are available to any plant growing in association with the legume. The vigorous development of a grass in a legume–grass mixture is evidence of this rapid release (Figure 13.25), as are the relatively high nitrate concentrations sometimes measured in groundwater under legume crops. Some crops, such as beans and peas, are such weak nitrogen fixers that most of the nitrogen they absorb must come from the soil. Consequently, it should not be assumed that the symbiotic systems

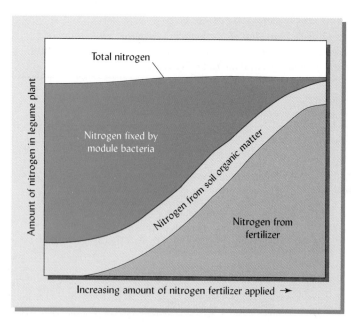

FIGURE 13.23 Influence of adding inorganic nitrogen on the nitrogen found in a representative legume plant. As more nitrogen fertilizer is added, the plant obtains less of its nitrogen by biological fixation. The inorganic nitrogen merely replaces biologically fixed N, saving the plant some energy that would have been allocated to the nodule bacteria. Plant growth (not shown) is also little affected by the fertilizer application. However, some legume species fix nitrogen so inefficiently (e.g., phaseolus bean) that they are more likely to respond positively to the use of nitrogen fertilizer. The uptake of inorganic nitrogen released by mineralization (middle pool in diagram) may or may not be much affected by the fertilizer. (Diagram courtesy of R. Weil)

TABLE 13.4 Typical Levels of Nitrogen Fixation from Different Systems

Crop or plant	Associated organism	Typical levels of nitrogen fixation, kg N/ha/yr
Symbiotic		
Legumes (nodulated)		
Ipil-ipil tree (*Leucaena leucocephala*)	Bacteria (*Rhizobium*)	100–500
Locust tree (*Robina* spp.)		75–200
Alfalfa (*Medicago sativa*)		150–250
Clover (*Trifolium pratense L.*)		100–150
Lupine (*Lupinus*)		50–100
Vetch (*Vicia vileosa*)		50–150
Bean (*Phaseolus vulgaris*)		30–50
Cowpea (*Vigna unguiculata*)	Bacteria (*Bradyrhizobium*)	50–100
Peanut (*Arachis*)		40–80
Soybean (*Glycine max L.*)		50–150
Pigeon pea (*Cajunus*)		150–280
Kudzu (*Pueraria*)		100–140
Nonlegumes (nodulated)		
Alders (*Alnus*)	Actinomycetes (*Frankia*)	50–150
Species of Gunnera	Cyanobacteria[a] (*Nostoc*)	10–20
Nonlegumes (nonnodulated)		
Pangola grass (*Digitaria decumbens*)	Bacteria (*Azospirillum*)	5–30
Bahia grass (*Paspalum notatum*)	Bacteria (*Azobacter*)	5–30
Azolla	Cyanobacteria[a] (*Anabena*)	150–300
Nonsymbiotic	Bacteria (*Azobacter, Clostridium*)	5–20
	Cyanobacteria[a] (various)	10–50

[a] Sometimes referred to as *blue-green algae*.

always increase soil nitrogen. Only in cases where the soil is low in available nitrogen and vegetation includes strong nitrogen fixers would this be likely to be true.

In the case of legume crops harvested for seed or hay, most of the nitrogen fixed is removed from the field with the harvest. Nitrogen additions from such crops should be considered as nitrogen *savers* for the soil rather than nitrogen builders. On the other hand, considerable buildup of soil nitrogen can be achieved by perennial legumes (such as alfalfa) and by annual legumes (such as hairy vetch) whose entire growth is returned to the soil as **green manure**. Such contributions can be useful in managing agricultural systems (Section 13.15) and should be taken into account when estimating nitrogen fertilizer needs for maximum plant production with minimal environmental pollution.

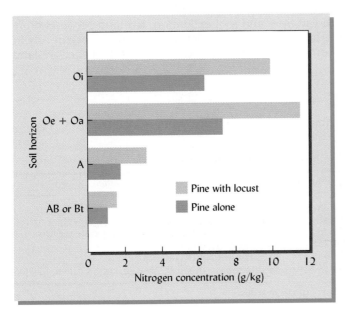

FIGURE 13.24 Nitrogen contents of forest soil horizons showing the effects of New Mexican locust trees (*Robinia neomexicana*) growing in association with ponderosa pine (*Pinus ponderosa*) in a region of Arizona receiving about 670 mm of rainfall per year. The data are means from 20 stands of pondersosa pine, half of them with the nitrogen-fixing legume trees (locust) in the understory. The soils are Eutrustalfs and Argiustolls with loam and clay loam textures. [Data from Klemmedson (1994)]

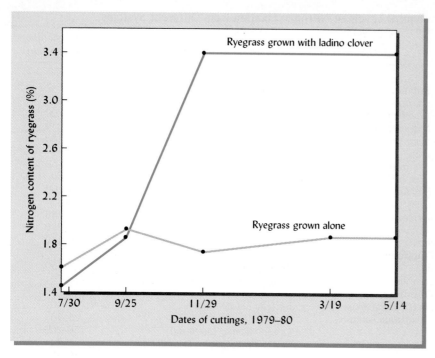

FIGURE 13.25 Nitrogen content of five field cuttings of ryegrass grown alone or with ladino clover. For the first two harvests, nitrogen fixed by the clover was not available to the ryegrass, and the nitrogen content of the ryegrass forage was low. In subsequent harvests, the fixed nitrogen apparently was available and was taken up by the ryegrass. This was probably due to the mineralization of dead ladino clover root tissue. [From Broadbent et al. (1982)]

13.12 SYMBIOTIC FIXATION WITH NONLEGUMES

Nodule-Forming Nonlegumes

Nearly 200 species from more than a dozen genera of nonlegume plants are known to develop nodules and to accommodate symbiotic nitrogen fixation. Included are several important groups of angiosperms, listed in Table 13.5. These plants, which are present in certain forested areas and wetlands, form distinctive nodules (*inset*, Figure 13.26) when their root hairs are invaded by soil actinomycetes of the genus *Frankia*.

The rates of nitrogen fixation per hectare compare favorably with those of the legume–*Rhizobium* associations (see Table 13.4). On a worldwide basis, the total nitrogen fixed in this way may even exceed that fixed by agricultural legumes. Because of their nitrogen-fixing ability, certain of the tree–actinomycete associations are able to colonize infertile soils and newly forming soils on disturbed lands, which may have extremely low fertility as well as other conditions that limit plant growth (Figure 13.26). Once nitrogen-fixing plants become established and begin to build up the soil nitrogen supply through leaf litter and root exudation, the land becomes more hospitable for colonization by other species. *Frankia* thus play a very important role in the nitrogen economy of areas undergoing succession, as well as in established wetland forests.

TABLE 13.5 **Number and Distribution of Major Actinomycete-Nodulated Nonlegume Angiosperms**

In comparison, there are about 13,000 legume species.

Genus	Family	Species[a] nodulated	Geographic distribution
Alnus	Betulaceae	33/35	Cool regions of the northern hemisphere
Ceanothus	Rhamnaceae	31/35	North America
Myrica	Myricaceae	26/35	Many tropical, subtropical, and temperate regions
Casuarina	Casuarinaceae	24/25	Tropics and subtropics
Elaeagnus	Elaeagnaceae	16/45	Asia, Europe, North America
Coriaria	Coriariaceae	13/15	Mediterranean to Japan, New Zealand, Chile to Mexico

[a] Number of species nodulated/total number of species in genus.
Selected from Torrey (1978).

FIGURE 13.26 Soil actinomycetes of the genus *Frankia* can nodulate the roots of certain woody plant species and form a nitrogen-fixing symbiosis that rivals the legume–*Rhizobia* partnership in efficiency. The actinomycete-filled root nodule (*a*) is the site of nitrogen fixation. The red alder tree (*b*) is among the first pioneer tree species to revegetate disturbed or badly eroded sites in high-rainfall areas of the Pacific Northwest in North America. This young alder is thriving despite the nitrogen-poor, eroded condition of the soil because it is not dependent on soil nitrogen for its needs. (Photos courtesy of R. Weil)

Certain cyanobacteria are known to develop nitrogen-fixing symbiotic relations with green plants. One involves nodule formation on the stems of *Gunnera*, an angiosperm common in marshy areas of the southern hemisphere. In this association, cyanobacteria of the genus *Nostoc* fix 10 to 20 kg N/ha/yr (see Table 13.4).

Symbiotic Nitrogen Fixation without Nodules

Among the most significant nonnodule nitrogen-fixing systems are those involving cyanobacteria. One system of considerable practical importance is the *Azolla–Anabaena* complex, which flourishes in certain rice paddies of tropical and semitropical areas. The *Anabaena* cyanobacteria inhabit cavities in the leaves of the floating fern *Azolla* and fix quantities of nitrogen comparable to those of the more efficient *Rhizobium*–legume complexes (see Table 13.4).

A more widespread but less intense nitrogen-fixing phenomenon is that which occurs in the *rhizosphere* of certain grasses and other nonlegume plants. The organisms responsible are bacteria, especially those of the *Spirillum* and *Azotobacter* genera (see Table 13.4). Plant root exudates supply these microorganisms with energy for their nitrogen-fixing activities.

Scientists have reported a wide range of rates for rhizosphere nitrogen fixation with the highest values observed in association with certain tropical grasses. Even if typical rates are only 5 to 30 kg N/ha/yr, the vast areas of tropical grasslands suggest that the total quantity of nitrogen fixed by rhizosphere organisms is likely very high (see Table 13.4).

13.13 NONSYMBIOTIC NITROGEN FIXATION[9]

Certain free-living microorganisms present in soils and water are able to fix nitrogen. Because these organisms are not directly associated with higher plants, the transformation is referred to as *nonsymbiotic* or *free-living*.

---

[9] For insights into the exploitation of nonsymbiotic N fixation for agriculture, see Kennedy et al. (2004).

Fixation by Heterotrophs

Video and more on Azotobacter:
http://www.microbiologybytes .com/video/Azotobacter.html

Several different groups of bacteria and cyanobacteria are able to fix nitrogen non-symbiotically. In upland mineral soils, the major fixation is brought about by species of several genera of heterotrophic aerobic bacteria, *Azotobacter* and *Azospirillum* (in temperate zones) and *Beijerinckia* (in tropical soils). Certain anaerobic bacteria of the genus *Clostridium* are also active in fixing nitrogen. Because pockets of low oxygen supply exist within aggregates even in well-drained soils (see Section 7.4), aerobic and anaerobic bacteria probably work side by side in many well-drained soils. These organisms obtain their carbon either from root exudates in the rhizosphere or by saprophytic decomposition of soil organic matter, and they operate best where soil nitrogen is limited.

The amount of nitrogen fixed by these heterotrophs varies greatly with the pH, soil nitrogen level, and sources of organic matter available. In some natural ecosystems these organisms undoubtedly make an important contribution to the nitrogen needs of the plant community. Because of limited carbon supplies, in conventional agricultural systems they probably fix only 5 to 20 kg N/ha/yr (see Table 13.4); however, with proper organic matter management, it is thought that the rates may be considerably higher. If agriculturalists are able to take advantage of these organisms, the benefits would go beyond increased crop yields to include reducing the need to manufacture nitrogen fertilizer (and therefore lowering the amount of reactive nitrogen circulating in the environment) as well as reductions in N_2O emissions.

Fixation by Autotrophs

In the presence of light, certain photosynthetic bacteria and cyanobacteria are able to fix carbon dioxide and nitrogen simultaneously. The contribution of the photosynthetic bacteria is uncertain, but that of cyanobacteria is thought to be of some significance, especially in wetlands (including in rice paddies). In some cases, cyanobacteria contribute a major part of the nitrogen needs of rice, but nonsymbiotic species rarely fix more than 20 to 30 kg N/ha/yr. Nitrogen fixation by cyanobacteria in upland soils also occurs (including in the desert microbiotic crusts discussed in Section 11.14), but at much lower levels than found under wetland conditions.

13.14 NITROGEN DEPOSITION FROM THE ATMOSPHERE

The atmosphere contains small quantities of ammonia and nitrogen oxide gases released from soils, plants, and fossil fuel combustion (especially in vehicle engines—see Section 9.6), as well as nitrates formed by lightning strikes. The term *nitrogen deposition* refers to the addition of these atmosphere-borne nitrogen compounds to soils (usually after transformation to the NH_4^+ or NO_3^- forms) through rain, snow, dust, and gaseous absorption.

The quantity of ammonia and nitrates in precipitation varies markedly with location (see Figure 13.27). Deposition is greatest in high-rainfall areas downwind from cities (nitrate from nitrogen oxides in car exhaust) and concentrated animal farming areas (ammonium volatilized from manure). The ratio of nitrate to ammonium nitrogen deposited varies with location, the nitrate share ranging from about one-third to two-thirds.

Although the nitrogen may stimulate greater plant growth in agricultural systems, the effects on forests, grasslands, and aquatic ecosystems are quite damaging. The nitrates in particular are associated with acidification of rain (as discussed in Section 9.6), but since the ammonium soon nitrifies, both forms lead to soil acidification. Nitrogen added by deposition as ammonium and nitrate from the atmosphere (or by fertilization) also impacts another soil process with important global change implications: methane oxidation (Figure 13.28). Methane is an important greenhouse gas affecting climate change, and its removal from the atmosphere by soil oxidation helps maintain its global balance (see also Section 12.9, Table 12.7). Forested soils have particularly high rates of methane oxidation, but also may be hardest hit by additions of mineral nitrogen. The methane oxidation capacities of grasslands and croplands are also significantly reduced by mineral nitrogen.

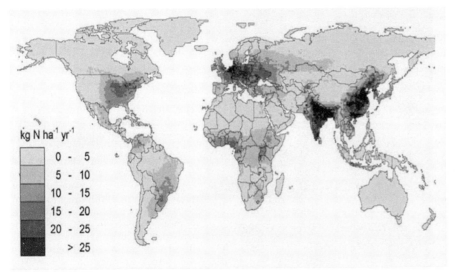

FIGURE 13.27 Global distribution of reactive nitrogen deposition to land from the atmosphere. Excessive N deposition can damage forests and other natural ecosystems. Highest deposition (darkest shading) occurs in high-rainfall regions downwind of intensive livestock feeding, rice production, and/or industrialized population centers. Ammonia from manure and nitrogen oxides from rice paddies and fossil fuel combustion are principal sources of the N that falls with rain, snow, and dust. Although variable, there is usually more ammonium than nitrate deposited. [From Eickhout et al. (2006)]

As the Earth's growing human population puts ever more reactive nitrogen into circulation, its unwanted deposition is becoming an increasingly serious global environmental problem.

Effects on Forest Ecosystems

Nitrogen in precipitation might be considered beneficial fertilizer when it falls on farmland, but it can be a serious pollutant when chronically added to some forested soils. Most forest soils are nitrogen limited—that is, they contain a surplus of carbon so that any nitrogen added is quickly tied up by microbial and chemical immobilization and very little nitrate is lost by leaching. However, a condition known as *nitrogen saturation* has been found to result from high levels of nitrogen deposition on certain mature forests in northern Europe and to a lesser extent in North America. Nitrogen saturation refers to the inability of the forest system to retain all or even most of the nitrogen received by deposition, leading to the leaching of nitrates and the associated soil acidification and loss of calcium and magnesium (as described in Sections 9.6 and 13.21). Nitrogen deposition can eventually reduce tree growth and disrupt the forest soil ecosystem in numerous ways, many of which are probably related to nitrogen saturation (Figure 13.29). Nitrogen deposition greater than about 8 kg N/ha/yr is expected to eventually cause damage to sensitive forest and aquatic ecosystems.

The amounts of nitrogen deposited; the type of forest (e.g., deciduous or evergreen); soil properties such as texture, mineralogy, and acidity; the history of land use and age

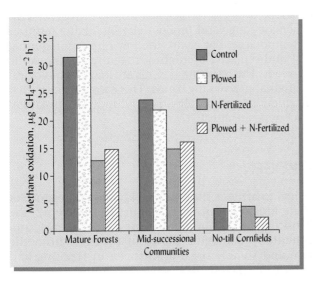

FIGURE 13.28 Mineral nitrogen reduced the capacity of soils to oxidize methane and thereby remove this potent greenhouse gas from the atmosphere. Forested soils exhibited the highest rates of methane oxidization and the greatest impairment due to the addition of N. A one-time physical disturbance (plowing) had little impact. Nitrogen was applied as a solution of ammonium nitrate (100 kg N ha^{-1}). The study was on sandy loam soils (Typic Hapludalfs) in southern Michigan. [Modified from Suwanwaree and Robertson (2005)]

FIGURE 13.29 Drastic effects of chronic high N additions on a pine forest ecosystem in Massachusetts. The experimental addition of 150 kg ha^{-1} yr^{-1} (=15 g m^{-2} yr^{-1}) of N as NH_4NO_3 since 1988 has decimated the tree canopy (*a*) and understory (*b*) compared to the unamended plot (*c* and *d*). By the 14th year of the study, 56% of the trees had died in the high N plot compared to 12% in the control plot. The high N treatment decreased soil microbial biomass by 40% and soil respiration by 35%. The fungal/bacterial ratio and microbial diversity also decreased. Much more N (inorganic and organic) was lost in leaching water from the high N plots. Foliage analyses showed significantly lower leaf calcium in the high N plots, suggesting that soil acidification and loss of calcium may have played a role in the trees' demise. The Montauk stony sandy loam soils (Typic Dystrochrepts) formed from glacial till. [Photos courtesy of R. Weil; data from Magill et al. (2004)]

of the forest; and climatic variables all seem to influence how chronic nitrogen additions will impact forest soils and ecosystems. For example, mature coniferous forests on poorly buffered soils may reach a state of nitrogen saturation more rapidly than young, rapidly growing forests such as those subject to frequent logging or recently converted from agriculture.

Effects on Rangeland Systems

Natural rangeland ecosystems contain a wide variety of native plant species that are known for their ability to conserve nitrogen and efficiently utilize low levels of deposited nitrogen (2 to 5 kg N/ha/yr). However, research using deliberate nitrogen additions suggests that as nitrogen deposition increases, certain plants—often exotic species—that are highly responsive to high levels of nitrogen quickly crowd out many of the desirable native species that are adapted to grow efficiently at low nitrogen levels. Attempts that have been made to increase rangeland productivity by adding fertilizer nitrogen may increase productivity temporarily, but again, the low-nitrogen-requiring native species are soon crowded out and replaced by high-nitrogen-requiring exotic species that are generally considered weeds. The resulting systems are lower in biological diversity and productivity than the original native rangelands. This case illustrates that well-meaning efforts to increase productivity of complex, poorly understood natural ecosystems can, in fact, have the exact opposite effect.

13.15 PRACTICAL MANAGEMENT OF SOIL NITROGEN

Sustainable nitrogen management aims to achieve three goals: (1) the maintenance of soil organic matter to ensure adequate long-term nitrogen supplies in the soil, (2) the regulation of the soluble forms of nitrogen to ensure that plant needs are met, and (3) the minimization of environmentally damaging losses of nitrogen from the soil–plant system, including nitrate and soluble organic nitrogen in leaching and runoff water, as well as ammonia and nitrogen oxide in gaseous emissions.

A particular type of soil in any given combination of climate and farming system tends to assume what may be called a *normal* or *equilibrium content* of nitrogen. Consequently, under ordinary methods of cropping and manuring, any attempt to permanently raise the nitrogen content to a level higher than this will result in unnecessary waste as nitrogen leaches, volatilizes, or is otherwise lost before it may be used. The equilibrium level of soil nitrogen is governed by the management practices employed and is likely to change as they are changed. For example, if cultivated soil that is initially low in organic matter and nitrogen is planted to perennial grass and fertilized or mixed with legumes, the cessation of tillage and high rate of root matter return will cause soil organic matter to build up quickly (see Section 12.8). If this long-term grassland is then plowed under, the reverse will take place, and the nitrogen stored over the years will be rapidly released—possibly causing a flush of nitrate leaching.

Using Nitrogen Fertilizers Wisely

Nitrogen management in field vegetables:
http://res2.agr.gc.ca/stjean/publication/bulletin/nitrogen 6-azote6_e.htm

Ammonium and nitrate ions from fertilizer are taken up by plants and participate in the nitrogen cycle in exactly the same way as ammonium and nitrate derived from organic matter mineralization or other sources. In fact, most of the N added as soluble fertilizers enters the biological cycle *before* the fertilized plants use it (see Figure 13.30). Nonetheless, the application of fertilizer usually results in a much larger concentration of these soluble nitrogen ions than would be found in unfertilized soils. Usually, plants and microorganisms cannot assimilate the applied nitrogen fast enough to prevent major losses by leaching, surface runoff, denitrification, and ammonia volatilization. The environmental consequences of these losses have already been discussed. Box 13.5 describes some technologies that can make nitrogen fertilizers less prone to cause pollution. However, even with the best fertilizer technologies, it is still important to avoid applying more nitrogen than is actually needed.

Making matters worse, managers may increase the amount of nitrogen applied to compensate for expected losses. Managers may also fail to take into account nitrogen deposition from the atmosphere, mineralization from soil organic matter, fixation by legumes, and release from current or past applications of animal manure or other organic materials. All these sources will contribute varying amounts of N to the pool of

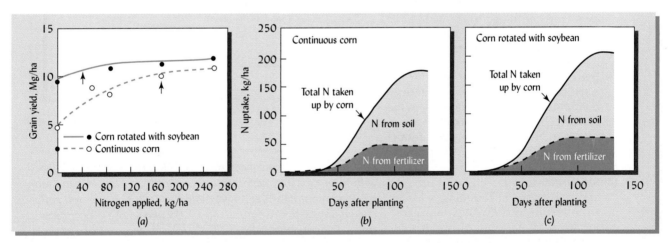

FIGURE 13.30 Crop rotation can increase yields and reduce fertilizer requirements. (*a*) The amount of fertilizer nitrogen required (↑) for high corn yields was less than 50 kg N/ha in the corn–soybean rotation (solid line), but over 150 kg N/ha in the continuous corn system (dashed line). These soils supported either corn every year or corn every second year alternating with soybeans for the 12 years. (*b*) The use of isotopically labeled fertilizer ($^{15}NH_4NO_3$) showed that even when a relatively high rate was applied (168 kg N/ha), most of the corn's N needs did not come directly from the fertilizer but from the mineralization of soil organic matter. This was especially true in the latter part of the growing season. By harvesttime, the fertilizer-derived nitrogen taken up by the corn represented less than one-third of the nitrogen taken up by the plants and accounted for less than one-third of the fertilizer nitrogen applied to the soil. (*c*) Corn grown in rotation with soybeans took up more nitrogen from both soil and fertilizer sources than did corn grown after corn. These data are averages for two years and two soils (a silt loam Ustoll and a loam Udoll) in Kansas, but the results are typical of those found in many parts of the world (compare to Table 16.15). The rotation corn maintained some yield advantage even when high amounts of nitrogen were applied, indicating that in addition to increased nitrogen availability after soybeans, the crop rotation conferred other benefits, such as pest reduction or improved soil microbial activity. Crop rotations involving more than two crops usually produce even greater benefits, especially if several years in the rotation are devoted to perennial grass-legume hay or pasture. [Data recalculated from Omay et al. (1998)]

BOX 13.5   REGULATION OF SOLUBLE NITROGEN WITH FERTILIZER TECHNOLOGY

A great challenge in managing nitrogen (N) fertilizers is to supply plant-available N at the proper time, in sufficient but not excessive amounts, and with a minimum of loss to the environment. Here we will discuss four types of fertilizer technologies that can help us meet this goal.

(1) **Split application** of N fertilizer refers to splitting the total N application for the growing season into several small doses, rather than applying the entire amount at once (usually at or before planting for annual crops). Application of the entire amount at the beginning of the season may result in much of the N leaching below the root zone before plants have had a chance to use it. This is especially true where soils are sandy or where precipitation (rainfall and irrigation) is high early in the season. In regions with wet winters, it is important to avoid leaving a large amount of soluble nitrogen in the soil profile after the crop has completed its growth in fall.

(2) **Chemical nitrification inhibitors**[a] are compounds that inhibit the activity of the *Nitrosomonas* bacteria, which convert NH_4^+ to NO_3^- in the first step of nitrification. (Note from eq. 13.5 that a chemical inhibiting *Nitrobacter* would *not* be useful, as this would cause toxic nitrite (NO_2^-) to accumulate.) As long as N remains as NH_4^+ it has little susceptibility to loss by leaching and does not transform into the greenhouse-forcing gases (N_2O and NO) by denitrification. Therefore, N fertilizers might be used more efficiently if the conversion of NH_4^+ to NO_3^- could be slowed down until the crop is ready to make use of the mobile NO_3^-. To this end chemical companies have developed such nitrification inhibitors as dicyandiamide (DCD), nitrapyrin (N-Serve®), and etridiazol (Dwell®). When mixed with nitrogen fertilizers, these compounds can temporarily prevent NO_3^- formation. The key word here is *temporarily,* for when conditions are favorable for nitrification, the inhibition usually lasts only a few weeks (less if soils are above 20 °C). Fertilizers treated with these chemicals are expensive and can be expected to pay for themselves by saving N only in rainy conditions on sandy soils (by reduced leaching) and on temporarily waterlogged soils (by reduced denitrification).

(3) **Biological inhibition of nitrification** may offer a new way of controlling nitrate formation without the limitations and expense of synthetic fertilizer additives. Researchers have discovered that root exudates produced by a tropical forage grass (*Brachiaria humidicola*) effectively inhibit *Nitrosomonas*, accounting for the high efficiency of N use in certain pasture systems. Scientists hope to transfer the nitrification inhibition gene from *B. humidicola* to other plants, with the aim of achieving widespread improvements in N use efficiency and reductions in N impacts on groundwater and greenhouse gases.

(4) **Slow-release fertilizers** provide another means of reducing N losses from fertilized soils. Most inorganic fertilizers contain entirely water soluble N, but the N in slow-release materials is formulated to dissolve slowly in the soil over a period of weeks or months so that N availability is better synchronized with plant uptake. Some stabilized organic materials, such as compost (see Section 12.10) and digested sewage sludge (e.g.,

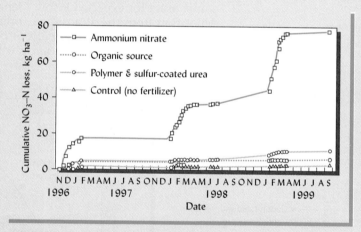

FIGURE 13.31   *Nitrate leaching from a New England lawn fertilized with 50 kg N/ha in November, May, and July of each year in the form of (1) ammonium nitrate, 100% soluble N, (2) polymer and sulfur-coated urea with dissolution rate of 25%/week, or (3) a composted turkey litter organic source with 5% total N. The control was unfertilized. The Paxton fine sandy loam soil (Oxyaquic Dystrudepts) with 57% sand, 12% clay, and 5.2% organic matter grew a mix of Kentucky bluegrass, perennial ryegrass, and creeping red fescue. No irrigation was used, and leaching occurred mainly in fall-winter. Funnels (gravity lysimeters) embedded 38 cm under the natural soil profile were used to collect the leachate. [From Guillard and Kopp (2004), with permission from ASA-CSSA-SSSA]*

(continued)

---

[a] For a review of chemical nitrification inhibitors, see Prasad and Power (1995). For information on biological nitrification inhibition in tropical pastures, see Ishikawa et al. (2003).

BOX 13.5 (Cont.) REGULATION OF SOLUBLE NITROGEN WITH FERTILIZER TECHNOLOGY

Milorganite®), are used for this purpose. However, most slow-release N fertilizers are made by treating urea with materials that slow its dissolution or inhibit its hydrolysis to ammonium. Urea-formaldehyde, isobutylidene diurea (IBDU), resin-coated fertilizers (e.g., Osmocote®), and polymer and/or sulfur-coated urea are all examples of slow-release N fertilizers. In the case of the latter, the sulfur content (10 to 20%) itself may be beneficial where sulfur is in low supply (Section 13.18) or problematic where the extra acidity generated by the sulfur (Section 13.22) would be undesirable. Sulfur-coated urea should not be used in flooded rice paddies as reduced iron will combine with the sulfur to form insoluble FeS, locking up the N in the fertilizer. Slow-release fertilizers cost from 1.5 (for sulfur-coated urea) to 5 times as much as plain urea per unit of N. Therefore, they are most widely used for high-value plant production (e.g., certain vegetables, turfgrass, and ornamentals) where the cost of fertilizers is not critical. They are commonly applied to turfgrass because they are not likely to cause fertilizer burn and because they provide the convenience of fewer applications. Figure 13.31 illustrates the dramatic reductions in N leaching under a turfgrass lawn that can be achieved by replacing water-soluble fertilizers with organic or slow-release materials.

nitrogen available to the plants, so the amount added in fertilizer should be reduced accordingly. That is, fertilizer should be considered only as a supplement to the nitrogen made available from organic matter mineralization, biological nitrogen fixation, atmospheric deposition, animal manure, and plant residues. This strategy is often referred to as taking "nitrogen credits" for these nonfertilizer sources.

Preventing N Losses

Fate of applied nitrogen: http://www.pmep.cce.cornell .edu/facts-slides-self/facts/ nit-el-grw89.html

Nitrogen losses by leaching and denitrification generally become a problem only when nitrogen fertilization exceeds the amount needed to fill the gap between crop uptake needs and the supply from these other sources. If losses could be perfectly controlled and "nitrogen credits" accurately accounted for, fertilizer N additions would be no more or less than needed to bring the inputs in balance with the intended output.

Regardless of the farming system, the goal should be to minimize the losses of nitrogen by all pathways except appropriate controlled harvest of a crop (this output is, after all, what the manager sells to make a profit or feed the family). The need for commercial nitrogen fertilizer will be governed by the degree to which the manager is able to integrate symbiotic nitrogen fixation, manure and residue recycling, and loss minimization into the farming system. Not every system needs to export a nitrogen-containing product. In the case of lawn grass, the aim of management is not to produce grass clippings, but rather a healthy lawn. In contrast to the heavy fertilization used on many lawns, the nitrogen-balance principles just discussed suggest that if the clippings are not removed after every mowing, then little or no nitrogen will be removed and little fertilizer need be applied (see Figure 13.32).

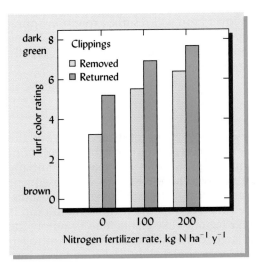

FIGURE 13.32 Turfgrass quality as influenced by nitrogen fertilizer and return or removal of grass clippings. Note that the practice of leaving the grass clippings on the lawn to recycle the nitrogen seemed to provide as much benefit as 100 kg/ha of fertilizer N, even providing acceptable quality turf with no N application. Modern rotary mowers cut grass leaves into tiny pieces that easily sift down between the grass blades and decompose rapidly without accumulating as undesirable thatch. If substituted for some or all N fertilizer, recycling of clippings can reduce the significant contribution that lawns make to N in runoff in urban and suburban watersheds. [Drawn from data in Heckman et al. (2000)]

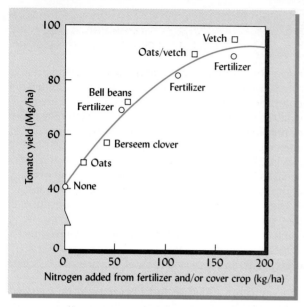

FIGURE 13.33 Effects of winter cover crops and inorganic fertilizer on the yield of the following main crop of processing tomatoes grown on a Xeralf soil in California. The amount of nitrogen shown on the x-axis was added either as inorganic fertilizer or as aboveground residues of the cover crops. Note that nitrogen from either source seemed to be equally effective. For the tomato crop, which requires nitrogen over a long period of time, vetch alone or vetch mixed with oats produced enough nitrogen for near-optimal yields. [Data abstracted from Stivers et al. (1993). © Lewis Publishers, an imprint of CRC Press, Boca Raton, Florida]

Legume Cover Crops Reduce the Need for Nitrogen Fertilizer

Calculating N credits for legumes:
http://www.extension.umn.edu/distribution/cropsystems/DC3769.html

A legume cover crop system may be able to replace part or all of the nitrogen fertilizer normally used to grow the main crop (Figure 13.33). The cover crop is usually killed mechanically or by a herbicide spray about a week before the main crop is planted. The cover crop residues can be left on the surface as a mulch in a no-till system (see Section 17.6) or plowed under in a conventional tillage system. When evaluating the feasibility of using cover crops, consideration should be given to the many other benefits of cover crops in addition to nitrogen supply (Section 16.2) as well as the costs and risks of growing the cover crop.

Nitrogen is supplied to the following main crop over a period of several months as the cover crop residues decay, rather than all at once like a soluble fertilizer application. In most cases the low C/N ratio legume residues decay rapidly enough to keep up with the nitrogen demands of the main crop. Legume cover crop systems have been adapted for vineyards, rice paddies, grain crops, vegetable fields, and home gardens wherever there is sufficient water to support both the cover crop and the main (summer) crop. Cover crops are increasingly important in organic farming systems because environmental restrictions on excessive phosphorus application (Sections 16.2 and 16.12) are forcing organic farmers to reduce their traditional dependence for nitrogen on purchased manure and compost (which contain high levels of phosphorus).

In temperate regions, winter annual legumes, such as vetch, clovers, and peas, can be sown in fall after the main crop harvest or, if the growing season is short, they can be seeded by airplane while the main crop is still in the field. After surviving the coldest part of the winter in a dormant state, the cover crop will resume growth in spring and associated microorganisms will fix as much as 3 kg/ha of nitrogen daily during the warmer spring weather. The amount of nitrogen provided is therefore partially determined by how long the cover crop is allowed to grow.

Legume Main Crops in Rotations

A detailed consideration of different cropping sequences (**crop rotations**) is beyond the scope of this book, but the nutrient-management aspects of rotating legumes with nonlegumes deserve mention here. A legume main crop grown one year may substantially reduce the amount of nitrogen fertilizer that needs to be used to grow a subsequent nonleguminous crop the next year. However, unlike for the legume cover crops just discussed, most of the biomass and accumulated nitrogen in the legume main crop is removed with the crop harvest. Perennial forage legumes like alfalfa tend to supply the greatest amounts of nitrogen because of their large root contribution, but the nitrogen contributions of grain legumes (e.g., soybeans or peanuts) should also be taken into account when planning fertilizer applications to nonlegumes that follow in the rotation. For example, many studies have shown that compared to corn grown after corn, corn grown after soybeans responds less dramatically to nitrogen fertilizer and requires less of it for optimal growth (Figure 13.30*a*).

Growing the same crop year after year on the same land generally produces lower yields of that crop and engenders more negative impacts on the soil and environment than if that crop is grown in rotation with other crops. This is true even when legumes are not part of the rotation, so more than nitrogen fixation must be involved. The improved plant productivity may result from interruption of weed, disease, and insect pest cycles; from complimentary soil exploitation by differing root systems; different types of residues and nutrient requirements; synergistic effects (see Section 11.14); and, possibly, positive effects on mycorrhizal diversity (see Section 11.9). These and other phenomena may explain why, even where pests and nutrient supply are optimally controlled, crops consistently yield 10 to 20% more in rotations than in continuous culture.

Bringing Nitrogen Inputs into Balance with Outputs

To understand the problem of nitrogen control, it is useful to consider the nitrogen inputs and outputs for the system in question. This is especially important for concentrated animal-feeding operations (CAFOs) where the raising of animals is not well integrated on the land with the production of their feed. CAFOs confine so many animals on a farm that much of the animal feed must be imported from other areas. The feed contains nitrogen (as protein) that mostly ends up in the animal manure. If much of the livestock feed was grown at distant places, the manure produced at the CAFO will contain far more nitrogen than can be used by crops in the local fields. If nitrogen brought onto the farm exceeds that exported from the farm in plant or animal products, the excess will likely be "exported" to the environment by leaching, runoff, erosion, and volatilization (see also Sections 16.1 and 16.4). Table 13.6 illustrates the magnitude of nitrogen excesses in several places with many CAFOs.

When we examine the nitrogen balance for a backyard, farm, region, or nation, it is often clear that more nitrogen is being applied than can be effectively used by the crops grown, creating a surplus that is susceptible to environmentally (and financially!) damaging losses. This excess will have to be reduced if nitrogen pollution is to be controlled.

As a detailed example, Figure 13.34 shows the nitrogen budget for the state of Illinois. The features that make a system dominated by agriculture different from a natural ecosystem are mainly: (1) the import of nitrogen in fertilizer and feed and (2) the removal of nitrogen in the harvested product. The principles discussed in this chapter suggest several basic strategies for achieving a rational reduction of excessive nitrogen inputs while maintaining or improving production levels and profitability in agricultural enterprises. These approaches include (1) taking into account the nitrogen contribution from *all* sources and reducing the amount of fertilizer applied accordingly; (2) improving the efficiency with which fertilizer and organic amendments (e.g., manure) are used; (3) avoiding overly optimistic yield goals that lead to fertilizer application rates designed to meet crop needs that are much higher than actually occur in most years; and (4) improving crop response knowledge, which identifies the lowest nitrogen application that is likely to produce optimum profit. These and other strategies of nutrient management will be discussed further in Chapter 16.

TABLE 13.6 **Regional Nitrogen Surpluses that Can Occur with Intensive Livestock Production**

The difference between the N supplied and that used by crops is listed as surplus. At first, some of the surplus N may be incorporated into soil organic matter, but most eventually will leach, run off, or volatilize into the environment.

Country or state	Nitrogen supply, 1000 Mg			Nitrogen use by crop uptake	Surplus nitrogen	
	Manure	*Fertilizer*	*Total*	*1000 Mg*	*Total, 1000 Mg*	*Per hectare, kg*
Belgium/Luxembourg	380	199	580	211	369	240
Denmark	434	381	816	287	529	187
The Netherlands	752	504	1,255	285	970	480
Delaware, U.S.	8	19	27	16	11	49

Data from Leuch et al. (1995) and Sims and Wolf (1994).

FIGURE 13.34 Annual inputs and outputs of nitrogen for the state of Illinois, averaged for 18 years (1979–96 for terrestrial fluxes and 1980–97 for riverine export). The units are 1000 Mg N/year. The major inputs of nitrogen are commercial fertilizers, nitrogen fixation, and deposition from the atmosphere. Fluxes inside the box are for nitrogen in the total grain harvested, its consumption by animals and humans, and the human sewage that moves into rivers from sewage plants. The major outputs are the nitrogen in the animals sold and the grain exported outside the state, the nitrogen that leaches or runs off into rivers (riverine export), and that lost by denitrification from the soil and from streams as they move toward the sea. Note in the box that a small amount of N (13,000 Mg/yr) remains in the system. This may contribute to the level of nitrogen in the soil. Note that animal manure and crop residues are not considered as either an input or an output since they represent recycling of nitrogen that comes from the soil and goes back into the soil. This diagram illustrates many of the principles covered in this chapter. [From David and Gentry (2000)]

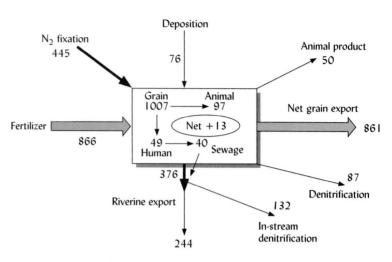

13.16 IMPORTANCE OF SULFUR[10]

Sulfur has long been recognized as indispensable for many reactions in living cells. In addition to its vital roles in plant and animal nutrition, sulfur is also responsible for several types of air, water, and soil pollution and is therefore of increasing environmental interest. The environmental problems associated with sulfur include acid precipitation, certain types of forest decline, acid mine drainage, acid sulfate soils, and even some toxic effects in drinking water used by humans and livestock.

Roles of Sulfur in Plants and Animals

Role of Sulfur:
http://www.soil.ncsu.edu/publications/Soilfacts/AG-439-15/

Sulfur is a constituent of the amino acids methionine, cysteine, and cystine, deficiencies of which result in serious human malnutrition. The vitamins biotin, thiamine, and B1 contain sulfur, as do many protein enzymes that regulate such activities as photosynthesis and nitrogen fixation. It is believed that sulfur-to-sulfur bonds link certain sites on long chains of amino acids, causing proteins to assume the specific three-dimensional shapes that are the key to their catalytic action. Sulfur is closely associated with nitrogen in the processes of protein and enzyme synthesis. Sulfur is also an essential ingredient of the aromatic oils that give the cabbage and onion families of plants their characteristic odors and flavors. It is not surprising that among the plants, the legume, cabbage, and onion families require especially large amounts of sulfur.

Deficiencies of Sulfur

Healthy plant foliage generally contains 0.15 to 0.45% sulfur, or approximately one-tenth as much sulfur as nitrogen. Plants deficient in sulfur tend to become spindly and to develop thin stems and petioles. Their growth is slow, and maturity may be delayed. They also have a chlorotic light green or yellow appearance (Plate 92, after page 656). Symptoms of sulfur deficiency are similar to those associated with nitrogen deficiency (see Section 13.1). However, unlike nitrogen, sulfur is relatively immobile in the plant, so the chlorosis develops first on the youngest leaves as sulfur supplies are depleted (in nitrogen-deficient plants, chlorosis develops first on the older leaves). Sulfur-deficient leaves on some plants show interveinal chlorosis or faint striping that distinguishes them from nitrogen-deficient leaves. Also, unlike nitrogen-deficient plants, sulfur-deficient plants tend to have low sugar but high nitrate contents in their sap.

---

[10] For a discussion of sulfur and agriculture, see Tabatabai (1986).

As a result of the following three independent trends, sulfur deficiencies in agricultural plants have become increasingly common during the past several decades:

1. Enforcement of clean air standards has led to reduction of sulfur dioxide (SO_2) emissions to the atmosphere from the burning of fossil coal and oil.
2. The sulfur contents of today's highly concentrated N-P-K fertilizers are far lower than those of a generation ago. Thus, less sulfur is being added.
3. As crop harvests have increased, larger amounts of sulfur are being removed from the soils. Consequently, the need for sulfur has increased just as the inputs from all sources have declined.

AREAS OF DEFICIENCY. Sulfur deficiencies have been reported in most areas of the world but are most prevalent in areas where soil parent materials are low in sulfur, where extreme weathering and leaching have removed this element, or where there is little replenishment of sulfur from the atmosphere. In many tropical countries, one or more of these conditions prevail and sulfur-deficient areas are common.

Burning of plant biomass results in a loss of sulfur to the atmosphere. In many parts of the world, crop residues and native vegetation are routinely burned as a means of clearing the land. Soils of the African savannas are particularly deficient in sulfur as a result of the annual burning of plant residues during the dry season. Fire converts much of the sulfur in the plant residues to sulfur gases, such as sulfur dioxide. Sulfur in these gases and in smoke particulates is subsequently carried by the wind hundreds of kilometers away to areas covered by rain forest, where some of the sulfur dioxide is absorbed by moist soils and foliage and some is deposited with rainfall. Thus, the soils of the savannas tend to export their sulfur to those of the rain forest (e.g., Oxisols). Consequently, the latter often contain significant accumulations of sulfur in their profiles (see Figure 13.35).

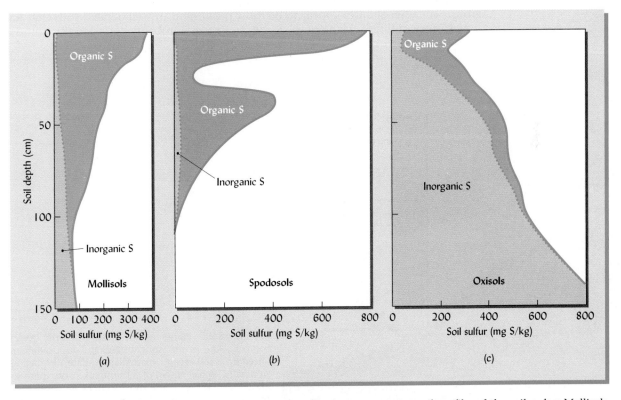

FIGURE 13.35 The distribution of organic and inorganic sulfur in representative soil profiles of the soil orders Mollisols, Spodosols, and Oxisols. In each, soil organic forms dominate the surface horizon. Considerable inorganic sulfur, both as adsorbed sulfate and calcium sulfate minerals, exists in the lower horizons of Mollisols. Relatively little inorganic sulfur exists in Spodosols. However, the bulk of the profile sulfur in the humid tropics (Oxisols) is present as sulfate adsorbed to colloidal surfaces in the subsoil. (Diagram courtesy of R. Weil)

In the United States, deficiencies of sulfur are most common in the Southeast, the Northwest, California, and the Great Plains. In the Northeast and in other areas with heavy industry and large cities, sulfur deficiencies are not yet widespread.

13.17 NATURAL SOURCES OF SULFUR

Sulfur in nature, see sulfur section in:
http://www..iitap.iastate.edu/gcp/chem/nitro/nitro_lecture.html

The three major natural sources of sulfur that can become available for plant uptake are (1) *organic matter*, (2) *soil minerals*, and (3) *sulfur gases in the atmosphere*. In natural ecosystems where most of the sulfur taken up by plants is eventually returned to the same soil, these three sources combined are usually sufficient to supply the needs of growing plants (Figure 13.36). These three natural sources of sulfur will be considered in order.

Organic Matter[11]

In surface soils in temperate, humid regions, 90 to 98% of the sulfur is usually present in organic forms (see Figure 13.35). As is the case for nitrogen, the exact forms of the sulfur in the organic matter are not known. Recent research using sophisticated spectroscopic analyses has shown that three principal groups of organic sulfur compounds exist in soil organic matter. The first group consists of the most reduced forms of S bonded to carbon in compounds such as sulfides, disulfides, thiols, and thiophenes. These include proteins with amino acids such as cysteine, cystine, and methionine. A second group of intermediate redox state includes sulfoxides and sulfonates in which the sulfur is bonded to carbon but also to oxygen (C—S—O). The third group consists of highly oxidized forms of S in ester sulfates (C—O—S), in which S is bound to oxygen rather than directly to carbon.

Typically, undisturbed soils under grasslands, wetlands, and moist forest vegetation have much less than half of the organic sulfur that is in the most oxidized ester-sulfate group. The ester-sulfates appear to be most prominent in well-aerated soils that have a long history of tillage. Many of the S-compounds are probably bound to the humus and

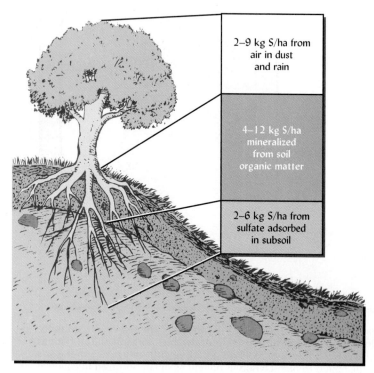

2–9 kg S/ha from air in dust and rain

4–12 kg S/ha mineralized from soil organic matter

2–6 kg S/ha from sulfate adsorbed in subsoil

FIGURE 13.36 Plants take up sulfur primarily from three sources: sulfur in atmospheric gases and dust, sulfate mineralized from soil organic matter, and sulfate adsorbed on soil minerals. Typical ranges of sulfur uptake from these sources are shown. Where these three sources are insufficient for optimal growth, the application of sulfur-containing fertilizers may be warranted. In areas downwind of coal-burning plants and metal smelters, the atmospheric contribution may be much larger than indicated here.

---

[11] For a study of organic sulfur forms in a range of soils, see Zhao et al. (2006).

clay fractions and are thereby somewhat protected from microbial attack. Examples of compounds in the three main fractions of organic sulfur are as follows:

Ester sulfate (glucose sulfate)   Sulfoxide compound   Carbon-bonded sulfur (cysteine)

Over time, soil microorganisms break down these organic sulfur compounds into soluble forms analogous to the release of ammonium and nitrate from organic matter, discussed in Section 13.3. As with nitrogen, most soil sulfur is organic with only a very small percentage in the mineralized (sulfate) form, even in the sandy Spodosols described in Table 13.7 (compare with Figure 13.35*b*).

In dry regions, less organic matter is present in the surface soils. However, gypsum ($CaSO_4 \cdot 2H_2O$), which supplies inorganic sulfur, is often present in the subsurface horizons. Therefore, the proportion of organic sulfur is not likely to be as high in arid- and semiarid-region soils as it is in humid-region soils. This is especially true in the subsoils, where organic sulfur may constitute only a small fraction of the sulfur present and where gypsum is prominent.

Soil Minerals

The inorganic forms of sulfur are not as plentiful as the organic forms, but they include the soluble and available compounds on which plants and microbes depend. The two most common inorganic sulfur compounds are sulfates and sulfides. The sulfate minerals are most easily solubilized, and the sulfate ion (SO_4^{2-}) is easily assimilated by plants. Sulfate minerals are most common in regions of low rainfall, where they accumulate in the lower horizons of some Mollisols and Aridisols (see Figure 13.35). They may also

TABLE 13.7 Concentration of Sulfur Compounds in Forested Spodosols from Several Locations

Note that the organic layers (Oa) have high S levels and that the two organic forms (C-bonded and ester sulfate) contain 84 to 99% of the soil sulfur.

Horizon	Totals, µg S/g	Proportion of soil S, %			
		C-bonded S	Ester sulfate	Sulfate S	Inorganic S[a]
Hubbard Brook, N.H.[b]					
Oa	1563	71	28	0.3	—
Bh	303	72	22	3.0	—
Bs1	452	64	26	8.1	—
Huntington Forest, N.Y.[c]					
Oa	1780	77	22	0.2	0.7
Bh	761	83	13	3.1	2.2
Bs1	527	70	22	4.2	4.4
Conifer site[d]					
Oa	2003	88	11	0.3	0.6
Bh	540	83	11	2.4	3.4
Bs1	515	57	27	13	3.0

[a] Inorganic sulfur compounds, such as sulfides.
[b] From Schindler et al. (1986).
[c] From David et al. (1982).
[d] From Mitchell et al. (1992).

accumulate as neutral salts in the surface horizons of saline soils in arid and semiarid regions.

Sulfides that are found in some humid-region soils with restricted drainage must be oxidized to the sulfate form before the sulfur can be assimilated by plants. When these soils are drained, oxidation can occur, and ample available sulfur is released. In some cases, so much sulfur is oxidized that problems of extreme acidity result (see Section 13.20).

Another mineral source of sulfur is the clay fraction of some soils high in Fe, Al oxides and kaolinite. These clays are able to strongly adsorb sulfate from soil solution and subsequently release it slowly by anion exchange, especially at low pH. Oxisols and other highly weathered soils of the humid tropics and subtropics may contain large stores of sulfate, especially in their subsoil horizons (see Figure 13.35). Considerable sulfate may also be bound by the metal oxides in the spodic horizons under certain temperate forests.

Atmospheric Sulfur[12]

The atmosphere contains varying quantities of carbonyl sulfide (COS), hydrogen sulfide (H_2S), sulfur dioxide (SO_2), and other sulfur gases, as well as sulfur-containing dust particles. These atmospheric forms of sulfur arise from volcanic eruptions, volatilization from soils, ocean spray, biomass fires, and industrial plants (such as electric-generation stations fired by high-sulfur coal, and metal smelters). In the past half century, the contribution from industrial sources has dominated sulfur deposition in certain locations.

In the atmosphere, most of the sulfur materials are eventually oxidized to sulfates, forming H_2SO_4 and sulfate salts, such as $CaSO_4$ and $MgSO_4$. The "acid rain" problem caused by this atmospheric sulfur (as well as nitrogen) was discussed in Section 9.6 (see Figure 9.16) and the global distribution of sulfur shown in Figure 9.18. When sulfur is returned to the Earth as dry particles and gases, it is called *dry deposition;* when it is brought down with precipitation, it is called *wet deposition.* Although the proportion of these two forms of deposition varies from one place to another, each typically supplies about half of the total (Figure 13.37).

After watching forests and lakes become seriously damaged by acid rain in the 1970s and 1980s, governments in North America and parts of Europe established a regulatory program to reduce sulfur emissions. As a result, sulfur emissions in these regions have declined by almost half since the late 1980s (although nitrogen oxide emissions have not been equally addressed). Today in eastern North America, annual sulfur deposition rarely exceeds 15 kg S/ha (= 37 kg sulfate) and is more commonly less than 8 to 10 kg S/ha. On the other hand, sulfur emissions are on the rise in China, India, and other newly industrializing regions, where burning of coal and oil cause as much as 50 to 75 kg S/ha to come down in a year. In other areas little affected by industrial emissions (e.g., most rural areas of the western United States), deposition is generally only 2 to 5 kg S/ha/yr. In rural Africa, as little as 1 to 4 kg S/ha/yr is deposited.

Atmospheric sulfur becomes part of the soil–plant system in three ways. The wet deposition materials, which are usually high in H_2SO_4, are mostly absorbed by soils, with some also absorbed through plant foliage. Part of the dry deposition is also absorbed directly by soils, while some is absorbed directly by plants. The quantity that plants can absorb directly is variable, but in some cases 25 to 35% of the plant sulfur can come from this source even if available soil sulfate is adequate. In sulfur-deficient soils, about half of the plant needs can come from the atmosphere (see Figure 13.36).

The acid precipitation caused partly by atmospheric sulfur (as well as nitrogen) is a serious threat to the health of lake, forest, and agroecosystems. In areas immediately downwind from industrial plants, sulfur deposition may be great enough to cause direct toxicity to trees and crops (not to mention respiratory problems in people). As far as 1000 km downwind, the deposited sulfate may mobilize toxic soil aluminum, acidify lakes, and deplete soils of calcium needed by forests. On the other hand, reduced sulfur emissions in recent years are resulting in plant deficiencies of this element in some areas, especially for high-yield-potential agricultural crops. In these cases, part of the crop's sulfur requirement may have to be supplied by increasing the sulfur content of the fertilizers used, thereby increasing crop production costs.

---

[12] For a discussion of global sulfur deposition trends, see Stern (2006).

FIGURE 13.37 This apparatus collects both wet and dry sulfur deposition. A sensor (*a*) triggers the small roof (*b*) to move over and cover the dry deposition collection chamber (*c*) at the first sign of precipitation. The wet deposition chamber (*d*) is then exposed to collect precipitation. When the precipitation ceases, the sensor triggers the roof to move back over the wet deposition collection chamber so that dry deposition can again be collected. The map shows the average geographic distribution of sulfate deposition from the atmosphere in eastern North America. The data are based on measurements of sulfate (SO_4^{2-}) in wet deposition only (average of 1996–2000). To estimate the *total* deposition as the element S, the values for wet deposition of sulfate can be multiplied by 2 (assuming dry deposition = wet deposition) and then by 0.2 (to convert mass of SO_4^{2-} into S). [Photo courtesy of R. Weil; map modified from International Joint Commission (2002)]

13.18   THE SULFUR CYCLE

Reactions in the sulfur cycle:
http://filebox.vt.edu/users/
chagedor/biol_4684/Cycles/
Scycle.html

The major transformations that sulfur undergoes in soils are shown in Figure 13.38. The inner circle shows the relationships among the four major forms of this element: (1) *sulfides*, (2) *sulfates*, (3) *organic sulfur*, and (4) *elemental sulfur*. The outer portions show the most important sources of sulfur and how this element is lost from the system.

Considerable similarity to the nitrogen cycle is evident (compare Figures 13.2 and 13.38). In each case, the atmosphere is an important source of the element in question. Both elements are held largely in the soil organic matter, both are subject to microbial oxidation and reduction, both can enter and leave the soil in gaseous forms, and both are subject to some degree of leaching in the anionic form. Microbial activities are responsible for many of the transformations that determine the fates of both nitrogen and sulfur.

Figure 13.38 should be referred to frequently in conjunction with the following more detailed examination of sulfur in plants and soils.

13.19   BEHAVIOR OF SULFUR COMPOUNDS IN SOILS

Mineralization

Sulfur behaves much like nitrogen as it is absorbed by plants and microorganisms and moves through the sulfur cycle. The organic forms of sulfur must be mineralized by soil organisms if the sulfur is to be used by plants. The rate at which this occurs depends on the same environmental factors that affect nitrogen mineralization, including moisture, aeration, temperature, and pH. When conditions are favorable for general microbial activity, sulfur mineralization occurs. Some of the more easily decomposed organic compounds in the soil are sulfate esters, from which microorganisms release sulfate ions directly. However, in much of the soil organic matter, sulfur in the reduced state is

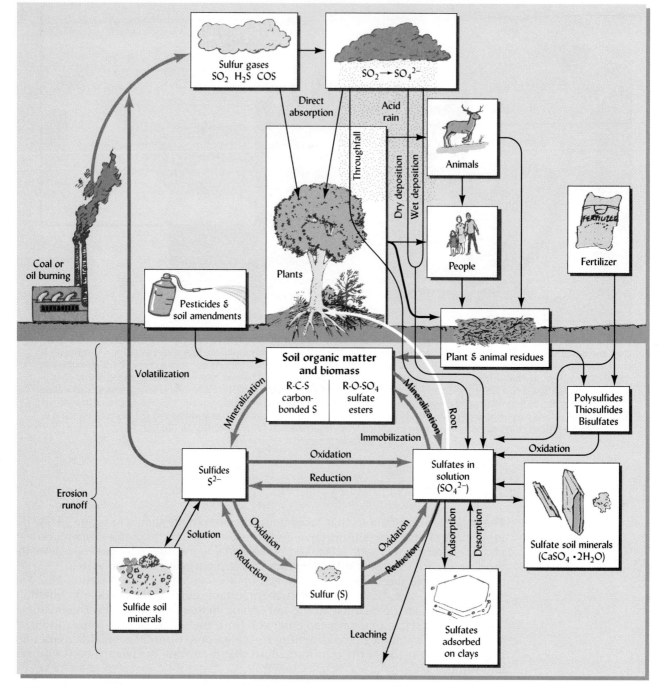

FIGURE 13.38 The sulfur cycle, showing some of the transformations that occur as this element is cycled through the soil–plant–animal–atmosphere system. In the surface horizons of all but a few types of arid-region soils, the great bulk of sulfur is in organic forms. However, in deeper horizons or in excavated soil materials, various inorganic forms may dominate. The oxidation and reduction reactions that transform sulfur from one form to another are mainly mediated by soil microorganisms.

bonded to carbon atoms in protein and amino acid compounds. In the latter case the mineralization reaction might be expressed as follows:

$$\underset{\substack{\text{Proteins and} \\ \text{other organic} \\ \text{combinations}}}{\text{Organic sulfur}} \longrightarrow \underset{\substack{\text{H}_2\text{S and other} \\ \text{sulfides are} \\ \text{simple examples}}}{\text{decay products}} \xrightarrow{\text{O}_2} \underset{\text{Sulfates}}{\text{SO}_4{}^{2-} + 2\text{H}^+} \tag{13.9}$$

Because this release of available sulfate is mainly dependent on microbial processes, the supply of available sulfate in soils fluctuates with seasonal, and sometimes daily, changes in environmental conditions (Figure 13.39). These fluctuations lead to the

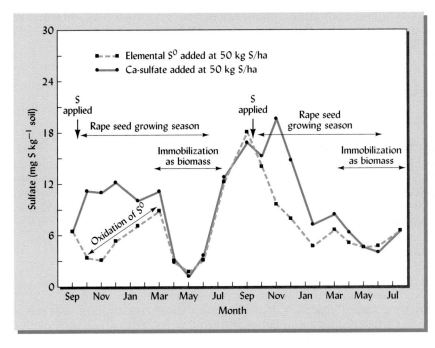

FIGURE 13.39 Seasonal changes in the sulfate form of sulfur available in the surface horizon of a soil (Argixeroll) in Oregon used to grow the oilseed crop, rape. This crop is sown in the fall, grows slowly during the winter, and then grows rapidly during the cool spring months. Data are shown for plots that were fertilized with either elemental S or calcium sulfate. The vertical arrows indicate dates on which these amendments were applied. Note that sulfate concentration was greater in the calcium sulfate-fertilized soils for the first few months after each application while the elemental S was slowly converted to sulfate by microbial oxidation. A distinct depression in sulfate concentration occurred each spring as the soil warmed up, stimulating both immobilization of sulfate into microbial biomass and uptake of sulfate into the rapeseed crop. Sulfate concentrations reached a peak in late summer and early fall, when crop uptake ceased after harvest and microbial mineralization was rapidly occurring. Movement of dissolved sulfate from the lower horizons up into the surface soil may also have occurred during hot, dry weather. [Modified from Castellano and Dick (1991)]

same difficulties in predicting and measuring the amount of sulfur available to plants as were discussed in the case of nitrogen.

Immobilization

Immobilization of inorganic forms of sulfur occurs when low-sulfur, energy-rich organic materials are added to soils. The immobilization mechanism is thought to be the same as for nitrogen—the energy-rich material stimulates microbial growth, and the inorganic sulfate is assimilated into microbial tissue. A C/S ratio greater than 400:1 generally leads to such immobilization of sulfur. When the microbial activity subsides, the inorganic sulfate reappears in the soil solution.

The pattern of S immobilization in soils suggests that, like nitrogen, sulfur in soil organic matter may be associated with organic carbon in a reasonably constant ratio. The ratio among carbon, nitrogen, and sulfur for a number of soils on three different continents is given in Table 13.8. The C/N/S ratio of 100:8:1 is reasonably representative.

During the microbial breakdown of organic materials, several sulfur-containing gases are formed, including hydrogen sulfide (H_2S), carbon disulfide (CS_2), carbonyl sulfide (COS), and methyl mercaptan (CH_3SH). All are more prominent in anaerobic soils. Hydrogen sulfide is commonly produced in waterlogged soils by reduction of sulfates by anaerobic bacteria. Most of the others are formed from the microbial decomposition

TABLE 13.8 Mean Carbon/Nitrogen/Sulfur Ratios in a Variety of Soils

Location	Description and number of soils	C/N/S ratio
North Scotland	Agricultural, noncalcareous (40)	104:7:1
Minnesota	Mollisols (6)	74:7:1
Minnesota	Spodosols (24)	108:8:1
Oregon	Agricultural, varied (16)	143:10:1
Eastern Australia	Acid soils (128)	126:8:1
Eastern Australia	Alkaline soils (27)	92:7:1
Sweden	Agricultural, Inceptisols	69:7:1
Highland Ethiopia	Cultivated soils (3)	90:7:1
Highland Ethiopia	Forested soils (3)	67:6:1

Data for Sweden from Kirchmann et al. (1996), for Ethiopia from Solomon et al. (2001), and for other soils from Whitehead (1964).

of sulfur-containing amino acids. Although these gases can be adsorbed by soil colloids, some escape to the atmosphere, where they undergo chemical changes and eventually return to the soil.

13.20 SULFUR OXIDATION AND REDUCTION

The Oxidation Process

Oxidation rates of commercial sulfur products: http://soil.scijournals.org/cgi/content/full/65/1/239

During the microbial decomposition of organic carbon-bonded sulfur compounds, sulfides are formed along with other incompletely oxidized substances, such as elemental sulfur (S^0), thiosulfates ($S_2O_3^{2-}$), and polythionates ($S_{2x}O_{3x}^{2-}$). These reduced substances are subject to oxidation, just as are the ammonium compounds formed when nitrogenous materials are decomposed. The oxidation reactions may be illustrated as follows, with hydrogen sulfide and elemental sulfur:

$$H_2S + 2O_2 \rightarrow H_2SO_4 \rightarrow 2H^+ + SO_4^{2-} \tag{13.10}$$

$$2S + 3O_2 + 2H_2O \rightarrow 2H_2SO_4 \rightarrow 4H^+ + SO_4^{2-} \tag{13.11}$$

The oxidation of some sulfur compounds, such as sulfites (SO_3^{2-}) and sulfides (S^{2-}), can occur by strictly chemical reactions. However, most sulfur oxidation in soils is *biochemical* in nature, carried out by a number of autotrophic bacteria, which include five species of the genus *Thiobacillus*. Since the environmental requirements and tolerances of these five species vary considerably, the process of sulfur oxidation occurs over a wide range of soil conditions. For example, sulfur oxidation may occur at pH values ranging from <2 to >9. This flexibility is in contrast to the comparable nitrogen oxidation process, nitrification, which requires a rather narrow pH range closer to neutral.

The Reduction Process

Like nitrate ions, sulfate ions tend to be unstable in anaerobic environments. They are reduced to sulfide ions by a number of bacteria of two genera, *Desulfovibrio* (five species) and *Desulfotomaculum* (three species). The organisms use the oxygen in sulfate to oxidize organic materials. A representative reaction showing the reduction of sulfur coupled with organic matter oxidation is as follows:

$$2R-CH_2OH + SO_4^{2-} \rightarrow 2R-COOH + 2H_2O + S^{2-} \tag{13.12}$$

Organic alcohol   Sulfate   Organic acid   Sulfide

In poorly drained soils, the sulfide ion reacts immediately with iron or manganese, which in anaerobic conditions are typically present in the reduced forms. By tying up the soluble reduced iron, the formation of iron sulfides helps prevent iron toxicity in rice paddies and marshes. This reaction may be expressed as follows:

$$Fe^{2+} + S^{2-} \rightarrow FeS \tag{13.13}$$

Dissolved ferrous iron   Sulfide   Iron sulfide (solid)

$$Mn^{2+} + S^{2-} \rightarrow MnS \tag{13.14}$$

Dissolved reduced manganese   Sulfide   Manganese sulfide (solid)

Sulfide ions will also undergo hydrolysis to form gaseous hydrogen sulfide, which causes the rotten-egg smell of swampy or marshy areas (Figure 13.40). Sulfur reduction may take place with sulfur-containing ions other than sulfates. For example, sulfites (SO_3^{2-}), thiosulfates ($S_2O_3^{2-}$), and elemental sulfur (S^0) are readily reduced to the sulfide form by bacteria and other organisms.

The oxidation and reduction reactions of inorganic sulfur compounds play an important role in determining the quantity of sulfate (the plant-available nutrient form of sulfur) present in soils at any one time. Also, the state of sulfur oxidation is an important factor in the acidity of soil and water draining from soils.

FIGURE 13.40 Using a handful of soil from a coastal marsh, soil scientist Robert Darmondy performs the "whiff test" he originated. The test checks for sulfidic material, which emits H_2S gas with an odor like that of a rotten egg. (Photo courtesy of R. Weil)

Acidity from Sulfur Oxidation

Reactions 13.10 and 13.11 show that, like nitrogen oxidation, sulfur oxidation is an acidifying process. These reactions explain why elemental sulfur can be applied to lower soil pH if it is higher than desired and why sulfur-containing fertilizers (e.g., ammonium sulfate) sometimes dramatically lower soil pH below desirable levels (see Sections 9.10 and 10.10).

Along with nitrogen, the sulfur in the atmosphere (as described in Section 13.17) forms strong acids that acidify rainwater to a pH of 4 or even lower from the normal pH of 5.6 or higher. Section 9.6 explains how "acid rain" forms and how it damages soils, forests, and lakes in many regions. As described in Section 13.21, part of the damage to these ecosystems stems from the leaching of sulfate anions, which can promote serious losses of calcium and magnesium.

EXTREME SOIL ACIDITY.   The acidifying effect of sulfur oxidation can bring about extremely acid soil conditions that cause serious soil management problems and broader environmental pollution. Certain soils and sedimentary geologic materials are termed *sulfidic* because they contain high levels of reduced sulfur, usually inherited from their present or past association with seawater (which is high in sulfur). A common form of reduced sulfur is the mineral pyrite (iron disulfide, FeS_2). The sulfides in these *potential acid-sulfate* materials are stable so long as oxygen is not present, but if submerged or buried materials are drained or excavated, the sulfides and/or elemental sulfur quickly oxidize and form sulfuric acid, driving pH levels as low as 1.5 (see Section 9.6, reactions 9.20–9.21). Plants cannot grow under these conditions (Figure 13.41). The quantity of limestone needed to neutralize the acidity is so high that it is impractical to remediate

FIGURE 13.41 Construction of this highway cut through several layers of sedimentary rock. One of these layers contained reduced sulfide materials. Now exposed to the air and water, this layer is producing copious quantities of sulfuric acid as the sulfide materials are oxidized. Note the failure of vegetation to grow below the zone from which the acid is draining. (Photo courtesy of R. Weil)

these soils by liming (for reactions, solutions, and other details see Section 9.8). If allowed to proceed unchecked, the acids may wash into nearby streams. Thousands of kilometers of streams have been seriously polluted in this manner, the water and rocks in such streams often exhibiting orange colors from the iron compounds in the acid drainage (see Plates 108 and 109, after page 656).

13.21 SULFUR RETENTION AND EXCHANGE

The sulfate ion is the form in which plants absorb most of their sulfur from soils. Since many sulfate compounds are quite soluble, the sulfate would be readily leached from the soil, especially in humid regions, were it not for its adsorption by the soil colloids. As was pointed out in Chapter 8, most soils have some anion exchange capacity that is associated with iron and aluminum oxide coatings and clays and, to a limited extent, with 1:1-type silicate clays. Sulfate ions are attracted by the positive charges that characterize acid soils containing these clays. They also react directly with hydroxy groups exposed on the surfaces of these clays. Figure 13.42 illustrates sulfate adsorption mechanisms on the surface of some Fe, Al oxides and 1:1-type clays. Note that adsorption increases at lower pH values as positive charges that become more prominent on the particle surfaces attract the sulfate ions. Some sulfate reacts with the clay particles, becoming tightly bound, and is only slowly available for plant uptake and leaching.

In warm, humid regions, surface soils are typically quite low in sulfur. However, much sulfate may be held by the iron and aluminum oxides and 1:1-type silicate clays that tend to accumulate in the subsoil horizons of the Ultisols and Oxisols of these regions (see Figure 13.35). Symptoms of sulfur deficiency commonly occur early in the growing season on Ultisols in the southeastern United States, especially those with sandy, low-organic-matter surface horizons. However, the symptoms may disappear as the crop matures and its roots reach the deeper horizons where sulfate is retained.

Sulfate Adsorption and Leaching of Nonacid Cations

When the sulfate ion leaches from the soil, it is usually accompanied by equivalent quantities of cations, including Ca and Mg and other nonacid cations. In soils with high sulfate adsorption capacities, sulfate leaching is low and the loss of companion cations is

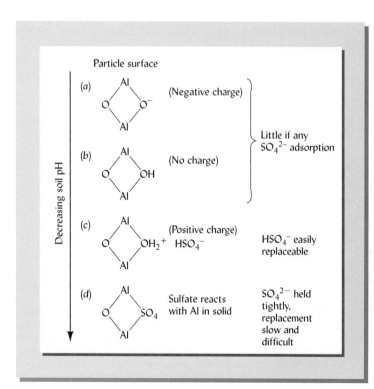

FIGURE 13.42 Effect of decreasing soil pH on the adsorption of sulfates by 1:1-type silicate clays and oxides of Fe and Al (reaction with a surface-layer Al is illustrated). At high pH levels (*a*), the particles are negatively charged, the cation exchange capacity is high, and base-forming cations are adsorbed. Sulfates are repelled by the negative charges. As acidity is increased (*b*), the H$^+$ ions are attracted to the particle surface and the negative charge is satisfied, but the SO$_4^{2-}$ ions are still not attracted. At still lower pH values (*c*), more H$^+$ ions are attracted to the particle surface, resulting in a positive charge that attracts the SO$_4^{2-}$ ion. This is easily exchanged with other anions. At still lower pH levels, the SO$_4^{2-}$ reacts directly with Al and becomes a part of the crystal structure. Such sulfate is tightly bound, and it is removed very slowly, if at all.

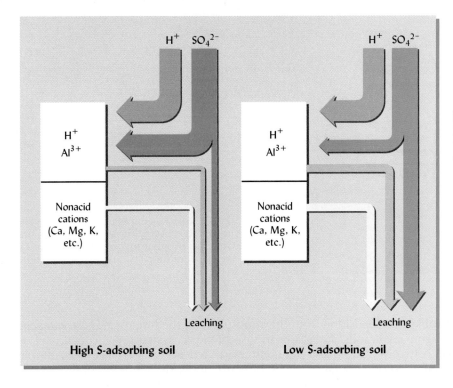

FIGURE 13.43 Diagrams illustrating cation leaching losses as influenced by sulfate adsorption capacities of forest soils. When acid rain containing SO_4^{2-} and H^+ ions falls on soils with high SO_4^{2-}-adsorbing capacities (*left*), the small quantities of SO_4^{2-} available for leaching are accompanied by correspondingly small amounts of cations such as Ca^{2+} and Mg^{2+}. Where acid rain falls on soils with low SO_4^{2-}- adsorbing capacity (*right*), most of the sulfate remains in the soil solution and is leached from the soil along with equivalent quantities of cations, including Ca^{2+} and Mg^{2+}. Al^{3+} ions that commonly replace the Ca^{2+} and Mg^{2+} lost from the soil exchange complex are toxic to many forest species. This likely accounts for at least part of the negative effects of acid rain on some forested areas. [Redrawn from Mitchell et al. (1992)]

also low (Figure 13.43). In contrast, sulfate leaching losses from low-sulfate-adsorbing soils are commonly high and take with them considerable quantities of nonacid cations. Sulfur is thus seen as an indirect conserver of these cations in the soil solution. This is of considerable importance in soils of forested areas that receive acid rain.

13.22 SULFUR AND SOIL FERTILITY MAINTENANCE

Figure 13.44 depicts the major gains and losses of sulfur from soils. The problem of maintaining adequate quantities of sulfur for mineral nutrition of plants is becoming increasingly important. Even though chances for widespread sulfur deficiencies are generally less than for nitrogen, phosphorus, and potassium, increasing crop removal of sulfur makes it essential that farmers be attentive to prevent deficiencies of this element. In some parts of the world (especially in certain semiarid grasslands), sulfur is already the next most limiting nutrient after nitrogen.

Crop residues and farmyard manures can help replenish the sulfur removed in crops, but these sources generally can help to recycle only those sulfur supplies that already exist within a farm. In regions with low-sulfur soils, greater dependence must be placed on fertilizer additions. Regular applications of sulfur-containing materials are now necessary for good crop yields in large areas far removed from industrial plants. There will certainly be an increased necessity for the use of sulfur in the future.

13.23 CONCLUSION

The cycles of sulfur and nitrogen have much in common, including many processes that operate in soils. Both elements are held by soil colloids in slowly available forms. In surface soil horizons, the bulk of both elements are found as constituents of soil organic matter. Their release to inorganic ions (SO_4^{2-}, NH_4^+, and NO_3^-) is accomplished by soil microorganisms and makes them available to plants. Anaerobic soil organisms change both elements into gaseous forms, which are emitted to the atmosphere. While in the atmosphere, some serve as greenhouse gases that accelerate climate change. Both

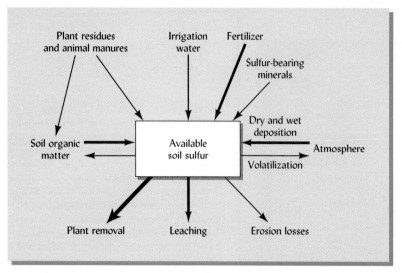

FIGURE 13.44 Major gains and losses of available soil sulfur. The thickness of the arrows indicates the relative amounts of sulfur involved in each process under average conditions. Considerable variation occurs in the field.

elements are subject to deposition from the atmosphere in the form of *acid precipitation*, seriously damaging lakes, soils, and plants in susceptible ecosystems.

Differences between nitrogen and sulfur also exist. In deeper soil layers much sulfur is found in gypsum and other sulfates (predominantly under dry climatic conditions) or pyrite and other sulfides (under low-oxygen conditions). No equivalent minerals commonly exist in soils with regard to nitrogen. Certain soil organisms have the ability to fix elemental N_2 gas into compounds usable by plants. No analogous process occurs for sulfur.

Plants remove about 10 to 20 times as much nitrogen as sulfur. If plant material is regularly removed from the system, it is usually more critical for nitrogen than for sulfur that the supply be regularly replenished. However, in areas away from industrial emissions, sulfur deficiency is increasingly common in plants so that this element will likely join nitrogen as a regular component of soil fertility programs.

Excessive amounts of nitrogen or sulfur cause serious environmental problems. Acid deposition and acid sulfate weathering are the main environmental concerns associated with sulfur cycling in soils. Progress has been greater in reducing excesses of sulfur than of nitrogen. Nitrogen also causes acidity, but eutrophication of coastal waters by nitrogen, especially in the highly mobile nitrate form, is probably its most ecologically damaging effect. The manufacture of fertilizer, along with combustion of fossil fuels, continues to magnify these problems by swelling the Earth's total pool of reactive nitrogen.

Current knowledge of the nitrogen and sulfur cycles can help us alleviate environmental degradation and enhance life-supporting productivity. However, much remains to be learned about both cycles, as well as about the cycles of other nutrient elements such as phosphorus and potassium, which we will consider in the next chapter.

STUDY QUESTIONS

1. The manager of a landscaping company is having a bit of an argument with the landscape architect about plans to fertilize and lime the soil in a new installation before planting turf and ornamental trees. The manager planned to use mostly urea for supplying nitrogen. The landscape architect says the urea will have an alkaline reaction and raise the soil pH. The manager says that urea will have an acid reaction and lower the soil pH. Who is correct? Explain, using chemical reactions to support your argument.

2. A sandy loam soil under a golf course fairway has an organic matter content of 3% by weight. Calculate the approximate amount of nitrogen (in kg N/ha) you would expect this soil to provide for plant uptake during a typical year. Show your work and state what assumptions or estimates you made to do this calculation.

3. The grass in the fairway referred to in question #2 is mowed weekly from May through October and produces an average of 200 kg/ha dry matter in clippings each time it is mowed. The clippings contain 2.5% N on average (dry weight basis). How much N from fertilizer would need to be applied to maintain this growth pattern? Show your calculations and state what assumptions or estimates you made.

4. Both sulfur and nitrogen are added to soils by atmospheric deposition. In what situations is this phenomenon beneficial and under what circumstances is it detrimental?

5. About 2000 kg of wheat straw was applied to 1 ha of land. Tests showed the soil to contain 25 kg nitrate-N per ha. The straw contained 0.4% N. How much N was applied in the straw? Explain why two weeks after the straw was applied, new tests showed no detectable nitrate N. Show your work and state what assumptions or estimates you made to do this calculation.

6. Why do CAFOs (concentrated animal-feeding operations) on industrial-style farms present some environmental and health problems relating to nitrogen? What are these problems and how can they be managed?

7. What differences would you expect in nitrate contents of streams from a forested watershed and one where agricultural crops are grown, and why?

8. What is *acid rain,* what are the sources of acidity in this precipitation, and how does this acidity damage natural ecosystems?

9. Nitrogen is "fixed" from the atmosphere and is also "fixed" by vermiculite clays and humus. Differentiate between these two processes and indicate the role of microbes, if any, in each process.

10. Chemical fertilizers and manures with high N contents are commonly added to agricultural soils. Yet these soils are often lower in total N than are nearby soils under natural forest or grassland vegetation. Explain why this is the case.

11. Why are S deficiencies in agricultural crops more widespread today than 20 years ago?

12. How do riparian forests help reduce nitrate contamination of streams and rivers?

13. What are *potential acid sulfate soils,* where would you find them, and under what circumstances are they likely to cause serious problems?

14. In some tropical regions, agroforestry systems that involve mixed cropping of trees and food crops are used. What advantages in nitrogen management do such systems have over monocropping systems that do not involve trees?

REFERENCES

Addiscott, T. M. 2005. *Nitrate, Agriculture and the Environment.* (Wallingford, UK: CABI Publishing), 279 pp.

Addiscott, T. M. 2006. "Is it nitrate that threatens life or the scare about nitrate?" *Journal of the Science of Food and Agriculture,* **86**:2005–2009.

Baethgen, W. E., and M. M. Alley. 1987. "Nonexchangeable ammonium nitrogen contributions to plant available nitrogen," *Soil Sci. Soc. Amer. J.,* **51**:110–115.

Bateman, E. J., and E. M. Baggs. 2005. "Contributions of nitrification and denitrification to N_2O emissions from soils at different water-filled pore space," *Biol. Fertil. Soils,* **41**:379–388.

Bernston, G. M., and J. D. Aber. 2000. "Fast nitrate immobilization in N saturated temperate forest soils," *Soil Biol. and Biochem.,* **32**:151–156.

Broadbent, F. E., T. Nakashima, and G. Y. Chang. 1982. "Estimation of nitrogen fixation by isotope dilution in field and greenhouse experiments," *Agron. J.,* **74**:625–628.

CAST. 1999. *Gulf of Mexico hypoxia: Land and sea interactions.* Task Force Report 134 (Ames, Iowa: Council for Agricultural Science and Technology).

Castellano, S. D., and R. P. Dick. 1991. "Cropping and sulfur fertilization influence on sulfur transformations in soil," *Soil Sci. Soc. Amer. J.,* **54**:114–121.

Colbourn, P. 1993. "Limits to denitrification in two pasture soils in a maritime temperate climate," *Agriculture, Ecosystems and Environment,* **43**:49–68.

David, M. B., and L. E. Gentry. 2000. "Anthropogenic inputs of nitrogens and phosphorus and riverine export for Illinois, USA," *J. Environ. Qual.,* **29**:494–508.

David, M. B., M. J. Mitchell, and J. P. Nakas. 1982. "Organic and inorganic sulfur constituents of a forest soil and their relationship to microbial activity," *Soil Sci. Soc. Amer. J.,* **46**:847–852.

Diaz, R. J. 2001. "Overview of hypoxia around the world," *J. Environ. Qual.,* **30**:275–281.

Eickhout, B., A. F. Bouwman, and H. Van Zeijts. 2006. "The role of nitrogen in world food production and environmental sustainability," *Agric. Ecosyst. Environ.,* **116**:4–14.

Emsley, J. 1991. "Metals trace the secrets of nitrogen fixation," *New Scientist,* **131** (1784):19.

Franzen, D. W. 2004. "Volatilization of urea affected by temperatures and soil pH." North Dakota State University Extension Service. http://www.ag.ndsu.edu/procrop/fer/ureavo05.htm.

Galloway, J. N. and E. B. Cowling. 2002. Reactive nitrogen and the world: 200 years of change. *Ambio,* **31**:64–71.

Galloway, J., J. Aber, J. Erisman, S. Seitzinger, R. Howarth, E. Cowling, and B. Cosby. 2003. "The nitrogen cascade," *Bioscience,* **53**:341–356.

Glibert, P., J. Harrison, C. Heil, and S. Seitzinger. 2006. "Escalating worldwide use of urea: A global change contributing to coastal eutrophication," *Biogeochemistry,* **77**:441–463.

Guillard, K., and K. L. Kopp. 2004. "Nitrogen fertilizer form and associated nitrate leaching from cool-season lawn turf," *J. Environ. Qual.,* **33**:1822–1827.

Hanson, G. C., P. M. Groffman, and A. J. Gold. 1994. "Denitrification in riparian wetlands receiving high and low groundwater nitrate inputs," *J. Environ. Qual.,* **23**:917–922.

Heckman, J. R., H. Lui, W. Hill, M. Demilia, and W. L. Anastasai. 2000. "Kentucky bluegrass responses to mowing practice and nitrogen fertility management," *J. Sustain. Agric.,* **15**:25–33.

International Joint Commission. 2002. "The Canada-United States air quality agreement 2002 progress report." International Joint Commission. http://www.epa.gov/airmarkets/usca/airus02.pdf (verified 15 August 2006).

Ishikawa, T., G. V. Subbarao, O. Ito, and K. Okada. 2003. "Suppression of nitrification and nitrous oxide emission by the tropical grass *brachiaria humidicola,*" *Plant Soil,* **255**:413–419.

Kennedy, I. R., A.T.M.A. Choudhury, and M. L. Kecskes. 2004. "Non-symbiotic bacterial diazotrophs in crop-farming systems: Can their potential for plant growth promotion be better exploited?" *Soil Biol. Biochem.,* **36**:1229–1244.

Khalil, K., B. Mary, and P. Renault. 2004. "Nitrous oxide production by nitrification and denitrification in soil aggregates as affected by O_2 concentration," *Soil Biol. Biochem.,* **36**:687–699.

Kirchmann, H., F. Pichlmayer, and M. H. Gerzabek. 1996. "Sulfur balances and sulfur-34 abundance in a long-term fertilizer experiment," *Soil Sci. Soc. Amer. J.,* **60**:174–178.

Klemmedson, J. O. 1994. "New Mexican locust and parent material: Influence on forest floor and soil macronutrients," *Soil Sci. Soc. Amer. J.,* **58**:974–980.

Lambert, K. F., and C. Driscoll. 2003. "Nitrogen pollution: From the sources to the sea," *Science Links Publication*, Vol. 1, No. 2. Hubbard Brook Research Foundation, Hanover, NH.29. http://www.hubbardbrook.org/hbrf/nitrogen/Nitrogen.pdf.

Leffelaar, P. A., and W. W. Wessel. 1988. "Denitrification in a homogeneous, closed system: Experimental and simulation," *Soil Sci.,* **146**:335–349.

Leuch, D., S. Haley, P. Liapis, and B. McDonald. 1995. *The EU Nitrate Directive and CAP Reform: Effects on Agricultural Production, Trade and Residual Soil Nitrogen.* Report 255 (Washington, D.C.: USDA Economic Research Service).

L'hirondel, J., and J.-L. L'hirondel. 2002. *Nitrate and Man: Toxic, Harmless or Beneficial?* (Wallingford, U.K.: CABI), 168 pp.

Magill, A. H., J. D. Aber, W. S. Currie, K. J. Nadelhoffer, M. E. Martin, W. H. Mcdowell, J. M. Melillo, and P. Steudler. 2004. "Ecosystem response to 15 years of chronic nitrogen additions at the Harvard forest LTER, Massachusetts, USA," *Forest Ecology and Management,* **196**:7–28.

Mallin, M. A., and T. L. Wheeler. 2000. "Nutrients and fecal coliform discharge from coastal North Carolina golf courses," *J. Environ. Qual.,* **29**:979–986.

Mekonnen, K., R. J. Buresh, and B. Jama. 1997. "Root and inorganic nitrogen distributions in *Sesbania* fallow, natural fallow, and maize fields," *Plant and Soil,* in press.

Mitchell, M. J., M. B. David, and R. B. Harrison. 1992. "Sulfur dynamics of forest ecosystems," in R. W. Howarth, J.W.B. Stewart, and M. V. Ivanov (eds.), *Sulfur Cycling on the Continents,* 1992 SCOPE (New York: Wiley).

Murphy, D. W. et al. 2000. "Soluble nitrogen in agricultural soils," *Biol. Fertil. Soils,* 30:374–387.

Neff, J. C., F. S. Chapin, and P. M. Vitousek. 2003. "Breaks in the cycle: Dissolved organic nitrogen in terrestrial ecosystems," *Frontiers of Ecology and the Environment,* 1:205–211.

Omay, A., C. Rice, D. Maddux, and W. Gordon. 1998. "Corn yield and nitrogen uptake in monoculture and in rotation with soybean," *Soil Sci. Soc. Amer. J.,* 62: 1596–1603.

Patrick, W. H., Jr. 1982. "Nitrogen transformations in submerged soils," in F. J. Stevenson (ed.), *Nitrogen in Agricultural Soils.* Agronomy Series No. 27 (Madison, Wis.: Amer. Soc. Agron., Crop Sci. Soc. Amer., Soil Sci. Soc. Amer.).

Prasad, R., and J. F. Power. 1995. "Nitrification inhibitors for agriculture, health and the environment," *Advances in Agronomy,* 54:233–280.

Sanchez, P. A., R. J. Buresh, and R.R.B. Leakey. 1997. "Trees, soils and food security," in *Philosophical Transactions of the Royal Society,* London, Series A, 355.

Santamaria, P. 2006. "Nitrate in vegetables: Toxicity, content, intake and EC regulation," *Journal of the Science of Food and Agriculture,* 86:10–17.

Schindler, S. C. et al. 1986. "Incorporation of ^{35}S-sulfate into inorganic and organic constituents of two forest soils," *Soil Sci. Soc. Amer. J.,* 150:457–462.

Sims, J. T., and D. C. Wolf. 1994. "Poultry waste management: Agricultural and environmental issues," *Adv. Agron.,* 52:1–83.

Solomon, D., J. Lehmann, M. Tekalign, F. Fritzsche, and W. Zech. 2001. "Sulfur fractions in particle-size separates of the sub-humid Ethiopian highlands as influenced by land use changes," *Geoderma,* 102:41–59.

Stern, D. I. 2006. "Reversal of the trend in global anthropogenic sulfur emissions," *Global Environmental Change,* 16:207–220.

Stivers, L. J., C. Shennen, E. Jackson, K. Groody, and C. J. Griffin. 1993. "Winter cover cropping in vegetable production systems in California," in M. G. Paoletti, et al. (eds.), *Soil Biota, Nutrient Cycling and Farming Systems,* (Boca Raton, FL: Lewis Press).

Suwanwaree, P., and G. P. Robertson. 2005. "Methane oxidation in forest, successional, and no-till agricultural ecosystems: Effects of nitrogen and soil disturbance," *Soil Sci. Soc. Amer. J.,* 69:1722–1729.

Tabatabai, S. J. 1986. *Sulfur in Agriculture.* Agronomy Series No. 27 (Madison, Wis.: Amer. Soc. Agron., Crop Sci. Soc. Amer., Soil Sci. Soc. Amer.).

Torrey, J. G. 1978. "Nitrogen fixation by actinomycete-induced angiosperms," *BioScience,* 28:586–592.

Walvoord, M. A., F. M. Phillips, D. A. Stonestrom, R. D. Evans, P. C. Hartsough, B. D. Newman, and R. G. Striegl. 2003. "A reservoir of nitrate beneath desert soils," *Science,* 302:1021–1024.

Whitehead, D. C. 1964. "Soil and plant nutrition aspects of the sulfur cycle," *Soils and Fertilizers,* 27:1–8.

Zhao, F. J., J. Lehmann, D. Solomon, M. A. Fox, and S. P. McGrath. 2006. "Sulphur speciation and turnover in soils: Evidence from sulphur K-edge xanes spectroscopy and isotope dilution studies," *Soil Biol. Biochem.,* 38:1000–1007.

Skeleton of Chincoteague pony returns phosphorus to soil. (R. Weil)

14

SOIL PHOSPHORUS AND POTASSIUM

*No sólo son raíces bajo las
piedras teñidas de sangre,
no sólo sus pobres huesos
derribados definitivamente
trabajan en la tierra . . .*

*(They are not only roots
beneath the bloodstained
stones, not only do their
poor demolished bones
definitely till the soil . . .)*
—PABLO NERUDA, CHILEAN POET, ESPAÑA EN CORAZON
(SPAIN IN OUR HEARTS)

Among the nutrient elements, phosphorus[1] is second only to nitrogen in its impact on the productivity and health of terrestrial and aquatic ecosystems. The total quantity of phosphorus in most native soils is low, with most of what is present in forms quite unavailable to plants. Phosphorus is so scarce in natural ecosytems that archaeologists often test soils for the presence of high phosphorus concentrations as an indication of prehistoric human habitations.

We shall see that natural ecosystems have evolved ways of maximizing the efficiency with which plants use and recycle limited phosphorus supplies. Historically, agricultural land uses have accelerated the loss and removal of phosphorus without encouraging its recycling and replacement. Low phosphorus availability in agricultural soils often leads to major social and environmental problems. Unproductive soils fail to produce adequate crops, forcing poor people to clear more land in order to produce enough food to survive. The cleared land supports little vegetative cover and so is subject to erosion that further degrades the soils and pollutes rivers and lakes with sediment-laden runoff. This scenario pertains today in much of sub-Saharan Africa, where phosphorus-deficient soils and inadequate supplies of phosphorus fertilizers have contributed to this being the main region of the world where hunger is more widespread today than 30 years ago.

Industrialized countries have for many years overcompensated for initial low phosphorus availability by adding far more phosphorus to agricultural soils than removed in crop harvests. Excessive application of phosphorus-containing fertilizers and concentration of livestock and their organic wastes have led to a phosphorus buildup in the surface soil of many agricultural watersheds. Phosphorus-laden runoff water from such watersheds, along with the runoff and sewage flows from urban watersheds, constitutes

---

[1] For a fascinating historical account about all aspects of this element, see Emsley (2002). For in-depth technical reviews of environmental, biogeochemical, and agricultural aspects of phosphorus, see Tiessen (1995).

one of the most serious types of water pollution. This pollution can jeopardize drinking water supplies and restrict the use of aquatic systems, especially freshwaters, for fisheries, industry, and recreation. Control of this *nutrient cum pollutant* is therefore a high priority for national and regional water quality programs.

Potassium does not play as direct a role in water quality as phosphorus, but inadequate supplies of this element commonly limit plant productivity, crop quality, and resistance to disease and insect pests. The latter may lead to greater pesticide use on potassium-deficient soils. Even though most soils have large total supplies of this element, most of that present is tied up as insoluble minerals and is unavailable for plant use. Plants require potassium in much larger amounts than phosphorus, so that careful management practices are necessary to ensure sufficient short- and long-term availability of this nutrient for vigorous plant growth.

In this chapter we will learn how to optimize plant availability of both potassium and phosphorus, as well as how to minimize the pollution potential of the latter.

14.1 PHOSPHORUS IN PLANT NUTRITION AND SOIL FERTILITY[2]

Neither plants nor animals can live without phosphorus. It is an essential component of the organic compound **adenosine triphosphate (ATP)**, which is the *energy currency* that drives most biochemical processes. For example, the uptake of nutrients and their transport within the plant, as well as their assimilation into different biomolecules, are energy-using plant processes that require ATP.

Phosphorus is an essential component of **deoxyribonucleic acid** (DNA), the seat of genetic inheritance, and of **ribonucleic acid** (RNA), which directs protein synthesis in both plants and animals. Phospholipids, which play critical roles in cellular membranes, are another class of universally important phosphorus-containing compounds. Bones and teeth are made of the calcium-phosphate compound apatite. In fact, ground bone (called bone meal) is often used as a phosphorus fertilizer. In healthy plants, leaf tissue phosphorus content is usually about 0.2 to 0.4% of the dry matter; comparable figures for nitrogen are about 10 times greater.

Phosphorus and Plant Growth

ASPECTS OF PLANT GROWTH ENHANCED. Adequate phosphorus nutrition enhances the fundamental processes of photosynthesis, nitrogen fixation, flowering, fruiting (including seed production), and maturation. Phosphorus is needed in especially large amounts in meristematic tissues. Root growth, particularly development of lateral roots and fibrous rootlets, is encouraged by phosphorus. In cereal crops, good phosphorus nutrition strengthens structural tissues such as those found in straw or stalks (Table 14.1), thus helping to prevent lodging (falling over). Improvement of crop quality, especially in forages and vegetables, is another benefit attributed to this nutrient.

SYMPTOMS OF PHOSPHORUS DEFICIENCY IN PLANTS. A phosphorus-deficient plant is usually stunted, thin-stemmed, and spindly, but its foliage, rather than being pale, is often dark, almost bluish-green. Thus, unless much larger, healthy plants are present to make a comparison, phosphorus-deficient plants often seem quite normal in appearance (Figure 14.1). Phosphorus-deficient plants are also characterized by delayed maturity, sparse flowering, and poor seed quality. In severe cases, phosphorus deficiency can cause yellowing and

---

[2] For a review of the plant availability of this element, see Sharpley (2000).

TABLE 14.1   **Influence of Phosphorus Nutrition on Cornstalk Rot**

Breeding line of corn	P in ear leaf, mg P/kg dry matter	Erect plants, % of all plants	Crushing strength of stalk, kg/cm²
High P accumulator	0.75	92	288
Low P accumulator	0.24	70	105

Data from Porter et al. (1981).

FIGURE 14.1 Garden radishes grown in low-phosphorus soil with (*right*) and without (*left*) P fertilizer. Dramatic stunting without obvious foliar symptoms is typical of P deficiency on many plants. Note the pronounced reduction in root carbohydrate storage. (Photo courtesy of R. Weil)

senescence of leaves. Some plants are stunted and develop purple colors (see Plates 90, 96, and 102, after page 656) in their leaves and stems as a result of phosphorus deficiency. Related stresses, such as cold temperatures, can also cause purple pigmentation. Phosphorus is very mobile within the plant, being transferred from older to younger leaves when in short supply. Therefore, older leaves show deficiency symptoms first.

The Phosphorus Problem in Soil Fertility

The phosphorus problem in soil fertility is threefold. *First,* the total phosphorus content of soils is relatively low, ranging from 200 to 2000 kg P in the upper 15 cm of 1 ha of soil. *Second,* the phosphorus compounds commonly found in soils are mostly unavailable for plant uptake, often because they are highly insoluble. *Third,* when soluble sources of phosphorus, such as those in fertilizers and manures, are added to soils, they are fixed (changed to unavailable forms) and in time form highly insoluble compounds (see Section 14.8).

Fixation[3] reactions in soils may allow only a small fraction (10 to 15%) of the phosphorus applied in fertilizers and manures to be taken up by plants in the year of application. Consequently, farmers in developed countries who could afford to do so typically applied two to four times as much phosphorus as was removed in the crop harvest. Repeated over many years, such practices saturated the phosphorus-fixation capacity and built up the level of available soil phosphorus. Such long-term buildup of phosphorus certainly improves soil fertility. However, to avoid the environmental damages discussed in Section 14.2, little or no additional phosphorus should be applied until soil phosphorus is drawn down to more moderate levels over a period of years. Fertilizer-use trends in developed countries reflect the fact that fertilizer applications can be reduced where soil phosphorus levels have been built up (Figure 14.2).

In many developing countries, especially in Africa, where poor farmers cannot afford to buy fertilizer, underuse rather than overuse of fertilizer phosphorus is the rule. In most of sub-Saharan Africa, where per-capita food production has declined in recent decades, fertilizer additions to food crops supply only a small fraction of the amount of phosphorus removed in the harvest. Soils there are depleted of phosphorus year after year, such that in many areas, lack of this element is the first limiting factor in food-crop production. The decline in per-capita food production in sub-Saharan Africa will not likely be reversed until the critical phosphorus deficiency problems are solved.

---

[3] Note that the term *fixation* as applied to phosphorus has the same general meaning as the chemical fixation of potassium or ammonium ions; that is, the chemical being fixed is bound, entrapped, or otherwise held tightly by soil solids in a form that is relatively unavailable to plants. In contrast, the fixation of gaseous nitrogen refers to the *biological* conversion of N_2 gas to reactive forms that plants can use.

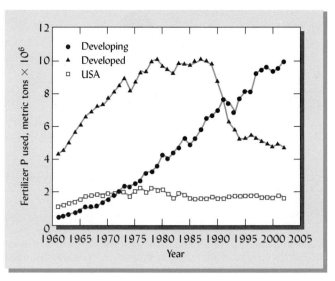

FIGURE 14.2 Fertilizer-phosphorus consumption trends. Developing countries continue to rapidly increase their fertilizer-phosphorus use as they raise their agricultural production on soils initially low in phosphorus. It can be expected that use in some Asian countries will soon level off as soils there begin to accumulate phosphorus. However, the need for more phosphorus fertilizer in sub-Saharan Africa is likely to remain great in the coming decades. Phosphorus use in developed countries increased even more rapidly during the period of 1960–1980. Use then leveled off during the 1980s, mainly because after decades of accumulating phosphorus, soils in these countries no longer produced a dramatic response to phosphorus fertilizer. The sharp decline in developed-country phosphorus use since the 1990s largely reflects rising concerns about excessive phosphorus in aquatic environments, but also has been affected by reduced use in the former Soviet Union for economic reasons starting in the mid-1980s. In the United States, phosphorus use has remained steady since the mid-1980s. [Based on data from FAO (2006)]

14.2 EFFECTS OF PHOSPHORUS ON ENVIRONMENTAL QUALITY[4]

Two major global environmental problems related to soil phosphorus are **land degradation** caused by too little available phosphorus and **accelerated eutrophication** caused by too much. Both problems are related to the role of phosphorus as a plant nutrient.

Land Degradation

Many highly weathered soils in the warm, humid, and subhumid regions of the world have very little capacity to supply phosphorus for plant growth. The low phosphorus availability is partly a result of extensive losses of phosphorus during long periods of intense weathering and partly due to the low availability of phosphorus in the aluminum and iron combinations that are the dominant forms of phosphorus in these soils. Undisturbed natural ecosystems in these regions usually contain enough phosphorus in the plant biomass and soil organic matter to maintain a substantial standing crop of trees or grasses. Most of the phosphorus taken up by the plants is that which was released from the decomposing residues of other plants. Very little is lost as long as the system remains undisturbed.

Once the forest is cleared (by timber harvest, by forest fires, or for agriculture), the losses of phosphorus in eroded soil particles, in runoff water, and in biomass removals (harvests) can be substantial. Within just a few years, the system may lose most of the phosphorus that had cycled between the plants and the soils. The remaining inorganic phosphorus in the soil is largely unavailable for plant uptake. In this manner, the phosphorus-supplying capacity of the disturbed soil rapidly becomes so low that regrowth of natural vegetation is sparse, and, on land cleared for agricultural use, crops soon fail to produce useful yields (Figure 14.3).

Leguminous plants that might be expected to replenish soil nitrogen supplies are particularly hard-hit by phosphorus deficiency, because low phosphorus supply inhibits effective nodulation and retards the biological nitrogen-fixation process. The spindly plants, deficient in both phosphorus and nitrogen, can provide little vegetative cover to prevent heavy rains from washing away the surface soil. The resulting erosion will further reduce soil fertility and water-holding capacity. The increasingly impoverished soils can support less and less vegetative cover, and so the degradation accelerates. Meanwhile, the soil particles lost by erosion become sediment farther down in the watershed, filling in reservoirs and increasing the turbidity of the rivers (see Chapter 17).

---

[4] For an excellent set of research and review papers on practical and innovative measures to control losses of phosphorus from agricultural lands, see *J. Environ. Qual.* **29**:1–181, 2000. For forward-thinking approaches to problems of soil phosphorus in Africa, see Buresh et al. (1997) and Sanchez (2002).

FIGURE 14.3   On highly weathered soils in the tropics and subtropics, low availability of phosphorus can limit the regrowth of natural vegetation or the production of crops after natural vegetation is cleared. (*Left*) Several years after this Caribbean rain forest was disturbed by a landslide, lack of sufficient soil phosphorus is still inhibiting the regrowth of vegetation within the demarcated unfertilized plot. The surrounding area was fertilized with phosphorus shortly after the landslide. (*Right*) Phosphorus has been so depleted in this African cornfield that the woman will have little or no grain to harvest from her severely phosphorus-deficient crop. In both cases, stunted plant growth leaves the soil susceptible to accelerated erosion and further degradation. (Photos courtesy of R. Weil)

There are probably 1 to 2 billion ha of land in the world where phosphorus deficiency limits growth of both crops and native vegetation. Much of this land lies in poor countries whose farmers have little money for fertilizers. Without properly managed inputs of phosphorus, there can be little hope of restoring these lands to productivity and their inhabitants to prosperity. Halting and reversing this type of land degradation will require managing the phosphorus cycle to make efficient use of scarce phosphorus resources.

Water Quality Degradation

Accelerated or cultural eutrophication (Box 14.1) is caused by phosphorus entering streams from both **point sources** and **nonpoint sources**. Point sources, such as outflows from sewage treatment plants and industrial factories, are relatively easy to identify, regulate, and clean up. During the past several decades many developed countries have greatly reduced phosphorus loading from point sources. Nonpoint sources, in contrast, are difficult to identify and control. Nonpoint sources of phosphorus are principally runoff water and eroded sediments from soils scattered throughout an affected watershed. These diffuse sources of phosphorus are now the main cause of eutrophication in many regions.

PHOSPHORUS ENRICHMENT OF SOILS.   Phosphorus losses from a watershed can be increased by a variety of human activities, including timber harvest, livestock overgrazing, soil tillage, and soil application of animal manures and phosphorus-containing fertilizers. Because of the phosphorus enrichment of agricultural surface soils, and because many agricultural operations increase the level of surface runoff and erosion, soils devoted to agriculture lose far more phosphorus to streams than do those covered by relatively undisturbed grasslands or forests (Figure 14.5).

Animal wastes from industrial-style farms often contain elevated levels of phosphorus because of excessive phosphorus in animal feeds. The resulting manures allow rainwater to dissolve large amounts of soluble phosphorus in both inorganic and organic forms (Table 14.2). Such manure is commonly overapplied with respect to the phosphorus needs of crops, especially on fields convenient to livestock facilities. These practices have resulted in dramatic increases in the phosphorus content of the surface soils. Runoff and erosion from such P-saturated soils are likely to be responsible for serious downstream eutrophication problems.

PHOSPHORUS LOSSES IN RUNOFF.   Agricultural management that involves disturbing the soil surface with tillage generally increases the amount of phosphorus carried away on eroded sediment (i.e., **particulate P**). On the other hand, fertilizer or manure that is left

BOX 14.1   PHOSPHORUS AND EUTROPHICATION

Abundant growth of plants in terrestrial systems is usually considered beneficial, but in aquatic systems too much growth can cause water quality problems. The unwanted growth of algae (floating single-celled plants) and aquatic weeds can make a lake unsuitable as a source of drinking water or as habitat for fish.

In unpolluted lakes and streams, the water is commonly clear, free of excess growth of algae and other aquatic plants, and is inhabited by highly diverse communities of organisms. When phosphorus is added to a phosphorus-limited lake, it stimulates a burst of algal growth (referred to as an *algal bloom*) and, often, a shift in the dominant algal species. The overfertilization is termed **eutrophication**. The word *eutrophication* comes from the Greek *eutruphos*, meaning well-nourished or nourishing. **Natural eutrophication**—the slow accumulation of nutrients over centuries—causes lakes to slowly fill in with dead plants, eventually forming Histosols (see Figure 2.22). Excessive input of nutrients under human influence, called **cultural eutrophication**, tremendously speeds this process. Critical levels of phosphorus in water, above which eutrophication is likely to be triggered, are approximately 0.03 mg/L of dissolved phosphorus and 0.1 mg/L of total phosphorus.

During eutrophication, phosphorus-stimulated algae and plants may suddenly cover the surface of the water with what resembles a mat of algal scum and floating plants (Plate 95). When these aquatic weeds and algal mats die, they sink to the bottom, where their decomposition by microorganisms uses up the oxygen dissolved in the water. The process is accelerated by warm water temperatures. The decrease in oxygen (anoxic conditions) severely limits the growth of many aquatic organisms, especially fish. Such eutrophic lakes often become turbid, limiting growth of beneficial submerged aquatic vegetation and benthic (bottom-feeding) organisms that serve as food for much of the fish community. In extreme cases eutrophication can lead to massive fish kills (Figure 14.4).

Eutrophic conditions favor the growth of *Cyanobacter*, blue-green algae, at the expense of zooplankton, a major food source for fish. These *Cyanobacter* produce toxins and bad-tasting and bad-smelling compounds that make the water unsuitable for human or animal consumption. Some filamentous algae can clog water treatment intake filters and thereby increase the cost of water remediation. Dense growth of both algae and aquatic weeds may make the water useless for boating and swimming. Furthermore, eutrophic waters generally have a reduced level of biological diversity (fewer species) and fewer fish of desirable species. (See also Box 13.3 on eutrophication and nitrogen.) Thus, eutrophication can transform clear, oxygen-rich, good-tasting water into cloudy, oxygen-poor, foul-smelling, bad-tasting, and possibly toxic water in which a healthy aquatic community cannot survive.

FIGURE 14.4   *In extreme cases of eutrophication, massive fish kills can occur in sensitive lakes and rivers. The kills result from anoxic conditions that are brought on by the decay of the masses of algae stimulated by elevated inputs of phosphorus (or sometimes of nitrogen). (Photo courtesy of R. Weil)*

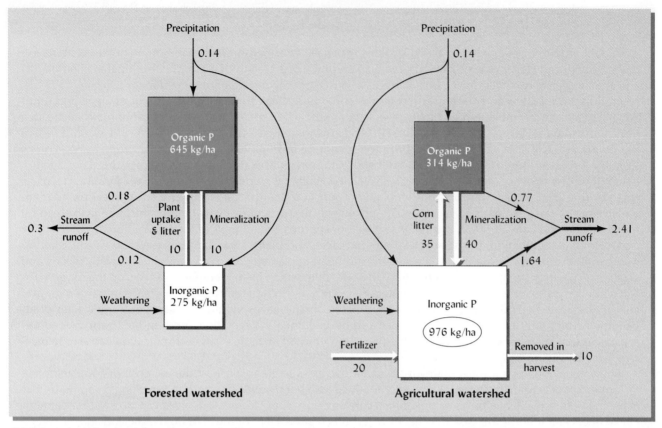

FIGURE 14.5 Phosphorus balance in surface soils (Ultisols) of adjacent forested and agricultural watersheds. The forest consisted primarily of mature hardwoods that had remained relatively undisturbed for 45 or more years. The agricultural land was producing row crops for more than 100 years. It appears that in the agricultural soil, about half of the organic phosphorus has been converted into inorganic forms or lost from the system since cultivation began. At the same time, substantial amounts of inorganic phosphorus accumulated from fertilizer inputs. Compared to the forested soil, mineralization of organic phosphorus was about 4 times as great in the agricultural soil, and the amount of phosphorus lost to the stream was 8 times as great. Flows of phosphorus, represented by arrows, are given as kg/ha/yr. Although not shown in the diagram, it is interesting to note that nearly all (95%) of the phosphorus lost from the agricultural soil was in particulate form, while losses from the forest soil were 33% dissolved and 77% particulate. [Data from Vaithiyanathan and Correll (1992)]

unincorporated on the surface of cropland or pastures usually leads to increased losses of phosphorus dissolved in the runoff water (i.e., **dissolved P**). These trends can be seen in Table 14.3 by comparing phosphorus losses from no-till and conventionally tilled wheat fields.

TABLE 14.2 Inorganic and Organic Phosphorus Dissolved from Several Manures and Composts by 5 Simulated Rainfalls

The high levels of dissolved phosphorus in these manures are largely a result of the excessive addition of phosphorus to animal feeds. While much of the dissolved phosphorus may react with the soil, some will flow with surface runoff or follow preferential channels to groundwater.

| | Phosphorus leached by 5 simulated rainfalls, mg P/kg material | |
Manure/Composts	Inorganic P	Organic P
Dairy manure	1925	375
Dairy compost	2232	232
Poultry manure	4380	1531
Poultry litter	2918	400
Poultry compost	1859	53
Swine slurry	3994	972

Data from Sharpley and Moyer (2000).

TABLE 14.3 Influence of Wheat Production and Tillage on Annual Losses of Phosphorus in Runoff Water and Eroded Sediments Coming from Soils in the Southern Great Plains

The total phosphorus lost includes the phosphorus dissolved in the runoff water and the phosphorus adsorbed to the eroded particles.[a] Although cattle grazing on the natural grasslands probably increased losses of phosphorus from these watersheds, the losses from the agricultural watersheds were about 10 times as great. The no-till wheat fields lost much less particulate phosphorus, but more dissolved phosphorus, than the conventionally tilled wheat fields.

| | | kg P/ha/yr | | |
Location and soil	Management	Dissolved P	Particulate P	Total P
El Reno, Okla. Paleustolls, 3% slope	Wheat with conventional plow and disk	0.21	3.51	3.72
	Wheat with no-till	1.04	0.43	1.42
	Native grass, heavily grazed	0.14	0.10	0.24
Woodward, Okla. Ustochrepts, 8% slope	Wheat with conventional sweep plow and disk	0.23	5.44	5.67
	Wheat with no-till	0.49	0.70	1.19
	Native grass, moderately grazed	0.02	0.07	0.09

[a] Wheat was fertilized with up to 23 kg/ha of P each fall.
Data from Smith et al. (1991).

The decision to use tillage to incorporate phosphorus-bearing soil amendments involves a trade-off between the advantages of incorporating phosphorus into the soil (less phosphorus dissolves in the runoff water and more is available for plant uptake) and the disadvantages associated with disturbing the surface soil (increased loss of soil particles by erosion and usually increased amounts of runoff water; see Table 14.3). In no-tillage systems, surface application of manure without incorporation may result in lower total loss of phosphorus, because this type of management achieves substantial reductions in soil erosion and total runoff. The effect of these reductions may outweigh the effect of the increased phosphorus concentration in the relatively small volume of water that does run off. If the equipment is available to do so, the best option might be to handle the high-phosphorus amendment as a liquid and inject it into the soil with a minimum of disturbance to the soil surface.

Disturbances to natural vegetation, such as timber harvest or wildfires, also increase the loss of phosphorus, primarily via eroded sediment (Figure 14.6). Erosion tends to transport predominantly the clay and organic matter fractions of the soil (which are relatively rich in phosphorus), leaving behind the coarser, lower-phosphorus fractions. Thus, compared to the original soil, eroded sediment is often more than twice as concentrated in phosphorus—an **enrichment ratio** of 2 or more being typical.

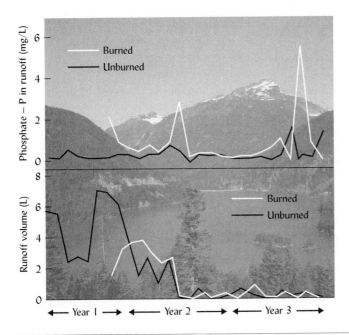

FIGURE 14.6 Effect of wildfires on nutrient runoff in the Sierra Nevada Mountains surrounding Lake Tahoe. The renowned clarity of Lake Tahoe's waters has suffered in recent years, and research suggests that runoff of phosphorus (as well as N) from the mountain slopes is partly responsible for the eutrophication of the lake. Note that although the wildfire had little effect on the runoff volumes generated, the runoff from burned areas sporadically contained much more reactive phosphorus (PO_4-P), even several years after the fire. [Redrawn from Miller et al. (2006); Photo courtesy of R. Weil]

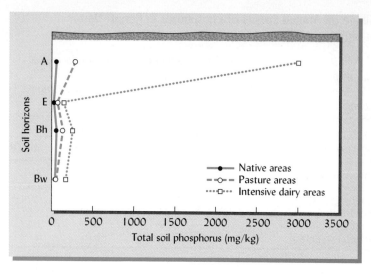

FIGURE 14.7 Average total soil phosphorus levels of intensively managed areas, and nearby pasture and native (forested) areas of three dairy farms near Lake Okeechobee in South Florida. The intensively used areas near the barns where cattle are held prior to milking are extremely high in phosphorus, some of this element having moved down into the subsoil. Runoff water from such areas of animal concentration into nearby streams and lakes is a major source of phosphorus that stimulates eutrophication. [Drawn from data in Nair et al. (1995)]

LIVESTOCK CONCENTRATION.   Runoff from facilities where thousands of cattle, hogs, or poultry are confined is an increasingly significant source of phosphorus that contributes to eutrophication. Manure from 10,000 beef cattle contains as much phosphorus and nitrogen as is found in human wastes from a city of 100,000 (see Section 16.4). Even if animal manure is spread on nearby fields, heavy rains may induce runoff and erosion that carry significant quantities of soluble and particulate phosphorus (Figure 14.7) into lakes, ponds, and streams where eutrophication occurs (Plate 95).

Pollution from intensive animal production, as well as from suburban septic drainfields, is thought to be responsible for blooms of *pfiesteria* and other toxic algae that have caused massive fish kills in estuaries and rivers of the eastern United States and elsewhere. Elevated phosphorus in streams is believed to encourage the life stage of *pfiesteria* that releases a toxin 1000 times more deadly than cyanide; it can kill small fish in a matter of minutes and larger ones in a few hours. These *pfiesteria* infestations have also poisoned a number of people, motivating the creation of programs designed to reduce phosphorus losses from industrial-type hog and poultry operations.

14.3   THE PHOSPHORUS CYCLE

In order to manage phosphorus for economic plant production and for environmental protection, we will have to understand the nature of the different forms of phosphorus found in soils and the manner in which these forms of phosphorus interact within the soil and in the larger environment. The cycling of phosphorus within the soil, from the soil to higher plants and back to the soil, is illustrated in Figure 14.8.

PHOSPHORUS IN SOIL SOLUTION.   Compared to other macronutrients, such as sulfur and calcium, the concentration of phosphorus in the soil solution is very low, generally ranging from 0.001 mg/L in very infertile soils to about 1 mg/L in rich, heavily fertilized soils. Plant roots absorb phosphorus dissolved in the soil solution, mainly as phosphate ions (HPO_4^{2-} and $H_2PO_4^-$), but some soluble organic phosphorus compounds are also taken up. The chemical species of phosphorus present in the soil solution is determined by the solution pH, as shown in Figure 14.9. In strongly acid soils (pH 4 to 5.5), the monovalent anion $H_2PO_4^-$ dominates, while alkaline solutions are characterized by the divalent anion HPO_4^{2-}. Both anions are important in near-neutral soils. Of the two anions, $H_2PO_4^-$ is thought to be slightly more available to plants, but effects of pH on phosphorus reactions with other soil constituents are more important than the particular phosphorus anion present.

UPTAKE BY ROOTS AND MYCORRHIZAE.   Plant uptake of phosphate ions from the soil solution is curtailed by the slow movement of these ions to root surfaces. This is overcome in part by root proliferation into zones where the ions are held. Also, phosphate ions move to

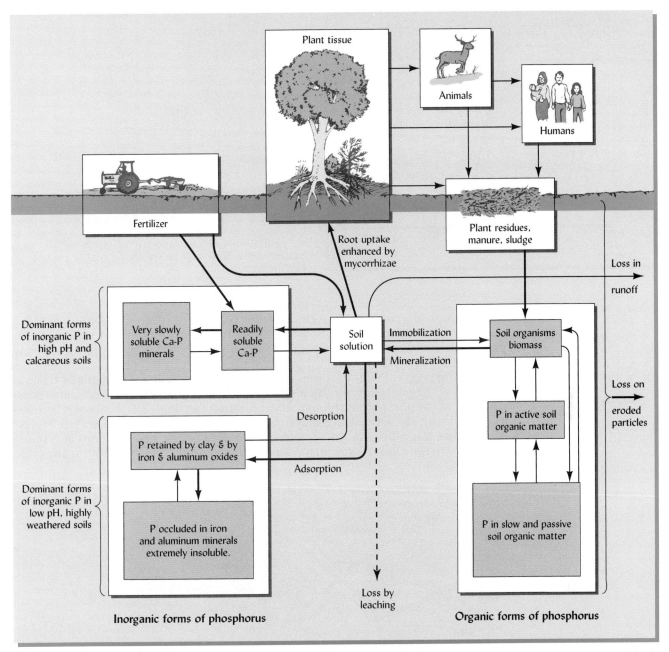

FIGURE 14.8 The phosphorus cycle in soils. The boxes represent pools of the various forms of phosphorus in the cycle, while the arrows represent translocations and transformations among these pools. The three largest white boxes indicate the principal groups of phosphorus-containing compounds found in soils. Within each of these groups, the less soluble, less available forms tend to dominate. Thick arrows represent the principal pathways. (Diagram courtesy of R. Weil)

the roots of many plants through symbiosis with mycorrhizal fungi (Figure 14.10). The microscopic, threadlike mycorrhizal hyphae extend out into the soil several centimeters from the root surfaces (see also Plate 54 and Section 11.9). The hyphae are able to absorb phosphorus ions as the ions enter the soil solution and may even be able to access some strongly bound forms of phosphorus. The hyphae then bring the phosphate to the root by transporting it inside the hyphal cells, where soil-retention mechanisms cannot interfere with phosphate movement. Generally this mycorrhizal association is best developed where host plants are growing undisturbed in soils with low phosphorus availability. However, the fungi have been observed to benefit early-season growth of annual plants even in soils testing high in available phosphorus if the soil is kept vegetated with plants that can serve as suitable hosts (for example, Table 14.4).

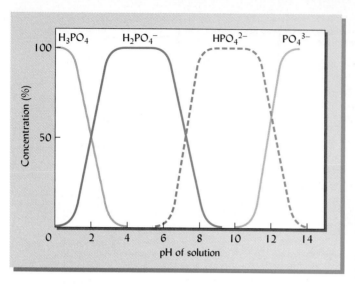

FIGURE 14.9 The effect of pH on the relative concentrations of the three species of phosphate ions. At lower pH values, more H^+ ions are available in the solution, and thus the phosphate ion species containing more hydrogen predominates. In near-neutral soils, HPO_4^{2-} and $H_2PO_4^-$ are found in nearly equal amounts. Both of these species are readily available for plant uptake.

DECOMPOSITION OF PLANT RESIDUES. Once in the plant, a portion of the phosphorus is translocated to the plant shoots, where it becomes part of the plant tissues. As the plants shed leaves and their roots die, or when they are eaten by people or animals, phosphorus returns to the soil in the form of plant residues, leaf litter, and wastes from animals and people. Microorganisms that decompose the residues temporarily tie up at least part of the phosphorus in their cells (microbial biomass-P), but eventually release a portion of the phosphorus through mineralization (Section 12.2). Some of it becomes associated with the active and passive fractions of the soil organic matter (see Section 12.6), where it is subject to storage and future release. These organic forms also slowly mineralize to the soluble forms that plant roots can absorb, thereby repeating the cycle.

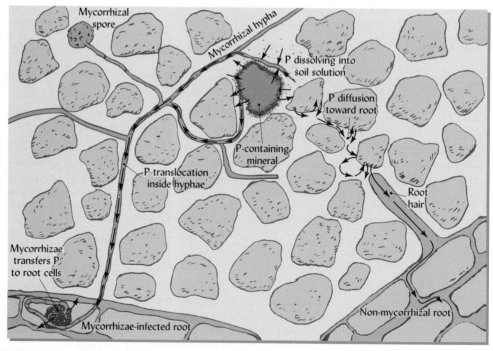

FIGURE 14.10 Roles of diffusion and mycorrhizal hyphae in the movement of phosphate ions to plant roots. In soils with low solution phosphorus concentration and high phosphorus fixation, slow diffusion may seriously limit the ability of roots to obtain sufficient phosphorus. The hyphae of symbiotic mycorrhizal fungi help overcome this problem. They penetrate the soil, absorb the phosphorus, and by cytoplasmic streaming inside the hyphae, transport phosphorus to the plant roots. This makes a plant much less dependent on the diffusion of phosphate ions through the soil. (Diagram courtesy of R. Weil)

TABLE 14.4 Effect of Previous Land Use (Fallow vs. Crop) on Early Season Mycorrhizal Colonization, Seedling Growth, and Grain Yield of Corn on a Soil Testing High in Available Phosphorus

In each year, the lack of a continuous host for mycorrhizal fungi due to previous fallow resulted in reduced corn root colonization at the 3-leaf stage, which in turn depressed growth and P uptake by the 6-leaf stage, leading to lower final grain yields. Practices that encourage mycorrhizae may help plants get off to a quick start with a minimum of starter fertilizer.

	Previous land use	Mycorrhizal root colonization on 3-leaf stage corn, %	Shoot dry wt. at 6-leaf stage, kg/ha	P concentration at 6-leaf stage, %	P uptake at 6-leaf stage, g/ha	Grain yield, kg/ha
Year 1	Crop	20.2	193	0.284	563	2903
	Fallow	11.0	142	0.228	337	2378
Year 2	Crop	46.9	103	0.262	273	7176
	Fallow	12.8	81	0.178	148	6677
Year 3	Crop	17.2	261	0.336	882	5495
	Fallow	8.0	158	0.293	469	4980

Data selected from Bittman et al. (2006).

CHEMICAL FORMS IN SOILS.   In most soils, the amount of phosphorus available to plants from the soil solution at any one time is very low, seldom exceeding about 0.01% of the total phosphorus in the soil. The bulk of the soil phosphorus exists in three general groups of compounds—namely, *organic phosphorus, calcium-bound inorganic phosphorus, and iron- or aluminum-bound inorganic phosphorus* (see Figure 14.8). The organic phosphorus is distributed among the *active, slow,* and *passive* fractions of soil organic matter (see Section 12.6). Of the inorganic phosphorus, the calcium compounds predominate in most alkaline soils, while the iron and aluminum forms are most important in acidic soils. All three groups of compounds slowly contribute phosphorus to the soil solution, but most of the phosphorus in each group is of very low solubility and not readily available for plant uptake.

Unlike nitrogen and sulfur, phosphorus is not generally lost from the soil in gaseous form.[5] Because soluble inorganic forms of phosphorus are strongly adsorbed by mineral surfaces (see Section 14.6), leaching losses of inorganic phosphorus are generally very low, but they still may be sufficient to stimulate eutrophication in downstream waters.

GAINS AND LOSSES.   The principal pathways by which phosphorus is lost from the soil system are plant removal (5 to 50 kg ha^{-1} yr^{-1} in harvested biomass), erosion of phosphorus-carrying soil particles (0.1 to 10 kg ha^{-1} yr^{-1} on organic and mineral particles), phosphorus dissolved in surface runoff water (0.01 to 3.0 kg ha^{-1} yr^{-1}), and leaching to groundwater (0.0001 to 0.4 kg ha^{-1} yr^{-1}). For each pathway, the higher figures cited for annual phosphorus loss would most likely apply to cultivated soils (see Section 16.12 and Figure 14.5).

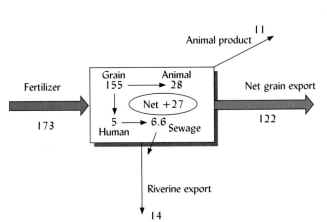

FIGURE 14.11 Diagram showing the annual inputs and outputs of phosphorus for the State of Illinois, averaged for 18 years (1979–1996 for terrestrial fluxes and 1980–1997 for riverine export). The units are 1000 Mg P/yr. The major input into the system is fertilizer phosphorus. Fluxes inside the box are for phosphorus in the total grain harvested, its consumption by animals and humans, and the sewage from animals and humans that moves into rivers. The major outputs are the net grain export outside the state, the phosphorus in the animals sold, and the phosphorus moved from the state in the river systems (riverine export). Note that within the box some 27,000 Mg of phosphorus remains in the system, mostly residing in the soils of the state. Phosphorus in manures is not considered as either an input or output, since it is merely recycled from the soil into the plants and back into the soil. This diagram illustrates some of the principles covered in this chapter. It emphasizes the buildup of phosphorus in the soil and the loss of a portion of this element in river waters, a portion that means little for plant nutrition but one that could have damaging effects on water quality downstream. [From David and Gentry (2000)]

---

[5] Some phosphorus may be lost as fishy-smelling phosphine gas (PH_3)—a phenomenon noted in certain graveyard soils.

The amount of phosphorus that enters the soil from the atmosphere (sorbed on dust particles) is quite small (0.05 to 0.5 kg ha^{-1} yr^{-1}), but may nearly balance the losses from the soil in undisturbed forest and grassland ecosystems. As already discussed, in an agroecosystem, optimal crop production may initially require the input from fertilizer to exceed the removal in crop harvest, but only until enough phosphorus accumulates to reduce the P-fixing capacity of the soil (see Section 14.8). The level of soil fertility and severity of environmental pollution are largely determined by the balance—or lack of balance—between inputs from fertilizer and feed and outputs as plant and animal products. As was done in Figure 13.34 for nitrogen, Figure 14.11 on the previous page illustrates a real-world, input–output budget for phosphorus in Illinois.

14.4 ORGANIC PHOSPHORUS IN SOILS

Both inorganic and organic forms of phosphorus occur in soils, and both are important to plants as sources of this element. The relative amounts in the two forms vary greatly from soil to soil, but the data in Table 14.5 give some idea of their relative proportions in a range of mineral soils. The organic fraction generally constitutes 20 to 80% of the total phosphorus in surface soil horizons (Figure 14.12). The deeper horizons may hold large amounts of inorganic phosphorus, especially in soils from arid and semiarid regions.

Organic Phosphorus Compounds

Until recently, scientists have focused more attention on the inorganic than on the organic phosphorus in soils, and our knowledge of the specific nature of most of the organic-bound phosphorus in soils is quite limited. However, three broad groups of organic phosphorus compounds are known to exist in soils: (1) inositol phosphates or phosphate esters of a sugarlike compound, inositol [$C_6H_6(OH)_6$]; (2) nucleic acids; and (3) phospholipids. While other organic phosphorus compounds are present in soils, the identity and amounts present are less well understood.

Inositol phosphates are the most abundant of the known organic phosphorus compounds, making up 10 to 50% of the total organic phosphorus. Their abundance in soils is probably related to their high stability in both acid and alkaline conditions, and their interaction with the fulvic and humic acid components of soil humus. One of the

TABLE 14.5 **Total Phosphorus Content of Surface Soils from Different Locations and the Percentage of Total Phosphorus in the Organic Form**

Soils	Number of samples	Total P, mg/kg	Organic fraction, %
New York			
Histosols (cultivated)	8	1491	52
Iowa			
Mollisols and Alfisols	6	561	44
Arizona	19	703	36
Australia	3	422	75
Texas			
Ustolls	2	369	34
Hawaii			
Andisol	1	4700	37
Oxisol	1	1414	19
Ethiopia highlands			
Vertisols	15	454	47
Zimbabwe			
Alfisols	22	899	56
Maryland			
Ultisols (silt loams)	6	650	59
Ultisols (forested sandy loams)	3	472	70
Ultisols (sandy loam, cropland)	4	647	25

Data for Iowa and Arizona from sources quoted by Brady (1974); Australia from Fares et al. (1974); New York from Cogger and Duxbury (1984); Hawaii from Soltanpour et al. (1988); Zimbabwe from R. Weil and F. Folle, unpublished; Texas from Raven and Hossner (1993); Maryland (silt loams) from Weil et al. (1988); Maryland (sandy loams) from Vaithiyanathan and Correll (1992); Ethiopian Vertisols from Tekalign et al. (1988).

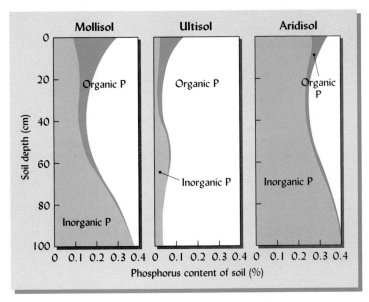

FIGURE 14.12 Phosphorus contents of representative soil profiles from three soil orders. All three soils contain a high proportion of organic phosphorus in their surface horizons. The Aridisol has a high inorganic phosphorus content throughout the profile because rainfall during soil formation was insufficient to leach much of the inorganic phosphorus compounds from the soil. The increased phosphorus in the subsoil of the Ultisol is due to adsorption of inorganic phosphorus by iron and aluminum oxides in the B horizon. In both the Mollisol and Aridisol, most of the subsoil phosphorus is in the form of inorganic calcium-phosphate compounds.

most common inositol phosphates in soils is *phytic acid*, a compound in which plants store P in their seeds (including grains like corn). The fact that nonruminant animals cannot digest phytic acid means that grain-fed swine and poultry require supplemental phosphorus in their feed and that their manure is artificially high in phosphorus (see Section 16.4).

Nucleic acids are adsorbed by humic compounds as well as by silicate clays. Adsorption on these soil colloids probably helps protect the phosphorus in nucleic acids from microbial attack. Still, the nucleic acids and phospholipids together probably make up only 1 to 2% of the organic phosphorus in most soils.

The other chemical compounds that contain most of the soil organic phosphorus have not yet been identified, but much of the organic phosphorus appears to be associated with the fulvic acid fraction of the soil organic matter. Our ignorance of the specific compounds involved does not detract from the importance of these compounds as suppliers of phosphorus through microbial breakdown.[6]

Much of the phosphorus in the soil solution and in leachates of areas that have received large quantities of animal wastes is present as **dissolved organic phosphorus** (DOP). DOP is generally more mobile than soluble inorganic phosphates, probably because it is not so readily adsorbed by organic matter clays and $CaCO_3$ in the soil. In the lower horizons of such soils, the DOP commonly makes up more than 50% of the total soil solution phosphorus. As a consequence, in heavily manured areas with sandy soils DOP can leach downward to nearly 2 m (Figure 14.13). In fields with high water tables, the phosphorus can move with the groundwater to nearby lakes or streams and thereby contribute significantly to eutrophication.

Mineralization of Organic P

Phosphorus held in organic forms can be mineralized and immobilized by the same general processes that release nitrogen and sulfur from soil organic matter (see Chapter 13):

$$\xleftarrow{\text{Immobilization}}$$

$$\text{Organic P forms} \underset{\text{Microbes}}{\overset{\text{Microbes}}{\rightleftharpoons}} \underset{\substack{\text{Soluble} \\ \text{phosphate}}}{H_2PO_4^-} \underset{}{\overset{Fe^{3+},\, Al^{3+},\, Ca^{2+}}{\rightleftharpoons}} \underset{\text{Insoluble fixed P}}{\text{Fe, Al, Ca phosphates}} \quad (14.1)$$

$$\xrightarrow{\text{Mineralization}}$$

Net immobilization of soluble phosphorus is most likely to occur if residues added to the soil have a C/P ratio greater than 300:1, while net mineralization is likely if the

---

[6] Although plants are able to absorb some organic phosphorus compounds directly, the level of such absorption is thought to be very low compared to that of inorganic phosphates.

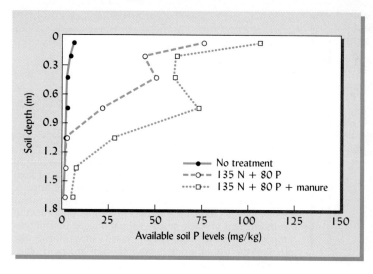

FIGURE 14.13 Effect of adding fertilizer with and without cow manure for a period of 42 years on the level of available phosphorus at different depths of a Mollisol in Nebraska on which continuous corn had been grown. Both treatments resulted in large increases in the available P level in the upper horizons, but only where manure was also applied did increases appear below a depth of 1 m. Apparently the organic forms of P in the manured plot were not adsorbed by the soil, thereby permitting deeper penetration of the phosphorus, which may then have been subject to movement through the groundwater to nearby waterways. [Redrawn from Eghball et al. (1996)]

ratio is below 200:1. Mineralization of organic phosphorus in soils is subject to many of the same influences that control the general decomposition of soil organic matter—such as temperature, moisture, and tillage (see Section 12.3). In temperate regions, mineralization of organic phosphorus in soils typically releases 5 to 20 kg P/ha/yr, most of which is readily absorbed by growing plants. These values can be compared to the annual uptake of phosphorus by most crops, trees, and grasses, which generally ranges from 5 to 30 kg P/ha. When forested soils are first brought under cultivation in tropical climates, the amount of phosphorus released by mineralization may exceed 50 kg/ha/yr, but unless phosphorus is added from outside sources these high rates of mineralization will soon decline due to the depletion of readily decomposable soil organic matter. In Florida, rapid mineralization of organic matter in Histosols (Saprists) drained for agricultural use is estimated to release about 80 kg P/ha/yr. Unlike most mineral soils, these organic soils possess little capacity to retain dissolved phosphorus, so water draining from them is quite concentrated in phosphorus (0.5 to 1.5 mg P/L) and is thought to be contributing to the degradation of the Everglades wetland system.

Contribution of Organic Phosphorus to Plant Needs

Recent evidence indicates that the readily decomposable or easily soluble fractions of soil *organic phosphorus* are often the most important factor in supplying phosphorus to plants in *highly weathered soils* (e.g., Ultisols and Oxisols), even though the total organic matter content of these soils may not be especially high. The inorganic phosphorus in the highly weathered soils is far too insoluble to contribute much to plant nutrition. Apparently plant roots and mycorrhizal hyphae are able to obtain some of the phosphorus released from organic forms before it forms inorganic compounds that quickly become insoluble. In contrast, it appears that the more soluble *inorganic forms* of phosphorus play the biggest role in phosphorus fertility of *less weathered soils* (e.g., Mollisols and Vertisols), even though these generally contain relatively high amounts of soil organic matter. Green manure crops and cover crop residues left on the soil surface as a mulch can improve phosphorus availability in both groups of soils. In temperate regions, freezing and thawing can lyse plant cells in growing plants and fresh residues, rapidly releasing soluble P from the cell cytoplasm. Unfortunately, recent research suggests that some of this soluble organic P may be lost in winter runoff water before the spring flush of plant growth can take it up.

14.5 INORGANIC PHOSPHORUS IN SOILS

Of all the macronutrients found in soils, phosphorus has by far the smallest quantities in solution or in readily soluble forms in mineral soils. Likewise, the relative immobility of inorganic phosphorus in mineral soils is well known. Two phenomena tend to control the concentration of phosphorus in the soil solution and the movement of phosphorus in soils: (1) the solubility of phosphorus-containing minerals, and (2) the fixation or adsorption of phosphate ions on the surface of soil particles. In practice, it is difficult to separate the influence of these two types of reactions or even determine the exact nature of inorganic phosphorus compounds present in a particular soil.

FIXATION AND RETENTION. Dissolved phosphate ions in mineral soils are subject to many types of reactions that tend to remove the ions from the soil solution and produce phosphorus-containing compounds of very low solubility. These reactions are sometimes collectively referred to by the general terms *phosphorus fixation* and *phosphorus retention*. *Phosphorus retention* is a somewhat more general term that includes both precipitation and fixation reactions.

The tendency for soils to fix phosphorus in relatively insoluble, unavailable forms has far-reaching consequences for phosphorus management. For example, phosphorus fixation may be viewed as troublesome if it prevents plants from using all but a small fraction of fertilizer phosphorus applied. On the other hand, phosphorus fixation can be viewed as a benefit if it causes most of the dissolved phosphorus to be removed from phosphorus-rich wastewater applied to a soil (Box 14.2). The fixation reactions responsible in both situations will be discussed as they apply to the availability of phosphorus under acidic and alkaline soil conditions. We will begin by describing the various inorganic compounds and their solubility.

Inorganic Phosphorus Compounds

As indicated by Figure 14.8, most inorganic phosphorus compounds in soils fall into one of two groups: (1) those containing calcium, and (2) those containing iron and aluminum (and, less frequently, manganese).

As a group, the calcium phosphate compounds become more soluble as soil pH decreases; hence, they tend to dissolve and disappear from acid soils. On the other hand, the calcium phosphates are quite stable and very insoluble at higher pH and so become the dominant forms of inorganic phosphorus present in neutral to alkaline soils.

Of the common calcium compounds containing phosphorus (Table 14.6), the **apatite** minerals are the least soluble and are therefore the least available source of phosphorus. Some apatite minerals (e.g., fluorapatite) are so insoluble that they persist even in weathered (acid) soils. The simpler mono- and dicalcium phosphates are readily available for plant uptake. Except on recently fertilized soils, however, these compounds are present in only extremely small quantities because they easily revert to the more insoluble forms.

In contrast to calcium phosphates, the iron and aluminum hydroxy phosphate minerals, **strengite** ($FePO_4 \cdot 2H_2O$) and **variscite** ($AlPO_4 \cdot 2H_2O$), have very low solubilities in strongly acid soils and become more soluble as soil pH rises. These minerals would therefore be quite unstable in alkaline soils, but are prominent in acid soils, in which they are quite insoluble and stable.

Other similar compounds, combining phosphorus with iron, aluminum, or manganese, are also found in acid soils. Some are products of surface reactions between phosphate ions and the somewhat amorphous hydroxy polymers that often exist as coatings on soil particles. Evidence suggests that phosphate ions even react with aluminum near the edges of silicate clay crystals, forming insoluble products similar to the aluminum phosphates described in the preceding paragraph.

Effect of Aging on Inorganic Phosphate Availability

In both acid and alkaline soils, phosphorus tends to undergo sequential reactions that produce phosphorus-containing compounds of lower and lower solubility. Therefore, the longer that phosphorus remains in the soil, the less soluble—and, therefore, less plant-available—it tends to become. Usually, when soluble phosphorus is added to a soil, a rapid reaction removes the phosphorus from solution (*fixes* the phosphorus) in the first few hours. Slower reactions then continue to gradually reduce phosphorus solubility for months or years as the phosphate compounds age. The freshly fixed phosphorus may be slightly soluble and of some value to plants. With time, the solubility of the fixed phosphorus tends to decrease to extremely low levels. The effect of aging appears to be due to such factors as the regularity and size of crystals in precipitated

BOX 14.2   PHOSPHORUS REMOVAL FROM WASTEWATER

Environmental soil scientists and engineers remove phosphorus from municipal wastes by taking advantage of some of the same reactions that bind phosphorus in soils. After primary and secondary sewage treatment that removes solids and oxidizes most of the organic matter, tertiary treatment in huge, specially designed tanks (Figure 14.14, *left*) causes phosphorus to precipitate through reactions with iron and aluminum compounds, such as the following:

$$Al_2(SO_4)_3 \cdot 14H_2O + 2PO_4^{3-} \rightarrow 2AlPO_4 + 3SO_4^{2-} + 14H_2O \qquad (14.2)$$

Alum   Soluble   Insoluble AlP
phosphate

$$FeCl_3 + PO_4^{3-} \rightarrow FePO_4 + 3Cl^- \qquad (14.3)$$

Ferric   Soluble   Insoluble FeP
chloride   phosphate

The insoluble aluminum and iron phosphates settle out of solution and are later mixed with other solids from the wastewater to form sewage sludge. The low-phosphorus water, after minor processing, is returned to the river.

FIGURE 14.14 *Increasingly, modern sewage plants (left) are required to include tertiary treatment facilities for phosphorus removal. The chemical reactions in the sewage treatment process, which are similar to those affecting phosphorus availability in soils, occur when the treatment plant effluent is passed through an artificial wetland (above right) to remove nitrogen by denitrification (see Chapter 13). In the final stage of pollutant removal, the effluent is applied to a grass covered field (lower right) using a spray irrigation system (see arrow), so that the wastewater can be further treated by chemical and biological processes as it percolates through the soil. (Photos courtesy of R. Weil)*

Other less-expensive tertiary treatment approaches involve the spraying of the wastewater on vegetated soils. Natural soil and plant processes clean the phosphorus and other constituents out of the waste water. In some *infiltration* systems, the water percolates through relatively permeable soils. Other systems, termed *overland flow* systems, use finer-textured, less-permeable soils, over which water slowly flows, permitting the upper few centimeters of soil and the vegetation to remove most of the soluble phosphorus and other contaminants. In both systems, advantage is taken of the soil's phosphate-fixing capacity (see Section 14.8).

As the following example illustrates, knowledge of soil properties and processes is essential for effective design of advanced land-based wastewater treatment systems. A large environmental engineering firm won a contract to build a new type of wastewater treatment facility that would look more like a park than a sewage plant because it used constructed wetlands and soils to clean the wastewater. After flowing through a number of artificial marshes and filtering systems (Figure 14.14, *upper right*), the wastewater was sprayed into a large

(continued)

BOX 14.2 *(Cont.)* PHOSPHORUS REMOVAL FROM WASTEWATER

field covered with a layer of artificial permeable "soil" several meters thick (Figure 14.14, *lower right*). The expectation was that the phosphorus would be fixed as it moved through the "soil," leaving the groundwater sufficiently low in this element that it could be released into a nearby estuary.

Unfortunately, the system didn't work. The water coming from the bottom of the artificial profile was higher in phosphorus than before the treatment. The problem? The designers had specified peat (very low in mineral colloids) rather than mineral soil as the artificial "soil" through which the water percolated. Consequently, insufficient iron, aluminum, or calcium compounds were available to fix the phosphorus. Soil scientists brought in to assist the municipality recognized the problem and recommended that a few trainloads of steel wool dust (metallic iron) be incorporated into the peat. As the steel wool rusted (oxidized), ferric iron formed and reacted strongly with the dissolved phosphorus in the wastewater. The phosphorus was precipitated without appreciably decreasing the desirable high permeability of the peat "soil."

phosphates, more permanent bonding of adsorbed phosphate into the calcium carbonate or metal oxide particles, and the extent to which sorbed phosphate is buried as surface precipitation reactions continue (Figure 14.15). The nature of these and other reactions of phosphorus in soils is discussed in the following sections.

14.6 SOLUBILITY OF INORGANIC PHOSPHORUS IN ACID SOILS

The particular types of reactions that fix phosphorus in relatively unavailable forms differ from soil to soil and are closely related to soil pH (Figure 14.16). In acid soils these reactions involve mostly Al, Fe, or Mn, either as dissolved ions, oxides, or hydrous oxides. Many soils contain such hydrous oxides as coatings on soil particles and as interlayer precipitates in silicate clays. In alkaline and calcareous soils, the reactions primarily involve precipitation as various calcium phosphate minerals (see Table 14.6) or adsorption to the iron impurities on the surfaces of carbonates and clays. At moderate pH values, adsorption on the edges of kaolinite or on the iron oxide coating on kaolinite clays plays an important role.

Precipitation by Iron, Aluminum, and Manganese Ions

Probably the easiest type of phosphorus-fixation reaction to visualize is the simple reaction of $H_2PO_4^-$ ions with dissolved Fe^{3+}, Al^{3+}, and Mn^{3+} ions to form insoluble hydroxy phosphate precipitates (Figure 14.17a). In strongly acid soils, enough soluble Al, Fe, or

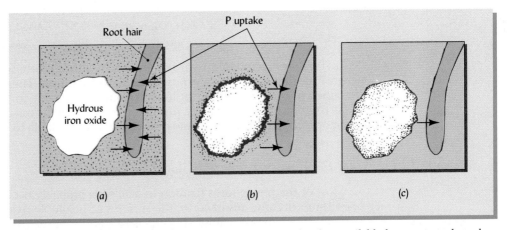

FIGURE 14.15 How relatively soluble phosphates are rendered unavailable by compounds such as hydrous oxides of Fe and Al. (*a*) The situation just after application of a soluble phosphate. The root hair and the hydrous iron oxide particle are surrounded by soluble phosphate ions (small dots). (*b*) Within a very short time most of the soluble phosphate has reacted with the surface of the iron oxide crystal. The phosphorus is still fairly readily available to the plant roots, since most of it is located at the surface of the particle where exudates from the plant can encourage exchange. (*c*) In time the phosphorus penetrates the crystal, and only a small portion is found near the surface. Under these conditions its availability is low.

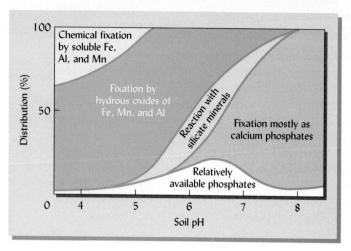

FIGURE 14.16 Inorganic fixation of added phosphates at various soil pH values. Average conditions are postulated, and it is not to be inferred that any particular soil would have exactly this distribution. The actual proportion of the phosphorus remaining in an available form will depend upon contact with the soil, time for reaction, and other factors. It should be kept in mind that some of the added phosphorus may be changed to an organic form in which it would be temporarily unavailable.

TABLE 14.6 **Inorganic Phosphorus-Containing Compounds Commonly Found in Soils**

In each group, the compounds are listed in order of increasing solubility.

Compound	Formula
Iron and aluminum compounds	
Strengite	$FePO_4 \cdot 2H_2O$
Variscite	$AlPO_4 \cdot 2H_2O$
Calcium compounds	
Fluorapatite	$[3Ca_3(PO_4)_2] \cdot CaF_2$
Carbonate apatite	$[3Ca_3(PO_4)_2] \cdot CaCO_3$
Hydroxy apatite	$[3Ca_3(PO_4)_2] \cdot Ca(OH)_2$
Oxyapatite	$[3Ca_3(PO_4)_2] \cdot CaO$
Tricalcium phosphate	$Ca_3(PO_4)_2$
Octacalcium phosphate	$Ca_8H_2(PO_4)_6 \cdot 5H_2O$
Dicalcium phosphate	$CaHPO_4 \cdot 2H_2O$
Monocalcium phosphate	$Ca(H_2PO_4)_2 \cdot H_2O$

Mn is usually present to cause the chemical precipitation of nearly all dissolved $H_2PO_4^-$ ions by reactions such as the following (using the aluminum cation as an example):

$$Al^{3+} + H_2PO_4^- + 2H_2O \rightleftharpoons 2H^+ + Al(OH)_2H_2PO_4$$
$$\text{(soluble)} \qquad\qquad\qquad\qquad \text{(insoluble)}$$

(14.4)

Freshly precipitated hydroxy phosphates are slightly soluble because they have a great deal of surface area exposed to the soil solution. Therefore, the phosphorus contained in them is, initially at least, somewhat available to plants. Over time, however, as the precipitated hydroxy phosphates age, they become less soluble and the phosphorus in them becomes almost completely unavailable to most plants.

Reaction with Hydrous Oxides and Silicate Clays

Most of the phosphorus fixation in acid soils probably occurs when $H_2PO_4^-$ ions react with, or become adsorbed to, the surfaces of insoluble oxides of iron, aluminum, and manganese, such as gibbsite ($Al_2O_3 \cdot 3H_2O$) and goethite ($Fe_2O_3 \cdot 3H_2O$; see Figure 14.15) and with 1:1 type silicate clays. These hydrous oxides occur as crystalline and noncrystalline particles and as coatings on the interlayer and external surfaces of clay particles. Fixation of phosphorus by clays probably takes place over a relatively wide pH range (see Figure 14.16). The large quantities of Fe, Al oxides and 1:1 clays present in many soils make possible the fixation of extremely large amounts of phosphorus by these reactions.

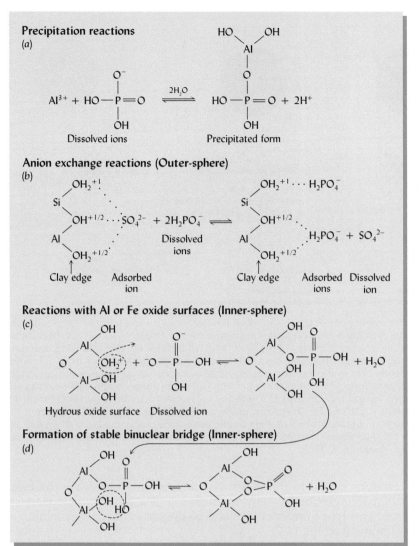

Precipitation reactions
(a)

Al³⁺ + HO—P=O ⇌(2H₂O) HO—P=O + 2H⁺

Dissolved ions   Precipitated form

Anion exchange reactions (Outer-sphere)
(b)

Clay edge   Adsorbed ion   + 2H₂PO₄⁻ ⇌ Clay edge   Adsorbed ions Dissolved ion

Dissolved ions

Reactions with Al or Fe oxide surfaces (Inner-sphere)
(c)

Hydrous oxide surface   Dissolved ion   ⇌   + H₂O

Formation of stable binuclear bridge (Inner-sphere)
(d)

⇌   + H₂O

FIGURE 14.17 Several of the reactions by which phosphate ions are removed from soil solution and fixed by reaction with iron and aluminum in various hydrous oxides. Freshly precipitated aluminum, iron, and manganese phosphates (a) are relatively available, though over time they become increasingly unavailable. In (b) the phosphate is reversibly adsorbed by anion exchange. In reactions of the type shown in (c) a phosphate ion replaces an —OH₂ or an —OH group in the surface structure of Al or Fe hydrous oxide minerals. In (d) the phosphate further penetrates the mineral surface by forming a stable binuclear bridge. The adsorption reactions (b, c, d) are shown in order— from those that bind phosphate with the least tenacity (relatively reversible and somewhat more plant-available) to those that bind phosphate most tightly (almost irreversible and least plant-available). It is probable that, over time, phosphate ions added to a soil may undergo an entire sequence of these reactions, becoming increasingly unavailable. Note that (b) illustrates an outer-sphere complex, while (c) and (d) are examples of inner-sphere complexes (see Figure 8.20).

Although all the exact mechanisms have not been identified, $H_2PO_4^-$ ions are known to react with iron and aluminum mineral surfaces in several different ways, resulting in different degrees of phosphorus fixation. Some of these reactions are shown diagrammatically in Figure 14.17.

The $H_2PO_4^-$ anion may be attracted to positive charges that develop under acid conditions on the surfaces of iron and aluminum oxides and the broken edges of kaolinite clays (see Figure 14.17b). The adsorbed $H_2PO_4^-$ anions form outer-sphere complexes and are subject to anion exchange with certain other anions, such as OH^-, SO_4^{2-}, MoO_4^{2-}, or organic acids ($R—COO^-$; see Section 8.7). Since this type of adsorption of $H_2PO_4^-$ ions is reversible, the phosphorus may slowly become available to plants. Availability of such adsorbed $H_2PO_4^-$ may be increased by (1) liming the soil to increase the hydroxyl ions, or (2) adding organic matter to increase organic acids (anions) capable of replacing $H_2PO_4^-$.

The phosphate ion may also replace a structural hydroxyl to form an inner-sphere complex with the oxide (or clay) surface (see Figure 14.17c). This reaction, while reversible, binds the phosphate too tightly to allow its ready replacement by other anions. The availability of phosphate bound in this manner is very low. Over time, a second oxygen of the phosphate ion may replace a second hydroxyl, so that the phosphate becomes chemically bound to two adjacent aluminum (or iron) atoms in the hydrous oxide surface (see Figure 14.17d). With this step, the phosphate becomes an integral part of the oxide mineral, and the likelihood of its release back to the soil solution is extremely small.

Finally, as more time passes, the precipitation of additional iron or aluminum hydrous oxide may bury the phosphate deep inside the oxide particle (see Figure 14.15). Such phosphate is termed *occluded* and is the least available form of phosphorus in most acid soils.

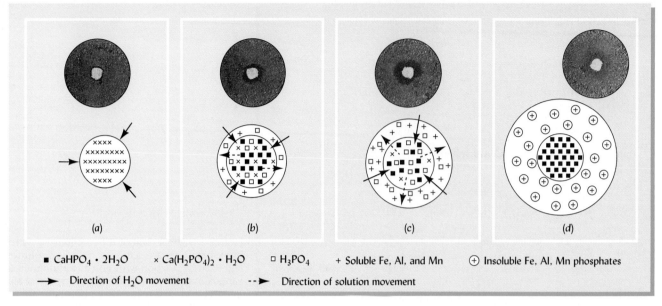

■ CaHPO₄ · 2H₂O   × Ca(H₂PO₄)₂ · H₂O   □ H₃PO₄   + Soluble Fe, Al, and Mn   ⊕ Insoluble Fe, Al, Mn phosphates

→ Direction of H₂O movement   --▶ Direction of solution movement

FIGURE 14.18   When a granule of soluble calcium monophosphate [Ca(H₂PO₄)₂ · H₂O] fertilizer is added to a moist soil, the following series of reactions rapidly and dramatically reduces the availability of the added phosphorus: (a) The Ca(H₂PO₄)₂ · H₂O in the fertilizer granules attracts water from the soil. (b) In the moistened granule, phosphoric acid is formed by the following reaction: Ca(H₂PO₄)₂·H₂O + H₂O → CaHPO₄ · 2H₂O + H₃PO₄. As more water is attracted, an H₃PO₄-laden solution with a pH of about 1.4 moves outward from the granule. (c) This solution is sufficiently acid to dissolve and displace large quantities of iron, aluminum, and manganese. These ions promptly react with the phosphate to form low-solubility compounds. (d) Later, these compounds revert to the hydroxy phosphates of iron, aluminum, and manganese in acid soils. In neutral to alkaline soils, equally insoluble calcium phosphates are formed. In both cases, insoluble dicalcium phosphate (CaHPO₄ · 2H₂O) remains in the granule. Fortunately, the phosphorus in the freshly precipitated compounds is slightly available for plant uptake. But when these freshly precipitated compounds are allowed to age or to revert to more insoluble forms, the phosphorus becomes almost completely unavailable to plants in the short term. (Photos courtesy of G. L. Terman and National Plant Food Institute, Washington, D.C.)

Precipitation reactions similar to those just described are responsible for the rapid reduction in availability of phosphorus added to soil as soluble $Ca(H_2PO_4)_2 \cdot H_2O$ in fertilizers (Figure 14.18). This type of reaction can also be used to control the solubility of phosphorus in wastewater (see Box 14.2).

Effect of Iron Reduction Under Wet Conditions

Phosphorus bound to iron oxides by the mechanisms just discussed is very insoluble under well-aerated conditions. However, prolonged anaerobic conditions can reduce the iron in these complexes from Fe^{3+} to Fe^{2+}, making the iron–phosphate complex much more soluble and causing it to release phosphorus into solution. The release of phosphorus from iron phosphates by means of the reduction and subsequent solubilization of iron improves the phosphorus availability in soils used for paddy rice.

These reactions are also of special relevance to water quality. Phosphorus bound to soil particles may accumulate in river- and lake-bottom sediments, along with organic matter and other debris. As the sediments become anoxic, the reducing environment may cause the gradual release of phosphorus held by hydrous iron oxides. Thus, the phosphorus eroded from soils today may aggravate the problem of eutrophication for years to come, even after the erosion and loss of phosphorus from the land has been brought under control.

14.7   INORGANIC PHOSPHORUS AVAILABILITY AT HIGH PH VALUES[7]

The availability of phosphorus in alkaline soils is determined principally by the solubility of the various calcium phosphate compounds present. In alkaline soils (e.g., pH = 8), soluble $H_2PO_4^-$ quickly reacts with calcium to form a sequence of products of decreasing

[7] See Sample et al. (1980) for a discussion of this subject.

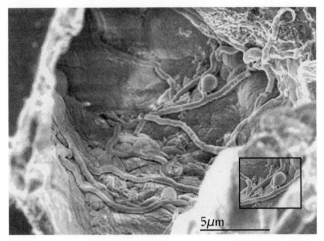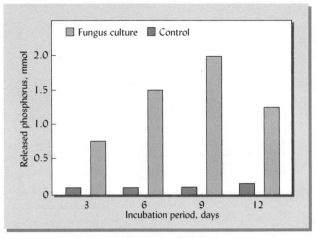

FIGURE 14.19 Certain soil microorganisms can increase the availability of phosphorus in minerals such as rock phosphate and aluminum phosphates that normally hold the phosphorus in very insoluble forms. (*Left*) A micrograph of a fungus growing on the surface of an aluminum phosphate found in soils. The fungus is thought to produce organic acids that help release some soluble phosphorus. (*Right*) In another experiment, phosphorus is released from phosphate rock by a culture of a fungus (*Aspergillus niger*) that had been isolated from a tropical soil. [Micrograph (*left*) courtesy of Dr. Anne Taunton, University of Wisconsin; graph (*right*) drawn from data in Goenadi et al. (2000)]

solubility. For instance, highly soluble monocalcium phosphate [$Ca(H_2PO_4)_2 \cdot H_2O$] added as concentrated superphosphate fertilizer rapidly reacts with calcium carbonate in the soil to form first dicalcium phosphate ($CaHPO_4 \cdot 2H_2O$) and then tricalcium phosphate [$Ca_3(PO_4)_2$], as follows:

$$Ca(H_2PO_4)_2 \cdot H_2O + 2H_2O \xrightarrow{CaCO_3} 2(CaHPO_4 \cdot 2H_2O) + CO_2\uparrow \xrightarrow{CaCO_3} Ca_3(PO_4)_2 + CO_2\uparrow + 5H_2O$$

Monocalcium phosphate (soluble) — Dicalcium phosphate (slightly soluble) — Tricalcium phosphate (very low solubility)

(14.5)

The solubility of these compounds and, in turn, the plant availability of the phosphorus they contain decrease as the phosphorus changes from the $H_2PO_4^-$ ion to tricalcium phosphate [$Ca_3(PO_4)_2$]. Although this compound is quite insoluble, it may undergo further reactions to form even more insoluble compounds, such as the hydroxy-, oxy-, carbonate-, and fluorapatite compounds (apatites) shown in Table 14.6. These compounds are thousands of times less soluble than freshly formed tricalcium phosphates. The extreme insolubility of apatites in neutral or alkaline soils generally makes powdered phosphate rock (which consists mainly of apatite minerals) not very effective as a source of phosphorus for plants unless it is ground very fine (to increase weathering surface) and applied to relatively acidic soils.

Reversion of soluble fertilizer phosphorus to extremely insoluble calcium phosphate forms is most serious in the calcareous soils of low-rainfall regions (e.g., western United States). Iron and aluminum impurities in calcite particles may also adsorb considerable amounts of phosphate in these soils. Because of the various reactions with $CaCO_3$, phosphorus availability tends to be nearly as low in the Aridisols, Inceptisols, and Mollisols of arid regions as in the highly acid Spodosols and Ultisols of humid regions, where iron, aluminum, and manganese limit phosphorus availability.

Bacteria and fungi can enhance the solubility of both calcium and aluminum phosphates by releasing citric and other organic acids that either dissolve the calcium phosphates or form metal complexes that release the P from iron and aluminum phosphates in acid soils (Figure 14.19). The released P is likely used first by the microorganisms themselves, but is eventually made available to plants as well.

14.8   PHOSPHORUS-FIXATION CAPACITY OF SOILS

pH and phosphorus fixation:
www.petrik.com/
PUBLIC/library/misc__/
phosphavailability.htm

Soils may be characterized by their capacity to fix phosphorus in unavailable, insoluble forms. The phosphorus-fixation capacity of a soil may be conceptualized as the total number of sites on soil particle surfaces capable of reacting with phosphate ions. Phosphorus fixation may also be due to reactive soluble iron, aluminum, or manganese. The different types of fixation mechanisms are illustrated schematically in Figure 14.20.

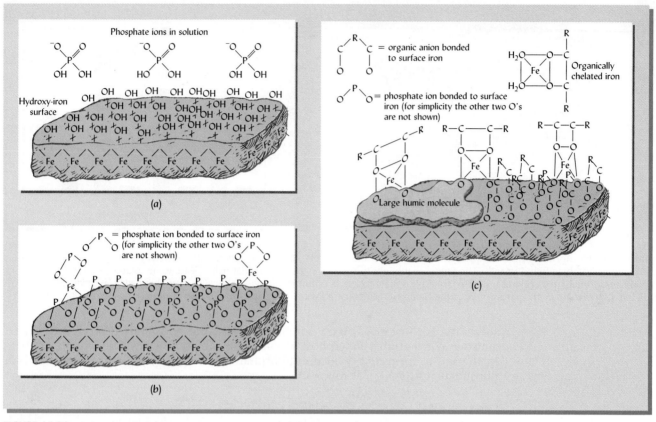

FIGURE 14.20 Schematic illustrations of phosphorus-fixation sites on a soil particle surface showing hydrous iron oxide as the primary fixing agent. In part (*a*) the sites are shown as + symbols, indicating positive charges or hydrous metal oxide sites, each capable of fixing a phosphate ion. In part (*b*) the fixation sites are all occupied by phosphate ions (the soil's fixation capacity is satisfied). Part (*c*) illustrates how organic anions, larger organic molecules, and certain strongly fixed inorganic anions can reduce the sites available for fixing phosphorus. Such mechanisms partially account for the reduced phosphorus fixation and greater phosphorus availability brought about when mulches and other organic materials are added to a soil. (Diagram courtesy of R. Weil)

One way of determining the phosphorus-fixing capacity of a particular soil is to shake a known quantity of the soil in a phosphorus solution of known concentration. After about 24 hours an equilibrium will be approached, and the concentration of phosphorus remaining in the solution (the **equilibrium phosphorus concentration** [EPC]) can be determined. The difference between the initial and final (*equilibrium*) solution phosphorus concentrations represents the amount of phosphorus fixed by the soil. If this procedure is repeated using a series of solutions with different initial phosphorus concentrations, the results can be plotted as a phosphorus-fixation curve (Figure 14.21), and the maximum phosphorus-fixation capacity can be extrapolated from the value at which the curve levels off.

Phosphorus fixation by soils is not easily reversible. However, if a portion of the fixed phosphorus is present in relatively soluble forms (see Section 14.6) and most of the fixation sites are already occupied by a phosphate ion, some release of phosphorus to solution is likely to occur when the soil is exposed to water with a very low phosphorus concentration. This release (often called *desorption*) of phosphorus is indicated in Figure 14.21 where the curve for soil A crosses the zero fixation line and becomes negative (negative fixation = release). The solution concentration (*x*-axis) at which zero fixation occurs (phosphorus is neither released nor retained) is called the EPC_0. Such release of previously fixed phosphorus helps to resupply soil solution phosphorus depleted by plant uptake. Release of fixed phosphorus is also very important in determining losses of dissolved phosphorus in the surface runoff from a watershed. The EPC_0 is an important parameter for both soil fertility and environmental assessment because it indicates (1) the capacity to replenish the soil solution as it is depleted of P by plant roots, and (2) the rate at which the soil will release phosphorus into runoff and leaching waters.

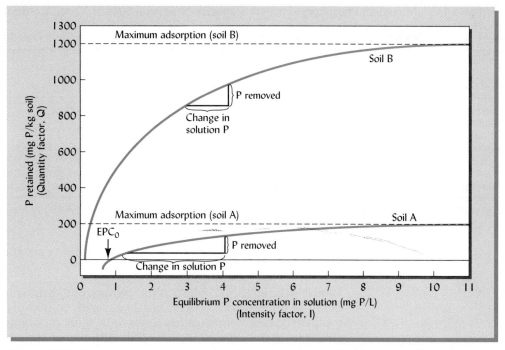

FIGURE 14.21 The relationship between phosphorus fixation and phosphorus in solution when two different soils (A and B) are shaken with solutions of various initial phosphorus concentrations. Initially, each soil removes nearly all of the phosphorus from solution, and as more and more concentrated solutions are used, the soil fixes greater amounts of phosphorus. However, eventually solutions are used that contain so much phosphorus that most of the phosphorus-fixation sites are satisfied, and much of the dissolved phosphorus remains in solution. The amount fixed by the soil levels off as the *maximum phosphorus-fixing capacity* of the soil is reached (see horizontal dashed lines: for soil A, 200 mg P/kg soil; for soil B, 1200 mg P/kg soil). If the initial phosphorus concentration of a solution is equal to the *equilibrium phosphorus concentration (EPC)* for a particular soil, that soil will neither remove phosphorus from nor release phosphorus to the solution (i.e., phosphorus fixation = 0 and EPC = EPC_0). If the solution phosphorus concentration is less than the EPC_0, the soil will release some phosphorus (i.e., the fixation will be negative). In this example, soil B has a much *higher* phosphorus-fixing capacity and a much *lower* EPC than does soil A. It can also be said that soil B is highly *buffered* because much phosphorus must be added to this soil to achieve a small increase in the equilibrium solution phosphorus concentration. On the other hand, if a plant root were to remove a relatively large amount of total phosphorus from soil B, only a small change would occur in the equilibrium solution concentration.

QUANTITY-INTENSITY RELATIONSHIPS.[8] The relationship between phosphorus in solution and phosphorus in slowly soluble or fixed forms is an example of the balance between *quantity* factors and *intensity* factors in soil fertility. The intensity factor is the amount of a nutrient dissolved in the soil solution. The quantity factor is the amount of that nutrient associated with the solid framework of the soil and in equilibrium with the nutrient ions in solution. In Figure 14.21, the intensity factor would be represented on the *x*-axis and the quantity factor on the *y*-axis. The slope of the curve that defines the relationship for each soil represents the amount of change in the quantity factor *Q* that results from a given change in the intensity factor *I*; that is, slope = $\Delta Q/\Delta I$. This is another way of expressing the *potential buffering capacity* (PBC) of the soil:

$$PBC = \frac{\Delta Q}{\Delta I}$$

This general relationship applies not only to phosphorus, but to potassium and any other substance whose solution concentration is controlled by retention reactions with the soil solids. In Section 9.4, for example, buffering of pH was discussed.

FACTORS AFFECTING THE EXTENT OF PHOSPHORUS FIXATION IN SOILS. Soils that remove more than 350 mg P/kg of soil (i.e., a phosphorus-fixing capacity of about 700 kg P/ha) from solution are generally considered to be high phosphorus-fixing soils. High phosphorus-fixing soils tend to maintain low phosphorus concentrations in the soil solution and in runoff water. Table 14.7 lists values of maximum phosphorus-fixing capacity for a range

---

[8] For a review of buffering capacity as it relates to the available phosphorus and potassium, as well as other nutrients, see Nair (1996).

TABLE 14.7 Maximum Phosphorus-Fixation Capacity of Several Soils of Varied Content and Kinds of Clays

Fe, Al oxides (especially amorphous types) fix the largest quantities and silicate clays (especially 2:1-type) fix the least.

Soil Great Group (and series, if known)	Location	Clay Percent	Clay Type	Maximum P fixation, mg P/kg soil
Evesboro (Quartzipsamment)	Maryland	6	Kaolinite, Fe, Al oxides	125
Kandiustalf	Zimbabwe	20	Kaolinite, Fe oxides	394
Kitsap (Xerept)	Washington	12	2:1 clays, allophane	453
Matapeake (Hapludult)	Maryland	15	Chlorite, kaolinite, Fe oxides	465
Rhodustalf	Zimbabwe	53	Kaolinite, Fe oxides	737
Newberg (Haploxeroll)	Washington	38	2:1 clays, Fe oxides	905
Tropohumults	Cameroon	46	Fe, Al oxides, kaolinite	2060

Cameroon data courtesy of V. Ngachie; Washington data from Kuo (1988); Maryland and Zimbabwe data courtesy of R. Weil and F. Folle.

of soils. The effects of clay content and type of clay are apparent. These and other factors will now be discussed.

AMOUNT OF CLAY PRESENT.   Most of the compounds with which phosphorus reacts are in the finer soil fractions. Therefore, if soils with similar pH values and mineralogy are compared, phosphorus fixation tends to be more pronounced, and ease of phosphorus release tends to be lowest in those soils with higher clay contents.

TYPE OF CLAY MINERALS PRESENT.   Some clay minerals are much more effective at phosphorus fixation than others. Generally, those clays that possess greater anion exchange capacity (due to positive surface charges) have a greater affinity for phosphate ions. For example, extremely high phosphorus fixation is characteristic of allophane clays typically found in Andisols and other soils associated with volcanic ash. Oxides of iron and aluminum, such as gibbsite and goethite, also strongly attract and hold phosphorus ions. Among the layer silicate clays, kaolinite has a greater phosphorus-fixation capacity than most. The 2:1 clays of less-weathered soils have relatively little capacity to bind phosphorus. Thus, the soil components responsible for phosphorus-fixing capacity are, in order of increasing extent and degree of fixation:

2:1 clays << 1:1 clays < carbonate crystals < crystalline Al, Fe, Mn oxides
< amorphous Al, Fe and Mn oxides, allophane

To some degree, the preceding phosphorus-fixing soil components are distributed among soils in relation to soil taxonomy. Vertisols and Mollisols generally are dominated by 2:1 clays and have low phosphorus-fixation capacities. Iron and aluminum oxides are prominent in Ultisols and Oxisols. Andisols, characterized by large quantities of amorphous oxides and allophane, have the greatest phosphorus-fixing capacity, and their productivity is often limited by this property (Figure 14.22).

EFFECT OF SOIL pH.   The greatest degree of phosphorus fixation occurs at very low and very high soil pH. As pH increases from below 5.0 to about 6.0, the iron and aluminum phosphates become somewhat more soluble. Also, as pH drops from greater than 8.0 to below 6.0, calcium phosphate compounds increase in solubility. Therefore, as a general rule in mineral soils, phosphate fixation is at its lowest (and plant availability is highest) when soil pH is maintained in the 6.0 to 7.0 range (see Figure 14.16).

Even if pH ranges from 6.0 to 7.0, phosphate availability may still be very low, and added soluble phosphates will be readily fixed by soils. The low recovery by plants of phosphates added to field mineral soils in a given season is partially due to this fixation. A much higher recovery would be expected in organic soils (see Histosol, Figure 14.22) and in many potting mixes where calcium, iron, and aluminum concentrations are not as high as in mineral soils.

EFFECT OF ORGANIC MATTER.   Organic matter has little capacity to strongly fix phosphate ions. To the contrary, amending soil with organic matter, especially decomposable material, is likely to reduce phosphorus fixation by several mechanisms (see Figure 14.20). First, large, humic molecules adhering to sorbing surfaces can mask fixation sites and prevent them from interacting with phosphorus ions in solution. Second,

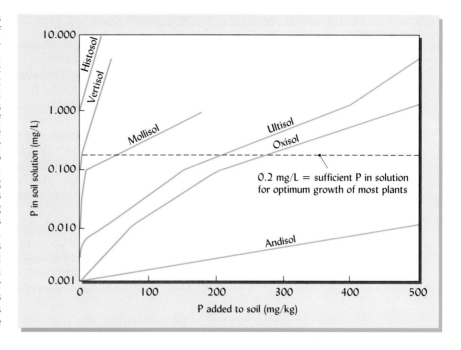

FIGURE 14.22 Typical soil solution levels of phosphorus when increasing amounts of fertilizer phosphorus are added to representative soils from several different orders. Without help from mycorrhizal fungi, most crop plants require a concentration of about 0.2 mg/L in the soil solution for optimum growth (see Table 14.8). The graph suggests very low phosphorus-fixing capacities of Histosols and Vertisols, making it possible for soils in these orders when intially cultivated to require little if any fertilizer to support good plant growth. In contrast, because of their high phosphorus-fixing capacities, unfertilized Ultisols and Oxisols would require a minimum of 200–300 kg P/ha for good plant growth, and Andisols would require many times this amount. Keep in mind that the above relationships would not necessarily apply for a given soil in each order, especially if the soil had been cultivated for a long time without fertilizer additions or with high rates of phosphorus fertilizers that would have satisfied the soil's P-fixing capacity. Curves are representative of data from many sources.

organic acids produced by plant roots and microbial decay can serve as organic anions, which compete with phosphorus ions for positively charged sites on the surfaces of clays and hydrous oxides. Third, certain organic compounds can entrap reactive Al and Fe in stable organic complexes called *chelates* (see Section 15.8). Once chelated, the metals are unavailable for reaction with phosphorus ions in solution. In addition, phosphorus fixation is also likely reduced by the release of phosphorus ions by microbial mineralization of many P-rich organic materials. Figure 14.23 illustrates that added organic material (treated sewage sludge) reduces the Q/I ratio (slope of sorption curve) and increases EPC_0 (level of phosphorus in solution).

14.9 PRACTICAL CONTROL OF PHOSPHORUS IN SOILS

The principles of soil phosphorus behavior discussed in this chapter suggest a number of approaches to ameliorate soil deficiencies and prevent excessive losses of this critical element.

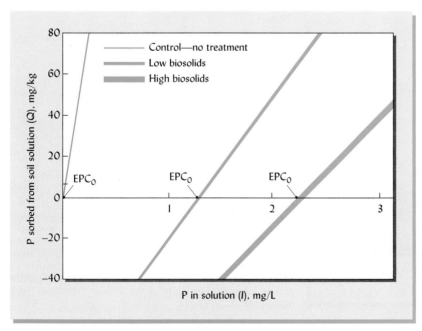

FIGURE 14.23 Phosphorus quantity/intensity (Q/I) relationships for a Mollisol without and with two rates of organic amendment (sewage sludge). Note that the phosphorus in solution for the untreated soil remains very low, even when as much as 80 mg P/kg soil has been sorbed. When sludge is added to the soil, however, much higher phosphorus is maintained in solution. At lower solution phosphorus values, the amount sorbed (Q) was negative, indicating release from (instead of sorption to) the soil. The lower P sorption is partly due to P release from the sewage sludge, but the organic compounds in the waste material may have replaced by anion exchange some sorbed phosphate, or may have formed complexes (chelates) with some of the P-sorbing cations (e.g., Fe^{3+}), thereby causing them to release to the soil solution some of the sorbed phosphorus. The high levels of solution phosphorus are beneficial for plant nutrition, but also indicate that phosphorus can be easily lost from this soil by leaching and runoff leading to downstream eutrophication. [From Sui and Thompson (2000)]

ADJUST APPLICATION TO SOIL STATUS. Where phosphorus-fixing capacity is grossly unsaturated, optimum crop yields will likely require additions that considerably exceed plant uptake. However, as the excess phosphorus begins to saturate fixation sites, rates of application should be lowered to supply no more than what plants take up so as to prevent excessive phosphorus accumulations.

LOCALIZED PLACEMENT. Placement of phosphorus fertilizer in a localized zone is less likely to undergo fixation reactions (Figure 14.18) than if it were mixed into the bulk soil, and thereby reduces the amount of fertilizer required. This end can be achieved by placement of fertilizer in narrow bands or in small holes and by the use of pellets instead of fine powders. In untilled systems, broadcasting on the surface effectively creates a horizontal "band."

COMBINE AMMONIUM WITH PHOSPHORUS. When ammonium and phosphorus fertilizers are mixed in a band, the acidity produced by oxidation of ammonium ions (see Section 13.7) and by uptake of excess cations as ammonium (see Section 9.1) keeps the phosphate in more soluble compounds and enhances plant P uptake.

CYCLING OF ORGANIC MATTER. During the microbial breakdown of organic materials, phosphorus is released slowly and can be taken up by plants or mycorrhizae before it can be fixed by the soil. In addition, organic compounds can reduce soil P fixation capacity (Figure 14.20). Residues or prunings from certain phosphorus-efficient plants (e.g., the African shrub *Tithonia diversifolia*, Figure 14.24) can be harvested from a donor site and transferred to a low-fertility receiving site where phosphorus will be supplied as the transferred plant material decays.

CONTROL OF SOIL pH. Phosphorus availability can be optimized in most soils by proper liming or acidification (see Sections 9.8 and 9.10) to a pH level between 6 and 7.

ENHANCE MYCORRHIZAL SYMBIOSIS. Practices that enhance mycorrhizal symbiosis usually improve the utilization of soil phosphorus. Such practices range from including good host plants, to reducing tillage, to inoculation with appropriate fungi (see Section 11.9).

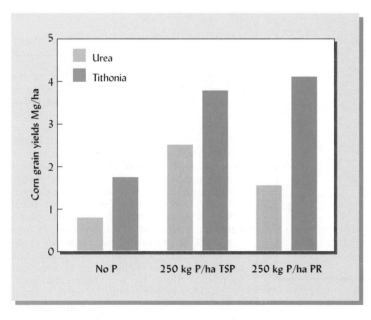

FIGURE 14.24 Residues of the common hedge tithonia (*Tithonia diversifolia*) enhanced the yields of corn growing on an Oxisol in Kenya equally well whether applied with relatively insoluble phosphate rock (PR) or with triple super phosphate (TSP), a carrier of relatively available phosphorus. Tithonia also supplied some nitrogen and potassium. Research suggests that the decomposition products of tithonia may increase soil microbial activity and slightly reduce P-sorption capacity of the Oxisol. [Modified from Sanchez et al. (1997)]

TABLE 14.8 Concentration of Phosphorus
in Soil Solution Required for Near-Optimal
Growth (95% of Maximum Yield) of Various
Plants

Plant	Approximate P in soil solution, mg/L
Cassava	0.005
Peanut	0.01
Corn	0.05
Sorghum	0.06
Cabbage	0.04
Soybean	0.20
Tomato	0.20
Head lettuce	0.30

Data from Fox (1981).

CHOOSE P-EFFICIENT PLANTS.   Some plant species require much less phosphorus in the soil solution than do others. Table 14.8 indicates the requirements of several agricultural species, but there are also wide differences among native plants. Although much remains to be learned, different species enhance phosphorus uptake by at least four strategies: (1) Monocots exhibit extensive fibrous root systems and mycorrhizal associations. (2) N-fixing legumes use little nitrate and take up an excess of cations over anions, leading to rhizosphere acidification (Section 9.1) and subsequent release of P from low-solubility Ca-phosphates. (3) Certain species excrete specific compounds (such as *piscidic acid* produced by pigeon pea) that complex with Fe to greatly increase the availability of iron-bound soil phosphorus. (4) Plants in the *Cruciferae* family (cabbage, mustard, etc.) compensate for their very poor mycorrhizal properties by excreting citric and malic acids, forming extensive fine root hairs, and taking up high amounts of Ca^{2+}. Knowledge of these plant characteristics can aid in choosing plants for restoration ecology, as well as for low-income farmers who can afford little fertilizer.

Reducing Phosphorus Losses to Water

AVOID EXCESS ACCUMULATION.   This goal requires carefully keeping the sum of all phosphorus inputs (deposition, fertilizers, organic amendments, plant residues, and animal feed) from consistently exceeding plant removals or accumulating beyond the lowest levels that will support near-optimum plant growth (see Sections 14.3 and 16.12).

MINIMIZE LOSS IN RUNOFF AND SEDIMENT.   Use of conservation tillage practices that minimize runoff and erosion, *especially from land already high in phosphorus,* is essential. Cover crops and plant residues can increase infiltration and reduce runoff (see Section 6.2). Application of manure or fertilizer to frozen soils should be avoided.

CAPTURE P FROM RUNOFF.   Natural or constructed wetlands (Section 7.7) and riparian (shoreline) buffer strips (Section 16.2) can tie up some phosphorus before runoff enters sensitive lakes or streams. These measures remove some dissolved P, but mainly P bound to sediment.

TIE UP P WITH INORGANIC AMENDMENTS.   Various iron-, aluminum-, or calcium-containing materials react with dissolved P from phosphorus saturated soils or P-rich organic waste products. Highly insoluble compounds are formed, similar to those described in Section 14.5 and Box 14.2. Treating poultry litter in the chicken house with aluminum sulfate (alum) (Figure 14.25) and spreading inorganic amendments on the surface of grass sod are practices that greatly reduce the loss of phosphorus dissolved from P-rich organic fertilizers such as manure or compost (Figure 14.26).

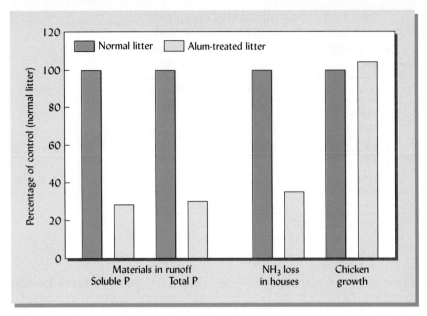

FIGURE 14.25 Effect of adding alum (aluminum sulfate) to chicken litter in poultry houses on the emission of ammonia in the houses and on the soluble and total phosphorus in the runoff and erosion from fields on which the chicken litter was spread. The growth rate of the chickens was increased slightly (*right*) in the alum-treated sheds. Such treatment may greatly decrease the movement of phosphorus into water bodies and reduce the health hazard for workers and chickens from the ammonia in the houses. [Calculated from Moore et al. (2000)]

14.10 POTASSIUM: NATURE AND ECOLOGICAL ROLES[9]

Of all the essential elements, potassium is the third most likely, after nitrogen and phosphorus, to limit plant productivity. For this reason it is commonly applied to soils as fertilizer and is a component of most mixed fertilizers.

The potassium story differs in many ways from that of phosphorus. Unlike phosphorus (or sulfur and, to a large extent, nitrogen), potassium is present in the soil solution only as a positively charged cation, K^+. Like phosphorus, potassium does not form any gases that could be lost to the atmosphere. Its behavior in the soil is influenced primarily by soil cation exchange properties (see Chapter 8) and mineral weathering (Chapter 2), rather than by microbiological processes. Unlike nitrogen and phosphorus, potassium causes no off-site environmental problems when it leaves the soil system. It is not toxic and does not cause eutrophication in aquatic systems.

---

[9] For further information on this topic, see Mengel and Kirkby (2001) and Munson (1985).

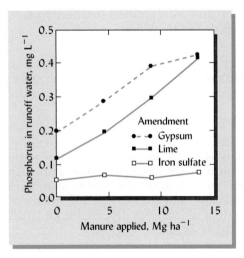

FIGURE 14.26 Effect of surface application of iron sulfate, gypsum, or lime on the concentration of phosphorus dissolved in runoff water. Bermuda grass sod was fertilized with composted dairy manure and subjected to two 30-minute rain events. The iron sulfate quickly reacted with phosphate ions as they were released from the compost, forming highly insoluble iron phosphate compounds. Gypsum and lime, which formed some calcium-phosphates, were far less effective in reducing phosphorus in the runoff water. [Drawn from data in Torbert et al. (2005)]

Although potassium plays numerous roles in plant and animal nutrition, it is not actually incorporated into the structures of organic compounds. Instead, potassium remains in the ionic form (K^+) in solution in the cell or acts as an activator for cellular enzymes. Potassium is known to activate over 80 different enzymes responsible for such plant and animal processes as energy metabolism, starch synthesis, nitrate reduction, photosynthesis, and sugar degradation. Certain plants, many of which evolved in sodium-rich semiarid environments, can substitute sodium or other monovalent ions to carry out some, but not all, of the functions of potassium.

As a component of the plant cytoplasmic solution, potassium plays a critical role in lowering cellular osmotic water potentials, thereby reducing the loss of water from leaf stomata and increasing the ability of root cells to take up water from the soil (see Section 5.3 for a discussion of osmotic water potential). Potassium is essential for photosynthesis, for protein synthesis, for nitrogen fixation in legumes, for starch formation, and for the translocation of sugars. As a result of several of these functions, a good supply of this element promotes the production of plump grains and large tubers. The potassium content of normal, healthy leaf tissue can be expected to be in the range of 1 to 4% in most plants, similar to that of nitrogen but an order of magnitude greater than that of phosphorus.

Potassium is especially important in helping plants adapt to environmental stresses. Good potassium nutrition is linked to improved drought tolerance, improved winter-hardiness, better resistance to certain fungal diseases, and greater tolerance to insect pests (Figure 14.27). In the latter role, potassium fertilization is often an important component of integrated pest-management programs designed to reduce the use of toxic pesticides. Potassium also enhances the quality of flowers, fruits, and vegetables by improving flavor and color and strengthening stems (thereby reducing lodging). In many of these respects, potassium seems to counteract some of the detrimental effects of excess nitrogen. Maintaining a balance between potassium and other nutrients (especially nitrogen, phosphorus, calcium, and magnesium) is an important goal in managing soil fertility.

In animals, including humans, potassium plays critical roles in regulating the nervous system and in the maintenance of healthy blood vessels. Diets that include such high-potassium foods as bananas, potatoes, orange juice, and leafy green

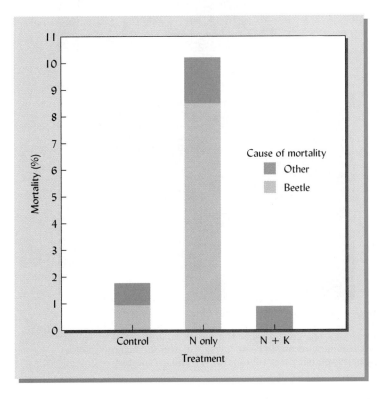

FIGURE 14.27   Influence of potassium and nitrogen fertilizer treatments on the percentage of Ponderosa pine trees dying from beetle damage and other causes in the first four years after planting in western Montana. Nitrogen used alone (224 kg N/ha) stimulated a large increase in tree mortality, but adding potassium (224 kg K/ha) completely counteracted this effect. Both fertilization treatments stimulated the growth of the surviving trees. [From Mandzak and Moore (1994)]

vegetables have been shown to lower human risk of stroke and heart disease. Maintaining a balance between potassium and sodium is especially important in human diets.

Deficiency Symptoms in Plants

Compared to deficiencies of phosphorus and many other nutrients, a deficiency of potassium is relatively easy to recognize in most plants. In addition to the characteristics previously mentioned (reduced drought tolerance, increased lodging, etc.), specific foliar symptoms are associated with potassium deficiency. Because potassium is very mobile within the plant, it is translocated from older tissues to younger ones if the supply becomes inadequate. The symptoms of deficiency therefore usually occur earliest and most severely on the oldest leaves.

In general, when potassium is deficient the tips and edges of the oldest leaves begin to yellow (chlorosis) and then die (necrosis), so that the leaves appear to have been burned on the edges (Figure 14.28 and Plate 87, after page 656). On some plants the

(a)

(b)

(c)

(d)

FIGURE 14.28 Potassium deficiency often produces easily recognized foliar symptoms, mainly on older leaves: (a) chlorotic margins on boxwood leaves; (b) chlorotic leaf margins on soybean; (c) ragged, necrotic margins of older banana leaves; and (d) small, white necrotic spots on hairy vetch leaflets. (Photos (a), (c), and (d) courtesy of R. Weil; photo (b) courtesy of Potash & Phosphate Institute)

necrotic leaf edges may tear, giving the leaf a ragged appearance (Figure 14.28c). In several important forage and cover-crop legume species, potassium deficiency produces small, white necrotic spots that form a unique pattern along the leaflet margins; this easily recognized symptom is one that people often mistake for insect damage (see Figure 14.28d).

Potassium deficiency should not be confused with damage from excess salinity, which can also produce brown, necrotic leaf margins. Salinity damage is more likely to affect the newer leaves (see Figures 10.25 and 10.27).

14.12 THE POTASSIUM CYCLE

Figure 14.29 shows the major forms in which potassium is held in soils and the changes it undergoes as it is cycled through the soil–plant system. The original sources of potassium are the primary minerals, such as micas (biotite and muscovite) and potassium feldspar (orthoclase and microcline). As these minerals weather, their rigid lattice structures become more pliable. For example, potassium held between the 2:1-type crystal layers of mica is in time made more available, first as nonexchangeable but slowly available forms near the weathered edges of minerals and, eventually, as the readily exchangeable forms and the soil solution forms from which it is absorbed by plant roots.

Potassium is taken up by plants in large quantities. Depending on the type of ecosystem under consideration, a portion of this potassium is leached from plant foliage by rainwater (throughfall) and returned to the soil, and a portion is returned to the soil with the plant residues. In natural ecosystems, most of the potassium taken up by plants is returned in these ways or as wastes (mainly urine) from animals feeding on the vegetation. Some potassium is lost with eroded soil particles and in runoff water, and some is lost to groundwater by leaching. In agroecosystems, from one-fifth (e.g., in cereal grains) to nearly all (e.g., in hay crops) of the potassium taken up by plants may be exported to distant markets, from which it is unlikely to return.

At any one time, most soil potassium is in primary minerals and nonexchangeable forms. In relatively fertile soils, the release of potassium from these forms to the exchangeable and soil solution forms that plants can use directly may be sufficiently rapid to keep plants supplied with enough potassium for optimum growth. On the other hand, where high yields of agricultural crops or timber are removed from the land, or where the content of weatherable potassium-containing minerals is low, the levels of exchangeable and solution potassium may have to be supplemented by outside sources, such as chemical fertilizers, poultry manure, or wood ashes. Without these additions, the supply of available potassium will likely be depleted over a period of years, and the productivity of the soil will likewise decline.

An example of depletion and restoration of available soil potassium is given in Figure 14.30. Farming without fertilizers for over a century depleted the exchangeable potassium in a sandy soil in New York. After the supply of available potassium was exhausted, the land was abandoned in the 1920s. In the 1920s and 1930s a red pine forest was planted. The trees in some plots were fertilized with potassium, causing the expected rapid recovery of exchangeable potassium to its preagricultural level. Even where the trees were not fertilized, the level of exchangeable potassium in the surface soil was replenished, but slowly, over a period of about 80 years.

The replenishment of exchangeable potassium in the unfertilized forest plots provides an example of the ability of unharvested perennial plants, such as forest trees and pasture legumes, to ameliorate soil fertility over time. Deep-rooted perennials often act as "nutrient pumps," taking potassium from deep subsoil horizons into their root systems, translocating it to their leaves, and then recycling it back to the surface of the soil via leaf fall and leaching (see Section 16.3 and Figure 16.15).

In most mature natural ecosystems the small (1 to 5 kg/ha) annual losses of potassium by leaching and erosion are more than balanced by weathering of potassium from primary minerals and nonexchangeable forms in the soil profile, followed by vegetative translocation to the surface of the soil. In most agricultural systems, leaching losses are far greater because much higher exchangeable K levels are generally maintained for crop production, but crop roots are active for only part of the year. The sections that follow give greater details on the reactions involved in the potassium cycle.

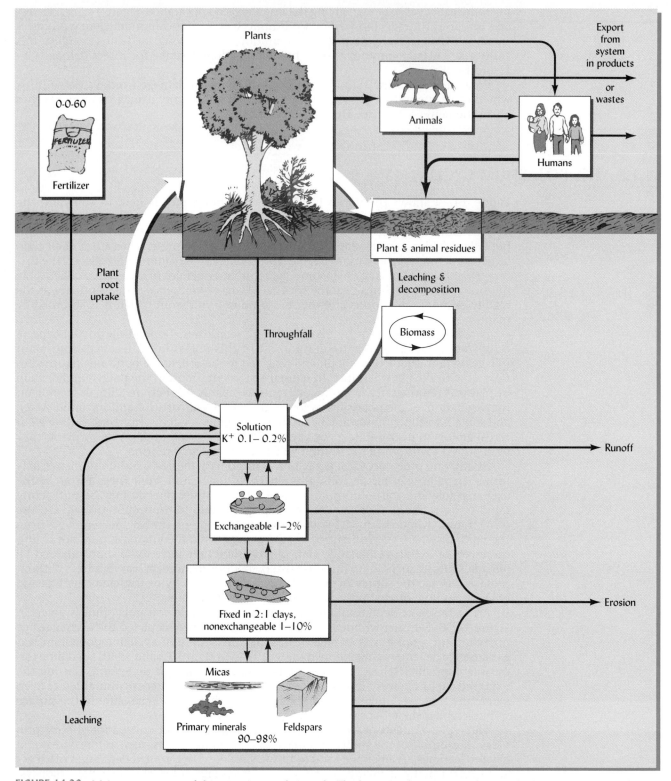

FIGURE 14.29 Major components of the potassium cycle in soils. The large circular arrow emphasizes the biological cycling of potassium from the soil solution to plants and back to the soil via plant residues or animal wastes. Primary and secondary minerals are the original sources of the element. Exchangeable potassium may include those ions held and released by both clay and humus colloids, but potassium is not a structural component of soil humus. The interactions among solution, exchangeable, nonexchangeable, and structural potassium in primary minerals is shown. The bulk of soil potassium occurs in the primary and secondary minerals and is released very slowly by weathering processes. (Diagram courtesy of R. Weil)

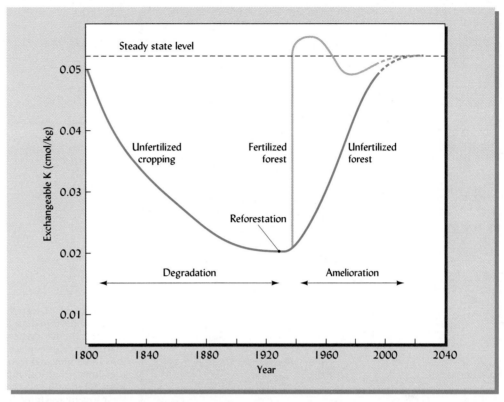

FIGURE 14.30 The general pattern of depletion of A-horizon exchangeable potassium by decades of exploitative farming, followed by its restoration under forest vegetation. The forest consisted of red pine trees planted on a Plainfield loamy sand (Udipsamment) in New York. This soil has a very low cation exchange capacity and low levels of exchangeable K+. [From Nowak et al. (1991)]

14.13 THE POTASSIUM PROBLEM IN SOIL FERTILITY

Availability of Potassium

In contrast to phosphorus, potassium is found in comparatively high levels in most mineral soils, except those consisting mostly of quartz sand. In fact, the total quantity of this element is generally greater than that of any other major nutrient element. Amounts as great as 30,000 to 50,000 kg potassium in the upper 15 cm of 1 ha of soil are not at all uncommon (see Table 1.3).

Yet the quantity of potassium held in an easily exchangeable condition at any one time often is very small. Most of this element is held rigidly as part of the primary minerals or is fixed in forms that are, at best, only moderately available to plants. Therefore, the situation with respect to potassium utilization parallels that of phosphorus and nitrogen in at least one way: A very large proportion of all three of these elements in the soil is insoluble and relatively unavailable to growing plants.

Leaching Losses

Potassium is much more readily lost by leaching than is phosphorus. Drainage waters from soils receiving liberal fertilizer applications usually contain considerable quantities of potassium. From representative humid-region soils growing annual crops and receiving only moderate rates of fertilizer, the annual loss of potassium by leaching is usually about 25 to 50 kg/ha, the greater values being typical of acid, sandy soils.

Losses would undoubtedly be much larger were the leaching of potassium not slowed by the attraction of the positively charged potassium ions to the negatively charged cation exchange sites on clay and humus surfaces. Liming an acid soil to raise its pH can reduce the leaching losses of potassium because of the *complementary ion effect* (see Chapter 8 and Figure 14.31). The ease with which ions may be removed from the exchange complex varies among different elements. Typically, trivalent ions (Al^{3+})

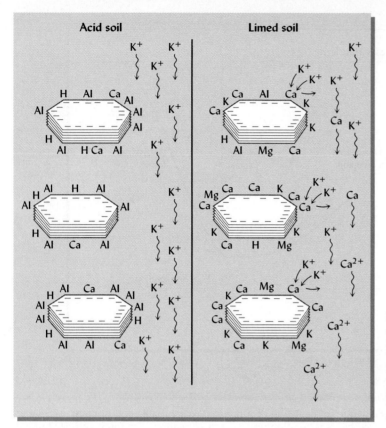

Acid soil | **Limed soil**

FIGURE 14.31 Diagrammatic illustration of how liming an acid soil can reduce leaching losses of potassium. The fact that the K^+ ions can more easily replace Ca^{2+} ions than they could replace Al^{3+} ions allows more of the K^+ ions to be removed from solution by cation exchange in the limed (high-calcium) soil. The removal of K^+ ions from solution by adsorption on the colloids will reduce their loss by leaching, but they will still be at least moderately available for plant uptake.

are more tightly held than divalent ions (Ca^{2+} and Mg^{2+}). In a limed soil, where higher levels of exchangeable calcium and magnesium are present, monovalent potassium ions are better able to replace them on the exchange complex. Where higher levels of exchangeable aluminum saturate the exchange complex, potassium is less likely to be adsorbed. Thus, in a limed soil, K^+ can be more readily retained on the exchange complex, and leaching of this element is reduced.

Plant Uptake and Removal

Plants take up very large amounts of potassium, often five to ten times as much as for phosphorus and about the same amount as for nitrogen. If most or all of the aboveground plant parts are removed in harvest, the drain on the soil supply of potassium can be very large. For example, a 60-Mg/ha yield of corn silage may remove 160 kg/ha of potassium. Conventional bolewood timber harvest typically removes about 100 kg/ha of potassium. If the entire tree is chipped and removed, as for paper pulp, the removal of potassium may be twice as great. A high-yielding legume hay crop may remove 400 kg/ha of potassium each year.

Luxury Consumption

Moreover, this situation is made even more critical by the tendency of plants to take up soluble potassium far in excess of their needs if sufficiently large quantities are present. This tendency is termed *luxury consumption,* because the excess potassium absorbed does not increase plant growth.

The principles involved in luxury consumption are shown in Figure 14.32. For many plants there is a direct relationship between the available potassium (soil plus fertilizer) and the removal of this element by the plants. However, only a certain amount of this element is needed for optimum growth, and this is termed *required potassium.* Potassium taken up by the plant above this critical required level is considered a *luxury.* If plant residues are not returned to the soil, the removal of this

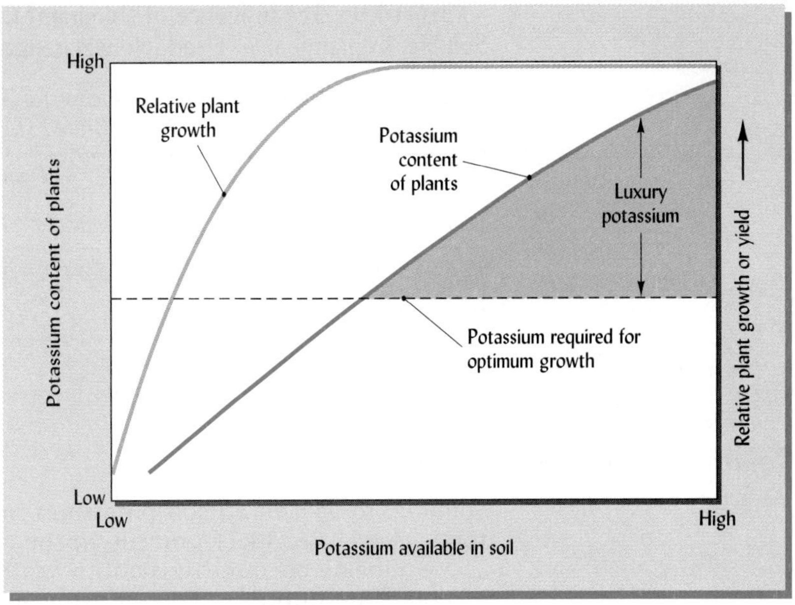

FIGURE 14.32 The general relationship between available potassium level in soil, plant growth, and plant uptake of potassium. If available soil potassium is raised above the level needed for maximum plant growth, many plants will continue to increase their uptake of potassium without any corresponding increase in growth. The potassium taken up in excess of that needed for optimum growth is termed *luxury consumption.* Such luxury consumption may be wasteful, especially if the plants are completely removed from the soil. It may also cause dietary imbalance in grazing animals. (Diagram courtesy of R. Weil)

excess potassium is decidedly wasteful. In addition, high levels of potassium may depress calcium and magnesium uptake and cause nutritional imbalances both in the plants and in animals that consume them. This calcium problem is of particular concern in forage crops for dairy cows.

In summary, then, the problem of potassium is at least threefold: (1) a very large proportion is relatively unavailable to higher plants; (2) it is subject to leaching losses; and (3) the removal of potassium by plants is high, especially when luxury quantities of this element are supplied. With these ideas as a background, the various forms and availabilities of potassium in soils will now be considered.

14.14 FORMS AND AVAILABILITY OF POTASSIUM IN SOILS

Potassium in soils:
www.soils.wisc.edu/~barak/
soilscience326/potassium.htm

Four forms of soil potassium are shown in the potassium-cycle diagram (Figure 14.29, from the bottom upward): (1) K in primary mineral crystal structures, (2) K in nonexchangeable positions in secondary minerals, (3) K in exchangeable form on soil colloid surfaces, and (4) potassium ions soluble in water. The total amount of potassium in a soil and the distribution of potassium among the four major pools shown in Figure 14.29 is largely a function of the kinds of clay minerals present in a soil. Generally, soils dominated by 2:1 clays contain the most potassium; those dominated by kaolinite contain the least (Table 14.9). In terms of availability for plant uptake, the following interpretation applies to the different forms of soil potassium:

K in primary mineral structure	*Unavailable*
Nonexchangeable K in secondary minerals	*Slowly available*
Exchangeable K on soil colloids ⎫ K soluble in water ⎭	*Readily available*

All plants can easily utilize the readily available forms, but the ability to obtain potassium held in the slowly available and unavailable forms differs greatly among plant species. Many grass plants with fine, fibrous root systems are able to exploit potassium held in clay interlayers and near the edges of mica and feldspar crystals of clay and silt size. A few plants adapted to low-fertility sandy soils, such as elephant grass (*Pennisetum purpureum* Schum.), have been shown to obtain potassium from even sand-sized primary minerals, a form of potassium usually considered to be unavailable.

TABLE 14.9 **The Influence of Dominant Clay Minerals on the Amounts of Water-Soluble, Exchangeable, Fixed (Nonexchangeable), and Total Potassium in Soils**

The values given are means for many soils in 10 soil orders sampled in the United States and Puerto Rico.

Potassium pool	Dominant clay mineralogy of soils, mg K/kg soil		
	Kaolinitic (26 soils)	Mixed (53 soils)	Smectitic (23 soils)
Total potassium	3340	8920	15780
Exchangeable potassium	45	224	183
Water-soluble potassium	2	5	4

Data from Sharpley (1990).

Relatively Unavailable Forms

Some 90 to 98% of all soil potassium in a mineral soil is in relatively unavailable forms (see Figure 14.29), mostly in the crystal structure of feldspars and the micas. These minerals are quite resistant to weathering and supply relatively small quantities of potassium during a given growing season. However, their cumulative release of potassium over a period of years undoubtedly is of some importance. This release is enhanced by the solvent action of carbonic acid and of stronger organic and inorganic acids, as well as by the presence of acidic clays and humus (see Section 2.1). As already mentioned, the roots of some plants can obtain a significant portion of their potassium supply from these minerals, apparently by depleting the potassium ions from the solution around the edges of these minerals, thereby favoring the dissolution of the mineral.

Readily Available Forms

Only 1 to 2% of the total soil potassium is readily available. Available potassium exists in soils in two forms: (1) in the soil solution, and (2) exchangeable potassium adsorbed on the soil colloidal surfaces. Although most of this available potassium (approximately 90%) is in the exchangeable form, soil solution potassium is most readily absorbed by higher plants (but, unfortunately, is subject to considerable leaching loss).

As represented in Figure 14.29, these two forms of readily available potassium are in dynamic equilibrium. This equilibrium is of extreme practical importance. When plants absorb potassium from the soil solution, some of the exchangeable potassium immediately moves into the soil solution until the equilibrium is again established. When water-soluble fertilizers are added, the soil solution becomes potassium enriched and the reverse of the described adjustment occurs—potassium from soil solution moves onto the exchange complex. The exchangeable potassium can be seen as an important buffer mechanism for soil solution potassium.

Slowly Available Forms

In the presence of vermiculite, smectite, and other 2:1-type minerals, the K^+ ions (as well as the similarly sized NH_4^+ ions) in the soil solution (or added as fertilizers) not only become adsorbed but also may become definitely fixed by the soil colloids (Figure 14.33). Potassium (and ammonium) ions fit in between layers in the crystals of these normally expanding clays and become an integral part of the crystal. These ions cannot be replaced by ordinary exchange processes and consequently are referred to as *nonexchangeable ions*. As such, these ions are not readily available to most higher plants. This form is in equilibrium, however, with the more available forms and consequently acts as an extremely important reservoir of slowly available nutrients (Figure 14.34).

Release of Fixed Potassium

The quantity of nonexchangeable or fixed potassium in some soils is quite large. The fixed potassium in such soils is continually released to the exchangeable form in amounts large enough to be of great practical importance. The data in Table 14.10

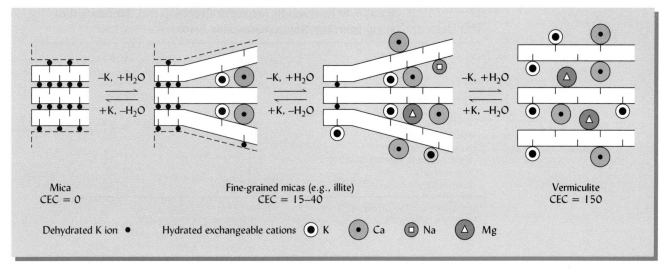

FIGURE 14.33 Diagrammatic illustration of the release and fixation of potassium between primary micas, fine-grained mica (illite clay), and vermiculite. In the diagram, the release of potassium proceeds to the right, while the fixation process proceeds to the left. Note that the dehydrated potassium ion is much smaller than the hydrated ions of Na^+, Ca^{2+}, Mg^{2+}, etc. Thus, when potassium is added to a soil containing 2:1-type minerals such as vermiculite, the reaction may go to the left and potassium ions will be tightly held (fixed) in between layers within the crystal, producing a fine-grained mica structure. Ammonium ions (NH_4^+, not shown) are of a similar size and charge to potassium ions and may be fixed by similar reactions. [Modified from McLean (1978)]

indicates the magnitude of the release of nonexchangeable potassium from certain soils. In these soils, the potassium removed by plants was supplied largely from nonexchangeable forms. The entire equilibrium may be represented for potassium as follows:

$$\text{Nonexchangeable K} \underset{}{\overset{\text{Slow}}{\rightleftharpoons}} \text{Exchangeable K} \underset{}{\overset{\text{Rapid}}{\rightleftharpoons}} \text{Soil solution K}$$

As a result of these relationships, very sandy soils with low CEC are poorly buffered with respect to potassium. In them, the potassium ion concentration may be quite high at the beginning of a growing season or just after fertilization, but the soils have little capacity to maintain the potassium concentration, as plants remove the dissolved potassium from the soil solution during the growing season. Late-season potassium deficiency may result. In finer-textured soils with a greater CEC (and therefore greater buffering capacity), the initial solution concentration of potassium may be somewhat

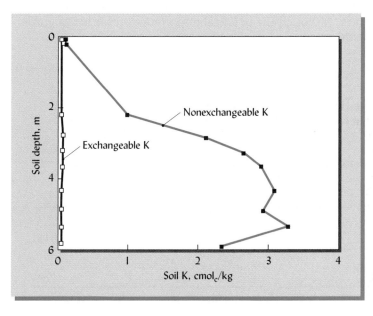

FIGURE 14.34 Exchangeable and nonexchangeable potassium levels in a Ultisol in South Carolina after 30 years growth of a loblolly pine following 150 years of cultivated crops. Although the exchangeable potassium level was quite low, tree growth was not adversely affected, and large quantities of this element were absorbed by the trees over the 30-year period. This was made possible by the conversion of nonexchangeable potassium to the exchangeable form, which was readily taken up by the trees. The upper horizons may have been depleted somewhat of nonexchangeable K, but the deep tree roots were able to use the potassium released from the nonexchangeable form in the lower horizons. [From Richter et al. (1994)]

TABLE 14.10   Potassium Removal by Intensive Cropping and the Amount of This Element Coming from the Nonexchangeable Form

Soil	Total K used by crops kg/ha	Percent derived from nonexchangeable form
Wisconsin soils[a]		
Carrington silt loam	133	75
Spencer silt loam	66	80
Plainfield sand	99	25
Mississippi soils[b]		
Robinsonville fine silt loam	121	33
Houston clay	64	47
Ruston sandy loam	47	24

[a]Average of six consecutive cuttings of ladino clover. [From Evans and Attoe (1948)]
[b]Average of eight consecutive crops of millet. [From Gholston and Hoover (1948)]

lower, but the soil is capable of maintaining a fairly constant supply of solution potassium ions throughout the growing season.

14.15   FACTORS AFFECTING POTASSIUM FIXATION IN SOILS

Four soil conditions markedly influence the amounts of potassium fixed: (1) the nature of the soil colloids, (2) wetting and drying, (3) freezing and thawing, and (4) the presence of excess lime.

Effects of Type of Clay and Moisture

The ability of the various soil colloids to fix potassium varies widely. Kaolinite and other 1:1-type clays fix little potassium. On the other hand, clays of the 2:1 type, such as vermiculite, fine-grained mica (illite), and smectite, fix potassium very readily and in large quantities. Even silt-sized fractions of some micaceous minerals fix and subsequently release potassium (Table 14.11).

The potassium and ammonium ions are attracted between layers in the negatively charged clay crystals. The tendency for fixation is greatest in minerals where the major source of negative charge is in the silica (tetrahedral) sheet. Consequently, vermiculite has a greater fixing capacity than montmorillonite.

Alternate wetting/drying and freezing/thawing has been shown to enhance both the fixation of potassium in nonexchangeable forms and the release of previously fixed potassium to the soil solution. Although the practical importance of this is recognized, its mechanism is not well understood.

TABLE 14.11   Average Uptake of Potassium by the Roots of Ryegrass Grown on 14 High Mica Arable Loess-Derived Soils (Alfisols) with and without the Clay Removed

The mica in the silt and sand fractions provided about ¾ as much potassium as did the whole soils. It is probable that the silt fraction was a major source of potassium in these soils.

	K uptake by ryegrass root (mg/pot)		
	First harvest	Second harvest	Total
Whole soils	103.2	41.3	144.5
Silt + sand fraction	73.9	23.7	97.6

Calculated from Mengel et al. (1998).

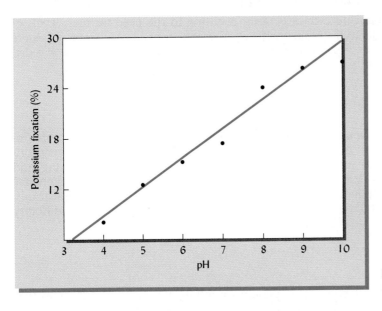

FIGURE 14.35 The effect of pH on the fixation of potassium in soils of India. [From Grewal and Kanwar (1976)]

Influence of pH

Applications of lime sometimes result in an increase in potassium fixation of soils (Figure 14.35). This is not surprising, since in strongly acid soils the tightly held H^+ and hydroxy aluminum ions prevent the potassium ions from being closely associated with the colloidal surfaces, which reduces their susceptibility to fixation. As the pH increases, the H^+ and hydroxyl aluminum ions are removed or neutralized, and it is easier for potassium ions to move closer to the colloidal surfaces, where they are more susceptible to fixation in 2:1 clays.

Furthermore, high calcium and magnesium levels in the soil solution may reduce potassium uptake by the plant because cations tend to compete against one another for uptake by roots. Since the absorption of the potassium ion by plant roots is affected by the activity of other ions in the soil solution, some authorities prefer to use the ratio

$$\frac{[K^+]}{\sqrt{[Ca^{2+}] + [Mg^{2+}]}}$$

rather than the potassium concentration to indicate the available potassium level in solution. Finally, potassium deficiency frequently occurs in calcareous soils even when the amount of exchangeable potassium present would be adequate for plant nutrition on other soils. Potassium fixation as well as cation ratios may be responsible for these adverse effects on calcium carbonate-rich soil.

14.16 PRACTICAL ASPECTS OF POTASSIUM MANAGEMENT

Except in very sandy soils, the problem of potassium fertility is rarely one of total supply, but rather one of adequate *rate* of transformation from nonavailable to available forms. Where little plant material is removed (e.g., in forests, rangeland, and some ornamental systems), cycling between plant and soil may be adequate for continued plant growth. However, where crops are removed, especially if little plant residue is returned, then the plant–soil cycle must be supplemented by release of potassium from less available mineral forms and, to some degree, by fertilization.

Vigorous growth of high-potassium-content plants places great demands on the soil supply of available potassium. Moreover, the rate of potassium uptake is not constant, but varies with plant growth stage and season. If high yields of forage legumes such as alfalfa are to be produced, the soil may have to be capable of supplying potassium for very high uptake rates during certain periods, resulting in the need for high levels of fertilization even on soils well supplied with weatherable minerals. However, excessive levels of potassium that depress calcium and magnesium in the forage must be avoided to maintain plant and animal health (see grass tetany, Section 15.2).

Frequency of Application

Although a heavy dressing applied every few years may be most convenient, more frequent light applications of potassium may offer the advantages of reduced luxury consumption of potassium by some plants, reduced losses of this element by leaching, and reduced opportunity for fixation in unavailable forms before plants have had a chance to use the potassium applied. Although such fixation has definite conserving features, in most cases these tend to be outweighed by the disadvantages of leaching and luxury consumption.

Potassium-Supplying Power of Soils

Full advantage should be taken of the potassium-supplying power of soils. The idea that each kilogram of potassium removed by plants or through leaching must be returned in fertilizers may not always be correct. In many soils, the large quantities of moderately available forms already present can be utilized so that only a part of the total amount removed by harvest need be replaced by fertilizer. Moreover, the importance of lime in reducing leaching losses of potassium should not be overlooked as a means of effectively utilizing the power of soils to furnish this element.

Soils of arid regions are often well supplied with weatherable potassium-containing minerals and can therefore supply adequate potassium for many years, even under irrigation where leaching is more important and plant removal is great. However, continued crop removal can deplete the available potassium pools even in these soils. Also, deep-rooted plants such as cotton and fruit trees may depend on the subsoil for much of their potassium. Increasing the availability of this element at depths below the plow layer is difficult.

Potassium Losses and Gains

The problem of maintaining soil potassium is diagrammed in Figure 14.36. Plant removal of potassium generally exceeds that of the other essential elements, with the possible exception of nitrogen. Annual losses from plant removal as great as 400 kg/ha or more of potassium are not uncommon, particularly if the plant is a legume and is cut several times for hay. As might be expected, therefore, the return of plant residues and manures is very important in maintaining soil potassium.

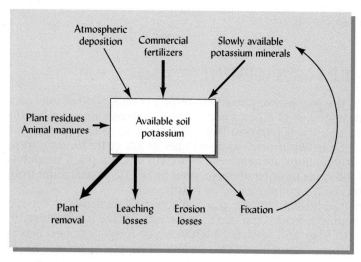

FIGURE 14.36 Gains and losses of *available* soil potassium under average field conditions. The approximate magnitude of the changes is represented by the width of the arrows. For any specific case, the actual amounts of potassium added or lost undoubtedly may vary considerably from this representation. As was the case with nitrogen and phosphorus, commercial fertilizers are important in meeting plant demands where intensive agriculture is practiced.

The annual losses of available potassium by leaching and erosion greatly exceed those of nitrogen and phosphorus. They are generally not as great, however, as the corresponding losses of available calcium and magnesium. Such losses of soil minerals have serious implications for sustainable soil productivity.

Increased Use of Potassium Fertilizers

The global use of potassium fertilizers will continue to increase as the pools of available soil potassium are depleted and increasing crop yields place demands on the soil that cannot be met by mineral weathering alone. This is especially true in row crop agriculture and in regions where sandy soils or highly weathered soils are prominent. Farmers in the United States increased their use of potassium fertilizer for many years until about 1980, by which time the potassium-supplying power of their soils (on average) had largely reached the level at which only maintenance additions were needed. This maintenance level in many regions is at least half the level of average potassium removal (the other part of removals being replaced by potassium released by mineral weathering). In less developed agricultural regions of the world, potassium fertilizer use will have to increase for many years to come if yields are to be increased or even maintained.

14.17   CONCLUSION

Soil phosphorus presents us with a double-edged sword; its management is crucially important from both an environmental standpoint and for soil fertility. On one hand, too little phosphorus commonly limits the productivity of natural and cultivated plants and is the cause of widespread soil and environmental degradation. On the other hand, industrialized agriculture has concentrated too much phosphorus in some cases, resulting in losses from soil that cause egregious eutrophication in surface waters. Some of these situations have been caused by excessive buildup of soil phosphorus with fertilizer. Others are the result of concentration of animal production, so that phosphorus in the manure produced at many livestock facilities far exceeds that required by the crops grown on the surrounding land.

Except in cases of extreme buildup, the availability of phosphorus to plant roots has a double constraint: the low total phosphorus level in soils and the small percentage of this level that is present in available forms. Furthermore, even when soluble phosphates are added to soils, they are quickly fixed into insoluble forms that in time become quite unavailable to growing plants. In acid soils, the phosphorus is fixed primarily by iron, aluminum, and manganese; in alkaline soils, by calcium and magnesium. This fixation greatly reduces the efficiency of phosphate fertilizers, with little of the added phosphorus being taken up by plants. In time, however, this unused phosphorus can build up and serve as a reserve pool for plant absorption.

Potassium is generally abundant in soils, but it, too, is present mostly in forms that are quite unavailable for plant absorption. Fortunately, however, some soils contain considerable nonexchangeable but slowly available forms of this element. Over time this potassium can be released to exchangeable and soil solution forms that can be quickly absorbed by plant roots. This is fortunate since the plant requirements for potassium are high—5 to 10 times that of phosphorus and similar to that of nitrogen.

Finally, in studying the use and management of these mineral nutrients, it is important to keep in mind the sobering truth that, while potassium supplies are quite large, the world's supply of useable phosphorus is predicted to become exhausted by the end of this century, if today's wasteful practices continue. While substitutes can be found for many finite natural resources mined from the Earth—copper in phone lines can be replaced by fiber optics, steel in car bodies by composite plastics, and petroleum in transportation fuels by biodiesel or hydrogen cells—this is *not* the case for phosphorus in food production. The U.S. Geological Survey[10] lists the following under the heading *Substitutes for Phosphate Rock*: "THERE ARE NO SUBSTITUTES FOR PHOSPHORUS . . . "

---

[10] See Jasinski (2006).

STUDY QUESTIONS

1. You have learned that nitrogen, potassium and phosphorus are all "fixed" in the soil. Compare the processes of these fixations and the benefits and constraints they each provide.

2. Assume you add a soluble phosphate fertilizer to an Oxisol and to an Aridisol. In each case, within a few months most of the phosphorus has been changed to insoluble forms. Indicate what these forms are and the respective compounds in each soil responsible for their formation.

3. How does the phosphorus content of cultivated soils in the United States compare with that of nearby forest soils that have never been cleared? What is the reason for this difference?

4. What is meant by *eutrophication,* and how is it influenced by farm practices involving phosphorus?

5. Which is likely to have the higher buffering capacity for phosphorus and potassium, a sandy loam or a clay? Explain.

6. In the spring a certain surface soil showed the following soil test: soil solution K = 20 kg/ha; exchangeable K = 200 kg/ha. After two crops of alfalfa hay that contained 250 kg/ha of potassium were harvested and removed, a second soil test showed soil solution K = 15 kg/ha and exchangeable K = 150 kg/ha. Explain why there was not a greater reduction in soil solution and exchangeable K levels.

7. What is the effect of soil pH on the availability of phosphorus, and what are the unavailable forms at the different pH levels?

8. What is *luxury consumption* of plant nutrients, and what are its advantages and disadvantages?

9. How does phosphorus that forms relatively insoluble inorganic compounds in soils find its way into streams and other waterways?

10. The incorporation of large amounts of wheat straw into a soil may bring about P deficiency in the following crop. What is the likely reason for this?

11. Compare the organic P levels in the upper horizons of a forested soil with those of a nearby soil that has been cultivated for 25 years. Explain the difference.

12. For establishment of new turf in a golf fairway, the recommendations from soil testing call for the application of 1000 kg of limestone powder, 200 kg of triple super phosphate fertilizer, and 200 kg of urea. The materials are to be incorporated by raking into the upper 4 cm of soil. The superintendent suggests that all three materials be mixed together and then applied in a single operation to save time and money. Is this a good or bad idea? Give a detailed explanation based on chemistry.

REFERENCES

Bittman, S., C. G. Kowalenko, D. E. Hunt, T. A. Forge, and X. Wu. 2006. "Starter phosphorus and broadcast nutrients on corn with contrasting colonization by mycorrhizae," *Agron. J.,* **98**:394–401.

Brady, N. C. 1974. *The Nature and Properties of Soils,* 8th ed. (New York: Macmillan).

Buresh, R. J., P. C. Smithson, and D. T. Hellums. 1997. "Building soil phosphorus capital in Africa," in R. J. Buresh, P. A. Sanchez, and F. Calhoun (eds.), *Replenishing Soil Fertility in Africa.* SSSA Special Publication No. 51 (Madison, Wis.: Soil Science Society of America).

Cogger, C., and J. M. Duxbury. 1984. "Factors affecting phosphorus loss from cultivated organic soils," *J. Environ. Qual.,* **13**:111–114.

David, M. B., and L. E. Gentry. 2000. "Anthropogenic inputs of nitrogen and phosphorus and riverine export for Illinois, USA," *J. Environ. Qual.,* **29**:494–508.

Eghball, B., G. D. Binford, and D. D. Baltensperger. 1996. "Phosphorus movement and adsorption in a soil receiving long-term manure and fertilizer application," *J. Environ. Qual.,* **25**:1339–1343.

Emsley, J. 2002. The 13th Element: The Sordid Tale of Murder, Fire, and Phosphorus (New York: John Wiley & Sons), 352 pp.

Evans, C. E., and O. J. Attoe. 1948. "Potassium supplying power of virgin and cropped soils," *Soil Sci.,* **66**:323–334.

FAO. 2006. FAOSTAT. Food and Agriculture Organization of the United Nations. http://faostat.fao.org/ (verified 20 August 2006).

Fares, F., J. C. Fardeau, and F. Jacquin, 1974. "Quantitative survey of organic phosphorus in different soil types," *Phosphorus and Agric.,* **63**:25-41.

Fox, R. L. 1981. "External phosphorus requirements of crops," in *Chemistry in the Soil Environment.* ASA Special Publication No. 40 (Madison, Wis.: American Society of Agronomy and Soil Science Society of America), pp. 223–239.

Gholston, L. E., and C. D. Hoover. 1948. "The release of exchangeable and nonexchangeable potassium from several Mississippi and Alabama soils upon continuous cropping," *Soil Sci. Soc. Amer. Proc.,* **13**:116–121.

Goenadi, D. H., Siswanto, and Y. Sugiarto. 2000. "Bioactivation of poorly soluble phosphate rocks with a phosphorus-solubilizing fungus," *Soil Sci. soc. Amer. J.,* **64**:927–932.

Grewal, J. S., and J. S. Kanwar. 1976. *Potassium and Ammonium Fixation in Indian Soils* (review) (New Delhi, India: Indian Council for Agricultural Research).

Jasinski, S. M. 2006. "Phosphate rock." U.S. Department of the Interior, U.S. Geological Survey, Reston, Va, pp. 124–125. http://minerals.usgs.gov/minerals/pubs/commodity/phosphate_rock/phospmcs06.pdf.

Kuo, S. 1988. "Application of modified Langmuir isotherm to phosphate sorption by some acid soils," *Soil Sci. Soc. Amer. J.,* **52**:97–102.

Mandzak, J. M., and J. A. Moore. 1994. "The role of nutrition in the health of inland Western forests," *J. Sustainable Forestry,* **2**:191–210.

McLean, E. O. 1978. "Influence of clay content and clay composition on potassium availability," pp. 1–19, in G. S. Sekhon (ed.), *Potassium in Soils and Crops* (New Delhi, India: Potash Research Institute of India).

Mengel, K., and E. A. Kirkby. 2001. *Principles of Plant Nutrition,* 5th ed. (Dordrecht, Netherlands: Kluwer Academic Publishers).

Mengel, K., Rahmatullah, and H. Dou. 1998. "Release of potassium from the silt and sand fractions of loess-derived soils," *Soil Science,* **163**:805–813.

Miller, W. W., D. W. Johnson, T. M. Loupe, J. S. Sedinger, E. M. Carroll, J. D. Murphy, R. F. Walker, and D. Glass. 2006. "Nutrients flow from runoff at burned forest site in Lake Tahoe basin," *Calif. Agric.,* **60**:65–71.

Moore, P. A., T. C. Daniel, and D. R. Edwards. 2000. "Reducing phosphorus runoff and inhibiting ammonia loss from poultry manure with aluminum sulfate," *J. Environ. Qual.,* **29**:37–49.

Munson, R. D. (ed.). 1985. *Potassium in Agriculture* (Madison, Wis.: American Society of Agronomy).

Nair, K.P.P. 1996. "The buffering power of plant nutrients and effects on availability," *Advances in Agronomy,* **57**:237–287.

Nair, V. D., A. A. Graetz, and K. M. Portier. 1995. "Forms of phosphorus in soil profiles from dairies of South Florida," *Soil Sci. Soc. Amer. J.,* **59**:1244–1249.

Nowak, C. A., R. B. Downard, Jr., and E. H. White. 1991. "Potassium trends in red pine plantations at Pack Forest, New York," *Soil Sci. Soc. Amer. J.,* **55**:847–850.

Porter, R. M., J. E. Ayers, M. W. Johnson, Jr., and P. E. Nelson. 1981. "Influence of differential phosphorus accumulation on corn stalk rot," *Agron. J.,* **73**:283–287.

Raven, K. P., and L. R. Hossner. 1993. "Phosphorus desorption quantity-intensity relationships in soils," *Soil Sci. Soc. Amer. J.,* **57**:1501–1508.

Richter, D. D., D. Markewitz, C. G. Wells, H. L. Allen, R. April, P. R. Heine, and B. Urrego. 1994. "Soil chemical change during three decades in an old-field Loblolly pine (*Pinus taeda* L.) ecosystem," *Ecology,* **75**:1463–1473.

Sample, E. C., F. E. Khasawneh, E. C. Sample, and E. J. Kamprath(e) Sanchez, P. A., K. D. Shepherd, M. J. Soule, F. M. Place, R. J. Buresh, and A-M. N. Izak. 1980. "Reactions of phosphate fertilizers in soils," in F. E. Khasawneh et al. (eds.), *The Role of Phosphorus in Agriculture* (Madison, Wis.: American Society of Agronomy).

Sanchez, P. A. 2002. "Soil fertility and hunger in Africa," *Science,* **295**:2019–2020.

Sanchez, P. A. et al. 1997. "Soil fertility replenishment in Africa: An investment in natural resource capital," in R. J. Buresh, P. A. Sanchez, and F. Calhoun (eds.), *Replenishing Soil Fertility in Africa.* SSSA Special Publication No. 51 (Madison, Wis.: Soil Science Society of America).

Sharpley, A. 2000. "Phosphorus availability," pp. D-18–D-38, in M. E. Summer (ed.), *Handbook of Soil Science* (New York: CRC Press).

Sharpley, A. N. 1990. "Reaction of fertilizer potassium in soils of differing mineralogy," *Soil Sci.*, **49**:44–51.

Sharpley, A., and B. Moyer. 2000. "Phosphorus forms in manure and compost and their release during simulated rainfall," *J. Environ. Qual.*, **29**:1462–1469.

Smith, S. J., A. N. Sharpley, J. W. Naney, W. A. Berg, and O. R. Jones. 1991. "Water quality impacts associated with wheat culture in the Southern Plains," *J. Environ. Qual.*, **20**:244–249.

Soltanpour, P. N., R. L. Fox, and R. C. Jones. 1988. "A quick method to extract organic phosphorus from soils," *Soil Sci. Soc. Amer. J.*, **51**:255–256.

Stevenson, F. J. 1986. *Cycles of Soil Carbon, Nitrogen, Phosphorus, Sulfur, and Micronutrients* (New York: Wiley).

Sui, Y., and M. L. Thompson. 2000. "Phosphorus sorption, desorption, and buffering capacity in a bio-solids-amended Mollisol," *Soil Sci. Soc. Amer. J.*, **64**:164–169.

Tekalign, M., I. Haque, and C. S. Kamara. 1988. "Phosphorus status of some Ethiopian highland Vertisols," pp. 232–252, in Jutzi S. C., I. Haque, J. McIntyre, and J. E. S. Stares (eds.), *Management of Vertisols in Sub-Saharan Africa*. Proceedings of a conference held at the International Livestock Research Center (ILCA), Addis Ababa, Ethiopia, 31 August–4 September, 1987. ILCA, Addis Ababa.

Tiessen, H. (ed.). 1995. *Phosphorus in the Global Environment—Transfers, Cycles and Management* (New York: John Wiley and Sons), 480 pp. http://www.icsu-scope.org/downloadpubs/scope54/TOC.htm.

Torbert, H. A., K. W. King, and R. D. Harmel. 2005. "Impact of soil amendments on reducing phosphorus losses from runoff in sod," *J. Environ Qual.*, **34**:1415–1421.

Vaithiyanathan, P., and D. L. Correll. 1992. "The Rhode River watershed: Phosphorus distribution and export in forest and agricultural soils," *J. Environ. Qual.*, **21**:280–288.

Weil, R. R., P. W. Benedetto, L. J. Sikora, and V. A. Bandell. 1988. "Influence of tillage practices on phosphorus distribution and forms in three Ultisols," *Agron. J.*, **80**:503–509.

Soils provide mountain goat with trace elements via plants. (R. Weil)

15

CALCIUM, MAGNESIUM, AND TRACE ELEMENTS

Look and you will find it . . . what is unsought will go undetected . . .
—SOPHOCLES

Calcium and magnesium are the two most abundant cations on the exchange complex in most soils. These elements in their exchangeable and weatherable forms influence all ecosystems through their critical role in counteracting soil and water acidification. As discussed in Chapter 9, these nonacid cations enhance pH buffering, lower the levels of acid cation saturation, and are the cations supplied in various ground limestones used to raise the pH of acid soils. Calcium and magnesium are also essential nutrients for plant, animal, and microbial life. They are especially important in helping plants overcome a wide range of environmental stresses.

Eight elements (iron, manganese, zinc, copper, boron, molybdenum, nickel, and chlorine) are essential for plant growth, but required in such small quantities that they are called *micronutrients*. This term must not be construed to imply that these nutrients are somehow less important than macronutrients. To the contrary, too little or too much of any one micronutrient can stimulate dramatic effects in terms of competitive shifts in plant species, stunted growth, low yields, dieback, and even plant death. When they are needed, very small applications of micronutrients can produce striking results.

Micronutrient deficiencies are becoming increasingly commonplace in agriculture as a consequence of higher levels their of removal by ever more productive crops, but also as a result of reduced inadvertent application of these elements in fertilizers and organic amendments. Low levels of these elements in food crops can have widespread negative impacts on human health.

Trace element is a more general term used to describe elements present in tissues and other environmental samples in very small concentrations. Trace elements include the micronutrients just mentioned, as well as such elements as cobalt and vanadium, which are not universally essential but can improve the growth of some plant species. Animals, including humans, also require most of these elements in their diets. The trace elements selenium, chromium, tin, iodine, and fluorine serve as essential micronutrients for animals but apparently not for plants. Some trace elements, such as arsenic and cadmium, are of environmental interest primarily because of their potentially toxic effects. Levels of trace elements toxic to plants or to animals may result from natural soil conditions,

639

from pollution, or from soil-management practices. Excess levels of certain trace elements can render plants unsafe to eat and can lead to toxic water pollution.

This chapter will provide some of the background and tools needed to recognize and deal effectively with the varied deficiencies and toxicities encountered in a wide range of soil–plant systems. The problems of trace elements in contaminated soils will be addressed in Chapter 18.

15.1 CALCIUM AS ESSENTIAL NUTRIENT[1]

Calcium (Ca) is a macronutrient essential for all plants. The ability of a soil to supply this element is intimately tied to soil acidity because this nonacid cation helps buffer soil pH and reduce aluminum saturation (Sections 9.3–9.4). The calcium status of soils has a major influence on the species composition and productivity of terrestrial ecosystems. For animals, the calcium content of the plants they eat is important because calcium is a major component of bones and teeth and plays important roles in many physiological processes. Before the application of calcium transformed the region into the thriving center of South America's agricultural economy, the acid soils of the Brazilian Cerrado region were so infertile that the area was considered a wasteland. The soils were so low in calcium that ranchers trying to graze herds on the Cerrado vegetation would lose cattle from broken bones due to calcium deficiency. It has even been suggested that the relatively higher calcium status of soils in Africa compared to those in South America may account for the occurrence of such large herbivores as elephants, zebras, and giraffes in semiarid savannas in Africa but not in South America.

Calcium in Plants

AMOUNTS TAKEN UP. Plants generally use Ca in amounts second only to nitrogen and potassium; however, the amount of calcium in plant foliage varies widely from as low as 0.1 to as high as 5% of the dry matter. Most monocots are considered to be *calcifuge* (calcium avoiding) plants and grow well with 0.15 to 0.5% Ca in leaf tissues. Many dicots are considered to be *calcicoles* (Latin: *chalk dwelling*) and need 1 to 3% Ca in their leaves for optimal growth. Trees store a great deal of calcium in their woody tissues; the net calcium uptake by many trees is close to that of nitrogen (see Table 16.6).

PHYSIOLOGICAL ROLES. Calcium is a major component of the middle lamella of cell walls, Ca–pectates giving the wall much of its stiffness. Calcium is also intimately involved with cell elongation and division, membrane permeability, and the activation of several critical enzymes. Through its role in maintaining the integrity of the cell membranes, calcium is critical for protecting the cell against toxicities of other elements.

In contrast to most nutrients, calcium is taken up almost exclusively by young root tips whose endodermis cells have not yet matured to the point of becoming impermeable (suberized). Redistribution of calcium within the plant occurs mainly with the transpiration water in the xylem, rather than in the phloem. These two facts account for the restricted movement of calcium to growing tissues that are not drawing water by transpiration (that is, to fruits rather than to leaves).

DEFICIENCY SYMPTOMS. Deficiencies of calcium are quite rare for most plants, except in very acid soils. When calcium deficiency does occur, it is usually associated with growing points (meristems) such as buds, unfolding leaves, fruits, and root tips (Figure 15.1). Under calcium deficiency, the root system is often shorter and denser than normal. When calcium is deficient, normally harmless levels of other metals can become toxic to the plant—including nutrient metals like magnesium, zinc, or manganese, as well as nonnutrient metals such as aluminum. Under such conditions, calcium-deficient roots become severely stunted and gelatinous (Figure 15.2). These root effects can cause sensitive forest trees to loose branches and eventually die. In very acid soils, Ca deficiency is often accompanied by Al or Mn toxicity, and the related effects on plants are difficult to tell apart.

---

[1] For a review of calcium in soils and plants, see Chapter 11 in Mengel and Kirby (2001). For a brief interview with renowned soil scientist Pedro Sanchez on the change that calcium made in the Cerrado, see Taylor (2005). For a summary of calcium depletion in acidified forests, see Lawrence and Huntington (1999).

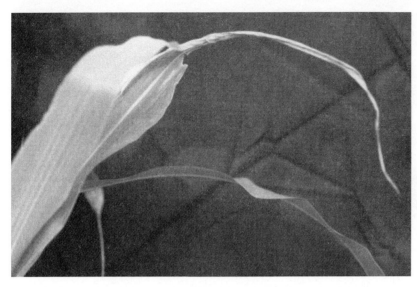

FIGURE 15.1 Young leaves that stick together and fail to properly unfold are a typical symptom of calcium deficiency in monocots such as shown here in sorghum–sudangrass, a hybrid forage species. Foliar symptoms of calcium deficiency are not commonly seen. However, they may occur in very acid soils where plants are likely to also suffer from aluminum toxicity and other problems. (Photo courtesy of R. Weil)

Certain plants show Ca deficiency, even when the soil pH is adequately maintained by liming. Such deficiencies are often related to the transport of Ca within the plant. Blossom-end rot is an exceptionally disheartening example. Imagine a gardener about to pick a delicious-appearing ripe tomato—only to find that the bottom (blossom-end) of the fruit is a soft, black, rotten mass (Figure 15.3). This disorder, especially common in melons and tomatoes, is caused by inadequate Ca for the cell walls of the expanding fruit. It is usually associated with unevenness in the water supply that interrupts the flow of Ca. Similarly, a peanut grower may have healthy-appearing plants, but find that the pods are empty shells (called pops) with no peanut inside. Again, the cause is inadequate Ca transport to the developing seed. The remedy is to supply lime or gypsum directly to the shallow layer of soil in which the pods develop for direct uptake through the shell rather than via the root system located deeper in the soil. Calcium applications, sometimes made directly to the tree foliage, are often required to avoid bitter pit symptoms on apples, which give the fruit brown, corky-textured and bitter-tasting flesh.

Soil Forms and Processes

FORMS IN SOILS. Calcium in the soil is found mainly in three pools that resupply the soil solution: (1) calcium-containing minerals (such as calcite or plagioclase), (2) calcium complexed with soil humus, and (3) calcium held by cation exchange on the clay and humus colloids. The cycling of calcium among these and other soil pools, and the gains and losses of calcium by such mechanisms as plant uptake, atmospheric deposition as dust and soot, liming, and leaching, comprise the calcium cycle, as illustrated in Figure 15.4. In most soils, the principal sources of Ca for plant uptake are (1) exchangeable Ca and (2) Ca in

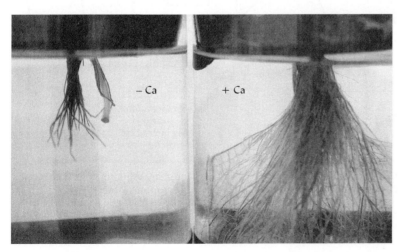

FIGURE 15.2 Root growth was almost completely inhibited by low calcium in the nutrient solution (*left*) compared to healthy roots in the same nutrient solution but with calcium added (*right*). If the ratio of calcium to all other cations in solution drops below 5:1, the integrity of root membranes is lost, causing many other elements to become toxic to the plants. (Photo courtesy of R. Weil)

FIGURE 15.3 Blossom-end rot on tomato caused by inadequate calcium supply to the fruit. Note the black, rotten bottom side of the two tomato fruits. The problem may be associated with low availability of calcium and is usually aggravated by irregular soil water availability. (Photo courtesy of R. Weil)

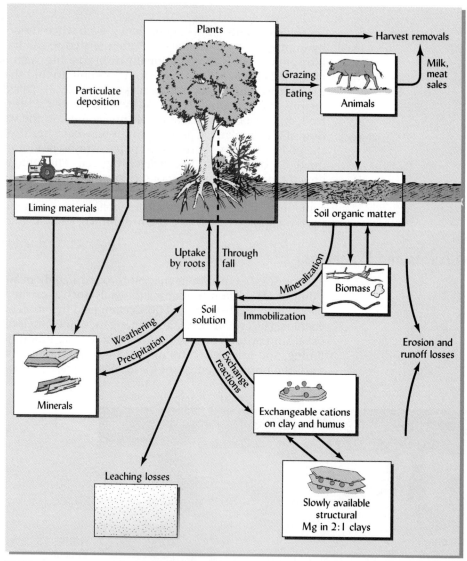

FIGURE 15.4 Simplified diagram of the cycling of calcium and magnesium in soils. The rectangular compartments represent pools of these elements in various forms, while the arrows represent processes by which the elements are transformed or transported from one pool to another. (Diagram courtesy of R. Weil)

readily weathered minerals (such as carbonates and apatite). In arid and semiarid regions, the high pH, high carbonate nature of the soil solution greatly diminishes the solubility of calcium-containing minerals.

Downwind from "dirty" coal-burning industries without effective environmental controls, considerable calcium is deposited in particulate air pollution. Deposition of dust from wind erosion of calcareous desert soils can make substantial contributions to soil calcium thousands of kilometers downwind. Such calcium deposition can partially offset acidification caused by nitrogen and sulfur deposition (see Sections 13.14 and 13.17). Central and southeastern China provide some of the most extreme examples of Ca deposition as environmental policies struggle to keep pace with rapid industrialization. For example, the combination of desert dust and industrial coal burning deposits as much as 120 kg Ca ha^{-1} yr^{-1} on soils near Tie Shan Ping in central China. Yet, with over 200 kg ha^{-1} yr^{-1} of combined nitrogen and sulfur deposition, the deposited calcium succeeds only in raising the mean pH of the rain from about 3.0 to about 4.1, leaving a very serious acidification problem.

Fungi may help trees mine Ca: http://news.bbc.co.uk/1/hi/sci/tech/2040623.stm

LEACHING LOSSES FROM CROPLAND.   The need for repeated applications of limestone in humid regions (Section 9.8) suggests significant losses of calcium and magnesium from the soil. Table 15.1 illustrates losses of these elements by leaching compared with those from crop removal and soil erosion. Note that the total loss from all three causes (leaching, erosion, and crop removal) in humid-region agricultural soils, expressed in the form of carbonates, approaches 1 Mg/ha per year. Typical recommendations in humid regions suggest about 1 Mg/ha should be applied every four years to maintain soil pH, once the desired level is achieved. This amount of lime may replace exchangeable Ca lost by leaching, but not that which is lost by crop removal and erosion. Presumably, in most soils the difference is made up by weathering of Ca from minerals.

LOSSES FROM FORESTS.   Similarly, timber harvest methods that leave the soil open to erosion and nutrient leaching lead to rapid losses of calcium (and magnesium) from forest ecosystems. Even relatively undisturbed forests in the humid eastern United States may lose calcium by leaching and removal in trees at rates that exceed inputs by atmospheric deposition (Table 15.1). Scientists are concerned that for some soils in the humid regions,

TABLE 15.1 Annual Calcium and Magnesium Losses from Cropland and Forestland in Humid Regions

Note that the annual combined loss of Ca and Mg from the cropland was nearly 1 Mg/ha, expressed as carbonates. For the eight eastern U.S. forested watersheds, the average annual input of Ca from atmospheric deposition (6 kg/ha) is enough to approximately offset only the leaching losses of Ca, but not the removal of Ca in timber harvest.

Ecosystem and manner of loss	Loss, kg/ha/yr	
	Calcium	Magnesium
Cropland in Missouri (silt loam, 4% slope)		
Erosion by water	95	33
Crop removal in standard rotation	50	25
Leaching	115	25
Total from cropland	260 (651 as CaCO$_3$)	83 (291 as MgCO$_3$)
	942 total carbonates	
Apple orchard in New York—leaching losses	170	84
Douglas fir watersheds, loss in streams		
From clear-cut and slash/burned	81	26
From undisturbed watershed	26	8
Average of 4 eastern U.S. mixed hardwood forests		
Net tree removal (uptake-litterfall)	13	NR[a]
Leaching loss in streams	5	NR
Average of 4 eastern U.S. pine forests		
Net tree removal (uptake-litterfall)	5	NR
Leaching loss in streams	7	NR

[a] NR = not reported.
Eastern U.S. forest data calculated from Lawrence and Huntington (1999); apple orchard from Richards et al. (1998); others from Brady and Weil (1996).

TABLE 15.2 **Cascade of Calcium-related Effects from Acid Deposition on Forests**

Biogeochemical response to acid deposition	Physiological response	Effect on forest function
Leaching of Ca from leaf membrane	Reduced cold tolerance of needles in red spruce	Loss of current-year needles in red spruce
Reduction of the ratios of Ca to Al and Ca to Mn in soil and in soil solutions	Dysfunction in fine roots of red spruce, blocking uptake of Ca	Decreased growth and increased susceptibility to stress in red spruce
Reduction of the ratios of Ca to Al and Ca to Mn in soil and in soil solutions	Greater energy use to acquire Ca in soils with low Ca:Al ratios	Decreased growth and increased photosynthate allocation to roots
Reduction of the availability of nutrient cations in marginal soils	Sugar maples on drought-prone or nutrient-poor soils are less able to withstand stresses	Episodic dieback and growth impairment in sugar maple

Modified from Fenn et al. (2006).

the release of Ca from mineral weathering will not be able to keep up with the losses and that acid rain combined with intensive timber harvesting may be depleting the Ca reserves in the more poorly buffered watersheds. The acidification affects a number of biogeochemical processes involving calcium. These processes in turn influence tree physiology, which ultimately impacts the ecological functioning of forests (Table 15.2). Nonetheless, research in this area is still inconclusive and up to now forests have rarely shown a positive growth response to applied calcium.

15.2 MAGNESIUM AS A PLANT NUTRIENT

Magnesium in Plants

ROLES IN PLANT. Plants generally take up Mg in similar or somewhat smaller amounts (0.15 to 0.75% of dry matter) than Ca. About one-fifth of the magnesium in plant tissue is found as the central component of the chlorophyll molecule and so is intimately involved with photosynthesis in plants. Magnesium also plays critical roles in the synthesis of oils and proteins and in the activation of enzymes involved in energy metabolism. The Mg^{2+} ion forms a bridge connecting ATP molecules (which provide energy for cellular reactions, see Section 14.1) to the enzymes that catalyze numerous physiological processes involving phosphorylation.

DEFICIENCY SYMPTOMS. The deficiency of Mg is much more common than that of Ca, at least when the soil pH is at an appropriate level. The most common symptom of Mg deficiency is interveinal chlorosis on the older leaves, which appears as a mottled green and yellow coloring in dicots (Plate 89) and a striping in monocots. Because Mg (unlike Ca) is readily translocated in most plants from the older to the younger, still-growing leaves, the oldest leaves are the first to be affected by low Mg supplies. Common on very sandy soils with low CEC, these symptoms are sometimes termed *sand drown* as they appear somewhat like those caused by oxygen-starvation in a waterlogged soil.

Forages with low contents of Mg compared to Ca and K can cause grazing animals to suffer from a sometimes-fatal Mg deficiency known as *grass tetany*. High levels of K can aggravate this problem by reducing Mg uptake.

Spruce and fir trees growing on soils low in exchangeable Mg have exhibited reduced Mg in needle tissue, stunted growth, and needles that turn yellow, especially on the tips. Broadleaf trees severely deficient in Mg exhibit interveinal chlorosis on the older leaves, which in citrus trees sometimes coalesces into large, brilliant yellow-colored patches on either side of the midrib.

Magnesium in Soil

FORMS IN SOIL. The main source of plant-available Mg in most soils is the pool of exchangeable Mg on the clay-humus complex (Figure 9.3). As plants and leaching remove this Mg, the easily exchangeable pool is replenished by Mg weathered from minerals (such as

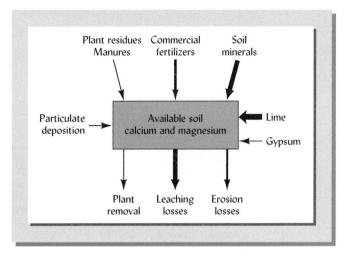

FIGURE 15.5 The principal ways by which calcium and magnesium are supplied to and removed from the plant-available pool in soils. The relative thickness of the arrows indicate that major losses in agricultural systems are usually through leaching and erosion and are usually replaced mainly by the addition of liming materials, with manures, mulches, and certain types of fertilizers providing secondary sources. In forest systems, losses by erosion and leaching are relatively small and may be equaled or exceeded by plant uptake and eventual loss in timber (especially whole tree) harvest. Calcium-rich particulate deposition (dust) provides inputs that are of considerable significance for many forest ecosystems.

dolomite, biotite, hornblende, and serpentine). In some soils, replenishment also takes place from a pool of slowly available Mg in the structure of certain 2:1 clays (see Section 8.3). Variable amounts of Mg are made available by the breakdown of plant residues and soil organic matter. In unpolluted forests, research suggests that atmospheric deposition, rather than rock weathering, may supply much of the magnesium used by trees. The inputs and outputs to the pools of plant-available calcium and magnesium in soils are summarized in Figure 15.5.

Ratio of Calcium to Magnesium[2]

Being less tightly held (more easily leached) than Ca, exchangeable Mg commonly saturates only 5 to 20% of the effective CEC, as compared to the 60 to 90% typical for Ca in neutral to moderately acid soils (see Figure 9.5). Some agriculturists believe that optimum plant growth and soil tilth require a ratio of exchangeable Ca:Mg very near 6:1 (65% Ca and 10% Mg saturation of the CEC). This belief can lead to the wasteful use of soil amendments in an effort to achieve this so-called "ideal" Ca:Mg ratio (see Figure 16.38). Numerous research studies have shown that, in fact, plants grow very well and meet their Ca and Mg needs in soils with Ca:Mg ratios anywhere from 1:1 to 15:1. Well-documented research shows that soil aggregation and biological activity are also largely unaffected by a similarly wide range in this ratio. However, even though plant and soil health are not likely to be affected by soil Ca:Mg ratios, the ratio of Ca to Mg content in plant tissue may be altered enough to influence the nutrition of grazing animals, and Ca or Mg mineral supplements may be needed in animal diets.

Soils formed on serpentine rock, which is rich in Mg but contains little or no Ca, offer an unusual but dramatic exception to the above statements. Exchangeable Mg in these soils is typically 3 to 9 times as plentiful as exchangeable Ca, giving a Ca:Mg ratio much smaller than 1.0. Serpentine-derived soils may, therefore, exhibit severe imbalances between Ca and Mg, causing severe deficiency of Ca and toxicity of Mg for all but the few plant species that evolved to grow in this unique soil environment.

15.3 DEFICIENCY VERSUS TOXICITY

It is axiomatic that anything can be toxic if taken in large enough amounts. At low levels of a nutrient, deficiency and reduced plant growth may occur (*deficiency range*). As the level of nutrient is increased, plants respond by taking up more of the nutrient and increasing their growth. If a level of nutrient availability has been reached that is sufficient to meet the plants' needs (*sufficiency range*), raising the level further will have little effect on plant growth, although the concentration of the nutrient may continue to increase in the plant tissue. At some level of availability, the plant will take up too

---

[2] For an objective review of how Ca/Mg ratios and cation balancing in general became widely, but wrongly, promoted for fertility management, see Kopittke and Menzies (2007).

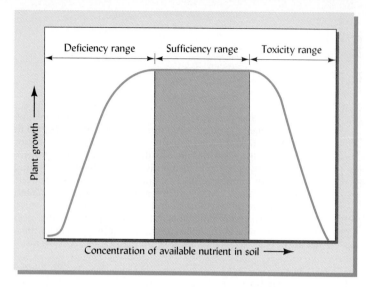

FIGURE 15.6 The relationship between the amount of a micronutrient available for plant uptake and the growth of the plant. Within the deficiency range, as nutrient availability increases so does plant growth (and uptake, which is not shown here). Within the sufficiency range, plants can get all of the nutrient they need, and so their growth is little affected by changes within this range. At higher levels of availability a threshold is crossed into the toxicity range, in which the amount of nutrient present is excessive and causes adverse physiological reactions that lead to reduced growth and even death of the plant. (Diagram courtesy of R. Weil)

much of the nutrient for its own good (*toxicity range*), causing adverse physiological reactions to take place. The relationship among deficient, sufficient, and toxic levels of nutrient availability is described in Figure 15.6.

The sufficiency range is very broad, for macronutrients and toxicity seldom occurs. However, for micronutrients the difference between deficient and toxic levels may be very narrow, making the possibility of toxicity quite real. For example, in the cases of boron and molybdenum, severe toxicity may result from applying as little as 3 to 4 kg/ha of available nutrient to a soil initially deficient in these elements. While the sufficiency range for other micronutrients is much wider and toxicities are not as likely from overfertilization, toxicities of copper, zinc, and nickel have been observed on soils contaminated by industrial sludges, manure from concentrated hog facilities, metal smelter wastes, and long-term application of copper sulfate fungicide. Toxicity of manganese is quite common in association with low soil pH (see Section 9.7). In certain forests it appears that the high ratios of manganese to magnesium or calcium are associated with the decline and death of sensitive tree species (Table 15.3).

High levels of molybdenum may occur naturally in certain poorly drained alkaline soils. In some cases, enough of this element may be taken up by plants to cause toxicity, not only to susceptible plants but also to livestock grazing forages on these soils. Boron, too, may occur naturally in alkaline soils at levels high enough to cause plant toxicity. Although somewhat larger amounts of most other micronutrients are required and can be tolerated by plants, great care needs to be exercised in applying micronutrients, especially for maintaining nutrient balance.

Deficiency and toxicity symptoms:
http://hort.ufl.edu/teach/orh3254/DefSymptoms.htm

TABLE 15.3 Magnesium and Manganese in Soil and Leaves in Relation to Declining Sugar Maples in Northern Pennsylvania Forests

Sugar maple death is thought to be related to soil acidification from nitrogen deposition. Ratios of Mg/Mn in soil and in tree foliage were much better indicators of adverse conditions on these soils than were corresponding ratios with aluminum.

Soil Great Group, parent material	Dead sugar maples %	A horizon pH$_w$	Ca	Mg	Mn	Mg/Mn ratio	Mg	Mn	Mg/Mn ratio
			–Exchangeable ions in soil –				— Leaf content —		
			— cmol$_c$ kg^{-1} —				— % —		
Oxyaquic Glossudalfs, glacial till	3	5	4.4	0.627	0.033	19	0.172	0.048	9.2
Aquic Hapludults, shale and sandstone	64	4.08	0.66	0.123	0.121	1.0	0.082	0.254	0.7

Data selected from Kogelmann and Sharpe (2006).

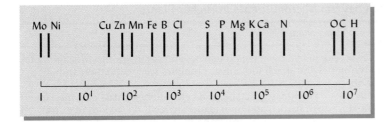

FIGURE 15.7 Relative numbers of atoms of the essential elements in alfalfa at bloom stage, expressed logarithmically. Note that there are more than 10 million hydrogen atoms for each molybdenum atom. Even so, normal plant growth would not occur without molybdenum. [Modified from Viets (1965)]

In addition, irrigation water in dry regions may contain enough dissolved boron, molybdenum, or selenium to damage sensitive crops, even if the original levels of these trace elements in the soil were not very high. Therefore, it is prudent to monitor the content of these elements in water used for irrigation (see Section 10.8). Selenium does not appear essential for plants, but it is required by animals and can be toxic to both (see Section 15.9).

15.4 MICRONUTRIENT ROLES IN PLANTS[3]

Micronutrients are required in very small quantities, their concentrations in plant tissue being one or more orders of magnitude lower than for the macronutrients (Figure 15.7). The ranges of plant tissue concentrations considered deficient, adequate, and toxic for several micronutrients are illustrated in Figure 15.8.

Increasing attention is being directed toward micronutrient deficiencies for several reasons:

1. Intensive plant production practices have increased crop yields, resulting in greater removal of micronutrients from soils.

2. The trend toward more concentrated, high analysis fertilizers has reduced the use of impure salts and organic manures, which formerly supplied significant amounts of micronutrients.

3. Increased knowledge of plant nutrition and improved methods of analysis in the laboratory are helping in the diagnosis of micronutrient deficiencies that might formerly have gone unnoticed.

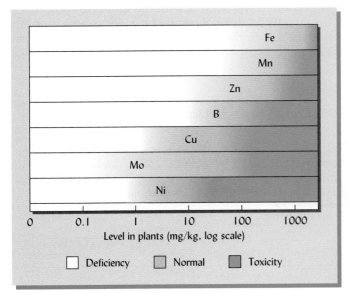

FIGURE 15.8 Deficiency, normal, and toxicity levels in plants for seven micronutrients. Note that the range is shown on a logarithmic scale and that the upper limit for manganese is about 10,000 times the lower range for molybdenum and nickel. In using this figure, keep in mind the remarkable differences in the ability of different plant species and cultivars to accumulate and tolerate different levels of micronutrients. (Based on data from many sources)

[3] For an excellent overview of micronutrients in global agriculture, see Fageria et al. (2002). For a detailed review of their physiological roles in plants, see Welch (1995). For a comprehensive reference for soil and plant aspects of micronutrients in agriculture, see Mortvedt et al. (1991).

TABLE 15.4   Functions of Several Micronutrients in Higher Plants

Micronutrient	Functions in higher plants
Zinc	Present in several dehydrogenase, proteinase, and peptidase enzymes; promotes growth hormones and starch formation; promotes seed maturation and production.
Iron	Present in several peroxidase, catalase, and cytochrome oxidase enzymes; found in ferredoxin, which participates in oxidation-reduction reactions (e.g., NO_3^- and SO_4^{2-} reduction and N fixation); important in chlorophyll formation.
Copper	Present in laccase and several other oxidase enzymes; important in photosynthesis, protein and carbohydrate metabolism, and probably nitrogen fixation.
Manganese	Activates decarboxylase, dehydrogenase, and oxidase enzymes; important in photosynthesis, nitrogen metabolism, and nitrogen assimilation.
Nickel	Essential for urease, hydrogenases, and methyl reductase; needed for grain filling, seed viability, iron absorption, and urea and ureide metabolism (to avoid toxic levels of these nitrogen-fixation products in legumes).
Boron	Activates certain dehydrogenase enzymes; facilitates sugar translocation and synthesis of nucleic acids and plant hormones; essential for cell division and development.
Molybdenum	Present in nitrogenase (nitrogen fixation) and nitrate reductase enzymes; essential for nitrogen fixation and nitrogen assimilation.
Cobalt	Essential for nitrogen fixation; found in vitamin B_{12}.
Chloride	Essential for photosynthesis and enzyme activation. Plays role in regulation of water uptake on salt-affected soils.

4. Increasing evidence indicates that food grown on soils with low levels of trace elements may provide insufficient human dietary levels of certain elements, even though the crop plants themselves show no signs of deficiency.

Physiological Roles in Plants

Micronutrients play many complex roles in plant nutrition. While most of the micronutrients participate in the functioning of a number of enzyme systems (Table 15.4), there is considerable variation in the specific functions of the various micronutrients in plant and microbial growth processes. For example, copper, iron, and molybdenum are capable of acting as electron carriers in the enzyme systems that bring about oxidation-reduction reactions in plants. Such reactions are essential steps in photosynthesis and many other metabolic processes. Zinc and manganese function in many plant enzyme systems as bridges to connect the enzyme with the substrate upon which it is meant to act. Apparently, through its catalysis of certain enzymes, a sufficient supply of manganese plays an important role in the mechanisms by which plants defend themselves from pathogen attack (Table 15.5).

TABLE 15.5   Effect of Manganese Levels on the Incidence of Root Rot and Root Peroxidase Activity in Cowpea[a]

Manganese catalyses peroxidase, an exo-enzyme that aids the synthesis of both lignin and monophenols. The lignin acts as a mechanical and chemical barrier against fungal invasion, while monophenols are fungal toxins that may also act in defense of the plant. Mn fertilizer controlled root rot almost as well as the chemical fungicide Carbendazim.

Treatment	Cowpea tissue conc.		Disease incidence 10 days after sowing (%)	Peroxidase activity index
	Mn (mg/kg)	N (%)		
Mn, 0 mg/kg soil	45	3.15	75	16
Mn, 5 mg/kg soil	78	2.54	47	24
Mn, 10 mg/kg soil	92	2.85	44	27
Carbendazim, 0.2% on seeds	45	3.43	31	31

[a] Grown in low-fertility, alkaline (pH 8.6) loamy sand inoculated with *Rhizoctonia bataticola*.
Data selected from Kalim et al. (2003).

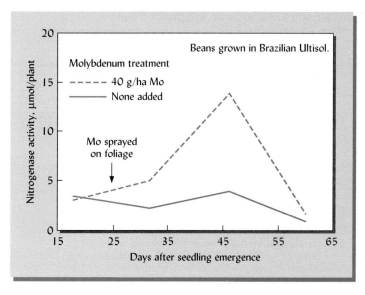

FIGURE 15.9 An example of a major benefit coming from a small application of a micronutrient. Note that the spraying of an extremely small quantity of molybdenum (only 40 g/ha) on bean leaves dramatically increased the plant level of the enzyme nitrogenase that is critical for the process of nitrogen fixation. Molybdenum is essential for this process, which makes available for plants and animals globally more than one hundred million metric tons of reactive nitrogen every year. [From Vieria et al. (1998); used with permission of Marcel Dekker, Inc., New York]

Molybdenum and manganese are essential for certain nitrogen transformations in microorganisms as well as in plants. Molybdenum and iron are components of the enzyme *nitrogenase*, which is essential for the processes of symbiotic and nonsymbiotic nitrogen fixation (Figure 15.9). Molybdenum is also present in the enzyme *nitrate reductase*, which is responsible for the reduction of nitrates in soils and plants.

Nickel has only recently been added to the list of elements shown to be essential to higher plants. It is essential for the function of several enzymes, including *urease*, the enzyme that breaks down urea into ammonia and carbon dioxide. Nickel-deficient legumes accumulate toxic levels of urea in their leaves; seeds of cereal plants deficient in nickel are not viable and fail to germinate.

Zinc plays a role in protein synthesis, in the formation of some growth hormones, and in the reproductive process of certain plants. Copper is involved in both photosynthesis and respiration and in the use of iron. It also stimulates lignification of cell walls. The roles of boron have yet to be clearly defined, but boron appears to be involved with cell division, water uptake, and sugar translocation in plants. Iron is involved in chlorophyll formation and degradation and in the synthesis of proteins and nucleic acids. Manganese seems to be essential for photosynthesis, respiration, and nitrogen metabolism.

The role of chlorine is still somewhat obscure; however, it is known to influence photosynthesis and root growth. Cobalt is essential for the symbiotic fixation of nitrogen. In addition, legumes and some other plants have a cobalt requirement independent of nitrogen fixation, although the amount required is small compared to that for the nitrogen-fixation process.

Managing micronutrients:
http://web.missouri.edu/
~umcsnrsoilwww/313_
W2004/microl_2004.htm

Deficiency Symptoms

Insufficient supply of a nutrient is commonly expressed by visible plant symptoms that can be used to diagnose the micronutrient deficiencies (see Table 15.6). Most of the micronutrients are relatively immobile in the plant (Plate 88). That is, the plant cannot efficiently transfer the nutrient from older leaves to newer ones. Therefore, the concentration of the nutrient tends to be lowest, and the symptoms of deficiency most pronounced, in the younger leaves that develop after the supply of the nutrient has run low.[4] The iron-deficient sorghum in Plate 99 and the zinc-deficient peach leaves in Plate 97 (following page 656) illustrate the pattern of pronounced deficiency symptoms on the younger leaves.

Photos of mineral
deficiencies in plants:
http://www.hbci.com/
~wenonah/min-def/list.htm

In the case of zinc deficiency in corn (Plate 98), broad white bands on both sides of the midrib are typical in young corn plants, but the symptoms may disappear as the soil warms up and the maturing plant's root system expands into a larger volume of

---

[4] This pattern is in contrast to most of the macronutrients (except sulfur), which are more easily translocated by the plants and so become most deficient in the older leaves.

Micro-nutrient	Symptoms of deficiency	Symptoms of toxicity
Fe	Young leaves show interveinal chlorosis, distinct green veins, entire leaf is white or yellow if severe. Roots grow profuse root hairs, bottle-brush effect.	Bronze or black discoloration of leaf margins, dark, slimy roots, mainly on submerged soils.
Mn	Young leaves, interveinal yellowing with greener veins, tissue near veins remains green, mottled appearance on dicots. Dead tissue in later stages. No lateral roots.	Dark green leaves with red flecks early, later bronze/yellow interveinal tissue. Patchy green colors. May induce Fe deficiency.
Zn	Young mature or old leaves depending on plant show yellowing, usually interveinal, rosetting and dwarfing of leaves, broad white band on either side of corn leaf. Dark, shriveled leaves on turfgrass.	Pale leaves with necrotic interveinal lesions. Leaf tips water-soaked. Increased lateral roots, dense root system, barbed-wire pattern.
Cu	Young leaves yellow, rolled or dead leaf tips, stunted leaves and plants. Drooping leaves on trees.	Stunted plants, pale leaves—may induce Fe deficiency. Red coloration on leaf margins. Roots short with barbed-wire pattern.
Ni	Newest leaves chlorotic, small "mouse ear" leaves on pinnate-leafed trees, death of meristems.	Young leaflets distorted, white interveinal banding, dark green veins, irregular oblique streaking or white stripes brown patches. Brown, stunted roots.
Mo	May mimic N deficiency in legumes, mottled yellowing in young leaves, narrow, whip-like leaves in brassicas, whitish color, leaf tip death.	Reddish colors along leaf margins, normal-appearing roots.
B	Growing points of shoots and roots die. Reddish young leaves. Malformed buds, necrosis of internal tissues in fleshy stems, tubers, seeds. Stubby, bushy root system.	Interveinal chlorosis, marginal necrosis with distinct boundaries. Relative normal root appearance.
Cl	Yellowing and reduced leaf size. Later bronzing and necrosis. In cereals elongated necrotic patches.	Leaf tip and margin burn (necrosis), resembles K deficiency, but on younger leaves. Death of root tips.

Sources: Fageria et al. (2002); Baligar et al. (1998); Rashid and Ryan (2004); and others.

soil. Interveinal chlorosis (darker green veins and light yellowish areas between the veins) on the younger leaves and the whorl of tiny leaves at the terminal end of the branch (a symptom known as *little leaf*) are characteristic of tree foliage suffering from too little zinc.

Nickel deficiency has recently been found to be the cause of the commonly seen "mouse ear" symptom of malformed, small-sized leaflets on pecan trees growing in sandy, low-nickel soils (Figure 15.10).

FIGURE 15.10 Nickel deficiency on a pecan tree growing on sandy, low-nickel soil in Georgia. The left side of the tree was sprayed with a solution of NiSO$_4$ · 6H$_2$O just before leafing out in spring, obviously correcting the deficiency. The right side was not sprayed with Ni. The inset is a close-up view of the characteristic Ni deficiency symptom on pecan trees, a malformation called "mouse-ear" due to the shape of the stunted, malformed leaves. In some older pecan orchards, accumulation of zinc in the soil from many years of applications to counter zinc deficiency seems to aggravate the nickel problem, possibly because Zn competes for the same uptake sites on the root membrane. [Photos courtesy of Bruce W. Wood, USDA/ARS Southeastern Fruit and Nut Tree Research Lab. For details, see Wood et al. (2004).]

(a)   (b)

FIGURE 15.11   (a) A misshapen rose, or *bullhead,* on a boron-deficient tea rose. (b) A well-formed bloom on the same plant after a solution containing boron was applied to the leaves and young buds. (Photos courtesy of R. Weil)

Boron deficiency generally affects the growing points of plants, such as their buds, fruits, flowers, and root tips. Plants low in boron may produce deformed flowers (Figure 15.11); aborted seeds; thickened, brittle, puckered leaves; or dead growing points. Young leaves may turn red on some plants (Plate 100). In all of these cases, a small dose of boron applied at the correct time could make the difference between a marketable or an unmarketable plant product.

15.5 SOURCE OF MICRONUTRIENTS

Deficiencies and toxicities of micronutrients may be related to the total contents of these elements in the soil. More often, however, these problems result from the chemical forms of the elements in the soil and, particularly their solubility and availability to plants.

Inorganic Forms

Sources of the eight micronutrients vary markedly from area to area. The wide variability in the content of these elements in soils and suggested contents in a representative soil is shown in Table 15.7.

All of the micronutrients have been found in varying quantities in igneous rocks. Two of them—iron and manganese—have prominent structural positions in primary silicate minerals, such as biotite and hornblende. Others, such as cobalt and zinc, also may occupy structural positions as minor replacements for the major constituents of silicate minerals, including clays.

The mineral forms of micronutrients are altered as mineral decomposition and soil formation occur. Oxides and, in some cases, sulfides of elements such as iron, manganese, and zinc are formed (see Table 15.7). Secondary silicates, including the clay minerals, may contain considerable quantities of iron and manganese and smaller quantities of zinc and cobalt. Ultramafic rocks, especially serpentinite, are high in nickel. The micronutrient cations released as weathering occurs are subject to colloidal adsorption, just as are the calcium or aluminum ions (see Section 8.7).

Anions such as borate and molybdate in soils may undergo adsorption or reactions similar to those of the phosphates. For boron, the adsorption may be represented as follows:

$$\boxed{\text{Soil colloid}}\begin{matrix}-\text{O}\\[-0.3em]\\-\text{O}\end{matrix}\!\!\!\!\diagdown\!\!\diagup\!\text{B}-\text{OH}$$

TABLE 15.7 **Major Sources of Plant Micronutrients, Ranges and Representative Contents of These Nutrients in Soils, Along with Their Representative Contents in Harvested Crops**

The ratio of soil to crop contents emphasizes the primary need to increase the efficiency of plants in absorbing these nutrients from the soil.

Element	Major sources	Range, kg/ha/15 cm	Representative contents of Soil, kg/ha/15 cm	Crop, kg/crop	Soil/crop ratio
Fe	Oxides, sulfides, silicates	20,000–220,000	56,000	2	28,000
Mn	Oxides, silicates, carbonates	45–9,000	2200	0.5	4,400
Zn	Sulfides, carbonates, silicates	25–700	110	0.3	366
Cu	Sulfides, hydroxy carbonates, oxides	4–2,000	45	0.1	450
Ni	Silicates (e.g., serpentine), (Fe Ni), Si	10–2,200	45	0.02	2,250
B	Borosilicates, borates	8–200	22	0.2	110
Mo	Sulfides, oxides, molybdates	0.4–10	5	0.02	250
Cl	Chlorides	15–100	22	2.5	9

Chlorine, by far the most soluble of the group, is added to soils in considerable quantities each year through rainwater. Its incidental addition to soils in fertilizers and in other ways helps prevent the deficiency of chlorine under field conditions. Chlorine is only very weakly adsorbed to soil colloids.

Organic Forms

Organic matter is an important secondary source of some of the trace elements. Several of them tend to be held as complex combinations by organic (humus) colloids. Copper is especially tightly held by organic matter—so much so that its availability can be very low in organic soils (Histosols). In uncultivated profiles, there is a somewhat greater concentration of micronutrients in the surface soil, much of it presumably in the organic fraction. Correlations between soil organic matter and contents of copper, molybdenum, and zinc have been noted. Although the elements thus held are not always readily available to plants, their release through decomposition is undoubtedly an important fertility factor. Animal manures are a good source of micronutrients, much of it present in organic forms.

Forms in Soil Solution

The dominant forms of micronutrients that occur in the soil solution are listed in Table 15.8. The specific forms present are determined largely by the pH and by soil aeration (i.e., redox potential). Note that the cations are present in the form of either simple cations or hydroxy metal cations. The simple cations tend to be dominant under highly acid conditions. The more complex hydroxy metal cations become more prominent as the soil pH is increased.

Molybdenum is present mainly as MoO_4^{2-}, an anionic form that reacts at low pH in ways similar to those of phosphorus (see Chapter 14). Although boron also may be present

TABLE 15.8 **Forms of Micronutrients Dominant in the Soil Solution**

Micronutrient	Dominant soil solution forms
Iron	Fe^{2+}, $Fe(OH)_2^+$, $Fe(OH)^{2+}$, Fe^{3+}
Manganese	Mn^{2+}
Zinc	Zn^{2+}, $Zn(OH)^+$
Copper	Cu^{2+}, $Cu(OH)^+$
Molybdenum	MoO_4^{2-}, $HMoO_4^-$
Boron	H_3BO_3
Cobalt	Co^{2+}
Chlorine	Cl^-
Nickel	Ni^{2+}, Ni^{3+}

From data in Lindsay (1972).

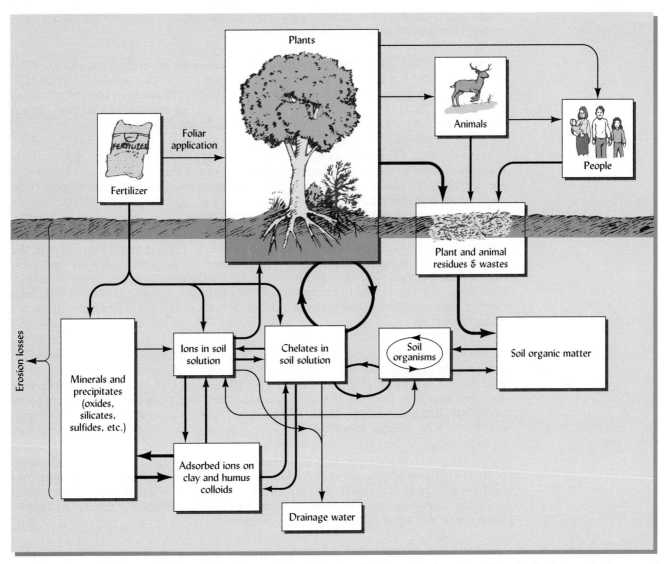

FIGURE 15.12 Cycling and transformations of micronutrients in the soil–plant–animal system. Although all micronutrients may not follow each of the pathways shown, most are involved in the major components of the cycle. The formation of chelates, which keep most of these elements in soluble forms, is a unique feature of this cycle. (Diagram courtesy of R. Weil.)

in anionic form at high pH levels, research suggests that undissociated boric acid (H_3BO_3) is the form that is dominant in the soil solution and is absorbed by plants.

The cycling of micronutrients through the soil–plant–animal system is illustrated in a generalized way by Figure 15.12. Although not every micronutrient will participate in every pathway shown in this figure, it can be seen that organic chelates, soil colloids, soil organic matter, and soil minerals all contribute micronutrients to the soil solution and, in turn, to growing plants. As we turn our attention to micronutrient availability, it will be helpful to refer back to Figure 15.12 to see the relationships among the processes involved.

15.6 GENERAL CONDITIONS CONDUCIVE TO TRACE ELEMENT DEFICIENCY/TOXICITY

Deficiencies

Micronutrients are most apt to limit plant growth in the following types of soil conditions.

LEACHED, ACID, SANDY SOILS. Strongly leached, acid, sandy soils are low in most micronutrients for the same reasons they are deficient in most of the macronutrients—their parent materials were initially low in the elements, and acid leaching has removed much of the small quantity of micronutrients originally present. In the case of molybdenum, acid soil conditions markedly decrease availability.

ORGANIC SOILS.   The micronutrient contents of organic soils depend on the extent of the washing or leaching of these elements into the bog area as the soils were formed. In most cases, this rate of movement is too slow to produce deposits as high in micronutrients as are the surrounding mineral soils. The ability of organic soils to bind certain elements, notably copper, also accentuates micronutrient deficiencies.

INTENSIVE CROPPING.   Intensive cropping, in which large amounts of plant nutrients are removed in the harvest, accelerates the depletion of micronutrient reserves in the soil and increases the likelihood of micronutrient deficiencies. Such depletion is most common where high yields are produced with the aid of chemical fertilizers that supply some of the macronutrients, but not the micronutrients.

EXTREMES OF pH.   Soil pH, especially in well-aerated soils, has a decided influence on the availability of all the micronutrients except chlorine. Under acid conditions, molybdenum is rendered unavailable, while most trace element cations are freely available, sometimes at toxic levels. In contrast, at high pH levels, the availability of molybdenum is high (sometimes too high), while deficiencies of the trace element cations are all too common.

ERODED SOILS.   Soil erosion can influence micronutrient availability. Erosion of topsoil carries away considerable soil organic matter, in which much of the potentially available micronutrients are held. Also, removal of the topsoil exposes subsoil horizons that are often higher in pH than the topsoil, a condition that leads to deficiencies of some micronutrients, such as zinc. Eroded ridges or hillsides are common sites of micronutrient deficiencies in some areas.

PARENT MATERIALS.   Trace element deficiencies and toxicities are often related to the level of these elements in the parent materials from which the soils form or to the minerals in the watersheds that might be transported by irrigation or floodwaters to the soil in question. Attention must be given to means of either supplying micronutrient supplements or to removing the soluble excess nutrients.

WASTE DISPOSAL.   Trace elements are common constituents of industrial and domestic wastes that are often applied to soils. The macronutrient contents of these wastes (particularly those of nitrogen and phosphorus) have long been appreciated, but only recently have we recognized the significance of their trace element contents. Small quantities of trace elements applied in these wastes can help alleviate nutrient deficiencies. However, repeated land applications of large quantities of wastes have increased the soil levels of some trace elements to their toxicity ranges. These levels are adversely affecting not only the plants, but the animals that consume them as well.

15.7 FACTORS INFLUENCING THE AVAILABILITY OF THE TRACE ELEMENT CATIONS

Trace element cations including iron, manganese, zinc, copper, and nickel are each influenced in a characteristic way by the soil environment. However, certain soil factors have the same general effects on the availability of all of them.

Soil pH

The micronutrient cations are most soluble and available under acid conditions (Figure 15.13). In fact, under these conditions, the soil solution concentrations or activities of one or more of these elements (most commonly manganese) are often sufficiently high as to be toxic to common plants. As indicated in Chapter 9, one of the primary reasons for liming acid soils is to reduce the solubility of manganese and aluminum.

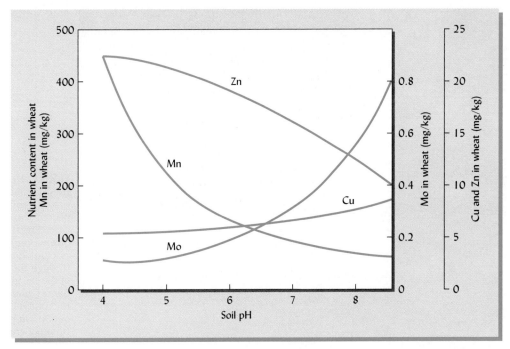

FIGURE 15.13 Effect of soil pH on the concentrations of manganese, zinc, copper, and molybdenum in wheat plants. The soils were from different countries around the world. The molybdenum levels are extremely low, but increase with increasing pH. Manganese and zinc levels decrease as the pH rises, while copper is little affected. [Redrawn from Sillanpaa (1982)]

As the pH is increased, the ionic forms of the micronutrient cations are changed first to the hydroxy ions and, finally, to the insoluble hydroxides or oxides of the elements. The following example uses the ferric ion as typical of the group:

$$Fe^{3+} \xrightarrow{OH^-} Fe(OH)^{2+} \xrightarrow{OH^-} Fe(OH)_2^+ \xrightarrow{OH^-} Fe(OH)_3 \qquad (15.1)$$

Simple cation (soluble)   Hydroxy metal cations (soluble)   Hydroxide (insoluble)

Iron fertilization:
http://web.missouri.edu/
~umcsnrsoilwww/313_W2004/
fe2_2004.htm

All of the hydroxides of the micronutrient cations are relatively insoluble—some more so than others. The exact pH at which precipitation occurs varies from element to element and between oxidation states of a given element. For example, the higher valence states of iron and manganese form hydroxides that are much more insoluble than their lower-valence counterparts. In any case, the principle is the same—at low pH values, the solubility of micronutrient cations is high, and as the pH is raised, their solubility and availability to plants decrease. Overliming of an acid soil often leads to deficiencies of iron, manganese, zinc, copper, and sometimes boron. Such deficiencies associated with high pH occur naturally in many of the calcareous soils of arid regions.

The general desirability of a slightly acid soil (with a pH between 6 and 7) largely stems from the fact that for most plants, this pH condition allows micronutrient cations to be soluble enough to satisfy plant needs without becoming so soluble as to be toxic (see Figure 9.22). Certain plants, especially those that are native to very acid soils, have only a poor ability to take up iron and other micronutrients unless the soil is quite acid (pH about 5). Such acid-loving plants therefore become deficient in these elements when the soil pH is such that iron solubility is lowered (usually above pH 5.5; see Figure 15.14). Section 9.7 gives more specific information on the pH preferences of various plants.

Zinc availability, like that of iron, is reduced when soil pH is raised by liming. However, in addition to the pH effects, the addition of high-magnesium liming materials (e.g., dolomitic limestone) further decreases zinc availability, because zinc is tightly adsorbed to dolomite and magnesium carbonate crystal surfaces. Zinc deficiency may also be aggravated by interaction between zinc and magnesium in the plant.

FIGURE 15.14 The chlorotic foliage of these azaleas is a sign of iron deficiency inadvertently induced by a too-high soil pH. In this case, the iron deficiency was caused by the use of marble gravel as a decorative mulch. Marble consists mostly of calcium carbonate. Rainwater percolating through the gravel mulch dissolved enough of the calcium to lime the soil below, raising the pH gradually from 5.2 to 6.0. At pH 6.0 the solubility of iron is too low for azaleas to obtain what they need. Calcium leaching from concrete walkways can have a similar effect on acid-loving vegetation growing in adjacent soil. (Photo courtesy of R. Weil)

Oxidation State and pH

The trace element cations iron, manganese, nickel, and copper occur in soils in more than one valence state. In the lower valence states, the elements are considered *reduced;* in the higher valence state they are *oxidized*. Metallic cations generally become reduced when the oxygen supply is low, as occurs in wet soils containing decomposable organic matter. Reduction can also be brought about by organic metabolic reducing agents, such as NADPH or caffeic acid, produced by plants and microorganisms in the soil. Oxidation and reduction reactions in relation to soil drainage are discussed in Sections 7.4 and 7.5.

The changes from one valence state to another are, in most cases, brought about by microorganisms and organic matter.[5] In some cases, the organisms may obtain their energy directly from the inorganic reaction. For example, the oxidation of manganese from Mn(II) in manganous oxides (MnO) to Mn(IV) in manganic oxides (MnO_2) can be carried out by certain bacteria and fungi. In other cases, organic compounds formed by microbes or plant roots may be responsible for the oxidation or reduction.

Diagnosing copper deficiency:

http://www.lagric.gov.ab.ca/$department/deptdocs.nsf/all/agdex3476?opendocument

INTERACTION OF SOIL REACTION AND AERATION. At pH values common in soils, the oxidized states of iron, manganese, and copper are generally much less soluble than are the reduced states. The hydroxides (or hydrous oxides) of these high-valent forms precipitate even at low pH values and are extremely insoluble. For example, the hydroxide of trivalent ferric iron precipitates at pH values of 3.0 to 4.0, whereas ferrous hydroxide does not precipitate until a pH of 6.0 or higher is reached.

The interaction of soil acidity and aeration in determining micronutrient availability is of great practical importance. Iron, manganese, and copper are generally more available under conditions of restricted drainage or in flooded soils (Figure 15.15). Very acid soils that are poorly drained may supply toxic quantities of iron and manganese. Manganese toxicity has been reported to occur when certain high-manganese acid soils are thoroughly wetted during irrigation. Andisols (volcanic soils) with high-organic-matter melanic epipedons are known to cause manganese toxicity problems when they are wet by heavy rains. Iron toxicity is common in flooded rice paddies. Such toxicity is much less apt to occur under well-drained conditions, unless soil pH is very low. The influence of aeration can be quite complex in the case of zinc (Box 15.1).

---

[5] For a review of microbial reduction of certain trace elements, see Lovley (1996).

PLATE 61   A mini-profile of urban turf reveals a thick thatch layer.

PLATE 62   Water-stressed soybeans show extent of oak tree root competition for water.

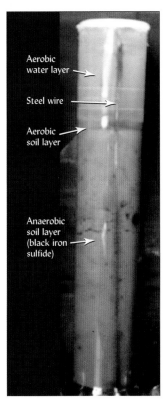

Aerobic water layer

Steel wire

Aerobic soil layer

Anaerobic soil layer (black iron sulfide)

PLATE 63   Oxidized (orange iron oxide) and reduced (black iron sulfide) zones along steel wire embedded in Winogradsky column.

Topsoil

Ap1

Ap2

25

Sand layer

'2C

50

'2Bs

Sulfuric horizon

'3Bw

75

Sulfidic clay layer

'3Cgd

PLATE 64   Profile of 6 year old landfill cap with layers of topsoil, sand and (sulfidic) clay.

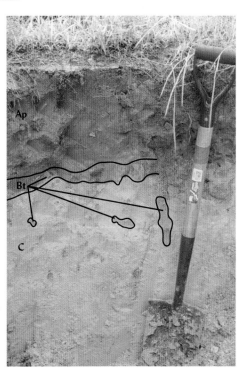

Ap

Bt

C

PLATE 65   Ultisol profile in Maryland truncated by 300 years of erosion under tillage.

PLATE 66   Grass turned green by septage, a sign of septic drain field malfunction.

PLATE 67   Dark green grass growing over septic drain lines. Texas.

PLATE 68   Soil quality kit tests for soil respiration, bulk density and infiltration.

PLATE 69   Finger flow- the uneven water infiltration and movement due to hydrophobic organic coatings on sand particles.

PLATE 70   Forest floor or O horizons in Vermont.

PLATE 71   The darker surface soil was brushed aside to expose a hydrophobic layer caused by burning the Chaparral vegetation. Water beads up rather than soaking into this layer (see p. 294).

PLATE 72   Contour strip fields and grassed waterway (arrow) in New York.

PLATE 73   Sod includes about 1 cm of soil removed from the sod farm.

PLATE 74   Efficient drip irrigation of a young apple orchard in Mexico.

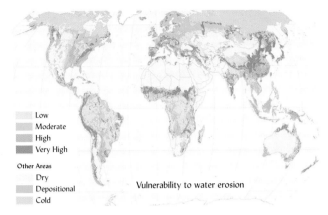

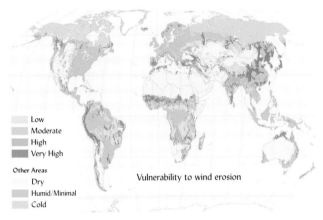

Low
Moderate
High
Very High

Other Areas
Dry
Depositional
Cold

Vulnerability to water erosion

PLATE 75   Global soil vulnerability to erosion by water.

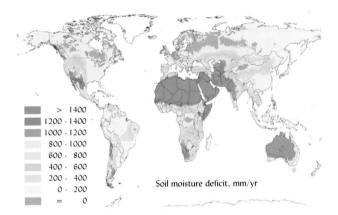

> 1400
1200 - 1400
1000 - 1200
800 - 1000
600 - 800
400 - 600
200 - 400
0 - 200
= 0

Soil moisture deficit, mm/yr

PLATE 76   Annual soil moisture deficit (in mm).

Low
Moderate
High
Very High

Other Areas
Dry
Humid/Minimal
Cold

Vulnerability to wind erosion

PLATE 77   Global soil vulnerability to erosion by wind.

PLATE 78 Constructed Entisols (world reference group, Technosols) in urban plaza, including layers of geotextile and sand.

PLATE 79   House in Africa made of several kinds of soil.

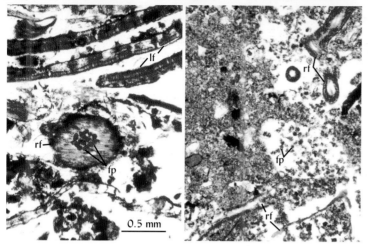

PLATE 80 Influence of soil fauna on microstructure in O (*left*) and A (*right*) horizons of a forested Ultisol in Tennessee. Particulate organic matter (POM) includes leaf fragments (lf), fecal pellets (fp) and root fragments (rf).

PLATE 81 Roots from sweet pepper plants follow organic matter-lined earthworm burrows through the compacted B subsoil of a Pennsylvania Inceptisol. Earthworm activity was encouraged by 15 years of no-till practices and cover crops.

PLATE 82 Plant roots grow along the gleyed coating of a fragipan prism whose reddish interior is too dense for roots to grow and the roots are squeezing flat between prisms.

PLATE 83 This cicada nymph nestled 60 cm deep in the B horizon of a forested Ultisol will feed by sucking sap from oak tree roots for several years before emerging as an adult. Free water in the macropore bathes the cicada whose burrowing promotes drainage and enhances root growth.

PLATE 84 Extension of tree roots far beyond the drip line can be seen by competition for water between trees and grass in the surface soil. Contrast this pattern of drought stress to that shown in Plate 85.

PLATE 85 Drought stressed grass shows importance of soil depth. The rectangular area of brown grass is underlain by a shallow (25 cm) layer of soil atop the roof of an underground library. The trees and green grass at right grow in deeper soil.

PLATE 86 The soil on the right of this hydrangea was limed, that on the left was acidified (with FeSO$_4$). After a year, blue flowers formed on the low pH side, pink on the high pH side.

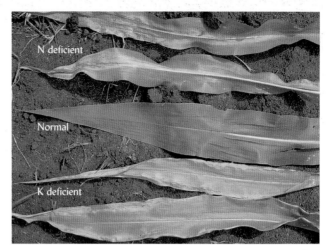

PLATE 87 Leaves from near the bottom of N deficient (yellow tip and mid-rib), K deficient (necrotic leaf edges) and normal corn plants. All the leaves came from the same field.

PLATE 88 This iron deficient azalea was sprayed with FeSO$_4$ on one side three days before being photographed. Soil pH higher than 5.5 can induce such Fe deficiency.

PLATE 89 Magnesium deficiency causes interveinal chlorosis on the older leaves. Poinsettia.

PLATE 90 Phosphorus deficiency causes severe stunting and purpling of older leaves. Tomato.

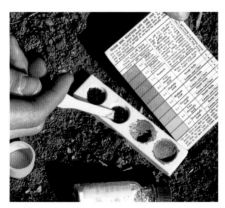

PLATE 91 After wetting soil with pH-sensitive dye, the color is compared to a chart (yellow at pH 4.0 to purple at pH 8.5) in order to estimate soil pH in the field. This soil has a pH of about 7. See page 372.

PLATE 92 Sulfur deficiency typically causes chlorosis (yellowing) on the *youngest* leaves first, or on all the foliage, as in the sorghum plant on the left.

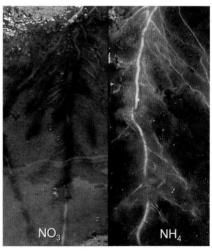

PLATE 93 Dye shows grass plant rhizosphere pH as affected by nitrogen form used. Color key as in Plate 91.

PLATE 94 Nitrogen-deficient corn on Udolls in central Illinois. Ponded water after heavy rains resulted in nitrogen loss by denitrification and leaching.

PLATE 95 Slow-moving coastal plain stream choked with algal bloom caused by nitrogen and phosphorus from upstream farmland.

PLATE 96 Normal (*left*) and phosphorus-deficient (*right*) corn plants. Note stunting and purple color.

PLATE 97 Zinc deficiency on peach tree. Note whorl of small, misshaped leaves.

PLATE 98 Zinc deficiency on sweet corn. Note broad whitish bands.

PLATE 99 Eroded calcareous soil (Ustolls) with iron-deficient sorghum.

PLATE 100 Boron deficiency on alfalfa. Note reddish foliage.

PLATE 101 Pink blooms belong to pioneering redbud (*Cercis canadensis* L.) trees, a nitrogen-fixing legume that enriches the soil for the other species (which eventually will take over as the forest matures).

PLATE 102 Phosphorus-deficient grape leaves.

PLATE 103 The large soybean root nodule was cut open to show its red interior indicative of active nitrogen fixation. The red comes from an iron coordinated compound very similar to the hemoglobin that makes human blood turn red when oxygenated.

PLATE 104 Looking like snow, the salt crust covering this soil formed when salt-laden groundwater in this salt marsh rose by capillarity and evaporated from the soil surface, leaving the dissolved salt behind. Near the Great Salt Lake in Utah.

PLATE 105 Sea spray has caused salt injury (brown leaves) despite the high salt tolerance of this Bermuda grass at Pebble Beach Golf Course in California.

PLATE 106 It's a good thing this homeowner readjusted his spreader before he finished fertilizing the lawn. Salt "burn" from too much N fertilizer.

PLATE 107 Iron deficiency causes yellowing with sharply contrasting green veins on the younger leaves. Rose growing in soil with pH 6.8.

PLATE 108 Stream polluted by acid drainage caused by sulfuricization of soils forming in coal mine spoil. $FeSO_4$ in the acid drainage oxides in the stream to causes the orange color. See p. 380.

PLATE 109 Early stages of soil formation in material dredged from Baltimore Harbor. Sulfidic materials (black), acid drainage (orange liquid), salt accumulations (whitish crust), and initiation of prismatic structure (cracks) are all evident.

PLATE 110 Landsat Thematic Mapper image of Washington, D.C. (*upper left*), and sediment-laden Potomac River (*center*). Composite image with natural colors. See page 857.

PLATE 111 Landsat Thematic Mapper image of Palo Verde Valley, California, irrigation scheme. Composite image using bands 2, 3, and 4. Lush vegetation appears bright red. See page 860.

Plates 1–4, 7, 10, 11, 13, 15–23, 25–36, 38–53, 55–63, 65–68, 70–74, 78–79, 81–92, 94–101 and 103–109 courtesy of Ray Weil; plates 8 and 12 courtesy of R.W. Simonson; plates 5 and 14 courtesy of Chien-Lu Ping, Agriculture and Forestry Experiment Station, University of Alaska—Fairbanks; plates 6 and 9 courtesy of Soil Science Society of America; plate 24 courtesy of Carlos F. Dorronsoro Fernandez, Univ. of Granada, Spain; plates 53, 54 and 102 courtesy of P.R. Schreiner, Oregon State Univ.; plate 64 courtesy of Chris Smith, USDA-NRCS; plate 69 courtesy of Stefan Doerr, Univ. of Swansea Wales; plates 75 and 77 courtesy of USDA/NRCS, plate 76 from Tao et al (2003) courtesy of Swedish Academy of Science, plate 80 courtesy of Debra Phillips, Oak Ridge National Laboratory, Tennessee, plate 93 courtesy of Joseph Heckman, Rutgers Univ.; plate 110 courtesy Space Imaging, Inc.; plate 111 courtesy of Earth Satellite Corp, Rockville, Md.

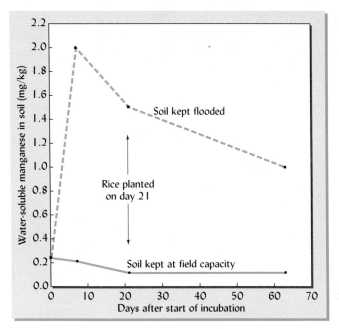

FIGURE 15.15 Effect of flooding on the amount of water-soluble manganese in soils. The data are the averages for 13 unamended Ultisol horizons with initial pH values ranging from 3.9 to 7.1. [Data from Weil and Holah (1989)]

At the high end of the soil pH range, good drainage and aeration often have the opposite effect. Well-oxidized calcareous soils are sometimes deficient in available iron, zinc, or manganese even though adequate total quantities of these trace elements are present. The hydroxides of the high-valence forms of these elements are too insoluble to supply the ions needed for plant growth. In contrast, at high soil pH values, molybdenum availability may be excessively high. **Molybdenosis** is a potentially fatal disorder caused by excessive molybdenum in the diet of livestock grazing plants grown on certain very high pH soils.

There are marked differences in the sensitivity of different plant varieties to iron deficiency in soils with high pH. This is apparently caused by differences in their ability to solubilize iron immediately around the roots. Efficient varieties respond to iron stress by acidifying the immediate vicinity of the roots and by excreting compounds capable of reducing the iron to a more soluble form, with a resultant increase in its availability (Figure 15.17). It appears that most of the iron-reducing activity is concentrated in the root hairs of actively growing young roots (Figure 15.18).

Other Inorganic Reactions

Micronutrient cations interact with silicate clays in two ways. First, they may be involved in cation exchange reactions much like those of calcium or aluminum. Second, they may be tightly bound or fixed to certain silicate clays, especially the 2:1 type. Zinc, manganese, and iron ions sometimes occur in the crystal structure of these clays. Depending on conditions, they may be released from the clays or fixed by them in a manner similar to that by which potassium is fixed (see Section 14.13). The fixation may cause serious deficiency in the case of zinc, because this element is present in soil in such small quantities (see Table 15.7).

The application of large quantities of phosphate fertilizers can adversely affect the supply of some of the micronutrients. The uptake of both iron and zinc may be reduced in the presence of excess phosphates. For environmental quality reasons, as well as from a practical standpoint, phosphate fertilizers should be used in only those quantities required for good plant growth.

Lime-Induced Chlorosis

Iron deficiency in fruit trees and many other plants is encouraged by the presence of the bicarbonate ion. Bicarbonate-containing irrigation waters increase the level of this ion in some soils. In other soils, especially poorly buffered sandy soils, the problem

BOX 15.1   COMPLEX SOIL CHANGES YIELD ZINC-DEFICIENT RICE

China's rice farmers, who are adapting new systems of growing rice in upland aerated conditions instead of traditional flooded rice paddies, are running into some unforeseen micronutrient problems. Although they saw no Zn deficiency in the traditional flooded rice, they are finding their rice to be Zn deficient on the same soils once they switch to the drained, aerated system. Many soils used for flooded rice cultivation in northern China have very low levels of available zinc to begin with, less than 1.0 mg Zn/kg soil extractable by the chelating agent DTPA. (Chelating agents are often used in testing soils for micronutrient availability, as they somewhat mimic the action of the plant root, as explained in Section 15.8).

The low Zn availability of these soils is related to the high pH, high carbonate levels, and low redox potential of the flooded soils. Yet, these soils were able to make available just enough Zn to avoid deficiencies in rice grown under flooded conditions. However, when the soils are drained for the aerated rice production system, at least five changes occur. Of these, two processes are likely to increase the availability of zinc and three are likely to decrease it:

Soil changes likely to increase zinc availability:
- Iron in the soils is oxidized, causing acidification that should increase Zn solubility.
- Decomposition of soil organic matter is accelerated, releasing the Zn previously adsorbed.

Soil changes likely to decrease zinc availability:
- The oxidized iron soon precipitates as $Fe[OH]_3$, which can strongly adsorb Zn from solution, decreasing its availability.
- Aeration stimulates nitrification, causing plants to take up more NO_3^- and less NH_4^+, in turn causing less rhizosphere acidification (see Section 9.1) and lower Zn solubility.
- The drained soil contains less water-filled pore space, thus providing few pathways through which Zn can move by diffusion and mass flow toward the roots.

With so many processes working in opposite directions, the net effect on Zn bioavailability is difficult to predict on theoretical grounds, but experiments with five strains (genotypes) of rice showed that Zn is consistently less available under the aerated conditions (Figure 15.16). For some rice genotypes, reduction in Zn availability keeps their uptake below the level considered sufficient for good growth. Fortunately, several other rice genotypes are able to obtain enough Zn for their needs, even under the aerated system. Soil scientists, in collaboration with plant breeders, are attempting to sort things out to allow China's rice farmers to take advantage of the new, more productive aerated system without being held back by a micronutrient deficiency.

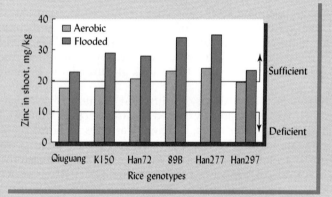

FIGURE 15.16 *Zinc concentration at tillering stage in rice plants under aerated and flooded production systems without Zn application. [From Gao et al. (2006)]*

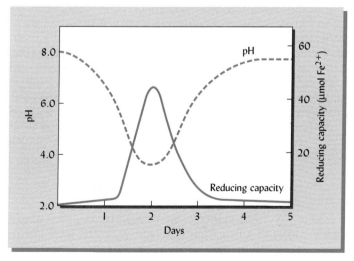

FIGURE 15.17   Response of one variety of sunflowers to iron deficiency. When the plant became stressed owing to iron deficiency, plant exudates lowered the pH and increased the reducing capacity immediately around the roots. Iron is solubilized and taken up by the plant, the stress is alleviated, and conditions return to normal. [From Marschner et al. (1974) as reported by Olsen et al. (1981); used with permission of *American Scientist*]

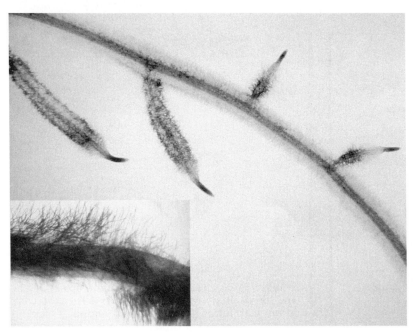

FIGURE 15.18 Reduction of iron by tomato roots is made visible by soaking live roots in a solution containing a stain that turns a dark blue color when the iron in the stain is reduced to the Fe(II) form. Note that the reduction reaction occurred only in the mature part of young branch rootlets (not in the zone just behind the root tip). The expanding root tips had not yet developed the reducing mechanisms. The close-up (*inset*) shows that the reduction reaction occurs only in the root hairs. The main part of the root remains undarkened. (Photos courtesy of Paul Bell, Louisiana State University Agricultural Center)

stems from application of more lime than is needed to reach the proper pH. The chlorosis apparently results from iron deficiency in soils with high pH because the bicarbonate ion interferes in some way with iron metabolism. This interference is coupled with the fact that iron solubility in the soil is greatly reduced with higher pH.

Organic Matter

Examples of zinc and copper:
http://www.extension.umn.edu/distribution/cropsystems/DC0720.html

Organic matter, organic residues, and manure applications affect the immediate and potential availability of micronutrient cations. Some organic compounds react with these cations to form water-insoluble complexes that can protect the nutrients from interactions with mineral particles that can bind them in even more insoluble forms. Other complexes provide slowly available nutrients as they undergo microbial breakdown. Organic complexes that enhance micronutrient availability are considered in Section 15.8. Figure 15.19 illustrates the large role that sorption on soil organic matter (and iron oxides) plays in the solubility of Cu and the much smaller role of organic matter with regard to Zn.

Deficiencies of copper and, to a lesser extent, manganese are often found on poorly drained soils high in organic matter (e.g., peats and marshes). Zinc is also retained by organic matter, but deficiencies stemming from this retention are not common.

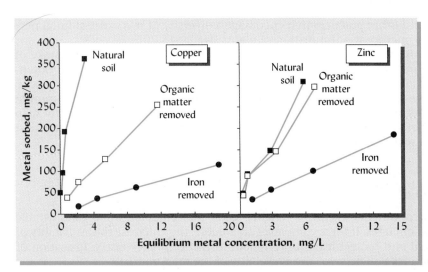

FIGURE 15.19 Sorption of Cu and Zn in a tropical Alfisol as affected by removal of the organic and iron components of the soil. Removal of organic matter greatly reduced the sorption of Cu but had little effect on Zn. The subsequent removal of "free" iron from soil particle surfaces reduced the sorption of both metals. On the iron coatings and on the organic matter the metal binding sites were more selective for Cu than for Zn. The data illustrate that reactions with organic matter and amorphous iron are the major controls on Cu sorption, whereas Zn is held mainly by the iron and probably by cation exchange reactions. [Redrawn from Agbenin and Olojo (2004)]

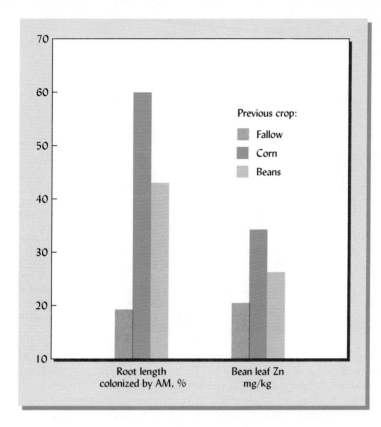

FIGURE 15.20 Effect of crop rotation on uptake of zinc and formation of mycorrhizae by beans. The bean crop, grown with furrow irrigation on an Aridisol (Calcid) in Idaho, was preceded by either a corn crop, a bean crop, or a year of bare fallow. The corn-followed-by-bean rotation favored both arbuscular mycorrhizae (AM) formation and zinc uptake by the second-year bean crop. The fallow period deprived the AM fungi of the host plants they need to survive (see Section 11.9). [Data from Hamilton et al. (1993)]

The microbial decomposition of organic plant residues and animal manures can result in the release of micronutrients by the same mechanisms that stimulate the release of macronutrient ions. As was the case for macronutrients such as nitrogen, however, temporary deficiencies of the trace elements may occur when the residues are added due to the assimilation of micronutrients in the bodies of the active microorganisms.

Micronutrient-enriched organic products have been used as a nutrient source on soils deficient in available trace elements. For example, composts of iron-enriched organic materials, such as forest by-products, peat, animal manures, and plant residues, have been found to be effective on iron-deficient soils.

Role of Mycorrhizae

A symbiosis between most higher plants and certain soil fungi produces mycorrhizae (fungus roots), which are far more efficient than normal plant roots in several respects. The nature of the mycorrhizal symbiosis and its importance in phosphorus nutrition were described in Sections 11.9 and 14.3, but it is worth mentioning here that mycorrhizae have also been shown to increase plant uptake of micronutrients (Table 11.8). Crop rotations and other practices that encourage a diversity of mycorrhizal fungi may thereby improve micronutrient nutrition of plants (Figure 15.20).

Surprisingly, mycorrhizae also appear to protect plants from excessive uptake of micronutrients and other trace elements where these elements are present in potentially toxic concentrations. Seedlings of such trees as birch, pine, and spruce are able to grow well on sites contaminated with high levels of zinc, copper, nickel, and aluminum only if their roots are sheathed by ectomycorrhizae. The mycorrhizae apparently help exclude these metallic cations from the root stele and prevent long-distance transport of metal cations within the plant.

15.8 ORGANIC COMPOUNDS AS CHELATES

Chelating agents:
http://scifun.chem.wise.edu/
chemweek/Chelates/Chelates.
html

The cationic micronutrients react with certain organic molecules to form organometallic complexes called **chelates**. If these complexes are soluble, they increase the availability of the micronutrient and protect it from precipitation reactions. Conversely, formation of an insoluble complex will decrease the availability of the micronutrient.

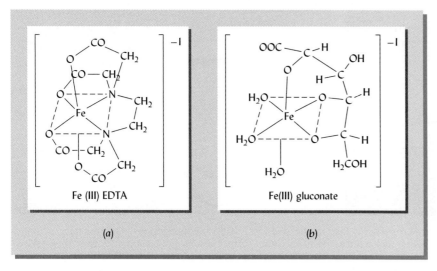

FIGURE 15.21 Structural formula for two common iron chelates, ferric ethylenediaminetetra-acetate (Fe-EDTA) (*a*) and ferric gluconate (*b*). In both chelates, the iron is protected and yet can be used by plants. [Diagrams from Clemens et al. (1990); reprinted by permission of Kluwer Academic Publishers]

A chelate (from the Greek *chele*, claw) is an organic compound in which two or more atoms are capable of bonding to the same metal atom, thus forming a ring. These organic molecules may be synthesized by plant roots and released to the surrounding soil, may be present in the soil humus, or may be synthetic compounds added to the soil to enhance micronutrient availability. In complexed form, the cations are protected from reaction with inorganic soil constituents that would make them unavailable for uptake by plants. Iron, zinc, copper, and manganese are among the cations that form chelate complexes. Two examples of an iron chelate ring structure are shown in Figure 15.21.

The effect of chelation can be illustrated with iron. In the absence of chelation, when an inorganic iron salt such as ferric sulfate is added to a calcareous soil, most of the iron is quickly rendered unavailable by reaction with hydroxide, as follows:

$$\underset{\text{(available)}}{Fe^{3+} + 3OH^-} \rightleftharpoons \underset{\text{(unavailable)}}{FeOOH + H_2O} \qquad (15.2)$$

In contrast, if the iron is chelated, it largely remains in the chelate form, which is available for uptake by plants. In this reaction, the available iron chelate reactant is favored:

$$\underset{\text{(available)}}{Fe \text{ chelate} + 3OH} \rightleftharpoons \underset{\text{(unavailable)}}{FeOOH + chelate^{3-} + H_2O} \qquad (15.3)$$

The mechanism by which micronutrients from chelates are absorbed by plants is different for different plants. Many dicots appear to remove the metallic cation from the chelate at the root surface, reducing (in the case of iron) and taking up the cation while releasing the organic chelating agent in the soil solution. Roots of certain grasses have been shown to take in the entire chelate–metal complex, reducing and removing the metallic cation inside the root cell, then releasing the organic chelate back to the soil solution (Figure 15.22). In both cases, it appears that the primary role of the chelate is to allow metallic cations to remain in solution so they can diffuse through the soil to the root. Once the micronutrient cations are inside the plant, other organic chelates (such as citrates) may be carriers of these cations to different parts of the plant.

Stability of Chelates

Some of the major synthetic chelating agents are listed in Table 15.9. Many similar chelating compounds occur naturally in the soil.

Chelates vary in their stability and therefore in their suitability as sources of micronutrients. The stability of a chelate is measured by its stability constant, which is related to the tenacity with which a metal ion is bound in the chelate. If the binding and release of a metal by a chelating agent is represented by the following reaction

$$Metal\text{-}chelate^- \rightleftharpoons metal^{2+} + chelate^{3-} \qquad (15.4)$$

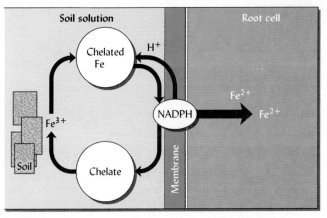 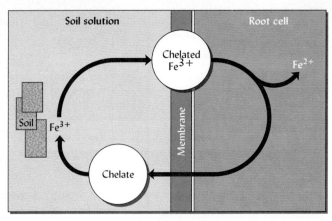

FIGURE 15.22 Two ways in which plants utilize micronutrients held in chelated form. (*Left*) Dicotyledonous plants such as cucumber and peanuts produce strong reducing agents (NADPH) that reduce iron at the outer surface of the root membrane. They then take in only the reduced iron, leaving the organic chelate in the soil solution where it can complex another iron atom. (*Right*) Grass plants such as wheat or corn apparently take the entire chelate-metal complex into their root cells. They then remove the iron, reduce it, and return the chelate to the soil solution. (Diagram courtesy of R. Weil)

then the stability constant K for the metal-chelate complex is calculated as follows, where the values in brackets are concentrations in solution:

$$K = \frac{[\text{metal-chelate}^-]}{[\text{metal}^{2+}][\text{chelate}^{3-}]} \quad (15.5)$$

Stability, chelation and the chelate effect:

http://wwwchem.uwimona.edu.jm:1104/courses/chelate.html

The larger the stability constant, the greater the tendency for the metal to remain chelated. The stability constant K for each metal-chelate complex is different and is usually expressed as the logarithm of K (see Table 15.9).

The stability constant is useful in predicting which chelate is best for supplying which micronutrient. An added metal chelate must be reasonably stable within the soil if it is to have lasting advantage. For example, the stability constant for EDDHA-Fe^{3+} is 33.9, but that for EDDHA-Zn^{2+} is only 16.8. We can therefore predict that if EDDHA-Zn were added to a soil, the Zn in the chelate would be rapidly and almost completely replaced by Fe^{3+} from the soil, leaving the Zn in the unchelated form and subject to precipitation:

$$\text{Zn chelate}^- + Fe^{3+} \rightleftharpoons \text{Fe chelate} + Zn^{2+} \quad (15.6)$$

Since the iron chelate is more stable than its zinc counterpart, the reaction goes to the right, and the released zinc ion is subject to reaction with the soil. Similarly, calcium can replace micronutrients from chelates. Even though the stability constant for Ca chelates is generally low, calcium often replaces micronutrients from chelates in

TABLE 15.9 **Stability Constants K for Selected Chelating Agents and Nutrient Cations**

The stability constants are given as logarithms, so a difference of 1.0 represents a 10-fold difference in stability. The macronutrient calcium is included because it is usually the metallic cation with by far the greatest activity in the soil solution. As such, calcium competes with micronutrient cations for binding sites in chelating agents. The relative cost of chelating agents is also given.

Chelating agent	Log K[a]						Relative cost[b]
	Fe^{3+}	Fe^{2+}	Zn^{2+}	Cu^{2+}	Mn^{2+}	Ca^{2+}	
EDTA	25.0	14.27	14.87	18.70	13.81	11.0	4.4
EDDHA	33.9	14.3	16.8	23.94	—	7.2	43
HEEDTA	19.6	12.2	14.5	17.4	10.7	8.0	5.5
Citrate	11.2	4.8	4.86	5.9	3.7	4.68	—
Gluconate	37.2	1.0	1.7	36.6	—	1.21	1.0

[a] $K = \dfrac{[\textit{metal-chelate}]}{[\textit{metal}][\textit{chelate}]}$

[b] Cost per kg of iron relative to the cost of iron gluconate. [From Clemens et al. (1990)]

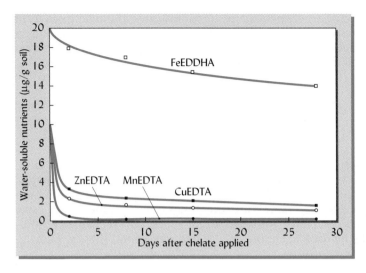

FIGURE 15.23 Average reduction in water solubility of four chelated micronutrients when incubated with four calcareous soils with pH higher than 8. The stability constants listed in Table 15.9 explain the behavior of the four micronutrient chelates shown here. The chelate with iron was most stable in these soils, that with manganese the least. Because of the small quantities needed by plants, even the copper and zinc chelates would likely provide adequate nutrients for plant absorption. [Drawn using data from Ryan and Hariq (1983)]

practice, because the concentration of Ca^{2+} in the soil solution is far greater than the concentrations of micronutrients.

It should not be inferred that only iron chelates are effective. The chelates of other micronutrients, including zinc, manganese, and copper, have been used successfully to supply these nutrients (Figure 15.23). Apparently, replacement of other micronutrients in the chelates by iron from the soil is sufficiently slow to permit absorption by plants of the other added micronutrients. Also, because foliar spray and banded applications are often used to supply zinc and manganese, the possibility of reaction of these elements in chelates with iron and calcium in the soil can be reduced or eliminated.

The use of synthetic chelates in industrial countries is substantial, in spite of the fact that they are quite expensive. They are used primarily to ameliorate micronutrient deficiencies of fruit trees and ornamentals. Although chelates may not replace the more conventional methods of supplying most micronutrients, they offer possibilities in special cases. Some of the chelators used in micronutrient fertilizers, such as gluconate, are naturally occurring and can supply certain micronutrients much more economically than can the more expensive aminopolycarboxylate compounds (e.g., EDDHA) listed in Table 15.9. Agricultural and chemical research will likely continue to increase the opportunities for the effective use of chelates.

In addition, selection of plant varieties whose roots produce their own chelating agents and practices that encourage the production of natural chelating agents from decomposing organic matter may take increasing advantage of chelation phenomena to improve micronutrient fertility of soils.

15.9 FACTORS INFLUENCING THE AVAILABILITY OF THE TRACE ELEMENT ANIONS

Unlike the cations needed in trace quantities by plants, the anion micronutrients seem to have relatively little in common with each other. Chlorine, molybdenum, and boron are quite different chemically, so little similarity would be expected in their reactions in soils.

Chlorine

Chlorine is absorbed in larger quantities by most crop plants than any of the micronutrients except iron. Most of the chlorine in soils is in the form of the chloride ion, which leaches rather freely from humid-region soils. In semiarid and arid regions, a higher concentration might be expected, with the amount reaching the point of salt toxicity in some of the poorly drained saline soils. In most well-drained areas, however, one would not expect a high chlorine content in the surface of arid-region soils.

Except where toxic quantities of chlorine are found in saline soils, there are no known natural soil conditions that reduce the availability and use of this element. Accretions of chlorine from the atmosphere, along with those from fertilizer salts such as potassium chloride, are sufficient to meet most crop needs. However, beneficial effects of chlorine on plant growth are common. This element helps to control several fungal diseases in plants, such as stalk rot in corn and "take all" in wheat. Chlorine also has an indirect effect on plant nutrition since it tends to suppress nitrification. This leads to a higher NH_4^+ to NO_3^- ratio in the soil solution, and, as the NH_4^+ ion is taken up by the plants, to decrease in rhizosphere pH. This greater acidity increases the availability and uptake of manganese that may, in turn, suppress the "take all" disease.

Chlorine seems to directly affect the nutrition of some plants. For example, tropical palms, adapted to growth in coastal soils where ocean spray contributes much chlorine, sometimes show chlorine deficiency if they are grown on inland soils with relatively low chlorine levels.

Toxic effects of chlorine have also been noted. This element is an obligatory component of the most widely used potassium fertilizer, potassium chloride. Consequently, when high rates of potassium are required for optimum plant growth, equally high rates of chlorine are applied. Such high levels are sometimes toxic to some plants, especially if they are of the *Solonaceae* family (tomatoes and the like). As a preventative measure, potassium sulfate may be used to supply the needed potassium, thereby alleviating any difficulties from excess chlorine.

Boron

Boron:
http://www.extension.umn.edu/
distribution/cropsystems/
DC0723.html

Boron is one of the most commonly deficient of all the micronutrients. The availability of boron is related to the soil pH, this element being most available in acid soils. While it is most available at low pH, boron is also rather easily leached from acid, sandy soils. Therefore, although deficiency of boron is relatively common on acid, sandy soils, it occurs because of the low supply of total boron rather than because of low availability of the boron present.

Soluble boron is present in soils mostly as boric acid [$B(OH)_3$] or as $B(OH)_4^-$. These compounds can exchange with the OH groups on the edges and surfaces of variable-charge clays such as kaolinite, and especially with the oxides of iron and aluminum. Reactions such as the following take place:

$$(15.7)$$

The boron so adsorbed is quite tightly bound, especially between pH 7 and 9, the range of lowest availability of this element. This probably accounts for the lime-induced boron deficiency noted on some soils as the pH is raised to pH 7 and above.

Boron is also adsorbed by humus, the strength of binding being even greater than for inorganic colloids. Boron is also a component of soil organic matter that is released by microbial mineralization. Consequently, organic matter serves as a major reservoir for boron in many soils and exerts considerable control over the availability of this nutrient. The adsorption of boron to organic and inorganic colloids increases with increasing pH (Figure 15.24).

Boron management by plant type:
http://www.borax.com/
agriculture/boronl.html

Boron availability is impaired by long dry spells, especially following periods of optimum moisture conditions. This may be related to the fact that boron is generally taken up with the transpiration stream of water rather than by active ion transport, as is the case for uptake of most other nutrients. Dry conditions may also reduce the mineralization of organically held boron during the dry periods. Boron deficiencies are also common in calcareous Aridisols and in neutral-to-alkaline soils with a high pH.

Molybdenum

Soil pH is the most important factor influencing the availability and plant uptake of molybdenum. The following equations show the forms of this element present at low and high soil pH:

$$H_2MoO_4 \underset{+H^+}{\overset{+OH^-}{\rightleftharpoons}} HMoO_4^- + H_2O$$

$$HMoO_4^- \underset{+H^+}{\overset{+OH^-}{\rightleftharpoons}} MoO_4^{2-} + H_2O$$

$$(15.8)$$

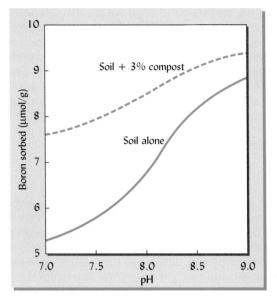

FIGURE 15.24 Effect of organic matter (added as compost in this experiment) and soil pH on adsorption of boron. Along with clays, humic compounds in decomposed organic matter adsorb boron. The adsorption capacities from both clays and humic compounds increase with rising pH in the neutral-to-alkaline range. While adsorption lowers the boron concentration in the soil solution, and hence its availability for plant uptake, it also protects boron from leaching loss. The soil in this study was a Calcic Haploxeralf. [Redrawn from Yermiyahu et al. (1995)]

At low pH values, the molybdenum is adsorbed by silicate clays and, more especially, by oxides of iron and aluminum through *ligand exchange* with hydroxide ions on the surface of the collodial particles. Reactions such as the following occur:

$$\begin{array}{c} \text{Al} \\ \diagdown \\ \diagup \\ \text{Al} \end{array} \text{OH} + \text{HMoO}_4^- \rightarrow \begin{array}{c} \text{Al} \\ \diagdown \\ \diagup \\ \text{Al} \end{array} \text{O} - \overset{\displaystyle \text{O}}{\underset{\displaystyle \text{O}}{\text{Mo}}} - \text{O} + \text{H}_2\text{O} \tag{15.9}$$

Surface hydroxyls　　Soluble Mo　　Adsorbed Mo

The liming of acid soils will usually increase the availability of molybdenum (Figure 15.25). The effect is so striking that some researchers, especially those in Australia and New Zealand, argue that the primary reason for liming very acid soils is to supply molybdenum. Furthermore, in some instances 30 g or so of molybdenum added to acid soils has given about the same increase in the yield of legumes as has the application of several megagrams of lime.

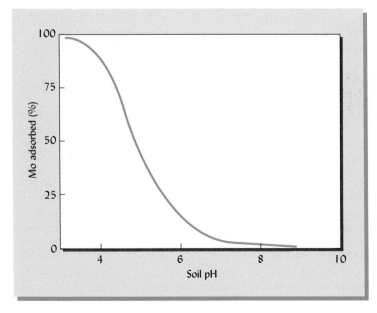

FIGURE 15.25 Effect of pH on the adsorption of molybdenum on a Hesperia coarse, loamy sand from California. The high adsorption at low pH values could result in molybdenum deficiency. Likewise, at high pH values the molybdenum is not adsorbed and is free to be taken up by plants, sometimes at toxic levels. [Redrawn from Goldberg et al. (1996)]

The phosphate anion seems to improve the availability of molybdenum by competing with the latter for sorption sites on soil surfaces. For this reason, molybdate salts are often applied along with phosphate carriers to molybdenum-deficient soils. This practice apparently encourages the uptake of both elements and is a convenient way to add the extremely small quantities of molybdenum required. Legume seeds coated in a mixture of superphosphate and sodium molybdate have also been used successfully to improve the grazing quality of acid-soil range and savanna lands.

A second common anion, the sulfate ion, seems to have the opposite effect on plant utilization of molybdenum. Sulfate reduces molybdenum uptake and seems to compete with molybdenum at functional sites on plant metabolic compounds.

Selenium

Selenium and livestock:

http://www.ansci.cornell.edu/plants/toxicagents/selenium/selenium.html

Selenium is not essential for plant growth, but is essential for human and animal health. This element also occurs in some soils at levels that are toxic both to plants and to animals that consume the plants. For example, excess selenium causes "blind staggers" in ruminants and embryonic deformities in waterfowl. Natural selenium toxicities are found notably in soils developed on marine sedimentary parent material that is high in selenium. But human activities such as irrigation and mining can significantly increase the levels of selenium in soils, sometimes to toxic levels.

Selenium is found in nature in four major solid forms and several volatile forms. The particular forms present determine the degree of toxicity much more than does the total amount of selenium in the soil. The relationship among these forms may be shown by the following reactions, which illustrate the microbiological reduction of the soluble and highly oxidized selenates to reduced and less soluble forms.[6]

$$SeO_4^{2-} \rightleftharpoons SeO_3^{2-} \rightleftharpoons Se \rightleftharpoons Se^{2-} \rightleftharpoons (CH_3)_2Se$$

[Se(VI)]	[Se(IV)]	[Se(0)]	[Se(-II)]	[Se(-II)organic]
Selenate	Selenite	Elemental selenium	Selenide	Dimethyl selenide

(15.10)

Similar to sulfur, selenium is found in four oxidation states in soils: 2−, 0, 4+ and 6+. The particular compounds present at any one time are determined largely by factors such as redox potential, pH, organic ligands present, and the activities of soil microorganisms. Selenates are most soluble and are prominent in well-aerated soils, especially if the pH is high (above 7). They seem to be responsible for most environmental selenium toxicity. Selenites are commonly dominant under acid (pH 4.5–6.5), poorly drained conditions, but are only slowly available since they are adsorbed by iron oxides. If added to soils to reduce selenium deficiencies they will not likely induce selenium toxicities.

Elemental selenium and selenides are quite insoluble and accumulate in wetland sediments, as do some Se-organic compounds. Some plants, in association with fungi and bacteria, absorb both organic and inorganic forms of selenium and produce volatile organics such as dimethyl selenide and dimethyl diselenide that can be released as gases to the atmosphere. These are relatively nontoxic compounds. As explained in Box 15.2, these reactions are used as a promising means of **bioremediation** to remove toxic levels of soluble selenium from soils and water (see also Section 18.5).

Other Trace Elements

A number of other trace elements found in the anionic form have been reported to be essential for either plants or animals, but we will consider only arsenic because of its toxic effects on humans and other animals.

ARSENIC.[7]   Some studies have shown arsenic to be essential for both plant and animal life. This element is a common natural constituent of soils, but its level is augmented in

---

[6] Nonbiological reduction of selenates also occurs in the presence of ferrous/ferric hydroxide $[Fe(II)_aFe(III)_b(OH)_{12}] X \cdot 3H_2O$; where $a = 1–4$, $b = 1–2$, and X = an anion, such as SO_4^{2-}, a compound known as *green rust*, commonly found in wet soils. See Myneni et al. (1997).

[7] For a comprehensive treatment of the chemistry and management of arsenic in the environment, see Naidu et al. (2006); for a review of the serious problem of arsenic in soils and groundwater in Bangladesh, see Hossain (2006).

BOX 15.2 SELENIUM—BOUND AND VAPORIZED[a]

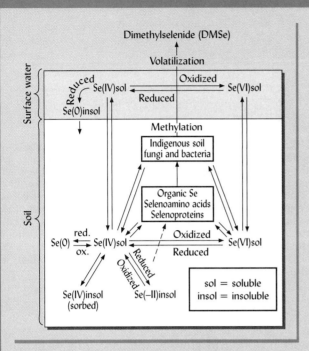

Dimethylselenide (DMSe)

Volatilization

sol = soluble
insol = insoluble

FIGURE 15.26 *Selenium transformation in wetland soils.*

Irrigation waters carry two relatively soluble forms of selenium, selenates [$Se(VI)O_4^{2-}$] and selenites [$Se(IV)O_3^{2-}$]. When selenium first moves into the soil, some of it is reduced quickly to very insoluble elemental selenium (Se^0), which is largely unavailable to plants and is non-toxic. Further transformations take place as both soluble forms move downward into the soil (Figure 15.26). Reducing conditions favor formation of selenites, which tend to be tightly sorbed by iron oxides. Further reduction induced by microbes leads to the formation of not only elemental selenium [Se^0] but to selenides [Se^{2-}], both of which are quite insoluble. Thus, reducing conditions encourage the formation of insoluble forms, thereby lowering the toxicity of the selenium present.

As microbes and plants metabolize selenium, it is assimilated into organic forms such as selenoamino acids and selenoproteins, most of which are also quite insoluble. Certain plant species such as rice and members of the *Brassica family* (generally in association with soil fungi and bacteria) are able to attach methyl groups (methylation) to organoselenium compounds, thereby forming volatile gases such as dimethylselenide (DMSe). DMSe is 700 times less toxic than the selenates and can be dispersed into the atmosphere without any environmental damage. The process seems to work best in soils that are moist but not flooded and that are well supplied with organic materials to provide metabolic energy for these reactions. To allow continued irrigated crop production without damaging the environment, soil scientists are working to harness both pathways for selenium detoxification—the process that changes selenium to insoluble forms and the process that releases the selenium into the atmosphere.

[a] Concepts for the diagram are from Hayes and Traina (1998) and Frankenberger and Losi (1995). An excellent review of research on phytovolatilization of selenium is found in Zayed et al. (2000).

some cases by human activities. For example, it has been widely used as an herbicide and can be leached from certain mine tailings. A few reports have shown beneficial effects of arsenic on plant growth, but its negative effect, particularly on humans, is most widely recognized.

For centuries, arsenic has been known and used as a human and animal poison. Its presence in soils, groundwater, and well water is of concern to people around the world, but especially in Bangladesh, India, China, Chile, and Slovakia. In Bangladesh, for example, more than 20 million of the country's 126 million people are believed to be drinking arsenic-contaminated water. Thousands are suffering from skin cancer that is caused by naturally occurring arsenic toxicity. The well water in the United States is generally safe, but the arsenic content of some wells exceeds the current maximum contaminant level (MCL), which has recently become even more stringent (Figure 15.27).

Arsenic in Drinking Water—
USEPA rule:
http://www.epa.gov/safewater/arsenic/index.html

Arsenic is found as a minor constituent of many minerals (especially sulfides). Upon their breakdown it becomes associated with the soil in two major forms, arsenite [AsO_3^{3-}, or three-valent As(III)] and arsenate (AsO_4^{3-}, or five-valent As(V)]. Both forms are sorbed by oxides and hydroxides of iron, but As(V) is generally more strongly sorbed, especially in acid soils. Consequently, arsenites [As(III)] are generally more mobile and move more easily into groundwater, the source of drinking water in many parts of the world. For this reason, wet and reduction-prone conditions are to be avoided to minimize dissolution and movement of the most toxic forms of arsenic.

Scientists are evolving methods for the remediation of arsenic-contaminated waters. For example, they are trying to use hydrous oxides of iron as sorbing agents for the arsenic in drinking water. Also, they have discovered that certain plants are *hyperaccumulators* of arsenic. For example, an uncultivated fern is known to have accumulated as

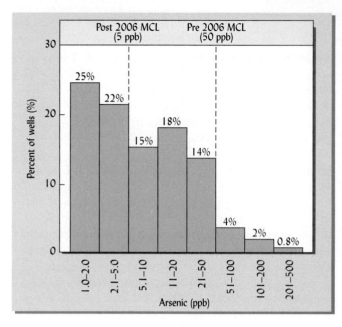

FIGURE 15.27 The percentage of 247 Arizona drinking water wells with increasing levels of arsenic in the water. Some 7% of the wells had water with arsenic exceeding 50 ppb, the maximum contaminant level (MCL) acceptable by the U.S. EPA prior to 2006 and a level typical in the regions of Bangladesh where arsenic poisoning is a common malady. More than half of the wells would not meet the 5 ppb standard promulgated by both the U.S. EPA (post-2006) and the World Health Organization as the level above which human health is at risk. [Data from the U.S. Geological Survey National Water Quality Assessment; see Spencer (2000)]

much as 2% arsenic in research plots in Florida. By growing and harvesting such plants, it may be possible to reduce the level of soluble arsenic in the soils and the groundwater.

15.10 NEED FOR NUTRIENT BALANCE

Nutrient balance among the trace elements is as essential as, but even more difficult to maintain than, macronutrient balance. Some of the plant enzyme systems that depend on micronutrients require more than one element. For example, both manganese and molybdenum are needed for the assimilation of nitrates by plants. The beneficial effects of combinations of phosphates and molybdenum have already been discussed. Apparently, some plants need zinc and phosphorus for optimum use of manganese. The use of boron and calcium depends on the proper balance between these two nutrients. A similar relationship exists between potassium and copper and between potassium and iron in the production of good-quality potatoes. Copper utilization is favored by adequate manganese, which in some plants is assimilated only if zinc is present in sufficient amounts. Ruminant animals fed plant tissue with low Cu/Mo ratios suffer molybdenum toxicity (see Figure 15.28). Of course, the effects of these and other nutrients will depend on the specific plant being grown, but the complexity of the situation can be seen from the examples cited.

Antagonism and Synergism

Some enzymatic and other biochemical reactions requiring a given micronutrient may be poisoned by the presence of a second trace element in toxic quantities. Other negative effects occur because one element competes with or otherwise reduces uptake of a second element by the plant root. On the other hand, a good supply of a certain nutrient may enhance the utilization of a second nutrient element in what is termed a *synergistic effect*. Some of these interactions are summarized in Table 15.10.

Some of the *antagonistic* effects may be used effectively in reducing toxicities of certain of the micronutrients. For example, copper toxicity of citrus groves caused by residual copper from fungicidal sprays may be reduced by adding iron and phosphate fertilizers. Sulfur additions to calcareous soils containing toxic quantities of soluble molybdenum may reduce the availability, and hence the toxicity, of molybdenum. The hyperaccumulation of phosphorus, manganese, and magnesium in zinc-deficient plants is an example of the complex interactions among essential elements as they influence plant nutrition.

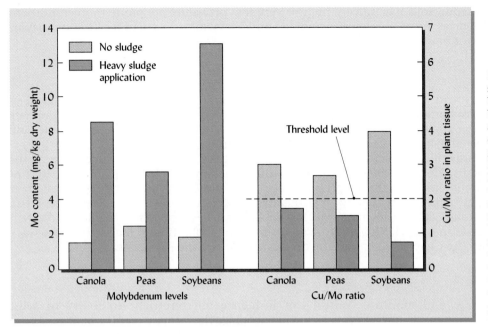

FIGURE 15.28 Heavy applications of sludges containing trace elements can bring about nutritional imbalances in both plants and animals. On the left are shown the molybdenum contents of the tops of three plant species grown on a soil with and without a heavy sewage sludge treatment that had been made 20 years previously. On the right is shown the ratio of the copper content to that of molybdenum (Cu/Mo ratio) in these plants. Unfortunately, ruminant animals suffer molybdenosis (molybdenum toxicity) if they eat feeds that have a Cu/Mo ratio less than 2 (dotted line). Great care must be used to prevent such imbalances when sludges are applied to soils. [Drawn from data in McBride et al. (2000)]

These examples of nutrient interactions, both beneficial and detrimental, emphasize the highly complicated nature of the biological transformations in which micronutrients are involved. The total land area on which unfavorable nutrient balances require special micronutrient treatment is increasing as soils are subjected to more intensive cropping methods.

Before leaving the topic of nutrient balance and imbalance, it should be noted that human activities from mining to industrial spills can contaminate soils with high levels of micronutrients (as well as non-nutrient metals) that threaten to be toxic to plants, animals, fish and/or people. In Chapter 18 (Sections 18.7 and 18.8) we will address the remediation of such contaminated land, including bioremediation using special plants that can accumulate the offending metals in their tissues for removal from the soil.

15.11 SOIL MANAGEMENT AND TRACE ELEMENT NEEDS

Although the characteristics of each micronutrient are quite specific, some generalizations with respect to management practices are possible.

In seeking the cause of plant abnormalities, one should keep in mind the conditions under which micronutrient deficiencies or toxicities are likely to occur. Sandy soils,

TABLE 15.10 **Some Antagonistic (Negative) and Synergistic (Positive) Effects of Other Nutrients on Micronutrient Utilization by Plants[a]**

The occurrence of so many interactions emphasizes the need for balance among all nutrients and avoidance of excess application of any particular nutrient.

	Elements decreasing utilization		Elements increasing utilization	
Micronutrient	Soil and root surface reactions	Plant metabolic reactions	Soil and root surface reactions	Plant metabolic reactions
Fe	B, Cu, Zn, Mo, Mn	Mn, Mo, P, S, Zn	B, Mo	
Mn	Fe, B	Fe	B	
Zn	Mg, Cu, B, Fe, P	Fe, N, Mg	N, B	Fe
Cu	B, Zn, Mo	P, N	B	
B	Ca, K			N
Mo	S, Cu	S	P	P
Ni	Ca, Fe	Fe, Zn		

[a] Summarized from many sources.

mucks, and soils having very high or very low pH values are prone to micronutrient deficiencies. Areas of intensive cropping and heavy macronutrient fertilization may be deficient in the micronutrients.

Changes in Soil Acidity

In very acid soils, one might expect toxicities of iron and manganese and deficiencies of phosphorus and molybdenum. These can be corrected by liming and by appropriate fertilizer additions. Alkaline soils may have deficiencies of iron, manganese, zinc, and copper and, in a few cases, a toxicity of molybdenum.

No specific statement can be made concerning the pH value most suitable for all the elements. However, medium-textured soils generally supply adequate quantities of micronutrients when the soil pH is held between 6 and 7. In sandy soils, a somewhat lower pH may be justified because the total quantity of micronutrients is low, and even at pH 6.0, some cation deficiencies may occur. It is important to be on guard to recognize inadvertent increases in soil pH, such as those occurring from application of lime-stabilized sewage sludge (see Chapter 16) or leaching of calcareous gravel and pavements.

Soil Moisture

Drainage and moisture control can influence micronutrient solubility in soils. Improving the drainage of acid soils will encourage the formation of the oxidized forms of iron and manganese. These are less soluble and, under acid conditions, less toxic than the reduced forms. Flooding a soil will favor the reduced forms of iron and manganese, which are more available to growing plants (Figure 15.29). Excessive moisture at high pH values can have the opposite effect since poorly drained soils have a high carbon dioxide concentration, encouraging the formation of bicarbonate ions, which reduces iron availability. Poor drainage also increases the availability of molybdenum in some alkaline soils to the point of producing plants with toxic levels of this element.

Fertilizer Applications

Micronutrient deficiencies are relatively unlikely to be a problem in most soils to which plant residues are returned and organic amendments, such as animal manure or sewage sludge, are regularly applied. Animal manures applied at normal rates sufficient to supply

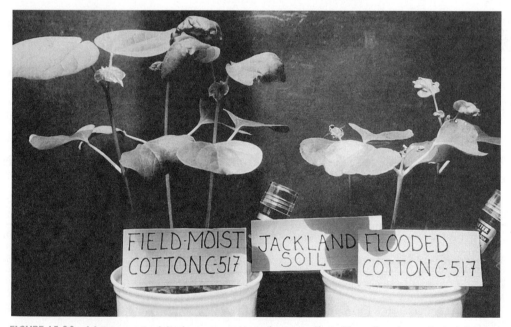

FIGURE 15.29 Manganese toxicity in young cotton plants as affected by soil moisture during the three weeks prior to planting the cotton seed. The crinkled-leaf symptom and stunted size resulted from toxic levels of manganese in the soil that had been flooded prior to planting. Both pots were maintained at field-capacity water content from sowing to harvest. The Jackland soil (Aquic Hapludult) used is high in manganese-containing minerals and has a pH of approximately 5.8. [From Weil et al. (1997)]

TABLE 15.11   A Few Commonly Used Fertilizer Materials That Supply Micronutrients

Micronutrient	Commonly used fertilizers		Nutrient content, %
Boron	Borax	$Na_2B_4O_7 \cdot 10H_2O$	11
	Sodium pentaborate	$Na_2B_{10}O_{16} \cdot H_2O$	18
Copper	Copper sulfate	$CuSO_4 \cdot 5H_2O$	25
Iron	Ferrous sulfate	$FeSO_4 \cdot 7H_2O$	19
	Iron chelates	NaFeEDDHA	6
Manganese	Manganese sulfate	$MnSO_4 \cdot 3H_2O$	26–28
	Manganese oxide	MnO	41–68
Molybdenum	Sodium molybdate	$Na_2MoO_4 \cdot 2H_2O$	39
	Ammonium molybdate	$(NH_4)_6Mo_7O_{24} \cdot 4H_2O$	54
Zinc	Zinc sulfate	$ZnSO_4 \cdot H_2O$	35
	Zinc oxide	ZnO	78
	Zinc chelate	$Na_2ZnEDTA$	14

Selected from Murphy and Walsh (1972).

macronutrient needs carry enough copper, zinc, manganese, and iron to supply a major portion of micronutrient needs as well (see Table 16.9). In addition, the chelates produced from manure enhance the availability of these micronutrients.

Nonetheless, the most common management practice to overcome micronutrient deficiencies (and some toxicities) is the application of commercial fertilizers. Examples of fertilizer materials applied to supply micronutrients are shown in Table 15.11. The materials are most commonly applied to the soil, although foliar sprays and even seed treatments can be used. Foliar sprays of dilute inorganic salts or organic chelates are more effective than soil treatments where high soil pH and other factors render the soil-applied nutrients unavailable.

For soil applications, about one-half to one-fourth as much fertilizer is needed if the application is banded rather than broadcast (see Figure 16.30). About one-fifth to one-tenth as much fertilizer is needed if the material is sprayed on the plant foliage. Foliar application may, however, require repeated applications in a single year. Treating seeds with small dosages (20 to 40 g/ha) of molybdenum has had satisfactory results on molybdenum-deficient acid soils. Typical rates of application to soil are given in Table 15.12.

The micronutrients can be applied to the soil either as separate materials or incorporated in standard macronutrient carriers. Unfortunately, the solubilities of copper,

TABLE 15.12   Plants Known to be Especially Susceptible or Tolerant to, and Soil Conditions Conducive to, Micronutrient Deficiencies

Plants which are most susceptible to deficiency of a micronutrient often have a relatively high requirement for that nutrient and may be relatively tolerant to levels of that nutrient that would be high enough to cause toxicity to other plants.

Micronutrient	Common range in rates recommended for soil application[a], kg/ha	Plants most commonly deficient (high requirement or low efficiency of uptake)	Plants rarely deficient (low requirement or high efficiency of uptake)	Soil conditions commonly associated with deficiency
Iron	0.5–10.0	Blueberries, azaleas, roses, holly, grapes, nut trees, maple, bean, sorghum, oaks	Wheat, alfalfa, sunflower, cotton	Calcareous, high pH, waterlogged alkaline soils
Manganese	2–20	Peas, oats, apple, sugar beet, raspberry, citrus	Cotton, soybean, rice, wheat	Calcareous, high pH, drained wetlands, low organic matter, sandy soils
Zinc	0.5–20	Corn, onion, pines, soybeans, beans, pecans, rice, peach, grapes	Carrots, asparagus, safflower, peas, oats, crucifers, grasses	Calcareous soils, acid, sandy soils, high phosphorus
Copper	0.5–15	Wheat, corn, onions, citrus, lettuce, carrots	Beans, potato, peas, pasture grasses, pines	Histosols, very acid, sandy soils
Boron	0.5–5	Alfalfa, cauliflower, celery, grapes, conifers, apples, peanut, beets, rapeseed, pines	Barley, corn, onion, turfgrass, blueberry, potato, soybean	Low organic matter, acid, sandy soils, recently limed soils, droughty soils, soils high in 2:1 clays
Molybdenum	0.05–0.5	Alfalfa, sweet clover, crucifers (broccoli, cabbage, etc.), citrus, most legumes	Most grasses	Acid sandy soils, highly weathered soils with amorphous Fe and Al

[a] The lower end of each range is typical for banded applications; the higher end is typical for broadcast applications.

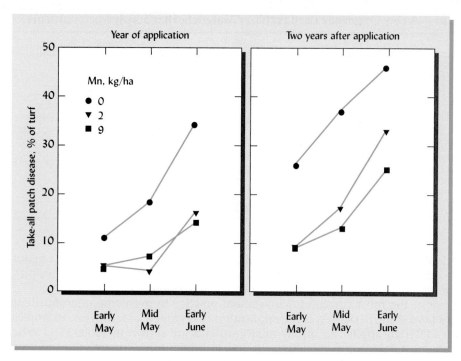

FIGURE 15.30 Application of manganese to control take-all patch disease in a bent-grass turf. Just 2 kg/ha of Mn significantly reduced the disease severity. The effects remained even 2 years after the application. However, applications 12 to 18 months apart are recommended for acceptable control. The disease is named "take-all" because of total destruction of the plant by the fungal pathogen *Gaeumannomyces gramini*. This pathogen also causes take-all disease of wheat, another member of the grass family. It is thought that improved Mn nutrition enhances lignin biosynthesis in the plant cell walls, making the plant cell walls more resistant to attack (see also Table 15.5). For both plant species, Mn application controls the disease, but only if the soil pH is kept low enough for Mn to remain readily available. [Drawn from data in Heckman et al. (2003)]

iron, manganese, and zinc can be reduced by such incorporation, but boron and molybdenum remain in reasonably soluble condition. Liquid macronutrient fertilizers containing polyphosphates encourage the formation of complexes that protect added micronutrients from adverse chemical reactions. In effect, polyphosphates in fertilizer solutions act as chelating agents for micronutrients. High-surface-area pitted glasslike beads, called **frits**, are manufactured with boron, copper, zinc, and other micronutrients incorporated into the glass. These fritted materials slowly release their micronutrients as the glass weathers in the soil. This slow release avoids some of the problems of precipitation and sorption that might otherwise occur, particularly in an alkaline soil.

Economic responses to micronutrients are becoming more widespread as intensity of plant production and our knowledge about micronutrients increases. In high-value ornamental horticulture and turfgrass management, a small application of a micronutrient such as manganese can sometimes serve as a very inexpensive and environmentally friendly means of controlling plant disease (Figure 15.30). Profitability is also seen in the yield and quality responses to micronutrients by fruits, vegetables, and field crops in areas of neutral to alkaline soils. Even on acid soils, deficiencies of these elements are increasingly encountered. Molybdenum, which has been used for some time for forage crops and for cauliflower and other vegetables, has received attention in recent years for forest nurseries and soybeans, especially on acid soils. Micronutrients have been used for decades in plant production on muck soils and on very sandy soils. To maintain optimal plant production, the use of micronutrients will be increasingly common under many other types of soil conditions, such as those previously mentioned.

A soil will often produce a micronutrient deficiency in some plants but not in others. Plant species, and varieties within species, differ widely in their susceptibility to micronutrient deficiency or toxicity. Table 15.12 provides examples of plant species known to be particularly susceptible or tolerant to several micronutrient deficiencies.

Marked differences in crop needs for micronutrients make fertilization a problem where rotations are being followed. On general-crop farms, vegetables are sometimes grown in rotation with small grains and forages. If the boron fertilization is adequate for a vegetable crop such as red beets, or even for alfalfa, the small-grain crop grown in the rotation may show toxicity damage. These facts emphasize the need for specificity in determining crop nutrient requirements and for care in meeting these needs.

Plant Selection and Breeding

Variation in plant abilities to accumulate micronutrients and other trace elements suggests that plant selection and breeding may hold potential for overcoming deficiencies and toxicities of these elements. For example, certain species in the *Brassicaceae* family are able to

TABLE 15.13 Effect of Soils at Two Kansas Locations on the Concentrations ($\mu g\ g^{-1}$) of Four Elements Important as Micronutrients for Humans in Hard Red Winter Wheat Used for Making Bread

For each location, the mean and the range of values for 14 different wheat cultivars is given. For the 3 plant nutrients (Fe, Zn, and Cu), the variations among cultivars and between soils were relatively small, but for Se (not a plant nutrient), values differed almost tenfold between soils and much less among plant cultivars. The newer, higher-yielding cultivars in this study tended to have the lower values within each range.

Element	Hutchinson, KS		Manhattan, KS	
	Mean	Range	Mean	Range
Fe	31.4	24.4–42.8	33.7	30.2–38.3
Zn	20.9	16.0–26.3	29.3	26.1–33.9
Cu	2.12	1.74–2.82	4.2	3.68–5.68
Se	0.36	0.28–0.48	0.05	0.04–0.06

From Garvin et al. (2007). Copyright of Soc. of Chemical Industry. Reproduced with permission granted by John Wiley & Sons, Ltd. on behalf of SCI.

accumulate 5 to 100 times as much of certain trace metals as found in most other plants. Similarly, black gum (*Nyassa sylvatica*) is an accumulator of cobalt. *Bioremediation* may take advantage of such accumulators in removing toxic quantities of some trace elements from polluted soil (see Section 18.5). Human nutrition and health are also affected by the trace element content of plants used as food. Some recent research suggests that plant breeding and biotechnology may be used to select and/or create plant varieties that more efficiently absorb trace elements or that tolerate their toxic levels in soils.

Research on micronutrients in human diets:
http://www.ars.usda.gov/research/projects/projects.htm?ACCN_NO=408299&fy=2004

FIGHTING MICRONUTRIENT HUNGER. Micronutrient deficiencies are a serious human health concern. Billions of people suffer from iron deficiency, and millions more suffer from deficiencies of zinc and copper. In large parts of Europe and Asia, the Se in peoples' diets is below the needed levels. In many cases these deficiencies may be prevalent because people lack diversity—especially green vegetables—in their diets and rely upon staple grain products made from rice, wheat, or corn for most of their nutrition.

Although plants exert considerable control over the uptake of plant nutrients such as Fe, Zn, and Cu (see Sections 15.7 and 15.8), the influence of soil properties is still significant. In the case of Se (see Section 15.9), soil chemistry seems to have an even greater influence, as plants exert little control over the uptake of this element, which though not needed by the plant, is essential for animals (Table 15.13). Plants may easily take up levels of Se ranging from so low that people and animals may suffer deficiency to so high that animals may succumb to toxicity (see Box 15.2). The Se content of wheat used to make bread is of major importance to human health, as Se is essential for all animals and at relatively high levels may even reduce the incidence of some cancers. Some species of *Astragalus* and *Stanleya* are used as dietary supplements as they can accumulate about 5 times as much selenium as do such cereal crops grown in similar soils.

Pro biotech view of enhancing micronutrients in food:
http://www.agbioworld.org/biotech-info/topics/goldenrice/bouis.html

World political leaders have turned to plant scientists to create high-yielding rice varieties that have high contents of iron that is readily assimilated by humans. Some progress has been made in improving the iron content of rice by using conventional plant-breeding techniques, and studies show the potential for even greater progress by using biotechnical methods. However, there is evidence that plant breeders have inadvertently selected for cultivars that, while high yielding, may be less effective in taking up trace elements important in human diets (Table 15.13). Similar differences exist in genetic tolerance of plants to nutrient deficiencies. The future calls for close collaboration between soil and plant scientists in exploring the long-term health and sustainable development implications of these findings.

Micronutrient Availability

Major sources of micronutrients and the general reactions that make them available to higher plants and microorganisms are summarized in Figure 15.31. Original and secondary minerals are the primary sources of these elements, while the breakdown of

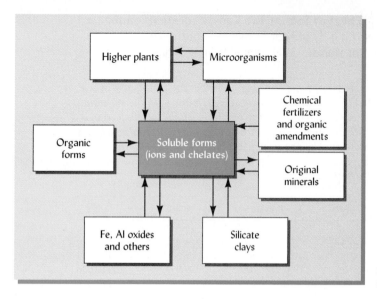

FIGURE 15.31 Diagram of soil sources of soluble forms of micronutrients and their utilization by plants and microorganisms.

organic forms releases ions to the soil solution. The micronutrients are used by higher plants and microorganisms in important life-supporting processes. Removal of nutrients in crop or timber harvest reduces the soluble ion pool, which may need to be replenished with manures or chemical fertilizers to avoid nutrient deficiencies.

Worldwide Management Problems

Micronutrients in global fertilizer practice:
http://www.fertilizer.org/ifa/publicat/PDF/2004_ag_new_delhi_brown_slides.pdf

Trace element deficiencies and toxicities have been diagnosed in most areas of crop production and some forest areas of the United States and Europe. However, in some developing countries, particularly in the tropics, the extent of these deficiencies is much less well known. Limited research suggests that there may be large areas with deficiencies or toxicities of one or more of these elements. Irrigation schemes that bring calcareous soils in desert areas under cultivation are often plagued with deficiencies of iron, zinc, copper, and manganese. Copper and zinc deficiency have been noted in cereal crops on highly weathered soils in the humid and subhumid tropics. As macronutrient deficiencies are addressed and yields are increased, more micronutrient deficiencies will undoubtedly come to the fore. One encouraging factor is the small quantity of micronutrients usually needed, making transportation of fertilizer materials to remote parts of the world much less of an expense than is the case for macronutrient fertilizers. The management principles established for the economically developed countries should also be helpful in alleviating micronutrient deficiencies in less-developed countries.

Equal attention must be given in developing countries to situations where trace element toxicities prevail or may prevail. Naturally occurring toxic levels must be dealt with using the same methods being tried in the industrialized countries, including the utilization of native plants as nutrient accumulators to lower the soil level of some of the offending elements. Application to the land of excess quantities of trace elements in industrial and domestic wastes must be avoided at all cost. The developing countries can learn from the mistakes of their more affluent neighbors by not using the land as an indiscriminate dumping ground.

15.12 CONCLUSION

Calcium and magnesium are macronutrients usually used in quantities similar to sulfur and phosphorus. However, for woody plants and for certain "calcicole" plants, calcium is used in amounts that rival nitrogen and potassium (1 to 3% Ca in the dry matter). Both Ca and Mg are intimately associated with the processes of soil acidification and thus linked to sulfur and nitrogen deposition. The cycling of these two nonacid cations is of major concern in biogeochemistry and soil Ca supplies greatly impact the nature of the ecology that will develop in an area. Calcium and magnesium also impact global carbon balance through the precipitation and dissolution of Ca and Mg carbonates.

Micronutrients are becoming increasingly important to world agriculture as crop removal of these essential elements increases. Soil and plant tissue tests confirm that these elements are limiting crop production over wide areas and suggest that attention to them will likely increase in the future.

Micronutrient deficiencies are due not only to low contents of these elements in soils but more often to their unavailability to growing plants. They are adsorbed by inorganic constituents such as Fe, Al, oxides and form complexes with organic matter, some of which are only sparingly available to plants. Other such organic complexes, known as *chelates*, protect some of the micronutrient cations from inorganic adsorption and make them available for plant uptake.

Toxicities of micronutrients retard both plant and animal growth. Removing these elements from soil and water, or rendering them unavailable for plant uptake, is one of the challenges facing soil and plant scientists. Adequate micronutrient supply is also a potential tool for helping to manage plant diseases, to which results using Mn attest.

In most cases, soil-management practices that avoid extremes in soil pH, that optimize the return of plant residues and animal manures, and that promote chelate production by actively decomposing organic matter will minimize the risk of micronutrient deficiencies or toxicities. But increasingly noted are situations where micronutrient problems can be most practically solved by the application of micronutrient fertilizers. Such materials are becoming common components of fertilizers for field and garden use and will likely become even more so in the future.

STUDY QUESTIONS

1. If a forest soil has 5 mg/L (ppm) of Al^{3+} ions and 5 mg/L of Ca^{2+} ions in the soil solution, what is the *molar* ratio of Ca/Al in that solution? (That is, the ratio of mol_c Ca^{2+}/mol_c Al^{3+}, as presented in Table 15.3.)

2. Which is likely to be a better indicator of acidification stress on forest trees, a molar ratio of Ca/Al in leaf tissue less than 1.0 or a molar ratio of Ca/Mn less than 1.0 in the leaf tissue? Why?

3. What portion of the plant would you look at to find symptoms of Ca and Mg deficiencies, respectively?

4. During a year's time, some 250 kg nitrogen and only 30 g molybdenum have been taken up by the trees growing on a hectare of land. Would you therefore conclude that the nitrogen was more essential for the tree growth? Explain.

5. Since only small quantities of micronutrients are needed annually for normal plant growth, would it be wise to add large quantities of these elements now to satisfy future plant needs? Explain.

6. Iron deficiency is common for peaches and other fruits grown on highly alkaline irrigated soils of arid regions, even though these soils are quite high in iron. How do you account for this situation, and what would you do to alleviate the difficulty?

7. How do Fe and Al oxides effect the availability of Mo and B in soils? Explain.

8. Give two examples of fungal-caused plant diseases that can be effectively reduced by fertilizing with a micronutrient. Name the micronutrient and explain why it helps control the disease.

9. Soybeans growing on a recently limed soil show evidence of a deficiency of a nutrient, thought by some to be molybdenum. Do you agree with this diagnosis? If not, what is your explanation?

10. What are *chelates*, how do they function, and what are their sources?

11. The addition of only 1 kg/ha of a nutrient to an acid soil on which lime-loving cauliflower was being grown gave considerable growth response. Which of the nutrients would it likely have been? Explain.

12. Two Aridisols, both at pH 8, were developed from the same parent material, one having restricted drainage, the other being well drained. Plants growing on the

well-drained soils showed iron deficiency symptoms while those on the less-well-drained soil did not. What is the likely explanation for this?

13. Animals, both domestic and wild, are adversely affected by deficiencies and toxicities of two of the micronutrients. Which elements are these, and what are the conditions responsible for their effects?

14. Discuss the role plant breeders and geneticists might play in managing micronutrient deficiencies and toxicities.

15. Since boron is required for the production of good-quality table beets, some companies purchase only beets that have been fertilized with specified amounts of this element. Unfortunately, an oat crop following the beets does very poorly compared to oats following unfertilized beets. Give possible explanations for this situation.

REFERENCES

Agbenin, J. O., and L. A. Olojo. 2004. "Competitive adsorption of copper and zinc by a Bt horizon of a savanna Alfisol as affected by pH and selective removal of hydrous oxides and organic matter," *Geoderma*, **119**:85–95.

Brady, N. and R. Weil. 1996. *The Nature and Properties of Soils*. 11th ed. (Upper Saddle River: Prentice Hall).

Baligar, V. C., N. K. Fageria, and M. A. Elrashidi. 1998. "Toxicity and nutrient constraints to root growth," *Hortscience*, **33**:960–965.

Brown, P. H., R. M. Welch, and E. E. Cary. 1987. "Nickel: A micronutrient essential for higher plants," *Plant Physiol.*, **85**:801–803.

Clemens, D. F., B. M. Whitehurst, and G. B. Whitehurst. 1990. "Chelates in agriculture," *Fertilizer Research*, **25**:127–131.

Fageria, N. K., V. C. Baligar, and R. B. Clark. 2002. "Micronutrients in crop production," *Adv. Agron*, **77**:185–268.

Fenn, M. E., T. G. Huntington, S. B. Mclaughlin, C. Eagar, and R. B. Cook. 2006. "Status of soil acidification in North America," *Journal of Forest Science*, **52**:3–13.

Frankenberger, W. T., and M. E. Losi. 1995. "Applications of bioremediation in the cleanup of heavy metals and metalloids," SSSA Special Publication 43 (Madison, Wis.: Soil Science Society of America).

Gao, X., C. Zou, X. Fan, F. Zhang, and E. Hoffland. 2006. "From flooded to aerobic conditions in rice cultivation: Consequences for zinc uptake," *Plant Soil*, **280**:41–47.

Garvin, D. F., R. M. Welch, and J. W. Finley. 2007. "Historical shifts in the seed mineral micronutrient concentration of U.S. hard red winter wheat germplasm," *J. Sci. Food Agric*. In press.

Goldberg, S., H. S. Forster, and C. L. Godfrey. 1996. "Molybdenum adsorption on oxides, clay minerals, and soils," *Soil Sci. Soc. Amer. J.*, **60**:425–432.

Hamilton, M. A., D. T. Westermann, and D. W. James. 1993. "Factors affecting zinc uptake in cropping systems," *Soil Sci. Soc. Amer. J.*, **57**:1310–1315.

Hayes, K. F., and S. J. Traina. 1998. "Metal ion speciation and its significance in ecosystem health," in P. M. Huang, D. C. Adriano, T. J. Logan, and R. T. Checkai (eds.), *Soil Chemistry and Ecosystem Health*. SSSA Spec. Publ. 52 (Madison, Wis.: Soil Science Society of America).

Heckman, J. R., B. B. Clarke, and J. A. Murphy. 2003. "Optimizing manganese fertilization for the suppression of take-all patch disease on creeping bentgrass," *Crop Sci.*, **43**:1395–1398.

Hossain, M. F. 2006. "Arsenic contamination in Bangladesh—An overview," *Agric. Ecosyst. Environ.*, **113**:1–16.

Kalim, S., Y. P. Luthra, and S. K. Gandhi. 2003. "Cowpea root rot severity and metabolic changes in relation to manganese application," *Journal of Phytopathology*, **151**:92–97.

Kogelmann, W. J., and W. E. Sharpe. 2006. "Soil acidity and manganese in declining and nondeclining sugar maple stands in Pennsylvania," *J. Environ. Qual.*, **35**:433–441.

Kopittke, P. M., and N. W. Menzies. 2007. "A review of the use of the basic cation saturation ratio and the "ideal" soil". *Soil Sci. Soc. Amer. J.*, **71**:259–265.

Lawrence, G., and T. G. Huntington. 1999. "Soil-calcium depletion linked to acid rain and forest growth in the eastern United States," *Science for a Changing World Report*, WRIR 98-4267. U.S. Geological Survey. bqs.usgs.gov/acidrain/WRIR984267.pdf.

Lindsay, W. L. 1972. "Inorganic phase equilibria of micronutrients in soils," in J. J. Mortvedt, P. M. Giordano, and W. L. Lindsay (eds.), *Micronutrients in Agriculture* (Madison, Wis.: Soil Science Society of America).

Lovley, D. R. 1996. "Microbial reduction of iron, manganese and other metals," *Advances in Agronomy*, **54**:175–231.

Marschner, H. 1995. *Mineral Nutrition of Higher Plants*, 2nd ed. (New York: Academic Press).

Marschner, H., A. Kalisch, and V. Romheld. 1974. "Mechanism of iron uptake in different plant species," *Proc. 7th Int. Colloquium on Plant Analysis and Fertilizer Problems*, Hanover, West Germany.

McBride, M. B. et al. 2000. "Molybdenum uptake by forage crops grown on sewage sludge-amended soils in the field and greenhouse," *J. Environ. Qual.*, **29**:848–854.

Mengel, K., and J. M. Kirby. 2001. *Principles of Plant Nutrition.* (Dordrecht, Netherlands: Kluwer Academic Publishers), 864 pp.

Mortvedt, J. J., F. R. Cox, L. M. Shuman, and R. M. Welch (eds.). 1991. *Micronutrients in Agriculture*. SSSA Book Series, No. 4 (Madison, Wis.: Soil Science Society of America).

Murphy, L. S., and L. M. Walsh. 1972. "Correction of micronutrient deficiencies with fertilizers," in J. J. Mortvedt, P. M. Giordano, and W. L. Lindsay (eds.), *Micronutrients in Agriculture* (Madison, Wis.: Soil Science Society of America).

Myneni, S.C.B., T. K. Tokunaga, and G. E. Brown. 1997. "Abiotic selenium redox transformations in the presence of Fe (II,III) oxides," *Science*, **278**:1106–1109.

Naidu, R., E. Smith, G. Owens, P. Bhattacharya, and P. Nadebaum (eds.). 2006. *Managing Arsenic in the Environment: From Soil to Human Health* (Collingwood, Australia: CSIRO Publishing), 656 pp.

Olsen, R. A., R. B. Clark, and J. H. Bennett. 1981. "The enhancement of soil fertility by plant roots," *Amer. Scientist*, **69**:378–384.

Rashid, A., and J. Ryan. 2004. "Micronutrient constraints to crop production in soils with mediterranean-type characteristics: A review," *Journal of Plant Nutrition*, **27**:959–975.

Richards, B. K., T. Steenhuis, J. Peverly, and M. B. McBride. 1998. "Metal mobility at an old heavily loaded sludge application site," *Environmental Pollution*, **99**:365–377.

Ryan, J., and S. N. Hariq. 1983. "Transformation of incubated micronutrients in calcareous soils," *Soil Sci. Soc. Amer. J.*, **47**:806–810.

Sillanpaa, M. 1982. *Micronutrients and the Nutrient Status of Soils: A Global Study* (Rome: U.N. Food and Agricultural Organization).

Spencer, J. E. 2000. "Arsenic in groundwater," *Arizona Geology*, **30**(3).

Stevenson, F. J. 1986. *Cycles of Soil Carbon, Nitrogen, Sulfur and Micronutrients* (New York: Wiley).

Taylor, M. Z. 2005. No bad soils. AgWeb.com division of Farm Journal, Inc. http://www.agweb.com/get_article.asp?sigcat=topproducer&pageid=117257 (posted 26 August 2005 verified 26 August 2006).

Terry, N., and G. Banuelos (eds.). 2000. *Phytoremediation of Contaminated Soil and Water* (New York: Lewis Publishers).

Vieira, R. F., E. J. B. N. Cardoso, C. Vieira, and S. T. A. Cassini. 1998. "Foliar application of molybdenum in common beans: I. Nitrogenase and reductase activities in a soil of high fertility," *J. Plant Nutrition*, **21**:169–180.

Viets, F. J., Jr. 1965. "The plants' need for and use of nitrogen," in *Soil Nitrogen* (*Agronomy*, No. 10) (Madison, Wis.: American Society of Agronomy).

Weil, R. R., C. D. Foy, and C. A. Coradetti. 1997. "Influence of soil moisture regimes on subsequent soil manganese availability and toxicity in two cotton genotypes," *Agron. J.*, **89**:1–8.

Weil, R. R., and S. S. Holah. 1989. "Effects of submergence on the availability of micronutrients in three Ultisols," *Plant and Soil*, **114**:147–157.

Welch, R. M. 1995. "Micronutrient nutrition of plants," *Critical Reviews in Plant Science*, **14**(1):49–82.

Wood, B. W., C. C. Reilly, and A. P. Nyczepir. 2004. "Mouse-ear of pecan: A nickel deficiency," *Hortscience*, **39**:1238–1242.

Yermiyahu, U., R. Keren, and Y. Chen. 1995. "Boron sorption by soil in the presence of composted organic matter," *Soil Sci. Soc. Amer. J.*, **59**:405–409.

Zayed A. E., E. Pilon-Smits, M. de Souza, Z. Q. Lin, and N. Terry. 2000. "Remediation of selenium-polluted soils and waters by phytovolatilization," Chapter 4, in N. Terry and G. Bañuelos (eds.), *Phytoremediation of Contaminated Soil and Water* (New York: Lewis Publishers), 61–83 pp.

Pastoral scene belies unmanaged nutrients. (R. Weil)

16

PRACTICAL NUTRIENT MANAGEMENT

For every atom lost to the sea, the prairie pulls another out of the decaying rocks. The only certain truth is that its creatures must suck hard, live fast, and die often, lest its losses exceed its gains.
—ALDO LEOPOLD, A SAND COUNTY ALMANAC *(1949)*

As stewards of the land, soil managers must keep nutrient cycles in balance. By doing so they maintain the soil's capacity to supply the nutritional needs of plants and, indirectly, of us all. While undisturbed ecosystems may need no intervention, few ecosystems are so undisturbed. More often, human hands have directed the output of the ecosystem for human ends. Forests, farms, fairways, and flower gardens are ecosystems modified to provide us with lumber, food, recreational opportunities, and aesthetic satisfaction. By their very nature, managed ecosystems need management.

In managed ecosystems, nutrient cycles can become unbalanced through increased removals (e.g., harvest of timber and crops), through increased system leakage (e.g., leaching and runoff), through simplification (e.g., monoculture, be it of pine tree or sugarcane), through increased demands for rapid plant growth (whether the soil is naturally fertile or not), and through increased animal density (especially if imported feed brings in nutrients from outside the ecosystem). Some of the greatest impacts of land management are felt not on the land but in the water, where excess nutrients play havoc with aquatic ecosystems. The land manager is, of necessity, also a nutrient and environmental manager.

In this chapter we will discuss methods of enhancing nutrient recycling, as well as sources of additional nutrients that can be applied to soils or plants. We will learn how to diagnose nutritional disorders of plants and correct soil fertility problems. Building on the principles set out earlier, this chapter contains much practical information on profitable production of abundant, high-quality plant products and on maintaining the quality of both the soil and the rest of the environment.

16.1 GOALS OF NUTRIENT MANAGEMENT[1]

Nutrient management is one aspect of a holistic approach to managing soils in the larger environment. It aims to achieve four broad, interrelated goals: (1) cost-effective production of high-quality plants, (2) efficient use and conservation of nutrient resources, (3) maintenance or enhancement of soil quality, and (4) protection of the environment beyond the soil.

Plant Production

150-yr study of soil fertility impacts on plant ecology:
http://news.bbc.co.uk/1/hi/sci/tech/4766081.stm

Three of the primary types of plant production in which people engage are: (1) *agriculture,* (2) *forestry,* and (3) *ornamental landscaping.* Agriculturists range from small-scale subsistence farm families or home gardeners who produce only for their own use to large-scale farmers whose primary goal is to make a profit from the plants and animals they produce. Regardless of the scale, the main nutrient-management goal for agriculture is to increase plant yield and quality thereby helping subsistent farmers to feed their families and commercial farmers to enhance their incomes. Unfortunately, farmers in both groups tend to judge the success or failure of their management schemes in only three to six months, the period of growth of the crops they are producing. This is nothing like the span of several human generations that may be needed to fully evaluate the effectiveness of the practices they choose to use.

In forestry, the principal plant product may be measured in terms of volume of lumber or paper produced. In these instances, nutrient management aims to enhance rate of growth so that the time between investment and payoff can be minimized. The survival rate of tree seedlings is also important. Wildlife habitat and recreational values may also be primary or secondary products. The time frame in forestry, measured in decades or even centuries, tends to limit the intensity of nutrient management interventions that can be profitably undertaken.

When soils are used for ornamental landscaping purposes, the principal objective is to produce quality, aesthetically pleasing plants. Whether the plants are produced for sale or not, relatively little attention is paid to yield of biomass produced. Hardiness, resistance to pests, color, and abundance of blooms are much more important. Labor costs and convenience are generally of more concern than fertilizer costs; hence, expensive, slow-release fertilizers are widely used.

Conservation of Nutrient Resources

Two concepts that are key to the goal of conserving nutrient resources are (1) renewal or reuse of the resources, and (2) nutrient budgeting that reflects a balance between system inputs and outputs.

The first law of thermodynamics suggests that all material resources are ultimately renewable, since the elements are not destroyed by use but are merely recombined and moved about in space. In practical terms, however, once a nutrient has been removed from a plot of land and dispersed into the larger environment, it may be difficult if not impossible to use it again as a nutrient for plant growth. For instance, phosphorus deposited in a lake bottom with eroded sediment and nitrogen buried in a landfill as a component of garbage are not available for reuse. In contrast, land application of composted municipal garbage and irrigation with sewage effluent are examples of practices that treat nutrients as reusable resources.

Recycling is a form of reuse in which nutrients are returned to the same land from which they were previously removed. Litter fall from perennial vegetation recycles nutrients to the soil naturally. Leaving crop residues in the field and spreading barnyard manure onto the land from which the cattle feed was harvested are both examples of managed nutrient recycling. The term *renewable resource* best applies to soil nitrogen, which can be replenished from the atmosphere by biological nitrogen fixation (see Sections 13.10 to 13.12). Manufacture of nitrogen fertilizers also fixes atmospheric nitrogen, but at a large cost in nonrenewable fossil fuel energy.

---

[1] For an overview of issues and advances in nutrient management for agriculture and environmental quality, see Magdoff et al. (1997). For a standard textbook on management of agricultural soil fertility and fertilizers, see Havlin et al. (2005).

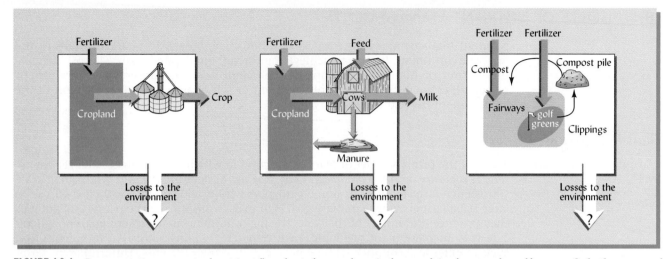

FIGURE 16.1 Representative conceptual nutrient flowcharts for a cash-grain farm, a dairy farm, and a golf course. Only the managed inputs, key recycling flows, and outputs are shown. Unmanaged inputs, such as nutrient deposition in rainfall, are not shown. Outputs that are difficult to manage, such as leaching and runoff losses to the environment, are shown as being variable. Although information on unmanaged inputs and outputs is not always readily available, it must be taken into consideration in developing a complete nutrient management plan. Such flowcharts are a starting point in identifying imbalances between inputs and outputs that could lead to wasted resources, reduced profitability, and environmental damage.

Farm nutrient budgeting made easy (New Zealand):
http://www.ew.govt.nz/enviroinfo/land/management/nutrients/index.htm

Other fertilizer nutrients, such as potassium and phosphorus, are mined or extracted from nonrenewable mineral deposits or from mineral-laden seawater. The size of known global reserves varies according to the nutrient. As farmers in Asia increase their use of phosphorus fertilizer to levels now common in Europe and the United States, the world's high-quality sources of this crucial and irreplaceable element are likely to be depleted within as little as a single century. Already the United States has gone from being the world's largest exporter of phosphate to a net importer, and is predicted to have depleted virtually all of its domestic supplies as early as 2035. Furthermore, as the best, most concentrated, and most accessible sources of these nutrients are depleted, the cost of producing fertilizer will likely rise in terms of money, energy, and environmental disruption. Careful husbandry of nutrient resources must be an integral part of any long-term nutrient management program.

Nutrient Budgets[2]

A useful step in planning nutrient management is to conceptualize the nutrient flows for the particular system under consideration. Such a flowchart should attempt to account for all the major inputs and outputs of nutrients. Simplified examples of such nutrient budgets are shown in Figure 16.1.

Addressing nutrient imbalances, shortages, and surpluses may call for analysis of nutrient flows, not just on a single farm or enterprise, but on a watershed, regional, or even national scale. For example, many countries in Africa are net exporters of nutrients. That is, exports of agricultural and forest products carry with them more nutrients than are imported into the country as fertilizers, food, or animal feed. During the past 30 years, an average of 22 kg/ha N, 2.5 kg/ha P, and 15 kg/ha K have been lost *annually* from about 200 million ha of cultivated land in sub-Saharan Africa (excluding South Africa). This net *negative* nutrient balance appears to be a contributing factor in the impoverishment of African soils, the reduction of agricultural productivity, and the stagnation or decline of national economies (see Section 20.10).

By contrast, in temperate regions, the cultivated land on average receives nutrients in excess of those removed in crops, runoff, and erosion. During the past 30 years, the nearly 300 million ha of cultivated temperate-region soils had a net *positive* nutrient balance of at least 60 kg/ha N, 20 kg/ha P, and 30 kg/ha K. While nutrient deficiencies still occur in some fields, the great majority of soils have experienced a nutrient buildup. Some of the excess nutrients move into streams, lakes, or the atmosphere, where they can contribute to environmental damage (Sections 16.2, 14.2, and 13.4).

---

[2] For discussions of the negative nutrient balance in African soils, see Sanchez et al. (1997) and Smaling et al. (1997).

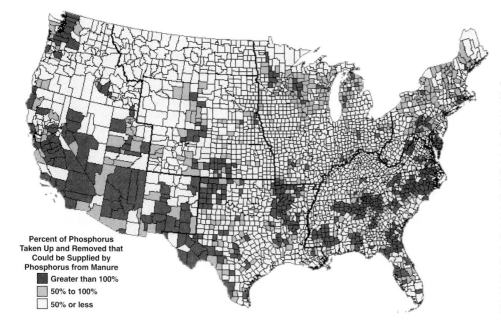

FIGURE 16.2 The phosphorus (P) content of manure produced in each region compared to the amount of P harvested in crops on all the non-legume cropland and hayland in each county of the United States. In the most darkly shaded counties, livestock feeding operations are producing more manure P than could be used by crops, even if all the manure could be applied to the fields that most need it. Agricultural land in these areas is therefore accumulating P, and losses of this element are likely to cause water-quality impairment. [Source: USDA Natural Resources Conservation Service, Resource Assessment and Strategic Planning Division, Washington, D.C. (2000)]

Percent of Phosphorus Taken Up and Removed that Could be Supplied by Phosphorus from Manure

- Greater than 100%
- 50% to 100%
- 50% or less

Changes in the structure of agricultural production in some countries have led to *regional* nutrient imbalances and concomitant serious water-pollution problems. The concentration of livestock production facilities in a region that must import feed from other areas is a case in point. The animal manure produced in these areas contains nutrients far in excess of the amounts that can be used in an efficient and environmentally safe manner by crops in nearby fields (see Figure 16.2 for areas where such concentrations are found). The concentrated poultry industry in Delaware is one example for which the imbalance between phosphorus sources and utilization potential has been documented (Figure 16.3). Similar excess nutrient supplies in several other areas were documented for nitrogen in Table 13.6.

Soil Quality and Productivity

The concept of using nutrient management to enhance soil quality goes far beyond simply supplying nutrients for the current year's plant growth. Rather, it includes the long-term nutrient-supplying and nutrient-cycling capacity of the soil, improvement of

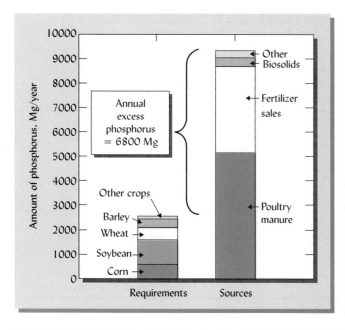

FIGURE 16.3 Phosphorus balance for the State of Delaware. Crop requirements estimated from recommended fertilizer rates based on soil tests and expected crop responses (see Section 16.11) totaled about 2600 Mg P. The manure generated by the state's poultry industry alone contained about twice as much P as required by all the crops in the state. The total amount of P applied in fertilizers *and* manures (plus some other P-containing wastes) totaled more than 9400 Mg P, leaving a surplus of more than 6800 Mg P. If the total P sources were spread on all 217,900 ha of cropland in the state, there would be 31 kg/ha of excess P each year. Most of this excess P came from poultry feed, and has resulted in P buildup in soils and increased P losses in runoff. [Data from Beegle et al. (2002)].

soil physical properties or tilth, maintenance of above- and belowground biological functions and diversity, and the avoidance of chemical toxicities. Likewise, the management tools employed go far beyond the application of various fertilizers (although this may be an important component of nutrient management). Nutrient management requires the integrated management of physical, chemical, and biological processes. The effects of tillage on organic matter accumulation (Chapter 12), the increase in nutrient availability brought about by earthworm activity (Chapter 11), the role of mycorrhizal fungi in phosphorus uptake by plants (Chapter 14), and the impact of fire on soil nutrient and water supplies (Chapter 7) are all examples of components of integrated nutrient management.

16.2 ENVIRONMENTAL QUALITY

Nutrients in the nation's waters—too much of a good thing? http://water.usgs.gov/ nawqa/circ-1136/ circ-1136main.html

Nutrient management impacts the environment most directly with water-quality problems caused by N and P. Together, these two nutrients are the most widespread cause of water-quality impairment in lakes and estuaries, and are second only to sediment (see Chapter 17) among pollutants impairing the water quality of rivers and streams. As explained in Chapters 13 and 14 (Boxes 13.3 and 14.1), the growth of aquatic plants (e.g., seaweeds, algae, and phytoplankton) often explodes when concentrations of N or P exceed critical levels, leading to numerous undesirable changes in the aquatic ecosystem (Table 16.1). In most freshwater (lakes and streams), P is the limiting nutrient that can set off eutrophication; as little as 0.02 mg/L of total P is thought to be potentially damaging in lakes and streams. In saltier waters (estuaries and coastal areas), N is the nutrient most likely to cause eutrophication (see Box 13.3). Levels of total dissolved N above 2 mg/L are often considered above normal and damaging to the ecosystem. In addition to stimulating eutrophication, N in the form of dissolved ammonia gas (NH_3) can be directly toxic to fish. For this reason, average levels of ammonium-N (which is in equilibrium with ammonia) should be kept below 2 mg/L. The nitrate form of N is also of concern, as levels above 10 mg nitrate-N/L are considered unfit for human drinking water (see Box 13.2).

Most industrialized countries have made great strides in reducing nutrient pollution from factory and municipal sewage outfalls (called **point sources** because these sources are clearly localized). However, much less has been accomplished with regard to controlling nutrients in the runoff water coming from the landscape (called **nonpoint sources** because these sources are diffuse and not easily identified). Streams draining forestland and rangeland are generally far lower in nutrients than those draining watersheds dominated by agricultural and urban land uses. Two examples are worth considering. Scientists estimate that agricultural activities account for some 40% of the N and

TABLE 16.1 Adverse Aquatic Ecosystems Effects of Eutrophication Stimulated by Excess N or P [a]

Eutrophication results from excessive nutrient inputs that stimulate increased algal growth and species changes that lead to the other adverse effects listed. Freshwater systems are usually most sensitive to inputs of P, while more saline estuarine and ocean systems are generally more sensitive to inputs of N.

• Increased phytoplankton growth	• Increased masses of algae covering bottom sediments and foliage of submerged plants	• Oxygen depletion
• Increases in bloom-forming, and in some cases toxic, phytoplankton species	• Increased water turbidity and decreased light transmission	• Fish kills
• Blooms of toxic dinoflagellates	• Decreased growth of submerged aquatic vegetation grasses that serve as fish habitat and underpin some aquatic food webs	• Death of coral reef communities
• Increased blooms of gelatinous zooplankton in coastal waters		• Loss of desirable fish species
• Shifts in species of macroscopic plants		• Reduced harvests of marketable fish and shellfish
		• Taste, odor, and water treatment problems

[a]Also see Boxes 13.3 and 14.1 for further background on eutrophication.

P loads in the 170,000 km^2 Chesapeake Bay watershed (on the U.S. East Coast), while point sources account for another 20%. In the more intensively farmed and less densely populated Mississippi River basin covering nearly 3 million km^2, some 65% of the N load comes from agriculture and only 6% from point sources. See Chapters 13 and 14 to review the many pathways by which N and P can be lost from soils.

Nutrient Management Plans

One tool for reducing nonpoint source N and P pollution is a nutrient management plan—a document that records an integrated strategy and specific practices for how nutrients will be used in plant production (Table 16.2). Increasingly, these plans are seen as legal documents used to implement government regulations on nonpoint source nutrient pollution. Generally, the document is prepared by a specially trained soil scientist who consults closely with the landowner to meet both environmental goals and the practical needs. The plan attempts to balance the inputs of N and P (and other nutrients) with their desirable outputs (i.e., removal in harvested products) to prevent undesirable outputs (runoff, leaching) that exceed **maximum allowable daily loadings (MDL)**, the largest amount of nutrient runoff and leaching (in g ha^{-1} day^{-1}) permitted from an area of land. An important component of many nutrient management plans, the *P site index*, is explained in Section 16.12.

Best Management Practices

The primary means of preventing nutrient pollution of waters is to avoid excessive applications of these nutrients on the landscape, and to manage soils and plants to reduce the transport of nutrients (and other pollutants) from soils to groundwater and surface waters. In the United States, practices officially sanctioned to implement these strategies are known as *best management practices* (BMPs). Four general types of practices will now be briefly considered: (1) buffer strips, (2) cover crops, (3) conservation tillage, and (4) forest stand management.

Riparian Buffer Strips[3]

Buffer strips of dense vegetation situated along the bank of a stream or other body of water (the **riparian zone**) are a simple and generally cost-effective method to protect water from the polluting effects of a nutrient-generating land use. Fertilized cropland, poultry or livestock operations, farmland that has been amended with organic wastes, forest harvest operations, and urban development are examples of land uses that have the potential to generate nutrient or sediment loadings. The vegetation in the buffer

---

[3] For details about riparian buffer strips in forested areas, see Belt and O'Laughlin (1994); for information related to buffers in agricultural settings, see Chesapeake Bay Program (1995).

TABLE 16.2 Typical Components of a Nutrient Management Plan

This chapter explains many of the tools used to plan for nutrient application to land in a manner that maximizes nutrient-use efficiency and minimizes water pollution risks.

• Aerial site photographs or maps and a soil map	• Realistic yield goals and a description of how they were determined	• Planned rates, methods, and timing of nutrient applications
• Current and/or planned plant production sequences or crop rotations	• A complete nutrient budget for N, P, and K in the production system	• Location of environmentally sensitive areas or resources, if present
• Soil test results and recommended nutrient application rates	• An accounting of all nutrient inputs such as fertilizers, animal manure, sewage sludge, irrigation water, compost, and atmospheric deposition	• The potential risk of N and/or P water pollution as assessed by a *N Leaching Index*, a *P Site Index*, or other acceptable assessment tools
• Plant tissue analysis results		
• Nutrient analysis of manure or other soil amendments		

FIGURE 16.4 The zone of grassy vegetation maintained along the left bank of this stream serves as a riparian buffer strip. The grass should substantially reduce the amount of nutrients washing into the stream from the uphill cropland. However, by retaining nutrients from the runoff water, the buffer strip itself may become nutrient-saturated and lose some of its effectiveness. In the forested areas in the background, a similar strip of land along the stream will be left undisturbed to act as a buffer during logging operations. (Photo courtesy of R. Weil)

strips may consist of natural or planted species, including grasses (Figure 16.4), shrubs, trees (Plate 60, after page 112), or a combination of these vegetation types (Figure 16.5).

HOW THEY WORK.   Water running off the surface of the nutrient-rich land passes through the riparian buffer strip before it reaches the stream. Trees (and their litter layer) or grass plants (and their thatch layer) reduce water velocity and increase the tortuosity of the water's travel paths. Under these conditions, most of the sediment and attached nutrients will settle out of the slowly flowing water. In addition, dissolved nutrients are adsorbed by the soil, immobilized by microorganisms, or are taken up by the buffer strip plants. Occasional mowing of grass or thinning of tree stands may help maintain high

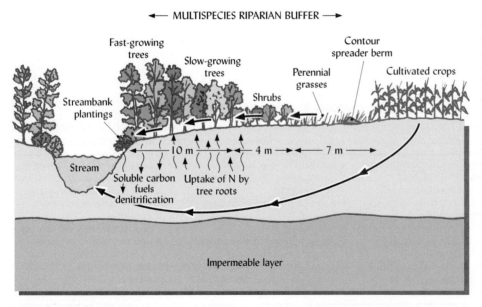

FIGURE 16.5 A general plan for a multispecies riparian buffer strip designed to protect the stream from nutrients and sediment in cropland runoff while also providing wildlife habitat benefits. A grass-covered level berm spreads runoff water evenly to avoid gullies. The perennial grasses filter out sediments and take up dissolved nutrients. Deep tree roots remove some nutrients from shallow groundwater. Soluble carbon from tree litter percolates downward to provide an energy source for anaerobic denitrifying bacteria that can remove additional nitrogen from shallow groundwater. The woody vegetation also provides wildlife habitat, shades the stream, and provides in-stream woody debris for fish habitat. A total buffer width of 10 to 20 m is usually sufficient to obtain most of the potential environmental benefits. [Diagram courtesy R. Weil]

rates of nutrient uptake. The decreased flow velocity also increases the retention time—the length of time during which microbial action can work to break down pesticides before they reach the stream. Under some circumstances, buffer strips along streams can also reduce the nitrate levels in the groundwater flowing under them (although emission of nitrous oxide to the atmosphere may be a consequence of nitrate reduction; see Figure 13.17).

DESIGN AND MANAGEMENT. To preserve the dense vegetation and litter cover, riparian zones should be kept off limits to timber harvest machinery, and if the adjacent land is grazed, cattle should be fenced out. The width needed for optimum cleanup may vary from 6 to 60 m, although a width of 10 to 20 m is usually sufficient to obtain most of the nutrient- and sediment-removal benefits on slopes of less than 8% (Figure 16.6). Setting land aside as a buffer often represents a significant reduction in harvestable cropland or timberland (See Figure 17.31, *left*). However, a well-designed buffer strip can compensate the landowner for this economic loss by providing some real benefits: turnaround space for field equipment, valuable hay from grass buffers, improved fishing by shading the stream and providing large woody debris for fish habitat, and enhanced recreational values associated with increased wildlife populations. Thus, installation of buffer strips can often be a win-win situation.

Cover Crops

Dynamic cover crop bibliography:
http://www.nal.usda.gov/wqic/Bibliographies/dynamic.html#cover

Instead of being harvested, a *cover crop* is grown to provide vegetative cover for the soil and then is killed and either left on the surface as a mulch, or tilled into the soil as a *green manure*. In climates with enough precipitation to allow for some water use by the cover crop, numerous benefits can be achieved in comparison to leaving the soil unvegetated for the off-season. The plants may provide habitat for wildlife and for beneficial insects; they protect the soil from the erosive forces of wind and rain (see Section 17.7); they add to the soil organic matter (see Section 12.7); and, if leguminous, they may increase the available nitrogen in the soil (see Section 13.15).

Cover crops can also reduce the loss of nutrients and sediment in surface runoff. First, the protective foliage and litter prevent the formation of a crust at the soil surface, thus maintaining a high rate of infiltration (see Figure 6.8, *right*) and reducing runoff. Second, for the runoff that does occur, the cover crop helps remove both sediment and nutrients by the same mechanisms that operate in a buffer strip, as previously described.

COVER CROPS REDUCE LEACHING LOSSES. Cover crops can also serve as important nutrient managment tools to reduce the leaching losses of nutrients, principally nitrogen (Figure 16.7). In many temperate humid regions, the greatest potential for leaching of nitrate from cropland occurs during the fall and winter, after harvest and before planting of the main crop in the spring. During this time of vulnerability, an actively growing cover crop will reduce percolation of water and remove much of the nitrogen from the water that does percolate, incorporating this nutrient into plant tissue. For this purpose, an ideal

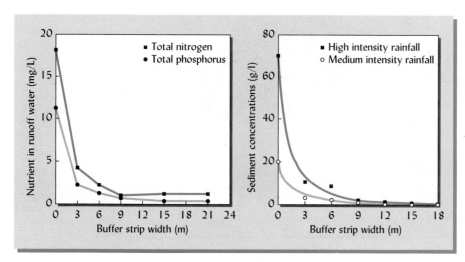

FIGURE 16.6 Removal of nutrients and sediments from runoff by vegetative buffer strips of various widths. Nitrogen and phosphorus were measured in the runoff from a field that had received swine manure (*left*). In another experiment, sediment was measured in the runoff from a field left fallow that had received different intensities of rainfall (*right*). Note that most of the nutrients and the sediment were removed in the first 9 m of the vegetative buffer strips. Depending on the type of vegetation and soil characteristics, the width of satisfactory buffers may vary from 6 to 60 m. [Nutrient data from Chauby et al. (1994); sediment data from Robinson et al. (1996)]

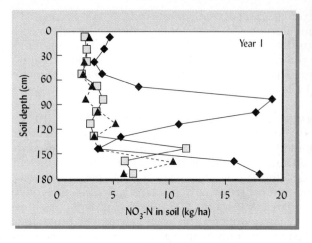

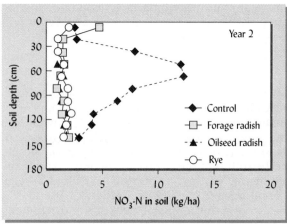

Forage radish   Rye

FIGURE 16.7 Cover crops can capture soluble nitrogen (N) left in the soil profile after the main cropping season. They thereby substantially reduce N leaching to groundwater during the winter. Forage radish and rye are among the temperature region cover crops (*photo*) capable of capturing more than 100 kg/ha of such residual N in fall, cleaning the soil profile of soluble N to considerable depths. The graphs show soil nitrate-N in November of two years, expressed as kg N/ha for each 15 cm depth increment. For comparison, the control plots had some weeds, but no cover crop. Oilseed radish and rye were grown in different years on the sandy Ultisol. [Unpublished data of Dean and Weil, Univ. of Maryland]

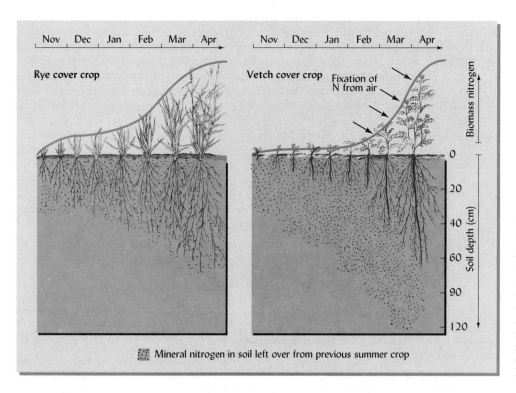

Mineral nitrogen in soil left over from previous summer crop

FIGURE 16.8 Relative effectiveness of a nonlegume (rye) and a legume (vetch) winter cover crop in mopping up the nitrate remaining in the soil after the harvest of a heavily fertilized summer crop. Rye grows very rapidly during mild fall weather just after harvest, while the legume grows very little until the soil warms again the following spring. These traits, and the facts that, unlike the legume, rye is dependent on soil nitrogen and has a fibrous root system, combine to make rye an excellent choice of cover crop to reduce winter leaching of soil nitrate.

FIGURE 16.9 As for any plants, vigorous growth of a cover crop requires adequate levels of available nutrients in the soil. The thin, slow-growing oat cover crop on the low-P and -K soil in the foreground is not able to help very much in preventing nitrate leaching on this coarse-textured Ultisol. The soil in the background with much more vigorously growing oats received phosphorus and potassium during the previous year at the rates of 36 and 70 kg/ha, respectively. (Photo courtesy of R. Weil)

cover crop should produce an extensive root system, as quickly as possible, once the main crop has ceased growth. Largely because of their more rapid root growth in fall, winter annual cereals (rye, wheat, oats) and Brassicas (rape, forage radish, mustards) have proven to be more efficient than legumes (vetch, clover, etc.) at mopping up left-over soluble nitrogen (Figure 16.8).

A cover crop, like any plant, requires a balanced nutrient supply and will not be capable of effectively reducing nitrate leaching if the soil is poorly supplied with nutrients (Figure 16.9). In this regard, P and K are especially important for cover crops growing during periods of cool temperatures. On sandy, low organic matter soils it may even be necessary to apply a small dressing of N in fall to promote vigorous growth that will enable the cover crop roots to catch up with N already moved deeply down the profile, as shown in Figure 16.7.

Conservation Tillage

The term **conservation tillage** applies to agricultural practices that keep at least 30% of the soil surface covered by plant residues. The effects of conservation tillage on soil properties and on the prevention of soil erosion are discussed elsewhere in this textbook (see Sections 6.4 and 17.6). Here, we emphasize the effects on nutrient losses.

Compared to plowed fields with little residue cover, conservation tillage usually reduces the total amount of water running off the land surface, and reduces even more the load of nutrients and sediment carried by that runoff (Tables 16.3 and 14.3). When combined with a cover crop, the reductions are greater still. In most situations, the less the soil surface is disturbed by tillage, even when manure or sewage sludge is spread on it, the smaller are the losses of nutrients in surface runoff. The relatively small amounts of nutrients lost from untilled land (no-till cropland, pastures, and forests) tend to be mostly dissolved in the water rather than attached to sediment particles, while the reverse is true for tilled land. Because of large, sediment-associated nutrient losses, the total nutrient loss in surface runoff from conventionally tilled land generally is far greater than that from land where no-till or conservation tillage methods are used.

On the other hand, the loss of nutrients by leaching can be somewhat greater with conservation tillage than conventional tillage. In conservation tillage systems, a higher percentage of the precipitation or irrigation water infiltrates into the soil, where it may carry nutrients downward. Over time, many no-till soils develop large macropores (such as worm burrows) that are open to the soil surface. Rain and irrigation water may move down rapidly through these large macropores. However, nutrients held in the finer pores of the soil matrix (as opposed to those on the soil surface) are bypassed by such

TABLE 16.3 Effect of Land Use and Tillage on the Loss of Nutrients in Runoff

The alfalfa was a three-year-old stand, and the cultivated plots were planted to corn after having been in grass for 10 years. Ridge tillage is a form of conservation tillage that leaves the soil covered by residues in winter and disturbs only part of the soil in spring.

Water/nutrient loss	Tillage system		
	Untilled alfalfa	Ridge-tilled corn	Conventionally tilled corn
Runoff, % of rainfall	18	33	40
Nutrient loss[a] kg/ha/yr			
Nitrogen	13	49	315
Phosphorus	0.21	1.12	2.65

[a] Sediment and runoff.
Data for a silty clay loam Hapludalf in Ohio, from Thomas et al. (1992).

flow and do not leach into the lower horizons (see Figure 6.29). In such situations, nutrient leaching may actually be less where conservation tillage is used.

Nutrient Losses Associated with Forest Management

Manage Forests for Healthy Ecosystems:
http://www.utextension.utk edu/publications/pbfiles/ pb1574.pdf

Undisturbed forests lose nutrients primarily by (1) leaching and runoff of dissolved ions and organic compounds, (2) erosion of nutrient-containing organic litter and mineral particles, and (3) volatilization of certain nutrients, especially during fires. Most often, the output of such elements as calcium, magnesium, and potassium (but usually not nitrogen) in streams is greater than the input from atmospheric deposition. However, weathering from soil and rock minerals, combined with atmospheric deposition, usually can maintain the plant-available supply of these elements. Nitrogen losses from forested ecosystems commonly range from about 1 to 5 kg/ha each year, while atmospheric inputs of nitrogen (not including biological nitrogen fixation) are usually two to three times as great. Annual losses of phosphorus from forest soils are typically very low (<0.1 kg/ha), and are closely balanced by inputs of this element from the atmosphere and by slow release from mineral weathering.

Management of forests to produce marketable wood products tends to increase losses by all three pathways just mentioned, plus it adds a fourth very significant pathway, namely, removal of nutrients in forest products (whole trees, logs, or pine straw). Forest management practices that physically disturb the soil (and therefore tend to increase nutrient losses mainly by erosion) include building roads for timber harvesting, dragging logs on the ground (skidding), and after-harvest site preparation for planting new trees (see Section 17.9).

SOIL DISTURBANCE. Disturbance of forest soil not only leaves it more vulnerable to erosion, but also can alter the nutrient balance in several other ways. Carefully planned tree harvesting and regeneration using methods that minimize soil disturbance and hasten revegetation can keep nutrient losses to low levels. Two practices should generally be avoided, as they can be particularly damaging to the forest nutrient cycle: First, extended suppression of unwanted vegetation with repeated use of herbicides is inadvisable because it delays the repopulation of the soil with active roots (Figure 16.10). Second, windrowing of stumps and slash (tree branches and tops left after harvesting logs) can be detrimental even if it clears land for easy replanting. Organic matter decomposition and mineral nutrient release are accelerated because of physical mixing of the O and A horizons, and a large proportion of site nutrients are concentrated in the area of the windrow, causing less efficient plant use and greater susceptibility to leaching.

HARVEST METHODS. *Clear-cutting* calls for the simultaneous harvest of all the trees in an area of forestland. This practice is part of *even-aged stand management*, as all the trees in a stand will be planted or will regenerate at the same time. Large openings in the canopy, such as are produced by a clear-cut, are necessary for the regeneration of shade-intolerant species, including many of the most economically important conifers. Having all trees in a stand be of the same age and species simplifies future management

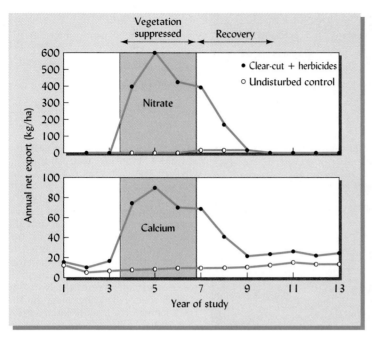

FIGURE 16.10 Greatly elevated export of nitrogen and calcium in stream water from an experimental forest watershed where clear-cut harvesting was followed by herbicide suppression of unwanted vegetation for three years. Although the losses shown are about 10 times greater than losses reported from other clear-cut watersheds, they amply demonstrate the potential for high nutrient losses when forest ecosystems are drastically disturbed. The extraordinarily high losses were partly due to the nearly nitrogen-saturated condition of this New Hampshire hardwood forest, and partly to the unusual extended period before living roots were allowed to repermeate the soils and take up nutrients released from litter and slash decomposition. [Data from Bormann and Likens (1979). © Springer-Verlag]

(fertilization, thinning, and harvesting can be uniform) and usually maximizes net productivity.

Selective cutting of only a few mature trees at a time confers several advantages. It allows the forest soil to continuously maintain its network of active roots and mycorrhizae, creates only modest disturbance of the forest canopy and landscape appearance, and allows for the maintenance of high species diversity. However, compared to even-aged management, selective cutting is often more difficult to manage, requires the stand to be disturbed with roads and logging equipment more frequently, and often tends to remove only the best trees, leaving behind the poorest-quality trees to populate and regenerate the stand.

Whichever harvest strategy is used, it is important to minimize disturbance to the forest floor, during both harvest and regeneration. Leaving slash and stumps in place after a clear-cut harvest encourages an even distribution of nutrients (Figure 16.11).

FIGURE 16.11 Clear-cut harvest of loblolly pine on Coastal Plain soils. A few mature trees were left standing as seed trees. The forest floor is covered with slash of all sizes, but little living vegetation has appeared (one month after harvest in spring). Removal of the tree canopy will allow the sun to warm the soil, accelerating mineralization of nutrients in the forest floor, but decomposition of the high-C/N-ratio slash may immobilize much of the nitrogen released. (Photo courtesy of R. Weil)

FIGURE 16.12 Clear-cut harvest of a red alder stand on steep Andisols in Oregon. The slash and root mass from this nitrogen-fixing species has a narrow C/N ratio, so mineralization is rapid and nitrogen losses after harvest may be high. In this particular case, the risk of nitrogen loss was increased further because the operator wanted to replant with Douglas fir and therefore planned to use herbicides to suppress weeds and alder regrowth. Note the crawler tractor skidding a log and the truck taking on a load. (Photo courtesy of R. Weil)

The slash in most forests has a very high C/N ratio (see Section 12.3), so its slow decay promotes immobilization and reduces the loss of nitrogen mineralized in the forest floor. On the other hand, when slash has a lower C/N ratio, it may release nitrogen faster than the newly establishing vegetation can use it. This is the case for the nitrogen-fixing red alder forests of the Pacific Northwest (Figure 16.12), and to a lesser degree for other hardwoods if logged in full summer leaf.

NUTRIENT LOSSES. Because of increased litter decomposition and reduced plant uptake, streams draining many clear-cut watersheds periodically carry somewhat elevated levels of nitrogen and other nutrients for many years after the timber harvest has occurred (Figure 16.13). The concentrations of nitrogen are usually (but not always)

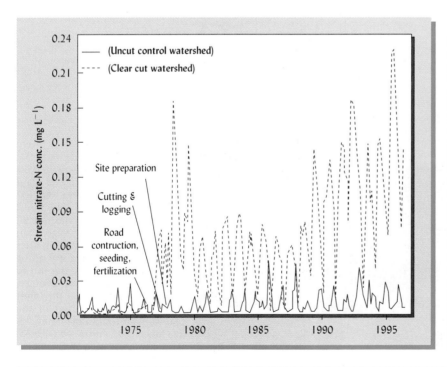

FIGURE 16.13 Mean monthly nitrate-N concentrations in stream water from two commercial forested watersheds in North Carolina. Data are shown for a calibration period when both watersheds were managed identically (1971–1976), a treatment period (1976–1977) when one watershed was clear-cut and replanted, and a post-harvest period (1978–1996) when both watersheds were left undisturbed. Nitrate-N concentrations increased in the logged watershed and continued to be higher even after 20 years of regrowth, suggesting that nutrient-recycling processes were disrupted in some fundamental way. However, when concentrations were multiplied by streamflow volume (not shown) to calculate nitrate-N mass losses, the differences between watersheds were relatively small, nitrate-N losses in the logged watershed ranging from 0.25 to 1.27 kg N ha^{-1} yr^{-1} more than the control watershed during the first 5 years after logging. These differences can be compared to 4.5 kg N ha^{-1} yr^{-1} received in atmospheric deposition. [Redrawn from Swank et al. (2001)]

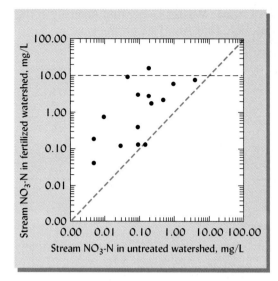

FIGURE 16.14 Short-term peak nitrate-N (NO_3–N) concentrations in streams draining forested watersheds with or without application of 200 to 250 kg N/ha as urea fertilizer. Each data point represents one pair of watersheds (fertilized versus untreated). The diagonal dashed line represents no treatment effect; points above the line indicate increased nitrate-N concentration compared to the control. Note that log scales are used because the range of concentrations among watersheds is very large (0.003 to 20). In all but two of the watersheds, fertilizing the forest substantially increased peak nitrate concentrations in the stream water, in some cases to more than 10 mg/L. As in agriculture, best management practices are necessary in forestry to protect environmental quality while increasing plant productivity. [Data from a survey of research results in Brown and Binkley (1994); see also Binkley et al. (1999)]

too small (<2 mg N/L) to immediately threaten stream water quality. However, this loss of nutrients, combined with nutrient removal in the harvested trees, raises concerns about soil depletion and site productivity, especially on sites that are nutrient-poor to begin with.

FERTILIZATION. Fertilizer application to forests is becoming an increasingly common practice. For example, every year about 1 million hectares of short-rotation slash pine and loblolly pine forests in the southern United States receive applications of N and/or P. These nutrients, along with some sulfur and boron, are also applied widely to forests in the Pacific Northwest. Foresters apply fertilizers for the same reasons that farmers do. Fertilizers prevent soil impoverishment in the face of the nutrient losses just discussed. Application of fertilizer can also raise the fertility level of a site, substantially accelerating plant growth and raising productivity. As might be expected, increases in peak nutrient exports can often be detected when forested watersheds undergo fertilization (Figure 16.14). Although the effects of forest fertilization on water quality do not yet approach those associated with fertilizer use in agriculture, foresters would be well served to study the lessons learned from fertilizer use on farms and so avoid making the nutrient management mistakes that have plagued agriculture.

In the remainder of this chapter we will concentrate on the properties and uses of various nutrient sources, measures designed to meet the goals of nutrient management outlined in the preceding, and methods of assessing plant and soil nutrient status in order to determine which nutrients need to be supplemented and in what quantities.

16.3 NUTRIENT RESOURCES AND CYCLES

Forest Nutrient Cycles:
http://www.umext.maine
.edu/onlinepubs/htmpubs/
7029.htm

Internal nutrient resources come from within the ecosystem, be it a forest, a watershed, a farm, or a home in the suburbs, and are generally preferred, since their financial and environmental costs are minimized. These resources include the process of mineral weathering within the soil profile, biological nitrogen fixation, acquisition of nutrients from atmospheric deposition, and various forms of internal recycling, such as animal manure application, utilization of cover crops, and plant-residue management.

If internal resources prove inadequate, nutrients must be imported from resources external to the system, (see Section 13.15). Such *external* resources are usually purchased inorganic or organic fertilizers. Some of the organic residues and (so-called) wastes available in the United States are listed in Table 16.4, along with the percentage of each that is used on the land.

Depending on the parent materials and climate, weathering of minerals (see Section 2.1) may release significant quantities of nutrients (Table 16.5). For timber production, most nutrients are released fast enough from either parent materials or decaying organic matter to supply adequate nutrients. In agricultural systems, however, some

TABLE 16.4 **Estimated Quantities of Major Organic Wastes Generated Annually in the United States, and Percentages of These Materials That Are Used on the Land**

Organic waste	Annual production, millions of dry metric tons	Used on land, %
Crop residues	450	75
Animal manures	175	90
Municipal refuse	145	10
Logging and wood manufacture	35	10
Industrial organics	9	5
Sewage sludge and septage	6	50
Food processing	3	15

Estimates from USDA (1980) and other sources.

nutrients generally must be added, since nutrients are removed from the land annually in harvested crops. Negligible amounts of nitrogen are released by weathering of mineral parent materials, so other mechanisms (including biological nitrogen fixation) are needed to resupply this important nutrient. Release of nutrients by mineralization of soil organic matter is important in short-term nutrient cycling, but in the long run, the organic matter and the nutrients it contains must be replenished or soil fertility will be depleted.

Nutrient Economy of Forests[4]

RECYCLING. In most forests, organic matter mineralization is the main source of nutrients for tree growth, and the rates of nutrient uptake from the soil by the trees closely match the rates of release by mineralization. Surface plant litter makes up but a small proportion of the nutrients cycled from trees to the soil decomposers each year. The biggest portion comes from a combination of fine roots and mycorrhizal hypha (see Sections 11.7 and 11.9). A second major source of nutrients for each season's tree growth is the recycling process within the trees themselves. Nutrients are translocated from the leaves to the twigs and branches just prior to litter fall. The translocated nutrients are then remobilized early in the next growing season to produce new leaf growth and wood increment. Additional sources are atmospheric deposition and weathering of soil minerals.

Forests have evolved several mechanisms to conserve nutrients in nutrient-poor sites. For instance, conifers in cold regions (e.g., black spruce) may retain their needles

---

[4] For in-depth treatments of this topic, see Perry (1994) and Likens and Bormann (1995). Aber et al. (2000) provide insights on ecological challenges in forestry.

TABLE 16.5 **Amounts of Selected Nutrients Released by Mineral Weathering in a Representative Humid, Temperate Climate, Compared with Amounts Removed by Silvicultural and Agricultural Harvests**

For the forest, weathering release and harvest removal are roughly balanced, but cropping removes much more of some nutrients than can be released by weathering. Leaching losses are not shown.

	Amount released or removed, kg/ha			
	P	K	Ca	Mg
Weathered from igneous parent material over 50 years	5–25	250–1000	150–1500	50–500
Removed in harvest of 50-year-old deciduous bole wood	10–20	60–150	175–250	25–100
Removed by 50 annual harvests of a corn–wheat–soybean rotation	1200	2000	550	500

Estimated from many sources.

TABLE 16.6   Representative Nutrient Distribution in Several Types of Forest Ecosystems[a]

	In vegetation, kg/ha	In forest floor, kg/ha	Residence time,[b] years
Nitrogen			
Boreal coniferous	300–500	600–1100	100–300
Temperate deciduous	100–1200	200–1000	5–7
Tropical rain forest	1000–4000	30–50	0.4–0.8
Phosphorus			
Boreal coniferous	30–60	75–150	150–450
Temperate deciduous	60–80	20–100	4–8
Tropical rain forest	200–300	1–5	0.4–0.8
Potassium			
Boreal coniferous	150–350	300–750	50–150
Temperate deciduous	300–600	50–150	0.8–1.5
Tropical rain forest	2000–3500	20–40	0.1–0.3
Calcium			
Boreal coniferous	200–600	150–500	100–200
Temperate deciduous	1000–1200	200–400	2–4
Tropical rain forest	3500–5000	100–200	0.2–0.4

[a] Data derived from many sources.
[b] Residence time in forest floor (O horizons).

for several decades rather than shed them more frequently, enabling these trees to grow on sites with very low levels of nitrogen availability (the needles are much higher in nitrogen than the woody tissues). Forests generally produce more aboveground biomass (100 to 200 kg) per kilogram of nitrogen taken up than other ecosystems (e.g., corn produces about 60 to 70 kg biomass per kg N taken up). In cold climates or in coniferous forests, slow rates of organic matter decomposition result in the immobilization of the system's nitrogen in the forest floor (Table 16.6).

SURFACE SOIL ENRICHMENT.   Trees may obtain most of their nutrients from the surface horizons, but are also well adapted to gathering nutrients from deep in the soil profile, where much of the nutrient release from parent material takes place. Overall nutrient-use efficiency can sometimes be increased by combining trees and agricultural crops into what are known as *agroforestry systems* (Figures 16.15 and 20.25).

Trees may improve fertility of the upper soil horizons in several ways. In Chapter 2 (Figure 2.27) we saw that trees can act as nutrient pumps, taking up nutrients that occur deep in the profile because of weathering or leaching, and depositing them at the soil surface as litter that will decompose and release the nutrients where they can be of use to relatively shallow-rooted agricultural crops. Nitrogen-fixing trees (mostly legumes) can also add nitrogen to the surface soil with their nitrogen-rich leaf litter. Trees may also enhance the fertility of the soil in their vicinity by trapping wind-blown dust, thus increasing the deposition of such nutrients as calcium, phosphorus, and sulfur.

Some Effects of Fire

Hot-burning forest wildfires convert a great deal of nitrogen and sulfur, and some phosphorus, to gaseous forms in which they are lost from the site. Burned-over land continues to lose more phosphorus in runoff for several years after a high-intensity burn (see Figure 14.6). Ashes, both from wildfires and from prescribed burns (low-intensity intentional fires), contain high levels of soluble K, Mg, Ca, and P, increasing the short-term availability of these nutrients, but also increasing the rate of loss of these nutrients from the forest ecosystem. Wildfires usually result in greater nutrient losses than do prescribed burns, because the high heat associated with wildfires destroys some of the soil organic matter as well as the aboveground biomass.

FIGURE 16.15 Two examples of agroforestry systems. (*Left*) The deep-rooted *Acacia albida* enriches the soil under its spreading branches (arrow points to a man). Conveniently, these trees leaf out in the dry season and lose their leaves during the rainy season when crops are grown. It is an African tradition to leave these trees standing when land is cleared for crop production. Crops growing under the trees yield more and have higher contents of sulfur, nitrogen, and other nutrients. (*Right*) Branches pruned from widely spaced rows of leguminous trees are spread as a mulch on the soil surface in the alleys between the tree rows, thus enriching the alleys with nutrients from the leaves as well as conserving soil moisture. The crops grown in this alley-cropping system may yield better than crops grown alone, but only if competition between trees and crop plants for light and water can be kept to a minimum. (Photos courtesy of R. Weil)

Rangeland Nutrient Cycling

The burning of rangeland grasses (or crop residues) usually produces much less volatilization of nitrogen and sulfur than do forest wildfires, as grass fires move quickly and burn at relatively low temperatures. While the loss of organic matter consumed by the fire is undeniable, the nutrients released may stimulate enough extra plant biomass production that soil organic matter may actually accumulate to higher levels under grasslands subject to occasional burns than under those where fire is completely controlled. Grazing by large mammals, if not too frequent and intense, can also stimulate increased plant production and quality (Box 16.1). Fire and grazing, both natural components in rangeland ecosystems, can be important tools in nutrient management.

16.4 RECYCLING NUTRIENTS THROUGH ANIMAL MANURES[5]

Maryland's manure-matching service helps manure do the most good: http://www.mda.state.md.us/resource_conservation/financial_assistance/manure_management/index.php

For centuries, the use of farm manure has been synonymous with a successful and stable agriculture. In this context, manure supplies organic matter and plant nutrients to the soil and is associated with the production of soil-conserving forage crops used to feed animals. About half of the solar energy captured by plants grown for animal feed ultimately is embodied in animal manure, which if returned to the soil can be a major driver of soil quality.

Huge quantities of farm manure are available each year for the recycling of essential elements to the land. For each kilogram of liveweight, farm animals produce about 4 kg dry weight of manure per year. In the United States, the farm animal population voids some 350 million Mg of manure solids per year, about 10 times as much as does the human population.

Concentrated Animal-Feeding Operations (CAFOs)

Unfortunately, in most industrialized countries, the advent of huge, concentrated animal-feeding operations (CAFOs) has changed the perception of animal manure from an *opportunity for recycling nutrients* as efficiently as possible to an *obligation for disposing of*

---

[5] Some of the challenges of such recycling are discussed in Gardner (1997).

BOX 16.1 GRAZING MAMMALS AND NUTRIENT CYCLING IN SOILS

In the Serengeti plains of East Africa, the spatial distribution of nonmigratory antelopes and gazelles (and by inference, the lions and cheetahs that prey on them) is influenced by variations in soil fertility. The graphs in Figure 16.16 describe a study in which animals were found grazing most frequently in areas where the soils were high in nitrogen and sodium, two mineral nutrients critical to the health and survival of pregnant or lactating females and their young. Not only did the animals seek out the more fertile soils, but their activities also actually enhanced the cycling of nutrients, making the soils they frequented more fertile. First, the animals left manure and urine that contained most of the nutrients they consumed, but in a more easily decomposable form than in the original plant material. Second, the animals' grazing action seems to have stimulated vigorous growth of the palatable, easily decomposed plant species, thus speeding the cycling of nutrients. Leaf nitrogen concentrations in plant regrowth after grazing generally were higher than in ungrazed plants, making the resulting plant residues more readily recyclable in the soil.

Overgrazing occurs when animal density exceeds the carrying capacity of the land. Improvements in the condition of vegetation from grazing by free-ranging wild animals contrast with the negative impacts caused by overgrazing. Continual grazing by livestock may kill off the most palatable species, so that the vegetation becomes relatively sparse and dominated by less palatable plants. In some cases, the residues of these plants are also less decomposable. The resulting impairment of the nutrient cycling processes accelerates the deterioration of the vegetative cover and exposes the soil to the erosive action of wind and water (see Section 17.11). On the other hand, a well-grazed pasture should not be fertilized as if it were a hayfield. Hay is mechanically removed several times a year from the hayfield, but no urine or feces are returned by grazing animals.

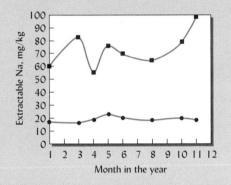

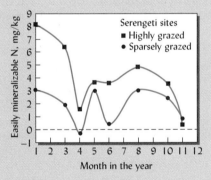

FIGURE 16.16 *Gazelle in the Savannah (top) Plant-available sodium (middle) and nitrogen (bottom) in soils where gazelles habitually graze heavily or almost not at all. Averages for two pairs of sites and two years of study in Tanzania's Serengeti National Park. [Data recalculated from McNaughton et al. (1997)]*

wastes with as little cost and environmental damage as possible (Figure 16.17). Due to inadequate manure-holding and disposal practices, the groundwater under such CAFOs as cattle feedlots is often polluted with nitrates and pathogens, and water from nearby wells may be unfit to drink (Table 16.7).

CATTLE FEEDLOTS. To visualize the enormity of the manure disposal problem, consider a 100,000-head beef feedlot. We can estimate that the feedlot produces 200,000 Mg of manure (dry matter) per year:

$$\frac{4 \text{ Mg } \textbf{manure}}{\text{Mg } \textbf{liveweight}} \times \frac{0.5 \text{ Mg } \textbf{liveweight}}{\textbf{animal}} \times 100,000 \textbf{ animals} \approx 200,000 \text{ Mg } \textbf{manure}$$

FIGURE 16.17 Aerial view of a large feedlot in Colorado where 100,000 cattle are fed on grain imported from distant farms. The inset is a closeup of a similar feedlot. It is difficult for such concentrated animal feeding operations to recycle the nutrients in the manure back to the land on which the cattle feed was grown. Instead of being seen as a valued resource, the manure in this situation may be considered a waste to be disposed of. Agriculture is challenged to structure itself in a more ecologically balanced manner that better integrates its animal and crop components. (Large photo courtesy ConAgra Feed Lots, Greeley, Colo.; inset courtesy of R. Weil)

If this manure contains 2% N and the corn silage (whole plants) grown to feed the cattle removes some 240 kg N/ha, then we can estimate that the manure should be applied at 12 Mg/ha (~6 tons/acre),[6]

$$\frac{1 \text{ Mg } \textbf{manure}}{0.02 \text{ Mg } \textbf{N}} \times \frac{240 \text{ kg } \textbf{N}}{\text{ha}} \times \frac{1 \text{ Mg}}{1000 \text{ kg}} = \frac{12 \text{ Mg } \textbf{manure}}{\text{ha}}$$

[6] In the unlikely event that the land had not been previously manured, some N fertilizer might be needed for the corn silage in the first year or two to supplement the N released from the manure, but soon N released from previous years' applications would make supplementary fertilizer unnecessary (see Section 16.9).

TABLE 16.7   The Effect of Soil and Site Characteristics on the Percentage of Wells in Five Midwestern States Having Nitrate-N Concentrations Greater than 10 mg/L, the Upper Limit Considered Suitable for Human Consumption

Sites with sandy soils, near cropland, near barnyards, and having shallow wells had the highest nitrate-N levels.

Characteristics	Texture of soils		Proximity to cropland		Proximity to feedlot or barnyard		Well depth		Shallow wells near barnyard or feedlot
	Sandy	Clayey	<6 m	Out of sight	<6 m	Out of sight	Deep >30 m	Shallow <15 m	
No. of wells	2412	6415	1684	3098	704	7520	5106	3467	158
Percent with nitrate-N levels >10 mg/L	7.2	3.1	6.4	1.8	12.2	2.8	1.1	9.7	25.3

Data from Richards et al. (1996).

and that utilization of the manure in this manner would require some 17,000 ha of land,

$$\frac{1 \text{ ha}}{12 \text{ Mg } \textbf{manure}} \times 200,000 \text{ Mg } \textbf{manure} \simeq 17,000 \text{ ha.}$$

If the corn were grown in rotation with soybean (a legume crop that does not need applied N), then the total amount of cropland required for manure utilization would double to 34,000 ha or 340 km². To find this much cropland, some of the manure would have to be hauled 20 km (12 miles) or more from the feedlot! Finally, if soil phosphorus is already at (or above) optimal levels from previous manuring, as is usually the case near CAFOs (Section 14.9), manure should be applied at a much lower rate tailored to meet the P (not N) needs of the crops, thus requiring an even larger land base.

In practice, to save transportation costs and time, manure is often applied to nearby fields at higher-than-needed rates. However, applications at these higher rates will likely result in the pollution of surface and groundwater by nitrogen and phosphorus and may cause salinity damage to crops and soils.

POULTRY AND SWINE MANURE. Even more concentration of nutrients exists in the poultry and swine industries. Nearly all chickens and most hogs produced in the United States are grown in large "factory farms" that are concentrated near meat-processing plants. Not only do such CAFOs import nutrients in feed grains, but they also import calcium-P mineral feed supplements to compensate for the inability of their nonruminant animals to digest *phytic acid*, the form of P found in most seeds (see Section 14.4). The manure produced therefore contains more nitrogen and far more phosphorus than the local cropland base can properly utilize (see Figures 16.1–16.3). As a result, farm fields near large CAFOs tend to have very high levels of N and P in both the soil and in the water draining from the land.

SOME STOP-GAP MEASURES. Although probably not long-term solutions to an unbalanced agricultural system, the public welfare may be served by such approaches as the following: (1) Discourage further manure applications to fields already saturated with nutrients; instead, facilitate transportation of manure to areas with low P soils. (2) Encourage the use of new corn varieties that contain less phytic acid P and more inorganic P, allowing better assimilation by nonruminant animals and making it less necessary to purchase P feed supplements, thereby reducing the amount of P excreted. (3) Promote composting of manure to reduce the volume of material and the solubility of the nutrients in it. (4) Eliminate the overfeeding of P supplements to all types of livestock, in order to reduce the concentration of P in the manure. (5) Mix iron or aluminum compounds with the manure to reduce the solubility of its phosphorus (see Section 14.5).

Nutrient Composition of Animal Manures

Generally, about 75% of the N, 80% of the P, and 90% of the K ingested by animals passes through the digestive system and appears in the manure. For this reason, animal manures are valuable sources of both macro- and micronutrients. For a particular type of animal, the actual water and nutrient content of a load of manure will depend on the nutritional quality of the animals' feed, how the manure was handled, and the conditions under which it has been stored (Table 16.8). The variability in nutrient content from one type of animal manure to another (e.g., poultry manure compared to horse manure) is even greater. Therefore, one has to be cautious in interpreting general statements about the value and use of manure.

Both the urine (except for poultry, which produce solid uric acid instead of urine) and feces are valuable components of animal manure. On the average, a little more than *one-half of the N*, almost *all of the P*, and about *two-fifths of the K* are found in the solid manure. Nevertheless, this higher nutrient content of the solid manure is offset by the more ready availability of the constituents carried by the urine. Effective nutrient conservation requires that manure handling and storage minimize the loss of the liquid portion.

The data in Table 16.9 show that manures and most other organic nutrient sources have a relatively low nutrient content in comparison with commercial fertilizer. On a

TABLE 16.8   Example of the Variable Composition of Farm Manure

The data are based on 28 samples of horse manure sent in to one lab over a period of five years.

	Percent of fresh weight				
	Total N	*Soluble N*	*P*	*K*	*Water*
Lowest analysis	0.21	0.0	0.04	0.07	39
Highest analysis	0.85	0.14	0.75	1.0	80
Average analysis	0.51	0.03	0.16	0.35	63

Courtesy of V. A. Bandel, University of Maryland.

dry-weight basis, animal manures contain from 2 to 5% N, 0.5 to 2% P, and 1 to 3% K. These values are one-half to one-tenth as great as are typical for commercial fertilizers.

Furthermore, manure is rarely spread in the dry form, but usually contains a great deal of water. As it comes from the animal, the water content is 30 to 50% for poultry to 70 or 85% for cattle (see Table 16.9). If the fresh manure is handled as a solid and spread directly on the land (Figure 16.18, *right*), the high water content is a nuisance that adds to the expense of hauling. If the manure is handled and digested in a liquid form or slurry and applied to the land as such, even more water is involved (Figure 16.18, *left*). All this water dilutes the nutrient content of manure, as normally spread in the field, to values much lower than those cited for dry manure in Table 16.9. The high content of water and low content of nutrients makes it difficult to economically justify transporting manure to distant fields where it might do the most good. However, the value of the micronutrients in manure (Table 16.9) and the nonnutrient benefits of its organic matter (see Section 12.5) may be even greater than that of its N-P-K content, and should be included in any economic evaluation of manure transport.

Storage, Treatment, and Management of Animal Manures

Manure woes on Michigan dairy farm with cracking soils:
http://www.pmac.net/AM/big_stink.html

INTEGRATED ANIMAL PRODUCTION.   Where animal and crop production are integrated on a farm, manure handling is not too much of a problem. The use of carefully managed pasture for cattle can be maximized so that the animals themselves spread much of the manure while grazing. The total amount of nutrients in the manure produced on the farm is likely to be somewhat less than that needed to grow the crops; thus, modest amounts of inorganic fertilizers may be needed to make up the difference.

Hog manure handling:
http://www.epa.gov/agriculture/ag101/porkmanure.html

MANURE HANDLING IN CONFINEMENT SYSTEMS.   Where animals are concentrated in large confinement systems, the problem of manure disposal takes precedence over its utilization. The manure can be collected and *spread daily* (Figure 16.18), *packed in piles* where it is allowed to partially decompose before spreading, stored in *aerated ponds* that promote oxidation of the organic materials, or stored in deep *anaerobic lagoons* in which the manure ferments in the absence of oxygen gas. The latter method produces methane (which can be captured to use as fuel) and ammonia (which contributes to acrid odors) as gaseous products of fermentation, while nitrogen gases are produced by denitrification. Most of the P and K and some of the N remain in the lagoon manure and can be applied to the soil as a liquid. The nutrient content of the manure can be markedly affected by the handling methods used (Table 16.10). Also, storage of manure in composting piles or in liquid slurry lagoons for several months before spreading on the land can greatly reduce the potential for contaminating fruits and vegetables with pathogens that can cause human illness.

Methods of manure handling that both prevent pollution and preserve nutrients in a form that can be easily transported and sold commercially would make a major contribution to ameliorating the manure problem for concentrated animal production enterprises. Options currently being developed include the following: (1) **Heat-dry** and **pelletize** technology transforms the sloppy manure into small pellets that handle like commercial fertilizer. Although expensive in terms of energy and capital, the product is popular in the landscaping and lawn industries as a slow-release fertilizer. (2) **Commercial composting** systems (see Section 12.10) represent a low-energy-use, low-cost way to produce an easy-to-handle, nonodiferous, relatively high-analysis, slow-release fertilizer. Composting of manure at large cattle feedlots (Figure 16.19) is

TABLE 16.9 Commonly Used Organic Nutrient Sources: Their Approximate Nutrient Contents and Other Characteristics

Along with nitrogen-fixing legumes grown in rotation and as cover crops, materials such as these (except sewage sludge and municipal solid wastes) provide the mainstay of nutrient supply in organic farming. The nutrient contents shown for animal manures are typical of well-fed livestock in confinement production systems. Manure from free-range animals not given feed supplements may be considerably lower in both nitrogen and phosphorus.

Material	Water,[a] %	Percent of dry weight						g/Mg of dry weight						
		Total N	P	K	Ca	Mg	S	Fe	Mn	Zn	Cu	B	Mo	
Activated sewage sludge	<10	6	1.5	0.5	—	—	—	—	—	450	—	—	—	Most common form is Milorganite, N available over 2 to 6 months.
Coffee grounds[d]	60	1.6	0.01	0.04	0.08	0.01	0.11	330	50	15	40	—	—	May acidify soil.
Cottonseed meal	<15	7	1.5	1.5	—	—	—	—	—	—	—	—	—	Acidifies the soil. Commonly used as livestock feed.
Dairy cow manure[b]	75	2.4	0.7	2.1	1.4	0.8	0.3	1,800	165	165	30	20	—	May contain high-C bedding.
Dried blood	<10	13	1	1	—	—	—	—	—	—	—	—	—	Slaughterhouse by-product, N is available quickly.
Dried fish meal	<15	10	3	3	—	—	—	—	—	—	—	—	—	Incorporate or compost because of bad odors. Can feed to livestock.
Feedlot cattle manure[c]	80	1.9	0.7	2.0	1.3	0.7	0.5	5,000	40	8	2	14	1	May contain soil and soluble salts.
Hardwood tree leaves[f]	20	1.0	0.1	0.4	1.6	0.2	0.1	1,500	550	80	10	38	—	High Pb for some street trees.
Horse manure[c]	63	1.4	0.4	1.0	1.6	0.6	0.3	—	200	125	25	—	—	May contain high-C bedding.
Municipal solid waste compost[e]	40	1.2	0.3	0.4	3.1	0.3	0.2	14,000	500	650	280	60	7	May have high C/N and contain heavy metals, plastic, and glass.
Poultry (broiler) manure[b]	35	4.4	2.1	2.6	2.3	1.0	0.6	1,000	413	480	172	40	0.7	May contain high-C bedding, high soluble salts, arsenic, and ammonia.
Sewage sludge[b]	80	4.5	2.0	0.3	1.5[g]	0.2	0.2	16,000[g]	200	700	500	100	15	May contain high soluble salts, toxic heavy metals.
Sheep manure[c]	68	3.5	0.6	1.0	0.5	0.2	0.2	—	150	175	30	30	—	May contain weed seeds.
Spoiled legume hay	40	2.5	0.2	1.8	0.2	0.2	0.2	100	100	50	10	1,500	3	May contain elevated Cu levels.
Swine manure[c]	72	2.1	0.8	1.2	1.6	0.3	0.3	1,100	182	390	150	75	0.6	Very high C/N ratio; must be supplemented by other N.
Wood wastes	—	—	0.2	0.2	0.2	1.1	0.2	2,000	8,000	500	50	30	—	
Young rye green manure	85	2.5	0.2	2.1	0.1	0.05	0.04	100	50	40	5	5	.05	Nutrient content decreases with advanced growth stage.

[a] Water content given for fresh materials. Processing and storage methods may alter water content to less than 5% (heat-dried) or to more than 93% (slurry).

[b] Broiler and dairy manure composition estimated from means of approximately 800 and 400 samples analyzed by the University of Maryland manure analysis program 1985–1990.

[c] Composition of swine, sheep, and horse manure calculated from North Carolina Cooperative Extension Service Soil Fact Sheets prepared by Zublena et al. (1993).

[d] Coffee grinds data from Krogmann et al. (2003).

[e] Composition of municipal solid waste compost based on mean values for the products of 10 composting facilities in the United States as reported by He et al. (1995). Sulfur as sulfate-S.

[f] Hardwood leaf data from Heckman and Kluchinski (1996).

[g] Sludge contents of Ca and Fe may vary 10-fold depending on the wastewater treatment processes used.

Data derived from many sources.

FIGURE 16.18 Spreading dairy manure as liquid slurry after lagoon storage (*left*) and as a solid after packing and storage in a pile (*right*). Such methods of manure spreading are effective means of recycling nutrients, but are labor intensive and time consuming. Many loads of manure will be hauled to fertilize each field. The manure should not be spread when the soil is frozen and should be incorporated as soon as possible after spreading. Calibration of the spreaders is important to prevent unintentional overapplication of nutrients. (Photos courtesy of R. Weil)

TABLE 16.10 Influence of Handling and Storage Methods on Nutrient Losses from Animal Manure Between Excretion and Application to Land

Manure handling and storage method	Percentage lost		
	N	P	K
Solid systems			
Daily scrape and haul	15–35	10–20	20–30
Manure pack or compost	20–50	5–10	5–10
Liquid systems			
Aerobic tank storage	5–25	5–10	0–5
Lagoon (anaerobic) storage	70–80	50–80[a]	50–80

[a] Most of the P and K in a lagoon system is recoverable only when the sludge is dredged out.
From Sutton (1994).

FIGURE 16.19 A water truck maintains moist conditions in a long compost windrow located in an Arizona cattle feedlot. Note the mixture of dark, low-C/N manure and lighter, high-C/N straw in the windrows. The piles will be turned once or twice to stimulate decomposition, producing an easy-to-handle product that takes up about half the volume of the original manure. (Photo courtesy of Dr. Mohammed A. Zerkoune, University of Arizona)

one way of reducing leaching and runoff losses of soluble nutrients and reducing the volume of manure that must be transported. (3) **Anaerobic Digestion** (discussed previously) can be enhanced with the collection of biogas, which contains about 80% methane and 20% carbon dioxide and can be burned much like commercial natural gas. Small-scale manure digesters can supply cooking and heating fuel for remote villages and large-scale digesters can generate electricity.

In the future, such technologies, along with more integrated animal farming systems, may help to redress the serious nutrient imbalances that have developed with regard to manure production in industrialized agriculture.

16.5 INDUSTRIAL AND MUNICIPAL BY-PRODUCTS[7]

In addition to farm manures, four major types of organic wastes are of significance in land application: (1) municipal garbage, (2) sewage effluents and sludges (see Sections 16.6 and 18.7), (3) food-processing wastes, and (4) wastes of the lumber industry. Because of their uncertain content of toxic chemicals, these and other industrial wastes may or may not be acceptable for land application.

Society's concern for environmental quality has forced waste generators to seek non-polluting, but still affordable, ways of disposing of these materials. Although once seen as mere waste products to be flushed into rivers and out to sea, these materials are increasingly seen as sources of nutrients and organic matter that can be used beneficially to promote soil productivity in agriculture, forestry, landscaping, and disturbed-land reclamation.

Garbage

Municipal garbage has been used for centuries to enhance soil fertility in Asian countries. Most municipal solid waste (MSW) in industrial countries is incinerated or land-filled (see Section 18.10), but growing concerns about air quality and scarcity of land-fill space are raising the level of interest in using soil application as a means of disposal. About 50 to 60% of MSW consists of decomposable materials (paper, food scraps, yard waste, street tree leaves, etc.). Once the inorganic glass, metals, and so forth are removed, MSW can be *composted* (Section 12.10), sometimes in conjunction with such nutrient-rich materials as sewage sludge or animal manure, to produce MSW compost that is then applied to the land. Because of its very low nutrient content (see Table 16.9), MSW compost is an expensive way to distribute nutrients. However, alternative disposal options are often even more expensive. Potentially, the entire annual production of organic-material MSW, some 80 million cubic meters in the United States, could be recycled as a soil amendment using less than 10% of the agricultural land in the country. Increasing numbers of compost operations of all sizes are being run by communities, small businesses, or even individuals.

Food-Processing Wastes

Land application of food-processing wastes is being practiced in selected locations, but the practice is focused almost entirely on pollution abatement and not on soil enhancement. Liquid wastes are commonly applied through sprinkle irrigation to permanently grassed fields.

Wood Wastes

Sawdust, wood chips, and shredded bark from the lumber industry have long been sources of soil amendments and mulches, especially for home gardeners and landscapers. Because of their high C/N ratios and high lignin contents, these materials decompose very slowly. They make good mulching material, but do not readily supply plant nutrients. In fact, sawdust incorporated into soils to improve soil physical properties may cause plants to become nitrogen deficient unless an additional source of nitrogen is applied (see Section 12.3).

---

[7] For a collection of technical papers discussing the potential benefits and problems associated with the land application of these by-products, see Powers and Dick (2000).

Wastewater Treatment By-Products

Sewage treatment has evolved over the past century to help society avoid polluting rivers and oceans with pathogens, oxygen-demanding organic debris, and eutrophying nutrients. The ever-more-stringent effort to clean up the wastewater before returning it to natural waters has two basic consequences. First, the amount of material *removed* from the wastewater during the treatment process has increased tremendously. This solid material, known as sewage *sludge,* must also be disposed of safely. Second, a goal of advanced wastewater treatment is to remove nutrients (mainly phosphorus, but increasingly also nitrogen) from the sewage **effluent** (the treated water that is returned to the stream), a job that can be accomplished safely and economically by allowing the partially treated effluent to interact with a soil–plant system. Therefore, there is a growing interest in using soils to assist with the sewage problem in two ways: (1) as a system of assimilating, recycling, or disposing of the solid sludge; and (2) as a means of carrying out the final removal of nutrients and organics from the liquid effluent.

Sewage Effluent

Some cities operate sewage farms on which they produce crops, usually animal feeds and forages that offset part of the expense of effluent disposal. Forest irrigation is a cost-effective method of final effluent cleanup and produces enhanced tree growth as a bonus (Figure 16.20). The rate of wood production is greatly increased as a result of both the additional water and the additional nutrients supplied therewith. This method of advanced wastewater treatment is used by a number of cities around the world.

In a carefully planned and managed effluent irrigation system, the combination of (1) nutrient uptake by the plants, (2) adsorption of inorganic and organic constituents by soil colloids, and (3) degradation of organic compounds by soil microorganisms results in the purification of the wastewater. Percolation of the purified water eventually replenishes the groundwater supply.

Sewage Sludge or Biosolids

Sewage sludge is the solid by-product of domestic and/or industrial wastewater treatment plants (Figure 16.21). It has been spread on the land for decades, and its use will likely increase in the future. If sewage sludge has been treated to meet certain land-application standards (low pathogen and contaminant levels), the term **biosolids** may be applied. The product Milorganite®, a dried, activated (oxygenated) sludge sold by the

FIGURE 16.20 Final treatment of sewage effluent and recharge of groundwater are being accomplished by natural soil and plant processes in this effluent-irrigated forest on Ultisols near Atlanta, Georgia. Nutrient flows, groundwater quality, and tree growth are carefully monitored. Some of the greatly increased production of wood is used as an energy source to run the sewage treatment plant. (Photo courtesy of R. Weil)

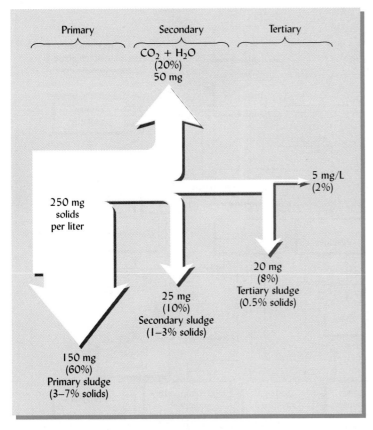

FIGURE 16.21 Diagram showing the removal of suspended solids from 1 liter of raw waste water. Primary treatment permits the separation of most of the solids from raw sewage. Secondary treatment encourages oxidation of much of the organic matter and separation of more solids. Tertiary treatment usually involves the use of calcium, aluminum, or iron compounds to remove phosphorus from the sewage water. [Redrawn from Loehr et al. (1979); used with permission of Van Nostrand Reinhold Company, New York]

Milwaukee Sewerage Commission, has been widely used as a slow-release fertilizer in North America since 1927, especially on turfgrass. Numerous other cities market composted sludge products to landscaping and other specialty users. However, the great bulk of sewage sludge used on land is applied as liquid slurry or as partially dried cake.

Biosolids to Fertilize Farmland (Un. Nebraska): http://lancaster.unl.edu/enviro/biosolids/overvew.htm

COMPOSITION OF SEWAGE SLUDGE. As might be expected, the composition of sludge varies from one sewage treatment plant to another. The variations depend on the nature of treatment the sewage receives, especially the degree to which the organic material is allowed to digest. Representative values for plant nutrients are given in Table 16.9. Like manure and other organic nutrient sources, sewage sludge contributes micronutrients as well as macronutrients. Levels of plant micronutrient metals (zinc, copper, iron, manganese, and nickel) as well as other heavy metals (cadmium, chromium, lead, etc.) are determined largely by the degree to which industrial wastes have been mixed in with domestic wastes. In the United States, the levels of metals in sewage are far lower than they were in the past, because of source-reduction programs that require industrial facilities to remove pollutants *before* sending their sewage to municipal treatment plants (see Section 18.7). Nonetheless, vigilance must be maintained to avoid sludges too contaminated for safe land application.

In comparison with inorganic fertilizers, sludges are generally low in nutrients, especially potassium (which is soluble and found mainly in the effluent). Representative levels of N, P, and K are 4, 2, and 0.3%, respectively (see Table 16.9). The phosphorus content is higher where advanced sewage treatment is designed to remove phosphorus from the effluent and deposit it in the sludge (see Box 14.2). If the sewage treatment precipitates phosphorus by reactions with iron or aluminum compounds, the phosphorus in the sludge will likely have a very low availability to plants.

Integrated Recycling of Wastes

For most of the industrialized countries, widespread recycling of organic wastes other than animal manures is a relatively recent phenomenon. In heavily populated areas of Asia, however, and particularly in China and Japan, such recycling has long been

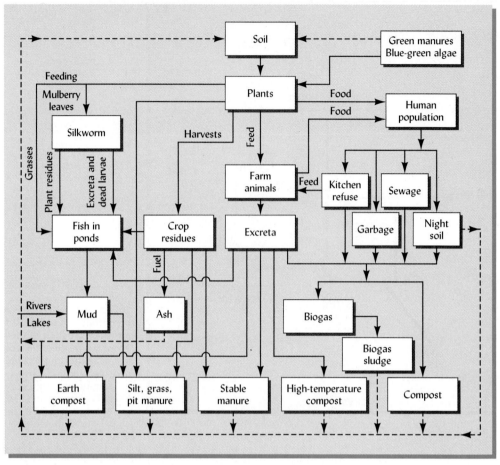

FIGURE 16.22 Recycling of organic wastes and nutrient elements in the People's Republic of China. Note the degree to which the soil is involved in the recycling processes. [Concepts from FAO (1977) and Yang (2006)]

practiced (Figure 16.22). There, organic "wastes" have traditionally been used for biogas production, as food for fish, and as a source of heat from compost piles. The plant nutrients and organic matter are recycled and returned to the soil. Despite China's increasing use of chemical fertilizers, the traditional respect for what others might see as wastes supports the complex recycling systems that continue to supply about 50% of the nutrients used in China's agriculture. As they look to achieving a more sustainable future, Western countries have much to learn from traditional Chinese attitudes and practices.

16.6 PRACTICAL UTILIZATION OF ORGANIC NUTRIENT SOURCES

In Section 12.5 we discussed the many beneficial effects on soil physical and chemical properties that can result from amendment of soils with decomposable organic materials, such as manure or sludge. Here we will focus on the principles of ecologically sound management of nutrients from sewage sludge, farm manure, MSW compost, and other organic materials. The first step (required by law in the case of sewage sludge) is usually to have a representative sample analyzed in a reputable laboratory so you know what you are working with.

The rate of application is generally governed by the amount of nitrogen or phosphorus that the organic material will make available to plants. Nitrogen usually is the first criterion because nitrogen is needed in the largest quantity by most plants, and because excess nitrogen can present a pollution problem (see Section 13.8). The ratio of P to N in most organic sources is higher than in plant tissue. Consequently, if organic materials supply sufficient N to meet plant needs, they probably supply excessive levels of P (see Section 14.2) and the build up of soil P must be taken into account in the long run. For soils already high in P levels, the application rate for an organic amendment

TABLE 16.11   Release of Mineral Nitrogen from Various Organic Materials Applied to Soils, as Percent of the Organic Nitrogen Originally Present[a]

For example, if 10 Mg of poultry floor litter initially contains 300 kg N in organic forms, 50% or 150 kg of N would be mineralized in year 1. Another 15% (0.15 × 300) or 45 kg of N would be released in the year 2.

Organic nitrogen source	Year 1	Year 2	Year 3	Year 4
Poultry floor litter	50	15	8	3
Dairy manure (fresh solid)	25	18	9	4
Swine manure lagoon liquid	45	12	6	2
Feedlot cattle manure	35	15	6	2
Composted feedlot manure	20	8	4	1
Lime-stabilized, aerobically digested sewage sludge	40	12	5	2
Anaerobically digested sewage sludge	20	8	4	1
Composted sewage sludge	10	5	3	2
Activated, unstabilized sewage sludge	45	15	4	2

[a]These values are approximate and may need to be increased for warm climates or sandy soils and decreased for cold or dry climates or heavy clay soils.
Sources of data: Eghball et al. (2002) and Brady and Weil (1996).

may be limited by the P supplied, rather than by the N content. Potentially, toxic heavy metals in some materials may also limit the rate of application (see Section 18.7).

A small fraction of the N in manure or sludge may be soluble (ammonium or nitrate) and immediately available, but the bulk of the N must be released by microbial mineralization of organic compounds. Table 16.11 indicates the N mineralization rates for various organic materials. Materials partially decomposed during treatment and handling (e.g., by composting or digestion) release a lower percentage of their nitrogen. For example, Figure 16.23 compares the rate of nitrate-N released from fresh and composted poultry litter.

If a field is treated annually with an organic material, the application rate needed will become progressively smaller because, after the first year, the amount of nitrogen released from material applied in previous years must be subtracted from the total to be applied afresh (see Box 16.2). This is especially true for composts for which the initial availability of the nitrogen is quite low. Instead of making progressively smaller applications, another practical strategy is to use a moderate application every year, but supplement the nitrogen from other sources in the first few years until nitrogen release from previous and current applications can supply the entire requirement (see also Section 16.4).

Special Uses

The organic matter component plays a dominant role in applications of organic nutrient sources to soil areas denuded from erosion, from land-leveling for irrigation, or from mining operations (Table 16.12). Improvements in water-holding capacity and

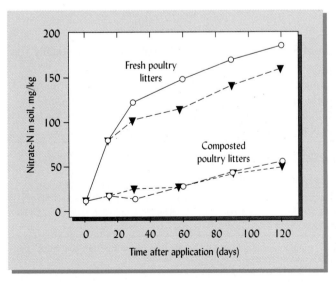

FIGURE 16.23 Nitrate nitrogen accumulation in a silt loam soil (Hapludalfs) incubated with composted or fresh poultry litter from two sources. Each of the litters was mixed with soil in amounts sufficient to provide 0.23 g of total N per kg of dry soil (approximately equivalent to 300 kg N/ha). Soil-litter mixtures were incubated at 25 °C for 120 days in controlled environment chambers. [Modified from Preusch et al. (2002)]

If the soil phosphorus level is not already above the optimal range, the rate of release of available nitrogen usually determines the proper amount of manure, sludge, or other organic nutrient source to apply. The amount of nitrogen made available in any year should meet, but not exceed, the amount of nitrogen that plants can use for optimum growth. Our example here is a field producing corn two years in a row. The goal is to produce 7000 kg/ha of grain each year. This yield normally requires the application of 120 kg/ha of available nitrogen (about 58 kg of grain per kg N applied; see Section 16.13). We expect to obtain this N from a lime-stabilized sewage sludge containing 4.5% total N and 0.2% mineral N (ammonium and nitrate).

Year 1

Calculation of amount of sludge to apply per hectare:

% organic N in sludge   = total N − mineral N = 4.5% − 0.2% = 4.3%.

Organic N in 1 Mg of sludge   = 0.043 × 1000 kg = 43 kg N.

Mineral N in 1 Mg of sludge   = 0.002 × 1000 kg = 2 kg N.

Mineralization rate for lime-stabilized sludge in first year (Table 16.11) = 40% of organic N.

Available N mineralized from 1 Mg sludge in first year = 0.40 × 43 kg N = 17.2 kg N.

Total available N from 1 Mg sludge = mineral N + mineralized N = 2.0 + 17.2 = 19.2 kg N.

Amount of (dry) sludge needed = 120 kg N/(19.2 kg available N/Mg dry sludge) = 6.25 Mg dry sludge.

Adjust for moisture content of sludge (e.g., assume sludge has 25% solids and 75% water).

Amount of wet sludge to apply: 6.25 Mg dry sludge/(0.25 Mg dry sludge/Mg wet sludge) = 6.25/0.25 = 25 Mg wet sludge.

Year 2

Calculate amount of N mineralized in year 2 from sludge applied in year 1:

Second year mineralization rate (Table 16.11) = 12% of original organic N.

N mineralized from sludge in year 2 = 0.12 × 43 Kg N/Mg × 6.25 Mg dry sludge = 32.25 kg N from sludge in year 2.

Calculate amount of sludge to apply in year 2:

N needed from sludge applied in year 2 = N needed by corn − N released from sludge applied in year 1
= 120 kg − 32.25 kg = 87.75 kg N needed/ha.

Amount of (dry) sludge needed per ha = 87.75 kg N/(19.2 kg available N/Mg dry sludge) = 87.75/19.2 = 4.57 Mg dry sludge/ha.

Adjust for moisture content of sludge (e.g., assume sludge has 25% solids and 75% water).

Mg of wet sludge to apply: 4.57 Mg dry sludge/(0.25 Mg dry sludge/Mg wet sludge) = 4.57/0.25
= 18.3 Mg wet sludge.

Note that the 10.82 Mg dry sludge (6.25 + 4.57) also provided plenty of P: 216 kg P/ha (assuming 2% P; see Table 16.9), an amount that greatly exceeds the crop requirement and will soon lead to excessive buildup of P.

TABLE 16.12 **Heavy Application of Sewage Sludge Helps Restoration of Drastically Disturbed Site[a]**

The revegetation effort benefited from both the nutrients and the organic matter in the sludge. Compared to fertilizer, sludge application increased the concentration of some potentially toxic metals in grass tissue, but not above normal levels. In the case of nickel, tissue levels were actually reduced because plant growth increased more than plant nickel uptake.

	Standing plant biomass		Metals in plant tissue in year 2, mg/kg		
	Year 2	Year 5	Cr	Ni	Cu
Control (no amendment)	1.55	1.49	0.9	5.7	12.1
Commercial fertilizer	4.79	3.10	0.5	1.6	14.8
Sewage sludge (368 Mg/ha)	5.68	3.07	1.8	0.6	24.8

[a]The sludge was compared to fertilizer or no amendment. Acid, infertile, shaley soils forming from coal mine spoil were seeded in year 1 with a mix of grass and broadleaf species.
Data from Haering et al. (2000).

soil structure brought about by decomposing organic materials may be just as important as the nutrient-supplying capacity of the amendment. For example, in the mine reclamation project featured in Table 16.12, the researchers noted that in addition to the effects shown, woody and herbaceous native perennial plants preferentially invaded the sludge-treated plots. Initial applications of 50 to more than 100 Mg/ha may be worked into the soil in the affected areas. These rates may be justified to supply organic matter as well as nutrients, provided the material is not so high in nitrogen as to create the potential for nitrate leaching.

Micronutrient deficiencies can be ameliorated with manure application. Such treatments are sometimes used when there is some uncertainty about which specific nutrient is lacking. Manure applications can usually be made with little concern for adding toxic quantities of the micronutrients. However, swine manure can cause the buildup of phytotoxic levels of copper in amended soils if the swine have been fed copper supplements to stimulate their growth.

16.7 INORGANIC COMMERCIAL FERTILIZERS

The worldwide use of fertilizers on farms increased dramatically since the middle of the 20th century (Figure 16.24), accounting for a significant part of the equally dramatic increases in crop yields during the same period. Improved soil fertility through the application of fertilizer nutrients is an essential factor in enabling the world to feed the billions of people that are being added to its population (see Section 20.4).

The need to supplement forest soil fertility is increasing as the demands for forest products increase the removal of nutrients and competing uses of land leave forestry with more infertile, marginal sites. Currently, most fertilization in forestry is concentrated on tree nurseries and on seed trees, where the benefits of fertilization are relatively short term and of high value, and the logistics of fertilizer application are not so difficult and expensive as for extensive forest stands. However, fertilization of new forest plantings and existing stands is increasingly practiced in Japan, northwestern and southeastern United States, the Scandinavian countries, and Australia, with the fertilizer spread by helicopter.

Regional Use of Fertilizers

Fertilizer use statistics within a region tell whether fertilizers are contributing to soil quality enhancement or to environmental degradation. For example, in Europe and East Asia where moisture is abundant and intensified cropping is common, fertilizer

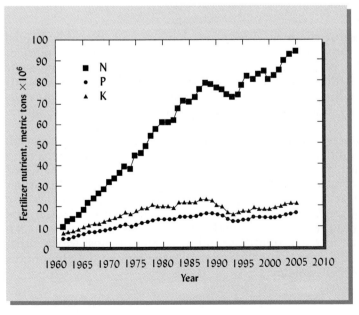

FIGURE 16.24 World fertilizer use since 1960 by major nutrient element. The use of nitrogen (N) has increased much faster than that of phosphorus (P) and potassium (K). The dip in world fertilizer use in the early 1990s was due mainly to drastic reductions in Russia and Ukraine after the collapse of the Soviet Union. In most industrialized countries fertilizer use has leveled off or declined, but it continues to increase in China, India, and much of the developing world. In the latter countries, the removal of nutrients in crop harvests may still exceed the amounts returned to the soil. [Data selected from FAO (2006)]

nutrient application rates are nearly triple the world average. In the Netherlands, nitrogen additions from fertilizers and manures are more than four times the removal of this element in harvested crops. In contrast, soils in sub-Saharan Africa are literally being mined—far less nutrients are being added from all sources than are being removed in the crops. The fertilizer nutrient rate in this region is only about 10% of the world average. We must be as site specific as possible in evaluating the role of fertilizers in meeting humanitarian and environmental goals.

Origin and Processing of Inorganic Fertilizers

Most fertilizers are inorganic salts containing readily available plant nutrient elements. Some are manufactured, but others, such as phosphorus and potassium, are found in natural geological deposits. Beds of solid salts located far beneath the Earth's surface are the primary sources of potassium. These underground deposits are found in many locations, including sites in Canada, France, Germany, and Russia, as well as in New Mexico. The salts are mined and then purified, yielding such compounds as potassium chloride and potassium sulfate. Apatite found in phosphate rock deposits is the primary source of phosphorus fertilizers. Since apatite is extremely insoluble, however, it is treated with sulfuric, phosphoric, or nitric acid to produce materials, such as triple superphosphate, that yield phosphorus in a readily available form. Phosphate rock deposits are located around the world; among the most extensive are found in Florida, North Carolina, and Morocco.

Under very high temperatures and pressures, nitrogen gas in the atmosphere is fixed with hydrogen from natural gas to produce ammonia gas, NH_3. This fixation consumes tremendous quantities of energy—usually from fossil fuel. When the resulting ammonia gas is then put under moderate pressure, it liquifies, forming anhydrous ammonia. Because of its low cost and ease of application, more nitrogen is applied directly to soils in anhydrous ammonia than in any other fertilizer. More important, it is the starting point for the manufacture of most of the other nitrogen carriers, including urea, ammonium nitrate, ammonium sulfate, sodium nitrate, and liquid mixtures known as "nitrogen solutions." The industrial process of nitrogen fixation has greatly altered the recycling of nitrogen on Earth (see Section 13.2).

Many commercial fertilizers contain two or more of the primary nutrient elements. In some cases, mixtures of the primary nutrient carriers are used. But in others, a given compound may carry two nutrients, examples being monoammonium phosphate, diammonium phosphate, and potassium nitrate. Care must be used in selecting the components of a mixed fertilizer since some compounds are not compatible with others, resulting in poor physical condition and reduced nutrient availability in the mixture.

Properties and Use of Inorganic Fertilizers

The composition of inorganic commercial fertilizers is much more precisely defined than is the case for the organic material discussed above. Table 16.13 lists the nutrient element contents and other properties of some of the more commonly used inorganic fertilizers. In most cases, fertilizers are used to supply plants with the macronutrients nitrogen, phosphorus, and/or potassium—sometimes called the *primary fertilizer elements*. Fertilizers that supply sulfur, magnesium, and the micronutrients are also manufactured.

It can be seen from the data in Table 16.13 that a particular nutrient (say, nitrogen) can be supplied by many different *carriers,* or fertilizer compounds. Decisions as to which fertilizers to use must take into account not only the nutrients they contain, but also a number of other characteristics of the individual carriers. Table 16.13 provides information about some of these characteristics, such as the salt hazard (see also Section 10.7), acid-forming tendency (see also Section 13.8), tendency to volatilize, ease of solubility, and content of nutrients other than the principal one. Of the nitrogen carriers, anhydrous ammonia, nitrogen solutions, and ureas are the most widely used. Diammonium phosphate and potassium chloride supply the bulk of the phosphorus and potassium used in the United States.

TABLE 16.13 Commonly Used Inorganic Fertilizer Materials: Their Nutrient Contents and Other Characteristics

Fertilizer	N	P	K	S	Salt hazard	Acid formation[b]	Other nutrients & comments
	Percent by weight						
Primarily sources of nitrogen							
Anhydrous ammonia (NH$_3$)	82				Low	−148	Pressurized equipment needed; toxic gas; must be injected into soil.
Urea [CO(NH$_2$)$_2$]	45				Moderate	−84	Soluble; hydrolyses to ammonium forms. Volatilizes if left on soil surface.
Ammonium nitrate (NH$_4$NO$_3$)	33				High	−59	Absorbs moisture from air; can be left on soil surface. Can explode if mixed with organic dust or S.
Sulfur-coated urea	30–40			13–16	Low	−110	Variable slow rate of release.
UF (ureaform-aldehyde)	30–40				Very low	−68	Slowly soluble; faster with warm temperatures.
UAN solution	30				Moderate	−52	Most commonly used liquid N.
IBDU (isobutylidene diurea)	30				Very low	—	Slowly soluble.
Ammonium sulfate [(NH$_4$)$_2$SO$_4$]	21			24	High	−110	Rapidly lowers soil pH; very easy to handle.
Sodium nitrate (NaNO$_3$)	16				Very high	+29	Hardens, disperses soil structure.
Potassium nitrate (KNO$_3$)	13		36	0.2	Very high	+26	Very rapid plant response.
Primarily sources of phosphorus							
Monoammonium phosphate (NH$_4$H$_2$PO$_4$)	11	21–23		1–2	Low	−65	Best as starter.
Diammonium phosphate [(NH$_4$)$_2$HPO$_4$]	18–21	20–23		0–1	Moderate	−70	Best as starter.
Triple superphosphate		19–22		1–3	Low	0	15% Ca.
Phosphate rock [Ca$_3$(PO$_4$)$_2$·CaX]		8–18[a]			Very low	Variable	Low to extremely low availability. Best as fine powder on acid soils. 30% Ca. Contains some Cd, F, etc.
Single superphosphate		7–9		11	Low	0	Nonburning, can place with seed. 20% Ca.
Bonemeal	1–3[a]	10[a]		0.4	Very low	—	Slow availability of N, P as for phosphate rock. 20% Ca.
Colloidal phosphate		8[a]			Very low	—	P availability as for phosphate rock. 20% Ca.
Primarily sources of potassium							
Potassium chloride (KCl)			50		High	0	47% Cl–may reduce some diseases.
Potassium sulfate (K$_2$SO$_4$)			42	17	Moderate	0	Use where Cl not desirable.
Wood ashes		0.5–1	1–4		Moderate to high	+40	About ½ the liming value of limestone; caustic. 10–20% Ca, 2–5% Mg, 0.2% Fe, 0.8% Mn.
Greensand			0.6	6	Very low	0	Very low availability.
Granite dust			4		Very low	0	Very slow availability.
Primarily sources of other nutrients							
Basic slag		1–7			Low	+70	10% Fe, 2% Mn, slow availability; best on acid soils. 3–30% Ca, 3% Mg.
Gypsum (CaSO$_4$·2H$_2$O)				19	Low	0	Stabilizes soil structure; no effect on pH; Ca and S readily available. 23% Ca.
Calcitic limestone (CaCO$_3$)					Very low	+95	Slow availability; raises pH. 36% Ca.
Dolomitic limestone [CaMg(CO$_3$)$_2$]					Very low	+95	Very slow availability; raises pH. ~24% Ca, ~12% Mg.
Epsom salts (MgSO$_4$·7H$_2$O)				13	Moderate	0	No effect on pH; water soluble. 2% Ca, 10% Mg.
Sulfur, flowers (S)				95	—	−300	Irritates eyes; very acidifying; slow acting; requires microbial oxidation.
Solubor					Moderate	—	Very soluble; compatible with foliar sprays. 20.5% B.
Borax (Na$_2$B$_4$O$_7$·10H$_2$O)					Moderate	—	Very soluble. 11% B; 9% Na.
EDTA chelates					—	—	See label. Usually 13% Cu or 10% Fe or 12% Mn or 12% Zn.
Cu, Fe, Mn, or Zn sulfates				13–20			25% Cu, 19% Fe, 27% Mn, or 35% Zn, very soluble.

[a] Highly variable contents.
[b] A negative number indicates that acidity is produced; a positive number indicates that alkalinity is produced; kg CaCO$_3$/100 kg material needed to neutralize acidity.

Physical Forms of Marketed Fertilizer

Commercial fertilizers in many countries are still sold in bags, transported to the field in trucks, then emptied by hand into fertilizer spreaders and applied to the land. However, less than 10% of all fertilizer in the United States is handled in this manner. Increases in fertilizer use rates and labor costs, along with improved means of transporting and handling the fertilizer and increased availability of custom applicators, have favored two alternative means of marketing fertilizer: (1) unbagged, dry solids handled in *bulk* form, and (2) *liquid* or fluid forms stored, transported, and applied from tanks. In both cases the costs, particularly in terms of labor, are reduced.

Bulk spreading, often done at times when trucks or large fertilizer spreaders can get on the land, is the favorite for applying multinutrient fertilizers. Liquid fertilizers comprise more than half of the single-nutrient carriers sold in the United States and about 40% of all fertilizers. Labor costs are low, since the fertilizer is transferred from one tank to another and applied to the field with the aid of mechanical pumps.

Fertilizer Grade

Commercial fertilizers were first manufactured in the late 19th and early 20th centuries. Analytical procedures and laws passed to regulate the new industry resulted in certain labeling and marketing conventions that are still in common use today.

Of these conventions it is most important to be familiar with the *fertilizer grade*. Every fertilizer label states the **grade** as a three-number code, such as 10-5-10 or 6-24-24. These numbers stand for percentages indicating the *total* nitrogen (N) content, the *available* phosphate (P_2O_5) content, and the *soluble* potash (K_2O) content. Plants do not take up phosphorus and potassium in these chemical forms, nor do any fertilizers actually contain P_2O_5 or K_2O.

The oxide expressions for phosphorus (P_2O_5) and potassium (K_2O) are relics of the days when geochemists reported the contents of rocks and minerals in terms of the oxides formed upon heating. Unfortunately, these expressions found their ways into laws governing the sale of fertilizers, and there is considerable resistance to changing them, although some progress is being made. In scientific work and in this textbook, the simple elemental contents are used (P and K) wherever possible. Box 16.3 and Figure 16.25 explain how to convert between the elemental and oxide forms of expression.

The grade is important from an economic standpoint because it conveys the analysis or concentration of the nutrient in a carrier. When properly applied, most fertilizer carriers give equally good results for a given amount of nutrient element. The more concentrated carriers are usually the most economical to use, because less weight of fertilizer must be transported to supply the needed quantity of a given nutrient. Hence, economic comparisons among different equally suitable fertilizers should be based on the price per kilogram of nutrient, not the price per kilogram of fertilizer.

Fate of Fertilizer Nutrients

A common myth about fertilizers suggests that inorganic fertilizers applied to soil directly feed the plant, and that therefore the biological cycling of nutrients, such as described by Figures 13.2 for nitrogen and 14.8 for phosphorus, are of little consequence where inorganic fertilizers are used. The reality is that nutrients added by normal application of fertilizers, whether organic or inorganic, are incorporated into the complex soil nutrient cycles, and that relatively little of the fertilizer nutrient (from 10 to 60%) actually winds up in the plant being fertilized during the year of application. Even when the application of fertilizer greatly increases both plant growth and nutrient uptake, the fertilizer stimulates increased cycling of the nutrients, and the nutrient ions taken up by the plant come largely from various pools in the soil and not directly from the fertilizer. For example, some of the added N may go to satisfy the needs of microorganisms, preventing them from competing with plants for other pools of N. This knowledge has been obtained by careful analysis of dozens of nutrient studies that used fertilizer with isotopically tagged nutrients. Results from such a study are summarized in Table 16.14, which shows somewhat more N uptake from fertilizer than is typically reported. Generally, as fertilizer rates are increased, the efficiency of fertilizer nutrient use decreases, leaving behind in the soil an increasing proportion of the added nutrient.

BOX 16.3   HOW MUCH NITROGEN, PHOSPHORUS, AND POTASSIUM IS IN A BAG OF 6-24-24?

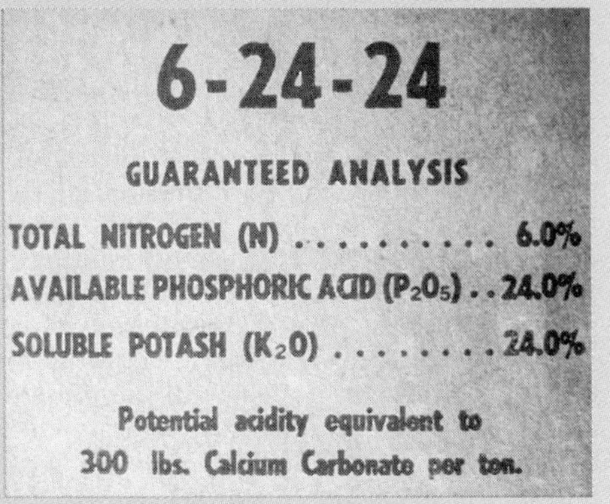

6-24-24

GUARANTEED ANALYSIS

TOTAL NITROGEN (N) 6.0%

AVAILABLE PHOSPHORIC ACID (P₂O₅) . . 24.0%

SOLUBLE POTASH (K₂O) 24.0%

Potential acidity equivalent to 300 lbs. Calcium Carbonate per ton.

FIGURE 16.25 *A typical commercial fertilizer label. Note that a calculation must be performed to determine the percentage of the nutrient elements P and K in the fertilizer since the contents are expressed as if the nutrients were in the forms of P_2O_5 and K_2O. Also note that after interacting with the plant and soil, this material would cause an increase in soil acidity that could be neutralized by 300 units of $CaCO_3$ per 2000 units (1 ton = 2000 lbs) of fertilizer material.*

Conventional labeling of fertilizer products reports percentage N, P_2O_5, and K_2O. Thus, a fertilizer package (Figure 16.25) labeled as 6-24-24 (6% nitrogen, 24% P_2O_5, 24% K_2O) actually contains 6% N, 10.5% P, and 19.9% K (see calculations below).

To determine the amount of fertilizer needed to supply the recommended amount of a given nutrient, first convert percent P_2O_5 and percent K_2O to percent P and K, by calculating the proportion of P_2O_5 that is P and the proportion of K_2O that is K. The following calculations may be used:

Given that the molecular weights of P, K, and O are 31, 39, and 16 g/mol, respectively:

Molecular weight of P_2O_5 = 2(31) + 5(16) = 142 g/mol

$$\text{Proportion P in } P_2O_5 = \frac{2P}{P_2O_5} = \frac{2(31)}{2(31) + 5(16)} = 0.44$$

To convert $P_2O_5 \rightarrow P$, multiply percent P_2O_5 by 0.44

Molecular weight of K_2O = 2(39) + 16 = 94

$$\text{Proportion K in } K_2O = \frac{2K}{K_2O} = \frac{2(39)}{2(39) + 16} = 0.83$$

To convert $K_2O \rightarrow K$, multiply percent K_2O by 0.83

Thus, if the bag in Figure 16.25 contains 25 kg of the 6-24-24 fertilizer, it will supply 1.5 kg N (0.06×25); 2.6 kg P($0.24 \times 0.44 \times 25$); and 5 kg K ($0.24 \times 0.83 \times 25$).

The Concept of the Limiting Factor

Two German chemists (Justus von Liebig and Carl Sprengel) are credited with first publishing, in the mid-1800s, "the law of the minimum" which holds that *plant production can be no greater than that level allowed by the growth factor present in the lowest amount relative to the optimum amount for that factor.* This growth factor, be it temperature, nitrogen, or water supply, will limit the amount of growth that can occur and is therefore called the **limiting factor** (Figure 16.26).

If a factor is not the limiting one, increasing it will do little or nothing to enhance plant growth. In fact, increasing the amount of a nonlimiting factor may actually reduce plant growth by throwing the system further out of balance. For example, if a plant is limited by lack of phosphorus, adding more nitrogen may only aggravate the phosphorus deficiency.

TABLE 16.14   **Source of Nitrogen in Corn Plants Grown in North Carolina on an Enon Sandy Loam Soil (Ultic Hapludalf) Fertilized with Three Rates of Nitrogen as Ammonium Nitrate**

The source of the nitrogen in the corn plant was determined by using fertilizer tagged with the isotope ^{15}N. Moderate fertilizer use increased the uptake of N already in the soil system as well as that derived from the fertilizer.

Fertilizer nitrogen applied, kg/ha	Corn grain yield, Mg/ha	Total N in corn plant, kg/ha	Fertilizer-derived N in corn, kg/ha	Soil-derived N in corn, kg/ha	Fertilizer-derived N in corn as percent of total N in corn	Fertilizer-derived N in corn as percent of N applied
50	3.9	85	28	60	33	56
100	4.6	146	55	91	38	55
200	5.5	157	86	71	55	43

Calculated from Reddy and Reddy (1993).

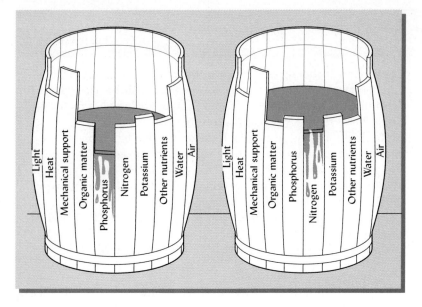

FIGURE 16.26 An illustration of the law of the minimum and the concept of the limiting factor. Plant growth is constrained by the essential element (or other factor) that is most limiting. The level of water in the barrel represents the level of plant production. (*Left*) Phosphorus is represented as being the factor that is most limiting. Even though the other elements are present in more than adequate amounts, plant growth can be no greater than that allowed by the level of phosphorus available. (*Right*) When phosphorus is added, the level of plant production is raised until another factor becomes most limiting—in this case, nitrogen.

Looked at another way, applying available phosphorus (the first limiting nutrient in this example) may allow the plant to respond positively to a subsequent addition of nitrogen. Thus, the increased growth obtained by applying two nutrients together often is much greater than the sum of the growth increases obtained by applying each of the two nutrients individually. Such an *interaction* or *synergy* between two nutrients can be seen in the left panel of Figure 16.27, which also illustrates a physical limiting growth factor, compaction.

16.8   FERTILIZER APPLICATION METHODS

Wise, effective fertilizer use involves making correct decisions regarding *which* nutrient element(s) to apply, *how much* of each needed nutrient to apply, *what type* of material or carrier to use (Tables 16.9 and 16.13 list some of the choices), *in what manner* to apply the material, and, finally, *when* to apply it. We will leave information on the first two decisions until Section 16.10. Here we will discuss the alternatives available with regard to the last two decisions.

There are three general approaches to applying fertilizers: (1) *broadcast application,* (2) *localized placement,* and (3) *foliar application.* Each method has some advantages and disadvantages and may be particularly suitable for different situations. Often some combination of the three methods is used.

Broadcasting

In many instances fertilizer is spread evenly over the entire field or area to be fertilized. This method is called *broadcasting.* Often the broadcast fertilizer is mixed into the plow layer by means of tillage, but in some situations it is left on the soil surface and allowed to be carried into the root zone by percolating rain or irrigation water. The broadcast method is most appropriate when a large amount of fertilizer is being applied with the aim of raising the fertility level of the soil over a long period of time. Broadcasting is the most economical way to spread large amounts of fertilizer over wide areas (Figure 16.28).

For close-growing vegetation, broadcasting provides an appropriate distribution of the nutrients. It is therefore the most commonly used method for rangeland, pastures, small grains, turf grass, and forests. Fertilizers are also broadcast on some row cropland, especially in the fall when it is most convenient, although certainly not most efficient. The broadcast fertilizer may or may not be incorporated into the soil (Figure 16.29*b* and *c*). Unfortunately for crops with wide row spacing or young tree seedlings in forest

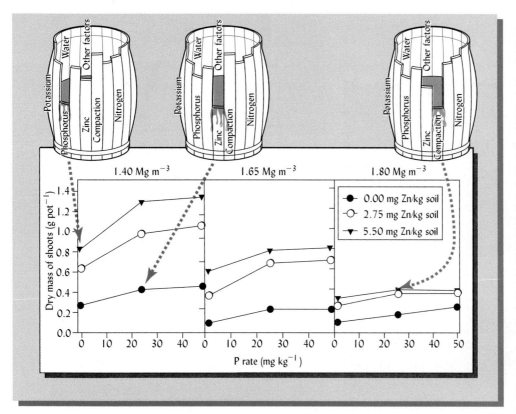

FIGURE 16.27 The law of the minimum illustrated by responses of wheat to increasing levels of soil compaction (1.40, 1.65, and 1.80 Mg/m³ bulk density), phosphorus fertilization (0, 25, and 50 mg P/kg soil) and zinc fertilization (0, 2.75, and 5.5 mg Zn/kg soil). The shortest (leaking) barrel stave shows which of these three factors is the most limiting in various situations. For example, the middle barrel shows that zinc is most limiting when compaction is low and a moderate level of P has been applied—that is, there is a large response to adding zinc but little response to adding more P. This panel also illustrates a *synergy* or *interaction* between P and Zn—namely, the response to P is greater once some Zn is added, and vise versa. In the right panel compaction is limiting (bulk density = 1.8 Mg m⁻³) and wheat shoots would respond to reduced compaction (compare to left panel) but not to adding more P or Zn. [Modified from Barzegar et al. (2006)].

plantings, broadcasting places the fertilizer where it is just as accessible to the *weeds* as to the target plants.

For phosphorus, zinc, manganese, and other nutrients that tend to be strongly retained by the soil, broadcast applications are usually much less efficient than localized placement. Often 2 to 3 kg of fertilizer must be broadcast to achieve the same response as from 1 kg that is placed in a localized area.

A heavy one-time application of phosphorus and potassium fertilizer, broadcast and worked into the soil, is a good preparation for establishing perennial plants such as lawns, pastures, and orchards. It may be necessary to broadcast a top dressing in subsequent years, being careful not to allow fertilizers with high salt hazards (see Table 16.13) to remain in contact with the foliage long enough to cause salt burn. Because of its mobility in the soil, nitrogen does not suffer from reduced availability when broadcast, but if left on the soil surface, much may be lost by volatilization. Volatilization losses are especially troublesome for urea and ammonium fertilizers applied to soils with a high pH. Nitrogen is commonly broadcast (sprayed) in a liquid form, often in a solution that also contains other nutrients or chemicals (see Figure 16.28 *Lower left*). Runoff studies have shown that most of the annual loss of nutrients (or herbicides, if surface broadcast) usually occurs during the first one or two heavy-rainfall events after the broadcast application. Where sprinkler irrigation is practiced, liquid fertilizers can be broadcast in the irrigation water, a practice sometimes called **fertigation**.

Localized Placement

Although it is commonly thought that nutrients must be thoroughly mixed throughout the root zone, research has clearly shown that a plant can easily obtain its entire supply of a nutrient from a concentrated localized source in contact with only a small fraction of its root system. In fact, a small portion of a plant's root system can grow and proliferate in a band of fertilizer even though the salinity level caused by the fertilizer would be fatal to a germinating seed or to a mature plant were a large part of its root system exposed. This finding allowed the development of techniques for localized fertilizer placement.

FIGURE 16.28 (*Upper left*) Broadcasting granular phosphorus and potassium fertilizer on a hayfield in Maryland. (*Lower left*) Broadcasting a nitrogen solution (mixed with fungicide) on a wheat field in France. *(Right)* Helicopter broadcasting special forestry-grade fertilizer pellets on pine plantations in southeastern U.S. (Helicopter photo courtesy of Westvaco, Inc.; others courtesy of R. Weil)

There are at least two reasons why fertilizer is often more effectively used by plants if it is placed in a localized concentration rather than mixed with soil throughout the root zone. First, localized placement reduces the amount of contact between soil particles and the fertilizer nutrient, thus minimizing the opportunity for adverse fixation reactions. Second, in the fertilized zone the concentration of the nutrient in the soil solution at the root surface will be very high, resulting in greatly enhanced uptake by the roots.

Localized placement is especially effective for young seedlings, in cool soils in early spring, and for plants that grow rapidly with a big demand for nutrients early in the season. For these reasons, **starter fertilizer** is often applied in bands on either side of the seed as the crop is planted. Since germinating seeds can be injured by fertilizer salts, and since these salts tend to move upward as water evaporates from the soil surface, the best placement for starter fertilizer is approximately 5 cm below and 5 cm off to the side from the seed row (see Figure 16.29d and Figure 16.30).

Liquid fertilizers and slurries of manure and sewage sludge can also be applied in bands rather than broadcast. Bands of these liquids are placed 10 to 30 cm deep in the soil by a process known as *knife injection* (Figures 16.29e and 16.31). In addition to the advantages mentioned for banding fertilizer, injection of these organic slurries reduces runoff losses and odor problems. Anhydrous ammonia and pressurized nitrogen solutions must be injected into the soil to prevent losses by volatilization. Injecting bands at depths of 15 and 5 cm, respectively, are considered adequate for these two materials.

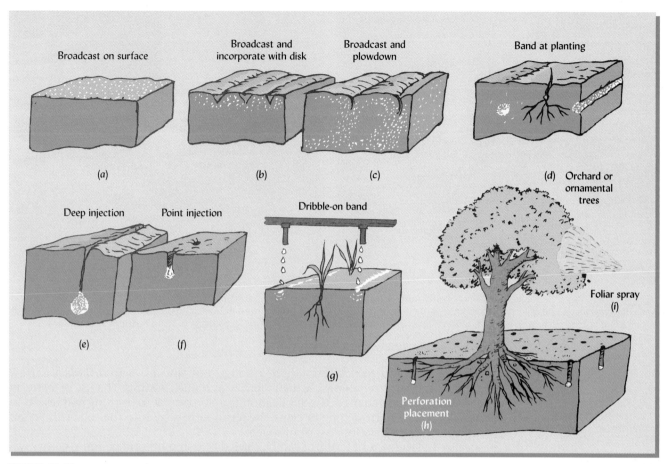

FIGURE 16.29 Fertilizers may be applied by many different methods, depending on the situation. Methods (*a*) to (*c*) represent broadcast fertilizer, with or without incorporation. Methods (*d*) to (*h*) are variations of localized placement. Method (*i*) is foliar application and has special advantages, but also limitations. Commonly, two or three of these methods may be used in sequence. For example, a field may be prepared with (*c*) before planting; (*d*) may be used during the planting operation; (*g*) may be used as a **side dressing** early in the growing season; and, finally, (*i*) may used to correct a micronutrient deficiency that shows up in the middle of the season.

FIGURE 16.30 Many mechanical planters are equipped with a fertilizer-banding attachment that places starter fertilizer for row crops about 5 cm below and 5 cm to the side of the seed. This placement eliminates the danger of fertilizer burn, yet concentrates the nutrients near the seed where the crop roots will encounter them shortly after the seed germinates. (Courtesy National Plant Food Institute, Washington, D.C.)

FIGURE 16.31 (*Left*) Sewage-sludge slurry being knife-injected into the soil before planting crops on the land. This injection method reduces runoff losses and objectionable odors. Lighter knives (not shown) are used to inject liquid fertilizers. (*Right*) A spike-wheel applicator designed for point injection of liquid fertilizer. [Photo (*left*) courtesy of R. Weil; photo (*right*) courtesy of Fluid Fertilizer Foundation]

Another approach to banding liquids (though not slurries) is to **dribble** a narrow stream of liquid fertilizer alongside the crop row as a side-dressing. The use of a stream instead of a fine spray changes the application from broadcast to banding and results in enough liquid in a narrow zone to cause the fertilizer to soak into the soil. This action greatly reduces volatilization loss of nitrogen.

Localized placement of fertilizer can be carried one step beyond banding with a system called **point injection**. With this system, small portions of liquid fertilizer can be applied next to every individual plant without significantly disturbing either the plant root or the surface residue cover left by conservation tillage. The point injection implement shown in Figure 16.31 (*right*) is a modern version of the age-old dibble stick with which peasant farmers in the lesser developed countries plant seeds and later apply a portion of fertilizer in the soil next to each plant, all with a minimum of disturbance of the surface mulch.

The use of **drip irrigation** systems (see Section 6.9) has greatly facilitated the localized application of nutrients in irrigation water. Because drip fertigation is applied at frequent intervals, the plants are essentially spoon-fed, and the efficiency of nutrient use is quite high.

PERFORATION METHOD FOR TREES. Trees in orchards and ornamental plantings are best treated individually, the fertilizer being applied around each tree within the spread of the branches but beginning approximately 1 m from the trunk (see Figure 16.29*h*). The fertilizer is best applied by what is called the *perforation* method. Numerous small holes are dug around each tree within the outer half of the branch-spread zone and extending down into the upper subsoil where the fertilizer is placed. Special large fertilizer pellets are available for this purpose. This method of application places the nutrients within the tree root zone and avoids an undesirable stimulation of the grass or cover that may be growing around the trees. If the cover crop or lawn around the trees needs fertilization, it is treated separately, the fertilizer being drilled in at the time of seeding or broadcast later.

Foliar Application

Plants are capable of absorbing nutrients through their leaves in limited quantities. Under certain circumstances, the best way to apply a nutrient is *foliar application*—spraying a dilute nutrient solution directly onto the plant leaves (Figure 16.29*i*). Diluted NPK fertilizers, micronutrients, or small quantities of urea can be used as foliar sprays, although care must be taken to avoid significant concentrations of Cl^- or NO_3^-, which can be toxic to some plants. Foliar fertilization may conveniently fit in with other field operations for horticultural crops, because the fertilizer is often applied simultaneously with pesticide sprays.

The amount of nutrients that can be sprayed on leaves in a single application is quite limited. Therefore, while a few spray applications may deliver the entire season's requirement for a micronutrient, only a small portion of the macronutrient needs can be supplied in this manner. The danger of leaf injury is especially high during dry, hot weather, when the solution quickly evaporates from the leaf surface, leaving behind the fertilizer salts. Spraying on cool, overcast days or during early morning or late evening hours reduces the risk of injury, as does the use of a dilute solution containing, for example, only 1 or 2% nitrogen.

16.9   TIMING OF FERTILIZER APPLICATION

The timing of nutrient applications in the field is governed by several basic considerations: (1) making the nutrient available when the plant needs it; (2) avoiding excess availability, especially of nitrogen, before and after the principal period of plant uptake; (3) making nutrients available when they will strengthen, not weaken, long-season and perennial plants; and (4) conducting field operations when conditions make them practical and feasible.

Availability When the Plants Need It

For mobile nutrients such as nitrogen (and to some degree potassium), the general rule is to make applications as close as possible to the period of rapid plant nutrient uptake. For rapid-growing summer annuals, such as corn, this means making only a small starter application at planting time and applying most of the needed nitrogen as a side dressing just before the plants enter the rapid nutrient accumulation phase, usually about four to six weeks after planting. For cool-season plants, such as winter wheat or certain turf grasses, most of the nitrogen should be applied about the time of spring "green-up," when the plants resume a rapid growth rate. For trees, the best time is when new leaves are forming. With slow-release organic sources, time should be allowed for mineralization to take place prior to the plants' period of maximum uptake.

Environmentally Sensitive Periods

In temperate (**Udic** and **Xeric**) climates, most leaching takes place in the winter and early spring when precipitation is high and evapotranspiration is low. Nitrates left over after plant uptake has ceased have the potential for leaching during this period. In this regard it should be noted that, for grain crops, the rate of nutrient uptake begins to decline during grain-filling stages and has virtually ceased long before the crop is ready for harvest. With inorganic nitrogen fertilizers, avoiding leftover nitrates is largely a matter of limiting the amount applied to what the plants are expected to take up. However, for slow-release organic sources applied in late spring or early summer, mineralization is likely to continue to release nitrates after the crop has matured and ceased taking them up. To the extent that this timing of nitrate release is unavoidable, non-leguminous cover crops should be planted in the fall to absorb the excess nitrate being released.

SPLIT APPLICATIONS.   In high-rainfall conditions and on permeable soils, dividing a large dose of fertilizer into two or more split applications may avoid leaching losses prior to the crop's establishment of a deep root system. In cold climates, another environmentally sensitive period occurs during early spring, when snowmelt over frozen or saturated soils results in torrents of runoff water. This runoff may pollute rivers and streams with soluble nutrients from manure or fertilizer that is at or near the soil surface.

Application of urea fertilizers to mature forests is usually carried out when rains can be expected to wash the nutrients into the soil and minimize volatilization losses. Furthermore, nitrogen fertilization of forests should occur just prior to the onset of the growing season (early spring, or in warm climates, winter) so that the tree roots will have the entire growing season to utilize the nitrogen. Fertilizer applications commonly result in a pulse of nitrate leaving the watershed for several weeks following the fertilizer application. The nitrogen loss can be reduced if a 10- to 15-m unfertilized buffer is maintained along all streams.

Physiologically Appropriate Timing

It is important to make nutrients available when they will strengthen plants and improve their quality. For example, too much nitrogen in the summer may stress a cool-season turf grass, while high nitrogen late in the season will reduce the sugar content of sugar crops. A good supply of potassium is particularly important in the fall to enable plants to improve their winter-hardiness. Planting-hole or broadcast application of phosphorus at the time of the tree planting brings good results on phosphorus-poor sites. However, broadcast application of nitrogen to trees soon after the seedlings are planted may benefit fast-growing weeds more than the desired trees. Later in the development of a forest stand, when the tree canopy has matured, application of fertilizer, usually from a helicopter, can be quite beneficial.

Practical Field Limitations

Sometimes it is simply not possible to apply fertilizers at the ideal time of the year. For example, although a crop may respond to a late-season side dressing, such an application will be difficult if the plants are too tall to drive over without damaging them. Using an airplane may allow more flexibility in fertilizer timing. Early spring applications may be limited by the need to avoid compacting wet soils. Economic costs or the time demands of other activities may also require that compromises be made in the timing of nutrient application.

16.10 DIAGNOSTIC TOOLS AND METHODS[8]

Three basic tools are available for diagnosing soil fertility problems: (1) *field observations*, (2) *plant tissue analysis*, and (3) *soil analysis* (soil testing). To effectively guide the application of nutrients, as well as to diagnose problems as they arise in the field, all three approaches should be integrated. There is no substitute for careful observation and *recording* of circumstantial evidence and symptoms in the field. Effective observation and interpretation requires skill and experience, as well as an open mind. It is not uncommon for a supposed soil fertility problem to actually be caused by soil compaction, weather conditions, pest damage, or human error. The task of the diagnostician is to use all the tools available in order to identify the factor that is limiting plant growth, and then devise a course of action to alleviate the limitation.

Plant Symptoms and Field Observations

Diagnosing field problems in rice. Int'l. Rice Res. Instit.: http://www.knowledgebank. irri.org/riceDoctor_MX/default. htm

This detectivelike work can be one of the more exciting and challenging aspects of nutrient management. To be an effective soil fertility diagnostician, several general guidelines are helpful.

1. Develop an organized way to *record* your observations. The information you collect may be needed to properly interpret soil and plant analytical results obtained at a later date.
2. Talk to the person who owns or manages the land. Ask when the problem was first observed and if any recent changes have taken place. Obtain records on plant growth or crop yield from previous years, and ascertain the history of management of the site for as many years as possible. It is often useful to sketch a map of the site showing features you have observed and the distribution of symptoms.

---

[8] For a detailed discussion of both plant analysis and soil testing, see Westerman (1990).

3. Look for *spatial patterns*—how the problem seems to be distributed in the landscape and in individual plants. Linear patterns across a field may indicate a problem related to tillage, drain tiles, or the incorrect spreading of lime or fertilizer. Poor growth concentrated in low-lying areas may relate to the effects of soil aeration. Poor growth on the high spots in a field may reflect the effects of erosion and possibly exposure of subsoil material with an unfavorable pH.

4. Closely examine individual plant leaves to characterize any foliar symptoms. Nutrient deficiencies can produce characteristic symptoms on leaves and other plant parts. Examples of such symptoms are shown in several figures in Chapters 13 to 15 and in Plates 87 to 107. Determine if the symptoms are most pronounced on the younger leaves (as is the case for most of the micronutrient cations) or on the older leaves (as is the case for nitrogen, potassium, and magnesium). Some nutrient deficiencies are quite reliably identified from foliar symptoms, while others produce symptoms that may be confused with herbicide damage, insect damage, or damage from poor aeration.

5. Observe and *measure* differences in plant growth and crop yield that may reflect different levels of soil fertility, even though no leaf symptoms are apparent. Check *both* aboveground and belowground growth. Are mycorrhizae associated with tree roots? Are legumes well nodulated? Is root growth restricted in any way?

Plant Tissue Analysis

NUTRIENT CONCENTRATIONS.   The concentration of essential elements in plant tissue is related to plant growth or crop yield, as shown in Figure 16.32. The range of tissue concentrations at which the supply of a nutrient is sufficient for optimal plant growth is termed the **sufficiency range**. At the upper end of the sufficiency range, plants may be participating in luxury consumption, as the additional nutrient uptake has not produced additional plant growth (see Section 14.13). At concentrations above the sufficiency range, plant growth may decline as nutrient elements reach concentrations that are toxic to plant cells or interfere with the use of other nutrients. If tissue concentrations are in the **critical range**, the supply is just marginal and growth is expected to decline if the nutrient becomes any less available, even though visible foliar symptoms may not be exhibited ("hidden hunger"). Plants with tissue concentrations below the sufficiency range for a nutrient are likely to respond to additions of that nutrient if no other factor is more limiting. The sufficiency range and critical range have been well

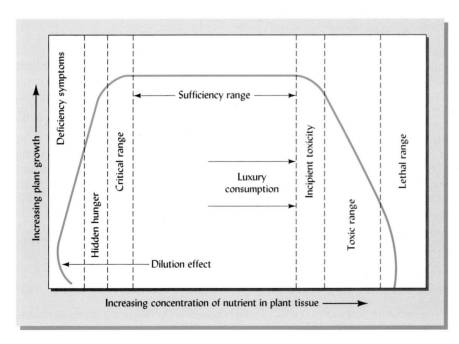

FIGURE 16.32   The relationship between plant growth or yield and the concentration of an essential element in the plant tissue. For most nutrients there is a relatively wide range of values associated with normal, healthy plants (the sufficiency range). Beyond this range, plant growth suffers from either too little or too much of the nutrient. The critical range (CR) is commonly used for the diagnosis of nutrient deficiency. Nutrient concentrations below the CR are likely to reduce plant growth even if no deficiency symptoms are visible. This moderate level of deficiency is sometimes called *hidden hunger.* The odd hook at the lower left of the curve is the result of the so-called dilution effect that is often observed when extremely stunted, deficient plants are given a small dose of the limiting nutrient. The growth response may be so great that even though somewhat more of the element is taken up, it is diluted in a much greater plant mass.

TABLE 16.15 A Guide to Sufficiency Ranges for Tissue Analysis of Selected Plant Species

Values apply only to the indicated plant parts and stage of growth. Normally, 6 to 20 plants should be sampled. Leaves should be washed briefly in distilled water to remove any soil or dust and then dried before submitting for analysis.

Plant species and Part to sample	Content, %						Content, µg/g				
	N	P	K	Ca	Mg	S	Fe	Mn	Zn	B	Cu
Pine trees (*Pinus* spp.) Current-year needles near terminal	1.2–1.4	0.10–0.18	0.3–0.5	0.13–0.16	0.05–0.09	0.08–0.12	20–100	50–600	20–50	3–9	2–6
Oak tree (*Quercus*) Mature leaves	1.9–3.0	0.15–0.30	1.0–1.5	0.3–0.5	0.15–0.30	–	50–150	35–200	15–30	15–40	6–12
Turfgrasses, warm season Clippings	2.7–3.5	0.25–0.55	1.3–3.0	0.50–1.2	0.15–0.60	0.15–0.6	35–500	25–150	15–55	6–60	5–30
Turfgrasses, cool season Clippings	3.0–5.0	0.3–0.4	2–4	0.3–0.8	0.2–0.4	0.25–0.8	40–500	20–100	20–50	5–20	6–30
Corn (*Zea mays*) Ear-leaf at tasseling	2.5–3.5	0.20–0.50	1.5–3.0	0.2–1.0	0.16–0.40	0.16–0.50	25–300	20–200	20–70	6–40	6–40
Soybean (*Glycine max*) Youngest mature leaf at flowering	4.0–5.0	0.31–0.50	2.0–3.0	0.45–2.0	0.25–0.55	0.25–0.55	50–250	30–200	25–50	25–60	8–20
Apple (*Malus* spp.) Leaf at base of nonfruiting shoots	1.8–2.4	0.15–0.30	1.2–2.0	1.0–1.5	0.25–0.50	0.13–0.30	50–250	35–100	20–50	20–50	5–20
Wheat (*Triticum* spp.) Youngest mature leaf at flowering	2.2–3.3	0.24–0.36	2.0–3.0	0.28–0.42	0.19–0.30	0.20–0.30	35–55	30–50	20–35	5–10	6–10
Rice (*Oryza sativa*) Youngest mature leaf at tillering	2.8–3.6	0.14–0.27	1.5–3.0	0.16–0.40	0.12–0.22	0.17–0.25	90–200	40–800	20–160	5–25	6–25
Tomato (*Solanum lycopersicum*) Youngest mature leaf at flowering	3.2–4.8	0.32–0.48	2.5–4.2	1.7–4.0	0.45–0.70	0.60–1.0	120–200	80–180	30–50	35–55	8–12
Alfalfa (*Medicago sativa*) Upper third of plant at first flower	3.0–4.5	0.25–0.50	2.5–3.8	1.0–2.5	0.3–0.8	0.3–0.5	50–250	25–100	25–70	6–20	30–80

Data derived from many sources.

Plant analysis handbook for agronomic and horticultural crops. Univ. of Georgia: http://aesl.ces.uga.edu/publications/plant/plant.html

Plant tissue testing in Asian agriculture: http://www.fftc.agnet.org/library/article/eb536.html#eb536t4

characterized for many plants, especially for agronomic and major horticultural crops. Less is known about forest trees and ornamentals.[9] Sufficiency ranges for 11 essential elements in a variety of plants are listed in Table 16.15.

TISSUE ANALYSIS. Tissue analysis can be a powerful tool for identifying plant nutrient problems if several simple precautions are taken. First, it is critical that the correct plant part be sampled. Second, the plant part must be sampled at the specified stage of growth, because the concentrations of most nutrients decrease considerably as the plant matures. Third, it must be recognized that the concentration of one nutrient may be affected by that of another nutrient, and that sometimes the ratio of one nutrient to another (e.g., Mg/K, N/S, or Fe/Mn) may be the most reliable guide to plant nutritional status (Figure 16.33). In fact, several elaborate mathematical systems for assessing the ratios or balance among nutrients have proven useful for certain plant species.[10] Because of the uncertainties and complexities in interpreting tissue concentration data, it is wise to sample plants from the best and worst areas in a field or stand. The difference between samples may provide valuable clues concerning the nature of the nutrient problem.

CORNSTALK NITRATE. The *end-of-season cornstalk nitrate test* (CSNT) measures the nitrate content of the lower portion of mature cornstalks to improve future nitrogen management

[9] Detailed information on tissue analysis for a large number of plant species can be found in Reuter and Robinson (1986).
[10] The best developed of the multinutrient ratio systems is known as the Diagnostic Recommendation Integrated System (DRIS). For details, see Walworth and Sumner (1987).

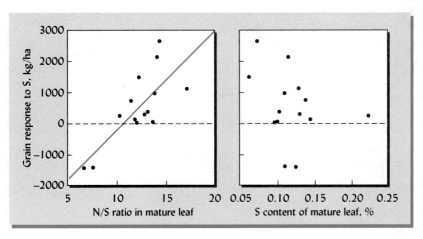

FIGURE 16.33 Relationship between sulfur content in leaves (*right*) or the N/S ratio in leaves (*left*) and the corn grain increases produced in response to application of sulfur. The graphs illustrate that the ratio of two interacting elements is often a better guide to plant nutrient status than the tissue contents of either element alone. Nitrogen and sulfur are both needed to synthesize plant proteins. As the N/S ratio of the unfertilized corn increased, so did the positive response to application of sulfur (as gypsum). There was no clear relationship between response to sulfur application and the leaf sulfur content by itself, even though most of the corn had sulfur levels below the sufficiency range indicated in Table 16.15. The data are from experiments on 14 small farms in Malawi. Sulfur deficiencies are widespread in central African soils. [Data from Weil and Mughogho (2000)]

for corn. This test is particularly helpful in identifying excessive levels of nitrogen in the plant at harvest time, which, in turn, is an indication of excessive levels in the soil at the end of the season. Such high cornstalk nitrate levels alert farmers to reduce their nitrogen applications to avoid excess nitrogen leaching in future years.

16.11 SOIL ANALYSIS

Since the total amount of an element in a soil tells us very little about the ability of that soil to supply that element to plants, more meaningful *partial soil analyses* have been developed. *Soil testing* is the routine partial analysis of soils for the purpose of guiding nutrient management.

The soil testing process consists of three critical phases: (1) *sampling* the soil, (2) chemically *analyzing* the sample, and (3) *interpreting* the analytical result to make a recommendation on the kind and amounts of nutrients to apply.

Sampling the Soil

Sampling turf for soil testing:
http://www.cropsoil.uga.edu/turf/index.html

Soil sampling is widely acknowledged to be one of the weakest links in the soil testing process. Part of the problem is that about a teaspoonful of soil (Figure 16.34) is eventually used to represent millions of kilograms of soil in the field. Since soils are highly variable, both horizontally and vertically, it is essential to carefully follow the sampling instructions from the soil testing laboratory.

Because of the variability in nutrient levels from spot to spot, it is always advisable to divide a given field or property into as many distinct areas as practical, taking soil samples from each to determine nutrient needs. For example, suppose a 20-ha field has a 2-ha low spot in the middle, and 5 ha at one end that used to be a permanent pasture. These two areas should be sampled, and later managed, separately from the remainder of the field. Similarly, a homeowner should sample flower beds separately from lawn areas, low spots separately from sloping areas, and so on. On the other hand, known areas of unusual soil that are too small or irregular to be managed separately should be avoided and not included in the composite sample from the whole field.

COMPOSITE SAMPLE. Usually, a soil probe is used to remove a thin cylindrical core of soil from at least 12 to 15 randomly scattered places within the land area to be represented (Figure 16.35). The 12 to 15 subsamples are thoroughly mixed in a plastic bucket, and about 0.5 L of the soil is placed in a labeled container and sent to the lab. If the soil is moist, it should be air-dried without sun or heat prior to packaging for routine soil tests. Heating the sample might cause a falsely high result for certain nutrients.[11]

---

[11] Samples for analysis of soil nitrates should be rapidly dried under a fan or in an oven at about 65 °C.

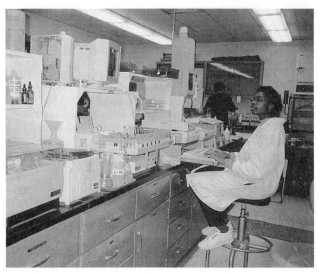

FIGURE 16.34 (*Right*) After a soil sample is received by the soil testing lab, the soil is ground and screened to make a homogenous powder. Then a small scooped or weighed sample is analyzed chemically. This small amount of soil must represent thousands of metric tons of soil in the field. (*Left*) After a portion of the nutrients have been extracted from the soil sample, the solution containing these nutrients undergoes elemental analysis. Since soil test labs must run hundreds or thousands of samples each day, the analysis is generally automated and the results are recorded and interpreted by computer. (Photos courtesy of R. Weil)

Two questions must be addressed when sampling a soil: (1) the depth to which the sample should be taken, and (2) the time of year when the soil should be sampled.

DEPTH TO SAMPLE.   The standard depth of sampling for a plowed soil is the depth of the plowed layer, about 15 to 20 cm, but various other depths are also used (see Figure 16.35). Because in many unplowed soils nutrients are stratified in contrasting layers, the depth of sampling can greatly alter the results obtained.

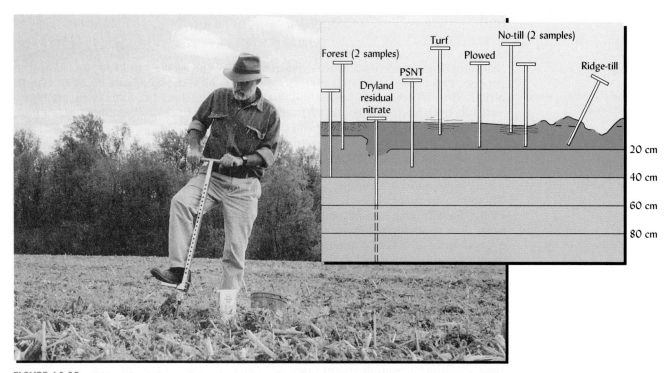

FIGURE 16.35   Taking the soil sample in the field is often the most error-prone step in the soil testing process because soil properties vary greatly both with depth and from place to place in even a uniform-appearing field. Areas that are obviously unusual (wet spots, places where manure was piled, eroded spots, etc.) should be avoided. (diagram) The proper depth to sample depends on the purpose of the soil test and the nature of the soil. Some suggested depths for different situations are shown. (Photo courtesy of R. Weil)

TIME OF YEAR. Seasonal changes are often observed in soil test results for a given area. For example, the potassium level is usually highest in early spring, after freezing and thawing has released some fixed K ions from clay interlayers, and lowest in late summer, after plants have removed much of the readily available supply. The time of sampling is especially important if year-to-year comparisons are to be made. A good practice is to sample each area every year or two (always at the same time of year), so that the soil test levels can be tracked over the years to determine whether nutrient levels are being maintained, increased, or depleted.

TIMING FOR SPECIAL NITROGEN TESTS. Timing is especially critical to determine the amount of mineralized nitrogen in the root zone. In relatively dry, cold regions (e.g., the Great Plains), the *residual nitrate* test is done on 60-cm-deep samples obtained sometime between fall and before planting in spring. In humid regions, where nitrate leaching is more pronounced, a special test has been developed to determine whether a soil will mineralize enough nitrogen for a corn crop. The samples for this *Presidedress Nitrate Test* (PSNT) are taken from the upper 30 cm when the corn is about 30 cm tall, just in time to determine how much nitrogen to apply as the crop enters its period of most rapid nitrogen uptake. In this case, the soil must be sampled during the narrow window of time when spring mineralization has peaked, but plant uptake has not yet begun to deplete the nitrate produced (see Figure 13.11).

SITE-SPECIFIC MANAGEMENT ZONES. The just-described standard procedure of compositing many small soil samples into a single mixed sample to represent the "average" soil in a large field must be recognized for the compromise that it is. Fertilizer recommendations based on the *average* soil condition in the field will likely be either too high or too low for almost any particular spot in that field. Using *geographic information systems* (GIS) computer technology, fertilizer rates can be much more precisely tailored to account for soil variations within a field. However, the benefits of doing this are not always worth the costs. The costs include field labor and fees for collecting and processing a large number of soil samples collected in a grid pattern within the field. Since each sample is geo-referenced as to its specific location (using space satellite–based *geo-positioning systems*, GPS), computer software can generate maps showing management zones with defined soil properties and fertilizer needs. Computer-controlled fertilizer-spreading equipment can be automatically adjusted "on the go" to spread more or less fertilizer as called for by the map of management zones. The development and use of this type of spatial information about soils will be discussed in more detail in Chapter 19.

Chemical Analysis of the Sample

Sampling for Site Specific Management (Manitoba):
http://www.gov.mb.ca/agriculture/soilwater/soilfert/fbd01s02.html

In general, soil tests attempt to extract from the soil amounts of essential elements that are correlated with the nutrients taken up by plants. Different extraction solutions are employed by various laboratories. Buffered salt solutions, such as sodium or ammonium acetate, or mixtures of dilute acids and chelating agents are the extracting agents most commonly used. The extractions are accomplished by placing a small measured quantity of soil in a bottle with the extracting agent and shaking the mixture for a certain number of minutes. The amount of the various nutrient elements brought into solution is then determined. The whole process is usually automated so that a modern laboratory can handle hundreds of samples each day (see Figure 16.34).

The most common and reliable tests are those for soil pH, potassium, phosphorus, and magnesium. Micronutrients are sometimes extracted using chelating agents, especially for calcareous soils in the more arid regions. While the nitrate and sulfate present in the soil at the time of sampling can be measured, predicting the availability of nitrogen and sulfur is considerably more difficult because of the many biological factors involved.

Because the methods used by different labs may be appropriate for different types of soils, it is advisable to send a soil sample to a lab in the same region from which the soil originated. Such a lab is likely to use procedures appropriate for the soils of the region and should have access to data correlating the analytical results to plant responses on soils of a similar nature.

MEHLICH 3 EXTRACT. Although many different extracting solutions are still in use by various soil-testing labs, a solution known as the "Mehlich 3 extractant" deserves special mention as this method has recently gained wide acceptance (especially in humid regions) as a nearly "universal" soil test extractant.[12] This complex extracting solution contains 0.2 M CH_3COOH + 0.25 M NH_4NO_3 + 0.15 M NH_4F + 0.013 M HNO_3 + 0.001 M EDTA. These components allow the extractant to remove "plant-available" nutrients from soils by the processes of dissolution (by the acids), desorption (by the NH_4F and NH_4NO_3), and chelation (by the EDTA). In this method, a measured volume (2.5 cm^3) of dry, sieved soil is shaken with 100 mL of the extracting solution for 5 minutes, and then filtered. The filtrate can be analyzed for almost all of the essential elements, usually by an automated inductively coupled plasma emission spectrograph (ICP), which is capable of determining dozens of elements in a single analysis.

Soil tests designed for use on soils or soil-based potting media generally do not give meaningful results when used on artificial peat-based soilless potting media. Special extraction procedures must be used for the latter, and the results must then be correlated to nutrient uptake and growth of plants grown in similar media.

These examples should further emphasize the importance of providing the soil testing lab with complete information concerning the nature of your soil, its management history, and your plans for its future use.

Interpreting the Results to Make a Recommendation

Problems with "cation balancing" U of WI-Madison:
http://www.soils.wisc.edu/extension/FAPM/approvedppt2004/Kelling1.pdf

This is, perhaps, the most controversial aspect of soil testing. The soil test values themselves are merely *indices* of nutrient-supplying power. They do not indicate the actual amount of nutrient that will be supplied. For this reason, it is best to think of soil test reports as more indicative than quantitative.

Many years of field experimentation at many sites are needed to determine which soil test level indicates a low, medium, or high capacity to supply the nutrient tested. Such categories are used to predict the likelihood of obtaining a profitable response from the application of the nutrient tested (Figure 16.36). Since the actual units of measurement (ppm, mg/L, or lb/acre) have little actual meaning to the end user, soil test results are increasingly reported as index values on a relative scale (often with 75 to 100

---

[12]*After* he had officially retired from North Carolina State University, Dr. A. Mehlich developed the Mehlich 3 extractant, the third and most successful of his soil test methods. See Mehlich (1984).

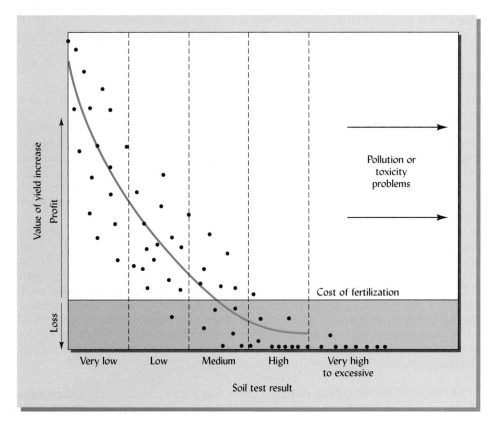

FIGURE 16.36 The relationship between soil test results for a nutrient and the extra yield obtained by fertilizing with that nutrient. Each data point represents the *difference* in plant yield between the fertilized and the unfertilized soil. Because many factors affect yield and because soil tests can only approximately predict nutrient availability, the relationship is not precise, but the data points are scattered about the trend line. If the point falls above the fertilizer cost line, the extra yield was worth more than the cost of the fertilizer and a profit would be made. For a soil testing in the very low and low categories, a profitable response to fertilizer is very likely. For a soil testing medium, a profitable response is a 50:50 proposition. For soils testing in the high category, a profitable response is unlikely.

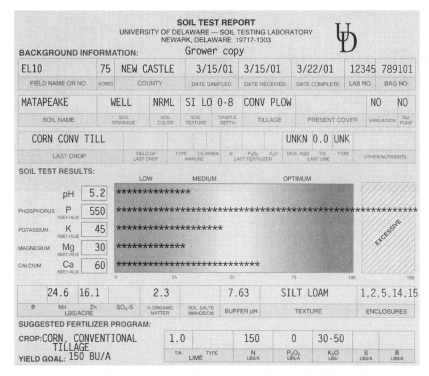

SOIL TEST REPORT
UNIVERSITY OF DELAWARE — SOIL TESTING LABORATORY
NEWARK, DELAWARE 19717-1303
Grower copy

BACKGROUND INFORMATION:

EL10	75	NEW CASTLE	3/15/01	3/15/01	3/22/01	12345	789101
FIELD NAME OR NO.	ACRES	COUNTY	DATE SAMPLED	DATE RECEIVED	DATE COMPLETE	LAB NO.	BAG NO.

MATAPEAKE	WELL	NRML	SI LO	0-8	CONV PLOW		NO	NO
SOIL NAME	SOIL DRAINAGE	SOIL COLOR	SOIL TEXTURE	SAMPLE DEPTH	TILLAGE	PRESENT COVER	IRRIGATION	INJ. PUMP

CORN CONV TILL						UNKN	0.0	UNK	
LAST CROP	YIELD OF LAST CROP	TYPE MANURE	T/A WHEN	N	P₂O₅ LAST FERTILIZER	K₂O	MOS. AGO LAST LIME	T/A TYPE	OTHER NUTRIENTS

SOIL TEST RESULTS:

		LOW	MEDIUM	OPTIMUM	EXCESSIVE
pH	5.2	*****************			
PHOSPHORUS P INDEX VALUE	550	**			
POTASSIUM K INDEX VALUE	45	*********************			
MAGNESIUM Mg INDEX VALUE	30	*************			
CALCIUM Ca INDEX VALUE	60	****************************			

scale: 0   25   50   75   100   150

24.6	16.1		2.3		7.63	SILT LOAM	1,2,5,14,15
B	Mn Zn LBS/ACRE		SO₄-S	% ORGANIC MATTER	SOL. SALTS MMHOS/CM	BUFFER pH	TEXTURE	ENCLOSURES

SUGGESTED FERTILIZER PROGRAM:

CROP: CORN, CONVENTIONAL TILLAGE
YIELD GOAL: 150 BU/A

1.0		150	0	30-50		
T/A LIME	TYPE	N LBS/A	P₂O₅ LBS/A	K₂O LBS/	S LBS/A	B LBS/A

FIGURE 16.37 A typical soil test report. Soil test results are relative indicators and many labs now report them as index values spanning such categories as low, medium, and optimum or high. The report records the nature and history of the field, reports the levels of certain nutrients and soil properties that were measured in the lab, and gives recommendations for the amounts and types of soil amendments to apply for a specific crop. For the particular field sampled here, the following interpretations can be made. The pH is low and should be raised to near 5.8 to 6.0. At least some of the liming material should be dolomitic limestone to add Mg, which is in low to medium supply. Moderate amounts of K are likely to give an economic response. Sulfur and B are not part of the standard analysis package in Delaware, but might be included if requested. The P level is highly excessive, so P additions of any kind should be avoided, and steps should be taken to manage transport of P to waterways. As Delaware is in a humid region where most soluble N may leach away over the winter, N application is recommended based on the yield goal and soil properties that affect mineralization, but not on an analytical test for available N. In an arid or semiarid region, a soil test lab would likely also measure and report nitrate and electrical conductivity in the soil and use these values in making recommendations concerning N fertilization and salt management, respectively. (Soil test report courtesy of K. L. Gartley, University of Delaware)

considered optimal). The report shown in Figure 16.37 uses both interpretative categories and a relative index scale.

Recommendations for nutrient applications take into consideration practical knowledge of the plants to be grown, the characteristics of the soil under study, and other environmental conditions. Management history and field observations can help relate soil test data to fertilizer needs.

The interpretation of soil test data is best accomplished by experienced and technically trained personnel who fully understand the scientific principles underlying the common field procedures. In most commercial or university soil test laboratories, the factors to be considered in making fertilizer recommendations are programmed into a computer, and the interpretation is printed out for the farmer's or gardener's use (Figure 16.37).

Merits of Soil Testing

It must not be inferred from the preceding discussion that the limitations of soil testing outweigh its advantages. When the precautions already described are observed, soil testing is an invaluable tool in making fertilizer recommendations. These tests are most useful when they are correlated with the results of field fertilizer experiments (Figure 16.38). Adding amendments to achieve some "ideal" balance of nutrients in the soil is often wasteful of money and resources. Rather, soil tests used correctly in conjunction with calibration experiments, can indicate what level of amendment needs to be added, if any, to allow the soil to supply sufficient nutrients for optimal plant growth (see also Section 15.2).

Dependability in predicting how plants will respond to soil amendments varies with the particular soil test, the tests for some parameters being much more reliable than others because of the consistency and breadth of field correlation data. In general, the tests for pH (need for liming or acidification), P, K, Mg, B, and Zn are quite reliable. Some soil test labs report additional soil properties such as other micronutrients (Cu, Mn, Fe, Mo, etc.), humus fractions (fulvic, humic, etc.), nonacid cation ("base") saturation, and even various microbial populations (fungi, bacteria, nonparasitic nematodes, etc.). However, the basis for making valid and practical recommendations for managing soil fertility based on such soil test parameters is currently very scant, at best.

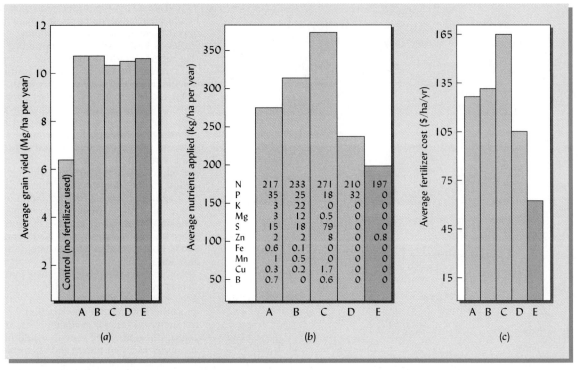

FIGURE 16.38 (a) The yield of corn from plots near North Platte, Nebraska, receiving five different rates of fertilizer as recommended by five soil testing laboratories (A to E). Data averaged over a six-year period. The soil was a Cozad silt loam (Fluventic Haplustoll). (b) The nutrient levels recommended by these laboratories and (c) the cost of the fertilizer applied each year. Note that the yields from all fertilized plots were about the same, even though rates of fertilizer application differed markedly. The fertilizer rates recommended by laboratory E utilized a sufficiency-level concept that was based on the calibration of soil tests with field yield responses. Those recommended by laboratories A to D were based either on a maintenance concept that required replacement of all nutrients removed by a crop or on the supposed need to maintain set cation ratios (Ca/Mg, Ca/K, and Mg/K). Obviously, the sufficiency-level concept provided more economical results in this trial. As a result of this and many other similar comparison studies, many soil test labs have adopted the sufficiency-level approach. [From Olson et al. (1982); used with permission of American Society of Agronomy]

Generally, soil testing has been most relied upon in agricultural systems, while foliar analysis has proved more widely useful in forestry. The limited use of soil testing in forestry is partly due to the fact that trees, with their extensive perennial root systems, integrate nutrient bioavailability throughout the profile. This, combined with the typically complex nutrient stratification in forested soils, creates a great deal of uncertainty about how to obtain a representative sample of soil for analysis. In addition, because of the comparably long time frame in forestry, limited information is available on the correlation of soil test levels with timber yields in the sense that such information is widely available for agronomic crops (see Figure 16.36). An exception may be the conifer plantations on Ultisols of the southeastern United States, where the wood volume responses to P fertilization at various soil test levels have been studied extensively. Even though the relationship between tree growth and soil test level is not well known for most other forest systems, standard agronomic soil testing can still be useful in distinguishing those soils whose ability to supply P or K is adequate from those with very low supplying power for these nutrients.

16.12 SITE-INDEX APPROACH TO PHOSPHORUS MANAGEMENT

As concerns have increased about nonpoint source water pollution, research has identified phosphorus movement from land to water as a major cause of aquatic ecosystem degradation, especially for lakes. In most industrial countries, municipal sewage systems are *the* major point source of phosphorus water pollution and agricultural land is *the* major nonpoint source of phosphorus water pollution. Society continues to invest in upgrading sewage treatment facilities to reduce phosphorus loading from those sources. The situation demands that the amounts of phosphorus contributed from agricultural land also be drastically reduced in some areas.

Phosphorus movement from land to water is determined by *phosphorus-transport, phosphorus-source*, and *phosphorus-management* parameters. Transport parameters include the mechanisms that govern how rain and irrigation water cause phosphorus to move across the landscape in runoff and sediment. Source parameters include the amount and forms of phosphorus on and in the soil. Management parameters include the method of application, timing, and placement of phosphorus and such disturbances as tillage.

Overenrichment of Soils

In Section 14.1 we learned how phosphorus, so scarce in nature, has accumulated over a period of decades to excessive levels in many agricultural soils as a result of two historical trends. In the case of phosphorus fertilizer, research in the early and mid-20th century suggested that farmers needed to add several times the amount of phosphorus that plants were likely to use, because soils were found to fix most of the added phosphorus in unavailable forms. Applications of 40 to 100 kg P/ha became standard practice, even though crop harvests removed only about one-third of that amount. The phosphorus not removed in harvest accumulated in the soil, and eventually satisfied a large fraction of the soil's phosphorus fixation capacity. It was not until the 1980s and 1990s that many farmers and researchers realized that soils with a long history of phosphorus fertilization were no longer in need of phosphorus application levels above what the crops would remove.

The second trend that contributed to the phosphorus pollution problem was the concentration of livestock and the subsequent heavy applications of animal manure to soils nearby (see Section 16.2). Much of the nitrogen added in the manure was lost to the environment by volatilization or leaching, but the phosphorus, being much less mobile, mostly remained in the soil. In recent years, farmers have begun to apply manure in amounts calculated to just meet the crop's nitrogen needs. Even so, the amount of phosphorus applied is still far more than plants can use (remember, from Section 16.6, that plants use about 10 times as much nitrogen as phosphorus, but manures—especially if not handled carefully to avoid nitrogen losses—contain only 1 to 4 times as much nitrogen as phosphorus).

Because of these trends, some farm fields have become so high in phosphorus that rain or irrigation water interacting with the soil carries away enough phosphorus to impair the ecology of receiving waters. It is imperative that sites with such a high potential to cause phosphorus pollution be identified and appropriately managed to reduce their environmental impact. It may take decades to lower soil test P levels back to the optimum range (Figure 16.39).

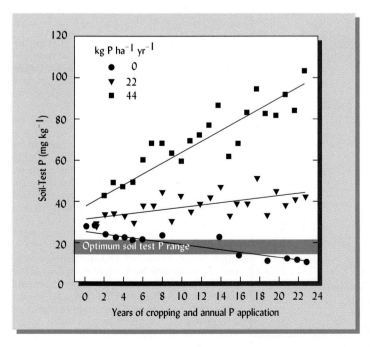

FIGURE 16.39 Changes in soil test P as a result of annual applications of 0, 22 or 44 kg P ha^{-1} to a corn-soybean rotation on an Iowa Mollisol. Data from this and other sites in the study suggest that application of 12 to 20 kg P ha^{-1} yr^{-1} is sufficient to balance P removed in crop harvests and maintain soil test P in the optimal range for many cropping systems. The soil tests were conducted using the Bray-1 method for which the optimal levels are 16 to 22 mg P extracted per kg soil. [Data from Dodd and Mallarino (2005) with permission of the Soil Sci. Soc. of America]

Transport of Phosphorus from Land to Water

Two fundamental conditions must be met for a soil to cause significant phosphorus pollution. First, the supply of phosphorus in the soil must be relatively large; second, the characteristics of the site and soil must allow significant transport of phosphorus from the soil in the field to a receiving water body such as a lake or stream. A review of the phosphorus cycle (Section 14.3) reminds us that phosphorus can move from land to water by three principal pathways:

1. Attached to eroded soil particles (the main P-loss pathway from tilled soils)
2. Dissolved in water running off the surface of the land (a major pathway for pastures, woodland, and no-till cropland)
3. Dissolved or attached to suspended colloidal particles in water percolating down the profile and through the groundwater aquifers that feed streams or lakes from below (a significant pathway in very sandy or poorly drained soils with shallow water tables and possibly in heavily manured soils with many continuous macropores).

Phosphorus Soil Test Level as Indicator of Potential Losses

PHOSPHORUS IN ERODED SEDIMENT. Phosphorus attached to eroded soil particles accounts for the greatest amount of phosphorus contributed to receiving waters from most agricultural land. As discussed in Section 14.2, the eroded fraction is usually more concentrated in phosphorus than the soil left behind. While phosphorus strongly adsorbed onto mineral particles may not be as available to aquatic phytoplankton and algae as that carried in solution, such sediments tend to accumulate on lake and river bottoms, where their phosphorus becomes increasingly soluble because of iron reduction reactions. Soils that are both highly erodible and highly enriched with phosphorus (as indicated by excessive phosphorus soil test levels) represent the most serious potential for phosphorus pollution.

PHOSPHORUS IN RUNOFF WATER. As described in Section 16.11, soil testing was designed to determine how well a soil can supply nutrients to plant roots. However, routinely used phosphorus soil tests have been shown to also provide a useful (though not perfect) indication of how readily phosphorus will desorb from a soil and dissolve in runoff water. The relationship between phosphorus soil test level in the upper centimeter or two of soil and phosphorus in runoff water is not linear (Figure 16.40). Rather, the relationship exhibits a threshold effect such that little phosphorus is lost to runoff if soils are near or

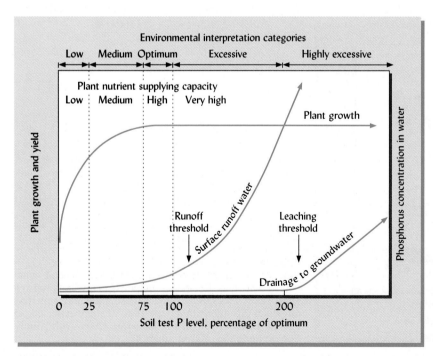

FIGURE 16.40 The generalized relationship between levels of plant-available phosphorus in soils (soil test P level) and environmental losses of P dissolved in surface runoff and subsurface drainage waters. The generalized relationship between traditional plant nutrient supply interpretation categories and environmental interpretation categories is also indicated. The diagram suggests that, fortunately, soil P levels can be achieved that are both conducive to optimum plant growth and protective of the environment. If P losses by soil erosion are controlled (although not shown, this is a big if), significant quantities of dissolved P would be lost only when soils contain P levels in excess of those needed for optimum plant growth. Losses of P by leaching in drainage water would be significant only at very high soil P levels, as this pathway rapidly increases only after the P-sorption capacity of the soil is substantially saturated. Note the threshold levels (vertical arrows) for P losses. The vertical axes are not to scale. [Figure based on data and concepts discussed in Sharpley (1997), Beegle et al. (2000), and Higgs et al. (2000)]

below the critical range for phosphorus, but the amount released from the soil to runoff water increases exponentially as phosphorus soil test levels rise much higher than this. Therefore, as suggested by Figure 16.40, it appears that soils can be maintained at phosphorus levels sufficient for optimal plant growth without undue losses of phosphorus in runoff water. As is also the case with nitrogen, significant environmental damages are associated with excess—the application of more nutrient than is really needed. Therefore, the most profitable level of phosphorus fertility should also be an environmentally sound level, and one that also conserves the world's finite stocks of this essential element.

PHOSPHORUS IN DRAINAGE WATER. In most soils under most circumstances, significant quantities of phosphorus are unlikely to be lost in drainage water. Drainage water is usually very low in phosphorus because as phosphorus-laden water percolates from the surface down through the subsurface soil horizons, dissolved $H_2PO_4^-$ is strongly sorbed by inner-sphere complexation onto Fe-, Al-, Mn-, and Ca-containing mineral surfaces (Sections 8.7 and 14.8). However, under certain circumstances this mechanism is not so effective in removing phosphorus from drainage water. These circumstances include the following:

1. The percolating water is flowing through macropores and has little contact with soil surfaces (Section 6.8). This may occur under long-term no-till management or in pastures.
2. The dissolved phosphorus is not in mineral ($H_2PO_4^-$ or HPO_4^{2-}) forms, but instead is part of soluble organic molecules, which are not strongly sorbed by mineral surfaces. This situation is common under forests or heavily manured soils.
3. The phosphorus is sorbed onto the surface of dispersed colloidal particles that are so small that they remain suspended in the percolating water.
4. The phosphorus-fixation capacity of the soil is so small (as in sands, Histosols, and some waterlogged soils with reduced iron) or already so saturated with phosphorus (as in soils overloaded with phosphorus after many years of excessive manure and fertilizer applications) that little more phosphorus can be sorbed (see Figure 16.40).

Site Characteristics Influencing Transport of Phosphorus

To cause environmental damage, the phosphorus-rich sites must also have characteristics conducive to the transport of phosphorus to sensitive bodies of water. Such characteristics might include close proximity to the water body, the absence of a protective vegetated buffer, erodible soils or high runoff rates caused by low permeability and steep slopes. Control of soil erosion by conservation tillage, mulches, and other practices discussed in Chapter 17 can greatly reduce the transport of sediment-associated phosphorus, but will have little effect on, or may even increase, the loss of phosphorus dissolved in runoff or drainage water. In fact, when no-till fields are fertilized, the phosphorus tends to accumulate in the upper 1 to 2 cm of soil, where it is most susceptible to desorption into runoff water. The beneficial influence of buffers in removing both dissolved and sediment-associated phosphorus from runoff was discussed in Section 16.2.

Phosphorus Site Index

Usually, most of the phosphorus entering a river or lake originates from only a small fraction of the watershed land area. Effective environmental management requires identification of the high-risk sites. As just discussed, these sites can be expected to be those with high-risk phosphorus source characteristics (large amounts of phosphorus-containing materials are applied and/or phosphorus soil test levels are excessively high) and high phosphorus transport characteristics. Researchers have attempted to integrate phosphorus source, transport, and management characteristics of a site into an index of phosphorus pollution risk, commonly referred to as the *phosphorus site index* (Table 16.16). Box 16.4 illustrates how such an index can be calculated for a specific farm field (or a separately identified part of a field). The phosphorus index is designed to identify sites where the risk of phosphorus movement may be relatively higher than that of other sites. Analysis of the individual index parameters should indicate which ones most influenced the index. One can then develop plans for corrective action to address problems associated with these identified parameters. An important part of such plans is the restriction of further phosphorus additions (Table 16.17).

TABLE 16.16 Phosphorus Site Index Based on Site Characteristics That Affect the Availability of Phosphorus at the Source and the Transport of Phosphorus to Sensitive Receiving Waters[a]

This example illustrates the types of parameters and relationships among parameters that might be used to rank individual farm fields or parts of fields as potential sources of phosphorus pollution. In adapting this type of phosphorus index tool, the site and management characteristics included, the weights used for each, and the method of calculation should reflect local conditions.

Parameter	Weight factor (A)	Phosphorus loss rating (B)					Score (A × B)
		None 0	Low 1	Medium 2	High 4	Very high 8	
P transport characteristics							
1. Annual soil erosion by rain[b]	1.5	<2	2–8	8–12	12–32	>32	☐
2. Erosion by irrigation water	1.0	Not irrigated	Tailwater return or no visible runoff	QS[c] >38 for erosion-resistant soils	QS >38 for erodible soils or visible runoff at field border	QS >38 for very erodible soils or rills visible	☐
3. Runoff class[d]	1	Very low	Low	Medium	High	Very high	☐
4. Distance from field edge to lake or stream	1	>30 m	15–30 m	8–15 m	4–8 m	<4 m	☐
5. Vegetated buffer between field and lake or stream[e]	0.5	>30 m	15–30 m	8–15 m	4–8 m	<4 m	☐
6. No P application zone (above any vegetated buffer)	0.5	>15 m	10–15 m	5–10 m	2–5 m	<2 m	☐
7. Receiving water priority rating[f]	1	Very low	Low	Medium	High	Very high	☐
Score 1 + score 2 + · · · + score 7 = **total score for P transport→**							
P source characteristics							
1. Soil test P category where >100 is considered excessive[g]	1	Low 0–25	Medium 26–75	Optimum 76–100	Excessive 101–200	Highly excessive >201	☐
2. P fertilizer application rate, kg P/ha	0.75	None	1–15	16–45	46–75	>75	☐
3. P fertilizer application method	0.5	None applied	Inject or band deeper than 5 cm	Incorporate within 5 days of application	Surface apply in summer or incorporate after >5 days	Surface apply during winter and not incorporate	☐
4. Organic P source application rate, kg P/ha[h]	1.0	None	1–15	16–30	31–45	>46	☐
5. Organic P source application method	1.0	None applied	Inject or band deeper than 5 cm	Incorporate within 5 days of application	Surface apply in summer or incorporate after >5 days	Surface apply during winter and not incorporate	☐
Score 1 + score 2 + · · · + score 5 = **total score for P source →**							☐
(P transport score) × (P source score) = **combined P site index for risk of water pollution →**							☐

Combined P site index:	0–10	11–50	51–250	251–1000	>1000
Risk of P pollution:	Negligible	Low	Medium	High	Very high

[a] Adapted from models and concepts discussed in USDA/NRCS (2006), Gburek et al. (2000) and Snyder et al. (2005). In any given region, the appropriate phosphorus index matrix table would differ somewhat from the general model shown here. English units are commonly used rather than metric for communication with farmers in the United States.

[b] Mg/ha as determined by RUSLE—see Section 17.4.

[c] Q = flow rate (L/min) of water applied to a furrow; S = slope (%). $Q \times S = QS$.

[d] Ranging from permeable soils on nearly level sites (none) to impermeable soils on slopes >20% (very high).

[e] See Section 16.2. Perennial grass nearest the field combined with forest vegetation nearest the stream is ideal.

[f] Not all areas have official priority rankings. Low priority would typically be an already-polluted artificial holding pond, while highest priority would be a pristine lake very sensitive to eutrophication.

[g] Not all soil test reports give results on this scale. "Excessive" may be termed "very high" by some labs (see Figure 16.37).

[h] The P from an organic source may be multiplied by an availability factor to estimate the fraction of total P in an organic material that will be released in a year.

BOX 16.4   HOW A PHOSPHORUS SITE INDEX WORKS

The phosphorus site index table assigns a rating of 0 to 8 for different degrees of risk from each parameter. In Table 16.16, the parameters are grouped into phosphorus source parameters and phosphorus transport parameters. The weights listed in the second column represent professional judgments as to how much each parameter is likely to influence phosphorus pollution. The weight multiplied by the rating gives a partial score for each parameter. These partial scores are then added together to give two total scores, one for phosphorus source and one for phosphorus transport. The product obtained by multiplying these two scores yields an overall combined score that indicates to which of five categories of risk the site belongs. Consider the following example:

A particular field is located only 12 m from the edge of a sensitive lake (priority = high) and there is no zone left at the field edge without phosphorus application. The field is bordered by a 12-m-wide vegetated buffer strip. The field is not irrigated and is moderately erodible (10 Mg/ha estimated annual erosion). The runoff class is medium because although the soil is moderately permeable, the site is steeply sloping. The upper layer of soil tests in the highly excessive range (300) because of a long history of phosphorus applications. No inorganic phosphorus fertilizer is now applied, but plans are to surface-apply, in winter, enough liquid manure to supply about 40 kg P/ha. The parameter scores in Table 16.16 are calculated here. For example, for the first parameter, the weight factor from Table 16.16 is 1.5 and the rating is 2, so $1.5 \times 2 = 3.0$, the partial score for this parameter.

Transport Parameter	Calculations (source parameters)	Partial score		P Source Parameter	Calculations (transport parameters)	Partial score
1	$1.5 \times 2 =$	3.0		1	$1.0 \times 8 =$	8.0
2	$1.0 \times 0 =$	0.0		2	$0.75 \times 0 =$	0
3	$1.0 \times 2 =$	2.0		3	$0.5 \times 0 =$	0
4	$1.0 \times 2 =$	2.0		4	$1.0 \times 4 =$	4.0
5	$0.5 \times 2 =$	1.0		5	$1.0 \times 8 =$	8.0
6	$0.5 \times 8 =$	4.0		Total P source score		20.0
7	$1.0 \times 4 =$	4.0				
Total P transport score		16.0				

Combined P pollution risk index: $16 \times 20 = 320$, indicates high risk

Based on the combined score of 320, this site has a high risk for phosphorus movement from the site (see rating chart at bottom of Table 16.16). There is a high probability for an adverse impact to surface-water resources unless remedial action is taken. Soil and water conservation as well as phosphorus management practices are necessary to reduce the risk of phosphorus movement and probable water-quality degradation. The partial scores suggest that the situation could be corrected by addressing transport parameter 6 (create a no-phosphorus application zone along the lower edge of the field), source parameter 4 (reduce the amount of organic nutrient source applied), and source parameter 5 (incorporate the organic nutrient source rather than applying it to the surface in winter). Transport parameter 7 (the priority ranking of the receiving water) and source parameter 1 (the current phosphorus soil test level) also contribute substantially to the combined index, but these factors are not subject to management change.

TABLE 16.17   Phosphorus Application in Accordance with Various Measures of Water Pollution Risk

The guidelines given here are generalized from USDA/NRCS national guidelines and should be modified as needed according to local soil, water, and agricultural conditions.

	Determine one of these measures of soil P status		Suitable application rates of P from organic **or** inorganic P sources	
Soil test P level[a]	Soil test P relative to runoff loss threshold	Site index for risk of P pollution[c]	Organic nutrient source[d]	Inorganic P fertilizer[b]
Low	<0.75 × threshold	Low	To supply needed N	20–50 kg/ha
Medium	<0.75 × threshold	Medium	To supply needed N	15–30 kg/ha
High	0.75 to 1.5 × threshold	Medium–high	To supply only the P in harvested crop	0–20 kg/ha
Very high (excessive)	1.5 to 2.0 × threshold	High	To supply ½ the P in harvested crop	0–10 kg/ha starter only
Highly excessive	>2 × threshold	Very high	No P application	

[a] Interpretive levels based on the high category being sufficient for optimum plant production. See Section 16.11.
[b] For use when soil specific phosphorus threshold values are available for the soils in question. See Figure 16.40 for an illustration of this concept.
[c] An index that integrates soil, landscape, and water resource characteristics such as Table 16.16.
[d] Such as manure, sewage sludge, compost, etc. If these materials are applied at rates to meet N needs, P will accumulate and eventually move the site into a higher soil P status category.

Because of the prominent response by most nonlegume plants, and because of its implications for environmental quality, the initial focus in most soil fertility plans is on nitrogen. Applications of phosphorus, potassium, sulfur, calcium, and other nutrients are usually made to balance and supplement the nitrogen supply whether it be from the soil, crop residues (especially legumes), organic wastes, or organic and inorganic fertilizers.

PLANT RESPONSE.   Because it is very difficult to predict the nitrogen-supplying ability of a soil from chemical tests, nitrogen fertilizer recommendations are usually based on field experiments that define the relationship between added nitrogen and plant growth or crop yield. Generally, these field studies are carried out on a variety of soils and under a range of different weather conditions (which cause the response to nitrogen to vary greatly from one year to the next). From the shape of the response curve and economic considerations (see following), the optimal level of nitrogen fertilizer is determined.

NITROGEN CREDITS.   This optimal fertilizer rate should be adjusted by the amount of any additional gains or losses not taken into account in the standard response curve. For example, nitrogen contributions from previous or current manure applications (see Box 16.2), legume cover crop (see Figure 13.33), previous legume in the rotation (see Figure 13.30), or nitrate in irrigation water should be subtracted from the amount of fertilizer recommended.

In arid and semiarid regions where overwinter leaching is minimal, the nitrate nitrogen found in the upper 60 to 120 cm of the profile in spring is often used along with data on soil organic matter levels to estimate the nitrogen that will be available. In humid regions the soil nitrate-N concentration in the upper 30 cm of soil just before side-dressing time is used to predict nitrogen fertilizer needs for certain crops. Several chemical tests and computer programs also have been developed that attempt to predict the amount of nitrogen that will be mineralized during a growing season.

PROFITABILITY.   A second aspect relates to economics. Farmers do not use fertilizers just to grow big crops or to increase the nutrient content of their soils. They do so to make a living. The most profitable rate is determined by the ratio of the value of the extra yield expected to the cost of the fertilizer applied. The law of diminishing returns applies. Therefore, the most profitable fertilizer rate will be somewhat less than the rate that would produce the very highest yield (Figure 16.41).

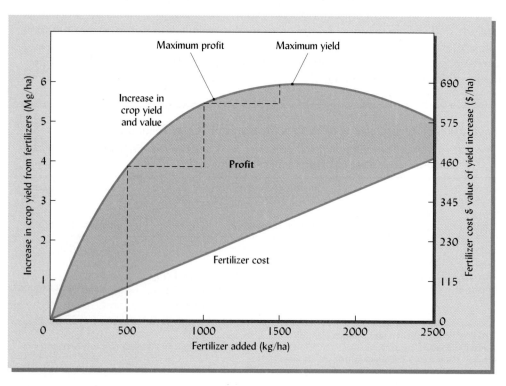

FIGURE 16.41 Relationships among the rate of fertilizer addition, crop yield increase, fertilizer costs, and profit from adding fertilizers. Note that the yield increase (and profit) from the first 500 kg of fertilizer is much greater than from the second and third 500 kg. Also note that the maximum profit is obtained at a lower fertilizer rate than that needed to give a maximum yield. The calculations assume that the yield response to added fertilizer takes the form of a smooth quadratic curve, an assumption which often leads to an overestimate of the amount of fertilizer needed to optimize profit.

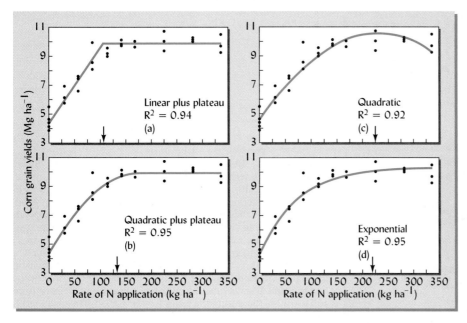

FIGURE 16.42 An example of how the mathematical function chosen to represent fertilizer-response data can affect the amount of fertilizer recommended. The data in all five graphs are exactly the same and represent the response of corn yields in Iowa to increasing levels of nitrogen fertilization. The vertical arrows indicate the recommended, most profitable rate of nitrogen to apply according to each mathematical model. Note that all models fit the data equally well (as indicated by the very similar R^2 values), but that the linear-plus-plateau model predicts an optimum nitrogen rate of 104 kg/ha, while the standard quadratic model suggests that 222 kg/ha is the optimum. Apparently the extra 118 kg/ha of nitrogen may have no effect on crop yield, but may greatly increase the risk of environmental damage. [Redrawn from Cerrato and Blackmer (1990)]

RESPONSE CURVES. Traditionally, economic analysis of optimum fertilizer rates has assumed that the plant response to fertilizer inputs was represented by a smooth curve following a quadratic function ($y = a + bx + cx^2$). In fact, actual data obtained can be just as well described by a number of other mathematical functions (Figure 16.42). This seemingly esoteric observation can have a great effect on the amount of fertilizer recommended and, in turn, on the likelihood of environmental harm from excessive fertilizer use (Figure 16.43). Among the various models studied, the linear-plateau approach (Figure 16.42a) usually leads to the lowest fertilizer recommendation, least wasted fertilizer, and least environmental damage. Nonetheless, the uncertainties related to nitrogen supply are still sufficiently high, and the cost of nitrogen fertilizer may be sufficiently low, that farmers tend to err on the side of oversupply, just for "insurance." In most years, this philosophy can be disadvantageous to both the farmer and the environment.

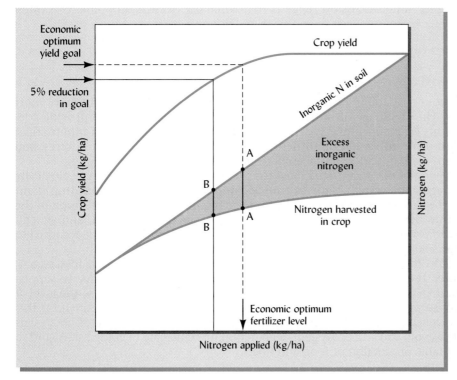

FIGURE 16.43 Influence of rate of nitrogen fertilization on crop yield, nitrogen removed in harvest, and the amount of excess inorganic soil nitrogen potentially available for loss into the environment. The fine lines show that a relatively small decrease (5%) in the yield goal for which nitrogen was applied would result in a relatively large reduction (about 30%) in the potential for nitrogen pollution. Line A-A represents the excess nitrogen present when fertilizer is applied at a rate designed to reach the economic optimum yield goal. Line B-B represents the excess nitrogen if the yield goal is 5% less than the economic optimum. Unfortunately, in many cases a yield reduction of 5% may reduce profits by a much larger percentage. On the other hand, studies show that most farmers set unrealistically high yield goals. [Adapted from National Research Council (1993)]

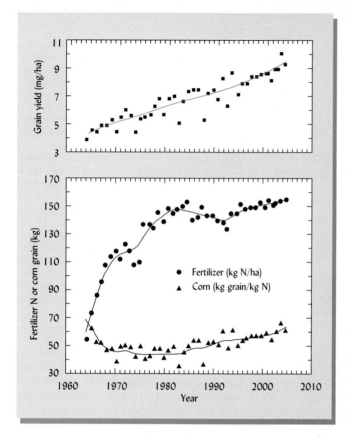

FIGURE 16.44 Trends in U.S. corn grain yields (kg/ha, upper curve, squares), fertilizer N rate used for corn (kg N/ha, middle, circles), and resulting N use efficiency (lower, triangles). Nitrogen-use efficiency, defined as kg corn grain produced per kg N applied, declined during the 1960s and early 1970s when energy and N fertilizer were cheap. That is, even though corn yields were increasing, fertilizer use rate was increasing faster, so more and more N was being used to grow each kg of grain. Corn yields continued to rise steadily (due to more productive corn varieties and better soil management) from the mid-1980s to the mid-1990s, despite the fact that the average rate of fertilizer N application declined from 150 to 130 kg/ha during the same 10-year period. The lower fertilizer rates were encouraged by the U.S. farm crisis, rising energy and fertilizer costs, and increasing environmental concerns. Efficiency continued to rise after the mid-1990s, as grain yield increases outpaced a slight rise in fertilizer rates. Future data will tell whether improvements in soil and water management and corn varieties will continue to allow greater yields with little or no increase in N fertilizer. [Data courtesy of USDA National Agricultural Statistics Service and Jim Porterfield, American Farm Bureau Federation]

Fertilizer use and corn yield data for the United States (Figure 16.44) suggest that N fertilizer was not always used very efficiently, much to the detriment of both water quality and farmer profitably. Although some advisors continue to recommend as much as 27 kg N per Mg (1.5 lbs. of N per bushel) of expected corn grain, most field data suggest that as little as 14 to 16 kg N/ Mg grain (0.8 or 0.9 lbs N/bushel) is all that is really required.

Anyone involved with the actual production of plants can testify to the enormous improvements that accrue from judicious use of organic and inorganic nutrient supplements. The preceding discussion suggests that while use of such suppliments is often necessary, their optimum levels are difficult to determine with precision.

16.14   CONCLUSION

The continuous availability of plant nutrients is critical for the sustainability of most ecosystems. The challenge of nutrient management is threefold: (1) to provide adequate nutrients for plants in the system; (2) to simultaneously ensure that inputs are in balance with plant utilization of nutrients, thereby conserving nutrient resources; and (3) to prevent contamination of the environment with unutilized nutrients.

The recycling of plant nutrients must receive primary attention in any ecologically sound management system. This can be accomplished in part by returning plant residues to the soil. These residues can be supplemented by judicious application of the organic wastes that are produced in abundance by municipal, industrial, and agricultural operations worldwide. The use of cover crops grown specifically to be returned to the soil is an additional organic means of recycling nutrients.

For sites from which crops or forest products are removed, nutrient losses commonly exceed the inputs from recycling. Inorganic fertilizers will continue to supplement natural and managed recycling to replace these losses and to increase the level of soil fertility so as to enable humankind to not only survive, but to flourish on this planet. In extensive areas of the world, fertilizer use will have to be increased above current levels to avoid soil and ecosystem degradation, to remediate degraded soils, and to enable profitable production of food and fiber.

The use of fertilizers, both inorganic and organic, should not be done in a simply habitual manner or for so-called insurance purposes. Rather, soil testing and other diagnostic tools should be used to determine the true need for added nutrients. In managing nitrogen and phosphorus, increasing attention will have to be paid to the potential for transport of these nutrients from soils where they are applied to waterways where they can become pollutants. If soils are low in available nutrients, fertilizers often return several dollars' worth of improved yield for every dollar invested. However, where the nutrient-supplying power of the soil is already sufficient, adding fertilizers is likely to be damaging both to the bottom line and to the environment.

STUDY QUESTIONS

1. The groundwater under a heavily manured field is high in nitrates, but by the time it reaches a stream bordering the field, the nitrate concentration has declined to acceptable levels. What are likely explanations for the reduction in nitrate?

2. You want to plant a cover crop in fall to minimize nitrate leaching after the harvest of your corn crop. What characteristics would you look for in choosing a cover crop to ameliorate this situation?

3. What management practices on forested sites can lead to significant nitrogen losses, and how can the losses be prevented?

4. What effect do forest fires have on nutrient availabilities and losses to streams?

5. Compare the resource-conservation and environmental-quality issues related to each of the three so-called fertilizer elements, N, P, and K.

6. A park manager wants to fertilize an area of turfgrass with nitrogen and phosphorus at the rates of 60 kg/ha of N and 20 kg/ha of P. He has stocks of two types of fertilizers: urea (45-0-0) and diammonium phosphate (18-46-0). How much of each should he blend together to fertilize a 10-ha area of turfgrass?

7. How much phosphorus (P) is there in a 25-kg bag of fertilizer labeled "20-20-10"?

8. Compare the relative advantages and disadvantages of organic and inorganic nutrient sources.

9. A certified organic grower plans to grow a crop that requires the application of 120 kg of plant-available N/ha and 20 kg/ha of P. She has a source of compost that contains 1.5% total N (with 10% of this available in the first year) and 1.1% total P (with 80% of this available in the first year). (a) Assuming her soil has a low P soil test level, how much compost should she apply to provide the needed N and P? (b) If her soil is already optimal in P, how can she provide the needed amounts of both N and P *without* causing further P buildup?

10. Discuss the concept of the *limiting factor* and indicate its importance in enhancing or constraining plant growth.

11. Consider Figure 16.27, which illustrates the response of wheat plants (shoot weight) to increasing soil P, Zn, and bulk density. (a) Explain, using specific data, how the graphs illustrate the principle of limiting factors (or "the law of the minimum"). (b) Describe the conditions in the experiment under which compaction (rather than zinc or phosphorus) is the most limiting factor. (c) Explain in detail, using specific data, how the graphs also illustrate exceptions to or limitations of the "law of the minimum."

12. Why are nutrient-cycling problems in agricultural systems more prominent than those in forested areas?

13. Discuss how GIS-based, site-specific nutrient-application technology might improve profitability and reduce environmental degradation.

14. Discuss the value and limitations of soil tests as indicators of plant nutrient needs and water pollution risks.

15. When might the use of plant tissue analyses have advantages over soil testing for correcting nutrient imbalances?

REFERENCES

Aber, J. et al. 2000. "Applying ecological principles to management of the U.S. National Forests," *Issues in Ecology*, **6** www.esa.org/science_resources/Issues/FileEnglish/ Issues6 .pdf (confirmed 12 September 2006) (Washington D.C.: Ecological Society of America).

Barzegar, A. R., H. Nadian, F. Heidari, S. J. Herbert, and A. M. Hashemi. 2006. "Interaction of soil compaction, phosphorus and zinc on clover growth and accumulation of phosphorus," *Soil Tillage Res.*, **87**:155–162.

Beegle, D. B., O. T. Carton, and J. S. Bailey. 2000. "Nutrient management planning: Justification, theory, practice," *J. Environ. Quality*, **29**:72–79.

Beegle, D. B., L. E. Lanyon, and J. T. Sims. 2002. "Nutrient balances," pp. 171–193, in P. M. Haygarth and S. C. Jarvis (eds.), *Agriculture, Hydrology and Water Quality*. (Wallingford, U.K.: CAB International).

Belt, G. H., and J. O'Laughlin. 1994. "Buffer strip design for protecting water quality and fish habitat," *Western J. Applied Forestry*, **9**(2):4145.

Binkley, D., H. Burnham, and H. L. Allen. 1999. "Water quality impacts of forest fertilization with nitrogen and phosphorus," *Forest Ecology and Management*, **121**: 191–213.

Bormann, F. H., and G. E. Likens. 1979. *Pattern and Process in a Forested Ecosystem* (New York: Springer-Verlag).

Brady, N. C., and R. R. Weil. 1996. *The Nature and Properties of Soils*, 11th ed. (Upper Saddle River, N.J.: Prentice Hall Inc.). xi + 740 pp.

Brown, T. C., and D. Binkley. 1994. "Effect of management on water quality in North American forests." USDA Forest Service General Technical Report RM-248 (Ft. Collins, Colo.).

Cerrato, M. E., and A. M. Blackmer. 1990. "Comparison of models from describing corn yield response to nitrogen fertilizer," *Agron. J.*, **98**:138–143.

Chauby, I., D. R. Edwards, T. C. Daniel, P. A. Moore, Jr., and D. J. Nichols. 1994. "Effectiveness of vegetative filter strips in retaining surface-applied swine manure constituents," *Trans. Am. Soc. Agric. Engineers*, **37**:845–850.

Chesapeake Bay Program. 1995. "Water quality functions of riparian forest buffer systems in the Chesapeake Bay watershed." EPA 903-R-95–004 (Washington, D.C.: USEPA).

Dodd, J. R., and A. P. Mallarino. 2005. "Soil-test phosphorus and crop grain yield responses to long-term phosphorus fertilization for corn-soybean rotations," *Soil Sci. Soc. Am. J.*, **69**:1118–1128.

Eghball, B., B. J. Wienbold, J. E. Gilley, and R. A. Eigenberg. 2002. "Mineralization of manure nutrients," *J. Soil Water Conserv.*, **57**:470–473.

FAO. 1977. *China: Recycling of Organic Wastes in Agriculture*. FAO Soils Bulletin 40 (Rome: U.N. Food and Agriculture Organization).

FAO. 2006. FAOSTAT. Food and Agriculture Organization of the United Nations. http://faostat.fao.org/ (verified 20 August 2006).

Gardner, G. 1997. *Recycling Organic Waste: From Urban Pollutant to Farm Resource*. Worldwatch Paper 135 (Washington, D.C.: Worldwatch Institute).

Gburek, W. J., A. N. Sharpley, L. Heatherwaite, and G. J. Folmar. 2000. "Phosphorus management at the watershed scale: A modification of the phosphorus index," *J. Environ. Quality*, **29**:130–144.

Haering, K. C., W. L. Daniels, and S. E. Feagley. 2000. "Reclaiming mined lands with biosolids, manures and papermill sludges," Chapter 24 (pp. 615–644), in R. I. Barnhisel, W. L. Daniels, and R. G. Darmody (eds.), *Reclamation of Drastically Disturbed Lands*. Agronomy Monograph 41 (Madison, Wis.: American Society of Agronomy).

Havlin, J. L., J. D. Beaton, S. L. Tisdale, and W. L. Nelson. 2005. *Soil Fertility and Fertilizers—An Introduction to Nutrient Management*, 7th ed. (Upper Saddle River, N.J.: Prentice Hall). 515 pp.

He, Xin-Tao, T. Logan, and S. Traina. 1995. "Physical and chemical characteristics of selected U. S. municipal solid waste composts," *J. Environ. Qual.*, **24**:543–552.

Heckman, J. R., and D. Kluchinski. 1996. "Chemical composition of municipal leaf waste and hand-collected urban leaf litter," *J. Environ. Qual.*, **25**:355–362.

Higgs, B., A. E. Johnston, J. L. Salter, and C. J. Dawson. 2000. "Some aspects of achieving sustainable phosphorus use in agriculture," *J. Environ. Quality*, **29**:80–87.

Krogmann, U., B. F. Rogers, L. S. Boyles, W. J. Bamka, and J. R. Heckman. 2003. "Guidelines for land application of non-traditional organic wastes (food processing by-products and municipal yard wastes) on farmlands in New Jersey," *Bulletin e281*. Rutgers Cooperative Extension, New Jersey Agricultural Experiment Station, Rutgers, The State University of New Jersey. http://www.rce.rutgers.edu/pubs/pdfs/e281.pdf (posted June 2003; verified 28 November 2004).

Likens, G. E., and F. H. Bormann. 1995. *Biogeochemistry of a Forested Ecosystem*, 2nd ed. (New York: Springer-Verlag).

Loehr, R. C. et al. 1979. *Land Application of Wastes*, Vol. I (New York: Van Nostrand Reinhold).

Magdoff, F., L. Lanyon, and B. Liebhardt. 1997. "Nutrient cycling, transformations, and flows: Implications for a more sustainable agriculture," *Advances in Agronomy*, **60**:2–73.

McNaughton, S. J., F. F. Banyikwa, and M. M. McNaughton. 1997. "Promotion of the cycling of diet-enhancing nutrients by African grazers," *Science*, **278**:1798–1800.

Mehlich, A. 1984. "Mehlich 3 soil test extractant: A modification of Mehlich 2 extractant," *Commun. Soil Sci. Plant Anal.*, **15**:1409–1416.

National Research Council. 1993. *Soil and Water Quality: An Agenda for Agriculture* (Washington, D.C.: National Academy of Sciences).

Olson, R. A., K. D. Frank, P. H. Graboushi, and G. W. Rehm. 1982. "Economic and agronomic impacts of varied philosophies of soil testing," *Agron. J.*, **74**:492–499.

Perry, D. A. 1994. *Forest Ecosystems* (Baltimore: Johns Hopkins University Press).

Powers, J. F., and W. P. Dick (eds.). 2000. *Land Application of Agricultural, Industrial, and Municipal By-Products*. Soil Science Society of America Book Series no. 6 (Madison, Wis.: Soil Sci. Soc. Amer.).

Preusch, P. L., P. R. Adler, L. J. Sikora, and T. J. Tworkoski. 2002. "Nitrogen and phosphorus availability in composted and uncomposted poultry litter," *J. Environ. Qual.*, **31**:2051–2057.

Reddy, G. B., and K. R. Reddy. 1993. "Fate of nitrogen-15 enriched ammonium nitrate applied to corn," *Soil Sci. Soc. Amer. J.*, **57**:111–115.

Reuter, D. J., and J. B. Robinson. 1986. *Plant Analysis: An Interpretation Manual* (Melbourne, Australia: Inkata Press).

Richards, R. P. et al. 1996. "Well water, well vulnerability, and agricultural contamination in the midwestern United States," *J. Environ. Quality*, **25**:389–402.

Robinson, C. A., M. Ghaffarzadeh, and R. M. Cruze. 1996. "Vegetative filter strip effects on sediment concentration in cropland runoff," *J. Soil and Water Conserv.*, **50**:227–230.

Sanchez, P. A. et al. 1997. "Soil fertility replenishment in Africa: An investment in natural resource capital," in R. J. Buresh, P. A. Sanchez, and F. Calhoun (eds.), *Replenishing Soil Fertility in Africa*. ASA/SSSA Publication (Madison, Wis.: Soil Sci. Soc. Amer.).

Sharpley, A. N. 1997. "Rainfall frequency and nitrogen and phosphorus runoff from soil amended with poultry litter," *J. Environ. Quality*, **26**:1127–1132.

Smaling, E. M. A., S. M. Nwanda, and B. H. Jensen. 1997. "Soil fertility in Africa is at stake," in R. J. Buresh, P. A. Sanchez, and F. Calhoun (eds.), *Replenishing Soil Fertility in Africa*. ASA/SSSA Publication (Madison, Wis.: Soil Sci. Soc. Amer.).

Snyder, C. S., T. W. Bruulsema, A. N. Sharpley, and D. B. Beegle. 2005. "Site-specific use of the environmental phosphorus index concept," p. 4. Potash & Phosphate Institute, Norcross, Ga.

Sutton, A. L. 1994. "Proper animal manure utilization," in *Nutrient Management*, supplement to *J. Soil Water Conserv.*, **49**(2): 65–70.

Swank, W. T., J. M. Vose, and K. J. Elliot. 2001. "Long-term hydrologic and water quality responses following commercial clear cutting of mixed hardwoods on a southern Appalachian catchment," *Forest Ecology and Management*, **143**:163–178.

Thomas, M. L., R. Lal, T. Logan, and N. R. Fausey. 1992. "Land use and management effects on non-point loading from Miamian soil," *Soil Sci. Soc. Amer. J.*, **56**: 1871–1875.

USDA Natural Resources Conservation Service. *The P Index: A Phosphorus Assessment Tool*. www.nrcs.usda.gov/TECHNICAL/ECS/nutrient/pindex.html (confirmed July 2007).

USDA. 1980. *Appraisal, 1980 Soil and Water Resources Conservation Act, Review Draft*, Part I (Washington, D.C.: USDA).

Walworth, J. L., and M. E. Sumner. 1987. "The diagnosis and recommendation integrated system (DRIS)," *Advances in Soil Science*, **6**:149–187.

Weil, R. R., and S. K. Mughogho. 2000. "Sulfur nutrition of maize in four regions of Malawi," *Agron. J*, **92**:649–656.

Westerman, R. L. (ed.). 1990. *Soil Testing and Plant Analaysis*, 3rd ed. (Madison, Wis.: Soil Sci. Soc. Amer.).

Yang, H. S. 2006. "Resource management, soil fertility and sustainable crop production: Experiences of China," *Agric. Ecosyst. Environ.*, **116**:27–33.

Zublena, J. P., J. C. Barker, and T. A. Carter. 1993. "Poultry manure as a fertilizer source," *Soil Facts* (Raleigh, N.C.: North Carolina Cooperative Extension Service, North Carolina State University).

Wind erosion degrades desert soil and vegetation. (R. Weil)

17

SOIL EROSION AND ITS CONTROL

The wind crosses the brown land, unheard . . .
—*T.S. ELIOT*, THE WASTE LAND

No soil phenomenon is more destructive worldwide than the erosion caused by wind and water. Since prehistoric times people have brought the scourge of soil erosion upon themselves, suffering impoverishment and hunger in its wake. Past civilizations have disintegrated as their soils, once deep and productive, washed away, leaving only thin, rocky relics of the past. It is hard to imagine that agricultural communities once flourished in the now nearly barren hills in parts of India, Greece, Lebanon, or Syria.

Since 1960, farmers have had to more than double world food output to feed the unprecedented numbers of people on Earth. As the ratio of people to land steadily rises, poor people see little choice but to clear and burn steep, forested slopes or plow up natural grasslands to plant their crops. Population pressures have also led to overgrazing of rangelands and overexploitation of timber resources. All these activities lead to a downward spiral of ecological deterioration, land degradation, and deepening poverty. The impoverished crops and rangelands leave little if any residues to protect the soil, leading to further erosion, driving ever more desperate people to clear and cultivate—and degrade—still more land. Add to this the intense, concentrated erosion on sites disturbed by construction or mining activity, as well as globalization pressures to expand cropping and logging to marginal lands, and it is clear that the current threat of soil erosion is more ominous than at any time in history.

The degraded productivity of farm, forest, range, and urban lands tells only part of the sad erosion story. Soil particles washed or blown from the eroding areas are subsequently deposited elsewhere—in nearby low-lying sites within the landscape or far away—even on other continents. Far downstream or downwind, the sediment and dust cause major water and air pollution and bring enormous economic and social costs to society.

Combating soil erosion is everybody's business. Fortunately, much has been learned about the mechanisms of erosion and techniques have been developed that can effectively and economically control soil loss in most situations. This chapter will equip you with some of the concepts and tools you will need to do your part in solving this pressing world problem.

17.1 SIGNIFICANCE OF SOIL EROSION AND LAND DEGRADATION[1]

Land Degradation

Soil erosion in Maine: http://www.greenworks.tv/waterquality/erosion.htm

During the past half century, human land use and associated activities have degraded some 5 billion ha (about 43%) of the Earth's vegetated land. Such **land degradation** results in a reduced productive potential and a diminished capacity to provide benefits to humanity. Much of this degradation (on about 3.6 billion ha) is linked to **desertification**, the spreading of desert conditions that disrupt semiarid and arid ecosystems (including agroecosystems). A major cause of desertification is overgrazing by cattle, sheep, and goats, a factor that likely accounts for about a third of all land degradation, mainly in such dry regions as the Sahel in northern Africa and the rangeland of the American Southwest. Likewise, the indiscriminate felling of rain forest trees has already degraded nearly 0.5 billion ha in the humid tropics. Additionally, inappropriate agricultural practices continue to degrade land in all climatic regions.

Soil-Vegetation Interdependency

Degraded lands may suffer from destruction of native vegetation communities, reduced agricultural yields, lowered animal production, and simplification of once-diverse natural ecosystems with or without accompanying degradation of the soil resource. On about 2 billion of the 5 billion ha of degraded lands in the world, soil degradation is a major part of the problem (Figure 17.1). In some cases the soil degradation occurs mainly as deterioration of physical properties by compaction or surface crusting (see Sections 4.7 and 4.6), or as deterioration of chemical properties by acidification (see Section 9.6) or salt accumulation (see Section 10.3). However, most (~85%) soil degradation stems from erosion—the destructive action of wind and water.

The two main components of land degradation—damage to plant communities and deterioration of soil—interact to cause a downward spiral of accelerating ecosystem damage and human poverty (Figure 17.2). Due to overgrazing, deforestation, or inappropriate

---

[1] For a readable account of historical degradation of land and water resources, see Hillel (1991). Overviews of the current extent of soil erosion and other forms of land degradation are given by Oldeman (1994), Daily (1997), and Rosensweig and Hillel (1998).

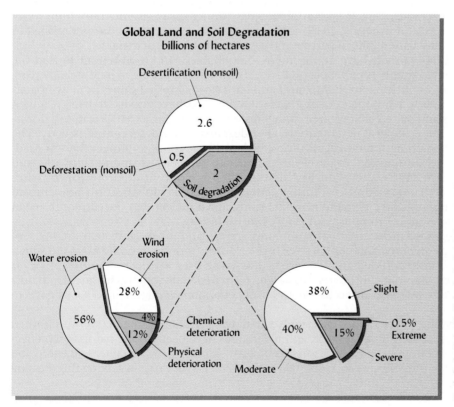

FIGURE 17.1 Soil degradation as a part of global land degradation caused by overgrazing, deforestation, inappropriate agricultural practices, fuel wood overexploitation, and other human activities. About 60% of degraded land has suffered vegetative, but not soil degradation. Of the 2 billion ha of land with degraded soils, most could be restored easily (*slight* degradation) or with considerable financial and technical investments (*moderate* degradation). *Severely* degraded soils are currently useless for agriculture and would require major international assistance for restoration. About 9 million ha (0.5% of degraded soils) are *extremely* degraded and incapable of restoration. About 85% of the soil degradation is caused by erosion by wind and water. [FAO data selected from Oldeman (1994) and Daily (1997)]

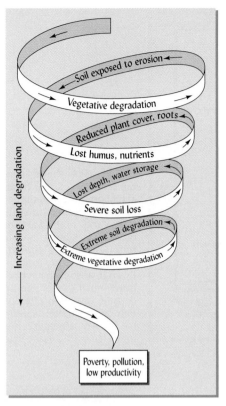

FIGURE 17.2 The downward spiral of land degradation resulting from the feedback loop between soil and vegetation. As the natural vegetation is disturbed, soil becomes exposed to raindrops and wind leading to erosion and loss of soil, including organic matter and nutrients. The now impoverished soil can support only stunted crops or other vegetation, which leaves the soil with even less protective cover and root mass than before. Soil loss becomes severe, such that the soil depth and the capacity to hold water are greatly reduced and vegetation can barely survive, leaving extremely degraded soil. Incapable of providing nutrients and water needed to support healthy growth of natural vegetation or crops, the site continues to erode, polluting rivers with sediment and impoverishing the people who attempt to grow their food on the land. (Diagram courtesy of R. Weil)

methods of crop production, vegetation becomes less dense and vigorous, and thus provides the soil with less and less protection from erosion. Simultaneously, as the soil is degraded by such processes as erosion and nutrient depletion, it becomes less and less capable of supporting a protective canopy of vegetation. Soil degradation weakens the vegetation through its effects on runoff and infiltration of rainwater (see Section 6.2). With as much as 50 to 60% of the rainfall lost as runoff, scarcity of soil water on eroded soils can become a serious impediment to plant growth. Improvements in both soil and vegetation management must go hand-in-hand if the productive potential of the land is to be protected—or even be restored by moving *up* rather than *down* the spiral—a prospect that is attainable.

Geological versus Accelerated Erosion

GEOLOGICAL EROSION.   Erosion is a process that transforms soil into **sediment.** Soil erosion that takes place naturally, without the influence of human activities, is termed **geological erosion.** It is a natural leveling process. It inexorably wears down hills and mountains, and through subsequent deposition of the eroded sediments, it fills in valleys, lakes, and bays. Many of the landforms we see around us—canyons, buttes, rounded hills, river valleys, deltas, plains, and pediments—are the result of geological erosion and deposition. The vast deposits that now appear as sedimentary rocks originated in this way.

In most settings, geological erosion wears down the land slowly enough that new soil forms from the underlying rock or regolith faster than the old soil is lost from the surface. The very existence of soil profiles bears witness to the net accumulation of soil and the effectiveness of undisturbed natural vegetation in protecting the land surface from erosion.

The rate of geological soil erosion varies greatly with both rainfall and type of material comprising the regolith. Geological erosion by water tends to be greatest in semiarid regions where rainfall is enough to be damaging, but not enough to support dense, protective vegetation (Figure 17.3). Areas blanketed by deep deposits of silts may have exceptionally high erosion rates under such conditions. The gullied, barren landscape of the North American badlands (Figure 17.4) is an extreme example of geological

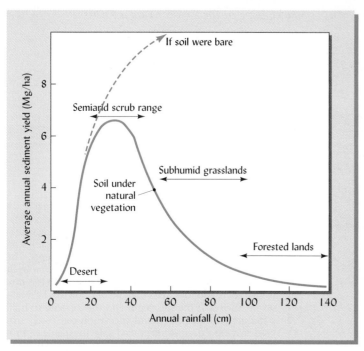

FIGURE 17.3 Generalized relationship between annual rainfall and soil loss from geologic erosion by water. The actual amount of sediment lost annually per hectare will depend on other climatic variables, topography, and the type of soils in the watershed. Note that sediment yields are greatest in semiarid regions. Here a number of severe runoff–generating storms occur in most years, but the total rainfall is too little to support much protective plant cover. By comparison, the very dry deserts have too little rain to cause much erosion, and the well-watered regions support dense forests that effectively protect the soil. Where the natural vegetation is destroyed by plowing, erosion from the bare soils is much higher with increasing rainfall, as indicated by the dashed curve.

FIGURE 17.4 The effects of geologic erosion and sedimentation can be dramatic, as in the semiarid badlands scene in South Dakota (*left*). There is little vegetation to protect the hills from ravages of sudden summer thunderstorms. Note the petrified log (*arrow*) becoming exposed as the hillside erodes in the foreground. Its presence indicates that deep layers of clayey sediment had been deposited on this site during ancient cycles of erosion and sedimentation. (*Right*) The extreme gully erosion in a site in central Tanzania highlights the cutting power of turbulent water, the flow of which was probably enhanced by the deforestation of the surrounding watershed. The scene also indicates the important role of raindrops in detaching soil particles. Note the tall, thin pedestals of soil that remain where a layer of rocklike ironstone protects the underlying soft material from the impact of raindrops (compare to the much smaller pedestals in Figure 17.11*a*). (Photos courtesy of R. Weil)

TABLE 17.1 Annual Sediment Loads for Nine of the World's Major Rivers

River	Countries	Annual sediment load, million Mg	Erosion, Mg/ha drained
Yangtze	China	1600	479
Ganges	India, Nepal	1455	270
Amazon	Brazil, Peru, etc.	363	13
Mississippi	United States	300	93
Irrawaddy	Burma	299	139
Kosi	India, Nepal	172	555
Mekong	Vietnam, Thailand, etc.	170	43
Red	China, Vietnam	130	217
Nile	Sudan, Egypt, etc.	111	8

Data from different sources compiled by El-Swaify and Dangler (1982).

erosion occurring where unstable clay and silt deposits are subjected to infrequent but intense rainstorms, yet the soil is usually too dry (partly because of high runoff losses) to support much vegetation.

SEDIMENT LOADS. Rainfall, geology, and other factors (including human activities) influence the sediment loads carried by the world's great rivers (Table 17.1). Although rivers like the Mississippi and Yangtze were muddy before humans disturbed their watersheds, current sediment loads are far greater than before. To gain some perspective on the enormous amount of soil transported to the sea by these rivers, consider the Mississippi's sediment load (only a fifth as great as that of the Yangtze or the Ganges). If the 300 million Mg of sediment were carried to the Gulf of Mexico by dump trucks, it would take a continuous, year-round caravan of more than 80,000 large trucks, stretching all the way from Wisconsin to New Orleans (1600 km) and back, with a 20-Mg load being dumped into the Gulf about every 2 seconds.

HUMAN-ACCELERATED EROSION. We stand in awe at the edge of the Grand Canyon, which was formed over millenia by geologic erosion—yet few realize that humankind has now become the preeminent force on the landscape, now moving nearly twice as much soil per year as global geologic process, and two-thirds of that *unintentionally* through erosion, mainly associated with agricultural activities (Figure 17.5).

Accelerated erosion occurs when people disturb the soil or the natural vegetation by grazing livestock, cutting forests for agricultural use (Figure 17.6), plowing hillsides, or tearing up land for construction of roads and buildings. Accelerated erosion is often

Soil erosion in U.S. agriculture:
http://www.epa.gov/
agriculture/ag101/cropsoil.
html#envconcerns

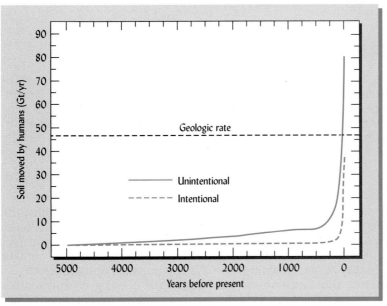

FIGURE 17.5 Estimates of the total amount of soil material moved annually by humans as a function of time. Intentional soil movement refers mainly to construction and excavation activities. Unintentional soil movement refers mainly to soil loss due to agricultural activities such as land clearing, tillage, overgrazing and long periods without vegetative cover. Humans now move more soil material than all natural processes combined. This may not mean that sediment loads of major rivers has increased this dramatically, as movement of soil within a landscape does not necessarily lead to sediment in rivers. [For comparisons of agricultural to geologic soil movement, see Wilkinson and McElroy (2007). [Graph redrawn from (Hooke, 2000)].

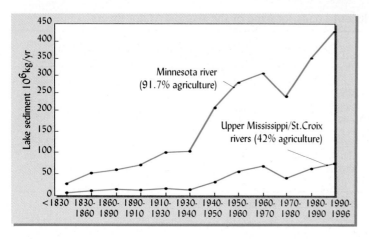

FIGURE 17.6 Sediment deposition rates into Lake Pepin (Minnesota-Wisconsin border) from watersheds of the Minnesota river (92% cultivated) and the upper Mississippi and St. Croix rivers (42% cultivated). Neither watershed deposited much sediment before about 1830 because little of the land had been cleared for farming. But as agriculture expanded, deposition rates increased, especially from the Minnesota watershed where intensive row-crop agriculture became dominant. [Redrawn from Kelley and Nater (2000)]

10 to 1000 times as destructive as geological erosion, especially on sloping lands in regions of high rainfall. Rates of erosion by wind and water on agricultural land in Africa, Asia, and South America are thought to average about 30 to 40 Mg/ha annually. In the United States, the average erosion rate on cropland is about 12 Mg/ha—7 Mg by water and 5 Mg by wind. Some cultivated soils are eroding at 10 times these average rates. In comparison, erosion on undisturbed humid-region grasslands and forests generally occurs at rates considerably below 0.1 Mg/ha.

About 4 billion Mg of soil is moved annually by soil erosion in the United States, some two-thirds by water and one-third by wind. More than half of the movement by water and about 60% of the movement by wind take place on croplands that produce most of the country's food. Much of the remainder comes from semiarid rangelands, from logging roads and timber harvest on forest lands, and from soils disturbed for highway and building construction. Although progress has been made in reducing erosion (see Section 17.14), current high losses are simply not acceptable for long-term sustainability and must be further reduced.

Under the influence of accelerated erosion, soil is commonly washed or blown away faster than new soil can form by weathering or deposition. As a result, the soil depth suitable for plant roots is often reduced. In severe cases, gently rolling terrain may become scarred by deep gullies, and once-forested hillsides may be stripped down to bare rock. Accelerated erosion often makes the soils in a landscape more heterogenous. An example can be seen in the striking differences in surface soil color that develop as ridgetop soils are truncated, exposing material from the B or C horizon at the land surface, while soils lower in the landscape are buried under organic-matter-enriched sediment (Figure 17.7 and Plates 36 and 65).

17.2 ON-SITE AND OFF-SITE EFFECTS OF ACCELERATED SOIL EROSION

Erosion damages the site on which it occurs and also has undesirable effects off-site in the larger environment. The off-site costs relate to the effects of excess water, sediment, and associated chemicals on downhill and downstream environments. While the costs associated with either or both of these types of damages may not be immediately apparent, they are real and grow with time. Landowners and society as a whole must eventually foot the bill.

Types of On-Site Damages

The most obviously damaging aspect of erosion is the loss of soil itself. In reality, the damage done to the soil is greater than the amount of soil lost would suggest, because the soil material eroded away is almost always more valuable than that left behind. Not only are surface horizons eroded while less fertile subsurface horizons remain untouched, but the quality of the remaining topsoil is also impaired. Erosion selectively removes organic matter and fine mineral particles, while leaving behind mainly relatively less active, coarser fractions. Experiments have shown organic matter and

FIGURE 17.7 Erosion and deposition occur simultaneously across a landscape. (*Left*) The soil on this ridgetop was worn down by erosion during nearly 300 years of cultivation. The surface soil exposed on the ridgetop consists mainly of light-colored C horizon material. At sites lower down the slope the surface horizon shows mainly A and B horizon material, some of which has been deposited after eroding from locations upslope. (*Right*) Erosion on the sloping wheat field in the background has deposited a thick layer of sediment in the foreground, burying the plants at the foot of the hill. [Photos courtesy of R. Weil (*left*) and USDA Natural Resources Conservation Service (*right*)]

nitrogen in the eroded material to be five times as high as in the original topsoil, giving an **enrichment ratio** of 5 for these soil components. Comparable enrichment ratios for phosphorus and potassium are commonly 2 and 3, respectively. The quantity of essential nutrients lost from the soil by erosion is quite high, although only a portion of these nutrients are lost in forms that would be available to plants in the short term. The soil left behind usually has lower water-holding and cation-exchange capacities, less biological activity, and a reduced capacity to supply nutrients for plant growth.

In addition to the just-mentioned reduction in soil-quality factors, soil movement during erosion can spread plant disease organisms from the soil to plant foliage and from a higher to a lower-lying field. The deterioration of soil structure often leaves a dense crust on the soil surface, which, in turn, greatly reduces water infiltration and increases water runoff. Newly planted seeds and seedlings may be washed downhill, trees may be uprooted, and small plants may be buried in sediment. In the case of wind erosion, fruits and foliage may be damaged by the sandblasting effect of blowing soil particles.

Finally, gullies that carve up badly eroded land may make the use of tractors impossible and may undercut pavements and building foundations, causing unsafe conditions and expensive repairs.

Types of Off-Site Damages

Ecosystem threats to Big Darby Creek watershed: http://www.nature.org/wherewework/northamerica/states/ohio/bigdarby/

Erosion moves sediment and nutrients off the land, creating the two most widespread water pollution problems in our rivers and lakes. The nutrients impact water quality largely through the process of eutrophication caused by excessive nitrogen and phosphorus, as was discussed in Sections 13.8 and 14.2. In addition to nutrients, sediment and runoff water may also carry toxic metals and organic compounds, such as pesticides. The sediment itself is a major water pollutant, causing a wide range of environmental damages.

DAMAGES FROM SEDIMENT. Sediment deposited on the land may smother crops and other low-growing vegetation (Figure 17.7). It fills in roadside drainage ditches and creates hazardous driving conditions where mud covers the roadway.

Sediment that washes into streams makes the water cloudy or turbid (Figure 17.8*a* and Plate 110). High **turbidity** prevents sunlight from penetrating the water and thus reduces photosynthesis and survival of the *submerged aquatic vegetation* (SAV). The demise of the SAV, in turn, degrades the fish habitat and upsets the aquatic food chain.

Sediment laden
water from tributary

Clear water
of river

(a)

(b)

FIGURE 17.8 Off-site damages caused by soil erosion include the effects of sediment on aquatic systems. (a) A sediment-laden tributary stream empties into the relatively clear waters of a larger river. The turbid water will foul fish gills, inhibit submerged aquatic vegetation, and clog water-purification systems. Part of the sediment will settle out on the river bottom, covering fish-spawning sites and raising the river bed enough to aggravate the severity of future flooding episodes. (b) Expensive dredging and excavation (by the dragline in the foreground) is being undertaken to restore the beauty, recreational value, and flood-control function of this pond in a neighborhood park after accumulated sediment had transformed it into a mere mudflat. The watershed upstream from the pond had undergone a period of rapid suburban development, during which adequate sediment-control practices were not used on the construction sites. [Photos courtesy of USDA Natural Resources Conservation Service (a) and R. Weil (b)]

The muddy water also fouls the gills of some fish. Sediment deposited on the stream bottom can have a disastrous effect on many freshwater fish by burying the pebbles and rocks among which they normally spawn. The buildup of bottom sediments can actually raise the level of the river, so that flooding becomes more frequent and more severe. For example, to counter the rising river bottom, flood-control levees along the Mississippi must be constantly enlarged.

A number of major problems occur when the sediment-laden rivers reach a lake, reservoir, or estuary. Here, the water slows down and drops its load of sediment. Eventually reservoirs—even those formed by giant dams—become mere mudflats, completely filled in with sediment (see Figure 17.8b). Prior to that, the capacity of the reservoir to store water for irrigation or municipal water systems is progressively reduced, as is the capacity for floodwater retention or hydroelectric generation. It is estimated that 1.5 billion Mg of sediment are deposited each year in the nation's reservoirs. Similarly, harbors and shipping channels fill in and become impassible. The loss of function and the costs of dredging, excavation, filtering, and construction activities necessary to remedy these situations run into the billions of dollars every year.

WINDBLOWN SAND AND DUST. Wind erosion also has its off-site effects. Blowing sands may bury roads and fill in drainage ditches, necessitating expensive maintenance. The sand-blasting effect of wind-borne soil particles may damage the fruits and foliage of crops in neighboring fields, as well as the paint on vehicles and buildings many kilometers downwind from the eroding site. Finer wind-blown dust with clay-size particles causes the most expensive and far-reaching damages. Much of this dust—especially particulate matter (PM) with diameters between 2.5 and 10 microns (PM_{10})—arises from wind erosion on cropland, rangelands, and construction sites (as well as from traffic on unpaved roads). Even more damaging are particles smaller than 2.5 microns ($PM_{2.5}$), which arise mainly from vehicle exhaust and smoke from fires and industrial plants. The off-site damages from these dust particles include the aesthetics-related costs of added house-cleaning, more frequent car washes, and lost tourist revenues when majestic views at recreational parks are obscured. Even more serious are the major health hazards presented by these very fine wind-blown particles.

HEALTH HAZARDS FROM PM_{10} AND $PM_{2.5}$.[2] While silt-sized particles are generally filtered by nose hairs or trapped in the mucous of the windpipe and bronchial tubes, smaller clay-sized particles often pass through these defenses and lodge in the alveoli (air sacs) of the lungs. The particles themselves cause inflammation of the lungs, and they may also often carry toxic substances that cause further lung damage. For example, airborne clay particles adsorb water vapor and may become coated with sulfuric or nitric acids found in the atmosphere (see Section 9.6). Human pathogens may also adhere to dust particles and travel with them to spread disease.

Epidemiological studies suggest that the number of deaths resulting from people inhaling this fine *fugitive dust* is in the thousands every year, and may even exceed the number of deaths from traffic accidents. The U.S. Environmental Protection Agency therefore has set standards that call for the 24-hour average concentrations in the air not to exceed 150 µg/m^3 and 35 µg/m^3 of PM_{10} and $PM_{2.5}$, respectively. However, wind-blown dust is a global problem such that wind erosion in the Sahara desert in Africa and Gobi desert in China has been implicated in respiratory diseases in North America.

ESTIMATED COSTS OF EROSION.[3] Although no precise data exist, national or regional average wind and water erosion rates have been used to estimate the total costs of erosion in the United States. Included in such calculations are the on-site costs of replacing nutrients and water lost through accelerated erosion, as well as crop yield reductions due to reduced soil depth. Depending mainly on assumptions about the value of nutrients lost in sediment and runoff (should only those in readily useable form be valued?), the total annual on-site costs have been estimated at between $8 and $40 billion.

The off-site costs of erosion are likely even greater, especially because of the health effects of windblown particles and the reduced recreational (fishing, swimming, and aesthetic) value of muddy waters. The total of these annual off-site costs has been estimated at $10 and $30 billion. The grand total annual cost of erosion in the United States is therefore likely between $15 and $60 billion. Such high costs are a sobering reminder of the burden that society bears as a result of poor land management and would seem to justify increasing the sums allocated to the battle against erosion.

MAINTENANCE OF SOIL PRODUCTIVITY.[4] Although extreme soil erosion can reduce soil productivity to almost zero, in most cases the effect is too subtle to notice between one year and the next. Where farmers can afford to do so, they compensate for the loss of nutrients by increasing the use of fertilizer. The losses of organic matter and water-holding capacity are much more difficult to overcome. Over the long term, accelerated soil erosion that exceeds the rate of soil formation leads to declining productivity on most soils. In the United States, crop yields on severely eroded soils are often 20 to 40% lower than on similar soils with only slight erosion.

Ultimately, the rate of decline of soil productivity, or the cost of maintaining constant crop-yield levels, is determined by such soil properties as *depth to a root-restricting layer* and *permeability of the subsoil*. As shown in Figure 17.9, a deep, well-drained, and well-managed soil may not decline much in productivity even though it suffers some erosion. In contrast, erosion on a shallow, low-permeability soil may bring about a rapid productivity decline.

---

[2] For details about PM_{10} and $PM_{2.5}$, see USEPA (2006).
[3] A detailed analysis giving estimates in 1995 dollars can be found in Pimental et al. (1995).
[4] For estimates of the effect of erosion on the productive potential of African soils, see Lal (1995), and for U.S. Mollisols and Alfisols, see Weesies et al. (1994).

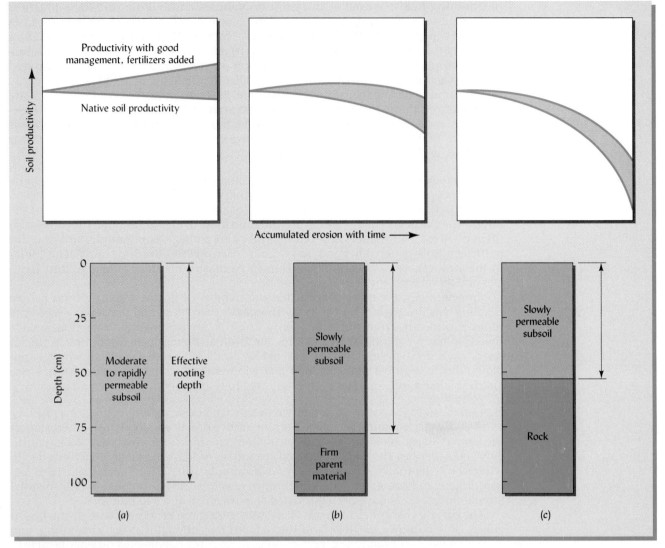

FIGURE 17.9 Effect of erosion over time on the productivity of three soils differing in depth and permeability. Productivity on soil (*a*) actually increases with time because of good management practices and fertilizer additions, even though the native soil productivity declines as a result of erosion. Because soil (*c*) is shallow and has restricted permeability, its productivity declines rapidly as a result of erosion, a decline that good management and fertilizers cannot prevent. Soil (*b*), which is intermediate in both depth and permeability, suffers only a slight decline in productivity due to erosion. Soil characteristics clearly influence the effect of erosion on soil productivity.

Soil-Loss Tolerance[5]

The loss of *any* amount of soil by erosion is detrimental, but years of field experience, as well as scientific research, indicate that some loss can be tolerated. Scientists of the USDA Natural Resources Conservation Service, working in cooperation with field personnel throughout the country, have developed tentative soil-loss tolerance limits for most cultivated soils in the United States.

A tolerable soil loss (***T* value**) is the maximum amount of soil that can be lost annually by the combination of water and wind erosion on a particular soil without degrading that soil's long-term productivity. Currently, *T* values are based on the best judgment of informed soil scientists, rather than on rigorous research data.

COMMON RANGE OF *T* VALUES. The *T* values for soils in the United States commonly range from 5 to 11 Mg/ha. They depend on a number of soil-quality and -management factors,

---

[5] For a discussion of how *T* values were derived, see Schertz (1983).

including soil depth, organic matter content, and the use of water-control practices. Soils with shallow layers of infertile, impermeable, or rocky material are generally assigned T values near the low end of the range. In some tropical areas, the low nutrient-supplying power of the subsoil suggests that appropriate T values may be considerably lower than those used in the United States.

The majority of agricultural soils in the United States are currently assigned the highest T value, 11 Mg/ha. This represents a maximum allowable loss of about 0.9 mm of soil depth annually, a rate at which it would take about 225 years to lose the equivalent of an entire Ap horizon. Under good agricultural management in a deep, permeable soil profile, this may be sufficient time to allow replacement of the lost material as new Ap horizon material forms from underlying subsoil material. Development of horizons in undisturbed soils under natural vegetation is most likely much slower than this, so the T value assigned to a particular type of soil probably would not provide sufficient protection for many rangeland or forest sites.

SIGNIFICANCE OF T VALUES. Even with its limitations, the concept of T values is useful in focusing attention on the soils where improved practices are needed to maintain long-term productivity. Because T values are used in determining compliance with various regulatory programs, there is considerable controversy as to whether the values should be increased or lowered. Concerns about long-term soil productivity and about the off-site effects of sediment from eroded fields suggest that the T values currently in use may be too high.

Clearly, much work remains to be done in controlling excessive soil loss.

17.3   MECHANICS OF WATER EROSION

Movie of raindrop impact:
http://www.public.asu.edu/
~mschmeec/rainsplash.html

Soil erosion by water is fundamentally a three-step process (Figure 17.10):

1. *Detachment* of soil particles from the soil mass
2. *Transportation* of the detached particles downhill by floating, rolling, dragging, and splashing
3. *Deposition* of the transported particles at some place lower in elevation

On comparatively smooth soil surfaces, the beating action of raindrops causes most of the detachment. Where water is concentrated into channels, the cutting action of turbulent, flowing water detaches soil particles. In some situations, freezing-thawing action also contributes to soil detachment.

Influence of Raindrops

A raindrop accelerates as it falls until it reaches *terminal velocity*—the speed at which the friction between the drop and the air balances the force of gravity. Larger raindrops fall faster, reaching a terminal velocity of about 30 km/h, or about as fast as a person can run. As the speeding raindrops impact the soil with explosive force, they transfer their high kinetic energy to the soil particles (see Figure 17.10).

Raindrop impact exerts three important detrimental effects: (1) it detaches soil; (2) it destroys granulation; and (3) its splash, under certain conditions, causes an appreciable transportation of soil. So great is the force exerted by raindrops that they not only loosen and detach soil granules, but may even beat the granules to pieces. As the dispersed material dries it may develop into a hard crust, which will prevent the emergence of seedlings and will encourage runoff from subsequent precipitation (see Section 4.6).

History may someday record that one of the truly significant scientific advances of the 20th century was the realization that most erosion is initiated by the impact of raindrops, rather than the flow of running water. For centuries prior to this realization, soil conservation efforts aimed at controlling the more visible flow of water across the land, rather than protecting the soil surface from the impact of raindrops.

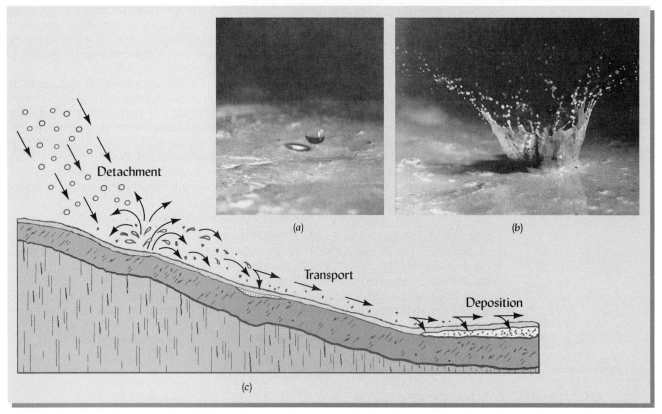

FIGURE 17.10 The three-step process of soil erosion by water begins with the impact of raindrops on wet soil. (*a*) A raindrop speeding toward the ground. (*b*) The splash that results when the drop strikes a wet, bare soil. Such raindrop impact destroys soil aggregates, encouraging sheet and interrill erosion. Also, considerable soil may be moved by the splashing process itself. The raindrop affects the detachment of soil particles, which are then transported and eventually deposited in locations downhill (*c*).

Transportation of Soil

RAINDROP SPLASH EFFECTS. When raindrops strike a wet soil surface, they detach soil particles and send them flying in all directions (see Figure 17.10). On a soil subject to easy detachment, a very heavy rain may splash as much as 225 Mg/ha of soil, some of the particles splashing as much as 0.7 m vertically and 2 m horizontally. If the land is sloping or if the wind is blowing, this splashing may be greater in one direction, leading to considerable net horizontal movement of soil.

ROLE OF RUNNING WATER. Runoff water plays the major role in the transportation step of soil erosion. If the rate of rainfall exceeds the soil's infiltration capacity, water will pond on the surface and begin running downslope. The soil particles sent flying by raindrop impact will then land in flowing water, which will carry them down the slope. So long as the water is flowing smoothly in a thin layer (sheet flow), it has little power to detach soil. However, in most cases the water is soon channeled by irregularities in the soil surface which cause it to increase in both velocity and turbulence. The channelized flow then not only carries along soil splashed by raindrops, but also begins to detach particles as it cuts into the soil mass. This is an accelerating process, for as a channel is cut deeper, it fills with greater and greater volumes of flowing water. So familiar is the power of runoff water to cut and carry that the public generally ascribes to it all the damage done by heavy rainfall.

Types of Water Erosion

Three types of water erosion are generally recognized: (1) *sheet*, (2) *rill*, and (3) *gully* (Figure 17.11). In **sheet erosion**, splashed soil is removed more or less uniformly, except that tiny columns of soil often remain where pebbles intercept the raindrops

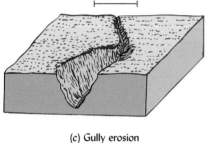

(a) Sheet erosion

(b) Rill erosion

(c) Gully erosion

0.1m

1 m

10 m

FIGURE 17.11 Three major types of soil erosion. *Sheet erosion* is relatively uniform erosion from the entire soil surface. Note that the perched stones and pebbles have protected the soil underneath from sheet erosion. The pencil gives a sense of scale. *Rill erosion* is initiated when the water concentrates in small channels (rills) as it runs off the soil. Subsequent cultivation may erase rills, but it does not replace the lost soil. *Gully erosion* creates deep channels that cannot be erased by cultivation. Although gully erosion looks the most catastrophic of the three, far more total soil is lost by the less obvious sheet and rill erosion. [Drawings from FAO (1987); photos courtesy USDA Natural Resources Conservation Service]

(see Figure 17.11*a*). However, as the sheet flow is concentrated into tiny channels (termed **rills**), **rill erosion** becomes dominant. Rills are especially common on bare land, whether newly planted or in fallow (see Figure 17.11*b*). Rills are channels small enough to be smoothed by normal tillage, but the damage is already done—the soil is lost. When sheet erosion takes place primarily between irregularly spaced rills, it is called **interrill erosion**.

Where the volume of runoff is further concentrated, the rushing water cuts deeper into the soil, deepening and coalescing the rills into larger channels termed **gullies** (see Figure 17.11*c*). This is **gully erosion**. Gullies on cropland are obstacles for tractors and cannot be removed by ordinary tillage practices. All three types may be serious, but sheet and rill erosion, although less noticeable than gully erosion, are responsible for most of the soil moved.

Deposition of Eroded Soil

Erosion may send soil particles on a journey of a thousand kilometers or more—off the hills, into creeks, and down great muddy rivers to the ocean. On the other hand, eroded soil may travel only a meter or two before coming to rest in a slight depression on a hillside or at the foot of a slope (as was shown in Figure 17.7). The amount of soil delivered to a stream, divided by the amount eroded, is termed the **delivery ratio**.

Soil from an eroding PA
streambank travels to the
Chesapeake Bay:
http://www.bayjournal.com/
article.cfm?article=699

As much as 60% of eroded soil may reach a stream (delivery ratio = 0.60) in certain watersheds where valley slopes are very steep. As little as 1% may reach the streams draining a gently sloping coastal plain. Typically, the delivery ratio is larger for small watersheds than for large ones, because the latter provide many more opportunities for deposition before a major stream is reached. It is estimated that about 5 to 10% of all eroded soil in North America is washed out to sea. The remainder is deposited in reservoirs, river beds, on flood plains, or on relatively level land farther up the watershed.

17.4   MODELS TO PREDICT THE EXTENT OF WATER-INDUCED EROSION[6]

Land managers and policymakers have many needs to predict the extent of soil erosion:

> To plan for the best management of a nation's soil resources
>
> To evaluate the consequences of alternative tillage practices
>
> To determine compliance with environmental regulations
>
> To develop sediment-control plans for construction projects
>
> To estimate the years it will take to silt-in a hydroelectric dam

The Water Erosion Prediction Project (WEPP)[7]

The WEPP model explained
and available for download:
http://topsoil.nserl.purdue.
edu/nserlweb/weppmain/
wepp.html

The detachment, transport, and deposition processes of soil erosion can be predicted mathematically by soil erosion *models*. These are equations—or sets of linked equations—that interrelate information about the rainfall, soil, topography, vegetation, and management of a site with the amount of soil likely to be lost by erosion. The most ambitious and sophisticated of the erosion models developed so far is a complex, process-based computer program called the Water Erosion Prediction Project (WEPP). It is based on an understanding of the fundamental mechanisms involved with each process leading to soil erosion.

WEPP is a *simulation* model that computes, on a daily basis, the rates of hydrologic, plant-growth, and even litter-decay processes. Theoretically, it can predict exactly how rainfall will interact with the soil on a site during a particular rainstorm or during the course of an entire year. If sufficient data are available to feed into the model, it can predict both on-site and off-site effects of raindrop impact, splash erosion, interrill flow, rill formation, channelization, gully formation, and sediment deposition. Currently, researchers with the USDA Forest Service and the Natural Resources Conservation Service, along with others throughout the world, are compiling the necessary databases, testing and improving the WEPP model, and making it accessible via the Internet.

The Universal Soil-Loss Equation (USLE)

In contrast to the process-based operation of WEPP, most predictions of soil erosion continue to rely on much simpler models that statistically relate soil erosion to a number of easily observed factors. Scientists can make such *empirical* models if they know that certain conditions are associated with soil erosion, even if they do not understand the details of *why* this is so. At the heart of these models is the realization that water-induced erosion results from the interaction of rain and soil. Decades of erosion

---

[6] For discussion of the original USLE, see Wischmeier and Smith (1978), and for the RUSLE, see Renard et al. (1997). In this textbook, we use the scientifically acceptable SI units for the R and K factors in our discussion of these erosion equations. However, since these soil-loss equations were published in the United States for use by landowners and the general public, most maps, tables, and computer programs available supply values for the R and K factors in customary English units, rather than in SI units. When using English units for the R and K factors, the soil loss A is expressed in tons (2000 lb) per acre, which can be easily converted to Mg/ha by multiplying by 2.24. For details on converting the customary English units to SI units, see Foster et al. (1981).

[7] The WEPP model and associated databases for use on personal computers can be downloaded from: http://topsoil.nserl.purdue.edu/nserlweb/weppmain/wepp.html. See Flanagan et al. (2001) for details about the availability and uses of the WEPP materials.

research have clearly identified the major factors affecting this interaction. These factors are quantified in the **universal soil-loss equation (USLE):**

$$A = R \times K \times LS \times C \times P \qquad \text{(eq 17.1)}$$

A, the predicted annual soil loss, is the product of

R = rainfall erosivity } Rain-related factor

K = soil erodibility
L = slope length
S = slope gradient or steepness } Soil-related factors

C = cover and management
P = erosion-control practices } Land-management factors

Working together, these factors determine how much water enters the soil, how much runs off, how much soil is transported, and when and where it is redeposited. Note that because the factors are multiplied together, *if any one factor could be reduced to zero, the resulting amount of erosion (A) would also be reduced to zero.* More details about the erosion factors are given in Table 17.2.

Unlike the WEPP program, the USLE was designed to predict only the amount of soil loss by sheet and rill erosion in an average year for a given location. It cannot predict erosion from a specific year or storm, nor can it predict the extent of gully erosion and sediment delivery to streams. It can, however, show how varying any combination of the soil- and land-management-related factors might be expected to influence soil erosion, and therefore can be used as a decision-making aid in choosing the most effective strategies to conserve soil.

TABLE 17.2 Summary of Major Differences between USLE and RUSLE

Factor	Universal soil-loss equation (USLE)	Revised universal soil-loss equation (RUSLE)
R	Based on long-term average rainfall conditions for specific geographic areas in the United States.	Generally the same as USLE in the eastern United States. Values for western states (Montana to New Mexico and west) are based on data from more weather stations and thus are more precise for any given location. RUSLE computes a correction to *R* to reflect, for flat land, the effect of raindrop impact on water ponded on the surface.
K	Based on soil texture, organic matter content, permeability, and other factors inherent to soil type.	Same as USLE but adjusted to account for seasonal changes, such as freezing and thawing, soil moisture, and soil consolidation.
LS	Based on length and steepness of slope, regardless of land use.	Refines USLE by assigning new equations based on the ratio of rill to interrill erosion, and accommodates complex slopes.
C	Based on cropping sequence, surface residue, surface roughness, and canopy cover, which are weighted by the percentage of erosive rainfall during the six crop stages. Lumps these factors into a table of soil-loss ratios, by crop and tillage scheme.	Uses these subfactors: prior land use, canopy cover, surface cover, surface roughness, and soil moisture. Refines USLE by dividing each year in the rotation into 15-day intervals, calculating the soil-loss ratio for each period. Recalculates a new soil-loss ratio every time a tillage operation changes one of the subfactors. RUSLE provides improved estimates of soil-loss changes as they occur throughout the year, especially relating to surface and near-surface residue and the effects of climate on residue decomposition.
P	Based on installation of practices that slow runoff and thus reduce soil movement. *P* factor values change according to slope ranges with some distinction for various ridge heights.	*P* factor values are based on hydrologic soil groups, slope, row grade, ridge height, and the 10-year single storm erosion index value. RUSLE computes the effect of strip-cropping based on the transport capacity of flow in dense strips relative to the amount of sediment reaching the strip. The *P* factor for conservation planning considers the amount and location of deposition.

Based on, Renard et al. (1994).

The Revised Universal Soil-Loss Equation (RUSLE)

(RUSLE-2) download and training:
http://fargo.nserl.purdue.edu/rusle2_dataweb/RUSLE2_Index.htm

The USLE has been used widely since the 1970s. In the early 1990s, the basic USLE was updated and computerized to create an erosion-prediction tool called the **revised universal soil-loss equation (RUSLE)**. The RUSLE uses the same basic factors of the USLE just shown, although some are better defined and interrelationships among them improve the accuracy of soil-loss prediction. The RUSLE is a computer software package that is constantly being improved and modified as experience is gained from its use around the world. The major differences between USLE and RUSLE are shown in Table 17.2.

As we are about to see, the five factors that comprise the USLE provide a useful framework for understanding soil erosion and its control.

17.5   FACTORS AFFECTING INTERRILL AND RILL EROSION

Rainfall Erosivity Factor R

The rainfall **erosivity** factor R represents the driving force for sheet and rill erosion. It takes into consideration the total rainfall and, more important, the intensity and seasonal distribution of the rain. Rain intensity is of great importance for two reasons: (1) intense rains have a large drop size, which results in much greater kinetic energy being available to detach soil particles; and (2) the higher the rate of rainfall, the more runoff that occurs, providing the means to transport detached particles. Gentle rains of low intensity may cause little erosion, even if the total annual precipitation is high. In contrast, a few torrential downpours may result in severe damage, even in areas of low annual rainfall. Likewise, soil losses are heavy if the rain falls when the soil is just thawing or is relatively bare because of recent disturbance.

An index of the kinetic energy of each storm is calculated from data related to the intensity and amount of rainfall. Then the indices for all storms occurring during a year are summed to give an annual index. An average of such indexes for many years is used as the R value in the universal soil-loss equation. The RUSLE includes more precise values, especially for the western part of the United States. It also gives special consideration to raindrops striking water ponded on the surface of flat slopes and to runoff from the thawing of frozen soils.

Rainfall index values for locations in the United States are shown in Figure 17.12. Note that they vary from less than 10 in areas of the west to more than 700 along the coasts of Louisiana [the map gives R values in English units that can be converted to SI units of $(MJ \cdot mm)/(ha \cdot h \cdot yr)$ if multiplied by 17.02]. Similar data has been generated in other parts of the world. Generally, rainfall tends to be more intense and more erosive in subtropical and tropical regions than in temperate regions.

Rainfall intensity in most locations is so highly variable that actual erosivity in any one year is commonly 2 to 5 times greater or smaller than the long-term average. In fact, a few unusually intense, heavy storms often account for most of the erosion that takes place (Figure 17.13). Conservation practices based on the predictions of the USLE or RUSLE using long-term average R factors may not be sufficient to limit erosion damages from these relatively rare, but extremely damaging, storms.

Soil Erodibility Factor K

The soil **erodibility** factor K indicates a soil's inherent susceptibility to erosion. The K value assigned to a particular type of soil indicates the amount of soil lost per unit of erosive energy in the rainfall, assuming a standard research plot (22 m long, 9% slope) on which the soil is kept continuously bare by tillage.

The two most significant and closely related soil characteristics influencing erodibility are (1) *infiltration capacity,* and (2) *structural stability.* High infiltration means that less water will be available for runoff, and the surface is less likely to be ponded (which would make it more susceptible to splashing). Stable soil aggregates resist the beating action of rain, and thereby save soil even though runoff may occur. Certain tropical clay soils high in hydrous oxides of iron and aluminum are known for their highly stable aggregates that resist the action of torrential rains. Downpours of a similar magnitude on swelling-type clays would be disastrous.

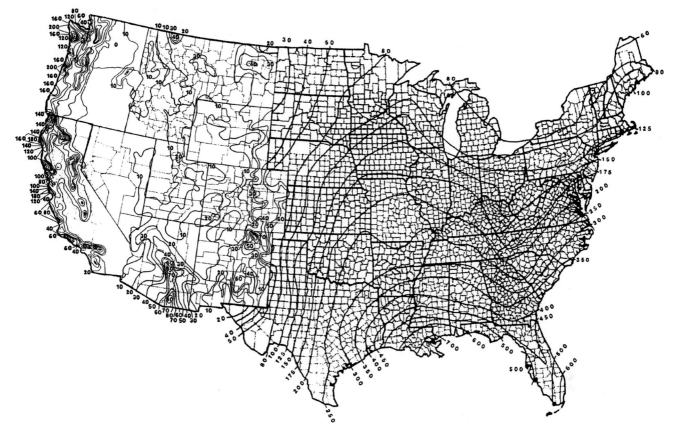

FIGURE 17.12 The geographic distribution of *R* values for rainfall erosivity in the continental United States. Note the very high values in the humid, subtropical Southeast, where annual rainfall is high and intense storms are common. Similar amounts of annual rainfall along the coast of Oregon and Washington in the Northwest result in much lower *R* values because there the rain mostly falls gently over long periods. The complex patterns in the West are mainly due to the effects of mountain ranges. Values on map are in units of 100 (ft·ton·in.)/(acre·yr). To convert to Sl units of (MJ·mm)/(ha·h·yr), multiply by 17.02. [Redrawn from USDA (1995)]

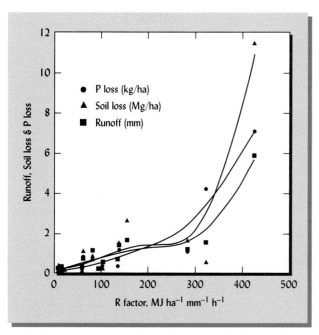

FIGURE 17.13 Influence of the R factor (rainfall energy) on runoff water, soil loss, and phosphorus runoff loss in a vineyard in northeastern Spain. The vineyard had been leveled prior to planting the grape vines and the soil exposed at the surface was very low in organic matter. The data are based on measurements during 17 natural rainfall events over a 3-year period. The study results illustrate that most soil and nutrient losses result from a small number of extreme events. [Data selected from Ramos and Martinez-Casasnovas (2006)]

Basic soil properties that tend to result in high K values include high contents of silt and very fine sand; expansive types of clay minerals; a tendency to form surface crusts; the presence of impervious soil layers; and blocky, platy, or massive soil structure. Soil properties that tend to make the soil more *resistant to erosion* (low K values) include high soil organic matter content, nonexpansive types of clays, and strong granular structure. Approximate K values for soils at selected locations are shown in Table 17.3.

Site-specific K values obtained with the RUSLE program will vary somewhat from these values. Note that the K factor in SI units normally varies from near zero to about 0.1. Soils with high rates of water infiltration commonly have K values of 0.025 or below, while more easily eroded soils with low infiltration rates have K factors of 0.04 or higher.

Unlike the USLE, the RUSLE takes into account the fact that K values vary seasonally. For example, in cold regions, thawing of frozen soil in spring results in higher K values, because the soil is supersaturated with water and is "fluffy" from freeze-thaw action. Also, the K factor in RUSLE may be reduced with time in stony soils, because they become protected by an armor of stone fragments left on the soil surface after the finer soil components erode away.

Topographic Factor LS

The topographic factor *LS* reflects the influence of length and steepness of slope on soil erosion. It is expressed as a unitless ratio with soil loss from the area in question in the numerator, and that from a standard plot (9% slope, 22 m long) in the denominator. The longer the slope, the greater the opportunity for concentration of the runoff water.

Figure 17.14 illustrates the increases in LS factors that occur as slope length and steepness increase. Three graphs are given for sites with low, moderate, and high ratios of rill to interrill (sheet) erosion. Most sites cultivated to row crops have moderate rill to interrill erosion ratios. On sites where this ratio is low, such as rangelands, more of the soil movement occurs by interrill erosion. On these sites, slope steepness (%) has a relatively greater influence on erosion, while the slope length has a relatively smaller influence. The opposite is true for freshly excavated construction areas and other highly disturbed sites, which have high rill to interrill erosion ratios. Here, where rill erosion predominates, slope length has a greater influence. While such generalized LS factor values for simple slopes can be used with the USLE, the RUSLE computer

TABLE 17.3 Computed K Values for Soils at Different Locations

The values listed are in SI units. The computerized RUSLE model may give more accurate values for some locations.

Soil	Location	Compounds[a] K
Udalf (Dunkirk silt loam)	Geneva, NY	0.091
Udalf (Keene silt loam)	Zanesville, OH	0.063
Udult (Lodi loam)	Blacksburg, VA	0.051
Udult (Cecil sandy clay loam)	Watkinsville, GA	0.048
Udoll (Marshall silt loam)	Clarinda, IA	0.044
Udalf (Hagerstown silty clay loam)	State College, PA	0.041
Ustoll (Austin silt)	Temple, TX	0.038
Aqualf (Mexico silt loam)	McCredie, MO	0.034
Udult (Cecil sandy loam)	Clemson, SC	0.034
Udult (Cecil sandy loam)	Watkinsville, GA	0.030
Alfisols	Indonesia	0.018
Alfisols	Benin	0.013
Oxisols	Ivory Coast	0.013
Udult (Tifton loamy sand)	Tifton, GA	0.013
Ultisols	Hawaii	0.012
Alfisols	Nigeria	0.008
Udept (Bath flaggy silt loam)	Arnot, NY	0.007
Ultisols	Nigeria	0.005
Oxisols	Puerto Rico	0.001

[a] To convert these K values from (Mg · ha · h)/(ha · MJ · mm) to English units of (ton · acre · h)/(100 acres · ft-ton · in.), simply multiply the values in this table by 7.6. From Wischmeier and Smith (1978); data for tropical soils cited by Cassel and Lal (1992).

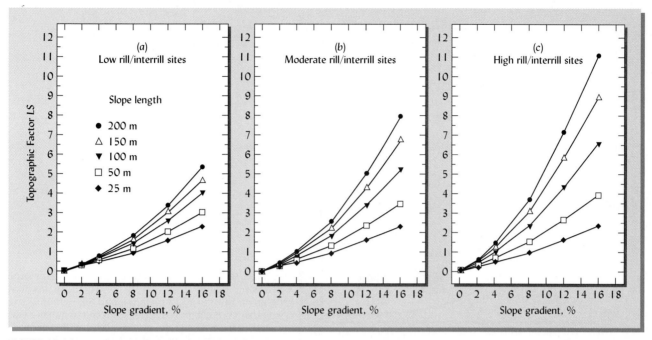

FIGURE 17.14 Relationship between values of the topographic factor *LS* and the slope gradient for several lengths of slope on three types of sites: (*a*) sites with low ratios of rill to interrill erosion, such as many rangelands; (*b*) sites with moderate ratios of rill to interrill erosion, such as most tilled row-crop land; and (*c*) sites with high ratios of rill to interrill erosion, such as freshly disturbed construction sites and new seedbeds. The LS values extrapolated from these graphs can be used in the Universal Soil-Loss Equation. [Graphs based on data in Renard et al. (1997)]

program calculates more location-specific values, including values for complex (nonuniform) slopes.

Cover and Management Factor C

Erosion and runoff are markedly affected by different types of vegetative cover and cropping systems (Table 17.4). Undisturbed forests and dense grass provide the best soil protection and are about equal in their effectiveness. Forage crops (both legumes and grasses) are next in effectiveness because of their relatively dense cover. Small grains, such as wheat and oats, are intermediate and offer considerable obstruction to surface wash. Row crops, such as corn, soybeans, and potatoes, offer relatively little living cover during the early growth stages and thereby leave the soil susceptible to erosion unless residues from previous crops cover the soil surface.

Cover crops consist of plants that are similar to the forage crops just mentioned. They can provide soil protection during the time of year between the growing seasons

TABLE 17.4 **Effect of Plant Cover on Soil Erosion by Water in the Humid Zone of West Africa**

The data are averaged over many nearby sites, all with similar slopes, soils, and amounts of rainfall. The erosive influence of cultivation (with various types of tillage), especially bare fallows, is illustrated, as is the protective effect of undisturbed forest vegetation.

Type of cover	Number of sites	Mean rainfall, mm/y	Runoff, % of rainfall	Erosion, Mg/ha
Forest protected from fire	11	1293	0.9	0.10
Forest with light fires	13	1289	1.1	0.27
Natural grass fallow	7	1203	16.6	4.88
Groundnut (peanut)	32	1329	20.7	7.70
Upland rice	17	946	23.3	5.52
Maize (corn)	17	1405	17.7	7.63
Failed crops and bare soil	11	1154	39.5	21.28

Data selected from that cited by Pierre (1992).

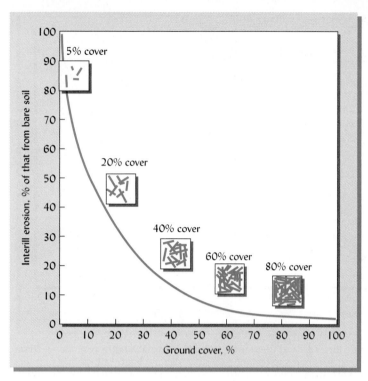

FIGURE 17.15 Reduction in interrill erosion achieved by increasing ground cover percentage. The diagrams above the graph illustrate 5, 20, 40, 60, and 80% ground cover. Note that even a light covering of mulch has a major effect on soil erosion. The graph applies to interrill erosion. On steep slopes, some rill erosion may occur even if the soil is well covered. [Generalized relationship based on results from many studies]

for annual crops. For widely spaced perennial plantings such as orchards and vineyards, cover crops can permanently protect the soil between rows of trees or vines. A mulch of plant residue or applied materials is also effective in protecting soils. Research on all continents has shown that a surface mulch does not have to be thick or cover the soil completely to make a major contribution to soil conservation. Even small increases in surface cover result in large reductions in soil erosion, particularly interrill erosion (Figure 17.15).

Regulation of grazing to maintain a dense vegetative cover on range- and pasture-land and the inclusion of close-growing hay crops in rotation with row crops on arable land will help control both erosion and runoff. Likewise, the use of conservation tillage systems, which leave most of the plant residues on the surface, greatly decreases erosion hazards.

The C factor in the USLE or RUSLE is the ratio of soil loss under the conditions in question to that which would occur under continuously bare soil. This ratio C will approach 1.0 where there is little soil cover (e.g., a bare seedbed in the spring or freshly graded bare soil on a construction site). It will be low (e.g., < 0.10) where large amounts of plant residues are left on the land or in areas of dense perennial vegetation.

Values of C are specific to each region and type of vegetation or soil management. Estimates based on experiment data and field experience are available from conservation officers and the RUSLE program. Examples of C values are given in Table 17.5.

Support Practice Factor P[8]

On some sites with long and/or steep slopes, erosion control achieved by management of vegetative cover, residues, and tillage must be augmented by the construction of physical structures or other steps aimed at guiding and slowing the flow of runoff water. These **support practices** determine the value of the P factor in the USLE. The P factor is

---

[8] Many of the erosion-control practices or management techniques discussed with regard to the C and P factors and in later sections of this chapter are considered to be **best management practices (BMPs)** (see also on Section 16.2) under provisions of the Clean Water Act in the United States. The act defines BMPs as "optimal operating methods and practices for reducing or eliminating water pollution" from land-use activities.

TABLE 17.5 Examples of C Values for the Cover and Vegetation Management Factor

The C values indicate the ratio of soil eroded from a particular vegetation system to that expected if the soil were kept completely bare. Note the effects of canopy cover, surface litter (residue) cover, tillage, and crop rotation. The C values are site- and situation-specific and must be calculated from local information on plant growth habits, climate, and so on. In the United States, specific values may be obtained from the RUSLE computer program or from local offices of the USDA Natural Resources Conservation Service.

Vegetation	Management/condition	C value
Range grasses and low (<1 m) shrubs	75% canopy cover, no surface litter	0.17
	75% canopy cover, 60% cover with decaying litter	0.032
Scrub brush about 2 m tall	25% canopy cover, no litter	0.40
	75% canopy cover, no litter	0.28
Trees with no understory, about 4 m drop fall	75% canopy cover, no litter	0.36
	75% canopy cover, 40% leaf litter cover	0.09
	75% canopy cover, 100% leaf litter cover	0.003
Woodland with understory	90% canopy cover, 100% litter cover	0.001
Permanent pasture	Dense stand of grass sod	0.003
Corn–soybean rotation	Fall plowing, conventional tillage, residues removed	0.53
	Spring chisel plow–plant conservation tillage, 2500 kg/ha surface residues after planting	0.22
	No-till planting, 5000 kg/ha surface residues after planting	0.06
Corn–soybean–wheat–hay rotation	Fall plowing, conventional tillage, residues removed	0.20
	Spring chisel plow–plant conservation tillage, 2500 kg/ha surface residues after planting	0.13
	No-till planting, 5000 kg/ha surface residues after planting	0.05
Corn–oats–hay–hay rotation	Spring conventional plowing before planting	0.05
	No-till planting	0.03

Values typical of midwestern United States. Based on Wischmeier and Smith (1978) and Schwab et al. (1996).

the ratio of soil loss with a given support practice to the corresponding loss if row crops were planted up and down the slope. If there are no support practices, the *P* factor is 1.0. The support practices include tillage on the contour, contour strip-cropping, terrace systems, and grassed waterways, all of which will tend to reduce the *P* factor.

CONTOUR CULTIVATION. Rows of plants slow the flow of runoff water if they follow the contours across the slope gradient (but the rows *encourage* channelization and gullies if they run up and down the slope; Figure 17.16). Even more effective is planting on ridges built up of soil along the contours. However, ridges must be designed to carry heavy runoff safely from the field (Figure 17.17).

On long slopes subject to sheet and rill erosion, the fields may be laid out in narrow strips across the incline, alternating the tilled crops, such as corn and potatoes, with hay and small grains. Water cannot achieve an undue velocity on the narrow strips of tilled land, and the hay and grain crops check the rate of runoff. Such a layout is called **strip-cropping** and is the basis for erosion control in many hilly agricultural areas (Figure 17.18 and Plate 72). This arrangement can be thought of as shortening the effective slope length.

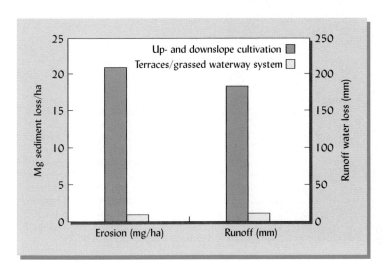

FIGURE 17.16 Erosion and water runoff losses from small watersheds where potatoes (a row crop) were grown either up and down the slope, or on the contour in a system with diversion terraces and a grassed waterway. The contour practices provided dramatic soil and water conservation effects. (Data are averages of three years.) [From Chow et al. (1999)]

FIGURE 17.17 Contour ridges must be carefully laid out with sufficient height to hold back water from even heavy rainfall. Here, surface retention is fast becoming surface runoff. (Photo courtesy of R. Weil)

When the cross strips are laid out rather definitely on the contours, the system is called **contour strip-cropping.** The width of the strips will depend primarily on the degree of slope, the permeability of the land, and the soil erodibility. Widths of 30 to 125 m are common. Contour strip-cropping is often augmented by diversion ditches and waterways between fields. Permanent sod established in the swales produces **grassed waterways** that can safely carry water off the land without the formation of gullies (see Figure 17.18).

TERRACES. Construction of various types of terraces reduces the effective length and gradient of a slope (Figure 17.19). **Bench terraces** are used where nearly complete control of the water runoff must be achieved, such as in rice paddies (see Figure 6.45). Where farmers use large machinery and need to farm all the land in a field, **broad-based terraces** are more common. Broad-based terraces waste little or no land and are quite effective if properly maintained. Water collected behind each terrace flows gently

Grassed waterway

FIGURE 17.18 An aerial photo of farmland in Kentucky where contour strip-cropping is practiced and grassed waterways (see arrow) control the flow of water off cropland and prevent gully erosion. (*Inset*) A grassed waterway in action, showing how the permanent grass sod can resist the scouring force of water and safely conduct water off the field. (Photos courtesy of USDA National Resources Conservation Service)

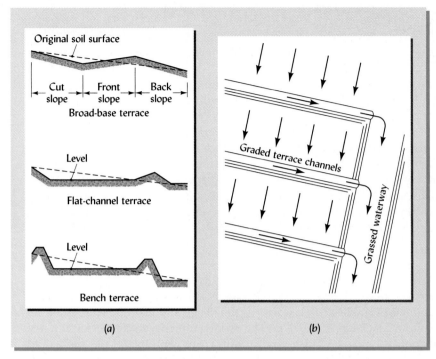

FIGURE 17.19   (a) Three types of terraces in use around the world. Broad-based terraces permit the entire surface to be cropped and are widely used in the United States. Flat-channel terraces allow large volumes of water to move off the soil without erosion. Bench terraces can keep water on the land and are used widely in rice production. (b) Diagram of controlled flow of runoff water from a field to terrace channels and onto a grassed waterway from which it can move to a canal or stream channel.

across (rather than down) the field in a terrace channel, which has a drop of only about 50 cm in 100 m (0.5%). The terrace channel usually guides the runoff water to a grassed waterway, through which the water moves downhill to a nearby canal, stream, or river.

Examples of *P* values for contour tillage and strip-cropping at different slope gradients are shown in Table 17.6. Note that *P* values increase with slope and that they are lower for strip-cropping, illustrating the importance of this practice for erosion control. Terracing also reduces the *P* values. Unlike the USLE, the RUSLE takes into account interactions between support practices and subfactors such as slope and soil water infiltration.

The five factors of the USLE (*R, K, LS, C,* and *P*) have suggested many approaches to the practical control of soil erosion. A sample calculation is shown in Box 17.1 to illustrate how the USLE can help evaluate erosion-control options.

More than half of the soil eroded in the United States comes from agricultural land, with the remainder coming from forests, rangeland, and construction sites. We will now focus on several specific erosion-control technologies appropriate for these various types of land uses.

TABLE 17.6   P Factors for Contour and Strip-Cropping at Different Slopes and the Terrace Subfactor at Different Terrace Intervals

The product of the contour or strip-cropping factors and the terrace subfactor gives the P value for terraced fields.

Slope, %	Contour P factor	Strip-cropping P factor	Terrace interval, m	Terrace subfactor	
				Closed outlets	Open outlets
1–2	0.60	0.30	33	0.5	0.7
3–8	0.50	0.25	33–44	0.6	0.8
9–12	0.60	0.30	43–54	0.7	0.8
13–16	0.70	0.35	55–68	0.8	0.9
17–20	0.80	0.40	69–60	0.9	0.9
21–25	0.90	0.45	90	1.0	1.0

Contour and strip-cropping factors from Wischmeier and Smith (1978); terrace subfactor from Foster and Highfill (1983).

BOX 17.1 CALCULATIONS OF EXPECTED SOIL LOSS USING USLE

The RUSLE computer software is designed to calculate the expected soil loss from specific cropping systems at a given location.

The principles involved in both USLE and RUSLE can be verified by making calculations using USLE and its associated factors. Note that the factors in the USLE are related to each other in a multiplicative fashion. Therefore, if any one factor can be made to be near zero, the amount of soil loss A will be near zero.

Assume, for example, a location in Iowa on a Marshall silt loam with an average slope of 6% and an average slope length of 100 m. Assume further that the land is clean-tilled and fallowed.

Figure 17.12 shows that the R factor for this location is about 150 in English units or (150×17) 2550 in SI units. The K factor for a Marshall silt loam in central Iowa is 0.044 (Table 17.3) and the topographic factor LS from Figure 17.14 is 1.7 (high rill to interrill ratio on soil kept bare). The C factor is 1.0, since there is no cover or other management practice to discourage erosion. If we assume the tillage is up and down the hill, the P value is also 1.0. Thus, the anticipated soil loss can be calculated by the USLE ($A = RKLSCP$):

$$A = (2550)(0.044)(1.7)(1.0)(1.0) = 191 \text{ Mg/ha or } 85.2 \text{ tons/acre}$$

If the crop rotation involved corn–soybean–wheat–hay and conservation tillage practices (e.g., spring chisel plow tillage) were used, a reasonable amount of residue would be left on the soil surface. Under these conditions, the C factor may be reduced to about 0.13 (Table 17.5). Likewise, if the tillage and planting were done on the contour, the P value would drop to about 0.5 (Table 17.6). P could be reduced further to 0.4 (0.5×0.8) if terraces with open outlets were installed about 40 m apart. Furthermore, with crops on the land the site would have a moderate rill to interrill erosion ratio, so the LS factor would be only 1.4 (middle of Figure 17.14). With these figures the soil loss becomes:

$$A = (2550)(0.044)(1.4)(0.13)(0.4) = 8.2 \text{ Mg/ha or } 3.4 \text{ tons/acre}$$

The units for the calculation are not normally shown, but they are:

$$\frac{MJ \cdot mm}{ha \cdot h \cdot yr} \times \frac{Mg \cdot ha \cdot h}{ha \cdot MJ \cdot mm} = \frac{Mg}{ha \cdot yr} \ (LS, C, \text{ and } P \text{ are unitless ratios}).$$

The benefits of good cover and management and support practices are obvious. The figures cited were chosen to provide an example of the utility of the universal soil-loss equation, but calculations can be made for any specific location. In the United States, pertinent factor values that can be used for erosion prediction in specific locations are generally available from state offices of the USDA Natural Resource Conservation Service. The necessary factors are also built into the RUSLE software.

17.6 CONSERVATION TILLAGE[9]

Conservation tillage pros, cons, and methods:
http://www.ncsu.edu/sustainable/tillage/tillage.html

For centuries, conventional agricultural practice around the world encouraged extensive soil tillage that leaves the soil bare and unprotected from the ravages of erosion. During the last three decades, two technological developments have allowed many farmers to avoid this problem by managing their soils with greatly reduced tillage—or no tillage at all. First came the development of herbicides that could kill weeds chemically rather than mechanically. Second, farmers and equipment manufacturers developed machinery that could plant crop seeds even if the soil was covered by plant residues. These developments obviated two of the main reasons that farmers tilled their soils. Farmer interest in reduced tillage heightened as it was shown that these systems produced equal or even higher crop yields in many regions while saving time, fuel, money—and soil. The latter attribute earned these systems the name of **conservation tillage**.

---

[9] For reviews of conservation tillage technology, see Blevins and Frye (1992) and Carter (1994). For a review of economic and policy considerations for conservation agriculture, see FAO (2001).

TABLE 17.7   General Classification of Different Conservation Tillage Systems

All systems maintain at least 30% of the crop residues on the surface.

Tillage system	Operation involved
No-till	Soil undisturbed prior to planting, which occurs in narrow seedbed, 2.5 to 7.5 cm wide. Weed control primarily by herbicides.
Ridge till (till, plant)	Soil undisturbed prior to planting, which is done on ridges 10 to 15 cm higher than row middles. Residues moved aside or incorporated on about one-third of soil surface. Herbicides and cultivation to control weeds.
Strip till	Soil undisturbed prior to planting. Narrow and shallow tillage in row using rotary tiller, in-row chisel, and so on. Up to one-third of soil surface is tilled at planting time. Herbicides and cultivation to control weeds.
Mulch till	Soil surface disturbed by tillage prior to planting, but at least 30% of residues left on or near soil surface. Tools such as chisels, field cultivators, disks, and sweeps are used (e.g., stubble mulch). Herbicides and cultivation to control weeds.
Reduced till	Any other tillage and planting system that keeps at least 30% of residues on the surface.

Definitions used by Conservation Technology Information Center, West Lafayette, Ind.

Conservation Tillage Systems

Pursuing conservation tillage in organic farming systems:
http://attra.ncat.org/attra-pub/organicmatters/conservationtillage.html

While there are numerous conservation tillage systems in use today (Table 17.7), all have in common that they leave significant amounts of organic residues on the soil surface after planting. Keep in mind that conventional tillage involves first moldboard plowing (Figure 17.20 *left*) to completely bury weeds and residues, followed by one to three passes with a harrow to break up large clods, then planting the crop, and subsequently several cultivations between crop rows to kill weeds. Every pass with a tillage implement bares the soil anew and also weakens the structure that helps soil resist water erosion.

Conservation tillage systems range from those that merely reduce excess tillage to the no-tillage system, which uses no tillage beyond the slight soil disturbance that occurs as the planter cuts a planting slit through the residues to a depth of several cm into the soil (Figure 17.21, *inset*). The conventional moldboard plow was designed to leave the field "clean"; that is, free of surface residues. In contrast, conservation tillage systems, such as **chisel plowing** (Figure 17.20 *right*), stir the soil but only partially incorporate surface residues, leaving more than 30% of the soil covered. **Stubble mulching**, whose water-conserving attributes were highlighted in Section 6.4, is another example. **Ridge tillage** is a conservation system in which crops are planted on top of permanent 15- to 20-cm-high ridges. About 30% soil coverage is maintained, even though the ridges are scraped off a bit for planting and then built up again by shallow tillage to control weeds.

FIGURE 17.20   Conventional inversion tillage and conservation tillage in action. (*Left*) In conventional tillage, a moldboard plow inverts the upper soil horizon, burying all plant residues and producing a bare soil surface. (*Right*) A chisel plow, one type of conservation tillage implement, stirs the soil but leaves a good deal of the crop residues on the soil surface. (Photos courtesy of R. Weil)

FIGURE 17.21 In no-till systems, one crop is planted directly into the residue of a cover crop or of a previous cash crop, with only a narrow band of soil disturbed. No-till systems leave virtually all of the residue on the soil surface, providing up to 100% cover and nearly eliminating erosion losses. Here corn was planted into a cover crop killed with a herbicide (weed-killing chemical) to form a surface mulch. The inset shows a closeup of the no-till planter in action (direction of travel is to the right). The rolling furrow openers cut a slot through the residue and soil into which the seed is placed at a depth set by the depth wheel. Snug seed-to-soil contact is ensured by the press wheel that closes the slot. (Photos courtesy of R. Weil)

With **no-tillage** systems we can expect 50 to 100% of the surface to remain covered (Table 17.8). Well-managed continuous no-till systems in humid regions include cover crops during the winter and high-residue-producing crops in the rotation. Such systems keep the soil completely covered at all times and build up organic surface layers somewhat like those found in forested soils.

No-till without herbicides. Rodale Research Institute: http://www.newfarm.org/depts/NFfield_trials/1103/notillroller.shtml

Conservation tillage systems generally provide yields equal to or greater than those from conventional tillage, provided the soil is not poorly drained and in a cool region. However, during the transition from conventional tillage to no-tillage, crop yields may decline slightly for several years for reasons associated with some of the effects outlined in the following subsections.

Adaptation by Farmers[10]

Video on no-till for sustainable rural development: http://info.worldbank.org/etools/bspan/PresentationView.asp?PID=665&EID=339

In recent years conservation tillage has become increasingly popular, being used on nearly two-fifths of the nation's cropland. Conservationists project that as much as 60% of the cropland in the United States will be managed with some kind of conservation

[10] For the inspiring story of one Chilean farmer's struggle to conquer the forces of erosion and degradation and restore the health of his soil, see Crovetto (1996).

TABLE 17.8 **The Effect of Tillage Systems in Nebraska on the Percentage of Land Surface Covered by Crop Residues**

Note that the conventional moldboard plow system provided essentially no cover, while no-till and planting on a ridge (ridge till) provided best cover.

Tillage system	Number of fields	Percentage of fields with residue cover greater than			
		15%	*20%*	*25%*	*30%*
Moldboard	33	3	0	0	0
Chisel	20	40	15	5	0
Disk	165	40	20	9	4
Field cultivate	13	46	23	0	0
Ridge till (till, plant)	2	100	50	50	0
No-till	3	100	100	100	100
All systems	236	36	18	9	4

From Dickey et al. (1987).

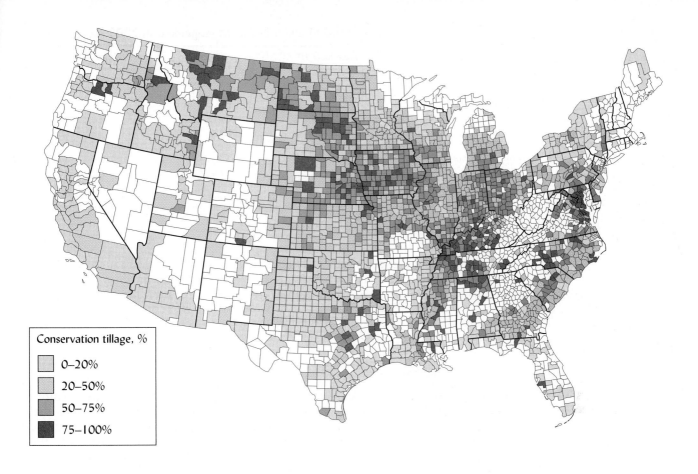

FIGURE 17.22 The percentage of cropland managed with conservation tillage in different counties (local jurisdictions) of the United States in 2004. Note that in some counties more than 3 out of 4 hectares is in conservation tillage. Overall, conservation tillage was used on about 41% of the cropland in the United States. Counties shown in white have less than 4,000 ha of cropland. [Courtesy of Conservation Information Center (CTIC), West Lafayette, IN].

tillage by about 2020. Already, 80% or more of the cropland in a few areas within the United States is managed with conservation tillage (Figure 17.22).

No-tillage systems, especially, have spread to nearly all regions of the USA and are now used in some form on almost half of all the conservation tillage hectares (see Figure 17.23). The no-till system has been used continuously on some farms in the

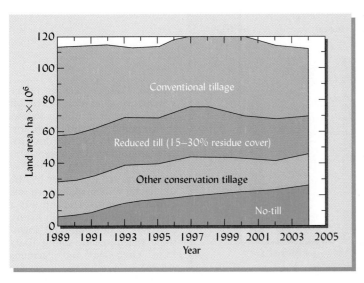

FIGURE 17.23 Adoption of conservation tillage by farmers in the United States. Since the late 1990s about half of U.S. cropland has been managed with conservation tillage practices that keep at least 30% of the soil surface covered by crop residues. About one quarter of U.S. cropland and about half of the conservation tillage area are managed with no-till techniques. Adoption of no-till continues to increase. [Graph courtesy of R. Weil using data from Conservation Technology Information Center]

Conservation tillage in
Zambia:
http://www.fao.org/ag/ags/
agse/agse_s/3ero/namibia1/
til_nam.htm

TABLE 17.9 **Global Area of Cropland under No-till Management in 2005, by Country**

Country	Land area, ha x 10^3	Country	Land area, ha x 10^3
USA	25,300	Venezuela	300
Brazil	21,863	Uruguay	288
Argentina	16,000	France	150
Canada	13,400	Chile	130
Australia	9000	Italy	80
Paraguay	1500	Colombia	70
India & Pakistan	1500	Mexico	50
Bolivia	417	Ghana	45
South Africa	300	Others	1000
Spain	300	Total	91,693

Data from Conservation Tillage Information Center.

eastern United States since about 1970 (more than 30 to 40 years without any tillage). No-tillage and other conservation systems are also being used in other parts of the world (Table 17.9). One of the most significant examples of no-tillage expansion has been in southern Brazil. Thousands of small-scale soybean and corn farmers there have successfully adapted cover-crop-based no-tillage systems using animal traction or small tractors.

Erosion Control by Conservation Tillage

Since conservation tillage systems were initiated, hundreds of field trials have demonstrated that these tillage systems allow much less soil erosion than do conventional tillage methods. Surface runoff is also decreased, although the differences are not as pronounced as with soil erosion (Figure 17.24). These differences are reflected in the much lower C factor values assigned to conservation tillage systems (see Table 17.5).

The erosion-control value of an undisturbed surface residue mulch was discussed in the previous section. Conservation tillage also significantly reduces the loss of nutrients dissolved in runoff water or attached to sediment (review Tables 14.3 and 16.2).

Effect on Soil Properties

When soil management is converted from plow tillage to conservation tillage (especially no-tillage), numerous soil properties are affected, mostly in favorable ways. The changes are most pronounced in the upper few centimeters of soil (Figure 12.25).

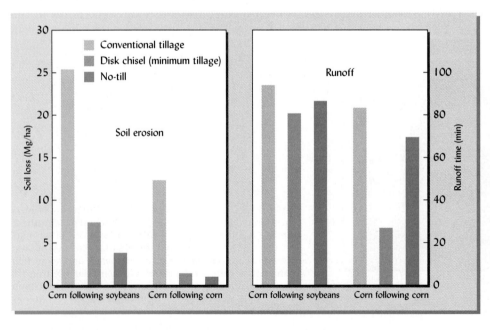

FIGURE 17.24 Effect of tillage systems on soil erosion and runoff from corn plots in Illinois following corn and following soybeans. Soil loss by erosion was dramatically reduced by the conservation tillage practices. The period of runoff was reduced most by the disk chisel system where corn was grown after corn. The soil was a Typic Argiudoll (Catlin silt loam), 5% slope, planted up- and down-slope, tested in early spring. [Data from Oschwald and Siemens (1976)]

Generally, the changes are greatest for systems that produce the most plant residue (especially corn and small grains in humid regions), retain the most residue coverage, and cause the least soil disturbance. Many of these changes are illustrated in other chapters of this textbook, so the discussion here will be brief.

PHYSICAL PROPERTIES. Macroporosity and aggregation (see Sections 4.5 and 4.6) are increased as active organic matter builds up and earthworms and other organisms establish themselves. Infiltration and internal drainage are generally improved, as is soil water-holding capacity. Some of these effects are illustrated in Figure 17.25. The enhanced infiltration capacity of no-till-managed soils is generally quite desirable, but in some cases it may lead to more rapid leaching of nitrates and other water-soluble chemicals. Residue-covered soils are generally cooler and more moist (see Sections 6.5 and 7.11). This is an advantage in the hot part of the year, but may be detrimental to early crop growth in the cool spring of temperate regions.

In cool regions, soils with restricted drainage may yield somewhat less using conservation tillage, because soil conditions are wetter and cooler than with conventional tillage. Reduced yields have discouraged the adoption of conservation tillage in these regions. However, limited preplanting tillage over the crop row (see Section 7.11), or ridge tillage are conservation tillage systems that allow at least part of the soil to warm faster and largely overcome these problems.

CHEMICAL PROPERTIES. No-tillage systems significantly increase the organic matter content of the upper few centimeters of soil (see Figures 17.25 and 12.25). During the initial four to six years of no-till management, the buildup of organic matter results in the immobilization of nutrients (see Section 12.3), especially nitrogen. This is in contrast to the mineralization of nutrients that is encouraged by the decline of soil organic matter under conventional tillage. Eventually, when soil organic matter stabilizes at a new higher level, nutrient mineralization rates under no-till increase. Higher moisture and lower oxygen levels may also stimulate denitrification (see Section 13.9). These processes sometimes result in the need for greater levels of nitrogen fertilization for optimum yields during the early years of no-till management.

In no-tillage systems, nutrient elements tend to accumulate in the upper few centimeters of soil as they are added to the soil surface in crop residues, animal manures, chemical fertilizers, and lime. However, research indicates that because of the surface mulch, crop roots (like those of trees in the untilled forest soil environment) have no trouble obtaining nutrients from the near-surface soil layers. The stratification must be taken into account in sampling soils for fertility testing (see Section 16.11).

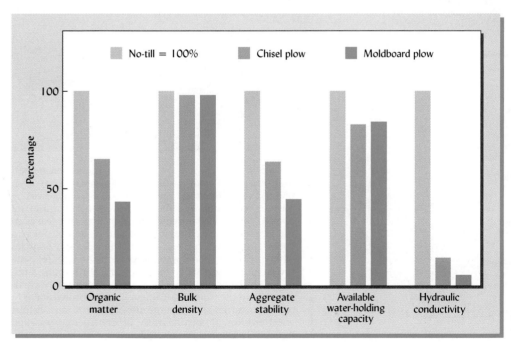

FIGURE 17.25 The comparative effects of 28 years of three tillage systems on soil organic matter content and a number of soil physical properties of an Alfisol surface soil in Ohio. Values for the no-till system were taken as 100, and the others are shown in comparison. Bulk density was about the same for each tillage system, but for all other properties the no-till system was decidedly more beneficial than either of the other two systems. The saturated hydraulic conductivity was especially high with no-till. [From Mahboubi et al. (1993); used with permission of the Soil Science Society of America]

Without tillage to mix the soil, the acidifying effects of nitrogen oxidation, residue decomposition, and rainfall are concentrated in the upper few centimeters of soil, the pH of which may drop more rapidly than that of the whole plow layer in conventional systems (see Section 9.6). In humid regions this acidity must be countered by application of liming materials to the soil.

BIOLOGICAL EFFECTS. The abundance, activity, and diversity of soil organisms tend to be greatest in conservation tillage systems characterized by high levels of surface residue and little physical soil disturbance (see Section 11.14). Earthworms and fungi, both important for soil structure, are especially favored (see Table 11.11). However, organic residues left on the surface in no-till are actually more slowly decomposed than those incorporated by conventional tillage. No-till residues are in less intimate contact with the soil particles so their breakdown is delayed, and they remain as a protective surface barrier for a longer period of time.

17.7 VEGETATIVE BARRIERS

Narrow rows of permanent vegetation (usually grasses or shrubs) planted on the contour can be used to slow down runoff, trap sediment, and eventually build up "natural" or "living" terraces (Figure 17.26). In some situations, tropical grasses (e.g., a deep-rooted, drought-tolerant species called *vetiver grass*) have shown considerable promise as an affordable alternative to the construction of terraces.

FIGURE 17.26 The use of vegetative barriers to create natural terraces. (*Photo*) A tropical grass (vetiver) has been planted vegetatively on the contour in a cassava field by pushing root sprigs into the soil. (*a*) The root cuttings are planted perpendicular to the slope direction. In a year or so the grass will be well established and its dense root and shoot growth will serve as a barrier to hold soil particles while permitting some water to pass on through. (*b*) Note the buildup of soil above the grass, basically forming a terrace wall. Perennial tall wheatgrass is being used experimentally in the Northern Plains area of the United States to serve as a barrier against snow movement and later against wind erosion. (Photo courtesy of Centro Internacional Agricultura Tropical in Cali, Colombia)

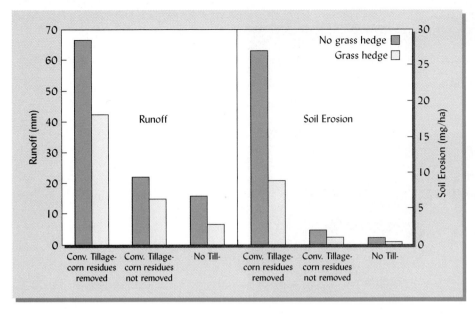

FIGURE 17.27 Narrow grass hedges can be effective tools to reduce losses of runoff water and soil. Switchgrass hedges 0.72 m wide were established across the slope every 16 m in a field of continuous corn that then received a total of 120 mm of simulated rainfall. The soil was a Typic Hapludoll in Iowa with an average slope of 12%. Corn residues left on the soil surface as well as no-till systems also drastically reduced losses, especially of the soil. [Data estimated from a figure in Gilley et al. (2000)]

The deep-rooted vetiver plants have dense, stiff stems that tend to filter out soil particles from the muddy runoff. This sediment builds up on the upslope side of the grass barrier and, in time, actually creates a terrace that may be more than 1 m above the soil surface on the downslope side of the plants. Vetiver grass is particularly well suited to survive under harsh conditions, as its root system can forage deeply for water and its foliage is not palatable to wandering cattle. Other grasses are used that can serve a dual purpose as cattle fodder, but these tend to take more water out of the root zone of the adjacent food crops.

Research has shown that narrow grass hedges can be effective in reducing runoff and erosion from soils in the U.S. Midwest (Figure 17.27). Also, research is being conducted to evaluate numerous systems of vegetative barriers that combine grasses with trees especially in some tropical countries. Such systems may provide many benefits (fruit, firewood, fodder, and nutrient-rich mulch) in addition to erosion control, making them more attractive to farmers who must invest their limited labor and resources in establishing such conservation practices. In temperate regions, conservation technologies (like the winter cover crops discussed in Section 16.3) will improve crop production in the short term as well as protect soil and water quality in the longer term. The benefits of reduced erosion and increased crop yields resulting from several vegetation erosion-control practices are summarized in Table 17.10.

17.8 CONTROL OF GULLY EROSION AND MASS WASTING

Gullies rarely form in soils protected by healthy, dense, forest or sod vegetation, but are common on deserts, rangeland, and open woodland in which the soil is only partially covered. Gullies also readily form in soils exposed by tillage or grading if small rills are allowed to coalesce so that running water eats into the land (Figure 17.28, top). Water concentrated by poorly designed roads and trails may cause gullies to form even in dense forests. In many cases, neglected gullies will continue to grow and after a few

TABLE 17.10 Range in the Effects of Mulching, Contour Cultivation, and Grass Contour Hedges on Soil Erosion and Crop Yields

Practice	Reduction in soil erosion, %	Increase in crop yield, %
Mulching	78–98	7–188
Contour cultivation	50–86	6–66
Grass contour hedges	40–70	38–73

From a review of more than 200 studies, from Doolette and Smyle (1990).

FIGURE 17.28 The devastation of gully erosion. (*Upper*) Gully erosion in action on a highly erodible soil in western Tennessee. The roots of the small wheat plants are powerless to prevent the cutting action of the concentrated water flow. (*Lower*) The legacy of neglect of human-induced accelerated erosion. Tillage of sloping soils during the days of the Roman Empire began a process of accelerated erosion that eventually turned swales into jagged gullies that continue to cut into this Italian landscape with each heavy rain. For a sense of scale, note the olive trees and houses on the grassy, gentle slopes of the relatively uneroded hilltops. (Upper photo courtesy of the USDA Natural Resources Conservation Service; lower photo courtesy of R. Weil)

years, devastate the landscape (see Figure 17.28, *bottom*). On the other hand, in some stony soils coarse fragments left behind in the channel bottom may protect it from further cutting action.

Remedial Treatment of Gullies

If small enough, gullies can be filled in, shaped for smooth water flow, sown to grass, and thereafter be left undisturbed to serve as grassed waterways. When the gully erosion is too active to be checked in this manner, more extensive treatment may be required. If the gully is still small, a series of check dams about 0.5 m high may be constructed at intervals of 4 to 9 m, depending on the slope. These small dams may be constructed from materials available on site, such as large rocks, rotted hay bales, brush, or logs. Wire netting may be used to stabilize these structures. Check dams, whether large or small, should be constructed with the general features illustrated in Figure 17.29. After a time, enough sediment may collect behind the dams to form a series of bench terraces and the ditch may be filled in and put into permanent sod.

With very large gullies, it may be necessary to divert the runoff away from the head of the channel and install more permanent dams of earth, concrete, or stone in the

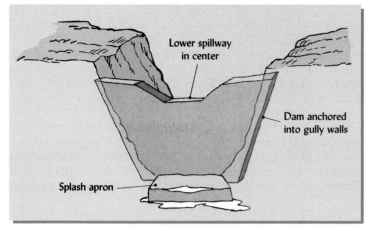

FIGURE 17.29 A schematic drawing of a check dam used to arrest gully erosion. Whether made from rock, brush, concrete, or other materials, a check dam should have the general features shown. The structure should be dug into the walls of the gully to prevent water from going around it. The center of the dam should be lower so that water will spill over there and not wash out the soil of the gully walls. An erosion-resistant apron made from densely bundled brush, concrete, large rocks, or similar material should be installed beneath the center of the dam to prevent the overflow from undercutting the structure. In contrast to the gully-healing effect of a well-designed check dam, the haphazard dumping of rocks, brush, or junked cars into a gully will make matters worse, not better. (Diagram courtesy of R. Weil)

channel itself. Again, sediment deposited above the dams will slowly fill in the gully. Semipermanent check dams, flumes, and riprap-lined channels are also used on construction sites, but are generally too expensive for extensive use on agricultural land.

Mass Wasting on Unstable Slopes

Liquefaction of saturated soils:

www.ce.washington.edu/~liquefaction/html/what/what1.html

The downhill movement of large masses of unstable soil (**mass wasting**) is quite different from the erosion of the soil surface, which is the main topic of this chapter. Mass wasting can be a problem on very steep slopes (usually greater than 60% slope). While this type of soil loss sometimes occurs on steep pastures, it is most common on nonagricultural land. Mass wasting can take several forms. **Soil creep** (Plate 46) is the slow deformation (without shear failure) of the soil profile as the upper layers move imperceptibly downhill. **Landslides** occur with the sudden shear failure and downhill movement of a mass of soil, usually under very wet conditions (Plates 41 and 42). **Mud flows** involve the partial liquefaction and flow of saturated soil due to loss of cohesion between particles (see also Section 4.9 and Figure 4.49).

Mass wasting is sometimes triggered by human activities that undermine natural stabilizing forces or cause the soil to become water saturated as a result of concentrated water flow. The rotting of large soil-anchoring tree roots several years after clear-cutting a forest or the construction of a road cut at the toe of a steep slope are all-too-common examples.

17.9 CONTROL OF ACCELERATED EROSION ON RANGE- AND FORESTLAND

Rangeland Problems

Many semiarid rangelands lose large amounts of soil under natural conditions (see Figure 17.4), but accelerated erosion can lead to even greater losses if human influences are not carefully managed. Overgrazing, which leads to the deterioration of the vegetative cover on rangelands, is a prime example. Grass cover generally protects the soil better than the scattered shrubs that usually replace it under the influence of poorly managed livestock grazing. In addition, cattle congregating around poorly distributed water sources and salt licks may completely denude the soil. Cattle trails, as well as ruts from off-road vehicles, can channelize runoff water and spawn gullies that eat into the landscape. Because of the prevalence of dry conditions, wind erosion (to be discussed in Sections 17.11–17.12), also plays a major role in the deterioration of rangeland soils.

Erosion on Forestlands

In contrast to deserts and rangelands, land under healthy, undisturbed forests lose very small amounts of soil. However, accelerated erosion can be a serious problem on forested land, both because the rates of soil loss may be quite high and because the amount of land involved is often enormous. The main cause of accelerated erosion in forested watersheds is usually the construction of logging roads, timber-harvest operations, and the trampling of trails and off-trail areas by large numbers of recreational users (or cattle, in some areas).

To understand and correct these problems, it is necessary to realize that the secret of low natural erosion from forested land is the undisturbed forest floor, the O horizons that protect the soil from the impact of raindrops and allow such high infiltration rates that surface runoff is very small or absent. Contrary to the common perception, it is the forest floor, rather than the tree canopy or roots, that protects the soil from erosion (Figure 17.30*a*). In fact, rainwater dripping from the leaves of tall trees often forms very large drops that reach terminal velocity and impact the ground with more energy than direct rain from even the most intense of storms. If the forest floor has been disturbed and mineral soil exposed, serious splash erosion can result (see Figure 17.30*b*). Gully erosion can also occur under the forest canopy if water is concentrated, as by poorly designed roads.

Practices to Reduce Soil Loss Caused by Timber Production[11]

Forest Service WEPP Interfaces for roads, forests, etc.:
http://forest.moscowfsl.wsu.edu/fswepp/

The main sources of eroded soil from timber production are *logging roads* (that are built to provide access to the area by trucks), *skid trails* (the paths along which logs are dragged), and *yarding areas* (the areas where collected logs are sized and loaded onto trucks). Relatively little erosion results directly from the mere felling of the trees (except where large tree roots are needed to anchor the soil against mass wasting, as discussed in Section 17.8). Strategies to control erosion should include consideration of (1) intensity of timber harvest, (2) methods used to remove logs, (3) scheduling of timber harvests, and (4) design and management of roads and trails. Soil disturbance in preparation for tree regeneration (such as tillage to eliminate weed competition or provide better seed-to-soil contact) must also be limited to sites with low susceptibility to erosion.

INTENSITY OF TIMBER HARVEST. On the steepest, most erodible sites, environmental stewardship may require that timber harvest be foregone and the land be given only protective management. On somewhat less susceptible sites, selective cutting (occasional removal of only the oldest trees) may be practiced without detrimental results. The shelterwood system (in which a substantial number of large trees are left standing after all others are harvested) is probably the next most intensive method and can be used on moderately

---

[11] For general introduction to this topic, see Nyland (1996). For an analysis of erosion causes on steep forest lands in the tropics, see (Sidle et al. 2006).

(a)

(b)

FIGURE 17.30 The leaf mulch on the forest floor, rather than the tree roots or canopy, provides most of the protection against erosion in a wooded ecosystem. (*a*) An undisturbed temperate deciduous forest floor (as seen through a rotten stump). The leafless canopy will do little to intercept rain during winter months. During the summer, rainwater dripping from the foliage of tall trees may impact the forest floor with as much energy as unimpeded rain. (*b*) Severe erosion has taken place under the tree canopy in a wooded area where the protective forest floor has been destroyed by foot traffic. The exposed tree roots indicate that nearly 25 cm of the soil profile has washed away. (Photos courtesy of R. Weil)

susceptible sites. Clear-cutting (removal of all trees from large blocks of forest) should be used only on gentle slopes with stable soils.

METHOD OF TREE REMOVAL. The least expensive and most commonly used method of tree removal is by wheeled tractors called *skidders* (see Figure 4.37). This method generally disrupts the forest floor, exposing the mineral soil on perhaps 30 to 50% of the harvested area. In contrast, more expensive methods using cables to lift one end of the log off the ground are likely to expose mineral soil on only 15 to 25% of the area. Occasionally, for very sensitive sites, logs are lifted to yarding areas by balloon or helicopter, practices which are very expensive but result in as little as 4 to 8% bare mineral soil.

SCHEDULING OF TIMBER HARVEST. Much erosion can be prevented by limiting entry into the forest by machinery to those periods when the soil is either dry or (in temperate regions) frozen and covered with snow. Damage to the forest floor (including both compaction and exposure of mineral soil) occurs much more readily if the soil is very wet. In addition, wheel ruts easily form in wet soils and channel runoff to initiate gully erosion.

DESIGN AND MANAGEMENT OF ROADS. Poorly built logging roads may lose as much as 100 Mg/ha of soil by erosion of the road surface, the drainage ditch walls, or the soil exposed by road cuts into the hillside. Roads also collect and channelize large volumes of water, which can cause severe gullying. Roads should be so aligned as to avoid these problems. Although expensive, placing gravel on the road surface, lining the ditches with rocks, and planting perennial vegetation on exposed road cuts can eliminate up to 99% of the soil loss. A much less expensive measure is to provide cross channels (shallow ditches or **water bars**, as shown in Figure 17.31 *right*) every 25 to 100 m to prevent excessive accumulation of water and safely spread it out onto areas protected by natural vegetation. After timber harvest is complete, the roads in an area should be grassed over and closed to traffic.

DESIGN OF SKIDDING TRAILS. Skidding trails that lead runoff water downhill toward a yarding area invite the formation of gullies. Repeated trips dragging logs over the same secondary trails also greatly increase the amount of mineral soil exposed to erosive forces. Both practices should be avoided, and yarding areas should be located on the highest-elevation, most level, and well-drained areas available. (Figure 17.31, *left*).

FIGURE 17.31 Two important forestry practices designed to minimize damage from erosion due to timber harvests. (*Left*) An aerial view of clear-cut and unharvested block of pine forest in Alabama. Narrow skid trails can be seen that lead *up* to a staging area located on high ground and spread apart going downhill toward the streams. This is in contrast to the common and easier practice of dragging logs downhill so that skid trails converge at a low point, inviting runoff water to concentrate into gully-cutting torrents. Also visible are several dark-colored buffer strips where the trees were left undisturbed along streams to provide protection for water quality. (*Right*) An open-top culvert (also called a water bar) in a well-designed logging road in Montana. This simple structure, along with proper road bed alignment, can greatly reduce gully erosion caused by water flowing unimpeded along roads and trails in forested areas. Water bars placed at frequent intervals lead runoff water, a little at a time, off the road and into densely vegetated areas. (Photos courtesy of R. Weil)

BUFFER STRIPS ALONG STREAM CHANNELS. When forests are harvested, buffer strips as wide as 1.5 times the height of the tallest trees should generally be left untouched along all streams (Figure 17.31, *left*). As discussed in Section 16.2, buffer strips of dense vegetation have a high capacity to remove sediment and nutrients from runoff water. Forested buffers also protect the stream from excessive logging debris. In addition, streamside trees shade the water, protecting it from the undesirable heating that would result from exposure to direct sunlight.

17.10 EROSION AND SEDIMENT CONTROL ON CONSTRUCTION SITES

Although active construction sites cover relatively little land in most watersheds, they may still be a major source of eroded sediment because the potential erosion per hectare on drastically disturbed land is commonly 100 times that on agricultural land. Heavy sediment loads are characteristic of rivers draining watersheds in which land use is changing from farm and forest to built-up land. Historically, once urbanization of a watershed is complete (all land being either paved over or covered by well-tended lawns), sedimentation rates return to levels as low (or lower) than before the development took place.

To prevent serious sediment pollution from construction sites, governments in the United States (e.g., through state laws and the federal Clean Water Act of 1992) and in many other industrialized countries require that contractors develop detailed erosion- or sediment-control plans before initiating construction projects that will disturb more than about 1 ha of land. The goals of erosion control on construction sites are (1) to avoid on-site damage, such as undercutting of foundations or finished grades and loss of topsoil needed for eventual landscaping; and (2) to retain eroded sediment on-site so as to avoid all the environmental damages (and liabilities) that would result from deposition of sediment on neighboring land and roads, and in ditches, reservoirs, and streams.

Principles of Erosion Control on Construction Sites

Erosion and sediment control planning for construction sites: http://www.civil.ryerson.ca/stormwater/menu_5/index.htm

Five basic steps are useful in developing plans to meet the aforementioned goals:

1. When possible, schedule the main excavation activities for low-rainfall periods of the year.
2. Divide the project into as many phases as possible, so that only a few small areas must be cleared of vegetation and graded at any one time.
3. Cover disturbed soils as completely as possible, using vegetation or other materials.
4. Control the flow of runoff to move the water safely off the site without destructive gully formation.
5. Trap the sediment before releasing the runoff water off-site.

The last three steps bear further elaboration. They are best implemented as specific practices integrated into an overall erosion-control plan for the site.

Keeping the Disturbed Soil Covered

Soils freshly disturbed by excavation or grading operations are characterized by very high erodibility (K values). This is especially true for low-organic-matter subsoil materials. Potential erosion can be extremely high (200 to 400 Mg/ha is not uncommon) unless the C value is made very low by providing good soil cover. This is best accomplished by allowing the natural vegetation to remain undisturbed for as long as possible, rather than clearing and grading the entire project area at the beginning of construction (see step 1, preceding). Once a section of the site is graded, any sloping areas not directly involved in the construction should be sodded or sown to fast-growing grass species adapted to the soil and climatic conditions (Plate 43 shows erosion on an unprotected road bank).

Seeded areas should be covered with **mulch** or specially manufactured **erosion blankets** (Figure 17.32, *bottom*). Erosion blankets, made of various biodegradable or nonbiodegradable materials, provide instant soil cover and protect the seed from being washed away (Figure 17.33).

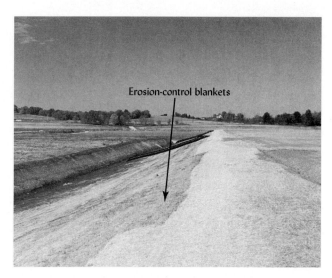

Erosion-control blankets

FIGURE 17.32 Two methods of establishing vegetative cover on steep, unstable slopes. (*Left*) A hydroseeder allows vegetative cover to be efficiently established on difficult-to-reach areas. The machine is spraying a mixture of water, chopped straw, grass seed, fertilizer, and sticky polymers that hold the mulch in place until the grass seed can take root. (*Right*) Erosion-control mats or blankets made of plastic netting or natural materials like jute are laid down over newly seeded grass to hold the seed and soil in place until the vegetative cover is established. (Photos courtesy of R. Weil)

A commonly used technology to protect steep slopes and areas difficult to access, such as road cuts, is the **hydroseeder** (Figure 17.32, *top*) that sprays out a mixture of seed, fertilizer, lime (if needed), mulching material, and sticky polymers. Good construction-site management includes removal and stockpiling of the A-horizon material before an area is graded (see Figure 1.16). This soil material is often quite high in fertility and is a potential source of sediment and nutrient pollution. The stockpile should therefore be given a grass cover to protect it from erosion until it is used to provide topsoil for landscaping around the finished structures.

Controlling the Runoff

Freshly exposed and disturbed subsoil material is highly susceptible to the cutting action of flowing water. The gullies so formed may ruin a grading job, undercut pavements and foundations, and produce enormous sediment loads. The flow of runoff

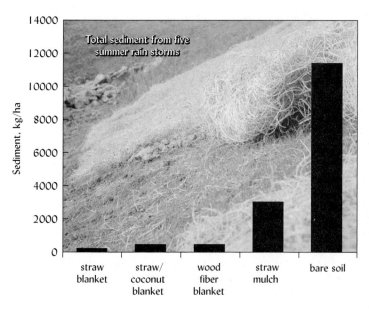

FIGURE 17.33 Total sediment generated by five summer storms on construction site soil left bare or protected by straw mulch or various types of commercial erosion blankets. The clayey soil met the State Department of Transportation specifications for topsoil. It was raked, fertilized, and seeded to grass mixture. The experimental plots had a 35% slope gradient and were 9.75 m long. Immediately after seeding, the soils were covered by the various erosion-control materials. The best grass vegetative cover and biomass was achieved on straw mulch plots in the first year. The best short-term sediment control was achieved by the commercial blankets, such as the wood-fiber blanket shown in the background photo. [Drawn from data in Benik et al. (2003); photo courtesy of R. Weil]

FIGURE 17.34 The flow of runoff from large areas of bare soil must be carefully controlled if off-site pollution is to be avoided. Here, a carefully designed channel with a grass sod bottom and sides lined with large rocks (riprap) prevents gully erosion, reduces soil loss, and guides runoff around the perimeter of a construction site. (Photo courtesy of R. Weil)

grass sod

rip rap

water must be controlled by carefully planned grading, terracing, and channel construction. Most construction sites require a perimeter waterway to catch runoff before it leaves the site and to channel it to a retention basin.

The sides and bottom of such channels must be covered with "armor" to withstand the cutting force of flowing water. Where high water velocities are expected, the soil must be protected with **hard armor** such as **riprap** (large angular rocks, such as are shown in Figure 17.34), **gabions** (rectangular wire-mesh containers filled with hand-sized stone), or interlocking concrete blocks. The soil is first covered with a **geotextile** filter cloth (a tough nonwoven material) to prevent mixing of the soil into the rock or stone.

In smaller channels, and on more gentle slopes where relatively low water velocities will be encountered, **soft armor**, such as grass sod or erosion blankets, can be used. Generally, soft armor is cheaper and more aesthetically appealing than hard armor. Newer approaches to erosion control often involve reinforced vegetation (e.g., trees or grasses planted in openings between concrete blocks or in tough erosion mats).

The term **bioengineering** describes techniques that use vegetation (locally native, noninvasive species are preferred) and natural biodegradable materials to protect channels subject to rather high water velocities. Examples include the use of **brush mattresses** to stabilize steep slopes. In this technique, live tree branches are tightly bundled together, staked down flat using long wooden pegs, and partially covered with soil. The so-called **live stake** technique (Figure 17.35) is another example of a bioengineering approach commonly used to stabilize soil along channels subject to high velocity water. In both cases, the soil is provided some immediate physical protection from scouring water, and eventually the dormant cuttings take root to provide permanent, deep-rooted vegetative protection.

Trapping the Sediment

For small areas of disturbed soils, several forms of sediment barriers can be used to filter the runoff before it is released. The most commonly used types of silt barriers are straw bales and woven fabric silt fences. If installed properly, both can effectively slow the water flow so that most of the sediment is deposited on the uphill side of the barrier (Figure 17.36), while relatively clear water passes through.

On large construction sites, a system of protected slopes and channels leads storm runoff water to one or more retention and sedimentation ponds located at the lowest elevation of the site. As the flowing water meets the still water in the pond it drops most of its sediment load (Figure 17.37), allowing the relatively clear water to be skimmed off the top and released to the next pond or off the site. Wetlands (Section 7.7) are often

FIGURE 17.35 An example of *bioengineering* along a stream bed that had been disturbed during the construction of a commercial airport in Illinois. Living willow branches are pounded into the soft erodible stream bank to anchor the soil and reduce the scouring power of the water during high flows (*right*). Eventually the willow branches will take root and sprout (*left*), providing trees that will permanently stabilize the stream bank and improve wildlife habitat. (Photos courtesy of R. Weil)

constructed to help purify the overflow from sedimentation ponds before the water is released into the natural stream or river.

Construction site erosion-control measures are commonly designed to retain the runoff from small storms on-site. The retention ponds must also be able to deal with runoff generated from intense rainstorms—the kind that may be expected to occur on a site only once in every 10 to even 100 years. In designing the capacity of sediment-retention ponds, the erosion models discussed in Section 17.4 are used to estimate the amount of sediment that is likely to be eroded from the site. While expensive to construct, well-designed sediment-retention ponds can be incorporated as permanent aesthetic water features that enhance the value of the final project.

FIGURE 17.36 Several types of sediment-control measures used around the periphery of a construction site. (*Left*) A line of straw bales pegged to the ground allows water to seep through, but filters out much of the sediment and slows down the flow so that the water drops its sediment load. (*Right*) A properly installed silt fence effectively removes sediment from runoff water leaving the edge of a construction site. The silt fencing material is a woven plastic that allows water to flow through at a much reduced rate. Note the light-colored sediment on the inside and the undisturbed forest floor on the outside of the silt fence. Improperly installed silt fencing is useless, as sediment-laden runoff passes underneath. A silt fence must be embedded in the soil to avoid this problem. (Photos courtesy of R. Weil)

FIGURE 17.37 The effect of a small sediment impoundment structure on the retention of suspended soil eroded from a construction site. Note the riprap-lined channel leading runoff water into the pond and the deltalike alluvial fan of sediment deposited as the water's velocity is reduced where it enters the pond. The standpipe in the background has holes that allow the clearest water near the top to overflow and leave the construction site. (Photo courtesy of R. Weil)

17.11 WIND EROSION: IMPORTANCE AND FACTORS AFFECTING IT

Wind erosion in the dustiest place on Earth:
http://www.nature.com/news/2005/050411/full/434816a.html

Desert Wind Erosion Dust Threatens Mountain Snow—(audio report):
http://www.npr.org/templates/story/story.php?storyId=5415308

Up to this point we have focused on soil erosion by water, but wind, too, causes much soil loss. Wind erosion is most common in arid and semiarid regions, and it is a problem on some soils in humid climates as well. It occurs when strong winds blow across soils with relatively dry surface layers. All kinds of soils and soil materials are affected. The finer soil particles may be carried to great heights and for thousands of kilometers—even from one continent, across the ocean, to another.

Wind erosion is a worldwide problem. Large areas of the former Soviet Union and sub-Saharan Africa have been badly damaged by wind erosion. Overgrazing and other misuses of the fragile lands of arid and semiarid areas have allowed wind erosion to depress soil productivity and bring starvation and misery to millions of people in these areas.

In May 1934, a huge dust storm originating in the southern Great Plains region of the United States sent great clouds of silt, clay, and organic matter eastward to the Atlantic seaboard and out over the ocean. It is said that the dust-darkened skies over Washington, D.C., helped convince Congress to fund new, far-reaching programs to combat soil erosion.

Wind erosion causes widespread damage, not only to the vegetation and soils of the eroding site, but also to anything that can be damaged by the abrasiveness of soil-laden wind, and finally to the offsite area where the eroded soil settles back to earth (Figure 17.38). Wind moves about 40% of the soil transported by erosion in the United States. About 12% of the continental United States is somewhat affected by wind erosion—8% moderately so and perhaps 2 to 3% greatly.

In six of the Great Plains states, annual wind erosion exceeds water erosion on cropland, averaging from 4 Mg/ha in Nebraska to 29 Mg/ha in New Mexico, where mismanagement of plowed lands and overgrazing of range grasses have greatly increased the susceptibility of the soils to wind action. In dry years, the results have been most deplorable.

Even in humid regions, certain soils suffer significant wind erosion when their surface layer dries out and wind velocity is high. The movement of sand dunes along the Atlantic coast and on the eastern shore of Lake Michigan are examples. Wind erosion also damages cultivated sandy or peaty soils when conditions are dry. The on-site and off-site damages caused by wind erosion were discussed in Section 17.2.

The erosive force of the wind tends to be much greater in some regions than in others. For example, the semiarid Great Plains region of the United States is subject to winds with 5 to 10 times the erosive force of the winds common in the humid East. In the Great Plains, the winds are most powerful in the winter season. In other regions, high winds occur most commonly during the hot summers.

FIGURE 17.38 Wind erosion in action. (*Upper left*) Much wind erosion comes from dust storms such as this one moving across the High Plains of Texas. The swirling black cloud consists of fine particles eroded from the soil by high winds sweeping across the flat farmland and range. Apparently, most of the land was not as well covered as the wheat field in the foreground. (*Upper right*) Soil eroded by wind during a single dust storm has piled up to a depth of nearly 1 m along a fencerow in Idaho. The existing soil and plants are covered by deposits that are quite unproductive because the soil structure has been destroyed. Also, these deposits are subject to further movement when the wind direction changes. (*Lower left*) Direct wind damage to a tomato crop in a sandy field in Delaware. The tomato plants are buried beneath windblown sandy soil. (*Lower right*) Young tomato fruits show damage incurred by sandblasting during a windstorm. (Upper left photo courtesy of Dr. Chen Weinan, USDA Agricultural Research Service, Warm Springs, TX; other photos courtesy R. Weil)

Mechanics of Wind Erosion

Animation of saltation in action:

http://plantandsoil.unl.edu/ croptechnology2005/soil_sci/ animationOut.cgi?anim_ name=saltation-modd.swf

Like water erosion, wind erosion involves three processes: (1) *detachment,* (2) *transportation,* and (3) *deposition.* The moving air, itself, results in some detachment of tiny soil grains from the granules or clods of which they are a part. However, when the moving air is laden with soil particles, its abrasive power is greatly increased. The impact of these rapidly moving grains dislodges other particles from soil clods and aggregates. These dislodged particles are now ready for one of the three modes of wind-induced transportation, depending mostly on their size.

SALTATION. The first and most important mode of particle transportation is that of **saltation,** or the movement of soil by a series of short bounces along the ground surface (Figure 17.39). The particles remain fairly close to the ground as they bounce, seldom rising more than 30 cm or so. Depending on conditions, this process may account for 50 to 90% of the total movement of soil.

SOIL CREEP. Saltation also encourages **soil creep,** or the rolling and sliding along the surface of the larger particles. The bouncing particles carried by saltation strike the large aggregates and speed up their movement along the surface. Soil creep accounts for the movement of particles up to about 1.0 mm in diameter, which may amount to 5 to 25% of the total movement.

SUSPENSION. The most spectacular method of transporting soil particles is by movement in **suspension.** Here, dust particles of a fine-sand size and smaller are moved parallel to the ground surface and upward. Although some of them are carried at a height no

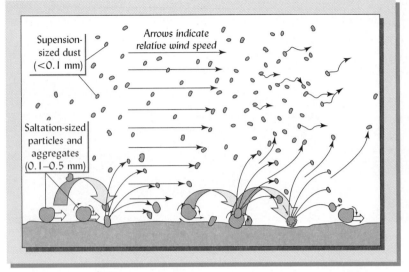

FIGURE 17.39 How particles move during wind erosion. As indicated by the straight arrows, the wind is blowing from left to right and is slowed somewhat by friction and obstructions near the soil surface. Fine particles are picked up from the soil surface and carried into the atmosphere, where they remain suspended until the wind velocity is reduced. Medium-sized particles or aggregates, being too large to be carried up in suspension, are bounced along the soil surface. When they strike larger soil aggregates, they break up to release particles of various sizes. The finer particles move in suspension up into the air above, and the medium-sized particles continue to bounce along the soil surface. This process of particle movement stimulated by medium-sized particles skipping along the surface is termed *saltation*. (Diagram courtesy of R. Weil)

greater than a few meters, the turbulent action of the wind results in others being carried kilometers upward into the atmosphere and many hundreds of kilometers horizontally. These particles return to the earth only when the wind subsides and/or when precipitation washes them down. Although it is the most striking manner of transportation, suspension seldom accounts for more than 40% of the total and is generally no more than about 15%.

Factors Affecting Wind Erosion

Wet soils do not blow because of the adhesion between water and soil particles. Dry winds generally lower the moisture content to below the wilting point before wind erosion takes place. Other factors that influence wind erosion are (1) wind velocity and turbulence, (2) soil surface conditions, (3) soil characteristics, and (4) the nature and orientation of the vegetation.

WIND VELOCITY AND TURBULENCE.   The rate of wind movement, especially gusts having greater than average velocity, will influence erosion. The *threshold velocity*—the wind speed required to initiate soil movement—is usually about 25 km/h (7 m/s) (Table 17.11). At higher wind speeds, soil movement is proportional to the cube of the wind velocity. Thus, the quantity of soil carried by wind increases dramatically as speeds above 30 km/h are reached. Wind turbulence also influences the capacity of the atmosphere to transport matter. Although the wind itself has some direct influence in picking up fine soil, the impact of wind-carried particles as they strike the soil is probably more important.

SURFACE ROUGHNESS.   Wind erosion is less severe where the soil surface is rough. This roughness can be obtained by proper tillage methods, which create large clods or ridges. Leaving a stubble mulch (see Section 6.4) is an even more effective way of reducing wind-borne soil losses.

SOIL PROPERTIES.   In addition to moisture content, wind erosion is also influenced by (1) mechanical stability of soil clods and aggregates, (2) stability of soil crusts, (3) bulk density and (4) size of erodible soil fractions. Some clods resist the abrasive action of wind-carried particles. If a soil crust resulting from a previous rain is present, it, too, may be able to withstand the wind's erosive power. The presence of clay, organic matter, and other cementing agents is also important in helping clods and aggregates resist abrasion. This is one reason why sandy soils, which are low in such agents, are so easily eroded by wind. Because they participate in saltation, soil particles or aggregates about 0.1 mm in diameter are more erodible than those larger or smaller in size.

TABLE 17.11   Threshold Wind Velocity, Height of Transition from Saltation to Suspension, and Mass of Soil Carried by Creep, Saltation, and Suspension During Four Wind Erosion Episodes on Cropland at Big Spring, Texas

Note that saltation accounted for most of the soil material moved, and that this movement mostly took place at heights of less than 30 cm above the soil surface. The location is semiarid, receiving an average of 470 mm of rainfall annually. The soil was a bare, smooth field of Amarillo fine sandy loam (Ustalfs).

Date	Threshold wind velocity, m/s	Height of transition from saltation to suspension, cm	Creep	Saltation	Suspension	Total
			Mass of soil material moved per width of wind path, kg/m			
22 January	6.8	22	0.87	13.33	0.72	14.92
5 February	7.0	22	7.68	142.9	13.43	164.01
14 March	7.2	28	0.43	15.9	3.98	20.31
21 March	7.8	27	0.10	2.41	0.30	2.80

Selected from Fryrear and Saleh (1993).

VEGETATION.   Vegetation or stubble mulch will reduce wind erosion hazards, especially if the rows run perpendicular to the prevailing wind direction. This effectively slows wind movement near the soil surface. In addition, plant roots help bind the soil and make it less susceptible to wind damage.

17.12   PREDICTING AND CONTROLLING WIND EROSION

A wind erosion prediction equation (WEQ) has been in use since the late 1960s:

$$E = f(I \times C \times K \times L \times V) \qquad \text{(eq. 17.2)}$$

The predicted wind erosion E is a function f of:

I = soil erodibility factor
C = climate factor
K = soil-ridge-roughness factor
L = width of field factor
V = vegetative cover factor

The WEQ involves the major factors that determine the severity of the erosion, but it also considers how these factors interact with each other. Consequently, it is not as simple as is the USLE for water erosion. Evidence of the interaction among factors is seen in Figure 17.40. The **soil erodibility factor** I relates to the properties of the soil and to the degree of slope of the site in question. The **soil-ridge-roughness** factor K takes into consideration the cloddiness of the soil surface, vegetative cover V, and ridges on the soil surface. The **climatic factor** C involves wind velocity, soil temperature, and precipitation (which helps control soil moisture). The **width of field factor** L is the width of a field in the downwind direction. Naturally, the width changes as the direction of the wind changes, so the prevailing wind direction is generally used. The **vegetative cover** V relates not only to the degree of soil surface covered with residues, but to the nature of the cover—whether it is living or dead, still standing, or flat on the ground.

A revised, more complex, and more accurate computer-based prediction model has been developed, and is known as the **revised wind erosion equation (RWEQ)**. It is still an empirical model based on many years of research to characterize the relationship between observable conditions and resulting wind erosion severity. Table 17.12 outlines the factors taken into consideration by RWEQ, which (like RUSLE) calculates the erosion hazard during 15-day intervals throughout the year. For each interval of time, the RWEQ makes adjustments in the residue, soil erodibility, and soil roughness parameters, based on the input information about management operations and weather conditions. For example, it assumes that residues decompose over time, that tillage operations flatten standing residues, and that rainfall shakes clods to reduce soil roughness.

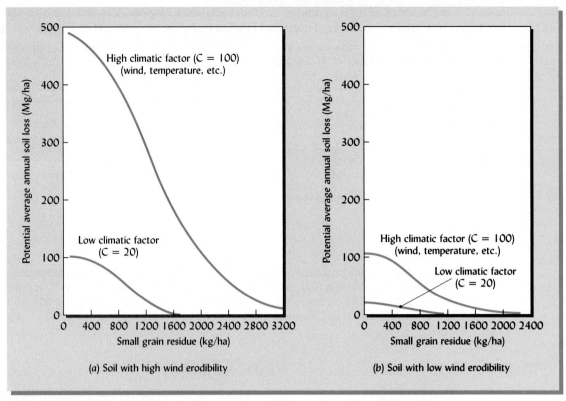

FIGURE 17.40 Effect of small grain residues on the potential wind erosion of soils with high (*a*) and low (*b*) erodibility. Strong, dry winds and high temperatures encourage erosion, especially on the soil with high wind-erodibility characteristics. Surface residues can be used to help control this wind erosion. [From Skidmore and Siddoway (1978); used with permission of the American Society of Agronomy]

TABLE 17.12 Some Factors Integrated in the Revised Wind Erosion Equation (RWEQ) Model

The RWEQ program calculates values for each factor for each 15-day period during the year. It also calculates interactions; that is, it uses the value of one factor to modify other factors.

Model factor	Subfactor terms in the model	Comments
Weather factor	Weather factor *WF*; includes terms for wind velocity, direction, air temperature, solar radiation, rainfall, and snow cover.	Modified by other factors such as soil wetness.
Soil factors	Erodible fraction *EF*; fraction smaller than 0.84 mm diameter.	Based on sand, silt, organic matter, and rock cover.
	Soil crust factor *SCF*.	Induced by rainfall, eliminated by tillage.
	Surface roughness *SR*; a random component due to clods and/or an oriented component due to ridges.	Interacts with rainfall and tillage.
	Soil wetness *SW*.	Computed from rainfall minus evapotranspiration.
Tillage factor	Tillage factor *TF*; depends on type of implement, soil conditions, timing, and so forth.	Modifies surface roughness, crust factor, and so forth.
Hill factor	Hill factor *HF*; slope gradients and length input.	Affects wind speed (high going upslope, lower going down).
Irrigation factor	Irrigation factor *IR*.	Equivalent to added rainfall.
Crops factor	Flat residue SLR_f factor.	Soil cover by residues lying on the surface is estimated for each crop, including changes over time due to decay, and so forth.
	Standing residue SLR_s; depends on crop, harvest height, plant density, and so forth.	Standing residues reduce the wind speed at the soil surface. Decay is slower than for flat residues.
	Crop canopy factor SLR_c.	Changes daily with crop growth.
Barriers	Barriers (e.g., tree windbreaks); includes orientation, density, height, spacing.	Reduce leeward wind velocity.

Based on information in Fryrear et al. (2000).

Scientists and engineers around the world are also cooperating in the development of a much more complex process-based model known as the **Wind Erosion Prediction System (WEPS)**. Like its water erosion sister, WEPP, this computer program simulates all the basic processes of wind interaction with soil. Scientists are continually improving the model and testing its predictions against data observed in the real world. The USDA has plans to make this model available in a version that can be integrated with WEPP in a user-friendly format. However, since to run both of these complex process-based models one must supply a great deal of information about every aspect of the site on which erosion is to be predicted, it remains to be seen whether they will ever become as widely used as the simpler, tried-and-true empirical models now in widespread use.

Control of Wind Erosion

SOIL MOISTURE.   The factors of the wind erosion equation give clues to methods of reducing wind erosion. For example, since soil moisture increases cohesiveness, the wind speed required to detach soil particles increases dramatically as soil moisture increases. Therefore, where irrigation water is available, it is common to moisten the soil surface when high winds are predicted (Figure 17.41). Unfortunately, most wind erosion occurs in dry regions without available irrigation. A vegetative cover also discourages soil blowing, especially if the plant roots are well established. In dry-farming areas, however, sound moisture-conserving practices require summer fallow on some of the land, and hot, dry winds reduce the moisture in the soil surface. Consequently, other means must be employed on cultivated lands of these areas.

TILLAGE.   Certain conservation tillage practices described in Section 17.6 were used for wind erosion control long before they became popular as water erosion control practices. Keeping the soil surface rough and maintaining some vegetative cover is accomplished by using appropriate tillage practices. However, the vegetation should be well anchored into the soil to prevent it from blowing away. Stubble mulch has proven to be effective for this purpose (see Section 6.4).

The effect of tillage depends not only on the type of implement used, but also on the timing of the tillage operation. Tillage can greatly reduce wind erosion if it is done while there is sufficient soil water to cause large clods to form. Tillage on a dry soil may produce a fine, dusty surface that aggravates the erosion problem. Tillage to provide for a cloddy surface condition should be at right angles to the prevailing winds. Likewise,

FIGURE 17.41 Wind erosion control in an area of productive Histosols (Saprists) in central Michigan. Prior to being cleared and drained, this area was a partially forested bog. When dry, cultivated Histosols are very light and fluffy and susceptible to wind erosion. The rows of trees (mainly willows) were planted perpendicular to the prevailing winds to slow the wind velocity and protect these valuable organic soils from erosion. Wetting the soil surface is another effective means of reducing wind erosion, as seen by the darker-colored field in the background (where the water table was raised) and the darker circles in the left foreground where sprinkler irrigation was used. Note that the photo was taken in early spring before most crops were planted and before the trees had fully leafed out. (Photo courtesy of USDA Natural Resources Conservation Service)

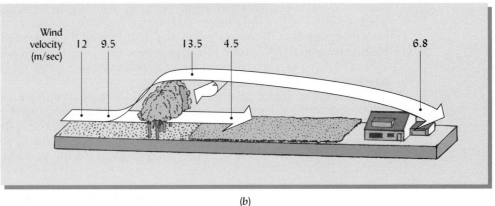

(b)

FIGURE 17.42 Wind breaks to reduce wind erosion. (a) Trees make good windbreaks and add beauty to a North Dakota farm home-stead. Note newly planted rows of trees in foreground. (b) The effect of a windbreak on wind velocity. The wind is deflected upward by the trees and is slowed down even before reaching them. On the leeward side further reduction occurs; the effect being felt as far as 20 times the height of the trees. (c) Narrow strips of cereal rye act as miniature windbreaks to protect watermelons from wind erosion on a loamy sand on the mid-Atlantic coastal plain. [Photo (a) courtesy of USDA Natural Resources Conservation Service; (b) from FAO (1987); (c) courtesy of R. Weil]

strip-cropping and alternate strips of cropped and fallowed land should be perpendicular to the wind.

BARRIERS. Barriers such as shelter belts (Figure 17.42a) are effective in reducing wind velocities for short distances and for trapping drifting soil. Various devices are used to control blowing of sands, sandy loams, and cultivated peat soils (even in humid regions). Windbreaks and tenacious grasses and shrubs are especially effective. Picket fences and burlap screens, though less efficient as windbreaks than such trees as willows, are often preferred because they can be moved from place to place as crops and cropping practices are varied. Rye, planted in narrow strips across the field, is some-times used on peat lands and on sandy soils (see Figure 17.42c). Narrow rows of such perennial grasses as tall wheatgrass are being evaluated for a combination of wind erosion control and capturing of winter snows in the Northern Plains states.

The land capability classification system devised by the U.S. Department of Agriculture has been used since the 1950s to assess the appropriate uses of various types of land. It is especially helpful in identifying land uses and management practices that can minimize soil erosion, especially that induced by rainfall.

The system uses eight **land capability classes** to indicate the *degree* of limitation imposed on land uses (Figure 17.43), with Class I the least limited and Class VIII the most limited. Each land use class may have four subclasses that indicate the *type* of limitation encountered: risks of erosion (e); wetness, drainage, or flooding (w); root-zone limitations, such as acidity, density, and shallowness (s); and climatic limitations, such as a short growing season (c). The erosion (e) subclasses are the most common, and they will be the focus of our attention here. For example, Class IIe land is slightly susceptible to erosion, while Class VIIIe is extremely susceptible. Figure 17.44 shows the appropriate intensity of use allowable for each of these land capability classes if erosion losses (or problems associated with the other subclasses) are to be avoided.

Table 17.13 provides a summary of areas with major limitations on land use in the United States. Overall, erosion and sedimentation problems are the most serious limitations for nearly 60% of the land. Excessive wetness and shallowness are each problems on about 20% of the land.

In the United States, Classes I, V, and VIII each account for less than 3% of the nonfederal land. The other classes are much more common, with about 20% of the land in each of Classes II, III, VI, and VII and about 14% in Class IV. This means that about 43% of the land in the United States (242 million ha) is suitable for regular cultivation (Classes I, II, and III). Another 14% (78 million ha) is marginal for growing cultivated crops. The remainder is suited primarily for grasslands and forests, and is used mostly for those purposes. A brief description of the land use capability classes follows.

Class I. These soils are deep and well drained, have nearly level topography (see Figure 17.43), and have few limitations on their use. They can be used continuously for row crops, tree nurseries, or for any of the less-intensive uses shown in Figure 17.44. For most purposes, these soils provide the greatest productivity at the least cost.

FIGURE 17.43 Several land capability classes in San Mateo County, California. A range is shown from the nearly level land in the foreground (Class I), which can be cropped intensively, to the badly eroded hillsides (Classes VII and VIII). Although topography and erosion hazards are emphasized here, it should be remembered that other factors—drainage, stoniness, droughtiness—also limit soil usage and help determine the land capability class. (Courtesy USDA Natural Resources Conservation Service)

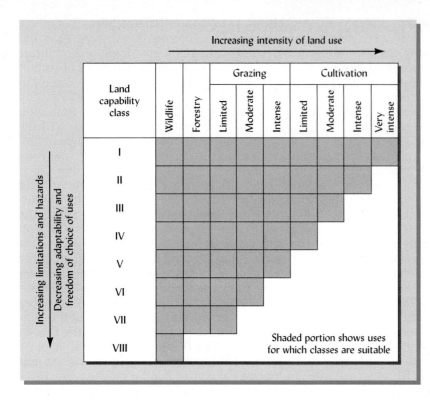

FIGURE 17.44 Intensity with which each land capability class can be used with safety. Note the increasing limitations on the safe uses of the land as one moves from Class I to Class VIII. [Modified from Hockensmith and Steele (1949)]

CLASS II. Soils in Class II have some limitations that reduce the choice of uses or require moderate conservation practices. Tillage and row-crop production should be restricted, or some conservation practices (conservation tillage, grassed waterways, contour strips, etc.) should be installed.

CLASS III. Soils in this class have severe limitations that reduce the choice of plants or require special conservation practices, or both. The same crops may be grown on Class III land as on Classes I and II land, but crops that provide soil cover, such as grasses and legumes, must be more prominent in the rotations used. The amount of land used for row crops is severely restricted. For the w subclass, drainage systems may be needed. Special care and scheduling for timber-harvest operation may be required.

CLASS IV. Soils in this class can be used for cultivation, but there are severe limitations on the choice of crops and management practices. If the land is to be cultivated, crops should be mostly close-growing types (e.g., wheat or barley) and sod or hay crops must be used extensively. Row crops cannot be grown safely, except in no-till systems. The choice of crops may be limited by excess moisture (subclass w), as well as by erosion

TABLE 17.13 Percentage of U.S. Land with Major Limitations in the Different Land Use Capability Classes

Erosion and sedimentation are the most significant limitations, but wetness and shallowness are also important

	Percentage of land with indicated major limitation			
LUC class	Erosion and sedimentation	Wetness	Shallowness	Climate hazards
II	51	34	8	7
III	67	22	9	2
IV	69	16	14	1
VI	62	8	26	3
VII	47	6	45	2
Total	59	18	21	3

Calculated from USDA (1994).

hazards. For tilled cropland, terracing, contour strips, or other special conservation practices are required. For timber harvest on forest land, special techniques may be required.

CLASS V.   These lands are generally not suited to crop production because of factors other than erosion hazards. Such limitations include (1) frequent stream overflow, (2) growing season too short for crop plants, (3) stony or rocky soils, and (4) ponded areas where drainage is not feasible. Often, pastures can be improved on this class of land.

CLASS VI.   Soils in this class have extreme limitations (typically, steep slopes and high erosion hazards) that restrict use largely to pasture, range, woodland, or wildlife. Forest entry should be limited, special care should be used in path construction, and alternative timber-harvest techniques should be used.

CLASS VII.   Severe limitations restrict the use of this land to controlled grazing, woodland, or wildlife. The physical limitations are the same as for Class VI, except they are so strict that pasture improvement is impractical. Cable or balloon timber-harvest methods may be necessary. Only small areas should be harvested, and entry should be strictly limited.

CLASS VIII.   In this land class are soils that should not be used for any kind of commercial plant production. Land use is restricted to recreation, wildlife, water supply, or aesthetic purposes. This land includes sand beaches, river wash, and rock outcrops.

The land-use capability classification scheme illustrates the practical use that can be made of soil surveys (see Section 19.7). The many soils delineated on a map by the soil surveyor are viewed in terms of their safest and best longtime use. The eight land capability classes have become the starting point in the development of land-use plans that promote wise use and conservation of the land resource by thousands of farmers, ranchers, and other landowners.

17.14   PROGRESS IN SOIL CONSERVATION

Soil Erosion

W. C. Lowdermilk's 1939 travelogue about soil erosion in Europe, N. Africa, and the Middle East:

http://www.soilandhealth.org/01aglibrary/010119lowdermilk.usda/cls/html

Soil erosion in the United States accelerated when the first European settlers chopped down trees and began to farm the sloping lands of the humid eastern part of the country. Soil erosion was a factor in the declining productivity of these lands that, in time, led to their abandonment and the westward migration of people in search of new farmlands.

It wasn't until the worldwide depression and widespread droughts of the early 1930s accentuated rural poverty and displaced millions of people that governments began to pay attention to the rapid deterioration of soils. In 1930, Dr. H. H. Bennett and associates recognized the damage being done and obtained U. S. government support for erosion-control efforts. Since then, considerable reductions in erosion have been achieved in the United States and elsewhere (for example, Figure 17.45).

During the 1940s and 1950s such physical practices as contour strips, terraces, and windbreaks were installed with much persuasion and assistance from government agencies. Some of this progress was reversed as the terraces and windbreaks appeared to stand in the way of the "fence row to fence row" all-out crop production policies of the 1970s. But, since 1982 rather remarkable progress has been achieved in reducing soil erosion (Figure 17.46), largely as a result of two factors: (1) the spread of conservation tillage (see Section 17.6), and (2) the implementation of land-use changes as part of the conservation reserve program. Progress continues to be made on both fronts.

However, about one-third of the cultivated cropland in the United States is still losing more than 11 Mg/ha/yr, the maximum loss that can be sustained without serious loss of productivity on most soils. After some 80 years of soil conservation efforts, soil erosion is still a major problem on about half the cropland of the United States, and in much of the world the problem has actually worsened.

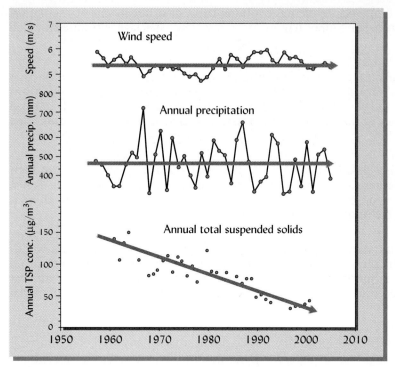

FIGURE 17.45 Wind-blown dust (total suspended solids, TSP) has declined significantly since 1960 in the Southern High Plains of Texas (lower trend arrow). Records show that neither average wind speed nor the average precipitation changed during this period (upper two trend arrows), so the decline in wind erosion was not due to less erosive climatic conditions. Minor changes occurred in land use with some cropland being converted to rangeland (not shown), but the area involved was far too small to account for the reductions in dust. Most likely it was the widespread adoption of conservation tillage and other improved agricultural practices that were responsible for the downward trends in wind erosion activity and dust emissions (TSP) during this period. [Compiled from figures in Stout and Lee (2003)]

The Conservation Reserve Program

A major part (about 60%) of the reduction in soil erosion experienced in the United States since 1982 is due to government programs that have paid farmers to shift some land from crops to grasses and forests. Establishing grass or trees on these former croplands reduced the sheet and rill erosion from an average of 19.3 to 1.3 Mg/ha and the wind erosion from 24 to 2.9 Mg/ha. Between 1982 and 2006, 14 million ha of cropland were diverted to such noncultivated uses through the **Conservation Reserve Program (CRP)**. In 1995, the CRP was reoriented to better target **highly erodible land (HEL)**[12] and other environmentally sensitive areas. The distribution of CRP land (Figure 17.47a) now closely reflects the distribution of highly erodible croplands (Figure 17.47b).

The CRP is basically an arrangement by which the U.S. taxpayers pay rent to farmers to forego cropping part of their farmland and instead plant grass or trees on it (more

---

[12] Highly erodible land is defined by the USDA Natural Resources Conservation Service as land for which the USLE predicts that soil loss on bare, unprotected soil would be eight times the T value for that soil: $R \times K \times LS \geq 8T$.

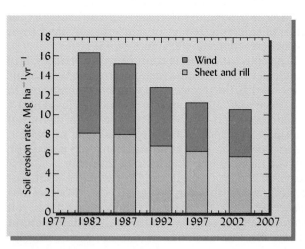

FIGURE 17.46 Average rates of soil loss by water and wind erosion in the United States from 1982 to 2003. Farmer adoption of conservation tillage practices along with the Conservation Reserve Program likely account for the nearly 40% reduction in rate of combined wind and water erosion. [Calculated from data in USDA-NRCS (2006)]

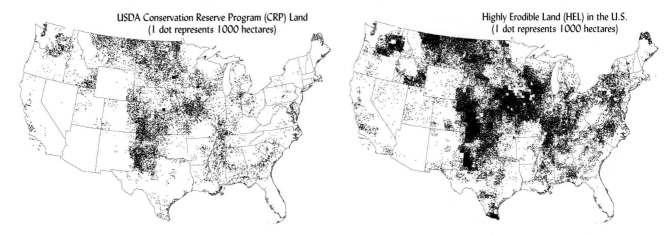

FIGURE 17.47 Distribution of land in the United States (*a*) enrolled in the Conservation Reserve Program (CRP) and (*b*) considered to be highly erodible land (HEL). Note that the distributions appear to be quite similar, indicating that the lands most sensitive to erosion are those being taken out of crop production and put into grasses or trees. (Maps courtesy of Conservation Technology Information Center, West Lafayette, Ind.)

rent is paid for trees). The rental agreements (leases) run for 10 to 15 years, during which time the land is undisturbed. The benefits to the nation have been evident, not only in greatly reduced soil losses and sediment pollution, but also in a dramatic upswing in wildlife as bird and animal populations take advantage of the newly restored habitat. Where strips of land along streams (riparian zone buffers) have been incorporated into the CRP, water-quality benefits have also been amplified.

Conservation Management to Enhance Soil Quality

In a broader sense, conservation management practices are those that improve soil quality in more ways than just by protecting the soil from erosion. Properties that indicate the level of soil quality, especially those associated with soil organic matter, can be enhanced by such conservation measures as minimizing tillage, maximizing residue cover of the soil surface, providing for diversity of plant types, keeping soil under grass sod vegetation for at least part of the time, adding organic amendments where practical, and maintaining balanced soil fertility. Improved soil quality, in turn, enhances the soil's capacity to support plants, resist erosion, prevent environmental contamination, and conserve water. Conversation management therefore can lead to the upward spiral of soil and environmental improvement referred to in Section 17.1.

Adapting Soil Conservation to the Needs of Resource-Poor Farmers

As satisfying as the progress of the past decades has been, soil losses by erosion are still much too high. Continued efforts must be made to protect the soil and to hold it in place. In the United States, some 30 million ha of highly erodible cropland continue to lose an average of more than 15 Mg/ha of soil each year from water erosion, and an equal amount from wind erosion. In spite of remarkable progress, conservation tillage systems have *not* been adopted for more than half of the nation's cropland. And no one knows what will happen to the CRP lands when the rental leases expire. The battle to bring erosion under control has just begun, not only in the United States but throughout the world.

In much of the world, so little land is available to each farmer for food production that nations cannot afford the luxury of following the land-use capability classification recommendations as outlined in Section 17.13. Many farmers must use *all* land capable of food production simply to stave off starvation and impoverishment. These farmers often realize that farming erodible land jeopardizes their future livelihood and that of their children, but they see no choice. It is imperative to either find nonagricultural employment for these people or to find farming systems that are sustainable on these erodible lands.

Fortunately, some farmers and scientists have developed, through long traditions of adaptation or through innovation and research, farming systems that *can* produce food and profits while conserving such erodible soil resources. Examples include the traditional Kandy Home Gardens of Sri Lanka's humid mountains, in which a rain forest–like mixed stand of tall fruit and nut trees is combined with an understory of pepper vines, coffee bushes, and spice plants to provide valuable harvests while keeping the soil under perennial vegetative protection. Another example comes from Central America, where farmers have learned to plant thick stands of velvet bean (*Mucuna*) or other viney legumes that can be chopped down by machete to leave a soil-protecting, water-conserving, weed-inhibiting mulch on steep farmlands. In Asia, steep lands have been carefully terraced in ways that allow production of food, even paddy rice, on very steep land without causing significant erosion.

Many examples from the United States and around the world make it clear that when governments cajole, pay, or force farmers into installing soil conservation measures on their land, the results are unlikely to be long-lasting. Usually, farmers will abandon the unwanted practices as soon as the pressure is off. On the other hand, if scientists and conservationists work *with* farmers to help them develop and adapt conservation systems that the farmers feel are of benefit to them and their land, then effective and lasting progress can be made. Experience with conservation tillage systems in the United States, mulch farming systems in Central America, and vegetative contour barriers in Asia have shown that farmers can help develop practices that are good for their land and for their profits: a win-win situation.

Interactive map depicts barriers posed by soil and climate conditions, including erosion risk: http://www.sciencemag.org/cgi/content/full/304/5677/1616/DC1#map

17.15 CONCLUSION

Accelerated soil erosion is one of the most critical environmental and social problems facing humanity today. Erosion degrades soils, making them less capable of producing the plants on which animals and people depend. Equally important, erosion causes great damage downstream in reservoirs, lakes, waterways, harbors, and municipal water supplies. Wind erosion also causes fugitive dust that may be very harmful to human health.

Nearly 4 billion Mg of soil is eroded each year on land in the United States alone. Half of this erosion occurs on the nation's croplands, and the remainder on harvested timber areas, rangelands, and construction sites. Some one-third of the cropland still suffers from erosion that exceeds levels thought to be tolerable.

Water carries away most of the sediment in humid areas by sheet and rill erosion. Gullies created by infrequent, but violent, storms account for much of the erosion in drier areas. Wind is the primary erosion agent in many drier areas, especially where the soil is bare and low in moisture during the season when strong winds blow.

Protecting soil from the ravages of wind or water is by far the most effective way to constrain erosion. In croplands and forests, such protection is due mainly to the cover of plants and their residues. Conservation tillage practices maintain vegetative cover on at least 30% of the soil surface, and the widening adoption of these practices has contributed to the significant reductions in soil erosion achieved over the past two decades. Crop rotations that include sod and close-growing crops, coupled with such practices as contour tillage, strip cropping, and terracing, also help combat erosion on farmland.

In forested areas, most erosion is associated with timber-harvesting practices and forest road construction. For the sake of future forest productivity and current water quality, foresters must become more selective in their harvest practices and invest more in proper road construction.

Construction sites for roads, buildings, and other engineering projects lay bare many scattered areas of soils that add up to a serious erosion problem. Control of sediment from construction sites requires carefully phased land clearing, along with vegetative and artificial soil covers, and installation of various barriers and sediment-holding ponds. These measures may be expensive to implement, but the costs to society that result when sediment is not controlled are too high to ignore. Once construction is completed, erosion rates on urban areas are commonly as low as those on areas under undisturbed native vegetation.

Erosion-control systems must be developed in collaboration with those who use the land, and especially the poor, for whom immediate needs must overshadow concerns for the future. As put succinctly in *The River,* a classic 1930s documentary film produced during the rebirth of American soil erosion consciousness, "Poor land makes poor people, and poor people make poor land."

STUDY QUESTIONS

1. Explain the distinction between *geologic erosion* and *accelerated erosion.* Is the difference between the two greater in humid or arid regions?

2. When erosion takes place by wind or water, what are three important types of damages that result on the land whose soils are eroding? What are five important types of damages that erosion causes in locations away from the eroding site?

3. What is a common *T* value, and what is meant by this term? Explain why certain soils have been assigned a higher *T* value than other soils.

4. Describe the three main steps in the water erosion process.

5. Many people assume that the amount of soil eroded on the land in a watershed (*A* in the universal soil-loss equation) is the same as the amount of sediment carried away by the stream draining that watershed. What factor is missing that makes this assumption incorrect? Do you think that this means the USLE should be renamed?

6. Why is the total annual rainfall in an area *not* a very good guide to the amount of erosion that will take place on a particular type of bare soil?

7. Contrast the properties you would expect in a soil with either a very high *K* value or a very low *K* value.

8. How much soil is likely to be eroded from a Keene silt loam in central Ohio, on a 12% slope, 100 m long, if it is in dense permanent pasture and has no support practices applied to the land? Use the information available in this chapter to calculate an answer.

9. What type of conservation tillage leaves the greatest amount of soil cover by crop residues? What are the advantages and disadvantages of this system?

10. Why are narrow strips of grass planted on the contour sometimes called a "living terrace"?

11. In most forests, which component of the ecosystem provides the primary protection against soil erosion by water, the *tree canopy, tree roots,* or *leaf litter?*

12. Certain soil properties generally make land susceptible to erosion by wind or erosion by water. List four properties that characterize soils highly susceptible to wind erosion. Indicate which two of these properties should also characterize soils highly susceptible to water erosion, and which two should not.

13. Which two factors in the wind erosion prediction equation (WEQ) can be affected by tillage? Explain.

14. Describe a soil in land capability Class IIw in comparison with one in Class IVe.

15. Why is it important that there be a close relationship between land in the CRP and that considered to be HEL?

REFERENCES

Benik, S. R., B. N. Wilson, D. D. Biesboer, B. Hansen, and D. Stenlund. 2003. "Evaluation of erosion control products using natural rainfall events," *J. Soil Water Conserv.,* **58:**98–104.

Blevins, R. L., and W. W. Frye. 1992. "Conservation tillage: An ecological approach to soil management," *Advances in Agronomy,* **51:**33–78.

Carter, M. R. 1994. *Conservation Tillage in Temperate Agroecosystems* (Boca Raton, Fla.: Lewis Publishers).

Cassel, D. K., and R. Lal. 1992. "Soil physical properties of the tropics: Common beliefs and management constraints," in R. Lal and P. A. Sanchez (eds.), *Myths and Science of Soils of the Tropics*. SSA Special Publication no. 29 (Madison, Wis.: Soil Sci. Soc. of Amer.), pp. 61–89.

Chow, T. L., H. W. Rees, and J. L. Daigle. 1999. "Effectiveness of terraces/grassed waterway for soil and water conservation: A field evaluation," *J. Soil Water Cons.*, **54**:577–583.

Crovetto, C. 1996. *Stubble Over the Soil* (Madison, Wis.: Amer. Soc. Agron.).

Daily, G. 1997. "Restoring value to the world's degraded lands," *Science*, **269**:350–354.

Daily G. C., T. Söderqvist, S. Aniyar, K. Arrow, P. Dasgupta, P. R. Ehrlich, C. Folke, A. Jansson, B-O. Jansson, N. Kautsky, S. Levin, J. Lubchenco, K-G. Mäler, D. Simpson, D. Starrett, D. Tilman, and B. Walker. 2000. "The value of nature and the nature of value," *Science*, **289**:395–396.

Dickey, E. C., P. T. Jasa, B. J. Dolesh, L. A. Brown, and S. K. Rockwell. 1987. "Conservation tillage: Perceived and actual use," *J. Soil Water Cons.*, **42**:431–434.

Doolette, J. B., and J. W. Smyle. 1990. "Soil and moisture conservation technologies: Review of literature," in J. B. Doolette and W. B. Magrath (eds.), *Watershed Development in Asia: Strategies and Technologies* (Washington, D.C.: World Book Technical Paper 127).

El-Swaify, S. A., and E. W. Dangler. 1982. "Rainfall erosion in the tropics: A state-of-the-art," *Soil Erosion and Conservation in the Tropics*. ASA Special Publication no. 43 (Madison, Wis.: Amer. Soc. Agron.).

FAO. 1987. *Protect and Produce* (Rome: U.N. Food and Agriculture Organization).

FAO. 2001. "The economics of conservation agriculture," FAO Y2781/E. Food and Agriculture Organization of the United Nations, Rome 73 pp. http://www.fao.org/docrep/004/Y2781E/y2781e00.htm#toc.

Flanagan, D. C., J. C. Ascough II, M. A. Nearing, and J. M. Laflen. 2001. "The water erosion prediction project (wepp) model," Chapter 7, in R. S. Harmon and W. W. Doe III (eds.), *Landscape Erosion and Evolution Modeling*, Norwell, Mass.: Kluwer Academic Publishers.

Foster, G. R., D. K. McCool, K. G. Renard, and W. C. Moldenhauer. 1981. "Conversion of the universal soil loss equation to SI metric units," *J. Soil Water Cons.*, **36**:355–359.

Foster, G. R., and R. E. Highfill. 1983. "Effect of terraces on soil loss: USLEP factor values for terraces," *J. Soil Water Cons.*, **38**:48–51.

Fryrear, D. W., and A. Saleh. 1993. "Field wind erosion: Vertical distribution," *Soil Sci.*, **155**:294–300.

Fryrear, D.W., J. D. Bilbro, A. Saleh, H. Schomberg, J. E. Stout, and T. M. Zobeck. 2000. "RWEQ: Improved wind erosion technology," *J. Soil Water Cons.*, **55**:183–189.

Gilley, J. E., B. Eghball, L. A. Kramer and T. B. Moorman. 2000. "Narrow grass hedge effects on runoff and soil loss," *J. Soil Water Cons.*, **55**:190–196.

Hillel, D. 1991. *Out of the Earth: Civilization and the Life of the Soil* (New York: The Free Press).

Hockensmith, R. D., and J. G. Steele. 1949. "Recent trends in the use of the land-capability classification," *Soil Sci. Soc. Amer. Proc.*, **14**:383–388.

Hooke, R. L. 2000. "On the history of humans as geomorphic agents," *Geology*, **28**:843–846.

Hudson, N. 1995. *Soil Conservation*, 3rd ed. (Ames, Iowa: Iowa State University Press).

Kelley, D. W., and E. A. Nater. 2000. "Historical sediment flux from three watersheds into Lake Pepin, Minnesota, USA," *J. Environ. Qual.*, **29**:561–568.

Lal, R. 1995. "Erosion-crop productivity relationships for the soils of Africa," *Soil Sci. Soc. Amer. J.*, **59**:661–667.

Mahboubi, A. A., R. Lal, and N. R. Faussey. 1993. "Twenty-eight years of tillage effects on two soils in Ohio," *Soil Sci. Soc. Amer. J.*, **57**:506–512.

Nyland, R. D. 1996. *Silviculture: Concepts and Applications* (New York: McGraw-Hill).

Oldeman, L. R. 1994. "The global extent of soil degradation," in D. J. Greenland and I. Szabolcs (eds.), *Soil Resilience and Sustainable Land Use* (Wallingford, U.K.: CAB International).

Oschwald, W. R., and J. C. Siemens. 1976. "Conservation tillage: A perspective," Agronomy Facts SM-30 (Urbana, Ill.: University of Illinois).

Pierre, C. J. M. G. 1992. *Fertility of Soils: A Future for Farming in the West African Savannah* (Berlin: Springer-Verlag).

Pimentel D., C. Harvey, P. Resosudarmo, K. Sinclair, D. Kurz, M. McNair, S. Crist, L. Shpritz, L. Fitton, R. Saffouri, and R. Blair. 1995. "Environmental and economic costs of soil erosion and conservation benefits," *Science,* **267**:1117–1122.

Ramos, M. C., and J. A. Martinez-Casasnovas. 2006. "Nutrient losses by runoff in vineyards of the Mediterranean Alt Penedes region (ne Spain)," *Agric. Ecosyst. Environ.,* **113**:356–363.

Renard, K. G., G. Foster, D. Yoder, and D. McCool. 1994. "RUSLE revisited: Status, questions, answers and the future," *J. Soil Water Cons.,* **49**:213–220.

Renard K. G., G. R. Foster, G. A. Weesies, D. K. McCool, and D. C. Yoder. 1997. *Predicting Soil Erosion by Water: A Guide to Conservation Planning with the Revised Universal Soil Loss Equation (RUSLE).* Agricultural Handbook no 703 (Washington, D.C.: USDA).

Rosensweig, C., and D. Hillel. 1998. *Climate Change and the Global Harvest: Potential Impacts of the Greenhouse Effect on Agriculture* (Cary, N.C.: Oxford University Press).

Schwab, G. O., D. D. Fangmeirer, and W. J. Elliot. 1996. *Soil and Water Management Systems,* 4th ed. (New York: Wiley).

Schertz, D. L. 1983. "The basis for soil loss tolerance," *J. Soil Water Cons.,* **30**:10–14.

Sidle, R. C., A. D. Ziegler, J. N. Negishi, A. R. Nik, R. Siew, and F. Turkelboom. 2006. "Erosion processes in steep terrain—Truths, myths, and uncertainties related to forest management in Southeast Asia," *Forest Ecology and Management,* **224**:199–225.

Skidmore, E. L., and F. H. Siddoway. 1978. "Crop residue requirements to control wind erosion," in W. R. Oschwalk (ed.), *Crop Residue Management Systems.* ASA Special Publication no. 31 (Madison, Wis.: Amer. Soc. Agron.; Crop Sci. Soc. Amer.; and Soil Sci. Soc. Amer.).

Stout, J. E., and J. A. Lee. 2003. "Indirect evidence of wind erosion trends on the southern high plains of North America," *Journal of Arid Environments,* **55**:43–61.

USDA. 1994. *Summary Report: 1992 National Resources Inventory* (Washington, D.C.: USDA Natural Resources Conservation Service).

USDA. 1995. Agricultural Handbook no. 703 (Washington, D.C.: U.S. Department of Agriculture).

USDA-NRCS. 2006. "Annual national resources inventory for 2003: Soil erosion." U.S. Department of Agriculture, Natural Resources Conservation Service. http://www .nrcs.usda.gov/technical/land/nri03/SoilErosion-mrb.pdf (posted May 2006; verified 31 May 2006).

USEPA. 2006. "Particulate matter: Basic information." U.S. Environmental Protection Agency. http://www.epa.gov/air/particlepollution/basic.html (verified 23 September 2006).

Weesies, G. A., S. J. Livingston, W. D. Hosteter, and D. L. Schertz. 1994. "Effect of soil erosion on crop yield in Indiana: Results of a 10 year study," *J. Soil Water Cons.,* **49**:597–600.

Wilkinson, B. H., and B. J. McElroy. 2007. The impact of humans on continental erosion and sedimentation. *Geological Society of America Bulletin* **119**:140–150.

Wischmeier, W. J., and D. D. Smith. 1978. *Predicting Rainfall Erosion Loss—A Guide to Conservation Planning.* Agricultural Handbook no. 537 (Washington, D.C.: USDA).

Polluted soil awaits remediation near abandoned oil refinery. (R. Weil)

18

SOILS AND CHEMICAL POLLUTION

*Black and portentous
this humor prove,
unless good counsel
may the cause remove . . .*
—*W. SHAKESPEARE*, ROMEO AND JULIET

The soil is a primary recipient by design or accident of a myriad of wastes, chemicals and products used in modern society, many of which we have conveniently "thrown away." Every year, millions of tons of industrial, domestic, and agricultural products find their way into the world's soils. Once there, they become part of biological cycles that affect all forms of life.

In previous chapters we highlighted the enormous capacity of soils to accommodate added organic and inorganic chemicals. Tons of organic residues are broken down by soil microbes each year (Chapter 12), and large quantities of inorganic chemicals are fixed or bound tightly by soil minerals (Chapter 14). But we also learned of the limits of the soil's capacity to sorb these chemicals, and how environmental quality suffers when these limits are exceeded (Chapters 8 and 16).

We have seen how soil processes affect the production and sequestering of greenhouse gases, such as nitrous oxide, methane, and carbon dioxide (see, e.g., Sections 12.9 and 13.8). Other nitrogen- and sulfur-containing gases come to earth in acid rain (see, e.g., Section 9.6). Mismanaged irrigation projects on arid-region soils result in the accumulation of salts (see Section 10.3), including toxic levels of sodium, selenium and arsenic (Chapter 15).

We have also seen how fertilizer and manure applications that leave excess quantities of nutrients in the soil can result in the contamination of ground and surface waters with nitrates (Section 13.8) and phosphates (Section 14.2). The eutrophication of lakes, estuaries and slow-moving rivers is evidence of these nutrient buildups. Huge "animal factories" for meat and poultry production produce mountains of manure that must be disposed of without loading the environment with unwanted chemicals and with pathogens that are harmful to humans and other animals (Section 16.4).

In this chapter we will focus on chemicals that contaminate and degrade soils, including some whose damage extends to water, air, and living things. The brief review of soil pollution is intended as an introduction to the nature of the major pollutants, their reactions in soils, and alternative means of managing, destroying, or inactivating them.

18.1 TOXIC ORGANIC CHEMICALS

Comprehensive information
on cleaning contaminated
soils at abandoned industrial
sites:
http://www.epa.gov/brownfields/

Industrialized societies have synthesized thousands of organic (carbon-containing) compounds for thousands of uses. An enormous quantity of organic chemicals is manufactured every year—more than 60 million Mg in the United States alone. Included are plastics and plasticizers, lubricants and refrigerants, fuels and solvents, pesticides and preservatives. Some are extremely toxic to humans and other life. Through accidental leakage and spills or through planned spraying or other treatments, synthetic organic chemicals can be found in virtually every corner of our environment—in the soil, in the groundwater, in the plants, and in our own bodies.

Environmental Damage from Organic Chemicals

Find out how to reduce
environmental impact of
your household hazardous
wastes:
http://www.klickitatcounty.org/
SolidWaste/default.asp?fCateg
oryIDSelected=-1671944469

These artificially synthesized compounds are termed **xenobiotics** because they are unfamiliar to the living world (Greek *xeno,* strange). Being nonnatural, many xenobiotics are both toxic to living organisms and resistant to biological decay. The chemical structures of xenobiotic compounds may be quite similar to those of naturally occurring compounds produced by microorganisms and plants. The difference is commonly the insertion of halogen atoms (Cl, F, Br) or multivalent nonmetal atoms (such as S and N) into the structure (see Figure 18.1).

Some xenobiotic compounds are relatively inert and harmless, but others are biologically damaging even in very small concentrations. Those that find their way into soils may inhibit or kill soil organisms, thereby undermining the balance of the soil community (see Section 11.14). Other chemicals may be transported from the soil to the air, water, or vegetation, where they may be contacted, inhaled, or ingested by any number of organisms, including people. It is imperative, therefore, that we control the release of organic chemicals and that we learn of their fate and effects once they enter the soil.

Organic chemicals may enter the soil as contaminants in industrial and municipal organic wastes applied to or spilled on soils, as components of discarded machinery, in large or small lubricant and fuel leaks, as military explosives, or as sprays applied to control pests in terrestrial ecosystems. Pesticides are probably the most widespread organic pollutants associated with soils. In the United States, pesticides are used on some 150 million ha of land, three-fourths of which is agricultural land. Soil contamination by other organic chemicals is usually much more localized. We will therefore emphasize the pesticide problem.

The Nature of the Pesticide Problem

Pest Management Practices
in U.S. Agriculture, 1990–97,
ERS/USDA:
http://www.ers.usda.gov/
publications/sb969/sb969d.
pdf

Pesticides are chemicals that are designed to kill pests (that is, any organism that the pesticide user perceives to be damaging). Some 600 chemicals in about 50,000 formulations are used to control pests. They are used extensively in all parts of the world. About 600,000 Mg of organic pesticide chemicals are used annually in the United States, with more than three times that amount used in the rest of the world. Although the total amount of pesticides used has remained relatively constant or even dropped since the 1980s, formulations in use today are generally more potent, so that smaller quantities are applied per hectare to achieve toxicity to the pest.

BENEFITS OF PESTICIDES. Pesticides have provided many benefits to society. They have helped control mosquitoes and other vectors of such human diseases as yellow fever and malaria. They have protected crops and livestock against insects and diseases. Without the control of weeds by chemicals called *herbicides,* conservation tillage (especially no-tillage) would be much more difficult to adopt; much of the progress made in controlling soil erosion probably would not have come about without herbicides. Also, pesticides reduce the spoilage of food as it moves from farm fields to distant dinner tables.

PROBLEMS WITH PESTICIDES. While the benefits to society from pesticides are great, so are the costs (Table 18.1). Widespread and heavy use of pesticides on agricultural soils and suburban and urban landscapes has led to contamination of both surface and groundwater (Table 18.2). Therefore, when pesticides are used, they should be chosen for low toxicity to humans and wildlife, low mobility on soils, and low persistence

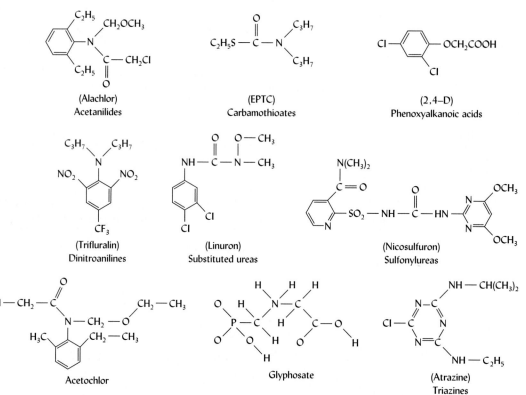

FIGURE 18.1 Structural formulae of representative compounds in 18 classes of widespread organic contaminants. Carbaryl, DDT, and parathion are insecticides; the lower nine compounds shown are herbicides. The widely differing structures result in a great variety of toxicological properties and reactions in the soil.

TABLE 18.1 Total Estimated Environmental and Social Costs from Pesticide Use in the United States

Type of impact[a]	Cost, $ million/yr
Public health impacts	1260
Domestic animal deaths and contamination	50
Loss of natural enemies	900
Cost of pesticide resistance	2250
Honeybee and pollination losses	500
Crop losses	1500
Fishery losses	38
Groundwater contamination and cleanup costs	2900
Cost of government regulations to prevent damage	320
Total	9720

[a]The death of an estimated 60 million wild birds may represent an additional substantial cost in lost revenues from hunters and bird watchers.
Data adjusted for inflation to 2006 dollars, from Pimental et al. (1992). © American Institute of Biological Sciences.

TABLE 18.2 Selected Pesticides Commonly Found in Water of Urban and Agricultural Watersheds in the United States

The distribution coefficients for organic carbon (K_{oc}), solubilities in water (S_w), and half-lives help determine the ease with which the compounds move to ground and surface water. Detections, maximum concentrations observed, and health advisory levels should be compared to suggest the seriousness of the groundwater contamination.

Pesticide compound	log K_{oc} (mL/g)	S_w (mg/L)	Half-life for transformation in aerobic soil (days)	Half-life for transformation in water (days)	Detections in wells (%)	Maximum concentration observed (µg/L)	Health criterion (µg/L)
Agricultural herbicides and degradates detected most frequently in water							
Atrazine	2.00	30	146	742	38	3.8	3[b]
Deethylatrazine	1.90	2700	170	NA	34	2.6	—
Metolachlor	2.26	430	26	410	15	5.4	100[a]
Alachlor	2.23	240	20.4	640	2.4	0.55	2[b]
Acetochlor	2.38	223	14	2300	—	—	—
Metribuzin	1.72	1000	172	>200	3.1	0.30	200[a]
EPTC	2.30	370	7	>200	1.5	0.45	—
Trifluralin	4.14	0.5	169	>32	0.5	0.014	5[a]
Urban herbicides detected most frequently in water							
Simazine	2.11	5	91	>32	1.4	1.3	4[b]
2,4-D	1.68	890	2.3	732	—	—	70[b]
Diuron	2.60	40	372	>500	—	—	10[a]
Bromacil	1.86	815	275	>30	—	—	90[a]
Insecticides detected most frequently in water							
Diazinon	2.76	60	39	140	0.7	0.077	0.6[a]
Chlorpyrifos	3.78	0.73	30.5	29	0.02	0.005	20[b]
Carbofuran	2.02	351	11	289	0.7	1.3	40[b]
Carbaryl	2.36	120	17	11	1.1	0.02	700[a]
Malathion	3.26	145	<1	6.3	0.2	0.004	100[a]
Dieldrin	4.08	0.17	NA	3830	1.4	0.45	0.002[c]
Organochlorine pesticide compounds detected frequently in fish tissue							
p,p'-DDE	5.0	0.04	NA	>44,000	3.9	0.006	0.1[c]
p,p'-DDT	5.4	0.006	NA	5,000	—	—	—

Source: Gilliom et al. (2006).
[a]Health advisory level (HA-L), the concentration expected to cause health problems (noncancer) with lifetime exposure.
[b]Maximum contaminate level (MCL) permissible as an annual average concentration in public-use water.
[c]Cancer risk concentration (CRC).

(see Section 18.3). Even then, the use of pesticides often has wide-ranging detrimental effects on the microbial and faunal communities. In fact, the harm done, though not always obvious, may outweigh the benefits. Examples include insecticides that kill natural enemies of pest species as well as the target pest (sometimes creating new major pests from species formerly controlled by natural enemies) and fungicides that kill both disease-causing and beneficial mycorrhizal fungi (see Section 11.9). Given these facts, it should not come as a surprise that despite the widespread use of pesticides, insects, diseases, and weeds still cause the loss of one-third of the crop production, about the same proportion of crops lost to these pests in the United States, before synthetic organic pesticides were in use.

Integrated Pest Management, information from University of California: http://www.ipm.ucdavis.edu/

ALTERNATIVES TO PESTICIDES.   Pesticides should not be seen as a panacea, or even as indispensable. Some farmers, most notably the small but increasing number who practice **organic farming**,[1] produce profitable, high-quality yields without the use of synthetic pesticides. In managing the effects of pests in any type of plant community (agricultural, ornamental, or forest), chemical pesticides should be used as a *last* resort, rather than as a *first* resort. Before resorting to the use of an insecticide or herbicide, every effort should be made to minimize the detrimental effects of insects and weeds by means of crop diversification, establishment of habitat for beneficial insects, application of organic soil amendments, implementation of cultural practices to reduce weed competition, and selection of pest-resistant plant cultivars. Too often, because pesticides are available as a convenient crutch, these more sophisticated approaches to plant management are not thoroughly explored.

Farmscaping: An alternative to pesticides (ATTRA): http://www.attra.org/attra-pub/farmscape.html

NONTARGET DAMAGES.   Although some pesticides are intentionally applied to soils, most reach the soil because they miss the insect or plant leaf that is the application target. When pesticides are sprayed in the field, most of the chemical misses the target organism. For pesticides aerially applied to forests, about 25% reaches the tree foliage, and far less than 1% reaches a target insect. About 30% may reach the soil, while about half of the chemical applied is likely to be lost into the atmosphere or in runoff water.

Designed to kill living things, many of these chemicals are potentially toxic to organisms other than the pests for which they are intended. Some are detrimental to nontarget organisms, such as beneficial insects and certain soil organisms. Those chemicals that do not quickly break down may be biologically magnified as they move up the food chain. For example, as earthworms ingest contaminated soil, the chemicals tend to concentrate in the earthworm bodies. When birds and fish eat the earthworms, the pesticides can build up further to lethal levels. The near extinction of certain birds of prey (including the American bald eagle) during the 1960s and 1970s called public attention to the sometimes-devastating environmental consequences of pesticide use. More recently, evidence is mounting to suggest that human endocrine (hormone) balance may be disrupted by the minute traces of some pesticides found in water, air, and food.

18.2 KINDS OF ORGANIC CONTAMINANTS

Industrial Organics

Industrial organics that often end up contaminating soils by accident or neglect include petroleum products used for fuel [gasoline components such as benzene, and more complex polycyclic aromatic hydrocarbons, (PAHs)], solvents used in manufacturing processes [such as trichloroethylene (TCE)], and military explosives such as trinitrotoluene (TNT). Several examples of their structures are shown in Figure 18.1. Polychlorinated biphenyls (PCBs) constitute a particularly troublesome class of widely dispersed compounds. These compounds can disrupt reproduction in birds and cause cancer and hormone effects in humans and other animals. Several hundred varieties of liquid or resinous PCBs were produced from 1930 to 1980 and used as specialized

---

[1] The term *organic farming* has little to do with the chemical definition of organic, which simply indicates that a compound contains carbon. Rather, it refers to a system and philosophy of farming that eschews the use of synthetic chemicals while it emphasizes soil organic matter and biological interactions to manage agroecosystems. See section 20.9.

Leaking underground
storage tanks—the threat to
public health and the
environment:
http://www.sierraclub.org/
toxics/Leaking_USTs/index.asp

lubricants, hydraulic fluids, and electrical transformer insulators, as well as in certain epoxy paints and many other industrial and commercial applications. Because of their extreme resistance to natural decay and their ability to enter food chains, even today soil and water all over the globe contain at least traces of PCBs.

The sites most intensely contaminated with organic pollutants are usually located near chemical manufacturing plants or oil storage facilities, but railway, shipping, and highway accidents also produce hot spots of contamination. Thousands of neighborhood gas stations represent potential or actual sites of soil and groundwater contamination as gasoline leaks from old, rusting underground storage tanks (Figure 18.2). However, as already mentioned, by far the most widely dispersed xenobiotics are those designed to kill unwanted organisms (i.e., pests).

Pesticides

Pesticides are commonly classified according to the group of pest organisms targeted: (1) *insecticides,* (2) *fungicides,* (3) *herbicides* (weed killers), (4) *rodenticides,* and (5) *nematocides.* In practice, all find their way into soils. Since the first three are used in the largest quantities and are therefore more likely to contaminate soils, they will be given primary consideration. Figure 18.1 shows that most pesticides contain aromatic rings of some kind, but that there is great variability in pesticide chemical structures.

The Future Role of Pesticides
in U.S. Agriculture:
http://books.nap.edu/
books/0309065267/html/
index.html

INSECTICIDES.   Most of these chemicals are included in three general groups. The *chlorinated hydrocarbons,* such as DDT, were the most extensively used until the early 1970s, when their use was banned or severely restricted in many countries due to their low biodegradability and persistence, as well as their toxicity to birds and fish. The *organophosphate* pesticides are generally biodegradable, and thus less likely to build up in soils and water. However, they are extremely toxic to humans, so great care must be used in handling and applying them. The *carbamates* are considered least dangerous because of their ready biodegradability and relatively low mammalian toxicity. However, they are highly toxic to honeybees and other beneficial insects and to earthworms.

FUNGICIDES.   Fungicides are used mainly to control diseases of fruit and vegetable crops and as seed coatings to protect against seed rots. Some are also used to protect harvested fruits and vegetables from decay, to prevent wood decay, and to protect clothing from mildew. Organic materials such as the thiocarbamates and triazoles are currently in use.

FIGURE 18.2 Leaking underground storage tank (LUST) replacement at a gas station in California. The old rusting steel tanks have been removed and replaced by more corrosion-resistant fiberglass tanks, which are set in the ground and covered with pea gravel. The soil and groundwater aquifer beneath the tanks were cleaned up using special techniques to stimulate soil microorganisms and to pump out volatile organics such as benzene vapors. Remediation and replacement typically costs $700,000 for a single gas station. (Photo courtesy of R. Weil)

HERBICIDES. The quantity of herbicides used in the United States exceeds that of the other types of pesticides combined. Starting with 2,4-D (a chlorinated phenoxyalkanoic acid), dozens of chemicals in literally hundreds of formulations have been placed on the market (see Figure 18.1). These include the *triazines,* used mainly for weed control in corn; *substituted ureas; some carbamates; the relatively new sulfonylureas,* which are potent at very low rates; *dinitroanilines;* and *acetanilides,* which have proved to be quite mobile in the environment. One of the most widely used herbicides, *glyphosate* (e.g. Roundup®), does not belong to any of the aforementioned chemical groups. Unlike most herbicides, it is nonselective, meaning that it will kill almost any plant, including crops. However, a gene that confers resistance to its effects has been discovered and engineered into several major crops. These genetically engineered crops can then be grown with a very simple, convenient method of weed control that usually consists of one or two sprayings of glyphosate that will kill all plants other than the resistant crop.

As one might expect, this wide variation in chemical makeup provides an equally wide variation in properties. Most herbicides are biodegradable, and most of them are relatively low in mammalian toxicity. However, some are quite toxic to fish, soil fauna, and perhaps to other wildlife. They can also have deleterious effects on beneficial aquatic vegetation that provides food and habitat for fish and shellfish.

NEMATOCIDES. Although nematocides are not as widely used as herbicides and insecticides, some of them are known to contaminate soils and the water draining from treated soils. For example, some carbamate nematocides dissolve readily in water, are not adsorbed onto soil surfaces, and consequently easily leach downward and into the groundwater. Other nematicidal chemicals are volatile soil fumigants that kill virtually all life in the soil, both the helpful and the harmful (Section 18.4). Methyl bromide, once the most commonly used of these fumigants, has been banned because of its adverse effects on the atmosphere and parts of the environment. Happily, the search for substitutes has led to the development of many nonchemical means to manage the pests once controlled by this highly toxic chemical (e.g., see Figure 11.15).

18.3 BEHAVIOR OF ORGANIC CHEMICALS IN SOIL[2]

Once they reach the soil, organic chemicals, such as pesticides or hydrocarbons, move in one or more of seven directions (Figure 18.3): (1) they may vaporize into the atmosphere without chemical change; (2) they may be absorbed by soils; (3) they may move downward through the soil in liquid or solution form and be lost from the soil by leaching; (4) they may undergo chemical reactions within or on the surface of the soil; (5) they may be broken down by soil microorganisms; (6) they may wash into streams and rivers in surface runoff; and (7) they may be taken up by plants or soil animals and move up the food chain. The specific fate of these chemicals will be determined at least in part by their chemical structures, which are highly variable.

Volatility

Organic chemicals vary greatly in their volatility and subsequent susceptibility to atmospheric loss. Some soil fumigants, such as methyl bromide (now banned from most uses), were selected because of their very high vapor pressure, which permits them to penetrate soil pores to contact the target organisms. This same characteristic encourages rapid loss to the atmosphere after treatment, unless the soil is covered or sealed.

---

[2] For reviews on organic chemicals in the soil environment, see Sawhney and Brown (1989) and Pierzynski et al. (2004); for pesticides, see Cheng (1990).

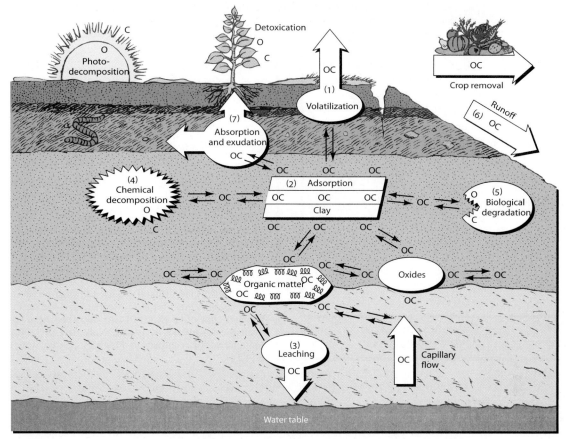

FIGURE 18.3 Processes affecting the dissipation of organic chemicals (OC) in soils. Note that the OC symbol is split up by decomposition (both by light and chemical reaction) and degradation by microorganisms, indicating that these processes alter or destroy the organic chemical. In transfer processes, the OC remains intact. [From Weber and Miller (1989)

A few herbicides (e.g., trifluralin) and fungicides (e.g., PCNB) are sufficiently volatile to make vaporization a primary means of their loss from soil. The lighter fractions of crude oil (e.g., gasoline and diesel) and many solvents vaporize to a large degree when spilled on the soil.

The assumption that disappearance of pesticides from soils is evidence of their breakdown is questionable. Some chemicals lost to the atmosphere are known to return to the soil or to surface waters with the rain.

Adsorption

The adsorption of organic chemicals by soil is determined largely by the characteristics of the compound and of the soils to which they are added. Soil organic matter and high-surface-area clays tend to be the strongest adsorbents for some compounds (Figure 18.4), while oxide coatings on soil particles strongly adsorb others. The presence of certain functional groups, such as —OH, —NH_2, —NHR, —$CONH_2$, —COOR, and —$^+NR_3$, in the chemical structure encourages adsorption, especially on the soil humus. Hydrogen bonding (see Sections 5.1 and 8.3) and protonation [adding of H^+ to a group such as an —NH_2 (amino) group] probably promotes some of the adsorption. Everything else being equal, larger organic molecules with many charged sites are more strongly adsorbed.

Some organic chemicals with positively charged groups, such as the herbicides diquat and paraquat, are strongly adsorbed by silicate clays. Adsorption by clays of some pesticides tends to be pH-dependent (Figure 18.5), with maximum adsorption occurring at low pH, which encourages protonation. Adding H^+ ions to functional groups (e.g., —NH_2) yields a positive charge on the herbicide, resulting in greater attraction to negatively charged soil colloids.

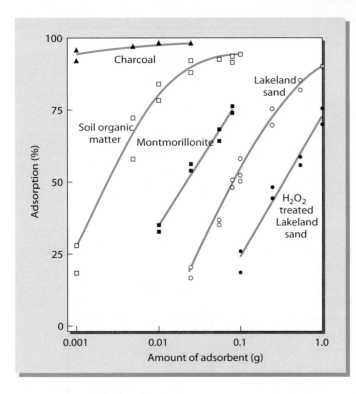

FIGURE 18.4 Adsorption of polychlorinated biphenyl (PCB) by different soil materials. The Lakeland sand (Typic Quartzipsamments) lost much of its adsorption capacity when treated with hydrogen peroxide H_2O_2 to remove its organic matter. The amount of soil material required to adsorb 50% of the PCB was approximately 10 times as great for montmorillonite (a 2:1 clay mineral) as for soil organic matter, and 10 times again as great for H_2O_2 treated Lakeland sand. Later tests showed that once the PCB was adsorbed, it was no longer available for uptake by plants. Note that the amount of soil material added is shown on a log scale. [From Strek and Weber (1982)]

Leaching and Runoff

Pesticide leaching potential by Watershed for 13 crops grown in the U.S. http://www.unl.edu/ nac/atlas/Map_Html/Clean_ Water/National/NRI_%20 Pesticide_Leaching_1992/ Pesticide_leaching.htm

The tendency of organic chemicals to leach from soils is closely related to their solubility in water and their potential for adsorption. Some compounds, such as chloroform and phenoxyacetic acid, are a million times more water-soluble than others, such as DDT and PCBs, which are quite soluble in oil but not in water. High water-solubility favors leaching losses.

Strongly adsorbed molecules are not likely to move down the profile (Table 18.3). Likewise, conditions that encourage such adsorption will discourage leaching. Leaching is apt to be favored by water movement, the greatest leaching hazard occurring in highly permeable, sandy soils that are also low in organic matter. Periods of high rainfall around the time of application of the chemical promote both leaching and runoff losses (Table 18.4). With some notable exceptions, herbicides seem to be somewhat more mobile than most fungicides or insecticides, and therefore are more likely to find their way to groundwater supplies and streams (Figure 18.6).

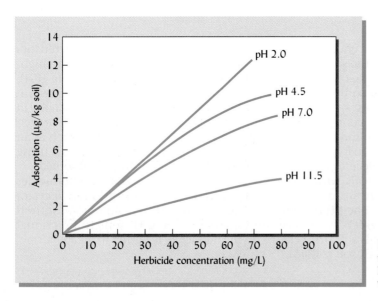

FIGURE 18.5 The effect of pH of kaolinite on the adsorption of glyphosate, a widely used herbicide. [From McConnell and Hossner (1985), copyright 1984; reprinted with permission of American Chemical Society]

TABLE 18.3   The Degree of Adsorption of Selected Herbicides

Weakly adsorbed herbicides are more susceptible to movement in the soil than those that are more tightly adsorbed.

Common name	Example trade name	Adsorptivity to soil colloids
Dalapon	Dowpon	None
Chloramben	Amiben	Weak
Bentazon	Basagran	Weak
2,4-D	Several	Moderate
Propachlor	Ramrod	Moderate
Atrazine	AAtrex	Strong
Alachlor	Lasso	Strong
EPTC	Eptam	Strong
Diuron	Karmex	Strong
Glyphosate	Roundup	Very strong
Paraquat	Paraquat	Very strong
Trifluralin	Treflan	Very strong
DCPA	Dacthal	Very strong

TABLE 18.4   Surface Runoff and Leaching Losses (Through Drain Tiles) of the Herbicide Atrazine from a Clay Loam Lacustrine Soil (Alfisols) in Ontario, Canada

The herbicide was applied at 1700 g/ha in late May. The data are the average of three tillage methods. Note that the rainfall for May and June is related to the amount of herbicide lost by both pathways.

Year of study	Surface runoff loss	Drainage water loss	Total dissolved loss	Percent of total applied, %	Rainfall, May–June, mm
	Atrazine loss, g/ha				
1	18	9	27	1.6	170
2	1	2	3	0.2	30
3	51	61	113	6.6	255
4	13	32	45	2.6	165

Data abstracted from Gaynor et al. (1995).

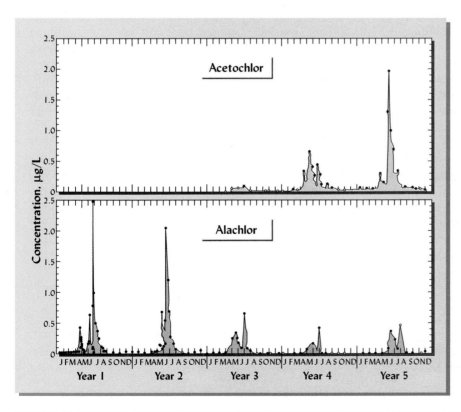

FIGURE 18.6   Herbicides in the main U.S. corn-growing region illustrate the direct and rapid connection between the use of a chemical on the land and its concentration in streams and rivers. The White River near Hazelton, Indiana, was monitored over a 5-year period. Note that the concentrations of the herbicide Alachlor peaked every year in June, about a month after most farmers in the watershed sprayed their corn and soybean fields. In year 3, a new compound, Acetochlor, partially replaced the older herbicide Alachlor. Within a year of the introduction of the newer compound, Acetochlor concentrations increased while Alachlor concentrations decreased. [From Gilliom et al. (2006)

Contamination of Groundwater

Experts once maintained that contamination of groundwater by pesticides occurred only from accidents such as spills, but it is now known that many pesticides reach the groundwater from normal agricultural use. Since many people (e.g., 40% of Americans) depend on groundwater for their drinking supply, leaching of pesticides is of wide concern. Table 18.2 lists some of the 40 pesticides found in a national survey of well waters in the United States. The concentrations are given in μg/L or parts per billion (see Box 18.1). In some cases, the amount of pesticide found in the drinking water has been high enough to raise long-term health concerns (for example, atrazine and dieldrin exceed the health safety criteria in Table 18.2).

Chemical Reactions

Upon contacting the soil, some pesticides undergo chemical modification independent of soil organisms. For example, iron cyanide compounds decompose within hours or days if exposed to bright sunlight. DDT, diquat, and the triazines are subject to

---

BOX 18.1   CONCENTRATIONS AND TOXICITY OF CONTAMINANTS IN THE ENVIRONMENT

As analytical instrumentation becomes more sophisticated, contaminants can be detected at much lower levels than was the case in the past. Since humans and other organisms can be harmed by almost any substance if large enough quantities are involved, the subject of toxicity and contamination must be looked at *quantitatively*. That is, we must ask *how much*, not simply *what*, is in the environment. Many highly toxic (meaning harmful in very small amounts) compounds are produced by natural processes and can be detected in the air, soil, and water—quite apart from any activities of humans.

The mere presence of a natural toxin or a synthetic contaminant may not be a problem. Toxicity depends on (1) the *concentration* of the contaminant, and (2) the level of *exposure* of the organism. Thus, low concentrations of certain chemicals that would cause no observable effect by a single *exposure* (e.g., one glass of drinking water) may cause harm (e.g., cancer, birth defects) to individuals exposed to these concentrations over a long period of time (e.g., three glasses of water a day for many years).

Regulatory agencies attempt to estimate the effects of long-term exposure when they set standards for no-observable-effect levels (NOEL) or health-advisory levels (see Table 18.2). Some species and individuals within a species will be much more sensitive than others to any given chemical. Regulators attempt to consider the risk to the most susceptible individual in any particular case. For nitrate in groundwater, this individual might be a human infant whose entire diet consists of infant formula made with the contaminated water. For DDT, the individual at greatest risk might be a bird of prey that eats fish that eat worms that ingest lake sediment contaminated with DDT. For a pesticide taken up by plants from the soil, the individual at greatest risk might be an avid gardener who eats vegetables and fruits mainly from the treated garden over the course of a lifetime.

It is important to get a feel for the meaning of the very small numbers used to express the concentration of contaminants in the environment. For instance, concentrations are often given in parts per billion (ppb). This is equivalent to micrograms per kilogram or μg/kg. In water this would be μg/L (Table 18.2). To comprehend the number 1 billion imagine a billion golf balls: lined up, they would stretch completely around the Earth. One bad ball out of a billion (1 ppb) seems like an extremely small number. On the other hand, 1 ppb can seem like a very large number. Consider water contaminated with 1 ppb of potassium cyanide, a very toxic substance consisting of a carbon, a potassium, and a nitrogen atom linked together (KCN). If you drank just one *drop* of this water, you would be ingesting almost 1 trillion molecules of potassium cyanide:

$$\frac{6.023 \times 10^{23}\ \text{molecules}}{1\ \text{mol}} \times \frac{1\ \text{mol}}{65\ \text{g KCN}} \times \frac{1\ \text{g KCN}}{10^6\ \mu\text{g KCN}} \times \frac{1\ \mu\text{g KCN}}{L} \times \frac{L}{10^3\ \text{cm}^3} \times \frac{\text{cm}^3}{10\ \text{drops}} = \frac{9.3 \times 10^{11}\ \text{molecules}}{\text{drop}}$$

In the case of potassium cyanide, the molecules in this drop of water would probably not cause any observable effect. However, for other compounds, this many molecules may be enough to trigger DNA mutations or the beginning of cancerous growth. Assessing these risks is still an uncertain business.

---

slow photodecomposition in sunlight. The triazine herbicides (e.g., atrazine) and organophosphate insecticides (e.g., malathion) are subject to hydrolysis and subsequent degradation. While the complexities of molecular structure of the pesticides suggest different mechanisms of breakdown, it is important to realize that degradation independent of soil organisms does in fact occur.

Microbial Metabolism

Biochemical degradation by soil organisms is the single most important method by which pesticides are removed from soils. Certain polar groups on the pesticide molecules, such as —OH, —COO$^-$, and —NH$_2$, provide points of attack for the organisms.

DDT and other chlorinated hydrocarbons, such as aldrin, dieldrin, and heptachlor, are very slowly broken down, persisting in soils for 20 or more years. In contrast, the organophosphate insecticides, such as parathion, are degraded quite rapidly in soils, apparently by a variety of organisms (Figure 18.7). Likewise, most herbicides (e.g., 2,4-D, the phenylureas, the aliphatic acids, and the carbamates) are readily attacked by a host of organisms. Exceptions are the triazines, which are slowly degraded, primarily by chemical action. Most organic fungicides are also subject to microbial decomposition, although the rate of breakdown of some is slow, causing troublesome residue problems.

Plant Absorption

Pesticides are commonly absorbed by higher plants. This is especially true for those pesticides (e.g., systemic insecticides and most herbicides) that must be taken up in order to perform their intended function. The absorbed chemicals may remain intact inside the plant, or they may be degraded. Some degradation products are harmless, but others are even more toxic to humans than the original chemical that was absorbed. Understandably, society is quite concerned about pesticide residues found in the parts of plants that people eat, whether as fresh fruits and vegetables or as processed foods. The use of pesticides and the amount of pesticide residues in food are strictly regulated by law to ensure human safety. Despite widespread concerns, there is little evidence that the small amounts of residues permissible in foods by law have had any ill effects on public health. However, routine testing by regulatory agencies has shown that about 1 to 2% of food samples tested contain pesticide residues above the levels permissible.

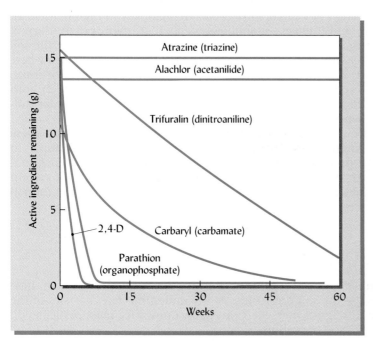

FIGURE 18.7 Degradation of four herbicides (alachlor, atrazine, 2,4-D, and trifuralin) and two insecticides (parathion and carbaryl), all of which are used extensively in the Midwest of the United States. Note that atrazine and alachlor are quite slowly degraded, whereas parathion and 2,4-D are quickly broken down. [From Krueger and Seiber (1984), copyright 1984; reprinted with permission from American Chemical Society]

TABLE 18.5 Common Range of Persistence of a Number of Organic Compounds

Risks of environmental pollution are highest with those chemicals with greatest persistence.

Organic chemical	Persistence in soils
Chlorinated hydrocarbon insecticides (e.g., DDT, chlordane, and dieldrin)	3–20 yr
PCBs	2–10 yr
Triazine herbicides (e.g., atrazine and simazine)	1–2 yr
Glyphosate herbicide	6–20 mo
Benzoic acid herbicides (e.g., amiben and dicamba)	2–12 mo
Urea herbicides (e.g., monuron and diuron)	2–10 mo
Vinyl chloride	1–5 mo
Phenoxy herbicides (2,4-D and 2,4,5-T)	1–5 mo
Organophosphate insecticides (e.g., malathion and diazinon)	1–12 wk
Carbamate insecticides	1–8 wk
Carbamate herbicides (e.g., barban and CIPC)	2–8 wk

Persistence in Soils

The persistence of chemicals in the soil is the net result of all their reactions, movements, and degradations. Marked differences in persistence are the rule (see Figure 18.7). For example, organophosphate insecticides may last only a few days in soils. The widely used herbicide 2,4-D persists in soils for only two to four weeks. PCBs, DDT, and other chlorinated hydrocarbons may persist for 3 to 20 years or longer (Table 18.5). The persistence times of other pesticides and industrial organics fall generally between the extremes cited. The majority of pesticides degrade rapidly enough to prevent buildup in soils receiving normal annual applications. Those that resist degradation have a greater potential to cause environmental damage.

Continued use of the same pesticide on the same land can increase the rate of microbial breakdown of that pesticide. Apparently, having a constant food source allows a population build up of those microbes equipped with the enzymes needed to break down the compound. This is an advantage with respect to environmental quality and is a principle sometimes applied in environmental cleanup of toxic organic compounds, but the breakdown may become sufficiently rapid to reduce a pesticide's effectiveness.

Regional Vulnerability to Pesticide Leaching

The vulnerability of groundwater to contamination by pesticide leaching varies greatly from one area to another. Highest vulnerability occurs in regions with high rainfall, an abundance of sandy soils, and intensive cropping systems that involve high usage of those types of pesticides that are most soluble and least strongly adsorbed by the soil colloids. For example, the southern Atlantic Coast of the United States is an area where sandy soils are prominent, and where pesticide-intensive cropping systems (for fruits, vegetables, peanuts, and cotton) are used. Likewise, vulnerability to leaching of both pesticides and nitrates is high in the Corn Belt, where much of the land is under continuous corn production with its high herbicide and nitrogen fertilizer use.

It should be pointed out that pesticide hazards are site-specific, and that these regional generalizations might mask localized areas of vulnerability. For example, in arid regions irrigated areas of intensive vegetable crop production may experience considerable leaching of both pesticides and nitrates. Likewise, application of certain water-soluble pesticides may result in groundwater contamination even where the soil may not be coarse in texture.

18.4 EFFECTS OF PESTICIDES ON SOIL ORGANISMS

Since pesticides are formulated to kill organisms, it is not surprising that some of these compounds are toxic to specific soil organisms. At the same time, the diversity of the soil organism population is so great that, excepting a few fumigants, most pesticides do not kill a broad spectrum of soil organisms.

Fumigants

Fumigants are compounds used to free a soil of a given pest, such as nematodes. These compounds have a more drastic effect on both the soil fauna and flora than do other pesticides. For example, 99% of the microarthropod population is usually killed by such fumigants as DD and vampam, and it may take as long as two years for the population to fully recover. Fortunately, the recovery time for the microflora is generally much less.

Fumigation reduces the number of species of both flora and fauna, especially if the treatment is repeated, as is often the case where nematode control is attempted. At the same time, the total number of bacteria is frequently much greater following fumigation than before. This increase is probably due to the relative absence of competitors and predators following fumigation and to the carbon and energy sources left by dead organisms for microbial utilization.

Effects on Soil Fauna

The effects of pesticides on soil animals varies greatly from chemical to chemical and from organism to organism. Nematodes are not generally affected, except by specific fumigants. Mites are generally sensitive to most organophosphates and to the chlorinated hydrocarbons, with the exception of aldrin. Springtails vary in their sensitivity to both chlorinated hydrocarbons and organophosphates, some chemicals being quite toxic to these organisms.

EARTHWORMS. Fortunately, many pesticides have only mildly depressing effects on earthworm numbers, but there are exceptions. Among insecticides, most of the carbamates (carbaryl, carbofuran, aldicarb, etc.) are highly toxic to earthworms. Among the herbicides, simazine is more toxic than most. Among the fungicides, benomyl is unusually toxic to earthworms. The concentrations of pesticides in the bodies of the earthworms are closely related to the levels found in the soil (Figure 18.8). Thus, earthworms can magnify the pesticide exposure of birds, rodents, and other creatures that prey upon them.

Pesticides have significant effects on the numbers of certain predators and, in turn, on the numbers of prey organisms. For example, an insecticide that reduces the

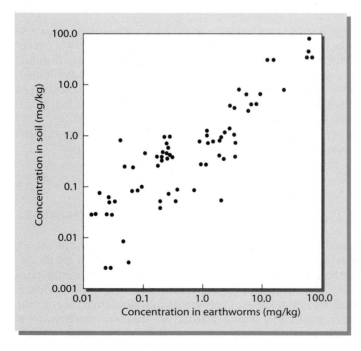

FIGURE 18.8 Effect of concentration of pesticides in soil on their concentration in earthworms. Birds or rodents eating the earthworms at any level of concentration would further concentrate the pesticides. [Data from several sources gathered by Thompson and Edwards (1974); used with permission of Soil Science Society of America]

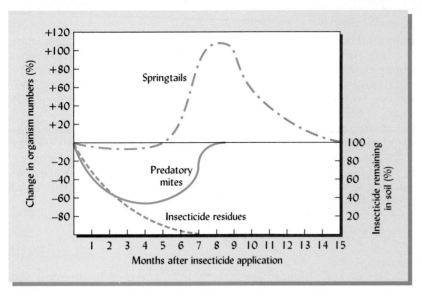

FIGURE 18.9 The direct effect of insecticide on predatory mites in a soil and the indirect effect of reducing mite numbers on the population of springtails (tiny insects) that serve as prey for the mites. [Replotted from Edwards (1978)]

numbers of predatory mites may stimulate numbers of springtails, which serve as prey for the mites (Figure 18.9). Such organism interaction is normal in most soils.

Effects on Soil Microorganisms

The overall levels of bacteria in the soil are generally not too seriously affected by pesticides. However, the organisms responsible for nitrification and nitrogen fixation are sometimes adversely affected. Insecticides and fungicides affect both processes more than do most herbicides, although some of the latter can reduce the numbers of organisms carrying out these two reactions. Recent evidence suggests that some pesticides can enhance biological nitrogen fixation by reducing the activity of protozoa and other organisms that are competitors or predators of the nitrogen-fixing bacteria. These findings illustrate the complexity of life in the soil.

Fungicides, especially those used as fumigants, can have marked adverse effects on soil fungi and actinomycetes, thereby slowing down the humus formation in soils. Interestingly, however, the process of ammonification is often stimulated by pesticide use.

The negative effects of most pesticides on soil microorganisms are temporary, and after a few days or weeks, organism numbers generally recover. But exceptions are common enough to dictate caution in the use of the chemicals. Care must be taken to apply them only when alternate means of pest management are not available.

This brief review of the behavior of organic chemicals in soils reemphasizes the complexity of the changes that take place when new and exotic substances are added to our environment. Our knowledge of the soil processes involved certainly reaffirms the necessity for a thorough evaluation of potential environmental impacts prior to approval and use of new chemicals for extensive use on the land.

18.5 REMEDIATION OF SOILS CONTAMINATED WITH ORGANIC CHEMICALS

U.S. Geological Survey bioremediation projects:
http://water.usgs.gov/wid/html/bioremed.html

Soils contaminated with organic pollutants are found throughout the world. The wide areas contaminated with organic pesticides are best addressed by modifying the agro-ecosystems to enable the use of pesticides to be reduced or eliminated, or by using less toxic, less mobile, and more rapidly degradable pesticide compounds. In many cases, the soil ecosystem should be able to recover its function and diversity through *natural attenuation* over a reasonable period. Natural attenuation may involve any or all of the chemical, physical, and biological processes that were illustrated in Figure 18.3.

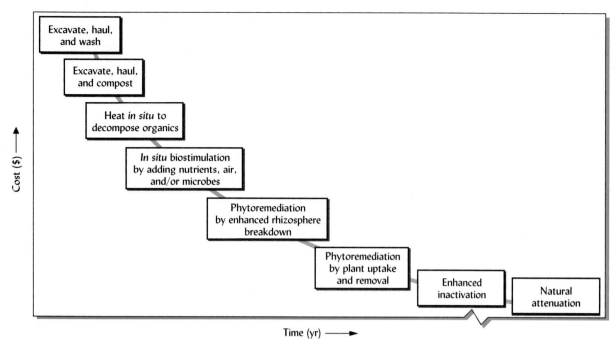

FIGURE 18.10 A wide range of methods is available to remediate (clean up) polluted soils. At one extreme are remediation techniques that are very expensive and disruptive, but usually quite rapid. Technologies at the other extreme may be quite inexpensive and nondisruptive, but usually take much more time to accomplish the cleanup. [Modified from Reynolds et al. (1999)]

Perhaps more problematic, however, are sites where accidental spills of toxic organic materials have occurred or where, through the decades, organic wastes from industrial and domestic processes have been dumped on soils. The levels of such *acute contamination* are often sufficiently high that plant growth is restrained or even prevented. Pollutants can move into the groundwater, making it unfit for human consumption. Fish and wildlife may be decimated. Because of public concerns, businesses and government are spending billions of dollars annually to clean up (**remediate**) these contaminated soils. We shall consider a few of the methods in use and under development in the rapidly evolving soil remediation industry. In general, efforts to remediate polluted soils face the need to compromise between speed and certainty that cleanup standards will be met on one hand, and expense and disruption of the site on the other (Figure 18.10).

Physical and Chemical Methods

The earliest and still most widely used methods of soil remediation involve physical and/or chemical treatment of the soil, either in place (*in situ*) or by moving the soil to a treatment site (*ex situ*).

Alternative Cleanup
Technologies for
Underground Storage
Tank Sites:
http://www.epa.gov/OUST/
pubs/turns.htm

Ex Situ Treatment. *Ex situ* treatment may involve excavating the soil to treatment bins where it may be incinerated to drive off volatile chemicals and to destroy other pollutants by high-temperature chemical decomposition. Water-soluble and volatile chemicals may also be removed by pushing or pulling air or water through the soil by vacuum extraction or leaching. Such treatments are usually quite effective in removing or destroying the contaminants, but are expensive, especially if large quantities of soil must be excavated and treated. And, of course, the treated soil is also destroyed as a living system and must be either replaced on the site or deposited in a landfill.

In Situ Treatment. *In situ* treatments are usually preferred if viable technologies are available. The soil is left in place, thereby reducing excavation, treatment, and disposal costs and providing greater flexibility in future land use. The contaminants are either removed from the soil (*decontamination*) or are sequestered (*bound up*) in the soil matrix (*stabilized*). Decontamination in situ involves some of the same techniques of water flushing, leaching, vacuum extraction, and heating used in ex situ processes. Water treatment is not effective, however, with nonpolar compounds that are repelled by water. To help remove such compounds, scientists and engineers have sprayed onto the soil surface or have injected into the soil compounds called *surfactants*. As these move downward in the soil, they dissolve organic contaminants, which can then be pumped out of the soil as in the water-washing systems.

ORGANOCLAYS.　Certain surfactants may also be used to immobilize or stabilize soil contaminants. They are positively charged and through cation exchange can replace metal cations on soil clays. For example, one group of such surfactants, quaternary ammonium compounds (QACs), has the general formula $(CH_3)_3NR^+$, where R is an organic alkyl or aromatic group. The positive charges on QACs stimulate cation exchange by reactions such as the following, using a monovalent exchangeable cation such as K^+ as an example:

$$\boxed{\text{Colloid}}\ K^+ + (CH_3)_3NR^+ \rightarrow \boxed{\text{Colloid}}\ (CH_3)_3NR^+ + K^+ \qquad \text{(eq. 18.1)}$$

Untreated clay　　　　QAC　　　　Organoclay

The resulting products, known as *organoclays*, have properties quite different from the untreated clays. They attract rather then repel nonpolar organic compounds. Thus, the injection of a QAC into the zone of groundwater flow can stimulate the formation of organoclays and thereby immobilize soluble organic groundwater contaminants, holding them until they can be degraded (Figure 18.11).

DISTRIBUTION COEFFICIENTS K_d.　As we learned in Section 8.12, the degree of sorption of organic compounds by soil colloids is commonly indicated by the coefficient of distribution K_d between the sorbed and solution portions of the organic compound.

$$K_d = \frac{(\text{mg contaminant/kg soil})}{(\text{mg contaminant/L solution})} \qquad \text{(eq. 18.2)}$$

The K_d for the adsorption of many nonpolar organic compounds on untreated clays is very low because the clays are hydrophilic (water-loving) and their adhering water films repel the hydrophobic, nonpolar organic compounds (Figure 18.12). Surface soil horizons containing significant quantities of humus often exhibit a much higher K_d because of the sorption of the organic contaminant into the organic matter coatings. This is the reason that K_{oc} is often a better measure of a compound's tendency to become immobilized in various surface soils. Deeper soil layers, especially near and

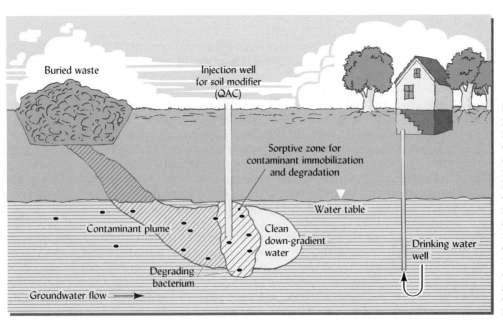

FIGURE 18.11　How a combination of a quaternary ammonium compound (QAC), hexadecyltrimethylammonium, and bioremediation by degrading bacteria could be used to hold and remove an organic contaminant. The pollutant is moving into groundwater from a buried waste site. The QAC reacts with soil clays to form organoclays and soil organic matter complexes that adsorb and stabilize the contaminant, giving microorganisms time to degrade or destroy it. [Redrawn from Xu et al. (1997)]

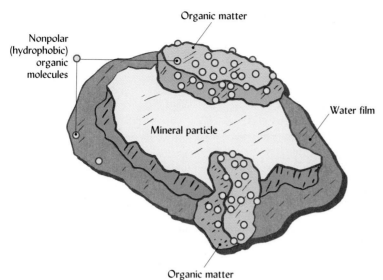

FIGURE 18.12 Because mineral colloids in soils are nearly always surrounded by at least a thin film of water, hydrophobic organic molecules tend to sorb onto humus more readily than clay. The nonpolar organic molecules cannot compete with the polar water molecule for a place on the charged mineral surfaces. This is one reason why, for a specific organic contaminant, the K_{oc} is more consistent than the K_d in characterizing the tendency to be held by various soils. [Diagram courtesy of R. Weil]

below the water table, generally contain little humus, and so have limited capacity to immobilize organic contaminants.

In contrast, organoclays effectively sorb organic contaminants, leaving little in the soil solution, thereby reducing their movement into the groundwater and eventually into streams or drinking water. Consequently, the K_d values of organic contaminants on organoclays are commonly 100 to 200 times those measured on the untreated clays. Table 18.6 shows K_d values for some common organic contaminants on organoclays and indicates the very high sorbing power of the newly created sorbants. Their tenacity is complemented by the very strong complexation of organic compounds by soil organic matter. Organoclays thus offer promising mechanisms for holding organic soil pollutants until they can be destroyed by biological or physicochemical processes.

Bioremediation[3]

Composting to clean up munitions-contaminated soils:

www.epa.gov/epaoswer/non-hw/compost/explos.pdf

For many heavily contaminated soils, there is a biological alternative to incineration, soil washing, and landfilling—namely, **bioremediation**. Simply put, this technology uses enhanced plant and/or microbial action to degrade organic contaminants into harmless metabolic products. Analysis of microbial DNA has shown that degradation of contaminants in soils is almost always the work of genetically diverse *consortia* of many organisms, rather than just one or two bacterial species. Petroleum constituents,

---

[3] For reviews of the theories and technologies regarding this topic, see Alexander (1994), Skipper and Turco (1995), Wise et al. (2000), and Eccles (2007).

TABLE 18.6 **The Organic Level After Treating Clays Varying in Cation Exchange Capacity with a Quaternary Ammonium Compound (QAC) to Form Organoclays, and the Sorption Coefficients K_d of Five Organic Compounds on these Organoclays**

High K_d values suggest high retention of the pollutants and low concentration in the soil solution. Note the low tendency for kaolinite and illite to form organoclays, and the variability in sorption coefficients of the different compounds.

Clay	CEC of untreated clay, $cmol_c/kg$	Organic C in organoclay, %	K_d of organic contaminants on organoclays				
			Benzene	Toluene	Ethylbenzene	Propylbenzene	Naphthalene
Illite	24	2.5	39	77	156	—	1270
Vermiculite	80	16.4	68	169	448	1618	1387
Smectite (high charge)	130	23.0	184	319	583	1412	4818
Kaolinite	4	1.0	3	7	21	—	—

Data from Jaynes and Boyd (1997).

FIGURE 18.13 Hot water vapor rises in cold winter air as windrows of high-temperature compost are mixed and aerated by a special compost-turning machine in order to accelerate the breakdown of organic compounds. The method can hasten the degradation of organic pollutants in soil material excavated from a contaminated site, mixed with decomposable organic materials, and made into windrows. (Photo courtesy of R. Weil)

including the more resistant polyacrylic aromatic hydrocarbons (PAHs), as well as several synthetic compounds, such as pentachlorophenol (PCP) and trichloroethylene (TCE), can be broken down, primarily by soil bacteria and so-called white-rot fungi. Bioremediation is usually accomplished *in situ,* but polluted soil may also be excavated and treated *ex-situ*—that is, hauled to a treatment site where such techniques as high-temperature composting may be used to destroy the organic contaminants in the soil (Figure 18.13).

BIOAUGMENTATION. In some cases, the remediation process depends on organisms native to the soil. In others, microbes specifically selected for their ability to remove the contaminants are introduced into the polluted soil zone to *augment* the natural microbial populations. This approach is called **bioaugmentation.** For example, certain bacteria have been identified that can detoxify perchloroethene (PCE), a common, highly toxic groundwater pollutant that is suspected of being a carcinogen. Scientists can inoculate with these organisms to expedite the step by step removal of the four chlorines from the PCE, producing ethylene, a gas that is relatively harmless to humans.

$$\underset{\substack{\text{PCE}\\\text{(Suspected carcinogen)}}}{\text{Cl}_2\text{C}=\text{CCl}_2} \quad \xrightarrow[\text{4HCl}]{8\text{H}} \quad \underset{\substack{\text{Ethylene}\\\text{(Harmless gas)}}}{\text{H}_2\text{C}=\text{CH}_2}$$

Global interest group for bioremediation—try "Biolinks":
http://www.bioremediationgroup.org/AboutUs/Home.htm

BIOSTIMULATION. The use of bioremediation technology that assists the naturally occurring microbial populations in breaking down chemicals is called **biostimulation.** Usually, the soil naturally contains some bacteria or other microorganisms that can degrade the specific contaminant. But the rate of natural degradation may be far too slow to be very effective. Both growth rate and metabolic rate of the organisms capable of using the contaminant as a carbon source are often limited by insufficient mineral nutrients, especially nitrogen and phosphorus (see Section 12.3 for a discussion of the C/N ratio in organic decomposition). Special fertilizers have been formulated and used successfully to greatly speed up the degradation process. One such fertilizer of French manufacture is an oil-in-water microemulsion of urea, lauryl phosphate, and an emulsion stabilizer. It acts not only as a supplier of nutrients, but also as a surfactant that can enhance interaction between microbes and the organic contaminants. It received its first major test in Alaska in 1989.

ALASKA OIL SPILL CLEANUP. The 30 or more different genera of bacteria and fungi known to degrade hydrocarbons are found in almost any soil or aquatic environment. But they may need help. The cleanup of crude oil contamination from the 1989 Exxon Valdez

FIGURE 18.14 Bioremediation of crude oil from the Exxon Valdez oil spill off the coast of Alaska. The oil contaminating the beach soils was degraded by indigenous bacteria when an oil-soluble fertilizer containing nitrogen and phosphorus was sprayed on the beach (o data points in graph on the left). The control sections of the beach (+ data points) were left unfertilized for 70 days. By then the effect of the fertilization was so dramatic that a decision was made to treat the control sections as well. The index of oil remaining is based on natural logarithms, so each whole number indicates more than doubling of oil remaining. The photo (*right*) shows the clear delineation between the oil-covered control section and the fertilized parts of the beach. [Data from Bragg et al. (1994); reprinted with permission from *Nature*, © 1994 Macmillan Magazines Limited; photo courtesy of P. H. Pritchard, USEPA, Gulf Breeze, Fla.; from Pritchard et al. (1992); reprinted by permission of Kluwer Academic Publishers]

oil spill in Alaskan waters was a spectacular case of successful bioremediation by fertilization (Figure 18.14). A special fertilizer was sprayed on the oil-soaked beaches (Entisols). The fertilizer was formulated to be oliophilic (soluble in oil but not in water) so that it would stay with the oil and not contribute to eutrophication of Prince William Sound. Within a few weeks, and despite the cold temperatures, most of the oil in the test area was degraded. The success of bioremediation was greatest where nitrogen was most available to the microorganisms.

IN SITU BIOSTIMULATION TECHNIQUES. Other situations call for the use of other bioremediation techniques in situ. In some cases, low soil porosity causes oxygen deficiency that limits microbial activity. Techniques are being developed that use *in situ* bioremediation to clean up oxygen-deficient soils and associated groundwater contamination. For example, organic-solvent-contaminated soils have been bioremediated (Figure 18.15) by piping in a mixture of air (for oxygen), methane (to act as a carbon source to stimulate specific bacteria), and phosphorus (a nutrient that is needed for bacteria growth).

Some success has been achieved by inoculating contaminated soils with improved organisms that can degrade the pollutant more readily than can the native population. Although genetic engineering may prove useful in making "superbacteria" in the future, most inoculation has been achieved with naturally occurring organisms. Organisms isolated from sites with a long history of the specific contamination or grown in laboratory culture on a diet rich in the pollutant in question tend to become acclimated to metabolizing the target chemical.

Phytoremediation

Higher plants can also participate in bioremediation, a process termed **phytoremediation**. For years, plant-based systems have been used for the removal of municipal wastewater contaminants (Sections 16.2 and 13.9). More recently, this concept has been

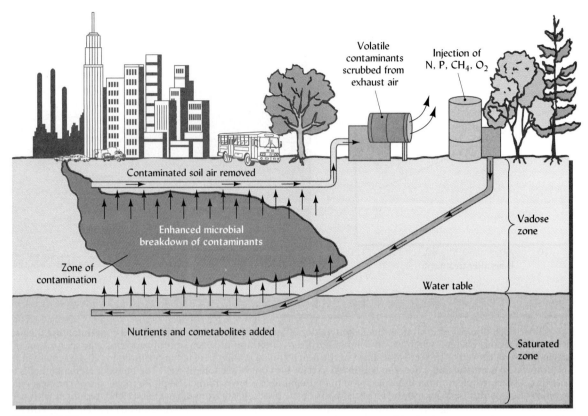

FIGURE 18.15 *In situ* bioremediation of soil and groundwater contaminated with volatile organic solvents. The scheme illustrated is typical of the *biostimulation* approach to soil remediation. Breakdown of the organic contaminant is stimulated by adding such components as nutrients, oxygen, and cometabolites that improve the soil environment for the growth of native bacteria capable of metabolizing the contaminant. In this instance, methane (CH_4) is added intermittently as a substrate for certain methane-oxidizing bacteria which multiply rapidly and turn to the solvent as a carbon source whenever methane is not available. The nutrients are added and the contaminated soil air is removed by perforated pipes inserted by horizontal well-drilling techniques. Such biostimulation schemes can significantly cut the time and cost for cleanup of contaminated soils. [Based on Hazen (1995)]

Links to many sites on phytoremediation: http://www.dsa.unipr.it/ phytonet/links.htm

extended to industrial pollutants and to the removal of shallow groundwater pollutants of all kinds, both organic and inorganic.

Phytoremediation uses plants in two fundamentally different ways (Figure 18.16). In the first, plant roots take up the pollutant from the soil. The plant may then either accumulate large amounts of the contaminant in aboveground biomass, or it may metabolize the contaminant into harmless by-products. The accumulation of unusually high concentrations of a contaminant in the plant biomass is called **hyperaccumulation.** Hyperaccumulating plants take up and tolerate very high concentrations of a contaminant, most commonly a toxic metal such as zinc or nickel, but also certain organics such as trinitrotoluene (TNT). Hyperaccumulation allows the contaminant to be removed by harvesting the plant tissue.

The second type of cleanup using plants is called **enhanced rhizosphere phytoremediation.** In this process, the plants do not take up the contaminant. Instead, the plant roots excrete into the soil carbon compounds that serve as microbial substrates and growth regulators (see Section 11.7). These compounds stimulate the growth of the rhizosphere bacteria that, in turn, degrade the organic contaminant. The transpiration of water by the plant causes soil water, with its load of dissolved contaminant molecules, to move toward the roots, thus increasing the efficiency of the rhizosphere reactions.

Research has shown that some plant species are better than others at stimulating the degradation of specific compounds in their rhizospheres. Many plant species, domesticated and wild, have been used in phytoremediation. Prairie grasses can stimulate the degradation of petroleum products, including PAHs, and spring wildflower plants in Kuwait were recently found to degrade the hydrocarbons in oil spills. Fast-growing hybrid poplars can remove ammunition compounds, such as TNT, as well as some

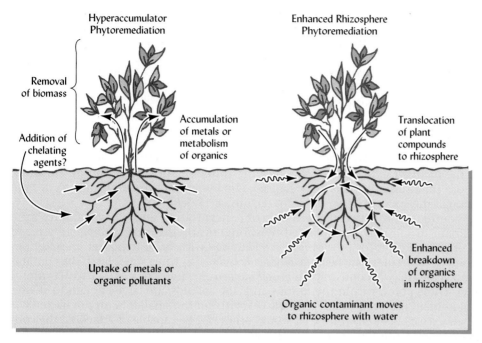

FIGURE 18.16 Two approaches to phytoremediation—the use of plants to help clean up contaminated soils. (*Left*) *Hyperaccumulating plants* take up and tolerate very high concentrations of an inorganic or organic contaminant. In the case of metal contaminants, the addition of chelating agents may increase the rate of metal uptake, but can add a major expense and may allow metals to migrate below the root zone. (*Right*) In *enhanced rhizosphere phytoremediation,* the plants do not take up the contaminant. Instead, the plant roots excrete substances that stimulate the microbes in the rhizosphere soil, speeding their degradation of organic contaminants. Transpiration-driven movement of water and dissolved contaminants to the enhanced rhizosphere zone improves the system's effectiveness. (Diagram courtesy of R. Weil)

pesticides and excess nitrates. Figure 18.17 illustrates how phytoremediation hastened the cleanup of two oil-contaminated sites under cold-weather conditions.

Phytoremediation is particularly advantageous where large areas of soil are contaminated with only moderate concentrations of organic pollutants. However, phytoremediation also commonly takes a longer time to remove large quantities of contaminants than do the more costly engineering procedures.

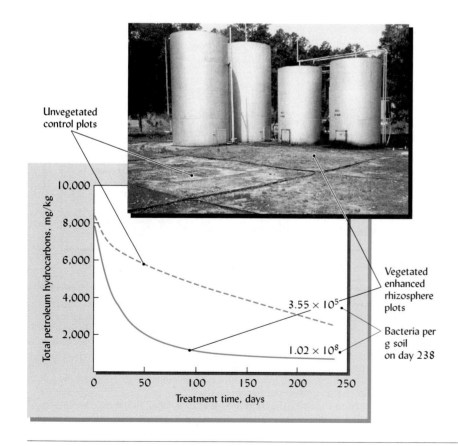

FIGURE 18.17 Rhizosphere-enhancement phytoremediation of oil contaminated soils. (*Photo*) Test plots adjacent to oil storage tanks in Arkansas where an accidental spill occurred. Two cool-season grasses (tall fescue and ryegrass) were seeded on the enhanced rhizosphere plots in January, and the photograph was taken 5 months later. (*Graph*) Attenuation of total petroleum hydrocarbons (TPHs) in another similar oil-contaminated soil near Fairbanks, Alaska. Where plant roots created rhizosphere conditions in much of the soil volume, an 80% reduction in TPH concentration was achieved in less than 100 days. In the unvegetated plots, natural attenuation had not achieved this level of cleanup even after 238 days. Fertilizer was used to stimulate the growth of plants, whose roots, in turn, enhanced the growth and activity of microorganisms, which finally provided the enzymatic activity to break down the petroleum compounds in the soil. Note that bacteria were nearly 300 times as numerous in the vegetated and fertilized *enhanced rhizosphere* plots as in the unvegetated *natural attenuation* control plots. [Photo courtesy of Duane Wolf, University of Arkansas; graph redrawn from Reynolds et al. (1999) with permission of Cambridge University Press.

As we have seen, when an organic contaminant is added to soil, biological and chemical processes begin to degrade the contaminant so its total concentration declines steadily for weeks, months, or years, depending on its half-life. Researchers have found, however, that many organic contaminants may undergo an aging process in the soil, whereby over time the contaminant becomes less and less subject to decomposition even though relatively high concentrations can still be detected by laboratory analyses. When soil scientists use bacteria, earthworms, plants, or animal feeding in **bioassays** to determine the effect of the compound on living organisms, the initial period of degradation is usually paralleled by a similar period of declining biological activity. In other words, the *bioavailability* at first parallels the contaminant concentration. However, after more time (usually years) has elapsed, the bioavailability continues to decline even though the concentration does not (Figure 18.18, *left*).

The contradiction between the trends of concentration and bioavailabiliity for a contaminant in soil may be explained by strong, nearly irreversible *sorption* by Fe, Al oxides or silicate clays, or by *chemical complexation* with organic matter. In addition, contaminant molecules may become *physically isolated* in three ways. First, the contaminant molecules may become *trapped in soil nanopores*, tiny pore spaces 1 to 100 nm in diameter. These pores, within the humus and clay colloids, are large enough to shelter the contaminant molecule (*a* in Figure 18.18 *right*), but too small for entry of bacteria or even their extracellular enzymes that would otherwise be capable of attacking the contaminant. Second, the contaminant molecules may diffuse into or be absorbed onto the solid structure of a humus particle, where, again, they would not be exposed to living cells or their enzymes (*b* in Figure 18.18 *right*). Third, contaminant molecules may become buried, or occluded, under precipitated mineral coatings, again isolating them from biological interactions (*c* in Figure 18.18 *right*). Pollutant molecules trapped by any of these mechanisms may offer little risk of environmental mobility or biological toxicity even if their total concentration in the soil remains high. Because of this aging process, some scientists believe that environmental cleanup standards should be based on a contaminant's measurable bioavailability rather than on its total concentration.

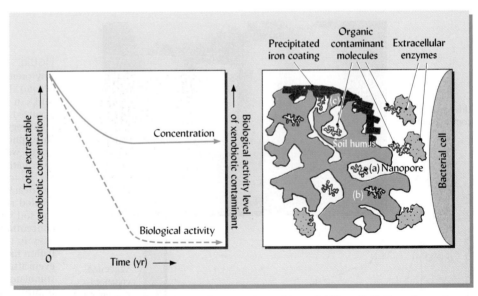

FIGURE 18.18   The effect of aging on soil contaminant bioavailability. (*Left*) Over time, chemical and physical "aging" interactions with soil colloids may slow microbial degradation of many organic contaminants, even while bioavailability continues to decline. (*Right*) The decline in bioavailability can be due to irreversible sorption by mineral colloids and chemical complexation by humus, but some contaminant molecules (*b*) may become physically isolated in nanopores (*a*) that are too small to give access to bacteria or even their large, extracellular enzyme molecules. Since pollutants so held are not likely to move into the groundwater or have an effect on organisms in the soil or in the food chain, some scientists argue that there is no need to attempt to destroy them or remove them from the soil. [Diagrams courtesy of R. Weil, based on concepts in Alexander (2000) and (Stokes et al., 2005)

Therefore, regulations that set standards for total soil pollutant concentrations may be more strict (and more expensive) than is actually necessary to achieve an acceptably low level of risk. The question of whether to base environmental standards on the total amount of a compound present or on the bioavailability of the compound also applies to contamination by inorganic substances, the topic of the following section.

18.6 CONTAMINATION WITH TOXIC INORGANIC SUBSTANCES[4]

The toxicity of inorganic contaminants released into the environment every year is now estimated to exceed that from organic and radioactive sources combined. A fair share of these inorganic substances ends up contaminating soils. The greatest problems most likely involve mercury, cadmium, lead, arsenic, nickel, copper, zinc, chromium, molybdenum, manganese, selenium, fluorine, and boron. To a greater or lesser degree, all of these elements are toxic to humans and other animals. Cadmium and arsenic are extremely poisonous; mercury, lead, nickel, and fluorine are moderately so; boron, copper, manganese, and zinc are relatively lower in mammalian toxicity. Table 18.7 provides background information on the uses, sources, and effects of some of these elements. Although the metallic elements (see periodic table, Appendix B) are not all, strictly speaking, "heavy" metals, for the sake of simplicity this term is often used in referring to them.

Sources and Accumulation

There are many sources of the inorganic chemical contaminants that can accumulate in soils. The burning of fossil fuels, smelting (Figure 18.19), and other processing techniques release into the atmosphere tons of these elements, which can be carried for miles and later deposited on the vegetation and soil. Lead, nickel, and boron are gasoline additives that are released into the atmosphere and carried to the soil through rain and snow. Boron as the mineral borax is used in detergents, fertilizers, and forest fire retardants, all of which commonly reach the soil. Superphosphate and limestone, two widely used soil amendments, usually contain small quantities of cadmium, copper, manganese, nickel, and zinc. Cadmium is used in plating metals and in the manufacture

---

[4] For an excellent in-depth review of all aspects of this topic, see Adriano (2001). Additional information is available in Ahmad et al. (2006). Heavy metal contamination of food grown in China is discussed in Jamiska and Spencer (2007).

TABLE 18.7 **Sources of Selected Inorganic Soil Pollutants**

Chemical	Major uses and sources of soil contamination	Organisms principally harmed[a]	Human health effects
Arsenic	Pesticides, plant desiccants, animal feed additives, coal and petroleum, mine tailings, detergents, and irrigation water	H, A, F, B	Cumulative poison, cancer, skin lesions
Cadmium	Electroplating, pigments for plastics and paints, plastic stabilizers, batteries, and phosphate fertilizers	H, A, F, B, P	Heart and kidney disease, bone embrittlement
Chromium	Stainless steel, chrome-plated metals, pigments, refractory brick manufacture, and leather tanning	H, A, F, B	Mutagenic; also essential nutrient
Copper	Mine tailings, fly ash, fertilizers, windblown copper-containing dust, and water pipes	F, P	Rare; mental problems, fatigue; essential nutrient
Lead	Combustion of oil, gasoline, and coal; iron and steel production; solder in water-pipes; paint pigment	H, A, F, B	Brain damage, convulsions
Mercury	Pesticides, catalysts for synthetic polymers, metallurgy, and thermometers	H, A, F, B	Nerve damage
Nickel	Combustion of coal, gasoline, and oil; alloy manufacture; electroplating; batteries; and mining	F, P	Lung cancer
Selenium	High Se geological formations and irrigation wastewater in which Se is concentrated	H, A, F, B	Rare; loss of hair; nail deformities; essential nutrient
Zinc	Galvanized iron and steel, alloys, batteries, brass, rubber manufacture, mining, and old tires	F, P	Rare; essential nutrient

[a]H = humans, A = animals, F = fish, B = birds, P = plants.

FIGURE 18.19 A partially denuded hillside just downwind from a copper smelter in Anaconda, Montana. The heavy metal-laden fumes have contaminated this area with copper, zinc, nickel, and other metals to levels that are highly toxic to most plants and many other organisms, decimating the ecosystems of the area. Note the serious erosion that has resulted from the devegetation, despite the recent invasion of the area by a few metal-tolerant plant species. (Photo courtesy of R. Weil)

of batteries. Arsenic was for many years used as a wood preservative as well as an insecticide on cotton, tobacco, fruit crops, lawns, and as a defoliant or vine killer. Some of these mentioned elements are found as constituents in specific organic pesticides and in domestic and industrial sewage sludge. Additional localized contamination of soils with metals results from ore-smelting fumes, industrial wastes, and air pollution.

Some of the toxic metals are being released to the environment in increasing amounts, while others (most notably lead, because of changes in gasoline formulation) are decreasing. All are daily ingested by humans, either through the air or through food, water, and—yes—soil (see Box 18.2).

Concentration in Organism Tissue

Irrespective of their sources, toxic elements can and do reach the soil, where they become part of the food chain: soil→plant→animal→human (Figure 18.20). Unfortunately, once the elements become part of this cycle, they may accumulate in animal and human body tissue to toxic levels. This situation is especially critical for fish and other wildlife and for humans at the top of the food chain. It has already resulted in restrictions on the use of certain fish and wildlife for human consumption. Because of the globalization of our food supply, crops grown on polluted soils in countries with weak environmental regulations (especially those with a history of "dirty" industrialization, such as China) may threaten the safety of food consumed in importing countries around the world. Western nations have had to closely regulate the release of these toxic elements in the form of industrial wastes.

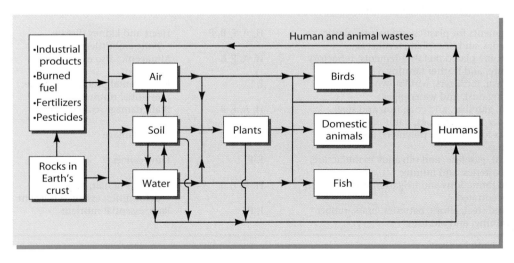

FIGURE 18.20 Sources of heavy metals and their cycling in the soil–water–air–organism ecosystem. It should be noted that the content of metals in tissue generally builds up from left to right, indicating the vulnerability of humans to heavy metal toxicity.

Lead contamination is a serious and widespread form of inorganic soil pollution. Long-term exposure to low levels of lead can profoundly affect a child's development and neurological function, including intelligence. Lead poisoning has been shown to contribute to mental retardation, poor academic performance, and juvenile delinquency. In the past (and, unfortunately, in the present in many developing countries that still use leaded gasoline), much of the lead exposure came from burning leaded fuels. The content of lead in soils commonly increases with proximity to major highways. The lead content of soils also usually increases as one approaches the center of a major city. Residents of inner cities generally live surrounded by lead-contaminated soils. The soil on the windward side of apartment buildings often shows the highest accumulations of lead, as it is there that the wind-carried particulates tend to settle out of the air. A second reason for high lead concentrations in urban soils is related to the lead-based pigments in paint from pre-1970 era buildings. Paint chips, flakes, and dust from sanding painted surfaces spread the lead around, and eventually much of it ends up in the

FIGURE 18.21 *Hand-to-mouth activity is a major pathway for lead poisoning in young children. Lead from car exhaust and old paint accumulates in urban soils. (Photo courtesy of R. Weil)*

soil. During dry weather, soil particles blow about, spreading the lead and contributing to the dust that settles on floors and windowsills. Although plants do not readily take up lead through their roots, lead-contaminated dust may stick to foliage and fruits.

Eating these garden products and breathing in lead-contaminated dust are two pathways for human lead exposure (see Figure 18.20). However, the most serious pathway, at least for young children, is thought to be hand-to-mouth activity—basically, eating dirt (see also Box 1.1). Anyone who has observed a toddler knows that a child's hands are continually in its mouth (Figure 18.21). Lead-contaminated dust on surfaces in the home can therefore be an important source of lead exposure for young children; so, too, can lead-contaminated soil in outdoor play areas. Having children wash their hands frequently can significantly cut down their exposure to this insidious toxin. The U.S. EPA has set standards for the cleanup of lead in soil around homes: 400 parts per million (ppm) of lead in bare soil in children's play areas or 1200 ppm average for bare soil in the rest of the yard. Soils with lead levels higher than these standards require some remediation.

Since 1970, major government programs have aimed to reduce lead exposure from paint (lead paints are banned and existing lead paint must be removed or sealed), drinking water, and food (lead has been banned

from solder used in pipe joints and food cans). As a result, median levels of lead in blood samples from U.S. children fell from 0.18 mg/L in 1970 to 0.03 mg/L in 1994. For children under age 6, blood levels above 10 mg Pb/L are considered elevated and a threat to health. Unfortunately, lead in soils has not been similarly addressed. The U.S. EPA reports that nearly 1 million children in America (especially in central cities) have dangerously elevated levels of lead in their blood (above 0.10 mg/L). Not only are children of poor, urban families exposed to lead-contaminated soils, but their diets are typically low in phosphorus and calcium, nutrients that could make them less susceptible to lead toxicity.

Current measures designed to protect children from lead in soil around the home include (1) excavation and removal of the soil, (2) dilution by mixing in large amounts of noncontaminated soil, or (3) stabilization of the lead away from the reach of children and dust-creating winds. Excavation of soil around homes is extremely expensive and, given the low mobility of lead in soils, probably not necessary. Instead, the contaminated soil area may be covered with a thick layer of uncontaminated topsoil, a

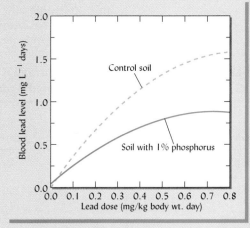

FIGURE 18.22 *In-situ treatment of highly lead-contaminated soil with 1% P by weight substantially reduced the bioavailability of the lead, as evidenced by lower lead levels in the blood of pigs fed the soil. [Graphed from data in Ryan et al. (2004)]*

BOX 18.2 *(Cont.)* LEAD CONTAMINATION AND POISONING

wooden deck, or pavement. Well-maintained turfgrass will prevent most dust formation and soil ingestion. Removal of lead-based paints is likewise very expensive and difficult, so isolating the lead-based paint under several coats of fresh paint can limit exposure.

Effective phytoremediation techniques have not yet been developed to get the lead out of soils; however, leaving the lead in the soil but transforming it into a form that is not bioavailable might be just as good. Recent research (Figure 18.22) shows that phosphorus (fertilizer) can effectively transform bioavailable lead compounds into relatively inert, nonbioavailable forms. Orthophosphate ions (PO_4^{3-}) efficiently react with Pb in the soil solution to precipitate very insoluble minerals, mainly lead pyromorphite [$Pb_5(PO_4)_3OH,Cl,F$]. This approach was tested near Joplin, Missouri where a lead smelter had contaminated local soils from the 1880s to the 1960s, resulting in average Pb levels of 2400 mg Pb/kg. The soil was amended with phosphorus equal to 1% of the soil by weight (500 to 1000 times normal fertilization rates). After 32 months, P-treated and untreated soils were sampled and fed at several dose-rates to young pigs (whose digestive systems are similar to humans'). Lead levels in the blood of pigs fed the phosphorus-treated soil were much lower than those fed the control soil (Figure 18.22), suggesting that although P-treatment leaves the lead in the soil, it could substantially reduce the hazard to children.

18.7 POTENTIAL HAZARDS OF CHEMICALS IN SEWAGE SLUDGE

The domestic and industrial sewage sludges considered as nutrient sources in Chapter 16 can be important sources of potentially toxic chemicals. Nearly half of the municipal sewage sludge produced in the United States is being applied to the soil, either on agricultural land or to remediate land disturbed by mining and industrial activities. Industrial sludges commonly carry significant quantities of inorganic as well as organic chemicals that can have harmful environmental effects.

SOURCE REDUCTION PROGRAMS. A great deal was learned during the 1970s and 1980s about the contents, behavior, and toxicity of metals in municipal sewage sludges. As a result of the research, source-reduction programs were implemented, which required industries to clean pollutants out of their wastewater *before* sending it to municipal wastewater treatment plants. In many cases, the recovery of valuable metal pollutants was actually profitable for industries. Because of these programs, municipal sewage sludges are much cleaner than in the past (Table 18.8). Note that the median levels of the most toxic industrial pollutants (Cd, Cr, Pb, and PCB) declined dramatically between the 1976 survey and the 1990 survey. Since much of the copper comes from the plumbing in homes (metallic copper is slightly solubilized in areas with acidic water supplies), that metal has been less affected by the source reduction regulations.

TABLE 18.8 Median Pollutant Concentrations Reported in Sewage Sludges Surveyed Across the United States in 1976 and 1990 and in Uncontaminated Agricultural Soils and Cow Manure

	Concentration, mg/kg dry weight			
Pollutant	Sludges surveyed in 1990[a]	Sludges surveyed in 1976[b]	Agricultural soils[d]	Typical values for cow manure
As	6	10	5.2	4
Cd	7	260	0.20	1
Cr	40	890	37	56
Cu	463	850	18.5	62
Hg	4	5	0.06	0.2
Mo	11	—	—	14
Ni	29	82	18.2	29
Pb	106	500	11.0	16
Zn	725	1740	53.0	71
PCB	0.21	9[c]	—	0

[a]Data from Chaney (1990).
[b]Data from Sommers (1977).
[c]1976 PCB value is median of cities in New York; from Furr et al. (1976).
[d]Median of 3045 surface soils reported by Holmgren et al. (1993).

TABLE 18.9 **Regulatory Limits on Inorganic Pollutants (Heavy Metals) in Sewage Sludge Applied to Agricultural Land**

Element	Maximum concentration in sludge, USEPA,[a] mg/kg	Annual pollutant loading rates, USEPA, kg/ha/yr	Cumulative allowable pollutant loading kg/ha		
			USEPA	Germany	Ontario
As	75	2.0	41	—	28
Cd	85	1.9	39	3.2	3.2
Cr	3000	150.0	3000	200	240
Cu	4300	75.0	1500	120	200
Hg	57	0.85	17	2	1.0
Mo	75	—	—	—	8
Ni	420	21	420	100	64
Pb	840	15	300	200	120
Se	100	5.0	100	—	3.2
Zn	7500	140	2800	400	440

[a]U.S. Environmental Protection Agency (1993).

Virginia guidelines on land application of sludge: http://www.ext.vt.edu/pubs/compost/452-303/452-303.html

REGULATION OF SLUDGE APPLICATION TO LAND. The lower levels of metals (and of organic pollutants) make municipal sewage sludges much more suitable for application to soils than in the past. Today, the amount of sludge that can be applied to agricultural land is more often limited by the potential for nitrate pollution from the nitrogen or phosphorus it contains, rather than by the metal content of the sludge. Nonetheless, application of sewage sludge to farmland is closely regulated to ensure that the metal concentrations in the sludge do not exceed the standards and that the total amount of metal applied to the soil over the years does not exceed the maximum accumulative loading limit listed in Table 18.9. The fact that metal-loading standards differ considerably between the United States and other countries (see Table 18.9) is an indication that the nature of the metal contamination threat is still somewhat controversial.

TOXIC EFFECTS FROM SLUDGE. The uncertainties as to the nature of many of the organic chemicals found in the sludge, as well as the cumulative nature of the metals problem, dictate continued caution in the regulations governing application of sludge to croplands. The effect of application of a high-metal sludge on heavy metal content of soils and of earthworms living in the soil is illustrated in Table 18.10. The sludge-treated soil areas, as well as the bodies of earthworms living in these soils, were higher in some of these elements than was the case in areas where sludge had not been applied. One would expect further concentration to take place in the tissue of birds and fish, many of which consume the earthworms.

Farmers must be assured that the levels of inorganic chemicals in sludge are not sufficiently high to be toxic to plants (a possibility mainly for zinc and copper) or to humans and other animals who consume the plants (a serious consideration for Cd, Cr, and Pb). For relatively low-metal municipal sludges, application at rates just high enough to supply needed nitrogen seems to be quite safe (Table 18.11).

TABLE 18.10 **The Effect of Sewage Sludge Treatment on the Content of Heavy Metals in Soil and in Earthworms Living in the Soil**

Note the high concentration of cadmium and zinc in the earthworms.

Metal	Concentration of metal, mg/kg			
	Soil		Earthworms	
	Control	Sludge-treated	Control	Sludge-treated
Cd	0.1	2.7	4.8	57
Zn	56	132	228	452
Cu	12	39	13	31
Ni	14	19	14	14
Pb	22	31	17	20

From Beyer et al. (1982).

SOILS AND CHEMICAL POLLUTION 821

Note that the metals show the typical pattern of less accumulation in the grain than in the leaves and stalks (stover). The annual sludge rate of about 10.5 Mg was designed to supply the nitrogen needs of the corn. The sludge had little effect on the metal content of the plants, except in the case of zinc (which increased, but not beyond the normal range for corn).

	Zn	Cu	Cd	Pb	Ni	Cr
Cumulative metal applied in sludge, kg/ha	175	135	1.2	49	4.9	1045
Treatment	Uptake in stover, mg/kg					
Fertilizer	18	8.4	0.16	0.9	0.7	0.9
Sludge	46.5	7.0	0.18	0.8	0.6	1.4
	Uptake in grain, mg/kg					
Fertilizer	20	3.2	0.29	0.4	0.4	0.2
Sludge	26	3.2	0.31	0.5	0.3	0.2

Data abstracted from Dowdy et al. (1994).

Direct ingestion of soils and sludge is also an important pathway for human and animal exposure. Animals should not be allowed to graze on sludge-treated pastures until rain or irrigation has washed the sludge from the forage. Children may eat soil while they play, and a considerable amount of soil eventually becomes dust in many households. Direct ingestion of soil and dust is particularly harmful in lead toxicity.

18.8   REACTIONS OF INORGANIC CONTAMINANTS IN SOILS

Heavy Metals in Sewage Sludge

Sewage sludge—the case for caution:
http://cwmi.css.cornell.edu/Sludge.html

Concern over the possible buildup of heavy metals in soils resulting from large land applications of sewage sludges has prompted research on the fate of these chemicals in soils. Most attention has been given to zinc, copper, nickel, cadmium, and lead, which are commonly present in significant levels in these sludges. Many studies have suggested that if only moderate amounts of sludge are added, and the soil is not very acid (pH >6.5), these elements are generally bound by soil constituents; they do not then easily leach from the soil, nor are they readily available to plants. Only in moderately to strongly acid soils have most studies shown significant movement down the profile from the layer of application of the sludge. Monitoring soil acidity and using judicious applications of lime have been widely recommended to prevent leaching into groundwaters and minimize uptake by plants.

More recently, studies using large amounts of sludge (up to and exceeding what is permitted by U.S. EPA regulations) have suggested that metals from sludge may initially be more mobile in soils than was previously thought. In fact, several studies have reported that from 20 to 80 percent of the metals applied with sludge at high rates have been leached from the root zone and, in all likelihood, were lost to the groundwater. The metals in these studies probably moved as soluble organic complexes while the sludge-soil mixture was still fresh. Over time, the metals remaining in the soil appear to be stabilized in various low-solubility soil fractions (Table 18.12).

FORMS FOUND IN SOILS TREATED WITH SLUDGE.   By using a sequence of chemical extractants, researchers have found that heavy metals are associated with soil solids in four major ways (Table 18.12). First, a very small proportion is held in *soluble* or *exchangeable forms,* which are available for plant uptake. Second, the elements are bound by the *soil organic matter* and by the *organic materials* in the sludge. High proportions of the copper and chromium are commonly found in this form, while lead is not so highly attracted. Organically bound elements are not readily available to plants, but may be released over a period of time.

TABLE 18.12 Forms of Six Heavy Metals Found in the Ap Horizon of a Metea Sandy Loam (Typic Hapludalfs) in Michigan That Received 870 Mg/ha (Dry Weight) of a "Dirty" Sewage Sludge Over 10 Years

The sludge application rate far exceeded that required to supply nitrogen to the crops grown, suggesting that the purpose was disposal rather than utilization. The sludge was incorporated into the soil between 1977 and 1986, prior to the implementation of source reduction programs to reduce the metal contents of most sewage sludges. The data are for soil samples taken 4 years after the last sludge application. The soil CEC was 7 cmol$_c$/kg, the organic matter content was 7%, and the pH was 6.9.

Forms in Soil	Solubility	Metal Content, mg/kg					
		Cd	Cr	Cu	Pb	Ni	Zn
Exchangeable and dissolved	Most	—	<1	4	<4	62	520
Acid soluble (carbonates, some organic)		3	38	140	19	170	1940
Organic matter		<1[a]	200	56	35	31	89
Fe and Mn oxides		<1	331	96	28	180	370
Residual (very insoluble sulfides, etc.)	Least	<1	48	11	99	24	56
Total of all forms		=4.5	617	307	=181	467	2975

Totals		Metal Content, kg/ha[b]					
Total measured in Ap horizon		=12	1728	859	=507	1308	8330
Total content in sludge applied		21	3000	1800	480	2100	11,300
Apparent recovery, %		=60	58	48	=106	62	74

[a]Numbers preceded by < indicate that the level present was less than the lowest concentration detectable by the analytical method used.
[b]The conversion from mg/kg to kg/ha assumes a bulk density of 1.4 Mg/m^3 and a sampling depth of 20 cm. Metal concentration data from Berti and Jacobs (1996). See also McBride et al. (1999) for further evidence of sludge-borne metal mobility in soils.

The third and fourth associations of heavy metals in soils are with *carbonates* and with *oxides of iron and manganese*. These forms are less available to plants than either the exchangeable or the organically bound forms, especially if the soils are not allowed to become too acid. The fifth association is commonly known as the *residual form*, which consists of sulfides and other very insoluble compounds that are less available to plants than any of the other forms.

It is fortunate that most soil-applied heavy metals are not readily absorbed by plants and that they are not easily leached from the soil. However, the immobility of the metals means that they will accumulate in soils if repeated sludge applications are made. Care must be taken not to add such large quantities that the capacity of the soil to react with a given element is exceeded. It is for this reason that regulations set maximum cumulative loading limits for each metal (see Table 18.9).

Other Inorganic Pollutants

Arsenic has accumulated in certain orchard soils following years of application of arsenic-containing pesticides. Being present in an anionic form (e.g., $H_2AsO_4^-$), this element is absorbed (as are phosphates) by hydrous iron and aluminum oxides, especially in acid soils. In spite of the capacity of most soils to tie up arsenates, long-term additions of arsenical sprays can lead to toxicities for sensitive plants and earthworms. The arsenic toxicity can be reduced by applications of sulfates of zinc, iron, and aluminum, which tie up the arsenic in insoluble forms. Toxicities from naturally occuring arsenic were discussed in Section 15.9.

Lead contaminates soils primarily from vehicle exhaust and from old lead-pigmented paints (paint chips and dust from painted woodwork). Most of the lead is tied up in the soil as low solubility carbonates, sulfides, and in combination with iron, aluminum, and manganese oxides (see Table 18.12). Consequently, the lead is largely unavailable to plants and not mobile enough to readily leach to groundwater. However, it can be absorbed by children who put contaminated soil in their mouths (Box 18.2).

Boron can contaminate soil via high-boron irrigation water, by excessive fertilizer application, or by the use of power plant fly ash as a soil amendment. Boron may be adsorbed by organic matter and clays but may still be available to plants, except at high soil pH. Boron is relatively soluble in soils, toxic quantities being leachable, especially from acid sandy soils. Boron toxicity in plants is usually a localized problem and is probably much less important than boron deficiency.

Fluorine toxicity is also generally localized. Drinking water for animals and fluoride fumes from industrial processes often contain toxic amounts of fluorine. The fumes can be ingested directly by animals or deposited on nearby plants. If the fluorides are adsorbed by the soil, their uptake by plants is restricted. The fluorides formed in soils are highly insoluble, the solubility being least if the soil is well supplied with lime.

Mercury is released mainly from burning coal to generate electricity. When it contaminates lake beds and swampy areas, the result is toxic levels of mercury among certain species of fish. Insoluble forms of mercury in soils, not normally available to plants or, in turn, to animals, are converted by microorganisms to an organic form, methylmercury, in which it is more soluble and available for plant and animal absorption. The methylmercury is concentrated in fatty tissue as it moves up the food chain, until it accumulates in some fish to levels that may be toxic to humans. This series of transformations illustrates how reactions in soil can influence human toxicities.

Chromium in trace amounts is essential for human life, but, like arsenic, it is a carcinogen when absorbed in larger doses. This element is widely used in steel, alloys, and paint pigments. Chromium is found in two major oxidation states in ordinary soils: a trivalent form [Cr(III)], and a hexavalent form [Cr(VI)]. In contrast to most metals, the more highly oxidized state [Cr(VI)] is the more soluble, and its solubility increases about pH 5.5. This behavior is opposite that of Cr(III), which forms insoluble oxides and hydroxides above that pH level.

To remediate Cr(VI)-contaminated soil and water, it is useful to reduce the chromium to Cr(III) (see also Section 7.5). This reduction process is enhanced by anerobic conditions [wet soil with an abundance of decomposable organic material to provide a large, biological oxygen demand (BOD)]. The organic matter serves as an electron donor and thereby hastens the reduction of Cr(VI) to the trivalent state [Cr(III)]. Provided the pH is maintained above 5.5, chromium in this reduced state will remain relatively stable, immobile, and non-toxic.

Selenium, which derives mainly from certain soil parent material, can accumulate in soils and plants to toxic levels, especially in arid regions (see Section 15.9 and Box 15.2 for details).

18.9 PREVENTION AND ELIMINATION OF INORGANIC CHEMICAL CONTAMINATION

Three primary methods of alleviating soil contamination by toxic inorganic compounds are (1) to eliminate or drastically reduce the soil application of the toxins; (2) to immobilize the toxin by means of soil management, to prevent it from moving into food or water supplies; and (3) in the case of severe contamination, to remove the toxin by chemical, physical, or biological remediation.

Reducing Soil Application

The first method requires action to reduce unintentional aerial contamination from industrial operations and from automobile, truck, and bus exhausts. Decision makers must recognize the soil as an important natural resource that can be seriously damaged if its contamination by unintended addition of inorganic toxins is not curtailed. Also, there must be judicious reductions in intended applications to soil of the toxins through pesticides, fertilizers, irrigation water, and solid wastes.

Immobilizing the Toxins

Soil and crop management can help reduce the continued cycling of these inorganic chemicals. This is done primarily by keeping the chemicals in the soil rather than encouraging their uptake by plants. The soil becomes a sink for the toxins, and thereby breaks the soil–plant–animal (humans) cycle through which the toxin exerts its effect. The soil breaks the cycle by immobilizing the toxins. For example, most of these elements are rendered less mobile and less available if the pH is kept near neutral or above (Figure 18.23). Liming of acid soils reduces metal mobility; hence, regulations require that the pH of sludge-treated land be maintained at 6.5 or higher.

Draining wet soils should be beneficial, since the oxidized forms of the several toxic elements are generally less soluble and less available for plant uptake than are the reduced forms. However, the opposite is true for chromium. The oxidized Cr(VI) is mobile and highly toxic to humans (see Section 18.8).

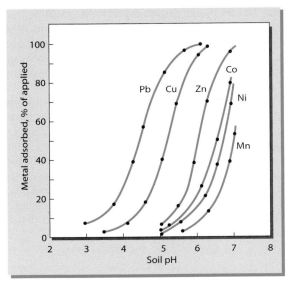

FIGURE 18.23 The effect of pH on the adsorption of six heavy metals. The metals were adsorbed by clay-sized goethite (an iron oxide mineral) that forms coatings on many soil particles. Maintaining the soil pH near 7 (neutral) is expected to maximize the sorption and thereby minimize the solution concentration of most heavy metals, especially of copper and lead [Modified from Basta et al. (2005)

Heavy phosphate applications reduce the availability of some metal cations (see Box 18.2) but may have the opposite effect on arsenic, which is found in the anionic form. Leaching may be effective in removing excess boron, although moving the toxin from the soil to water may not be of any real benefit.

Care should be taken in selecting plants to be grown on metal-contaminated soil. Generally, plants translocate much larger quantities of metals to their leaves than to their fruits or seeds (see Table 18.11). The greatest risk for food-chain contamination with metals is therefore through leafy vegetables, such as lettuce and spinach, or through forage crops eaten by livestock.

Bioremediation by Metal Hyperaccumulating Plants

Certain plants that have evolved in soils naturally very high in metals are able to take up and accumulate extremely high concentrations of metals without suffering from toxicity. Plants have been found that accumulate more than 20,000 mg/kg nickel, 40,000 mg/kg zinc, and 1000 mg/kg cadmium. While such **hyperaccumulator** plants would pose a serious health hazard if eaten by animals or people, they may facilitate a new kind of bioremediation for metal-contaminated soils.

If sufficiently vigorously growing genotypes of such plants can be found, it may be possible to use them to remove metals from contaminated soils. For example, several plants in the genus *Thlaspi* have been grown in soils contaminated by smelter fumes (Figure 18.24). These soils are so contaminated that they are virtually barren. Accumulating nearly 40,000

FIGURE 18.24 *Thlaspi caerulescens*, a zinc and cadmium hyperaccumulator plant growing in smelter-contaminated soil near Palmerton, Pa. This plant has been reported to accumulate up to 4% zinc in its tissue (dry weight basis). Research with such plants aims at developing technology to biologically remove and recover metals from heavily contaminated soils. (Photo by H. Witham; courtesy of R. Chaney, USDA)

mg/kg (about 4%) zinc in their tissues, the *Thlaspi* plants grown on this site could be harvested to remove large quantities of the metals from the soil. The plant tissue is so concentrated that it could be used as an "ore" for smelting new metal. This and other bioremediation technologies for metals (e.g., the bioreduction of chromium and selenium discussed earlier) hold promise for cleaning up badly contaminated soils without resorting to expensive and destructive excavation and soil-washing methods.

Genetic and bioengineering techniques are being utilized to develop high-yielding hyperaccumulating plants that can remove larger quantities of heavy metal contaminants from soils. For example, wide genetic variation in heavy metal accumulation by different strains of Alpine pennycress suggests the potential for breeding improved accumulating plants. Also, research to insert genes responsible for contaminant accumulation into other higher-yielding plants, such as canola and Indian mustard, is underway.

A combination of chelates and phytoremediation has been used to remove lead from contaminated soil. This element is sparingly available to plants, being strongly bound by both mineral and organic matter. The chelates solubilize the lead, and plants such as Indian mustard are used to remove it.

18.10 LANDFILLS[5]

A visit to the local landfill would convince anyone of the wastefulness of modern societies. Roughly 300 million Mg of municipal wastes are generated each year by people in the United States. Most (about 70%) of this waste material is organic in nature, largely paper, cardboard, and yard wastes (e.g., grass clippings, leaves, and tree prunings). The other 30% consists mainly of such nonbiodegradeables as glass, metals, and plastic. Currently, despite an upsurge in recycling efforts, the great majority of these materials are buried in the ground (Figure 18.25).

The Solid Waste Problem

First, it is important to understand that the entire waste disposal problem could be greatly reduced by creating less waste in the first place. Second, it is possible to eliminate most problems associated with waste disposal by two simple measures: (1) keeping the metals, glass, plastics, and paper separate in the household for easy recycling, and (2) composting the yard wastes, food wastes, and some of the paper products. The composted product from a number of municipalities is successfully used as a beneficial soil amendment (see Section 16.5). The small fraction of more hazardous wastes remaining can then be detoxified or concentrated and immobilized.

---

[5] For details on the soil and geotechnical aspects of landfill design, see Qian et al. (2002).

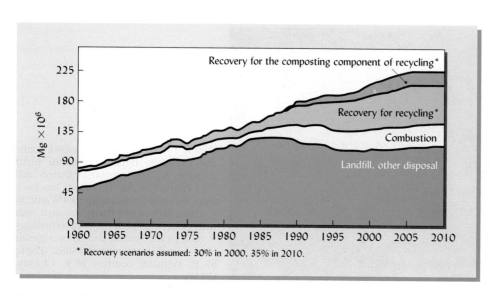

FIGURE 18.25 Historical and predicted trends in municipal solid-waste management in the United States. Soils play a central role in the composting and landfilling options (shown in dark green). [Data from U.S. EPA (2005a)].

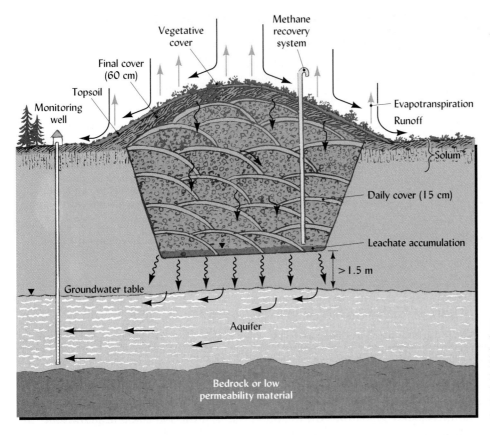

FIGURE 18.26 A natural attenuation landfill depends largely on soil processes to attenuate the contaminants in the leachate before they reach the groundwater. Compare to the containment landfill design illustrated in Figure 18.29. (Diagram courtesy of R. Weil)

The present reality is that most municipal solid wastes are buried in the ground and will probably continue to be disposed of in this manner for some time to come. In the past, wastes were merely placed in open dumps and, often, set afire. The term *landfill* came into use because wastes were often dumped in swampy lowland areas where, eventually, their accumulation filled up the lowland, creating upland areas for such uses as city parks and other facilities. Locating landfills on wetlands is no longer an acceptable practice.

Two Basic Types of Landfill Design

Although landfill designs vary with the characteristics of both the site and the wastes, two basic types of landfills can be distinguished: (1) the natural attenuation or unsecured landfill, and (2) the containment or secured landfill. We will briefly discuss the main features of each.

Natural Attenuation Landfills

Leaking landfill (USGS): http://pubs.usgs.gov/fs/ fs-040-03/

The purpose of a natural attenuation landfill is to contain nonhazardous municipal wastes in a sanitary manner, protect them from animals and wind dispersal, and, finally, to cover them sufficiently to allow revegetation and possible reuse of the site. Although the landfill is engineered to reduce water infiltration, some rainwater is allowed to percolate through the waste and down to the groundwater (Figure 18.26). Natural processes are relied upon to attenuate the leachate contaminants before the leachate reaches the groundwater. Soils play a major role in these natural attenuation processes through physical filtering, adsorption, biodegradation, and chemical precipitation (Table 18.13).

SOIL REQUIREMENTS. Finding a site with suitable soil characteristics is critical for a natural attenuation landfill. There must be at least 1.5 m of soil material between the bottom of the landfill and the highest groundwater level. This layer of soil should be only moderately permeable. If too permeable (sandy, gravelly, or highly structured), it will allow the leachate to pass through so quickly that little attenuation of contaminants will take place. The soil must have sufficient cation exchange capacity to adsorb NH_4^+, K^+, Na^+,

TABLE 18.13   Some Organic and Inorganic Contaminants in Untreated Leachate from Municipal Landfills

Range of concentrations and typical sources of the contaminants and mechanisms by which soils can attenuate the contaminants are also given. The ranges show that leachates vary greatly among landfills.

Chemical	Concentration, μg/L	Common sources	Mechanisms of attenuation
		Organics	
Dissolved organic matter, as Chemical Oxygen Demand (COD)	140,000 – 150,000,000	Rotting yard wastes, paper, and garbage	Biological degradation
Benzene	0.2–1630	Adhesives, deodorants, oven cleaner, solvents, paint thinner, and medicines	Filtration, biodegradation, and methanogenesis
Trans 1,2-Dichloroethane	1.6–6500	Adhesives and degreasers	Biodegradation and dilution
Toluene	1–12,300	Glues, paint cleaners and strippers, adhesives, paints, dandruff shampoo, and carburetor cleaners	Biodegradation and dilution
Xylene	0.8–3500	Oil and fuel additives, paints, and carburetor cleaners	Biodegradation and dilution
		Metals	
Nickel	15–1300	Batteries, electrodes, and spark plugs	Adsorption and precipitation
Chromium	20–1500	Cleaners, paint, linoleum, and batteries	Precipitation, adsorption, and exchange
Cadmium	0.1–40	Paint, batteries, and plastics	Precipitation and adsorption

Leachate concentration ranges from a review of hundreds of landfills built since 1965, in Kjeldsen et al. (2002).

Cd^{2+}, Ni^{2+}, and other metallic cations that the wastes are expected to release. The soil should also adsorb and retard organic contaminants long enough to allow a high degree of microbial degradation. On the other hand, if the soil is too impermeable, the leachate will build up, flood the landfill, and seep out laterally.

DAILY AND FINAL SOIL COVER.   The site for a natural attenuation landfill should also provide soils suitable for daily and final cover materials. At the end of every workday, the waste must be covered by a layer of relatively impermeable soil material (Figure 18.27). The final cover for the landfill is much thicker than the daily covers, and includes a 60- to 100-cm-thick layer of low-permeability, clay soil material designed to minimize percolation of water into the landfill. This impermeable layer of compacted clay is usually

FIGURE 18.27 A bulldozer compacts and buries trash in a natural attenuation landfill in deep, well-drained soils. The trash is covered daily by 15 cm of soil to prevent it from being carried away by wind or infested with rats. Once the entire trench is filled, a much thicker final cover of soil will be applied. Note the lack of any geomembrane liner or leachate collection system in this landfill. (Photo by R. Weil)

FIGURE 18.28 A soil pit dug in an engineered soil used to cap a completed section of a landfill in New Jersey. The pit excavation has revealed the three basic layers of the landfill cap. A compacted clay layer serves to seal the underlying garbage and prevent infiltration of water so as to minimize the formation of landfill leachate. A layer of loose, clean, medium sand was installed above the clay to serve as a drainage layer, which guides water horizontally over the clay and off the landfill cell to a collection pond (not shown). On top of the sand, a layer of sandy loam topsoil was installed from Ap horizon material hauled in from a distant site. The topsoil serves as growing medium for the grassy vegetative cover that protects and stabilizes the landfill cap. See also Plate 64. (Photo courtesy of Chris Smith, USDA/NRCS)

then covered with a 30 to 45 cm layer of highly permeable medium to coarse sand. This sand layer is designed to allow water to drain laterally off the landfill to a collection area. On top of the sand, a thinner layer of loamy "topsoil" is installed. The moderately permeable topsoil layer is meant to support a vigorous plant cover that will prevent erosion and use up water by evapotranspiration. The whole system is designed to limit the amount of water percolating through the waste, so that the amounts of contaminated leachate generated will not overwhelm the attenuating capacity of the soil between the landfill bottom and the groundwater (Figure 18.28 and Plate 64).

Containment or Secured Landfills

The second main type of landfill is much more complex and expensive to construct, but its construction and function is much less dependent on the nature of the soils at the site. The design (see Figure 18.29) is intended to contain, pump, and treat all leachate from the landfill, rather than to depend on soil processes for cleansing the leachate on its way to the groundwater. To accomplish the containment, one or more impermeable liners are set in place around the sides and bottom of the landfill. These are often made of expanding clays (e.g., bentonite) that swell to a very low permeability when wet. Plastic, watertight geomembranes are also used in making the liners. The membranes are covered with a tough, nonwoven, synthetic fabric (geotextiles) and then covered with a thick layer of fine gravel or sand to protect the liner from accidental punctures. A system of slotted pipes and pumps is installed to collect all the leachate from the bottom of the landfill (Figure 18.30, *left*). The collected leachate is then treated on or off the site. The principal soil-related concerns are the requirement for suitable sources of sand and gravel, of soil for daily cover, for clayey material to form the final cover, and for topsoil to support protective vegetation.

Environmental Impacts of Landfills

Today, regulations require that wastes be buried in carefully located and designed sanitary landfills. As a result, the number of landfill sites in the United States was reduced from about 16,000 in 1970 to fewer than 1700 in 2006. The remaining landfills are mostly very large, highly engineered containment-type systems. A major concern with regard to landfills is the potential water pollution from the rainwater that percolates through the wastes, dissolving and carrying away all manner of organic and inorganic contaminants (see Table 18.13). In addition to the general load of oxygen-demanding dissolved organic carbon

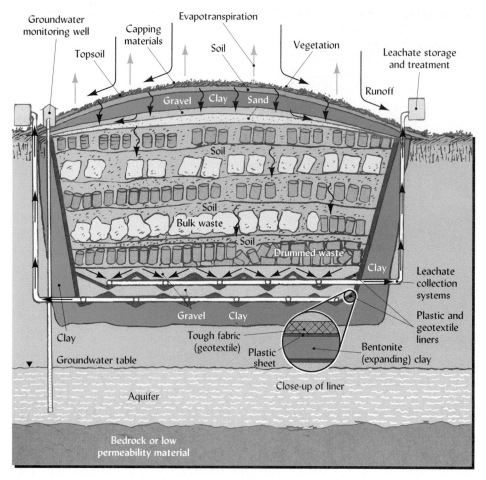

Groundwater monitoring well
Topsoil
Capping materials
Evapotranspiration
Soil
Vegetation
Leachate storage and treatment
Runoff
Gravel   Clay   Sand
Soil
Soil
Bulk waste
Soil
Drummed waste
Clay
Leachate collection systems
Plastic and geotextile liners
Gravel   Clay
Tough fabric (geotextile)
Plastic sheet
Bentonite (expanding) clay
Close-up of liner
Clay
Groundwater table
Aquifer
Bedrock or low permeability material

FIGURE 18.29 A containment-type landfill is designed to collect all the leachate and pump it out for storage and treatment. The bottom of the landfill cell is sealed with a waterproof geomembrane which is protected by a covering of geotextile and gravel. Versions of this basic design are used for more hazardous wastes or when soil conditions on the site are unsuitable for a natural attenuation landfill design, illustrated in Figure 18.26. (Diagram courtesy of R. Weil)

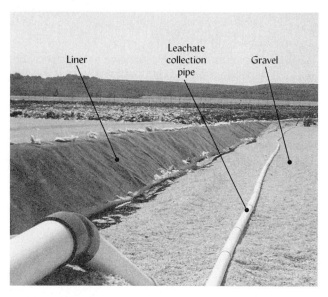

FIGURE 18.30 Engineered systems for the collection of leachate and gas emissions in a containment landfill. (*Left*) A black geomembrane liner covered with white pea gravel and a leachate collection pipe in a new cell being prepared in a containment-type landfill. The low hills in the background are completed cells blanketed with a vegetated final cover. The pollutant-laden leachate will be piped to a treatment facility. (*Right*) Gas wells collecting landfill gas (a mixture of mainly methane and carbon dioxide) from anaerobic decomposition in a completed landfill cell. The methane is used to fuel turbines that generate electricity that is used for the waste-disposal operation or sold to the local electric utility company. (Photos courtesy of R. Weil).

compounds, many of the contaminants in landfill leachate are highly toxic and would create a serious pollution problem if they reached the groundwater under the landfill.

In addition to efficiency of resource use, avoidance of particular landfill management problems is another reason that the organic components of refuse (mainly paper, yard trimmings, and food waste) should be composted to produce a soil amendment rather than landfilled. First, as these materials decompose in a finished landfill, they lose volume and cause the landfill to settle and the landfill surface to subside. This physical instability severely limits the uses that can be made of the land once a landfill is completed.

Second, decomposition of the organic refuse produces undesirable liquid and gaseous products. Within a few weeks, decomposition uses up the oxygen in the landfill, and the processes of anaerobic metabolism take over, changing the cellulose in paper wastes into butyric, propionic, and other volatile organic acids, as well as hydrogen and carbon dioxide. After a month or so, methane-producing bacteria become dominant, and for several years (or even decades) a gaseous mixture of about one-third carbon dioxide and two-thirds methane (known as *landfill gas*) is generated in quantity.

Gas problems in apartments built on former landfill: http://www.eti-geochemistry.com/walnut/index.html

The production of methane gas by the anaerobic decomposition of organic wastes in a landfill can present a very serious explosion hazard if this gas is not collected (and possibly burned as an energy source; see Figure 18.30, *right*). Where the soil is rather permeable, the gas may diffuse into basements up to several hundred meters away from the landfill. A number of fatal explosions have occurred by this process. Anaerobic decomposition in landfills also emits other harmful gases, the effects of which are less well known.

18.11 RADIONUCLIDES IN SOIL

U.S. EPA division of radiation: http://www.epa.gov/radiation/

Soils contain small quantities of ^{238}U, ^{40}K, ^{87}Rb, ^{14}C, and a number of other naturally occurring radioactive isotopes (radionuclides) that are characterized by long half-lives and give off minute amounts of radiation in the form of alpha particles (bundles of two neutrons and two protons) and beta particles (positive or negatively charged particles). As a radionuclide decays, its nucleus discharges these particles, transforming the atom into a different isotope or element with a lighter nucleus. The time it takes for one-half of the atoms of a particular radioactive isotope to undergo such decay is termed the *half-life* of the isotope. After 10 half-lives, 99.1% of the original atoms will have decayed. The intensity of radioactivity present—or, more precisely, the rate of radioactive decay—is expressed using the SI unit *becquerel* (Bq), which represents one decay per second. An older metric unit, still in wide use, is the *curie* (Ci), which equals 3.7×10^{10} Bq.[6]

Radioactivity from Nuclear Fission

The process of nuclear fission, in connection with atomic weapons testing and nuclear power generation, has contaminated soils with a number of additional radionuclides. However, only two of these are sufficiently long-lived to be of significance in soils: strontium 90 (half-life = 28 yr) and cesium 137 (half-life = 30 yr). The average level of ^{90}Sr in soil in the United States is about 14.4 kilobecquerels per square meter (kBq/m^2) or 388 millicuries per square kilometer (mCi)/km^2. The average level for ^{137}Cs is about 22.9 kBq/m^2 (620 mCi/km^2). The levels of radioactivity caused by nuclear fallout are quite small compared to that for naturally occurring radionuclides. For example, for naturally occurring ^{40}K the average level is about 1900 kBq/m^2 (51,800 mCi/km^2).

Partly because of the cation-exchange properties of soils, the levels of these fission radionuclides found in most soils are not high enough to be hazardous. Most of the ^{90}Sr and ^{137}Cs reaching the soil is adsorbed by the soil colloids in exchange for other cations previously adsorbed, with the result that plants take up mainly the replaced cations rather than the added radionuclides (see Section 8.8). Even during the peak periods of weapons testing in the early 1960s, soils did not contribute significantly to the level of these nuclides in plants. Atmospheric fallout directly onto foliage was the primary source of radionuclides in the food chain. Consequently, only in the event of a catastrophic supply of fission products could toxic soil levels of ^{90}Sr and ^{137}Cs be expected. Such high levels of

---

[6] The curie was named after Marie and Pierre Curie, Polish scientists who discovered radium, an element that decays at the rate of 3.7×10^{10} Bq/g. The becquerel was named after Antoine Henry Becquerel, a French scientist who discovered radioactivity in uranium.

^{90}Sr and ^{137}Cs, as well as ^{131}I, did contaminate soils in Ukraine, Scandinavia, and Eastern Europe in the wake of the 1986 reactor meltdown at Chernobyl in Ukraine (then part of the Soviet Union). The accident deposited more than 200 k Bq/m^2 of radioactivity on soils as far away as the United Kingdom. Fortunately, considerable research has been accomplished on the behavior of these nuclides in the soil–plant system.

STRONTIUM 90.   In the soil–plant–animal system, ^{90}Sr behaves very much like calcium, to which it is closely related chemically (see periodic table, Appendix B). It enters the soil from the atmosphere in soluble forms and is quickly adsorbed by the colloidal fraction, both organic and inorganic. It is taken up by plants and assimilated much like calcium. Contamination of forages and, ultimately, of milk by this radionuclide is of concern, as the ^{90}Sr could potentially be assimilated into the bones of the human body. Fortunately, when it exchanges with aluminum or hydrogen ions adsorbed on the colloids in an acid soil, it comprises such a minute fraction of the exchangeable cations that its availability is quite low. However, should these soils be limed, the large quantities of added calcium are likely to cause the desorption of the strontium from the exchange sites, making it more available for leaching and plant update. However, the preponderance of calcium in the limed soil solution would compete with strontium for uptake by plant roots, and so reduce the amount of strontium entering the food chain.

CESIUM 137.   Although chemically similar to potassium, cesium tends to be less readily available in many soils. Apparently, ^{137}Cs is firmly fixed by vermiculite and related interstratified minerals. The fixed nuclide is nonexchangeable, much as is fixed potassium in some interlayers of clay (see Box 8.3 and Section 14.15). Plant uptake of ^{137}Cs from vermiculitic soils is very limited. Where vermiculite and related clays are absent, as in some tropical soils, ^{137}Cs uptake is more rapid. In any case, the soil tends to dampen the movement of ^{137}Cs into the food chain of animals, including humans.

IODINE 131.   When it partially melted down, the nuclear reactor at Chernobyl emitted significant quantities of ^{131}I, which accumulates in the human thyroid. People in the area around the reactor have since suffered an increased incidence of thyroid cancer. Because of its short half-life (8.1 days), reactions in the soil and movement through the soil-plant-animal food chain are less significant than contamination of drinking water, inhalation of contaminated dust particles, and contamination of edible plant foliage.

Research is underway to take advantage of plant uptake of radionuclides in phytoremediation exercises. Certain plants, such as sunflowers, are being used to remove ^{90}Sr and ^{137}Cs from ponds and soils near the site of the Chernobyl nuclear disaster. Indian mustard is also being used in nearby sites to remove such nuclide contaminants.

Radioactive Wastes[7]

In addition to radionuclides added to soils because of weapons testing and nuclear power plant accidents, soils may interact with radioactive waste materials that have leaked from their holding tanks or have been intentionally buried for disposal. Plutonium, uranium, americium, neptunium, curium, and cesium are among the elements whose nuclides occur in radioactive wastes. These wastes are generated by research and medical facilities (where the radionuclides are used in cancer therapy and the like), and at power plants and weapons manufacturing sites.

Because of the secrecy and lack of regulation associated with the latter, they constitute some of the most polluted locations on Earth. For example, the U.S. Defense Department's now-abandoned plutonium-production complex at Hanford, Idaho, represents one of the biggest environmental cleanup challenges in the world. Among the hazards plaguing that site are hundreds of huge, in many cases leaking, underground tanks, in which high-level radioactive wastes have been stored for decades. Billions of cubic meters of soil and water have been contaminated with radioactive wastes at U.S. weapons manufacturing sites and at similar, equally polluted sites in the former Soviet Union.

PLUTONIUM TOXICITY.   Plutonium 239, a major pollutant at these sites, is dangerous both because of its intense radioactivity and because of its high level of toxicity to humans.

---

[7] For a description of the environmental challenges at the Hanford site, see Zorpette (1996).

TABLE 18.14   Concentrations of Several Breakdown Products of Uranium 238 and Thorium 232 (Nucleotides) in Six Different Soil Suborders in Louisiana

Note marked differences among levels in the different soils.

Soil suborder	No. of samples	^{238}U breakdown products, Bq/kg			^{232}Th breakdown products, Bq/kg		
		^{226}Ra	^{214}Pb	^{214}Bi	^{212}Pb	^{137}Cs	^{40}K
Udults	22	37.3	27.7	28.9	27.4	16.7	136
Aquults	24	30.4	36.7	38.1	50.0	10.9	100
Aqualfs	37	51.1	38.3	36.6	59.7	13.5	263
Aquepts	93	92.2	47.6	45.2	63.8	16.1	636
Aquolls	57	90.4	45.8	44.7	59.5	8.7	608
Hemists	18	136.3	49.4	49.0	74.9	19.4	783

From Meriwether et al. (1988).

The ^{239}Pu itself is quite immobile in soils, having a K_d estimated at about 1000. Nor is it taken up readily by plants, so it does not accumulate along terrestrial food chains. It does, however, accumulate in algae. Furthermore, oily liquid wastes carrying ^{239}Pu seep into the groundwater and nearby rivers, and contaminated surface soil blows in the desert wind, spreading the radionuclides for many kilometers. Cleanup may be impossible at some of these former weapons sites; the agencies responsible are struggling merely to stabilize and contain the contamination. With a half-life of 24,400 years, ^{239}Pu contamination is a problem that will not go away.

LOW-LEVEL WASTES.   Low-level radioactive wastes also present some environmental challenges. Even though the waste materials may be solidified before being placed in shallow land burial pits, some dissolution and subsequent movement in the soil are possible. Nuclides in wastes vary greatly in water solubility, uranium compounds being quite soluble, compounds of plutonium and americium being relatively insoluble, and cesium compounds being intermediate in solubility. Cesium, a positively charged ion, is adsorbed by soil colloids. Uranium is thought to occur as a UO_2^{2+} ion that is also adsorbed by soil. The charge on plutonium and americium appears to vary, depending on the nature of the complexes these elements form in the soil.

There is considerable variability in the actual uptake by plants of these nuclides from soils, depending on such properties as pH and organic matter content. The uptake from soils by plants is generally lowest for plutonium, highest for neptunium, and intermediate for americium and curium. Fruits and seeds are generally much lower in these nuclides than are leaves, suggesting that grains may be less contaminated by nuclides than forage crops and leafy vegetables.

Since soils are being used as burial sites for low-level radioactive wastes, care should be taken that soils are chosen whose properties discourage leaching or significant plant uptake of the chemicals. Data in Table 18.14 illustrate differences in the ability of different soils to hold breakdown products of two radionuclides. It is evident that monitoring of nuclear waste sites will likely be needed to assure minimum transfer of the nuclides to other parts of the environment.

18.12   RADON GAS FROM SOILS[8]

The Health Hazard

The soil is the primary source of the colorless, odorless, tasteless radioactive gas *radon,* which has been shown to cause lung cancer. Although landfills containing radioactive wastes have been known to emit radon gas at elevated concentrations, most concern regarding this potential toxin is directed toward radon that occurs naturally in soils.

---

[8] For a straightforward discussion of the hazards of indoor radon and what you can do to protect yourself, see U.S. EPA (2005b). For a review of the radon in soils and geologic material with a focus on mapping radon hazard areas, see Appleton (2007).

Therefore, radon is not usually considered a soil pollutant because it has not been introduced into the soil by human activity. Nonetheless, radon gas is thought to be a serious environmental health hazard when it moves from the soils and accumulates inside buildings. Deaths from breathing in radon are estimated at about 20,000 per year in the United States, some 10 to 50 times more numerous than deaths caused by contaminants in drinking water. The gas may be a causal factor in about 10% of lung cancer cases.

The health hazard from this gas stems from its transformation to radioactive polonium isotopes, which are solids that tend to attach to dust particles. The polonium-contaminated dust particles may lodge in the lungs, where alpha-particle radiation emitted by the polonium can penetrate the lung tissue and cause cancer. The principal concern is with radon accumulation in homes, offices, and schools where people breathe the air in basement or ground-level rooms for extended periods.

How Radon Accumulates in Buildings

GEOLOGIC FACTORS. Radon originates from uranium (^{238}U) found in minerals, sorbed on soil colloids, or dissolved in groundwater. Over billions of years, the uranium undergoes radioactive decay forming radium, which in turn gives off radiation over thousands of years and transforms into radon (Figure 18.31). Both uranium and radium are solids; however, radon is a gas that can diffuse through pores and cracks and emerge into the atmosphere. Soils and rocks that contain high concentrations of uranium will likely produce large amounts of radon gas. Soils formed from certain highly deformed metamorphic rocks and from marine sediments, limestones, and coal or oil-bearing shales tend to have the highest potential levels for radon production. However, nearby houses built on soils formed from the same parent material may differ widely in their indoor radon concentrations (Figure 18.32, *right*), the difference being due to variations in soil properties and/or house construction.

Geology of radon and its indoor risk potential:
http://energy.cr.usgs.gov/radon/georadon/4.html

SOIL PROPERTIES. To become a hazard, radon must travel from its source in the underlying rock or soil, up through overlying soil layers, and finally into an enclosed building where it might accumulate to unhealthful concentrations. It must also make this trip quite rapidly, because the half-life of radon is only 3.8 days. Within several weeks, radon completely decays to polonium, lead, and bismuth, radioactive solids that last for only minutes. Whether significant quantities of radon reach a building foundation depends mainly on two factors: (1) the distance that the radon must travel from its source and (2) the permeability of the soil through which it travels. Since radon is an inert gas, the soil does not react with it, but merely serves as a channel through which the gas moves.

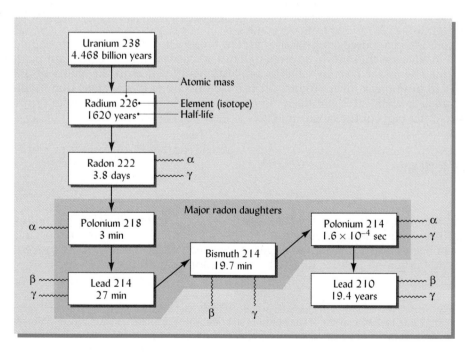

FIGURE 18.31 Radioactive decay of uranium 238 in soils that results in the formation of inert but radioactive radon. This gas emits alpha (α) particles and gamma (γ) rays and forms *radon daughters* that are capable of emitting alpha (α) and beta (β) particles and gamma (γ) rays. The alpha particles damage lung tissue and cause cancer. Radon gas may account for about 20,000 deaths annually in the United States. [Modified from Boyle (1988)

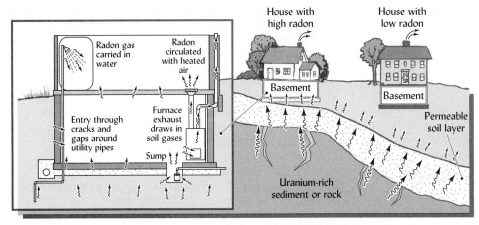

FIGURE 18.32 The foundation of a house, as well as the nature of the soil and rocks in the vicinity, help determine whether dangerous levels of radioactive radon gas are likely to accumulate inside. (*Right*) Levels of indoor radon can vary greatly from one house to another. On a regional basis, radon is likely to be highest where the soils are formed from uranium-bearing parent materials. The soil that underlies a house plays a role in the movement of radon gas from its source in rock and soil minerals to the air inside the house. Dry, coarse-textured permeable soil layers allow much faster diffusion of radon gas than does a wet or fine-textured soil. If soils underlying a house are relatively impermeable, radon movement will be so slow that nearly all of the radon emitted will have decayed before it can reach the house foundation. (*Left*) Once it arrives at a house foundation, radon may enter through a variety of openings, such as cracks in the foundation blocks, joints between the walls and concrete floor, and gaps where utility pipes enter the house. Radon risks are greatest during cold weather. Heated air escapes up the furnace chimney but windows are sealed, creating negative air pressure inside the house, which, in turn, draws in air from the soil. (Diagrams courtesy of R. Weil)

As explained in Sections 7.1 and 7.2, gases move through soil both by diffusion and by mass flow (convection). The rate of radon diffusion through a soil depends on the total soil porosity (more so than on pore size) and the degree to which the pores are filled with water. Radon diffuses through air-filled pores about 10,000 times faster than through water-filled pores. Movement is most rapid through sandy or gravelly soil layers that tend to hold little water. Therefore, some of the highest indoor radon concentrations have been found in houses where only a thin, well-drained, gravelly soil separates the house foundation from uranium-rich rock. At the other extreme, a thick, wet clay layer would provide an excellent barrier against radon diffusion. Convective airflow, which is mainly stimulated by rainwater entering the soil and by changes in atmospheric pressure, may play a significant role in radon movement during stormy weather.

BUILDING FACTORS. Because concrete is an effective barrier against diffusion, radon enters the building primarily by convective flow, which is driven by the difference between the air pressure inside the building and in the soil air around the foundation. Open upper-story windows, exhaust pipes, and chimneys tend to exhaust air from a building, lowering the internal air pressure and drawing in air from outside (Figure 18.32, *left*). If the basement has no open windows, air will be drawn into the building through cracks in the basement walls and floors, and around openings where utility pipes enter the basement. Since modern airtight buildings permit little exchange of air with the outside, radon can accumulate to harmful levels. A building with a crawl space between the first floor and the soil surface will draw in fresh outside air from the atmosphere, but one with a basement or a slab foundation directly on the soil will draw in soil air that may be enriched in radon.

Radon Testing and Remediation

TESTING. Since the occurrence of high radon levels cannot be accurately predicted, the only sure way to determine the risk of radon is to test for its presence. Testing is usually carried out in two stages. The first uses an inexpensive (about $15) charcoal canister, which is placed in the test area, unsealed, and left to absorb radon for the specified period (usually 3 days). The canister should be placed in the most high-risk location in the building—for example, in a basement bedroom without cross ventilation—during a time when the building is heated. After the test period, the canister is resealed and sent to a lab where the amount of radon absorbed can be measured and related to the radon concentration that was in the air. If the results suggest a radon level above 4 piCu/L (148 Bq/m³), the U.S. EPA advises that a long-term test be conducted using a somewhat more expensive alpha-track detector for a period of 3 to 12 months. If the long-term test also suggests levels above 4 piCi/L, modifications should be made to the building to reduce the accumulation of radon inside.

REMEDIATION. Depending on the levels of radon and the condition of the building, the modification may be as simple as caulking cracks in the floor and walls and filling gaps around utility-pipe entrances. Remediation of higher radon levels may require alterations that are more extensive. Ventilation of the room with outside air can prevent unhealthful radon buildup, but a more energy-efficient solution is a subslab ventilation system. For the latter, perforated pipes are installed in a layer of gravel under the foundation slab, and the air pressure there is lowered either by a mechanical fan or by convective draw from a special chimney. In this way, gas coming from the soil is intercepted and redirected to the atmosphere before it can enter the building. Installation of a subslab ventilation system is much less expensive during new construction than as a retrofit, and is now standard practice in many areas with high-uranium soils.

18.13 CONCLUSION

Three major conclusions may be drawn about soils in relation to environmental quality. First, since soils are valuable resources, they should be protected from environmental contamination, especially that which does permanent damage. Second, because of their vastness and remarkable capacities to absorb, bind, and break down added materials, soils offer promising mechanisms for the disposal and utilization of many wastes that otherwise may contaminate the environment. Third, soil contaminants and the products of their breakdown in soil reactions can be toxic to humans and other animals if the soil is ingested or the contaminants move from the soil into plants, soil fauna, the air, and—particularly—into water supplies.

To gain a better understanding of how soils might be used and yet protected in waste-management efforts, soil scientists devote a considerable share of their research efforts to environmental-quality problems. Furthermore, soil scientists have much to contribute to the research teams that search for better ways to clean up environmental contamination. Some of the most promising technological advances have been in the field of bioremediation, in which the biological processes of the soil are harnessed to effect soil cleanup. Finding appropriate sites where soils can be safely used to clean up or store hazardous wastes involves geographic information about soils, the topic of the next chapter.

STUDY QUESTIONS

1. What agricultural practices contribute to soil and water pollution, and what steps must be taken to reduce or eliminate such pollution?

2. Discuss the types of reactions pesticides undergo in soils, and indicate what we can do to encourage or prevent such reactions.

3. Discuss the environmental problems associated with the disposal of large quantities of sewage sludge on agricultural lands, and indicate how the problems could be alleviated.

4. What is *bioremediation,* and what are its advantages and disadvantages compared with physical and chemical methods of handling organic wastes?

5. Even though large quantities of the so-called heavy metals are applied to soils each year, relatively small quantities find their way into human food. Why?

6. Compare the design, operation, and management of today's containment land-fills with the natural attenuation type most common 30 years ago, and indicate how the changes affect soil and water pollution.

7. What are *organoclays,* and how can they be used to help remediate soils polluted with nonpolar organic compounds?

8. Soil organic matter and some silicate clays chemically sorb some organic pollutants and protect them from microbial attack and leaching from the soil. What are the implications (positive and negative) of such protection for efforts to reduce soil and water pollution?

9. What radionuclides are of greatest concern in soil and water pollution, and why are they not more readily taken up by plants?

10. What are the comparative advantages and disadvantages of *in situ* and *ex situ* means of remediating soils polluted with organic compounds?

11. What are two approaches to *phytoremediation,* and for what kinds of pollutants are they useful? Explain.

12. Suppose a nickel-contaminated soil 15 cm deep contained 800 mg/kg Ni. Vegetation was planted to remove the nickel by phytoremediation. The above-ground plant parts average 1% Ni on a dry-weight basis and produce 4000 kg/ha of harvestable dry matter. If two harvests are possible per year, how many years will it take to reduce the Ni level in the soil to a target of 80 mg/kg?

REFERENCES

Adriano, D. C. 2001. *Trace Elements in Terrestrial Environments: Biogeochemistry, Bioavailability, and Risks of Metals* (New York: Springer). 880 pp.

Ahmad, I., S. Hayat, and J. Pichtel (eds.). 2006. *Heavy Metal Contamination of Soil: Problems and Remedies* (Enfield, N.H.: Science Publishers Inc.). 252 pp.

Alexander, M. 1994. *Biodegradation and Bioremediation* (San Diego: Academic Press).

Alexander, M. 2000. "Aging, bioavailability, and overestimation of risk from environmental pollutants," *Environ. Sci. Tech.,* **34**:4259–4265.

Appleton, J. D. 2007. Radon: Sources, health risks, and hazard mapping. *AMBIO: A Journal of the Human Environment*: 85–89.

Basta, N. T., J. A. Ryan, and R. L. Chaney. 2005. Trace element chemistry in residual-treated soil: Key concepts and metal bioavailability. *J. Environ. Qual.,* **34**:49–63.

Berti, W. R., and L. W. Jacobs. 1996. "Chemistry and phytotoxicity of soil trace elements from repeated sewage sludge applications," *J. Environ. Qual.,* **25**:1025–1032.

Beyer, W. N., R. L. Chaney, and B. M. Mulhern. 1982. "Heavy metal concentration in earthworms from soil amended with sewage sludge," *J. Environ. Qual.,* **11**:381–385.

Boyle, M. 1988. "Radon testing of soils," *Environ. Sci. Tech.,* **22**:1397–1399.

Bragg, J. R., R. C. Prince, E. J. Harner, and R. M. Atlas. 1994. "Effectiveness of bioremediation for the *Exxon Valdez* oil spill," *Nature,* **368**:413–418.

Chaney, R. L. 1990. "Public health and sludge utilization," Part II, *Biocycle,* **31**(10): 68–73.

Cheng, H. H. (ed.). 1990. *Pesticides in the Soil Environment: Processes, Impacts, and Modeling* (Madison, Wis.: Soil Sci. Soc. Amer.).

Dowdy, R. H., C. E. Clapp, D. R. Linden, W. E. Larson, T. R. Halbach, and R. C. Polta. 1994. "Twenty years of trace metal partitioning on the Rosemount sewage sludge watershed," pp. 149–155, in C. E. Clapp, W. E. Larson, and R. H. Dowdy (eds.), *Sewage Sludge: Land Utilization and the Environment* (Madison, Wis.: Soil Sci. Soc. Amer.).

Eccles, H. 2007. *Bioremediation* (New York: Taylor & Francis). 372 pp.

Edwards, C. A. 1978. "Pesticides and the micro-fauna of soil and water," pp. 603–622, in I. R. Hill and S. J. Wright (eds.), *Pesticide Microbiology* (London: Academic Press).

Furr, A. K., A. W. Lawerence, S. S. C. Tong, M. C. Grandolfo, R. A. Hofstader, C. A. Bache, W. H. Gutemann, and D. J. Lisk. 1976. "Multielement and chlorinated hydrocarbon analysis of municipal sewage sludges of American cities," *Environ. Sci. Tech.,* **10**:683–687.

Gaynor, J. D., D. C. MacTavish, and W. I. Findlay. 1995. "Atrazine and metolachlor loss in surface and subsurface runoff from three tillage treatments in corn," *J. Environ. Qual.,* **24**:246–256.

Gilliom, R. J., J. E. Barbash, C. G. Crawford, P. A. Hamilton, J. D. Martin, N. Nakagaki, L. H. Nowell, J. C. Scott, P. E. Stackelberg, G. P. Thelin, and D. M. Wolock. 2006. "The quality of our nation's waters: Pesticides in the nation's streams and ground water, 1992–2001," USGS Circular 1291 (U.S. Geological Survey, Reston, Virginia). 172 pp. http://pubs.usgs.gov/circ/2005/1291/.

Hazen, Terry C. 1995. "Savannah river site—a test bed for cleanup technologies," *Environ. Protection* (April): 10–16.

Holmgren, G. G. S., M. W. Meyer, R. L. Chaney, and R. B. Daniels. 1993. "Cadmium, lead, zinc, copper, and nickel in agricultural soils of the United States of America," *J. Environ. Qual.,* **22**:335–348.

Jaynes, W. F., and S. A. Boyd. 1991. "Clay mineral type and organic compound sorption by hexadecyltrimethylammonium-exchanged clays," *Soil Sci. Soc. Amer. J.,* **55**:43–48.

Kabata-Pendias, A., and H. Pendias. 1992. *Trace Elements in Soils and Plants* (Boca Raton, Fla.: CRC Press).

Kjeldsen, P., M. Barlaz, A. Rooker, A. Baun, A. Ledin, and T. Christensen. 2002. "Present and long-term composition of msw landfill leachate: A review," *Critical Reviews in Environmental Science and Technology,* **32**:297–336.

Kreuger, R. F., and J. N. Seiber (eds.). 1984. *Treatment and Disposal of Pesticide Wastes* (Washington, D.C.: Amer. Chem. Soc.).

McBride, M. B., B. K. Richards, T. Steenhuis, and G. Spiers. 1999. "Long-term leaching of trace elements in a heavily sludge-amended silty clay loam soil," *Soil Science,* **164**:613–623.

McConnell, J. S., and L. R. Hossner. 1985. "pH-dependent adsorption isotherm of glyphosate," *J. Agric. Food Chem.,* **33**:1075–1078.

Meriwether, J. R., J. N. Beck, D. F. Keeley, M. P. Langley, R. N. Thompson, and J. C. Young. 1988. "Radionuclides in Louisiana soils," *J. Environ. Qual.,* **17**:562–568.

Pierzynski, G. M., J. T. Sims, and G. F. Vance. 2004. *Soils and Environmental Quality,* 3rd ed. (Boca Raton, Fla.: CRC Press/Lewis Publishers).

Pimental, D., H. Acquay, M. Biltonen, P. Rice, M. Silva, J. Nelson, V. Lipner, S. Giordano, A. Horowitz, and M. D'Amore. 1992. "Environmental and economic costs of pesticide use," *Bioscience,* **42**:750–760.

Pritchard, P. H., J. G. Mueller, J. C. Rogers, F. V. Kremer, and J. A. Glaser. 1992. "Oil spill bioremediation: Experiences, lessons and results from the *Exxon Valdez* oil spill in Alaska," *Biodegradation,* **3**:315–335.

Qian, X., R. M. Koerner, and D. H. Gray. 2002. *Geotechnical Aspects of Landfill Design and Construction* (Upper Saddle River, N.J.: Prentice Hall). 716 pp.

Reynolds, C. M., D. C. Wolf, T. J. Gentry, L. B. Perry, C. S. Pidgeon, B. A. Koenen, H. B. Rogers, and C. A. Beyrouty. 1999. "Plant enhancement of indigenous soil microorganisms: A low cost treatment of contaminated soils," *Polar Record,* **35**(192): 33–40.

Ryan, J. A., K. G. Scheckel, W. R. Berti, S. L. Brown, S. W. Casteel, R. L. Chaney, J. Hallfrisch, M. Doolan, P. Grevatt, M. Maddaloni, and D. Mosby. 2004. "Reducing children's risk from lead in soil," *Environ. Sci. Technol.,* A-Pages, **38**:18A–24A.

Sawhney, B. L., and K. Brown (eds.). 1989. *Reactions and movement of organic chemicals in soils* (Madison, Wis.: Soil Sci. Soc. Amer.).

Skipper, H. D., and R. F. Turco. 1995. *Bioremediation: Science and Applications.* Special Publication no. 43 (Madison, Wis.: Soil Sci. Soc. Amer.).

Sommers, L. E. 1977. "Chemical composition of sewage sludges and analysis of their potential use as fertilizer," *J. Environ. Qual.,* **6**:225–232.

Stokes, J. D., G. I. Paton, and K. T. Semple. 2005. "Behaviour and assessment of bioavailability of organic contaminants in soil: Relevance for risk assessment and remediation," *Soil Use Manage.,* **21**:475–486.

Strek, H. J., and J. B. Weber. 1982. "Adsorption and reduction in bioactivity of polychlorinated biphenyl (Aroclor 1254) to redroot pigweed by soil organic matter and montmorillonite clay," *Soil Sci. Soc. Amer. J.,* **46**:318–322.

Thompson, A. R., and C. A. Edwards. 1974. "Effects of pesticides on nontarget invertebrates in freshwater and soil," pp. 341–386, in W. D. Guenzi (ed.), *Pesticides in Soil and Water* (Madison, Wis.: Soil Sci. Soc. Amer.).

U.S. EPA. 1993. *Clean Water Act,* sec. 503, vol. 58, no. 32 (Washington, D.C.: U.S. Environmental Protection Agency).

U.S. EPA. 2005a. "Municipal solid waste generation, recycling, and disposal in the United States: Facts and figures for 2003." A530-F-05-003. Environmental Protection Agency, Washington, D.C. http://www.epa.gov/msw/pubs/msw05rpt.pdf.

U.S. EPA. 2005b. "A citizen's guide to radon: The guide to protecting yourself and your family from radon. U.S. EPA 402-K-02-006, Revised. U.S. EPA, Indoor Environments Division, Washington, D.C.:16 http://www.epa.gov/radon/pubs/citguide.html# howdoes.

Weber, J. B., and C. T. Miller. 1989. "Organic chemical movement over and through soil," in B. L. Sawhney and K. Prown (eds.), *Reactions and Movement of Organic Chemicals in Soils.* SSSA Special Publication no. 22 (Madison, Wis.: Soil Sci. Soc. Amer.).

Wise, D. L., D. J. Trantolo, E. J. Cichon, H. I. Inyang, and U. Stottmeister (eds.). 2000. *Bioremediation of Contaminated Soils* (New York: Marcel Dekker). 920 pp.

Xu, S., G. Sheng, and S. A. Boyd. 1997. "Use of organoclays in pollution abatement," *Advances in Agronomy,* **59**:25–62.

Zamisk, N. and J. Spencer. 2007. "China faces new worry: heavy metals in the food," page A1 July 2, 2007 issue. *Wall Street Journal,* New York.

Zorpette, G. 1996. "Hanford's nuclear wasteland," *Scientific American* (May): 88–97.

19

GEOGRAPHIC SOILS
INFORMATION

*The soil/landscape portrait
thus evolved is an artwork
of the soil scientist . . .*
—L. P. WILDING

The one great constant concerning soils is their variability. Anyone who works intimately with soils soon realizes that soils are anything but uniform. In our discussion of soil profile development (Chapter 2) we focused on vertical variability in soils—the differences among soil horizons. The subject of this chapter is horizontal variability—how soils differ from place to place across the landscape.

In order to make practical use of soil science principles, the land resource manager must know not only the "what" and "why" of soils, but must also know the "where." If builders of an airport runway are to avoid the hazards of swelling clay soils, they must know *where* these troublesome soils are located. An irrigation expert probably knows what soil properties will be necessary for cost-efficient irrigation; but for the project to be a success, he or she must also know *where* soils with these properties can be found. Almost any project involving soils, from planning a game park to fertilizing a farm field, can take advantage of geographic information about soils and soil properties. This chapter provides an introduction to some of the tools that tell us *what is where.*

19.1   SOIL SPATIAL VARIABILITY IN THE FIELD

A quick review of Figure 7.9 (changing aeration from the surface of a soil granule to its center) and Figure 12.21 (changing soil organic matter across North America) reminds us that soil properties are variable at all scales. In this chapter we will consider soil variations occurring across distances having geographic meaning for land management—from a few meters to many kilometers.

When attempting to understand geographic variation of soils and how best to use each part of the land, it is often useful to analyze each site in terms of the five factors responsible for soil formation: *climate, parent material, organisms, topography,* and *time* (see Sections 2.3 and 2.7).

Climate usually influences soil variability at very large scales (regional differences), but where the landscape includes large water bodies or significant hills and mountains, rainfall and temperature may differ greatly over distances of 1 km or less. For example, the microclimate on north-facing slopes may differ in many ways from that on south-facing slopes. Likewise, parent materials often vary in large-scale regional patterns (for example, loess plateaus versus residual rock), but very small scale differences may also occur. The soil scientist should always be alert for the possible presence of such localized parent materials as colluvial deposits at the foot of a slope or alluvium along stream courses. The field soil scientist also needs some training in biology so as to be able to recognize changes in botanical composition that might result from localized variations in soil properties such as saturated conditions in depressions or calcareous outcroppings. As these examples suggest, most small-scale soil variations involve changes in topography, so an awareness of even subtle changes in slope is critical to understanding how soils change across a landscape.

Small-Scale Soil Variability

3D Soil-Landscape Models in Florida:
http://grunwald.ifas.ufl.edu/Projects/3D_Florida/3D_FL.htm

Soil properties are likely to change markedly across small distances: within a farm field, within a suburban house lot, and even within a single soil individual (as defined in Section 3.1). At this scale, variations most often relate to small changes in topography and thickness of parent material layers (Figure 19.1) or to the effects of organisms (e.g., the effects of individual trees or past human management). Plate 94 shows dramatic small-scale variability in both surface soil color and plant vigor, resulting from the exposure of a calcareous horizon on the hillocks by wind erosion.

This small-scale variability may be difficult to measure and not readily apparent to the casual observer. In some cases, the height or vigor of vegetation reflects the subsurface variability (see Plate 45). In other cases, the changes in soil properties are detected by analyzing soil samples taken from many evenly spaced borings made throughout the plot of land in question. This and other techniques will be discussed in Section 19.2 with regard to the practical use of such observations in making soil maps.

Variability in soil fertility often reflects past soil-management practices (Figure 19.2), as well as differences in soil profile characteristics. Analysis of small-scale variability has practical uses in managing soil fertility for a given field or nursery. As described in Section 16.11, soil tests for fertility management are traditionally performed on one composite sample (a mixture of small cores from 15 to 20 randomly scattered spots) that represents an entire field, or an area as large as 10 to 20 ha.

Where small-scale variability exists, the actual level of fertility at most spots in the field is likely to be either considerably higher or lower than the average soil test value for the field. To visualize the effect of the unaccounted-for variability, consider an analogy with shoe sizes. Fertilizer recommendations based on a field average may fit the

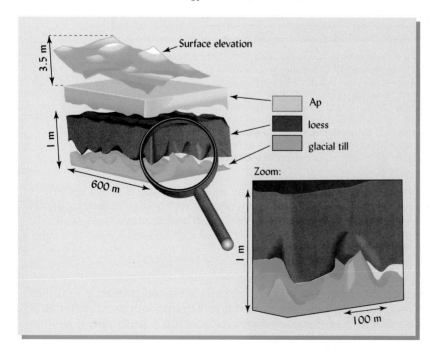

FIGURE 19.1 Three-dimensional model of a small area in a Mollisols landscape (about 600 × 400 m) in southern Wisconsin. The three layers are separated in the model to emphasize the variability in the thickness and boundaries of the horizons. The Ap horizon and two layers of underlying glacial parent materials (loess and glacial till) are depicted. The topmost layer represents the surface topography of the land as a kind of 3-D contour map showing several small knolls. The surface contours were generated from data in a digital elevation model (DEM) database. The double headed arrows indicate the different scales used in different parts of the model. [Modified from Grunwald et al. (2000). See also Grunwald (2006)]

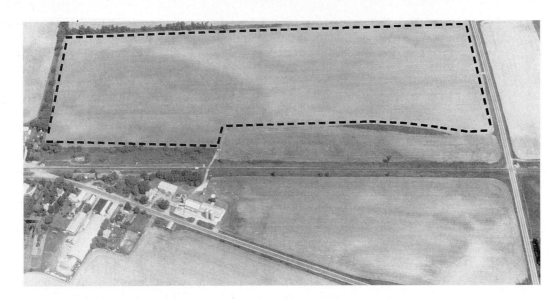

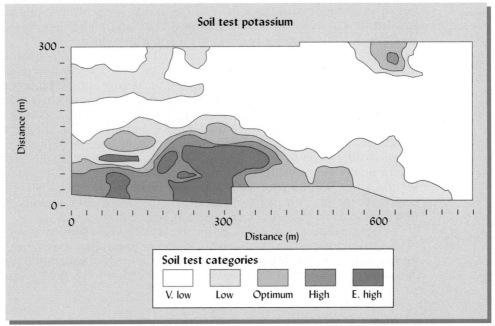

FIGURE 19.2 An oblique air photo (*top*) of a farm in central Wisconsin. The 22-ha field outlined by the heavy dashed line was studied in detail to make a map (*bottom*) of the spatial variability of available (soil test) potassium. The map was computer-generated using soil test values from 199 samples (each made up of five subsamples) taken at 32-m intervals in a grid pattern across the entire field. Note that the very high potassium levels correspond with a section of the field closest to the farmstead (lower left in photo). In the past, the farmer had managed that section as a separate field and had used it to dispose of manure from the nearby barnyard. The high-potassium spot in the upper right of the field marks the location of a manure pile that existed a number of years prior to the study. The spatial variation in potassium levels appears to be unrelated to the soil boundaries (light dashed lines) mapped in the county soil survey report. [From Wollenhaupt et al. (1994)]

needs of the soil about as well as buying everyone in a family size 6 shoes because it happens to be the average shoe size for that family. Several technological developments have made it much easier than in the past to detect, map and manage small-scale soil variability (see Sections 19.3 and 19.10).

Medium-Scale Soil Variability

For many soil properties, variability across a landscape is related primarily to differences in a particular soil-forming factor, such as soil topography or parent material. If one understands the influences of these soil-forming factors in a landscape, it is often possible to define sets of individual soils that tend to occur together in sequence across the land. Identifying one member of the set often makes it possible to predict soil properties in the landscape positions occupied by other members of the set. Such sets of soils include *lithosequences* (occurring across a sequence of parent materials), *chronosequences* (occurring across similar parent materials of varying age), and *toposequences* (with soils arranged according to changes in relief).

As one moves downslope in a particular landscape, the soils often interact with water and the water table in a systematic, predictable way that defines a **catena** of soils (see Section 2.6). The catena concept is similar to that of a toposequence, except that in a catena the member soils may or may not share a common parent material. A catena of soils usually consist of a well-drained member near the slope summit, sometimes an excessively drained member on the shoulder slope, a moderately well-drained member near the toe slope, and somewhat to very poorly drained members at the bottom. The concept of a soil catena is helpful in relating the soils to the landscape in a given region. The various soils in a catena can often be distinguished by the colors of the surface soil, and even more clearly by the colors of the B horizons. For example, in the tropical catena shown in Plate 16 (after page 112), the colors vary from dark gray at the bottom of the slope to dusky red at the top.

The relationship can also be seen by referring to Figure 19.3, where the Bath–Mardin–Volusia–Alden catena is shown. Although all four or five members of the catena are not always found together in a given area, the diagram illustrates the spatial relationship among the soils with respect to their drainage status. As shown, the drainage status of each catena member gives rise to distinct profile characteristics that affect plant rooting depth, species adaptation, and engineering uses of the soils.

Although different features may change across catenas in different regions, several trends are commonly observed moving from upslope to downslope positions, the changes being more pronounced in humid than in dry regions. Because of the relationship to the groundwater and the anaerobic conditions that accompany waterlogging, the chroma of the subsurface soil colors tends to be lower in down-slope soils, and mottles (iron depletion spots) or other redoximorphic features become increasingly numerous, more pronounced, and occur closer to the soil surface. In addition, the A horizons usually increase in thickness and darken in color due to denser vegetation and slower decomposition in the wetter soils (see Section 12.8), and deposition of organic-enriched surface material eroded from upslope soils (see Section 17.3). The texture of the surface horizon is also likely to differ. For example, in deep loess, erosion may preferentially move the silt and clay fractions downhill, leaving sandier material on the summit and shoulder and deposits of silt and clay in the lower positions. On the other hand, in catenas with well-developed argillic horizons, the upper catena members may have the finer-textured surfaces because erosion has removed most of their coarse-textured A and E horizons, exposing the clayey material from the B horizons.

A **soil association** is a more general grouping of individual soils that occur together in a landscape. Soil associations are named after the two or three dominant soils in the group, but may contain several additional, less extensive soils. The soils may be from the same soil order or they may be from different orders (Figure 19.4a and b). They may have formed in the same or in different parent materials. The only requirement is that the soils occur together in the same area. Identifying soil associations is of practical importance, because they enable us to characterize landscapes across large areas and they assist in planning general patterns of land use. A given soil association represents a defined range of soil properties and landscape relationships, even though the range of conditions included may be quite large (see Figure 19.4b).

Large-Scale Soil Variability

At a very large scale, soil patterns are principally the result of climate and vegetation patterns and secondarily related to parent material differences. Although it is often useful to refer to general regional soil characteristics, it must be remembered that much localized variation exists within each regional grouping. A study of the World Soils Map printed on the frontpapers of this textbook and the U.S. Soils Map on the endpapers will reveal important regional patterns. These patterns are also highlighted in the series

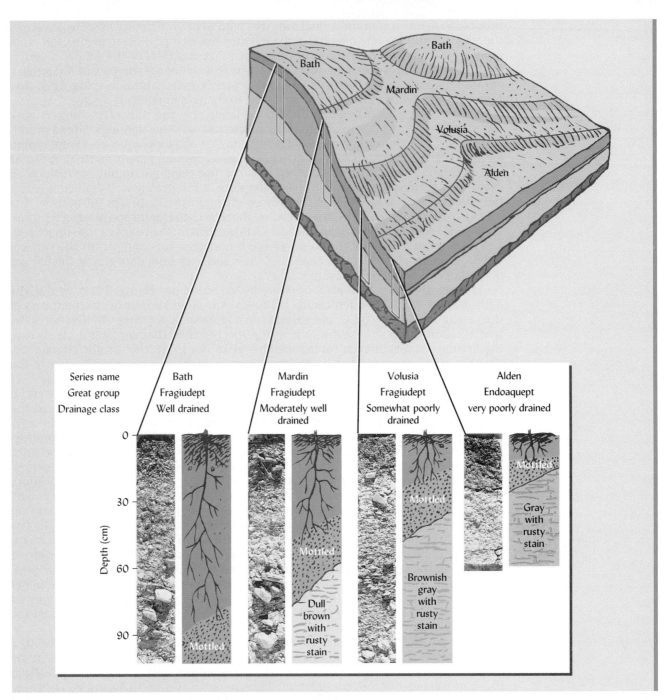

Series name	Bath	Mardin	Volusia	Alden
Great group	Fragiudept	Fragiudept	Fragiudept	Endoaquept
Drainage class	Well drained	Moderately well drained	Somewhat poorly drained	very poorly drained

FIGURE 19.3 Profile monoliths of four soils of a drainage catena (*below*) and a block diagram showing their topographic association in a landscape (*above*). Note the decrease in the depth of the well-aerated zone (above the mottled layers) from the Bath (well-drained, upslope) to the Alden (poorly drained, downslope). The Alden soil remains wet throughout the growing season. These soils are all developed from the same parent material and differ only in drainage and topography. All four soils belong to the Inceptisols order. With the exception of the cultivated Volusia, the monoliths shown were taken from forested sites. [Based on Cline and Marshall (1977)]

of small soil-order maps shown in Chapter 3. From these maps it can be seen that highly weathered Oxisols can be found principally in the hot, humid regions of South America and Africa drained by the Amazon and Congo rivers, respectively. Mollisols can be seen to characterize the semiarid grasslands of the world; Aridisols are located in the desert regions. The great expanse of loess deposits in the central United States (see Figure 2.16) is an example of the regional influence of parent materials. Soils information on this scale can make an important contribution to inventorying the natural resources of a state, province, or nation.

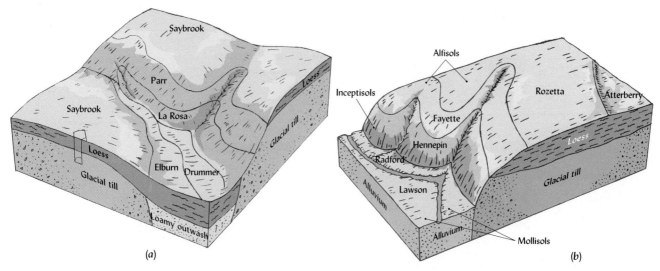

FIGURE 19.4 Two soil associations from Bureau County, Illinois. The soils of the Saybrook-Parr-La Rosa association (*a*) are all Mollisols. They differ principally with regard to topography and parent material (Parr and La Rosa developed in glacial till, the others mainly in loess). The Rozetta-Fayette-Hennepin association (*b*) includes a wider range of soil conditions and includes soils from three soil orders. [Based on Zwicker (1992)]

19.2 TECHNIQUES AND TOOLS FOR MAPPING SOILS[1]

Geographic information about soils is often best communicated to land managers by means of a soil map. Soil maps are in great demand as tools for practical land planning and management. Many soil scientists therefore specialize in mapping soils. Before beginning the actual mapping process, a soil scientist must learn as much as possible about the soils, landforms, and vegetation in the survey area. Therefore, the first step in mapping soils is to collect and study older or smaller-scale soil maps, geological and topographic maps, previous soil descriptions, and any other information available on the area. Once the soil survey begins, the soil scientist's task is threefold: (1) to define each soil unit to be mapped (see Section 19.3); (2) to compile information about the nature of each soil; and (3) to delineate the boundaries where each soil unit occurs in the landscape. We will now discuss some of the procedures and tools that soil scientists use to delineate soils in the field.

Soil Description

USDA Soil survey division field book for describing and sampling soils:
http://soils.usda.gov/technical/fieldbook/

Soil scientists may use computers and satellites, but they also use spades and augers. Despite all the technological advances of recent years, the heart of soil mapping is still the soil pit. A soil pit, whether dug by hand or with a backhoe, is basically a rectangular hole large enough and deep enough to allow one or more people to enter and study a typical pedon (see Section 3.1) as exposed on the pit face. Plates 1 to 12 are photographs taken of such pit faces. After cleaning away loose debris from the pit face, the soil scientist will examine the colors, texture, consistency, structure, plant rooting patterns, and other soil features to determine which horizons are present and at what depths their boundaries occur (Figure 19.5). Often the horizon boundaries are etched with a trowel or soil knife, as can be seen on the right side of Plate 9.

A soil description is then written in a standard format (see Table 19.1 for an example) that facilitates communication with other soil scientists and comparison with

---

[1] For an internationally oriented guide to all aspects of the process of making soil maps, see Legros (2006). For official procedures for making soil surveys in the United States, see USDA-NRCS (2006).

FIGURE 19.5 A soil pit allows detailed observations to be made of the soil in place. Here, several Natural Resources Conservation Service soil scientists describe a typical pedon for a mapping unit (Thorndale series) as revealed on the wall of a soil pit. The soil scientist standing in the pit is comparing the colors and textures of soil samples he has removed from several horizons while another soil scientist (*upper right*), records the observations in a notebook to make a soil profile description (such as the one in Table 19.1). Later when these soil scientists are boring transects of auger holes to map soils in a landscape (see Figures 19.6 and 19.7), they can determine the identity of the soils they encounter by comparing the properties of samples brought up in their augers to the properties listed in detailed written descriptions of the soil pit (Photo courtesy of R. Weil).

TABLE 19.1   Soil Profile Description (with Soil Taxonomy Diagnostic Horizons) for the Thorndale Soil Series[a]

Figure 19.5 shows soil scientists describing a Thorndale soil in Pennsylvania.

Horizon designation	Diagnostic horizon	Horizon boundaries	Description of horizon in typical pedon
Ap	Ochric epipedon	0–20 cm	Dark grayish brown (2.5Y 4/2) silt loam; weak medium granular structure; friable, slightly sticky, slightly plastic; many fine roots; neutral; clear smooth boundary.
Btg1		20–43 cm	Light olive gray (5Y 6/2) silty clay loam; moderate coarse subangular blocky structure; slightly firm, sticky, plastic; few medium and fine roots; many prominent dark grayish brown (10YR 4/2) clay films on faces of peds; common medium prominent reddish brown (5YR 4/3) masses of iron accumulations; slightly acid; clear smooth boundary.
Btg2	Argillic horizon	43–65 cm	Light olive gray (5Y 6/2) silty clay loam; weak coarse prismatic structure parting to moderate medium subangular blocky; firm, sticky, plastic; few fine roots; many prominent dark grayish brown (10YR 4/2) clay films on prisms and faces of peds; common medium and fine prominent reddish brown (5YR 4/4) masses of iron accumulations in the matrix; slightly acid; gradual smooth boundary.
Btxg	Fragipan	65–103 cm	Grayish brown (10YR 5/2) silty clay loam; weak very coarse prismatic structure parting to weak medium subangular blocky; firm, brittle, moderately sticky, moderately plastic; few fine roots; many faint dark grayish brown (10YR 4/2) clay films on prism faces and few faint dark grayish brown (10YR 4/2) clay films on faces of peds; common fine to medium prominent reddish brown (5YR 4/4) and brown (7.5YR 5/4) masses of iron accumulations in the matrix; slightly acid; abrupt smooth boundary.
C		103–163 cm	Strong brown (7.5YR 5/6) and reddish yellow (7.5YR 7/6) silt loam; massive; friable, slightly sticky, slightly plastic; common medium prominent grayish brown (2.5Y 5/2) iron depletions in the matrix; slightly acid.

[a] Thorndale soils are very deep, poorly drained soils formed in medium-textured colluvium derived from limestone, calcareous shale, and siltstone. Slopes are 0 to 8%. Permeability is slow. Mean annual precipitation and temperature are about 100 cm and 12 °C. Taxonomic class is Fine-silty, mixed, active, mesic Typic Fragiaqualfs.
Adapted from USDA-NRCS (2002).

European digital archive of soil maps of the world: http://eusoils.jrc.it/esdb_archive/EuDASM/indexes/access.htm

other soils. Sometimes the soil scientist will use field kits to make chemical tests, for pH and free carbonates (effervescence of carbon dioxide when dilute hydrochloric acid is added) (see Plate 91).

As far as possible at this stage, the soil horizons will be given master (A, E, B, etc.) and subordinate (2Bt, Ap, etc.) designations (see Table 2.6). Finally, samples of soil material will be obtained from each horizon. These will be used for detailed laboratory analyses and for archiving. The laboratory analyses will provide information for the chemical, physical, and mineralogical characterization of each soil.

Using these techniques, the soil scientists assigned to map an area will familiarize themselves with the soils they expect to find, learning certain unique characteristics that they can look for to quickly identify each soil and distinguish it from other soils in the area.

Delineating Soil Boundaries

For obvious reasons, a soil scientist cannot dig pits at many locations on the landscape to determine which soils are present and their boundaries. Instead, he or she will bring up soil material from numerous small boreholes made with a hand auger or hydraulic probe (Figure 19.6*a–b*). The texture, color, and other properties of the soil material from various depths can be compared mentally to characteristics of the known soils in the region.

With hundreds of different soils in many regions, this might seem to be a hopeless task. However, the job is not as daunting as one might suppose, for the soil scientist is not blindly or randomly boring holes. Rather, he or she is working from an understanding of the soil associations and how the five soil-forming factors determine which soils are likely to be found in which landscape positions. Usually there are only a few soils likely to occupy a particular location, so only a few characteristics must be checked. The soil auger is used primarily to confirm that the type of soil predicted to occur in a particular landscape position is the type actually there.

The nature of soil units and the locations of the boundary lines surrounding them are inferred from information obtained by auger borings at numerous locations across a landscape. A simple but laborious and time-consuming approach to obtaining soils information is to make auger borings at regular intervals (say, every 50 m) in a grid pattern across the landscape (Figure 19.7, *left*). Points with similar properties can then be connected to form soil boundaries. This approach is sometimes used in developing countries where labor to survey the sampling points and auger the soils is quite inexpensive.

Knowledge of the interplay of the soil-forming factors in a landscape can greatly expedite the soil scientist's work. An efficient soil mapper will use clues from changes in topography, vegetation, and soil surface colors as a guide in locating sites to make borings. A typical approach is to traverse the landscape along selected transects (straight-line paths), augering only at enough points to confirm expected soil properties and boundaries. In order to pinpoint soil boundaries, more frequent borings are made near the places where breaks in slope or other landscape clues suggest that soil boundaries will occur (see Figure 19.7 *right*). In Figure 19.8 a soils map made using transect borings is compared to an aerial photo of the bare soil in the same area and also to a map of one surface soil property measured in a large number of samples taken in a grid pattern covering the same area.

It is often useful for a soil mapper to visualize the landscape model in the form of a block diagram, which gives a stylized depiction of the main landforms and associated soils. A block diagram can communicate a deeper understanding of the existing landscape and aid in engineering applications if it accurately depicts the underlying substrata (parent materials, C and R horizons) as well as the A and B horizons on which most soil classification is focused. Figure 19.9 illustrates such a diagram.

19.3 MODERN TECHNOLOGY FOR SOIL INVESTIGATIONS

While it is still the mainstay of soil investigations and mapping, soil augering is intrusive (i.e., it makes holes) and labor-intensive. Several nonintrusive methods of soil investigation are finding increasing use in helping to identify subsurface features and locate soil boundaries. These technologies include (1) ground-penetrating radar, (2) electromagnetic induction, (3) geopositioning systems, and (4) remote sensing of surface features.

FIGURE 19.6 Soil maps are prepared by soil scientists who examine the soils in the field using such tools as a hand-powered soil auger (*a*) or a truck-mounted hydraulic soil probe (*b*). Most soil maps are initially prepared by outlining the boundaries of soil mapping units on an air photo (*c*). A finished map is then made by superimposing soil boundaries, roads, and other features on an aerial photo map base (*d*). Each soil mapping unit outlined is identified by a three-part code. For example, in the two small areas (arrows) labeled "145B2," the "145" codes for the soil series and surface texture, in this case a Saybrook silt loam. (On many soil survey maps a two-letter code is used in place of this three-digit one.) The capital letter "B" indicates slopes of 2 to 5%, and the final "2" indicates that the soil is moderately eroded. Thus the areas so marked contain the 2 to 5% slope, slightly eroded, silt loam surface soil phase of the Saybrook soil series. (Photos and maps courtesy of USDA/NRCS)

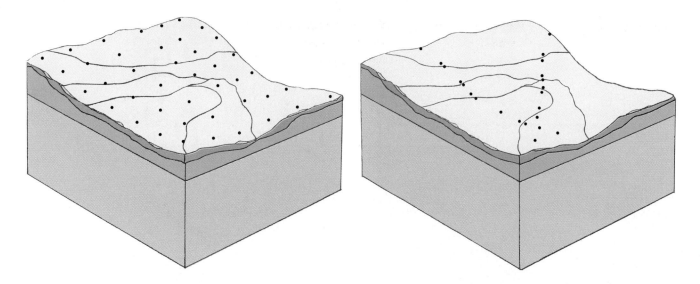

FIGURE 19.7 Two approaches to collecting information on soil properties and boundaries by soil auger borings (indicated by dots). The regularly spaced grid pattern (*left*) of borings is simple in concept, but very labor-intensive to carry out. In a much more efficient approach (*right*), the soil scientist traverses the landscape along selected transects (straight-line paths), augering only at enough points to confirm soil properties and boundaries predicted on the basis of soil–landscape relationships. Note that in order to pinpoint soil boundaries, extra borings are made near the places these boundaries are expected to occur. (Diagram Courtesy of R. Weil)

Ground-Penetrating Radar

Ground-penetrating radar (GPR) can be used to investigate subsurface layers while increasing the quality and reducing the cost of making soil maps. Ground-penetrating radar (GPR) is a noninvasive geophysical tool (no digging is required) that can produce images of contrasting layers in the upper regolith (0 to 30 m). The GPR transmits oscillating waves of electromagnetic energy that move through the soil until they hit a contrasting layer

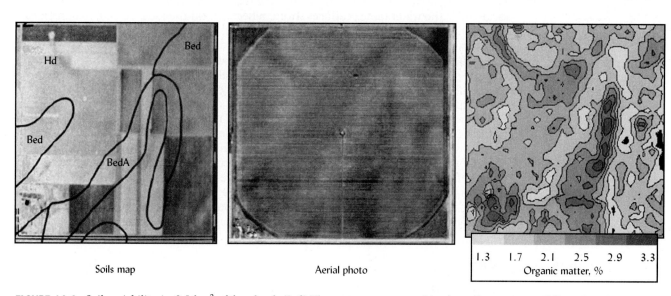

Soils map   Aerial photo

| 1.3 | 1.7 | 2.1 | 2.5 | 2.9 | 3.3 |

Organic matter, %

FIGURE 19.8 Soil variability in 2.5 km² of farmland. (*Left*) The area as represented in the soil survey map delineating six mapping units (and two soil series). The black-and-white air photo background was taken while crops covered all but the three dark-colored fields (where the bare soil can be seen). (*Middle*) An air photo taken some years after the land was converted to center pivot irrigation (Section 6.9) and the many rectangular fields were combined into one large, rounded field. The soil has just been plowed and variations in soil color (mainly related to the effect of topography on organic matter content) are easily distinguished. (*Right*) Spatial distribution of surface soil organic matter content (ranging from 1.3 to 3.3%) presented on a map created from analysis of many soil samples obtained in a grid pattern across the entire area. The spatial correlation between the soil organic matter data (*right*) and the soil colors (*middle*) are very close. The delineation of soil mapping units (*left*) also approximates the organic matter/drainage differences, the mapping units in the swales coded as "Hd" being higher in soil organic matter than the mapping units on the hills coded as "Bed". (Images courtesy of USDA/ARS Agroecosystems Management Unit, Nebraska.)

Landscape

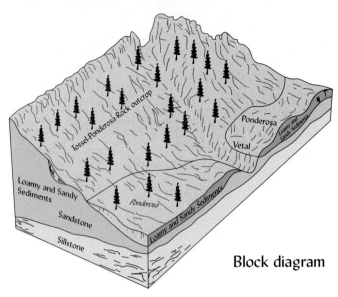

Block diagram

FIGURE 19.9   A block diagram can represent the spatial arrangement of landscape elements including soil series, landforms, and underlying parent materials. Soil scientists often observe geologic layers, but usually describe only the surface layers to a depth of about 2 m in soil surveys. Block diagrams can be used to associate information on substrata with the surface soil mapping units to help land users understand and manage both soils and underlying parent materials. [Photo and diagram adapted from Wysocki et al. (2005)]

Radar to map soil moisture in vineyard:
http://www.berkeley.edu/news/media/releases/2003/10/16_wine.shtml

which reflects back some of the energy. This reflected energy is measured and displayed on a radar record as the instrument is pulled across the land surface (Figure 19.10a). The amount of energy reflected depends on the degree of contrast in dielectric permittivity of the two adjoining layers. The dielectric permittivity of different soil layers is a property largely determined by the moisture content, salinity, and density of the material. For example, in many soils, a subsurface clay accumulation (Bt horizon) is more dense and moist than the layers above it and so causes an identifiable reflection front that allows the upper boundary of this subsoil horizon to be seen in the radar record (Figure 19.10b). In sandy soils, the water table produces strong reflections on the radar record, making it easy to map the depth to the water table (Figure 19.10c). The vertical scale on the record actually shows the time (in nanoseconds, ns) it took for the energy pulse to travel down to the reflecting layer and back to the instrument. Calibrating the instrument for the particular type of soil material encountered determines the velocity (between about 3 and 20 m/ns) at which the energy traverses the material, allowing the vertical scale on the radar record to be expressed as depth (m) rather than travel times (ns). Figure 19.10c shows the radar record from a single pass (transect) of the GPR. If many parallel transects are made the resulting records can be statistically combined into a three-dimensional map of the groundwater surface, enabling the investigator to predict the direction and rate of groundwater flow. Such a map provides an invaluable resource for investigating the movement of pollutants. The GPR is not suitable for all soils, because the combination of moisture, salt content, and type of clay may make the reflectance record difficult to interpret. Where it is effective, however, the cost of GPR is only about one-third that of detecting soil or groundwater boundaries by making multiple borings with a soil auger.

Electromagnetic Induction[2]

Electromagnetic induction (EM) techniques offer another noninvasive, rapid method of investigating subsurface features. Using a handheld instrument approximately the size and shape of a carpenter's level, this method measures the apparent conductivity of the soil for electromagnetic energy. The conductivity measured is influenced by the moisture content, salinity of the soil (see Section 10.4), and the amount and type of clays in the soil. This technique has been successfully used to map out the depth and thickness of claypan horizons in humid regions and to investigate groundwater contamination and salinity in arid regions.

---

[2] For an example of the use of this technique, see Doolittle et al. (1994).

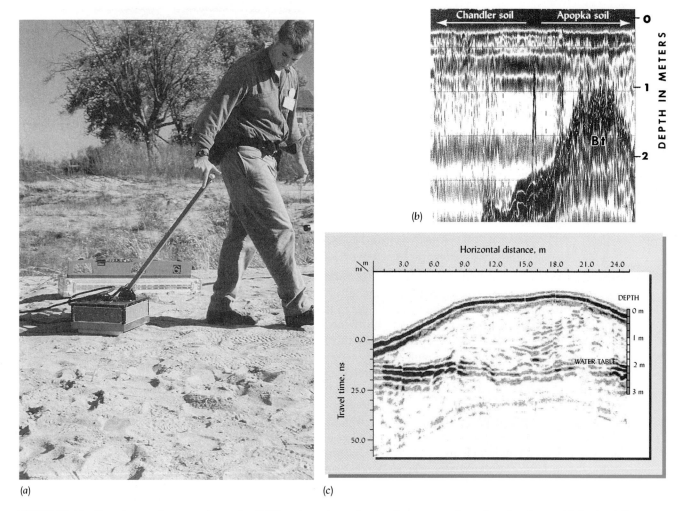

(a)

(b)

(c)

FIGURE 19.10 Use of ground-penetrating radar (GPR) to investigate the depth to contrasting subsurface layers. (*a*) As the GPR instrument is pulled over the soil surface, it emits an oscillating radar signal that penetrates the soil to as much as 30 m depth under favorable conditions. (*b*) An example of a radar record output from a GPR instrument, showing the depth to the Bt horizon in two Florida soils. The GPR signals are sensitive to the dramatic change in soil properties between the coarse sands of the upper horizons and the clay enriched Bt horizon. On the left side of the graph are the deep sands of the Chandler soil, a Quartzipsamment without an argillic horizon in the upper 2 m. On the right side of the graph, the Bt horizon of an Apopka soil (a Paleudult) can be clearly seen. The boundary between the two soils is gradual, but could be delineated at about the middle of the graph. The distance across the entire graph is about 140 m. (*c*) In sandy soils, GPR can be used to chart the depth to the groundwater table. [Photo courtesy of R. Weil; graphs from Doolittle (1987) and Doolittle et al. (2006)]

It should be noted that these electronic methods of soil investigation are not yet available as off-the-shelf technology, but must be adapted to each situation by the user. The user must initially quantify the relationship between the soil properties of interest and the electronic signals recorded by the instrument. A suitable computer program may be needed to analyze the data. Once so adapted, these technologies can provide detailed information about subsurface features and can be a great aid in determining the location of soil boundaries on detailed soil maps (Figure 19.11).

Global Positioning Systems[3]

An obvious prerequisite for delineating the location of soil bodies in the field is that the soil mappers know their location as they traverse a landscape. Traditionally, soil mappers have used large-scale base maps, air photos (see Section 19.5) and a compass to ascertain location. However, in nearly featureless terrain or heavily vegetated areas these aids are of limited use. Fortunately, soil mappers can now take advantage of satellite technology to identify precise locations anywhere in the world (Box 19.1).

---

[3] For a readable discussion of the principles behind the GPS, see Herring (1996).

FIGURE 19.11 Map of a 7 ha field showing the apparent electrical conductivity (EC$_a$) in the surface 90 cm of soil as measured by an electromagnetic induction [EM] meter (see Section 10.4) pulled back and forth (north–south) across the field. A differential global positioning system (GPS) receiver mounted on the vehicle geo-referenced each EC$_a$ reading. The data were grouped into four EC$_a$ classes, shown as black (highest EC$_a$), dark gray, light gray, and white (lowest EC$_a$). Soil cores were also collected (locations shown by 24 white or black dots) to confirm the EC$_a$ readings with laboratory EC$_s$ analysis. Other analyses on these soil samples showed that the EC$_a$ readings were closely correlated with depth of A horizon and soil contents of total organic C, clay, and salts, all properties that influence soil water-holding capacity. [Modified from Grigera et al. (2006) with permission from the Soil Science Society of America]

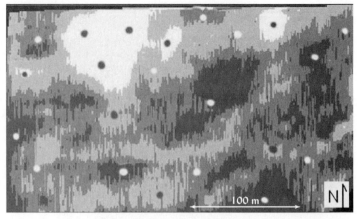

BOX 19.1   USING THE GLOBAL POSITIONING SYSTEM TO MAP SOILS INFORMATION

A network of several dozen Global Positioning System (GPS) satellites was developed by the United States Defense Department to aid in the navigation of military planes and ships during the 1980s. The system can now be used by civilians almost anywhere in the world to determine their precise location almost instantaneously. The GPS satellites are orbiting some 20,000 km above the Earth in precisely known patterns arranged so that at least four satellites are broadcasting simultaneously to any point on Earth's surface. The group of satellites in communication with a GPS receiver is called the *satellite constellation*.

An electronic clock in the receiver measures how long it takes for a pattern of radio signals to travel to it from each satellite. Given that the signals travel at the velocity of light, the receiver can calculate the distance to each satellite (distance = velocity × time). A receiver located at, say, 20,200 km from a particular satellite must be located someplace on the surface of an imaginary sphere that has its center at the satellite transmitter and has a diameter of 20,200 km. If the same receiver is also located at 20,700 km from a second satellite, it must also be located someplace along the surface of a second sphere 20,700 km in diameter. Simple geometry tells us that four such spheres can intersect at only one point in space (Figure 19.12, *left*), that point being the exact location of the receiver.

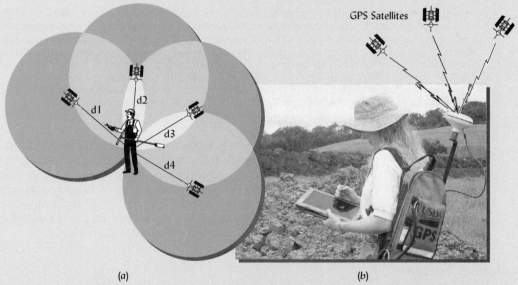

(a)   (b)

FIGURE 19.12 The Global Positioning System of satellites and receivers. (*a*) A GPS receiver uses radio signals to determine its distance (d1, d2, etc.) from each of four orbiting satellites. The spheres of distance around the four individual satellites intersect at only one position in space, the location of the receiver. (*b*) A soil scientist maps soils in the field using a computer tablet to record observed soil properties and a backpack GPS receiver to geo-reference the location of each observation. Three of the satellites in contact with the GPS receiver are schematically illustrated. The computerized GPS system can also be used to guide the mapper to a set of pre-determined soil sampling locations. (Diagram and photo courtesy of R. Weil).

(continued)

BOX 19.1 (Cont.) USING THE GLOBAL POSITIONING SYSTEM TO MAP SOILS INFORMATION

Because the setting of the satellite clock may not be known exactly, accuracy is not always optimal. One corrective measure involves installing a stationary receiver at an exactly known location on a tall building or tower. Mobile receivers then use broadcasts from the stationary receiver to calculate the "errors" in each satellite clock. Other *differential correction* signals are available from government and commercial sources broadcasting from FM radio towers. Without these corrections, simple, hand-held receivers can determine locations to within 2 to 6 m. With these corrective measures in use, even relatively small and inexpensive GPS receivers can determine locations to within 1 to 3 m, and larger, more sophisticated receivers (Figure 19.12, *right*) can determine locations to within a few centimeters.

This technology can be applied in making soil maps, allowing the soil scientist in the field to record the precise geographic coordinates of each soil observation made. The soil information mapped may be the name of the taxonomic map unit (as for soil surveys) or it may be a measured soil property, such as the organic matter content or nutrient availability. In the latter case, special maps can be prepared by *geostatistical* computer programs that estimate values for the soil property in question at all points between the points actually sampled (such as in Figure 19.8, *right*).

19.4 REMOTE SENSING TOOLS FOR SOILS INVESTIGATIONS[4]

Remote sensing describes the gathering of information from a distance. In this general sense of the term, we are remote sensing any time we use our eyes to see an object from a distance rather than picking up that object in our hands and feeling it. When we perceive an object, our brains are forming a mental image in response to light energy that has reflected off the object into our eyes. In an analogous manner, a photographic or digital image can be formed by sensors (e.g., cameras) mounted on a platform (e.g., an airplane or space satellite), providing a suitable vantage point for observing a particular area of land. While our eyes respond only to reflected energy with wavelengths in the visible range, other sensors can form images from additional wavelengths of energy, such as infrared. We will briefly describe several types of imagery and their uses in making geographic soils investigations. Air photos and other imagery covering most locations in the United States and the world are available from a number of government agencies and private companies (Table 19.2).

19.5 AIR PHOTOS

Most air photos are made with panchromatic black-and-white film, which "sees" a range of light wavelengths that closely corresponds to the spectrum visible to our eyes. The photograph produced is black and white or, more correctly, many shades of gray. Black-and-white air photos can reveal a wealth of information about landforms, vegetation, human influences, and, yes, soils. But experience is required to recognize the various gray tones and patterns as different types of vegetation, drainage patterns, and soil bodies.

Other films, such as natural color or infrared film, are also used for aerial photography. Black-and-white prints from infrared film are commonly used by forest managers because conifer needles absorb infrared energy much more completely than do hardwood leaves, allowing the conifers to be easily distinguished by their darker gray tones on the photograph.

Air photos have been used since 1935 to increase the speed and accuracy of making soil maps. Now often used in digital form, they can be used to assist soil investigations in at least three ways: (1) in providing a base map, (2) as a source of proxy information, and (3) in directly sensing soil properties.

---

[4] For an excellent, comprehensive Web-based tutorial on remote sensing, see Short (2006).

TABLE 19.2 **Partial Listing of Sources for Remote Sensing Imagery[a]**

Imagery	Source	Web site/ Contact numbers	Address
Air photos	Bureau of Land Management-National Science and Tech. Ctr	http://www.blm.gov/nstc/aerial/index.html Phone: (303) 236-7991	P.O. Box 25047 Denver, Co 80225-0047, USA
Air photos	U.S. Geological Survey (EROS)	http://www.usgs.gov/pubprod/aerialsatellite.html http://edc.usgs.gov/about/featurelinks.html Phone: 800-252-4547	Earth Resources Observation & Science 47914 252nd Street Sioux Falls, SD 57198-0001, USA
Air photos	U.S. Dept. of Agriculture Aerial-Photo Field Office	http://www.fsa.usda.gov/FSA/apfoapp?area=apfohome&subject=landing&topic=landing Phone: (801) 975-3503	User Services ASCS-USDA P.O. Box 30010 Salt Lake City, UT 84130, USA
Landsat Imagery	GeoEye	http://www.geoeye.com/ Phone: (703) 480-7500	21700 Atlantic Boulevard Dulles, VA 20166, USA
Landsat Imagery	MDA Federal, Inc.	http://www.mdafederal.com/geocover Phone: (301) 231-0660	6011 Exec. Blvd. Suite 400 Rockville, MD 20852, USA
Quick Bird, World View Imagery	Digital Globe, Inc.	http://www.digitalglobe.com/ Phone: (800) 496-1225	1601 Dry Creek Drive Suite 260 Longmont, CO 80503, USA
SPOT Imagery	SPOT Image Corporation	http://www.spot.com/html/SICORP/_401_.php Phone: (703) 715-3100	14595 Avion Parkway, Suite 500 Chantilly, VA 20151, USA
Radar Imagery	MDA (MacDonald, Dettwiler & Associates Ltd.)	http://gs.mdacorporation.com/ Phone: (604) 278-3411	13800 Commerce Parkway Richmond, BC, Canada V6V 2J3
Digital terrain elevations	NGA (National Geospatial-Intelligence Agency) Office of Corporate Relations	http://geoengine.nga.mil/geospatial/SW_TOOLS/NIMAMUSE/webinter/rast_roam.html Phone: 800-455-0899	Public Affairs Division, MS D-54 4600 Sangamore Road Bethesda, MD 20816-5003, USA

[a] Neither the authors nor the publishers promote the listed examples of sources in preference to other imagery providers.

Air Photographs as Base Maps

A sufficiently detailed air photo allows the soil scientist to determine his or her location in the field in relation to such features as buildings, roads, and streams that are visible on both the photograph and on the ground. Rather than using surveying equipment and a plane table to make a map from blank paper, the soil scientist can draw soil boundaries directly on the air photo. In this way the photograph serves as a *base map*.

It should be noted that uncorrected air photos are severely distorted, because the land areas shown near the edges of the photograph were considerably farther from the camera than were the areas directly beneath the passing airplane. Also, in hilly or mountainous terrain the hilltops would have been closer to the camera than the valleys. An *ortho photograph* is one that has been corrected for both types of distortion. In most modern soil surveys published, ortho photographs are used as base maps on which soil boundaries are drawn. New soil surveys use *digital ortho quads*, electronic versions of ortho photographs that are compatible with other computerized geographic information (see Section 19.9). However, uncorrected air photos are still used to allow stereoscopic coverage, as illustrated in Figure 19.13.

Air Photos for Proxy Information

U.S. Geological Survey satellite images:
http://earthshots.usgs.gov/tableofcontents

Soil investigations are usually concerned with features of soil profiles and other subsurface information. However, an air photo records radiant energy reflected off surfaces aboveground or, at best, a few centimeters belowground. Nonetheless, the surface tones, patterns, and features shown on the photograph often are related to conditions belowground. Once the soil scientist has learned what these relationships are, air photos can be used as a source of *proxy* information about soil conditions.

For example, the photographs do not directly record the depth to the water table; but dark tones indicate moist, high-organic-matter surface soil that may correlate with a seasonally shallow water table. If the soil scientist knows that a certain type of soil occurs in the drainageways, then drainageways visible by patterns of dark gray tones on the photograph serve as a proxy for digging a soil pit or boring an auger hole. Stereo pairs of photographs (Figure 19.13) give the viewer a three-dimensional (but vertically exaggerated) view of the land surface, and are especially useful in locating drainage-ways, slope breaks, and other features of relief. The features, in turn, help to locate the soils that are members of known soil associations and drainage catenas.

Vegetation often provides clues about the underlying soils. For example, a certain type of vegetation, recognizable on air photos, may grow only in areas of sodic soils with a natric horizon. The patches of this vegetation seen on the photograph then indicate the locations of the sodic soils.

Drainage patterns visible on air photos usually reflect the nature of soils and parent materials. For instance, many closely spaced (less than 1 cm apart on a 1:20,000 air photo) gullies and streams indicate the presence of relatively impervious bedrock and

"Earth as Art" satellite image gallery:
http://eros.usgs.gov/imagegallery/

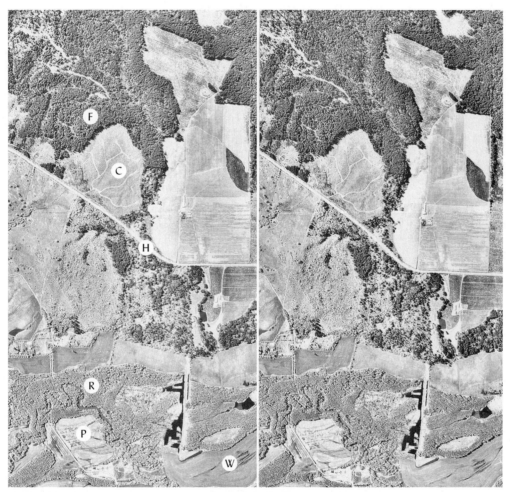

FIGURE 19.13 A stereo pair of air photos providing overlapping coverage of a scene in the Willamette valley of western Oregon. The soils are mostly Humults and Xerolls. The left- and right-hand photos show the same ground area but were taken when the airplane was in two different positions. If you place a pocket stereoscope over these photos, the two views of the same scene will trick your eyes into seeing the photos three-dimensionally. Note the coniferous forest (F) in the steeper terrain of the upper third of the scene above the highway (H), including a roundish clear-cut area [medium-gray zone, (C)] with a network of logging roads (white lines with "nodes" indicating leveled areas where logs were loaded onto trucks). In the lower third of the scene, the land is a nearly level river valley, and waterlogged spots (W) appear in the agricultural fields. A meandering river (R) can be seen in the lower quarter of the scene, and the patterns of alluvial soils can be discerned on the river flood plain (P). Soil differences are visible as different gray tones in fields where plowing has exposed bare soils. The scale of these photos is 1:10,000, so 1 cm on the photos represents 0.1 km. (Photos courtesy of WAC Corporation, Eugene, Ore.)

clayey, low-permeability soils. As another example, silty soils developed in loess usually produce a pinnate drainage pattern in which many small gullies and streams branch off from fairly straight larger streams at angles only slightly less than 90°.

A number of computerized interpretation tools have been developed to enhance the infromation obtainable from black-and-white air photos. One example is video image analysis (VIA), by which air photos are examined to distinguish up to 256 shades of gray (compared to only 32 that the human eye can distinguish). These gray shades or tones are related to soil and vegetation variations. Using appropriate calculations, the computer assists in detection of soil differences and delineation of soil boundaries on the air photos. Again, such interpretations of remotely sensed data must be checked against *ground truth* by going to the site with an auger and verifying soil types and boundaries.

Such interpretations can greatly speed the soil investigation by eliminating the need to investigate every mapping unit on the ground. However, the relationships between patterns on the photos and soil properties must be established afresh by ground investigation for each soil association or landscape type. The relationships so determined are likely to hold true only for landscapes in a limited area, typically 500 to 1500 km².

Direct Sensing of Soil Properties by Reflectance Spectroscopy[5]

Certain properties of the upper 2 to 20 mm of soil alter the manner in which the soil surface reflects various wavelengths of radiant energy (see Figure 19.14). The reflected energy can be scanned over the near infrared radiation spectra (wavelengths in the range of about 700 to 2500 nm) using a special monochromater instrument. At these wavelengths, each part of a complex soil organic or mineral substance produces a special absorption pattern related to particular vibrations caused by stretching and bending of molecular bonds. Depending on the wavelengths being monitored, a number of soil properties can be directly sensed and recorded. Direct sensing of soil properties usually requires a remote sensor tuned in to specific bands of wavelengths and a computer program designed to interpret the complex data using sophisticated statistical manipulation and correlations. Soil properties that have been successfully determined by such remote sensing include the soil contents of clay, sand, organic matter, water, iron oxide, copper, zinc, titanium, nitrogen, and phosphorus as well as albedo (overall reflectivity), temperature, cation exchange capacity, salinity, and soil structure.

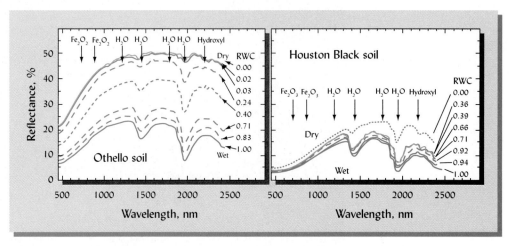

FIGURE 19.14 Reflectance spectra of two contrasting soils (Othello, a sandy Ultisol and Houston Black, a clayey Vertisol) at various relative water contents (RWC) ranging from oven dry (RWC = 0.0) to water saturated (RWC = 1.0). High iron content correlates with low reflectance in the two iron oxide (Fe₂O₃) bands. High organic matter correlates with low reflectance in the hydroxyl band. Comparing the dry soil, note the much lower overall reflectance for the Houston Black, which is, as its name suggests, a very black colored soil. Two "valleys" (inverted peaks of low reflectance) caused by water at 1450 and 1950 nm are especially obvious. The clear differences between the two soils and among water contents within a soil suggest the usefulness of such remotely sensed spectral data in mapping soils, soil moisture levels and other soil properties. Downward arrows indicate the reflectance wavelength attributed to Fe₂O₃, H₂O or hydroxyl. [Modified from Daughtry (2001)]

---

[5] For a comparison of soil analyses by chemical laboratory methods and reflectance spectral methods, see Nanni and Dematte (2006).

New uses for near infrared spectroscopy are being rapidly developed. Not only can the technology be used to interpret reflectance data remotely sensed from airplanes and space satellites (see Section 19.6), but variations of the technology have been developed that allow rapid determination of many soil properties in the lab or on the ground. Soil samples in the laboratory can now be analyzed for numerous nutrients and constituents without having to perform time-consuming and pollutant-generating wet chemistry procedures. Portable reflectance systems are being developed that can sense changes in soil organic matter, texture or water content "on the go"—that is, as the instrument is being pulled by a tractor through the surface soil across a landscape. Such technologies promise to revolutionize not only soil mapping, but also the fields of soil testing (Section 16.11) and precision farming (see Section 19.10).

19.6 SATELLITE IMAGERY

Many of the principles and techniques just mentioned with regard to air photos apply to satellite imagery as well. However, the imagery available from the sophisticated satellites now in orbit may be more complex and versatile than the types of air photos just discussed. Most satellite images are produced from computer-analyzed digital data obtained by multispectral scanners rather than by film cameras. The images are usually computer-enhanced by classification of each cell (pixel) of the image with regard to the type of vegetation, soil cover, land use, or similar theme. This classification is based on computer identification of each land cover's spectral fingerprint or pattern of reflectance over different wavelengths (as shown in Figure 19.14). An image may combine data sensed by different instruments, sometimes on different dates. In this manner, combinations of spectra can be used to identify vegetation types, soil properties, cultural features, water, and so forth.

The resolution of an image is said to be high if very small objects can be discerned. The resolution of satellite imagery available today is not as high as that obtained with low-altitude air photos, but it has steadily improved over the years. Images and multispectral data with resolutions as good as 2.5 m are available (usually for a substantial fee) from various satellites and sources (Table 19.2). Internet software such as Google Earth® has made imagery from a mix of satellites easily and freely available to the general public. The free software allows anyone with internet access to obtain a natural color image of virtually any location on the planet, although there is little choice of image date (important for accessing soils and vegetation) or resolution (varies about 80 m to 2.5 m).

Figure 19.15 illustrates resolution improvements over time between two types of Landsat satellite technologies. The older image (made in 1973) has a relatively low

Landsat Multispectral Image   28 June 73   Landsat Thematic Mapper Image   27 April 90
Bands 1-2-4   Bands 2-3-4

FIGURE 19.15 A black-and-white reproduction of two false-color Landsat satellite images showing a portion of an irrigation project in western Marja, Afghanistan. The upper photo is a Landsat multispectral image obtained in 1973; the lower photo is a Landsat Thematic Mapper digital image obtained in 1990. Each image used three different spectral bands. The highly reflective white tones indicate barren, salinized land. Black areas indicate irrigated cropland that has become waterlogged. Medium-gray tones (red on the original image) indicate well-drained, irrigated cropland. In 1973 an area of approximately 1000 ha (outlined) appeared to be cultivated but waterlogged, probably as a result of blocked drainage outlets. By 1990 the same area appears to be abandoned and salinized. (Photos courtesy of Earth Satellite Corporation, Rockville, Md.)

Landsat 7 characteristics, science and education applications, technical documentation, program policy, and history: http://landsat.gsfc.nasa.gov/

resolution of 80 m, while the newer image (from 1990) has a higher resolution that can resolve objects as small as 30 m. Note that the higher resolution results in a clearer, more detailed image. The difference in resolution is most obvious in distinguishing the drainage canals (arrow) in this arid region of Afghanistan.

This pair of Landsat images highlights the capability of multispectral imagery to portray soil conditions. The soil parent materials in the area are naturally high in soluble salts. Where irrigation without proper drainage raised the water table (evidenced by the dark tones of waterlogged soil in 1973), evaporation from groundwater caused salts to accumulate in the surface soils (as seen by the white, barren soils in 1990). The lack of proper maintenance of the drainage channels may have been related to the political upheaval and war experienced in Afghanistan between image dates.

The capability of satellite imagery to clearly portray landforms and vegetation over a wide area is illustrated by Figure 19.16. Examine this image closely. It is a black-and-white reproduction of a false-color Landsat Thematic Mapper image of the Palo Verde valley where the Colorado River forms the border between California and Arizona. The image is a composite made from three different spectral bands. Rugged mountainous terrain and alluvial fans are visible surrounding a nearly level, irrigated valley. The arid landscape is virtually bare of vegetation, except in rectangular fields where irrigation water has been applied. The small circles in the upper left are center-pivot irrigation systems, which result in lush green vegetation (red on the original image) in circular fields almost 1 km in diameter. Note the grid of streets that is the town of Palo Verde, near the center of the irrigation project.

The outlined rectangular portion of the Palo Verde valley image is enlarged as Plate 111. Much can be learned by closely examining this color plate. Notice the different blue-gray colors indicating soil differences visible in bare fields. Bright red areas are indicative of dense, green crops. The wavy lines of red to the east of the Colorado River indicate natural vegetation and weeds growing where gullies and drainageways have collected the little rainwater that falls in this arid region. Many bare fields are highlighted by yellow outlines, which were added to the image by a GIS program (see Section 19.9). The yellow-outlined fields are those that the City of Los Angeles paid the owners to keep bare, so that the city could obtain the water that would otherwise have been used for irrigating crops. The satellite image was used to verify that the owners of these fields were complying with this agreement.

A final example of satellite imagery is shown in Plate 110, which is a Landsat Thematic Mapper natural color image of the Potomac River basin from Washington, D.C. south, including parts of southern Maryland and northern Virginia. Major land-use patterns are clearly visible. The urbanized land of Washington and its suburbs is shown in blue-gray tones (the United States Capitol building grounds are visible as a white spot surrounded by a dark square of lawn). An enormous amount of sediment (yellow and brown colors) appears to be entering the Potomac River from streams draining the southern suburbs, possibly because of poor sediment control at construction sites. Forests appear green; most farm fields appear light yellow (mature corn and soybeans). The dark green-colored areas along the main river indicate the tidal marshes along the Potomac River and the associated Histosols and other hydric soils. An image such as this is very useful in resource inventory and general regional land planning.

Clearly the range of imagery and technology available for soil investigations is rapidly growing, presenting challenges and opportunities for those specializing in geographic information about soils.

19.7 MAKING A SOIL SURVEY

NRCS National Cooperative Soil Survey: http://soils.usda.gov/partnerships/ncss/

All of the technologies discussed in previous sections can be used to help make a soil survey. A **soil survey** is more than simply a soil map. The glossary describes a *soil survey* as "a systematic examination, description, classification, and mapping of the soils in a given area." Under some circumstances, soil scientists make special-purpose soil surveys which do not attempt to delineate and describe natural soil bodies, but merely aim to map the geographic distribution of selected soil properties, such as suitability for an irrigation project or conformation to a legal definition of wetlands.

FIGURE 19.16 A black-and-white reproduction of a Landsat Thematic Mapper image of the irrigated Palo Verde valley on the border between California and Arizona. The image is a composite made from bands 2 (green), 3 (red), and 4 (near infrared). The outlined rectangle is reproduced in color as Plate 111. (Photo courtesy of Earth Satellite Corporation, Rockville, Md.)

However, soil surveys are most valuable if they characterize and delineate natural soil bodies. Once the natural bodies are delineated and their properties are described, the soil survey can aid in making interpretations for all kinds of soil uses, not just those uses that were intended at the time the soil survey was conducted.

The basic steps in making a soil survey are:

1. *Mapping of the soils.* Described in Sections 19.1 to 19.3.
2. *Characterization of the mapping units.* See Section 19.2.
3. *Classification of the mapping units.* See Chapter 3 and Section 19.2.
4. *Correlation with other soil surveys.* Once a team of soil scientists completes a soil map, the map is reviewed by other soil scientists to verify that the soil boundaries match those mapped for adjacent areas and that the characterization and classification of mapping units are consistent with other soil surveys.
5. *Interpretation of soil suitability for various land uses.* A report is written to accompany the soil map in order to describe the suitability of each mapping unit for various land uses. The interpretive tables in the report often reflect many person-years of experience and observation, as well as standard interpretations of measured soil properties.
6. *Organization* of all this soils information to make it available to users in the form of maps, tabular information, and electronic data files.

Soil surveys may be conducted at different *orders* or levels of detail, ranging from very detailed surveys that attempt to delineate virtually every soil body in the landscape (first order) to general reconnaissance surveys of very large regions or entire continents (fifth order). Different kinds of *mapping units* and remotely sensed data sources are used to produce different orders of soil surveys (Table 19.3).

Map Scales

The *scale* of a map is the ratio of length on the map to actual length on the ground. A scale of 1:20,000 is commonly used for detailed soil maps and indicates that 1 cm on

TABLE 19.3 Different Orders of Soil Surveys

Soil surveys may be conducted at various scales or orders, ranging from very detailed surveys of small parcels of land to general surveys of very large regions. Different kinds of mapping units and remotely sensed data sources are used in producing different orders of soil surveys. The guidelines given in this table should be considered flexible and approximate.

	Order of Soil Survey				
	5th order	4th order	3rd order	2nd order	1st order
Type of survey	Reconnaissance	Reconnaissance	Semidetailed	Detailed	Intensive
Survey scale	1:250,000–1:10,000,000	1:50,000–1:300,000	1:20,000–1:65,000	1:12,000–1:32,000	1:1000–1:15,000
Size of mapping unit	2.5–500 km^2	15–250 ha	1.5–15 ha	0.5–4 ha	Smaller than 0.5 ha
Typical components of map units	Orders, suborders, and great groups	Great groups, subgroups, and families	Families, series, and phases of series	Soil series; phases of series	Phases of soil series
Kind of map unit	Associations, some consociations, and undifferentiated groups	Associations, and some complexes, consociations	Associations or complexes; some consociations	Consociations and complexes; few associations	Mostly consociations; some complexes
Remote sensing sources	←——————— Landsat Thematic Mapper digitized data ———————→				
	←——————— SPOT image digital data ———————→				
		←——————— High-altitude aerial photography ———————→			
			←——————— Low-altitude aerial photography ———————→		
Use of soil survey in land planning	←——— Resource inventory ———→				
		←——— Project location ———→			
			←——— Feasibility surveys ———→		
				←——— Management surveys ———→	

Adapted from Soil Survey Staff (2006).

the map represents a distance of 20,000 cm (or 0.2 km) on the ground. The ratio is unitless; therefore, 1 in. on the same map represents 20,000 in. (0.316 miles) on the ground.

The terms *large scale* and *small scale* sometimes confuse people. A *small-scale* map is one with a small *scale ratio* (e.g., 1:1,000,000 = 0.000001), on which a given object, such as a 100-ha lake, occupies only a tiny spot on the map. By contrast, a *large-scale* map has a large *scale ratio* (e.g., 1:10,000 = 0.0001), and a 100-ha lake would occupy a relatively large part of the map.

Mapping Units

Soil Taxonomy (or some other classification system; see Chapter 3) is usually the basis for preparing a soil survey. Because local features and requirements will dictate the nature of the soil maps and, in turn, the specific soil units that are mapped, the field *mapping units* may be somewhat different from the *classification units* found in *Soil Taxonomy*. The mapping units may represent some further differentiation below the soil series level—namely, *phases* of soil series; or the soil mappers may choose to group together similar or associated soils into conglomerate mapping units. Examples of such soil mapping units follow.

SOIL PHASE. Although technically not included as a class in *Soil Taxonomy*, a *phase* is a subdivision based on some important deviation that influences the use of the soil, such as surface texture, degree of erosion, slope, stoniness, or soluble salt content. Thus, a Cecil sandy loam, 3 to 5% slope, and a Hagerstown silt loam, stony phase, are examples of phases of soil series.

CONSOCIATIONS. The smallest practical mapping unit for most detailed soil surveys is an area that contains primarily one soil series and usually only one phase of that soil series. For example, a mapping unit may be labeled as the consociation "Saybrook silt loam, 2 to 5% slopes, moderately eroded." Quality-control standards for county soil surveys may indicate that a consociation mapping unit should be 50% "pure" and that the "impurities" should be so similar to the named phase that the differences do not affect land management. That is to say, if you walk out on the land to an area mapped as the just-mentioned consociation and make 20 auger borings, at least 10 of the borings should reveal properties within the range defined for the Saybrook silt loam. However, it would be expected that a few of the borings would indicate a similar soil, such as the Catlina silt loam, in which the loess layer is somewhat thicker than for the Saybrook soil but for which land-use interpretations are the same. Inclusions of *contrasting* soils should occupy less than 15% of the consociation.

SOIL COMPLEXES AND SOIL ASSOCIATIONS. Sometimes *contrasting* soils occur adjacent to each other in a pattern so intricate that the delineation of each kind of soil on a soil map becomes difficult, if not impossible. In such cases, a soil *complex* is indicated on a soil map, and an explanation of the soils present in the complex is contained in the soil survey report. A complex often contains two or three distinctly different soil series. As described in Section 19.1, relatively large-scale soil maps (e.g., third order) may display only soil *associations*—general groupings of soils that typically occur together in a landscape and could be mapped separately.

UNDIFFERENTIATED GROUPS. These units consist of soils that are *not* consistently found together, but are grouped and mapped together because their suitabilities and management are very similar for common land uses.

19.8 USING SOIL SURVEYS

Online soil survey reports and data:
http://soils.usda.gov/survey/

In the United States, soil scientists have worked for over 100 years to complete a detailed soil survey of the entire country. This effort, known as the *National Cooperative Soil Survey*, is an ongoing collaboration of federal, state, and local governments. The principal federal role is played by the U.S. Department of Agriculture Natural Resources Conservation Service (formerly the Soil Conservation Service). For most of its history, the soil survey proceeded on the basis of local political jurisdictions (counties). While soils in most counties in the U.S. have been mapped, there are still some counties without any soil map. Other counties were mapped many decades ago and their soil surveys need to be updated.

In 2006, the U.S. National Cooperative Soil Survey was reorganized to focus on 273 Major Land Resource Areas (MLRAs), zones defined by ecological characteristics, instead of the 3000+ politically defined counties formerly used to organize soil survey activities. Each MLRA is a geographically contiguous land area characterized by a common pattern of soils, land uses, water resources and climate. For example, two MLRAs are the "Central Nebraska Loess Hills" and the "Alabama and Mississippi Blackland Prairies."

The U.S. National Soil Information System (NASIS): http://nasis.usda.gov/intro

At about the same time, the National Cooperative Soil Survey began entering all its soils information into a digital database called the National Soil Information System (NASIS) for easy electronic retrieval and integration with other digitized and geo-referenced information such as geology, land use, vegetative cover, and the like. The National Soil Information System (NASIS) is central to the effort to replace static paper soil survey reports with a dynamic resource for soils information adapted to many types of users. However, the paper soil survey reports are still in wide use and we will now consider them to explain how soils information can be used effectively.

The County Soil Survey Report

A modern paper soil survey report consists of two major parts: (1) the *soil map*, and (2) accompanying *descriptive information* on the soil mapping units and their suitability for various land uses. In effect, the descriptive part of the report serves as a very elaborate key to explain the soil map.

GENERAL SOIL MAP.   The soil map usually consists of two parts. The first part is a fourth-order *General Soil Map* of the entire county, printed in color at a scale of approximately 1:200,000 (Figure 19.17*a*). This map shows the soil associations grouped into the main physiographic regions of the county. The General Soil Map is useful in providing an overview of the county land resources, but is too general to be used for site planning. Associated with the General Soil Map is an *index map* for the detailed map sheets. This index map is printed at the same scale as the General Soil Map and is divided into numerous rectangular sections, each representing a detailed map sheet. The index map shows enough nonsoil features, such as roads, towns and streams, to enable a user to locate the map sheet covering the area of interest.

DETAILED MAP SHEETS.   The second part of the county soils map is a detailed (2nd order) soil map consisting of many individual map sheets (Figure 19.17*b*), usually folded and bound into the back of the Soil Survey Report. Each map sheet covers an area of approximately 20 to 30 km^2, typically at a scale of 1:12,000 (1 cm = 0.12 km or 1 in. = 1000 ft), 1:15,840 (4 in. = 1 mile), 1:20,000 (1 cm = 0.2 km), or 1:24,000 (1 cm = 0.24 km or 1 in. = 2000 ft). To facilitate digitization using feet and inches, most soil survey updates in the United States are being produced at scales of 1:12,000 or 1:24,000. At these scales, individual trees and houses are discernable on the air photo map base, and areas of soil as small as 1 ha can be delineated (Figure 19.17*c*). The mapping units are mostly consociations consisting of soil phases. As such, these detailed maps are extremely useful for site planning.

INTERPRETIVE INFORMATION.   Use of the county soil survey for land use or site planning requires that the geographic information on the map be integrated with the descriptive information in the report. The report usually contains detailed soil profile descriptions for all of the mapping units, as well as tables that provide soil characterization data and interpretive rankings for the mapping units. Older soil surveys offer interpretive information on yield potentials for various locally important crops, on the suitability of soils for different irrigation methods, drainage requirements for soils, land capability classification of each mapping unit, and other information useful in making farm plans. Newer soil surveys also provide interpretive information on many nonagricultural land uses, including wildlife habitat, forestry, landscaping, waste disposal, building construction, and sources of roadbed materials.

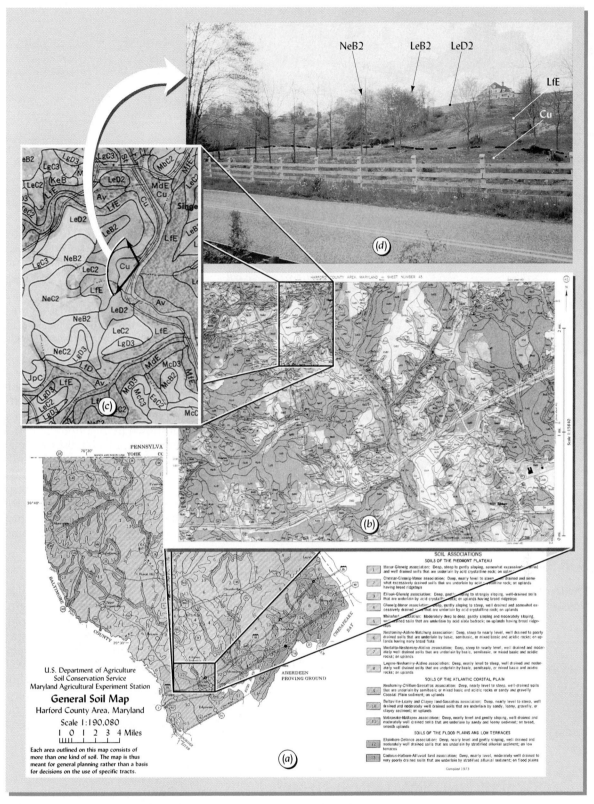

FIGURE 19.17 The soil map of Harford County, Maryland, an example of the soil map in a modern county soil survey. (*a*) One part of the soil map is the small-scale General Soil Map (originally at a scale of 1:190,080, in this case) reproduced here at much-reduced scale. (*b*) The detailed soil map consists of the collection of many large-scale map sheets (originally at a scale of 1:15,840, in this case), one of which (map sheet 45) is shown here at reduced scale. (*c*) A rectangular area of approximately 1.1 km² is outlined on the detailed map sheet and reproduced at the original scale, so that the mapping units and features on the air photo map base are visible (compare to Figure 19.19). (*d*) The scene showing the road in the foreground and the house on the hill in the background was photographed from the location indicated where the two arrows originate in (*c*). [Maps from Smith and Matthews (1975); photo courtesy of R. Weil]

FIGURE 19.18 A stereo pair of air photos including the area shown on the soil map in part (c) of Figure 19.14. Major features to notice are the dam, reservoir and small housing development in the upper left, the stream (Winters Run) flowing below the reservoir, and the hill and flood plain in the center of the photo. The latter features are the same as those shown from ground level in Figure 19.17d. (Air photos courtesy of USDA Agricultural Stabilization and Conservation Service)

EXAMPLE OF USE. As an example of how the soil survey report may be used, consider the parcel of land shown in the photograph in Figure 19.17d. This area is also included in the air photo stereo pair shown in Figure 19.18. Study Figure 19.18, preferably with a stereo-scope, and compare the topography observed with that shown in the photograph in Figure 19.17d. Locating this scene on the detailed map sheet (Figure 19.17c), we can determine that the parcel contains five mapping units encoded as Cu, LeB2, LeD2, LfE, and NeB2 (see Figure 19.6 for an explanation of this coding system). These codes correspond to the five mapping units described in Table 19.4, which also lists a few of the many interpretive ratings given in the soil survey report. If you were planning to develop a park on this site, the information in Table 19.4 would suggest that the Codorus silt loam (Cu) mapping unit would be suitable for a nature trail, but not for a visitor center with rest rooms.

Web Soil Survey

Use Web Soil Survey here: http://websoilsurvey.nrcs.usda .gov/app/

The traditional printed and bound soil survey maps and reports are being replaced by the interactive electronic "web soil survey" available on the Internet (see margin). The user first *defines* the *area of interest* using an address or by delineating a geographic area on a large-scale map. The second step is to *view* the chosen area of interest at a larger scale on a display that includes the soil boundaries and map unit names over an air photo background. Figure 19.19 illustrates such a map as displayed on the Web site. Finally, the Web site presents many options to *explore* the information by requesting maps that group soil series into suitability classes, slope classes, or other interpretive groupings. The user then can choose to save or print the map and related information or download the data for use in a geographic information system or GIS, a topic which we will now discuss in some detail.

19.9 GEOGRAPHIC INFORMATION SYSTEMS[6]

List of GIS-related websites within the Natural Resources Conservation Service: http://www.nrcs.usda.gov/ Technical/land/nrcsdata.html

For complex land planning analyses, the enormous quantity of information stored in digital databases or in a county soil survey report can perhaps be used to best advantage with the help of computerized geographic information systems (GIS).

An information system is a series of organized steps used to handle information or data, including the plan for the collection of data, actual collection of the data, manipulation of

---

[6] For a comprehensive introduction to GIS, see Halls et al. (2007).

TABLE 19.4   Examples of Interpretive Information Provided by a Soil Survey Report

The four mapping units described are those in the site shown in Figure 19.19d. The interpretations included are some of those that would be of interest for developing the site as a county park.

Mapping unit code	Name of mapping unit	Land use Capability Classification[a]	Limitations for use of soil			Suitability for open land wildlife
			Camp areas	Septic filter fields	Paths and trails	
Cu	Codorus silt loam	IIw-7	Severe—flood hazard	Severe—high water table, flood hazard	Moderate—flood hazard	Good
LeB2	Legore silt loam, 3 to 8% slopes, moderately eroded	IIe-10	Slight	Slight	Slight	Good
LeD2	Legore silt loam, 15 to 25% slopes, moderately eroded	IVe-10	Severe—slope	Severe—slope	Moderate—slope	Fair
LfE	Legore very stony silt loam, 25 to 45% slopes	VIIs-3	Severe—slope	Severe—slope	Severe—slope	Poor
NeB2	Neshaminy silt loam, 3 to 8% slopes, moderately eroded	IIe-4	Slight	Moderate—mod. permeability	Slight	Good

[a] See Section 17.13 for an explanation of the USDA Land Capability Classification System.
Data abstracted from Smith and Matthews (1975).

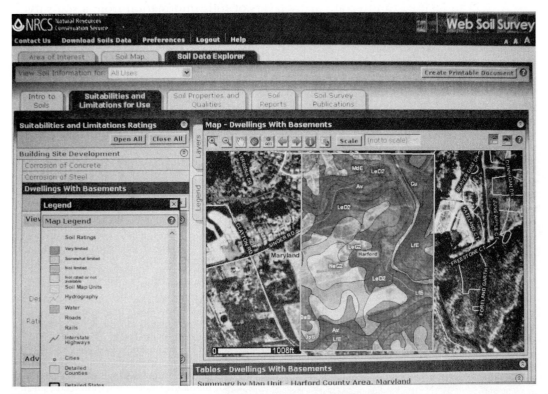

FIGURE 19.19   A screen shot from Web Soil Survey showing a defined *area of interest* (AOI, outlined rectangle in middle of map) in which soils information is superimposed over an air photo background. The map scale can be made large enough to show individual houses and drive ways. In the screen shot the "soil data explorer" option has been used to color-code the soils with regard to their suitability rating as sites for dwellings with basements. A legend of soil units, suitability ratings and geographic features is given in the panel at left. A tabular data (not shown) is also provided on the percent of the AOI occupied by each soil. The AOI shown here is the same area as in Figure 19.17c (note stream with a sharp bend in both figures). (Photo courtesy R. Weil)

Get Soil Surveys of Canada at:
http://sis.agr.gc.ca/cansis/
nsdb/detailed/intro.html

the data, and, finally, use of the data in making a product, such as a report or map. In a geographic information system (GIS), the data is tied to spatial locations. For example, the properties of a soil profile are associated with the location of the appropriate soil mapping units. A GIS includes the following five steps: (1) data acquisition, (2) preprocessing, (3) data management, (4) manipulation and analysis of data, and (5) product generation.

The soil mapping process described in Section 19.3 could be part of the data acquisition step, as could the collection of existing soil maps, reports, and zoning ordinances. This data will have to be preprocessed to be put into a form that can be integrated with other data. Data acquisition and preprocessing can be very time-consuming and expensive. An example of preprocessing is the digitization of an existing older paper soil map, entering all the soil boundaries and other spatial features into a computer database as *points, lines,* and geometrically defined areas called *polygons.* This process may be accomplished using a scanner, but often must be done by hand using a digitizing palette. The descriptive information, such as that illustrated in Table 19.4 would also have to be entered into the database or downloaded from NASIS. These steps are already accomplished in Web Soil Survey.

National Cartography and Geospatial Center—source for most GIS datasets maintained by NRCS:
http://www.ncgc.nrcs.usda.gov/

Once all the necessary information has been downloaded or entered into the database, it can be managed with computer programs and manipulated to create new types of information and insights. Finally, a product such as a map or report is produced. A number of complex computer programs have been designed to carry out the last three steps in a coordinated manner. Figure 19.20 shows a simple example of a product of such a system. The map in this figure is based on information taken from the detailed soil map sheet shown in Figure 19.17b. Although digitizing and entering the descriptive and geographic information for this single map sheet required more than 100 hours, subsequent computer generation of the map of campground limitations took only a few minutes. The map in Figure 19.20 has simplified a set of more than 90 different mapping units by grouping them into three simple, spatially displayed categories. The same information is already available in the soil survey report tables, but not in a spatially displayed form. Such maps are easy to generate using Web Soil Survey.

The just-mentioned map (Figure 19.20) comprises a single *layer* of information. A GIS can be used to integrate many layers of soils information from a soil survey report. As a hypothetical example of the use of a GIS, consider the following scenario. A local agency wished to find a suitable site for a public park in one of 20 large tracts of land delineated in the area covered by detailed map sheet 45 (from Figure 19.17b). The planners decided that the site for the park must meet the following four criteria:

1. Only slight limitations for picnic areas
2. Only slight limitations for paths and trails

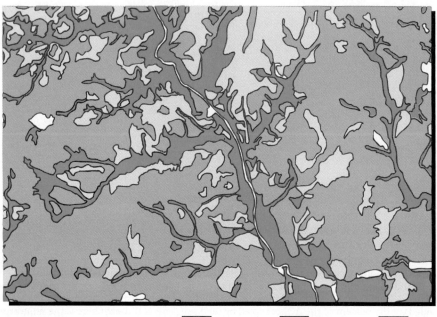

FIGURE 19.20 A simple interpretive soil map covering the same area as Figure 19.17b, map sheet 45 from Harford County, Maryland. The ratings of limitations from tables in the soil survey report have been associated with each of the thousands of soil polygons on the map sheet. The dark gray areas along the major streams are either too steep or too wet to be favorable for campgrounds and, therefore, are in the severe limitations category. This map is a single layer in that it contains only information on spatial distribution of campground limitations. It does not show roads, streams, or other data (nonsoil areas are left blank). Compare this map to that in Figure 19.17b. Note the principal river-tributary system running diagonally approximately through the center of both maps. (Courtesy of Margaret Mayers, University of Maryland)

Limitation for use as campgrounds: Moderate   Severe   Slight

3. Land capability class greater than II (to avoid using prime agricultural land that should be preserved for food production)

4. Woodland subclass 20, indicating a site highly productive for hardwood trees

The boundaries of every mapping unit polygon on map sheet 45 were entered into the GIS database, and each polygon was associated with the appropriate interpretive and descriptive data from the soil survey report. The GIS computer program searched for mapping unit polygons that met all four of the criteria and printed a map with shading used only for the qualifying polygons. Finally, the boundaries of the 20 tracts of land, as well as the locations of streams, were plotted on the map. The resulting map is shown in Figure 19.21. The uppermost tract just right of center appears to contain the largest area of land meeting the criteria for the proposed park.

In many land planning projects, soil properties comprise just one of several types of geographic information that must be integrated in order to take best advantage of a given site or to find the site best suited for the proposed land use. Planning the best use of different farm fields, finding a site suitable for a sanitary landfill, and designing a wildlife refuge are some examples of projects that could make good use of both soils and nonsoils geographic information. Nonsoils information to be considered might include topography, streams, vegetation, and present land use. A GIS can be used to combine all these types of information, giving greatest weight to those factors that the planner deems most critical.

An example of such an integrated GIS model is shown in Figure 19.22. Here, the objective was to develop a plan for conserving a rare and endangered wildflower, northern blazing star. The nature of the land, including soil properties such as drainage and water-holding capacity, as well as existing vegetation and legal conservation easements, determines which areas should be given priority in the effort to protect the wildflower habitat. A number of maps showing the spatial distribution of such soils and nonsoils factors were overlaid or combined in order to produce a final map that differentiated zones on the island according to their priority in a conservation action plan.

As more soils information becomes available in digital form, its use in making land planning decisions based on GIS analyses will undoubtedly become increasingly

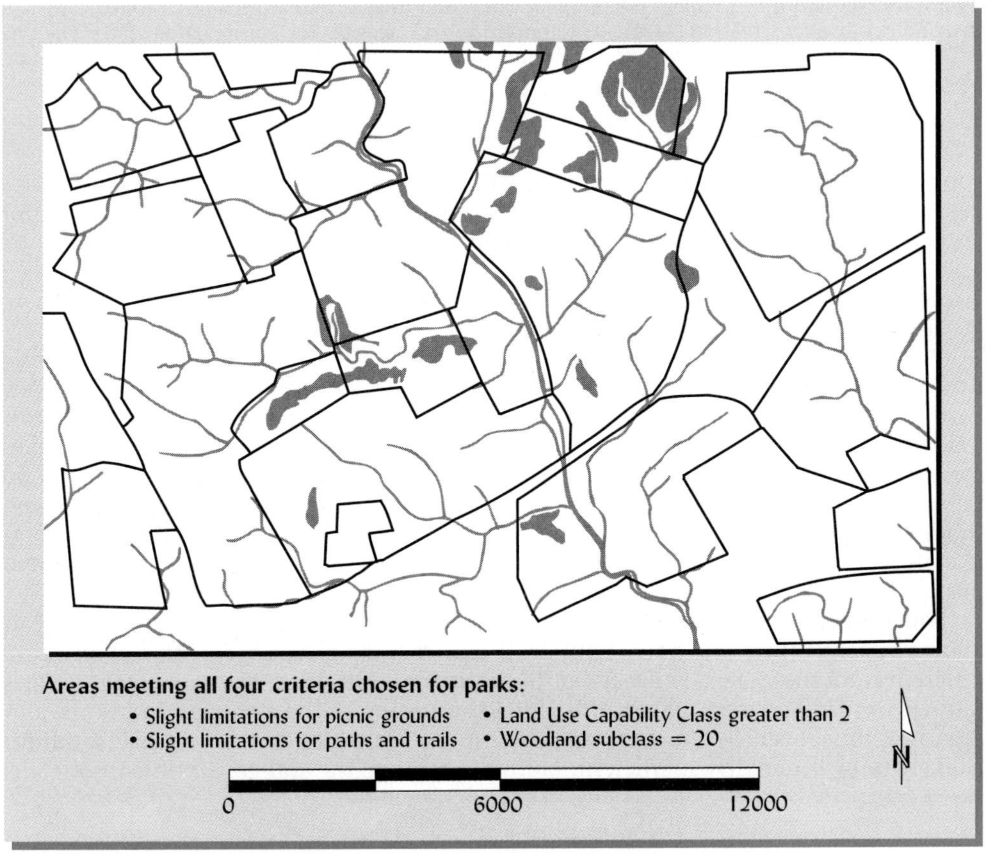

Interactive GIS maps of Canada (don't miss link in lower right corner, 'read more about this map'): http://nlwis-snite1.agr.gc.ca/slc-ppc22/index.phtml

FIGURE 19.21 A map of a portion of Harford County, Maryland (map sheet 45), with shaded areas determined to be suitable sites for a new park according to the four criteria listed. The land use capability class restriction (LUCC > 2) avoids taking prime farmland out of production (see Section 17.13). The property of "park suitability" is a new attribute created by the geographic information system. In addition to "park suitability," two other layers of geographic information are displayed: (1) streams (light, branched lines), and (2) boundaries of 20 hypothetical tracts of land (heavy, straighter lines). Which tract of land should the county government purchase for a new park? (Courtesy of Margaret Mayers, University of Maryland)

Areas meeting all four criteria chosen for parks:

- Slight limitations for picnic grounds
- Slight limitations for paths and trails
- Land Use Capability Class greater than 2
- Woodland subclass = 20

0   6000   12000

N

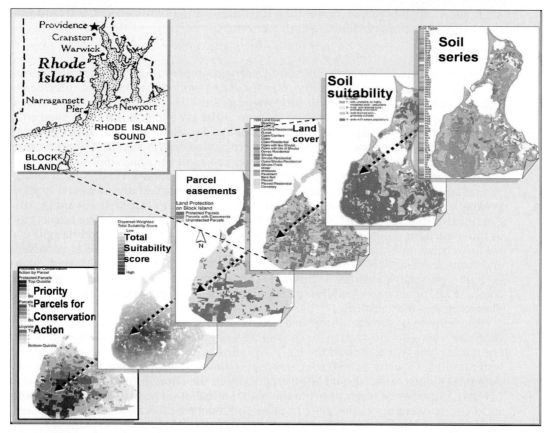

FIGURE 19.22 Using a geographic information system (GIS) to integrate spatial information on soils with physical, biological, and social land attributes to identify new land categories for a particular purpose. In this example, the goal was to save the rare wildflower, northern blazing star (*Liatris scariosa* var. *novae-angliae*). The plan prioritizes habitat patches for conservation action, based on habitat suitability and the spatial relationship of each patch to the entire population of the flower on Block Island, R.I. The soil series (first map layer) were grouped by suitability for northern blazing star habitat (second map layer) and used as a criterion along with existing land cover (forest, shrubs, wetlands, etc.) and existing conservation easements to prioritize land areas for the conservation effort. The final product (leftmost map) shows which parcels of land should be given high, medium or low priority in the conservation program. [Modified from maps courtesy of Vadeboncoeur (2003)]

common. The quality of those decisions will continue to depend on the quality of the geographic soils information provided by soil scientists and the quality of the decision criteria developed by planners and their clients.

19.10 GIS, GPS, AND SITE-SPECIFIC AGRICULTURE[7]

Advances in technology have combined (1) the Global Positioning System (GPS) of Earth-orbiting satellites (see Section 19.3), (2) computerized Geographic Information Systems (GIS) capable of making detailed maps that integrate information about many soil properties, and (3) variable rate technologies that allow farm equipment to alter the rate of fertilizer, seed, or chemical delivery on the go. The combined technology, sometimes referred to as "precision farming" or "site-specific agriculture," allows farmers to make the application of nutrients and other inputs more site-specific than was previously practical for large farming operations.

Soil properties in one part of a farm field may differ markedly from those in another part of the same field, with each part requiring management systems quite different from the others. However, until the advent of the just discussed technologies, farmers were forced to manage a whole field according to the average conditions, aiming for optimal average returns from their inputs. Managing for the average condition is likely to be inherently inefficient, failing to remove the soil constraints in one area

---

[7] For an assessment of the future of site-specific agriculture, see Dobermann et al. (2004).

of a field, while in another area providing excessive inputs that may lead to environmental contamination.

One approach to precision farming is to divide a large field into cells in a grid pattern, each cell usually being about 1 ha in area. For example, 18 separate geo-referenced soil samples (each being a composite of about 15 to 20 soil cores) may be collected from a single 18-ha field (Figure 19.23). These soil samples can then be analyzed for such properties as texture, organic matter content, pH, and soil test levels of phosphorus and potassium. A computer program can then produce a map of the spatial distribution of each soil property measured. Thus, one map might show areas of low, medium, and high soil test phosphorus levels. Another map might show areas of high, medium, and low clay content, and so on. Information on other spatial variables, such as soil classification, drainage class, past management, weed density, and so forth, can be fed into the program to generate additional maps. A sophisticated computer program then integrates the information from these individual maps to create a new combined map showing the different application rates of fertilizer (or other material) that are recommended for different parts of the field (see Figure 19.24).

For example, areas mapped as testing low in phosphorus might be slated to receive higher-than-average application rates of phosphorus fertilizer, while areas mapped as being high in phosphorus might receive no phosphorus fertilizer at all. Similarly, reduced rates of nitrogen fertilizer application might be mapped for areas of sandy soils with high leaching potential but low yield potential. The application-rate maps are then programmed into a computer on board the machine that spreads the fertilizer.

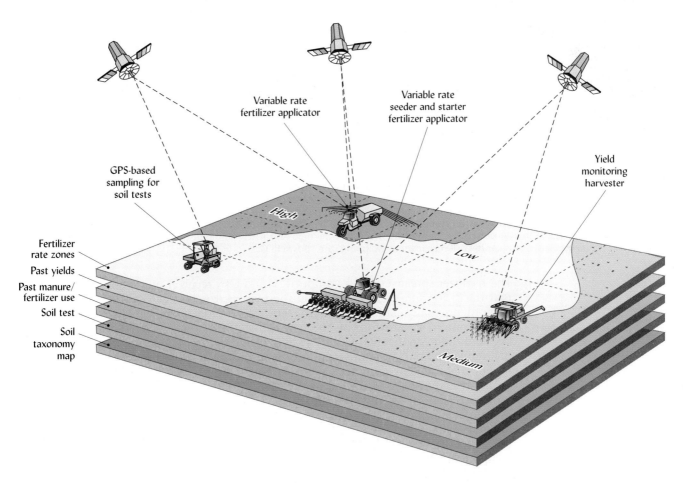

FIGURE 19.23 High-technology used to facilitate site-specific nutrient-management systems. The Global Positioning System (GPS) can plot the location of many soil sampling and plant production sites on a grid basis within a field. A soil sample (composite of 20 cores) is taken from each cell (about 1 ha) and is analyzed. With soil analysis and other data from these cells, computers can help create maps such as the one shown here for a 18-ha field. The top map combines data from the other layers to define "fertilizer rate zones." Satellite/computer systems can then be used to control variable-rate fertilizer applicators that apply only the amounts of nutrients that the soil tests and past soil management suggest are needed. At harvest time, similar satellite/computer connections make possible the monitoring of crop yields on the same grid basis when the harvest machine traverses the field. The yield data are used to create yield maps, which can then be used to further refine the nutrient-management system. [Modified from PPI (1996)]

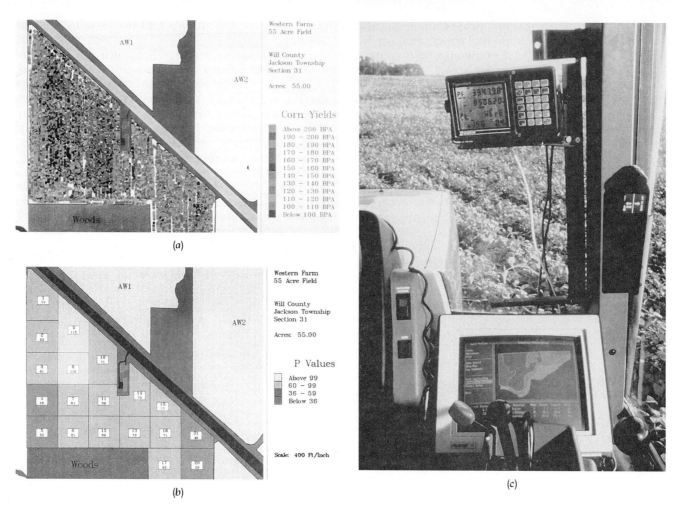

FIGURE 19.24 Use of GPS technology in *precision* or *site-specific* farming. (*a*) This map of corn yields throughout a 22-ha field was generated by a harvesting machine equipped with a yield monitor and a GPS receiver. Each dot on the map was generated as the monitor recorded the yield of corn every few meters while the machine worked up and down the field. Each yield measurement was associated with a map position determined by the GPS. (*b*) A map of the same field showing the available soil phosphorus as determined from samples collected in 1-ha cells. (*c*) A GPS-linked computer system installed in a tractor spreading fertilizer in a different field. The computer screen shows a map indicating the current location of the tractor in the field and where different rates of fertilizer should be applied. Compare to Figure 19.23. (Photos courtesy of Applications Mapping, Inc., Boswell, Ga.)

The fertilizer spreader also is equipped with a GPS receiver and associated computer that tracks the spreader's location on the map as it moves across the field. The computer then signals small motors on the fertilizer spreader to increase or decrease the rate of application, as called for on the map. Thus, the system is designed to reduce both overapplication on the high-fertility areas (which should save money and reduce nutrient-pollution potential) and underapplication on the low-fertility sites (which should make money by producing a higher yield). The total amount of fertilizer added may not be greatly different from that used in the past, but application rates should be more in tune with plant needs and environmental cautions.

At harvest time, similar computerized satellite linkages are used to monitor yields in different parts of the field and to create maps showing the yield differences. The yield maps produced by GPS-equipped harvest combines are proving to be the most popular component of the entire system, as they help farmers pinpoint problem areas for investigation. Yield levels in some parts of a field, even in one that is uniform in appearance, commonly are 2 to 3 times as high as in other parts of the field. By overlaying the yield map on the soil nutrient maps, it is possible to determine the extent to which nutrient-deficiencies are constraining yields. However, other problems such as poor drainage, low organic matter, or even feeding by wildlife may also come to light.

As we have just seen for fertilizer, site-specific management systems can also control insects and weeds and modify plant seeding rates and depths, allowing precision farming to manage land on a site-specific basis, rather than on a whole field basis.

Such computerized GPS-based systems are expensive in terms of equipment costs and costs to collect all the requisite information, especially the many grid sample analyses needed. The technology may not be economically viable for all farms, but its use is spreading, particularly among custom operators with computer expertise and the appropriate equipment. Precision farming using the GPS-GIS combination may help assure that fertilizers, manure, and pesticides are used only where they are needed to enhance plant production. The development of portable spectral sensors (Section 19.5) may lead to an even more efficient and affordable precision farming system that eliminates the expense of GPS, GIS, and grid soil sampling. Such sensors may be able to respond to changes in soil properties and control application equipment in real time and on the go.

19.11 CONCLUSION

Making soil surveys is both a science and an art by which many soil scientists apply their understanding of soils and landscapes to the real world. Mapping soils is not only a profession; many would say that it is a way of life. Working alone outdoors in all kinds of terrain and carrying all the necessary equipment, the soil scientist collects ground truth to be integrated with data from satellites and laboratories. The resulting soil maps and descriptive information in the soil survey reports and databases are used in countless practical ways by soil scientists and nonscientists alike. The soil survey, combined with powerful geographic information systems, enables planners to make rational decisions about what should go where. The challenge for soil scientists and concerned citizens is to develop the foresight and fortitude to use criteria in the GIS planning process that will help preserve our most valuable soils—not hasten their destruction under shopping malls and landfills.

STUDY QUESTIONS

1. What is the principal difference between a *soil association* on one hand, and a *catena* on the other? What is the difference between a *lithosequence* and a *toposequence?*

2. What is the main purpose of digging soil pits as part of making a soil survey?

3. Describe the kinds of information a soil mapper may use in deciding where to drill into the soil with an auger to bring up subsurface samples for study.

4. How can the GPS be used in a GIS?

5. What are the advantages of using aerial photos as a map base in making a soil survey?

6. For the region in which you live, describe some kinds of information that you would expect to be able to obtain from satellite imagery.

7. Assume that you have two wall maps, each approximately 1 m wide by 0.70 m tall. One is a map of Canada, and the other is of a ranch in California. Which map is the large-scale map and which is the small-scale map? If the ranch is roughly 20 km from east to west, what might be the approximate scale for its map?

8. A soil mapper drew a boundary around an area in which he made six randomly located auger borings, two in soil *A*, with an argillic horizon more than 60 cm thick and strong brown in color, and the other four in soil *B*, with a somewhat lighter brown argillic horizon between 50 and 70 cm thick. Other soil properties, as well as management considerations, were similar for the two types of soils. Would the map unit delineated likely be a *soil association*, a *soil consociation*, or a *soil complex?* Explain.

9. Assume you are planning to buy a 4-ha site on which to start a small orchard. Explain, step by step, how you could use the county soil survey to help determine if the prospective site was suitable for your intended use.

10. If you were hired by a state government to produce a GIS-based map showing where investments should be made to protect farmland from suburban development, what "layers" of information would you want to include in the GIS?

11. Try to produce the map shown in Figure 19.19 by visiting websoilsurvey.nrcs .usda.gov/app/. Hints: the location is about 4 miles due south of the town of Bel Air, Maryland (enter these as the city and state under the "Navigate by Address" tab). Be sure to study the instructions available on the website.

REFERENCES

Cline, M. G., and R. L. Marshall. 1977. "Soils of New York Landscapes," Inf. Bull. 119, Physical Sciences, Agronomy 6, New York State College of Agriculture and Life Sciences, Cornell University.

Daughtry, C. S. T. 2001. "Discriminating crop residues from soil by shortwave infrared reflectance," *Agron. J.*, **93**:125–131.

Dobermann, A., S. Blackmore, S. E. Cook, and V. I. Adamchuk. 2004. Precision farming: Challenges and future directions, in T. Fischer et al. (eds.), *New Directions for a Diverse Planet: Proceedings for the 4th International Crop Science Congress, Brisbane, Australia*, 26 September–1 October 2004. www.Cropscience.Org.Au.

Doolittle, J. A. 1987. "Using ground-penetrating radar to increase the quality and efficiency of soil surveys," in W. U. Reybold and G. W. Peterson (eds.), *Soil Survey Techniques*. SSSA Special Publication no. 20 (Madison, Wis.: Soil Sci. Soc. Amer.).

Doolittle, J. A., B. Jenkinson, D. Hopkins, M. Ulmer, and W. Tuttle. 2006. "Hydropedological investigations with ground-penetrating radar (GPR): Estimating water-table depths and local ground-water flow pattern in areas of coarse-textured soils," *Geoderma*, **131**:317–329.

Doolittle, J. A., K. A. Sudduth, N. R. Kitchen, and S. J. Indorante. 1994. "Estimating depths to claypans using electromagnetic induction methods," *J. Soil Water Cons.*, **49**:572–575.

Grigera, M. S., R. A. Drijber, K. M. Eskridge, and B. J. Wienhold. 2006. "Soil microbial biomass relationships with organic matter fractions in a Nebraska corn field mapped using apparent electrical conductivity," *Soil Sci. Soc. Am. J.*, **70**:1480–1488.

Grunwald, S. (ed.) 2006. *Environmental Soil-Landscape Modeling.* (New York: CRC Press). 496 pp.

Grunwald, S., P. Barak, K. McSweeney, and B. Lowery. 2000. "Soil landscape models at different scales portrayed in virtual reality modeling language," *Soil Sci.*, **165**:598–615.

Halls, P., C. McClean, and G. Newton-Cross. 2007. *GIS for Ecologists and Environmental Scientists* (New York: John Wiley & Sons). 320 pp.

Herring, T. A. 1996. "The global positioning system," *Scientific American*, **270**(2):44–50.

Legros, J.-P. 2006. *Mapping of the Soil* (Enfield, N. H.: *Science Publishers*). 422 pp.

Nanni, M. R., and J. A. M. Dematte. 2006. "Spectral reflectance methodology in comparison to traditional soil analysis," *Soil Sci. Soc. Am. J.*, **70**:393–407.

PPI. 1996. "Site-specific nutrient management systems for the 1990's." Pamphlet. (Norcross, Ga.: Potash and Phosphate Institute and Foundation for Agronomic Research).

Short, N. M., Sr. 2006. Remote sensing tutorial. NASA/Goddard Space Flight Center. http://rst.gsfc.nasa.gov/ (posted May 19, 2006; verified 15 June 2006).

Smith, H., and E. Matthews, 1975. *Soil Survey of Harford County Area, Maryland* (Washington, D.C.: U.S. Soil Conservation Service).

USDA-NRCS. 2002. Official series description—Thorndale series. National Cooperative Soil Survey. http://www2.ftw.nrcs.usda.gov/osd/dat/T/THORNDALE.html (posted December 2002; verified 04 October 2006).

USDA-NRCS. 2006. National soil survey handbook, title 430-vi. U.S. Department of Agriculture, Natural Resources Conservation Service. http://soils.usda.gov/technical/handbook/ (posted 03/22/2006; verified 20 October 2006).

Vadeboncoeur, M. 2003. Using GIS to prioritize land for management in the conservation of a rare species. Brown University. http://envstudies.brown.edu/thesis/2003/matthew_vadeboncoeur/ (posted November 16, 2004; verified 15 June 2006).

Wollenhaupt, N. C., R. P. Wolkowski, and M. K. Clayton. 1994. "Mapping soil test phosphorus and potassium for variable rate fertilizer application," *J. Prod. Agric.*, **7**:441–448.

Wysocki, D. A., P. J. Schoeneberger, and H. E. Lagarry. 2005. "Soil surveys: A window to the subsurface," *Geoderma*, **126**:167–180.

Zwicker, S. E. 1992. *Soil Survey of Bureau County, Illinois* (Washington, D.C.: USDA Natural Resources Conservation Service).

An African mother nurtures the future. (R. Weil)

20

PROSPECTS FOR GLOBAL SOIL QUALITY AS AFFECTED BY HUMAN ACTIVITIES

"… it is impossible to care for each other more or differently than we care for the earth."
—WENDELL BERRY

During the past few decades, we have begun to recognize that the actions of each country, each community, or even each individual can have global implications, whether those actions relate to politics, to markets, or to soils. Changes in soil productivity in one area affect food security and prices, as well as biodiversity and water quality, in both nearby and distant places. For example, soil particles picked up by the spring winds in China's loess plateau can soon be detected in the rainfall in the western United States. Likewise, excess salts, nutrients, or pesticides in the drainage water from soils in one nation can make the water unfit for use in another nation downstream.

This growing awareness of global interconnectedness is paralleled by the growing acceptance of the *ecosystem* concept as the prime basis for decisions regarding natural-resource management. This concept recognizes that biological communities interact with each other and with the environment at all scales, from the global terrestrial ecosystem to the ecosystem of a garden or a compost pile. Furthermore, components of one ecosystem may be impacted by processes in other ecosystems. For example, chemicals from an urban ecosystem's sewage plant or an agroecosystem's overfertilized field may adversely impact an estuary ecosystem many kilometers downstream.

Soils are integral components of all terrestrial ecosystems, whether urban, agricultural, forest, marsh, or grassland. The ecosystem approach continually reminds us of the interactions among physical, chemical, and biological components, as well as between the aboveground and belowground portions of our environment. We cannot clear forest or rangeland, lime a soil, add a new irrigation scheme, or apply domestic or industrial wastes to a soil without influencing that soil and its compliment of microorganisms, plants, and animals. Likewise, how we manage plant communities influences the long-term stability and quality of the soils in which they grow.

In previous chapters we explored soil processes that occur in various ecosystems, and actions that individual land users might take to influence these processes. We now turn to the global implications of local land-use decisions, and how these decisions affect the quality or health of the soil—which in turn, through various ecosystems, affects the well-being of all of us and our co-inhabitants on this planet.

873

20.1 THE CONCEPTS OF SOIL QUALITY AND SOIL HEALTH[1]

Intro. slide show on soil quality or health: http://southcenters.osu.edu/soil/slides/

People have traditionally evaluated soils using adjectives such as "good," "bad," "worn-out," "fertile," or "infertile." In recent years, scientists and land managers have needed new tools with which to better understand and evaluate the processes that improve or degrade soils, including the impacts of alternative managment practices on the overall fitness of soils. They also recognized the need to find simple, integrative means to communicate these impacts to farmers, foresters, and other land users.

To assist in taking this holistic view of soil improvement or degradation, soil scientists have developed the concepts of **soil health** and **soil quality**. Although these terms have often been used synonymously, they actually involve two distinct concepts. The term *soil health* refers to self-regulation, stability, resilience, and lack of stress symptoms in a soil *as an ecosystem*. Soil health describes the biological integrity of the soil community—the balance among organisms within a soil and between soil organisms and their environment. Chapter 11 (Sections 11.11–11.14) presented many concepts relevant to soil health.

The more utilitarian concept of *soil quality* is best applied to a soil *as a component of a larger ecosystem* that supports plant growth, regulates water flows, and so forth. Soil quality therefore describes the properties that make a soil fit to perform particular functions in support of the six broad ecological roles of soils introduced in Sections 1.1–1.7. Exactly what constitutes a high-quality soil may depend on which of these roles is under consideration—in other words, the intended use or the goal of soil management must be taken into account. For example, a soil well suited to serve as an engineering medium (say, as a building foundation) may not be well suited to support plant growth. In this chapter we will emphasize three broad functions for which soils are commonly managed: (1) plant productivity, (2) assimilation and recycling of waste materials, (3) and environmental protection of water and air quality.

Soil Quality In Relation to Management Goals

The management goal of plant productivity aims to maximize the growth and quality of desired plants. For a farmer the desired plant production might be marketable crops, for a rancher it might be palatable forage, and for a wildlife manager it might be habitat vegetation. The management goal of waste recycling aims to use the soil efficiently as a means to safely and beneficially deal with manure, sewage sludge, or other "wastes." The goal of environmental protection aims to detoxify, immobilize, or isolate potential contaminants so as to protect food webs and enhance (or at least maintain) the quality of soil, water, and air resources. The relationship between these broad management goals and more specific soil functions is illustrated in Figure 20.1. In this context, soil quality describes the capacity of a soil to perform a particular set of functions that are required to meet a given management goal.

---

[1] The term *soil quality* was first used by Alexander (1971). For reviews of the soil quality concept, practical means to use this concept in the field, and approaches to developing soil quality indices, see Shukla et al. (2006), Schjønning et al. (2004), Sanchez et al. (2003). For a contrary view, see Sojka et al. (2003).

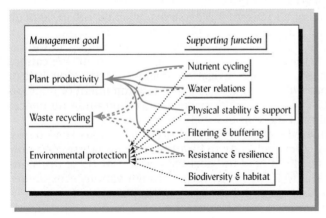

FIGURE 20.1 Examples of broad soil mangement goals and more specific soil functions that support these goals. The plant productivity management goal aims to maximize the quantity and quality of desired plant production (crops, forage, timber, or native vegetation). The waste recycling goal aims to use the soil efficiently as a means to safely and beneficially deal with manure, sewage sludge, effluent or other "wastes." The environmental protection goal aims to detoxify, immobilize, or isolate potential contaminants to protect air and water resources and the terrestrial food web. For example, a forest manager aiming to maximize timber production (plant productivity) would assess soil quality using indicator properties associated with nutrient cycling, water relations (ability to absorb, store, and release water), physical stability and support, and resistance and resilience functions. [Modified from Andrews et al. (2004)]

For example, from Figure 20.1 we can surmise that a forest manager aiming to maximize timber production would assess soil quality with regard to the capacity of the soil to perform four functions: (1) cycle nutrients—provide essential plant nutrients at appropriate times and amounts; (2) regulate water—absorb, store, and release water to plant roots; (3) provide physical support and maintain stability; (4) exhibit resistance and resilience functions (see Section 20.2) that minimize the degree and duration of impacts from major disturbances such as fire or compaction.

Each of the six specific functions shown in Figure 20.1 is associated with certain biological, chemical, and physical soil processes. Examples of such processes include the maintainance of plant-available nutrients in the soil solution, the leaching of pollutants through the soil to groundwater, the erosion of soil by wind or water, and the exchange of gases with the atmosphere that influence the soil's ability to perform. It is not always possible to measure directly the rates of these processes, but we can readily measure specific soil properties that are indicative of these rates. We can also use these measurements in simulation models to predict future changes in process rates and, in turn, soil quality. The properties measured are termed *soil quality indicators*.

Soil Quality Indicators

Table 20.1 illustrates a suggested minimum data set of indicator properties for the determination of soil quality in relation to the six functions described in Figure 20.1. For example, in the case of the nutrient-cycling function (that supports the plant productivity management goal), the soil quality indicators would include such measurable properties as potentially mineralizable nitrogen (PMN), soil test phosphorus level, soil pH, cation exchange capacity (CEC), and soil depth, among others.

TABLE 20.1 Selected Biological, Chemical and Physical Indicator Properties for Determining the Quality of Health of a Soil

The soil ecological functions are further explained in Figure 20.1 and some of the listed indicator properties are evaluated in terms of soil quality score in Figure 20.2.

Nutrient cycling	Water relations	Stability-support	Buffering-filtering	Resilience-resistance	Biodiversity habitat	Property symbol	Property Description	Related book sections
							Biological Properties	
X				X	X	MI	Maturity index of nematode trophic levels	Sections 11.2, 11.6
				X	X	qCO_2	Respiration (per unit microbial biomass per day)	Section 20.8
			X		X	MBC	Microbial biomass carbon	Sections 12.2, 20.8
X						MBP	Microbial biomass phosphorus	Sections 14.4, 20.8
X			X			PMN	Potentially mineralizable nitrogen	Section 13.3
X	X			X	X	Active C	Organic carbon oxidized by 0.02 M KMnO$_4$	Sections 12.2, 12.6
							Chemical Properties	
X			X		X	Soil test P	Available phosphorus by soil test (e.g. Mehlich 3)	Section 16.11
X			X	X		P_{sat}	Saturation of P fixing capacity ((Al+Fe)/P)	Section 16.12
X			X		X	Soil test K	Available potassium by soil test (e.g. Mehlich 3)	Section 16.11
X			X	X	X	Soil pH	Soil pH (in 1:1 water solution)	Section 9.5
	X	X		X	X	EC	Electrical conductivity	Section 10.4
X	X	X	X	X	X	SAR	Sodium absorption ration	Section 10.4
X	X	X	X	X	X	TOC	Total organic carbon	Section 12.6
X			X	X		CEC	Cation exchange capacity	Section 8.9
							Physical Properties	
X	X	X	X	X	X	AGG	Aggregate stability to slaking when wetted	Section 4.5
X	X	X		X	X	D_b	Bulk density	Section 4.7
X	X	X	X	X	X	Depth	Depth to root limiting layer	Sections 5.9, 17.2
	X		X	X	X	AWC	Plant-available water-holding capacity	Sections 5.4, 5.8
	X	X	X	X		S	Infiltration capacity (sorptivity)	Section 5.6
	X	X		X	X	Sand	Percentage of sand in the mineral fraction	Section 4.3
X	X	X	X	X	X	Clay+Silt	Percentage of clay + silt	Section 4.3

Most Closely Associated Soil Ecological Functions (heading spanning the first six columns)

Based on concepts in Andrews et al. (2004); Eigenberg et al. (2006); Karlen et al. (2006); Weil et al. (2003). For many of the methods, see Doran and Jones (1996).

A particular management goal or ecological role involves several soil functions; each function may involve several processes; and each process may be associated with several biological, chemical, and physical indicator properties. Therefore, the number of properties measured to assess soil quality may be a dozen or more, many of which can be measured on-the-spot in the field (Plate 68). Each of these measurements must be *interpreted* with regard to its implications for the soil functions under consideration. For example, if the bulk density is measured as 1.5 Mg/m^3, we must ask "is this too high, too low, or just right?" to facilitate the functions of regulating water, providing habitat, and so on. For some indicators (e.g., soil organic matter) higher values may be more desirable, for others (e.g., bulk density) lower values are better, and for still others (e.g., soil pH) there may be an optimum value, above or below which soil function suffers.

Assessing Soil Quality

Visual field assessment of soil quality. Approach from New Zealand:
http://www.landcareresearch.co.nz/research/soil/vsa/

To provide an overall assessment that will facilitate the comparison of soil quality from one soil to another, some way is needed to integrate many indicator properties. This task is made more complex by the fact that the measurements are expressed in many different kinds of units such as mg/kg, dS/m, $cmol_c/kg$ or Mg/m^3. To arrive at a single number that can be used to compare one soil to another, we can calculate a soil quality index (SQI) that integrates the many kinds of data collected. Soil scientists have created scoring curves that are used to give each indicator measurement a unitless score between 0 (nonfunctioning) to 10 (optimal functioning). Some examples of typical scoring curves are shown in Figure 20.2. It should be emphasized that many of these scoring curves depend on the types of soils, plants, or conditions involved. For example, the scoring curve shown for soil pH implies an optimum between pH 6–7. While this may be applicable to productivity for most agronomic crops, the optimum for crops on Histosols, or for certain forest plants, might be closer to pH 5 (see Section 9.7).

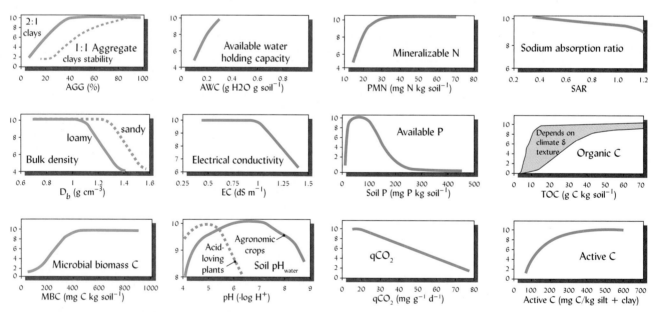

FIGURE 20.2 Scoring algorithms for 12 soil quality indicator properties that can be used to calculate a soil quality index (SQI). Once the 12 soil properties have been measured on soil samples (values on the X-axes) then curves like those shown are used to extrapolate or compute a unitless scoring value (Y-axes) for each property. For example, using the scoring curve for electrical conductivity (EC), a soil with EC measurement less than 1 dS m^{-1} would be given a score of 10 (best quality). Adding these scores together to create an index value integrates a wide range of chemical, physical, and biological soil properties. This SQI can then be used to assess or predict the capacity of different soils, or similar soils under differing management, to perform ecological functions. The scoring curves shown here are generalized to illustrate typical conditions. The relationships may differ somewhat among locations, and in some cases among plant species. [Modified from Andrews et al. (2004)]

BOX 20.1   CALCULATING A SOIL-QUALITY INDEX
FOR PLANT PRODUCTIVITY

Table 20.2 provides an example of how, if suitable measurements of soil properties were made, a soil quality index (SQI) for the plant productivity management goal could be derived using the information in Figures 20.1 and 20.2 and Table 20.1. Four functions of the soil supporting plant productivity are depicted: (1) *nutrient cycling*, (2) *water regulation*, (3) *physical stability*, and (4) *resilience and resistance*. In calculating the SQI, we will at first assume that each function is given equal weight. For simplicity in this example, only two indicator properties are shown for each function, therefore $n = 4 \times 2 = 8$. Others, such as those listed in Table 20.1, could have been added. Using interpretive scoring curves such as those in Figure 20.2, the measured values for the indicator properties are standardized on a common scoring scale (S) with a maximum = 10. Then the average score is calculated (eq. 20.1) by dividing the sum (Σ) of all the scores by the number of scores (n). This average is multiplied by 10 to provide a soil quality index that conveniently ranges between 0 and 100:

$$SQI = \frac{1}{n} \times \sum_{t=1}^{n} S_t \times 10 \qquad (20.1)$$

In some cases, there is good reason to believe that certain functions or indicators have a greater influence than others on the suitability of the soil to meet the management goal. In such a case, the indicator scores can be multiplied by appropriate weighting factors and a weighted soil quality index (SQI_w) can be calculated. For instance, assume that *nutrient cycling* was considered to be twice as important as the other functions in determining the soil's capacity to support plant growth. In that case, the indicator scores associated with nutrient cycling would be multiplied by a weighting factor (w) = 2. The SQI_w could then be calculated by summing all the weighted scores (S_{wt}) and dividing by n_w, the sum of all the weighting factors (n_w = 10 in this example), as shown in the last two columns of Table 20.2 using (equation 20.2):

$$SQI_w = \frac{1}{n} \times \sum_{t=1}^{n_w} S_{wt} \times 10 \qquad (20.2)$$

TABLE 20.2   Simplified Example of How Indicator Properties Can Be Combined to Give Soil Quality Indices[a]

For simplicity, only two examples of indicator properties are given for each ecological function. Note that an indicator (AGG in this example) may be used for more than one function. In the SQI all indicators and functions are given equal weight. In the SQI_w nutrient cycling is weighted to be 2 times as important as the other functions.

Management Goal	Supporting ecological function	Indicator property	Measured value	Indicator score (S)	Weighting factor (w)	Weighted indicator score (S · w = S_w)
Plant production	Nutrient cycling	Soil test P	80 mg P kg soil^{-1}	10	2	20
		PMN	20 mg N kg soil^{-1}	8	2	16
		etc.	...	...	...	...
	Water relations	AGG	30%	8	1	8
		AWC	20 g H$_2$O g soil^{-1}	8	1	8
		etc.	...	...	...	...
	Physical stability	AGG	30%	8	1	8
		Db	1.4 Mg m^{-3}	6	1	6
		etc.	...	...	...	...
	Resilience	TOC	25 g C kg soil^{-1}	4	1	4
		SAR	1.0	9.5	1	9.5
		etc.	...	...	...	...
		Sum of scores or factors =		61.5	n_w = 10	79.5

Average of S = 61.5/8 = 7.7   Average of S_w = 79.5/10 = 8.0
Unweighted Soil quality index   **Weighted Soil quality index**
(SQI) = 7.7 × 10 = 77   **(SQI_w) = 8.0 × 10 = 80**

[a] Indicator abbreviations are explained in Table 20.1. See Box text for equations to calculate SQI and SQI_w. See Figure 20.2 for typical curves used to convert measurements to indicator scores.

In any case, assuming appropriate scoring curves are available for use, scores for the various indicators are then averaged, thus integrating all the measurements into a single number—the SQI. If it is known that some indicators have a greater influence than others on the evaluated soil functions, the scores can be multiplied by a weighting factor before they are averaged. The example in Box 20.1 illustrates the process of determining a soil quality index by this method. Figure 20.3 shows how long term land-use can influence the soil quality index.

Time- and Place-Sensitive Functions

The relative importance of different soil functions and the weights given them may change as society's perceptions evolve. This is illustrated in Table 20.3, which shows how in 1900 food and fiber production was paramount in people's minds and would have been highly weighted compared to the five other nonproduction functions listed. Currently, more and more emphasis is being placed on environmental quality, especially in industrialized countries where food security is reasonably assured. The broader ecological roles of soils are becoming more widely recognized. In developing countries, however, where poverty, hunger, and even famine are still common, food and fiber production remains the soil quality issue of prime importance, as reflected in the high weight given to this function in Table 20.3.

Management-Sensitive Indicators

Some soil quality indicator properties are much more susceptible to change by soil management than are others. As shown in Figure 20.4, properties such as soil texture, mineralogy, steepness of slope, and stoniness are inherent characteristics of the soil and are not subject to change through land management practices. While these properties are important in determining the most appropriate system to use, they will not be altered by whatever system is chosen.

At the other extreme are properties that may be subject to almost daily control so that their effect on soil quality is immediate. Examples are the soil water content as affected by irrigation and rainfall, and levels of plant-available nutrient elements that change rapidly as chemical fertilizers are applied. Also, soil density can be radically reduced by a single pass of a tillage implement or increased by a single pass of a heavy vehicle. These properties are significant for soil function, but their use in soil quality assessment is problematic since they can change so readily from day to day.

Intermediate between these two extremes we find properties that are subject to change only through long-term management efforts. Soil organic matter content and active carbon levels, along with microbial biomass and soil aggregation, are examples of this intermediate class of soil quality indicators. It takes years of careful management to raise the level of these properties in soils, but once they are raised, they tend

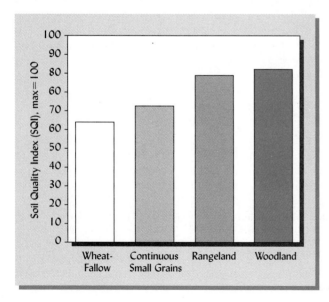

FIGURE 20.3 The effect of long-term land use on soil quality indices (SQI). Large numbers of surface horizon samples from soils under various types of land use were analyzed and their SQI values determined much as described in Table 20.2. The samples were collected across the semiarid Palouse and Nez Perce Prairies in Washington, Idaho, and Oregon from soils mainly in the suborders Albolls, Xerolls, and Xeralfs. The SQI was calculated using a wide range of chemical, biological, and physical indicator properties associated with the environmental protection management goal. The more continuous vegetative cover appears to be associated with higher SQI values. [Redrawn from data in Andrews et al. (2004)]

TABLE 20.3 Importance Assigned to Various Soil Functions in Ascertaining Soil Quality in Different Times and Circumstances

Note the very high weights for the food- and fiber-production function in 1900 worldwide, and in developing countries today. Other functions concerned with environmental and habitat issues are much more prominent today in industrialized countries.

	Probable Weights		
Soil function	Worldwide, 1900	Industrialized countries, 2010	Developing countries, 2010
1. Food and fiber production	85	40	70
2. Resistance to erosion	3	15	10
3. Water and air quality	1	10	5
4. Food quality	5	10	5
5. Wildlife habitat	1	15	5
6. Construction and transport base	5	10	5

to remain high for an extended period of time. These properties are important because of their major influence on soil processes such as water and air movement, soil erosion, and the generation of biodiversity. But they can be developed only if we as soil managers have at least a general understanding of the complex processes that generate them.

Farmers' Perceptions of Soil Quality

Although they may not use the term *soil quality*, farmers who work daily with soils usually note that some of their fields perform better than others. They tend to judge the quality or condition of their soils by such observable factors as the performance of crop plants, the colors associated with accumulation of organic matter, the ease of tillage, the presence of standing water after rain storms, and even the smell of the soil. In addition, most farmers in developed countries have their soils tested for nutrient availability as a guide to what types and amounts of fertilizer to use. During the last 50 years of the 20th century, much of the scientific effort toward improving soil management was directed at such soil testing and improving the supply and availability of mineral nutrients essential for plant growth. As a result, average soil test nutrient levels rose dramatically in intensively farmed areas such that it eventually became more common to find excessive amounts of the main fertilizer nutrients than to find deficiencies (see Sections 14.2 and 16.2).

Until recently, the importance of organic matter management was underplayed compared to the use of fertilizers and lime. Table 20.4 presents a comparison of soils from paired fields rated as being in "good" or "poor" condition or quality based on the experience of the farmers who worked with them. Note that in most cases the farmers' perceived difference in soil quality could be explained by the soil organic-matter-related

Ephemeral	Intermediate	Permanent
Changes within days or routinely managed	Subject to management over several years	Inherent to profile or site
• Water content	• Aggregation	• Soil depth
• Field soil respiration	• Microbial biomass	• Slope
• pH	• Basal respiration	• Climate
• Mineral N	• Specific respiration quotient	• Restrictive layers
• Available K		• Texture
• Available P	• Active C	• Stoniness
• Bulk density	• Organic matter content	• Mineralogy

FIGURE 20.4 Classification of soil properties contributing to soil quality based on their permanence and sensitivity to management. Some soil properties are quite ephemeral and change readily from day to day as a result of routine management practices or weather. Others are permanent properties inherent to the soil profile or site and are little affected by management. A management-oriented soil quality assessment would focus on properties that are intermediate, but all properties tend to be mutually reinforcing. [From Islam and Weil (2000)]

TABLE 20.4 Relationship Between Farmer Perception of Soil Quality and Measured Soil Properties from Pairs of Soils of the Same or Similar Soil Series on a Diversity of Farms in the Mid-Atlantic Region

All of the properties related to soil organic matter were significantly[a] better in the soils rated by farmers as having "good" soil quality, while this was true for only one of the routine soil fertility test properties.

Indicator soil property	N^b	Units	Mean for soils rated "good"	Significance of difference between good and poor	Mean for soils rated "poor"
Indicators related to soil organic matter					
TOC, total organic carbon	44	mg g^{-1}	17.6	***	14.6
MBC, microbial biomass carbon	45	mg g^{-1}	0.486	***	0.342
C_{MW}, carbon extracted after microwave irradiation	45	mg g^{-1}	0.189	***	0.141
C_{RS}, carbon in reducing sugars	45	mg g^{-1}	0.078	***	0.059
AGG, soil in aggregates stable when wetted	41	%	52.4	**	38.2
D_b, bulk density	45	Mg m^{-3}	1.42	***	1.48
Indicators from routine soil fertility testing					
pH, soil pH in 1:1 water: soil slurry	41	$-\log (H^+)$	6.5	ns	6.3
Ca, calcium soil test level (Mehlich 1)	40	mg kg^{-1}	1182	ns	989
Mg, magnesium soil test level (Mehlich 1)	36	mg kg^{-1}	152	*	140
Ca:Mg ratio	36	-na-	4.9	ns	4.2
K, potassium soil test level (Mehlich 1)	39	mg kg^{-1}	190	ns	182
P, phosphorus soil test level (Mehlich 1)	41	mg kg^{-1}	184	ns	154

[a] The symbols ***, **, and * indicate 99.9, 99 and 95% confidence that the adjacent means are truly different by statistical analysis; ns indicates that adjacent means are not truly different according to statistical analysis.
[b] N = number of soil pairs rated by farmers.
Modified from Gruver and Weil (2007).

properties, but not the routine fertility soil test properties. These results illustrate the dominant role that soil organic-matter plays in determining the quality and performance of soils in many intensive agricultural areas in developed countries.

20.2   SOIL RESISTANCE AND RESILIENCE

Before turning to specific agroecosystems that affect soil quality, two other concepts relating to soil quality should receive attention, namely, resistance and resilience. These terms were already used in relation to the soil functions shown in Figure 20.1. *Soil resistance* is the capacity of a soil to resist change when confronted with any kind of force or disturbance. A soil's capacity to resist change is an important component of soil quality, especially as it applies to resistance to soil erosion, compaction, acidification, as well as nutrient depletion. Resistance to various types of chemical changes is analogous to the concept of buffering discussed in Sections 9.4 and 14.8. For example, the soil solution levels of potassium in some fine-textured soils high in hydrous micas are not seriously affected by the removal of this element in harvested crops. The potassium extracted from the soil solution by plant roots is quickly replenished from exchangeable and nonexchangeable forms found in the clay and silt fractions of these soils. In other words, the soil resists change, a characteristic not found in most sandy soils that lack significant levels of exchangeable and nonexchangeable potassium.

A second important concept bearing on soil quality is that of *soil resilience*, or the capacity of a soil to rebound from changes stimulated by disturbances or external forces. A soil under natural forest or grassland vegetation is disturbed when the land is cleared for cultivation, and properties such as organic matter content, organic matter quality, and aggregate stability all decline, thereby reducing soil quality. If, however, the land is turned back to nature, or if other sustainable conservation systems of soil management are utilized, the soil will begin to recover and regain some of its lost properties.

The degree to which recovery takes place and its speed in doing so are measures of soil resilience, a vital component of soil quality. Figure 20.5 illustrates how two different soils might respond to a disturbance such as compaction during timber harvest. A soil with high resistance may suffer only a small degree of compaction and loss of functions such as promotion of root growth and water infiltration. Another soil with low resistance may suffer severely, but if it also has high resilience, it may rebound quickly and its functional impairment may be quite temporary.

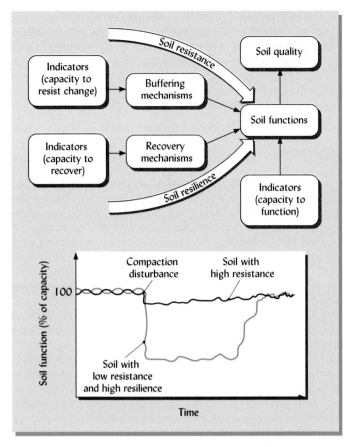

FIGURE 20.5 (*Upper*) The concept of how soil resistance and soil resilience relate to soil quality through soil functions. Resistance acts as a buffer in slowing down change stimulated by a disturbance, while resilience mechanisms help the soil recover from the negative effects of the disturbance. (*Lower*) The effect of a disturbance such as compaction on the functioning capacity of two soils differing in their resistance and resilience. One soil with low resistance to change, functions very poorly after the disturbance. In contrast, the disturbance only modestly affects the function of the second soil with its high resistance. Fortunately, the first soil has high resilience, so in a matter of time, its function recovers to the original level. Resistance assured the second soil's function in spite of the disturbance, while resilience brought the first soil back up to its original function level. [*Upper* modified from Seybold et al. (1999); used with permission of Lippincott, Williams, and Wilkins, Baltimore]

Factors Affecting Soil Resistance and Resilience

Soil resistance and resilience are affected by both inherited and dynamic or management-oriented characteristics. For example, inherited characteristics such as texture, type of clay minerals, slope, and climate largely determine soil resistance, and have significant effects on soil resilience. These factors are the main ones accounting for the differences exhibited by certain broad soil groups in various parts of the world, with respect to their levels of resistance to productivity decline in the face of soil loss by erosion (Figure 20.6).

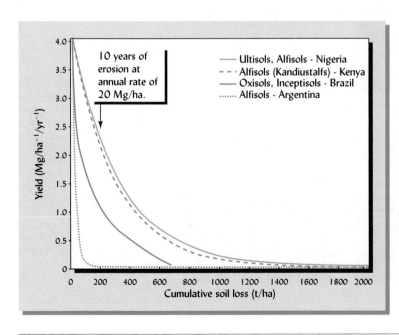

FIGURE 20.6 Examples of productivity (crop yield) reductions expected for different soils suffering from varying levels of cumulative soil loss by erosion. Note that the Alfisols in Argentina are far less resistant to degradation than are the Alfisols and Ultisols in Kenya and Nigeria. The Oxisols of Brazil appear to be intermediate in their level of resistance. The loss of productivity is partly related to the severity of the climate in the different locations and partly to the presence of toxic aluminum or other inhospitable conditions for root growth in the deeper soil layers. These deeper layers progressively become more important for plant roots as the surface soil is lost by erosion. Compare this resistance concept with the erosion tolerance concept presented in Figure 17.9. [Modified from Stocking (2003)]

Dynamic properties such as those associated with the type of vegetation, nutrient cycling, water and land management, as well as the underground community of organisms, play vital roles, especially for soil resilience. For example, properly managed cropping systems can speed up the rate of organic matter buildup in a degraded soil. In other words, these systems can enhance soil resilience, an important component of soil quality. The significance of both soil resistance and soil resilience will be seen later on as we focus on more specific ecosystems that are affecting soil quality.

20.3 SOILS AND GLOBAL ECOSYSTEM SERVICES

The state of the U.S. ecosystems, croplands, and soils:

http://www.us-ecosystems .org/ecosystems/report.html

Some practices that could enhance soil quality and **ecosystem services** are not adopted because they are considered to be uneconomical, that is, they fail to pay the landowner enough to justify their use. This is due in part to our economic system, which considers income as being only from products and services that can be bought or sold in the traditional marketplace. There is growing recognition, however, that such reasoning may not be valid in assessing the economic importance of ecological measures. It ignores the true value to society of many soil management practices and systems. It overlooks such services as carbon sequestration, protection of downstream water quality, enhancement of water supplies, fostering of biodiversity, protection of air quality (from dust, odors, and gaseous emissions), human disease mitigation, and even scenic beauty. Each of these services has intrinsic value, but its value may not be easily measured, nor are there always simple ways of rewarding those who enhance it.

Just as soil quality management influences the capacity of soils to perform certain ecosystem functions, so too some types of land use and ecosystems provide more or better services than others. Soil management is a key part of land managment that can maximize the value of the ecosystem services provided. Figure 20.7 illustrates the relative value of eight basic ecological functions as performed by five different ecosystems. Too often in the past, only one or two of these functions have been properly valued or even acknowledged as management goals. The examples illustrate that not only is the choice of land use important (e.g., forest versus farm), but the type of management, including soil management, is also critical (compare the conventional farm to the ecological farm). The ecological services provided can be greatly increased with small, if

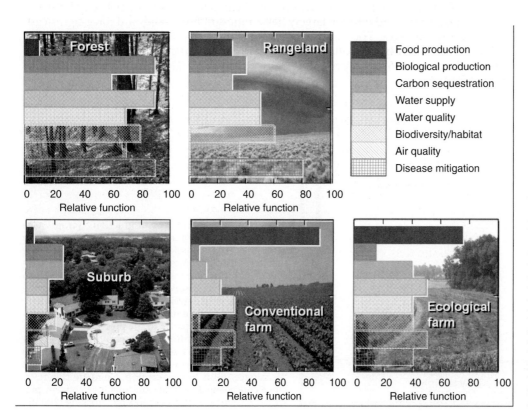

FIGURE 20.7 Relative performance of ecological functions by soils in five types of representative ecosystems. For humanity to meet its needs for all the listed ecosystem services, diverse types of ecosystems are required. Soil quality interacts with ecosystem characteristics to provide these essential services to humans and other species. The graphs for the conventional and ecological farms suggest that much can be done to increase ecosystem services with only a small sacrifice in food production. (Diagram and photos courtesy of R. Weil)

TABLE 20.5   Proportion of the Potential Income from Ecosystem Goods and Services Provided by Five Hypothetical Land-Based Businesses that Vary Greatly in Their Potential Impacts on Soil Quality

*Note that only 30 to 60 percent of the value is derived from traditional sources of income (**in bold**). Society will have to develop innovative political, social, and marketing means of appropriately rewarding each business for the total revenue or value it generates.*

Commodity or service	Proportion of net income (%)				
	Forestry business	Dairy farm business	Ranch business	Golf course	Grain farm business
Milk	0	**20**	0	0	0
Grains	0	**5**	0	0	**60**
Hay/forage	5	**5**	10	5	0
Cattle	0	**5**	**30**	0	0
Timber	**30**	5	5	0	0
Carbon sequestration	15	10	5	5	12
Water filtration/collection	15	10	10	10	8
Air purification	5	5	5	5	5
Soil fertility	5	5	5	5	3
Scenic beauty	5	10	5	15	5
Recreation	**5**	5	5	**45**	5
Biodiversity	10	10	10	5	2
Salinity control	0	0	5	5	0
Medicinal plants	5	5	5	0	0

For a discussion of ecosystem values and services, see Daily et al. (2000).

any, reductions in the amount of food produced if the farm adopts such practices as no-till (Section 17.6), reduced pesticide use (Section 18.2), riparian buffers, cover crops, diverse rotations (Section 16.2), and special areas devoted to habitat for wildlife and beneficial insects. Likewise, the ecosystem services provided by suburban and urban ecosystems (not shown) can be increased by such measures as green rooftops, rain gardens, permeable pavements (Section 6.2), on-site composting (Section 12.10), and increased plantings of native vegetation.

Ecological economists are attempting to measure and to express the value of ecosystem services in economic terms that have meaning to society. For example, in determining the total potential income from a forest ecosystem we should add to the timber sales the income equivalent attributable to such values as scenic beauty, watershed protection, and downstream water quality. Such methodology clearly demonstrates that the monetary income from ecosystems is often less than the "income" or societal benefit derived from other values of the ecosystem (see Table 20.5). It also suggests that the total income to be derived from soil management practices may far exceed income received from the sale of commercial products generated by these practices.

Society can recognize the benefits of various ecosystem management practices by providing incentives for using the beneficial practices or penalties for not doing so. Both penalties and incentives (often termed *green payments*) can effectively ensure that it is to the economic advantage of the landowner that appropriate soil management practices are followed, and that the loss of soil and water quality is minimized. Early examples of green payments include Conservation Reserve Program rentals in the United States (see Section 17.14), fees paid by sportsmen for private hunting rights on land managed to foster large wildlife populations, and payments made to farmers by utility companies in some countries for each unit of atmospheric carbon sequestered by improved soil management.

Endangered Soils[2]

Some scientists have proposed that certain soils are so unique and so rapidly succumbing to erosion, urbanization, and other destructive processes that they need society's protection. They propose protecting these soils by designating them as *rare* and

---

[2] For one proposal for protecting endangered soils, see Drohan and Farnham (2006).

Calculate your "ecological footprint"—Redefining Progress: http://www.earthday.net/footprint/index.asp

threatened, much as certain plant and animal species are protected by such designations. A soil would be designated *rare* if it exists over only a very limited area, while a soil might be designated as *threatened* if it is suffering from some form of degradation that is compromising its ability to function in the ecosystem. Protection from further degradation might be provided based on the value of the particular soil with regard to (1) economic productivity, (2) ecosystem services, (3) scientific value for the study of unique soil processes and properties, (4) historic or cultural value (such as the inclusion of fossils or artifacts within the profile), and (5) simple rarity.

These and other creative political, social, and economic actions are likely to be needed in increasing measure to ensure that scientific findings such as those discussed in this book are widely utilized to maximize soil quality and ecosystem services for the benefit of all.

20.4 SUSTAINING THE HUMAN POPULATION

The survival of the human species continues to depend directly on soil functions that sustain the growth of plants used for food, feed (for domesticated animals), fiber (for everything from cotton shirts to wooden houses) and, increasingly, fuel (biofuels). At the same time, the soil's capacities to support biological productivity sustain the interconnected diversity of life above- and belowground. It is important to recognize that the survival of other organisms, from bears to bacteria, is often determined by how we humans go about satisfying our basic needs.

The First 10,000 Years of Agriculture

Regional trends in world agriculture: http://www.ifpri.org/pubs/agm05/jvbagm2005.asp

Since the dawn of agriculture some 10,000 years ago, people have cleared forests and prairies so that the land could be used to grow food and fiber for their growing families. Initially, the transformation of natural ecosystems to the less stable agricultural systems had only local effects on soil quality because there was an abundance of land and relatively few people—only about 1 million people lived on Earth.

While 1000 ha of productive land could support perhaps a single family of humans gathering wild plants and hunting wild animals, the practice of agriculture allowed the same land area to support more than 100 families. As a result, humans survived in greater numbers and by 2000 years ago, their global population had reached some 200 million. Soil productivity began to suffer over wide areas. The salinization of the once very productive irrigated valleys of ancient Mesopotamia in the Middle East (see Section 10.3) and the severe water erosion of the hilly lands farmed by Greeks and Romans (see Figure 17.28) forced these peoples to expand into the less densely populated lands of North Africa and Europe. The rice cultures of Asia had less negative impacts on soil quality as their crop production was centered on riverine floodplains less subject to erosion or salinization, and periodically replenished with new sediments.

The Population Explosion

As human populations increased, food production kept pace primarily by expanding the area of land cleared for cultivation rather than by increasing the amount of food produced per hectare. One of the greatest expansions of farmland occurred after Europeans "discovered" the Western Hemisphere, on whose virgin soils they soon produced food not only for the new colonies, but for export to the food-deficient parts of the globe. As industrialization began to concentrate growing numbers of people in cities, farmland expansion was required to feed the urbanites. By the year 1800 there were some 1 billion human mouths to feed. By 1940 the number had grown to 2 billion, and by the end of the 20th century the human population surpassed 6 billion. Thus, while the first billion took some 10,000 years to accrue, the sixth billion took only 10 years.

This explosion of the human population during the 20th century was stimulated largely by the advent of such basic public health measures as purified drinking water, sanitary facilities, and drugs that prevented or controlled many common infectious diseases.

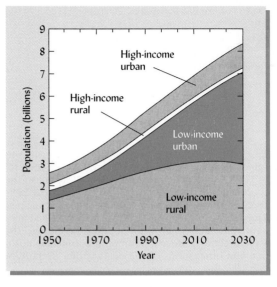

FIGURE 20.8 The world's population is expected to rise to about 8.5 billion by the year 2030. Essentially all the population increase will occur in the low income countries, most of which are already pressed to provide food for their current populations. Note the rapidly increasing proportion of developing country populations that live in urban areas. While considerable quantities of vegetables and other food crops are grown in and around cities, most of the food required by the urban populations must be produced in often remote, rural areas. [Graph based on the author's calculations using data from United Nations Secretariat (2007)]

State of food insecurity in the world, 2006:
http://www.fao.org/docrep/009/a0750e/a0750e00.htm

Whereas humans had always lived in a world in which most of their children died before reaching five years of age, they now found that most of their children survived to have children of their own. The life experiences and child survival changed so rapidly that it often took several generations for people to adapt their reproductive behaviors to the new realities—a process still ongoing in the poorer countries of the world.

Current projections estimate that the human population will level off at about 9 to 10 billion near the end of this century. Nearly all of the increase will occur in the world's poorer developing countries, and mainly in cities (Figure 20.8). To feed this exploding population, farmers have already had to produce more food in the past half century than had been produced in the previous 10,000 years of the history of agriculture. To achieve this vast increase in food production it was necessary to **(1) clear and cultivate more land** in the remaining forests (which often occurred on steep, erosion-prone terrain) or natural grasslands (many of which are semiarid, subject to drought and erosion by wind); or **(2) greatly increase the cropping intensity** and the yields per hectare on the more productive lands already under cultivation. Most of the needed food increases came (and will continue to come) from enhanced production on existing farmlands. As we shall see, both of these approaches to increasing food production can have serious consequences for the soil quality.

20.5 INTENSIFIED AGRICULTURE—THE GREEN REVOLUTION

When the human population explosion became evident after World War II, many experts predicted widespread starvation. Their predictions were based primarily on the assumption that, as in the past, expansion of cultivated land would be the primary means of increasing food production. They wrongly discounted possibilities for increased production intensity on land already in cultivation. Instead of mass starvation, by the 1970s the world actually saw food production increase more rapidly than population—resulting in rising per-capita food production in all major regions with the notable exception of sub-Saharan Africa (Figure 20.9).

Total world grain production more than doubled between 1960 and 2004 (Figure 20.10). Since the area of land under cultivation was increased only slightly during this period, it is clear that most of the increase in food production came from greater production per unit of land area. Indeed, the world average grain yield per ha also more than doubled, from about 1300 kg/ha in 1960 to over 3300 kg/ha in 2004. As a result, the threat of massive starvation was averted, and prices for staple foods like

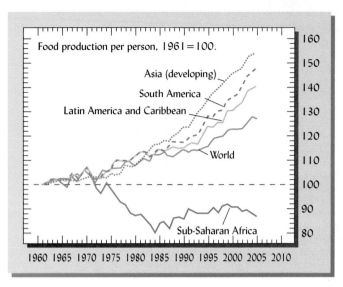

FIGURE 20.9 Relative changes in per-capita food production in different regions of the world between 1961 and 2005. Food production per person worldwide increased about 25%, but in the developing countries of Asia, the increase was more than 55%. Only in sub-Saharan Africa (excluding South Africa) did the per-capita food production decline. Most of the increases resulted from agricultural intensification begun with the Green Revolution. The reasons for the decline in Africa are complex, but include nutrient depletion and other forms of soil degradation. [Data from FAO (2006b)]

cereals actually fell by as much as 75% in some countries, making food much more affordable for low-income city dwellers in particular.

Increasing population will continue to expand world demand for food and fiber, a trend magnified by rising incomes and changing diets in developing countries. The first thing most poor people buy when their incomes rise is more food, in particular, more animal products. The production of 1 kg of chicken, pork, or beef requires at least 3, 5, or 8 kg of grain equivalent, respectively. During the coming century the overall food demand is therefore expected to increase by the interactive effects of increasing numbers of people, growing consumption per person, and a larger proportion of animal products in the diets.

COMPONENTS OF THE GREEN REVOLUTION.   How did such good fortune befall the peoples of the world? Through a concerted international effort, scientists and their farmer collaborators developed and put to use new farming systems that integrated the use of newly created *high-yielding varieties* of cereals (wheat, corn, and rice) with dramatic increases in nutrient inputs from *chemical fertilizers* (Figure 20.11) and increased water availability through *irrigation*. The new crop varieties were specially bred to respond to fertilizer nitrogen by producing large grain yields, but also stiff, short stems that prevented the plants from falling over as traditional varieties did (for an example, see Figure 13.1*d*). Monoculture systems were intensively used and new *synthetic pesticides* were sprayed to

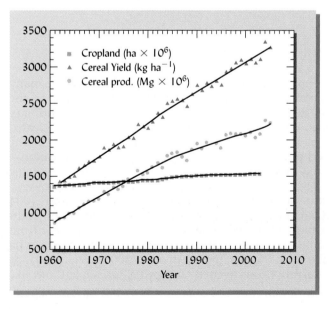

FIGURE 20.10 Between 1960 and 2005, world food production (•) increased by 250%. This rate of increase surpassed the rate of population growth, allowing a significant increase in the amount of food available per person. Nearly all the increase in food production was due to enhanced yields per unit of land area (▲) since the area of land under cultivation for all crops (•) increased relatively little and the proportion devoted to cereal cropland declined slightly (not shown). [Data from FAO (2006b) and author's calculations]

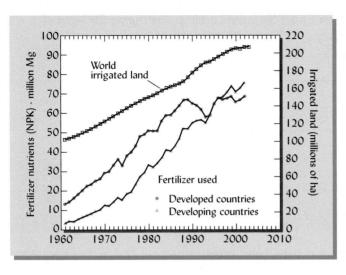

FIGURE 20.11 Increases in fertilizer use in industrial and developing countries and in world irrigated area since 1960. Note the 25-fold increase in fertilizer use in developing countries and the worldwide doubling of land under irrigation. The drop in developed-country fertilizer use during the early 1990s was due primarily to decreases in the former Soviet Union, along with a leveling-off in the United States and Europe. [From FAO (2006b) data and author's calculations]

keep insects at bay (though insect resistant crop varieties were also soon introduced). In some systems, *multiple cropping* was used with two or three crops being harvested annually from the same piece of land.

BENEFITS OF THE GREEN REVOLUTION. The results were most spectacular in Asia and Latin America, where the term *green revolution* was coined to describe the process. Wheat yields in India, for example, increased by nearly 400% from 1960 to 1985, and yields of rice in Indonesia and China more than doubled. The global average caloric intake increased to about 2700 kilocalories, about 16% above minimum needs. Although millions still remained hungry, human nutrition among the poor was greatly enhanced since rural farmers had greater harvests and urban poor paid lower prices for their food staples.

DOWNSIDE OF THE GREEN REVOLUTION. It should be noted that while the green revolution technologies certainly saved large parts of the world from starvation, the process was not entirely benign. It was accompanied by traumatic social upheavals as competition from farmers who could afford the new inputs drove from the land others who could not. Landless peasant families often migrated to already overcrowded cities in search of the means of survival. Serious environmental toxicities resulted from poorly controlled, poorly chosen, and overused pesticides. In some cases the pesticides actually created more pest problems than they cured because they killed natural enemies (such as predaceous insects and spiders) that had previously kept plant pests in check. With this mixed blessing in mind, we will now discuss some of the ways in which this new era of intensive agriculture has affected soil quality for good and for ill.

20.6 EFFECTS OF INTENSIFIED AGRICULTURE ON SOIL QUALITY

Direct and indirect evidence suggests that intensification of agricultural production systems has had both positive and negative effects on soil quality.

Positive Effects

SOIL NUTRIENT CONTENTS. On the positive side, intensified agriculture has generally maintained or even increased the levels of some **macronutrients** in soil, since these elements are commonly supplied from outside sources, such as manures, lime, or fertilizers. Where appropriate applications of chemical fertilizers have been used, the N, P, and K components of soil quality have often been enhanced.

SOIL ORGANIC MATTER. Intensified agriculture has also increased the level of plant production, permitting a corresponding increase in the amount of **crop residues** that can be returned to the soil to enhance soil quality. Such residues provide soil cover, reduce soil erosion, and can help maintain or increase soil organic matter levels (Table 20.6).

TABLE 20.6 **Organic Carbon and Total Nitrogen Levels in Soil During 29 Years of Continuous Rice Cropping (3 Crops per Year) with and without Nitrogen Fertilizer**

Organic C and N levels built up more where N was heavily applied. Phosphorus and potassium were applied to all plots.

Year	Organic carbon in soil, g/kg		Total N in soil, g/kg	
	No N applied	330 kg N/ha/yr applied	No N applied	330 kg N/ha/yr applied
1963	18.3	18.3	1.94	1.94
1978	18.8	21.4	1.97	2.22
1985	20.4	23.9	2.07	2.38
1992	20.7	23.0	2.09	2.30

Modified from Cassman et al. (1997).

LAND-USE PRESSURE. Agriculture has been intensified mostly on the more productive, relatively level soils, where risks from erosion are not too high. By producing most of the additional food on these soils, the need for expanding onto more fragile lands has been minimized, preserving the soil quality (and other environmental benefits) on the natural lands spared from cultivation. Figure 20.12 illustrates this point for India. Were it not for the wheat yield gains from the Green Revolution, the country would have been forced to plow an additional 42 million ha of fragile lands, mostly in forests, an area equivalent in size to the state of California. By this reasoning, worldwide, more than 600 million ha—equal to the area of the great Amazon basin—have been "saved" due to increased yields of all cereal crops in already cultivated land.

Negative Effects

When cleared lands are intensively cultivated and nutrients removed with harvests, deficiencies of nitrogen and phosphorus are usually first to appear. On some soils, especially those that are highly weathered and low in 2:1-type clays, crop removals soon also lower the potassium levels. The application of chemical fertilizers generally provides ample quantities of nitrogen, phosphorus, and potassium. However, the micronutrients removed in the bumper harvests are usually not replaced by standard N-P-K fertilizers, so *micronutrient deficiencies* may appear (Section 15.11). Heavy use of nitrogen fertilizers may also lead to increased soil acidification (Section 9.6). Both effects could lower soil quality.

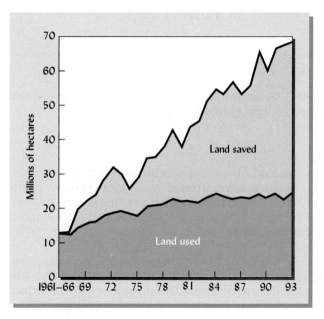

FIGURE 20.12 In the 1990s, if India had been forced to produce its wheat with technologies and varieties of the 1960s, farmers would have needed about 40 million more hectares of farmland. Most of this extra farmland would have to come from easily erodible forestlands that are characterized by steep slopes. [From *The Economist*, June 10, 1995]

EXCESS NUTRIENTS. In many areas of the world (particularly East Asia and Western Europe) nitrogen and phosphorus have been added in quantities far in excess of plant uptake. With time, as the levels of these nutrients built up in the soil, they moved as *pollutants* into the runoff or drainage waters or into the atmosphere (see Sections 13.8 and 14.2), seriously compromising the environmental protection goal of soil quality.

SALINIZATION. Irrigation-induced *salinization* is another negative effect of agricultural intensification on soil quality (see Section 10.3). For example, each year the salt added in irrigation water to the soils of Arizona is equivalent to about 350 kg for each of the 4 million people living in the state. Worldwide, some 30% of irrigated soils are significantly affected by salinization, some so seriously that the land has been abandoned.

INSECTICIDES. Chemical *insecticides* that are commonly used in intensified agriculture systems can adversely affect soil quality. While some organochemicals adversely affect a broad spectrum of soil organisms, others are selective, reducing biological diversity more than overall abundance (Sections 11.11, 18.4). Some soils treated decades ago with high levels of arsenic- or copper-containing insecticides still contain toxic levels of these chemicals. The use of insecticides, especially those with residual action or high mobility in the soil, can decrease soil quality with regard to the environmental protection and biodiversity roles of soils. Because of the uncertain effects of today's pesticides on soil quality, *integrated pest management* (IPM) systems that minimize the use of pesticides, or *organic farming* methods (see Section 20.8) that eliminate the use of synthetic pesticides should be emphasized.

HERBICIDES. Herbicides are chemicals that are meant to kill plants—presumably the unwanted plants that we call *weeds*. The use of herbicides has been associated less closely than insecticides with Green Revolution farming, partly because mechanical cultivation and hand-pulling alternatives have often been less costly where labor is cheap, and partly because wheat and rice are close-growing crops that on their own can often outcompete weeds without much effort by the farmer. Nonetheless, herbicides are very widely used in intensive production of such row crops as corn, soybeans, cotton, and the like, where the wider spacing between crop plants gives weeds an early opening to compete.

Different herbicidal chemicals vary widely with regard to their mobility, persistence, and toxicity to animals and people. Those that are long lasting and relatively mobile in soils have created major water pollution problems (Section 18.3). Others that are immobile and show no soil activity have played a very positive role in enhancing soil quality by facilitating conservation tillage (Section 17.6). To maximize soil and water quality, herbicides should be used very selectively and with restraint. Research shows that low to moderate weed populations can often be tolerated without sacrificing crop yields. Furthermore, the intensive use of herbicides to strictly enforce a single species monoculture negatively impacts soil quality because the lack of plant biodiversity can lead to reduced diversity and resilience of the above- and belowground communities associated with the shoots and roots of these plants.

HEALTHY DIET. Intensive agricultural systems have focused primarily on cereal crops, such as wheat, corn, and rice, which provide about half the world's calories and are quite responsive to external inputs, such as water and fertilizers. Unfortunately, less attention has been paid to the pulses (N-fixing grain legumes such as beans, peas, and lentils), fruits, and vegetables. As a result, the area planted to these crops actually decreased in some countries. For example, in India, the area of land devoted to pulses decreased by 13% from 1970 to 1995. In addition, the world average yield for pulses increased only 50% (from 600 to 900 kg/ha) since 1960 while that of the cereals increased 250%. This has implications for soil quality because integration of vegetables and legumes into farming systems provides for more flexible and diverse rotations and higher plant residue quality for the soil food web. There are also implications for *human health* because, compared to cereals, pulse crops and leafy vegetables are generally higher in proteins and micronutrients, or are richer in certain essential vitamins (although whole cereal grains are important for their B vitamins). Human diseases associated with *deficiencies* of iron, zinc, copper, and vitamin A are widespread in tropical countries (Section 15.11). Also, the residues from pulse crops provide organic nitrogen for the soil food web and for subsequent crop uptake, allowing reduction in N fertilizer use. Excessive emphasis on cereals has thus reduced several aspects of soil quality in many countries of the world.

PLANT DISEASE.   Similarly, the Green Revolution has had some negative impacts on soil quality because the improved cereals have commonly been grown in *monoculture* season after season. In some areas, research has shown a decline in the biological productivity of monoculture systems. This may be due to the buildup of *pathogens* or of *allelochemicals* that are toxic to the crop, or to declining levels of micronutrients in the soil. For any of these reasons, when cropping systems do not take advantage of the benefits of crop rotation, soil quality usually suffers.

REDUCED BIODIVERSITY.   High input, intensified agriculture using monoculture systems generally adversely affects *biodiversity*, including the genetic diversity within crop species. For example, before intensification of agriculture in China, farmers were growing 10,000 varieties of wheat. Today, more than 90% of these varieties have disappeared with only a handful of high-yielding varieties accounting for nearly all the wheat production.

Chemical intensive, monoculture systems also significantly affect the abundance and biodiversity of soil organisms as they provide little diversity in the organic residues and in the associated organisms that take part in their decay. We know that the clearing and cultivation of forested lands reduces the number of fungi and increases the relative numbers of bacteria. The ratio of fungal biomass to that of bacteria may be about 1:1 in intensively tilled soils, about 3:1 with minimum tillage, and more than 100:1 under forested vegetation. Modern DNA analysis techniques suggest that bacterial diversity may also be adversely affected. Monoculture systems, especially those where the crop residues are removed or burned, also reduce the number of earthworms and other macroorganisms, compared to their numbers in systems with crop rotation.

CONCENTRATED ANIMAL-FEEDING OPERATIONS.   The great bulk of pork, beef, poultry, and dairy production in industrial countries, and increasingly in the developing countries of Asia, is carried out in "factories" that are quite separated from the plant part of the agricultural cycle. Confinement of livestock in these concentrated animal-feeding operations (CAFOs) may be efficient in terms of labor costs and feed conversion to animal protein, but they have serious adverse effects on soil and environmental quality (see Section 16.4), as well as on the welfare of farmers and the animals themselves.

The CAFOs separate animals from the land on which their feed is grown, preventing the ecological integration of plants and animals, and oversimplifying the agroecosystems in both the animal and crop production areas. They remove plant products (and the nutrients they contain) from wide areas and concentrate them into a production factory, the wastes from which often pollute the surrounding soil and water systems with nitrogen, phosphorus, pathogens, hormones, and antibiotics. Adopting new, efficient but more integrative approaches to animal production is a major challenge that needs to be seriously taken up for the long-term sustainability of agriculture, soil quality, and the environment.

20.7   IMPACTS OF VASTLY INCREASED RATIOS OF PEOPLE TO LAND

The population explosion of the past 60 years has dramatically reduced the amount of land available per person, enormously increasing the intensity of land use—often beyond the bounds of soil resilience. We will discuss briefly the effects on several land-use systems that have suffered major losses in soil quality as a result of these population pressures.

Shifting Cultivation

Indigenous peoples and their ancestors around the world have developed traditional pastoral and crop-production systems that take advantage of soil resilience. These systems rely on the ability of soils under natural vegetation to recover from the temporary degradation caused by losses of nutrients, soil organic matter, and soil structure during brief periods of agricultural use. One of these indigenous systems is a type of shifting cultivation, which has been practiced for generations in tropical forests around the world. It involves the slashing of native vegetation to make small clearings in the forest and then burning the cut material (Figure 20.13 and Plate 34) to prepare the land for

FIGURE 20.13 (*Left*) Aerial view of a slash-and-burn area in the Amazon, where the trees have been cut down and are ready for burning. Note the second area in the background where similar cutting is under way. (*Right*) In another field that has been slashed and burned, a row of corn has been planted (arrow) and other crops will follow. There are no weeds at this point, but in a few years, weed infestation and soil fertility depletion will force the cultivator to move on to another site. (Photos courtesy of the International Center for Research on Agroforestry)

producing food crops for a few years. Hence the system is sometimes termed "slash-and-burn" agriculture. Crops are grown for only a few years until the supply of nutrients from the ashes and soil organic matter is depleted, or until weeds become intolerable. The farmer then moves on to slash and burn another small plot, cultivates it for a few years, and then abandons it, too. This process is repeated, shifting from one site to another, until after some 15 to 20 years, the initial site is once again cleared of the now semimature natural vegetation, and the process is started over. During the 15 to 20 years of forest regrowth, the natural vegetation at least partially rejuvenates the quality of the soil by adding organic residues, recycling nutrients from deep in the profile, and fixing nitrogen from the air.

During the past half century the number of people dependent on shifting cultivation has increased dramatically. More than 300 million people, most of them desperately poor, practice some form of shifting cultivation on enormous land areas, primarily in the tropics and subtropics. Their activities account for a major part of tropical deforestation. Once quite sustainable, the shifting cultivation system now fosters soil and environmental degradation and is nearing collapse in many regions. The reason is simple math—the amount of land needed per person for sustainable shifting cultivation is very large. Attempts to practice shifting cultivation with less land undercut the ability of the system to sustain soil quality.

For example, consider a shifting cultivator who can produce enough food for his family by cultivating 1 ha of land. In the sustainable traditional system, this farmer might cultivate the 1 ha for 3 years and then abandon it to forest revegetation for 15 years so that 18 years elapse between repeated clearings of a given plot of land. This means that 6 ha of land (18 y/(1 ha×3 y)) are required to produce the needed food—1 under cultivation plus 5 under some stage of natural vegetation fallow for recovery. As the human population in an area doubled and quadrupled during the past half century, the amount of land available per person was halved and quartered. Shifting cultivators either had to slash new, virgin forest lands (a difficult task and usually not an available option) or they had to reduce the number of years between abandoning a plot and returning to it. As fallow periods declined from nearly 15 to 20 down to only 3 to 5 years, the time was insufficient for regenerative processes to replenish soil nutrient and organic matter levels and to constrain soil erosion (Figure 20.14). As a consequence, soil quality has deteriorated, along with yields of food crops and family well-being in most areas where shifting cultivation is practiced.

Slash-and-burn systems also adversely affect the atmosphere. The CO_2 and other gases released in the burning process are thought to make a significant contribution to

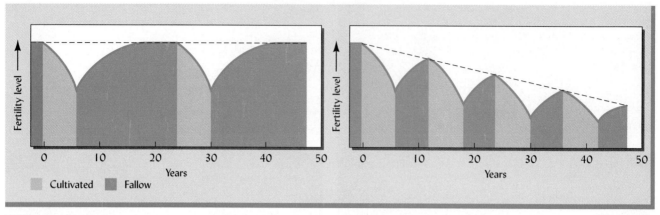

FIGURE 20.14   Changes in soil quality under two shifting-cultivation systems. (*Left*) In the system with a long fallow period, the natural vegetation is able to help regenerate the level of fertility and soil quality after cropping. (*Right*) With a shorter fallow period so common in areas of high population pressure, insufficient regeneration time is available, and the soil quality level declines rapidly. These figures illustrate the concept of soil resilience.

total global warming. To prevent the deterioration of air and soil quality, cultivators in these areas need viable alternatives to the intensified shifting-cultivation systems that they have been forced to use.

Nomadic Pastoral Systems

Nomadic pastoral systems on common lands in arid to semiarid areas, combined with increased human populations and periods of drought, have adversely affected soil quality, especially in the Sahel region of Africa. Much as in the case of shifting cultivation just described, when the numbers of nomads and their livestock were low, the range-land had many years to recover between visits of the grazing herds. Increased animal numbers, or even constant numbers with less forage to consume because of drought, result in little or no plant recovery between grazing episodes—in a word, *overgrazing*. The more nutritious grasses are mostly eliminated, and only less-palatable shrubs remain. Overall plant production is reduced, less plant residue is returned to the soil, and soil organic matter is not replenished. Wind erosion increases (see Section 17.11), and overall biological productivity declines, along with soil quality. Nearly 100 million people now live in areas where such effects are occurring.

Marginal Lands

Human impacts and conservation in the Cerrado:
http://www.biodiversityhotspots .org/xp/Hotspots/cerrado/ Pages/impacts.aspx

Human population pressures have also resulted in expansion of cultivation into areas that are marginal for agriculture, such as erosion-prone, steep hillsides. Many landless families in Central and South America, Africa, and Asia are forced to use these areas to survive. Including farm families that practice shifting cultivation, some 800 million people produce their food on forested hillsides. Removal of plant nutrients in the crops they harvest, along with increased soil erosion, decreases biological productivity and lowers soil quality.

Urbanization and Sprawl

The increased use of land for urban, industrial, and transportation purposes also has its impacts on soil quality. The problem of urban sprawl has resulted in losses of cropland, pasture, and forest lands around the world (Figure 20.15), especially in the developing countries, where population pressures are greatest and economic growth is most pronounced. Global urbanization alone is thought to claim well over 1 million ha annually. Furthermore, this land is usually prime land for plant productivity, since the growing cities were originally located to take advantage of good cropland nearby.

Before we look to the future, we should remind ourselves of what we have done in the past half century, during which time some 2 billion ha of land have suffered soil quality degradation (see Figure 17.1). Data in Table 20.7 suggest that by 1990 (the latest

FIGURE 20.15 Urban sprawl can devastate the quality of soils in forests and agroecosystems that once provided food, fiber, and other valuable ecosystem services. (*Left*) Poor people haphazardly expand Tegucigalpa (Honduras) onto steep, erodible forest lands. (*Right*) Wealthy people haphazardly expand Phoenix (Arizona) onto highly productive irrigated farmland. The soils will likely never again be available for forestry or agricultural. These soils become part of new ecosystems involving foundations for roads and buildings, and environments for lawns and gardens. If the soils are protected from erosion (too late for the scene at left), amply supplied with organic matter, and not treated with excess fertilizer and pesticide chemicals, soil quality for these new functions may well be quite satisfactory. (Photos courtesy of R. Weil)

Loss of U.S. farmland:
http://www.farmland.org/
resources/fote/default.asp

figure available) soil quality had declined on 38% of the world's agricultural land, on 21% of permanent pastures, and on 18% of forests and woodlands. Soil degradation on agricultural lands has been most drastic in Africa and Central America, but the problem extends to all continents. On about half of the land suffering soil degradation, erosion by water is the main cause, with another quarter of the area suffering from erosion by wind. Also widespread is degradation due to compaction, salinization, and depletion of nutrients. In more limited areas (but still tens of millions of ha worldwide) soils have suffered acidification, pollution, and waterlogging. Since rising demands for food, fiber, biomass, and living space are sure to continue to increase pressures on soil quality, we now need to join in aggressive efforts aimed at reversing the degradation trends.

TABLE 20.7 Land Areas Used for Agriculture, Permanent Pasture or Forest and the Percentages of Each Suffering Degradation (Loss in Soil Quality) as a Result of Human Activities. Millions of ha.

Reductions in soil quality are most severe in agricultural lands, particularly in Africa and Central America, where more than two-thirds of the cultivated lands are adversely affected.

Land use	Africa	Asia	South America	Central America	North America	Europe	Oceania	World
Agricultural land								
Area	187	536	142	38	236	287	49	1475
Percentage degraded	**65**	**38**	**45**	**74**	**26**	**25**	**16**	**38**
Permanent pasture								
Area	793	978	478	94	274	156	439	3212
Percentage degraded	**31**	**20**	**14**	**11**	**11**	**35**	**19**	**21**
Forest and woodlands								
Area	683	1273	896	66	621	353	156	4048
Percentage degraded	**19**	**27**	**13**	**38**	**1**	**26**	**8**	**18**
All lands								
Area	1663	2787	1516	198	1131	796	644	8735
Percentage degraded	**30**	**27**	**16**	**32**	**8**	**27**	**16**	**23**

Reorganized and calculated from Oldeman et al. (1990).

A twofold challenge lies ahead: (1) to satisfy the enormous and growing needs of society for food, fiber, and biomass produced using soils, while (2) maintaining or even enhancing the quality of the soil resource and the larger environment. To meet this challenge, a new paradigm of global agriculture is beginning to unfold.

A New Paradigm of Plant Production

USDA Sustainable Agriculture Research & Education (SARE):
www.sare.org

In the past, the limits of *technology* and *short-term economic feasibility* have been the primary bases underlying food-production systems. While these considerations must still be accommodated, there is a growing recognition that we must give greater emphasis to an *ecological basis* for our agricultural systems. This new sense of direction will use science and technology to understand how agroecosystems interact with other ecosystems. It will help us focus on establishing food-production systems that minimize losses and maximize the stewardship of natural resources for food production, for natural habitats, and for the other ecosystem services and soil quality goals discussed earlier in this chapter. We will start with some significant changes needed in intensified commercial agricultural systems.

Lectures and speeches on sustainable agriculture:
http://www.leopold.iastate
.edu/pubs/speech/speech.htm

Food and fiber production on prime agricultural lands will continue to be highly intensified, in terms of high yields and continuous cropping, but not necessarily in terms of chemical inputs. To sustain this intense production, we must employ new soil and crop management systems that will *reduce pollution* of soil, water, and the atmosphere, decrease *soil erosion*, increase the *efficiency of nutrient and water use*, maintain or increase the *quantity and quality of soil organic matter*, and increase *biological diversity*. Since maintenance of soil cover and nutrient cycling can affect several of these objectives, we will discuss them first.

Keeping the Soil Covered

Perhaps the most important requirement for maintaining and improving soil quality in intensive agriculture is to keep the soil surface covered with growing plants, residues, or mulches at all times. Proper cover can greatly reduce soil loss by erosion (Section 17.5) and water loss by runoff and surface evaporation (Section 6.4). A diversity and generous supply of plant residues on the soil surface also stimulates the soil food web organisms (Chapter 11), enhances aggregate stability (Sections 4.5–4.6), increases the infiltration of rain and irrigation water (Section 5.6), and increases both soil organic matter and biodiversity. In short, keeping the soil covered is one of the best paths to improved soil quality.

Conservation tillage, including no-till, not only ensures organic residue coverage on most soils, but costs less in terms of both money and energy inputs than do most tilled systems. Combining conservation tillage with cover crops or green manures provides synergistic benefits to soil quality and environmental protection (see Sections 16.3 and 17.5) in all but very dry climates.

Biomass for Biofuel Production[3]

On food and oil:
http://www.globalpublicmedia
.com/articles/507

An emerging challenge for soil managers is the new and growing demand for plant biomass, produced not for food or fiber, but as a renewable feed stock to replace dwindling supplies of petroleum for the energy industry. Projections suggest that soil resources may be increasingly devoted to energy crops as the United States and other countries attempt to shift their energy sectors away from reliance on nonrenewable and greenhouse gas-generating fossil fuels. In particular, dramatic future growth is projected for the production of biofuels from plant biomass (Figure 20.16, *center*). Advances are being made in fermentation processes used to produce ethanol, both from starchy material such as corn grain, and from cellulose-based materials such as crop residues and leafy tissues. If only the burning of the fuel itself is considered, biofuel may serve as a carbon-neutral

---

[3] For an example of biofuels production planning that appears to ignore soil quality, see Ragauskas et al. (2006) and for rebuttals recognizing the soils and sustainability aspects, see letters edited by Kavanagh (2006).

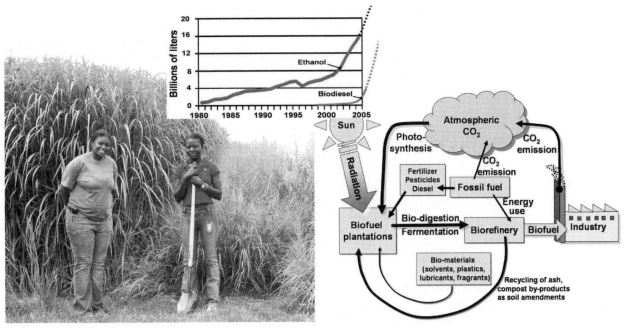

FIGURE 20.16 Soil resources are used to grow plant biomass for production of such biofuels as ethanol and biodiesel. Burning biofuels may be carbon-neutral, but much of the energy value of the biofuels produced is offset by the fossil fuel energy inputs used to grow the plants and run the biorefineries (*right*). The fossil fuel inputs for the biofuel plantations are mainly in the form of nitrogen fertilizer, pesticides and tractor fuel, the use of which can be minimized by ecological farming practices. Production of bio-fuels in the U.S. (*graph*) is expected to increase exponentially between 2005 and 2010 (dotted lines are projected trends) so that their contribution to the overall energy use in the U.S. may rise from the current 6% to as high as 20%. The photo (left) shows research assistants in front of University of Illinois plots in which two grass species are being studied as sources of biomass for cellulosic fermentation to ethanol. The taller species is *Miscanthus gianteous* and the shorter is *Panicum virgatum* (switchgrass). The goal is to use soil, water and nutrient resources to most efficiently produce maximum biomass. Such grasses present several advantages over corn grown for this purpose, including 1) a dense root system that should protect soil quality and very effectively utilize soil nitrogen and 2) a dense growth habit that may preclude the need for herbicides. It is important to remember that biofuel production can be sustainable only if adequate biomass is left in the field to maintain soil quality. [Production data from USDA (2006); diagram redrawn from Lal (2006) copyright Society of Chemical Industry and reproduced with permission granted by John Wiley & Sons, Ltd. on behalf of SCI; photo courtesy of R. Weil]

More on oil and food:
http://www.earthpolicy.org/
Updates/2005/Update48.htm

energy source. That is, the CO_2 emitted in burning the biofuel is balanced by the CO_2 just recently removed in growing the biomass. However, most studies show that large amounts of fossil fuel energy must be used to grow the biomass and run the biofuel refineries. In fact, the energy input into the biofuels industry may nearly equal—or even exceed—the energy output in the form of usable biofuels (Figure 20.16, *right*). Therefore, even with the recent technological advances, growth in the biofuels industry is—and will probably remain—quite dependent on large government subsidies.

The environmental, as well as economic, sustainability of biofuels must be carefully considered. Engineers and system designers have suggested plans to harvest as much fermentable biomass as possible from each ha of land. They have even called for plant breeders to create plant varieties that funnel larger amounts of energy to harvestable shoot biomass and devote less to root production. These suggestions run counter to the requirements of soil quality. It is essential, therefore, that soil scientists become vocal in reminding planners that biofuel production can be sustainable only if the crops are sustainably produced. It is also essential that an adequate portion of the biomass is left as surface residue in the field to protect against erosion and maintain soil quality (see Table 12.6). This is one reason why perennial grasses (Figure 20.16, *left*)—especially mixed species stands—may be a more sustainable option than corn residues as a biofuels feedstock.

Nutrient Management

Nutrient management that applies the principles outlined in Chapter 16 will continue to be essential for maintenance and improvement of soil quality and protection of the environment. Using the soil food web and chemical buffering processes to supply the right balance and amounts of nutrients for optimal plant growth and quality remains the central focus. Such management must significantly reduce the rates of nutrient application in parts of Europe, the United States, and Asia where nutrient additions currently far exceed plant uptake. In other areas, such as sub-Saharan Africa additional nutrients will need to be supplied. Increasing attention will have to be paid to micronutrients in some regions. In all regions, greater attention needs to be given to improving the efficiency of nutrient use, and to more effective recycling of all nutrients in the agroecosystem.

Combining Organic and Inorganic Sources

Sustainable nutrient management module: http://www.montana.edu/wwwpb/pubs/mt444915.pdf

Research has repeatedly shown that combinations of inorganic fertilizers and organic manures or residues from a variety of crops in the rotation can maintain (or even increase) yields and at the same time improve soil quality. This is due in part to enhanced nutrient cycling and to the influence of diversified crop residues on both the quality of soil organic matter and the microbial biomass. Figure 20.17 provides some insights into how the combination of organic and inorganic nutrient sources can employ the microbial biomass to enhance nutrient availability and plant growth.

In conventional systems with only one or two crops in the rotation, the level of the *active pool* of soil organic matter is usually low. In more diversified systems with several crops in the rotation, and with somewhat lower rates of fertilizer, much higher levels of the active organic matter fraction have been found. The soil in such high-crop-diversity systems usually also has higher levels of microbial biomass N and higher rates of N mineralization. The interaction among the diversity of crops being grown, the quality of soil organic matter, and the microbial numbers and their ability to release nitrogen and other nutrients for plant uptake are illustrated in Table 20.8.

Crop Rotations

Crop rotations provide numerous advantages over the monoculture systems often favored in intensified agriculture. Yields of two crops are higher if they alternate with each other on a given field, or if they are grown together simultaneously as an *intercrop*. If legumes are included in the rotation, they provide additional nitrogen to the system, and improve soil quality by enhancing the incorporation of crop residues into soil organic matter (Figure 20.18). Forage crops encourage animal production on a sustainable, decentralized basis and assure the availability of animal manures that supply nutrients and long-lasting organic matter benefits. Close-growing crops in the rotation also help control soil erosion, and, with the help of crop residues, markedly increase the soil's water-infiltration capacity. Even growers of such high-value crops as vegetables and ornamentals are beginning to see that they can profit more and do a better job of soil quality management if they practice crop rotations that use one-third or less of their land for the high-value (usually soil-depleting) crops while growing lower-value but better soil-building crops (such as hay or grains) on the remaining land. If fall or winter cover

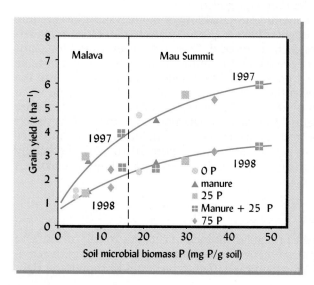

FIGURE 20.17 The relationship between soil microbial biomass P and maize grain yield in response to fertilization with inorganic P and manure and the combination of the two. The soil amendments tested were (•) a no P added control, (▲) 1.9 Mg/ha of local cattle manure (containing 5 kg P/ha), (•) 25 kg/ha of commercial fertilizer P, (▪) the combination of manure + 25 kg/ha fertilizer P, and (♦) 75 kg/ha of fertilizer P. The latter amendment was meant to partially saturate the soil P-fixing sites and provide P for several years of cropping. The combination amendment (▪) was designed to encourage the fertilizer P to cycle via the microbial biomass into organic forms, and thereby avoid fixation on the colloid surfaces. The experiment was repeated at two sites on village farms in Kenya. At Malava the soils were acid (pH 4.5) sandy clay Kandiudults with a high P-fixing capacity. At Mau Summit the soils were more fertile, less acid (pH 5.5) silty clay Melanudands with lower P-fixing capacity. The treatments were applied in 1997 and 1998 and corn was grown each year for grain. All plots also received 100 kg inorganic N ha^{-1}. Note the close relationship of grain yield to soil microbial biomass P across both sites. The generally higher yields in 1997 were due to better rainfall that year. [From Ayaga et al. (2006)]

TABLE 20.8 Effects of Conservation Practices on Some Soil Quality Factors Related to Organic Matter

In each region, soil was analyzed from six pairs of adjacent fields, one on which conservation practices (reduced tillage, greater crop diversity, more sod crops in rotation, and/or use of organic nutrient sources) were used, while conventional practices (more tillage, less diversity, etc.) were used on the other.

Properties	Coastal plain soils		Piedmont soils	
	Conservation management	Conventional management	Conservation management	Conventional management
Total organic C, g/kg	12.5	8.3	19.6	15.5
Active organic C,[a] mg/kg	121	75	134	112
Microbial biomass C, % of total organic	2.4	1.3	2.6	2.3
Nitrogen mineralization rate constant[b]	38	33	42	36
Aggregate stability, %	73	58	74	66
Specific maintenance respiration[c] (qCO_2), mg CO_2 g microbial biomass C^{-1} day^{-1}	41	72	18	32

[a] Mainly sugars extracted from soils after disruption with microwaves.
[b] The rate constant k, day^{-1}, in the first-order decay equation $N_p = N_o e^{kt}$.
[c] Higher numbers indicate ecosystem stress and more energy expended just to survive.
Data from Islam and Weil (2000).

crops are included (Section 16.2), end-of-season nutrients left by the main crop can be saved, water contamination can be averted, and additional residues are added to the soil. Cropping system studies have clearly established that rotations involving close-growing crops, including legumes, are well suited to maintain good soil quality.

Water Management

Scarcity of water is likely to challenge our ability to meet future food demands. Greater efficiency in using both precipitation and irrigation water must be achieved. Diversified crop rotations that include close-growing crops, along with conservation tillage, can

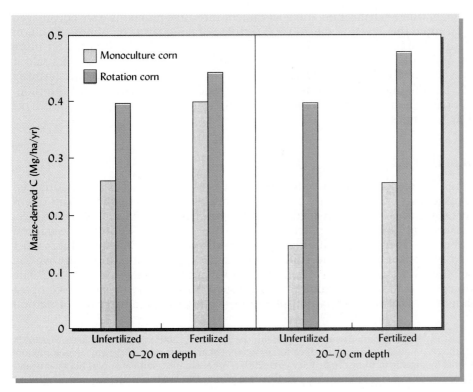

FIGURE 20.18 Growing corn in a legume-based rotation rather than in monoculture enhanced soil quality as evidenced by the amount of soil carbon that was derived from the corn residues. The benefit of using a rotation was especially evident in the subsoil (20 to 70 cm deep). These data are from a long-term (35 year) experiment in Ontario, Canada, on a Humaquept (Humic Gleysol, Canadian system). [Drawn from data in Gregorich et al. (2001)]

provide greater soil cover, increased water penetration, and decreased water runoff and soil erosion. This leads to more efficient water use for plant productivity. Likewise, such irrigation management systems as drip irrigation that reduce the quantity of water needed to produce plants should be more widely used. Improved drainage systems serving irrigated areas must reduce the buildup of salts, thereby maintaining the quality of the soil (Sections 10.7–10.9).

Biological Diversity

Intensified cropping systems that involve a diversity of crops in the rotation, especially if both legumes and nonlegumes are included, have been shown to increase the biological activity and diversity in the soil. Not only will there be more microorganisms, but the diversity of organic food for them will encourage greater microbial diversity. Likewise, the cycling of nutrients among organic and inorganic forms requires a diversity of soil organisms. Such diversity will enhance essential physical, chemical, and biological processes on which soil quality depends.

20.9 ORGANIC FARMING SYSTEMS[4]

Canadian organic-conventional farming comparision—Long term results:
http://www.umanitoba.ca/faculties/afs/plant_science/glenlea/glenlearesresults.html

There is a growing movement to view the farm as a living system whose many component parts (plants, animals, microbes, soils, people, etc.) are integrated in an *organic* manner, that is, they act in concert as a whole organism. Hence, the term *"organic"* in **organic farming**. Although organic farming systems are quite varied, they share a central focus on soil quality, as well as on the avoidance of synthetic chemicals (whether or not these chemicals are "organic" in the chemistry sense pertaining to carbon in their makeup). The trend toward organic farming has been stimulated largely by the concerns of food consumers about the safety and nutritional quality of their food, as well as about the environmental and social impacts of its production. To assure these consumers that they are getting what they pay extra for, official organic farming certification and inspection programs have been developed in many countries, including a program run by the U.S. Department of Agriculture. Still, the many scientific studies that have compared organically and conventionally produced foods have been able to document little consistent difference in nutritional quality or safety except that the former generally contains much lower (but not zero) levels of pesticide residues.

Western Canada organic-conventional cropping systems comparison. Click on "Results":
http://www.umanitoba.ca/faculties/afs/plant_science/glenlea/glenlea.html

In Europe, where certification and government incentive programs have encouraged organic farming, there were some 200,000 certified organic farms on almost 7 million ha of certified organic farmland by 2005. In the U.S., almost 9000 farms on more than 1.6 million ha were certified organic by 2005. A generation ago, skeptics said it would be "impossible" to grow food on a commercial scale using organic farming methods, but the organic food now common in modern supermarkets and the thousands of organic farms that produce reasonably high yields and make good profits have proved such skeptics wrong.

Organic Farming Philosophy and Practice

Organic farming effects on productivity and soil quality:
http://www.isofar.org/sections/wg-long-term-experiments.html

At its core, organic farming is a system that strives to integrate ecologically-oriented good management practices, many of which are discussed in this textbook. This orientation is expressed in the USDA regulation that governs organic farming in the U.S., which states that "Production practices . . . must maintain or improve the natural resources of the operation, including soil and water quality. . . . The producer must select and implement tillage and cultivation practices that maintain or improve the physical, chemical, and biological condition of soil and minimize soil erosion." The regulation goes on to spell out cultural practices such as rotations, cover crops, compost and manure application, companion plantings, timing of operations, use of natural

---

[4] For the history of the organic farming movement, see Heckman (2006) and (2007). See USDA/AMS (2006) for the USDA organic farming certification program. For an assessment of the global food production potential by organic farming see Badgley et al (2007). For a scientific comparison of the quality and safety of organically and conventionally grown food, see Winter and Davis (2007).

http://www.rodaleinstitute.org/science/home.html

enemies to manage pests, selection of crop species and varieties that are favored for the management of plant nutrition, weeds and insects. Only when such approaches are insufficient does the regulation provide for the use of certain approved, naturally-occuring chemicals. Unfortunately, regulators, consumers and farmers too often pay more attention to the list of allowed and forbidden materials than to the broader principles of ecological farming.

No hormones, synthetic drugs or antibiotics feed additives (see Box 11.2) are allowed in raising organic livestock. The feed for organic livestock must be from organically grown plants. Many official certifiers call for the seeds planted to be from organically grown seed crops and for the manures spread to be from organically raised livestock. The organic farming *philosophy* prohibits the use of synthetic chemicals, including most commercial fertilizers and pesticides. While the potential for adverse effects on both people and non-target organisms by the use of hormones or pesticides is well known, the *prohibition* (as opposed to the *de-emphasis*) of processed mineral fertilizers appears to have little basis in science. The aversion to mineral fertilizers in organic farming may have originated because of a belief in the "humus theory" of plant nutrition (Section 12.5) and also because of concerns that farmers were adopting cheap, easy-to-use mineral fertilizers *instead of* traditional nutrient sources such as manure and compost, rather than *in conjunction* with these organic materials. The anti-fertilizers strictures may also have grown out of bad experiences in the early days of fertilizers, when sodium nitrate was a popular nitrogen source. With heavy use, the sodium in this fertilizer can cause deterioration of soil structure (see Section 10.6), accounting for the description of soils so treated as "hard and lifeless." Ironically, naturally occuring (but equally damaging) sodium nitrate materials are among the few soluble mineral fertilizers allowed by organic certification programs.

Organic and conventional farming systems compared since 1978 in Switzerland: http://www.fibl.org/english/research/soil-sciences/dok/index.php

To maintain soil fertility, organic farmers (who do not speak of fertilizing crops, but rather of "feeding the soil") utilize organic materials and naturally-occuring, unprocessed mineral sources of plant nutrients. While animal manures are commonly used on both organic and non-organic farms, on the certified organic farms their use is closely regulated to avoid both over application of nutrients and problems with human pathogens contaminating food products. For example, organic certification prohibits application of raw manure to fields that have vegetables growing in them. Organic farms also may use the nutrients in such sources as crop residues (particularly of legumes), and powdered rocks (such as phosphate rock and greensand).

Organic versus Sustainable

A growing trend—small, local and organic: http://www.washingtonpost.com/wp-dyn/content/article/2006/11/05/AR2006110500887_pf.html

Researchers have conducted many studies in which organically managed farms or plots are compared to those that are conventionally managed. In some cases, an organic farm has been observed to produce declining yields, deplete soil nutrients or even allow greater nutrient leaching than its conventional counterpart. However, other studies show the opposite: on farms that convert to organic management yields increase over time, long-term soil fertility is maintained and soil quality is enhanced.

As with conventional farms, some organic farms perform better than others. It is difficult to objectively compare whole farming systems because they differ in so many inputs, outputs and processes. For example, a conventional 500 ha grain farm may sell just two crops, say corn and soybeans. A neighboring 500 ha organic farm would be likely to grow wheat, barely, oats, alfalfa/grass hay, beef cattle and sheep in addition to some corn and soybeans. Even when more limited comparisons are made on research station plots, objectivity can be compromised if the managers of the research farm are themselves more experienced and skillful in managing one system or the other. With these limitations in mind, most research studies comparing systems with similar tillage practices show that organic farms produce far less pollution and adverse environmental effects than do their conventional counterparts. Since no feed additives, antibiotics, or synthetic pesticides are used in the organic systems, the difference for environmental protection is most obvious with regard to residues of such chemicals on produce and in runoff and groundwater. In some comparisons, organic farming also reduces erosion, nutrient leaching and runoff. Organic farming usually has positive effects on soil quality, as shown in Figure 20.19 and Table 20.9.

In research plot comparisons, organically grown crops often yield 10 to 20% less than those grown conventionally. However, studies comparing commercial farms usually show that farmers well experienced in using organic or conventional methods generally produce similar yields per hectare of most crops. Moreover, the organic farmers often

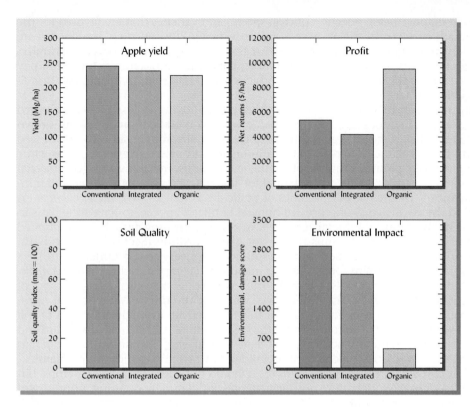

FIGURE 20.19 Sustainability of three apple production systems in Washington State. Cumulative apple yields during the six-year study did not differ among systems. However, the soil quality index (determined as in Box 20.1) was higher in the integrated and organic systems, which received organic amendments, compared to the conventional system, in which fertility was maintained by mineral fertilizers only. The profit in the final study year (which saw full-quality production for all systems) was much higher for the organic system because the price received for certified organic apples was higher and the yields were good. The researchers reported that a price premium of about 12% would be needed to allow the organic growers to recoup their extra costs. This price premium may be seen as a way that the markets pay organic producers for protecting the quality of soil and water resources. The environmental impact of the conventional apple orchard was potentially far more damaging than that of the other two systems, even though the conventional system in this study used all recommended practices and chemicals. [Data selected from tables and graphs in Reganold et al. (2001)].

make a greater profit. The enhanced profit is sometimes due to lower costs for such inputs as fertilizers and pesticides, but more often the greater profits result from premium prices that consumers are currently willing to pay for organically produced food (Figure 20.19).

However, it should be understood that *organic* does not always mean *sustainable*. For example, organic farms that rely on off-farm feed for their animals or use off-farm manure and compost for soil fertility may be challenged to avoid the nutrient pollution described in Sections 16.1 and 16.2. Perhaps a more serious limitation is that organic farmers cannot easily adopt conservation tillage practices that minimize soil disturbance and improve soil quality (Section 17.6), because such practices depend on prohibited herbicides for weed control. Therefore, some organic farms suffer more soil erosion and degradation than farms that use the best available no-till and other conservation tillage practices. To overcome this problem, researchers are attempting to develop high-residue "organic no-till" cropping systems for use on erodible soils. One approach is to use special roller-crimper machines to kill tall cover crops without herbicides. No-till equipment then plants into the rolled cover crop residues, which are meant to form a heavy mulch that controls weeds without herbicides.

TABLE 20.9 **Effects of Conventional and Organic-Based Cropping Systems on Selected Soil Quality Indicators at Three Locations in Nebraska**

The conventional systems involved continuous corn or corn-soybean rotation with inorganic fertilizers. The organic systems involved 4–6 crops in rotations that included 2–5 years of alfalfa, and crop residues and manure (from livestock fed the alfalfa) as the main sources of nutrients. The organic systems generally resulted in higher indicators of soil quality.

Soil Quality Indicator	Giltner NE (Udic Argiustoll)		Valley NE (Udic Argiustoll)		Deweese NE (Typic Haplaquoll)	
	Conv.	Organic	Conv.	Organic	Conv.	Organic
Bulk density, Mg/m³	1.34	1.26	1.35	1.08	1.22	1.18
Water-holding capacity, m³/m³	0.15	0.17	0.15	0.21	0.18	0.20
pH	5.93	6.25	4.96	6.46	6.12	6.59
Organic C, Mg/ha	53.4	55.6	42.34	57.92	55.57	70.28
Microbial biomass C, kg/ha	651	816	464	775	654	1192

Selected data from Liebig and Doran (1999).

Another sustainability problem is the negative nutrient budget (Section 16.1) maintained by some organic farms, especially with respect to phosphorus. Some organic farms that were established on land that had previously accumulated high levels of phosphorus, have been able to maintain high yields for years or even decades despite the fact that they export more P in products sold than they import in phosphate rock or other materials. However, eventually this imbalance results in P deficient soil that limits productivity. Furthermore, because available supplies of phosphorus and other nutrients are finite (Section 16.4), we must eventually learn to achieve a high level of nutrient recycling in our entire food system, whether or not it is managed with organic farming methods.

20.10 SUSTAINABLE AGRICULTURE SYSTEMS FOR RESOURCE-POOR FARMERS

Hundreds of millions of poor people in developing countries practice low-yield subsistence farming by which—in a *good* year—they barely eke out sufficient food to keep their families alive. Most have little or no extra production that they can sell to earn the cash that could enable them to pay school fees for their children or purchase fertilizers, improved seeds, or livestock to enhance their farm. Many are shifting cultivators (Section 20.7) trapped in the downward cycle of soil degradation (see Figure 17.2) because they no longer have enough land to allow the fallow periods needed for natural soil rejuvenation. Some who can afford small amounts of fertilizer use it to grow cash crops like cotton or coffee that will repay the cost of the fertilizer and yield a small cash profit, but such crops will not feed their families. Typically, to produce enough calories to stave off hunger, these farmers devote nearly all their land to the production of a starchy staple crop such as corn, sorghum, cassava or rice. This leaves little land for more protein-rich pulses such as beans, cowpeas, lentils, or pigeon peas. Commonly, women may tend a small kitchen garden area immediately next to the house where leafy vegetables, root crops, herbs, and sometimes pulses are grown to provide some balance and variety to the family diet. Innovative systems that are more productive and sustainable but that require very little cash investment are needed to lift the lives of hundreds of millions of people trapped in such circumstances around the world, and especially in Africa.

To illustrate how food production, income generation, and soil quality might all be enhanced in currently low-yield, resource-poor areas, we will focus on the situation in sub-Saharan Africa. However, the principles discussed here also apply to low-productivity agricultural systems in many other tropical areas.

Improving Soil Quality in Sub-Saharan Africa[5]

About 55% of the land area of the African continent is considered unsuitable for any kind of agriculture except nomadic grazing, yet some 30% of the continent's people eke out an existence in these areas. Soils of low quality are found on an additional 16% of the land where 23% of the people live. These soils are limited by such constraints as extreme acidity, impermeable layers in the subsoil, and accumulation of salts. Fortunately, medium- to high-quality soils are found on about 30% of the land where 47% of the people live. These soils are free of major constraints and are found in regions where rainfall is stable and sufficient for at least one crop a year. They offer great potential for increased production of food and fiber if sustainable, intensified systems can be utilized.

Nutrient Balances and Fertilizers[6]

Solutions to Africa's food woes remain elusive: http://www.npr.org/templates/story/story.php?storyId =6226828

As mentioned in Section 20.5, the high-yield, intensive agriculture of the Green Revolution increased food production and thereby the nutritional status of people in most regions of the world, except sub-Saharan Africa (Figure 20.20). The reasons that

---

[5] Our use of the term *sub-Saharan Africa* does not include South Africa, which has a large, modern agricultural sector. For background on soil quality and food production systems in Africa, see Buresh and Sanchez (1997) and Eswaran et al. (1997).

[6] For a summary of the need for replenishing Africa's soils to alleviate poverty, see Sanchez and Swaminathan (2005).

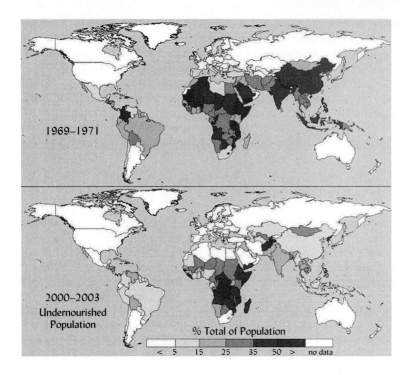

FIGURE 20.20 The percent of people in various countries suffering from too little to eat (undernutrition) around 1970 and 2002. The Green Revolution of the 1970s and the Asian economic boom of the 1990s greatly reduced hunger in Asia and South America (fewer dark-shaded countries). However, in sub-Saharan Africa the proportion of the population suffering from undernutrition has actually increased during the past three decades. [Adapted from FAO (2006a)]

Green Revolution technologies failed to take hold in Africa are complex (including lack of effective political support, poor infrastructure, and difficult climatic and soil conditions), but an important factor was the lack of effective programs to encourage the use of fertilizer (Figure 20.21). In most of sub-Saharan Africa, as in many of the more remote, marginal lands of Asia and South America, a primary problem is deficiency of plant nutrients. Soils are simply being mined of plant nutrients, because removal in harvested crops and by leaching far exceeds the amounts being returned from all sources. Nutrient input/output studies on farm lands of sub-Saharan Africa show an alarming *negative balance*, suggesting a decline in soil quality (Table 20.10). Fertilizer applications average less than 10 kg/ha, and in many cases are zero. Overall each year, cultivated lands in sub-Saharan Africa are losing an estimated 6.1 million Mg N, 0.74 million Mg P, and 4.6 million Mg K. During the last 30 years this has resulted in a cumulative loss of some 660 kg N, 75 kg P, and 450 kg K from the average ha of cultivated land in sub-Saharan Africa.

Soil acidity, deficiency of nutrients, and other soil constraints are prime reasons for low productivity in Africa and many other regions as well (Table 20.11). The quality of soils is declining annually, and this decline will continue if improvements cannot be

Is fertilizer a public or private good in Africa?
http://web.africa.ufl.edu/asq/v6/v6i1a13.htm#_edn2

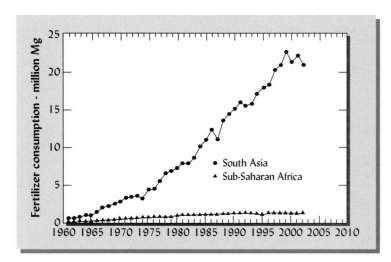

FIGURE 20.21 Changes in fertilizer consumption in South Asia (India, Bangladesh, and Pakistan) compared to that in sub-Saharan Africa. The rapid rise in the use of fertilizer in South Asia reflects the influence of the Green Revolution that took hold beginning in the late 1960s and solved that region's overall food shortage problem. Unfortunately, no such progress was made in sub-Saharan Africa. [Data from FAO (2006b)]

African fertilizer use and marketing references: http://www.aec.msu.edu/agecon/fs2/inputs/index.htm

TABLE 20.10 Average Annual N, P, and K Balances (Inputs Minus Outputs) for the Arable Land in Several African Countries

	Balance, kg/ha/yr		
Country	Nitrogen (N)	Phosphorus (P)	Potassium (K)
Malawi	−67	−10	−48
Ethiopia	−47	−7	−32
Senegal	−16	−2	−14
Average of 13 African countries	−39	−5	−30

From Stoorvogel et al. (1993).

made in soil management. Farming systems that improve nutrient cycling, maximize biological N fixation, and protect soils from erosion are an important part of the solution. However, in many areas of the tropics, the addition of nutrients from external inputs (mainly fertilizers) will be needed to jump-start severely nutrient-depleted systems.

Unfortunately, even though West Africa has natural gas resources (needed to make nitrogen fertilizers) and a number of African countries have significant phosphate rock deposits (the raw material for phosphate fertilizers), the development of African fertilizer manufacture capacity lags far behind the need. Because of the high costs of importing fertilizers from distant factories in Europe and North America, and because of the poor transportation infrastructure (many roads are too muddy for trucks to pass during the rainy season when crops are planted), the price of fertilizer in Africa is double or triple the world price by the time it is delivered to farmers. Furthermore, the fertilizer often arrives too late for use on that season's crop. All this discourages farmers from making nutrient additions, especially in remote areas such as those where shifting agriculture is practiced. Clearly there is a need for developing an indigenous fertilizer manufacturing industry and delivery infrastructure, as well as for innovative practices that combine organic and inorganic nutrient sources to help stretch fertilizer supplies, to increase nutrient inputs, and to cycle the nutrients once they reach the soil.

Improved Internal Cycling of Nutrients

The need to nurture sub-Saharan Africa's soil: http://www.ifpri.org/2020/newslet/nv_0702/nv_0702a.htm

While nutrient losses from the whole farm or agroecosystem are very important, nutrient transfers *within* the farm or agroecosystem can also have critical impacts (Figure 20.22). Studies of African village farming systems show that nutrients tend to move out of the common grazing lands, forested fallows, and main crop fields (which are usually some distance from the farmers' houses) toward the house compounds in the form of firewood brought home for cooking fires, residues left from grain crops brought to the home for threshing, crop residues and tall grasses brought home for thatching roofs, manure from livestock confined near the home compound overnight, food preparation wastes, and of course, human wastes. In contrast there are almost no transfers going in the opposite direction to replenish nutrients in the source areas. As a result, the soil immediately surrounding the house is usually highly enriched in nutrients.

Many villagers take advantage of this island of fertility by growing a productive kitchen garden. However, this enriched area is very small and the soil in it is so overenriched in certain nutrients that adding more is just a waste. A few farmers attempt to move some nutrients from the home to the main outlying fields by scraping up the manure from the animal confinement area and carrying it (in head baskets or in ox

TABLE 20.11 Percentage of Soil Areas in Five Agroecological Regions of the Tropics Having Different Chemical Constraints to Agricultural Plant Growth

Soil constraint	Humid tropics, %	Acid savannas, %	Semiarid tropics, %	Tropical steeplands, %	Tropical wetlands, %	Total, %
Acidity with aluminum toxicity	56	50	13	29	4	32
Acidity without Al toxicity	18	50	29	16	29	25
High P fixation by Fe oxides	37	32	9	20	0	22
Low CEC	11	4	6	—	—	5
Salinity	1	0	2	0	7	1

From Sanchez and Logan (1992).

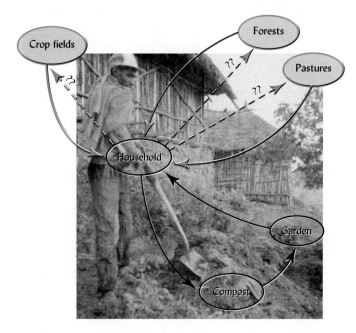

FIGURE 20.22 Recycling of nutrients is an essential component of sustainable farming. It is especially critical in developing countries for poor farmers who cannot afford to import enough fertilizer to their farms to replace all the nutrients removed from their soils by cropping, grazing, and forest harvest. The background photo shows a village in Wolayta, Ethiopia, where by local traditions people compost household wastes and reuse the nutrients in their kitchen gardens. However, most of the nutrients in the household originally came from outlying cropland (via harvested food and crop residues), forests (via firewood collected for cooking), and pastures (via manure from cattle kept overnight in the home compound). Over time, the tiny home gardens flourish while soil in the more distant croplands, pastures, and forests become depleted and degraded. (Diagram and photo courtesy of R. Weil)

carts) out to the field. However, it must be remembered that cattle manure in general has low concentrations of nutrients, especially phosphorus. Manure from African cattle grazing a low-quality diet of P-deficient forages and crop residues typically contains far lower concentrations of P and other nutrients compared to textbook values for cattle manure generated in the feedlots of developed countries. This means that many bulky, heavy loads of manure are needed to supply a modest quantity of nutrients. Moving the manure is so laborious that, in reality, very little manure ever gets transferred to the depleted outlying fields, and little or none to the depleted pastures areas where the bulk of the nutrients probably originated.

Composting the manure along with any excess crop residues has the potential to reduce the mass that needs to be carried back to the outlying fields, but is also laborious and only rarely practiced. An important but generally overlooked pool of nutrients indigenous to resource-poor village agroecosystems is the ash from cooking fires. Since it is very lightweight and relatively concentrated in nutrients, enough ash to replenish at least some of the depleted soils could easily be transported from the home to outlying fields (see Box 20.2).

Crop Residues and Livestock in Farming Systems

Management of organic inputs in soils of the tropics: http://www.plantpath.cornell.edu/mba_project/moist/home2.html

The dry stalks and husks (crop residues) left over after the removal of crop harvests might seem quite valueless, and indeed are often referred to as "trash" or "wastes." But in reality, these materials constitute the *plant litter* of the agroecosystem, and it is critically important that they be managed so they can help enhance soil organic matter, nutrient cycling, and erosion control. Just as they do in natural ecosystems, large herbivore animals can also play critical roles in agroecosystems, by recycling nutrients and organic matter in the very beneficial form of animal manure and by encouraging land uses that include stands of grasses and legumes, which also help build soil quality. In developing countries, livestock (oxen, donkeys, horses, and buffalo) also provide traction power—playing the roles that tractors do in developed-country agriculture. For animal agriculture to be integrated with crop production in a manner that builds rather than degrades soil quality and farm productivity there must be a balanced relationship among such major system components as livestock grazing, animal traction, tillage intensity, pasture grazing, forage crops, food crop residues, animal manures, and fuel wood trees (see Figure 20.24).

BOX 20.2 NUTRIENTS TO BURN? AN OVERLOOKED SOURCE
OF FREE FERTILIZER

Wood cut from trees, bundled crop residues, and cakes of dried animal manure are the principal sources of energy for cooking (and heating) in sub-Saharan Africa and other developing areas (as illustrated for Ethiopia in Table 20.12). Most of this biomass fuel is burned in open cooking fires that produce food flavored by smoky fragrances, as well as warmth and light to sit around in the evening. But the open fires are notoriously inefficient and require many hours of labor—usually by the women and girls of a village—to gather the fuel from distant forests or fields and carry it (often for several kilometers) to the home. Except for N, most of the nutrients in the firewood end up in the ashes.

By early morning, when the fire has burned out, the women can be seen sweeping the now cool ashes just beyond the living compound into the kitchen garden where the soil is already overenriched from countless previous ash additions. In some cases, ashes may be used to reduce odors in the family pit latrine. But it is very rare to find a villager who collects the ashes in a place where they are protected from rain and then spreads them evenly over the most distant fields where soil fertility is the lowest.

What is the potential significance of this nutrient resource? The statistics for Ethiopia are representative. Rural Ethiopians annually burn an average of about 0.52, 0.41, and 0.21 Mg dry weight of wood, crop residue, and cattle manure (for a total of 1.14 Mg dry biomass used as fuel per person). Ethiopia has about 1/6 ha arable land per person in the farming population. Thus, for each hectare of cropland there are 6 people burning approximately 6840 kg (6 × 1.14) of total dry biomass per year.

Depending somewhat on the tree species, animal feed, and fertility of cropland, this dry matter will produce ashes weighing about 2% of the initial dry mass, or about 137 kg (0.02 x 6840 kg) of ash per 6 people or per ha of available cropland. Using the nutrient concentrations from Figure 20.23, for every ha of available cropland this ash would contain 1.3 kg P (137 kg ash × 0.0095 kg P/kg ash), 0.75 kg S, 6.7 kg K, and 13 kg Ca, plus small but significant quantities of micronutrients. If protected from rain (the nutrients in ash are water soluble) and spread where most needed on 1/5 of the cropland, the application rates would be 5 times the rates just cited, or 6.5 kg P/ha, 3.8 kg S/ha, 33.6 kg K/ha, and 65 kg Ca/ha. This rate of ash application would supply sufficient nutrients to at least arrest the acidification process and replace nutrients removed by crop harvest and leaching (compare to negative balances given in Table 20.10).

The use of ash might produce substantial gains in crop productivity as suggested by the responses of faba bean and wheat shown in the bar graph (Figure 20.23). The ash might be best utilized first on leguminous crops or trees, in order to leverage the P, S, K, and Ca fertility into added nitrogen via enhanced biological N fixation. The value of the ashes might be further enhanced if modest amounts of N and P fertilizers could also be applied (but mixing the high pH ashes with these fertilizers should be avoided! See Sections 13.6 and 14.7). Another option to supply N would be the collection of human urine, which could supply as much as 60 to 90 kg N/ha on 1/5 of the available cropland each year.

In this example, the nutrients do not represent *additions* to the agroecosystem. Rather, the use of ashes represents mainly a *transfer* of nutrients within the agroecosystem: to the croplands from woodlots, hedges, and common grazing areas. The fields farthest from the house are usually the most depleted in all nutrients and so they are the fields where the greatest response to ash application might be expected. The cost to the farmer of utilizing this currently wasted resource would be very low in terms of both cash and labor, so even small, short-term crop responses could make the practice attractive.

TABLE 20.12 National and Per-Capita Use of Organic Materials as Fuel for Cooking in Ethiopia

Burning this biomass will produce ashes weighing about 2% of the initial dry mass.

Type of fuel	$Mg \times 10^6$	Mg/person[a]
Wood	21.4	0.52
Manure (Dung)	15.4	0.41
Crop residues	9.6	0.21
Total	46.4	1.14

[a] National fuel use / national population in 1990.
Fuel demand estimates for 1990 from Girma, (2001)

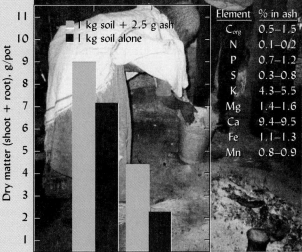

FIGURE 20.23 Nutrient concentrations (ranges upper right) found in ashes from several villages in the Ethiopian highlands and responses to the ash by faba bean and wheat plants (graph) grown in an acid (pH 5.5) soil from a degraded outlying crop field. In the background an Oromo woman collects ashes from her cooking fire to give to researchers. (Courtesy of R. Weil and Asgellil Dibabe)

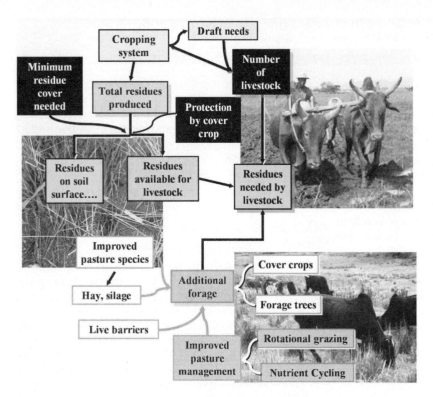

FIGURE 20.24 For most farmers in Africa, livestock are an integral part of farming. In many areas oxen are used for draft power in plowing the soil to prepare the fields for sowing seed. A cropping system that uses large amounts of tillage will require many oxen (and breeder cattle), which eat large amounts of fodder, often in the form of crop residues, thus depriving the soil of the vegetative cover it needs to counter erosion and organic matter depletion. A reduced-tillage system could reduce the number of oxen required for tillage and therefore allow more crop residue to be left as mulch to prevent erosion and build soil quality. More livestock feed from improved management of pasture areas could also help support the cattle without removal of crop residues from the soil. The situation is often complicated by additional competing demands for crop residues as fuel and as material for roofing and fencing. [Many of these concepts are reviewed in FAO (2000); diagram and photos courtesy of R. Weil]

For example, if a farming system involves large amounts of tilled land with many sequential tillage passes performed each year, farmers will desire a relatively large ratio of draft animals to land area. The draft animals (and associated livestock for breeding replacements), require large amounts of fodder, often in the form of crop residues. Farmers either collect the crop residues for use as dry season livestock feed, or they allow the livestock to graze the residues in the fields after harvest (the latter practice at least returns some nutrients and organic matter as manure dropped during grazing). In either case, the soil is deprived of the residue cover it needs to counter erosion and organic matter depletion.

Introduction of a reduced-tillage farming system might contribute to the solution from the demand side by reducing the number of oxen required for tillage and therefore allowing more crop residue to be left as mulch to prevent erosion and build soil quality. From the supply side, improved pastures could provide much more fodder than typically unmanaged and overgrazed common areas. Pasture productivity can be vastly increased by interseeding forage legumes in pastures and systematically rotating grazing animals from one area to another to allow forage plants in each area time to recover after periods of grazing. Pastures can be further enhanced by relatively small applications of nutrients, especially phosphorus, and on some soils, potassium. These nutrients are known to stimulate greater biological N fixation by pasture legumes, the fixed N in turn stimulating growth of both the legumes themselves and associated grasses. In these ways, the steeper, more marginal lands could support the needed livestock without having to sacrifice residue coverage on croplands. Additional competing demands for crop residues also exist—such as their use for cooking fuel and building material for thatched roofs and fencing. To meet these needs, fast-growing (preferably N fixing) trees can be planted in small woodlots or as field borders to provide the necessary fuel and building materials without harming—in fact often enhancing—soil quality. Such integration of trees into farming systems is known as *agroforestry*.

Agroforestry Systems

Agroforestry systems that integrate fast-growing, nitrogen-fixing woody species into small-scale farming systems show great promise in enhancing the supply and cycling of plant nutrients in the soil–plant system, while simultaneously increasing plant productivity and soil quality. We have already noted (see Section 13.8) the ability of deep-rooted,

TABLE 20.13   Yield of Corn in Kenya Following One or Two Years of Fallow in Which Various Species Were Grown

Three years of corn following 2 years of the nitrogen-fixing Sesbania tree produced more grain with less labor than the currently used systems (grass fallow and unfertilized corn).

Treatment	Maize grain yield, Mg/ha
2 years *Sesbania sesban*	5.4
Maize with fertilizer	4.0
1 year *Sesbania sesban*	3.4
Ground nut–maize rotation	1.9
Grass fallow	1.8
Maize without fertilizer	1.1

From Kwesiga et al. (1997).

fast-growing trees to recycle nitrates that have leached to depths of several meters in some tropical soils. The nitrogen so salvaged (often as much as 100 kg N/ha) can be taken up by food crops when the tree residues are spread on the soil and decompose. This retrieved nitrogen represents a real input to the food-producing system, since otherwise the nitrogen would have been lost in the drainage waters.

The growth of fast-growing, nitrogen-fixing trees during the fallow period after the food crops have been harvested is another source of nitrogen for the production systems in subhumid areas (Table 20.13 and Figure 20.25). Nitrogen-fixation rates are equivalent to those provided by leguminous forages in the developed countries. When the legume foliage is added to the soil and allowed to decompose, some of the nitrogen is released for plant uptake. Part of the N incorporated into labile forms of soil organic matter and into the microbial biomass will be available for later utilization by plants. Biological fixation is a significant source of nitrogen that must be more fully utilized.

FIGURE 20.25   Scientists and farmers have discovered the remarkably beneficial effects of including rapidly growing, nitrogen-fixing trees in rotations with food crops in Africa. (*Left*) The luxuriant growth of one such leguminous tree, *Sesbania sesban*, is shown in a farmer's field in Zambia. (*Right*) The very deep rooting system of this tree encourages the recycling of other nutrients as well as nitrogen. The remarkable yield response of corn following *Sesbania* is shown in Table 20.13. [Photo courtesy Dr. Steven Franzel, International Center for Research in Agroforestry; graph from Mekonnen et al. (1997), with permission of Kluwer Academic Publishers]

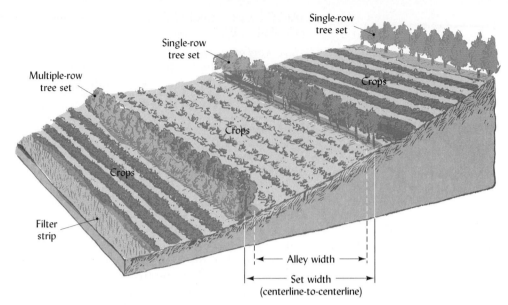

FIGURE 20.26 Alley cropping across the slope, showing alleys in which crops of various kinds are produced in between rows (single or multiple) of fast-growing shrubs or trees, preferably of the nitrogen-fixing variety. Before planting the cultivated crop, the trees can be trimmed and pruned back, and the pruned vegetation can be spread over the cropped area. This will provide organic matter, nutrients, and protection of the soil from erosion and will minimize shading by the trees during the growing season. Such systems have shown some promise where the rainfall is sufficient for both the planted crop and the border trees, and where there is enough labor for cutting and spreading the tree trimmings. [Modified from USDA (1997)]

Agroforestry Systems as Alternatives to Slash-and-Burn

There are several agroforestry systems that can provide viable alternatives to the destructive slash-and-burn methods in use today in Africa and elsewhere in the tropics. We will discuss two of these briefly.

MIXED TREE CROPS. First is the *mixed tree crop* system, in which the domesticated crop (often a tree species) is planted among existing native species (usually trees). In time, a number of the cultivated crop species may become dominant. The essential feature of the system, however, is that the soil will at no time be free of vegetation. Building on the experience of traditional systems, some scientists recommend the propagation of trees that produce high-valued products (fruits, nuts, and medicines) that can provide the producer with added income. The Kandy home gardens in Sri Lanka and the Telum forests of northeast Mexico are encouraging examples of successful systems developed by local people. The mixed tree crop system, which requires detailed knowledge of the species to be managed, offers economic returns while simultaneously assuring that soil quality will be maintained.

ALLEY CROPPING. An agroforestry system known as *alley cropping* involves growing food crops in alleys, the borders of which are formed by fast-growing trees or shrubs, usually legumes. The hedgerows are pruned regularly to prevent shading of the food crop, and

TABLE 20.14 The Influence of Conservation Tillage and Alley Cropping (with Leguminous Tree Hedgerows) on Runoff and Soil Erosion Losses in Nigeria

System	Runoff, % of rainfall		Soil erosion, Mg/ha	
	Corn	Cowpeas	Corn	Cowpeas
Plow-till, no hedgerows	17.0	4.3	4.3	0.63
No-till, no hedgerows	1.3	0.8	0.1	0.03
Leucaena Hedgerows	4.9	1.1	0.6	0.13
Gliricidia Hedgerows	4.3	0.7	0.6	0.07

From Lal (1989).

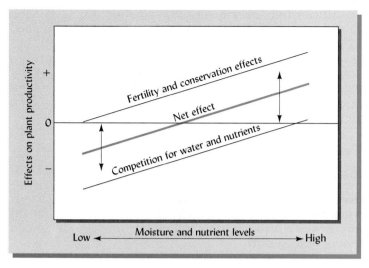

FIGURE 20.27 Benefits and constraints on plant productivity in alley cropping systems. The upper line illustrates the positive effects from nutrient cycling and soil conservation. The lower line shows the negative effects of competition from the border row species for water and nutrients. The heavy center line shows the net effect, which is generally positive in more humid areas with soils that have moderate plant nutrient levels. In drier, less-fertile areas, the competition for water and nutrients often results in a net reduction in crop yields.

the prunings are used as a mulch and a source of nutrients for the food crop. The mulch helps control weeds, prevents runoff and erosion, and reduces evaporation from the soil surface. Figure 20.26 illustrates how the system works.

In areas with ample rainfall and moderately fertile soils, alley cropping systems have proven to be successful and crop yields can be maintained without having to leave the land fallow for years. The soil is protected from excess erosion and soil quality is improved (Table 20.14). But in other, quite infertile subhumid to semiarid areas, the hedgerow species compete with the food crops for nutrients and moisture, and yields suffer. Figure 20.27 illustrates how competition reduces the effectiveness of alley cropping in some drier and more infertile regions. Thus, alley cropping may be beneficial in some cases, but is not a viable alternative to slash-and-burn in others.

Organic/Inorganic Combinations

In all parts of the world farmers and researchers are finding that by combining such organic materials as crop residues, agroforestry litter, animal wastes, and compost with modest levels of inorganic fertilizers, they can increase yields and simultaneously maintain or improve soil quality. The combination of organic materials and chemical fertilizers seems to provide a synergistic effect by which yield increases are greater from the combination than when the same inputs are supplied separately (Figure 20.28).

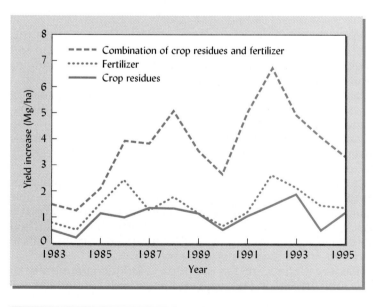

FIGURE 20.28 Increased yields of pearl millet from applications of crop residues and inorganic fertilizers supplied separately and in combination in 13 years of cropping in West Africa. The yield increases from fields receiving both organic and inorganic materials generally exceeded the sum of the yield increases from fields receiving the two materials separately. The beneficial effects of combined application of organic and inorganic materials were clearly demonstrated. [Calculated and redrawn from ICRISAT (1997)]

Research shows that most African soils have a very low potential for sustainable agricultural use as they are managed today. In most cases, the replenishment of phosphorus in African soils will have to come primarily from inorganic sources, with supplementation from biological sources. In contrast, nitrogen replenishment can likely come primarily from biological sources such as leguminous plants, with inorganic sources providing the supplementation. The synergistic benefits from using combinations of organic and inorganic inputs may greatly enhance the quality of soils in Africa, Asia, Latin America, North America, or elsewhere in the world.

20.11 IMPROVING SOIL QUALITY IN ASIA AND LATIN AMERICA

Land-use practices that keep the soil covered with either natural forests or grasslands, or with the residues from harvested crops, must be high on the list of criteria for any sustainable land-use system. Next we consider some soil quality problems in specific tropical and subtropical regions of the world.

South and Southeast Asia

The Impact of High-Yielding Rice Varieties In South India: http://www.ifpri.org/pubs/books/hazell91.htm

The remarkable food-production achievements of the last four decades in Asia might suggest that soil quality decline is not a major problem in this area. Unfortunately, such decline has taken place, but it has been masked by the increased use of fertilizers and irrigation water. Intensive agricultural practices, coupled with the widespread burning of crop residues, have had negative impacts on both the quantity and quality of soil organic matter. In South Asia, as in the parts of Africa just discussed, crop residues as well as animal manures are used as fuel for domestic cooking, or the residues are consumed by animals. In Southeast Asia, the rice and wheat straw that has been gathered to central threshing areas is burned just to get rid of it. In any case, the land is left mostly bare during intercropped periods, thereby encouraging erosion, a decline in soil organic matter, and a breakdown of stable soil structure. The protruding bedrock in once fertile and productive lands is a grim reminder of past soil destruction in some areas. Such degradation simply cannot be allowed to continue if the enormous and increasing populations of Asia are to be fed.

Research has shown that biological production and soil quality can be improved by adding a combination of organic and inorganic inputs into soils in this region. Steps must be taken, however, to assure that farmers have sources of organic materials to complement the inorganic fertilizers that they can purchase. In some areas, community tree lots are providing fuel for cooking, making it possible for the crop residues to remain on the land. In other areas, green manures and grass hedges (such as shown in Figure 17.26) are providing the organic matter needed.

South and Central America

The soil quality outlook in South and Central America is quite mixed. In much of this region, soil quality has been degraded by both agricultural intensification and soil "mining" processes. However the Amazon rainforest region of Brazil also contains many small patches of land on which centuries of farming has dramatically increased—rather than decreased—the fertility of the soil (Box 20.3). In addition, South America contains some of the largest remaining land areas in the world where there is still significant potential to expand cultivation.

For example, a woodland-savannah region of central Brazil known as the *Cerrado* offers such promise. Covering nearly 2 million km^2, the Cerrado has until recently escaped most conversion to agriculture because its soils are extremely acid, low in calcium, high in P-fixing capacity, and often aluminum-toxic. The recent development of high-yielding, aluminum-tolerant varieties of wheat, corn, and soybeans makes possible for the first time the production of crops on millions of hectares of level to rolling land in the Cerrado. The region has become one of the main centers of soybean production in the world. With suitable rotation systems and modest fertilizer and calcium applications (see Section 15.1), soils in the Cerrado can be quite productive.

It should be noted that the Cerrado region supports a unique community of plant species adapted to its long dry season and frequent fires, as well as to its acid soil

BOX 20.3   TERRA PRETA: THE CHARCOAL PATH TO SOIL QUALITY?[a]

In the Amazon basin, a region of highly weathered Oxisols and Ultisols, most soils are dominated by Fe and Al oxide clays, possess very low levels of fertility, and have little capacity to hold nutrients. So soil scientists exploring this region were both puzzled and astonished when they found small patches (averaging 20 ha) of dark colored, apparently highly fertile soils along the Amazon river and some of its major tributaries. Analyses indicated that the organic matter content of these soils was several times as great as that of the surrounding soils. Nutrient levels also proved to be much higher.

Isotope analyses indicated that most of the carbon in these soils accumulated several thousand years ago. It is believed that these patches of fertile soils were created by ancient peoples who once lived in small agricultural settlements carved out of the rainforest. Apparently they farmed these soils continuously for many centuries in such a manner that soil quality was greatly enhanced—rather than degraded—by agricultural use. Even today, some of these soils, which are called *Terra Preta* (= *dark soil* in Portuguese), are dug up and sold locally for their high fertility.

Investigations have shown that Terra Preta soils are much higher in calcium and phosphorus than the surrounding soil because of amendment with large amounts of human excrement as well as with the bones of small animals eaten by the ancient farmers. But the really unique aspect of these soils is that much of the carbon in them is present as *charcoal*. Charcoal is produced by *charring* plant materials in smoldering, low temperature fires (as opposed to *burning* them in high temperature fires that leave only ash). The tiny bits of charcoal found in Terra Preta soils have a very porous structure that greatly aids in holding water and provides a very high capacity to retain nutrients in the form of dissolved organic compounds. A complex aromatic chemical structure makes the charcoal remarkably resistant to microbial breakdown. These properties of the charcoal result in very high and stable accumulations of carbon as well as high nutrient availability for plant growth. In experimental plantings, crops grown in the Terra Preta soils have produced more than double the yield of those grown on nearby non-Terra Preta soils.

Soil scientists studying these unique human-influenced soils report that adding charcoal to soils may greatly improve soil quality and make soils resistant to degradation in agricultural use. Charcoal making produces an important fuel for people in the humid tropics (Figure 20.29) and can be sustainable if the wood required comes from forest plantations, rather than from the remaining natural forests. Waste bits of charcoal too small to sell for cooking fuel might provide a significant quantity of charcoal for soil amendment. Some researchers even suggest that charcoal could help reverse the degradation that soils now suffer under intensified "slash and burn" agriculture (see Section 20.7). To this end, they recommend developing what might be called a "slash and char" agriculture, in which the biomass from land clearing is smoldered rather than burned and soils are amended with food wastes and human and animal manure. The properties and potentials of charcoal in soils are just beginning to be explored.

FIGURE 20.29   *Making charcoal by smoldering wood in a baked mud kiln. [Photo courtesy of R. Weil]*

[a] For an overview of Terra Preta soils and the potential of charcoal to enhance soil quality, see Glaser (2007).

conditions. The unique biome is home to many endemic bird species and such large mammals as the maned wolf, jaguar and giant anteater. With very little (<6%) of their Cerrado habitat under conservation protection, the survival of the wildlife is threatened by the expanding corn, soybean and ranching operations. Fortunately, the expansion of farming into this region has taken advantage of no-till and other conservation technologies, so declining soil quality, erosion and loss of wildlife habitat may not be as

Tilled fields

FIGURE 20.30 A typical hillside farm in the mountainous terrain of Central America (Honduras). The forest vegetation has been almost entirely cleared away. Small (0.2 to 1 ha) fields are tilled and planted on 80 to 100% slopes using hand implements. The patches in the distance are tilled fields that are part of an intensified, unsustainable shifting cultivation pattern. Only rudimentary attempts to conserve the soil have been made by this farmer, such as planting on the contour and using contour bunds made from piled plant residues. (Photo courtesy of R. Weil)

The expert and the farmer in rural development cooperation:
http://www.geocities.com/eswaranpadma/Farmer.html

great a problem as in some earlier major agricultural expansions. If appropriate conservation practices are used to protect the soils, wildlife habitat and biodiversity, these previously uncultivated lands may well be partly utilized for agriculture with minimum negative effect on the overall quality of the soils and the ecosystem.

In potentially productive areas, such as the Cerrado of Brazil, the major challenge is to avoid the errors made in the past by farmers in North America and Europe. Native grasslands will likely be brought into cultivation, but care must be taken to use diverse rotations and organic materials, not just inorganic fertilizers and monocultures.

The clearing of forest lands for agricultural purposes, not only in the Amazon Basin but in most other regions of South and Central America, must be avoided to the extent possible. Erosion rates in Central America are among the highest in the world. This is largely due to the combination of high rainfall and the cultivation of very steep slopes by desperately poor farmers claiming patches of forest land to grow their corn, beans, and often cash crops such as coffee (Figure 20.30). Even where such conservation measures as vegetation barriers, mulching, cover crops, and no-till cropping are adopted, soil erosion is a constant threat. If soil quality is to be maintained or improved, the net change in land use in most areas should be toward increased forest lands, not more agricultural land. Judicious use of organic/inorganic combinations as well as an expansion of agroforestry systems should increase food production on land already in agriculture. Maintaining hilly lands and naturally infertile areas in forests would sustain ecological systems that provide valuable services and products, maintain or improve soil quality, and increase the potential for ecotourism earnings.

20.12 CONCLUSION

Soil scientists, in collaboration with colleagues in other disciplines, have developed the concept of *soil quality* to help quantify factors that are influencing the ability of the soil to function effectively in a variety of ecosystem roles. Although engineering uses are locally important, the primary measures of soil quality are enhanced plant productivity, waste recycling and protection of environmental quality leading to human and animal health, and biological diversity. Human population pressures are forcing increased emphasis on the soil's function in enhancing plant productivity, but if soil quality is to be improved, we must simultaneously achieve the other goals as well.

Farmers of the world have performed near miracles in the past half century. They have increased food production at higher rates than at any time in history, more than matching the unprecedented increases in human population numbers. The increased food prodcution has come primarily from intensification of agriculture on existing farmlands. While soil quality has been improved by some of the intensified systems, others have degraded soil, water, and atmospheric resources.

A major challenge facing humankind is the production of more food and fiber in the next 40 to 50 years than has been produced since the dawn of agriculture some 10,000 years ago. A new paradigm is needed to do this without degrading the environment, endangering human and animal health, or decreasing biological diversity. This paradigm envisages an ecosystem approach for farming—an approach that emphasizes the interaction of plant productivity with the productivity and well-being of all living organisms and with the quality of the environment. Organic farming will likely have a role to play in the new era of soil quality management, as will new, more effective farming systems for the resource poor farmers of the world. A major lesson of the past few decades is that organic materials and nutrient cycling must not be forgotten even where inorganic fertilizers are needed to replenish nutrients lost by previous poor management. Much evidence suggests that combinations of organic and inorganic nutrients perform better than either alone. Such interactions and other soil management steps based on sound science can stimulate marked improvements in soil quality, which, in turn, can enhance the provision of critical ecosystem services that benefit us all.

STUDY QUESTIONS

1. What is soil quality, how is it measured, and of what importance is it to all organisms that live in or on the soil?

2. Why is soil quality likely of greater societal concern today than it was 100 years ago?

3. Why does the biological productivity aspect of soil quality tend to receive more attention than do the aspects dealing with the environment and animal and human health? Is this comparative priority changing? Why?

4. Compare the provision of ecosystem services by soils in a mature forest; a mono-cropped, chemical-intensive vegetable farm; and a diverse, integrated vegetable, grain, and livestock farm.

5. What is meant by *soil resistance* and *soil resilience* as they relate to soil quality?

6. What technological inputs were largely responsible for the remarkable food- and fiber-production increases of the past three decades in the developing countries of Asia and Latin America? Will these inputs likely be significant in increasing food and fiber production in the next 30 years? Why or why not?

7. What are the positive and negative effects on soil quality of the intensive agricultural practices of the past half century?

8. How does the evolving ecosystem approach to plant production differ from the short-term economic return approach that has pertained in the past? Which approach will likely have a more positive effect on soil quality or health, and why?

9. What are some of the changes in nutrient management of intensified plant-production systems that are needed to improve soil productivity and agricultural sustainability and to enhance soil quality?

10. Africa is the only major region of the world where per-capita food production has declined in the past three decades. What are the reasons for this situation, and what major steps might be taken to change it?

11. What is *shifting cultivation*; why is its use today in many areas resulting in a decline in soil quality; and what are some promising alternatives?

12. In what areas of the world has the decline in soil quality been most extensive? Why has this decline taken place?

13. How can the proper integration of livestock with crop farming enhance soil quality in (a) developed-country intensive agriculture and (b) developing-country low-input agriculture?

Alexander, M. 1971. "Agriculture's responsibility in establishing soil quality criteria," pp. 66–71, in *Environmental Improvement—Agriculture's Challenge in the Seventies* (Washington, D.C.: National Academy of Sciences).

Andrews, S. S., D. L. Karlen, and C. A. Cambardella. 2004. "The soil management assessment framework: A quantitative soil quality evaluation method," *Soil Sci. Soc. Am. J.,* **68**:1945–1962.

Ayaga, G., A. Todd, and P. C. Brookes. 2006. "Enhanced biological cycling of phosphorus increases its availability to crops in low-input sub-Saharan farming systems," *Soil Biol. Biochem.,* **38**:81–90.

Badgley, C., J. Moghtader, E. Quintero, E. Zakem, M. J. Chappell, K. Avilés-Vásquez, A. Samulon, and I. Perfecto. 2007. Organic agriculture and the global food supply. Renewable Agriculture and Food Systems **22**:86–108.

Buresh, R. J., and P. A. Sanchez (eds.). 1997. *Replenishing Soil Fertility in Africa.* SSSA Special Publication no. 51 (Madison, Wis.: Soil Sci. Soc. Amer.).

Cassman, K. G., S. Peng, D. C. Olk, W. Reicharat, A. Doberman, and U. Singh. 1997. "Opportunities for increased nitrogen use efficiency from improved resource management in irrigated rice systems," *Field Crop Res.,* in press.

Daily, G. C., T. Soderqvist, S. Aniyar, K. Arrow, P. Dasgupta, P. R. Ehrlich, C. Folke, A. Jansson, B.-O. Jansson, N. Kautsky, S. Levin, J. Lubchenco, K.-G. Maler, D. Simpson, D. Starrett, D. Tilman, and B. Walker. 2000. The value of nature and the nature of value. Science **289**:395–396.

Doran, J. W., and A. J. Jones (eds.). 1996. *Methods for Assessing Soil Quality.* SSSA Special Publication no. 49. (Madison, Wis.: Soil Sci. Soc. Amer.).

Drohan, P. J., and T. J. Farnham. 2006. "Protecting life's foundation: A proposal for recognizing rare and threatened soils," *Soil Sci. Soc. Am. J.,* **70**:2086–2096.

Eigenberg, R. A., J. A. Nienaber, B. L. Woodbury, and R. B. Ferguson. 2006. "Soil conductivity as a measure of soil and crop status—A four-year summary," *Soil Sci. Soc. Am. J.,* **70**:1600–1611.

Eswaran H., R. Almaraz, E. vanden Berg, and P. Reich. 1997. "An assessment of the soil resources of Africa in relation to productivity," *Geoderma,* **77**:1–18.

FAO. 2000. "Manual on integrated soil management and conservation practices," FAO land and water bulletin No. 8. Food and Agriculture Organization of the United Nations, Rome. 215 pp.

FAO. 2006a. "Hunger map: Undernourished population." Food and Agriculture. Organization of the United Nations. http://www.fao.org/es/ess/faostat/foodsecurity/FSMap/map14.htm (posted 01 September 2006; verified 10 October 2006).

FAO. 2006b. FAOSTAT. Food and Agriculture Organization of the United Nations. http://faostat.fao.org/ (verified 20 November 2006).

Girma, T. 2001. "Land degradation: A challenge to Ethiopia," *Environmental Management,* **27**:815–824.

Glaser, B. 2007. "Prehistorically modified soils of central amazonia: A model for sustainable agriculture in the twenty-first century." *Phil. Trans Royal Society B,* **362**:187–196.

Gregorich, E. G., C. F. Drury, and J. A. Baldock. 2001. "Changes in soil carbon under long-term maize in monoculture and legume-based rotation," *Canadian J. Soil Sci.,* **81**:21–31.

Gruver, J. B., and R. R. Weil. 2007. "Farmer perceptions of soil quality and their relationship to management-sensitive soil parameters," *Renewable Agriculture and Food Systems,* **22**:271–281.

Heckman, J. R. 2006. "A history of organic farming: Transitions from Sir Albert Howard's war in the soil to USDA National Organic Program," *Renewable Agriculture and Food Systems,* **21**:143–150.

Heckman, J. R. 2007. Soil fertility in organic agriculture. The Soil Profile, Vol. 17.8 p. http://www.njaes.rutgers.edu/pubs/soilprofile/sp-v17.pdf

ICRISAT. 1997. ICRISAT Report 1996 (Patancheru, India: International Crops Research Institute for the Semi-Arid Tropics).

Islam, K. R., and R. R. Weil. 2000. "Soil quality indicator properties in Mid-Atlantic soils as influenced by conservation management," *J. Soil Water Cons.,* **55**:69–78.

Kang, B. T., L. Richards, and A. N. Atta-Krah. 1990. "Alley Farming," *Advances in Agronomy,* **43**:316–360.

Karlen, D. L., E. G. Hurley, S. S. Andrews, C. A. Cambardella, D. W. Meek, M. D. Duffy, and A. P. Mallarino. 2006. "Crop rotation effects on soil quality at three northern corn/soybean belt locations." *Agron. J.,* **98**:484–495.

Kavanagh, E. 2006. "Commentary letters—Looking at biofuels and bioenergy," *Science,* **312**:1743–1749.

Kwesiga, F. R., S. Franzel, F. Place, D. Phiri, and C. P. Simwanza. 1997. "Sesbania improved fallows in Eastern Zambia: Their inception, development and farmer enthusiasm," in R. J. Buresh and P. J. Cooper (eds.), *The Science and Practice of Short Term Improved Fallows* (Dordrecht, Netherlands: Kluwer Academic Publishers).

Lal, R. 1989. "Agroforestry systems and soil surface management of a tropical Alfisol: II: Water runoff, erosion and nutrient loss," *Agroforestry Systems,* **8**:97–111.

Lal, R. 2006. "Managing soils for feeding a global population of 10 billion," *Journal of the Science of Food and Agriculture,* **86**:2273–2284.

Liebig, M. A., and J. W. Doran. 1999. "Impact of organic production practices on soil quality indicators," *J. Environ. Qual.,* **28**:1601–1609.

Mekonnen, K., R. S. Buresh, and B. Jama. 1997. "Root and inorganic nitrogen distributions in Sesbania fallow, natural fallow and maize fields," *Plant and Soil* **188**:319–327.

Oldeman, L. R., R. T. A. Hakkeling, and W. G. Sombroek. 1990. *World Map of the Status of Human-Induced Soil Degradation: An Explanatory Note* (Wageningen, Netherlands: ISRIC; Kenya: UNEP).

Ragauskas, A. J., C. K. Williams, B. H. Davison, G. Britovsek, J. Cairney, C. A. Eckert, W. J. Frederick, Jr., J. P. Hallett, D. J. Leak, C. L. Liotta, J. R. Mielenz, R. Murphy, R. Templer, and T. Tschaplinski. 2006. "The path forward for biofuels and biomaterials," *Science,* **311**:484–489.

Reganold, J. P., J. D. Glover, P. K. Andrews, and H. R. Hinman. 2001. "Sustainability of three apple production systems," *Nature,* **410**:927–930.

Sanchez, P. A., C. A. Palm, and S. W. Buol. 2003. "Fertility capability soil classification: A tool to help assess soil quality in the tropics," *Geoderma,* **114**:157–185.

Sanchez, P. A., and M. S. Swaminathan. 2005. "Hunger in Africa: The link between unhealthy people and unhealthy soils," *Lancet,* **365**:442–444.

Sanchez, P. A., and T. J. Logan. 1992. "Myths and science about the chemistry and fertility of soils of the tropics," pp. 35–46, in P. A. Sanchez and R. Lal (eds.), *Myths and Science of Soils of the Tropics.* SSSA Special Publication no. 29 (Madison Wis.: Soil Sci. Soc. Amer.).

Schjønning, P., S. Elmholt, and B. T. Christensen (eds.). 2004. *Managing Soil Quality: Challenges in Modern Agriculture* (Wallingford, Oxon: CAB International). 671 pp.

Seybold, C. A., J. E. Herrick, and J. J. Brejda. 1999. "Soil resilience: A fundamental component of soil quality," *Soil Science,* **164**:224–234.

Shukla, M. K., R. Lal, and M. Ebinger. 2006. "Determining soil quality indicators by factor analysis," *Soil Tillage Res.,* **87**:194–204.

Sojka, R. E., D. R. Upchurch, and N. E. Borlaug. 2003. "Quality soil management or soil quality management: Performance versus semantics," *Adv. Agron.,* **79**:1–68.

Stocking, M. A. 2003. "Tropical soils and food security: The next 50 years," *Science,* **302**:1356–1359.

Stoorvogel, J. J., E. M. A. Smaling, and B. H. Janssen. 1993. "Calculating soil nutrient balances in Africa at different scales," *Fertilizer Research,* **35**:227–235.

United Nations Secretariat. 2007. World population prospects: The 2006 revision. Population Division of the Department of Economic and Social Affairs of the United Nations Secretariat http://esa.un.org//unpp, (verified 27 April 2007)

USDA. 1997. *Alley Cropping Conservation Practice.* Job Sheet 311 (Washington, D. C.: USDA Natural Resources Conservation Service).

USDA. 2006. "2007 farm bill theme paper-energy and agriculture," United States Department of Agriculture, Washington, D.C. http://www.usda.gov/documents/Farmbill07energy.pdf.

USDA/AMS. 2006. "The organic foods productioné act of 1990-regulatory text," USDA Agric. Marketing Service. http://www.ams.usda.gov/nop/NOP/standards/FullRegTextOnly.html.

Weil, R. R., K. R. Islam, M. A. Stine, J. B. Gruver, and S. E. Samson-Liebig. 2003. "Estimating active carbon for soil quality assessment: A simplified method for lab and field use," *Amer. J. of Alternative Agric.* **18**:3–17.

Winter, C.K., and S.F. Davis. 2007. Are organic foods healthier? Industry perspectives. Crops Soils Agronomy News 52:2–13, available online https://www.soils.org/news.html.

WORLD REFERENCE BASE, CANADIAN, AND AUSTRALIAN SOIL CLASSIFICATION SYSTEMS

TABLE A.1 Soil Reference Groups in the World Reference Base (WRB) for Soil Resources[a]

The World Reference Base provides a global vocabulary for communicating about different kinds of soils and a reference by which various national soil classification systems (such as U.S. Soil Taxonomy and the Canadian Soil Classification System discussed in this text) can be compared and correlated. The 32 Reference Soil Groups are differentiated mainly according to the primary pedogenesis process that has produced the characteristic soil features, except where "special" soil parent materials are of overriding importance. Each Reference Soil Group can be subdivided using a unique list of possible prefix and suffix qualifiers (not shown here[b]). These qualifiers indicate secondary soil-forming processes that significantly affected the primary soil characteristics especially important to soil use. To avoid making the classification of soils dependent on the availability of climatic data, separations are not based on specific climatic characteristics (as is the case in U.S. Soil Taxonomy).

Reference Soil Group[c]	Major Soil Characteristics	Approximate equivalents in U.S. Soil Taxonomy[d]
Organic Soils		
Histosols (HS)	Composed of organic materials	Most Histosols and Histels
Mineral Soils Dominantly Influenced by Human Activity		
Anthrosols (AT)	Soils with long and intensive agriculture use	Anthrepts and Anthropic, Plaggic great groups and subgroups
Technosols (TC)	Soils containing many artifacts	Entisols, such *proposed* suborders as Urbents, Garbents
Soils with Limited Rooting Due to Shallow Permafrost or Stoniness		
Cryosols (CR)	Ice-affected soils: Cryosols	Gelisols
Leptosols (LP)	Shallow or extremely gravelly soils	Lithic subgroups of Inceptisols and Entisols
Soils Influenced by Water		
Vertisols (VR)	Alternating wet-dry conditions, rich in swelling clays	Vertisols
Fluvisols (FL)	Young soils in alluvial deposits	Fluvents and Fluvaquents
Solonchaks (SC)	Strongly saline soils	Salids and salic or halic great groups of other orders
Solonetz (SN)	Soils with subsurface clay accumulation, rich in sodium	Natric great groups of Alfisols, Aridisols, and Mollisols
Gleysols (GL)	Groundwater-affected soils	Endoaquic great groups (e.g., Endoaqualfs, Endoaquolls, Endoaquults, Endoaquents, Endoaquepts)
Soils in Which Aluminium (Al) Chemistry Plays a Major Role in Their Formation		
Andosols (AN)	Young soils from volcanic ash and tuff deposits	Andisols
Podzols (PZ)	Acid soils with a subsurface accumulation of iron-aluminium-organic compounds	Spodosols
Plinthosols (PT)	Wet soils with an irreversibly hardening mixture of iron, clay, and quartz in the subsoil	Plinthic great groups of Aqualfs, Aquox, and Ultisols
Nitisols (NT)	Deep, dark red, brown, or yellow clayey soils having a pronounced shiny, nut-shaped structure	Parasesquic Inceptisols and some Oxisols and Ultisols
Ferralsols (FR)	Deep, strongly weathered soils with a chemically poor, but physically stable subsoil	Oxisols

(Continued)

Reference Soil Group[c]	Major Soil Characteristics	Approximate equivalents in US Soil Taxonomy[d]
Soils with Stagnant Water		
Planosols (PL)	Soils with a bleached, temporarily water-saturated topsoil on a slowly permeable subsoil	Albaqualfs and Albaquults and some albaquic subgroups of Alfisols and Ultisols
Stagnosols (ST)	Soil with temporarily water-saturated topsoil on structural or moderate textural discontinuity	Epiaquic great groups
Mineral Soils with Humus-rich Topsoils and a High Base Saturation Typically in Grasslands		
Chernozems (CH)	Soils with a thick, dark topsoil, rich in organic matter with a calcareous subsoil	Calciudolls
Kastanozems (KS)	Soils with a thick, dark brown topsoil, rich in organic matter and a calcareous or gypsum-rich subsoil	Many Calciustolls and Calcixerolls
Phaeozems (PH)	Soils with a thick, dark topsoil, rich in organic matter and evidence of removal of carbonates	Many Cryolls, Udolls, and Albolls
Soils with Accumulation of Nonsaline Substances and Influenced by Aridity		
Gypsisols (GY)	Soils with accumulation of secondary gypsum	Gypsids and some gypsic great groups of other orders
Durisols (DU)	Soils with accumulation of secondary silica	Durids and some Duric great groups of other orders
Calcisols (CL)	Soils with accumulation of secondary calcium carbonates	Calcids and Calcic great groups of Inceptisols
Mineral Soils with a Clay-Enriched Subsoil		
Albeluvisols (AB)	Acid soils with a bleached horizon penetrating into a clay-rich subsurface horizon	Some Glossudalfs
Alisols (AL)	Soils with subsurface accumulation of high activity clays, and low base saturation	Ultisols and ultic Alfisols
Acrisols (AC)	Soils with subsurface accumulation of low activity clays and low base saturation	Kandic great groups of Alfisols and Ultisols
Luvisols (LV)	Soils with subsurface accumulation of high activity clays and high base saturation	Haplo and pale great groups of Alfisols
Lixisols (LX)	Soils with subsurface accumulation of low activity clays and high base saturation	Kandic great groups of Alfisols with high base saturation
Relatively Young Soils or Soils with Little or No Profile Development		
Umbrisols (UM)	Acid soils with a thick, dark topsoil, rich in organic matter	Many Umbric great groups of Inceptisols
Arenosols (AR)	Very sandy soils featuring no or only very weak B horizon development	Psamments, grossarenic subgroups of other orders
Cambisols (CM)	Soil with only weakly to moderately developed B horizons	Cambids and many Inceptisols
Regosols (RG)	Soils with very limited soil development, often shallow to rock	Orthents, some Psamments and other Entisols

[a]Based on FAO (2006). World reference base for soil resources 2006: A framework for international classification correlation and communication. World Soil Resources Reports 103. Food and Agriculture Organization of the United Nations and United Nations Environmental Program, Rome. 128 pp. and on personal communication from Bob Engel (USDA/NRCS) and Michéli Erika (Univ. Agric. Sci., Hungary).

[b]As one example, the Kastanozems reference group can be subdivided using the prefix modifiers Vertic, Gypsic, Calcic, Luvic, Hyposodic, Siltic, Chromic, Anthric, and Haplic.

[c]Abbreviations (often used as symbols on soil maps) in parentheses.

[d]As discussed in Chapter 3 of this text.

TABLE A.2 The Australian Soil Classification System and Approximate Correlations with U.S. Soil Taxonomy

Order	Main Characteristics	Soil Taxonomy Order and Suborders
Anthroposols	"Man-made" soils	Some are Entisols (e.g., *proposed* Urbents, Spoilents)
Calcarosols	B horizon calcareous and lacking a marked clay accumulation	Aridisols, Alfisols (Ustalfs, Xeralfs)
Chromosols	Strong clay accumulation and pH > 5.5 in B horizon	Alfisols, some Aridisols
Dermosols	B horizon well structured but lacking a marked clay accumulation	Mollisols, Alfisols, Ultisols
Ferrosols	B horizon high in Fe and lacking a marked clay accumulation	Oxisols, some Alfisols
Hydrosols	Prolonged seasonal water saturation	Aquic subgroups of Alfisols, Ultisols, Inceptisols, salic Aridisols, and some Histosols
Kandosols	B horizon massive and lacking a marked clay accumulation	Alfisols, Ultisols, and Aridisols with massive B horizon structure
Kurosols	Strong clay accumulation and pH < 5.5 in B horizon	Ultisols, some Alfisols
Organosols	Organic materials	Histosols
Podosols	Acid soils with subsurface accumulation of Fe, Al-organic compounds	Spodosols, some Entisols
Rudosols	Negligible (rudimentary) horizon differentiation	Entisols, salic Aridisols
Sodosols	Strong clay accumulation in B horizon, with high sodium saturation	Natric subgroups of Alfisols, Aridisols
Tenosols	Weak horizon differentiation	Inceptisols, Aridisols, Entisols
Vertosols	High clay (>35%), deep cracks, slickensides	Vertisols

Modified from: CSIRO Land and Water (2003); http://www.clw.csiro.au/aclep/asc_re_on_line/soilhome.htm

TABLE A.3 Summary with Brief Descriptions of the Soil Orders in the Soil Classification System of Canada

The Canadian Soil Classification System is one of many national soil classification systems used in various countries around the world. Of these, it is perhaps the most closely aligned with the U.S. Soil Taxonomy. It includes five hierarchical categories: order, great group, subgroup, family, and series. The system is designed to apply principally to the soils of Canada. The soil orders of the Canadian System of Soil Classification are described in this Table and soil orders and some great groups are compared to the U.S. Soil Taxonomy in Table A.4.

Brunisolic	Soils with sufficient development to exclude them from the Regosolic order, but lacking the degree or kind of horizon development specified for other soil orders.
Chernozemic	Soils with high base saturation and surface horizons darkened by the accumulation of organic matter from the decomposition of plants from grassland or grassland-forest ecosystems.
Cryosolic	Soils formed in either mineral or organic materials that have permafrost either within 1 m of the surface or within 2 m if more than one-third of the pedon has been strongly cryoturbated, as indicated by disrupted, mixed, or broken horizons.
Gleysolic	Gleysolic soils have features indicative of periodic or prolonged saturation (i.e., gleying, mottling) with water and reducing conditions.
Luvisolic	Soils with light-colored, eluvial horizons that have illuvial B horizons in which silicate clay has accumulated.
Organic	Organic soils developed on well- to undecomposed peat or leaf litter.
Podzolic	Soils with a B horizon in which the dominant accumulation product is amorphous material composed mainly of humified organic matter combined in varying degrees with Al and Fe.
Regosolic	Weakly developed soils that lack development of genetic horizons.
Solonetzic	Soils that occur on saline (often high in sodium) parent materials, which have B horizons that are very hard when dry and swell to a sticky mass of very low permeability when wet. Typically the solonetzic B horizon has prismatic or columnar macrostructure that breaks to hard to extremely hard, blocky peds with dark coatings.
Vertisolic	Soils with high contents of expanding clays that have large cracks during the dry parts of the year and show evidence of swelling, such as gilgae and slickensides.

TABLE A.4 Comparison of U.S. Soil Taxonomy and the Canadian Soil Classification System

Note that because the boundary criteria differ between the two systems, certain U.S. Soil Taxonomy soil orders have equivalent members in more than one Canadian Soil Classification System soil order.[a]

U.S. Soil Taxonomy soil order	Canadian system soil order	Canadian system great group	Equivalent lower-level taxa in U.S. Soil Taxonomy
Alfisols	Luvisolic	Gray Brown Luvisols	Hapludalfs
		Gray Luvisols	Haplocryalfs, Eutrocryalfs, Fragudalfs, Glossocryalfs, Palecryalfs, and some subgroups of Ustalfs and Udalfs
	Solonetzic	Solonetz	Natrudalfs and Natrustalfs
		Solod	Glossic subgroups of Natraqualfs, Natrudalfs, and Natrustalfs
Andisols	Components of Brunisolic and Cryosolic		
Aridisols	Solonetzic		Frigid families of Natrargids
Entisols	Regosolic		Cryic great groups and frigid families of Entisols, except Aquents
		Regosol	Cryic great groups and frigid families of Folists, Fluvents, Orthents, and Psamments
Gelisols	Cryosolic	Turbic Cryosol	Turbels
		Organic Cryosol	Histels
		Stagnic Cryosol	Orthels
Histosols	Organic	Fibrisol	Cryofibrists, Sphagnofibrists
		Mesisol	Cryohemists
		Humisol	Cryosaprists
Inceptisols	Brunisolic	Melanic Brunisol	Some Eutrustepts
		Eutric Brunisol	Subgroups of Cryepts; frigid and mesic families of Haplustepts
		Sombric Brunisol	Frigid and mesic families of Udepts, and Ustept and Humic Dystrudepts
		Dystric Brunisol	Frigid families of Dystrudepts and Dystrocryepts
	Gleysolic		Cryic subgroups and frigid families of Aqualfs, Aquolls, Aquepts, Aquents, and Aquods
		Humic Gleysol	Humaquepts
		Gleysol	Cryaquepts and frigid families of Fragaquepts, Epiaquepts, and Endoaquepts
Mollisols	Chernozemic	Brown	Xeric and Ustic subgroups of Argicryolls and Haplocryolls
		Dark Brown	Subgroups of Argicryolls and Haplocryolls
		Black	Typic subgroups of Argicryolls and Haplocryolls
		Dark Gray	Alfic subgroups of Argicryolls
	Solonetzic	Solonetz	Natricryolls and frigid families of Natraquolls and Natralbolls
		Solod	Glossic subgroups of Natricryolls
Oxisols	Not relevant in Canada		
Spodosols	Podzolic	Humic Podzol	Humicryods, Humic Placocryods, Placohumods, and frigid families of other Humods
		Ferro-Humic Podzol	Humic Haplocryods, some Placorthods, and frigid families of humic subgroups of other Orthods
		Humo-Ferric Podzol	Haplorthods, Placorthods, and frigid families of other Orthods and Cryods except humic subgroups
Ultisols	Not relevant in Canada		
Vertisols	Vertisolic	Vertisol	Haplocryerts
		Humic Vertisol	Humicryerts

[a] Based on information in Soil Classification Working Group, 1998, *The Canadian System of Soil Classification*, 3rd ed. (Ottawa: Agriculture and Agri-Food Canada). Publication No. A53-1646/1997E.

BASIC SI UNITS OF MEASUREMENT

Parameter	Basic unit	Symbol
Amount of substance	mole	mol
Electrical current	ampere	A
Length	meter	m
Luminous intensity	candela	cd
Mass	gram (kilogram)	g (kg)
Temperature	kelvin	K
Time	second	s

PREFIXES USED TO INDICATE ORDER OF MAGNITUDE

Prefix	Multiple	Abbreviation	Multiplication factor
exa	10^{18}	E	1,000,000,000,000,000,000
peta	10^{15}	P	1,000,000,000,000,000
tera	10^{12}	T	1,000,000,000,000
giga	10^{9}	G	1,000,000,000
mega	10^{6}	M	1,000,000
kilo	10^{3}	k	1,000
hecto	10^{2}	h	100
deca	10	da	10
deci	10^{-1}	d	0.1
centi	10^{-2}	c	0.01
milli	10^{-3}	m	0.001
micro	10^{-6}	μ	0.000 001
nano	10^{-9}	n	0.000 000 001
pico	10^{-12}	p	0.000 000 000 001
femto	10^{-15}	f	0.000 000 000 000 001
atto	10^{-18}	a	0.000 000 000 000 000 001

FACTORS FOR CONVERTING NON-SI UNITS TO SI UNITS

Non-SI Unit	Multiply by[a]	To obtain SI Unit
Length		
inch, in.	2.54	centimeters, cm (10^{-2} m)
foot, ft	0.304	meter, m
mile,	1.609	kilometer, km (10^3 m)
micron, μ	1.0	micrometer, μm (10^{-6} m)
Ångstrom unit, Å	0.1	nanometer, nm (10^{-9} m)
Area		
acre, ac	0.405	hectare, ha (10^4 m^2)
square foot, ft^2	9.29×10^{-2}	square meter, m^2
square inch, in^2	645	square millimeter, mm^2
square mile, mi^2	2.59	square kilometer, km^2
Volume		
bushel, bu	35.24	liter, L
cubic foot, ft^3	2.83×10^{-2}	cubic meter, m^3
cubic inch, in.^{3}	1.64×10^{-5}	cubic meter, m^3
gallon (U.S.), gal	3.78	liter, L
quart, qt	0.946	liter, L
acre-foot, ac-ft	12.33	hectare-centimeter, ha-cm
acre-inch, ac-in.	1.03×10^{-2}	hectare-meters, ha-m
ounce (fluid), oz	2.96×10^{-2}	liter, L
pint, pt	0.473	liter, L
Mass		
ounce (avdp), oz	28.4	gram, g
pound, lb	0.454	kilogram, kg (10^3 g)
ton (2000 lb)	0.907	megagram, Mg (10^6 g)
tonne (metric), t	1000	kilogram, kg
Radioactivity		
curie, Ci	3.7×10^{10}	becquerel, Bq
picocurie per gram, pCi/g	37	becquerel per kilogram, Bq/kg
Yield and Rate		
pound per acre, lb/ac	1.121	kilogram per hectare, kg/ha
pounds per 1000 ft^2	48.8	kilogram per hectare, kg/ha
bushel per acre (60 lb), bu/ac	67.19	kilogram per hectare, kg/ha
bushel per acre (56 lb), bu/ac	62.71	kilogram per hectare, kg/ha
bushel per acre (48 lb), bu/ac	53.75	kilogram per hectare, kg/ha
gallon per acre (U.S.), gal/ac	9.35	liter per hectare, L/ha
ton (2000 lb) per acre	2.24	megagram per hectare, Mg/ha
miles per hour, mph	0.447	meter per second, m/s
gallon per minute (U.S.), gpm	0.227	cubic meter per hour, m^3/h
cubic feet per second, cfs	101.9	cubic meter per hour, m^3/h
Pressure		
atmosphere, atm	0.101	megapascal, MPa (10^6 Pa)
bar	0.1	megapascal, MPa
pound per square foot, lb/ft^2	47.9	pascal, Pa
pound per square inch, lb/in^2	6.9×10^3	pascal, Pa
Temperature		
degrees Fahrenheit (°F − 32)	0.556	degrees, °C
degrees Celsius (°C + 273)	1	Kelvin, K
Energy		
British thermal unit, Btu	1.05×10^3	joule, J
calorie, cal	4.19	joule, J
dyne, dyn	10^{-5}	newton, N
erg	10^{-7}	joule, J
foot-pound, ft-lb	1.36	joule, J
Concentrations		
percent, %	10	gram per kilogram, g/kg
part per million, ppm	1	milligram per kilogram, mg/kg
milliequivalents per 100 grams	1	centimole per kilogram, cmol/kg

[a] To convert from SI to non-SI units, *divide* by the factor given.

Periodic Table of the Elements with Notes Concerning Relevance to Soil Science

Based on atomic mass of $^{12}C = 12.0$. Numbers in parentheses are the mass numbers of the most stable isotopes of radioactive elements.

Group IA	Group IIA	Group IIIB	Group IVB	Group VB	Group VIB	Group VIIB	Group VIIIB	Group VIIIB	Group VIIIB	Group IB	Group IIB	Group IIIA	Group IVA	Group VA	Group VIA	Group VIIA	Group VIIIA
1 H 1.01 Hydrogen																	2 He 4.00 Helium
3 Li 6.94 Lithium	4 Be 9.01 Beryllium											5 B 10.81 Boron	6 C 12.01 Carbon	7 N 14.01 Nitrogen	8 O 16.00 Oxygen	9 F 19.00 Fluorine	10 Ne 20.18 Neon
11 Na 22.99 Sodium	12 Mg 24.30 Magnesium											13 Al 26.98 Aluminum	14 Si 28.09 Silicon	15 P 30.97 Phosphorus	16 S 32.07 Sulfur	17 Cl 35.45 Chlorine	18 Ar 39.95 Argon
19 K 39.10 Potassium	20 Ca 40.08 Calcium	21 Sc 44.96 Scandium	22 Ti 47.88 Titanium	23 V 50.94 Vanadium	24 Cr 52.00 Chromium	25 Mn 54.94 Manganese	26 Fe 55.85 Iron	27 Co 58.93 Cobalt	28 Ni 58.69 Nickel	29 Cu 63.55 Copper	30 Zn 65.38 Zinc	31 Ga 69.72 Gallium	32 Ge 72.59 Germanium	33 As 74.92 Arsenic	34 Se 78.96 Selenium	35 Br 79.90 Bromine	36 Kr 83.80 Krypton
37 Rb 85.47 Rubidium	38 Sr 87.62 Strontium	39 Y 88.91 Yttrium	40 Zr 91.22 Zirconium	41 Nb 92.91 Niobium	42 Mo 95.94 Molybdenum	43 Tc (98) Technetium	44 Ru 101.07 Ruthenium	45 Rh 102.91 Rhodium	46 Pd 106.42 Palladium	47 Ag 107.87 Silver	48 Cd 112.41 Cadmium	49 In 114.82 Indium	50 Sn 118.71 Tin	51 Sb 121.75 Antimony	52 Te 127.60 Tellurium	53 I 126.90 Iodine	54 Xe 131.29 Xenon
55 Cs 132.91 Cesium	56 Ba 137.33 Barium	57 La 138.91 Lanthanum	72 Hf 178.49 Hafnium	73 Ta 180.95 Tantalum	74 W 183.85 Tungsten	75 Re 186.21 Rhenium	76 Os 190.2 Osmium	77 Ir 192.22 Iridium	78 Pt 195.08 Platinum	79 Au 196.97 Gold	80 Hg 200.59 Mercury	81 Tl 204.38 Thallium	82 Pb 207.2 Lead	83 Bi 208.98 Bismuth	84 Po (209) Polonium	85 At (210) Astatine	86 Rn (222) Radon
87 Fr (223) Francium	88 Ra (226) Radium	89 Ac (227) Actinium	104 Unq (261) Unnilquadium	105 Unp (262) Unnilpentium	106 Unh (263) Unnilhexium	107 Uns (262) Unnilseptium	108 Uno (265) Unniloctium										

58 Ce 140.12 Cerium	59 Pr 140.91 Praseodymium	60 Nd 144.24 Neodymium	61 Pm (145) Promethium	62 Sm 150.36 Samarium	63 Eu 151.96 Europium	64 Gd 157.25 Gadolinium	65 Tb 158.93 Terbium	66 Dy 162.50 Dysprosium	67 Ho 164.93 Holmium	68 Er 167.26 Erbium	69 Tm 168.93 Thulium	70 Yb 173.04 Ytterbium	71 Lu 174.97 Lutetium
90 Th (232) Thorium	91 Pa (231) Protactinium	92 U (238) Uranium	93 Np (237) Neptunium	94 Pu (244) Plutonium	95 Am (243) Americium	96 Cm (247) Curium	97 Bk (247) Berkelium	98 Cf (251) Californium	99 Es (252) Einsteinium	100 Fm (257) Fermium	101 Md (258) Mendelevium	102 No (259) Nobelium	103 Lr (260) Lawrencium

Metals / Nonmetals

Atomic number
Symbol
Atomic mass

87
Fr
(223)
Francium

Elements known to be nutrients for animals or plants. Some are also toxic in excessive amounts.

Elements toxic to organisms in small amounts, and not known to serve as nutrients.

Other elements commonly studied in Soil Science because of soil-environmental impacts or because of their use as tracers or electrodes. (Br is used to trace anionic solutes such as nitrate. Isotopes of Rb and Sr are used to trace K and Ca in plants and soils. Cs and Ti are used to trace geological processes such as soil erosion. Pt and Ag are used in electrodes for measuring soil redox potential and pH, respectively.)

These 22 elements are needed as essential mineral nutrients by humans: macronutrients (calcium, chloride, magnesium, phosphorus, potassium, sodium, and sulfur) and micronutrients (chromium, cobalt, copper, fluoride, iodine, iron, manganese, molybdenum, nickel, selenium, silicon, tin, vanadium, zinc).

acacia, apple ring	*Faidherbia albida* (Del.) A. Chev. [syn. *Acacia albida*]	coffee	*Coffea spp.* L.
acacia, catclaw	*Acacia greggii* Gray	corn	*Zea maiz* L.
alder	*Alnus spp.* P. Mill.	cotton	*Gossypium hirsutum* L.
alder, red	*Alnus rubra* Bong.	cottonwood	*Populus deltoidies* Bartr. Ex. Marsh
alfalfa	*Medicago sativa* L.	cowpea	*Vigna unguiculata* (L.) Walp.
alkali grass, Nutall's	*Puccinellia nuttalliana* (J.A. Schultes) A.S. Hitchc.	cranberry	*Vaccinium macrocarpon* Ait.
		cranberry, small	*Vaccinium oxycoccos* L.
alkali sacaton	*Sporobolus airoides* (Torr.) Torr.	cucumber	*Cucumis sativus* L.
almond	*Prunus dulcis* (P. Mill.) D.A. Webber	currant	*Ribes spp.* L.
andromeda (bog rosemary)	*Andromeda polifolia* L.	cypress, bald	*Taxodium distichum* (L.) L.C. Rich.
apple	*Malus spp.* P. Mill.	dallisgrass	*Paspalum dilatatum* Poir
apricot	*Prunus armeniaca* L.	dogwood	*Cornus spp.* L.
arborvitae	*Thuja occidentalis* L.	dogwood, grey	*Cornus racemosa* Lam.
ash	*Fraxinus spp.* L.	elaeagnus	*Elaeagnus spp.* L.
ash, white	*Fraxinus americana* L.	elm	*Ulmus spp.* L.
asparagus	*Asparagus officinalis.* L.	elm, American	*Ulmus americana* L.
aspen	*Populus spp.* L.	eucalyptus	*Eucalyptus spp.*
aspen, quaking	*Populus tremuloides* Michx.	eucalyptus (jarrah)	*Eucalyptus marginata* Donn ex Sm.
autumn olive	*Elaeagnus umbellata* Thunb.	fescue	*Festuca spp.* L.
azalea	*Rhododendron spp.* L.	fescue, meadow	*Festuca pratensis* Huds.
azolla	*Azolla spp.* L.	fescue, red	*Festuca rubra* L.
bahia grass	*Paspalum notatum* Flueggé	fescue, sheep	*Festuca ovina* L.
banana	*Musa acuminata* Colla	fescue, tall	*Festuca elatior* L.
barley, forage	*Hordeum vulgare* L.	fig	*Ficus carica* L.
bean, broad (faba)	*Vicia faba* L.	filbert	*Corylus spp.* L.
bean, common	*Phaseolus vulgaris* L.	fir, Douglas	*Pseudotsuga menziesii* (Mirbel) Franco
bean, winged	*Psophocarpus tetragonobus* L. D.C.	gamagrass, eastern	*Tripsacum dactyloides* (L.) L.
beech, American	*Fagus grandifolia* Ehrh.	gliricidia (quickstick)	*Gliricidia sepium* (Jacq.) Kunth ex Walp.
beet, garden	*Beta procumbens* L.		
beet, sugar	*Beta vulgaris* L.	grape	*Vitus spp.* L.
bentgrass	*Agrostis stolonifera* L.	grapefruit	*Citrus paradisi* Macfad. (pro sp.) [*maxima sinensis*]
bermudagrass	*Cynodon dactylon* (L.) Pers.		
birch	*Betula spp.* L.	grevillea	*Grevillea spp.* R. Br. ex Knight
birch, black	*Betula lenta* L.	groundnut (peanut)	*Arachis hypogaea* L.
black cherry	*Prunus serotina* Ehrh.	guayule	*Parthenium argentatum* Gray
black locust	*Robinia pseudoacacia* L.	gunnera	*Gunnera spp.* L.
blackberry	*Rubus spp.* L.	harding grass	*Phalaris tuberosa* L. var. *stenoptera* (Hack) A.S. Hitchc.
blueberry	*Vaccinium spp.* L.		
bluegrass, Kentucky	*Poa pratensis* L. ssp. *pratensis*	hemlock, Canadian	*Tsuga canadensis* (L.) Carr.
bog rosemary	*Andromeda polifolia* L.	hemlock, Carolina	*Tsuga caroliniana* Engelm.
bougainvillea	*Bougainvillea spp.* Comm. ex Juss.	hibiscus	*Hibiscus spp.* L.
boxwood	*Buxus spp.* L.	hickory, bitternut	*Carya cordiformis* (Wangenh.) K. Koch
broccoli	*Brassica oleracea* L. var. *botrytis* L.		
brome grass	*Bromus spp.* L.	hickory, shagbark	*Carya ovata* (P. Mill.) K. Koch
buckwheat	*Eriogonum spp.* Michx.	holly, American	*Ilex opaca* Ait.
buffalo grass	*Buchloe dactyloides* (Nutt.) Engelm.	holly, burford	*Ilex cornuta* Lindl. & Paxton
cabbage	*Brassica oleracea* L.	honeysuckle	*Lonicera spp.* L.
canola (rapeseed)	*Brassica napus* L.	hydrangea	*Hydrangea spp.* L.
cantaloupe	*Cucumis melo* L.	ipil ipil tree	*Leucaena leucocephala* Benth.
carrot	*Daucus carota* L. ssp. *sativus* (Hoffm.) Arcang.	jarrah	*Eucalyptus marginata* Donn ex Sm.
		jojoba	*Simmondsia chinensis* (Link) Schneid.
cassava	*Manihot esculenta* Crantz	juniper	*Juniperus spp.* L.
casuarina (sheoak)	*Casuarina spp.* Rumph. ex L.	kale	*Brassica oleracea* L. (Acephala group)
cattail, common	*Typa latifolia* L.	kallargrass	*Leptochloa fusca* (L.) Kunth [syn. *Diplachne fusca* Beauv.]
cauliflower	*Brassica oleracea* L. (Botrytis group)		
ceanothus	*Ceanothus spp.* L.	kenaf	*Hibiscus cannabinus* L.
celery	*Apium graveolens* L. var. dulce (Mill.) Pers.	kochia, prostrate	*Kochia prostrata* (L.) Schrad.
		kudzu	*Pueraria montana* (Lour.) Merr. Var. lobata (Wild.)
cherry, flowering	*Prunus serrulata* Lindl.		
citrus	*Citrus spp.* L.	larch	*Larix spp.* P. Mill.
clover, alsike	*Trifolium hybridum* L.	lemon	*Citrus limon* (L.) Burm. F.
clover, berseem	*Trifolium alexandrinum* L.	lespedeza	*Lespedeza spp.* Michx.
clover, crimson	*Trifolium incarnatum* L.	lettuce	*Lactuca sativa* L.
clover, ladino	*Trifolium repens* L.	leucaena (lead tree)	*Leucaena spp.* Benth.
clover, red	*Trifolium pratense* L.	lilac	*Syringa spp.* L.
clover, strawberry	*Trifolium fragiferum* L.	linden	*Tillia spp.* L.
clover, sweet	*Melilotus indica* All	locust, black	*Robinia pseudoacacia* L.
clover, white	*Trifolium repens* L.	locust, honey	*Gleditsia triacanthos* L.
		lovegrass, weeping	*Eragrostis curvula* (Schrad.) Nees

lupine	*Lupinus spp.* L.	rice (paddy)	*Oryza sativa* L.
magnolia	*Magnolia spp.* L.	riverhemp, Egyptian	*Sesbania sesban* (L.) Merr.
maiden cane	*Panicum hemitomon* J.A. Schultes	rose-mallow, swamp	*Hibiscus moscheutos* L.
maize (corn)	*Zea mays* L.	rosemary	*Rosmarinus officinalis* L.
mandarin orange	*Citrus reticulata* Blanco	roses	*Rosa spp.* L.
maple	*Acer spp.* L.	rye (grain, forage)	*Secale cereale* L.
maple, red	*Acer rubrum* L.	rye, wild	*Elymus spp.*
maple, sugar	*Acer saccharum* Marsh.	ryegrass, perennial	*Lolium perenne* L.
mosquito fern	*Azolla spp.* L.	safflower	*Carthamus tinctorius* L.
mountain laurel	*Kalmia latifolia* L.	saltgrass, desert	*Distichlis spicta* L. var. *stricta* (Torr.) Bettle
mulberry	*Morus spp.* L.		
myrica	*Myrica spp.* L.	sesbania	*Sesbania sesban* (L.) Merr.
nut trees (e.g., almonds, hazelnuts)	*Prunus dulcis* (P. Mill.) D.A. Webber	skunk cabbage	*Symplocarpus foetidus* (L.) Salisb. Ex Nutt.
	Corylus spp. L.	sorghum	*Sorghum bicolor* (L.) Moench
oak	*Quercus spp.* L.	soy beans	*Glycine max* (L.) Merr.
oak, blackjack	*Quercus marilandica* Muenchh.	spartina (cordgrass)	*Spartina spp.* Schreb.
oak, chestnut	*Quercus prinus* L.	spinach	*Spinacia oleracea* L.
oak, northern red	*Quercus rubra* L.	spruce, black	*Picea mariana* (Mill.) B.S.P.
oak, pin	*Quercus palustris* Muenchh.	spruce, Norway	*Picea abies* (L.) Karst.
oak, southern red	*Quercus falcata* Michx.	spruce, red	*Picea rubens* Sarg.
oak, swamp white	*Quercus bicolor* Wild.	spruce, white	*Picea glauca* (Moench) Voss
oak, white	*Quercus alba* L.	squash	*Cucurbita pepo* L.
oak, willow	*Quercus pellos* L.	squash (zucchini)	*Cucurbita pepo* L. var. *melopepo* (L.) Alef.
oats	*Avena sativa* L.		
olive	*Olea europaea* L.	star jasmine	*Jasminum multiflorum* (Burm. f.) Andr
onion	*Allium cepa* L.	strawberry	*Fragaria x ananassa* Duch.
orange	*Citrus sinensis* (L.) Osbeck	sudan grass	*Sorghum sudanense* (Piper) Stapf
orchard grass	*Dactylis glomerata* L.	sugar beet	*Beta vulgaris* L.
pangola grass	*Digitaria eriantha* Steud.	sugar cane	*Saccharum officinarum* L.
pueraria, kudzu	*Pueraria phaseoloides* (Roxb.) Benth.	sumac	*Rhus spp.* L.
		sunflower	*Helianthus annuus* L.
pea	*Pisum sativa* L.	sycamore	*Plantus occidentalis* L.
pea, pigeon	*Cajanus cajan* (L.) Millsp.	tamarix (tamarisk)	*Tamarix gallica* L.
peach	*Prunus persica* (L.) Batsch	tea	*Camellia sinensis* (L.) O. Kuntze
peanut	*Arachis hypogaea* L.	teaberry	*Gaultheria procumbens* L.
pear	*Pyrus communis* L.	timothy	*Phleum pratense* L.
pecan	*Carya illinoinensis* (Wangenh.) K. Koch	tithonia	*Tithonia diversifolia* (Hemsl.) Gray
		tobacco	*Nicotiana spp.* L.
phragmities reed	*Phragmities australis* (Cav.) Trin. Ex Steud.	tomato	*Solanum lycopersicum* L.
		tree marigold	*Tithonia diversifolia* (Hemsl.) Gray
pine, loblolly	*Pinus taeda* L.	trefoil, birdsfoot	*Lotus corniculatus* L.
pine, Monterey	*Pinus radiata* D. Don	tulip poplar	*Liriodendron tulipifera* L.
pine, ponderosa	*Pinus ponderosa* Dougl. Ex P. & C. Laws.	turnip	*Brassica rapa* L. (Rapifera group)
		velvetleaf	*Abutilon theophrasti* Medik.
pine, red	*Pinus resinosa* Ait.	vetch	*Vicia spp.* L.
pine, white	*Pinus strobus* L.	vetch, common	*Vicia angustifolia* L.
pine, white Scotch	*Pinus sylvestris* L.	vetch, hairy	*Vicia villosa* Roth
pineapple	*Ananas comosus* (L.) Merrill	vetiver grass	*Vetiveria zizanioides* (L.) Nash ex Small
pitcher plant	*Sarracenia spp.* L.	viburnum	*Viburnum spp.* L.
plum (prune)	*Prunus domestica* L.	walnut	*Juglans spp.* L.
poinsettia	*Euphorbia pulcherrima* Willd. ex Klotzsch	water melon	*Citrullus lanatus* (Thunb.) Matsumura & Nakai
pomegranate	*Punica granatum* L.	water tupelo	*Nyssa aquatica* L.
poplar	*Populus spp.* L.	wheat	*Triticum aestivum* L.
potato	*Solanum tuberosum* L.	wheatgrass, crested	*Agropyron sibiricum* (Willd.) Beauvois
potato, sweet	*Ipomoea batatas* (L.) Lam.		
povertygrass	*Danthonia spicata* (L.) Beauv. Ex Roem. & Schult.	wheatgrass, fairway	*Agropyron cristatum* (L.) Gaertn.
		wheatgrass, tall	*Agropyron elongatum* (Hort) Beauvois
privet	*Ligustrum spp.* L.		
quickstick	*Gliricidia sepium* (Jacq.) Kunth ex Walp.	wheatgrass, western	*Pascopyrum smithii* (Rydb.) A. Löve
		wild rye, altai	*Leymus angustus* (Trin.) Pilger
radish	*Raphanus sativus* L.	wild rye, Russian	*Psathyrostachys juncea* (Fisch.) Nevski
rapeseed (see also canola)	*Brassica campestris* L. [syn. *B. rapa* L.]		
		willow	*Salix spp.* L.
raspberry	*Rubus idaeus* L.	willow, black	*Salix nigra* L.
red top	*Agrostis alba* L.	winged bean	*Psophocarpus tetragonolobus* (L.) DC
red top	*Agrostis gigantea* Roth	yew	*Taxus spp.* L.
reed canarygrass	*Phalaris arundinacea* L.		
rescuegrass	*Bromus catharticus* Vahl		
rhododendron	*Rhododendron spp.* L.		
rice	*Oryza spp.* L.		

GLOSSARY OF SOIL SCIENCE TERMS[1]

A horizon The surface horizon of a mineral soil having maximum organic matter accumulation, maximum biological activity, and/or eluviation of materials such as iron and aluminum oxides and silicate clays.

abiotic Nonliving basic elements of the environment, such as rainfall, temperature, wind, and minerals.

accelerated erosion Erosion much more rapid than normal, natural, geological erosion; primarily as a result of the activities of humans or, in some cases, of animals.

acid cations Cations, principally Al^{3+}, Fe^{3+}, and H^+, that contribute to H^+ ion activity either directly or through hydrolysis reactions with water. *See also* nonacid cations.

acid rain Atmospheric precipitation with pH values less than about 5.6, the acidity being due to inorganic acids (such as nitric and sulfuric) that are formed when oxides of nitrogen and sulfur are emitted into the atmosphere.

acid saturation The proportion or percentage of a cation-exchange site occupied by acid cations.

acid soil A soil with a pH value <7.0. Usually applied to surface layer or root zone, but may be used to characterize any horizon. *See also* reaction, soil.

acid sulfate soils Soils that are potentially extremely acid (pH < 3.5) because of the presence of large amounts of reduced forms of sulfur that are oxidized to sulfuric acid if the soils are exposed to oxygen when they are drained or excavated. A sulfuric horizon containing the yellow mineral jarosite is often present. *See also* cat clays.

acidity, active The activity of hydrogen ions in the aqueous phase of a soil. It is measured and expressed as a pH value.

acidity, residual Soil acidity that can be neutralized by lime or other alkaline materials but cannot be replaced by an unbuffered salt solution.

acidity, salt replaceable Exchangeable hydrogen and aluminum that can be replaced from an acid soil by an unbuffered salt solution such as KCl or NaCl.

acidity, total The total acidity in a soil. It is approximated by the sum of the salt-replaceable acidity plus the residual acidity.

Actinomycetes A group of bacteria that form branched mycelia that are thinner, but somewhat similar in appearance, to fungal hyphae. Includes many members of the order Actinomycetales.

activated sludge Sludge that has been aerated and subjected to bacterial action.

active layer The upper portion of a Gelisol that is subject to freezing and thawing and is underlain by permafrost.

active organic matter A portion of the soil organic matter that is relatively easily metabolized by microorganisms and cycles with a half-life in the soil of a few days to a few years.

adhesion Molecular attraction that holds the surfaces of two substances (e.g., water and sand particles) in contact.

adsorption The attraction of ions or compounds to the surface of a solid. Soil colloids adsorb large amounts of ions and water.

adsorption complex The group of organic and inorganic substances in soil capable of adsorbing ions and molecules.

aerate To impregnate with gas, usually air.

aeration, soil The process by which air in the soil is replaced by air from the atmosphere. In a well-aerated soil, the soil air is similar in composition to the atmosphere above the soil. Poorly aerated soils usually contain more carbon dioxide and correspondingly less oxygen than the atmosphere above the soil.

aerobic (1) Having molecular oxygen as a part of the environment. (2) Growing only in the presence of molecular oxygen, as aerobic organisms. (3) Occurring only in the presence of molecular oxygen (said of certain chemical or biochemical processes, such as aerobic decomposition).

---

[1] This glossary was compiled and modified from several sources, including *Glossary of Soil Science Terms* [Madison, Wis.: Soil Sci. Soc. Amer. (1997)] *Resource Conservation Glossary* [Anheny, Iowa: Soil Cons. Soc. Amer. (1982)], and *Soil Taxonomy* [Washington, D.C.: U.S. Department of Agriculture (1999)].

aerosolic dust A type of eolian material that is very fine (about 1 to 10 μm) and may remain suspended in the air over distances of thousands of kilometers. Finer than most *loess*.

aggregate (soil) Many soil particles held in a single mass or cluster, such as a clod, crumb, block, or prism.

agric horizon A diagnostic subsurface horizon in which clay, silt, and humus derived from an overlying cultivated and fertilized layer have accumulated. Wormholes and illuvial clay, silt, and humus occupy at least 5% of the horizon by volume.

agroforestry Any type of multiple cropping land-use that entails complementary relations between trees and agricultural crops.

agronomy A specialization of agriculture concerned with the theory and practice of field-crop production and soil management. The scientific management of land.

air-dry (1) The state of dryness (of a soil) at equilibrium with the moisture content in the surrounding atmosphere. The actual moisture content will depend upon the relative humidity and the temperature of the surrounding atmosphere. (2) To allow to reach equilibrium in moisture content with the surrounding atmosphere.

air porosity The proportion of the bulk volume of soil that is filled with air at any given time or under a given condition, such as a specified moisture potential; usually the large pores.

albic horizon A diagnostic subsurface horizon from which clay and free iron oxides have been removed or in which the oxides have been segregated to the extent that the color of the horizon is determined primarily by the color of the primary sand and silt particles rather than by coatings on these particles.

Alfisols An order in *Soil Taxonomy*. Soils with gray to brown surface horizons, medium to high supply of bases, and B horizons of illuvial clay accumulation. These soils form mostly under forest or savanna vegetation in climates with slight to pronounced seasonal moisture deficit.

algal bloom A population explosion of algae in surface waters, such as lakes and streams, often resulting in high turbidity and green- or red-colored water, and commonly stimulated by nutrient enrichment with phosphorus and nitrogen.

alkaline soil Any soil that has pH > 7. Usually applied to the surface layer or root zone but may be used to characterize any horizon or a sample thereof. *See also* reaction, soil.

allelochemical An organic chemical by which one plant can influence another. *See* allelopathy.

allelopathy The process by which one plant may affect other plants by biologically active chemicals introduced into the soil, either directly by leaching or exudation from the source plant, or as a result of the decay of the plant residues. The effects, though usually negative, may also be positive.

allophane A poorly defined aluminosilicate mineral whose structural framework consists of short runs of three-dimensional crystals interspersed with amorphous noncrystalline materials. Along with its more weathered companion, it is prevalent in volcanic ash materials.

alluvial fan Fan-shaped alluvium deposited at the mouth of a canyon or ravine where debris-laden waters fan out, slow down, and deposit their burden.

alluvium A general term for all detrital material deposited or in transit by streams, including gravel, sand, silt, clay, and all variations and mixtures of these. Unless otherwise noted, alluvium is unconsolidated.

alpha particle A positively charged particle (consisting of two protons and two neutrons) that is emitted by certain radioactive compounds.

aluminosilicates Compounds containing aluminum, silicon, and oxygen as main constituents. An example is microcline, $KAlSi_3O_8$.

amendment, soil Any substance other than fertilizers, such as lime, sulfur, gypsum, and sawdust, used to alter the chemical or physical properties of a soil, generally to make it more productive.

amino acids Nitrogen-containing organic acids that couple together to form proteins. Each acid molecule contains one or more amino groups ($—NH_2$) and at least one carboxyl group ($—COOH$). In addition, some amino acids contain sulfur.

Ammanox A biochemical process in the N cycle by which certain anaerobic bacteria or archaea oxidize ammonium ions using nitrite ions as the electron acceptor, the main product being N_2 gas.

ammonification The biochemical process whereby ammoniacal nitrogen is released from nitrogen-containing organic compounds.

ammonium fixation The entrapment of ammonium ions by the mineral or organic fractions of the soil in forms that are insoluble in water and are at least temporarily nonexchangeable.

amorphous material Noncrystalline constituents of soils.

anaerobic (i) The absence of molecular oxygen. (ii) Growing or occurring in the absence of molecular oxygen (e.g., anaerobic bacteria or biochemical reduction reaction).

anaerobic respiration The metabolic process whereby electrons are transferred from a reduced compound (usually organic) to an inorganic acceptor molecule other than oxygen.

andic properties Soil properties related to volcanic origin of materials, including high organic carbon content, low bulk density, high phosphate retention, and extractable iron and aluminum.

Andisols An order in *Soil Taxonomy*. Soils developed from volcanic ejecta. The colloidal fraction is dominated by allophane and/or Al-humus compounds.

angle of repose The maximum slope steepness at which loose, cohesionless material will come to rest.

anion Negatively charged ion; during electrolysis it is attracted to the positively charged anode.

anion exchange Exchange of anions in the soil solution for anions adsorbed on the surface of clay and humus particles.

anion exchange capacity The sum total of exchangeable anions that a soil can adsorb. Expressed as centimoles of charge per kilogram ($cmol_c/kg$) of soil (or of other adsorbing material, such as clay).

anoxic *See* anaerobic.

anthropic epipedon A diagnostic surface horizon of mineral soil that has the same requirements as the mollic epipedon but that has more than 250 mg/kg of P_2O_5 soluble in 1% citric acid, or is dry more than 10 months (cumulative) during the period when not irrigated. The anthropic epipedon forms under long-continued cultivation and fertilization.

antibiotic A substance produced by one species of organism that, in low concentrations, will kill or inhibit growth of certain other organisms.

Ap The surface layer of a soil disturbed by cultivation or pasturing.

apatite A naturally occurring complex calcium phosphate that is the original source of most of the phosphate fertilizers. Formulas such as $[3Ca_3(PO_4)_2] \cdot CaF_2$ illustrate the complex compounds that make up apatite.

aquic conditions Continuous or periodic saturation (with water) and reduction, commonly indicated by redoximorphic features.

aquiclude A saturated body of rock or sediment that is incapable of transmitting significant quantities of water under ordinary water pressures.

aquifer A saturated, permeable layer of sediment or rock that can transmit significant quantities of water under normal pressure conditions.

arbuscular mycorrhiza A common endomycorrhizal association produced by phycomycetous fungi and characterized by the development, within root cells, of small structures known as *arbuscules*. Some also form, between root cells, storage organs known as *vesicles*. Host range includes many agricultural and horticultural crops. Formerly called vesicular arbuscular mycorrhiza (VAM). *See also* endotrophic mycorrhiza.

arbuscule Specialized branched structure formed within a root cortical cell by endotrophic mycorrhizal fungi.

Archaea One of the two domains of single-celled prokaryote microorganisms. Includes organisms adapted to extremes of salinity and heat, and those that subsist on methane. Similar appearing, but evolutionarily distinct from bacteria.

argillan A thin coating of well-oriented clay particles on the surface of a soil aggregate, particle, or pore. A clay film.

argillic horizon A diagnostic subsurface horizon characterized by the illuvial accumulation of layer-lattice silicate clays.

arid climate Climate in regions that lack sufficient moisture for crop production without irrigation. In cool regions annual precipitation is usually less than 25 cm. It may be as high as 50 cm in tropical regions. Natural vegetation is desert shrubs.

Aridisols An order in *Soil Taxonomy*. Soils of dry climates. They have pedogenic horizons, low in organic matter, that are never moist for as long as three consecutive months. They have an ochric epipedon and one or more of the following diagnostic horizons: argillic, natric, cambic, calcic, petrocalcic, gypsic, petrogypsic, salic, or a duripan.

aspect (of slopes) The direction (e.g., south or north) that a slope faces with respect to the sun.

association, soil *See* soil association.

Atterberg limits Water contents of fine-grained soils at different states of consistency.
>**liquid limit (LL)** The water content corresponding to the arbitrary limit between the liquid and plastic states of consistency of a soil.
>**plastic limit (PL)** The water content corresponding to an arbitrary limit between the plastic and semisolid states of consistency of a soil.

autochthonous organisms Those microorganisms thought to subsist on the more resistant soil organic matter and little affected by the addition of fresh organic materials. *Contrast with* zymogenous organisms. *See also* k-strategist.

autotroph An organism capable of utilizing carbon dioxide or carbonates as the sole source of carbon and obtaining energy for life processes from the oxidation of inorganic elements or compounds such as iron, sulfur, hydrogen, ammonium, and nitrites, or from radiant energy. *Contrast with* heterotroph.

available nutrient That portion of any element or compound in the soil that can be readily absorbed and assimilated by growing plants. ("Available" should not be confused with "exchangeable.")

available water The portion of water in a soil that can be readily absorbed by plant roots. The amount of water released between the field capacity and the permanent wilting point.

B horizon A soil horizon, usually beneath the A or E horizon, that is characterized by one or more of the following: (1) a concentration of soluble salts, silicate clays, iron and aluminum oxides, and humus, alone or in combination; (2) a blocky or prismatic structure; and (3) coatings of iron and aluminum oxides that give darker, stronger, or redder color.

Bacteria One of two domains of single-celled prokaryote microorganisms. Includes all that are not Archaea.

bar A unit of pressure equal to 1 million dynes per square centimeter (10^6 dynes/cm^2). It approximates the pressure of a standard atmosphere.

base-forming cations (Obsolete) Those cations that form strong (strongly dissociated) bases by reaction with hydroxyl; e.g., K^+ forms potassium hydroxide (K^+ + OH). *See* nonacid cations.

base saturation percentage The extent to which the adsorption complex of a soil is saturated with exchangeable cations other than hydrogen and aluminum. It is expressed as a percentage of the total cation exchange capacity. *See* nonacid saturation.

bedding (Engineering) Arranging the surface of fields by plowing and grading into a series of elevated beds separated by shallow depressions or ditches for drainage.

bedrock The solid rock underlying soils and the regolith in depths ranging from zero (where exposed by erosion) to several hundred feet.

bench terrace An embankment constructed across sloping fields with a steep drop on the downslope side.

beta particle A high-speed electron emitted in radioactive decay.

bioaccumulation A buildup within an organism of specific compounds due to biological processes. Commonly applied to heavy metals, pesticides, or metabolites.

bioaugmentation The cleanup of contaminated soils by adding exotic microorganisms that are especially efficient at breaking down an organic contaminant. A form of *bioremediation.*

biodegradable Subject to degradation by biochemical processes.

biological nitrogen fixation Occurs at ordinary temperatures and pressures. It is commonly carried out by certain bacteria, algae, and actinomycetes, which may or may not be associated with higher plants.

biomass The total mass of living material of a specified type (e.g., microbial biomass) in a given environment (e.g., in a cubic meter of soil).

biopores Soil pores, usually of relatively large diameter, created by plant roots, earthworms, or other soil organisms.

bioremediation The decontamination or restoration of polluted or degraded soils by means of enhancing the chemical degradation or other activities of soil organisms.

biosequence A group of related soils that differ, one from the other, primarily because of differences in kinds and numbers of plants and soil organisms as a soil-forming factor.

biosolids Sewage sludge that meets certain regulatory standards, making it suitable for land application. *See* sewage sludge.

biostimulation The cleanup of contaminated soils through the manipulation of nutrients or other soil environmental factors to enhance the activity of naturally occurring soil microorganisms. A form of *bioremediation.*

blocky soil structure Soil aggregates with blocklike shapes; common in B horizons of soils in humid regions.

broad-base terrace A low embankment with such gentle slopes that it can be farmed, constructed across sloping fields to reduce erosion and runoff.

broadcast Scatter seed or fertilizer on the surface of the soil.

brownfields Abandoned, idled, or underused industrial and commercial facilities where expansion or redevelopment is complicated by real or perceived environmental contamination.

buffering capacity The ability of a soil to resist changes in pH. Commonly determined by presence of clay, humus, and other colloidal materials.

bulk blended fertilizers Solid fertilizer materials blended together in small blending plants, delivered to the farm in bulk, and usually spread directly on the fields by truck or other special applicator.

bulk blending Mixing dry individual granulated fertilizer materials to form a mixed fertilizer that is applied promptly to the soil.

bulk density, soil The mass of dry soil per unit of bulk volume, including the air space. The bulk volume is determined before drying to constant weight at 105 °C.

buried soil Soil covered by an alluvial, loessal, or other deposit, usually to a depth greater than the thickness of the solum.

by-pass flow *See* preferential flow.

C horizon A mineral horizon, generally beneath the solum, that is relatively unaffected by biological activity and pedogenesis and is lacking properties diagnostic of an A or B horizon. It may or may not be like the material from which the A and B have formed.

calcareous soil Soil containing sufficient calcium carbonate (often with magnesium carbonate) to effervesce visibly when treated with cold 0.1 N hydrochloric acid.

calcic horizon A diagnostic subsurface horizon of secondary carbonate enrichment that is more than 15 cm thick, has a calcium carbonate equivalent of more than 15%, and has at least 5% more calcium carbonate equivalent than the underlying C horizon.

caliche A layer near the surface, more or less cemented by secondary carbonates of calcium or magnesium precipitated from the soil solution. It may occur as a soft, thin soil horizon; as a hard, thick bed just beneath the solum; or as a surface layer exposed by erosion.

cambic horizon A diagnostic subsurface horizon that has a texture of loamy very fine sand or finer, contains some weatherable minerals, and is characterized by the alteration or removal of mineral material. The cambic horizon lacks cementation or induration and has too few evidences of illuviation to meet the requirements of the argillic or spodic horizon.

capillary conductivity (Obsolete) *See* hydraulic conductivity.

capillary fringe A zone in the soil just above the plane of zero water pressure (water table) that remains saturated or almost saturated with water.

capillary water The water held in the capillary or *small* pores of a soil, usually with a tension >60 cm of water. *See also* soil water potential.

carbon cycle The sequence of transformations whereby carbon dioxide is fixed in living organisms by photosynthesis or by chemosynthesis, liberated by respiration and by the death and decomposition of the fixing organism, used by heterotrophic species, and ultimately returned to its original state.

carbon/nitrogen ratio The ratio of the weight of organic carbon (C) to the weight of total nitrogen (N) in a soil or in organic material.

carnivore An organism that feeds on animals.

casts, earthworm Rounded, water-stable aggregates of soil that have passed through the gut of an earthworm.

cat clays Wet clay soils high in reduced forms of sulfur that, upon being drained, become extremely acid because of the oxidation of the sulfur compounds and the formation of sulfuric acid. Usually found in tidal marshes. *See* acid sulfate soils.

catena A group of soils that commonly occur together in a landscape, each characterized by a different slope position and resulting set of drainage-related proprieties. *See also* toposequence.

cation A positively charged ion; during electrolysis it is attracted to the negatively charged cathode.

cation exchange The interchange between a cation in solution and another cation on the surface of any surface-active material, such as clay or organic matter.

cation exchange capacity The sum total of exchangeable cations that a soil can adsorb. Sometimes called *total-exchange capacity, base-exchange capacity,* or *cation-adsorption capacity.* Expressed in centimoles of charge per kilogram ($cmol_c/kg$) of soil (or of other adsorbing material, such as clay).

cemented Indurated; having a hard, brittle consistency because the particles are held together by cementing substances, such as humus, calcium carbonate, or the oxides of silicon, iron, and aluminum.

channery Thin, flat fragments of limestone, sandstone, or schist up to 15 cm (6 in.) in major diameter.

chelate (Greek, claw) A type of chemical compound in which a metallic ion is firmly combined with an organic molecule by means of multiple chemical bonds.

chert A structureless form of silica, closely related to flint, that breaks into angular fragments.

chisel, subsoil A tillage implement with one or more cultivator-type feet to which are attached strong knife-like units used to shatter or loosen hard, compact layers, usually in the subsoil, to depths below normal plow depth. *See also* subsoiling.

chlorite A 2:1:1-type layer-structured silicate mineral having 2:1 layers alternating with a magnesium-dominated octahedral sheet.

chlorosis A condition in plants relating to the failure of chlorophyll (the green coloring matter) to develop. Chlorotic leaves range from light green through yellow to almost white.

chroma (color) *See* Munsell color system.

chronosequence A sequence of related soils that differ, one from the other, in certain properties primarily as a result of time as a soil-forming factor.

classification, soil *See* soil classification.

clay (1) A soil separate consisting of particles <0.002 mm in equivalent diameter. (2) A soil textural class containing >40% clay, <45% sand, and <40% silt.

clay mineral Naturally occurring inorganic material (usually crystalline) found in soils and other earthy deposits, the particles being of clay size, that is, <0.002 mm in diameter.

claypan A dense, compact, slowly permeable layer in the subsoil having a much higher clay content than the overlying material, from which it is separated by a sharply defined boundary. Claypans are usually hard when dry and plastic and sticky when wet. *See also* hardpan.

climosequence A group of related soils that differ, one from another, primarily because of differences in climate as a soil-forming factor.

clod A compact, coherent mass of soil produced artificially, usually by such human activities as plowing and digging, especially when these operations are performed on soils that are either too wet or too dry for normal tillage operations.

coarse fragments Mineral (rock) soil particles larger than 2 mm in diameter. *Compare to* fine earth fraction.

coarse texture The texture exhibited by sands, loamy sands, and sandy loams (except very fine sandy loam).

cobblestone Rounded or partially rounded rock or mineral fragments 7.5 to 25 cm (3 to 10 in.) in diameter.

co-composting A method of composting in which two materials of differing but complementary nature are mingled together and enhance each other's decomposition in a compost system.

cohesion Holding together: force holding a solid or liquid together, owing to attraction between like molecules. Decreases with rise in temperature.

collapsible soil Certain soil that may undergo a sudden loss in strength when wetted.

colloid, soil (Greek, gluelike) Organic and inorganic matter with very small particle size and a correspondingly large surface area per unit of mass.

colluvium A deposit of rock fragments and soil material accumulated at the base of steep slopes as a result of gravitational action.

color The property of an object that depends on the wavelength of light it reflects or emits.

columnar soil structure *See* soil structure types.

Compaction *See* Soil compaction.

companion planting The practice of growing certain species of plants in close proximity because one species has the effect of improving the growth of the other, sometimes by positive *allelopathic* effects.

compost Organic residues, or a mixture of organic residues and soil, that have been piled, moistened, and allowed to undergo biological decomposition. Mineral fertilizers are sometimes added. Often called *artificial manure* or *synthetic manure* if produced primarily from plant residues.

concretion A local concentration of a chemical compound, such as calcium carbonate or iron oxide, in the form of grains or nodules of varying size, shape, hardness, and color.

conduction The transfer of heat by physical contact between two or more objects.

conductivity, hydraulic *See* hydraulic conductivity.

conservation tillage *See* tillage, conservation.

consistence The combination of properties of soil material that determine its resistance to crushing and its ability to be molded or changed in shape. Such terms as *loose, friable, firm, soft, plastic,* and *sticky* describe soil consistence.

consistency The interaction of adhesive and cohesive forces within a soil at various moisture contents as expressed by the relative ease with which the soil can be deformed or ruptured.

consociation *See* soil consociation.

consolidation test A laboratory test in which a soil mass is laterally confined within a ring and is compressed with a known force between two porous plates.

constant charge The net surface charge of mineral particles, the magnitude of which depends only on the chemical and structural composition of the mineral. The charge arises from isomorphous substitution and is not affected by soil pH.

consumptive use The water used by plants in transpiration and growth, plus water vapor loss from adjacent soil or snow, or from intercepted precipitation in any specified time. Usually expressed as equivalent depth of free water per unit of time.

contour An imaginary line connecting points of equal elevation on the surface of the soil. A contour terrace is laid out on a sloping soil at right angles to the direction of the slope and nearly level throughout its course.

contour strip-cropping Layout of crops in comparatively narrow strips in which the farming operations are performed approximately on the contour. Usually strips of grass, close-growing crops, or fallow are alternated with those of cultivated crops.

controlled traffic A farming system in which all wheeled traffic is confined to fixed paths so that repeated compaction of the soil does not occur outside the selected paths.

convection The transfer of heat through a gas or solution because of molecular movement.

cover crop A close-growing crop grown primarily for the purpose of protecting and improving soil between periods of regular crop production or between trees and vines in orchards and vineyards.

creep Slow mass movement of soil and soil material down relatively steep slopes, primarily under the influence of gravity, but facilitated by saturation with water and by alternate freezing and thawing.

crop rotation A planned sequence of crops growing in a regularly recurring succession on the same area of land, as contrasted to continuous culture of one crop or growing different crops in haphazard order.

crotovina A former animal burrow in one soil horizon that has been filled with organic matter or material from another horizon (also spelled *krotovina*).

crumb A soft, porous, more or less rounded natural unit of structure from 1 to 5 mm in diameter. *See also* soil structure types.

crushing strength The force required to crush a mass of dry soil or, conversely, the resistance of the dry soil mass to crushing. Expressed in units of force per unit area (pressure).

crust (soil) (i) physical A surface layer on soils, ranging in thickness from a few millimeters to as much as 3 cm, that physical-chemical processes have caused to be much more compact, hard, and brittle when dry than the material immediately beneath it.

  (ii) microbiotic An assemblage of cyanobacteria, algae, lichens, liverworts, and mosses that commonly forms an irregular crust on the soil surface, especially on otherwise barren, arid-region soils. Also referred to as cryptogamic, cryptobiotic, or biological crusts.

cryophilic Pertaining to low temperatures in the range of 5 to 15 °C, the range in which cryophilic organisms grow best.

cryoturbation Physical disruption and displacement of soil material within the profile by the forces of freezing and thawing. Sometimes called *frost churning*, it results in irregular, broken horizons, involutions, oriented rock fragments, and accumulation of organic matter on the permafrost table.

cryptogam *See* crust (ii) microbiotic.

crystal A homogeneous inorganic substance of definite chemical composition bounded by planar surfaces that form definite angles with each other, thus giving the substance a regular geometrical form.

crystal structure The orderly arrangement of atoms in a crystalline material.

cultivation   A tillage operation used in preparing land for seeding or transplanting or later for weed control and for loosening the soil.

cutans   A modification of the texture, structure, or fabric at natural surfaces in soil materials due to concentration of particular soil constituents; e.g. "clay skins."

cyanobacteria   Chlorophyll-containing bacteria that accommodate both photosynthesis and nitrogen fixation. Formerly called blue-green algae.

deciduous plant   A plant that sheds all its leaves every year at a certain season.

decomposition   Chemical breakdown of a compound (e.g., a mineral or organic compound) into simpler compounds, often accomplished with the aid of microorganisms.

deflocculate   (1) To separate the individual components of compound particles by chemical and/or physical means. (2) To cause the particles of the *disperse phase* of a colloidal system to become suspended in the *dispersion medium*.

delineation   An individual polygon shown by a closed boundary on a soil map that defines the area, shape, and location of a map unit within a landscape.

delivery ratio   The ratio of eroded sediment carried out of a drainage basin to the total amount of sediment moved within the basin by erosion processes.

delta   An alluvial deposit formed where a stream or river drops its sediment load upon entering a quieter body of water.

denitrification   The biochemical reduction of nitrate or nitrite to gaseous nitrogen, either as molecular nitrogen or as an oxide of nitrogen.

density   *See* particle density; bulk density.

desalinization   Removal of salts from saline soil, usually by leaching.

desert crust   A hard layer, containing calcium carbonate, gypsum, or other binding material, exposed at the surface in desert regions.

desert pavement   A natural residual concentration of closely packed pebbles, boulders, and other rock fragments on a desert surface where wind and water action has removed all smaller particles.

desert varnish   A thin, dark, shiny film or coating of iron oxide and lesser amounts of manganese oxide and silica formed on the surfaces of pebbles, boulders, rock fragments, and rock outcrops in arid regions.

desorption   The removal of sorbed material from surfaces.

detritivore   An organism that subsists on detritus.

detritus   Debris from dead plants and animals.

diagnostic horizons   (As used in *Soil Taxonomy*): Horizons having specific soil characteristics that are indicative of certain classes of soils. Horizons that occur at the soil surface are called *epipedons;* those below the surface, *diagnostic subsurface horizons*.

diatomaceous earth   A geologic deposit of fine, grayish, siliceous material composed chiefly or wholly of the remains of diatoms. It may occur as a powder or as a porous, rigid material.

diatoms   Algae having siliceous cell walls that persist as a skeleton after death; any of the microscopic unicellular or colonial algae constituting the class Bacillariaceae. They occur abundantly in fresh and salt waters and their remains are widely distributed in soils.

diffusion   The movement of atoms in a gaseous mixture or of ions in a solution, primarily as a result of their own random motion.

dioctahedral sheet   An octahedral sheet of silicate clays in which the sites for the six-coordinated metallic atoms are mostly filled with trivalent atoms, such as Al^{3+}.

disintegration   Physical or mechanical breakup or separation of a substance into its component parts (e.g., a rock breaking into its mineral components).

disperse   (1) To break up compound particles, such as aggregates, into the individual component particles. (2) To distribute or suspend fine particles, such as clay, in or throughout a dispersion medium, such as water.

Dissimilatory nitrate reduction to ammonium (DNRA)   A bacterial process by which nitrate is converted to ammonium under a wide range of oxygen and carbon levels. Compare to dentrification (a different type of dissimilatory nitrate reduction) which is strictly anaerobic and requires an energy source.

dissolution   Process by which molecules of a gas, solid, or another liquid dissolve in a liquid, thereby becoming completely and uniformly dispersed throughout the liquid's volume.

distribution coefficient (K_d)   The distribution of a chemical between soil and water.

diversion terrace   *See* terrace.

drain   (1) To provide channels, such as open ditches or drain tile, so that excess water can be removed by surface or by internal flow. (2) To lose water (from the soil) by percolation.

drain field, septic tank   An area of soil into which the effluent from a septic tank is piped so that it will drain through the lower part of the soil profile for disposal and purification.

drainage, soil   The frequency and duration of periods when the soil is free from saturation with water.

drift   Material of any sort deposited by geological processes in one place after having been removed from another. Glacial drift includes material moved by the glaciers and by the streams and lakes associated with them.

drumlin   Long, smooth, cigar-shaped low hills of glacial till, with their long axes parallel to the direction of ice movement.

dryland farming The practice of crop production in low-rainfall areas without irrigation.

duff The matted, partly decomposed organic surface layer of forest soils.

duripan A diagnostic subsurface horizon that is cemented by silica, to the point that air-dry fragments will not slake in water or HCL. Hardpan.

dust mulch A loose, finely granular or powdery condition on the surface of the soil, usually produced by shallow cultivation.

E horizon Horizon characterized by maximum eluviation (washing out) of silicate clays and iron and aluminum oxides; commonly occurs above the B horizon and below the A horizon.

earthworms Animals of the Lumbricidae family that burrow into and live in the soil. They mix plant residues into the soil and improve soil aeration.

ecosystem A dynamic and interacting combination of all the living organisms and nonliving elements (matter and energy) of an area.

ecosystem services Products of natural ecosystems that support and fulfill the needs of human beings. Provision of clean water and unpolluted air are examples.

ectotrophic mycorrhiza (ectomycorrhiza) A symbiotic association of the mycelium of fungi and the roots of certain plants in which the fungal hyphae form a compact mantle on the surface of the roots and extend into the surrounding soil and inward between cortical cells, but not into these cells. Associated primarily with certain trees. *See also* endotrophic mycorrhiza.

edaphology The science that deals with the influence of soils on living things, particularly plants, including human use of land for plant growth.

effective cation exchange capacity The amount of cation charges that a material (usually soil or soil colloids) can hold at the pH of the material, measured as the sum of the exchangeable Al^{3+}, Ca^{2+}, Mg^{2+}, K^+, and Na^+, and expressed as moles or cmol of charge per kg of material. *See* cation exchange capacity.

effective precipitation That portion of the total precipitation that becomes available for plant growth or for the promotion of soil formation.

E_h In soils, it is the potential created by oxidation-reduction reactions that take place on the surface of a platinum electrode measured against a reference electrode, minus the Eh of the reference electrode. This is a measure of the oxidation-reduction potential of electrode-reactive components in the soil. *See also* pe.

electrical conductivity (EC) The capacity of a substance to conduct or transmit electrical current. In soils or water, measured in siemens/meter (or often dS/m), and related to dissolved solutes.

eluviation The removal of soil material in suspension (or in solution) from a layer or layers of a soil. Usually, the loss of material in solution is described by the term "leaching." *See also* illuviation and leaching.

endoaquic (endosaturation) A condition or moisture regime in which the soil is saturated with water in all layers from the upper boundary of saturation (water table) to a depth of 200 cm or more from the mineral soil surface. *See also* epiaquic.

endotrophic mycorrhiza (endomycorrhiza) A symbiotic association of the mycelium of fungi and roots of a variety of plants in which the fungal hyphae penetrate directly into root hairs, other epidermal cells, and occasionally into cortical cells. Individual hyphae also extend from the root surface outward into the surrounding soil. *See also* arbuscular mycorrhiza.

enrichment ratio The concentration of a substance (e.g., phosphorus) in eroded sediment divided by its concentration in the source soil prior to being eroded.

Entisols An order in *Soil Taxonomy*. Soils that have no diagnostic pedogenic horizons. They may be found in virtually any climate on very recent geomorphic surfaces.

eolian soil material Soil material accumulated through wind action. The most extensive areas in the United States are silty deposits (loess), but large areas of sandy deposits also occur.

epiaquic (episaturation) A condition in which the soil is saturated with water due to a perched water table in one or more layers within 200 cm of the mineral soil surface, implying that there are also one or more unsaturated layers within 200 cm below the saturated layer. *See also* endoaquic.

epipedon A diagnostic surface horizon that includes the upper part of the soil that is darkened by organic matter, or the upper eluvial horizons, or both. (*Soil Taxonomy.*)

equilibrium phosphorus concentration The concentration of phosphorus in a solution in equilibrium with a soil, the EPC_0 being the concentration of phosphorus achieved by desorption of phosphorus from a soil to phosphorus-free distilled water.

erosion (1) The wearing away of the land surface by running water, wind, ice, or other geological agents, including such processes as gravitational creep. (2) Detachment and movement of soil or rock by water, wind, ice, or gravity.

esker A narrow ridge of gravelly or sandy glacial material deposited by a stream in an ice-walled valley or tunnel in a receding glacier.

essential element A chemical element required for the normal growth of plants.

eukaryote An organism whose cells each have a visibly evident nucleus.

eutrophic Having concentrations of nutrients optimal (or nearly so) for plant or animal growth. (Said of algal-enriched bodies of water)

eutrophication Nutrient enrichment of lakes, ponds, and other such waters that stimulates the growth of aquatic organisms, which leads to a deficiency of oxygen in the water body.

evapotranspiration The combined loss of water from a given area, and during a specified period of time, by evaporation from the soil surface and by transpiration from plants.

exchange capacity The total ionic charge of the adsorption complex active in the adsorption of ions. *See also* anion exchange capacity; cation exchange capacity.

exchangeable ions Positively or negatively charged atoms or groups of atoms that are held on or near the surface of a solid particle by attraction to charges of the opposite sign, and which may be replaced by other like-charged ions in the soil solution.

exchangeable sodium percentage The extent to which the adsorption complex of a soil is occupied by sodium. It is expressed as follows:

$$ESP = \frac{\text{exchangeable sodium } (cmol_c/kg \text{ soil})}{\text{cation exchange capacity } (cmol_c/kg \text{ soil})} \times 100$$

exfoliation Peeling away of layers of a rock from the surface inward, usually as the result of expansion and contraction that accompany changes in temperature.

expansive soil Soil that undergoes significant volume change upon wetting and drying, usually because of a high content of swelling-type clay minerals.

external surface The area of surface exposed on the top, bottom, and sides of a clay crystal.

facultative organism An organism capable of both aerobic and anaerobic metabolism.

fallow Cropland left idle in order to restore productivity, mainly through accumulation of nutrients, water, and/or organic matter. Preceding a cereal grain crop in semiarid regions, land may be left in *summer fallow* for a period during which weeds are controlled by chemicals or tillage and water is allowed to accumulate in the soil profile. In humid regions, fallow land may be allowed to grow up in natural vegetation for a period ranging from a few months to many years. *Improved fallow* involves the purposeful establishment of plant species capable of restoring soil productivity more rapidly than a natural plant succession.

family, soil In *Soil Taxonomy*, one of the categories intermediate between the great group and the soil series. Families are defined largely on the basis of physical and mineralogical properties of importance to plant growth.

fauna The animal life of a region or ecosystem.

fen A calcium-rich, peat-accumulating wetland with relatively stagnant water.

ferrihydrite, $Fe_5HO_8 \cdot 4H_2O$ A dark reddish brown poorly crystalline iron oxide that forms in wet soils.

fertigation The application of fertilizers in irrigation waters, commonly through sprinkler systems.

fertility, soil The quality of a soil that enables it to provide essential chemical elements in quantities and proportions for the growth of specified plants.

fertilizer Any organic or inorganic material of natural or synthetic origin added to a soil to supply certain elements essential to the growth of plants.

fibric materials *See* organic soil materials.

field capacity (field moisture capacity) The percentage of water remaining in a soil two or three days after its having been saturated and after free drainage has practically ceased.

fine earth fraction That portion of the soil that passes through a 2 mm diameter sieve opening. *Compare to* coarse fragments.

fine texture Consisting of or containing large quantities of the fine fractions, particularly of silt and clay. (Includes clay loam, sandy clay loam, silty clay loam, sandy clay, silty clay, and clay textural classes.)

fine-grained mica A silicate clay having a 2:1-type lattice structure with much of the silicon in the tetrahedral sheet having been replaced by aluminum and with considerable interlayer potassium, which binds the layers together, prevents interlayer expansion and swelling, and limits interlayer cation exchange capacity.

fixation (1) For other than elemental nitrogen: the process or processes in a soil by which certain chemical elements are converted from a soluble or exchangeable form to a much less soluble or to a nonexchangeable form; for example, potassium, ammonium, and phosphorus fixation. (2) For elemental nitrogen: process by which gaseous elemental nitrogen is chemically combined with hydrogen to form ammonia. *See* biological nitrogen fixation.

flagstone A relatively thin rock or mineral fragment 15 to 38 cm in length commonly composed of shale, slate, limestone, or sandstone.

flocculate To aggregate or clump together individual, tiny soil particles, especially fine clay, into small clumps or floccules. Opposite of *deflocculate* or *disperse*.

floodplain The land bordering a stream, built up of sediments from overflow of the stream and subject to inundation when the stream is at flood stage. Sometimes called *bottomland*.

flora The sum total of the kinds of plants in an area at one time. The organisms loosely considered to be of the plant kingdom.

fluorapatite A member of the apatite group of minerals containing fluorine. Most common mineral in phosphate rock.

fluvial deposits Deposits of parent materials laid down by rivers or streams.

fluvioglacial *See* glaciofluvial deposits.

foliar diagnosis An estimation of mineral nutrient deficiencies (excesses) of plants based on examination of the chemical composition of selected plant parts, and the color and growth characteristics of the foliage of the plants.

food web   The community of organisms that relate to one another by sharing and passing on food substances. They are organized into trophic levels such as producers that create organic substances from sunlight and inorganic matter, to consumers and predators that eat the producers, dead organisms, waste products and each other.

forest floor   The forest soil O horizons, including litter and unincorporated humus, on the mineral soil surface.

fraction   A portion of a larger store of a substance operationally defined by a particular analysis or separation method. For example, the fulvic acid fraction of soil organic matter is defined by a series of laboratory procedures by which it is solubilized. *Compare* to pool.

fragipan   Dense and brittle pan or subsurface layer in soils that owes its hardness mainly to extreme density or compactness rather than high clay content or cementation. Removed fragments are friable, but the material in place is so dense that roots penetrate and water moves through it very slowly.

friable   A soil consistency term pertaining to soils that crumble with ease.

frigid   A soil temperature class with mean annual temperature below 8 °C.

fritted micronutrients   Sintered silicates having total guaranteed analyses of micronutrients with controlled (relatively slow) release characteristics.

fulvic acid   A term of varied usage but usually referring to the mixture of organic substances remaining in solution upon acidification of a dilute alkali extract from the soil.

functional diversity   The characteristic of an ecosystem exemplified by the capacity to carry out a large number of biochemical transformations and other functions.

functional group   An atom, or group of atoms, attached to a large molecule. Each functional group (e.g., —OH, —CH_3, —COOH, etc.) has a characteristic chemical reactivity.

fungi   Eukaryote microorganisms with a rigid cell wall. Some form long filaments of cells called *hyphae* that may grow together to form a visible body.

furrow slice   The uppermost layer of an arable soil to the depth of primary tillage; the layer of soil sliced away from the rest of the profile and inverted by a moldboard plow.

gabion   Partitioned, wire fabric containers, filled with stone at the site of use, to form flexible, permeable, and monolithic structures for earth retention.

gamma ray   A high-energy ray (photon) emitted during radioactive decay of certain elements.

Gelisols   An order in *Soil Taxonomy*. Soils that have permafrost within the upper 1 m, or upper 2 m if cryoturbation is also present. They may have an ochric, histic, mollic, or other epipedon.

gellic materials   Mineral or organic soil materials that have *cryoturbation* and/or ice in the form of lenses, veins, or wedges and the like.

genesis, soil   The mode of origin of the soil, with special reference to the processes responsible for the development of the solum, or true soil, from the unconsolidated parent material.

genetic horizon   Soil layers that resulted from soil-forming (pedogenic) processes, as opposed to sedimentation or other geologic processes.

geographic information system (GIS)   A method of overlaying, statistically analyzing, and integrating large volumes of spatial data of different kinds. The data are referenced to geographical coordinates and encoded in a form suitable for handling by computer.

geological erosion   Wearing away of the Earth's surface by water, ice, or other natural agents under natural environmental conditions of climate, vegetation, and so on, undisturbed by man. Synonymous with *natural erosion*.

gibbsite, Al(OH)$_3$ An aluminum trihydroxide mineral most common in highly weathered soils, such as Oxisols.

gilgai   The microrelief of soils produced by expansion and contraction with changes in moisture. Found in soils that contain large amounts of clay that swells and shrinks considerably with wetting and drying. Usually a succession of microbasins and microknolls in nearly level areas or of microvalleys and microridges parallel to the direction of the slope.

glacial drift   Rock debris that has been transported by glaciers and deposited, either directly from the ice or from the meltwater. The debris may or may not be heterogeneous.

glacial till   *See* till.

glaciofluvial deposits   Material moved by glaciers and subsequently sorted and deposited by streams flowing from the melting ice. The deposits are stratified and may occur in the form of outwash plains, deltas, kames, eskers, and kame terraces.

gleyed   A soil condition resulting from prolonged saturation with water and reducing conditions that manifest themselves in greenish or bluish colors throughout the soil mass or in mottles.

glomalin   A protein-sugar group of molecules secreted by certain fungi resulting in a sticky hyphal surface thought to contribute to aggregate stability.

goethite, FeOOH   A yellow-brown iron oxide mineral that accounts for the brown color in many soils.

granular structure   Soil structure in which the individual grains are grouped into spherical aggregates with indistinct sides. Highly porous granules are commonly called *crumbs*. A well-granulated soil has the best structure for most ordinary crop plants. *See also* soil structure types.

granulation   The process of producing granular materials. Commonly used to refer to the formation of soil structural granules, but also used to refer to the processing of powdery fertilizer materials into granules.

grassed waterway Broad and shallow channel, planted with grass (usually perennial species) that is designed to move surface water downslope without causing soil erosion.

gravitational potential That portion of the total *soil water potential* due to differences in elevation of the reference pool of pure water and that of the soil water. Since the soil water elevation is usually chosen to be higher than that of the reference pool, the gravitational potential is usually positive.

gravitational water Water that moves into, through, or out of the soil under the influence of gravity.

great group A category in *Soil Taxonomy*. The classes in this category contain soils that have the same kind of horizons in the same sequence and have similar moisture and temperature regimes.

green manure Plant material incorporated with the soil while green, or soon after maturity, for improving the soil.

greenhouse effect The entrapment of heat by upper atmosphere gases, such as carbon dioxide, water vapor, and methane, just as glass traps heat for a greenhouse. Increases in the quantities of these gases in the atmosphere will likely result in global warming that may have serious consequences for humankind.

groundwater Subsurface water in the zone of saturation that is free to move under the influence of gravity, often horizontally to stream channels.

grus A sediment or soil material comprised of loose grains of coarse sand and fine gravel size composed of quartz, feldspar and rock fragments. Produced from rocks by physical weathering or selectively transported by borrowing insects.

gully erosion The erosion process whereby water accumulates in narrow channels and, over short periods, removes the soil from this narrow area to considerable depths, ranging from 1 to 2 ft to as much as 23 to 30 m (75 to 100 ft).

gypsic horizon A diagnostic subsurface horizon of secondary calcium sulfate enrichment that is more than 15 cm thick.

gypsum requirement The quantity of gypsum required to reduce the exchangeable sodium percentage in a soil to an acceptable level.

halophyte A plant that requires or tolerates a saline (high salt) environment.

hard armor Pertains to the use of hard materials (such as large stones or concrete) to prevent soil and stream bank erosion by reducing the erosive force of flowing water. *See* soft armor.

hardpan A hardened soil layer, in the lower A or in the B horizon, caused by cementation of soil particles with organic matter or with such materials as silica, sesquioxides, or calcium carbonate. The hardness does not change appreciably with changes in moisture content and pieces of the hard layer do not slake in water. *See also* caliche; claypan.

harrowing A secondary broadcast tillage operation that pulverizes, smooths, and firms the soil in seedbed preparation, controls weeds, or incorporates material spread on the surface.

heaving The partial lifting of plants, buildings, roadways, fenceposts, etc., out of the ground, as a result of freezing and thawing of the surface soil during the winter.

heavy metals Those metals that have densities of 5.0 Mg/m or greater. Elements in soils include Cd, Co, Cr, Cu, Fe, Hg, Mn, Mo, Pb, and Zn.

heavy soil (Obsolete in scientific use) A soil with a high content of clay, and a high drawbar pull, hence difficult to cultivate.

hematite, Fe_2O_3 A red iron oxide mineral that contributes red color to many soils.

hemic material *See* organic materials.

herbicide A chemical that kills plants or inhibits their growth; intended for weed control.

herbivore A plant-eating animal.

heterotroph An organism capable of deriving energy for life processes only from the decomposition of organic compounds and incapable of using inorganic compounds as sole sources of energy or for organic synthesis. *Contrast with* autotroph.

histic epipedon A diagnostic surface horizon consisting of a thin layer of organic soil material that is saturated with water at some period of the year unless artificially drained and that is at or near the surface of a mineral soil.

Histosols An order in *Soil Taxonomy*. Soils formed from materials high in organic matter. Histosols with essentially no clay must have at least 20% organic matter by weight (about 78% by volume). This minimum organic matter content rises with increasing clay content to 30% (85% by volume) in soils with at least 60% clay.

horizon, soil A layer of soil, approximately parallel to the soil surface, differing in properties and characteristics from adjacent layers below or above it. *See also* diagnostic horizons.

horticulture The art and science of growing fruits, vegetables, and ornamental plants.

hue (color) *See* Munsell color system.

humic acid A mixture of variable or indefinite composition of dark organic substances, precipitated upon acidification of a dilute alkali extract from soil.

humic substances A series of complex, relatively high molecular weight, brown- to black-colored organic substances that make up 60 to 80% of the soil organic matter and are generally quite resistant to ready microbial attack.

humid climate Climate in regions where moisture, when distributed normally throughout the year, should not limit crop production. In cool climates annual precipitation may be as little as 25 cm; in hot climates, 150 cm or even more. Natural vegetation in uncultivated areas is forests.

humification The processes involved in the decomposition of organic matter and leading to the formation of humus.

humin The fraction of the soil organic matter that is not dissolved upon extraction of the soil with dilute alkali.

humus That more or less stable fraction of the soil organic matter remaining after the major portions of added plant and animal residues have decomposed. Usually it is dark in color.

hydration Chemical union between an ion or compound and one or more water molecules, the reaction being stimulated by the attraction of the ion or compound for either the hydrogen or the unshared electrons of the oxygen in the water.

hydraulic conductivity An expression of the readiness with which a liquid, such as water, flows through a solid, such as soil, in response to a given potential gradient.

hydric soils Soils that are water-saturated for long enough periods to produce reduced conditions and affect the growth of plants.

hydrogen bonding Relatively low energy bonding exhibited by a hydrogen atom located between two highly electronegative atoms, such as nitrogen or oxygen.

hydrologic cycle The circuit of water movement from the atmosphere to the Earth and back to the atmosphere through various stages or processes, as precipitation, interception, runoff, infiltration, percolation, storage, evaporation, and transpiration.

hydrolysis A reaction with water that splits the water molecule into H^+ and OH^- ions. Molecules or atoms participating in such reactions are said to *hydrolyze*.

hydronium A hydrated hydrogen ion (H_3O^+), the form of the hydrogen ion usually found in an aqueous system.

hydroperiod The duration of the presence of surface water in seasonal wetlands.

hydroponics Plant-production systems that use nutrient solutions and no solid medium to grow plants.

hydrostatic potential *See* submergence potential.

hydrous mica *See* fine-grained mica.

hydroxyapatite A member of the apatite group of minerals rich in hydroxyl groups. A nearly insoluble calcium phosphate.

hygroscopic coefficient The amount of moisture in a dry soil when it is in equilibrium with some standard relative humidity near a saturated atmosphere (about 98%), expressed in terms of percentage on the basis of oven-dry soil.

hyperaccumulator A plant with unusually high capacity to take up certain elements from soil resulting in very high concentrations of these elements in the plant's tissues. Often pertaining to concentrations of heavy metals to 1% or more of the tissue dry matter.

hyperthermic A soil temperature class with mean annual temperatures >22 °C.

hypha (pl. hyphae) Filament of fungal cells. Actinomycetes also produce similar, but thinner, filaments of cells.

hypoxia State of oxygen deficiency in an environment so low as to restrict biological respiration (in water, typically less than 2 to 3 mg O_2/L).

hysteresis A relationship between two variables that changes depending on the sequences or starting point. An example is the relationship between soil water content and water potential, for which different curves describe the relationship when a soil is gaining water or losing it.

igneous rock Rock formed from the cooling and solidification of magma that has not been changed appreciably since its formation.

illite *See* fine-grained mica.

illuvial horizon A soil layer or horizon in which material carried from an overlying layer has been precipitated from solution or deposited from suspension. The layer of accumulation.

illuviation The process of deposition of soil material removed from one horizon to another in the soil; usually from an upper to a lower horizon in the soil profile. *See also* eluviation.

immature soil A soil with indistinct or only slightly developed horizons because of the relatively short time it has been subjected to the various soil-forming processes. A soil that has not reached equilibrium with its environment.

immobilization The conversion of an element from the inorganic to the organic form in microbial tissues or in plant tissues, thus rendering the element not readily available to other organisms or to plants.

imogolite A poorly crystalline aluminosilicate mineral with an approximate formula $SiO_2Al_2O_3 \cdot 2.5H_2O$; occurs mostly in soils formed from volcanic ash.

impervious Resistant to penetration by fluids or by roots.

improved fallow *See* fallow.

Inceptisols An order in *Soil Taxonomy*. Soils that are usually moist with pedogenic horizons of alteration of parent materials but not of illuviation. Generally, the direction of soil development is not yet evident from the marks left by various soil-forming processes or the marks are too weak to classify in another order.

induced systemic resistance Plant defense mechanisms activated by a chemical signal produced by a rhizosphere bacteria. Although the process begins in the soil, it may confer disease resistance to leaves or other aboveground tissues.

indurated (soil) Soil material cemented into a hard mass that will not soften on wetting. *See also* consistence; hardpan.

infiltration The downward entry of water into the soil.

infiltration capacity A soil characteristic determining or describing the *maximum* rate at which water *can* enter the soil under specified conditions, including the presence of an excess of water.

inner-sphere complex A relatively strong (not easily reversed) chemical association or bonding directly between a specific ion and specific atoms or groups of atoms in the surface structure of a soil colloid.

inoculation The process of introducing pure or mixed cultures of microorganisms into natural or artificial culture media.

inorganic compounds All chemical compounds in nature except compounds of carbon other than carbon monoxide, carbon dioxide, and carbonates.

insecticide A chemical that kills insects.

intergrade A soil that possesses moderately well-developed distinguishing characteristics of two or more genetically related great soil groups.

interlayer (mineralogy) Materials between layers within a given crystal, including cations, hydrated cations, organic molecules, and hydroxide groups or sheets.

internal surface The area of surface exposed within a clay crystal between the individual crystal layers. *Compare with* external surface.

interstratification Mixing of silicate layers within the structural framework of a given silicate clay.

ionic double layer The distribution of cations in the soil solution resulting from the simultaneous attraction toward colloid particles by the particle's negative charge and the tendency of diffusion and thermal forces to move the cations away from the colloid surfaces. Also described as a diffuse double layer or a diffuse electrical double layer.

ions Atoms, groups of atoms, or compounds that are electrically charged as a result of the loss of electrons (cations) or the gain of electrons (anions).

iron-pan An indurated soil horizon in which iron oxide is the principal cementing agent.

irrigation efficiency The ratio of the water actually consumed by crops on an irrigated area to the amount of water diverted from the source onto the area.

isomorphous substitution The replacement of one atom by another of similar size in a crystal lattice without disrupting or changing the crystal structure of the mineral.

isotopes Two or more atoms of the same element that have different atomic masses because of different numbers of neutrons in the nucleus.

joule The SI energy unit defined as a force of 1 newton applied over a distance of 1 meter; 1 joule = 0.239 calorie.

K_d *See* distribution coefficient, K_d.

K_{oc} The distribution coefficient, K_d, calculated based on organic carbon content. $K_{oc} = K_d/foc$ where foc is the fraction of organic carbon.

kame A conical hill or ridge of sand or gravel deposited in contact with glacial ice.

kandic horizon A subsurface diagnostic horizon having a sharp clay increase relative to overlying horizons and having low-activity clays.

kaolinite An aluminosilicate mineral of the 1:1 crystal lattice group; that is, consisting of single silicon tetrahedral sheets alternating with single aluminum octahedral sheets.

K_{sat} Hydraulic conductivity when the soil is water saturated. *See also* hydraulic conductivity.

k-strategist An organism that maintains a relatively stable population by specializing in metabolism of resistant compounds that most other organisms cannot utilize. *Contrast with* r-strategist. *See also* autochthonous organisms.

labile A substance that is readily transformed by microorganisms or is readily available for uptake by plants.

lacustrine deposit Material deposited in lake water and later exposed either by lowering of the water level or by the elevation of the land.

land A broad term embodying the total natural environment of the areas of the Earth not covered by water. In addition to soil, its attributes include other physical conditions, such as mineral deposits and water supply; location in relation to centers of commerce, populations, and other land; the size of the individual tracts or holdings; and existing plant cover, works of improvement, and the like.

land capability classification A grouping of kinds of soil into special units, subclasses, and classes according to their capability for intensive use and the treatments required for sustained use. One such system has been prepared by the USDA Natural Resources Conservation Service.

land classification The arrangement of land units into various categories based upon the properties of the land or its suitability for some particular purpose.

land forming Shaping the surface of the land by scraping off the high spots and filling in the low spots with precision grading machinery to create a uniform, smooth slope, often for irrigation purposes. Also called *land smoothing*.

land-use planning The development of plans for the uses of land that, over long periods, will best serve the general welfare, together with the formulation of ways and means for achieving such uses.

laterite An iron-rich subsoil layer found in some highly weathered humid tropical soils that, when exposed and allowed to dry, becomes very hard and will not soften when rewetted. When erosion removes the overlying layers, the laterite is exposed and a virtual pavement results. *See also* plinthite.

layer (Clay mineralogy) A combination in silicate clays of (tetrahedral and octahedral) sheets in a 1:1, 2:1, or 2:1:1 combination.

leaching The removal of materials in solution from the soil by percolating waters. *See also* eluviation.

leaching requirement The leaching fraction of irrigation water necessary to keep soil salinity from exceeding a tolerance level of the crop to be grown.

leaf area index The ratio of the area of the total upper leaf surface of a plant canopy and the unit area on which the canopy is grown.

legume A pod-bearing member of the Leguminosae family, one of the most important and widely distributed plant families. Includes many valuable food and forage species, such as peas, beans, peanuts, clovers, alfalfas, sweet clovers, lespedezas, vetches, and kudzu. Nearly all legumes are associated with nitrogen-fixing organisms.

lichen A symbiotic relationship between fungi and cyanobacteria (blue-green algae) that enhances colonization of bare minerals and rocks. The fungi supply water and nutrients, the cyanobacteria the fixed nitrogen and carbohydrates from photosynthesis.

Liebig's law The growth and reproduction of an organism are determined by the nutrient substance (oxygen, carbon dioxide, calcium, etc.) that is available in minimum quantity with respect to organic needs; the *limiting factor*. Also attributed to Sprengel.

light soil (Obsolete in scientific use) A coarse-textured soil; a soil with a low drawbar pull and hence easy to cultivate. *See also* coarse texture; soil texture.

lignin The complex organic constituent of woody fibers in plant tissue that, along with cellulose, cements the cells together and provides strength. Lignins resist microbial attack and after some modification may become part of the soil organic matter.

lime (agricultural) In strict chemical terms, calcium oxide. In practical terms, a material containing the carbonates, oxides, and/or hydroxides of calcium and/or magnesium used to neutralize soil acidity.

lime requirement The mass of agricultural limestone, or the equivalent of other specified liming material, required to raise the pH of the soil to a desired value under field conditions.

limestone A sedimentary rock composed primarily of calcite ($CaCO_3$). If dolomite ($CaCO_3 \cdot MgCO_3$) is present in appreciable quantities, it is called a *dolomitic limestone*.

limiting factor *See* Liebig's law.

liquid limit (LL) *See* Atterberg limits.

lithosequence A group of related soils that differ, one from the other, in certain properties primarily as a result of parent material as a soil-forming factor.

loam The textural-class name for soil having a moderate amount of sand, silt, and clay. Loam soils contain 7 to 27% clay, 28 to 50% silt, and 23 to 52% sand.

loamy Intermediate in texture and properties between fine-textured and coarse-textured soils. Includes all textural classes with the words *loam* or *loamy* as a part of the class name, such as clay loam or loamy sand. *See also* loam; soil texture.

lodging Falling over of plants, either by uprooting or stem breakage.

loess Material transported and deposited by wind and consisting of predominantly silt-sized particles.

luxury consumption The intake by a plant of an essential nutrient in amounts exceeding what it needs. For example, if potassium is abundant in the soil, alfalfa may take in more than it requires.

lysimeter A device for measuring percolation (leaching) and evapotranspiration losses from a column of soil under controlled conditions.

macronutrient A chemical element necessary in large amounts (usually 50 mg/kg in the plant) for the growth of plants. Includes C, H, O, N, P, K, Ca, Mg, and S. (*Macro* refers to quantity and not to the essentiality of the element.) *See also* micronutrient.

macropores Larger soil pores, generally having a diameter greater than 0.06 mm, from which water drains readily by gravity.

map unit (mapping unit), soil A conceptual group of one to many component soils, delineated or identified by the same name in a soil survey, that represent similar landscape areas. *See also* delineation, soil consociation, soil complex, soil association, and undifferentiated group.

marl Soft and unconsolidated calcium carbonate, usually mixed with varying amounts of clay or other impurities.

marsh Periodically wet or continually flooded area with the surface not deeply submerged. Covered dominantly with sedges, cattails, rushes, or other hydrophytic plants. Subclasses include freshwater and saltwater marshes.

mass flow Movement of nutrients with the flow of water to plant roots.

matric potential That portion of the total *soil water potential* due to the attractive forces between water and soil solids as represented through adsorption and capillarity. It will always be negative.

mature soil A soil with well-developed soil horizons produced by the natural processes of soil formation and essentially in equilibrium with its present environment.

maximum retentive capacity The average moisture content of a disturbed sample of soil, 1 cm high, which is at equilibrium with a water table at its lower surface.

mechanical analysis (Obsolete) *See* particle size analysis; particle size distribution.

medium texture Intermediate between fine-textured and coarse-textured (soils). It includes the following textural classes: very fine sandy loam, loam, silt loam, and silt.

melanic epipedon A diagnostic surface horizon formed in volcanic parent material that contains more than 6% organic carbon, is dark in color, and has a very low bulk density and high anion adsorption capacity.

mellow soil A very soft, very friable, porous soil without any tendency toward hardness or harshness. *See also* consistence.

mesic A soil temperature class with mean annual temperature 8 to 15 °C.

mesofauna Animals of medium size, between approximately 2 and 0.2 mm in diameter.

mesophilic Pertaining to moderate temperatures in the range of 15 to 35 °C, the range in which mesophilic organisms grow best and in which mesophilic composting takes place.

metamorphic rock A rock that has been greatly altered from its previous condition through the combined action of heat and pressure. For example, marble is a metamorphic rock produced from limestone, gneiss is produced from granite, and slate is produced from shale.

methane, CH_4 An odorless, colorless gas commonly produced under anaerobic conditions. When released to the upper atmosphere, methane contributes to global warming. *See also* greenhouse effect.

micas Primary aluminosilicate minerals in which two silica tetrahedral sheets alternate with one alumina/magnesia octahedral sheet with entrapped potassium atoms fitting between sheets. They separate readily into visible sheets or flakes.

microfauna That part of the animal population which consists of individuals too small to be clearly distinguished without the use of a microscope. Includes protozoans and nematodes.

microflora That part of the plant population which consists of individuals too small to be clearly distinguished without the use of a microscope. Includes actinomycetes, algae, bacteria, and fungi.

micronutrient A chemical element necessary in only extremely small amounts (<50 mg/kg in the plant) for the growth of plants. Examples are B, Cl, Cu, Fe, Mn, and Zn. (*Micro* refers to the amount used rather than to its essentiality.) *See also* macronutrient.

micropores Relatively small soil pores, generally found within structural aggregates and having a diameter less than 0.06 mm. *Contrast to* macropores.

microrelief Small-scale local differences in topography, including mounds, swales, or pits that are only 1 m or so in diameter and with elevation differences of up to 2 m. *See also* gilgai.

mineral (i) An inorganic compound of defined composition found in rocks. (ii) An adjective meaning inorganic.

mineral nutrient An element in inorganic form used by plants or animals.

mineral soil A soil consisting predominantly of, and having its properties determined predominantly by, mineral matter. Usually contains <20% organic matter, but may contain an organic surface layer up to 30 cm thick.

mineralization The conversion of an element from an organic form to an inorganic state as a result of microbial decomposition.

minimum tillage *See* tillage, conservation.

minor element (Obsolete) *See* micronutrient.

moderately coarse texture Consisting predominantly of coarse particles. In soil textural classification, it includes all the sandy loams except the very fine sandy loam. *See also* coarse texture.

moderately fine texture Consisting predominantly of intermediate-sized (soil) particles or with relatively small amounts of fine or coarse particles. In soil textural classification, it includes clay loam, sandy loam, sandy clay loam, and silty clay loam. *See also* fine texture.

moisture potential *See* soil water potential.

mole drain Unlined drain formed by pulling a bullet-shaped cylinder through the soil.

mollic epipedon A diagnostic surface horizon of mineral soil that is dark colored and relatively thick, contains at least 0.6% organic carbon, is not massive and hard when dry, has a base saturation of more than 50%, has less than 250 mg/kg P_2O_5 soluble in 1% citric acid, and is dominantly saturated with bivalent cations.

Mollisols An order in *Soil Taxonomy*. Soils with nearly black, organic-rich surface horizons and high supply of bases. They have mollic epipedons and base saturation greater than 50% in any cambic or argillic horizon. They lack the characteristics of Vertisols and must not have oxic or spodic horizons.

molybdenosis A nutritional disease of ruminant animals in which high Mo in the forage interferes with copper absorption.

montmorillonite An aluminosilicate clay mineral in the smectite group with a 2:1 expanding crystal lattice, with two silicon tetrahedral sheets enclosing an aluminum octahedral sheet. Isomorphous substitution of magnesium for some of the aluminum has occurred in the octahedral sheet. Considerable expansion may be caused by water moving between silica sheets of contiguous layers.

mor Raw humus; type of forest humus layer of unincorporated organic material, usually matted or compacted or both; distinct from the mineral soil, unless the latter has been blackened by washing in organic matter.

moraine An accumulation of drift, with an initial topographic expression of its own, built within a glaciated region chiefly by the direct action of glacial ice. Examples are ground, lateral, recessional, and terminal moraines.

morphology, soil The constitution of the soil, including the texture, structure, consistence, color, and other physical, chemical, and biological properties of the various soil horizons that make up the soil profile.

mottling Spots or blotches of different color or shades of color interspersed with the dominant color.

mucigel   The gelatinous material at the surface of roots grown in unsterilized soil.

muck   Highly decomposed organic material in which the original plant parts are not recognizable. Contains more mineral matter and is usually darker in color than peat. *See also* muck soil; peat.

muck soil   (1) A soil containing 20 to 50% organic matter. (2) An organic soil in which the organic matter is well decomposed.

mulch   Any material such as straw, sawdust, leaves, plastic film, and loose soil that is spread upon the surface of the soil to protect the soil and plant roots from the effects of raindrops, soil crusting, freezing, evaporation, etc.

mulch tillage   *See* tillage, conservation.

mull   A humus-rich layer of forested soils consisting of mixed organic and mineral matter. A mull blends into the upper mineral layers without an abrupt change in soil characteristics.

Munsell color system   A color designation system that specifies the relative degrees of the three simple variables of color:
>   **chroma**   The relative purity, strength, or saturation of a color.
>   **hue**   The chromatic gradation (rainbow) of light that reaches the eye.
>   **value**   The degree of lightness or darkness of the color.

mycelium   A stringlike mass of individual fungal or actinomycetes hyphae.

myco   Prefix designating an association or relationship with a fungus (e.g., mycotoxins are toxins produced by a fungus).

mycorrhiza   The association, usually symbiotic, of fungi with the roots of seed plants. *See also* ectotrophic mycorrhiza; endotrophic mycorrhiza; arbuscular mycorrhiza.

natric horizon   A diagnostic subsurface horizon that satisfies the requirements of an argillic horizon, but that also has prismatic, columnar, or blocky structure and a subhorizon having more than 15% saturation with exchangeable sodium.

necrosis   Death associated with discoloration and dehydration of all or parts of plant organs, such as leaves.

nematodes   Very small (most are microscopic) unsegmented round worms. In soils they are abundant and perform many important functions in the soil food web. Some are plant parasites and considered pests.

neutral soil   A soil in which the surface layer, at least to normal plow depth, is neither acid nor alkaline in reaction. In practice this means the soil is within the pH range of 6.6 to 7.3. *See also* acid soil; alkaline soil; pH; reaction, soil.

nitrate depression period   A period of time, beginning shortly after the addition of fresh, highly carbonaceous organic materials to a soil, during which decomposer microorganisms have removed most of the soluble nitrate from the soil solution.

nitrification   The biochemical oxidation of ammonium to nitrate, predominantly by autotrophic bacteria.

nitrogen assimilation   The incorporation of nitrogen into organic cell substances by living organisms.

nitrogen cycle   The sequence of chemical and biological changes undergone by nitrogen as it moves from the atmosphere into water, soil, and living organisms, and upon death of these organisms (plants and animals) is recycled through a part or all of the entire process.

nitrogen fixation   The biological conversion of elemental nitrogen (N_2) to organic combinations or to forms readily utilized in biological processes.

nodule bacteria   *See* rhizobia.

nonacid cations   Those cations that do not react with water by hydrolysis to release H^+ ions to the soil solution. These cations do not remove hydroxyl ions from solution, but form strongly dissociated bases such as potassium hydroxide ($K^+ + OH$). Formerly called *base cations* or *base-forming cations* in soil science literature.

nonacid saturation   The proportion or percentage of a cation-exchange site occupied by nonacid cations. Formerly termed *base saturation*.

nonhumic substances   The portion of soil organic matter comprised of relatively low molecular weight organic substances; mostly identifiable biomolecules.

nonlimiting water range   The region bounded by the upper and lower soil water content over which water, oxygen, and mechanical resistance are not limiting to plant growth. *Compare with* available water.

nonpoint source   A pollution source that cannot be traced back to a single origin or source. Examples include water runoff from urban areas and leaching from croplands.

no-tillage   *See* tillage, conservation.

nucleic acids   Complex organic acids found in the nuclei of plant and animal cells; may be combined with proteins as nucleoproteins.

O horizon   Organic horizon of mineral soils.

ochric epipedon   A diagnostic surface horizon of mineral soil that is too light in color, too high in chroma, too low in organic carbon, or too thin to be a plaggen, mollic, umbric, anthropic, or histic epipedon, or that is both hard and massive when dry.

octahedral sheet   Sheet of horizontally linked, octahedral-shaped units that serve as the basic structural components of silicate (clay) minerals. Each unit consists of a central, six-coordinated metallic atom (e.g., Al, Mg, or Fe) surrounded by six hydroxyl groups that, in turn, are linked with other nearby metal atoms, thereby serving as interunit linkages that hold the sheet together.

oligotrophic   Environments, such as soils or lakes, which are poor in nutrients.

order, soil The category at the highest level of generalization in *Soil Taxonomy*. The properties selected to distinguish the orders are reflections of the degree of horizon development and the kinds of horizons present.

organic farming A system/philosophy of agriculture that does not allow the use of synthetic chemicals to produce plant and animal products, but instead emphasizes the management of soil organic matter and biological processes. In many countries, products are officially certified as being organic if inspections confirm that they were grown by these methods.

organic fertilizer By-product from the processing of animal or vegetable substances that contain sufficient plant nutrients to be of value as fertilizers.

organic soil A soil in which more than half of the profile thickness is comprised of organic soil materials.

organic soil materials (As used in *Soil Taxonomy*): (1) Saturated with water for prolonged periods unless artificially drained and having 18% or more organic carbon (by weight) if the mineral fraction is more than 60% clay, more than 12% organic carbon if the mineral fraction has no clay, or between 12 and 18% carbon if the clay content of the mineral fraction is between 0 and 60%. (2) Never saturated with water for more than a few days and having more than 20% organic carbon. Histosols develop on these organic soil materials. There are three kinds of organic materials:

> **fibric materials** The least decomposed of all the organic soil materials, containing very high amounts of fiber that are well preserved and readily identifiable as to botanical origin; with very low bulk density.
> **hemic materials** Intermediate in degree of decomposition of organic materials between the less decomposed fibric and the more decomposed sapric materials.
> **sapric materials** The most highly decomposed of the organic materials, having the highest bulk density, least amount of plant fiber, and lowest water content at saturation.

orographic Influenced by mountains (Greek *oros*). Used in reference to increased precipitation on the windward side of a mountain range induced as clouds rise over the mountain, leaving a *rain shadow* of reduced precipitation on the leeward side.

ortstein An indurated layer in the B horizon of Spodosols in which the cementing material consists of illuviated sesquioxides (mostly iron) and organic matter.

osmotic potential That portion of the total *soil water potential* due to the presence of solutes in soil water. It will generally be negative.

osmotic pressure Pressure exerted in living bodies as a result of unequal concentrations of salts on both sides of a cell wall or membrane. Water moves from the area having the lower salt concentration through the membrane into the area having the higher salt concentration and, therefore, exerts additional pressure on the side with higher salt concentration.

outer-sphere complex A relatively weak (easily reversed) chemical association or general attraction between an ion and an oppositely charged soil colloid via mutual attraction for intervening water molecules.

outwash plain A deposit of coarse-textured materials (e.g., sands and gravels) left by streams of meltwater flowing from receding glaciers.

oven-dry soil Soil that has been dried at 105 °C until it reaches constant weight.

oxic horizon A diagnostic subsurface horizon that is at least 30 cm thick and is characterized by the virtual *absence* of weatherable primary minerals or 2:1 lattice clays and the *presence* of 1:1 lattice clays and highly insoluble minerals, such as quartz sand, hydrated oxides of iron and aluminum, low cation exchange capacity, and small amounts of exchangeable bases.

oxidation The loss of electrons by a substance; therefore, a gain in positive valence charge and, in some cases, the chemical combination with oxygen gas.

oxidation ditch An artificial open channel for partial digestion of liquid organic wastes in which the wastes are circulated and aerated by a mechanical device.

oxidation-reduction potential *See* E_h and pe.

Oxisols An order in *Soil Taxonomy*. Soils with residual accumulations of low-activity clays, free oxides, kaolin, and quartz. They are mostly in tropical climates.

pans Horizons or layers in soils that are strongly compacted, indurated, or very high in clay content. *See also* caliche; claypan; fragipan; hardpan.

parent material The unconsolidated and more or less chemically weathered mineral or organic matter from which the solum of soils is developed by pedogenic processes.

particle density The mass per unit volume of the soil particles. In technical work, usually expressed as metric tons per cubic meter (Mg/m^3) or grams per cubic centimeter (g/cm^3).

particle size The effective diameter of a particle measured by sedimentation, sieving, or micrometric methods.

particle size analysis Determination of the various amounts of the different separates in a soil sample, usually by sedimentation, sieving, micrometry, or combinations of these methods.

particle size distribution The amounts of the various soil separates in a soil sample, usually expressed as weight percentages.

particulate organic matter A microbially active fraction of soil organic matter consisting largely of fine particles of partially decomposed plant tissue.

partitioning The distribution of organic chemicals (such as pollutants) into a portion that dissolves in the soil organic matter and a portion that remains undissolved in the soil solution.

pascal An SI unit of pressure equal to 1 newton per square meter.

peat Unconsolidated soil material consisting largely of undecomposed, or only slightly decomposed, organic matter accumulated under conditions of excessive moisture. *See also* organic soil materials; peat soil.

peat soil An organic soil containing more than 50% organic matter. Used in the United States to refer to the stage of decomposition of the organic matter, *peat* referring to the slightly decomposed or undecomposed deposits and *muck* to the highly decomposed materials. *See also* muck; muck soil; peat.

ped A unit of soil structure such as an aggregate, crumb, prism, block, or granule, formed by natural processes (in contrast to a *clod*, which is formed artificially).

pedology The science that deals with the formation, morphology, and classification of soil bodies as landscape components.

pedon The smallest volume that can be called *a soil*. It has three dimensions. It extends downward to the depth of plant roots or to the lower limit of the genetic soil horizons. Its lateral cross section is roughly hexagonal and ranges from 1 to 10 m^2 in size, depending on the variability in the horizons.

pedosphere The conceptual zone within the ecosystem consisting of soil bodies or directly influenced by them. A zone or sphere of activity in which mineral, water, air, and biological components come together to form soils. Usage is parallel to that for "atmosphere" or "biosphere."

pedoturbation Physical disturbance and mixing of soil horizons by such forces as burrowing animals (faunal pedoturbation) or frost churning (cryoturbation).

peneplain A once high, rugged area that has been reduced by erosion to a lower, gently rolling surface resembling a plain.

penetrability The ease with which a probe can be pushed into the soil. May be expressed in units of distance, speed, force, or work depending on the type of penetrometer used.

penetrometer An instrument consisting of a rod with a cone-shaped tip and a means of measuring the force required to push the rod into a specified increment of soil.

perc test *See* percolation test.

percolation, soil water The downward movement of water through soil. Especially, the downward flow of water in saturated or nearly saturated soil at hydraulic gradients of the order of 1.0 or less.

percolation test A measurement of the rate of percolation of water in a soil profile, usually to determine the suitability of a soil for use as a septic tank drain field.

perforated plastic pipe Pipe, sometimes flexible, with holes or slits in it that allow the entrance and exit of air and water. Used for soil drainage and for septic effluent spreading into soil.

permafrost (1) Permanently frozen material underlying the solum. (2) A perennially frozen soil horizon.

permanent charge *See* constant charge.

permanent wilting point *See* wilting point.

permeability, soil The ease with which gases, liquids, or plant roots penetrate or pass through a bulk mass of soil or a layer of soil.

petrocalcic horizon A diagnostic subsurface horizon that is a continuous, indurated calcic horizon cemented by calcium carbonate and, in some places, with magnesium carbonate. It cannot be penetrated with a spade or auger when dry; dry fragments do not slake in water; and it is impenetrable by roots.

petrogypsic horizon A diagnostic subsurface horizon that is a continuous, strongly cemented, massive gypsic horizon that is cemented by calcium sulfate. It can be chipped with a spade when dry. Dry fragments do not slake in water and it is impenetrable by roots.

pH, soil The negative logarithm of the hydrogen ion activity (concentration) of a soil. The degree of acidity (or alkalinity) of a soil as determined by means of a glass or other suitable electrode or indicator at a specified moisture content or soil-to-water ratio, and expressed in terms of the pH scale.

pH-dependent charge That portion of the total charge of the soil particles that is affected by, and varies with, changes in pH.

phase, soil A subdivision of a soil series or other unit of classification having characteristics that affect the use and management of the soil but do not vary sufficiently to differentiate it as a separate series. Included are such characteristics as degree of slope, degree of erosion, and content of stones.

photomap A mosaic map made from aerial photographs to which place names, marginal data, and other map information have been added.

phyllosphere The leaf surface.

physical properties (of soils) Those characteristics, processes, or reactions of a soil that are caused by physical forces and that can be described by, or expressed in, physical terms or equations. Examples of physical properties are bulk density, water-holding capacity, hydraulic conductivity, porosity, pore-size distribution, and so on.

physical weathering The breakdown of rock and mineral particles into smaller particles by physical forces such as frost action. *See also* weathering.

phytotoxic substances Chemicals that are toxic to plants.

placic horizon A diagnostic subsurface horizon of a black to dark reddish mineral soil that is usually thin but that may range from 1 to 25 mm in thickness. The placic horizon is commonly cemented with iron and is slowly permeable or impenetrable to water and roots.

plaggen epipedon   A diagnostic surface horizon that is human-made and more than 50 cm thick. Formed by long-continued manuring and mixing.

plant nutrients   *See* essential element.

plastic limit (PL)   *See* Atterberg limits.

plastic soil   A soil capable of being molded or deformed continuously and permanently, by relatively moderate pressure, into various shapes. *See also* consistence.

platy   Consisting of soil aggregates that are developed predominantly along the horizontal axes; laminated; flaky.

plinthite (brick)   A highly weathered mixture of sesquioxides of iron and aluminum with quartz and other diluents that occurs as red mottles and that changes irreversibly to hardpan upon alternate wetting and drying.

plow layer   The soil ordinarily moved when land is plowed; equivalent to *surface soil*.

plow pan   A subsurface soil layer having a higher bulk density and lower total porosity than layers above or below it, as a result of pressure applied by normal plowing and other tillage operations.

plowing   A primary broad-base tillage operation that is performed to shatter soil uniformly with partial to complete inversion.

point of zero charge   The pH value of a solution in equilibrium with a particle whose net charge, from all sources, is zero.

point source   A pollution source that can be traced back to its origin, which is usually an effluent discharge pipe. Examples are a wastewater treatment plant or a factory. *Opposite of* nonpoint source.

polypedon   (As used in *Soil Taxonomy*) Two or more contiguous pedons, all of which are within the defined limits of a single soil series; commonly referred to as a *soil individual*.

pool   A portion of a larger store of a substance defined by kinetic or theoretical properties. For example, the passive pool organic matter is defined by its very slow rate of microbial turnover. *Compare to* fraction.

pore size distribution   The volume of the various sizes of pores in a soil. Expressed as percentages of the bulk volume (soil plus pore space).

porosity, soil   The volume percentage of the total soil bulk not occupied by solid particles.

potential acidity   The acidity that could potentially be formed if reduced sulfur compounds in a potential acid sulfate soil were to become oxidized.

precision farming   The spatially variable management of a field or farm based on information specific to the soil or crop characteristics of many very small subunits of land. This technique commonly uses variable rate equipment, geo positioning systems and computer controls.

preferential flow   Nonuniform movement of water and its solutes through a soil along certain pathways, which are often macropores.

primary consumer   An organism that subsists on plant material.

primary mineral   A mineral that has not been altered chemically since deposition and crystallization from molten lava.

primary producer   An organism (usually a photosynthetic plant) that creates organic, energy-rich material from inorganic chemicals, solar energy, and water.

primary tillage   *See* tillage, primary.

priming effect   The increased decomposition of relatively stable soil humus under the influence of much enhanced, generally biological, activity resulting from the addition of fresh organic materials to a soil.

prismatic soil structure   A soil structure type with prismlike aggregates that have a vertical axis much longer than the horizontal axes.

Proctor test   A laboratory procedure that indicates the maximum achievable bulk density for a soil and the optimum water content for compacting a soil.

productivity, soil   The capacity of a soil for producing a specified plant or sequence of plants under a specified system of management. Productivity emphasizes the capacity of soil to produce crops and should be expressed in terms of yields.

profile, soil   A vertical section of the soil through all its horizons and extending into the parent material.

prokaryote   An organism whose cells do not have a distinct nucleus.

protein   Any of a group of nitrogen-containing organic compounds formed by the polymerization of a large number of amino acid molecules and that, upon hydrolysis, yield these amino acids. They are essential parts of living matter and are one of the essential food substances of animals.

protonation   Attachment of protons (H^+ ions) to exposed OH groups on the surface of soil particles, resulting in an overall positive charge on the particle surface.

protozoa   One-celled eukaryotic organisms, such as amoeba.

puddled soil   Dense, massive soil artificially compacted when wet and having no aggregated structure. The condition commonly results from the tillage of a clayey soil when it is wet.

rain, acid   *See* acid rain.

reaction, soil   (No longer used in soil science) The degree of acidity or alkalinity of a soil, usually expressed as a pH value or by terms ranging from extremely acid for pH values <4.5 to very strongly alkaline for pH values >9.0.

reactive nitrogen All forms of nitrogen that are readily available to biota (mainly ammonia, ammonium, and nitrate with smaller quantities of other compounds including nitrogen oxide gases) as opposed to unreactive nitrogen that exists mostly as inert N_2 gas.

recharge area A geographic area in which an otherwise confined aquifer is exposed to surficial percolation of water to recharge the groundwater in the aquifer.

redox concentrations Zones of apparent accumulations of Fe-Mn oxides in soils.

redox depletions Zones of low chroma (<2) where Fe-Mn oxides, and in some cases clay, have been stripped from the soil.

redox potential The electrical potential (measured in volts or millivolts) of a system due to the tendency of the substances in it to give up or acquire electrons.

redoximorphic features Soil properties associated with wetness that result from reduction and oxidation of iron and manganese compounds after saturation and desaturation with water. *See also* redox concentrations; redox depletions.

reduction The gain of electrons, and therefore the loss of positive valence charge, by a substance. In some cases, a loss of oxygen or a gain of hydrogen is also involved.

regolith The unconsolidated mantle of weathered rock and soil material on the Earth's surface; loose earth materials above solid rock. (Approximately equivalent to the term *soil* as used by many engineers.)

relief The relative differences in elevation between the upland summits and the lowlands or valleys of a given region.

residual material Unconsolidated and partly weathered mineral materials accumulated by disintegration of consolidated rock in place.

resilience The capacity of a soil (or other ecosystem) to return to its original state after a disturbance.

rhizobacteria Bacteria specially adapted to colonizing the surface of plant roots and the soil immediately around plant roots. Some have effects that promote plant growth, while others have effects that are deleterious to plants.

rhizobia Bacteria capable of living symbiotically with higher plants, usually in nodules on the roots of legumes, from which they receive their energy, and capable of converting atmospheric nitrogen to combined organic forms; hence the term *symbiotic nitrogen-fixing bacteria*. (Derived from the generic name *Rhizobium*.)

rhizoplane The root surface–soil interface. Used to describe the habitat of root-surface-dwelling microorganisms.

rhizosphere That portion of the soil in the immediate vicinity of plant roots in which the abundance and composition of the microbial population are influenced by the presence of roots.

rill A small, intermittent water course with steep sides; usually only a few centimeters deep and hence no obstacle to tillage operations.

rill erosion An erosion process in which numerous small channels of only several centimeters in depth are formed; occurs mainly on recently cultivated soils. *See also* rill.

riparian zone The area, both above and below the ground surface, that borders a river.

riprap Coarse rock fragments, stones, or boulders placed along a waterway or hillside to prevent erosion.

rock The material that forms the essential part of the earth's solid crust, including loose incoherent masses such as sand and gravel, as well as solid masses of granite and limestone.

root interception Acquisition of nutrients by a root as a result of the root growing into the vicinity of the nutrient source.

root nodules Swollen growths on plant roots. Often in reference to those in which symbiotic microorganisms live.

rotary tillage *See* tillage, rotary.

r-strategist Opportunistic organisms with short reproductive times that allow them to respond rapidly to the presence of easily metabolized food sources. *Contrast with* k-strategist. *See also* zymogenous organisms.

runoff The portion of the precipitation on an area that is discharged from the area through stream channels. That which is lost without entering the soil is called *surface runoff* and that which enters the soil before reaching the stream is called *groundwater runoff* or *seepage flow* from groundwater. (In soil science *runoff* usually refers to the water lost by surface flow; in geology and hydraulics *runoff* usually includes both surface and subsurface flow.)

salic horizon A diagnostic subsurface horizon of enrichment with secondary salts more soluble in cold water than gypsum. A salic horizon is 15 cm or more in thickness.

saline seep An area of land in which saline water seeps to the surface, leaving a high salt concentration behind as the water evaporates.

saline soil A nonsodic soil containing sufficient soluble salts to impair its productivity. The conductivity of a saturated extract is >4 dS/m, the exchangeable sodium adsorption ratio is less than about 13, and the pH is <8.5.

saline-sodic soil A soil containing sufficient exchangeable sodium to interfere with the growth of most crop plants and containing appreciable quantities of soluble salts. The exchangeable sodium adsorption ratio is >13, the conductivity of the saturation extract is >4 dS/m (at 25 °C), and the pH is usually 8.5 or less in the saturated soil.

salinization The process of accumulation of salts in soil.

saltation Particle movement in water or wind where particles skip or bounce along the stream bed or soil surface.

sand A soil particle between 0.05 and 2.0 mm in diameter; a soil textural class.

sapric materials *See* organic soil materials.

saprolite Soft, friable, weathered bedrock that retains the fabric and structure of the parent rock but is porous and can be dug with a spade.

saprophyte An organism that lives on dead organic material.

saturated paste extract The extract from a saturated soil paste, the electrical conductivity E_c of which gives an indirect measure of salt content in a soil.

saturation extract The solution extracted from a saturated soil paste.

saturation percentage The water content of a saturated soil paste, expressed as a dry weight percentage.

savanna (savannah) A grassland with scattered trees, either as individuals or clumps. Often a transitional type between true grassland and forest.

second bottom The first terrace above the normal floodplain of a stream.

secondary mineral A mineral resulting from the decomposition of a primary mineral or from the reprecipitation of the products of decomposition of a primary mineral. *See also* primary mineral.

sediment Transported and deposited particles or aggregates derived from soils, rocks, or biological materials.

sedimentary rock A rock formed from materials deposited from suspension or precipitated from solution and usually being more or less consolidated. The principal sedimentary rocks are sandstones, shales, limestones, and conglomerates.

seedbed The soil prepared to promote the germination of seed and the growth of seedlings.

self-mulching soil A soil in which the surface layer becomes so well aggregated that it does not crust and seal under the impact of rain but instead serves as a surface mulch upon drying.

semiarid Term applied to regions or climates where moisture is more plentiful than in arid regions but still definitely limits the growth of most crop plants. Natural vegetation in uncultivated areas is short grasses.

separate, soil One of the individual-sized groups of mineral soil particles—sand, silt, or clay.

septic tank An underground tank used in the deposition of domestic wastes. Organic matter decomposes in the tank, and the effluent is drained into the surrounding soil.

series, soil The soil series is a subdivision of a family in *Soil Taxonomy* and consists of soils that are similar in all major profile characteristics.

sewage effluent The liquid part of sewage or wastewater; it is usually treated to remove some portion of the dissolved organic compounds and nutrients present from the original sewage.

sewage sludge Settled sewage solids combined with varying amounts of water and dissolved materials, removed from sewage by screening, sedimentation, chemical precipitation, or bacterial digestion. Also called *biosolids* if certain quality standards are met.

shear Force, as of a tillage implement, acting at right angles to the direction of movement.

sheet (Mineralogy) A flat array of more than one atomic thickness and composed of one or more levels of linked coordination polyhedra. A sheet is thicker than a plane and thinner than a layer. Examples: tetrahedral sheet, octahedral sheet.

sheet erosion The removal of a fairly uniform layer of soil from the land surface by runoff water.

shelterbelt A wind barrier of living trees and shrubs established and maintained for protection of farm fields. Syn. *windbreak.*

shifting cultivation A farming system in which land is cleared, the debris burned, and crops grown for 2 to 3 years. When the farmer moves on to another plot, the land is then left idle for 5 to 15 years; then the burning and planting process is repeated.

short-range order minerals Minerals, such as allophane, whose structural framework consists of short distances of well-ordered crystalline structure interspersed with distances of noncrystalline amorphous materials.

shrinkage limit (SL) The water content above which a mass of soil material will swell in volume, but below which it will shrink no further.

side-dressing The application of fertilizer alongside row-crop plants, usually on the soil surface. Nitrogen materials are most commonly side-dressed.

siderophore A nonporphyrin metabolite secreted by certain microorganisms that forms a highly stable coordination compound with iron.

silica/alumina ratio The molecules of silicon dioxide (SiO_2) per molecule of aluminum oxide (Al_2O_3) in clay minerals or in soils.

silica/sesquioxide ratio The molecules of silicon dioxide (SiO_2) per molecule of aluminum oxide (Al_2O_3) plus ferric oxide (Fe_2O_3) in clay minerals or in soils.

silt (1) A soil separate consisting of particles between 0.05 and 0.002 mm in equivalent diameter. (2) A soil textural class.

silting The deposition of waterborne sediments in stream channels, lakes, reservoirs, or on floodplains, usually resulting from a decrease in the velocity of the water.

site index A quantitative evaluation of the productivity of a soil for forest growth under the existing or specified environment.

slag A product of smelting, containing mostly silicates; the substances not sought to be produced as matte or metal and having a lower specific gravity.

slash-and-burn *See* shifting cultivation.

slick spots Small areas in a field that are slick when wet because of a high content of alkali or exchangeable sodium.

slickensides Stress surfaces that are polished and striated and are produced by one mass sliding past another.

slope The degree of deviation of a surface from horizontal, measured in a numerical ratio, percent, or degrees.

slow fraction (of soil organic matter) That portion of soil organic matter that can be metabolized with great difficulty by the microorganisms in the soil and therefore has a slow turnover rate with a half-life in the soil ranging from a few years to a few decades. Often this fraction is the product of some previous decomposition.

smectite A group of silicate clays having a 2:1-type lattice structure with sufficient isomorphous substitution in either or both the tetrahedral and octahedral sheets to give a high interlayer negative charge and high cation exchange capacity and to permit significant interlayer expansion and consequent shrinking and swelling of the clay. Montmorillonite, beidellite, and saponite are in the smectite group.

sodic soil A soil that contains sufficient sodium to interfere with the growth of most crop plants, and in which the sodium adsorption ratio is 13 or greater.

sodium adsorption ratio (SAR)

$$SAR = \frac{[Na^+]}{\sqrt{1/2([Ca^{2+}] + [Mg^{2+}])}}$$

where the cation concentrations are in millimoles of charge per liter ($mmol_c/L$).

soft armor The bioengineering use of organic and/or inorganic materials combined with plants to create a living vegetation barrier of protection against erosion.

soil (1) A dynamic natural body composed of mineral and organic solids, gases, liquids and living organisms which can serve as a medium for plant growth. (2) The collection of natural bodies occupying parts of the Earth's surface that is capable of supporting plant growth and that has properties resulting from the integrated effects of climate and living organisms acting upon parent material, as conditioned by topography, over periods of time.

soil air The soil atmosphere; the gaseous phase of the soil, being that volume not occupied by soil or liquid.

soil alkalinity The degree or intensity of alkalinity of a soil, expressed by a value >7.0 on the pH scale.

soil amendment Any material, such as lime, gypsum, sawdust, or synthetic conditioner, that is worked into the soil to make it more amenable to plant growth.

soil association A group of defined and named taxonomic soil units occurring together in an individual and characteristic pattern over a geographic region, comparable to plant associations in many ways.

soil auger A tool used to bore small holes up to several meters deep in soils in order to bring up samples of material from various soil layers. It consists of a long T-handle attached to either a cylinder with twisted teeth or a screwlike bit.

soil classification (*Soil Taxonomy*) The systematic arrangement of soils into groups or categories on the basis of their characteristics. *See* order; suborder; great group; subgroup; family; and series.

Soil compaction The process or state of consolidation brought about by the application of mechanical forces to the soil which increase soil bulk density, and concomitantly decrease soil porosity. Often detrimental for plant growth and hydrologic soil functions.

soil complex A mapping unit used in detailed soil surveys where two or more defined taxonomic units are so intimately intermixed geographically that it is undesirable or impractical, because of the scale being used, to separate them. A more intimate mixing of smaller areas of individual taxonomic units than that described under *soil association*.

soil compressibility The property of a soil pertaining to its capacity to decrease in bulk volume when subjected to a load.

soil conditioner Any material added to a soil for the purpose of improving its physical condition.

soil conservation A combination of all management and land-use methods that safeguard the soil against depletion or deterioration caused by nature and/or humans.

soil consociation A kind of soil map unit that is named for the dominant soil taxon in the delineation, and in which at least half of the pedons are of the named soil taxon, and most of the remaining pedons are so similar as to not affect most interpretations.

soil correlation The process of defining, mapping, naming, and classifying the kinds of soils in a specific soil survey area, the purpose being to ensure that soils are adequately defined, accurately mapped, and uniformly named.

soil erosion *See* erosion.

soil fertility *See* fertility, soil.

soil genesis *See* genesis, soil.

soil geography A subspecialization of physical geography concerned with the areal distributions of soil types.

soil horizon *See* horizon, soil.

soil loss tolerance (T value) **(i)** The maximum average annual soil loss that will allow continuous cropping and maintain soil productivity without requiring additional management inputs. **(ii)** The maximum soil erosion loss that is offset by the theoretical maximum rate of soil development, which will maintain an equilibrium between soil losses and gains.

soil management The sum total of all tillage operations, cropping practices, fertilizer, lime, and other treatments conducted on or applied to a soil for the production of plants.

soil map A map showing the distribution of soil types or other soil mapping units in relation to the prominent physical and cultural features of the Earth's surface.

soil mechanics and engineering A subspecialization of soil science concerned with the effect of forces on the soil and the application of engineering principles to problems involving the soil.

soil moisture potential *See* soil water potential.

soil monolith A vertical section of a soil profile removed from the soil and mounted for display or study.

soil morphology The physical constitution, particularly the structural properties, of a soil profile as exhibited by the kinds, thicknesses, and arrangement of the horizons in the profile, and by the texture, structure, consistence, and porosity of each horizon.

soil order *See* order, soil.

soil organic matter The organic fraction of the soil that includes plant and animal residues at various stages of decomposition, cells and tissues of soil organisms, and substances synthesized by the soil population. Commonly determined as the amount of organic material contained in a soil sample passed through a 2-mm sieve.

soil porosity *See* porosity, soil.

soil productivity *See* productivity, soil.

soil profile *See* profile, soil.

soil quality The capacity of a specific kind of soil to function, within natural or managed ecosystem boundaries, to sustain plant and animal productivity, maintain or enhance water and air quality, and support human health and habitation. Sometimes considered in relation to this capacity in the undisturbed, natural state.

soil reaction *See* reaction, soil; pH, soil.

soil salinity The amount of soluble salts in a soil, expressed in terms of percentage, milligrams per kilogram, parts per million (ppm), or other convenient ratios.

soil separates *See* separate, soil.

soil series *See* series, soil.

soil solution The aqueous liquid phase of the soil and its solutes, consisting of ions dissociated from the surfaces of the soil particles and of other soluble materials.

soil strength A transient soil property related to the soil's solid phase cohesion and adhesion.

soil structure The combination or arrangement of primary soil particles into secondary particles, units, or peds. These secondary units may be, but usually are not, arranged in the profile in such a manner as to give a distinctive characteristic pattern. The secondary units

are characterized and classified on the basis of size, shape, and degree of distinctness into classes, types, and grades, respectively.

soil structure classes A grouping of soil structural units or peds on the basis of size from the very fine to very coarse.

soil structure grades A grouping or classification of soil structure on the basis of inter- and intraaggregate adhesion, cohesion, or stability within the profile. Four grades of structure, designated from 0 to 3, are recognized: *structureless, weak, moderate,* and *strong.*

soil structure types A classification of soil structure based on the shape of the aggregates or peds and their arrangement in the profile, including platy, prismatic, columnar, blocky, subangular blocky, granulated, and crumb.

soil survey The systematic examination, description, classification, and mapping of soils in an area. Soil surveys are classified according to the kind and intensity of field examination.

soil temperature classes A criterion used to differentiate soil in *Soil Taxonomy,* mainly at the family level. Classes are based on mean annual soil temperature and on differences between summer and winter temperatures at a depth of 50 cm.

soil textural class A grouping of soil textural units based on the relative proportions of the various soil separates (sand, silt, and clay). These textural classes, listed from the coarsest to the finest in texture, are sand, loamy sand, sandy loam, loam, silt loam, silt, sandy clay loam, clay loam, silty clay loam, sandy clay, silty clay, and clay. There are several subclasses of the sand, loamy sand, and sandy loam classes based on the dominant particle size of the sand fraction (e.g., loamy fine sand, coarse sandy loam).

soil texture The relative proportions of the various soil separates in a soil.

soil water deficit The difference between PET and ET, representing the gap between the amount of evapotranspiration water atmospheric conditions "demand" and the amount the soil can actually supply. A measure of the limitation that water supply places on plant productivity.

soil water potential (total) A measure of the difference between the free energy state of soil water and that of pure water. Technically it is defined as "that amount of work that must be done per unit quantity of pure water in order to transport reversibly and isothermically an infinitesimal quantity of water from a pool of pure water, at a specified elevation and at atmospheric pressure, to the soil water (at the point under consideration)." This *total* potential consists of *gravitational, matric,* and *osmotic* potentials.

solarization The process of heating a soil in the field by covering it with clear plastic sheeting during sunny conditions. The heat is meant to partially sterilize the upper 5 to 15 cm of soil to reduce pest and pathogen populations.

solum (pl. sola) The upper and most weathered part of the soil profile; the A, E, and B horizons.

sombric horizon A diagnostic subsurface horizon that contains illuvial humus but has a low cation exchange capacity and low percentage base saturation. Mostly restricted to cool, moist soils of high plateaus and mountainous areas of tropical and subtropical regions.

sorption The removal from the soil solution of an ion or molecule by adsorption and absorption. This term is often used when the exact mechanism of removal is not known.

species diversity The variety of different biological species present in an ecosystem. Generally, high diversity is marked by many species with few individuals in each.

species richness The number of different species present in an ecosystem, without regard to the distribution of individuals among those species.

specific gravity The ratio of the density of a mineral to the density of water at standard temperature and pressure.

specific heat capacity The amount of kinetic (heat) energy required to raise the temperature of 1 g of a substance (usually in reference to soil or soil components).

specific surface The solid particle surface area per unit mass or volume of the solid particles.

splash erosion The spattering of small soil particles caused by the impact of raindrops on very wet soils. The loosened and separated particles may or may not be subsequently removed by surface runoff.

spodic horizon A diagnostic subsurface horizon characterized by the illuvial accumulation of amorphous materials composed of aluminum and organic carbon with or without iron.

Spodosols An order in *Soil Taxonomy*. Soils with subsurface illuvial accumulations of organic matter and compounds of aluminum and usually iron. These soils are formed in acid, mainly coarse-textured materials in humid and mostly cool or temperate climates.

stem flow The process by which rain or irrigation water is directed by a plant canopy toward the plant stem so as to wet the soil unevenly under the plant canopy.

stratified Arranged in or composed of strata or layers.

strip-cropping The practice of growing crops that require different types of tillage, such as row and sod, in alternate strips along contours or across the prevailing direction of wind.

structure, soil *See* soil structure.

stubble mulch The stubble of crops or crop residues left essentially in place on the land as a surface cover before and during the preparation of the seedbed and at least partly during the growing of a succeeding crop.

subgroup, soil The *great groups* in *Soil Taxonomy* are subdivided into central concept subgroups that show the central properties of the great group, intergrade subgroups that show properties of more than one great group, and other subgroups for soils with atypical properties that are not characteristic of any great group.

submergence potential The positive hydrostatic pressure that occurs below the water table.

suborder, soil A category in *Soil Taxonomy* that narrows the ranges in soil moisture and temperature regimes, kinds of horizons, and composition, according to which of these is most important.

subsoil That part of the soil below the plow layer.

subsoiling Breaking of compact subsoils, without inverting them, with a special knifelike instrument (chisel), which is pulled through the soil at depths usually of 30 to 60 cm and at spacings usually of 1 to 2 m.

sulfidic Adjective used to describe sulfide-containing soil materials that initially have a pH > 4.0 and exhibit a drop of at least 0.5 pH unit within 8 weeks of aerated, moist incubation. Found in potential acid sulfate soils.

sulfuric horizon A diagnostic subsurface horizon in either mineral or organic soils that has a pH < 3.5 and fresh straw-colored mottles (called *jarosite mottles*). Forms by oxidation of sulfide-rich materials and is highly toxic to plants.

summer fallow *See* fallow.

surface runoff *See* runoff.

surface seal A thin layer of fine particles deposited on the surface of a soil that greatly reduces the permeability of the soil surface to water.

surface soil The uppermost part of the soil, ordinarily moved in tillage, or its equivalent in uncultivated soils. Ranges in depth from 7 to 25 cm. Frequently designated as the *plow layer*, the *Ap layer*, or the *Ap horizon*.

surface tension The elasticlike phenomenon resulting from the unbalanced attractions among liquid molecules (usually water) and between liquid and gaseous molecules (usually air) at the liquid–gas interface.

swamp An area of land that is usually wet or submerged under shallow fresh water and typically supports hydrophilic trees and shrubs.

symbiosis The living together in intimate association of two dissimilar organisms, the cohabitation being mutually beneficial.

synergism (i) The nonobligatory association between organisms that is mutually beneficial. Both populations can survive in their natural environment on their own, although, when formed, the association offers mutual advantages. **(ii)** The simultaneous actions of two or more factors that have a greater total effect together than the sum of their individual effects.

talus Fragments of rock and other soil material accumulated by gravity at the foot of cliffs or steep slopes.

taxonomy, soil The science of classification of soils; laws and principles governing the classifying of soil.

Also a specific *soil classification* system developed by the U.S. Department of Agriculture.

tensiometer   A device for measuring the negative pressure (or tension) of water in soil *in situ*; a porous, permeable ceramic cup connected through a tube to a manometer or vacuum gauge.

tension, soil-moisture   *See* soil water potential.

terrace   (1) A level, usually narrow, plain bordering a river, lake, or the sea. Rivers sometimes are bordered by terraces at different levels. (2) A raised, more or less level or horizontal strip of earth usually constructed on or nearly on a contour and designed to make the land suitable for tillage and to prevent accelerated erosion by diverting water from undesirable channels of concentration; sometimes called *diversion terrace*.

tetrahedral sheet   Sheet of horizontally linked, tetrahedron-shaped units that serve as one of the basic structural components of silicate (clay) minerals. Each unit consists of a central four-coordinated atom (e.g., Si, Al, Fe) surrounded by four oxygen atoms that, in turn, are linked with other nearby atoms (e.g., Si, Al, Fe), thereby serving as interunit linkages to hold the sheet together.

texture   *See* soil texture.

thermal analysis (differential thermal analysis)   A method of analyzing a soil sample for constituents, based on a differential rate of heating of the unknown and standard samples when a uniform source of heat is applied.

thermic   A soil temperature class with mean annual temperature 15 to 22 °C.

thermophilic   Pertaining to temperatures in the range of 45 to 90 °C, the range in which thermophilic organisms grow best and in which thermophilic composting takes place.

thermophilic organisms   Organisms that grow readily at temperatures above 45 °C.

thixotrophy   The property of certain clay soils of becoming fluid when jarred or agitated and then setting again when at rest. Similar to *quick*, as in quick clays or quicksand.

tile, drain   Pipe made of burned clay, concrete, or ceramic material, in short lengths, usually laid with open joints to collect and carry excess water from the soil.

till   (1) Unstratified glacial drift deposited directly by the ice and consisting of clay, sand, gravel, and boulders intermingled in any proportion. (2) To plow and prepare for seeding; to seed or cultivate the soil.

tillage   The mechanical manipulation of soil for any purpose; but in agriculture it is usually restricted to the modifying of soil conditions for crop production.

tillage, conservation   Any tillage sequence that reduces loss of soil or water relative to conventional tillage, which generally leaves at least 30% of the soil surface covered by residues, including the following systems:

> **minimum tillage**   The minimum soil manipulation necessary for crop production or meeting tillage requirements under the existing soil and climatic conditions.
>
> **mulch tillage**   Tillage or preparation of the soil in such a way that plant residues or other materials are left to cover the surface; also called *mulch farming, trash farming, stubble mulch tillage*, and *plowless farming*.
>
> **no-tillage system**   A procedure whereby a crop is planted directly into a seedbed not tilled since harvest of the previous crop; also called *zero tillage*.
>
> **ridge till**   Planting on ridges formed by cultivation during the previous growing period.
>
> **strip till**   Planting is done in a narrow strip that has been tilled and mixed, leaving the remainder of the soil surface undisturbed.

tillage, conventional   The combined primary and secondary tillage operations normally performed in preparing a seedbed for a given crop grown in a given geographic area. Usually said of non-conservation tillage.

tillage, primary   Tillage that contributes to the major soil manipulation, commonly with a plow.

tillage, rotary   An operation using a power-driven rotary tillage tool to loosen and mix soil.

tillage, secondary   Any tillage operations following primary tillage designed to prepare a satisfactory seedbed for planting.

tilth   The physical condition of soil as related to its ease of tillage, fitness as a seedbed, and its impedance to seedling emergence and root penetration.

topdressing   An application of fertilizer to a soil after the crop stand has been established.

toposequence   A sequence of related soils that differ, one from the other, primarily because of *topography* as a soil-formation factor, with other factors constant.

topsoil   (1) The layer of soil moved in cultivation. *See also* surface soil. (2) Presumably fertile soil material used to top-dress roadbanks, gardens, and lawns.

trace elements   Elements present in the Earth's crust in concentrations less than 1000 mg/kg. When referring to plant nutrients, the term *micronutrients* is preferred.

trioctahedral   An octahedral sheet of silicate clays in which the sites for the six-coordinated metallic atoms are mostly filled with divalent cations, such as Mg^{2+}.

trophic level   Levels in a food chain that pass nutrients and energy from one group of organisms to another.

truncated   Having lost all or part of the upper soil horizon or horizons.

tuff   Volcanic ash usually more or less stratified and in various states of consolidation.

tundra   A level or undulating treeless plain characteristic of arctic regions.

Ultisols   An order in *Soil Taxonomy*. Soils that are low in bases and have subsurface horizons of illuvial clay accumulations. They are usually moist, but during the warm season of the year some are dry part of the time.

umbric epipedon   A diagnostic surface horizon of mineral soil that has the same requirements as the mollic epipedon with respect to color, thickness, organic carbon content, consistence, structure, and P_2O_5 content, but that has a base saturation of less than 50%.

universal soil loss equation (USLE)   An equation for predicting the average annual soil loss per unit area per year; $A = RKLSPC$, where R is the climatic erosivity factor (rainfall plus runoff), K is the soil erodibility factor, L is the length of slope, S is the percent slope, P is the soil erosion practice factor, and C is the cropping and management factor.

unsaturated flow   The movement of water in a soil that is not filled to capacity with water.

vadose zone   The aerated region of soil above the permanent water table.

value (color)   *See* Munsell color system.

variable charge   *See* pH-dependent charge.

varnish, desert   A glossy sheen or coating on stones and gravel in arid regions.

vermicompost   Compost made by earthworms eating raw organic materials in moist aerated piles, which are kept shallow to avoid heat buildup that could kill the worms.

vermiculite   A 2:1-type silicate clay, usually formed from mica, that has a high net negative charge stemming mostly from extensive isomorphous substitution of aluminum for silicon in the tetrahedral sheet.

Vertisols   An order in *Soil Taxonomy*. Clayey soils with high shrink–swell potential that have wide, deep cracks when dry. Most of these soils have distinct wet and dry periods throughout the year.

vesicles   (1) Unconnected voids with smooth walls. (2) Spherical structures formed inside root cortical cells by vesicular arbuscular mycorrhizal fungi.

virgin soil   A soil that has not been significantly disturbed from its natural condition.

water deficit (soil)   The amount of available water removed from the soil within the vegetation's active rooting depth, or the amount of water required to bring the soil to field capacity.

water potential, soil   *See* soil water potential.

water table   The upper surface of groundwater or that level below which the soil is saturated with water.

water table, perched   The surface of a local zone of saturation held above the main body of groundwater by an impermeable layer of stratum, usually clay, and separated from the main body of groundwater by an unsaturated zone.

water use efficiency   Dry matter or harvested portion of crop produced per unit of water consumed.

waterlogged   Saturated with water.

watershed   All the land and water within the geographical confines of a drainage divide or surrounding ridges that separate the area from neighboring watersheds.

water-stable aggregate   A soil aggregate stable to the action of water, such as falling drops or agitation, as in wet-sieving analysis.

weathering   All physical and chemical changes produced in rocks, at or near the Earth's surface, by atmospheric agents.

wetland   An area of land that has hydric soil and hydrophytic vegetation, typically flooded for part of the year, and forming a transition zone between aquatic and terrestrial systems.

wetting front   The boundary between the wetted soil and dry soil during infiltration of water.

wilting point (permanent wilting point)   The moisture content of soil, on an oven-dry basis, at which plants wilt and fail to recover their turgidity when placed in a dark, humid atmosphere.

windbreak   Planting of trees, shrubs, or other vegetation perpendicular, or nearly so, to the principal wind direction to protect soils, crops, homesteads, etc., from wind and snow.

xenobiotic   Compounds foreign to biological systems. Often refers to compounds resistant to decomposition.

xerophytes   Plants that grow in or on extremely dry soils or soil materials.

zero tillage   *See* tillage, conservation.

zymogenous organisms   So-called opportunist organisms found in soils in large numbers immediately following addition of readily decomposable organic materials. *Contrast with* autochthonous organisms. *See also* r-strategist.

Calcium (*Continued*)
  and pH, 377, 378, 387, 393, 639, 640, 643, 644
  and plant nutrition, 409, 640–41
  and radionuclides in soil, 832
  soil forms and processes, 641–44
  in subsurface horizons, 84
Calcium carbonate, 369, 405–6, 410–11, 615
Calcium phosphate, 609, 612, 614–15
Calcium saturation percentage, 344
Calcium sulfate. *See* Gypsum
Cambids, 100, 101
Cambric horizon, 81, 83
Capacitance method for soil water measurement, 188, 189
Capillary action, 176–79, 202, 212, 221
Capillary fringe, 240, 241
Capillary water, 205, 221
Carbamates, 799
Carbon
  balance of, 495, 519–22
  carbon cycle, 495, 496–98, 531
  chlorinated hydrocarbons, 799, 805, 806
  FACE, 531
  global soil mass of, 498
  increase in soil, 888
  pools of organic, 517–19
  SOC, 342, 510, 522, 525, 527, 528
  soil organisms' sources of, 444, 447–48
Carbonates
  bicarbonate, 409–11, 657, 659
  calcium, 369, 405–6, 410–11, 615
  and pH, 371, 409–11
  and soil associations, 823
Carbon dioxide
  in carbon cycle, 497
  and carbonic acid, 359
  and gaseous exchange, 268
  and landfill decomposition, 831
  organic matter's regulation of, 530–35
  and pH, 387–88, 410–11
  and plant respiration, 274
  seasonal variations in soil release of, 279
  and soil air composition, 269
  soils' trapping of, 21, 25
Carbonic acid, 359, 497
Carbon/nitrogen (C/N) ratio, 504–5, 506–10, 524, 536, 548
Carbon/nitrogen/sulfur (C/N/S) ratio, 585
Carbon/phosphorus (C/P) ratio, 607–8
Carbon/sulfur (C/S) ratio, 585
Carbonyl sulfide, 582
Carboxylic chemical group, 325, 359
Carnivores, soil organism, 449, 450
Carriers, fertilizer, 708
Casts, earthworm, 455–56
Catchments, 8, 220
Catena, 61, 843, 844

Cation exchange capacity (CEC)
  and calcium/magnesium ratio, 645
  in dry-region soils, 407
  fundamentals, 337–45
  in Histols, 98
  and humus, 513, 515, 517
  in landfills, 827–28
  in Mollisols, 104–5
  and organoclay treatment of toxins, 811
  and pH, 365, 366–68, 371, 378, 389
  and potassium, 631–32
  and weathering, 347
Cations
  basic processes, 311, 312, 333–45
  and calcium, 641, 643
  and chelates, 660–63
  and colloidal adsorption, 311–12, 331–33
  cycling by trees, 57
  in dry-region soils, 407
  magnesium, 644–45
  micronutrient availability, 654–60
  and nitrification, 552
  and pH, 344, 358, 359, 364–65, 366–69, 371, 409, 411
  potassium, 622
  and radionuclides, 831
  selectivity of, 335
  in sodic soil dispersion, 422, 423
  and sulfate leaching, 588–89
  and weathering, 347, 362
Cattle feedlots, 695–97
CEC (cation exchange capacity). *See* Cation exchange capacity (CEC)
Cellulose and decomposition, 501
Central America, improving soil quality in, 910
Cerrado region in Brazil, 910–12
Cesium, 137, 831–32, 833
CFCs (chlorofluorocarbons), 530, 558
Channers, 128
Charcoal, 911
Charged colloidal surfaces, 317, 325, 328–33. *See also* Anion exchange capacity (AEC); Anions; Cation exchange capacity (CEC); Cations
Check dams, 770, 771
Chelates, 407, 517, 619, 660–63, 826
Chemical complexation, 816
Chemical decomposition, 35
Chemical equivalency principle, 338
Chemical nitrification inhibitors, 574
Chemical weathering, 38, 53, 56
Chemoautotrophs, 448, 477
Chemoheterotrophs, 448
Chenopodiaceae, 473
Chisel plowing, 763
Chlorinated hydrocarbons, 799, 805, 806
Chlorine, 648, 649, 650, 652, 663–64
Chlorite, 319
Chlorites, 321–22
Chlorofluorocarbons (CFCs), 530, 558
Chlorophyll, 543

Chlorosis, 385, 543
C horizons
  definition, 16
  development of, 62–63, 66
  land use impact of, 16–17, 18
  and regolith, 13
  role in soil profile, 70
Chroma, soil color, 122
Chromium, 278, 817, 822, 823, 824
Chronosequences, 41, 63, 843
Ciliates, 464, 465
Classes, soil texture, 127–32
Classification, soil. *See* Soil classification
Clay domain, 137
Clays. *See also* Kaolinite clays; Silicate clays; Smectite clays; Vertisols
  in Alfisols, 107
  allophane, 51, 80, 93, 314, 327, 552, 618
  binding of biomolecules to, 349–51
  and CEC estimation, 342
  in cohesive soils, 164
  compaction of, 166–67
  definition, 125, 127
  density of, 150
  dispersion of, 407, 421–24
  distribution of, 327, 328
  in dry-region soils, 407
  and erosion, 780
  feel method for classifying, 130
  flocculation of, 137, 138, 424
  formation of, 325–27
  friability of, 143–44
  function of, 22–23
  and humus, 513
  and hysteresis, 186
  identifying, 127–32
  and loam classes, 128
  and magnesium, 645
  micaceious, 424
  and nitrification, 552
  and nitrogen fixation, 549–50
  nonsilicated, 322, 324–25, 326, 327
  organic carbon levels in, 525
  organic matter's influence on, 515
  and organic toxic chemicals, 801, 816
  organoclays, 348, 349, 810, 811
  in Oxisols, 112
  particle size class, 116
  and pH, 367, 370–71
  and phosphorus availability, 618
  and potassium availability, 549, 630, 631
  and radionuclides in soil, 832
  and shale, 34
  and silt properties, 125
  as soil constituent, 19, 82
  and soil properties, 127
  structures dominated by, 135–36
  sulfur in, 582
  swelling-type, 167, 319–21, 351–54, 420, 422
  and tillage damage, 144

Moraines, 48
Morrow plots, 528
Moss peat, 53
Mound drain field system, 254
MSW (municipal solid waste), 701
Mucigel, 468
Muck, 53, 97
Mud flows, 771
Mulching
  and erosion control, 758, 769, 772, 774
  and evaporation, 235, 236, 237
  influence on organic matter, 526
  legumes and nitrogen management, 576
  and nutrient management, 684
  and phosphorus availability, 608
  and soil temperature, 303–6
  stubble mulching, 236, 763, 783–84
  wood wastes, 701
Mulch till system, 763
Multiple cropping, 887
Multispectral imagery, 858
Municipal solid waste (MSW), 701
Munsell color charts, 122
Mushrooms, 469, 471
Mutualistic relationships among soil organisms, 460, 487–90, 491
Mycelia, fungal, 469–70
Mycorrhizae
  aggregate stability role of, 139
  and micronutrient availability, 660
  and phosphorus, 602–3, 605, 620
  symbiotic relationship with plant roots, 472–75
Mycorrhizal inoculum, 475
Mycotoxins, fungal, 471–72
Nanopores, 162
National Soil Information System (NASIS), 861
Natric horizon, 82, 107–8
Natural attenuation
  in landfills, 827–29
  of toxic chemicals, 808–9
Natural bodies, soils as, 77, 860. *See also* Soil classification
Natural ecosystems vs. agroecosystems, 522
Natural tile drainage system, 249
Natural vs. cultural eutrophication, 599
Nematicides, 463, 800
Nematodes, 462–63, 471, 508, 807
Neptunium, 833
Net charge, colloids, 328
Neutral salts, 411, 419
Neutron scattering, 187
New Orleans, levee failure, 169
Newtons/m2, 183
Nickel
  and pH, 656
  plant nutrition role of, 648, 649, 650
  representative soil content, 652
  in sewage sludge, 822, 823

sources of, 817
  as toxic element, 817
"Night crawlers," 455, 456, 457
Nitrate depression period, 506, 536
Nitrate reductase, 649
Nitrates
  agricultural accumulation, 543
  and C/N ratios, 506–7
  leaching of, 552–57, 575
  and nitrification, 551
  and nitrogen cycle, 545
  and nutrient uptake, 543
  in organic waste, 705
  predominance in soils, 552, 553
  testing for, 720–21, 723
  and water contamination, 696
Nitric acid, 558
Nitric oxide, 557
Nitrification, 360, 551–52, 664
Nitrites, 543, 551
Nitrobacter, 550
Nitrogen. *See also* Ammonia; Ammonium; Nitrates
  C/N ratio, 504–5, 506–10, 524, 536, 548
  C/N/S ratio, 585
  and composting, 536–38
  decomposition release of, 509–10
  denitrification, 552, 557–62, 575, 767
  deposition from atmosphere, 570–72
  earthworm distribution of, 455
  erosion's removal of, 745
  in fertilizers, 375, 573–74, 708, 732
  fixation of, 478, 479, 482, 549–50, 562–70, 693, 808
  humus's protection of, 513
  immobilization of, 545–48, 549, 767
  increases in, 888, 889
  introduction, 542–43
  leaching of, 549, 685–86
  losses from forests, 688, 690–91
  and methane oxidation, 534
  micronutrient role in transformation of, 649
  mineralization of, 507–8, 545–48, 767
  and moisture content of soil, 524
  nematodes as releasers of, 462
  nitrification, 360, 551–52, 664
  nitrites, 543, 551
  nitrogen cycle, 544–45
  and organic matter, 529, 545, 548–49, 570
  organic vs. inorganic, 544–49
  in organic waste, 705, 706
  oxidation of, in acidic soils, 360
  and plant growth, 543–44
  prokaryotes' fixation role, 478, 479, 482
  and soil management, 556, 572–77, 578
  soluble organic nitrogen, 548–49
  and sulfur, 578
  testing methods, 723

Nitrogenase, 562, 563, 649
Nitrosomonas, 550
Nitrous oxide, 278, 279, 557
Nodulated legumes and nitrogen fixation, 565–66, 567
Nomadic pastoral systems, 892
Nonacid cations, 344, 359
Noncohesive soils, 165
Noncrystalline silicate clays, 314. *See also* Silicate clays
Nonexpansive clays, 321–22
Nonhumic substances, 511, 536
Nonnodulated systems and nitrogen fixation, 566, 567, 569
Nonsilicated clays, 322, 324–25, 326, 327
Nonsymbiotic nitrogen fixation, 569–70
Normal (non-salt-affected) soils, 418
No-tillage systems
  and disease control, 484
  and erosion control, 763, 764–66
  influence on organic matter, 526, 527
  and leaching, 687–88
  legumes and nitrogen management, 576
  liming challenges, 392, 393
  penetration resistance, 158
  and phosphorus losses, 601
  and soil moisture control, 236
  and soil properties, 767–68
  and soil temperature, 304
Nuclear fallout contamination, 336
Nuclear fission and radioactive waste, 831–32
Nucleic acids, 446, 595, 606
Nutrient balance, 668–69, 680, 681, 901–3
Nutrient cycling
  dung beetle's role in, 452
  fungi's role in, 471
  nutrient management, 694–97
  and organic/inorganic amendment combination, 896
  process of, 57
  sub-Saharan African challenges, 903–4
Nutrient management
  and amendment toxicity, 554, 682
  cover crops, 576, 608, 621, 684–87
  fertilizers, 707–18, 732–34
  goals of, 679–82
  introduction, 678
  methods of improving, 682–91
  nutrient resources and cycles, 691–94, 904, 905
  organic amendments, 694–707
  phosphorus site-index method, 726–31
  and soil diagnostics, 718–26
  and sustainable agriculture, 895–96
  technology tools for, 869–71
Nutrient management plans, 683

Plant-available water, 208
Plant growth. *See also* Nutrient
    management; Plant roots
    and acidic soils, 381–87
    aeration effects, 279–80
    and anion exchange, 346
    calcium's role in, 409, 640–41
    and cation saturation, 344
    earthworms' role in, 456
    and genetics, 426, 595, 886
    and glomalin, 139
    in islands of fertility, 403
    magnesium's role in, 644–45
    and micronutrients, 647–51
    nitrogen's role in, 543–44
    organic matter's role in, 22, 513–17
    organisms' effects on, 481–87
    overview, 25–29
    phosphorus's role in, 595–96, 608
    potassium's role in, 623–25
    in salt-affected soils, 420, 424–28,
        435
    soils as medium for, 3–7
    sulfur's role in, 578–80
    termites' effect on, 461
    water availability, 205, 207–14, 221,
        228
Plant growth-promoting rhizobacteria,
    482, 483
Plant production, 245–47, 679, 874,
    877. *See also* Agriculture
Plant residues (litter)
    composition of, 499–500
    and conservation tillage, 767
    crop residues, 235–36, 887–88, 904,
        906, 909, 910
    and decomposition process, 507–9
    and erosion control, 758
    in forest ecology, 70, 490, 508,
        692–93
    and humus formation, 511–12
    need for continuous supply of, 529
    phosphorus in, 604
    quality of, 504
    riparian buffer strips, 683–84
Plant roots. *See also* Nutrient uptake by
    plants
    and aeration, 274, 276, 281
    aerial roots, 268
    aggregation role, 138–39
    and calcium carbonate concentra-
        tions, 406
    and chelates, 662
    compaction and water supply, 156,
        208–9, 210
    effect on microorganisms, 468–69
    fungi/root symbiosis, 472–75
    and macropores, 160, 161
    nitrogen as essential for, 543
    nutrient uptake process, 27–29
    as organisms in soil, 466–69
    and phosphorus, 604
    physical disintegration by, 38
    and salt-affected soils, 432, 434
    soil as medium for, 3–7

and soil strength, 157
and soil temperature, 291
and spacial variation in pH, 374
and vapor movement, 203
and volumetric water content, 186
water supply mechanisms, 211–14
Plants. *See also* Plant growth; Plant
    residues (litter); Plant roots
    absorption of toxic chemicals, 805
    and aeration, 280–83
    breeding and amendments, 672–73
    C/N ratio, 504
    container-grown, 280–81, 426, 428,
        438
    diseases of, 483–84, 486, 487, 538,
        890
    interception of water precipitation,
        221
    metal hyperaccumulating, 825–26
    as nutrient sources for animals, 5
    pests of, 483, 486–87
    pH effects on, 381–87
    and photosynthesis, 7, 499
    phytoremediation, 813–17, 820, 826
    as primary carbon source, 498
    as primary producers, 448
    respiration by, 4, 274, 276
    soil-plant-atmosphere continuum,
        227–32, 519–22
    and soil temperature, 288–91
Plant tissue analysis, 719–20
Plasticity, soil, 125, 162–63, 167–68,
    352
Plasticity index (PI), 167–68, 352
Plastic limit (PL), 167–68, 205
Plastic mulch, 235, 236, 305–6
Platy structure, 135
Pleistocene epoch, 47
Plinthite, 109
Plowing. *See* Tillage
Plow layer, 72
Plow pans, 154
Plutonium, 239, 832–33, 833
PM (particulate matter) and wind
    erosion, 747
Point injection fertilizer application
    method, 715, 716
Point vs. nonpoint sources for water
    flow, 598, 682
Polarity in water molecule, 174
Pollutants. *See also* Toxins
    calcium deposition, 643
    colloidal sorption of contaminants,
        347–48
    from denitrification, 558–59
    and earthworms' leaching role,
        457–58
    and erosion, 745
    groundwater, 193–94, 195, 242–43,
        347–48, 804, 806
    herbicides as, 233, 795, 802, 803,
        806, 889
    human sources for water pollution,
        219
    and landfills, 829, 831

micronutrients as, 667–68, 823, 824
nitrogen excess, 542, 552–57, 682
and pH, 358
phosphorus excess, 597, 598, 599,
    600, 602, 607, 614, 682, 726–31
and preferential flow, 195
prokaryote remediation of, 477–78
and septic tank drain fields, 252–53
soil contaminant removal, 8–9, 295,
    296
and water movement, 7
Polonium, 834
Polyacrylamide (PAM), 147
Polychlorinated biphenyls (PCBs),
    798–99
Polypedon, 77
Polyphenols, 500, 508–10
Polysaccharides, 469, 488, 501, 512–13
POM (particulate organic matter), 518
Pools of organic carbon, 517–19
Pore space, 158–62. *See also* Soil pores
Positive feedback loop from plant
    growth, 403
Positively charged ions. *See* Cations
Potassium
    in fertilizers, 623, 634, 635, 664, 708
    fixation of, 549, 630–33
    forms and soil availability of, 629–33
    management of, 633–35
    and plant growth, 623–25
    potassium cycle, 625–26
    role of, 622
    and soil fertility issues, 625, 627–29
Potential acidity pool, 364, 379, 381
Potential acid-sulfate, 587
Potential buffering capacity (PBC), 617
Potential cation exchange capacity,
    368
Potential energy, 179–91
Potential evapotranspiration (PET),
    229, 234, 235, 239
Potential gradient, 192
Potential vs. active acid sulfate soils,
    380
Potentiometric methods for pH
    determination, 372
Potted plants, 280–81, 426, 428, 438
Poultry manure, 697
Prairie ecosystem, 106
Prairie pothole, 222
Precipitation. *See also* Acid rain
    and buffering of pH, 371
    effective, 53, 54–55
    in hydrologic cycle, 220
    intensity and leaching, 243
    and irrigation, 221
    nitrogen deposition, 570–72
    rainfall's erosion role, 741, 749–50,
        753, 754, 755
    seasonal distribution of, 54
    and solar energy, 297
    sulfur, 582
Precision agriculture, 418, 435, 868–71
Predators, soil organism, 444, 449, 450,
    462

Strong acids, 360
Strontium, 90, 831–32
Structural condition, 133
Structure, soil. *See* Soil structure
Stubble mulching, 236, 763, 783–84
Subangular blocky structure, 134, 135
Subgrouping *Soil Taxonomy*, 85
Submerged aquatic vegetation (SAV), 68, 745
Submicroaggregates, 137
Suborders in *Soil Taxonomy*, 85, 112–13
Sub-Saharan Africa, soil quality challenges in, 901–10
Subsidence, 99–100
Subsoil
    air vs. water in, 274
    and clays' ammonium fixation, 549
    compaction of, 154–56
    and lime incorporation, 392
    in soil profile, 16–17
    termites' transfer of material from, 460–61
Subsoiling, 154
Subsurface drainage, 247–49
Subsurface horizons, 82, 92, 105, 109
Suburbanization, 882, 883
Sufficiency range for nutrients, 645–46, 647, 719
Sulfates. *See also* Gypsum
    and acid rain, 378–81
    of aluminum, 396, 397–98, 621
    ester-sulfates, 580
    of iron, 396, 397–98
    leaching of, 588–89
    and molybdenum availability, 666
    potential acid-sulfate, 587
Sulfidic materials, 379, 587
Sulfur
    deposition from atmosphere, 378–81, 582
    immobilization of, 578, 585–86
    introduction, 542
    mineralization, 583–85
    and minerals, 581–82
    organic vs. inorganic, 580–82
    oxidation and reduction, 361, 378–81, 586–88
    and plant growth, 578–80
    retention and exchange, 588–89
    for sodic soil reclamation, 436–37
    and soil fertility, 589
    sulfur cycle, 583
Sulfur dioxide, 582
Sulfuric acid, 378–81, 582
Sulfuric horizon, 81, 83
Superphosphate, 817
Support practice factor P, 753, 758–61
Surface area, colloidal, 311
Surface drainage systems, 247, 250
Surface evaporation (E), 233–36
Surface horizons, 80–82, 105
Surface irrigation, 258–60
Surface runoff, 221
Surface seal, 145
Surface tension, 175, 177

Surfactants, 810
Survey, soil, 858, 860–64, 865
Suspension, wind erosion, 779–80, 781, 788
Sustainable agriculture, 894–99, 901–12
Swamps. *See* Wetlands
Swelling-type clays, 167, 319–21, 351–54, 420, 422
Swine manure, 697
Symbiosis, 472–75, 564–69. *See also* Mycorrhizae
Synergism and micronutrients, 668–69
Synthetic polymers, 438, 439
Tardigrades, 464, 466
TDR (time-domain reflectometry), 187
TDS (total dissolved solids), 415
T efficiency, 231
Temperature, atmospheric. *See also* Soil temperature
    and ammonia volatilization, 550
    and evaporation, 54
    and organic matter, 53
    and runoff, 222
    and slope aspect, 61
    soils as moderators of, 4–5
    and weathering, 37, 56
Tensile strength, 143, 144
Tensiometer, 189, 190
Tension plates, 189
Termites, 59, 458–61, 534
Terraces, 44, 760–61
Terra Preta soils, 911
Tertiary consumers, soil organism, 450–51
Testing, soil, 725–26
Teton Dam failure, 126
Tetrahedral sheets, 315, 328
Textural triangle, 132
Texture, soil. *See* Soil texture
Thermal properties of soils, 85, 299–303
Thermic temperature regime, 85, 116
Thermocouple psychrometer, 189–90
Thermophilic composting, 535–36
Thixotropy, 165
Thlaspi caerulescens, 825–26
Thorium, 232, 833
Threadworms (nematodes), 462–63, 471, 508, 807
Tile drains, 249, 250
Till, glacial, 48, 151
Tillage. *See also* Conservation tillage
    and aggregation, 142, 143–48, 767
    and density of soil, 154
    and disease control, 484
    effects on organisms, 458, 474, 489
    and heterogeneity of soil, 274
    influence on organic matter, 526, 527
    legumes and nitrogen management, 576
    and phosphorus losses, 601
    restrictions on, 529
    and salt-affected soils, 432, 434

and soil properties, 143–48, 766–68
and wind erosion, 782, 783–84
Tilth, soil, 143–48, 471
Timber harvesting, 152–53, 688–91, 740, 772–74
Time
    and formation of soil, 62–64
    and inorganic phosphorus availability, 609, 611
    residence time of water, 219, 285
    and soil temperature, 302–3
    and spacial variation in pH, 374
Time-domain reflectometry (TDR), 187
Titration curves, 369
Tonic composition of soil solution, 311–25
Topographic factor LS, 753, 756–57
Topography
    and effective precipitation, 54
    and formation of soils, 60–62
    and soil variability, 841, 843
Toposequences, 41, 61, 843
Topsoil, 16. *See also* A horizons
Torrands, 93, 94
Torrerts, 102
Torric moisture regime, 85
Torrox, 111
Total acidity pool, 364
Total dissolved solids (TDS), 415
Total porosity of soil, 158
Total soil water potential, 181
Total suspended solids (TSP), 788
Toxicity range for micronutrients, 645–46, 647
Toxins. *See also* Heavy metals; Inorganic chemicals; Organic toxins
    and aeration, 278
    in eroded materials, 745
    fungal, 471–72
    introduction, 794
    in irrigation drainage water, 430
    micronutrient excesses, 645–47, 653–54, 660
    nitrites, 543
    prokaryote breakdown of, 481, 482
    soils as protection from, 5
    sulfuric, 582
    and temperature, 295, 296
Trace elements, 639–40, 666–67. *See also* Micronutrients
Traffic pans, 154
Trails, forest, and erosion, 772
Transformations in soil formation, 64–65, 66, 69, 93
Transition horizons, 71
Translocations in soil formation, 65, 66–67, 69
Transpiration, 220, 229, 232–33, 257
Trees. *See also* Forests
    and aeration, 281–83
    perforation fertilizer application method, 715, 716
Triazines, 805

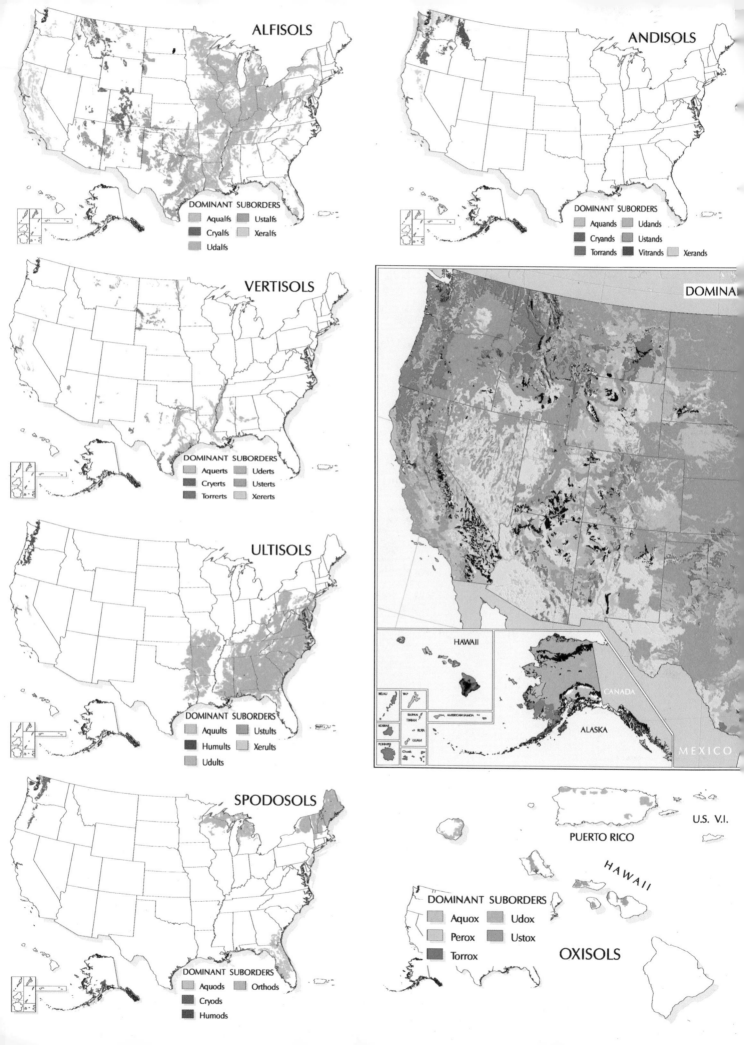

ALFISOLS

DOMINANT SUBORDERS
- Aqualfs
- Cryalfs
- Udalfs
- Ustalfs
- Xeralfs

ANDISOLS

DOMINANT SUBORDERS
- Aquands
- Cryands
- Torrands
- Udands
- Ustands
- Vitrands
- Xerands

VERTISOLS

DOMINANT SUBORDERS
- Aquerts
- Cryerts
- Torrerts
- Uderts
- Usterts
- Xererts

ULTISOLS

DOMINANT SUBORDERS
- Aquults
- Humults
- Udults
- Ustults
- Xerults

DOMINA

HAWAII

CANADA

ALASKA

MEXICO

SPODOSOLS

DOMINANT SUBORDERS
- Aquods
- Cryods
- Humods
- Orthods

PUERTO RICO

U.S. V.I.

HAWAII

DOMINANT SUBORDERS
- Aquox
- Perox
- Torrox
- Udox
- Ustox

OXISOLS